Prealgebra & Introductory Algebra

Fifth Edition

Elayn Martin-Gay
University of New Orleans

330 Hudson Street, NY NY 10013

Director, Portfolio Management: *Michael Hirsch*
Courseware Portfolio Manager: *Mary Beckwith*
Courseware Portfolio Management Assistant: *Alison Oehman*
Managing Producer: *Karen Wernholm*
Content Producer: *Patty Bergin*
Media Producer: *Audra Walsh*
Manager, Courseware QA: *Mary Durnwald*
Manager Content Development, Math: *Eric Gregg*
Product Marketing Manager: *Alicia Frankel*
Field Marketing Manager: *Jennifer Crum and Lauren Schur*
Product Marketing Assistant: *Hanna Lafferty*
Senior Author Support/Technology Specialist: *Joe Vetere*
Manager, Rights and Permissions: *Gina Cheselka*
Manufacturing Buyer: *Carol Melville, LSC Communications*
Text Design: *Tamara Newnam*
Composition and Production Coordination: *Integra*
Illustrations: *Scientific Illustrators*
Senior Designer: *Barbara T. Atkinson*
Cover Design: *Tamara Newman*
Cover Image: *Tamara Newman*

Copyright © 2019, 2015, 2011 by Pearson Education, Inc. All Rights Reserved. Printed in the United States of America. This publication is protected by copyright, and permission should be obtained from the publisher prior to any prohibited reproduction, storage in a retrieval system, or transmission in any form or by any means, electronic, mechanical, photocopying, recording, or otherwise. For information regarding permissions, request forms and the appropriate contacts within the Pearson Education Global Rights & Permissions department, please visit www.pearsoned.com/permissions/.

Attributions of third party content appear on page P1, which constitutes an extension of this copyright page.

PEARSON, ALWAYS LEARNING, and MYLAB MATH are exclusive trademarks owned by Pearson Education, Inc. or its affiliates in the U.S. and/or other countries.

Unless otherwise indicated herein, any third-party trademarks that may appear in this work are the property of their respective owners and any references to third-party trademarks, logos or other trade dress are for demonstrative or descriptive purposes only. Such references are not intended to imply any sponsorship, endorsement, authorization, or promotion of Pearson's products by the owners of such marks, or any relationship between the owner and Pearson Education, Inc. or its affiliates, authors, licensees or distributors.

Library of Congress Cataloging-in-Publication Data

Names: Martin-Gay, K. Elayn,- author.
Title: Prealgebra & introductory algebra / by Elayn Martin-Gay.
Other titles: Prealgebra and introductory algebra
Description: 5th edition. | New York, NY : Pearson, [2019] | Includes index.
Identifiers: LCCN 2017044022| ISBN 9780134707631 (se : alk. paper) | ISBN 9780134708485 (aie : alk. paper) | ISBN 9780134708515 (alc : alk. paper) | ISBN 9780134708478 (epub)
Subjects: LCSH: Arithmetic–Textbooks. | Algebra–Textbooks.
Classification: LCC QA107.2 .M39 2019 | DDC 512–dc23
LC record available at https://lccn.loc.gov/2017044022

2 18

ISBN-13: 978-0-13-470763-1 (Student Edition)
ISBN-10: 0-13-470763-X

ISBN-13: 978-0-13-470845-4 (Student Hardcover Edition)
ISBN-10: 0-13-470845-8

This book is dedicated to students everywhere—
and we should all be students. After all, is there anyone among
us who truly knows too much? Take that hint and continue
to learn something new every day of your life.

Best wishes from a fellow student:
Elayn Martin-Gay

Contents

Preface xiii

Applications Index xxiii

1 The Whole Numbers 1

1.1 Study Skill Tips for Success in Mathematics 2
1.2 Place Value, Names for Numbers, and Reading Tables 8
1.3 Adding and Subtracting Whole Numbers, and Perimeter 17
1.4 Rounding and Estimating 32
1.5 Multiplying Whole Numbers and Area 40
1.6 Dividing Whole Numbers 52
   Integrated Review—Operations on Whole Numbers 66
1.7 Exponents and Order of Operations 68
1.8 Introduction to Variables, Algebraic Expressions, and Equations 75
   Group Activity 84
   Vocabulary Check 85
   Chapter Highlights 85
   Chapter Review 89
   Getting Ready for the Test 95
   Chapter Test 96

2 Integers and Introduction to Solving Equations 98

2.1 Introduction to Integers 99
2.2 Adding Integers 108
2.3 Subtracting Integers 116
2.4 Multiplying and Dividing Integers 124
   Integrated Review—Integers 133
2.5 Order of Operations 135
2.6 Solving Equations: The Addition and Multiplication Properties 142
   Group Activity 151
   Vocabulary Check 152
   Chapter Highlights 152
   Chapter Review 154
   Getting Ready for the Test 159
   Chapter Test 160
   Cumulative Review 162

3 Solving Equations and Problem Solving 164

- **3.1** Simplifying Algebraic Expressions 165
- **3.2** Solving Equations: Review of the Addition and Multiplication Properties 175
  Integrated Review—Expressions and Equations 184
- **3.3** Solving Linear Equations in One Variable 186
- **3.4** Linear Equations in One Variable and Problem Solving 193
  Group Activity 202
  Vocabulary Check 202
  Chapter Highlights 203
  Chapter Review 205
  Getting Ready for the Test 209
  Chapter Test 210
  Cumulative Review 212

4 Fractions and Mixed Numbers 214

- **4.1** Introduction to Fractions and Mixed Numbers 215
- **4.2** Factors and Simplest Form 229
- **4.3** Multiplying and Dividing Fractions 242
- **4.4** Adding and Subtracting Like Fractions, Least Common Denominator, and Equivalent Fractions 254
- **4.5** Adding and Subtracting Unlike Fractions 268
  Integrated Review—Summary on Fractions and Operations on Fractions 280
- **4.6** Complex Fractions and Review of Order of Operations 282
- **4.7** Operations on Mixed Numbers 290
- **4.8** Solving Equations Containing Fractions 307
  Group Activity 316
  Vocabulary Check 316
  Chapter Highlights 317
  Chapter Review 321
  Getting Ready for the Test 326
  Chapter Test 327
  Cumulative Review 329

5 Decimals 331

- **5.1** Introduction to Decimals 332
- **5.2** Adding and Subtracting Decimals 344
- **5.3** Multiplying Decimals and Circumference of a Circle 357
- **5.4** Dividing Decimals 366
  Integrated Review—Operations on Decimals 376
- **5.5** Fractions, Decimals, and Order of Operations 378
- **5.6** Solving Equations Containing Decimals 387
- **5.7** Decimal Applications: Mean, Median, and Mode 392

Group Activity **400**

Vocabulary Check **401**

Chapter Highlights **401**

Chapter Review **404**

Getting Ready for the Test **410**

Chapter Test **411**

Cumulative Review **413**

6 Ratio, Proportion, and Percent **416**

6.1 Ratio and Proportion **417**

6.2 Percents, Decimals, and Fractions **427**

6.3 Solving Percent Problems with Equations **438**

6.4 Solving Percent Problems with Proportions **445**

Integrated Review—Ratio, Proportion, and Percent **453**

6.5 Applications of Percent **455**

6.6 Percent and Problem Solving: Sales Tax, Commission, and Discount **467**

6.7 Percent and Problem Solving: Interest **473**

Group Activity **479**

Vocabulary Check **479**

Chapter Highlights **480**

Chapter Review **483**

Getting Ready for the Test **487**

Chapter Test **488**

Cumulative Review **490**

7 Graphs, Triangle Applications, and Introduction to Statistics and Probability **492**

7.1 Pictographs, Bar Graphs, Histograms, Line Graphs, and Introduction to Statistics **493**

7.2 Circle Graphs **509**

Integrated Review—Reading Graphs **517**

7.3 Square Roots and the Pythagorean Theorem **519**

7.4 Congruent and Similar Triangles **526**

7.5 Counting and Introduction to Probability **535**

Group Activity **542**

Vocabulary Check **543**

Chapter Highlights **543**

Chapter Review **546**

Getting Ready for the Test **553**

Chapter Test **555**

Cumulative Review **560**

8 Geometry and Measurement 562

- 8.1 Lines and Angles 563
- 8.2 Perimeter 574
- 8.3 Area, Volume, and Surface Area 584
  - Integrated Review—Geometry Concepts 600
- 8.4 Linear Measurement 601
- 8.5 Weight and Mass 614
- 8.6 Capacity 624
- 8.7 Temperature and Conversions Between the U.S. and Metric Systems 632
  - Group Activity 641
  - Vocabulary Check 642
  - Chapter Highlights 642
  - Chapter Review 646
  - Getting Ready for the Test 652
  - Chapter Test 653
  - Cumulative Review 655

9 Equations, Inequalities, and Problem Solving 658

- 9.1 Symbols and Sets of Numbers 659
- 9.2 Properties of Real Numbers 668
- 9.3 Further Solving Linear Equations 677
  - Integrated Review—Real Numbers and Solving Linear Equations 686
- 9.4 Further Problem Solving 688
- 9.5 Formulas and Problem Solving 702
- 9.6 Linear Inequalities and Problem Solving 714
  - Group Activity 725
  - Vocabulary Check 725
  - Chapter Highlights 726
  - Chapter Review 729
  - Getting Ready for the Test 733
  - Chapter Test 734
  - Cumulative Review 736

10 Exponents and Polynomials 738

- 10.1 Exponents 739
- 10.2 Negative Exponents and Scientific Notation 751
- 10.3 Introduction to Polynomials 761
- 10.4 Adding and Subtracting Polynomials 771
- 10.5 Multiplying Polynomials 778
- 10.6 Special Products 785
  - Integrated Review—Exponents and Operations on Polynomials 793

10.7 Dividing Polynomials **795**
Group Activity **802**
Vocabulary Check **802**
Chapter Highlights **803**
Chapter Review **806**
Getting Ready for the Test **811**
Chapter Test **812**
Cumulative Review **814**

11 Factoring Polynomials **817**

11.1 The Greatest Common Factor and Factoring by Grouping **818**
11.2 Factoring Trinomials of the Form $x^2 + bx + c$ **828**
11.3 Factoring Trinomials of the Form $ax^2 + bx + c$ **835**
11.4 Factoring Trinomials of the Form $ax^2 + bx + c$ by Grouping **842**
11.5 Factoring Perfect Square Trinomials and the Difference of Two Squares **847**
Integrated Review—Choosing a Factoring Strategy **856**
11.6 Solving Quadratic Equations by Factoring **858**
11.7 Quadratic Equations and Problem Solving **866**
Group Activity **875**
Vocabulary Check **876**
Chapter Highlights **876**
Chapter Review **879**
Getting Ready for the Test **883**
Chapter Test **884**
Cumulative Review **886**

12 Rational Expressions **888**

12.1 Simplifying Rational Expressions **889**
12.2 Multiplying and Dividing Rational Expressions **900**
12.3 Adding and Subtracting Rational Expressions with the Same Denominator and Least Common Denominator **910**
12.4 Adding and Subtracting Rational Expressions with Different Denominators **918**
12.5 Solving Equations Containing Rational Expressions **925**
Integrated Review—Summary on Rational Expressions **934**
12.6 Rational Equations and Problem Solving **936**
12.7 Simplifying Complex Fractions **944**
Group Activity **952**
Vocabulary Check **952**
Chapter Highlights **953**
Chapter Review **957**
Getting Ready for the Test **961**
Chapter Test **962**
Cumulative Review **964**

13 Graphing Equations and Inequalities 967

- 13.1 The Rectangular Coordinate System 968
- 13.2 Graphing Linear Equations 981
- 13.3 Intercepts 991
- 13.4 Slope and Rate of Change 1001
- 13.5 Equations of Lines 1018
  Integrated Review—Summary on Linear Equations 1030
- 13.6 Introduction to Functions 1032
- 13.7 Graphing Linear Inequalities in Two Variables 1044
- 13.8 Direct and Inverse Variation 1053
  Group Activity 1063
  Vocabulary Check 1064
  Chapter Highlights 1064
  Chapter Review 1068
  Getting Ready for the Test 1075
  Chapter Test 1076
  Cumulative Review 1079

14 Systems of Equations 1082

- 14.1 Solving Systems of Linear Equations by Graphing 1083
- 14.2 Solving Systems of Linear Equations by Substitution 1094
- 14.3 Solving Systems of Linear Equations by Addition 1102
  Integrated Review—Summary on Solving Systems of Equations 1110
- 14.4 Systems of Linear Equations and Problem Solving 1111
  Group Activity 1123
  Vocabulary Check 1124
  Chapter Highlights 1124
  Chapter Review 1127
  Getting Ready for the Test 1130
  Chapter Test 1131
  Cumulative Review 1133

15 Roots and Radicals 1136

- 15.1 Introduction to Radicals 1137
- 15.2 Simplifying Radicals 1145
- 15.3 Adding and Subtracting Radicals 1153
- 15.4 Multiplying and Dividing Radicals 1157
  Integrated Review—Simplifying Radicals 1166
- 15.5 Solving Equations Containing Radicals 1168
- 15.6 Radical Equations and Problem Solving 1174
  Group Activity 1181
  Vocabulary Check 1182

Chapter Highlights **1182**
Chapter Review **1185**
Getting Ready for the Test **1188**
Chapter Test **1189**
Cumulative Review **1191**

16 Quadratic Equations **1193**

16.1 Solving Quadratic Equations by the Square Root Property **1194**
16.2 Solving Quadratic Equations by Completing the Square **1201**
16.3 Solving Quadratic Equations by the Quadratic Formula **1206**
Integrated Review—Summary on Solving Quadratic Equations **1215**
16.4 Graphing Quadratic Equations in Two Variables **1218**
Group Activity **1227**
Vocabulary Check **1228**
Chapter Highlights **1228**
Chapter Review **1230**
Getting Ready for the Test **1234**
Chapter Test **1235**
Cumulative Review **1237**

Appendices

Appendix A Tables **1240**
A.1 Table of Geometric Figures **1240**
A.2 Table of Percents, Decimals, and Fraction Equivalents **1242**
A.3 Table on Finding Common Percents of a Number **1243**
A.4 Table of Squares and Square Roots **1244**

Appendix B Factoring Sums and Differences of Cubes **1245**

Appendix C Mixture and Uniform Motion Problem Solving **1248**

Appendix D Systems of Linear Inequalities **1256**

Appendix E Geometric Formulas **1260**

Student Resources **1263**
  Study Skills Builders **1263**
  Bigger Picture—Study Guide Outline **1272**
  Practice Final Exam **1278**

  Answers to Selected Exercises **A1**

Subject Index **I-1**
Photo Credits **P1**

Preface

***Prealgebra & Introductory Algebra,* Fifth Edition**, was written in response to the needs of those teaching combined courses. My goals were to help students make the transition from arithmetic to algebra and to provide a solid foundation in algebra. To help students accomplish this, my goals for this text are:

- Most importantly, to write an organized, student-friendly text that is keyed to objectives and contains many worked-out examples.
- To introduce algebraic concepts early and repeat them often as I cover traditional arithmetic topics, thus laying the groundwork for the next algebra course your students will take. Specific care was taken to ensure that all core topics of an introductory algebra course are covered and that students have the most up-to-date, relevant text preparation for future courses that require an understanding of algebraic fundamentals.
- To show students the relevancy of mathematics in everyday life and in the workplace by emphasizing and integrating the following throughout this text: real-life and real-data applications, data interpretation, conceptual understanding, problem solving, writing, cooperative learning, number sense, estimation, critical thinking and geometric concepts.

The many factors that contributed to the success of the previous editions have been retained. In preparing the Fifth Edition, I considered comments and suggestions of colleagues, students, and many users of the prior edition throughout the country.

What's New in the Fifth Edition?

- **The Martin-Gay Program** has been revised and enhanced with a new design in the text and MyLab Math to actively encourage students to use the text, video program, and Video Organizer as an integrated learning system.
- **New Getting Ready for the Test** can be found before each Chapter Test. These exercises can increase student success by helping students prepare for their Chapter Test. The purpose of these exercises is to check students' conceptual understanding of the topics in the chapter as well as common student errors. It is suggested that students complete and check these exercises before taking a practice Chapter Test. All Getting Ready for the Test exercises are either Multiple Choice or Matching, and all answers can be found in the answer section of this text.

   Video Solutions of all exercises can be found in MyLab Math. These video solutions contain brief explanations and reminders of material in the chapter. Where applicable, incorrect choices contain explanations.

   Getting Ready for the Test exercise numbers marked in blue indicate that the exercise is available in **Learning Catalytics**. LC

- **New Learning Catalytics** is an interactive student response tool that uses students' smartphones, tablets, or laptops to engage them in more sophisticated tasks and thinking. Generate class discussion, guide your lecture, and promote peer-to-peer learning with real-time analytics. Accessible through MyLab Math, instructors can use Learning Catalytics to:
  - Pose a variety of open-ended questions that help your students develop critical thinking skills.
  - Monitor responses to find out where students are struggling.
  - Use real-time data to adjust your instructional strategy and try other ways of engaging your students during class.

- Manage student interactions by automatically grouping students for discussion, teamwork, and peer-to-peer learning.
- Pearson-created questions for developmental math topics are available to allow you to take advantage of this exciting technology. Additionally, "Getting Ready for the Test" exercises (marked in blue) are available in Learning Catalytics. Search the question library for "MGP&I" and the chapter number, for example, MGP&I7 would be the questions from Chapter 7.

- **Revised and updated Key Concept Activity Lab Workbook** includes Extension Exercises, Exploration Activities, Conceptual Exercises, and Group Activities. These activities are a great way to engage students in conceptual projects and exploration as well as group work. This workbook is available in MyLab Math, or can be packaged with a text or MyLab code.

- **Exercise Sets** have been carefully examined and revised. Special focus was placed on making sure that even- and odd-numbered exercises are carefully paired and that real-life applications are updated.

- **The Martin-Gay MyLab Math** course has been updated and revised to provide more exercise coverage, including assignable Video Check questions and an expanded video program. There are Lecture Videos for every section, which students can also access at the specific objective level; Student Success Tips videos; and an increased number of video clips at the exercise level to help students while doing homework in MyLab Math. Suggested homework assignments have been premade for assignment at the instructor's discretion.

Key Continuing Resources and Pedagogical Features

- **Vocabulary, Readiness & Video Check Questions** continue to be available in the text and for assignment in MyLab Math. The **Readiness** exercises center on a student's understanding of a concept that is necessary in order to continue to the exercise set. The **Video Check questions** are included in every section for every learning objective. These exercises are a great way to assess whether students have viewed and understood the key concepts presented in the videos. Answers to all Video Check questions are available in an answer section at the back of the text.

- **Interactive Lecture Series in MyLab Math**, featuring author Elayn Martin-Gay, provides students with active learning at their own pace. The videos offer the following resources and more:

    A complete lecture for each section of the text highlights key examples and exercises from the text. Pop-ups reinforce key terms, definitions, and concepts.

    An interface with menu navigation features allows students to quickly find and focus on the examples and exercises they need to review.

    Interactive Concept Check exercises measure students' understanding of key concepts and common trouble spots.

    Student Success Tips Videos are 3-5 minute videos designed to be daily reminders to students to continue practicing and maintaining good organizational and study habits. They include student success tips for general college success, tips specific to success in math courses, and content-specific tips to avoid common mathematical mistakes.

- **The Interactive Lecture Series** also includes the following resources for test prep:

    New Getting Ready for the Test Videos

    The Chapter Test Prep Videos help students during their most teachable moment—when they are preparing for a test. This innovation provides step-by-step solutions for the exercises found in each Chapter Test. For the Fifth Edition, the Chapter Test Prep Videos are also available on YouTube™. The videos are captioned in English and Spanish.

The Practice Final Exam Videos help students prepare for an end-of-course final. Students can watch full video solutions to each exercise in the Practice Final Exam at the end of this text.

- **The Video Organizer** helps students take notes and work practice exercises while watching the Interactive Lecture Series videos in their MyLab Math course. All content in the Video Organizer is presented in the same order as it is presented in the videos, making it easy for students to create a course notebook and build good study habits.
  - Covers all of the video examples in order.
  - Provides prompts with ample space for students to write down key definitions and properties.
  - Includes Play and Pause button icons to prompt students to follow along with the author for some exercises while they try others on their own.

The Video Organizer is available in a loose-leaf, notebook-ready format. It is also available for download in MyLab Math.

Key Pedagogical Features

The following key features have been retained and/or updated for the Fifth Edition of the text:

- **Problem-Solving Process** This is formally introduced in Chapter 3 with a four-step process that is integrated throughout the text. The four steps are **Understand, Translate, Solve, and Interpret.** The repeated use of these steps in a variety of examples shows their wide applicability. Reinforcing the steps can increase students' comfort level and confidence in tackling problems.

- **Exercise Sets Revised and Updated** The exercise sets have been carefully examined and extensively revised. Special focus was placed on making sure that even- and odd-numbered exercises are paired and that real-life applications were updated.

- **Examples** Detailed, step-by-step examples were added, deleted, replaced, or updated as needed. Many examples reflect real life. Additional instructional support is provided in the annotated examples.

- **Practice Exercises** Throughout the text, each worked-out example has a parallel Practice exercise. These invite students to be actively involved in the learning process. Students should try each Practice exercise after finishing the corresponding example. Learning by doing will help students grasp ideas before moving on to other concepts. Answers to the Practice exercises are provided at the bottom of each page.

- **Helpful Hints** Helpful Hints contain practical advice on applying mathematical concepts. Strategically placed where students are most likely to need immediate reinforcement, Helpful Hints help students avoid common trouble areas and mistakes.

- **Concept Checks** This feature allows students to gauge their grasp of an idea as it is being presented in the text. Concept Checks stress conceptual understanding at the point-of-use and help suppress misconceived notions before they start. Answers appear at the bottom of the page. Exercises related to Concept Checks are included in the exercise sets.

- **Mixed Practice Exercises** In the section exercise sets, these exercises require students to determine the problem type and strategy needed to solve it just as they would need to do on a test.

- **Integrated Reviews** This unique, mid-chapter exercise set helps students assimilate new skills and concepts that they have learned separately over several sections. These reviews provide yet another opportunity for students to work with "mixed" exercises as they master the topics.

- **Vocabulary Check** This feature provides an opportunity for students to become more familiar with the use of mathematical terms as they strengthen their verbal skills. These appear at the end of each chapter before the Chapter Highlights. Vocabulary, Readiness & Video exercises provide practice at the section level.

- **Chapter Highlights** Found at the end of every chapter, these contain key definitions and concepts with examples to help students understand and retain what they have learned and help them organize their notes and study for tests.

- **Chapter Review** The end of every chapter contains a comprehensive review of topics introduced in the chapter. The Chapter Review offers exercises keyed to every section in the chapter, as well as Mixed Review exercises that are not keyed to sections.

- **Chapter Test and Chapter Test Prep Videos** The Chapter Test is structured to include those problems that involve common student errors. The **Chapter Test Prep Videos** gives students instant access to a step-by-step video solution of each exercise in the Chapter Test.

- **Cumulative Review** This review follows every chapter in the text (except Chapter 1). Each odd-numbered exercise contained in the Cumulative Review is an earlier worked example in the text that is referenced in the back of the book along with the answer.

- **Writing Exercises** These exercises occur in almost every exercise set and require students to provide a written response to explain concepts or justify their thinking.

- **Applications** Real-world and real-data applications have been thoroughly updated, and many new applications are included. These exercises occur in almost every exercise set and show the relevance of mathematics and help students gradually and continuously develop their problem-solving skills.

- **Review Exercises** These exercises occur in each exercise set (except in Chapter 1) and are keyed to earlier sections. They review concepts learned earlier in the text that will be needed in the next section or chapter.

- **Exercise Set Resource Icons** Located at the opening of each exercise set, these icons remind students of the resources available for extra practice and support:

MyLab Math

See Student Resources descriptions on page xvii for details on the individual resources available.

Exercise Icons These icons facilitate the assignment of specialized exercises and let students know what resources can support them.

- ▷ Video icon: exercise worked in the Interactive Lecture Series found in MyLab Math.
- △ Triangle icon: identifies exercises involving geometric concepts.
- ✎ Pencil icon: indicates a written response is needed.
- 🖩 Calculator icon: optional exercises intended to be solved using a scientific or graphing calculator.

Group Activities Found at the end of each chapter, these activities are for individual or group completion, and are usually hands-on or data-based activities that extend the concepts found in the chapter, allowing students to make decisions and interpretations and to think and write about algebra.

Optional: Calculator Exploration Boxes and Calculator Exercises The optional Calculator Explorations provide keystrokes and exercises at appropriate places to give students an opportunity to become familiar with these tools. Section exercises that are best completed by using a calculator are identified by 🖩 for ease of assignment.

Student and Instructor Resources

STUDENT RESOURCES

Video Organizer

Designed to help students take notes and work practice exercises while watching the Interactive Lecture Series videos.

- Covers all of the video examples in order.
- Provides prompts with ample space for students to write down key definitions and rules.
- Includes "Play" and "Pause" button icons to prompt students to follow along with the author for some exercises while they try others on their own.
- Includes Student Success Tips Outline and Questions

Available in loose-leaf, notebook-ready format and in MyLab Math.

Key Concept Activity Lab Workbook

Includes Extension Exercises, Exploration Activities, Conceptual Exercises, and Group Activities. This workbook is available in MyLab Math, or can be packaged in printed form with a text or MyLab Math code.

Student Solutions Manual

Provides completely worked-out solutions to the odd-numbered section exercises; all exercises in the Integrated Reviews, Chapter Reviews, Chapter Tests, and Cumulative Reviews.

INSTRUCTOR RESOURCES

Annotated Instructor's Edition Contains all the content found in the student edition, plus the following: - Answers to even and odd exercises on the same text page - Teaching Tips throughout the text placed at key points	**Instructor's Resource Manual with Tests and Mini-Lectures** This resource includes: - Mini-lectures for each text section - Additional practice worksheets for each section - Several forms of tests per chapter—free response and multiple choice - Answers to all items **Instructor's Solutions Manual** **TestGen®** (These resources are available for download from MyLab Math or from the Instructor's Resource Center on pearson.com.)
Instructor-to-Instructor Videos—available in the Instructor Resources section of the MyLab Math course.	**Online Resources** **MyLab Math** (access code required) **MathXL®** (access code required)

Resources for Success

Get the Most Out of MyLab Math for *Prealgebra & Introductory Algebra,* Fifth Edition by Elayn Martin-Gay

Elayn Martin-Gay believes that every student can succeed, and each MyLab course that accompanies her texts is infused with her student-centric approach. The seamless integration of Elayn's award-winning content with the #1 choice in digital learning for developmental math gives students a completely consistent experience from print to MyLab.

A Comprehensive and Dynamic Video Program

The **Martin-Gay video program** is 100% presented by Elayn Martin-Gay to ensure consistency with the text. The video program includes full section lectures and shorter objective level videos, and an intuitive navigation menu and pop-ups that reinforce key definitions.

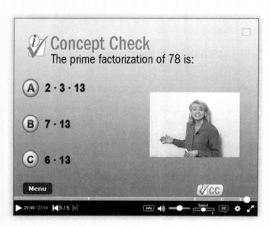

Within the section lecture videos, **Interactive Concept Checks** measure a student's understanding of key concepts and common trouble spots. Concept Checks ask students to try a question on their own within the video, after which Elayn Martin-Gay explains why they were correct or incorrect.

All videos can be assigned as a **media assignment** in the Assignment Manager, to ensure that students are getting the most out of their MyLab resources. Additionally, **Video Check questions** ensure that students have viewed and understood the key concepts from the section lecture videos.

Additional hallmark Martin-Gay video types include Student Success Tip videos and Chapter Test Prep videos. **Student Success Tip videos** are in short segments designed to be daily reminders to stay organized and to study. **Chapter Test Prep videos,** a Martin-Gay innovation, help students during their most teachable moment–when they are preparing for a test–with step-by-step solutions for the exercises in the Chapter Test.

New Tools Improve Preparedness and Personalize Learning

New! Getting Ready for the Test video solutions cover every Getting Ready for the Test exercise. These come at the end of each chapter to give students an opportunity to assess if they understand the big picture concepts of the chapter, and help them focus on avoiding common errors.

New! Skill Builder exercises offer just-in-time additional adaptive practice. The adaptive engine tracks student performance and delivers questions to each individual that adapt to his or her level of understanding. This new feature allows instructors to assign fewer questions for homework, allowing students to complete as many or as few questions needed.

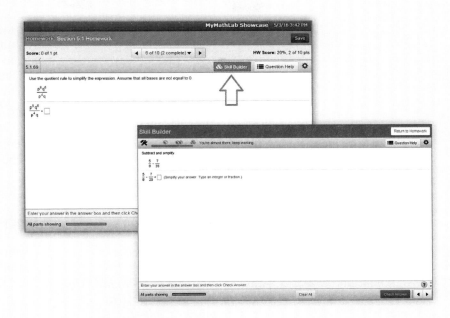

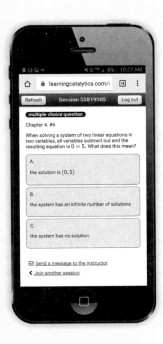

New Ways to Engage Students

New! Learning Catalytics
Martin-Gay-specific questions are pre-built and available through MyLab Math. Learning Catalytics is an interactive student response tool that uses students' smartphones, tablets, or laptops to engage them in more sophisticated tasks and thinking. **Getting Ready for the Test** exercises marked in blue in the text are pre-built in Learning Catalytics to use in class. These questions can be found in Learning Catalytics by searching for "MGP&I".

New! Vocab and Readiness questions in MyLab Math have been expanded to 100% coverage, and are now available with a new **Drag and Drop functionality!** Drag and Drop exercises allow students to manually select elements of the question, such as expressions, words, graphs, or images, and place them into a designated target area.

Easier Start-Up for Instructors

Enhanced Sample Assignments make course set-up easier by giving instructors a starting point for each section. Each assignment has been carefully curated for this specific text, and includes a thoughtful mix of question types.

pearson.com/mylab/math

Acknowledgments

There are many people who helped me develop this text, and I will attempt to thank some of them here. Courtney Slade and Cindy Trimble were *invaluable* for contributing to the overall accuracy of the text. Gina Linko and Patty Bergin provided guidance throughout the production process and Suellen Robinson provided many suggestions for updating applications during the writing of this Fifth Edition.

A very special thank you goes to my editor, Mary Beckwith, for being there 24/7/365, as my students say. And, my thanks to the staff at Pearson for all their support: Barbara Atkinson, Alicia Frankel, Michael Hirsch, Chris Hoag, Paul Corey, Michelle Renda, Jenny Crum and Lauren Schur among many others.

I would like to thank the following reviewers for their input and suggestions that have affected this and previous editions:

Lisa Angelo, *Bucks Community College*
Victoria Baker, *Nicholls State College*
Teri Barnes, *McLennan Community College*
Laurel Berry, *Bryant & Stratton*
Thomas Blackburn, *Northeastern Illinois University*
Gail Burkett, *Palm Beach Community College*
Anita Collins, *Mesa Community College*
Lois Colpo, *Harrisburg Area Community College*
Fay Dang, *Joliet Junior College*
Robert Diaz, *Fullerton College*
Tamie Dickson, *Reading Area Community College*
Latonya Ellis, *Gulf Coast Community College*
Sonia Ford, *Midland College*
Cheryl Gibby, *Cypress College*
Kathryn Gunderson, *Three Rivers Community College*
Elizabeth Hamman, *Cypress College*
Craig Hardesty, *Hillsborough Community College*
Lloyd Harris, *Gulf Coast Community College*
Teresa Hasenauer, *Indian River College*
Julia Hassett, *Oakton Community College*
Jeff Koleno, *Lorain County Community College*
Judy Langer, *Westchester Community College*
Sandy Lofstock, *St. Petersburg College*
Stan Mattoon, *Merced College*
Dr. Kris Mudunuri, *Long Beach City College*
Carol Murphy, *San Diego Miramar College*
Greg Nguyen, *Fullerton College*
Jean Olsen, *Pikes Peak Community College*
Darlene Ornelas, *Fullerton College*
Warren Powell, *Tyler Junior College*
Jeanette Shea, *Central Texas College*
Katerina Vishnyakova, *Collin County Community College*
Corey Wadlington, *West Kentucky Community and Technical College*
Edward Wagner, *Central Texas College*
Jenny Wilson, *Tyler Junior College*

I would also like to thank the following dedicated group of instructors who participated in our focus groups, Martin-Gay Summits, and our design review for the series. Their feedback and insights have helped to strengthen this edition of the text. These instructors include:

Billie Anderson, *Tyler Junior College*
Cedric Atkins, *Mott Community College*
Lois Beardon, *Schoolcraft College*
Laurel Berry, *Bryant & Stratton*
John Beyers, *University of Maryland*
Bob Brown, *Community College of Baltimore County–Essex*
Lisa Brown, *Community College of Baltimore County–Essex*
NeKeith Brown, *Richland College*
Gail Burkett, *Palm Beach Community College*
Cheryl Cantwell, *Seminole Community College*
Jackie Cohen, *Augusta State College*
Julie Dewan, *Mohawk Valley Community College*
Janice Ervin, *Central Piedmont Community College*
Richard Fielding, *Southwestern College*
Cindy Gaddis, *Tyler Junior College*
Nita Graham, *St. Louis Community College*
Pauline Hall, *Iowa State College*
Pat Hussey, *Triton College*

Dorothy Johnson, *Lorain County Community College*
Sonya Johnson, *Central Piedmont Community College*
Irene Jones, *Fullerton College*
Paul Jones, *University of Cincinnati*
Kathy Kopelousous, *Lewis and Clark Community College*
Nancy Lange, *Inver Hills Community College*
Judy Langer, *Westchester Community College*
Lisa Lindloff, *McLinnan Community College*
Sandy Lofstock, *St. Petersburg College*
Kathy Lovelle, *Westchester Community College*
Jean McArthur, *Joliet Junior College*
Kevin McCandless, *Evergreen Valley College*
Daniel Miller, *Niagra County Community College*
Marica Molle, *Metropolitan Community College*
Carol Murphy, *San Diego Miramar College*
Greg Nguyen, *Fullerton College*
Eric Oilila, *Jackson Community College*
Linda Padilla, *Joliet Junior College*
Davidson Pierre, *State College of Florida*
Marilyn Platt, *Gaston College*
Ena Salter, *Manatee Community College*
Carole Shapero, *Oakton Community College*
Janet Sibol, *Hillsborough Community College*
Anne Smallen, *Mohawk Valley Community College*
Barbara Stoner, *Reading Area Community College*
Jennifer Strehler, *Oakton Community College*
Ellen Stutes, *Louisiana State University Elinice*
Tanomo Taguchi, *Fullerton College*
MaryAnn Tuerk, *Elsin Community College*
Walter Wang, *Baruch College*
Leigh Ann Wheeler, *Greenville Technical Community College*
Valerie Wright, *Central Piedmont Community College*

A special thank you to those students who participated in our design review: Katherine Browne, Mike Bulfin, Nancy Canipe, Ashley Carpenter, Jeff Chojnachi, Roxanne Davis, Mike Dieter, Amy Dombrowski, Kay Herring, Todd Jaycox, Kaleena Levan, Matt Montgomery, Tony Plese, Abigail Polkinghorn, Harley Price, Eli Robinson, Avery Rosen, Robyn Schott, Cynthia Thomas, and Sherry Ward.

Elayn Martin-Gay

Personal Acknowledgements

I would like to personally thank my extended family. Although this list has grown throughout the years, it still warrants mentioning in my texts as each of these family members has contributed to my work in one way or another – from suggesting application exercises with data and updating/upgrading my computer to understanding that I usually work on "Vacations." I am deeply grateful to them all:

Clayton, Bryan (in heaven), Eric, Celeste, and Tové Gay; Leo and Barbara Miller; Mark and Madison Martin and Carrie Howard; Stuart and Earline Martin; Karen Martin Callac Pasch (in heaven); Michael, Christopher, Matthew, Nicole, and Jessica Callac; Dan Kirk; Keith, Mandy, Erin, and Clayton McQueen, Bailey Martin, Ethan, Avery, and Mia Barnes; Melissa and Belle Landrum.

About the Author

Elayn Martin-Gay has taught mathematics at the University of New Orleans for more than 25 years. Her numerous teaching awards include the local University Alumni Association's Award for Excellence in Teaching, and Outstanding Developmental Educator at University of New Orleans, presented by the Louisiana Association of Developmental Educators.

Prior to writing textbooks, Elayn Martin-Gay developed an acclaimed series of lecture videos to support developmental mathematics students in their quest for success. These highly successful videos originally served as the foundation material for her texts. Today, the videos are specific to each book in the Martin-Gay series.

The author has also created Chapter Test Prep Videos to help students during their most "teachable moment"—as they prepare for a test—along with Instructor-to-Instructor videos that provide teaching tips, hints, and suggestions for each developmental mathematics course, including basic mathematics, prealgebra, beginning algebra, and intermediate algebra.

Elayn is the author of 12 published textbooks as well as multimedia, interactive mathematics, all specializing in developmental mathematics courses. She has also published series in Algebra 1, Algebra 2, and Geometry. She has participated as an author across the broadest range of educational materials: textbooks, videos, tutorial software, and courseware. This provides an opportunity of various combinations for an integrated teaching and learning package offering great consistency for the student.

Applications Index

Advertising and marketing
 area of advertisement sign at Fenway Park, 613
 area of Coca-Cola sign, 612, 709
 Internet advertising, 435, 1109
 largest indoor illuminated sign, 596
 money spent on TV advertising, 65

Agriculture
 acres of wheat harvested in selected states, 502, 508
 annual orange production, 827
 apple production in selected states and in U.S., 695, 802
 average price per pound for chickens, 373
 bushels of oranges picked, 553
 corn production, 37, 667
 cropland prices per acre, 1017
 diameter of General Sherman sequoia tree, 638
 farm size increase, 465
 fencing needed for enclosure, 173
 fertilizer amounts needed, 711, 1121
 number of farms, 978, 1132
 number of trees in each row in orchard, 1079
 oxygen supply provided by lawns, 424
 pineapple sauce sales and profits, 1028
 rotten apples in shipment, 1133
 soil needed to fill hole, 654
 total value of potential crop, 489
 types of apples grown in Washington State, 514
 weights of heaviest zucchini ever grown, 622

Animals
 amount of grain eaten by cow in a year, 50
 amount of water in aquarium, 625–626
 area of wood needed for triangular water trough, 1157
 condor population changes, 131
 cricket chirping rates, 701, 710–711, 713
 dimensions of animal pens, 712
 distance bees chase fleeing human, 277
 diving speeds of birds, 198, 908
 dosage of medicine for dogs, 1043
 endangered and threatened species, 24–25, 84, 494
 height of bamboo at National Zoo, 611
 legal lobster size, 295
 length of pelican bill, 602
 life expectancies, 199
 lobster classification by weights, 316
 migrating distance of Monarch butterfly, 638
 mosquito control pesticide mixture, 424
 mourning dove population changes, 131
 number of fish in tank, 534, 711
 number of households owning reptiles, 461
 pet-related expenditures, 976
 sheep population, 30
 speed of cheetah, 905
 speed of cockroach, 342
 speed of sloth in tree, 277
 termite nest height, 638
 time for flying fish to travel certain distance, 714
 time for hyena to overtake giraffe, 944

Astronomy and space
 alignment of Mercury and Jupiter, 917
 amount of gamma rays produced by Sun, 758
 apparent magnitudes of stars, 668
 commercial space launches, 504
 day of week a past date fell on, 951
 days for Neptune to orbit Sun, 37
 Deep Space Network distance and weight conversions, 638
 degrees apart of Deep Space Network sites, 572
 degrees around Earth at the equator, 572
 deviation of mirrors on Hubble Space Telescope, 436
 diameter of largest crater on Moon, 638
 diameter of Milky Way, 807
 distance from Earth to Moon, 356
 distance from Earth to Sun, 366
 distance from Pluto to Sun, 412
 distance light travels over time, 759
 eclipse durations, 305
 elevation of Thirty Meter Telescope, 758
 length of day on Mars, 342
 orbit day lengths of planets around Sun, 342
 planetary radius, 24
 planets with days longer than Earth's, 225
 radio wave distances, 366
 radius of Earth at equator, 364
 radius of Saturn, 162
 rope length needed to encircle Earth, 711
 satellites for each planet, 698
 surface area of planetarium, 599
 surface temperature of planets, 103, 122
 time for space plane to travel around Earth, 710
 viewing distance of Thirty Meter Telescope, 758
 volume of Hayden Planetarium dome, 599
 volume of Jupiter, 807
 wavelengths observed by Thirty Meter Telescope, 760
 weight of elephant on Earth and Jupiter, 424
 weight of human above surface of Earth, 1059–1060
 weight of person on Earth, 63
 weight of satellite on Mars, 424
 weights of meteorites, 699

Automotive
 automobile theft numbers in U.S., 432
 average fuel economy for automobiles, 1016
 car rental fees and charges, 1119
 car sales volume by size, 30
 cost of owning and operating SUVs, 1015
 crossover utility vehicle sales in U.S., 425
 distance traveled on gasoline amount, 324, 328
 Ford vehicle sales, 465
 gas/alcohol mixture, 421
 gas/oil mixture, 426, 631
 gasoline mileage, 374, 552, 1279
 hybrid sales by make of car, 515
 hybrid vehicles sold in U.S., 1028
 interior space of cars, 30
 length of longest truck in the world, 612
 median automobile age, 1015
 monthly cost of owning and operating autos, 348–349
 number of cars manufactured in U.S. and Germany, 200
 number of not-blue cars on lot, 321
 number of not-white cars on lot, 322
 passenger vehicle production in U.S., 436
 percentage of people driving blue cars, 427
 percentage of white trucks sold in U.S., 436
 perimeter of stop sign, 580
 price per liter of gasoline, 631
 speed limits, 637
 speed of car and truck, 200
 top speed of dragsters, 200
 vehicles of each category sold in U.S., 699
 volume of gasoline in gas tanks, 631

Aviation
 air cargo and mail tonnage estimates, 35
 cruising speed of Boeing 747, 638
 flight time remaining before landing, 273–274
 runway length, 712
 speed of airplane in still air, 1119
 speed of hang glider, 710
 speed of plane in still air, 941, 1119
 time for one jet plane to overtake another, 942
 time for space plane to travel around Earth, 710

Business
 amount of money received from 3-D movies, 251
 area of parking lot, 1143
 Barnes & Noble stores in operation, 1001
 billable hours, 463
 billboard measurements, 423, 712, 731
 brand value estimates, 37
 break-even point, 1123
 car rental fees and charges, 1119
 chair production per month, 461
 commission, 468, 471–472, 485, 486, 489
 commission rate, 469, 471
 consumer spending for each category, 328
 cost of each item, 97, 353, 409
 cost to operate business depending on deliveries made per day, 1152
 Costco Gold Star Memberships, 980
 credit card late fees, 354
 CVS store numbers by state, 30
 daily sales rate, 1029
 decrease in number of employees, 464
 defective products, 460, 463, 489, 1279
 dimensions of huge chocolate bar, 1213
 dimensions of stacks of ovens in warehouse, 612
 discount amount, 469–470, 472, 473, 483, 485, 486, 489, 1279
 employee work shift length, 239
 fax machine production, 899
 fraction of Gap Corporation brand stores, 228
 fraction of goods types sold online, 266–267
 fraction of Hallmark employees in Kansas City, 239
 fraction of states in U.S. with Ritz-Carlton hotels, 239
 fresh salmon imports and exports, 1093
 Gap Inc. stores owned, 30
 global Internet use, 89, 90, 466
 gross profit margins, 899
 hourly minimum wage, 1041
 hourly pay increase, 485
 hourly wage, 978
 hours charged for completing a job, 696
 hours needed to complete a job, 695
 hours worked on each job, 691–692
 hours worked per week, 547
 IKEA's annual revenue, 989
 inventory of art dealer's shop, 226
 labor estimates, 941
 losses, 105, 131
 magazine sales in U.S., 1027
 manufacturing costs, 978, 1062, 1073
 manufacturing rate according to costs, 1001
 manufacturing rate per minutes, 483
 measurements of Tesla Giga factory, 600
 money change from purchase prices, 352
 net income, 115
 net sales, 970
 number of cars manufactured in U.S. and Germany, 200
 number of restaurants in U.S., 1028
 online spending per month, 516
 original price, 1120
 pay before taxes, 364
 paycheck according to hours worked, 1062
 percent decrease, 459, 463–464
 percent increase, 463–464, 656
 percent of McDonald's restaurants in U.S., 461
 percent off sales, 432
 perimeter of large buildings, 23
 pet-related expenditures, 976
 pie preparation for restaurant, 943
 postage for large envelopes in U.S., 1041
 price difference estimates, 38, 97
 price increase, 463
 price of each item, 213

xxiii

Business (*continued*)
  price rounding, 342
  pricing, 1022, 1112–1113, 1117, 1120
  profits per year, 496
  purchase price, 471
  purchase price not covered by trade-in, 239
  quantity pricing, 976
  range of staff salaries, 500
  ratio of Apple Inc.'s assets to debts, 453
  reduced pricing, 1120
  restaurant sales annually, 990
  restaurant-industry employment, 1009
  restaurant-industry sales, 1214
  revenue from downloaded singles, 343
  revenue from music sales, 465, 1078
  revenue from product sales, 1206
  salary increase, 463
  sale of bike and accessories, 200
  sale of tractor and plow, 200
  sale price, 29, 252, 469, 470, 472, 489, 1279
  sales needed to make to earn certain amount, 720, 732
  sales tax, 467, 470, 471, 473, 485, 486, 657
  sales tax rate, 468, 470, 471
  selling price, 472
  state energy consumption, 208
  time for worker to complete job alone, 959
  time for workers together to complete job, 937–938, 940, 942, 943, 960
  tipping, 473
  total cost estimates, 38, 51, 91, 94, 97, 365
  total price, 467, 470, 471, 472, 485, 486, 490, 657
  total revenue, 1214
  toy expenditures per child, 67
  trade balances, 123
  traveler spending per day, 183
  units manufactured at certain cost, 874
  value of manufactured products, 973
  values of global brands, 199
  vinyl album sales, 768
  weekly sales amounts, 1022
  wholesale and markup cost, 201
  work rate, 937–938, 942

Chemistry and physics
  Avogadro's number, 758
  brine solution mixture, 425
  copper amount in alloy, 489
  decibel levels, 30
  diameter of a DNA strand, 760
  diameter of a red blood cell, 760
  distance spring stretches with weights attached, 1062, 1073
  excess amount of water in lab mixture, 631
  melting points of elements, 131
  mixture/solution amounts, 650, 1115–1116, 1117, 1120, 1121, 1122, 1128, 1132, 1248–1249, 1254
  period of pendulum, 1063, 1144
  rock salt and ice mixture, 425
  rope lengths for experiment, 64
  size of angstrom, 807
  two resistors in parallel, 951

Demographics
  age distribution in U.S., 557
  areas/monuments maintained by Park Service, 241, 253, 279
  audiologist jobs prediction, 1109
  automobile theft numbers in U.S., 432
  average heights of humans in various countries, 349
  California population estimate, 253
  consumer spending on DVD and Blu-ray formats, 1101
  correctional officers employed in U.S., 464
  deaths from boating accidents, 465
  distribution of blood types, 239, 324, 384
  fastest-growing occupations, 437, 952
  favorite sports, 433
  fraction of employees being men/women, 225
  fraction of Habitat for Humanity affiliates in U.S., 228

  fraction of states containing Indian reservations, 226
  frequency of retirement ages, 501
  growth in nurse employment, 456
  height conversions, 641–642
  householders in selected age ranges, 505
  households with television in U.S., 1014
  increase in elementary and secondary teachers in U.S., 464
  increase in registered nurses, 464, 985
  international tourist arrivals to U.S., 1063
  Internet use by students, 433
  jobs with projected highest numerical increase, 517
  largest population of Native Americans, 200
  magazines in print in U.S., 1071
  number of female runners in race, 211
  number of girls on school bus, 253
  number of households owning reptiles, 461
  number of occupational therapy assistants in U.S., 461
  number of people in human chain, 40
  number of restaurants in U.S., 1028
  numbers working in auto industry, 425
  numbers working in service industries, 424
  nursing school applications accepted, 456
  percent decrease in crimes in New York City, 489
  percent decrease in population, 459, 464, 481
  percent of employees with no retirement plan, 488
  percent of female veterinarians in U.S., 461
  percentage of executives in their forties, 428
  population decrease, 461
  population density, 342
  population differences, 91
  population in U.S. with AB blood type, 436
  population increase, 463, 489
  population of U.S., 37
  population per square mile in U.S., 1028
  population projections, 29, 39, 51
  populations of largest cities, 503
  restaurant-industry employment, 1009
  smoked tobacco product usage in U.S., 435
  spoken languages, 493
  students receiving MMR vaccine, 425
  tourist destinations for selected countries, 979
  tourist numbers projections, 200
  violent crime percent decrease, 485
  visitation numbers to national natural sites, 724, 768
  visitors to Acadia and Grand Teton National Parks, 874
  visitors to Redwoods National park, 827
  visitors to U.S. by region, 509, 1237
  visitors to U.S. from Mexico, 510
  visitors to Yosemite National Park, 1231
  weight conversions, 641–642
  world population by continent, 435

Education
  class registration, 516
  college costs spent on books, 460
  college expenses, 60, 91
  college freshmen majors, 241, 251
  college tuition and fees, 989
  e-books read on cell phones, 419
  enrollment decrease in college, 461
  enrollment increase estimates at community college, 39
  four or more years of college by persons 25 or older, 547
  fraction of class being sophomores, 225
  fraction of students freshmen/not freshmen, 225
  fraction of students making an A on test, 239
  fraction of students with math/science as favorite subjects, 278
  fraction of two-year colleges, 253
  freshmen enrolled in prealgebra, 485
  grade point averages, 394–395, 398, 408, 412

  Head Start enrollment increase estimates, 39
  heights of students, 557
  hours of sleep per night by students, 548
  hours spent studying, 976
  increase in number of scholarship applications, 458–459, 656
  Internet use by students, 433
  living arrangements of college students, 513
  math problems completed in 30 minutes, 491
  number of children in day care center, 460
  number of freshmen at a high school, 463, 1133
  number of high school graduates, 826
  number of pages read by college classes, 31
  number of students in a university, 456
  number of students in class, 63
  numbers of female and male MIT students, 699
  numbers of graduate and undergraduate students, 666
  numbers of students admitted and applied, 666
  percentage of graduates returning home to live, 428
  percentage of students as freshmen, 427
  scores needed on final exams, 724
  students receiving MMR vaccine, 425
  students taking the SAT, 989
  test scores, 38, 61, 496–497, 507, 518, 557–558
  textbook costs, 50
  times for completing mazes by students, 393
  tuition increase, 463
  typing words per minute at end of course, 464

Electronics and computers
  area of faces of smartphones, 364
  cost of disks and notebooks, 1117
  decrease in number of cable TV systems, 464
  diameter of hard drives, 423
  digital and analog cinema screens in U.S., 461, 1092
  digital downloads of singles, 464
  DVD pricing, 1117, 1118
  DVD production, 899
  e-books read on cell phones, 419
  global Internet use, 89, 90, 466
  height of stack of compact discs, 650
  households with high-definition televisions in U.S., 436
  households without landlines, 435
  Internet advertising, 201
  Internet use in U.S., 385
  measurement conversion for compact discs, 633, 638
  megabytes held by DVDs/CDs, 46, 50, 329
  number of characters in line of print, 50
  number of download cards given to friends, 60, 163
  number of pixels on a screen, 50
  perimeter of smartphone, 353
  perimeter of top of compact disc, 581
  price of each item, 196, 213
  prices of Xbox and games, 199
  ratios of disk diameters, 423
  revenue from downloaded singles, 343
  smartphone dimensions, 698
  smartphone user numbers, 37
  surface area and volume of the Space Cube computer, 598
  telephone switchboard connections made simultaneously, 874
  text messages sent/received per day, 353, 841
  thickness of MacBook, 342
  two resistors in parallel, 951
  users of Google search, 759
  virtual reality device increase, 464
  wireless data usage, 1200
  wireless subscriber connections, 759

Entertainment and recreation
  admission pricing, 1112–1113
  admission total cost, 46, 51
  amount of money received from 3-D movies, 251
  area of movie screen, 173
  artificial wall climbing, 463

Applications Index   **XXV**

attendance at play, 458
Blu-ray DVD sales changes, 377
card game scores, 115, 127
CD pricing, 1117, 1118
choosing a number, 559
choosing card from deck, 541
choosing colored marble, 537, 540, 551, 554
cinema admission price, 978
circumference of Ferris wheels, 364
coin tossing, 535, 536–537, 541, 559
die rolling, 535–536, 551
digital and analog cinema screens in U.S., 461, 1092
digital movie screens in/not in U.S. and Canada, 280
dimensions of triangular sail, 550, 711, 867–868, 881
DVD sales changes, 377
favorite music types, 556
Frisbee prices, 1022
Fun Noodle sales rate, 1029
group rates for wall climbing, 463
height of kite, 1178
indoor cinema sites in U.S., 1028
length from area of Ping-Pong table, 801
lottery win amounts per person, 63
miles hiked on trail, 252
money earned by top-rated movies, 343, 354
movie industry box-office profits, 975
movie industry box-office sales, 1078
music download price increase, 465
netting length around trampoline, 582
number of DVDs vs. Blu-ray discs, 208
number of movie screens in U.S., 131, 1009–1010
number of moviegoers in U.S. and Canada, 251
number of roller coasters in amusement parks, 248
number of seats in theater, 1128
perimeter of Monopoly board, 30
perimeter of puzzle, 30
roller coaster heights and depths, 156
sale price of cruise, 252
shooting balls into bucket, 725
spinner spinning, 538–540, 551, 559
swim against current across river, 1151
ticket costs, 163
times for completing mazes, 393
toy prices, 1022
triangular mainsail measurements, 422
vinyl album sales, 768
visitors to Pecos National Historic Park, 724

Finance
annual inflation rate in U.S., 556
compound interest, 475–476, 485, 489
credit card balance transfer charges, 375, 497
credit card late fees, 354
currency exchange rates, 365
foreign currency exchange, 909
fraction of states without efile, 240
interest rate, 873
money received from estate division, 698
money spent on world tourism, 1014
national debts of selected countries, 758
number of electronically filed income tax returns, 192
price of gold, 1206
ratio of assets to debts, 453
rental car daily budgeting amounts, 1053
simple interest, 474, 485, 486, 489
stock gains and losses over time, 729
stock market losses, 130
stock prices, 1118
stock share changes, 353
total amount after interest over time, 476, 485, 486
total amount of investment, 475
total amount of loan, 489
tourism budgets for selected states, 700, 729
types of stocks owned, 1118

Food and nutrition
actual weight of cocoa in boxes, 623
actual weight of ham in cartons, 623
actual weight of oatmeal in carton, 623
actual weight of pineapple in cartons, 623
amount of rice from combining two containers, 622
annual food sales in U.S., 504
area of top of a pizza, 597
areas of two pizza sizes, 598
average price per pound for chickens, 373
better buy on pizzas, 598
caffeine content, 495
calories from fat, 251, 462
calories in food items, 49, 51, 97, 424, 464
candy mixture and price, 1122
chocolate consumption of selected countries, 355
chocolate mixture and costs, 1254
coffee bean blending and costs, 1120
color distribution of M&Ms, 514
cost of each food item, 1128
fiber content of selected foods, 504
fluid ounces of Jell-O in each dish, 631
food preferences, 484
fresh salmon imports and exports, 1093
fruit punch and ginger ale mixture, 631
grams of fat in food items, 49, 51, 324
iced tea remaining at party, 650
lobster classification by weights, 316
number of hamburgers from total pounds, 303
number of hot dogs and buns to purchase, 917
nut mixture and costs, 1120, 1254
pie preparation for restaurant, 943
pizza sizes and cost, 711
preservatives in box of dried fruit, 623
recipe ingredient amounts, 425, 650, 656
remaining amount of Mountain Dew in bottle, 631
saturated fat in food items, 364
tea bag production daily at plant, 51
total weight of candies, 649
types of milk consumed in U.S., 518
volume of a waffle ice cream cone, 597
volume of Coca-Cola shared with each person, 631
volume of soup in containers, 631
weight of 6 bottles of root beer, 622
weight of 24 cans of 7-Up, 622
weight of batch of cookies, 616
weight of food on pallet, 50
weight of meats in packages, 623
weight of one serving of sunflower seeds, 622
yogurt production in U.S., 1031

Geography and geology
angle of Khafre's Pyramid, Egypt, 573
angle of Monk's Mound, Illinois, 573
angle of pyramid at Chichen Itza, Mexico, 573
area of state of Colorado, 1079
area of state of Utah, 598
areas of selected deserts, 699
circumference of Meteor Crater, 582
counties for selected states, 698
deep-sea diving depths, 666
depth of cave, 329
depth of ocean trench, 104, 115, 122, 611
depths of Grand Canyon and Yellowstone, 611
distance glacier moves in one year, 714
diving depths, 105, 114, 130, 156, 161
elevation differences between points, 122, 158, 161
elevation of deepest point in lake, 161
elevations above/below sea level, 38, 104, 105, 106, 119, 122, 156
flow rate of water discharge into pond, 909
fraction of Earth's water taken up by oceans, 278
fraction of mountain peaks in India, 236
fraction of mountains in Colorado, 384
fraction of states adjacent to other states, 280
geothermal sources in Iceland, 435
Greenland and Antarctic Ice sheets, 435
high and low elevations, 666
highest U.S. dams, 517
highest/lowest elevations, 119
lake elevation differences, 122
land area drained by river basins, 29
land areas of continents, 513
large dams by continent, 515
lengths of logs cut from trees, 612
mean/median of longest rivers, 399
mountain elevations, 64
percentage of nuclear-generated electricity in France, 436
sizes of oceans, 515–516
sunrise times for Indianapolis, 1035–1036
sunset times for Seward, Alaska, 1040
surface area of Arctic and Pacific Oceans, 759
surface land area of each continent, 265–266
surface temperatures of planets, 122
thickness of ice on pond, 612
thickness of sediment in creek, 612
time for glacier to reach lake, 702–703, 737
time for lava flow to reach sea, 712
tornadoes in U.S. for selected years, 970
total electricity generated by solar and wind power, 777
total electricity generated by wind power only in U.S., 1231
volcano heights, 727–728
volume of Mount Fuji, 597
volume of water flowing over Niagara Falls, 759
widths of Grand Canyon and Yellowstone, 611
wildfires in U.S., 502–503

Geometry
angle measures, 567–568, 570–571, 600, 646, 650–651, 652, 653, 665, 692, 696, 697, 699, 700, 1126, 1279
area and circumference of circle, 1280
area and perimeter of geometric figures, 1279, 1280
area and perimeter of rectangle, 924
area of circle, 588, 651, 748, 827
area of geometric figures, 67, 206, 304, 407, 586–587, 593–594, 599, 647, 651, 784
area of parallelogram, 586, 748, 801
area of rectangle, 49, 50, 87, 92, 97, 170, 174, 184, 210, 252, 305, 315, 322, 325, 328, 386, 409, 596, 598, 748, 769, 783, 784, 791, 809, 827, 909, 1164
area of rectangular solids, 599
area of square, 74, 93, 97, 322, 613, 748, 769, 783, 784, 791, 809, 881, 909
area of trapezoid, 586
area of triangle, 383, 386, 412, 585, 758, 783, 1237
base height of triangle, 1235
base length of triangle, 873, 881
circumference of circle, 365, 406, 412, 578, 581–582, 583, 647
circumference of geometric shapes, 582, 600
circumference of telescope, 385
complementary angles, 566, 570, 573, 646, 925, 933, 1120, 1279
consecutive integers, 696, 697
consecutive numbers, 872
diameter of circle, 252, 278, 600, 653
dimensions of geometric shapes, 869, 872
dimensions of rectangle, 872, 875, 882, 1117, 1122, 1213
golden rectangles, 700, 701
height and width of signs, 709
length and width of rectangle from area, 827, 867
length of a side from the area, 1143
length of a side from the perimeter, 801, 810
length of diagonals, 522, 525, 534, 1157, 1178
length of long side of rectangle, 1184, 1186
length of side of cube from its surface area, 1151, 1152
length of side of cube from its volume, 1143, 1144
length of side of square-based pyramid, 1173

Applications Index

Geometry (continued)
- length of sides of geometric shapes, 315, 324, 375, 712, 881
- lengths of composite figures, 355
- perimeter of geometric figures, 28, 40, 49, 67, 75, 86, 90, 141, 173, 206, 210, 264, 277, 304, 323, 413, 478, 576–577, 579, 580, 581, 582, 584, 600, 651, 684, 685, 769, 776, 834, 842, 846, 881, 1156
- perimeter of octagon, 580
- perimeter of parallelogram, 579, 646, 1077, 1237
- perimeter of pentagon, 580
- perimeter of polygon, 22, 86, 576, 580
- perimeter of rectangle, 30, 40, 49, 97, 257, 315, 323, 328, 406, 574, 579, 600, 653, 834, 1156
- perimeter of square, 30, 74, 97, 257, 305, 353, 575, 580, 581, 917, 1043
- perimeter of trapezoid, 917
- perimeter of triangle, 40, 170, 184, 323, 353, 406, 575, 580, 646, 1043, 1278
- Pythagorean theorem, 521–522, 1174–1175, 1177, 1184
- radius of circle, 24, 162, 252, 583, 600, 653, 1190, 1200
- radius of cylinder, 1173
- radius of round ball from volume, 1164
- radius of sphere, 1187
- radius of sphere from volume, 1173
- ratio of rectangle width to length, 418, 453, 656
- ratio of rectangle width/length to perimeter, 418, 656
- ratio of square area to perimeter, 656
- ratio of triangle base to side, 418
- ratio of triangle side to perimeter, 418
- rectangle dimensions, 701, 705
- side lengths in congruent triangles, 527
- side lengths in similar triangles, 528–529, 532, 534
- side lengths of cube, 1151
- side lengths of geometric shapes, 711, 723, 1199
- side lengths of original square, 873
- side lengths of quadrilateral, 881
- side lengths of square, 874, 881, 1199
- side lengths of triangle, 867–868, 881, 882, 885, 1117, 1178, 1180, 1186, 1187
- sides of polygon, 1236
- supplementary angles, 567, 570–571, 573, 646, 925, 933, 1120, 1279
- surface area of box, 651
- surface area of cone, 595
- surface area of cube, 595, 750
- surface area of rectangular box, 808
- surface area of sphere, 1187
- total width/length of geometric object, 278
- unknown lengths in triangles, 524–525, 550, 552, 558, 561
- volume and surface area of box, 590, 596
- volume and surface area of geometric shapes, 592, 594–595, 648
- volume and surface area of sphere, 590–591, 597
- volume of box, 654
- volume of cone, 592, 597, 651
- volume of cube, 600, 648, 748, 750, 758, 783
- volume of cylinder, 591, 648, 654, 748
- volume of Hoberman Sphere, 713
- volume of rectangular box, 600
- volume of shipping crate, 712
- volume of sphere, 597, 598, 600, 648
- volume of squared-based pyramid, 592, 595, 596, 598, 600, 648, 1178
- width from area of rectangle, 816
- width of a side from the area, 810

Health and medicine
- administering medicine liquid dosages, 426
- amount of medication in decongestant tablet, 622
- aspirin usage, 265, 435
- blinking rate of human eye, 701
- blood pressure drug testing, 540
- body surface area of human, 1152
- body surface calculations, 641–642
- body-mass index, 899
- cephalic index, 899
- cholesterol in food items, 425
- components of bone, 436
- crutch lengths, 304
- disease cases estimates, 35
- distribution of blood types, 239, 324, 384
- dosage of medicine for a child, 899, 924
- dosage of medicine for dogs, 1043
- dosage of medicine per body weight, 421, 656
- doses of medicine in bottle, 374
- fraction of persons getting fewer than 8 hours sleep, 280
- grade of wheelchair ramp, 1014
- heart transplants in U.S., 1017
- human body temperature conversion, 636
- lung transplants in U.S., 1071
- lung weights in human, 638
- median/mean/mode pulse rates, 398
- medication received in 3.5 hours by patient, 628
- medication regime by ounces, 639
- most common blood types, 556
- number of days medicine will last, 374
- number of teaspoons in medicine bottle, 374
- ounces in tablets and caplets, 637
- population in U.S. with AB blood type, 436
- ratio of red blood cells to platelet cells, 423
- ratio of white blood cells to red blood cells, 423
- syringe measurements, 632
- weight of a child over time, 616
- weight of a person after being sick, 622
- weight of skin of average adult, 638
- weights of heaviest and lightest babies born, 622
- weights of human liver and heart, 633

Home construction and improvement
- amount of paint needed for wall, 370–371, 373
- area of computer chip, 304
- area of concrete block wall, 598
- area of glass in picture frame, 709
- area of lawn, 409
- area of lawn in square feet, 412
- area of roof, 1142
- area of table top, 813
- area of window, 599
- blueprint measurements, 420, 424
- border material needed for garden, 354
- bricks needed for house, 597
- carpet needed to cover floor, 647
- circumference of spa, 578
- circumference of watered area of lawn, 578
- deck area, 304
- deck length, 703, 885
- dimensions of swimming pool, 875
- driveway sealer needed, 647
- fencing cost and length needed, 577, 816, 1122
- fencing cost for rectangular garden, 580
- fencing materials needed, 29, 323, 375
- gallons of water seal needed for deck, 735
- garden dimensions, 710, 867, 875
- grass seed for yard, 596
- gutters around house, 29, 75
- insecticide amounts needed, 412
- installation of baseboard and carpet, 709
- insulation needed for attic, 598
- labor estimates, 941
- ladder length, 873, 1184
- length of connecting pipe, 1177
- length of swimming pool from volume, 801
- measuring between points on board, 951
- metal strip around workbench, 580
- painting of room border, 709
- perimeter of a ceramic tile, 581
- perimeter of a picture frame, 173, 709
- perimeter of a room, 173, 576–577
- perimeter of garden, 173
- perimeter of plot of land, 1156
- pitch of a roof, 1013, 1016
- railing amount needed for deck, 354
- scale drawings of plans, 479, 534
- sewer pipe slope, 1014
- shingles needed for roof, 597
- square feet of land plot, 596
- triangular flower bed, 710
- volume of swimming pool water, 750
- wallpaper cost, 577

Miscellaneous
- allowable weight of each passenger on elevator, 619
- amount collected by charity drive, 485
- amount of coal delivered by weight, 622
- amount of oil in drum, 625, 654
- angle measures of flag designs, 696
- angle measures of walls of Vietnam Veterans Memorial, 573, 692
- area codes of selected states and countries, 693, 697, 699
- area of a flag, 596
- area of an illuminated sign, 596
- area of circular watch face, 597
- area of Coca-Cola sign, 612, 709
- area of curtain wall of building, 908
- area of office and storage space in Pentagon, 908
- area of Uniroyal Giant Tyre sculpture, 599
- armed forces personnel per branch, 511
- average of several numbers, 92
- best-selling albums in U.S., 504
- boards for bookcase, 698
- bottled water consumption, 488, 826, 1101
- circumference of a can, 578
- circumference of circular barn, 582
- coin combinations, 356
- coin types in collection, 1117, 1129, 1255
- consecutive page numbers in book, 697, 872
- consecutive room numbers, 872
- depth of screw in wood after turning, 252
- dimensions of flag, 881
- dividend from spending at food cooperative, 460
- elevator shaft heights and depths, 155, 156
- fabric needed for sashes, 648
- fencing materials needed, 375
- formats of commercial radio stations, 386
- fraction of legal fireworks in states, 226
- fraction of marbles in certain colors, 226
- fraction of national monuments in New Mexico, 239
- fraction of national parks by state, 228
- fraction of wall as concrete, 239
- fundraiser food pricing, 1120
- gambling area size, 905
- garden dimensions, 700, 703–704, 709
- guy wire length, 873
- height of child since last birthday, 608
- heights in inches from meters, 365
- inner diameter of tubing/pipe, 278, 281
- inner diameter of washer, 278
- ladder length, 873, 1184
- length of diagonal of city block/park, 525
- length of diagonal path through park, 522
- length of piece of rope, 604–605
- length of piece remaining after cutting off shorter pieces, 325, 328
- lengths of two scarves, 324
- mail categories delivered by Postal Service, 279
- Marine Corps training centers in California, 228
- material remaining on bolt, 648
- median of list of numbers, 657
- newspaper circulation, 199, 1001
- No Smoking sign dimensions, 649
- number of apartments in building, 50
- number of apartments on one floor, 50
- number of books sold per hour, 374
- number of boxes filled, 92, 94
- number of boxes of crayons, 373
- number of boxes of envelopes office needs, 453
- number of boxes on pallet, 50
- number of download cards given to friends, 60, 163
- number of dresses from bolt of material, 296

number of items needed to fill box, 373
number of libraries in Mississippi, 252
number of meters in feet, 406
number of meters in inches, 373
number of pages in book, 29
number of seats in lecture hall, 50
number of shoe polish bottles in boxes, 650
number of windows in building, 52
number of words on pages of book, 47
numbers of each denomination of bills, 1250, 1254, 1255
numbers of each national park unit in U.S., 730
numbers owning microwave ovens, 485
percent of shoppers paying cash, 656
percent of soft drinks sold, 486
percentage of nuclear-generated electricity in France, 436
percentage of world's mail volume handled by U.S., 436
perimeter of photo, 654
perimeter of piece of land, 323
postage costs, 377
pottery styles sold at original and reduced prices, 1120
public libraries in selected states, 735, 1282
repeat contestants on television show, 252
scarf length, 608, 654
self-tanning lotion mixture and costs, 1254
square feet of material for drapery, 596
stamp purchases, 1118
stamp types in collection, 1129
strips of metal from stock piece, 296
tall buildings completed by continent, 549
tree girth comparisons, 295
types of books at a library, 514
types of coins in jar, 1118, 1129
unknown numbers, 213, 439–440, 447–449, 488, 561, 689, 723, 731, 735, 737, 810, 875, 936, 941, 959, 960, 1111–1112, 1118, 1128, 1129, 1237
volume of a birdbath, 596
volume of a building, 905
volume of a paperweight, 596
volume of a snow globe, 597
volume of all drawers in chest, 648
volume of microwave oven, 1164
volume of water pumped over time, 628
volume of water storage tank, 350, 595
waste disposal budget increase, 432
waste disposed daily per person, 507
waste dumping charge, 485
wedding budget, 720
weekly fundraising sales, 555
weight conversions on Australian postage stamps, 634, 638
weight of one bag of cement, 619

Personal finance
budget items for family, 549
charge account balances, 121
check writing, 333–334
checking account balances, 156, 161, 400, 506
credit card balance transfer charges, 375, 497
debt repayment amounts, 157, 158, 406
dividend from spending at food cooperative, 460
earnings shared by three persons, 60
monthly income on rent, 432
purchase price not covered by trade-in, 239
retirement party budget, 723
savings account, 1117
savings account balance, 29, 90, 162, 666
total earnings during three years, 90
total pay after withholdings, 462
wedding anniversary budget, 723
wedding budget, 1053

Politics and government
ballots cast in presidential election, 807
electoral votes for president, 198
electoral votes for selected states, 495
floor space of Pentagon, 697

fraction of presidents born in Ohio, 225
governors who are Democrats and Republicans, 698
Marine Corps training centers in California, 228
number of representatives of each political party, 690–691, 737
number of rulers in each country, 198
votes for candidates in election, 208

Real estate
commission, 468, 472
home sales commissions, 201, 252
loss in value of home, 486
lots in certain number of acres, 281
new home construction by region in U.S., 546
perimeter of triangular lot, 684
price of home from down payment figures, 462
value of a building, 1029

Sports
admission costs to baseball games, 51
area of basketball court, 170, 173
area of skating rink, 599
attendance at MLB games, 1132
average heights of basketball players, 723
average speeds of Daytona winners, 354
balance beam width, 637
baseball average salaries, 37
baseball caps sold at U.S. Open Golf Tournament, 239
baseball slugging averages, 898
basketball court measurements, 422
basketball free throws made, 211, 321, 433, 484
basketball points scored, 37, 201
betting losses, 157
bowling averages, 723
college football game scores, 698
cycling speed, 941
distance between golf course holes, 63
distance from home plate to first base, 1142
distance run by baseball home runner, 580
favorite sports, 433
football average salaries, 37
football yards lost, 130, 157
fraction of sports team being boys, 225
goals per game in Beach Soccer World Cup, 506–507
golf scores, 114, 121, 127, 141, 156, 506, 732
golf wages earned by professionals, 374
height of men's gymnastics rings from floor, 637
highest dive from diving board, 1236
ice hockey penalty killing percentage, 924
jogging speed, 941
length and width of soccer fields for children according to ages, 583
length of diagonal of football/soccer fields, 522, 525, 534
length of soccer field, 200
lime powder needed to mark foul lines, 580
medals awarded during Summer Olympics, 547, 695, 699
number of female runners in race, 211, 1279
perimeter of baseball diamond, 580
perimeter of football field, 580
points scored during basketball season, 52, 1118
races won by driver, 278
ratio of adults preferring selected sports, 509
runs batted in by two players, 1118
size of Olympic medals, 423
ski run ratings, 461
soccer field length per player age, 164
stadium capacities, 200
time for IndyCar to travel between cities, 712
touchdowns made during season, 64
volume of a Zorb, 597
volume of shipping crate for IndyCar, 712
weight resistance per body weight at gym, 460
women's tennis prize money winners, 950

Temperature and weather
average precipitation in California, 548
average rainfall, 304, 353, 354

average snowfall, 354
average temperature in Omaha, 498–499
average temperatures, 64, 138
boiling temperature, 107
daily high temperatures, 518
drops in temperature, 130, 157, 158
fraction of tropical storms as hurricanes, 225
high and low temperatures, 105, 111–112, 115, 121, 133, 329, 508, 518, 548
hurricane wind speeds per hour, 375
hurricanes making landfall by month, 503
monthly high temperatures, 548
monthly precipitation in Chicago, 555–556
surface temperature of planets, 103, 122
temperature at certain time of day, 122, 156, 158
temperature conversion, 82, 636, 639–640, 650, 704, 710, 712, 713
water freezing and boiling points, 665
wind speeds, 353, 375

Time and distance
average speed on level roads of trip, 1255
days for Neptune to orbit Sun, 37
Deep Space Network distance and weight conversions, 638
diagonal of field, 655
dimensions of city park, 534
distance between cities, 195–196, 199, 377, 406, 412, 639
distance between golf course holes, 63
distance between Romeo and Juliet, 1186
distance between two points, 1180
distance differences, 257
distance estimates, 35, 38, 39, 90, 91
distance from home after losing watch, 268
distance from home plate to first base, 1142
distance from starting point when plane overtakes another plane, 1255
distance glacier moves in one year, 714
distance light travels over time, 759
distance needed for car to stop, 1062
distance object falls, 82
distance of falling object over time, 1062
distance of skid, 1178
distance remaining to inspect tracks, 265
distance remaining to run, 265
distance shared by each driver, 649
distance traveled in 3 days, 329
diving speed of peregrine falcon, 908
driving time, 710, 712, 942, 1255
feet in one rod, 64
flight time remaining before landing, 273–274
fraction of centimeters in one meter, 239
fraction of days in month, 225
fraction of feet in one mile, 239
fraction of inches in one foot, 322
height of antenna, 525
height of average two-year-old child, 638
height of building, 529, 533, 549, 550
height of climbing wall, 463
height of Empire State Building, 425
height of falling cliff diver over time, 866, 887
height of falling stunt performer r over time, 855
height of fireworks at maximum, 1226
height of free-falling object over time, 763–764, 808, 813, 834, 855
height of launch gantry, 534
height of rocket fired upward over time, 768, 770, 865, 881
height of Space Needle deck, 533
height of Statue of Liberty, 424
height of students, 557
height of tallest fountain, 533
height of tower, 559
height of trees, 525, 529, 533, 654
height of Washington Monument, 730
height of woman given femur bone length, 1043
highest dive from diving board, 1236
highway mileage by state, 31, 37
hiking times/speeds, 1121, 1132, 1251–1252, 1255
how far you can see from top of building, 1179

Time and distance *(continued)*
  inches as fraction of a foot, 225
  land-speed record of solar-powered car, 908
  length differences of two scarves, 324
  length of concrete sound barriers along highway, 612
  length of connecting pipe, 1177
  length of diagonal of city block/park, 525, 655
  length of human index finger, 425
  length of longest truck in the world, 612
  length of pipe in feet, 304
  length of sides of Beijing Water Cube, 1199
  length of two ropes tied together, 611
  length of wire on flag pole, 1177
  lengths of pieces cut from total length, 604, 689–690, 695, 697, 730, 732, 776, 924
  map inches corresponding to miles, 420, 491
  map reading, 641
  maximum height of rocket fired upward, 865
  mileage of Appalachian Trail in selected states, 455
  miles hiked on trail, 252
  miles in certain number of yards, 92
  miles traveled on trip, 29
  miles walked on treadmill, 303
  proofreading copy, 942
  rate of travel over time, 1062, 1063
  remaining length of cut board, 304
  rope lengths, 64
  rowing distance, 1255
  rowing speed on river, 941
  runway length, 712
  speed of airplane in still air, 1119
  speed of boat in still water, 941, 942, 959, 1128, 1199, 1282
  speed of car before skidding, 1178
  speed of cheetah, 905
  speed of current, 1119, 1128
  speed of each train, 1252, 1254
  speed of falling object over time, 1028
  speed of fast sneeze, 638
  speed of plane in still air, 942
  speed of rowing a boat in still water, 1119
  speed of runner in track event, 905
  speed of two boats traveling apart at right angle, 874
  speed of vehicle 8 seconds after braking, 534
  speeds driven during trip, 942
  speeds of cars during test drive, 942
  speeds of cars over time, 942
  speeds of race cars, 944
  speeds of vehicle on different terrains, 941, 943
  speeds of vehicles traveling opposite directions, 943
  speeds of vehicles traveling same distance, 938–939, 943, 955–956, 959
  speeds of vehicles traveling toward each other, 1114, 1119, 1121
  time for conveyer belts to move a product, 940
  time for each of pumps to fill tank, 943
  time for falling object to reach ground, 855, 865, 885, 1178, 1197, 1199, 1200, 1214, 1226, 1230
  time for fireworks to reach maximum height, 1226
  time for flying fish to travel certain distance, 714
  time for glacier to reach lake, 702–703, 737
  time for inlet pipes to fill pond/pool, 939, 943, 959
  time for lava flow to reach sea, 712
  time for one vehicle to overtake another, 942, 1255
  time for rocket to reach heights, 1214
  time for trip through Channel Tunnel, 710
  time for vehicles to be certain distance apart, 1254
  time needed to bicycle/walk to burn calories, 723
  time of free fall, 1230
  time spent driving, 702
  time spent on bicycle during trip, 1119
  time to paint house, 941
  time to run/walk in charity race, 731
  time when falling object should hit ground, 834, 855, 872, 873, 881, 1028
  times for completing mazes, 393
  times spent jogging and walking during exercise, 1128
  total distance of trip from driving time, 1255
  travel times by same person on 2 vehicles, 942
  200-meter time in Olympic swimming, 342
  units of length conversion, 607–608
  velocity of falling object, 1176
  velocity of vehicle traveling on curved road, 1179
  viewing distance of Thirty Meter Telescope, 758
  walking rate in exercise program, 1114–1115
  walking rate of hikers, 942
  width of lake/pond by measuring distances across points, 1175
  wind speed, 941, 942, 1119
  yards in one mile, 64

Transportation
  amount hauled by truck each trip, 63
  bridge length, 67
  circumference of bridge caisson, 364
  deaths from boating accidents, 465
  distance remaining to inspect train tracks, 265
  distance shared by each driver, 649
  distance traveled on taxi charges, 696
  driving time, 702, 712
  freight truck weight, 273
  grade of a road, 1009, 1014
  grade of train tracks, 1014
  grade of wheelchair ramp, 1014
  height of bulge in railroad tracks, 1180
  high-speed train route, 710
  highway mileage by state, 31, 37
  highway speed limits, 265
  length of concrete sound barriers along highway, 612
  miles driven, 253, 374, 691, 695, 698
  miles traveled on trip, 29, 38, 91
  number of adults driving selected number of miles per week, 505
  number of bridges, 63, 1200
  number of lane dividers on highway, 63
  number of licensed drivers in U.S., 457–458
  number of light poles on highway, 64
  number of miles driven by adults per week, 505
  number of registered vehicles in U.S., 457
  number of roadway miles in selected states, 208
  parking lot dimensions, 710
  pricing of train tickets, 1118
  radius of curvature of road from velocity of vehicle, 1179
  railroad standard gauges in U.S. and Spain/Portugal, 303
  rate of travel over time, 1062, 1063
  sign dimensions, 1121
  span length of bridge, 654
  speed limits, 637
  weight of cinders spread on roads, 649

World records
  deepest bat colony, 100
  driest place in world, 354
  fastest train speeds, 697
  fastest tropical cockroach, 342
  heaviest and smallest babies born, 622
  heaviest zucchini ever grown, 622
  highest and lowest temperatures, 712, 731
  highest cliff diving in Mexico, 1199
  highest dive into lake, 1197
  highest U.S. dams, 517
  highest wind speed, 353
  largest American flag, 596
  largest building, 905
  largest casino, 905
  largest commercial building, 50
  largest hotel lobby, 50
  largest indoor illuminated sign, 596
  largest meteorite, 699
  largest optical telescope, 758
  largest round barn, 582
  largest suspension bridge, 364
  longest truck, 612
  Pearl of Lao-tze, 582
  record breaking weekend movie revenue, 14
  slowest mammal, 277
  smallest computer, 598
  snowiest city in U.S., 354
  steepest street, 1014
  tallest and shortest men, 611
  tallest building, 397, 855, 1179
  tallest fountain, 533
  tallest tree, 533

The Whole Numbers

1

A Selection of Resources for Success in This Mathematics Course

Whole numbers are the basic building blocks of mathematics. The whole numbers answer the question "How many?"

This chapter covers basic operations on whole numbers. Knowledge of these operations provides a good foundation on which to build further mathematical skills.

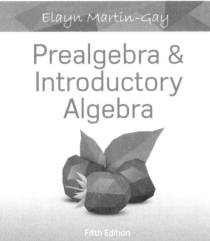

Textbook

Instructor

MyLab Math and MathXL

Video Organizer

Interactive Lecture Series

For more information about the resources illustrated above, read Section 1.1.

Sections

1.1 Study Skill Tips for Success in Mathematics

1.2 Place Value, Names for Numbers, and Reading Tables

1.3 Adding and Subtracting Whole Numbers, and Perimeter

1.4 Rounding and Estimating

1.5 Multiplying Whole Numbers and Area

1.6 Dividing Whole Numbers

Integrated Review— Operations on Whole Numbers

1.7 Exponents and Order of Operations

1.8 Introduction to Variables, Algebraic Expressions, and Equations

Check Your Progress

Vocabulary Check

Chapter Highlights

Chapter Review

Getting Ready for the Test

Chapter Test

1.1 Study Skill Tips for Success in Mathematics

Objectives

A Get Ready for This Course.

B Understand Some General Tips for Success.

C Know How to Use This Text.

D Know How to Use Text Resources.

E Get Help as Soon as You Need It.

F Learn How to Prepare for and Take an Exam.

G Develop Good Time Management.

Before reading this section, ask yourself a few questions.

1. Were you satisfied—really satisfied—with your performance in your last math course? In other words, do you feel that your outcome represented your best effort?
2. When you took your last math course, were your notes and materials from that course organized and easy to find, or were they disorganized and hard to find—if you saved them at all?

If the answer is "no" to these questions, then it is time to make a change. To begin, continue reading this section.

Objective A Let's Get Ready for This Course

1. *Start With a Positive Attitude.*

Now that you have decided to take this course, remember that a *positive attitude* will make all the difference in the world. Your belief that you can succeed is just as important as your commitment to this course. Make sure you are ready for this course by having the time and positive attitude that it takes to succeed.

2. *Understand How Your Course Material Is Presented—Lecture by Instructor, Online With Computer, or Both?*

Make sure that you are familiar with the way that this course is being taught. Is it a traditional course, in which you have a printed textbook and meet with an instructor? Is it taught totally online, and your textbook is electronic and you e-mail your instructor? Or is your course structured somewhere in between these two methods? (Not all of the tips that follow will apply to all forms of instruction.)

3. *Schedule Your Class So That It Does Not Interfere With Other Commitments.*

Make sure that you have scheduled your math course for a time that will give you the best chance for success. For example, if you are also working, you may want to check with your employer to make sure that your work hours will not conflict with your course schedule.

Objective B Here are a Few General Tips for Success

Below are some general tips that will increase your chance for success in a mathematics class. Many of these tips will also help you in other courses you may be taking.

1. *Most Important! Organize Your Class Materials. Unless Told Otherwise, Use a 3-Ring Binder Solely for Your Mathematics Class.*

In the next couple pages, many ideas will be presented to help you organize your class materials—notes, any handouts, completed homework, previous tests, etc. In general, you MUST have these materials organized. All of them will be valuable references throughout your course and when studying for upcoming tests and the final exam. One way to make sure you can locate these materials when you need them is to use a three-ring binder. This binder should be used solely for your mathematics class and should be brought to each and every class or lab. This way, any material can be immediately inserted in a section of this binder and will be there when you need it.

Helpful Hint

MyLab Math and MathXL When assignments are turned in online, keep a hard copy of your complete written work. You will need to refer to your written work to be able to ask questions and to study for tests later.

2. *Choose to attend all class periods.*

If possible, sit near the front of the classroom. This way, you will see and hear the presentation better. It may also be easier for you to participate in classroom activities.

3. *Complete Your Homework. This Means: Attempt All of It, Check All of It, Correct Any Mistakes, and Ask for Help if Needed.*

You've probably heard the phrase "practice makes perfect" in relation to music and sports. It also applies to mathematics. You will find that the more time you spend solving mathematics exercises, the easier the process becomes. Be sure to schedule enough time to complete your assignments before the due date assigned by your instructor.

Review the steps you took while working a problem. Learn to check your answers in the original exercises. You may also compare your answers with the "Answers to Selected Exercises" section in the back of the book. If you have made a mistake, try to figure out what went wrong. Then correct your mistake. If you can't find what went wrong, **don't** erase your work or throw it away. Show your work to your instructor, a tutor in a math lab, or a classmate. It is easier for someone to find where you had trouble if he or she looks at your original work.

It's all right to ask for help. In fact, it's a good idea to ask for help whenever there is something that you don't understand. Make sure you know when your instructor has office hours and how to find his or her office. Find out whether math tutoring services are available on your campus. Check on the hours, location, and requirements of the tutoring service.

> **Helpful Hint**
> **MyLab Math and MathXL**
> If you are doing your homework online, you can work and re-work those exercises that you struggle with until you master them. Try working through all the assigned exercises twice before the due date.

4. *Learn from your mistakes and be patient with yourself.*

Everyone, even your instructor, makes mistakes. (That definitely includes me—Elayn Martin-Gay.) Use your errors to learn and to become a better math student. The key is finding and understanding your errors.

Was your mistake a careless one, or did you make it because you can't read your own math writing? If so, try to work more slowly or write more neatly and make a conscious effort to carefully check your work.

Did you make a mistake because you don't understand a concept? Take the time to review the concept or ask questions to better understand it.

Did you skip too many steps? Skipping steps or trying to do too many steps mentally may lead to preventable mistakes.

> **Helpful Hint**
> **MyLab Math and MathXL**
> If you are completing your homework online, it's important to work each exercise on paper before submitting the answer. That way, you can check your work and follow your steps to find and correct any mistakes.

5. *Turn in assignments on time.*

This way, you can be sure that you will not lose points for being late. Show every step of a problem and be neat and organized. Also be sure that you understand which problems are assigned for homework. If allowed, you can always double-check the assignment with another student in your class.

> **Helpful Hint**
> **MyLab Math and MathXL**
> Be aware of assignments and due dates set by your instructor. Don't wait until the last minute to submit work online.

Objective C Knowing and Using Your Text or e-Text

Flip through the pages of this text or view the e-text pages on a computer screen. Start noticing examples, exercise sets, end-of-chapter material, and so on. Learn the way this text is organized by finding an example in your text of each type of resource listed below. Finding and using these resources throughout your course will increase your chance of success.

- *Practice Exercises.* Each example in every section has a parallel Practice exercise. Work each Practice exercise after you've finished the corresponding example. Answers are at the bottom of the page. This "learn-by-doing" approach will help you grasp ideas before you move on to other concepts.
- *Objectives.* Every section of this text is divided into objectives, such as **A** or **B**. They are listed at the beginning of the section and noted in that section. The main section of exercises in each exercise set is also referenced by an objective, such as **A** or **B**, and also an example(s). There is also often a section of exercises entitled "Mixed Practice," which is referenced by two or more objectives or sections. These are mixed exercises written to prepare you for your next exam. Use all of this referencing if you have trouble completing an assignment from the exercise set.

- *Icons (Symbols).* Make sure that you understand the meaning of the icons that are beside many exercises. ▶ tells you that the corresponding exercise may be viewed on the video Lecture Series that corresponds to that section. ✎ tells you that this exercise is a writing exercise in which you should answer in complete sentences. △ tells you that the exercise involves geometry.
- *Integrated Reviews.* Found in the middle of each chapter, these reviews offer you a chance to practice—in one place—the many concepts that you have learned separately over several sections.
- *End-of-Chapter Opportunities.* There are many opportunities at the end of each chapter to help you understand the concepts of the chapter.

    Vocabulary Checks contain key vocabulary terms introduced in the chapter.
    Chapter Highlights contain chapter summaries and examples.
    Chapter Reviews contain review problems. The first part is organized section by section and the second part contains a set of mixed exercises.
    Getting Ready for the Tests are multiple choice or matching exercises designed to check your knowledge of chapter concepts, before you attempt the chapter test. Video solutions are available for all these exercises.
    Chapter Tests are sample tests to help you prepare for an exam. The Chapter Test Prep Videos found in MyLab Math and YouTube provide the video solution to each question on each Chapter Test.
    Cumulative Reviews start at Chapter 2 and are reviews consisting of material from the beginning of the book to the end of that particular chapter.

- *Student Resources in Your Textbook.* You will find a **Student Resources** section at the back of this textbook. It contains the following to help you study and prepare for tests:

    Study Skill Builders contain study skills advice. To increase your chance for success in the course, read these study tips, and answer the questions.
    Bigger Picture—Study Guide Outline provides you with a study guide outline of the course, with examples.
    Practice Final provides you with a Practice Final Exam to help you prepare for a final.

- *Resources to Check Your Work.* The **Answers to Selected Exercises** section provides answers to all odd-numbered section exercises and to all integrated review, chapter review, getting ready for the test, chapter test, and cumulative review exercises. Use the **Solutions to Selected Exercises** to see the worked-out solution to every other odd-numbered exercise in the section exercises and chapter tests.

Objective D Knowing and Using Video and Notebook Organizer Resources ▶

Video Resources

Below is a list of video resources that are all made by me—the author of your text, Elayn Martin-Gay. By making these videos, I can be sure that the methods presented are consistent with those in the text. All video resources may be found in MyLab Math and some also on YouTube.

- *Interactive Video Lecture Series.* Exercises marked with a ▶ are fully worked out by the author. The lecture series provides approximately 20 minutes of instruction per section and is organized by Objective.
- *Getting Ready for the Test Videos.* These videos provide solutions to all of the Getting Ready for the Test exercises.

> **Helpful Hint**
>
> **MyLab Math**
> In MyLab Math, you have access to the following video resources:
> - Lecture Videos for each section
> - Getting Ready for the Test Videos
> - Chapter Test Prep Videos
> - Final Exam Videos
>
> Use these videos provided by the author to prepare for class, review, and study for tests.

- *Chapter Test Prep Videos.* These videos provide solutions to all of the Chapter Test exercises worked out by the author. They can be found in MyLab Math and YouTube. This supplement is very helpful before a test or exam.
- *Tips for Success in Mathematics.* These video segments are about 3 minutes long and are daily reminders to help you continue practicing and maintaining good organizational and study habits.
- *Final Exam Videos.* These video segments provide solutions to each question.

Video Organizer

This organizer is in three-ring notebook ready form. It is to be inserted in a three-ring binder and completed. This organizer is numbered according to the sections in your text to which it refers.

It is closely tied to the Interactive (Video) Lecture Series. Each section should be completed while watching the lecture video on the same section. Once completed, you will have a set of notes to accompany the (Video) Lecture Series section by section.

Objective E Getting Help

If you have trouble completing assignments or understanding the mathematics, get help as soon as you need it! This tip is presented as an objective on its own because it is so important. In mathematics, usually the material presented in one section builds on your understanding of the previous section. This means that if you don't understand the concepts covered during a class period, there is a good chance that you will not understand the concepts covered during the next class period. If this happens to you, get help as soon as you can.

Where can you get help? Try your instructor, a tutoring center, or a math lab, or you may want to form a study group with fellow classmates. If you do decide to see your instructor or go to a tutoring center, make sure that you have a neat notebook and are ready with your questions.

> **Helpful Hint**
> **MyLab Math and MathXL**
> - Use the **Help Me Solve This** button to get step-by-step help for the exercise you are working. You will need to work an additional exercise of the same type before you can get credit for having worked it correctly.
> - Use the **Video** button to view a video clip of the author working a similar exercise.

Objective F Preparing for and Taking an Exam

Make sure that you allow yourself plenty of time to prepare for a test. If you think that you are a little "math anxious," it may be that you are not preparing for a test in a way that will ensure success. The way that you prepare for a test in mathematics is important. To prepare for a test:

1. Review your previous homework assignments.
2. Review any notes from class and section-level quizzes you have taken. (If this is a final exam, also review chapter tests you have taken.)
3. Review concepts and definitions by reading the Chapter Highlights at the end of each chapter.
4. Practice working out exercises by completing the Chapter Review found at the end of each chapter. (If this is a final exam, go through a Cumulative Review. There is one found at the end of each chapter except Chapter 1. Choose the review found at the end of the latest chapter that you have covered in your course.) *Don't stop here!*
5. Take the Chapter Getting Ready for the Test. All answers to these exercises are available to you as well as video solutions.
6. Take a sample test with no notes, etc, available for help. It is important that you place yourself in conditions similar to test conditions to find out how you

> **Helpful Hint**
> **MyLab Math and MathXL** Review your written work for previous assignments. Then, go back and re-work previous assignments. Open a previous assignment, and click **Similar Exercise** to generate new exercises. Re-work the exercises until you fully understand them and can work them without help features.

will perform. There is a Chapter Test available at the end of each chapter, or you can work selected problems from the Chapter Review. Your instructor may also provide you with a review sheet. Then check your sample test. If your sample test is the Chapter Test in the text, don't forget that the video solutions are in MyLab Math and YouTube.

7. On the day of the test, allow yourself plenty of time to arrive at where you will be taking your exam.

When taking your test:

1. Read the directions on the test carefully.
2. Read each problem carefully as you take the test. Make sure that you answer the question asked.
3. Watch your time and pace yourself so that you can attempt each problem on your test.
4. If you have time, check your work and answers.
5. Do not turn your test in early. If you have extra time, spend it double-checking your work.

Objective G Managing Your Time

As a college student, you know the demands that classes, homework, work, and family place on your time. Some days you probably wonder how you'll ever get everything done. One key to managing your time is developing a schedule. Here are some hints for making a schedule:

1. Make a list of all of your weekly commitments for the term. Include classes, work, regular meetings, extracurricular activities, etc. You may also find it helpful to list such things as laundry, regular workouts, grocery shopping, etc.
2. Next, estimate the time needed for each item on the list. Also make a note of how often you will need to do each item. Don't forget to include time estimates for the reading, studying, and homework you do outside of your classes. You may want to ask your instructor for help estimating the time needed.
3. In the exercise set that follows, you are asked to block out a typical week on the schedule grid given. Start with items with fixed time slots like classes and work.
4. Next, include the items on your list with flexible time slots. Think carefully about how best to schedule items such as study time.
5. Don't fill up every time slot on the schedule. Remember that you need to allow time for eating, sleeping, and relaxing! You should also allow a little extra time in case some items take longer than planned.
6. If you find that your weekly schedule is too full for you to handle, you may need to make some changes in your workload, classload, or other areas of your life. You may want to talk to your advisor, manager or supervisor at work, or someone in your college's academic counseling center for help with such decisions.

1.1 Exercise Set MyLab Math

1. What is your instructor's name?

2. What are your instructor's office location and office hours?

3. What is the best way to contact your instructor?

4. Do you have the name and contact information of at least one other student in class?

5. Will your instructor allow you to use a calculator in this class?

6. Why is it important that you write step-by-step solutions to homework exercises and keep a hard copy of all work submitted?

7. Is there a tutoring service available on campus? If so, what are its hours? What services are available?

8. Have you attempted this course before? If so, write down ways that you might improve your chances of success during this attempt.

9. List some steps that you can take if you begin having trouble understanding the material or completing an assignment. If you are completing your homework in MyLab Math and MathXL, list the resources you can use for help.

10. How many hours of studying does your instructor advise for each hour of instruction?

11. What does the ✎ icon in this text mean?

12. What does the △ icon in this text mean?

13. What does the ▶ icon in this text mean?

14. Search the minor columns in your text. What are Practice exercises?

15. When might be the best time to work a Practice exercise?

16. Where are the answers to Practice exercises?

17. What answers are contained in this text and where are they?

18. What are Tips for Success in Mathematics and where are they located?

19. What and where are Integrated Reviews?

20. How many times is it suggested that you work through the homework exercises in MyLab Math or MathXL before the submission deadline?

21. How far in advance of the assigned due date is it suggested that homework be submitted online? Why?

22. Chapter Highlights are found at the end of each chapter. Find the Chapter 1 Highlights and explain how you might use it and how it might be helpful.

23. Chapter Reviews are found at the end of each chapter. Find the Chapter 1 Review and explain how you might use it and how it might be helpful.

24. Chapter Tests are found at the end of each chapter. Find the Chapter 1 Test and explain how you might use it and how it might be helpful when preparing for an exam on Chapter 1. Include how the Chapter Test Prep Videos may help. If you are working in MyLab Math and MathXL, how can you use previous homework assignments to study?

25. What is the Video Organizer? Explain the contents and how it might be used.

26. Read or reread objective G and fill out the schedule grid on the next page.

	Monday	Tuesday	Wednesday	Thursday	Friday	Saturday	Sunday
4:00 a.m.							
5:00 a.m.							
6:00 a.m.							
7:00 a.m.							
8:00 a.m.							
9:00 a.m.							
10:00 a.m.							
11:00 a.m.							
12:00 p.m.							
1:00 p.m.							
2:00 p.m.							
3:00 p.m.							
4:00 p.m.							
5:00 p.m.							
6:00 p.m.							
7:00 p.m.							
8:00 p.m.							
9:00 p.m.							
10:00 p.m.							
11:00 p.m.							
Midnight							
1:00 a.m.							
2:00 a.m.							
3:00 a.m.							

1.2 Place Value, Names for Numbers, and Reading Tables

Objectives

A Find the Place Value of a Digit in a Whole Number.

B Write a Whole Number in Words and in Standard Form.

C Write a Whole Number in Expanded Form.

D Read Tables.

The **digits** 0, 1, 2, 3, 4, 5, 6, 7, 8, and 9 can be used to write numbers. For example, the **whole numbers** are

0, 1, 2, 3, 4, 5, 6, 7, 8, 9, 10, 11, . . .

and the **natural numbers** are 1, 2, 3, 4, 5, 6, 7, 8, 9, 10, 11, . . .

The three dots (. . .) after the 11 mean that both lists continue indefinitely. That is, there is no largest whole number and there is no largest natural number. Also, the smallest whole number is 0 and the smallest natural number is 1.

Objective A Finding the Place Value of a Digit in a Whole Number

The position of each digit in a number determines its **place value.** For example, the average distance (in miles) between the planet Mercury and the planet Earth can be represented by the whole number 48,337,000. A place-value chart for this whole number is on the next page.

Section 1.2 | Place Value, Names for Numbers, and Reading Tables

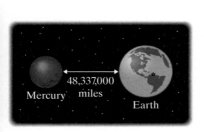

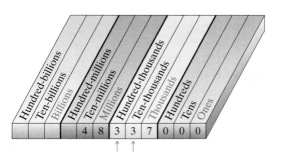

The two 3s in 48,337,000, shown above, represent different amounts because of their different placements. The place value of the 3 on the left is hundred-thousands. The place value of the 3 on the right is ten-thousands.

Examples Find the place value of the digit 3 in each whole number.

1. 396,418
↑
hundred-thousands

2. 93,192
↑
thousands

3. 534,275,866
↑
ten-millions

■ Work Practice 1–3

Practice 1–3

Find the place value of the digit 8 in each whole number.
1. 38,760,005
2. 67,890
3. 481,922

Objective B Writing a Whole Number in Words and in Standard Form ▶

A whole number such as 1,083,664,500 is written in **standard form.** Notice that commas separate the digits into groups of three, starting from the right. Each group of three digits is called a **period.** The names of the first four periods are shown below in red.

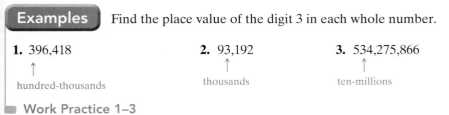

Writing a Whole Number in Words

To write a whole number in words, write the number in each period followed by the name of the period. (The ones period name is usually not written.) This same procedure can be used to read a whole number.

Fox example, we write 1,083,664,500 as

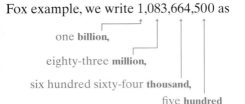

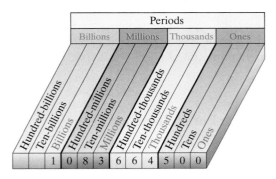

Helpful Hint Notice the commas after the name of each period.

Answers
1. millions 2. hundreds
3. ten-thousands

Practice 4–6

Write each whole number in words.

4. 54
5. 678
6. 93,205

Helpful Hint

The name of the ones period is not used when reading and writing whole numbers. For example,

9,265

is read as

"nine **thousand**, two hundred sixty-five."

Examples Write each whole number in words.

4. 72   seventy-two
5. 546   five hundred forty-six
6. 27,034   twenty-seven thousand, thirty-four

Work Practice 4–6

Helpful Hint The word "and" is *not* used when reading and writing whole numbers. It is used only when reading and writing mixed numbers and some decimal values, as shown later in this text.

Practice 7

Write 679,430,105 in words.

Example 7 Write 308,063,557 in words.

Solution: 308,063,557 is written as

three hundred eight **million**, sixty-three **thousand**, five hundred fifty-seven

Work Practice 7

✓ **Concept Check** True or false? When writing a check for $2600, the word name we write for the dollar amount of the check is "two thousand sixty." Explain your answer.

Practice 8–11

Write each whole number in standard form.

8. thirty-seven
9. two hundred twelve
10. eight thousand, two hundred seventy-four
11. five million, fifty-seven thousand, twenty-six

Writing a Whole Number in Standard Form

To write a whole number in standard form, write the number in each period, followed by a comma.

Examples Write each whole number in standard form.

8. forty-one   41
9. seven hundred eight   708
10. six thousand, four hundred ninety-three

   6,493   or   6493

11. three million, seven hundred forty-six thousand, five hundred twenty-two

   3,746,522

Work Practice 8–11

Answers

4. fifty-four
5. six hundred seventy-eight
6. ninety-three thousand, two hundred five
7. six hundred seventy-nine million, four hundred thirty thousand, one hundred five
8. 37
9. 212
10. 8,274 or 8274
11. 5,057,026

✓ **Concept Check Answer**
false

Section 1.2 | Place Value, Names for Numbers, and Reading Tables

Helpful Hint
A comma may or may not be inserted in a four-digit number. For example, both

6,493   and   6493

are acceptable ways of writing six thousand, four hundred ninety-three.

Objective C Writing a Whole Number in Expanded Form

The place value of a digit can be used to write a number in expanded form. The **expanded form** of a number shows each digit of the number with its place value. For example, 5672 is written in expanded form as

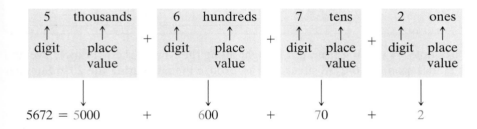

Example 12 Write 5,207,034 in expanded form.

Solution: 5,000,000 + 200,000 + 7000 + 30 + 4

Work Practice 12

Practice 12
Write 4,026,301 in expanded form.

We can visualize whole numbers by points on a line. The line below is called a **number line.** This number line has equally spaced marks for each whole number. The arrow to the right simply means that the whole numbers continue indefinitely. In other words, there is no largest whole number.

Number Line

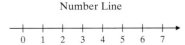

We will study number lines further in Section 1.4.

Objective D Reading Tables

Now that we know about place value and names for whole numbers, we introduce one way that whole number data may be presented. **Tables** are often used to organize and display facts that contain numbers. The following table shows the countries that won the most medals during the Summer Olympic Games in Rio de Janeiro, Brazil in 2016. (Although the medals are truly won by athletes from the various countries, for simplicity we will state that countries have won the medals.)

Answer
12. 4,000,000 + 20,000 + 6000 + 300 + 1

2016 Rio de Janeiro Summer Olympics Medal Count

Country	Gold	Silver	Bronze	Total	Country	Gold	Silver	Bronze	Total
United States	46	37	38	121	Japan	12	8	21	41
China	26	18	26	70	Australia	8	11	10	29
Great Britain	27	23	17	67	Italy	8	12	8	28
Russian Federation	19	18	19	56	Canada	4	3	15	22
Germany	17	10	15	42	South Korea	9	3	9	21
France	10	18	14	42	Netherlands	8	7	4	19
					Brazil	7	6	6	19

(*Source:* International Olympic Committee)

For example, by reading from left to right along the row marked "France," we find that France won 10 gold, 18 silver, and 14 bronze medals during the 2016 Summer Olympic Games.

Practice 13

Use the 2016 Summer Games table to answer each question.

a. How many bronze medals did Australia win during the 2016 Summer Olympic Games?

b. Which countries won more than 60 medals?

Answers

13. a. 10 b. United States, China, Great Britain

Example 13

Use the 2016 Summer Games table to answer each question.

a. How many silver medals did Germany win during the 2016 Summer Olympic Games?

b. Which countries shown won fewer gold medals than Japan?

Solution:

a. Find "Germany" in the left-hand column. Then read from left to right until the "silver" column is reached. We find that Germany won 10 silver medals.

b. Japan won 12 gold medals. Of the countries shown, France, Australia, Italy, Canada, South Korea, Netherlands, and Brazil each won fewer than 12 gold medals.

Work Practice 13

Vocabulary, Readiness & Video Check

Use the choices below to fill in each blank.

standard form   period   whole
expanded form   place value   words

1. The numbers 0, 1, 2, 3, 4, 5, 6, 7, 8, 9, 10, 11, 12, . . . are called _____ numbers.
2. The number 1286 is written in _____.
3. The number "twenty-one" is written in _____.
4. The number 900 + 60 + 5 is written in _____.
5. In a whole number, each group of three digits is called a(n) _____.
6. The _____ of the digit 4 in the whole number 264 is ones.

Section 1.2 | Place Value, Names for Numbers, and Reading Tables

Martin-Gay Interactive Videos Watch the section lecture video and answer the following questions.

See Video 1.2

Objective A 7. In Example 1, what is the place value of the digit 6?

Objective B 8. Complete this statement based on Example 3. To read (or write) a number, read from _____ to _____.

Objective C 9. In Example 5, what is the expanded form value of the digit 8?

Objective D 10. Use the table given in Example 6 to determine which breed shown has the fewest American Kennel Club registrations.

1.2 Exercise Set MyLab Math

Objective A *Determine the place value of the digit 5 in each whole number. See Examples 1 through 3.*

1. 657
2. 905
3. 5423
4. 6527

5. 43,526,000
6. 79,050,000
7. 5,408,092
8. 51,682,700

Objective B *Write each whole number in words. See Examples 4 through 7.*

9. 354
10. 316
11. 8279
12. 5445

13. 26,990
14. 42,009
15. 2,388,000
16. 3,204,000

17. 24,350,185
18. 47,033,107

Write each number in the sentence in words. (Do not write the years in words.) See Examples 4 through 7.

19. In 2016, the population of Monaco was 37,917. (*Source:* CIA World Factbook)

20. The state of Texas is divided into 254 counties. (*Source:* Texas Association of counties)

21. The Burj Khalifa, in Dubai, United Arab Emirates, a hotel and office building, is currently the tallest in the world at a height of 2720 feet. (*Source:* Council on Tall Buildings and Urban Habitat)

22. In 2016, there were 119,926 patients in the United States waiting for an organ transplant. (*Source:* United Network for Organ Sharing)

23. Each day, UPS delivers an average of 18,300,000 packages and documents worldwide. (*Source:* UPS)

24. The annual revenue for FedEx during fiscal year 2016 was $50,400,000,000. (*Source:* FedEx)

25. The highest point in Colorado is Mount Elbert, at an elevation of 14,433 feet. (*Source:* U.S. Geological Survey)

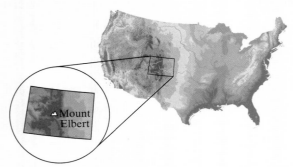

26. The highest point in Oregon is Mount Hood, at an elevation of 11,239 feet. (*Source:* U.S. Geological Survey)

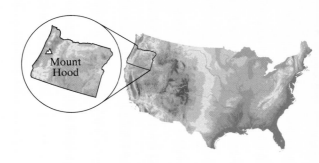

27. The average low price for a 2017 Honda Civic sedan was $17,410. (*Source:* Truecar.com)

28. The Goodyear blimp *Eagle* holds 202,700 cubic feet of helium. (*Source:* The Goodyear Tire & Rubber Company)

Write each whole number in standard form. See Examples 8 through 11.

29. Six thousand, five hundred eighty-seven

30. Four thousand, four hundred sixty-eight

31. Fifty-nine thousand, eight hundred

32. Seventy-three thousand, two

33. Thirteen million, six hundred one thousand, eleven

34. Sixteen million, four hundred five thousand, sixteen

35. Seven million, seventeen

36. Two million, twelve

37. Two hundred sixty thousand, nine hundred ninety-seven

38. Six hundred forty thousand, eight hundred eighty-one

Write the whole number in each sentence in standard form. See Examples 8 through 11.

39. The International Space Station orbits above Earth at an average altitude of two hundred fifty kilometers. (*Source:* space.com)

40. The average distance between the surfaces of Earth and the moon is about two hundred thirty-four thousand miles.

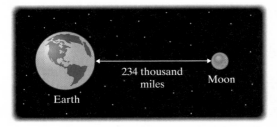

41. By 2016, there were one thousand, twenty-four buildings worldwide that were over 200 meters tall. (*Source:* Council on Tall Buildings and Urban Habitat)

42. The CN Tower in Toronto, Canada, is the tallest telecommunications/observation tower in the world. Its height is one thousand, eight hundred fifteen feet. (*Source:* Council on Tall Buildings and Urban Habitat)

43. The film *Star Wars Episode VII: The Force Awakens* shattered the world record for opening weekend income when it took in it two hundred forty-seven million, nine hundred sixty-six thousand, six hundred seventy-five dollars in 2015. (*Source:* Nash Information Services)

44. The film *Jurassic World* holds the record for the second-highest opening weekend income: it took in two hundred eight million, eight hundred six thousand, two hundred seventy dollars in 2015. (*Source:* Nash Information Service)

45. Morten Anderson, who played in the National Football League in 1982–2007, holds the record for most career field goals at five hundred sixty-five. (*Source:* NFL)

46. Morten Anderson also holds the record for the most field goals attempted in a career at seven hundred nine. (*Source:* NFL)

Objective **C** *Write each whole number in expanded form. See Example 12.*

47. 209   **48.** 789   **49.** 3470   **50.** 6040

51. 80,774   **52.** 20,215   **53.** 66,049   **54.** 99,032

55. 39,680,000   **56.** 47,703,029

Objective **D** *The table shows the beginning year of recent eruptions of major volcanoes in the Cascade Mountains. Use this table to answer Exercises 57 through 62. See Example 13.*

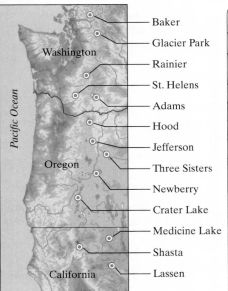

Recent Eruptions of Major Cascade Mountain Volcanoes (1750–present) *							
		Year of Eruption (beginning year)					
Volcano	State Location	1750–1799	1800–1849	1850–1899	1900–1949	1950–1999	2000–present
Mt. Baker	Washington	1792	1843	1870, 1880			
Glacier Peak	Washington	1750(?)					
Mt. Rainier	Washington		1841, 1843	1854			
Mt. St. Helens	Washington		1800			1980	
Mt. Hood	Oregon			1854, 1859, 1865			
Three Sisters	Oregon			1853(?)			
Medicine Lake	California				1910		
Mt. Shasta	California	1786		1855			
Cinder Cone*	California			1850			
Lassen Peak	California				1914		
Chaos Crags*	California			1854			

Other major volcanoes in the Cascades have had no eruptions from 1750 to present; *Source:* Harris
*Cinder Cone and Chaos Crags are located by Lassen Peak.

57. Mount Shasta erupted in the 1700s. Locate, then write this eruption year in standard form.

58. Mount Baker erupted in the 1700s. Locate, then write this eruption year in standard form.

59. Which volcano in the table has had the most eruptions?

60. Which volcano(es) in the table has had two eruptions?

61. Which volcano in the table had the earliest eruption?

62. Which volcano in the table had the most recent eruption?

The table shows the top ten popular breeds of dogs in 2016 according to the American Kennel Club. The breeds are listed in the order of number of registrations with the most popular breed listed first. Use this table to answer Exercises 63 through 68. See Example 13.

Top Ten American Kennel Club Registrations in 2016		
Breed	Average Dog Maximum Height (in inches)	Average Dog Maximum Weight (in pounds)
Labrador retriever	25	75
German shepherd	26	95
Golden retriever	24	80
Bulldog	26	90
Beagle	15	30
French bulldog	12	24
Yorkshire terrier	9	7
Poodle (standard, miniature, and toy)	standard: 26	standard: 70
Rottweiler	26	115
Boxer	25	70

(*Source:* American Kennel Club)

63. Which breed has a greater average dog maximum weight, the Bulldog or the German shepherd?

64. Which breed has more dogs registered, Golden retriever or German shepherd?

65. Which breed is the most popular dog? Write the average dog maximum weight for this breed in words.

66. Which breed is second in popularity? Write the average dog maximum height for this breed in words.

67. What is the maximum weight of an average-size Boxer?

68. What is the maximum height of an average-size standard poodle?

Concept Extensions

69. Write the largest four-digit number that can be made from the digits 1, 9, 8, and 6 if each digit must be used once. _____ _____ _____ _____

70. Write the largest five-digit number that can be made using the digits 5, 3, and 7 if each digit must be used at least once. _____ _____ _____ _____ _____

Check to see whether each number written in standard form matches the number written in words. If not, correct the number in words. See the Concept Check in this section.

71. Pay to the order of: $105.00
One Hundred Fifty and 00/100 Dollars

72. Pay to the order of: $7030.00
Seven Thousand Thirty and 00/100 Dollars

73. If a number is given in words, describe the process used to write this number in standard form.

74. If a number is written in standard form, describe the process used to write this number in expanded form.

75. How large is a trillion? To get an idea, one trillion seconds is over 31 thousand years. Look up "trillion" in a dictionary and use the definition to write this number in standard form.

76. How large is a quadrillion? To get an idea, if one quadrillion dollars were divided among the population of the United States, we would each receive over 9 million dollars. Look up "quadrillion" (in the American system) and write this number in standard form.

1.3 Adding and Subtracting Whole Numbers, and Perimeter

Objective A Adding Whole Numbers

An iPod is a hard drive–based portable audio player. In 2017, the iPod is still the most popular digital music player in the United States.

Suppose that an electronics store received a shipment of two boxes of iPods one day and an additional four boxes of iPods the next day. The **total** shipment in the two days can be found by adding 2 and 4.

2 boxes of iPods + 4 boxes of iPods = 6 boxes of iPods

The **sum** (or total) is 6 boxes of iPods. Each of the numbers 2 and 4 is called an **addend,** and the process of finding the sum is called **addition.**

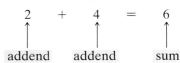

Objectives

A Add Whole Numbers.
B Subtract Whole Numbers.
C Find the Perimeter of a Polygon.
D Solve Problems by Adding or Subtracting Whole Numbers.

To add whole numbers, we add the digits in the ones place, then the tens place, then the hundreds place, and so on. For example, let's add 2236 + 160.

```
  2 2 3 6      Line up numbers vertically so that the place values correspond. Then
+   1 6 0      add digits in corresponding place values, starting with the ones place.
  2 3 9 6
```
↑ sum of ones
↑ sum of tens
↑ sum of hundreds
↑ sum of thousands

Example 1 Add: 46 + 713

Solution:
```
   46
+ 713
  759
```

■ Work Practice 1

Practice 1
Add: 4135 + 252

Adding by Carrying

When the sum of digits in corresponding place values is more than 9, **carrying** is necessary. For example, to add 365 + 89, add the ones-place digits first.

Carrying
```
    1
  3 6 5
+   8 9    5 ones + 9 ones = 14 ones or 1 ten + 4 ones
      4    Write the 4 ones in the ones place and carry the 1 ten to the tens place.
```

Next, add the tens-place digits.

```
  1 1
  3 6 5
+   8 9    1 ten + 6 tens + 8 tens = 15 tens or 1 hundred + 5 tens
    5 4    Write the 5 tens in the tens place and carry the 1 hundred to the hundreds place.
```

Answer
1. 4387

Next, add the hundreds-place digits.

$$\begin{array}{r} \overset{1\ 1}{3\ 6\ 5} \\ +\ \ \ 8\ 9 \\ \hline 4\ 5\ 4 \end{array}$$ 1 hundred + 3 hundreds = 4 hundreds
Write the 4 hundreds in the hundreds place.

Practice 2
Add: 47,364 + 135,898

Example 2 Add: 46,278 + 124,931

Solution:
$$\begin{array}{r} \overset{1\ 1\ 1}{46{,}278} \\ +\ 124{,}931 \\ \hline 171{,}209 \end{array}$$

■ Work Practice 2

 Concept Check What is wrong with the following computation?

$$\begin{array}{r} 394 \\ +\ 283 \\ \hline 577 \end{array}$$

Before we continue adding whole numbers, let's review some properties of addition that you may have already discovered. The first property that we will review is the **addition property of 0.** This property reminds us that the sum of 0 and any number is that same number.

Addition Property of 0

The sum of 0 and any number is that number. For example,

$7 + 0 = 7$
$0 + 7 = 7$

Next, notice that we can add any two whole numbers in any order and the sum is the same. For example,

$4 + 5 = 9$ and $5 + 4 = 9$

We call this special property of addition the **commutative property of addition.**

Commutative Property of Addition

Changing the **order** of two addends does not change their sum. For example,

$2 + 3 = 5$ and $3 + 2 = 5$

Another property that can help us when adding numbers is the **associative property of addition.** This property states that when adding numbers, the grouping of the numbers can be changed without changing the sum. We use parentheses to group numbers. They indicate which numbers to add first. For example, let's use two different groupings to find the sum of 2 + 1 + 5.

$(2 + 1) + 5 = 3 + 5 = 8$

Also,

$2 + (1 + 5) = 2 + 6 = 8$

Both groupings give a sum of 8.

Answer
2. 183,262

✓ **Concept Check Answer**
forgot to carry 1 hundred to the hundreds place

Section 1.3 | Adding and Subtracting Whole Numbers, and Perimeter

Associative Property of Addition

Changing the **grouping** of addends does not change their sum. For example,

$3 + (5 + 7) = 3 + 12 = 15$ and $(3 + 5) + 7 = 8 + 7 = 15$

The commutative and associative properties tell us that we can add whole numbers using any order and grouping that we want.

When adding several numbers, it is often helpful to look for two or three numbers whose sum is 10, 20, and so on. Why? Adding multiples of 10 such as 10 and 20 is easier.

 Example 3 Add: $13 + 2 + 7 + 8 + 9$

Solution: $13 + 2 + 7 + 8 + 9 = 39$

$20 + 10 + 9$

39

■ Work Practice 3

Practice 3
Add: $12 + 4 + 8 + 6 + 5$

Feel free to use the process of Example 3 any time when adding.

 Example 4 Add: $1647 + 246 + 32 + 85$

Solution:
```
  1 2 2
  1647
   246
    32
 +  85
 ─────
  2010
```

■ Work Practice 4

Practice 4
Add: $6432 + 789 + 54 + 28$

Objective B Subtracting Whole Numbers

If you have $5 and someone gives you $3, you have a total of $8, since $5 + 3 = 8$. Similarly, if you have $8 and then someone borrows $3, you have $5 left. **Subtraction** is finding the **difference** of two numbers.

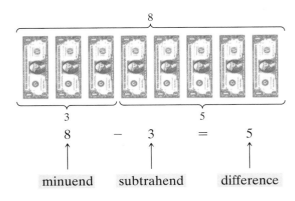

$8 - 3 = 5$

minuend   subtrahend   difference

Answers
3. 35   **4.** 7303

Chapter 1 | The Whole Numbers

In this example, 8 is the **minuend,** and 3 is the **subtrahend.** The **difference** between these two numbers, 8 and 3, is 5.

Notice that addition and subtraction are very closely related. In fact, subtraction is defined in terms of addition.

$8 - 3 = 5$ because $5 + 3 = 8$

This means that subtraction can be *checked* by addition, and we say that addition and subtraction are reverse operations.

Practice 5

Subtract. Check each answer by adding.
a. $14 - 6$
b. $20 - 8$
c. $93 - 93$
d. $42 - 0$

Example 5 Subtract. Check each answer by adding.

a. $12 - 9$
b. $22 - 7$
c. $35 - 35$
d. $70 - 0$

Solution:
a. $12 - 9 = 3$ because $3 + 9 = 12$
b. $22 - 7 = 15$ because $15 + 7 = 22$
c. $35 - 35 = 0$ because $0 + 35 = 35$
d. $70 - 0 = 70$ because $70 + 0 = 70$

■ Work Practice 5

Look again at Examples 5(c) and 5(d).

5(c) $35 - 35 = 0$ 5(d) $70 - 0 = 70$
      same   difference   a number   difference is the
      number   is 0   minus 0   same number

These two examples illustrate the subtraction properties of 0.

Subtraction Properties of 0

The difference of any number and that same number is 0. For example,

$11 - 11 = 0$

The difference of any number and 0 is that same number. For example,

$45 - 0 = 45$

To subtract whole numbers we subtract the digits in the ones place, then the tens place, then the hundreds place, and so on. When subtraction involves numbers of two or more digits, it is more convenient to subtract vertically. For example, to subtract $893 - 52$,

```
  8 9 3   ← minuend
-   5 2   ← subtrahend
  8 4 1   ← difference
      │ │ │
      │ │ 3 - 2
      │ 9 - 5
      8 - 0
```

Line up the numbers vertically so that the minuend is on top and the place values correspond. Subtract in corresponding place values, starting with the ones place.

To check, add.

```
  difference     or      841
+ subtrahend           +  52
  minuend               893   ← Since this is the original minuend,
                               the problem checks.
```

Answers
5. a. 8   b. 12   c. 0   d. 42

Section 1.3 | Adding and Subtracting Whole Numbers, and Perimeter

Example 6 Subtract: 7826 − 505. Check by adding.

Solution:
```
   7826          Check:    7321
 −  505                  +  505
   7321                    7826
```

■ Work Practice 6

Practice 6
Subtract. Check by adding.

a. 9143 − 122
b. 978 − 851

Subtracting by Borrowing

When subtracting vertically, if a digit in the second number (subtrahend) is larger than the corresponding digit in the first number (minuend), **borrowing** is necessary. For example, consider

```
   8|1
 − 6|3
```

Since the 3 in the ones place of 63 is larger than the 1 in the ones place of 81, borrowing is necessary. We borrow 1 ten from the tens place and add it to the ones place.

Borrowing

$$8 \text{ tens} - 1 \text{ ten} = 7 \text{ tens} \rightarrow \begin{array}{r} 7\ 11 \\ \cancel{8}\ \cancel{1} \\ -\ 6\ 3 \end{array} \leftarrow 1 \text{ ten} + 1 \text{ one} = 11 \text{ ones}$$

Now we subtract the ones-place digits and then the tens-place digits.

```
  7 11
  8 1                       18
 −6 3         Check:       +63
  1 8 ← 11 − 3 = 8          81    The original minuend.
     7 − 6 = 1
```

Example 7 Subtract: 543 − 29. Check by adding.

Solution:
```
   3 13
   5 4 3      Check:    514
 −   2 9              +  29
   5 1 4                543
```

■ Work Practice 7

Practice 7
Subtract. Check by adding.

a.
```
   697
 −  49
```

b.
```
   326
 − 245
```

c.
```
  1234
 − 822
```

Sometimes we may have to borrow from more than one place. For example, to subtract 7631 − 152, we first borrow from the tens place.

```
    2 11
  76 3 1
 −  1 5 2
        9 ← 11 − 2 = 9
```

In the tens place, 5 is greater than 2, so we borrow again. This time we borrow from the hundreds place.

6 hundreds − 1 hundred = 5 hundreds

```
     12
   5 2 11
   7 6 3 1              Check:    7479
 −   1 5 2                       + 152
   7 4 7 9                         7631  The original minuend.
```

1 hundred + 2 tens or 10 tens + 2 tens = 12 tens

Answers
6. a. 9021 b. 127
7. a. 648 b. 81 c. 412

Practice 8

Subtract. Check by adding.

a. 400
   −164

b. 1000
    − 762

Example 8
Subtract: 900 − 174. Check by adding.

Solution: In the ones place, 4 is larger than 0, so we borrow from the tens place. But the tens place of 900 is 0, so to borrow from the tens place, we must first borrow from the hundreds place.

$$\begin{array}{r} \overset{8}{\cancel{9}}\overset{10}{\cancel{0}}0 \\ -174 \end{array}$$

Now borrow from the tens place.

$$\begin{array}{r} \overset{8}{\cancel{9}}\overset{9}{\cancel{\overset{10}{\cancel{0}}}}\overset{10}{\cancel{0}} \\ -174 \\ \hline 726 \end{array}$$

Check: 726
           +174
           ─────
            900

■ Work Practice 8

Objective C Finding the Perimeter of a Polygon

In geometry, addition is used to find the perimeter of a polygon. A **polygon** can be described as a flat figure formed by line segments connected at their ends. (For more review, see Appendix A.1.) Geometric figures such as triangles, squares, and rectangles are called polygons.

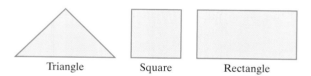

The **perimeter** of a polygon is the *distance around* the polygon, shown in red above. This means that the perimeter of a polygon is the sum of the lengths of its sides.

Practice 9

Find the perimeter of the polygon shown. (A centimeter is a unit of length in the metric system.)

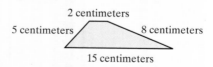

Example 9
Find the perimeter of the polygon shown.

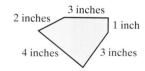

Solution: To find the perimeter (distance around), we add the lengths of the sides.

2 in. + 3 in. + 1 in. + 3 in. + 4 in. = 13 in.

The perimeter is 13 inches.

■ Work Practice 9

To make the addition appear simpler, we will often not include units with the addends. If you do this, make sure units are included in the final answer.

Answers
8. a. 236 b. 238
9. 30 cm

Example 10   Calculating the Perimeter of a Building

One way to compare large buildings is to calculate their footprint, or ground area. The second-largest building in the world by footprint is the Aalsmeer Flower Auction Building in Aalsmeer, Netherlands. The floor plan is a rectangle that measures 776 meters by 639 meters. Find the perimeter of this building. (A meter is a unit of length in the metric system.) (*Source: The Handy Science Answer Book,* Visible Ink Press and Internet research)

Practice 10
A park is in the shape of a triangle. Each of the park's three sides is 647 feet. Find the perimeter of the park.

Solution: Recall that opposite sides of a rectangle have the same length. To find the perimeter of this building, we add the lengths of the sides. The sum of the lengths of its sides is

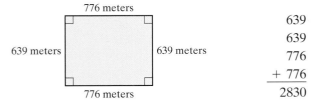

```
   639
   639
   776
 + 776
  2830
```

The perimeter of the building is 2830 meters.

(Note: upon completion, the largest footprint in the world will be the factory for Tesla vehicles, called the Gigafactory. It will have a footprint of about 100 football fields.)

■ Work Practice 10

Objective D Solving Problems by Adding or Subtracting

Often, real-life problems occur that can be solved by adding or subtracting. The first step in solving any word problem is to *understand* the problem by reading it carefully.

Descriptions in problems solved through addition or subtraction *may* include any of these key words or phrases:

Addition		
Key Words or Phrases	**Examples**	**Symbols**
added to	5 added to 7	7 + 5
plus	0 plus 78	0 + 78
increased by	12 increased by 6	12 + 6
more than	11 more than 25	25 + 11
total	the total of 8 and 1	8 + 1
sum	the sum of 4 and 133	4 + 133

Subtraction		
Key Words or Phrases	**Examples**	**Symbols**
subtract	subtract 5 from 8	8 − 5
difference	the difference of 10 and 2	10 − 2
less	17 less 3	17 − 3
less than	2 less than 20	20 − 2
take away	14 take away 9	14 − 9
decreased by	7 decreased by 5	7 − 5
subtracted from	9 subtracted from 12	12 − 9

Helpful Hint Be careful when solving applications that suggest subtraction. Although order *does not* matter when adding, order *does* matter when subtracting. For example, 20 − 15 and 15 − 20 do not simplify to the same number.

 Concept Check In each of the following problems, identify which number is the minuend and which number is the subtrahend.
a. What is the result when 6 is subtracted from 40?
b. What is the difference of 15 and 8?
c. Find a number that is 15 fewer than 23.

To solve a word problem that involves addition or subtraction, we first use the facts given to write an addition or subtraction statement. Next, we compute the addition or subtraction. Then we write the corresponding solution of the real-life problem. It is sometimes helpful to write the statement in words (brief phrases) and then translate to numbers.

Answer
10. 1941 ft

✓ **Concept Check Answers**
a. minuend: 40; subtrahend: 6
b. minuend: 15; subtrahend: 8
c. minuend: 23; subtrahend: 15

Practice 11

The radius of Uranus is 15,759 miles. The radius of Neptune is 458 miles less than the radius of Uranus. What is the radius of Neptune? (*Source:* National Space Science Data Center)

Example 11 — Finding the Radius of a Planet

The radius of Jupiter is 43,441 miles. The radius of Saturn is 7257 miles less than the radius of Jupiter. Find the radius of Saturn. (*Source:* National Space Science Data Center)

Solution:

In Words	Translate to Numbers
radius of Jupiter	43,441
− 7257	− 7257
radius of Saturn	36,184

The radius of Saturn is 36,184 miles.

Work Practice 11

Helpful Hint: Since subtraction and addition are reverse operations, don't forget that a subtraction problem can be checked by adding.

Graphs can be used to visualize data. The graph shown next is called a **bar graph**. For this bar graph, the height of each bar is labeled above the bar. To check this height, follow the top of each bar to the vertical line to the left. For example, the first bar is labeled 216. Follow the top of that bar to the left until the vertical line is reached, a bit more than halfway between 200 and 225, or 216.

Practice 12

Use the graph in Example 12 to answer the following:

a. Which country shown has the fewest threatened amphibians?

b. Find the total number of threatened amphibians for Madagascar, Peru, and Mexico.

Example 12 — Reading a Bar Graph

The bar graph below contains the number of threatened amphibians for selected countries. Each bar represents a country and the height of each bar represents the number of endangered amphibians identified in that country.

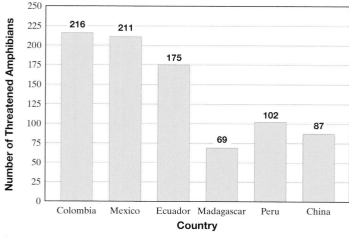

Source: theguardian.com

a. Which country shown has the greatest number of threatened amphibians?

b. Find the total number of threatened amphibians for Ecuador, China, and Colombia.

Answers
11. 15,301 miles
12. a. Madagascar   b. 382

Section 1.3 | Adding and Subtracting Whole Numbers, and Perimeter

Solution:

a. The country with the greatest number of threatened amphibians corresponds to the tallest bar, which is Colombia.

b. The key word here is "total." To find the total number of threatened amphibians for Ecuador, China, and Colombia, we add.

In Words		Translate to Numbers
Ecuador	⟶	175
China	⟶	87
Colombia	⟶	+ 216
	Total	478

The total number of threatened amphibians for Ecuador, China, and Colombia is 478.

Work Practice 12

Calculator Explorations Adding and Subtracting Numbers

Adding Numbers

To add numbers on a calculator, find the keys marked $+$ and $=$ or ENTER.

For example, to add 5 and 7 on a calculator, press the keys 5 $+$ 7 then $=$ or ENTER. The display will read 12.

Thus, $5 + 7 = 12$.

To add 687, 981, and 49 on a calculator, press the keys 687 $+$ 981 $+$ 49 then $=$ or ENTER. The display will read 1717.

Thus, $687 + 981 + 49 = 1717$. (Although entering 687, for example, requires pressing more than one key, here numbers are grouped together for easier reading.)

Use a calculator to add.

1. $89 + 45$
2. $76 + 97$
3. $285 + 55$
4. $8773 + 652$
5. 985
   1210
   562
   + 77
6. 465
   9888
   620
   + 1550

Subtracting Numbers

To subtract numbers on a calculator, find the keys marked $-$ and $=$ or ENTER.

For example, to find $83 - 49$ on a calculator, press the keys 83 $-$ 49 then $=$ or ENTER. The display will read 34. Thus, $83 - 49 = 34$.

Use a calculator to subtract.

7. $865 - 95$
8. $76 - 27$
9. $147 - 38$
10. $366 - 87$
11. $9625 - 647$
12. $10,711 - 8925$

Vocabulary, Readiness & Video Check

Use the choices below to fill in each blank. Some choices may be used more than once and some may not be used at all.

0	order	addend	associative
sum	number	grouping	commutative
perimeter	minuend	subtrahend	difference

1. The sum of 0 and any number is the same _____.
2. In 35 + 20 = 55, the number 55 is called the _____ and 35 and 20 are each called a(n) _____.
3. The difference of any number and that same number is _____.
4. The difference of any number and 0 is the same _____.
5. In 37 − 19 = 18, the number 37 is the _____, the 19 is the _____, and the 18 is the _____.
6. The distance around a polygon is called its _____.
7. Since 7 + 10 = 10 + 7, we say that changing the _____ in addition does not change the sum. This property is called the _____ property of addition.
8. Since (3 + 1) + 20 = 3 + (1 + 20), we say that changing the _____ in addition does not change the sum. This property is called the _____ property of addition.

Martin-Gay Interactive Videos *Watch the section lecture video and answer the following questions.*

See Video 1.3

Objective A 9. Complete this statement based on the lecture before Example 1. To add whole numbers, we line up _____ values and add from _____ to _____.

Objective B 10. In Example 5, explain how we end up subtracting 7 from 12 in the ones place.

Objective C 11. In Example 7, the perimeter of what type of polygon is found? How many addends are in the resulting addition problem?

Objective D 12. Complete this statement based on Example 8. To find the sale price, subtract the _____ from the _____ price.

1.3 Exercise Set MyLab Math

Objective A *Add. See Examples 1 through 4.*

1. 14 + 22
2. 27 + 31
3.   62
  +230
4.   37
  +542
5.   12
  13
+24

6.   23
  45
+30
7.   5267
+ 132
8.   236
+6243
9. 22,781 + 186,297
10. 17,427 + 821,059

Section 1.3 | Adding and Subtracting Whole Numbers, and Perimeter

11. $\begin{array}{r} 8 \\ 9 \\ 2 \\ 5 \\ +1 \\ \hline \end{array}$

12. $\begin{array}{r} 3 \\ 5 \\ 8 \\ 5 \\ +7 \\ \hline \end{array}$

13. $\begin{array}{r} 81 \\ 17 \\ 23 \\ 79 \\ +12 \\ \hline \end{array}$

14. $\begin{array}{r} 64 \\ 28 \\ 56 \\ 25 \\ +32 \\ \hline \end{array}$

15. $24 + 9006 + 489 + 2407$

16. $16 + 1056 + 748 + 7770$

17. $\begin{array}{r} 6820 \\ 4271 \\ +5626 \\ \hline \end{array}$

18. $\begin{array}{r} 6789 \\ 4321 \\ +5555 \\ \hline \end{array}$

19. $\begin{array}{r} 49 \\ 628 \\ 5\,762 \\ +29{,}462 \\ \hline \end{array}$

20. $\begin{array}{r} 26 \\ 582 \\ 4\,763 \\ +62{,}511 \\ \hline \end{array}$

21. $\begin{array}{r} 121{,}742 \\ 57{,}279 \\ 26{,}586 \\ +426{,}782 \\ \hline \end{array}$

22. $\begin{array}{r} 504{,}218 \\ 321{,}920 \\ 38{,}507 \\ +594{,}687 \\ \hline \end{array}$

Objective B *Subtract. Check by adding. See Examples 5 through 8.*

23. $\begin{array}{r} 749 \\ -149 \\ \hline \end{array}$

24. $\begin{array}{r} 957 \\ -257 \\ \hline \end{array}$

25. $\begin{array}{r} 62 \\ -37 \\ \hline \end{array}$

26. $\begin{array}{r} 55 \\ -29 \\ \hline \end{array}$

27. $\begin{array}{r} 922 \\ -634 \\ \hline \end{array}$

28. $\begin{array}{r} 674 \\ -299 \\ \hline \end{array}$

29. $\begin{array}{r} 600 \\ -432 \\ \hline \end{array}$

30. $\begin{array}{r} 300 \\ -149 \\ \hline \end{array}$

31. $\begin{array}{r} 6283 \\ -560 \\ \hline \end{array}$

32. $\begin{array}{r} 5349 \\ -720 \\ \hline \end{array}$

33. $\begin{array}{r} 533 \\ -29 \\ \hline \end{array}$

34. $\begin{array}{r} 724 \\ -16 \\ \hline \end{array}$

35. $1983 - 1904$

36. $1983 - 1914$

37. $50{,}000 - 17{,}289$

38. $40{,}000 - 23{,}582$

39. $7020 - 1979$

40. $6050 - 1878$

41. $51{,}111 - 19{,}898$

42. $62{,}222 - 39{,}898$

Objectives A B Mixed Practice *Add or subtract as indicated. See Examples 1 through 8.*

43. $\begin{array}{r} 986 \\ +48 \\ \hline \end{array}$

44. $\begin{array}{r} 986 \\ -48 \\ \hline \end{array}$

45. $76 - 67$

46. $80 + 93 + 17 + 9 + 2$

47. $\begin{array}{r} 9000 \\ -482 \\ \hline \end{array}$

48. $\begin{array}{r} 10{,}000 \\ -1\,786 \\ \hline \end{array}$

49. $\begin{array}{r} 10{,}962 \\ 4\,851 \\ +7\,063 \\ \hline \end{array}$

50. $\begin{array}{r} 12{,}468 \\ 3\,211 \\ +1\,988 \\ \hline \end{array}$

Objective C *Find the perimeter of each figure. See Examples 9 and 10.*

51.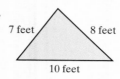

52.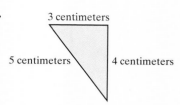

53.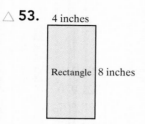

54.

55.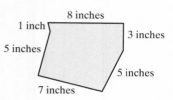

56.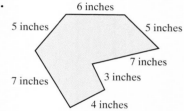

57.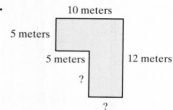

58.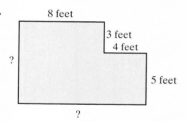

Objectives A B C D Mixed Practice–Translating *Solve. See Examples 9 through 12.*

59. Find the sum of 297 and 1796.

60. Find the sum of 802 and 6487.

61. Find the total of 76, 39, 8, 17, and 126.

62. Find the total of 89, 45, 2, 19, and 341.

63. Find the difference of 41 and 21.

64. Find the difference of 16 and 5.

65. What is 452 increased by 92?

66. What is 712 increased by 38?

67. Find 108 less 36.

68. Find 25 less 12.

69. Find 12 subtracted from 100.

70. Find 86 subtracted from 90.

Solve.

71. The population of Florida is projected to grow from 19,308 thousand in 2010 to 22,478 thousand in 2020. What is Florida's projected population increase over this time period?

72. The population of California is projected to grow from 39,136 thousand in 2010 to 44,126 thousand in 2020. What is California's projected population increase over this time period?

73. A new DVD player with remote control costs $295. A college student has $914 in her savings account. How much will she have left in her savings account after she buys the DVD player?

74. A stereo that regularly sells for $547 is discounted by $99 in a sale. What is the sale price?

A river basin is the geographic area drained by a river and its tributaries. The Mississippi River Basin is the third largest in the world and is divided into six sub-basins, whose areas are shown in the following bar graph. Use this graph for Exercises 75 through 78.

75. Find the total U.S. land area drained by the Upper Mississippi and Lower Mississippi sub-basins.

76. Find the total U.S. land area drained by the Ohio and Tennessee sub-basins.

77. How many more square miles of land are drained by the Missouri sub-basin than the Arkansas Red-White sub-basin?

78. How many more square miles of land are drained by the Upper Mississippi sub-basin than the Lower Mississippi sub-basin?

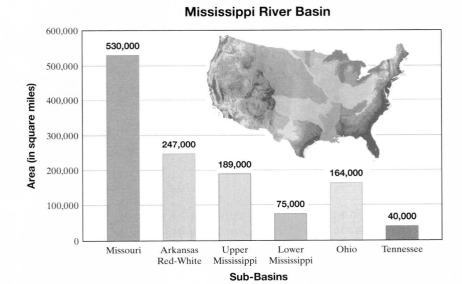

79. A homeowner is installing a fence in his backyard. How many feet of fencing are needed to enclose the yard below?

80. A homeowner is considering installing gutters around her home. Find the perimeter of her rectangular home.

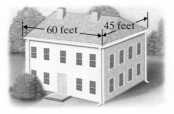

81. Professor Graham is reading a 503-page book. If she has just finished reading page 239, how many more pages must she read to finish the book?

82. When a couple began a trip, the odometer read 55,492. When the trip was over, the odometer read 59,320. How many miles did they drive on their trip?

83. In 2016, the country of New Zealand had 22,867,835 more sheep than people. If the human population of New Zealand in 2016 was 4,573,567, what was the sheep population? (*Source:* www.stats.govt.nz)

84. During one month in 2016, the two top-selling vehicles in the United States were the Ford F-Series and the Chevrolet Silverado, both trucks. There were 65,542 F-Series trucks and 49,768 Silverados sold that month. What was the total number of these trucks sold in that month? (*Source:* www.goodcarbadcar.com)

The decibel (dB) is a unit of measurement for sound. Every increase of 10 dB is a tenfold increase in sound intensity. The bar graph shows the decibel levels for some common sounds. Use this graph for Exercises 85 through 88.

85. What is the dB level for live rock music?

86. Which is the quietest of all the sounds shown in the graph?

87. How much louder is the sound of snoring than normal conversation?

88. What is the difference in sound intensity between live rock music and loud television?

Decibel Levels for Common Sounds

Sound Example	Decibels (dB)
Leaves Rustling	10
Normal Conversation	30
Live Rock Music	100
Snoring	88
Loud Television	70

89. In 2013, there were 2410 Gap Inc. (Gap, Banana Republic, Old Navy) stores located in the United States and 1034 located outside the United States. How many Gap Inc. stores were located worldwide? (*Source:* Gap, Inc.)

90. Automobile classes are defined by the amount of interior room. A subcompact car is defined as a car with a maximum interior space of 99 cubic feet. A midsize car is defined as a car with a maximum interior space of 119 cubic feet. What is the difference in volume between a midsize and a subcompact car?

91. The largest permanent Monopoly board is made of granite and is located in San Jose, California. It is in the shape of a square with side lengths of 31 ft. Find the perimeter of the square playing board.

92. The smallest commercially available jigsaw puzzle is a 1000-piece puzzle manufactured in Spain. It is in the shape of a rectangle with length of 18 inches and width of 12 inches. Find the perimeter of this rectangular-shaped puzzle.

The table shows the number of CVS pharmacies in ten states. Use this table to answer Exercises 93 through 98.

The Top Ten States for CVS Pharmacies in 2016	
State	**Number of CVS Pharmacies**
Massachusetts	356
California	867
Florida	756
Georgia	313
Indiana	301
New York	486
North Carolina	313
Ohio	309
Pennsylvania	408
Texas	659
(*Source:* CVS Pharmacy Inc.)	

93. Which state has the most CVS pharmacies?

94. Which of the states listed in the table has the fewest CVS pharmacies?

95. What is the total number of CVS pharmacies located in the three states with the most CVS pharmacies?

96. How many CVS pharmacies are located in the 10 states listed in the table?

97. Which pair of neighboring states have more CVS pharmacies combined, Pennsylvania and New York or Florida and Georgia?

98. There are 3048 CVS pharmacies located in the states not listed in the table. How many CVS pharmacies are there in the 50 states?

Section 1.3 | Adding and Subtracting Whole Numbers, and Perimeter

99. The state of Delaware has 2029 miles of urban highways and 3865 miles of rural highways. Find the total highway mileage in Delaware. (*Source:* U.S. Federal Highway Administration)

100. The state of Rhode Island has 5193 miles of urban highways and 1222 miles of rural highways. Find the total highway mileage in Rhode Island. (*Source:* U.S. Federal Highway Administration)

Concept Extensions

For Exercises 101–104, identify which number is the minuend and which number is the subtrahend. See the second Concept Check in this section.

101.
$$\begin{array}{r} 48 \\ -\ 1 \\ \hline \end{array}$$

102.
$$\begin{array}{r} 2863 \\ -1904 \\ \hline \end{array}$$

103. Subtract 7 from 70.

104. Find 86 decreased by 25.

105. In your own words, explain the commutative property of addition.

106. In your own words, explain the associative property of addition.

Check each addition below. If it is incorrect, find the correct answer. See the first Concept Check in this section.

107.
$$\begin{array}{r} 566 \\ 932 \\ +\ 871 \\ \hline 2369 \end{array}$$

108.
$$\begin{array}{r} 773 \\ 659 \\ +\ 481 \\ \hline 1913 \end{array}$$

109.
$$\begin{array}{r} 14 \\ 173 \\ 86 \\ +\ 257 \\ \hline 520 \end{array}$$

110.
$$\begin{array}{r} 19 \\ 214 \\ 49 \\ +\ 651 \\ \hline 923 \end{array}$$

Identify each answer as correct or incorrect. Use addition to check. If the answer is incorrect, write the correct answer.

111.
$$\begin{array}{r} 741 \\ -\ 56 \\ \hline 675 \end{array}$$

112.
$$\begin{array}{r} 478 \\ -\ 89 \\ \hline 389 \end{array}$$

113.
$$\begin{array}{r} 1029 \\ -\ 888 \\ \hline 141 \end{array}$$

114.
$$\begin{array}{r} 7615 \\ -\ 547 \\ \hline 7168 \end{array}$$

Fill in the missing digits in each problem.

115.
$$\begin{array}{r} 526\_ \\ -\ 2\_85 \\ \hline 28\_4 \end{array}$$

116.
$$\begin{array}{r} 10{,}4\_ \\ -\ 8\ 5\ 4 \\ \hline \_710 \end{array}$$

117. Is there a commutative property of subtraction? In other words, does order matter when subtracting? Why or why not?

118. Explain why the phrase "Subtract 7 from 10" translates to "10 − 7."

119. The local college library is having a Million Pages of Reading promotion. The freshmen have read a total of 289,462 pages; the sophomores have read a total of 369,477 pages; the juniors have read a total of 218,287 pages; and the seniors have read a total of 121,685 pages. Have they reached the goal of one million pages? If not, how many more pages need to be read?

1.4 Rounding and Estimating

Objectives

A Round Whole Numbers.

B Use Rounding to Estimate Sums and Differences.

C Solve Problems by Estimating.

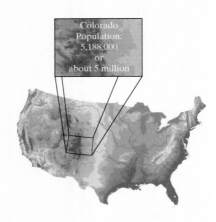

Objective A Rounding Whole Numbers

Rounding a whole number means approximating it. A rounded whole number is often easier to use, understand, and remember than the precise whole number. For example, instead of trying to remember the Colorado state population as 5,188,000, it is much easier to remember it rounded to the nearest million: 5,000,000, or 5 million people. (*Source:* U.S. census)

Recall from Section 1.2 that the line below is called a number line. To **graph** a whole number on this number line, we darken the point representing the location of the whole number. For example, the number 4 is graphed below.

On the number line, the whole number 36 is closer to 40 than 30, so 36 rounded to the nearest ten is 40.

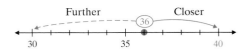

The whole number 52 is closer to 50 than 60, so 52 rounded to the nearest ten is 50.

In trying to round 25 to the nearest ten, we see that 25 is halfway between 20 and 30. It is not closer to either number. In such a case, we round to the larger ten, that is, to 30.

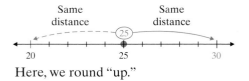

Here, we round "up."

To round a whole number without using a number line, follow these steps:

Rounding a Whole Number to a Given Place Value

Step 1: Locate the digit to the right of the given place value.

Step 2: If this digit is 5 or greater, add 1 to the digit in the given place value and replace each digit to its right by 0.

Step 3: If this digit is less than 5, replace it and each digit to its right by 0.

Section 1.4 | Rounding and Estimating

Example 1
Round 568 to the nearest ten.

Solution: 5 6 ⑧
↑
tens place

The digit to the right of the tens place is the ones place, which is circled.

5 6 ⑧
↑ ↖
Add 1. Replace with 0.

Since the circled digit is 5 or greater, add 1 to the 6 in the tens place and replace the digit to the right by 0.

We find that 568 rounded to the nearest ten is 570.

■ Work Practice 1

Practice 1
Round to the nearest ten.
a. 57
b. 641
c. 325

Example 2
Round 278,362 to the nearest thousand.

Solution:
Thousands place
│ ┌─ 3 is less than 5.
278,③62
↑   ↑
Do not add 1.   Replace with zeros.

The number 278,362 rounded to the nearest thousand is 278,000.

■ Work Practice 2

Practice 2
Round to the nearest thousand.
a. 72,304
b. 9222
c. 671,800

Example 3
Round 248,982 to the nearest hundred.

Solution:
Hundreds place
│ ┌─ 8 is greater than or equal to 5.
248,9⑧2
↑

Add 1. 9 + 1 = 10, so replace the digit 9 by 0 and carry 1 to the place value to the left.

$$2\ 4\ \overset{8+1}{8},\ \overset{0}{9}\ 8\ 2$$
↑   ↑
Add 1.   Replace with zeros.

The number 248,982 rounded to the nearest hundred is 249,000.

■ Work Practice 3

Practice 3
Round to the nearest hundred.
a. 3474
b. 76,243
c. 978,965

✓ **Concept Check** Round each of the following numbers to the nearest *hundred*. Explain your reasoning.
a. 59   **b.** 29

Objective B Estimating Sums and Differences

By rounding addends, minuends, and subtrahends, we can **estimate** sums and differences. An estimated sum or difference is appropriate when the exact number is not necessary. Also, an estimated sum or difference can help us determine if we made

Answers
1. **a.** 60 **b.** 640 **c.** 330
2. **a.** 72,000 **b.** 9000 **c.** 672,000
3. **a.** 3500 **b.** 76,200 **c.** 979,000

✓ **Concept Check Answers**
a. 100 **b.** 0

a mistake in calculating an exact amount. To estimate the sum below, round each number to the nearest hundred and then add.

$$
\begin{array}{rcl}
768 & \text{rounds to} & 800 \\
1952 & \text{rounds to} & 2000 \\
225 & \text{rounds to} & 200 \\
+\ 149 & \text{rounds to} & +\ 100 \\
\hline
& & 3100
\end{array}
$$

The estimated sum is 3100, which is close to the **exact** sum of 3094.

Practice 4

Round each number to the nearest ten to find an estimated sum.

$$
\begin{array}{r}
49 \\
25 \\
32 \\
51 \\
+\ 98 \\
\hline
\end{array}
$$

Example 4 Round each number to the nearest hundred to find an estimated sum.

$$
\begin{array}{r}
294 \\
625 \\
1071 \\
+\ 349 \\
\hline
\end{array}
$$

Solution:

Exact:		Estimate:
294	rounds to	300
625	rounds to	600
1071	rounds to	1100
+ 349	rounds to	+ 300
		2300

The estimated sum is 2300. (The exact sum is 2339.)

▪ Work Practice 4

Practice 5

Round each number to the nearest thousand to find an estimated difference.

$$
\begin{array}{r}
3785 \\
-\ 2479 \\
\hline
\end{array}
$$

Example 5 Round each number to the nearest hundred to find an estimated difference.

$$
\begin{array}{r}
4725 \\
-\ 2879 \\
\hline
\end{array}
$$

Solution:

Exact:		Estimate:
4725	rounds to	4700
−2879	rounds to	−2900
		1800

The estimated difference is 1800. (The exact difference is 1846.)

▪ Work Practice 5

Objective C Solving Problems by Estimating

Making estimates is often the quickest way to solve real-life problems when solutions do not need to be exact.

Answers
4. 260
5. 2000

Example 6 Estimating Distances

A driver is trying to quickly estimate the distance from Temple, Texas, to Brenham, Texas. Round each distance given on the map to the nearest ten to estimate the total distance.

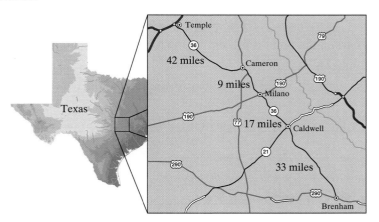

Solution:

Exact Distance:		Estimate:
42	rounds to	40
9	rounds to	10
17	rounds to	20
+33	rounds to	+30
		100

It is approximately 100 miles from Temple to Brenham. (The exact distance is 101 miles.)

■ Work Practice 6

Practice 6

Tasha Kilbey is trying to estimate how far it is from Gove, Kansas, to Hays, Kansas. Round each given distance on the map to the nearest ten to estimate the total distance.

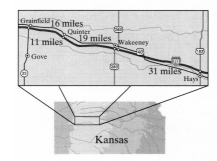

Example 7 Estimating Data

In three recent years the numbers of tons of air cargo and mail that went through Hartsfield-Jackson Atlanta International Airport were 629,700, 685,550, and 737, 655. Round each number to the nearest thousand to estimate the tons of mail that passed through this airport.

Solution:

Exact Tons of Cargo Mail:		Estimate:
629,700	rounds to	630,000
685,550	rounds to	686,000
+737,655	rounds to	+738,000
		2,054,000

The approximate tonnage of mail that moved through Atlanta's airport over this 3-year period was 2,054,000 tons. (The exact tonnage was 2,052,905 tons.)

■ Work Practice 7

Practice 7

In 2015, there were 2930 reported cases of mumps, 18,166 reported cases of pertussis (whooping cough), and 189 reported cases of measles. Round each number to the nearest thousand to estimate the total number of cases reported for these preventable diseases. (*Source:* Centers for Disease Control)

Answers
6. 80 mi
7. 21,000 total cases

Vocabulary, Readiness & Video Check

Use the choices below to fill in each blank.

60   rounding   exact
70   estimate   graph

1. To _____ a number on a number line, darken the point representing the location of the number.
2. Another word for approximating a whole number is _____.
3. The number 65 rounded to the nearest ten is _____, but the number 61 rounded to the nearest ten is _____.
4. A(n) _____ number of products is 1265, but a(n) _____ is 1000.

Martin-Gay Interactive Videos Watch the section lecture video and answer the following questions.

Objective A 5. In Example 1, when rounding the number to the nearest ten, why do we replace the digit 3 with a 4?

Objective B 6. As discussed in Example 3, explain how a number line can help us understand how to round 22 to the nearest ten.

Objective C 7. What is the significance of the circled digit in each height value in Example 5?

See Video 1.4

1.4 Exercise Set MyLab Math

Objective A *Round each whole number to the given place value. See Examples 1 through 3.*

1. 423 to the nearest ten
2. 273 to the nearest ten
3. 635 to the nearest ten

4. 846 to the nearest ten
5. 2791 to the nearest hundred
6. 8494 to the nearest hundred

7. 495 to the nearest ten
8. 898 to the nearest ten
9. 21,094 to the nearest thousand

10. 82,198 to the nearest thousand
11. 33,762 to the nearest thousand
12. 42,682 to the nearest ten-thousand

13. 328,495 to the nearest hundred
14. 179,406 to the nearest hundred
15. 36,499 to the nearest thousand

16. 96,501 to the nearest thousand
17. 39,994 to the nearest ten
18. 99,995 to the nearest ten

19. 29,834,235 to the nearest ten-million
20. 39,523,698 to the nearest million

Section 1.4 | Rounding and Estimating

Complete the table by estimating the given number to the given place value.

		Tens	Hundreds	Thousands
21.	5281			
22.	7619			
23.	9444			
24.	7777			
25.	14,876			
26.	85,049			

Round each number to the indicated place value.

27. The state of Texas contains 310,850 miles of urban and rural highways. Round this number to the nearest thousand. (*Source:* U.S. Federal Highway Administration)

28. The state of California contains 171,874 miles of urban and rural highways. Round this number to the nearest thousand. (*Source:* U.S. Federal Highway Administration)

29. It takes 60,149 days for Neptune to make a complete orbit around the Sun. Round this number to the nearest hundred. (*Source:* National Space Science Data Center)

30. Kareem Abdul-Jabbar holds the NBA record for points scored, a total of 38,387 over his NBA career. Round this number to the nearest thousand. (*Source:* National Basketball Association)

31. In 2016, the most valuable brand in the world was Apple, Inc. The estimated brand value of Apple at this time was $154,118,000,000. Round this to the nearest ten-billion. (*Source:* Forbes)

32. According to the U.S. Population Clock, the population of the United States was 322,962,019 in November 2016. Round this population figure to the nearest million. (*Source:* U.S. Census Population Clock)

33. The average salary for a professional baseball player in 2016 was $4,155,907. Round this average salary to the nearest hundred-thousand. (*Source:* ESPN)

34. The average salary for a professional football player in 2016 was $2,110,000. Round this average salary to the nearest million. (*Source:* ESPN)

35. The United States currently has 219,600,000 smartphone users, while India has 292,300,000 users. Round each user number to the nearest million. (*Source:* Pew Internet Research)

36. U.S. farms produced 15,226,000,000 bushels of corn in 2016. Round the corn production figure to the nearest ten-million. (*Source:* U.S. Department of Agriculture)

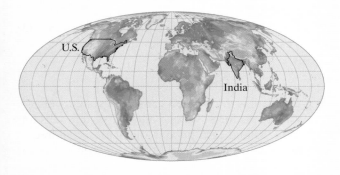

Objective B *Estimate the sum or difference by rounding each number to the nearest ten. See Examples 4 and 5.*

37. 39
     45
     22
    +17

38. 52
     33
     15
    +29

39. 449
    −373

40. 555
    −235

Estimate the sum or difference by rounding each number to the nearest hundred. See Examples 4 and 5.

41. 1913
     1886
    +1925

42. 4050
     3133
    +1220

43. 1774
    −1492

44. 1989
    −1870

45. 3995
     2549
    +4944

46. 799
     1655
    +271

Three of the given calculator answers below are incorrect. Find them by estimating each sum.

47. 463 + 219   600

48. 522 + 785   1307

49. 229 + 443 + 606   1278

50. 542 + 789 + 198   2139

51. 7806 + 5150   12,956

52. 5233 + 4988   9011

Helpful Hint Estimation is useful to check for incorrect answers when using a calculator. For example, pressing a key too hard may result in a double digit, while pressing a key too softly may result in the digit not appearing in the display.

Objective C *Solve each problem by estimating. See Examples 6 and 7.*

53. An appliance store advertises three refrigerators on sale at $899, $1499, and $999. Round each cost to the nearest hundred to estimate the total cost.

54. Suppose you scored 89, 97, 100, 79, 75, and 82 on your biology tests. Round each score to the nearest ten to estimate your total score.

55. The distance from Kansas City to Boston is 1429 miles and from Kansas City to Chicago is 530 miles. Round each distance to the nearest hundred to estimate how much farther Boston is from Kansas City than Chicago is.

56. The Gonzales family took a trip and traveled 588, 689, 277, 143, 59, and 802 miles on six consecutive days. Round each distance to the nearest hundred to estimate the distance they traveled.

57. The peak of Denali, in Alaska, is 20,320 feet above sea level. The top of Mt. Rainier, in Washington, is 14,410 feet above sea level. Round each height to the nearest thousand to estimate the difference in elevation of these two peaks. (*Source:* U.S. Geological Survey) (Note: Denali was formerly known as Mt. McKinley.)

58. A student is pricing new car stereo systems. One system sells for $1895 and another system sells for $1524. Round each price to the nearest hundred dollars to estimate the difference in price of these systems.

59. In 2016, the population of Springfield, Illinois, was 117,006, and the population of Champaign, Illinois, was 83,424. Round each population to the nearest ten-thousand to estimate how much larger Springfield was than Champaign. (*Source:* U.S. Census Bureau)

60. Round each distance given on the map to the nearest ten to estimate the total distance from North Platte, Nebraska, to Lincoln, Nebraska.

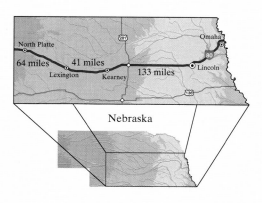

61. Head Start is a national program that provides developmental and social services for America's low-income preschool children ages three to five. Enrollment figures in Head Start programs showed a decrease from 1,128,030 in 2012 to 946,357 in 2016. Round each number of children to the nearest thousand to estimate this decrease. (*Source:* Office of Head Start)

62. Enrollment figures at a local community college showed an increase from 49,713 credit hours in 2005 to 51,746 credit hours in 2006. Round each number to the nearest thousand to estimate the increase.

Mixed Practice (Sections 1.2 and 1.4)
The following table shows a few of the airports in the United States with the largest volumes of passengers. Complete this table. The first line is completed for you. (*Source:* Airports Council International)

	City Location of Airport	Total passengers in 2015 (in hundred-thousands of passengers)	Amount Written in Standard Form	Standard Form Rounded to the Nearest Million	Standard Form Rounded to the Nearest Ten-Million
	Atlanta, GA	1015	101,500,000	102,000,000	100,000,000
63.	Los Angeles, CA	749			
64.	Chicago, IL	769			
65.	Dallas/Fort Worth, TX	641			
66.	New York JFK	568			

Concept Extensions

67. Find one number that when rounded to the nearest hundred is 5700.

68. Find one number that when rounded to the nearest ten is 5700.

69. A number rounded to the nearest hundred is 8600.
    a. Determine the smallest possible number.
    b. Determine the largest possible number.

70. On August 23, 1989, it was estimated that 1,500,000 people joined hands in a human chain stretching 370 miles to protest the fiftieth anniversary of the pact that allowed what was then the Soviet Union to annex the Baltic nations in 1939. If the estimate of the number of people is to the nearest hundred-thousand, determine the largest possible number of people in the chain.

71. In your own words, explain how to round a number to the nearest thousand.

72. In your own words, explain how to round 9660 to the nearest thousand.

73. Estimate the perimeter of the rectangle by first rounding the length of each side to the nearest ten.

74. Estimate the perimeter of the triangle by first rounding the length of each side to the nearest hundred.

1.5 Multiplying Whole Numbers and Area

Objectives

A Use the Properties of Multiplication.

B Multiply Whole Numbers.

C Find the Area of a Rectangle.

D Solve Problems by Multiplying Whole Numbers.

Multiplication Shown as Repeated Addition Suppose that we wish to count the number of laptops provided in a computer class. The laptops are arranged in 5 rows, and each row has 6 laptops.

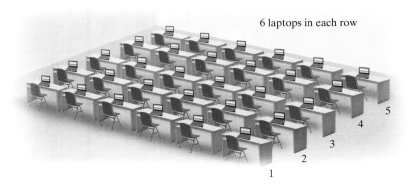

Adding 5 sixes gives the total number of laptops. We can write this as 6 + 6 + 6 + 6 + 6 = 30 laptops. When each addend is the same, we refer to this as **repeated addition.**

Multiplication is repeated addition but with different notation.

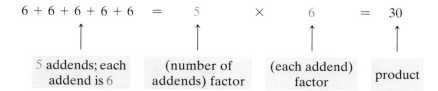

The × is called a **multiplication sign**. The numbers 5 and 6 are called **factors**. The number 30 is called the **product**. The notation 5 × 6 is read as "five times six." The symbols · and ( ) can also be used to indicate multiplication.

$$5 \times 6 = 30, \quad 5 \cdot 6 = 30, \quad (5)(6) = 30, \quad \text{and} \quad 5(6) = 30$$

✓ Concept Check

a. Rewrite $5 + 5 + 5 + 5 + 5 + 5 + 5$ using multiplication.
b. Rewrite 3×16 as repeated addition. Is there more than one way to do this? If so, show all ways.

Objective A Using the Properties of Multiplication

As with addition, we memorize products of one-digit whole numbers and then use certain properties of multiplication to multiply larger numbers. (If necessary, review the multiplication of one-digit numbers.)

Notice that when any number is multiplied by 0, the result is always 0. This is called the **multiplication property of 0.**

Multiplication Property of 0

The product of 0 and any number is 0. For example,

$$5 \cdot 0 = 0 \quad \text{and} \quad 0 \cdot 8 = 0$$

Also notice that when any number is multiplied by 1, the result is always the original number. We call this result the **multiplication property of 1.**

Multiplication Property of 1

The product of 1 and any number is that same number. For example,

$$1 \cdot 9 = 9 \quad \text{and} \quad 6 \cdot 1 = 6$$

Example 1 Multiply.

a. 4×1 **b.** $0(3)$ **c.** $1 \cdot 64$ **d.** $(48)(0)$

Solution:

a. $4 \times 1 = 4$ **b.** $0(3) = 0$
c. $1 \cdot 64 = 64$ **d.** $(48)(0) = 0$

Work Practice 1

Practice 1
Multiply.
a. 6×0
b. $(1)8$
c. $(50)(0)$
d. $75 \cdot 1$

✓ **Concept Check Answers**
a. 7×5
b. $16 + 16 + 16$; yes,
$3 + 3 + 3 + 3 + 3 + 3 + 3 + 3 +$
$3 + 3 + 3 + 3 + 3 + 3 + 3 + 3$

Answers
1. a. 0 **b.** 8 **c.** 0 **d.** 75

Like addition, multiplication is commutative and associative. Notice that when multiplying two numbers, the order of these numbers can be changed without changing the product. For example,

$$3 \cdot 5 = 15 \quad \text{and} \quad 5 \cdot 3 = 15$$

This property is called the **commutative property of multiplication.**

Commutative Property of Multiplication

Changing the **order** of two factors does not change their product. For example,

$$9 \cdot 2 = 18 \quad \text{and} \quad 2 \cdot 9 = 18$$

Another property that can help us when multiplying is the **associative property of multiplication.** This property states that when multiplying numbers, the grouping of the numbers can be changed without changing the product. For example,

$$(2 \cdot 3) \cdot 4 = 6 \cdot 4 = 24$$

Also,

$$2 \cdot (3 \cdot 4) = 2 \cdot 12 = 24$$

Both groupings give a product of 24.

Associative Property of Multiplication

Changing the **grouping** of factors does not change their product. From the previous work, we know that, for example,

$$(2 \cdot 3) \cdot 4 = 2 \cdot (3 \cdot 4)$$

With these properties, along with the **distributive property,** we can find the product of any whole numbers. The distributive property says that multiplication **distributes** over addition. For example, notice that $3(2 + 5)$ simplifies to the same number as $3 \cdot 2 + 3 \cdot 5$.

$$3(2 + 5) = 3(7) = 21$$

$$3 \cdot 2 + 3 \cdot 5 = 6 + 15 = 21$$

Since $3(2 + 5)$ and $3 \cdot 2 + 3 \cdot 5$ both simplify to 21, then

$$3(2 + 5) = 3 \cdot 2 + 3 \cdot 5$$

Notice in $3(2 + 5) = 3 \cdot 2 + 3 \cdot 5$ that each number inside the parentheses is multiplied by 3.

Section 1.5 | Multiplying Whole Numbers and Area   43

Distributive Property

Multiplication distributes over addition. For example,

$$2(3 + 4) = 2 \cdot 3 + 2 \cdot 4$$

Example 2 Rewrite each using the distributive property.

a. $5(6 + 5)$ **b.** $20(4 + 7)$ **c.** $2(7 + 9)$

Solution: Using the distributive property, we have

a. $5(6 + 5) = 5 \cdot 6 + 5 \cdot 5$
b. $20(4 + 7) = 20 \cdot 4 + 20 \cdot 7$
c. $2(7 + 9) = 2 \cdot 7 + 2 \cdot 9$

Work Practice 2

Practice 2

Rewrite each using the distributive property.

a. $6(4 + 5)$
b. $30(2 + 3)$
c. $7(2 + 8)$

Objective B Multiplying Whole Numbers

Let's use the distributive property to multiply 7(48). To do so, we begin by writing the expanded form of 48 (see Section 1.2) and then applying the distributive property.

$$7(48) = 7(40 + 8) \quad \text{Write 48 in expanded form.}$$
$$= 7 \cdot 40 + 7 \cdot 8 \quad \text{Apply the distributive property.}$$
$$= 280 + 56 \quad \text{Multiply.}$$
$$= 336 \quad \text{Add.}$$

This is how we multiply whole numbers. When multiplying whole numbers, we will use the following notation.

First:

$$\begin{array}{r} \overset{5}{4}8 \\ \times\ 7 \\ \hline 336 \end{array} \longleftarrow 7 \cdot 8 = 56 \quad \text{Write 6 in the ones place and carry 5 to the tens place.}$$

Next:

$$\begin{array}{r} \overset{5}{4}8 \\ \times\ 7 \\ \hline 336 \end{array} \quad 7 \cdot 4 + 5 = 28 + 5 = 33$$

The product of 48 and 7 is 336.

Example 3 Multiply:

a. $\begin{array}{r} 25 \\ \times\ 8 \end{array}$ **b.** $\begin{array}{r} 246 \\ \times\ 5 \end{array}$

Solution:

a. $\begin{array}{r} \overset{4}{2}5 \\ \times\ 8 \\ \hline 200 \end{array}$ **b.** $\begin{array}{r} \overset{2\,3}{2}46 \\ \times\ 5 \\ \hline 1230 \end{array}$

Work Practice 3

Practice 3

Multiply.

a. $\begin{array}{r} 29 \\ \times\ 6 \end{array}$ **b.** $\begin{array}{r} 648 \\ \times\ 5 \end{array}$

Answers

2. a. $6(4 + 5) = 6 \cdot 4 + 6 \cdot 5$
   c. $30(2 + 3) = 30 \cdot 2 + 30 \cdot 3$
   d. $7(2 + 8) = 7 \cdot 2 + 7 \cdot 8$
3. a. 174   **b.** 3240

Chapter 1 | The Whole Numbers

To multiply larger whole numbers, use the following similar notation. Multiply 89×52.

Step 1
$$\begin{array}{r} \overset{1}{8}9 \\ \times\ 52 \\ \hline 178 \end{array} \leftarrow \text{Multiply } 89 \times 2.$$

Step 2
$$\begin{array}{r} \overset{4}{8}9 \\ \times\ 52 \\ \hline 178 \\ \times 4450 \end{array} \leftarrow \text{Multiply } 89 \times 50.$$

Step 3
$$\begin{array}{r} 89 \\ \times\ 52 \\ \hline 178 \\ 4450 \\ \hline 4628 \end{array} \text{ Add.}$$

The numbers 178 and 4450 are called **partial products.** The sum of the partial products, 4628, is the product of 89 and 52.

Practice 4
Multiply.

$$\begin{array}{r} 306 \\ \times\ 81 \\ \hline \end{array}$$

Example 4 Multiply: 236×86

Solution:
$$\begin{array}{r} 236 \\ \times\ 86 \\ \hline 1\,416 \\ 18\,880 \\ \hline 20{,}296 \end{array} \begin{array}{l} \leftarrow 6(236) \\ \leftarrow 80(236) \\ \text{Add.} \end{array}$$

■ Work Practice 4

Practice 5
Multiply.

$$\begin{array}{r} 726 \\ \times 142 \\ \hline \end{array}$$

Example 5 Multiply: 631×125

Solution:
$$\begin{array}{r} 631 \\ \times\ 125 \\ \hline 3\,155 \\ 12\,620 \\ 63\,100 \\ \hline 78{,}875 \end{array} \begin{array}{l} \leftarrow 5(631) \\ \leftarrow 20(631) \\ \leftarrow 100(631) \\ \text{Add.} \end{array}$$

■ Work Practice 5

✓ **Concept Check** Find and explain the error in the following multiplication problem.

$$\begin{array}{r} 102 \\ \times\ 33 \\ \hline 306 \\ 306 \\ \hline 612 \end{array}$$

Answers
4. 24,786 5. 103,092

✓ **Concept Check Answer**
$$\begin{array}{r} 102 \\ \times\ 33 \\ \hline 306 \\ 3060 \\ \hline 3366 \end{array}$$

Objective C Finding the Area of a Rectangle

A special application of multiplication is finding the **area** of a region. Area measures the amount of surface of a region. For example, we measure a plot of land or the living space of a home by its area. The figures on the next page show two examples of units of area measure. (A centimeter is a unit of length in the metric system.)

Section 1.5 | Multiplying Whole Numbers and Area

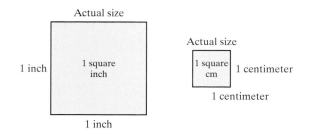

For example, to measure the area of a geometric figure such as the rectangle below, count the number of square units that cover the region.

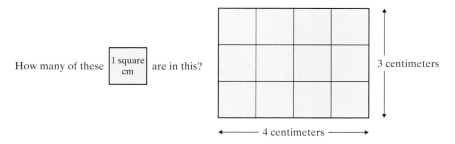

This rectangular region contains 12 square units, each 1 square centimeter. Thus, the area is 12 square centimeters. This total number of squares can be found by counting or by multiplying **4 · 3** (length · width).

Area of a rectangle = length · width
= (4 centimeters)(3 centimeters)
= 12 square centimeters

In this section, we find the areas of rectangles only. In later sections, we will find the areas of other geometric regions.

Helpful Hint

Notice that area is measured in **square** units while perimeter is measured in units.

Example 6 Finding the Area of a State

The state of Colorado is in the shape of a rectangle whose length is 380 miles and whose width is 280 miles. Find its area.

Solution: The area of a rectangle is the product of its length and its width.

Area = length · width
= (380 miles)(280 miles)
= 106,400 square miles

The area of Colorado is 106,400 square miles.

Work Practice 6

Practice 6

The state of Wyoming is in the shape of a rectangle whose length is 360 miles and whose width is 280 miles. Find its area.

Answer
6. 100,800 sq mi

Objective D Solving Problems by Multiplying

There are several words or phrases that indicate the operation of multiplication. Some of these are as follows:

Multiplication		
Key Words or Phrases	Examples	Symbols
multiply	multiply 5 by 7	$5 \cdot 7$
product	the product of 3 and 2	$3 \cdot 2$
times	10 times 13	$10 \cdot 13$

Many key words or phrases describing real-life problems that suggest addition might be better solved by multiplication instead. For example, to find the **total** cost of 8 shirts, each selling for $27, we can either add

$$27 + 27 + 27 + 27 + 27 + 27 + 27 + 27$$

or we can multiply 8(27).

Practice 7

A particular computer printer can print 16 pages per minute in color. How many pages can it print in 45 minutes?

Example 7 Finding DVD Space

A digital video disc (DVD) can hold about 4800 megabytes (MB) of information. How many megabytes can 12 DVDs hold?

Solution: Twelve DVDs will hold 12×4800 megabytes.

In Words	Translate to Numbers
megabytes per DVD $\rightarrow$	4800
$\times$ number of DVDs $\rightarrow$	$\times$ 12
	9600
	48000
total megabytes	57,600

Twelve DVDs will hold 57,600 megabytes.

■ Work Practice 7

Practice 8

A professor of history purchased DVDs and CDs through a club. Each DVD was priced at $11 and each CD cost $9. He bought eight DVDs and five CDs. Find the total cost of the order.

Example 8 Budgeting Money

A woman and her friend plan to take their children to the Georgia Aquarium in Atlanta, the world's largest aquarium. The ticket price for each child is $30 and for each adult, $36. If five children and two adults plan to go, how much money is needed for admission? (*Source:* GeorgiaAquarium.org)

Solution: If the price of one child's ticket is $30, the price for 5 children is $5 \times \$30 = \$150$. The price of one adult ticket is $36, so the price for two adults is $2 \times \$36 = \$72$. The total cost is:

In Words	Translate to Numbers
cost for 5 children $\rightarrow$	150
+ cost for 2 adults $\rightarrow$	+ 72
total cost	222

The total cost is $222.

■ Work Practice 8

Answers
7. 720 pages **8.** $133

Example 9 Estimating Word Count

The average page of a book contains 259 words. Estimate, by rounding each number to the nearest hundred, the total number of words contained on 212 pages.

Solution: The exact number of words is 259×212. Estimate this product by rounding each factor to the nearest hundred.

$$\begin{array}{rl} 259 & \text{rounds to} \quad 300 \\ \times 212 & \text{rounds to} \quad \times 200 \end{array}$$

$300 \times 200 = 60{,}000$

$3 \cdot 2 = 6$

There are approximately 60,000 words contained on 212 pages.

Work Practice 9

Practice 9
If an average page in a book contains 163 words, estimate, by rounding each number to the nearest hundred, the total number of words contained on 391 pages.

Answer
9. 80,000 words

Calculator Explorations Multiplying Numbers

To multiply numbers on a calculator, find the keys marked $\boxed{\times}$ and $\boxed{=}$ or $\boxed{\text{ENTER}}$. For example, to find $31 \cdot 66$ on a calculator, press the keys $\boxed{31}$ $\boxed{\times}$ $\boxed{66}$ then $\boxed{=}$ or $\boxed{\text{ENTER}}$. The display will read $\boxed{2046}$. Thus, $31 \cdot 66 = 2046$.

Use a calculator to multiply.

1. 72×48
2. 81×92
3. $163 \cdot 94$
4. $285 \cdot 144$
5. $983(277)$
6. $1562(843)$

Vocabulary, Readiness & Video Check

Use the choices below to fill in each blank.

| area | grouping | commutative | 1 | product | length |
| factor | order | associative | 0 | distributive | number |

1. The product of 0 and any number is _____.
2. The product of 1 and any number is the _____.
3. In $8 \cdot 12 = 96$, the 96 is called the _____ and 8 and 12 are each called a(n) _____.
4. Since $9 \cdot 10 = 10 \cdot 9$, we say that changing the _____ in multiplication does not change the product. This property is called the _____ property of multiplication.
5. Since $(3 \cdot 4) \cdot 6 = 3 \cdot (4 \cdot 6)$, we say that changing the _____ in multiplication does not change the product. This property is called the _____ property of multiplication.
6. _____ measures the amount of surface of a region.
7. Area of a rectangle = _____ $\cdot$ width.
8. We know $9(10 + 8) = 9 \cdot 10 + 9 \cdot 8$ by the _____ property.

Martin-Gay Interactive Videos

See Video 1.5

Watch the section lecture video and answer the following questions.

Objective A 9. The expression in Example 3 is rewritten using what property?

Objective B 10. During the multiplication process for Example 5, why is a single zero placed at the end of the second partial product?

Objective C 11. Why are the units to the answer to Example 6 not just meters? What are the correct units?

Objective D 12. In Example 7, why can "total" imply multiplication as well as addition?

1.5 Exercise Set MyLab Math

Objective A *Multiply. See Example 1.*

1. $1 \cdot 24$
2. $55 \cdot 1$
3. $0 \cdot 19$
4. $27 \cdot 0$

5. $8 \cdot 0 \cdot 9$
6. $7 \cdot 6 \cdot 0$
7. $87 \cdot 1$
8. $1 \cdot 41$

Use the distributive property to rewrite each expression. See Example 2.

9. $6(3 + 8)$
10. $5(8 + 2)$
11. $4(3 + 9)$

12. $6(1 + 4)$
13. $20(14 + 6)$
14. $12(12 + 3)$

Objective B *Multiply. See Example 3.*

15. 64×8
16. 79×3
17. 613×6
18. 638×5

19. 277×6
20. 882×2
21. 1074×6
22. 9021×3

Objective A B Mixed Practice *Multiply. See Examples 1 through 5.*

23. 89×13
24. 91×72
25. 421×58
26. 526×23
27. 306×81
28. 708×21

29. $(780)(20)$
30. $(720)(80)$
31. $(495)(13)(0)$
32. $(593)(47)(0)$
33. $(640)(1)(10)$

34. (240)(1)(20)   **35.** 1234 × 39   **36.** 1357 × 79   **37.** 609 × 234   **38.** 807 × 127

39. 8649   **40.** 1234   **41.**   589   **42.**   426   **43.** 1941   **44.** 1876
       × 274   × 567   × 110   × 110   × 2035   × 1407

Objective C Mixed Practice (Sections 1.3 and 1.5) *Find the area and the perimeter of each rectangle. See Example 6.*

45.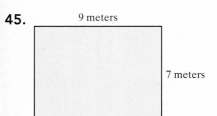
9 meters, 7 meters

△ **46.**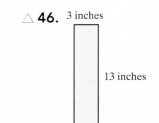
3 inches, 13 inches

47.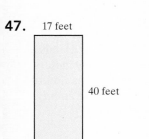
17 feet, 40 feet

△ **48.**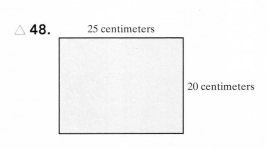
25 centimeters, 20 centimeters

Objective D *Estimate the products by rounding each factor to the nearest hundred. See Example 9.*

49. 576 × 354   **50.** 982 × 650   **51.** 604 × 451   **52.** 111 × 999

Without actually calculating, mentally round, multiply, and choose the best estimate.

53. 38 × 42 =
  a. 16
  b. 160
  c. 1600
  d. 16,000

54. 2872 × 12 =
  a. 2872
  b. 28,720
  c. 287,200
  d. 2,872,000

55. 612 × 29 =
  a. 180
  b. 1800
  c. 18,000
  d. 180,000

56. 706 × 409 =
  a. 280
  b. 2800
  c. 28,000
  d. 280,000

Objectives C D Mixed Practice–Translating *Solve. See Examples 6 through 9.*

57. Multiply 80 by 11.   **58.** Multiply 70 by 12.   **59.** Find the product of 6 and 700.

60. Find the product of 9 and 900.   **61.** Find 2 times 2240.   **62.** Find 3 times 3310.

63. One tablespoon of olive oil contains 125 calories. How many calories are in 3 tablespoons of olive oil? (*Source: Home and Garden Bulletin No. 72, U.S. Department of Agriculture*).

64. One ounce of hulled sunflower seeds contains 14 grams of fat. How many grams of fat are in 8 ounces of hulled sunflower seeds? (*Source: Home and Garden Bulletin No. 72, U.S. Department of Agriculture*).

65. The textbook for a course in biology costs $94. There are 35 students in the class. Find the total cost of the biology books for the class.

66. The seats in a large lecture hall are arranged in 14 rows with 34 seats in each row. Find how many seats are in this room.

67. Cabot Creamery is packing a pallet of 20-lb boxes of cheddar cheese to send to a local restaurant. There are five layers of boxes on the pallet, and each layer is four boxes wide by five boxes deep.
   a. How many boxes are in one layer?
   b. How many boxes are on the pallet?
   c. What is the weight of the cheese on the pallet?

68. An apartment building has *three floors*. Each floor has five rows of apartments with four apartments in each row.
   a. How many apartments are on 1 floor?
   b. How many apartments are in the building?

△ **69.** A plot of land measures 80 feet by 110 feet. Find its area.

△ **70.** A house measures 45 feet by 60 feet. Find the floor area of the house.

△ **71.** The largest hotel lobby can be found at the Hyatt Regency in San Francisco, CA. It is in the shape of a rectangle that measures 350 feet by 160 feet. Find its area.

△ **72.** Recall from an earlier section that the second-largest commercial building in the world under one roof is the flower auction building of the cooperative VBA in Aalsmeer, Netherlands. The floor plan is a rectangle that measures 776 meters by 639 meters. Find the area of this building. (A meter is a unit of length in the metric system.) (*Source: The Handy Science Answer Book,* Visible Ink Press)

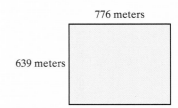

73. A pixel is a rectangular dot on a graphing calculator screen. If a graphing calculator screen contains 62 pixels in a row and 94 pixels in a column, find the total number of pixels on a screen.

74. A certain compact disc (CD) can hold 700 megabytes (MB) of information. How many MBs can 17 discs hold?

75. A line of print on a computer contains 60 characters (letters, spaces, punctuation marks). Find how many characters there are in 35 lines.

76. An average cow eats 3 pounds of grain per day. Find how much grain a cow eats in a year. (Assume 365 days in 1 year.)

77. One ounce of Planters Dry Roasted Peanuts has 160 calories. How many calories are in 8 ounces? (*Source:* RJR Nabisco, Inc.)

78. One ounce of Planters Dry Roasted Peanuts has 13 grams of fat. How many grams of fat are in 16 ounces? (*Source:* RJR Nabisco, Inc.)

79. The Thespian club at a local community college is ordering T-shirts. T-shirts size S, M, or L cost $10 each and T-shirts size XL or XXL cost $12 each. Complete the table below and use it to find the total cost. (The first row is filled in for you.)

T-Shirt Size	Number of Shirts Ordered	Cost per Shirt	Cost per Size Ordered
S	4	$10	$40
M	6		
L	20		
XL	3		
XXL	3		

80. The student activities group at North Shore Community College is planning a trip to see the local minor league baseball team. Tickets cost $5 for students, $7 for nonstudents, and $2 for children under 12. Complete the following table and use it to find the total cost.

Person	Number of Persons	Cost per Person	Cost per Category
Student	24	$5	$120
Nonstudent	4		
Children under 12	5		

81. Celestial Seasonings of Boulder, Colorado, is a tea company that specializes in herbal teas. Their plant in Boulder has bagging machines capable of bagging over 1000 bags of tea per minute. If the plant runs 24 hours day, how many tea bags are produced in one day? (*Source:* Celestial Seasonings)

82. There were about 3 million "older" Americans (ages 65 and older) in 1900. By 2020, this number is projected to increase eighteen times. Find this projected number of "older" Americans in 2020. (*Source:* Administration on Aging, U.S. Census Bureau)

Mixed Practice (Sections 1.3 and 1.5) *Perform each indicated operation.*

83. 128
     + 7

84. 126
     − 8

85. 134
     × 16

86. 47 + 26 + 10 + 231 + 50

87. Find the sum of 19 and 4.

88. Find the product of 19 and 4.

89. Find the difference of 19 and 4.

90. Find the total of 14 and 9.

Concept Extensions

Solve. See the first Concept Check in this section.

91. Rewrite 6 + 6 + 6 + 6 + 6 using multiplication.

92. Rewrite 11 + 11 + 11 + 11 + 11 + 11 using multiplication.

93. a. Rewrite 3 · 5 as repeated addition.
   b. Explain why there is more than one way to do this.

94. a. Rewrite 4 · 5 as repeated addition.
   b. Explain why there is more than one way to do this.

Find and explain the error in each multiplication problem. See the second Concept Check in this section.

95.
```
   203
 ×  14
   812
   203
  1015
```

96.
```
   31
 × 50
  155
```

Fill in the missing digits in each problem.

97.
```
     4_
 ×  _3
    126
   3780
   3906
```

98.
```
     _7
 ×  6_
    171
   3420
   3591
```

99. Explain how to multiply two 2-digit numbers using partial products.

100. In your own words, explain the meaning of the area of a rectangle and how this area is measured.

101. A window washer in New York City is bidding for a contract to wash the windows of a 23-story building. To write a bid, the number of windows in the building is needed. If there are 7 windows in each row of windows on 2 sides of the building and 4 windows per row on the other 2 sides of the building, find the total number of windows.

102. During the NBA's 2015–2016 regular season, LeBron James of the Cleveland Cavaliers scored 87 three-point field goals, 650 two-point field goals, and 359 free throws (worth one point each). How many points did LeBron James score during the 2015–2016 regular season? (*Source:* NBA)

1.6 Dividing Whole Numbers

Objectives

A Divide Whole Numbers.

B Perform Long Division.

C Solve Problems That Require Dividing by Whole Numbers.

D Find the Average of a List of Numbers.

Suppose three people pooled their money and bought a raffle ticket at a local fundraiser. Their ticket was the winner and they won a $75 cash prize. They then divided the prize into three equal parts so that each person received $25.

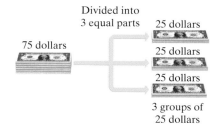

Objective A Dividing Whole Numbers

The process of separating a quantity into equal parts is called **division**. The division above can be symbolized by several notations.

Section 1.6 | Dividing Whole Numbers   53

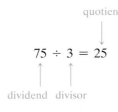

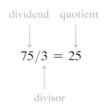

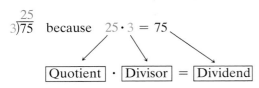

(In the notation $\frac{75}{3}$, the bar separating 75 and 3 is called a **fraction bar**.) Just as subtraction is the reverse of addition, division is the reverse of multiplication. This means that division can be checked by multiplication.

$3\overline{)75}^{25}$ because $25 \cdot 3 = 75$

$\boxed{\text{Quotient}} \cdot \boxed{\text{Divisor}} = \boxed{\text{Dividend}}$

Since multiplication and division are related in this way, you can use your knowledge of multiplication facts to review quotients of one-digit divisors if necessary.

Example 1 Find each quotient. Check by multiplying.

a. $42 \div 7$ **b.** $\frac{64}{8}$ **c.** $3\overline{)21}$

Solution:

a. $42 \div 7 = 6$ because $6 \cdot 7 = 42$

b. $\frac{64}{8} = 8$ because $8 \cdot 8 = 64$

c. $3\overline{)21}^{\,7}$ because $7 \cdot 3 = 21$

■ Work Practice 1

Practice 1

Find each quotient. Check by multiplying.

a. $9\overline{)72}$

b. $40 \div 5$

c. $\frac{24}{6}$

Example 2 Find each quotient. Check by multiplying.

a. $1\overline{)7}$ **b.** $25 \div 1$ **c.** $\frac{6}{6}$ **d.** $12 \div 12$ **e.** $\frac{20}{1}$ **f.** $18\overline{)18}$

Solution:

a. $1\overline{)7}^{\,7}$ because $7 \cdot 1 = 7$ **b.** $25 \div 1 = 25$ because $25 \cdot 1 = 25$

c. $\frac{6}{6} = 1$ because $1 \cdot 6 = 6$ **d.** $12 \div 12 = 1$ because $1 \cdot 12 = 12$

e. $\frac{20}{1} = 20$ because $20 \cdot 1 = 20$ **f.** $18\overline{)18}^{\,1}$ because $1 \cdot 18 = 18$

■ Work Practice 2

Practice 2

Find each quotient. Check by multiplying.

a. $\frac{7}{7}$ **b.** $5 \div 1$

c. $1\overline{)11}$ **d.** $4 \div 1$

e. $\frac{10}{1}$ **f.** $21 \div 21$

Answers
1. a. 8 **b.** 8 **c.** 4 **2. a.** 1 **b.** 5
c. 11 **d.** 4 **e.** 10 **f.** 1

Example 2 illustrates the important properties of division described next:

Division Properties of 1

The quotient of any number (except 0) and that same number is 1. For example,

$$8 \div 8 = 1 \qquad \frac{5}{5} = 1 \qquad 4\overline{)4} = 1$$

The quotient of any number and 1 is that same number. For example,

$$9 \div 1 = 9 \qquad \frac{6}{1} = 6 \qquad 1\overline{)3} = 3 \qquad \frac{0}{1} = 0$$

Practice 3

Find each quotient. Check by multiplying.

a. $\dfrac{0}{7}$ b. $8\overline{)0}$

c. $7 \div 0$ d. $0 \div 14$

Example 3 Find each quotient. Check by multiplying.

a. $9\overline{)0}$ b. $0 \div 12$ c. $\dfrac{0}{5}$ d. $\dfrac{3}{0}$

Solution:

a. $9\overline{)0} = 0$ because $0 \cdot 9 = 0$

b. $0 \div 12 = 0$ because $0 \cdot 12 = 0$

c. $\dfrac{0}{5} = 0$ because $0 \cdot 5 = 0$

d. If $\dfrac{3}{0} = $ a *number*, then the *number* times $0 = 3$. Recall from Section 1.5 that any number multiplied by 0 is 0 and not 3. We say, then, that $\dfrac{3}{0}$ is **undefined.**

■ Work Practice 3

Example 3 illustrates important division properties of 0.

Division Properties of 0

The quotient of 0 and any number (except 0) is 0. For example,

$$0 \div 9 = 0 \qquad \frac{0}{5} = 0 \qquad 14\overline{)0} = 0$$

The quotient of any number and 0 is not a number. We say that

$$\frac{3}{0}, \qquad 0\overline{)3}, \qquad \text{and} \qquad 3 \div 0$$

are **undefined.**

Objective B Performing Long Division

When dividends are larger, the quotient can be found by a process called **long division.** For example, let's divide 2541 by 3.

$$\text{divisor} \rightarrow 3\overline{)2541}$$
$$\uparrow$$
$$\text{dividend}$$

We can't divide 3 into 2, so we try dividing 3 into the first two digits.

$$3\overline{)2541}^{\,8} \qquad 25 \div 3 = 8 \text{ with 1 left, so our best estimate is 8.}$$
We place 8 over the 5 in 25.

Answers
3. a. 0 **b.** 0 **c.** undefined **d.** 0

Section 1.6 | Dividing Whole Numbers

Next, multiply 8 and 3 and subtract this product from 25. Make sure that this difference is less than the divisor.

$$\begin{array}{r} 8 \\ 3\overline{)2541} \\ -24 \\ \hline 1 \end{array}$$

$8(3) = 24$

$25 - 24 = 1$, and 1 is less than the divisor 3.

Bring down the next digit and go through the process again.

$$\begin{array}{r} 84 \\ 3\overline{)2541} \\ -24\downarrow \\ \hline 14 \\ -12 \\ \hline 2 \end{array}$$

$14 \div 3 = 4$ with 2 left

$4(3) = 12$

$14 - 12 = 2$

Once more, bring down the next digit and go through the process.

$$\begin{array}{r} 847 \\ 3\overline{)2541} \\ -24 \\ \hline 14 \\ -12\downarrow \\ \hline 21 \\ -21 \\ \hline 0 \end{array}$$

$21 \div 3 = 7$

$7(3) = 21$

$21 - 21 = 0$

The quotient is 847. To check, see that $847 \times 3 = 2541$.

Example 4
Divide: $3705 \div 5$. Check by multiplying.

Solution:

$$\begin{array}{r} 7 \\ 5\overline{)3705} \\ -35\downarrow \\ \hline 20 \end{array}$$

$37 \div 5 = 7$ with 2 left. Place this estimate, 7, over the 7 in 37.

$7(5) = 35$

$37 - 35 = 2$, and 2 is less than the divisor 5.

↳ Bring down the 0.

$$\begin{array}{r} 74 \\ 5\overline{)3705} \\ -35 \\ \hline 20 \\ -20\downarrow \\ \hline 05 \end{array}$$

$20 \div 5 = 4$

$4(5) = 20$

$20 - 20 = 0$, and 0 is less than the divisor 5.

↳ Bring down the 5.

$$\begin{array}{r} 741 \\ 5\overline{)3705} \\ -35 \\ \hline 20 \\ -20\downarrow \\ \hline 5 \\ -5 \\ \hline 0 \end{array}$$

$5 \div 5 = 1$

$1(5) = 5$

$5 - 5 = 0$

Practice 4
Divide. Check by multiplying.

a. $4908 \div 6$

b. $2212 \div 4$

c. $753 \div 3$

Answers
4. a. 818 **b.** 553 **c.** 251

(Continued on next page)

Helpful Hint Since division and multiplication are reverse operations, don't forget that a division problem can be checked by multiplying.

Practice 5

Divide and check by multiplying.

a. $7\overline{)2128}$

b. $9\overline{)45,900}$

Check:

$$\begin{array}{r} 741 \\ \times5 \\ \hline 3705 \end{array}$$

■ Work Practice 4

Example 5 Divide and check: $1872 \div 9$

Solution:

$$\begin{array}{r} 208 \\ 9\overline{)1872} \\ -18\downarrow \\ \hline 07 \\ -0\downarrow \\ \hline 72 \\ -72 \\ \hline 0 \end{array}$$

$2(9) = 18$
$18 - 18 = 0$; bring down the 7.
$0(9) = 0$
$7 - 0 = 7$; bring down the 2.
$8(9) = 72$
$72 - 72 = 0$

Check: $208 \cdot 9 = 1872$

■ Work Practice 5

Naturally, quotients don't always "come out even." Making 4 rows out of 26 chairs, for example, isn't possible if each row is supposed to have exactly the same number of chairs. Each of 4 rows can have 6 chairs, but 2 chairs are left over.

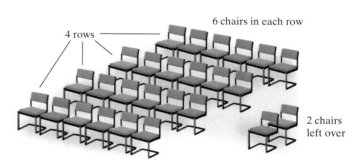

We signify "leftovers" or **remainders** in this way:

$$4\overline{)26}^{6\text{ R }2}$$

The **whole number part of the quotient** is 6; the **remainder part of the quotient** is 2. Checking by multiplying,

whole number part	·	divisor	+	remainder part	=	dividend
↓		↓		↓		↓
6	·	4	+	2		
		24	+	2	=	26

Answers
5. a. 304 b. 5100

Section 1.6 | Dividing Whole Numbers

Example 6 Divide and check: 2557 ÷ 7

Solution:

```
      365  R 2
   7)2557
    -21↓
      45
     -42↓
      37
     -35
       2
```

3(7) = 21
25 − 21 = 4; bring down the 5.
6(7) = 42
45 − 42 = 3; bring down the 7.
5(7) = 35
37 − 35 = 2; the remainder is 2.

Check: 365 · 7 + 2 = 2557

whole number part · divisor + remainder part = dividend

■ Work Practice 6

Practice 6
Divide and check.
a. 4)939
b. 5)3287

Example 7 Divide and check: 56,717 ÷ 8

Solution:

```
       7089  R 5
   8)56717
    -56↓
      07
      -0↓
      71
     -64↓
      77
     -72
       5
```

7(8) = 56
Subtract and bring down the 7.
0(8) = 0
Subtract and bring down the 1.
8(8) = 64
Subtract and bring down the 7.
9(8) = 72
Subtract. The remainder is 5.

Check: 7089 · 8 + 5 = 56,717

whole number part · divisor + remainder part = dividend

■ Work Practice 7

Practice 7
Divide and check.
a. 9)81,605
b. 4)23,310

When the divisor has more than one digit, the same pattern applies. For example, let's find 1358 ÷ 23.

```
         5
   23)1358
     -115↓
       208
```

135 ÷ 23 = 5 with 20 left over. Our estimate is 5.
5(23) = 115
135 − 115 = 20. Bring down the 8.

Answers
6. a. 234 R 3 **b.** 657 R 2
7. a. 9067 R 2 **b.** 5827 R 2

(Continued on next page)

Chapter 1 | The Whole Numbers

Now we continue estimating.

$$\begin{array}{r} 59 \text{ R } 1 \\ 23\overline{)1358} \\ -115 \\ \hline 208 \\ -207 \\ \hline 1 \end{array}$$

$208 \div 23 = 9$ with 1 left over.

$9(23) = 207$

$208 - 207 = 1$. The remainder is 1.

To check, see that $59 \cdot 23 + 1 = 1358$.

Practice 8

Divide: $8920 \div 17$

Example 8 Divide: $6819 \div 17$

Solution:

$$\begin{array}{r} 401 \text{ R } 2 \\ 17\overline{)6819} \\ -68 \\ \hline 01 \\ -0 \\ \hline 19 \\ -17 \\ \hline 2 \end{array}$$

$4(17) = 68$
Subtract and bring down the 1.
$0(17) = 0$
Subtract and bring down the 9.
$1(17) = 17$
Subtract. The remainder is 2.

To check, see that $401 \cdot 17 + 2 = 6819$.

■ Work Practice 8

Practice 9

Divide: $33{,}282 \div 678$

Example 9 Divide: $51{,}600 \div 403$

Solution:

$$\begin{array}{r} 128 \text{ R } 16 \\ 403\overline{)51600} \\ -403 \\ \hline 1130 \\ -806 \\ \hline 3240 \\ -3224 \\ \hline 16 \end{array}$$

$1(403) = 403$
Subtract and bring down the 0.
$2(403) = 806$
Subtract and bring down the 0.
$8(403) = 3224$
Subtract. The remainder is 16.

To check, see that $128 \cdot 403 + 16 = 51{,}600$.

■ Work Practice 9

Division Shown as Repeated Subtraction To further understand division, recall from Section 1.5 that addition and multiplication are related in the following manner:

$$\underbrace{3 + 3 + 3 + 3}_{\text{4 addends: each addend is 3}} = 4 \times 3 = 12$$

In other words, multiplication is repeated addition. Likewise, division is repeated subtraction.

For example, let's find

$35 \div 8$

Answers

8. 524 R 12 **9.** 49 R 60

by repeated subtraction. Keep track of the number of times 8 is subtracted from 35. We are through when we can subtract no more because the difference is less than 8.

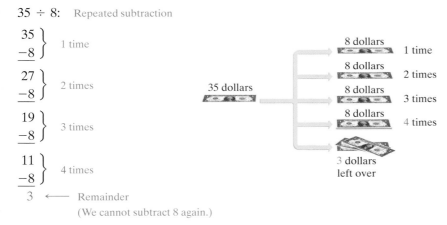

Thus, $35 \div 8 = 4 \text{ R } 3$.

To check, perform the same multiplication as usual, but finish by adding in the remainder.

$$\underbrace{4}_{\text{whole number part of quotient}} \cdot \underbrace{8}_{\text{divisor}} + \underbrace{3}_{\text{remainder}} = \underbrace{35}_{\text{dividend}}$$

Objective C Solving Problems by Dividing

Below are some key words and phrases that may indicate the operation of division:

Division		
Key Words or Phrases	**Examples**	**Symbols**
divide	divide 10 by 5	$10 \div 5$ or $\frac{10}{5}$
quotient	the quotient of 64 and 4	$64 \div 4$ or $\frac{64}{4}$
divided by	9 divided by 3	$9 \div 3$ or $\frac{9}{3}$
divided or shared equally among	$100 divided equally among five people	$100 \div 5$ or $\frac{100}{5}$
per	100 miles per 2 hours	$\frac{100 \text{ miles}}{2 \text{ hours}}$

✓ **Concept Check** Which of the following is the correct way to represent "the quotient of 60 and 12"? Or are both correct? Explain your answer.
a. $12 \div 60$
b. $60 \div 12$

✓ **Concept Check Answers**
a. incorrect
b. correct

Practice 10

Three students bought 171 blank CDs to share equally. How many CDs did each person get?

Example 10 — Finding Shared Earnings

Three college freshmen share a paper route to earn money for expenses. The total in their fund after expenses is $2895. How much is each person's equal share?

Solution:

In words: Each person's share = total money ÷ number of persons

Translate: Each person's share = 2895 ÷ 3

Then
```
       965
   3)2895
    -27
     ───
      19
     -18
      ──
      15
     -15
      ──
       0
```

Each person's share is $965.

■ Work Practice 10

Practice 11

Printers can be packed 12 to a box. If 532 printers are to be packed but only full boxes are shipped, how many full boxes will be shipped? How many printers are left over and not shipped?

Example 11 — Dividing Number of Downloads

As part of a promotion, an executive receives 238 cards, each good for one free song download. If she wants to share them evenly with 19 friends, how many download cards will each friend receive? How many will be left over?

Solution:

In words: Number of cards for each person = number of cards ÷ number of friends

Translate: Number of cards for each person = 238 ÷ 19

```
       12 R 10
   19)238
    -19
     ──
      48
     -38
      ──
      10
```

Each friend will receive 12 download cards. The cards cannot be divided equally among her friends since there is a nonzero remainder. There will be 10 download cards left over.

■ Work Practice 11

Answers
10. 57 CDs
11. 44 full boxes; 4 printers left over

Objective D Finding Averages

A special application of division (and addition) is finding the average of a list of numbers. The **average** of a list of numbers is the sum of the numbers divided by the *number* of numbers.

$$\text{average} = \frac{\text{sum of numbers}}{\textit{number} \text{ of numbers}}$$

Example 12 Averaging Scores

A mathematics instructor is checking a simple program she wrote for averaging the scores of her students. To do so, she averages a student's scores of 75, 96, 81, and 88 by hand. Find this average score.

Solution: To find the average score, we find the sum of the student's scores and divide by 4, the number of scores.

```
  75
  96
  81
+ 88
 340  sum
```

$$\text{average} = \frac{340}{4} = 85$$

```
      85
   4)340
     -32
      20
     -20
       0
```

The average score is 85.

■ Work Practice 12

Practice 12

To compute a safe time to wait for reactions to occur after allergy shots are administered, a lab technician is given a list of elapsed times between administered shots and reactions. Find the average of the times 4 minutes, 7 minutes, 35 minutes, 16 minutes, 9 minutes, 3 minutes, and 52 minutes.

Answer
12. 18 min

Calculator Explorations Dividing Numbers

To divide numbers on a calculator, find the keys marked ÷ and = or ENTER. For example, to find 435 ÷ 5 on a calculator, press the keys 435 ÷ 5 then = or ENTER. The display will read 87. Thus, 435 ÷ 5 = 87.

Use a calculator to divide.

1. 848 ÷ 16
2. 564 ÷ 12
3. 95)5890
4. 27)1053
5. $\dfrac{32{,}886}{126}$
6. $\dfrac{143{,}088}{264}$
7. 0 ÷ 315
8. 315 ÷ 0

Vocabulary, Readiness & Video Check

Use the choices below to fill in each blank. Some choices may be used more than once.

1	number	divisor	dividend
0	undefined	average	quotient

1. In 90 ÷ 2 = 45, the answer 45 is called the _____, 90 is called the _____, and 2 is called the _____.
2. The quotient of any number and 1 is the same _____.
3. The quotient of any number (except 0) and the same number is _____.
4. The quotient of 0 and any number (except 0) is _____.
5. The quotient of any number and 0 is _____.
6. The _____ of a list of numbers is the sum of the numbers divided by the _____ of numbers.

Martin-Gay Interactive Videos

Watch the section lecture video and answer the following questions.

Objective A 7. Look at Examples 6–8. What number can never be the divisor in division?

Objective B 8. In Example 10, how many 102s are in 21? How does this result affect the quotient?

9. What calculation would you use to check the answer in Example 10?

Objective C 10. In Example 11, what is the importance of knowing that the distance to each hole is the same?

Objective D 11. As shown in Example 12, what two operations are used when finding an average?

See Video 1.6

1.6 Exercise Set MyLab Math

Objective A *Find each quotient. See Examples 1 through 3.*

1. $54 \div 9$
2. $72 \div 9$
3. $36 \div 3$
4. $24 \div 3$
5. $0 \div 8$
6. $0 \div 4$
7. $31 \div 1$
8. $38 \div 1$
9. $\dfrac{18}{18}$
10. $\dfrac{49}{49}$
11. $\dfrac{24}{3}$
12. $\dfrac{45}{9}$
13. $26 \div 0$
14. $\dfrac{12}{0}$
15. $26 \div 26$
16. $6 \div 6$
17. $0 \div 14$
18. $7 \div 0$
19. $18 \div 2$
20. $18 \div 3$

Objectives A B Mixed Practice *Divide and then check by multiplying. See Examples 1 through 5.*

21. $3\overline{)87}$
22. $5\overline{)85}$
23. $3\overline{)222}$
24. $8\overline{)640}$
25. $3\overline{)1014}$
26. $4\overline{)2104}$
27. $\dfrac{30}{0}$
28. $\dfrac{0}{30}$
29. $63 \div 7$
30. $56 \div 8$
31. $150 \div 6$
32. $121 \div 11$

Divide and then check by multiplying. See Examples 6 and 7.

33. $7\overline{)479}$
34. $7\overline{)426}$
35. $6\overline{)1421}$
36. $3\overline{)1240}$
37. $305 \div 8$
38. $167 \div 3$
39. $2286 \div 7$
40. $3333 \div 4$

Divide and then check by multiplying. See Examples 8 and 9.

41. $55\overline{)715}$ **42.** $23\overline{)736}$ **43.** $23\overline{)1127}$ **44.** $42\overline{)2016}$ **45.** $97\overline{)9417}$

46. $44\overline{)1938}$ **47.** $3146 \div 15$ **48.** $7354 \div 12$ **49.** $6578 \div 13$ **50.** $5670 \div 14$

51. $9299 \div 46$ **52.** $2505 \div 64$ **53.** $\dfrac{12{,}744}{236}$ **54.** $\dfrac{5781}{123}$ **55.** $\dfrac{10{,}297}{103}$

56. $\dfrac{23{,}092}{240}$ **57.** $20{,}619 \div 102$ **58.** $40{,}853 \div 203$ **59.** $244{,}989 \div 423$ **60.** $164{,}592 \div 543$

Divide. See Examples 1 through 9.

61. $7\overline{)119}$ **62.** $8\overline{)104}$ **63.** $7\overline{)3580}$ **64.** $5\overline{)3017}$

65. $40\overline{)85{,}312}$ **66.** $50\overline{)85{,}747}$ **67.** $142\overline{)863{,}360}$ **68.** $214\overline{)650{,}560}$

Objective C Translating *Solve. See Examples 10 and 11.*

69. Find the quotient of 117 and 5.

70. Find the quotient of 94 and 7.

71. Find 200 divided by 35.

72. Find 116 divided by 32.

73. Find the quotient of 62 and 3.

74. Find the quotient of 78 and 5.

Solve.

75. Amad Ford teaches American Sign Language classes for $65 per student for a 7-week session. He collects $2145 from the group of students. Find how many students are in the group.

76. Kathy Gomez teaches Spanish lessons for $85 per student for a 5-week session. From one group of students, she collects $4930. Find how many students are in the group.

77. The gravity of Jupiter is 318 times as strong as the gravity of Earth, so objects on Jupiter weigh 318 times as much as they weigh on Earth. If a person would weigh 52,470 pounds on Jupiter, find how much the person weighs on Earth.

78. Twenty-one people pooled their money and bought lottery tickets. One ticket won a prize of $5,292,000. Find how many dollars each person received.

79. An 18-hole golf course is 5580 yards long. If the distance to each hole is the same, find the distance between holes.

80. A truck hauls wheat to a storage granary. It carries a total of 5768 bushels of wheat in 14 trips. How much does the truck haul each trip if each trip it hauls the same amount?

81. There is a bridge over highway I-35 every three miles. The first bridge is at the beginning of a 265-mile stretch of highway. Find how many bridges there are over 265 miles of I-35.

82. The white stripes dividing the lanes on a highway are 25 feet long, and the spaces between them are 25 feet long. Let's call a "lane divider" a stripe followed by a space. Find how many whole "lane dividers" there are in 1 mile of highway. (A mile is 5280 feet.)

83. Light poles along a highway are placed 492 feet apart. The first light pole is at the beginning of a 1-mile strip. Find how many poles should be ordered for the 1-mile strip of highway. (A mile is 5280 feet.)

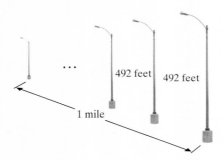

84. A piece of rope 185 feet long is to be cut into pieces for an experiment in a physics class. Each piece of rope is to be 8 feet long. Determine whether there is enough rope for 22 students. Determine the amount extra or the amount short.

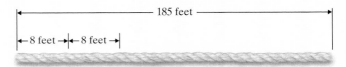

85. Broad Peak in Pakistan is the twelfth-tallest mountain in the world. Its elevation is 26,400 feet. A mile is 5280 feet. How many miles tall is Broad Peak? (*Source:* National Geographic Society)

86. Doug Baldwin of the Seattle Seahawks, Brandon Marshall of the New York Jets, and Allen Robinson of the Jacksonville Jaguars tied to lead the NFL in scoring touchdowns during the 2015 regular football season, scoring a total of 84 points each. If a touchdown is worth 6 points, how many touchdowns did each of these players make during the 2015 season? (*Source:* NFL)

87. Find how many yards are in 1 mile. (A mile is 5280 feet; a yard is 3 feet.)

88. Find how many whole feet are in 1 rod. (A mile is 5280 feet; 1 mile is 320 rods.)

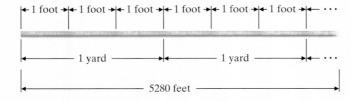

Objective D *Find the average of each list of numbers. See Example 12.*

89. 10, 24, 35, 22, 17, 12

90. 37, 26, 15, 29, 51, 22

91. 205, 972, 210, 161

92. 121, 200, 185, 176, 163

93. 86, 79, 81, 69, 80

94. 92, 96, 90, 85, 92, 79

The normal monthly temperatures in degrees Fahrenheit for Salt Lake City, Utah, are given in the graph. Use this graph to answer Exercises 95 and 96. (Source: National Climatic Data Center)

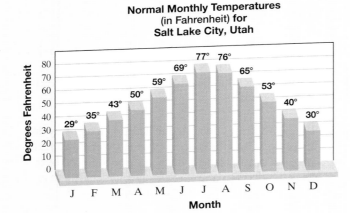

95. Find the average temperature for June, July, and August.

96. Find the average temperature for October, November, and December.

Section 1.6 | Dividing Whole Numbers   65

Mixed Practice (Sections 1.3, 1.5, and 1.6) *Perform each indicated operation. Watch the operation symbol.*

97. 82 + 463 + 29 + 8704

98. 23 + 407 + 92 + 7011

99. 546
× 28

100. 712
× 54

101. 722
− 43

102. 712
− 54

103. $\frac{45}{0}$

104. $\frac{0}{23}$

105. 228 ÷ 24

106. 304 ÷ 31

Concept Extensions

Match each word phrase to the correct translation. (Not all letter choices will be used.) See the Concept Check in this section.

107. The quotient of 40 and 8

108. The quotient of 200 and 20

a. 20 ÷ 200   **b.** 200 ÷ 20
c. 40 ÷ 8   **d.** 8 ÷ 40

109. 200 divided by 20

110. 40 divided by 8

The following table shows the top five automakers and the amount of money they spent on U.S. television advertising during a recent year. Use this table to answer Exercises 111 and 112. (Source: Automotive News)

Automotive Company	Amount Spent on Television Advertising During a Recent Year
General Motors	$3,500,000,000
Ford	$2,680,000,000
Fiat Chrysler Automobiles	$2,250,000,000
Toyota	$1,800,000,000
Nissan	$1,140,000,000

111. Find the average amount of money spent on television ads by the top two advertisers.

112. Find the average amount of money spent on television ads by the top four advertisers.

In Example 12 in this section, we found that the average of 75, 96, 81, and 88 is 85. Use this information to answer Exercises 113 and 114.

113. If the number 75 is removed from the list of numbers, does the average increase or decrease? Explain why.

114. If the number 96 is removed from the list of numbers, does the average increase or decrease? Explain why.

115. Without computing it, tell whether the average of 126, 135, 198, and 113 is 86. Explain why it is possible or why it is not.

116. Without computing it, tell whether the average of 38, 27, 58, and 43 is 17. Explain why it is possible or why it is not.

117. If the area of a rectangle is 60 square feet and its width is 5 feet, what is its length?

118. If the area of a rectangle is 84 square inches and its length is 21 inches, what is its width?

119. Write down any two numbers whose quotient is 25.

120. Write down any two numbers whose quotient is 1.

121. Find 26 ÷ 5 using the process of repeated subtraction.

122. Find 86 ÷ 10 using the process of repeated subtraction.

Integrated Review Sections 1.2–1.6

Operations on Whole Numbers

Answers

1. ______
2. ______
3. ______
4. ______
5. ______
6. ______
7. ______
8. ______
9. ______
10. ______
11. ______
12. ______
13. ______
14. ______
15. ______
16. ______
17. ______
18. ______
19. ______
20. ______
21. ______
22. ______
23. ______
24. ______
25. ______
26. ______
27. ______
28. ______
29. ______
30. ______

Perform each indicated operation.

1. $\begin{array}{r} 42 \\ 63 \\ +\ 89 \\ \hline \end{array}$

2. $\begin{array}{r} 7006 \\ -\ 451 \\ \hline \end{array}$

3. $\begin{array}{r} 87 \\ \times 52 \\ \hline \end{array}$

4. $8\overline{)4496}$

5. $1 \cdot 67$

6. $\frac{36}{0}$

7. $16 \div 16$

8. $5 \div 1$

9. $0 \cdot 21$

10. $7 \cdot 0 \cdot 8$

11. $0 \div 7$

12. $12 \div 4$

13. $9 \cdot 7$

14. $45 \div 5$

15. $\begin{array}{r} 207 \\ -\ 69 \\ \hline \end{array}$

16. $\begin{array}{r} 207 \\ +\ 69 \\ \hline \end{array}$

17. $3718 - 2549$

18. $1861 + 7965$

19. $7\overline{)1278}$

20. $\begin{array}{r} 1259 \\ \times\ \ 63 \\ \hline \end{array}$

21. $7\overline{)7695}$

22. $9\overline{)1000}$

23. $32\overline{)21{,}240}$

24. $65\overline{)70{,}000}$

25. $4000 - 2963$

26. $10{,}000 - 101$

27. $\begin{array}{r} 303 \\ \times 101 \\ \hline \end{array}$

28. $(475)(100)$

29. Find the total of 62 and 9.

30. Find the product of 62 and 9.

31. Find the quotient of 62 and 9.

32. Find the difference of 62 and 9.

33. Subtract 17 from 200.

34. Find the difference of 432 and 201.

Complete the table by rounding the given number to the given place value.

		Tens	Hundrds	Thousands
35.	9735			
36.	1429			
37.	20,801			
38.	432,198			

Find the perimeter and area of each figure.

39.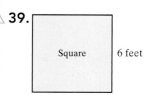

40.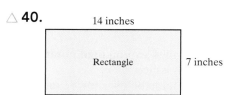

Find the perimeter of each figure.

41.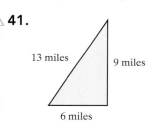

42.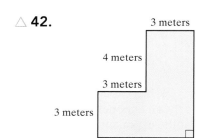

Find the average of each list of numbers.

43. 19, 15, 25, 37, 24

44. 108, 131, 98, 159

45. The Mackinac Bridge is a suspension bridge that connects the lower and upper peninsulas of Michigan across the Straits of Mackinac. Its total length is 26,372 feet. The Lake Pontchartrain Bridge is a twin concrete trestle bridge in Slidell, Louisiana. Its total length is 28,547 feet. Which bridge is longer and by how much? (*Sources:* Mackinac Bridge Authority and Federal Highway Administration, Bridge Division)

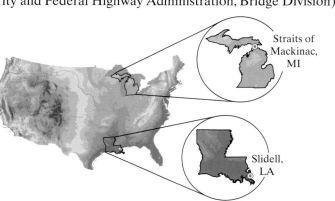

46. In the United States, the average toy expenditure per child is $309 per year. On average, how much is spent on toys for a child by the time he or she reaches age 18? (*Source:* statista)

1.7 Exponents and Order of Operations

Objectives
A Write Repeated Factors Using Exponential Notation.

B Evaluate Expressions Containing Exponents.

C Use the Order of Operations.

D Find the Area of a Square.

Objective A Using Exponential Notation

In the product $3 \cdot 3 \cdot 3 \cdot 3 \cdot 3$, notice that 3 is a factor several times. When this happens, we can use a shorthand notation, called an **exponent,** to write the repeated multiplication.

$3 \cdot 3 \cdot 3 \cdot 3 \cdot 3$ can be written as

3 is a factor 5 times

3^5 Read as "three to the fifth power."

This is called **exponential notation.** The **exponent,** 5, indicates how many times the **base,** 3, is a factor.

The table below shows examples of reading exponential notation in words.

Expression	In Words
5^2	"five to the second power" or "five squared"
5^3	"five to the third power" or "five cubed"
5^4	"five to the fourth power"

Usually, an exponent of 1 is not written, so when no exponent appears, we assume that the exponent is 1. For example, $2 = 2^1$ and $7 = 7^1$.

Practice 1–4

Write using exponential notation.
1. $8 \cdot 8 \cdot 8 \cdot 8$
2. $3 \cdot 3 \cdot 3$
3. $10 \cdot 10 \cdot 10 \cdot 10 \cdot 10$
4. $5 \cdot 5 \cdot 4 \cdot 4 \cdot 4 \cdot 4 \cdot 4 \cdot 4$

Examples Write using exponential notation.

1. $7 \cdot 7 \cdot 7 = 7^3$
2. $3 \cdot 3 = 3^2$
3. $6 \cdot 6 \cdot 6 \cdot 6 \cdot 6 = 6^5$
4. $3 \cdot 3 \cdot 3 \cdot 3 \cdot 9 \cdot 9 \cdot 9 = 3^4 \cdot 9^3$

■ Work Practice 1–4

Objective B Evaluating Exponential Expressions

To **evaluate** an exponential expression, we write the expression as a product and then find the value of the product.

Practice 5–8

Evaluate.
5. 4^2
6. 7^3
7. 11^1
8. $2 \cdot 3^2$

Answers
1. 8^4 2. 3^3 3. 10^5 4. $5^2 \cdot 4^6$
5. 16 6. 343 7. 11 8. 18

Examples Evaluate.

5. $9^2 = 9 \cdot 9 = 81$
6. $6^1 = 6$
7. $3^4 = 3 \cdot 3 \cdot 3 \cdot 3 = 81$
8. $5 \cdot 6^2 = 5 \cdot 6 \cdot 6 = 180$

■ Work Practice 5–8

Example 8 illustrates an important property: An exponent applies only to its base. The exponent 2, in $5 \cdot 6^2$, applies only to its base, 6.

Helpful Hint

An exponent applies only to its base. For example, $4 \cdot 2^3$ means $4 \cdot 2 \cdot 2 \cdot 2$.

Helpful Hint

Don't forget that 2^4, for example, is *not* $2 \cdot 4$. The expression 2^4 means repeated multiplication of the same factor.

$$2^4 = 2 \cdot 2 \cdot 2 \cdot 2 = 16, \quad \text{whereas } 2 \cdot 4 = 8$$

✓ **Concept Check** Which of the following statements is correct?
a. 3^5 is the same as $5 \cdot 5 \cdot 5$.
b. "Ten cubed" is the same as 10^2.
c. "Six to the fourth power" is the same as 6^4.
d. 12^2 is the same as $12 \cdot 2$.

Objective C Using the Order of Operations

Suppose that you are in charge of taking inventory at a local cell phone store. An employee has given you the number of a certain cell phone in stock as the expression

$$6 + 2 \cdot 30$$

To calculate the value of this expression, do you add first or multiply first? If you add first, the answer is 240. If you multiply first, the answer is 66.

Mathematical symbols wouldn't be very useful if two values were possible for one expression. Thus, mathematicians have agreed that, given a choice, we multiply first.

$$6 + 2 \cdot 30 = 6 + 60 \quad \text{Multiply.}$$
$$= 66 \quad \text{Add.}$$

This agreement is one of several **order of operations** agreements.

Order of Operations

1. Perform all operations within parentheses ( ), brackets [ ], or other grouping symbols such as fraction bars, starting with the innermost set.
2. Evaluate any expressions with exponents.
3. Multiply or divide in order from left to right.
4. Add or subtract in order from left to right.

✓ Concept Check Answer
c

Below we practice using order of operations to simplify expressions.

Practice 9
Simplify: $9 \cdot 3 - 8 \div 4$

Example 9 Simplify: $2 \cdot 4 - 3 \div 3$

Solution: There are no parentheses and no exponents, so we start by multiplying and dividing, from left to right.

$2 \cdot 4 - 3 \div 3 = 8 - 3 \div 3$ Multiply.
$ = 8 - 1$ Divide.
$ = 7$ Subtract.

Work Practice 9

Practice 10
Simplify: $48 \div 3 \cdot 2^2$

Example 10 Simplify: $4^2 \div 2 \cdot 4$

Solution: We start by evaluating 4^2.

$4^2 \div 2 \cdot 4 = 16 \div 2 \cdot 4$ Write 4^2 as 16.

Next we multiply or divide *in order* from left to right. Since division appears before multiplication from left to right, we divide first, then multiply.

$16 \div 2 \cdot 4 = 8 \cdot 4$ Divide.
$ = 32$ Multiply.

Work Practice 10

Practice 11
Simplify: $(10 - 7)^4 + 2 \cdot 3^2$

Example 11 Simplify: $(8 - 6)^2 + 2^3 \cdot 3$

Solution: $(8 - 6)^2 + 2^3 \cdot 3 = 2^2 + 2^3 \cdot 3$ Simplify inside parentheses.
$ = 4 + 8 \cdot 3$ Write 2^2 as 4 and 2^3 as 8.
$ = 4 + 24$ Multiply.
$ = 28$ Add.

Work Practice 11

Practice 12
Simplify:
$36 \div [20 - (4 \cdot 2)] + 4^3 - 6$

Example 12 Simplify: $4^3 + [3^2 - (10 \div 2)] - 7 \cdot 3$

Solution: Here we begin with the innermost set of parentheses.

$4^3 + [3^2 - (10 \div 2)] - 7 \cdot 3 = 4^3 + [3^2 - 5] - 7 \cdot 3$ Simplify inside parentheses.
$ = 4^3 + [9 - 5] - 7 \cdot 3$ Write 3^2 as 9.
$ = 4^3 + 4 - 7 \cdot 3$ Simplify inside brackets.
$ = 64 + 4 - 7 \cdot 3$ Write 4^3 as 64.
$ = 64 + 4 - 21$ Multiply.
$ = 47$ Add and subtract from left to right.

Work Practice 12

Answers
9. 25 **10.** 64 **11.** 99 **12.** 61

Section 1.7 | Exponents and Order of Operations

Example 13 Simplify: $\dfrac{7 - 2 \cdot 3 + 3^2}{5(2 - 1)}$

Solution: Here, the fraction bar is a grouping symbol. We simplify above and below the fraction bar separately.

$\dfrac{7 - 2 \cdot 3 + 3^2}{5(2 - 1)} = \dfrac{7 - 2 \cdot 3 + 9}{5(1)}$ Evaluate 3^2 and $(2 - 1)$.

$= \dfrac{7 - 6 + 9}{5}$ Multiply $2 \cdot 3$ in the numerator and multiply 5 and 1 in the denominator.

$= \dfrac{10}{5}$ Add and subtract from left to right.

$= 2$ Divide.

■ Work Practice 13

Practice 13

Simplify: $\dfrac{25 + 8 \cdot 2 - 3^3}{2(3 - 2)}$

Example 14 Simplify: $64 \div 8 \cdot 2 + 4$

Solution: $64 \div 8 \cdot 2 + 4 = \underline{8 \cdot 2} + 4$ Divide.
$= 16 + 4$ Multiply.
$= 20$ Add.

■ Work Practice 14

Practice 14

Simplify: $36 \div 6 \cdot 3 + 5$

Objective D Finding the Area of a Square

Since a square is a special rectangle, we can find its area by finding the product of its length and its width.

Area of a rectangle = length · width

By recalling that each side of a square has the same measurement, we can use the following procedure to find its area:

Area of a square = length · width
= side · side
= (side)^2

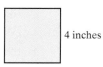
Square — Side / Side

Helpful Hint

Recall from Section 1.5 that area is measured in **square** units while perimeter is measured in units.

Example 15 Find the area of a square whose side measures 4 inches.

Solution: Area of a square = (side)^2
= (4 inches)^2
= 16 square inches

4 inches

The area of the square is 16 square inches.

■ Work Practice 15

Practice 15

Find the area of a square whose side measures 12 centimeters.

Answers
13. 7   **14.** 23   **15.** 144 sq cm

Calculator Explorations **Exponents**

To evaluate an exponential expression such as 4^7 on a calculator, find the keys marked $\boxed{y^x}$ or $\boxed{\wedge}$ and $\boxed{=}$ or $\boxed{\text{ENTER}}$. To evaluate 4^7, press the keys $\boxed{4}\,\boxed{y^x}$ (or $\boxed{\wedge}$) $\boxed{7}$ then $\boxed{=}$ or $\boxed{\text{ENTER}}$. The display will read $\boxed{16384}$. Thus, $4^7 = 16,384$.

Use a calculator to evaluate.

1. 4^6
2. 5^6
3. 5^5
4. 7^6
5. 2^{11}
6. 6^8

Order of Operations

To see whether your calculator has the order of operations built in, evaluate $5 + 2 \cdot 3$ by pressing the keys $\boxed{5}\,\boxed{+}\,\boxed{2}\,\boxed{\times}\,\boxed{3}$ then $\boxed{=}$ or $\boxed{\text{ENTER}}$. If the display reads $\boxed{11}$, your calculator does have the order of operations built in. This means that most of the time, you can key in a problem exactly as it is written, and the calculator will perform operations in the proper order. When evaluating an expression containing parentheses, key in the parentheses. (If an expression contains brackets, key in parentheses.) For example, to evaluate $2[25 - (8 + 4)] - 11$, press the keys $\boxed{2}\,\boxed{\times}\,\boxed{(}\,\boxed{25}\,\boxed{-}\,\boxed{(}\,\boxed{8}\,\boxed{+}\,\boxed{4}\,\boxed{)}\,\boxed{)}\,\boxed{-}\,\boxed{11}$ then $\boxed{=}$ or $\boxed{\text{ENTER}}$. The display will read $\boxed{15}$.

Use a calculator to evaluate.

7. $7^4 + 5^3$
8. $12^4 - 8^4$
9. $63 \cdot 75 - 43 \cdot 10$
10. $8 \cdot 22 + 7 \cdot 16$
11. $4(15 \div 3 + 2) - 10 \cdot 2$
12. $155 - 2(17 + 3) + 185$

Vocabulary, Readiness & Video Check

Use the choices below to fill in each blank.

addition	multiplication	exponent
subtraction	division	base

1. In $2^5 = 32$, the 2 is called the _____ and the 5 is called the _____.
2. To simplify $8 + 2 \cdot 6$, which operation should be performed first? _____
3. To simplify $(8 + 2) \cdot 6$, which operation should be performed first? _____
4. To simplify $9(3 - 2) \div 3 + 6$, which operation should be performed first? _____
5. To simplify $8 \div 2 \cdot 6$, which operation should be performed first? _____

Martin-Gay Interactive Videos Watch the section lecture video and answer the following questions.

See Video 1.7

Objective **A** 6. In the Example 1 expression, what is the 3 called and what is the 12 called?

Objective **B** 7. As mentioned in Example 4, what "understood exponent" does any number we've worked with before have?

Objective **C** 8. List the three operations needed to evaluate Example 7 in the order they should be performed.

Objective **D** 9. As explained in the lecture before Example 10, why does the area of a square involve an exponent whereas the area of a rectangle usually does not?

1.7 Exercise Set MyLab Math

Objective A *Write using exponential notation. See Examples 1 through 4.*

1. $4 \cdot 4 \cdot 4$
2. $5 \cdot 5 \cdot 5 \cdot 5$
3. $7 \cdot 7 \cdot 7 \cdot 7 \cdot 7 \cdot 7$
4. $6 \cdot 6 \cdot 6 \cdot 6 \cdot 6 \cdot 6 \cdot 6$
5. $12 \cdot 12 \cdot 12$
6. $10 \cdot 10 \cdot 10$
7. $6 \cdot 6 \cdot 5 \cdot 5 \cdot 5$
8. $4 \cdot 4 \cdot 3 \cdot 3 \cdot 3$
9. $9 \cdot 8 \cdot 8$
10. $7 \cdot 4 \cdot 4 \cdot 4$
11. $3 \cdot 2 \cdot 2 \cdot 2 \cdot 2$
12. $4 \cdot 6 \cdot 6 \cdot 6 \cdot 6$
13. $3 \cdot 2 \cdot 2 \cdot 2 \cdot 2 \cdot 5 \cdot 5 \cdot 5 \cdot 5 \cdot 5$
14. $6 \cdot 6 \cdot 2 \cdot 9 \cdot 9 \cdot 9 \cdot 9$

Objective B *Evaluate. See Examples 5 through 8.*

15. 8^2
16. 6^2
17. 5^3
18. 6^3
19. 2^5
20. 3^5
21. 1^{10}
22. 1^{12}
23. 7^1
24. 8^1
25. 2^7
26. 5^4
27. 2^8
28. 3^3
29. 4^4
30. 4^3
31. 9^3
32. 8^3
33. 12^2
34. 11^2
35. 10^2
36. 10^3
37. 20^1
38. 14^1
39. 3^6
40. 4^5
41. $3 \cdot 2^6$
42. $5 \cdot 3^2$
43. $2 \cdot 3^4$
44. $2 \cdot 7^2$

Objective C *Simplify. See Examples 9 through 14.*

45. $15 + 3 \cdot 2$
46. $24 + 6 \cdot 3$
47. $14 \div 7 \cdot 2 + 3$
48. $100 \div 10 \cdot 5 + 4$
49. $32 \div 4 - 3$
50. $42 \div 7 - 6$
51. $13 + \dfrac{24}{8}$
52. $32 + \dfrac{8}{2}$
53. $6 \cdot 5 + 8 \cdot 2$
54. $3 \cdot 4 + 9 \cdot 1$
55. $\dfrac{5 + 12 \div 4}{1^7}$
56. $\dfrac{6 + 9 \div 3}{3^2}$
57. $(7 + 5^2) \div 4 \cdot 2^3$
58. $6^2 \cdot (10 - 8)$
59. $5^2 \cdot (10 - 8) + 2^3 + 5^2$
60. $5^3 \div (10 + 15) + 9^2 + 3^3$
61. $\dfrac{18 + 6}{2^4 - 2^2}$
62. $\dfrac{40 + 8}{5^2 - 3^2}$
63. $(3 + 5) \cdot (9 - 3)$
64. $(9 - 7) \cdot (12 + 18)$
65. $\dfrac{7(9 - 6) + 3}{3^2 - 3}$

66. $\dfrac{5(12-7)-4}{5^2-18}$

67. $8 \div 0 + 37$

68. $18 - 7 \div 0$

69. $2^4 \cdot 4 - (25 \div 5)$

70. $2^3 \cdot 3 - (100 \div 10)$

71. $3^4 - [35 - (12 - 6)]$

72. $[40 - (8 - 2)] - 2^5$

73. $(7 \cdot 5) + [9 \div (3 \div 3)]$

74. $(18 \div 6) + [(3 + 5) \cdot 2]$

75. $8 \cdot [2^2 + (6 - 1) \cdot 2] - 50 \cdot 2$

76. $35 \div [3^2 + (9 - 7) - 2^2] + 10 \cdot 3$

77. $\dfrac{9^2 + 2^2 - 1^2}{8 \div 2 \cdot 3 \cdot 1 \div 3}$

78. $\dfrac{5^2 - 2^3 + 1^4}{10 \div 5 \cdot 4 \cdot 1 \div 4}$

79. $\dfrac{2 + 4^2}{5(20 - 16) - 3^2 - 5}$

80. $\dfrac{3 + 9^2}{3(10 - 6) - 2^2 - 1}$

81. $9 \div 3 + 5^2 \cdot 2 - 10$

82. $10 \div 2 + 3^3 \cdot 2 - 20$

83. $[13 \div (20 - 7) + 2^5] - (2 + 3)^2$

84. $[15 \div (11 - 6) + 2^2] + (5 - 1)^2$

85. $7^2 - \{18 - [40 \div (5 \cdot 1) + 2] + 5^2\}$

86. $29 - \{5 + 3[8 \cdot (10 - 8)] - 50\}$

△ Objective D **Mixed Practice (Sections 1.3 and 1.7)** *Find the area and perimeter of each square. See Example 15.*

87.
7 meters

△ 88.
9 centimeters

△ 89.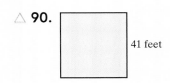
23 miles

△ 90.
41 feet

Concept Extensions

Answer the following true or false. See the Concept Check in this section.

91. "Six to the fifth power" is the same as 6^5.

92. "Seven squared" is the same as 7^2.

93. 2^5 is the same as $5 \cdot 5$.

94. 4^9 is the same as $4 \cdot 9$.

Section 1.8 | Introduction to Variables, Algebraic Expressions, and Equations 75

Insert grouping symbols (parentheses) so that each given expression evaluates to the given number.

95. $2 + 3 \cdot 6 - 2$; evaluates to 28

96. $2 + 3 \cdot 6 - 2$; evaluates to 20

97. $24 \div 3 \cdot 2 + 2 \cdot 5$; evaluates to 14

98. $24 \div 3 \cdot 2 + 2 \cdot 5$; evaluates to 15

99. A building contractor is bidding on a contract to install gutters on seven homes in a retirement community, all in the shape shown. To estimate the cost of materials, she needs to know the total perimeter of all seven homes. Find the total perimeter.

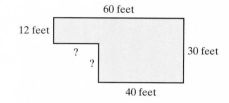

100. The building contractor from Exercise 99 plans to charge $4 per foot for installing vinyl gutters. Find the total charge for the seven homes given the total perimeter answer to Exercise 99.

Simplify.

101. $(7 + 2^4)^5 - (3^5 - 2^4)^2$

102. $25^3 \cdot (45 - 7 \cdot 5) \cdot 5$

103. Write an expression that simplifies to 5. Use multiplication, division, addition, subtraction, and at least one set of parentheses. Explain the process you would use to simplify the expression.

104. Explain why $2 \cdot 3^2$ is not the same as $(2 \cdot 3)^2$.

1.8 Introduction to Variables, Algebraic Expressions, and Equations

Objective A Evaluating Algebraic Expressions

Perhaps the most important quality of mathematics is that it is a science of patterns. Communicating about patterns is often made easier by using a letter to represent all the numbers fitting a pattern. We call such a letter a **variable.** For example, in Section 1.3 we presented the addition property of 0, which states that the sum of 0 and any number is that number. We might write

$0 + 1 = 1$
$0 + 2 = 2$
$0 + 3 = 3$
$0 + 4 = 4$
$0 + 5 = 5$
$0 + 6 = 6$

Objectives

A Evaluate Algebraic Expressions Given Replacement Values.

B Identify Solutions of Equations.

C Translate Phrases into Variable Expressions.

(Continued on next page)

continuing indefinitely. This is a pattern, and all whole numbers fit the pattern. We can communicate this pattern for all whole numbers by letting a letter, such as a, represent all whole numbers. We can then write

$$0 + a = a$$

Using variable notation is a primary goal of learning **algebra.** We now take some important first steps in beginning to use variable notation.

A combination of operations on letters (variables) and numbers is called an **algebraic expression** or simply an **expression.**

Algebraic Expressions

$$3 + x \quad 5 \cdot y \quad 2 \cdot z - 1 + x$$

If two variables or a number and a variable are next to each other, with no operation sign between them, the operation is multiplication. For example,

$$2x \quad \text{means} \quad 2 \cdot x$$

and

$$xy \text{ or } x(y) \quad \text{means} \quad x \cdot y$$

Also, the meaning of an exponent remains the same when the base is a variable. For example,

$$x^2 = \underbrace{x \cdot x}_{\text{2 factors of } x} \quad \text{and} \quad y^5 = \underbrace{y \cdot y \cdot y \cdot y \cdot y}_{\text{5 factors of } y}$$

Algebraic expressions such as $3x$ have different values depending on replacement values for x. For example, if x is 2, then $3x$ becomes

$$3x = 3 \cdot 2$$
$$= 6$$

If x is 7, then $3x$ becomes

$$3x = 3 \cdot 7$$
$$= 21$$

Replacing a variable in an expression by a number and then finding the value of the expression is called **evaluating the expression.** When finding the value of an expression, remember to follow the order of operations given in Section 1.7.

Practice 1

Evaluate $x - 2$ if x is 7.

Example 1 Evaluate $x + 6$ if x is 8.

Solution: Replace x with 8 in the expression $x + 6$.

$$x + 6 = 8 + 6 \quad \text{Replace } x \text{ with 8.}$$
$$= 14 \quad \text{Add.}$$

■ Work Practice 1

When we write a statement such as "x is 5," we can use an equal sign ($=$) to represent "is" so that

x is 5   can be written as $x = 5$.

Answer
1. 5

Section 1.8 | Introduction to Variables, Algebraic Expressions, and Equations

Example 2 Evaluate $2(x - y)$ for $x = 6$ and $y = 3$.

Solution: $2(x - y) = 2(6 - 3)$ Replace x with 6 and y with 3.

$ = 2(3)$ Subtract.

$ = 6$ Multiply.

■ Work Practice 2

Practice 2
Evaluate $y(x - 3)$ for $x = 8$ and $y = 4$.

Example 3 Evaluate $\dfrac{x - 5y}{y}$ for $x = 35$ and $y = 5$.

Solution: $\dfrac{x - 5y}{y} = \dfrac{35 - 5(5)}{5}$ Replace x with 35 and y with 5.

$\phantom{\dfrac{x-5y}{y}} = \dfrac{35 - 25}{5}$ Multiply.

$\phantom{\dfrac{x-5y}{y}} = \dfrac{10}{5}$ Subtract.

$\phantom{\dfrac{x-5y}{y}} = 2$ Divide.

■ Work Practice 3

Practice 3
Evaluate $\dfrac{y + 6}{x}$ for $x = 6$ and $y = 18$.

Example 4 Evaluate $x^2 + z - 3$ for $x = 5$ and $z = 4$.

Solution: $x^2 + z - 3 = 5^2 + 4 - 3$ Replace x with 5 and z with 4.

$ = 25 + 4 - 3$ Evaluate 5^2.

$ = 26$ Add and subtract from left to right.

■ Work Practice 4

Practice 4
Evaluate $25 - z^3 + x$ for $z = 2$ and $x = 1$.

Helpful Hint

If you are having difficulty replacing variables with numbers, first replace each variable with a set of parentheses, then insert the replacement number within the parentheses.

Example:

$x^2 + z - 3 = ()^2 + () - 3$

$ = (5)^2 + (4) - 3$

$ = 25 + 4 - 3$

$ = 26$

✓ **Concept Check** What's wrong with the solution to the following problem?
Evaluate $3x + 2y$ for $x = 2$ and $y = 3$.

Solution: $3x + 2y = 3(3) + 2(2)$

$ = 9 + 4$

$ = 13$

Answers
2. 20   **3.** 4   **4.** 18

✓ **Concept Check Answer**
$3x + 2y = 3(2) + 2(3)$
$ = 6 + 6$
$ = 12$

Practice 5

Evaluate $\dfrac{5(F-32)}{9}$ for $F = 41$.

Example 5 The expression $\dfrac{5(F-32)}{9}$ can be used to write degrees Fahrenheit F as degrees Celsius C. Find the value of this expression for $F = 86$.

Solution:
$$\dfrac{5(F-32)}{9} = \dfrac{5(86-32)}{9}$$
$$= \dfrac{5(54)}{9}$$
$$= \dfrac{270}{9}$$
$$= 30$$

Thus 86°F is the same temperature as 30°C.

■ Work Practice 5

Objective B Identifying Solutions of Equations

In Objective A, we learned that a combination of operations on variables and numbers is called an algebraic expression or simply an expression. Frequently in this book, we have written statements like $7 + 4 = 11$ or area = length · width. Each of these statements is called an **equation**. An equation is of the form

Helpful Hint An equation contains "=," while an expression does not.

An equation can be labeled as

$$\underbrace{x + 7}_{\text{left side}} \overset{\text{equal sign}}{=} \underbrace{10}_{\text{right side}}$$

When an equation contains a variable, deciding which values of the variable make the equation a true statement is called **solving** the equation for the variable. A **solution** of an equation is a value for the variable that makes the equation a true statement. For example, 2 is a solution of the equation $x + 5 = 7$, since replacing x with 2 results in the *true* statement $2 + 5 = 7$. Similarly, 3 is not a solution of $x + 5 = 7$, since replacing x with 3 results in the *false* statement $3 + 5 = 7$.

Practice 6

Determine whether 8 is a solution of the equation $3(y - 6) = 6$.

Example 6 Determine whether 6 is a solution of the equation $4(x - 3) = 12$.

Solution: We replace x with 6 in the equation.

$$4(x - 3) = 12$$
$$4(6 - 3) \overset{?}{=} 12 \quad \text{Replace } x \text{ with 6.}$$
$$4(3) \overset{?}{=} 12$$
$$12 = 12 \quad \text{True}$$

Since $12 = 12$ is a true statement, 6 *is* a solution of the equation.

■ Work Practice 6

Answers
5. 5 **6.** yes

Section 1.8 | Introduction to Variables, Algebraic Expressions, and Equations

A collection of numbers enclosed by braces is called a set. For example,

$\{0, 1, 2, 3, \ldots\}$

is the set of whole numbers that we are studying about in this chapter. The three dots after the number 3 in the set mean that this list of numbers continues in the same manner indefinitely.

The next example contains set notation.

Example 7 Determine which numbers in the set $\{26, 40, 20\}$ are solutions of the equation $2n - 30 = 10$.

Solution: Replace n with each number from the set to see if a true statement results.

Let n be 26.	Let n be 40.	Let n be 20.
$2n - 30 = 10$	$2n - 30 = 10$	$2n - 30 = 10$
$2 \cdot 26 - 30 \stackrel{?}{=} 10$	$2 \cdot 40 - 30 \stackrel{?}{=} 10$	$2 \cdot 20 - 30 \stackrel{?}{=} 10$
$52 - 30 \stackrel{?}{=} 10$	$80 - 30 \stackrel{?}{=} 10$	$40 - 30 \stackrel{?}{=} 10$
$22 = 10$ False	$50 = 10$ False	$10 = 10$ True ✓

Thus, 20 is a solution while 26 and 40 are not solutions.

Work Practice 7

Practice 7
Determine which numbers in the set $\{10, 6, 8\}$ are solutions of the equation $5n + 4 = 34$.

Objective C Translating Phrases into Variable Expressions

To aid us in solving problems later, we practice translating verbal phrases into algebraic expressions. Certain key words and phrases suggesting addition, subtraction, multiplication, or division are reviewed next.

Addition (+)	Subtraction (−)	Multiplication (·)	Division (÷)
sum	difference	product	quotient
plus	minus	times	divide
added to	subtract	multiply	shared equally among
more than	less than	multiply by	per
increased by	decreased by	of	divided by
total	less	double/triple	divided into

Example 8 Write as an algebraic expression. Use x to represent "a number."

a. 7 increased by a number
b. 15 decreased by a number
c. The product of 2 and a number
d. The quotient of a number and 5
e. 2 subtracted from a number

Solution:

a. In words: 7   increased by   a number
     Translate: 7   + x

(Continued on next page)

Practice 8
Write as an algebraic expression. Use x to represent "a number."

a. Twice a number
b. 8 increased by a number
c. 10 minus a number
d. 10 subtracted from a number
e. The quotient of 6 and a number

Answers
7. 6 is a solution.
8. a. $2x$ b. $8 + x$ c. $10 - x$
d. $x - 10$ e. $6 \div x$ or $\dfrac{6}{x}$

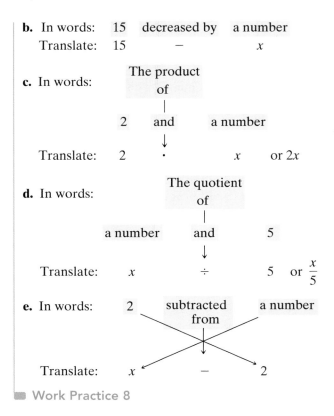

■ Work Practice 8

Helpful Hint

Remember that order is important when subtracting. Study the order of numbers and variables below.

Phrase	Translation
a number *decreased by* 5	$x - 5$
a number *subtracted from* 5	$5 - x$

Vocabulary, Readiness & Video Check

Use the choices below to fill in each blank. You may use each choice more than once.

 evaluating the expression variable(s) expression equation solution

1. A combination of operations on letters (variables) and numbers is a(n) _____.
2. A letter that represents a number is a(n) _____.
3. $3x - 2y$ is called a(n) _____ and the letters x and y are _____.
4. Replacing a variable in an expression by a number and then finding the value of the expression is called _____.
5. A statement of the form "expression = expression" is called a(n) _____.
6. A value for the variable that makes an equation a true statement is called a(n) _____.

Section 1.8 | Introduction to Variables, Algebraic Expressions, and Equations

Martin-Gay Interactive Videos  Watch the section lecture video and answer the following questions.

Objective A 7. Complete this statement based on the lecture before Example 1: When a letter and a variable are next to each other, the operation is an understood _____.

Objective B 8. In Example 5, why is a question mark written over the equal sign?

Objective C 9. In Example 6, what phrase translates to subtraction?

See Video 1.8

1.8 Exercise Set MyLab Math

Objective A Complete the table. The first row has been done for you. See Examples 1 through 5.

	a	b	$a+b$	$a-b$	$a \cdot b$	$a \div b$
	45	9	54	36	405	5
1.	21	7				
2.	24	6				
3.	152	0				
4.	298	0				
5.	56	1				
6.	82	1				

Evaluate each following expression for $x = 2$, $y = 5$, and $z = 3$. See Examples 1 through 5.

7. $3 + 2z$

8. $7 + 3z$

9. $3xz - 5x$

10. $4yz + 2x$

11. $z - x + y$

12. $x + 5y - z$

13. $4x - z$

14. $2y + 5z$

15. $y^3 - 4x$

16. $y^3 - z$

17. $2xy^2 - 6$

18. $3yz^2 + 1$

19. $8 - (y - x)$

20. $3 + (2y - 4)$

21. $x^5 + (y - z)$

22. $x^4 - (y - z)$

23. $\dfrac{6xy}{z}$

24. $\dfrac{8yz}{15}$

25. $\dfrac{2y - 2}{x}$

26. $\dfrac{6 + 3x}{z}$

27. $\dfrac{x + 2y}{z}$

28. $\dfrac{2z + 6}{3}$

29. $\dfrac{5x}{y} - \dfrac{10}{y}$

30. $\dfrac{70}{2y} - \dfrac{15}{z}$

31. $2y^2 - 4y + 3$

32. $3x^2 + 2x - 5$

33. $(4y - 5z)^3$

34. $(4y + 3z)^2$

35. $(xy + 1)^2$

36. $(xz - 5)^4$

 37. $2y(4z - x)$

38. $3x(y + z)$

39. $xy(5 + z - x)$

40. $xz(2y + x - z)$

41. $\dfrac{7x + 2y}{3x}$

42. $\dfrac{6z + 2y}{4}$

43. The expression $16t^2$ gives the distance in feet that an object falls after t seconds. Complete the table by evaluating $16t^2$ for each given value of t.

t	1	2	3	4
$16t^2$				

44. The expression $\dfrac{5(F - 32)}{9}$ gives the equivalent degrees Celsius for F degrees Fahrenheit. Complete the table by evaluating this expression for each given value of F.

F	50	59	68	77
$\dfrac{5(F - 32)}{9}$				

Objective B *Decide whether the given number is a solution of the given equation. See Example 6.*

45. Is 10 a solution of $n - 8 = 2$?

46. Is 9 a solution of $n - 2 = 7$?

47. Is 3 a solution of $24 = 80n$?

48. Is 50 a solution of $250 = 5n$?

49. Is 7 a solution of $3n - 5 = 10$?

50. Is 8 a solution of $11n + 3 = 91$?

51. Is 20 a solution of $2(n - 17) = 6$?

52. Is 0 a solution of $5(n + 9) = 40$?

53. Is 0 a solution of $5x + 3 = 4x + 13$?

54. Is 2 a solution of $3x - 6 = 5x - 10$?

▶ **55.** Is 8 a solution of $7f = 64 - f$?

56. Is 5 a solution of $8x - 30 = 2x$?

Determine which numbers in each set are solutions to the corresponding equations. See Example 7.

57. $n - 2 = 10$; $\{10, 12, 14\}$

58. $n + 3 = 16$; $\{9, 11, 13\}$

59. $5n = 30$; $\{6, 25, 30\}$

60. $3n = 45$; $\{15, 30, 45\}$

61. $6n + 2 = 26$; $\{0, 2, 4\}$

62. $4n - 14 = 6$; $\{0, 5, 10\}$

63. $3(n - 4) = 10$; $\{5, 7, 10\}$

64. $6(n + 2) = 23$; $\{1, 3, 5\}$

▶ **65.** $7x - 9 = 5x + 13$; $\{3, 7, 11\}$

66. $9x - 15 = 5x + 1$; $\{2, 4, 11\}$

Objective C Translating *Write each phrase as a variable expression. Use x to represent "a number." See Example 8.*

67. Eight more than a number

68. The sum of three and a number

69. The total of a number and eight

70. The difference of a number and five hundred

Section 1.8 | Introduction to Variables, Algebraic Expressions, and Equations

71. Twenty decreased by a number

72. A number less thirty

73. The product of 512 and a number

74. A number times twenty

75. The quotient of eight and a number

76. A number divided by 11

77. The sum of seventeen and a number added to the product of five and the number

78. The quotient of twenty and a number, decreased by three

79. The product of five and a number

80. The difference of twice a number, and four

81. A number subtracted from 11

82. Twelve subtracted from a number

83. A number less 5

84. The sum of a number and 7

85. 6 divided by a number

86. The product of a number and 7

87. Fifty decreased by eight times a number

88. Twenty decreased by twice a number

Concept Extensions

For Exercises 89 through 92, find and correct the error when evaluating each expression for $x = 3$ and $y = 5$. See the Concept Check in this section.

89. $2x + 3y = 2(5) + 3(3)$
$= 10 + 9$
$= 19$

90. $6y + 3x = 6(3) + 3(5)$
$= 18 + 15$
$= 33$

91. $7x + y^2 = 7(3) + 5^2$
$= 21 + 10$
$= 31$

92. $x^3 + 4y = 3^3 + 4(5)$
$= 9 + 20$
$= 29$

For Exercises 93 through 100, use a calculator to evaluate each expression for $x = 23$ and $y = 72$.

93. $x^4 - y^2$

94. $2(x + y)^2$

95. $x^2 + 5y - 112$

96. $16y - 20x + x^3$

97. If x is a whole number, which expression is the largest: $2x, 5x,$ or $\frac{x}{3}$?
Explain your answer.

98. If x is a whole number, which expression is the smallest: $2x, 5x,$ or $\frac{x}{3}$?
Explain your answer.

99. In Exercise 43, what do you notice about the value of $16t^2$ as t gets larger?

100. In Exercise 44, what do you notice about the value of $\frac{5(F - 32)}{9}$ as F gets larger?

Chapter 1 Group Activity

Investigating Endangered and Threatened Species

An **endangered** species is one that is thought to be in danger of becoming extinct throughout all or a major part of its habitat. A **threatened** species is one that may become endangered. The Division of Endangered Species at the U.S. Fish and Wildlife Service keeps close tabs on the state of threatened and endangered wildlife in the United States and around the world. The table below was compiled from early 2017 data in the Division of Endangered Species' box score. The "Total Species" column gives the total number of endangered and threatened species for each group.

1. Round each number of *endangered animal species* to the nearest ten to estimate the Animal Total.
2. Round each number of *endangered plant species* to the nearest ten to estimate the Plant Total.
3. Add the exact numbers of endangered animal species to find the exact Animal Total and record it in the table in the Endangered Species column. Add the exact numbers of endangered plant species to find the Plant Total and record it in the table in the Endangered Species column. Then find the total number of endangered species (animals and plants combined) and record this number in the table as the Grand Total in the Endangered Species column.
4. Find the Animal Total, Plant Total, and Grand Total for the Total Species column. Record these values in the table.
5. Use the data in the table to complete the Threatened Species column.
6. Write a paragraph discussing the conclusions that can be drawn from the table.

Endangered and Threatened Species Worldwide

	Group	Endangered Species	Threatened Species	Total Species
Animals	Mammals	329		377
	Birds	299		337
	Reptiles	87		139
	Amphibians	28		44
	Fishes	114		191
	Snails	41		53
	Clams	77		91
	Crustaceans	23		27
	Insects	69		80
	Arachnids	12		12
	Corals	1		23
	Animal Total			
Plants	Flowering Plants	704		867
	Conifers	1		6
	Ferns and others	29		31
	Lichens	2		2
	Plant Total			
	Grand Total			

Chapter 1 Vocabulary Check

Fill in each blank with one of the words or phrases listed below.

difference	factor	perimeter	dividend	minuend
place value	whole numbers	equation	divisor	variable
sum	set	addend	exponent	expression
solution	quotient	subtrahend	product	digits	area

1. The _____ are 0, 1, 2, 3, . . .
2. The _____ of a polygon is its distance around or the sum of the lengths of its sides.
3. The position of each digit in a number determines its _____.
4. A(n) _____ is a shorthand notation for repeated multiplication of the same factor.
5. To find the _____ of a rectangle, multiply length times width.
6. The _____ used to write numbers are 0, 1, 2, 3, 4, 5, 6, 7, 8, and 9.
7. A letter used to represent a number is called a(n) _____.
8. A(n) _____ can be written in the form "expression = expression."
9. A combination of operations on variables and numbers is called a(n) _____.
10. A(n) _____ of an equation is a value of the variable that makes the equation a true statement.
11. A collection of numbers (or objects) enclosed by braces is called a(n) _____.

Use the facts below for Exercises 12 through 21.

$2 \cdot 3 = 6$ $4 + 17 = 21$ $20 - 9 = 11$

12. The 21 above is called the _____.
13. The 5 above is called the _____.
14. The 35 above is called the _____.
15. The 7 above is called the _____.
16. The 3 above is called a(n) _____.
17. The 6 above is called the _____.
18. The 20 above is called the _____.
19. The 9 above is called the _____.
20. The 11 above is called the _____.
21. The 4 above is called a(n) _____.

> **Helpful Hint**
>
> ▶ Are you preparing for your test? To help, don't forget to take these:
> - Chapter 1 Getting Ready for the Test on page 95
> - Chapter 1 Test on page 96
>
> Then check all of your answers at the back of this text. For further review, the step-by-step video solutions to any of these exercises are located in MyLab Math.

1 Chapter Highlights

Definitions and Concepts	Examples
Section 1.2 Place Value, Names for Numbers, and Reading Tables	
The **whole numbers** are 0, 1, 2, 3, 4, 5, ... The position of each digit in a number determines its **place value**. A place-value chart is shown next with the names of the periods given. 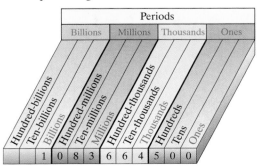	Examples of whole numbers: 0, 14, 968, 5, 268, 619

(continued)

Definitions and Concepts	Examples
Section 1.2 Place Value, Names for Numbers, and Reading Tables (*continued*)	
To write a whole number in words, write the number in each period followed by the name of the period. (The name of the ones period is not included.)	9,078,651,002 is written as nine billion, seventy-eight million, six hundred fifty-one thousand, two.
To write a whole number in standard form, write the number in each period, followed by a comma.	Four million, seven hundred six thousand, twenty-eight is written as 4,706,028.
Section 1.3 Adding and Subtracting Whole Numbers, and Perimeter	
To add whole numbers, add the digits in the ones place, then the tens place, then the hundreds place, and so on, carrying when necessary.	Find the sum: $\overset{2\,1\,1}{2689}$ ← addend 1735 ← addend $\underline{+\ \ 662}$ ← addend 5086 ← sum
To subtract whole numbers, subtract the digits in the ones place, then the tens place, then the hundreds place, and so on, borrowing when necessary.	Subtract: $\overset{8\,15}{7\,9\,5\,4}$ ← minuend $\underline{-5\,6\,7\,3}$ ← subtrahend $2\,2\,8\,1$ ← difference
The **perimeter** of a polygon is its distance around or the sum of the lengths of its sides.	△ Find the perimeter of the polygon shown. $\quad\quad\quad\quad\quad\quad$ 5 feet $\quad$ 2 feet $\diagup\diagdown$ 3 feet $\quad\quad\quad\quad\quad\quad$ 9 feet The perimeter is 5 feet + 3 feet + 9 feet + 2 feet = 19 feet.
Section 1.4 Rounding and Estimating	
Rounding Whole Numbers to a Given Place Value **Step 1:** Locate the digit to the right of the given place value. **Step 2:** If this digit is 5 or greater, add 1 to the digit in the given place value and replace each digit to its right with 0. **Step 3:** If this digit is less than 5, replace it and each digit to its right with 0.	Round 15,721 to the nearest thousand. $\quad\quad\quad$ 15, ⑦ 21 $\quad\quad\ $ Add 1 ↑↑ $\quad\quad\quad\quad\quad$ Replace $\ \ $ Since the circled digit is 5 or greater, add 1 to $\quad\quad\quad\quad\quad$ with $\quad\quad\ $ the given place value and replace digits to its $\quad\quad\quad\quad\quad$ zeros. $\quad\ $ right with zeros. 15,721 rounded to the nearest thousand is 16,000.

Chapter 1 Highlights

Definitions and Concepts	Examples
Section 1.5 Multiplying Whole Numbers and Area	
To multiply 73 and 58, for example, multiply 73 and 8, then 73 and 50. The sum of these partial products is the product of 73 and 58. Use the notation to the right.	$73 \leftarrow$ factor $\underline{\times 58} \leftarrow$ factor $584 \leftarrow 73 \times 8$ $\underline{3650} \leftarrow 73 \times 50$ $4234 \leftarrow$ product
To find the **area** of a rectangle, multiply length times width.	Find the area of the rectangle shown. 11 meters 7 meters area of rectangle = length · width = (11 meters)(7 meters) = 77 square meters
Section 1.6 Dividing Whole Numbers	
Division Properties of 0 The quotient of 0 and any number (except 0) is 0. The quotient of any number and 0 is not a number. We say that this quotient is **undefined**.	$\dfrac{0}{5} = 0$ $\dfrac{7}{0}$ is undefined
To divide larger whole numbers, use the process called **long division** as shown to the right.	$507 \text{ R } 2 \leftarrow$ quotient divisor $\rightarrow 14\overline{)7100} \leftarrow$ dividend $\underline{-70}\downarrow 5(14) = 70$ $10 $ Subtract and bring down the 0. $\underline{-0}\downarrow 0(14) = 0$ $100 $ Subtract and bring down the 0. $\underline{-98} 7(14) = 98$ $2 $ Subtract. The remainder is 2. To check, see that $507 \cdot 14 + 2 = 7100$.
The **average** of a list of numbers is $\text{average} = \dfrac{\textit{sum} \text{ of numbers}}{\textit{number} \text{ of numbers}}$	Find the average of 23, 35, and 38. $\text{average} = \dfrac{23 + 35 + 38}{3} = \dfrac{96}{3} = 32$

Definitions and Concepts	Examples
Section 1.7 Exponents and Order of Operations	

An **exponent** is a shorthand notation for repeated multiplication of the same factor.

$$3^4 = \underbrace{3 \cdot 3 \cdot 3 \cdot 3}_{\text{4 factors of 3}} = 81$$

where 3 is the base and 4 is the exponent.

Order of Operations

1. Perform all operations within parentheses ( ), brackets [ ], or other grouping symbols such as fraction bars, starting with the innermost set.
2. Evaluate any expressions with exponents.
3. Multiply or divide in order from left to right.
4. Add or subtract in order from left to right.

Simplify: $\dfrac{5 + 3^2}{2(7 - 6)}$

Simplify above and below the fraction bar separately.

$\dfrac{5 + 3^2}{2(7 - 6)} = \dfrac{5 + 9}{2(1)}$ Evaluate 3^2 above the fraction bar. Subtract $7 - 6$ below the fraction bar.

$= \dfrac{14}{2}$ Add. Multiply.

$= 7$ Divide.

The **area of a square** is $(\text{side})^2$.

Find the area of a square with side length 9 inches.

$$\text{Area of the square} = (\text{side})^2$$
$$= (9 \text{ inches})^2$$
$$= 81 \text{ square inches}$$

Section 1.8 Introduction to Variables, Algebraic Expressions, and Equations	

A letter used to represent a number is called a **variable**.

Variables:

$$x, \ y, \ z, \ a, \ b$$

A combination of operations on variables and numbers is called an **algebraic expression**.

Algebraic expressions:

$$3 + x, \quad 7y, \quad x^3 + y - 10$$

Replacing a variable in an expression by a number, and then finding the value of the expression, is called **evaluating the expression**.

Evaluate $2x + y$ for $x = 22$ and $y = 4$.

$2x + y = 2 \cdot 22 + 4$ Replace x with 22 and y with 4.
$= 44 + 4$ Multiply.
$= 48$ Add.

A statement written in the form "expression = expression" is an **equation**.

Equations:

$$n - 8 = 12$$
$$2(20 - 7n) = 32$$
$$\text{Area} = \text{length} \cdot \text{width}$$

A **solution** of an equation is a value for the variable that makes the equation a true statement.

Determine whether 2 is a solution of the equation $4(x - 1) = 7$.

$4(2 - 1) \stackrel{?}{=} 7$ Replace x with 2.
$4(1) \stackrel{?}{=} 7$ Subtract.
$4 = 7$ False

No, 2 is not a solution.

Chapter 1 Review

(1.2) *Determine the place value of the digit 4 in each whole number.*

1. 7640

2. 46,200,120

Write each whole number in words.

3. 7640

4. 46,200,120

Write each whole number in expanded form.

5. 3158

6. 403,225,000

Write each whole number in standard form.

7. Eighty-one thousand, nine hundred

8. Six billion, three hundred four million

The following table shows the Internet use by world regions. Use this table to answer Exercises 9 through 12 and other exercises throughout this review. (Source: International Telecommunications Union and Internet World Stats)

Internet Use by World Regions (in millions)			
World Region	**2016**	**2013**	**2008**
Africa	341	140	51
Asia	1846	1268	579
Europe	615	467	385
Middle East	141	141	42
North America	320	302	248
Latin America/Caribbean	385	280	139
Oceania/Australia	28	24	20

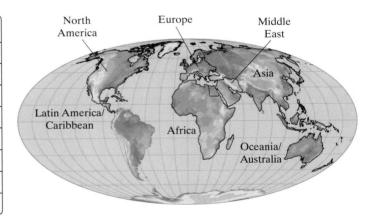

9. Find the number of Internet users in 2016 in Europe. Write your answer in standard form.

10. Find the number of Internet users in Oceania/Australia in 2016. Write your answer in standard form.

11. Which world region had the smallest number of Internet users in 2013?

12. Which world region had the greatest number of Internet users in 2008?

(1.3) *Add or subtract as indicated.*

13. 18 + 49

14. 28 + 39

15. 462 − 397

16. 583 − 279

17. 428 + 21

18. 819 + 21

19. 4000 − 86

20. 8000 − 92

21. 91 + 3623 + 497

22. 82 + 1647 + 238

Translating *Solve.*

23. Find the sum of 74, 342, and 918.

24. Find the sum of 49, 529, and 308.

25. Subtract 7965 from 25,862.

26. Subtract 4349 from 39,007.

27. The distance from Washington, DC, to New York City is 205 miles. The distance from New York City to New Delhi, India, is 7318 miles. Find the total distance from Washington, DC, to New Delhi if traveling by air through New York City.

28. Susan Summerline earned salaries of $62,589, $65,340, and $69,770 during the years 2004, 2005, and 2006, respectively. Find her total earnings during those three years.

Find the perimeter of each figure.

 29.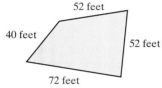

 30.

Use the Internet Use by World Regions table on page 89 for Exercises 31 and 32.

31. Find the increase in Internet users in Europe from 2008 to 2013.

32. Find the difference in the number of Internet users in 2016 between the Middle East and Oceania/Australia.

The following bar graph shows the monthly savings account balances for a freshman attending a local community college. Use this graph to answer Exercises 33 through 36.

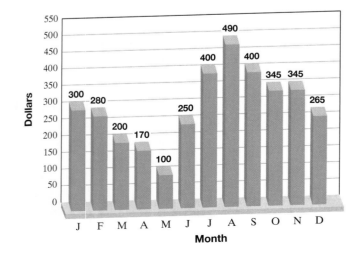

33. During what month was the balance the least?

34. During what month was the balance the greatest?

35. By how much did the balance decrease from February to April?

36. By how much did the balance increase from June to August?

Chapter 1 Review

(1.4) *Round to the given place value.*

37. 43 to the nearest ten

38. 45 to the nearest ten

39. 876 to the nearest ten

40. 493 to the nearest hundred

41. 3829 to the nearest hundred

42. 57,534 to the nearest thousand

43. 39,583,819 to the nearest million

44. 768,542 to the nearest hundred-thousand

Estimate the sum or difference by rounding each number to the nearest hundred.

45. 3785 + 648 + 2866

46. 5925 − 1787

47. A group of students took a week-long driving trip and traveled 630, 192, 271, 56, 703, 454, and 329 miles on seven consecutive days. Round each distance to the nearest hundred to estimate the distance they traveled.

48. In 2016, the population of Europe was 832,073,224 while the population of Latin America/Caribbean was 626,119,788. Round each number to the nearest million, and estimate the difference in population in 2016. (*Source:* Internet World Stats)

(1.5) *Multiply.*

49. 276 × 8

50. 349 × 4

51. 57 × 40

52. 69 × 42

53. 20(7)(4)

54. 25(9)(4)

55. 26 · 34 · 0

56. 62 · 88 · 0

57. 586 × 29

58. 242 × 37

59. 642 × 177

60. 347 × 129

61. 1026 × 401

62. 2107 × 302

Translating *Solve.*

63. Find the product of 6 and 250.

64. Find the product of 6 and 820.

65. A golf pro orders shirts for the company sponsoring a local charity golfing event. Shirts size large cost $32 while shirts size extra large cost $38. If 15 large shirts and 11 extra-large shirts are ordered, find the cost.

66. The cost for a South Dakota resident to attend Black Hills State University full-time is $20,199 per year. Determine the cost for 20 students to attend full-time. (*Source:* Black Hills State University)

Find the area of each rectangle.

△ **67.**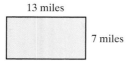
13 miles
7 miles

△ **68.**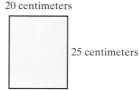
20 centimeters
25 centimeters

(1.6) *Divide and then check.*

69. $\dfrac{49}{7}$

70. $\dfrac{36}{9}$

71. $27 \div 5$

72. $18 \div 4$

73. $918 \div 0$

74. $0 \div 668$

75. $5\overline{)167}$

76. $8\overline{)159}$

77. $26\overline{)626}$

78. $19\overline{)680}$

79. $47\overline{)23{,}792}$

80. $53\overline{)48{,}111}$

81. $207\overline{)578{,}291}$

82. $306\overline{)615{,}732}$

Translating *Solve.*

83. Find the quotient of 92 and 5.

84. Find the quotient of 86 and 4.

85. A box can hold 24 cans of corn. How many boxes can be filled with 648 cans of corn?

86. One mile is 1760 yards. Find how many miles there are in 22,880 yards.

87. Find the average of the numbers 76, 49, 32, and 47.

88. Find the average of the numbers 23, 85, 62, and 66.

(1.7) *Simplify.*

89. 8^2

90. 5^3

91. $5 \cdot 9^2$

92. $4 \cdot 10^2$

93. $18 \div 2 + 7$

94. $12 - 8 \div 4$

95. $\dfrac{5(6^2 - 3)}{3^2 + 2}$

96. $\dfrac{7(16 - 8)}{2^3}$

97. $48 \div 8 \cdot 2$

98. $27 \div 9 \cdot 3$

99. $2 + 3[1^5 + (20 - 17) \cdot 3] + 5 \cdot 2$

100. $21 - [2^4 - (7 - 5) - 10] + 8 \cdot 2$

101. $19 - 2(3^2 - 2^2)$

102. $16 - 2(4^2 - 3^2)$

103. $4 \cdot 5 - 2 \cdot 7$

104. $8 \cdot 7 - 3 \cdot 9$

105. $(6 - 4)^3 \cdot [10^2 \div (3 + 17)]$

106. $(7 - 5)^3 \cdot [9^2 \div (2 + 7)]$

107. $\dfrac{5 \cdot 7 - 3 \cdot 5}{2(11 - 3^2)}$

108. $\dfrac{4 \cdot 8 - 1 \cdot 11}{3(9 - 2^3)}$

Find the area of each square.

109. A square with side length of 7 meters.

110. 3 inches

(1.8) *Evaluate each expression for $x = 5$, $y = 0$, and $z = 2$.*

111. $\dfrac{2x}{z}$

112. $4x - 3$

113. $\dfrac{x + 7}{y}$

114. $\dfrac{y}{5x}$

115. $x^3 - 2z$

116. $\dfrac{7 + x}{3z}$

117. $(y + z)^2$

118. $\dfrac{100}{x} + \dfrac{y}{3}$

Translating *Translate each phrase into a variable expression. Use x to represent a number.*

119. Five subtracted from a number

120. Seven more than a number

121. Ten divided by a number

122. The product of 5 and a number

Decide whether the given number is a solution of the given equation.

123. Is 5 a solution of $n + 12 = 20 - 3$?

124. Is 23 a solution of $n - 8 = 10 + 6$?

125. Is 14 a solution of $30 = 3(n - 3)$?

126. Is 20 a solution of $5(n - 7) = 65$?

Determine which numbers in each set are solutions to the corresponding equations.

127. $7n = 77$; $\{6, 11, 20\}$

128. $n - 25 = 150$; $\{125, 145, 175\}$

129. $5(n + 4) = 90$; $\{14, 16, 26\}$

130. $3n - 8 = 28$; $\{3, 7, 15\}$

Mixed Review

Perform the indicated operations.

131. $485 - 68$

132. $729 - 47$

133. 732×3

134. 629×4

135. $374 + 29 + 698$

136. $593 + 52 + 766$

137. $13\overline{)5962}$

138. $18\overline{)4267}$

139. 1968×36

140. 5324×18

141. $2000 - 356$

142. $9000 - 519$

Round to the given place value.

143. 842 to the nearest ten

144. 258,371 to the nearest hundred-thousand

Simplify.

145. $24 \div 4 \cdot 2$

146. $\dfrac{(15 + 3) \cdot (8 - 5)}{2^3 + 1}$

Solve.

147. Is 9 a solution of $5n - 6 = 40$?

148. Is 3 a solution of $2n - 6 = 5n - 15$?

149. A manufacturer of drinking glasses ships his delicate stock in special boxes that can hold 32 glasses. If 1714 glasses are manufactured, how many full boxes are filled? Are there any glasses left over?

150. A teacher orders 2 small whiteboards for $27 each and 8 boxes of dry erase pens for $4 each. What is her total bill before taxes?

Chapter 1 — Getting Ready for the Test

MULTIPLE CHOICE All the exercises below are **Multiple Choice**. Choose the correct letter. Some letters may be used more than once or not at all.

For Exercises 1 through 4, name the place value for the given digit. The number is 28,690,357,004 and choices are:

   A. tens   **B.** millions   **C.** ten-millions   **D.** ten-thousands   **E.** billions   **F.** hundred-millions

1. the digit 5

2. the digit 8

3. the digit 6

4. the digit 0 to the far left

For Exercises 5 through 8, identify the first operation to be performed to simplify the expression.

   A. add   **B.** subtract   **C.** multiply   **D.** divide

5. $6 - 3 \cdot 2$

6. $(6 - 3) \cdot 2$

7. $6 \div 3 \cdot 2$

8. $6 + 3 - 2$

9. The expression $5 \cdot 2^3$ simplifies to:

   A. 1000   **B.** 30   **C.** 40   **D.** 13

For Exercises 10 through 13, let $a = 35$ and $b = 5$. Choose the expression that gives each answer.

   A. $a - b$ **B.** $a \div b$ **C.** $a + b$ **D.** ab

10. answer: 7

11. answer: 175

12. answer: 30

13. answer: 40

Chapter 1

Test MyLab Math or YouTube For additional practice go to your study plan in MyLab Math.

Answers

Simplify.

1. Write 82,426 in words.

2. Write "four hundred two thousand, five hundred fifty" in standard form.

3. $59 + 82$

4. $600 - 487$

5. 496×30

6. $52,896 \div 69$

7. $2^3 \cdot 5^2$

8. $98 \div 1$

9. $0 \div 49$

10. $62 \div 0$

11. $(2^4 - 5) \cdot 3$

12. $16 + 9 \div 3 \cdot 4 - 7$

13. $6^1 \cdot 2^3$

14. $2[(6 - 4)^2 + (22 - 19)^2] + 10$

15. $5698 \cdot 1000$

16. Find the average of 62, 79, 84, 90, and 95.

17. Round 52,369 to the nearest thousand.

Estimate each sum or difference by rounding each number to the nearest hundred.

18. $6289 + 5403 + 1957$

19. $4267 - 2738$

Solve.

20. Subtract 15 from 107.

21. Find the sum of 15 and 107.

22. Find the product of 15 and 107.

23. Find the quotient of 107 and 15.

24. Twenty-nine cans of Sherwin-Williams paint cost $493. How much was each can?

25. Jo McElory is looking at two new refrigerators for her apartment. One costs $599 and the other costs $725. How much more expensive is the higher-priced one?

26. One tablespoon of white granulated sugar contains 45 calories. How many calories are in 8 tablespoons of white granulated sugar? (*Source: Home and Garden Bulletin No. 72,* U.S. Department of Agriculture)

27. A small business owner recently ordered 16 digital cameras that cost $430 each and 5 printers that cost $205 each. Find the total cost for these items.

Find the perimeter and the area of each figure.

28. Square, 5 centimeters

29. Rectangle, 20 yards by 10 yards

30. Evaluate $5(x^3 - 2)$ for $x = 2$.

31. Evaluate $\dfrac{3x - 5}{2y}$ for $x = 7$ and $y = 8$.

32. Translate the following phrases into mathematical expressions. Use x to represent "a number."
   a. The quotient of a number and 17
   b. Twice a number, decreased by 20

33. Is 6 a solution of the equation $5n - 11 = 19$?

34. Determine which number in the set is a solution to the given equation.
$n + 20 = 4n - 10$; $\{0, 10, 20\}$

24. _____

25. _____

26. _____

27. _____

28. _____

29. _____

30. _____

31. _____

32. a. _____

   b. _____

33. _____

34. _____

2 Integers and Introduction to Solving Equations

Thus far, we have studied whole numbers, but these numbers are not sufficient for representing many situations in real life. For example, to express 5 degrees below zero or $100 in debt, numbers less than 0 are needed. This chapter is devoted to integers, which include numbers less than 0, and operations on these numbers.

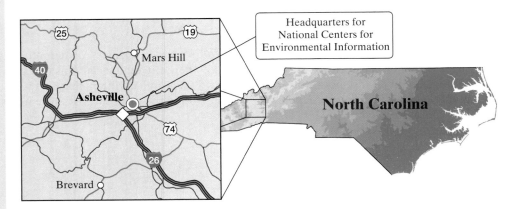

Sections

2.1 Introduction to Integers
2.2 Adding Integers
2.3 Subtracting Integers
2.4 Multiplying and Dividing Integers

Integrated Review—Integers

2.5 Order of Operations
2.6 Solving Equations: The Addition and Multiplication Properties

Check Your Progress

Vocabulary Check
Chapter Highlights
Chapter Review
Getting Ready for the Test
Chapter Test
Cumulative Review

Where Is U.S. Weather Data Stored?

To express whole degrees below zero, we need the set of integers. Below, we show the states of the United States with record minimum temperatures per state on one map and maximum temperatures on the other. How long have these temperatures been recorded and stored, and who stores this data? Much of temperature data recording in the U.S. started in 1895. In 1934, a unit was established in New Orleans, Louisiana, to store and process past weather records. This organization was moved to Asheville, North Carolina, in 1951 and was eventually called the National Climatic Data Center (NCDC). In 2015, the NCDC merged with two other centers to form the National Centers for Environmental Information (NCEI), headquartered in Asheville, North Carolina. See temperature exercises throughout this chapter.

2.1 Introduction to Integers

Objective A Representing Real-Life Situations

Thus far in this text, all numbers have been 0 or greater than 0. Numbers greater than 0 are called **positive numbers.** However, sometimes situations exist that cannot be represented by a number greater than 0. For example,

To represent these situations, we need numbers less than 0.
Extending the number line to the left of 0 allows us to picture **negative numbers,** which are numbers that are less than 0.

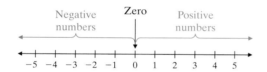

When a single + sign or no sign is in front of a number, the number is a positive number. When a single − sign is in front of a number, the number is a negative number. Together, we call positive numbers, negative numbers, and zero the **signed numbers.**

−5 indicates "negative five."
5 and +5 both indicate "positive five."
The number 0 is neither positive nor negative.

Some signed numbers are integers. The **integers** consist of the numbers labeled on the number line above. The integers are

..., −3, −2, −1, 0, 1, 2, 3, ...

Now we have numbers to represent the situations previously mentioned.

5 degrees below 0   −5°
20 feet below sea level   −20 feet

Objectives

A Represent Real-Life Situations with Integers.
B Graph Integers on a Number Line.
C Compare Integers.
D Find the Absolute Value of a Number.
E Find the Opposite of a Number.
F Read Bar Graphs Containing Integers.

> **Helpful Hint**
> Notice that 0 is neither positive nor negative.

> **Helpful Hint**
> A − sign, such as the one in −1, tells us that the number is to the left of 0 on the number line. −1 is read "negative one."
> A + sign or no sign tells us that a number lies to the right of 0 on the number line. For example, 3 and +3 both mean "positive three."

Practice 1

a. The world's deepest bat colony spends each winter in a New York zinc mine at a depth of 3805 feet. Represent this position with an integer. (*Source: Guinness Book of World Records*)

b. The tamarack tree, a type of conifer, commonly grows at the edge of the arctic tundra and survives winter temperatures of 85 degrees below zero Fahrenheit. Represent this temperature with an integer in degrees Fahrenheit.

Practice 2

Graph $-4, -1, 2,$ and -2 on the number line.

Example 1 Representing Depth with an Integer

The world's deepest cave is Krubera (or Voronja), in the country of Georgia, located by the Black Sea in Asia. It has been explored to a depth of 7188 feet below the surface of Earth. Represent this position using an integer.
(*Source:* MessagetoEagle.com and Wikipedia)

Solution: If 0 represents the surface of Earth, then 7188 feet below the surface can be represented by -7188.

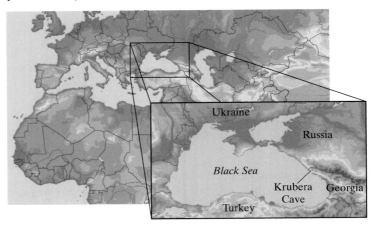

■ Work Practice 1

Objective B Graphing Integers

Example 2 Graph $0, -3, 5,$ and -5 on the number line.

Solution:

■ Work Practice 2

Objective C Comparing Integers

We can compare integers by using a number line. For any two numbers graphed on a number line, the number to the **right** is the **greater number** and the number to the **left** is the **smaller number**. Also, the symbols $<$ and $>$ are called **inequality symbols**.

The inequality symbol $>$ means "**is greater than**" and

the inequality symbol $<$ means "**is less than.**"

For example, both -5 and -7 are graphed on the number line below.

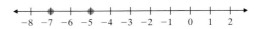

On the graph, -7 is **to the left of** -5, so -7 **is less than** -5, written as

$$-7 < -5$$

We can also write

$$-5 > -7$$

since -5 is **to the right of** -7, so -5 **is greater than** -7.

✓ **Concept Check** Is there a largest positive number? Is there a smallest negative number? Explain.

Answers
1. **a.** -3805 **b.** $-85°F$
2.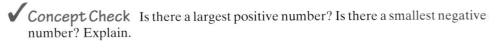

✓ Concept Check Answer
no

Example 3
Insert < or > between each pair of numbers to make a true statement.

a. −7 7   **b.** 0 −4   **c.** −9 −11

Solution:

a. −7 is to the left of 7 on a number line, so −7 < 7.
b. 0 is to the right of −4 on a number line, so 0 > −4.
c. −9 is to the right of −11 on a number line, so −9 > −11.

Work Practice 3

Practice 3
Insert < or > between each pair of numbers to make a true statement.

a. 0 −5   **b.** −3 3
c. −7 −12

Helpful Hint
If you think of < and > as arrowheads, notice that in a true statement the arrow always points to the smaller number.

5 > −4   −3 < −1
   ↑   ↑
smaller   smaller
number   number

Objective D Finding the Absolute Value of a Number

The **absolute value** of a number is the number's distance from 0 on the number line. The symbol for absolute value is | |. For example, |3| is read as "the absolute value of 3."

|3| = 3 because 3 is 3 units from 0.

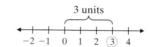

|−3| = 3 because −3 is 3 units from 0.

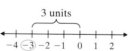

Example 4 Simplify.

a. |−9| **b.** |8| **c.** |0|

Solution:

a. |−9| = 9 because −9 is 9 units from 0.
b. |8| = 8 because 8 is 8 units from 0.
c. |0| = 0 because 0 is 0 units from 0.

Work Practice 4

Practice 4
Simplify.

a. |−6|
b. |4|
c. |−12|

Helpful Hint
Since the absolute value of a number is that number's *distance* from 0, the absolute value of a number is always 0 or positive. It is never negative.

|0| = 0 |−6| = 6
  ↑   ↑
 zero   a positive number

Answers

3. a. >   **b.** <   **c.** >
4. a. 6   **b.** 4   **c.** 12

Objective E Finding Opposites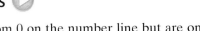

Two numbers that are the same distance from 0 on the number line but are on opposite sides of 0 are called **opposites.**

4 and −4 are opposites.

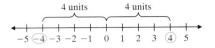

When two numbers are opposites, we say that each is the opposite of the other. Thus **4 is the opposite of −4** and **−4 is the opposite of 4.**

The phrase "the opposite" of is written in symbols as "−". For example,

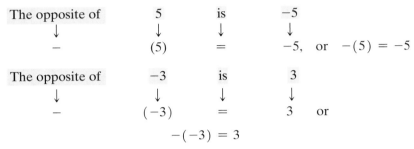

In general, we have the following:

Opposites

If a is a number, then $-(-a) = a$.

Notice that because "the opposite of" is written as "−", to find the opposite of a number we place a "−" sign in front of the number.

Practice 5

Find the opposite of each number.
a. 14   b. −9

Example 5 Find the opposite of each number.

a. 13   b. −2   c. 0

Solution:

a. The opposite of 13 is −13.
b. The opposite of −2 is $-(-2)$ or 2.
c. The opposite of 0 is 0.

Helpful Hint
Remember that 0 is neither positive nor negative.

■ Work Practice 5

✓ **Concept Check** True or false? The number 0 is the only number that is its own opposite.

Practice 6

Simplify.
a. $-|-7|$
b. $-|4|$
c. $-(-12)$

Example 6 Simplify.

a. $-(-4)$ b. $-|-5|$ c. $-|6|$

Solution:

a. $-(-4) = 4$ The opposite of negative 4 is 4.
b. $-|-5| = -5$ The opposite of the absolute value of −5 is the opposite of 5, or −5.
c. $-|6| = -6$ The opposite of the absolute value of 6 is the opposite of 6, or −6.

■ Work Practice 6

Answers
5. a. −14   b. 9
6. a. −7   b. −4   c. 12

✓ **Concept Check Answer**
true

Example 7
Evaluate $-|-x|$ if $x = -2$.

Solution: Carefully replace x with -2; then simplify.

$-|-x| = -|-(-2)|$ Replace x with -2.

Then $-|-(-2)| = -|2| = -2$.

Work Practice 7

Practice 7
Evaluate $-|x|$ if $x = -6$.

Objective F Reading Bar Graphs Containing Integers

The bar graph below shows the average daytime surface temperatures (in degrees Fahrenheit) of the eight planets, excluding the newly classified "dwarf planet," Pluto. Notice that a negative temperature is illustrated by a bar below the horizontal line representing 0°F, and a positive temperature is illustrated by a bar above the horizontal line representing 0°F.

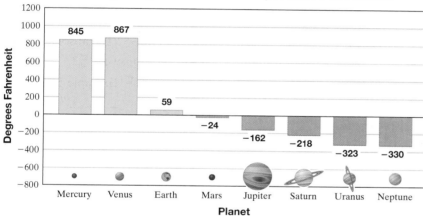

Average Daytime Surface Temperatures of Planets*

Source: The World Almanac
*For some planets, the temperature given is the temperature where the atmospheric pressure equals 1 Earth atmosphere.

Example 8
Which planet has the lowest average daytime surface temperature?

Solution: The planet with the lowest average daytime surface temperature is the one that corresponds to the bar that extends the farthest in the negative direction (downward). Neptune has the lowest average daytime surface temperature, −330°F.

Work Practice 8

Practice 8
Which planet has the highest average daytime surface temperature?

Answers
7. −6   8. Venus

Vocabulary, Readiness & Video Check

Use the choices below to fill in each blank. Not all choices will be used.

opposites	absolute value	right	is less than
inequality symbols	negative	positive	left
signed	integers	is greater than	

1. The numbers . . . −3, −2, −1, 0, 1, 2, 3, . . . are called _____.
2. Positive numbers, negative numbers, and zero together are called _____ numbers.
3. The symbols "<" and ">" are called _____.

4. Numbers greater than 0 are called _____ numbers while numbers less than 0 are called _____ numbers.
5. The sign "<" means _____ and ">" means _____.
6. On a number line, the greater number is to the _____ of the lesser number.
7. A number's distance from 0 on a number line is the number's _____.
8. The numbers −5 and 5 are called _____.

Martin-Gay Interactive Videos Watch the section lecture video and answer the following questions.

See Video 2.1

Objective A 9. In Example 1, what application is used to represent a negative number?

Objective B 10. In Example 2, the tick marks are labelled with what numbers on the number line?

Objective C 11. From Example 3 and your knowledge of a number line, complete this statement: 0 will always be greater than any of the _____ integers.

Objective D 12. What is the answer to Example 5? The absolute value of what other integer has this same answer?

Objective E 13. Complete this statement based on Example 10: A negative sign can be translated to the phrase "_____."

Objective F 14. In Examples 13 and 14, what other lake has a negative integer elevation?

2.1 Exercise Set MyLab Math

Objective A *Represent each quantity by an integer. See Example 1.*

1. A worker in a silver mine in Nevada works 1235 feet underground.

2. A scuba diver is swimming 25 feet below the surface of the water in the Gulf of Mexico.

3. The Challenger Deep in the Mariana Trench (the deepest-known part of the Pacific Ocean) is the lowest known point in the Earth's seabed hydrosphere. The bottom is estimated to be 35,814 feet below sea level.

4. The lowest elevation in the United States is found at Death Valley, California, at an elevation of 282 feet below sea level. (*Source:* U.S. Geological Survey)

Section 2.1 | Introduction to Integers

5. The record high temperature in Arkansas is 120 degrees above zero Fahrenheit. (*Source:* National Centers for Environmental Information)

6. The record high temperature in California is 134 degrees above zero Fahrenheit. (*Source:* National Centers for Environmental Information.)

7. The average depth of the Atlantic Ocean is 11,810 feet below its surface. (*Source: The World Almanac*, 2013)

8. The average depth of the Pacific Ocean is 14,040 feet below its surface. (*Source: The World Almanac*, 2013)

9. Uber lost $1270 million in the first half of 2016. (*Source:* Bloomberg)

10. Occidental Petroleum reported a loss of $241 million for the third quarter of 2016. (*Source:* Occidental Petroleum Corp.)

11. Two divers are exploring the wreck of the *Andrea Doria*, south of Nantucket Island, Massachusetts. Guillermo is 160 feet below the surface of the ocean and Luigi is 147 feet below the surface. Represent each quantity by an integer and determine who is deeper.

12. The temperature on one January day in Minneapolis was 10° below 0° Celsius. Represent this quantity by an integer and tell whether this temperature is cooler or warmer than 5° below 0° Celsius.

Use the Chapter 2 Opener for Exercises 13 and 14. First, locate the state. Then use the blue map to locate the coldest temperature on record for that state.

13. Find the coldest temperature on record for the state of Idaho.

14. Find the coldest temperature on record for the state of Arkansas.

Objective B *Graph each integer in the list on the same number line. See Example 2.*

15. 0, 3, 4, 6

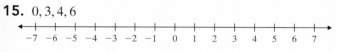

16. 7, 5, 2, 0

17. 1, −1, 2, −2, −4

18. 3, −3, 5, −5, 6

19. 0, 1, 9, 14

20. 0, 3, 10, 11

21. 0, −2, −7, −5

22. 0, −7, 3, −6

Objective C *Insert < or > between each pair of integers to make a true statement. See Example 3.*

23. 0   −7
24. −8   0
25. −7   −5
26. −12   −10
27. −30   −35
28. −27   −29
29. −26   26
30. 13   −13

Objective D *Simplify. See Example 4.*

31. |5|
32. |7|
33. |−8|
34. |−19|
35. |0|
36. |100|
37. |−55|
38. |−10|

Objective E *Find the opposite of each integer. See Example 5.*

39. 5
40. 8
41. −4
42. −6
43. 23
44. 123
45. −85
46. −13

Objectives D E Mixed Practice *Simplify. See Example 6.*

47. $|-7|$
48. $|-11|$
49. $-|20|$
50. $-|43|$
51. $-|-3|$
52. $-|-18|$
53. $-(-43)$
54. $-(-27)$
55. $|-15|$
56. $-(-14)$
57. $-(-33)$
58. $-|-29|$

Evaluate. See Example 7.

59. $|-x|$ if $x = -6$
60. $-|x|$ if $x = -8$
61. $-|-x|$ if $x = 2$
62. $-|-x|$ if $x = 10$
63. $|x|$ if $x = -32$
64. $|x|$ if $x = 32$
65. $-|x|$ if $x = 7$
66. $|-x|$ if $x = 1$

Insert <, >, or = between each pair of numbers to make a true statement. See Examples 3 through 6.

67. −12   −6
68. −4   −17
69. $|-8|$ $|-11|$
70. $|-8|$ $|-4|$
71. $|-47|$ $-(-47)$
72. $-|17|$ $-(-17)$
73. $-|-12|$ $-(-12)$
74. $|-24|$ $-(-24)$
75. 0   −9
76. −45   0
77. $|0|$ $|-9|$
78. $|-45|$ $|0|$
79. $-|-2|$ $-|-10|$
80. $-|-8|$ $-|-4|$
81. $-(-12)$ $-(-18)$
82. −22 $-(-38)$

Objectives D E Mixed Practice *Fill in the chart. See Examples 4 through 6.*

	Number	Absolute Value of Number	Opposite of Number
83.	31		
85.			−28

	Number	Absolute Value of Number	Opposite of Number
84.	−13		
86.			90

Objective F *The bar graph shows the elevations of selected lakes. Use this graph For Exercises 87 through 90. (Source: U.S. Geological Survey) See Example 8.*

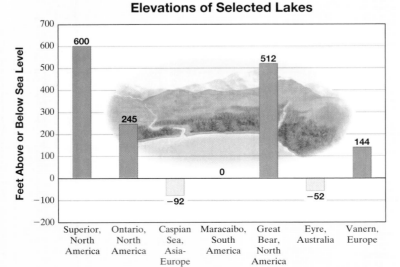

Elevations of Selected Lakes

87. Which lake shown has the lowest elevation?

88. Which lake has an elevation at sea level?

89. Which lake shown has the highest elevation?

90. Which lake shown has the second-lowest elevation?

The following bar graph represents the boiling temperature, the temperature at which a substance changes from liquid to gas at standard atmospheric pressure. Use this graph to answer Exercises 91 through 94.

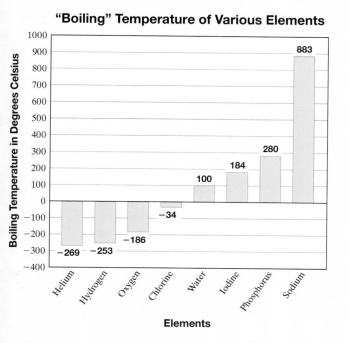

91. Which element has a positive boiling temperature closest to that of water?

92. Which element has the lowest boiling temperature?

93. Which element has a boiling temperature closest to $-200°C$?

94. Which element has an average boiling temperature closest to $+300°C$?

Review

Add. See Section 1.3.

95. $0 + 13$

96. $9 + 0$

97. $15 + 20$

98. $20 + 15$

99. $47 + 236 + 77$

100. $362 + 37 + 90$

Concept Extensions

Write the given numbers in order from least to greatest.

101. $2^2, -|3|, -(-5), -|-8|$

102. $|10|, 2^3, -|-5|, -(-4)$

103. $|-1|, -|-6|, -(-6), -|1|$

104. $1^4, -(-3), -|7|, |-20|$

105. $-(-2), 5^2, -10, -|-9|, |-12|$

106. $3^3, -|-11|, -(-10), -4, -|2|$

Choose all numbers for x from each given list that make each statement true.

107. $|x| > 8$
    a. -9 **b.** -5 **c.** 8 **d.** -12

108. $|x| > 4$
    a. 0 **b.** -4 **c.** 5 **d.** -100

109. Evaluate: $-(-|-8|)$

110. Evaluate: $(-|-(-7)|)$

Answer true or false for Exercises 111 through 115.

111. If $a > b$, then a must be a positive number.

112. The absolute value of a number is *always* a positive number.

113. A positive number is always greater than a negative number.

114. Zero is always less than a positive number.

115. The number $-a$ is always a negative number. (*Hint:* Read "$-$" as "the opposite of.")

116. Given the number line is it true that $b < a$?

117. Write in your own words how to find the absolute value of a signed number.

118. Explain how to determine which of two signed numbers is larger.

For Exercises 119 and 120, see the first Concept Check in this section.

119. Is there a largest negative number? If so, what is it?

120. Is there a smallest positive number? If so, what is it?

2.2 Adding Integers

Objectives

A Add Integers.

B Evaluate an Algebraic Expression by Adding.

C Solve Problems by Adding Integers.

Objective A Adding Integers

Adding integers can be visualized using a number line. A positive number can be represented on the number line by an arrow of appropriate length pointing to the right, and a negative number by an arrow of appropriate length pointing to the left.

Both arrows represent 2 or +2. They both point to the right and they are both 2 units long.

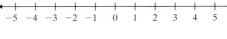

Both arrows represent -3. They both point to the left and they are both 3 units long.

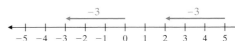

Practice 1

Add using a number line: $5 + (-1)$

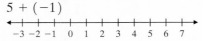

Answer

1.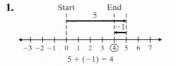

Example 1 Add using a number line: $5 + (-2)$

Solution: To add integers on a number line, such as $5 + (-2)$, we start at 0 on the number line and draw an arrow representing 5. From the tip of this arrow, we draw another arrow representing -2. The tip of the second arrow ends at their sum, 3.

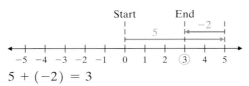

$5 + (-2) = 3$

■ Work Practice 1

Example 2
Add using a number line: $-1 + (-4)$

Start at 0 and draw an arrow representing -1. From the tip of this arrow, we draw another arrow representing -4. The tip of the second arrow ends at their sum, -5.

Solution:

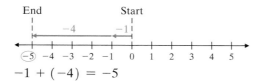

$-1 + (-4) = -5$

Work Practice 2

Practice 2
Add using a number line: $-6 + (-2)$

Example 3
Add using a number line: $-7 + 3$

Solution:

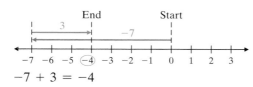

$-7 + 3 = -4$

Work Practice 3

Practice 3
Add using a number line: $-8 + 3$

Using a number line each time we add two numbers can be time consuming. Instead, we can notice patterns in the previous examples and write rules for adding signed numbers.

Rules for adding signed numbers depend on whether we are adding numbers with the same sign or different signs. When adding two numbers with the same sign, as in Example 2, notice that the sign of the sum is the same as the sign of the addends.

Adding Two Numbers with the Same Sign
Step 1: Add their absolute values.
Step 2: Use their common sign as the sign of the sum.

Example 4
Add: $-2 + (-21)$

Solution:
Step 1: $|-2| = 2$, $|-21| = 21$, and $2 + 21 = 23$.
Step 2: Their common sign is negative, so the sum is negative:
$-2 + (-21) = -23$

Work Practice 4

Practice 4
Add: $(-3) + (-19)$

Practice 5–6
Add.
5. $-12 + (-30)$ **6.** $9 + 4$

Examples
Add.
5. $-15 + (-10) = -25$
6. $2 + 6 = 8$

Work Practice 5–6

When adding two numbers with different signs, as in Examples 1 and 3, the sign of the result may be positive or negative, or the result may be 0.

Answers

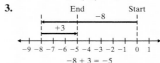

4. -22 **5.** -42 **6.** 13

Adding Two Numbers with Different Signs

Step 1: Find the larger absolute value minus the smaller absolute value.
Step 2: Use the sign of the number with the larger absolute value as the sign of the sum.

Practice 7
Add: $-1 + 26$

Example 7   Add: $-2 + 25$

Solution:
Step 1: $|-2| = 2$, $|25| = 25$, and $25 - 2 = 23$.
Step 2: 25 has the larger absolute value and its sign is an understood +:
$$-2 + 25 = +23 \text{ or } 23$$

■ Work Practice 7

Practice 8
Add: $2 + (-18)$

Example 8   Add: $3 + (-17)$

Solution:
Step 1: $|3| = 3$, $|-17| = 17$, and $17 - 3 = 14$.
Step 2: -17 has the larger absolute value and its sign is $-$:
$$3 + (-17) = -14$$

■ Work Practice 8

Practice 9–11
Add.
9. $-54 + 20$
10. $7 + (-2)$
11. $-3 + 0$

Examples   Add.

9. $-18 + 10 = -8$
10. $12 + (-8) = 4$
11. $0 + (-5) = -5$ The sum of 0 and any number is the number.

■ Work Practice 9–11

Recall that numbers such as 7 and -7 are called opposites. In general, the sum of a number and its opposite is always 0.

$$7 + (-7) = 0 \qquad -26 + 26 = 0 \qquad 1008 + (-1008) = 0$$
$\quad\;\;$ opposites $\qquad\qquad$ opposites $\qquad\qquad\quad$ opposites

If a is a number, then
$-a$ is its opposite. Also,
$$\left.\begin{array}{l} a + (-a) = 0 \\ -a + a = 0 \end{array}\right\} \text{ The sum of a number and its opposite is 0.}$$

Practice 12–13
Add.
12. $18 + (-18)$
13. $-64 + 64$

Examples   Add.

12. $-21 + 21 = 0$
13. $36 + (-36) = 0$

■ Work Practice 12–13

Answers
7. 25   **8.** -16 **9.** -34 **10.** 5
11. -3 **12.** 0   **13.** 0

✓**Concept Check Answer**
$5 + (-22) = -17$

✓**Concept Check** What is wrong with the following calculation?
$$5 + (-22) = 17$$

Section 2.2 | Adding Integers

In the following examples, we add three or more integers. Remember that by the associative and commutative properties for addition, we may add numbers in any order that we wish. In Examples 14 and 15, let's add the numbers from left to right.

Example 14 Add: $(-3) + 4 + (-11)$

Solution: $(-3) + 4 + (-11) = 1 + (-11)$
$= -10$

Work Practice 14

Practice 14
Add: $6 + (-2) + (-15)$

Example 15 Add: $1 + (-10) + (-8) + 9$

Solution: $1 + (-10) + (-8) + 9 = -9 + (-8) + 9$
$= -17 + 9$
$= -8$

Work Practice 15

Practice 15
Add: $5 + (-3) + 12 + (-14)$

A sum is the same if we add the numbers in any order. To see this, let's add the numbers in Example 15 by first adding the positive numbers together and the negative numbers together.

$1 + (-10) + (-8) + 9 = 10 + (-18)$ Add the positive numbers: $1 + 9 = 10$.
$= -8$ Add the negative numbers: $(-10) + (-8) = -18$.
  Add these results.

The sum is -8.

Helpful Hint Don't forget that addition is commutative and associative. In other words, numbers may be added in any order.

Objective B Evaluating Algebraic Expressions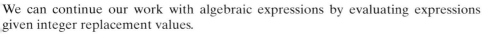

We can continue our work with algebraic expressions by evaluating expressions given integer replacement values.

Example 16 Evaluate $2x + y$ for $x = 3$ and $y = -5$.

Solution: Replace x with 3 and y with -5 in $2x + y$.
$2x + y = 2 \cdot 3 + (-5)$
$= 6 + (-5)$
$= 1$

Work Practice 16

Practice 16
Evaluate $x + 3y$ for $x = -6$ and $y = 2$.

Example 17 Evaluate $x + y$ for $x = -2$ and $y = -10$.

Solution: $x + y = (-2) + (-10)$ Replace x with -2 and y with -10.
$= -12$

Work Practice 17

Practice 17
Evaluate $x + y$ for $x = -13$ and $y = -9$.

Objective C Solving Problems by Adding Integers

Next, we practice solving problems that require adding integers.

Example 18 Calculating Temperature

In Pennsylvania, the record extreme high temperature is 111°F. Add −153° to this temperature and the result is the record extreme low temperature. Find this extreme low temperature. (*Source:* National Centers for Environmental Information)
(Continued on next page)

Practice 18
If the temperature was −7° Fahrenheit at 6 a.m., and it rose 4 degrees by 7 a.m. and then rose another 7 degrees in the hour from 7 a.m. to 8 a.m., what was the temperature at 8 a.m.?

Answers
14. −11 **15.** 0 **16.** 0 **17.** −22
18. 4°F

Solution:

In words:

extreme low temperature	=	extreme high temperature	+	decrease of of $-153°$
↓		↓		↓

Translate: extreme low temperature $= 111 + (-153)$

$= -42$

The record extreme low temperature in Pennsylvania, is $-42°F$.

■ Work Practice 18

Calculator Explorations Entering Negative Numbers

To enter a negative number on a calculator, find the key marked $\boxed{+/-}$. (Some calculators have a key marked $\boxed{CHS}$ and some calculators have a special key $\boxed{(-)}$ for entering a negative sign.) To enter the number -2, for example, press the keys $\boxed{2}$ $\boxed{+/-}$. The display will read $\boxed{-2}$.

To find $-32 + (-131)$, press the keys
$\boxed{32}$ $\boxed{+/-}$ $\boxed{+}$ $\boxed{131}$ $\boxed{+/-}$ $\boxed{=}$ or
$\boxed{(-)}$ $\boxed{32}$ $\boxed{+}$ $\boxed{(-)}$ $\boxed{131}$ $\boxed{ENTER}$

The display will read $\boxed{-163}$.
Thus $-32 + (-131) = -163$.

Use a calculator to perform each indicated operation.

1. $-256 + 97$
2. $811 + (-1058)$
3. $6(15) + (-46)$
4. $-129 + 10(48)$
5. $-108,650 + (-786,205)$
6. $-196,662 + (-129,856)$

Vocabulary, Readiness & Video Check

Use the choices below to fill in each blank. Not all choices will be used.

$-a$ a 0 commutative   associative

1. If n is a number, then $-n + n =$ _____.
2. Since $x + n = n + x$, we say that addition is _____.
3. If a is a number, then $-(-a) =$ _____.
4. Since $n + (x + a) = (n + x) + a$, we say that addition is _____.

Martin-Gay Interactive Videos Watch the section lecture video and answer the following questions.

See Video 2.2

Objective A 5. What is the sign of the sum in Example 6 and why?

Objective B 6. What is the sign of the sum in Example 8 and why?

Objective C 7. What does the answer to Example 10, -231, mean in the context of the application?

2.2 Exercise Set MyLab Math

Objective A *Add using a number line. See Examples 1 through 3.*

1. $-1 + (-6)$

2. $-6 + (-5)$

3. $-4 + 7$

4. $10 + (-3)$

5. $-13 + 7$

6. $9 + (-4)$

Add. See Examples 4 through 13.

7. $46 + 21$
8. $15 + 42$
9. $-8 + (-2)$
10. $-5 + (-4)$
11. $-43 + 43$
12. $-62 + 62$
13. $6 + (-2)$
14. $8 + (-3)$
15. $-6 + 0$
16. $-8 + 0$
17. $3 + (-5)$
18. $5 + (-9)$
19. $-2 + (-7)$
20. $-6 + (-1)$
21. $-12 + (-12)$
22. $-23 + (-23)$
23. $-640 + (-200)$
24. $-400 + (-256)$
25. $12 + (-5)$
26. $24 + (-10)$
27. $-6 + 3$
28. $-8 + 4$
29. $-56 + 26$
30. $-89 + 37$
31. $-45 + 85$
32. $-32 + 62$
33. $124 + (-144)$
34. $325 + (-375)$
35. $-82 + (-43)$
36. $-56 + (-33)$

Add. See Examples 14 and 15.

37. $-4 + 2 + (-5)$
38. $-1 + 5 + (-8)$
39. $-52 + (-77) + (-117)$
40. $-103 + (-32) + (-27)$
41. $12 + (-4) + (-4) + 12$
42. $18 + (-9) + 5 + (-2)$
43. $(-10) + 14 + 25 + (-16)$
44. $34 + (-12) + (-11) + 213$

Objective A Mixed Practice *Add. See Examples 1 through 15.*

45. $-6 + (-15) + (-7)$

46. $-12 + (-3) + (-5)$

47. $-26 + 15$

48. $-35 + (-12)$

49. $5 + (-2) + 17$

50. $3 + (-23) + 6$

51. $-13 + (-21)$

52. $-100 + 70$

53. $3 + 14 + (-18)$

54. $(-45) + 22 + 20$

55. $-92 + 92$

56. $-87 + 0$

57. $-13 + 8 + (-10) + (-27)$

58. $-16 + 6 + (-14) + (-20)$

Objective B *Evaluate $x + y$ for the given replacement values. See Examples 16 and 17.*

59. $x = -20$ and $y = -50$

60. $x = -1$ and $y = -29$

Evaluate $3x + y$ for the given replacement values. See Examples 16 and 17.

61. $x = 2$ and $y = -3$

62. $x = 7$ and $y = -11$

63. $x = 3$ and $y = -30$

64. $x = 13$ and $y = -17$

Objective C Translating *Translate each phrase; then simplify. See Example 18.*

65. Find the sum of -6 and 25.

66. Find the sum of -30 and 15.

67. Find the sum of -31, -9, and 30.

68. Find the sum of -49, -2, and 40.

Solve. See Example 18.

69. Suppose a deep-sea diver dives from the surface to 215 feet below the surface. He then dives down 16 more feet. Use signed numbers to represent this situation. Then find the diver's present depth.

70. Suppose a diver dives from the surface to 248 meters below the surface and then swims up 8 meters, down 16 meters, down another 28 meters, and then up 32 meters. Use signed numbers to represent this situation. Then find the diver's depth after these movements.

In golf, it is possible to have positive and negative scores. The following table shows the results of the eighteen-hole Round 4 for Jimmy Walker and Hideki Matsuyama at the 2016 PGA Championship in Baltusrol Golf Club in Springfield, New Jersey. Use the table to answer Exercises 71 and 72.

Player/Hole	1	2	3	4	5	6	7	8	9	10	11	12	13	14	15	16	17	18
Walker	0	0	0	0	0	0	0	0	0	−1	−1	0	0	0	0	0	−1	0
Matsuyama	0	0	0	0	0	0	0	0	0	0	−1	0	0	0	0	0	−1	0

71. Find the total score for each of the athletes in the round.

72. In golf, the lower score is the winner. Use the result of Exercise 71 to determine who won Round 4.

The following bar graph shows the yearly net income for Apple, Inc. Net income is one indication of a company's health. It measures revenue (money taken in) minus cost (money spent). Use this graph to answer Exercises 73 through 76. (Source: Apple, Inc.)

73. What was the net income (in dollars) for Apple, Inc. in 2016?

74. What was the net income (in dollars) for Apple, Inc. in 2015?

75. What was the difference in net income (in dollars) for Apple, Inc. from 2015 to 2016?

76. Find the total net income for the years 2014, 2015, and 2016.

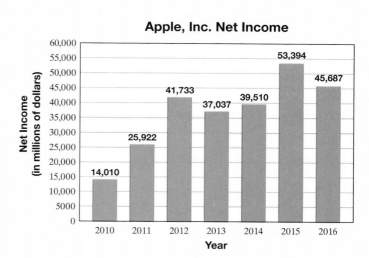

77. The temperature at 4 p.m. on February 2 was −10° Celsius. By 11 p.m. the temperature had risen 12 degrees. Find the temperature at 11 p.m.

78. In some card games, it is possible to have both positive and negative scores. After four rounds of play, Michelle had scores of 14, −5, −8, and 7. What was her total score for the game?

A small business reports the following net incomes. Use this table to answer Exercises 79 and 80.

Year	Net Income (in dollars)
2013	−$10,412
2014	−$1786
2015	$15,395
2016	$31,418

79. Find the sum of the net incomes for 2014 and 2015.

80. Find the sum of the net incomes for all four years shown.

81. The all-time record low temperature for Colorado is −61°F. Oklahoma's all-time record low temperature is 30°F higher than Colorado's record low. What is Oklahoma's record low temperature? (*Source:* National Centers for Environmental Information)

82. The all-time record low temperature for Minnesota is −60°F. In Georgia, the lowest temperature ever recorded is 43°F higher than Minnesota's all-time low temperature. What is the all-time record low temperature for Georgia? (*Source:* National Centers for Environmental Information)

83. The deepest spot in the Atlantic Ocean is the Puerto Rico Trench, which has an elevation of 8605 meters below sea level. The bottom of the Atlantic's Cayman Trench has an elevation 1070 meters above the level of the Puerto Rico Trench. Use a negative number to represent the depth of the Cayman Trench. (*Source:* Defense Mapping Agency)

84. The deepest spot in the Pacific Ocean is the Mariana Trench, which has an elevation of 10,924 meters below sea level. The bottom of the Pacific's Aleutian Trench has an elevation 3245 meters higher than that of the Mariana Trench. Use a negative number to represent the depth of the Aleutian Trench. (*Source:* Defense Mapping Agency)

116 Chapter 2 | Integers and Introduction to Solving Equations

Review

Subtract. See Section 1.3.

85. $44 - 0$ **86.** $91 - 0$ **87.** $200 - 59$ **88.** $400 - 18$

Concept Extensions

89. Name 2 numbers whose sum is -17.

90. Name 2 numbers whose sum is -30.

Each calculation below is incorrect. Find the error and correct it. See the Concept Check in this section.

91. $7 + (-10) \stackrel{?}{=} 17$

92. $-4 + 14 \stackrel{?}{=} -18$

93. $-10 + (-12) \stackrel{?}{=} -120$

94. $-15 + (-17) \stackrel{?}{=} 32$

For Exercises 95 through 98, determine whether each statement is true or false.

95. The sum of two negative numbers is always a negative number.

96. The sum of two positive numbers is always a positive number.

97. The sum of a positive number and a negative number is always a negative number.

98. The sum of zero and a negative number is always a negative number.

99. In your own words, explain how to add two negative numbers.

100. In your own words, explain how to add a positive number and a negative number.

2.3 Subtracting Integers

Objectives

A Subtract Integers.

B Add and Subtract Integers.

C Evaluate an Algebraic Expression by Subtracting.

D Solve Problems by Subtracting Integers.

In Section 2.1, we discussed the opposite of an integer.

The opposite of 3 is -3.
The opposite of -6 is 6.

In this section, we use opposites to subtract integers.

Objective A Subtracting Integers

To subtract integers, we will write the subtraction problem as an addition problem. To see how to do this, study the examples below.

$$10 - 4 = 6$$
$$10 + (-4) = 6$$

Since both expressions simplify to 6, this means that

$$10 - 4 = 10 + (-4) = 6$$

Also,

$$3 - 2 = 3 + (-2) = 1$$
$$15 - 1 = 15 + (-1) = 14$$

Thus, to subtract two numbers, we add the first number to the opposite of the second number. (The opposite of a number is also known as its **additive inverse**.)

Subtracting Two Numbers

If a and b are numbers, then $a - b = a + (-b)$.

Examples Subtract.

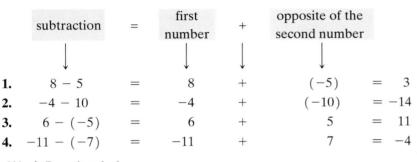

	subtraction	=	first number	+	opposite of the second number	
1.	$8 - 5$	=	8	+	(-5)	$= 3$
2.	$-4 - 10$	=	-4	+	(-10)	$= -14$
3.	$6 - (-5)$	=	6	+	5	$= 11$
4.	$-11 - (-7)$	=	-11	+	7	$= -4$

Work Practice 1–4

Practice 1–4

Subtract.
1. $13 - 4$
2. $-8 - 2$
3. $11 - (-15)$
4. $-9 - (-1)$

Examples Subtract.

5. $-10 - 5 = -10 + (-5) = -15$
6. $8 - 15 = 8 + (-15) = -7$
7. $-4 - (-5) = -4 + 5 = 1$

Work Practice 5–7

Practice 5–7

Subtract.
5. $6 - 9$
6. $-14 - 5$
7. $-3 - (-4)$

Helpful Hint

To visualize subtraction, try the following:
The difference between $5°F$ and $-2°F$ can be found by subtracting. That is,
$$5 - (-2) = 5 + 2 = 7$$
Can you visually see from the thermometer on the right that there are actually 7 degrees between $5°F$ and $-2°F$?

✓ **Concept Check** What is wrong with the following calculation?
$$-9 - (-5) = -14$$

Example 8 Subtract 7 from -3.

Solution: To subtract 7 *from* -3, we find

$$-3 - 7 = -3 + (-7) = -10$$

Work Practice 8

Practice 8

Subtract 6 from -15.

Answers
1. 9   2. -10 3. 26   4. -8
5. -3 6. -19 7. 1   8. -21

✓ **Concept Check Answer**
$-9 - (-5) = -9 + 5 = -4$

Objective B Adding and Subtracting Integers

If a problem involves adding or subtracting more than two integers, we rewrite differences as sums and add. Recall that by associative and commutative properties, we may add numbers in any order. In Examples 9 and 10, we will add from left to right.

Example 9 Simplify: $7 - 8 - (-5) - 1$

Solution:
$$7 - 8 - (-5) - 1 = \underbrace{7 + (-8)} + 5 + (-1)$$
$$= \underbrace{-1 + 5} + (-1)$$
$$= \underbrace{4 + (-1)}$$
$$= 3$$

■ Work Practice 9

Practice 9
Simplify: $-6 - 5 - 2 - (-3)$

Example 10 Simplify: $7 + (-12) - 3 - (-8)$

Solution:
$$7 + (-12) - 3 - (-8) = \underbrace{7 + (-12)} + (-3) + 8$$
$$= \underbrace{-5 + (-3)} + 8$$
$$= \underbrace{-8 + 8}$$
$$= 0$$

■ Work Practice 10

Practice 10
Simplify: $8 + (-2) - 9 - (-7)$

Objective C Evaluating Expressions

Now let's practice evaluating expressions when the replacement values are integers.

Example 11 Evaluate $x - y$ for $x = -3$ and $y = 9$.

Solution: Replace x with -3 and y with 9 in $x - y$.

$$\begin{array}{ccc} x & - & y \\ \downarrow & \downarrow & \downarrow \end{array}$$
$$= (-3) - 9$$
$$= (-3) + (-9)$$
$$= -12$$

■ Work Practice 11

Practice 11
Evaluate $x - y$ for $x = -5$ and $y = 13$.

Example 12 Evaluate $2a - b$ for $a = 8$ and $b = -6$.

Solution: Watch your signs carefully!

$$\begin{array}{ccc} 2a & - & b \\ \downarrow\downarrow & & \downarrow \end{array}$$
$= 2 \cdot 8 - (-6)$ Replace a with 8 and b with -6.
$= 16 + 6$ Multiply. Write subtraction as equivalent addition.
$= 22$ Add.

■ Work Practice 12

Practice 12
Evaluate $3y - z$ for $y = 9$ and $z = -4$.

> **Helpful Hint** Watch carefully when replacing variables in the expression $2a - b$. Make sure that all symbols are inserted and accounted for.

Answers
9. -10 10. 4 11. -18 12. 31

Objective D Solving Problems by Subtracting Integers

Solving problems often requires subtraction of integers. For Example 13 below, notice we begin to organize our solving process by numbering and labelling steps.

Example 13 Finding a Change in Elevation

The highest point in the United States is the top of Denali, at a height of 20,320 feet above sea level. The lowest point is Death Valley, California, which is 282 feet below sea level. How much higher is Denali than Death Valley? (*Source:* U.S. Geological Survey)

Solution:

1. UNDERSTAND. Read and reread the problem. To find "how much higher," we subtract. Don't forget that since Death Valley is 282 feet *below* sea level, we represent its height by -282. Draw a diagram to help visualize the problem.

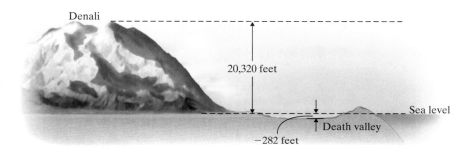

2. TRANSLATE.

   In words: how much higher is Denali = height of Denali minus height of Death Valley

   Translate: how much higher is Denali = $20{,}320 - (-282)$

3. SOLVE:

   $20{,}320 - (-282) = 20{,}320 + 282 = 20{,}602$

4. INTERPRET. Check and state your conclusion: Denali is 20,602 feet higher than Death Valley.

■ Work Practice 13

Practice 13

The highest point in Asia is the top of Mount Everest, at a height of 29,028 feet above sea level. The lowest point is the Dead Sea, which is 1312 feet below sea level. How much higher is Mount Everest than the Dead Sea? (*Source:* National Geographic Society)

Answer
13. 30,340 ft

Vocabulary, Readiness & Video Check

Multiple choice: Select the correct lettered response following each exercise.

1. It is true that $a - b =$ _____.
   a. $b - a$ b. $a + (-b)$ c. $a + b$

2. The opposite of n is _____.
   a. $-n$ b. $-(-n)$ c. n

3. To evaluate $x - y$ for $x = -10$ and $y = -14$, we replace x with -10 and y with -14 and evaluate _____.
   a. $10 - 14$ b. $-10 - 14$ c. $-14 - 10$ d. $-10 - (-14)$

4. The expression $-5 - 10$ equals _____.
   a. $5 - 10$ b. $5 + 10$ c. $-5 + (-10)$ d. $10 - 5$

120 Chapter 2 | Integers and Introduction to Solving Equations

Martin-Gay Interactive Videos Watch the section lecture video and answer the following questions.

See Video 2.3

Objective A 5. In the lecture before ▦ Example 1, what can the "opposite" of a number also be called? ▶

Objective B 6. In ▦ Example 7, how is the example rewritten in the first step of simplifying and why? ▶

Objective C 7. In ▦ Example 8, why do we multiply first? ▶

Objective D 8. What does the answer to ▦ Example 9, 263, mean in the context of the application? ▶

2.3 Exercise Set MyLab Math

Objective A Subtract. See Examples 1 through 7.

1. $-8 - (-8)$
2. $-6 - (-6)$
3. $19 - 16$
4. $15 - 12$
5. $3 - 8$
6. $2 - 5$
7. $11 - (-11)$
8. $12 - (-12)$
9. $-4 - (-7)$
10. $-25 - (-25)$
11. $-16 - 4$
12. $-2 - 42$
13. $3 - 15$
14. $8 - 9$
15. $42 - 55$
16. $17 - 63$
17. $478 - (-30)$
18. $844 - (-20)$
19. $-4 - 10$
20. $-5 - 8$
21. $-7 - (-3)$
22. $-12 - (-5)$
23. $17 - 29$
24. $16 - 45$

Translating Translate each phrase; then simplify. See Example 8.

25. Subtract 17 from -25.
26. Subtract 10 from -22.
27. Find the difference of -22 and -3.
28. Find the difference of -8 and -13.
29. Subtract -12 from 2.
30. Subtract -50 from -50.

Mixed Practice (Sections 2.2 and 2.3) Add or subtract as indicated.

31. $-37 + (-19)$
32. $-35 + (-11)$
33. $8 - 13$
34. $4 - 21$
35. $-56 - 89$
36. $-105 - 68$
37. $30 - 67$
38. $86 - 98$

Objective B Simplify. See Examples 9 and 10.

39. $8 - 3 - 2$
40. $8 - 4 - 1$
41. $13 - 5 - 7$
42. $30 - 18 - 12$
43. $-5 - 8 - (-12)$
44. $-10 - 6 - (-9)$
45. $-11 + (-6) - 14$
46. $-15 + (-8) - 4$
47. $18 - (-32) + (-6)$

48. 23 − (−17) + (−9) **49.** −(−5) − 21 + (−16) **50.** −(−9) − 14 + (−23)

51. −10 − (−12) + (−7) − 4 **52.** −6 − (−8) + (−12) − 7 **◐ 53.** −3 + 4 − (−23) − 10

54. 5 + (−18) − (−21) − 2

Objective **C** Evaluate $x - y$ for the given replacement values. See Examples 11 and 12.

55. $x = -4$ and $y = 7$ **56.** $x = -7$ and $y = 1$

57. $x = 8$ and $y = -23$ **58.** $x = 9$ and $y = -2$

Evaluate $2x - y$ for the given replacement values. See Examples 11 and 12.

◐ 59. $x = 4$ and $y = -4$ **60.** $x = 8$ and $y = -10$

61. $x = 1$ and $y = -18$ **62.** $x = 14$ and $y = -12$

Objective **D** Solve. See Example 13.

The bar graph shows the minimum and maximum temperatures for selected states. A temperature extreme is found by calculating maximum minus minimum. Find the temperature extreme for the states in Exercises 63 through 66.

63. Find the temperature extreme (difference in max and min) for the state of Florida.

64. Find the temperature extreme for the state of Texas.

65. Find the temperature extreme for the state with the coldest minimum.

66. Find the temperature extreme for the state with the hottest maximum.

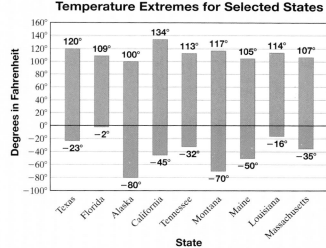

Source: National Centers for Environmental Information

Solve.

◐ 67. The coldest temperature ever recorded on Earth was −129°F in Antarctica. The warmest temperature ever recorded was 134°F in Death Valley, California. How many degrees warmer is 134°F than −129°F? (*Source: The World Almanac*)

68. The coldest temperature ever recorded in the United States was −80°F in Alaska. The warmest temperature ever recorded was 134°F in California. How many degrees warmer is 134°F than −80°F? (*Source: The World Almanac*)

69. Charley Hull from England finished first in the 2016 LPGA CME Group Tour Championship at the Tiburon Golf Club in Naples, Florida, with a score of −19, or nineteen strokes under par. In 63rd place with a +7 was Christina Kim of the United States. What was the difference in scores between Hull and Kim? (*Source:* LPGA)

70. A woman received a statement of her charge account at Old Navy. She spent $93 on purchases last month. She returned an $18 blouse because she didn't like the color. She also returned a $26 pajama set because it was damaged. What does she actually owe on her account?

71. The temperature on a February morning was −4° Celsius at 6 a.m. If the temperature drops 3 degrees by 7 a.m., rises 4 degrees between 7 a.m. and 8 a.m., and then drops 7 degrees between 8 a.m. and 9 a.m., find the temperature at 9 a.m.

72. Mauna Kea in Hawaii has an elevation of 13,796 feet above sea level. The Mid-America Trench in the Pacific Ocean has an elevation of 21,857 feet below sea level. Find the difference in elevation between those two points. (*Source:* National Geographic Society and Defense Mapping Agency)

Some places on Earth lie below sea level, which is the average level of the surface of the oceans. Use this diagram to answer Exercises 73 through 76. (Source: Fantastic Book of Comparisons, Russell Ash)

73. Find the difference in elevation between Death Valley and Qattâra Depression.

74. Find the difference in elevation between the Danakil and Turfan Depressions.

75. Find the difference in elevation between the two lowest elevations shown.

76. Find the difference in elevation between the highest elevation shown and the lowest elevation shown.

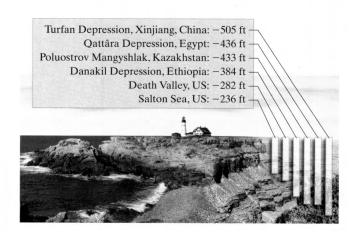

Turfan Depression, Xinjiang, China: −505 ft
Qattâra Depression, Egypt: −436 ft
Poluostrov Mangyshlak, Kazakhstan: −433 ft
Danakil Depression, Ethiopia: −384 ft
Death Valley, US: −282 ft
Salton Sea, US: −236 ft

The bar graph from Section 2.1 shows elevations of selected lakes. For Exercises 77 through 80, find the difference in elevation for the lakes listed. (Source: U.S. Geological Survey)

77. Lake Superior and Lake Eyre

78. Great Bear Lake and Caspian Sea

79. Lake Maracaibo and Lake Vanern

80. Lake Eyre and Caspian Sea

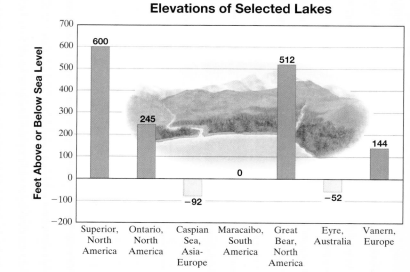

Elevations of Selected Lakes

Solve.

81. The average daytime surface temperature of the hottest planet, Venus, is 867°F, while the average daytime surface temperature of the coldest planet, Neptune, is −330°F. Find the difference in temperatures.

82. The average daytime surface temperature of Mercury is 845°F, while the average daytime surface temperature of Jupiter is −162°F. Find the difference in temperatures.

A country's exports minus its imports is called the country's trade balance. Use this for Exercises 83 and 84.

83. In November 2016, the United States had $186 billion in exports and $231 billion in imports. What was the U.S. trade balance in November 2016? (*Source:* U.S. Department of Commerce)

84. In 2016, the United States exported an average of 5101 thousand barrels of petroleum products per day and imported an average of 10,044 thousand barrels of petroleum products per day. What was the average U.S. trade balance for petroleum products per day in 2016? (*Source:* U.S. Energy Information Administration)

Mixed Practice–Translating (Sections 2.2 and 2.3) *Translate each phrase to an algebraic expression. Use "x" to represent "a number."*

85. The sum of −5 and a number.

86. The difference of −3 and a number.

87. Subtract a number from −20.

88. Add a number and −36.

Review

Multiply or divide as indicated. See Sections 1.5 and 1.6.

89. $\dfrac{100}{20}$

90. $\dfrac{96}{3}$

91. $\begin{array}{r} 23 \\ \times\ 46 \\ \hline \end{array}$

92. $\begin{array}{r} 51 \\ \times\ 89 \\ \hline \end{array}$

Concept Extensions

93. Name two numbers whose difference is −3.

94. Name two numbers whose difference is −10.

*Each calculation below is **incorrect**. Find the error and correct it. See the Concept Check in this section.*

95. $9 - (-7) \stackrel{?}{=} 2$

96. $-4 - 8 \stackrel{?}{=} 4$

97. $10 - 30 \stackrel{?}{=} 20$

98. $-3 - (-10) \stackrel{?}{=} -13$

Simplify. (Hint: Find the absolute values first.)

99. $|-3| - |-7|$

100. $|-12| - |-5|$

101. $|-5| - |5|$

102. $|-8| - |8|$

103. $|-15| - |-29|$

104. $|-23| - |-42|$

For Exercises 105 and 106, determine whether each statement is true or false.

105. $|-8 - 3| = 8 - 3$

106. $|-2 - (-6)| = |-2| - |-6|$

107. In your own words, explain how to subtract one signed number from another.

108. A student explains to you that the first step to simplify $8 + 12 \cdot 5 - 100$ is to add 8 and 12. Is the student correct? Explain why or why not.

2.4 Multiplying and Dividing Integers

Objectives

A Multiply Integers.

B Divide Integers.

C Evaluate an Algebraic Expression by Multiplying or Dividing.

D Solve Problems by Multiplying or Dividing Integers.

Multiplying and dividing integers is similar to multiplying and dividing whole numbers. One difference is that we need to determine whether the result is a positive number or a negative number.

Objective A Multiplying Integers

Consider the following pattern of products.

First factor decreases by 1 each time.
$$3 \cdot 2 = 6$$
$$2 \cdot 2 = 4$$
$$1 \cdot 2 = 2$$
$$0 \cdot 2 = 0$$

Product decreases by 2 each time.

This pattern can be continued, as follows.

$$-1 \cdot 2 = -2$$
$$-2 \cdot 2 = -4$$
$$-3 \cdot 2 = -6$$

This suggests that the product of a negative number and a positive number is a negative number.

What is the sign of the product of two negative numbers? To find out, we form another pattern of products. Again, we decrease the first factor by 1 each time, but this time the second factor is negative.

$$2 \cdot (-3) = -6$$
$$1 \cdot (-3) = -3$$
$$0 \cdot (-3) = 0$$

Product increases by 3 each time.

This pattern continues as:

$$-1 \cdot (-3) = 3$$
$$-2 \cdot (-3) = 6$$
$$-3 \cdot (-3) = 9$$

This suggests that the product of two negative numbers is a positive number. Thus we can determine the sign of a product when we know the signs of the factors.

Multiplying Numbers

The product of two numbers having the same sign is a positive number.

Product of Like Signs
$(+)(+) = +$
$(-)(-) = +$

The product of two numbers having different signs is a negative number.

Product of Different Signs
$(-)(+) = -$
$(+)(-) = -$

Practice 1–4

Multiply.

1. $-3 \cdot 8$ **2.** $-5(-2)$
3. $0 \cdot (-20)$ **4.** $10(-5)$

Answers
1. -24 **2.** 10 **3.** 0 **4.** -50

Examples Multiply.

1. $-7 \cdot 3 = -21$ **2.** $-3(-5) = 15$
3. $0 \cdot (-4) = 0$ **4.** $10(-8) = -80$

■ Work Practice 1–4

Section 2.4 | Multiplying and Dividing Integers

Recall that by the associative and commutative properties for multiplication, we may multiply numbers in any order that we wish. In Examples 5 and 6, we multiply from left to right.

Examples Multiply.

5. $7(-6)(-2) = -42(-2)$
$ = 84$

6. $(-2)(-3)(-4) = 6(-4)$
$ = -24$

7. $(-1)(-2)(-3)(-4) = -1(-24)$ We have -24 from Example 6.
$ = 24$

Work Practice 5–7

Practice 5–7

Multiply.

5. $8(-6)(-2)$
6. $(-9)(-2)(-1)$
7. $(-3)(-4)(-5)(-1)$

Helpful Hint

Have you noticed a pattern when multiplying signed numbers?
If we let $(-)$ represent a negative number and $(+)$ represent a positive number, then

The product of an even number of negative numbers is a positive result.
$(-)(-) = (+)$
$(-)(-)(-)(-) = (+)$

The product of an odd number of negative numbers is a negative result.
$(-)(-)(-) = (-)$
$(-)(-)(-)(-)(-) = (-)$

✓ **Concept Check** What is the sign of the product of five negative numbers? Explain.

Recall from our study of exponents that $2^3 = 2 \cdot 2 \cdot 2 = 8$. We can now work with bases that are negative numbers. For example,

$(-2)^3 = (-2)(-2)(-2) = -8$

Example 8 Evaluate: $(-5)^2$

Solution: Remember that $(-5)^2$ means 2 factors of -5.

$(-5)^2 = (-5)(-5) = 25$

Work Practice 8

Practice 8

Evaluate $(-2)^4$.

Notice in Example 8 the parentheses around -5 in $(-5)^2$. With these parentheses, -5 is the base that is squared. Without parentheses, such as -5^2, only the 5 is squared. In other words, $-5^2 = -(5 \cdot 5) = -25$.

Example 9 Evaluate: -7^2

Solution: Remember that without parentheses, only the 7 is squared.

$-7^2 = -(7 \cdot 7) = -49$

Work Practice 9

Practice 9

Evaluate: -8^2

Answers
5. 96 **6.** -18 **7.** 60 **8.** 16 **9.** -64

✓**Concept Check Answer**
negative; answers may vary

Helpful Hint

Make sure you understand the difference between Examples 8 and 9.

parentheses, so -5 is squared
$$(-5)^2 = (-5)(-5) = 25$$

no parentheses, so only the 7 is squared
$$-7^2 = -(7 \cdot 7) = -49$$

Objective B Dividing Integers

Division of integers is related to multiplication of integers. The sign rules for division can be discovered by writing a related multiplication problem. For example,

$$\frac{6}{2} = 3 \quad \text{because } 3 \cdot 2 = 6$$

$$\frac{-6}{2} = -3 \quad \text{because } -3 \cdot 2 = -6$$

$$\frac{6}{-2} = -3 \quad \text{because } -3 \cdot (-2) = 6$$

$$\frac{-6}{-2} = 3 \quad \text{because } 3 \cdot (-2) = -6$$

Helpful Hint Just as for whole numbers, division can be checked by multiplication.

Dividing Numbers

The quotient of two numbers having the same sign is a positive number.

The quotient of two numbers having different signs is a negative number.

Quotient of Like Signs

$$\frac{(+)}{(+)} = + \qquad \frac{(-)}{(-)} = +$$

Quotient of Different Signs

$$\frac{(+)}{(-)} = - \qquad \frac{(-)}{(+)} = -$$

Practice 10–12

Divide.

10. $\dfrac{42}{-7}$

11. $-16 \div (-2)$

12. $\dfrac{-80}{10}$

Examples Divide.

10. $\dfrac{-12}{6} = -2$

11. $-20 \div (-4) = 5$

12. $\dfrac{48}{-3} = -16$

Work Practice 10–12

✓ **Concept Check** What is wrong with the following calculation?

$$\frac{-36}{-9} = -4$$

Answers

10. -6 11. 8 12. -8

✓ **Concept Check Answer**

$\dfrac{-36}{-9} = 4$

Examples
Divide, if possible.

13. $\dfrac{0}{-5} = 0$ because $0 \cdot -5 = 0$

14. $\dfrac{-7}{0}$ is undefined because there is no number that gives a product of -7 when multiplied by 0.

Work Practice 13–14

Practice 13–14
Divide, if possible.

13. $\dfrac{-6}{0}$ 14. $\dfrac{0}{-7}$

Objective C Evaluating Expressions

Next, we practice evaluating expressions given integer replacement values.

Example 15
Evaluate xy for $x = -2$ and $y = 7$.

Solution: Recall that xy means $x \cdot y$.

Replace x with -2 and y with 7.

$xy = -2 \cdot 7$
$ = -14$

Work Practice 15

Practice 15
Evaluate xy for $x = 5$ and $y = -8$.

Example 16
Evaluate $\dfrac{x}{y}$ for $x = -24$ and $y = 6$.

Solution: $\dfrac{x}{y} = \dfrac{-24}{6}$ Replace x with -24 and y with 6.

$\phantom{\dfrac{x}{y}} = -4$

Work Practice 16

Practice 16
Evaluate $\dfrac{x}{y}$ for $x = -12$ and $y = -3$.

Objective D Solving Problems by Multiplying and Dividing Integers

Many real-life problems involve multiplication and division of signed numbers.

Example 17 Calculating a Total Golf Score

A professional golfer finished seven strokes under par (-7) for each of three days of a tournament. What was his total score for the tournament?

Solution:

1. UNDERSTAND. Read and reread the problem. Although the key word is "total," since this is repeated addition of the same number, we multiply.

2. TRANSLATE.

   In words: golfer's total score = number of days · score each day

   Translate: golfer's total = 3 · (−7)

3. SOLVE: $3 \cdot (-7) = -21$

4. INTERPRET. Check and state your conclusion: The golfer's total score was -21, or 21 strokes under par.

Work Practice 17

Practice 17
A card player had a score of -13 for each of four games. Find the total score.

Answers
13. undefined   14. 0   15. −40
16. 4   17. −52

Vocabulary, Readiness & Video Check

Use the choices below to fill in each blank. Each choice may be used more than once.

negative   0
positive   undefined

1. The product of a negative number and a positive number is a(n) _____ number.
2. The product of two negative numbers is a(n) _____ number.
3. The quotient of two negative numbers is a(n) _____ number.
4. The quotient of a negative number and a positive number is a(n) _____ number.
5. The product of a negative number and zero is _____.
6. The quotient of 0 and a negative number is _____.
7. The quotient of a negative number and 0 is _____.

See Video 2.4

Martin-Gay Interactive Videos *Watch the section lecture video and answer the following questions.*

Objective A 8. Explain the role of parentheses when comparing Examples 3 and 4.

Objective B 9. Complete this statement based on the lecture before Example 6: We can find out about sign rules for division because we know sign rules for _____.

Objective C 10. In Example 10, what are you asked to remember about the algebraic expression ab?

Objective D 11. In Example 12, how do we know the example will involve a negative number?

2.4 Exercise Set MyLab Math

Objective A *Multiply. See Examples 1 through 4.*

1. $-6(-2)$
2. $5(-3)$
3. $-4(9)$
4. $-7(-2)$
5. $9(-9)$
6. $-9(7)$
7. $0(-11)$
8. $-6(0)$

Multiply. See Examples 5 through 7.

9. $6(-2)(-4)$
10. $-2(3)(-7)$
11. $-1(-3)(-4)$
12. $-8(-3)(-3)$
13. $-4(4)(-5)$
14. $2(-5)(-4)$
15. $10(-5)(0)(-7)$
16. $3(0)(-4)(-8)$
17. $-5(3)(-1)(-1)$
18. $-2(-1)(3)(-2)$

Section 2.4 | Multiplying and Dividing Integers

Evaluate. See Examples 8 and 9.

19. -3^2
20. -2^4
21. $(-3)^3$
22. $(-1)^4$
23. -6^2
24. -4^3
25. $(-4)^3$
26. $(-3)^2$

Objective B *Find each quotient. See Examples 10 through 14.*

27. $-24 \div 3$
28. $90 \div (-9)$
29. $\dfrac{-30}{6}$
30. $\dfrac{56}{-8}$
31. $\dfrac{-77}{-11}$
32. $\dfrac{-32}{4}$
33. $\dfrac{0}{-21}$
34. $\dfrac{-13}{0}$
35. $\dfrac{-10}{0}$
36. $\dfrac{0}{-15}$
37. $\dfrac{56}{-4}$
38. $\dfrac{-24}{-12}$

Objectives A B Mixed Practice *Multiply or divide as indicated. See Examples 1 through 14.*

39. $-14(0)$
40. $0(-100)$
41. $-5(3)$
42. $-6 \cdot 2$
43. $-9 \cdot 7$
44. $-12(13)$
45. $-7(-6)$
46. $-9(-5)$
47. $-3(-4)(-2)$
48. $-7(-5)(-3)$
49. $(-7)^2$
50. $(-5)^2$
51. $-\dfrac{25}{5}$
52. $-\dfrac{30}{5}$
53. $-\dfrac{72}{8}$
54. $-\dfrac{49}{7}$
55. $-18 \div 3$
56. $-15 \div 3$
57. $4(-10)(-3)$
58. $6(-5)(-2)$
59. $-30(6)(-2)(-3)$
60. $-20 \cdot 5 \cdot (-5) \cdot (-3)$
61. $\dfrac{-25}{0}$
62. $\dfrac{0}{-14}$
63. $\dfrac{120}{-20}$
64. $\dfrac{63}{-9}$
65. $280 \div (-40)$
66. $480 \div (-8)$
67. $\dfrac{-12}{-4}$
68. $\dfrac{-36}{-3}$
69. -1^4
70. -2^3
71. $(-2)^5$
72. $(-11)^2$
73. $-2(3)(5)(-6)$
74. $-1(2)(7)(-3)$
75. $(-1)^{32}$
76. $(-1)^{33}$
77. $-2(-3)(-5)$
78. $-2(-2)(-3)(-2)$
79. $-48 \cdot 23$
80. $-56 \cdot 43$
81. $35 \cdot (-82)$
82. $70 \cdot (-23)$

Objective C *Evaluate ab for the given replacement values. See Example 15.*

83. $a = -8$ and $b = 7$
84. $a = 5$ and $b = -1$
85. $a = 9$ and $b = -2$

86. $a = -8$ and $b = 8$
87. $a = -7$ and $b = -5$
88. $a = -9$ and $b = -6$

Evaluate $\dfrac{x}{y}$ for the given replacement values. See Example 16.

89. $x = 5$ and $y = -5$
90. $x = 9$ and $y = -3$
91. $x = -15$ and $y = 0$

92. $x = 0$ and $y = -5$
93. $x = -36$ and $y = -6$
94. $x = -10$ and $y = -10$

Evaluate xy and also $\dfrac{x}{y}$ for the given replacement values. See Examples 15 and 16.

95. $x = -8$ and $y = -2$
96. $x = 20$ and $y = -5$
97. $x = 0$ and $y = -8$
98. $x = -3$ and $y = 0$

Objective D **Translating** *Translate each phrase; then simplify. See Example 17.*

99. Find the quotient of -54 and 9.
100. Find the quotient of -63 and -3.

101. Find the product of -42 and -6.
102. Find the product of -49 and 5.

Translating *Translate each phrase to an expression. Use x to represent "a number." See Example 17.*

103. The product of -71 and a number
104. The quotient of -8 and a number

105. Subtract a number from -16.
106. The sum of a number and -12

107. -29 increased by a number
108. The difference of a number and -10

109. Divide a number by -33.
110. Multiply a number by -17.

Solve. See Example 17.

111. A football team lost four yards on each of three consecutive plays. Represent the total loss as a product of signed numbers and find the total loss.

112. An investor lost $400 on each of seven consecutive days in the stock market. Represent his total loss as a product of signed numbers and find his total loss.

113. A deep-sea diver must move up or down in the water in short steps in order to keep from getting a physical condition called the "bends." Suppose a diver moves down from the surface in five steps of 20 feet each. Represent his total movement as a product of signed numbers and find the product.

114. A weather forecaster predicts that the temperature will drop five degrees each hour for the next six hours. Represent this drop as a product of signed numbers and find the total drop in temperature.

The graph shows melting points in degrees Celsius of selected elements. Use this graph to answer Exercises 115 through 118.

115. The melting point of nitrogen is 3 times the melting point of radon. Find the melting point of nitrogen.

116. The melting point of rubidium is −1 times the melting point of mercury. Find the melting point of rubidium.

117. The melting point of argon is −3 times the melting point of potassium. Find the melting point of argon.

118. The melting point of strontium is −11 times the melting point of radon. Find the melting point of strontium.

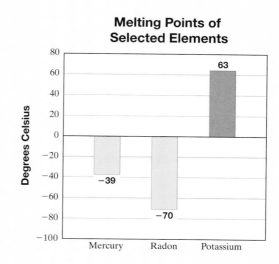

Solve. See Example 17.

119. For the second quarter of 2016, Sears Holding Corporation posted a loss of $396 million due to a decline in sales. If this trend were consistent for each month of the quarter, how much would you expect this loss to have been for each month? (*Source:* Sears Holdings Corporation)

120. From 2003 through 2015, the estimated population of mourning doves in the United States dropped by 73,023,000 birds. If the loss in birds were evenly spaced over these years, by how many birds would you expect the population to drop each year? (*Source:* U.S. Fish and Wildlife Service)

121. In 2010 there were 24,814 analog (nondigital) U.S. movie screens. In 2015, this number of screens dropped to 1109. (*Source:* Motion Picture Association of America)
   a. Find the change in the number of U.S. analog screens from 2010 to 2015.
   b. Find the average change per year in the number of analog movie screens over this period.

122. In 1987, the California condor was all but extinct in the wild, with about 30 condors in the world. The U.S. Fish and Wildlife Service captured the condors in the wild in an aggressive move to rebuild the population by breeding them in captivity and releasing the chicks into the wild. The condor population increased to approximately 436 by 2016. (*Source:* U.S. Fish and Wildlife Service)
   a. Find the change in the number of California condors from 1987 to 2016.
   b. Find the average change per year in the California condor population over the period in part **a**.

Review

Perform each indicated operation. See Section 1.7.

123. $90 + 12^2 - 5^3$ **124.** $3 \cdot (7 - 4) + 2 \cdot 5^2$ **125.** $12 \div 4 - 2 + 7$ **126.** $12 \div (4 - 2) + 7$

Concept Extensions

Mixed Practice (Sections 2.2, 2.3, and 2.4) *Perform the indicated operations.*

127. $-57 \div 3$

128. $-9(-11)$

129. $-8 - 20$

130. $-4 + (-3) + 21$

131. $-4 - 15 - (-11)$

132. $-16 - (-2)$

Solve. For Exercises 133 and 134, see the first Concept Check in this section.

133. What is the sign of the product of seven negative numbers?

134. What is the sign of the product of ten negative numbers?

Without actually finding the product, write the list of numbers in Exercises 135 and 136 in order from least to greatest. For help, see a helpful hint box in this section.

135. $(-2)^{12}, (-2)^{17}, (-5)^{12}, (-5)^{17}$

136. $(-1)^{50}, (-1)^{55}, 0^{15}, (-7)^{20}, (-7)^{23}$

137. In your own words, explain how to divide two integers.

138. In your own words, explain how to multiply two integers.

Sections 2.1–2.4 Integrated Review

Integers

1. The record low temperature in New Mexico is 50 degrees Fahrenheit below zero. The highest temperature in that state is 122 degrees above zero. Represent each quantity by an integer.

2. Graph the signed numbers on the given number line. $-4, 0, -1, 3$

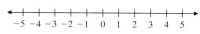

New Mexico

Insert $<$ or $>$ between each pair of numbers to make a true statement.

3. 0 -10
4. -4 4
5. -15 -5
6. -2 -7

Simplify.

7. $|-3|$
8. $|-9|$
9. $-|-4|$
10. $-(-5)$

Find the opposite of each number.

11. 11
12. -3
13. 64
14. 0

Perform the indicated operation.

15. $-3 + 15$
16. $-9 + (-11)$
17. $-8(-6)(-1)$
18. $-18 \div 2$

19. $65 + (-55)$
20. $1000 - 1002$
21. $53 - (-53)$
22. $-2 - 1$

Answers

1. _____
2. _____
3. _____
4. _____
5. _____
6. _____
7. _____
8. _____
9. _____
10. _____
11. _____
12. _____
13. _____
14. _____
15. _____
16. _____
17. _____
18. _____
19. _____
20. _____
21. _____
22. _____

23. $\dfrac{0}{-47}$

24. $\dfrac{-36}{-9}$

25. $-17 - (-59)$

26. $-8 + (-6) + 20$

27. $\dfrac{-95}{-5}$

28. $-9(100)$

29. $-12 - 6 - (-6)$

30. $-4 + (-8) - 16 - (-9)$

31. $\dfrac{-105}{0}$

32. $7(-16)(0)(-3)$

Translating *Translate each phrase; then simplify.*

33. Subtract -8 from -12.

34. Find the sum of -17 and -27.

35. Find the product of -5 and -25.

36. Find the quotient of -100 and -5.

Translating *Translate each phrase to an expression. Use x to represent "a number."*

37. Divide a number by -17

38. The sum of -3 and a number

39. A number decreased by -18

40. The product of -7 and a number

Evaluate the expressions below for $x = -3$ and $y = 12$.

41. $x + y$

42. $x - y$

43. $2y - x$

44. $3y + x$

45. $5x$

46. $\dfrac{y}{x}$

2.5 Order of Operations

Objective A Simplifying Expressions

We first discussed the order of operations in Chapter 1. In this section, you are given an opportunity to practice using the order of operations when expressions contain signed numbers. The rules for the order of operations from Section 1.7 are repeated here.

Objectives

A Simplify Expressions by Using the Order of Operations.

B Evaluate an Algebraic Expression.

C Find the Average of a List of Numbers.

Order of Operations
1. Perform all operations within parentheses ( ), brackets [ ], or other grouping symbols such as fraction bars, starting with the innermost set.
2. Evaluate any expressions with exponents.
3. Multiply or divide in order from left to right.
4. Add or subtract in order from left to right.

Before simplifying other expressions, make sure you are confident simplifying Examples 1 through 3.

Examples Find the value of each expression.

1. $(-3)^2 = (-3)(-3) = 9$ The base of the exponent is -3.
2. $-3^2 = -(3)(3) = -9$ The base of the exponent is 3.
3. $2 \cdot 5^2 = 2 \cdot (5 \cdot 5) = 2 \cdot 25 = 50$ The base of the exponent is 5.

Work Practice 1–3

Practice 1–3

Find the value of each expression.
1. $(-2)^4$
2. -2^4
3. $3 \cdot 6^2$

Helpful Hint

When simplifying expressions with exponents, remember that parentheses make an important difference.

$(-3)^2$ and -3^2 **do not** mean the same thing.
$(-3)^2$ means $(-3)(-3) = 9$.
-3^2 means the opposite of $3 \cdot 3$, or -9.

Only with parentheses around it is the -3 squared.

Example 4 Simplify: $\dfrac{-6(2)}{-3}$

Solution: First we multiply -6 and 2. Then we divide.

$$\dfrac{-6(2)}{-3} = \dfrac{-12}{-3}$$
$$= 4$$

Work Practice 4

Practice 4

Simplify: $\dfrac{-25}{5(-1)}$

Answers
1. 16   2. -16 3. 108   4. 5

Practice 5

Simplify: $\dfrac{-18 + 6}{-3 - 1}$

Example 5 Simplify: $\dfrac{12 - 16}{-1 + 3}$

Solution: We simplify above and below the fraction bar separately. Then we divide.

$$\dfrac{12 - 16}{-1 + 3} = \dfrac{-4}{2}$$
$$= -2$$

■ Work Practice 5

Practice 6

Simplify: $30 + 50 \div (-4)^3$

Example 6 Simplify: $60 + 30 \div (-2)^3$

Solution: $60 + 30 \div (-2)^3 = 60 + 30 \div (-8)$ Write $(-2)^3$ as -8.
$$= 90 + (-8)$$ Add from left to right.
$$= 82$$

■ Work Practice 6

Practice 7

Simplify: $-2^3 + (-4)^2 + 1^5$

Example 7 Simplify: $-4^2 + (-3)^2 - 1^3$

Solution:

$-4^2 + (-3)^2 - 1^3 = -16 + 9 - 1$ Simplify expressions with exponents.
$$= -7 - 1$$ Add or subtract from left to right.
$$= -8$$

■ Work Practice 7

Practice 8

Simplify:
$2(2 - 9) + (-12) - 3$

Example 8 Simplify: $3(4 - 7) + (-2) - 5$

Solution:

$3(4 - 7) + (-2) - 5 = 3(-3) + (-2) - 5$ Simplify inside parentheses.
$$= -9 + (-2) - 5$$ Multiply.
$$= -11 - 5$$ Add or subtract from left to right.
$$= -16$$

■ Work Practice 8

Practice 9

Simplify:
$(-5) \cdot |-8| + (-3) + 2^3$

Example 9 Simplify: $(-3) \cdot |-5| - (-2) + 4^2$

Solution:

$(-3) \cdot |-5| - (-2) + 4^2 = (-3) \cdot 5 - (-2) + 4^2$ Write $|-5|$ as 5.
$$= (-3) \cdot 5 - (-2) + 16$$ Write 4^2 as 16.
$$= -15 - (-2) + 16$$ Multiply.
$$= -13 + 16$$ Add or subtract from left to right.
$$= 3$$

■ Work Practice 9

Answers
5. 3 **6.** 16 **7.** 9 **8.** -29 **9.** -35

Example 10
Simplify: $-2[-3 + 2(-1 + 6)] - 5$

Solution: Here we begin with the innermost set of parentheses.

$-2[-3 + 2(-1 + 6)] - 5 = -2[-3 + 2(5)] - 5$ Write $-1 + 6$ as 5.
$= -2[-3 + 10] - 5$ Multiply.
$= -2(7) - 5$ Add.
$= -14 - 5$ Multiply.
$= -19$ Subtract.

Work Practice 10

Practice 10
Simplify:
$-4[-6 + 5(-3 + 5)] - 7$

✓ **Concept Check** True or false? Explain your answer. The result of
$-4(3 - 7) - 8(9 - 6)$
is positive because there are four negative signs.

Objective B Evaluating Expressions
Now we practice evaluating expressions.

Example 11
Evaluate x^2 and $-x^2$ for $x = -11$.

Solution: $x^2 = (-11)^2 = (-11)(-11) = 121$
$-x^2 = -(-11)^2 = -(-11)(-11) = -121$

Work Practice 11

Practice 11
Evaluate x^2 and $-x^2$ for $x = -15$.

Example 12
Evaluate $6z^2$ for $z = 2$ and $z = -2$.

Solution: $6z^2 = 6(2)^2 = 6(4) = 24$
$6z^2 = 6(-2)^2 = 6(4) = 24$

Work Practice 12

Practice 12
Evaluate $5y^2$ for $y = 4$ and $y = -4$.

Example 13
Evaluate $x + 2y - z$ for $x = 3, y = -5,$ and $z = -4$.

Solution: Replace x with 3, y with -5, z with -4, and simplify.

Helpful Hint
Remember to rewrite the subtraction sign.

$x + 2y - z = 3 + 2(-5) - (-4)$ Let $x = 3, y = -5,$ and $z = -4$.
$= 3 + (-10) + 4$ Replace $2(-5)$ with its product, -10.
$= -3$ Add.

Work Practice 13

Practice 13
Evaluate $x^2 + y$ for $x = -6$ and $y = -3$.

Answers
10. -23 **11.** $225; -225$ **12.** $80; 80$
13. 33

✓ **Concept Check Answer**
false; $-4(3 - 7) - 8(9 - 6) = -8$

Practice 14
Evaluate $4 - x^2$ for $x = -8$.

Example 14 Evaluate $7 - x^2$ for $x = -4$.

Solution: Replace x with -4 and simplify carefully!
$$7 - x^2 = 7 - (-4)^2$$
$$= 7 - 16 \qquad (-4)^2 = (-4)(-4) = 16$$
$$= -9 \qquad \text{Subtract.}$$

■ Work Practice 14

Objective C Finding Averages

Recall from Chapter 1 that the average of a list of numbers is

$$\text{average} = \frac{\text{sum of numbers}}{\text{number of numbers}}$$

Practice 15
Find the average of the temperatures for the months October through April.

Example 15 The graph shows some monthly normal temperatures for Barrow, Alaska. Use this graph to find the average of the temperatures for the months January through April.

Monthly Normal Temperatures for Barrow, Alaska

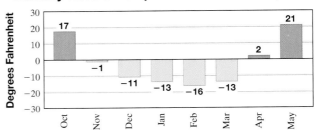

Solution: By reading the graph, we have

$$\text{average} = \frac{-13 + (-16) + (-13) + 2}{4} \qquad \text{There are 4 months from January through April.}$$
$$= \frac{-40}{4}$$
$$= -10$$

The average of the temperatures is $-10°$F.

■ Work Practice 15

Answers
14. -60 15. $-5°$F

🖩 Calculator Explorations   Simplifying an Expression Containing a Fraction Bar

Recall that even though most calculators follow the order of operations, parentheses must sometimes be inserted. For example, to simplify $\frac{-8 + 6}{-2}$ on a calculator, enter parentheses around the expression above the fraction bar so that it is simplified separately.

To simplify $\frac{-8 + 6}{-2}$, press the keys

(8 +/− + 6) ÷ 2 +/− = or
((−) 8 + 6) ÷ (−) 2 ENTER

The display will read ⬚ 1 ⬚.

Thus, $\frac{-8 + 6}{-2} = 1$.

Use a calculator to simplify.

1. $\dfrac{-120 - 360}{-10}$

2. $\dfrac{4750}{-2 + (-17)}$

3. $\dfrac{-316 + (-458)}{28 + (-25)}$

4. $\dfrac{-234 + 86}{-18 + 16}$

Vocabulary, Readiness & Video Check

Use the choices below to fill in each blank. Not all choices will be used.

average	subtraction	division	$-7 - 3(1)$
addition	multiplication		$-7 - 3(-1)$

1. To simplify $-2 \div 2 \cdot (3)$, which operation should be performed first? _____
2. To simplify $-9 - 3 \cdot 4$, which operation should be performed first? _____
3. The _____ of a list of numbers is $\dfrac{\text{sum of numbers}}{\text{number of numbers}}$.
4. To simplify $5[-9 + (-3)] \div 4$, which operation should be performed first? _____
5. To simplify $-2 + 3(10 - 12) \cdot (-8)$, which operation should be performed first? _____
6. To evaluate $x - 3y$ for $x = -7$ and $y = -1$, replace x with -7 and y with -1 and evaluate _____.

Martin-Gay Interactive Videos Watch the section lecture video and answer the following questions.

See Video 2.5

Objective A 7. In Example 1, what two things about the fraction bar are we reminded of?

Objective B 8. In Example 5, why is it important to place the replacement value for x within parentheses?

Objective C 9. From the lecture before Example 6, explain why finding the average is a good example of an application for this section.

2.5 Exercise Set MyLab Math

Objective A Simplify. See Examples 1 through 10.

1. $(-5)^3$
2. -2^4
3. -4^3
4. $(-2)^4$

5. $8 \cdot 2^2$
6. $5 \cdot 2^3$
7. $8 - 12 - 4$
8. $10 - 23 - 12$

9. $7 + 3(-6)$
10. $-8 + 4(3)$
11. $5(-9) + 2$
12. $7(-6) + 3$

13. $-10 + 4 \div 2$
14. $-12 + 6 \div 3$
15. $6 + 7 \cdot 3 - 10$
16. $5 + 9 \cdot 4 - 20$

17. $\dfrac{16 - 13}{-3}$
18. $\dfrac{20 - 15}{-1}$
19. $\dfrac{24}{10 + (-4)}$
20. $\dfrac{88}{-8 - 3}$

21. $5(-3) - (-12)$
22. $7(-4) - (-6)$
23. $[8 + (-4)]^2$
24. $[9 + (-2)]^3$

25. $8 \cdot 6 - 3 \cdot 5 + (-20)$ 26. $7 \cdot 6 - 6 \cdot 5 + (-10)$ 27. $4 - (-3)^4$ 28. $7 - (-5)^2$

29. $|7 + 3| \cdot 2^3$ 30. $|-3 + 7| \cdot 7^2$ 31. $7 \cdot 6^2 + 4$ 32. $10 \cdot 5^3 + 7$

33. $7^2 - (4 - 2^3)$ 34. $8^2 - (5 - 2)^4$ 35. $|3 - 15| \div 3$ 36. $|12 - 19| \div 7$

37. $-(-2)^6$ 38. $-(-2)^3$ 39. $(5 - 9)^2 \div (4 - 2)^2$ 40. $(2 - 7)^2 \div (4 - 3)^4$

41. $|8 - 24| \cdot (-2) \div (-2)$ 42. $|3 - 15| \cdot (-4) \div (-16)$ 43. $(-12 - 20) \div 16 - 25$

44. $(-20 - 5) \div 5 - 15$ 45. $5(5 - 2) + (-5)^2 - 6$ 46. $3 \cdot (8 - 3) + (-4) - 10$

47. $(2 - 7) \cdot (6 - 19)$ 48. $(4 - 12) \cdot (8 - 17)$ 49. $(-36 \div 6) - (4 \div 4)$

50. $(-4 \div 4) - (8 \div 8)$ 51. $(10 - 4^2)^2$ 52. $(11 - 3^2)^3$

53. $2(8 - 10)^2 - 5(1 - 6)^2$ 54. $-3(4 - 8)^2 + 5(14 - 16)^3$ 55. $3(-10) \div [5(-3) - 7(-2)]$

56. $12 - [7 - (3 - 6)] + (2 - 3)^3$ 57. $\dfrac{(-7)(-3) - (4)(3)}{3[7 \div (3 - 10)]}$ 58. $\dfrac{10(-1) - (-2)(-3)}{2[-8 \div (-2 - 2)]}$

59. $-3[5 + 2(-4 + 9)] + 15$ 60. $-2[6 + 4(2 - 8)] - 25$

Objective B Evaluate each expression for $x = -2, y = 4,$ and $z = -1$. See Examples 11 through 14.

61. $x + y + z$ 62. $x - y - z$ 63. $2x - 3y - 4z$ 64. $5x - y + 4z$

65. $x^2 - y$ 66. $x^2 + z$ 67. $\dfrac{5y}{z}$ 68. $\dfrac{4x}{y}$

Evaluate each expression for $x = -3$ and $z = -4$. See Examples 11 through 14.

69. x^2 70. z^2 71. $-z^2$ 72. $-x^2$

73. $2z^3$ 74. $3x^2$ 75. $10 - x^2$ 76. $3 - z^2$

77. $2x^3 - z$ 78. $3z^2 - x$

Objective C *Find the average of each list of numbers. See Example 15.*

79. $-10, 8, -4, 2, 7, -5, -12$

80. $-18, -8, -1, -1, 0, 4$

81. $-17, -26, -20, -13$

82. $-40, -20, -10, -15, -5$

Scores in golf can be 0 (also called par), a positive integer (also called above par), or a negative integer (also called below par). The bar graph shows final scores of selected golfers from a 2016 tournament. Use this graph for Exercises 83 through 88. (*Source:* PGA)

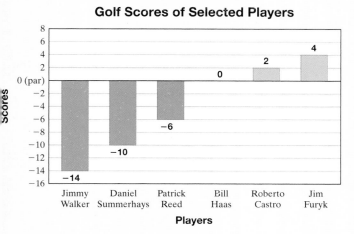

83. Find the difference between the lowest score shown and the highest score shown.

84. Find the difference between the two lowest scores.

85. Find the average of the scores for Reed, Haas, Castro, and Furyk. (*Hint:* Here, the average is the sum of the scores divided by the number of players).

86. Find the average of the scores for Walker, Summerhays, and Reed.

87. Can the average for the scores in Exercise 85 be greater than the highest score, 4? Explain why or why not.

88. Can the average of the scores in Exercise 86 be less than the lowest score, -14? Explain why or why not.

Review

Perform each indicated operation. See Sections 1.3, 1.5, and 1.6.

89. $45 \cdot 90$

90. $90 \div 45$

91. $90 - 45$

92. $45 + 90$

Find the perimeter of each figure. See Section 1.3.

93. Square, 8 in.

94. Parallelogram (opposite sides are equal in measure), 5 cm, 3 cm

95. Rectangle, 6 ft, 9 ft

96. Triangle, 17 m, 23 m, 32 m

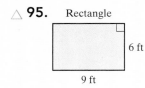

Concept Extensions

Insert parentheses where needed so that each expression evaluates to the given number.

97. $2 \cdot 7 - 5 \cdot 3$; evaluates to 12

98. $7 \cdot 3 - 4 \cdot 2$; evaluates to 34

99. $-6 \cdot 10 - 4$; evaluates to -36

100. $2 \cdot 8 \div 4 - 20$; evaluates to -36

101. Are parentheses necessary in the expression $3 + (4 \cdot 5)$? Explain your answer.

102. Are parentheses necessary in the expression $(3 + 4) \cdot 5$? Explain your answer.

103. Discuss the effect parentheses have in an exponential expression. For example, what is the difference between $(-6)^2$ and -6^2?

104. Discuss the effect parentheses have in an exponential expression. For example, what is the difference between $(2 \cdot 4)^2$ and $2 \cdot 4^2$?

Evaluate.

105. $(-12)^4$

106. $(-17)^6$

107. $x^3 - y^2$ for $x = 21$ and $y = -19$

108. $3x^2 + 2x - y$ for $x = -18$ and $y = 2868$

109. $(xy + z)^x$ for $x = 2, y = -5$, and $z = 7$

110. $5(ab + 3)^b$ for $a = -2, b = 3$

2.6 Solving Equations: The Addition and Multiplication Properties

Objectives

A Identify Solutions of Equations.

B Use the Addition Property of Equality to Solve Equations.

C Use the Multiplication Property of Equality to Solve Equations.

In this section, we introduce properties of equations and we use these properties to begin solving equations. Now that we know how to perform operations on integers, this is an excellent way to practice these operations.

First, let's recall the difference between an equation and an expression. From Section 1.8, a combination of operations on variables and numbers is an expression, and an equation is of the form "expression = expression."

Equations	Expressions
$3x - 1 = -17$	$3x - 1$
area = length · width	$5(20 - 3) + 10$
$8 + 16 = 16 + 8$	y^3
$-9a + 11b = 14b + 3$	$-x^2 + y - 2$

Section 2.6 | Solving Equations: The Addition and Multiplication Properties

> **Helpful Hint**
>
> Simply stated, an equation contains "=" while an expression does not. Also, we *simplify* expressions and *solve* equations.

Objective A Identifying Solutions of Equations

Let's practice identifying solutions of equations. Recall from Section 1.8 that a solution of an equation is a number that when substituted for the variable makes the equation a true statement.

For example,

-8 is a solution of

$\frac{x}{2} = -4$, because

$\frac{-8}{2} = -4$, or

$-4 = -4$ is true.

Also,

-8 is not a solution of

$x + 6 = 2$, because

$-8 + 6 = 2$ is false.

Let's practice determining whether a number is a solution of an equation. In this section, we will be performing operations on integers.

Example 1 Determine whether -1 is a solution of the equation $3y + 1 = 3$.

Solution:

$$3y + 1 = 3$$
$$3(-1) + 1 \stackrel{?}{=} 3$$
$$-3 + 1 \stackrel{?}{=} 3$$
$$-2 = 3 \quad \text{False}$$

Since $-2 = 3$ is false, -1 is *not* a solution of the equation.

■ Work Practice 1

Practice 1

Determine whether -2 is a solution of the equation $-4x - 3 = 5$.

Now we know how to check whether a number is a solution. But, given an equation, how do we find its **solution**? In other words, how do we find a number that makes the equation true? How do we solve an equation?

Objective B Using the Addition Property to Solve Equations

To solve an equation, we use properties of equality to write simpler equations, all equivalent to the original equation, until the final equation has the form

$x =$ **number** or **number** $= x$

Equivalent equations have the same solution, so the word "number" above represents the solution of the original equation. The first property of equality to help us write simpler, equivalent equations is the **addition property of equality.**

Answer
1. yes

Addition Property of Equality

Let a, b, and c represent numbers. Then

$a = b$	Also, $a = b$
and $a + c = b + c$	and $a - c = b - c$
are equivalent equations.	are equivalent equations.

In other words, the same number may be added to or subtracted from both sides of an equation without changing the solution of the equation. (Recall from Section 2.3 that we defined subtraction as addition of the first number and the opposite of the second number. Because of this, the addition property of equality also allows us to subtract the same number from both sides.)

A good way to visualize a true equation is to picture a balanced scale. Since it is balanced, the objects on each side of the scale weigh the same amount. Similarly, in a true equation the expressions on each side have the same value. Picturing our balanced scale, if we add the same weight to each side, the scale remains balanced.

Practice 2

Solve the equation for y: $y - 6 = -2$

Example 2 Solve: $x - 2 = -1$ for x.

Solution: To solve the equation for x, we need to rewrite the equation in the form $x = $ number. In other words, our goal is to get x alone on one side of the equation. To do so, we add 2 to both sides of the equation.

$$x - 2 = -1$$
$$x - 2 + 2 = -1 + 2 \quad \text{Add 2 to both sides of the equation.}$$
$$x + 0 = 1 \quad \text{Replace } -2 + 2 \text{ with 0.}$$
$$x = 1 \quad \text{Simplify by replacing } x + 0 \text{ with } x.$$

Check: To check, we replace x with 1 in the *original* equation.

$$x - 2 = -1 \quad \text{Original equation}$$
$$1 - 2 \stackrel{?}{=} -1 \quad \text{Replace } x \text{ with 1.}$$
$$-1 = -1 \quad \text{True}$$

Since $-1 = -1$ is a true statement, 1 is the solution of the equation.

■ Work Practice 2

Helpful Hint

Note that it is always a good idea to check the solution in the *original* equation to see that it makes the equation a true statement.

Let's visualize how we used the addition property of equality to solve an equation. Picture the equation $x - 2 = 1$ as a balanced scale. The left side of the equation has the same value as the right side.

Answer
2. 4

Section 2.6 | Solving Equations: The Addition and Multiplication Properties

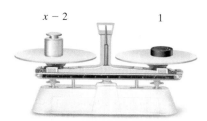

If the same weight is added to each side of a scale, the scale remains balanced. Likewise, if the same number is added to each side of an equation, the left side continues to have the same value as the right side.

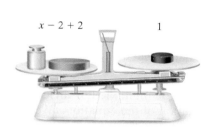
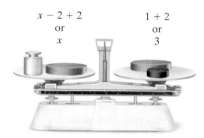

Example 3 Solve: $-8 = n + 1$

Solution: To get n alone on one side of the equation, we subtract 1 from both sides of the equation.

$$-8 = n + 1$$
$-8 - 1 = n + 1 - 1$ Subtract 1 from both sides.
$-9 = n + 0$ Replace $1 - 1$ with 0.
$-9 = n$ Simplify.

Check:
$-8 = n + 1$
$-8 \stackrel{?}{=} -9 + 1$ Replace n with -9.
$-8 = -8$ True

The solution is -9.

Work Practice 3

Helpful Hint Remember that we can get the variable alone on either side of the equation. For example, the equations $-9 = n$ and $n = -9$ both have the solution of -9.

Practice 3
Solve: $-2 = z + 8$

✓ **Concept Check** What number should be added to or subtracted from both sides of the equation in order to solve the equation $-3 = y + 2$?

Example 4 Solve: $x = -60 + 4 + 10$

Solution: Study this equation for a moment. Notice that our variable x is alone on the left side. Thus, we only need to add on the right side to find the value of x.

$x = -60 + 4 + 10$
$x = -56 + 10$ Add -60 and 4.
$x = -46$ Add -56 and 10.

Check to see that -46 is the solution.

Work Practice 4

Practice 4
Solve: $x = -2 + 90 + (-100)$

Answers
3. -10 4. -12

✓ **Concept Check Answer**
Subtract 2 from both sides.

Objective C Using the Multiplication Property to Solve Equations

Although the addition property of equality is a powerful tool for helping us solve equations, it cannot help us solve all types of equations. For example, it cannot help us solve an equation such as $2x = 6$. To solve this equation, we use a second property of equality called the **multiplication property of equality.**

> **Multiplication Property of Equality**
>
> Let a, b, and c represent numbers and let $c \neq 0$. Then
>
> | $a = b$ | Also, $a = b$ |
> |---|---|
> | and $a \cdot c = b \cdot c$ | and $\dfrac{a}{c} = \dfrac{b}{c}$ |
> | are equivalent equations. | are equivalent equations. |

In other words, both sides of an equation may be multiplied or divided by the same nonzero number without changing the solution of the equation. (We will see in Chapter 4 how the multiplication property allows us to divide both sides of an equation by the same nonzero number.)

To solve an equation like $2x = 6$ for x, notice that 2 is *multiplied* by x. To get x alone, we use the multiplication property of equality to *divide* both sides of the equation by 2, and simplify as follows:

$$2x = 6$$
$$\frac{2 \cdot x}{2} = \frac{6}{2} \quad \text{Divide both sides by 2.}$$

Then it can be shown that an expression such as $\dfrac{2 \cdot x}{2}$ is equivalent to $\dfrac{2}{2} \cdot x$, so

$$\frac{2 \cdot x}{2} = \frac{6}{2} \quad \text{can be written as} \quad \frac{2}{2} \cdot x = \frac{6}{2}$$
$$1 \cdot x = 3 \quad \text{or} \quad x = 3$$

Picturing again our balanced scale, if we multiply or divide the weight on each side by the same nonzero number, the scale (or equation) remains balanced.

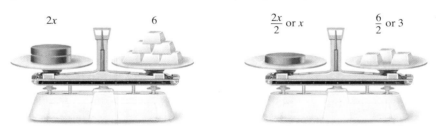

Practice 5

Solve: $3y = -18$

Example 5 Solve: $-5x = 15$

Solution: To get x alone, divide both sides by -5.

$$-5x = 15 \quad \text{Original equation}$$
$$\frac{-5x}{-5} = \frac{15}{-5} \quad \text{Divide both sides by } -5.$$
$$\frac{-5}{-5} \cdot x = \frac{15}{-5}$$
$$1x = -3 \quad \text{or} \quad x = -3 \quad \text{Simplify.}$$

Answer
5. -6

Section 2.6 | Solving Equations: The Addition and Multiplication Properties   147

Check: To check, replace x with -3 in the original equation.

$-5x = 15$ Original equation
$-5(-3) \stackrel{?}{=} 15$ Let $x = -3$.
$15 = 15$ True

The solution is -3.

Work Practice 5

Example 6   Solve: $-27 = 3y$

Solution: To get y alone, divide both sides of the equation by 3.

$-27 = 3y$

$\dfrac{-27}{3} = \dfrac{3y}{3}$ Divide both sides by 3.

$\dfrac{-27}{3} = \dfrac{3}{3} \cdot y$

$-9 = 1y$ or $y = -9$

Check to see that -9 is the solution.

Work Practice 6

Practice 6
Solve: $-32 = 8x$

Example 7   Solve: $-12x = -36$

Solution: To get x alone, divide both sides of the equation by -12.

$-12x = -36$

$\dfrac{-12x}{-12} = \dfrac{-36}{-12}$

$\dfrac{-12}{-12} \cdot x = \dfrac{-36}{-12}$

$x = 3$

Check: To check, replace x with 3 in the original equation.

$-12x = -36$
$-12(3) \stackrel{?}{=} -36$ Let $x = 3$.
$-36 = -36$ True

Since $-36 = -36$ is a true statement, the solution is 3.

Work Practice 7

Practice 7
Solve: $-3y = -27$

✓ **Concept Check**   Which operation is appropriate for solving each of the following equations, addition or division?
  a. $12 = x - 3$
  b. $12 = 3x$

The multiplication property also allows us to solve equations like

$\dfrac{x}{5} = 2$

Answers
6. -4 7. 9

✓ **Concept Check Answers**
a. addition
b. division

Here, x is *divided* by 5. To get x alone, we use the multiplication property to *multiply* both sides by 5.

$$5 \cdot \frac{x}{5} = 5 \cdot 2 \quad \text{Multiply both sides by 5.}$$

Then it can be shown that

$$5 \cdot \frac{x}{5} = 5 \cdot 2 \text{ can be written as } \frac{5}{5} \cdot x = 5 \cdot 2$$

$$1 \cdot x = 10 \quad \text{or} \quad x = 10$$

Practice 8

Solve: $\dfrac{x}{-4} = 7$

Example 8 Solve: $\dfrac{x}{3} = -2$

Solution: To get x alone, multiply both sides by 3.

$$\frac{x}{3} = -2$$

$$3 \cdot \frac{x}{3} = 3 \cdot (-2) \quad \text{Multiply both sides by 3.}$$

$$\frac{3}{3} \cdot x = 3 \cdot (-2)$$

$$1x = -6 \quad \text{or} \quad x = -6 \quad \text{Simplify.}$$

Check: Replace x with -6 in the original equation.

$$\frac{x}{3} = -2 \quad \text{Original equation}$$

$$\frac{-6}{3} \stackrel{?}{=} -2 \quad \text{Let } x = -6.$$

$$-2 = -2 \quad \text{True}$$

The solution is -6.

■ **Work Practice 8**

Answer
8. -28

Vocabulary, Readiness & Video Check

Use the choices below to fill in each blank. Some choices may be used more than once.

| equation | multiplication | addition |
| expression | solution | equivalent |

1. A combination of operations on variables and numbers is called a(n) _____.
2. A statement of the form "expression = expression" is called a(n) _____.
3. A(n) _____ contains an equal sign (=) while a(n) _____ does not.
4. A(n) _____ may be simplified and evaluated while a(n) _____ may be solved.
5. A(n) _____ of an equation is a number that when substituted for the variable makes the equation a true statement.
6. _____ equations have the same solution.

7. By the _____ property of equality, the same number may be added to or subtracted from both sides of an equation without changing the solution of the equation.

8. By the _____ property of equality, both sides of an equation may be multiplied or divided by the same nonzero number without changing the solution of the equation.

Martin-Gay Interactive Videos Watch the section lecture video and answer the following questions.

Objective A 9. From the lecture before Example 1, what does an equation have that an expression does not?

Objective B 10. In the lecture before Example 2, what does the addition property of equality mean in words?

Objective C 11. Complete this statement based on Example 7: To check a solution, we go back to the _____ equation, replace the variable with the proposed solution, and see if we get a _____ statement.

See Video 2.6

2.6 Exercise Set MyLab Math

Objective A Determine whether the given number is a solution of the given equation. See Example 1.

1. Is 6 a solution of $x - 8 = -2$?
2. Is 9 a solution of $y - 16 = -7$?
3. Is -5 a solution of $x + 12 = 17$?
4. Is -7 a solution of $a + 23 = -16$?
5. Is -8 a solution of $-9f = 64 - f$?
6. Is -6 a solution of $-3k = 12 - k$?
7. Is 3 a solution of $5(c - 5) = -10$?
8. Is 1 a solution of $2(b - 3) = 10$?

Objective B Solve. Check each solution. See Examples 2 through 4.

9. $a + 5 = 23$
10. $f + 4 = -6$
11. $d - 9 = -21$
12. $s - 7 = -15$
13. $7 = y - 2$
14. $1 = y + 7$
15. $-7 + 10 - 20 = x$
16. $-50 + 40 - 5 = z$

Objective C Solve. Check each solution. See Examples 5 through 8.

17. $5x = 20$
18. $6y = 48$
19. $-3z = 12$
20. $-2x = 26$

21. $\dfrac{n}{7} = -2$ **22.** $\dfrac{n}{11} = -5$ **23.** $2z = -34$ **24.** $7y = -21$

25. $-4y = 0$ **26.** $-9x = 0$ **27.** $-10x = -10$ **28.** $-31x = -31$

Objectives B C Mixed Practice *Solve. See Examples 2 through 8.*

29. $5x = -35$ **30.** $3y = -27$ **31.** $n - 5 = -55$ **32.** $n - 4 = -48$

33. $-15 = y + 10$ **34.** $-36 = y + 12$ **35.** $\dfrac{x}{-6} = -6$ **36.** $\dfrac{x}{-9} = -9$

37. $n = -10 + 31$ **38.** $z = -28 + 36$ **39.** $-12y = -144$ **40.** $-11x = -121$

41. $\dfrac{n}{4} = -20$ **42.** $\dfrac{n}{5} = -20$ **43.** $-64 = 32y$ **44.** $-81 = 27x$

Review

Translate each phrase to an algebraic expression. Use x to represent "a number." See Section 1.8.

45. A number decreased by -2

46. A number increased by -5

47. The product of -6 and a number

48. The quotient of a number and -20

49. The sum of -15 and a number

50. -32 multiplied by a number

51. -8 divided by a number

52. Subtract a number from -18.

Concept Extensions

Solve.

53. $n - 42{,}860 = -1286$

54. $n + 961 = 120$

55. $-38x = 15{,}542$

56. $\dfrac{y}{-18} = 1098$

57. Explain the differences between an equation and an expression.

58. Explain the differences between the addition property of equality and the multiplication property of equality.

59. Write an equation that can be solved using the addition property of equality.

60. Write an equation that can be solved using the multiplication property of equality.

Chapter 2 Group Activity

Magic Squares

Sections 2.1–2.3

A magic square is a set of numbers arranged in a square table so that the sum of the numbers in each column, row, and diagonal is the same. For instance, in the magic square below, the sum of each column, row, and diagonal is 15. Notice that no number is used more than once in the magic square.

2	9	4
7	5	3
6	1	8

The properties of magic squares have been known for a very long time and once were thought to be good luck charms. The ancient Egyptians and Greeks understood their patterns. A magic square even made it into a famous work of art. The engraving titled *Melencolia I*, created by German artist Albrecht Dürer in 1514, features the following four-by-four magic square on the building behind the central figure.

16	3	2	13
5	10	11	8
9	6	7	12
4	15	14	1

Exercises

1. Verify that what is shown in the Dürer engraving is, in fact, a magic square. What is the common sum of the columns, rows, and diagonals?

2. Negative numbers can also be used in magic squares. Complete the following magic square:

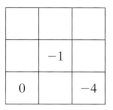

3. Use the numbers $-16, -12, -8, -4, 0, 4, 8, 12,$ and 16 to form a magic square:

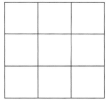

Chapter 2 Vocabulary Check

Fill in each blank with one of the words or phrases listed below.

inequality symbols   addition   solution   is less than   integers
expression   average   negative   absolute value   equation
positive   opposites   is greater than   multiplication

1. Two numbers that are the same distance from 0 on the number line but are on opposite sides of 0 are called _____.
2. The _____ of a number is that number's distance from 0 on a number line.
3. The _____ are ... , −3, −2, −1, 0, 1, 2, 3, ...
4. The _____ numbers are numbers less than zero.
5. The _____ numbers are numbers greater than zero.
6. The symbols "<" and ">" are called _____.
7. A(n) _____ of an equation is a number that when substituted for the variable makes the equation a true statement.
8. The _____ of a list of numbers is $\frac{\text{sum of numbers}}{\text{number of numbers}}$.
9. A combination of operations on variables and numbers is called a(n) _____.
10. A statement of the form "expression = expression" is called a(n) _____.
11. The sign "<" means _____ and ">" means _____.
12. By the _____ property of equality, the same number may be added to or subtracted from both sides of an equation without changing the solution of the equation.
13. By the _____ property of equality, both sides of an equation may be multiplied or divided by the same nonzero number without changing the solution of the equation.

> **Helpful Hint**
> Are you preparing for your test? To help, don't forget to take these:
> - Chapter 2 Getting Ready for the Test on page 159
> - Chapter 2 Test on page 160
>
> Then check all of your answers at the back of this text. For further review, the step-by-step video solutions to any of these exercises are located in MyLab Math.

2 Chapter Highlights

Definitions and Concepts	Examples
Section 2.1 Introduction to Integers	
Together, positive numbers, negative numbers, and 0 are called **signed numbers**.	−432, −10, 0, 15
The **integers** are ... , −3, −2, −1, 0, 1, 2, 3, ...	
The **absolute value** of a number is that number's distance from 0 on a number line. The symbol for absolute value is $|\ |$.	$\|-2\| = 2$ $\|2\| = 2$

Chapter 2 Highlights

Definitions and Concepts	Examples
Section 2.1 Introduction to Integers *(continued)*	
Two numbers that are the same distance from 0 on the number line but are on opposite sides of 0 are called **opposites**.	5 and −5 are opposites. 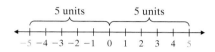
If a is a number, then $-(-a) = a$.	$-(-11) = 11$. Do not confuse with $-\|-3\| = -3$
Section 2.2 Adding Integers	
Adding Two Numbers with the Same Sign **Step 1:** Add their absolute values. **Step 2:** Use their common sign as the sign of the sum. **Adding Two Numbers with Different Signs** **Step 1:** Find the larger absolute value minus the smaller absolute value. **Step 2:** Use the sign of the number with the larger absolute value as the sign of the sum.	Add: $-3 + (-2) = -5$ $-7 + (-15) = -22$ $-6 + 4 = -2$ $17 + (-12) = 5$ $-32 + (-2) + 14 = -34 + 14$ $= -20$
Section 2.3 Subtracting Integers	
Subtracting Two Numbers If a and b are numbers, then $a - b = a + (-b)$.	Subtract: $-35 - 4 = -35 + (-4) = -39$ $3 - 8 = 3 + (-8) = -5$ $-10 - (-12) = -10 + 12 = 2$ $7 - 20 - 18 - (-3) = 7 + (-20) + (-18) + (+3)$ $= -13 + (-18) + 3$ $= -31 + 3$ $= -28$
Section 2.4 Multiplying and Dividing Integers	
Multiplying Numbers The product of two numbers having the same sign is a positive number. The product of two numbers having different signs is a negative number. **Dividing Numbers** The quotient of two numbers having the same sign is a positive number. The quotient of two numbers having different signs is a negative number.	Multiply: $(-7)(-6) = 42$ $9(-4) = -36$ Evaluate: $(-3)^2 = (-3)(-3) = 9$ Divide: $-100 \div (-10) = 10$ $\dfrac{14}{-2} = -7, \quad \dfrac{0}{-3} = 0, \quad \dfrac{22}{0}$ is undefined.

Definitions and Concepts	Examples
Section 2.5 Order of Operations	

Order of Operations

1. Perform all operations within parentheses ( ), brackets [ ], or other grouping symbols such as fraction bars, starting with the innermost set.
2. Evaluate any expressions with exponents.
3. Multiply or divide in order from left to right.
4. Add or subtract in order from left to right.

Simplify:

$$3 + 2 \cdot (-5) = 3 + (-10)$$
$$= -7$$

$$\frac{-2(5-7)}{-7 + |-3|} = \frac{-2(-2)}{-7 + 3}$$
$$= \frac{4}{-4}$$
$$= -1$$

Section 2.6   Solving Equations: The Addition and Multiplication Properties

Addition Property of Equality

Let $a, b,$ and c represent numbers.

If $a = b$, then

$a + c = b + c$ and $a - c = b - c$

In other words, the same number may be added to or subtracted from both sides of an equation without changing the solution of the equation.

Solve:

$$x + 8 = 1$$
$$x + 8 - 8 = 1 - 8 \quad \text{Subtract 8 from both sides.}$$
$$x = -7 \quad \text{Simplify.}$$

The solution is -7.

Multiplication Property of Equality

Let $a, b,$ and c represent numbers and let $c \neq 0$.

If $a = b$, then

$a \cdot c = b \cdot c$ and $\dfrac{a}{c} = \dfrac{b}{c}$

In other words, both sides of an equation may be multiplied or divided by the same nonzero number without changing the solution of the equation.

Solve:

$$-6y = 30$$
$$\frac{-6y}{-6} = \frac{30}{-6} \quad \text{Divide both sides by } -6.$$
$$\frac{-6}{-6} \cdot y = \frac{30}{-6}$$
$$y = -5 \quad \text{Simplify.}$$

The solution is -5.

Chapter 2   Review

(2.1) *Represent each quantity by an integer.*

1. A gold miner is working 1572 feet down in a mine.

2. Mount Hood, in Oregon, has an elevation of 11,239 feet.

Graph each integer in the list on the same number line.

3. −3, −5, 0, 7

4. −6, −1, 0, 5

Simplify.

5. $|-11|$

6. $|0|$

7. $-|8|$

8. $-(-9)$

9. $-|-16|$

10. $-(-2)$

Insert < or > between each pair of integers to make a true statement.

11. −18   −20

12. −5   5

13. $|-123|$ $-|-198|$

14. $|-12|$ $-|-16|$

Find the opposite of each integer.

15. −18

16. 42

Answer true or false for each statement.

17. If $a < b$, then a must be a negative number.

18. The absolute value of an integer is always 0 or a positive number.

19. A negative number is always less than a positive number.

20. If a is a negative number, then $-a$ is a positive number.

Evaluate.

21. $|y|$ if $y = -2$

22. $|-x|$ if $x = -3$

23. $-|z|$ if $z = -5$

24. $-|-n|$ if $n = -10$

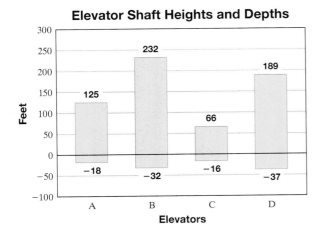

Elevator shafts in some buildings extend not only above ground, but in many cases below ground to accommodate basements, underground parking, etc. The bar graph shows four such elevators and their shaft distance above and below ground. Use the bar graph to answer Exercises 25 and 26.

25. Which elevator shaft extends the farthest below ground?

26. Which elevator shaft extends the highest above ground?

(2.2) *Add.*

27. $5 + (-3)$

28. $18 + (-4)$

29. $-12 + 16$

30. $-23 + 40$

31. $-8 + (-15)$

32. $-5 + (-17)$

33. $-24 + 3$

34. $-89 + 19$

35. $15 + (-15)$

36. $-24 + 24$

37. $-43 + (-108)$

38. $-100 + (-506)$

Solve.

39. The temperature at 5 a.m. on a day in January was $-15°$ Celsius. By 6 a.m. the temperature had fallen 5 degrees. Use a signed number to represent the temperature at 6 a.m.

40. A diver starts out at 127 feet below the surface and then swims downward another 23 feet. Use a signed number to represent the diver's current depth.

41. During the LPGA 2016 CME Group Tour Championship tournament, the winner, Charley Hull, had scores of -5, -2, -6, and -6. What was her total score for the tournament? (*Source:* LPGA)

42. The Ryder Cup, a biennial (every two years) pro men's golf tournament between an American team and a European team, scores holes won. During the 2016 Ryder Cup, the winners, the American team, had a score of 17. The losing team, the Europeans, had a score 6 less than the Americans' score. What was the European team's score? (*Source:* Ryder Cup 2016)

(2.3) *Subtract.*

43. $12 - 4$

44. $-12 - 4$

45. $-7 - 17$

46. $7 - 17$

47. $7 - (-13)$

48. $-6 - (-14)$

49. $16 - 16$

50. $-16 - 16$

51. $-12 - (-12)$

52. $-5 - (-12)$

53. $-(-5) - 12 + (-3)$

54. $-8 + (-12) - 10 - (-3)$

Solve.

55. If the elevation of Lake Superior is 600 feet above sea level and the elevation of the Caspian Sea is 92 feet below sea level, find the difference of the elevations.

56. Josh Weidner has $142 in his checking account. He writes a check for $125, makes a deposit of $43, and then writes another check for $85. Represent the balance in his account by an integer.

57. Some roller coasters travel above and below ground. One such roller coaster is Tremors, located in Silverwood Theme Park, Athol, Idaho. If this coaster rises to a height of 85 feet above ground, then drops 99 feet, how many feet below ground are you at the end of the drop? (*Source:* ultimaterollercoaster.com)

58. Go to the bar graph for Review Exercises **25** and **26** and find the total length of the elevator shaft for Elevator C.

Answer true or false for each statement.

59. $|-5| - |-6| = 5 - 6$

60. $|-5 - (-6)| = 5 + 6$

(2.4) *Multiply.*

61. $-3(-7)$ **62.** $-6(3)$ **63.** $-4(16)$ **64.** $-5(-12)$

65. $(-5)^2$ **66.** $(-1)^5$ **67.** $12(-3)(0)$ **68.** $-1(6)(2)(-2)$

Divide.

69. $-15 \div 3$ **70.** $\dfrac{-24}{-8}$ **71.** $\dfrac{0}{-3}$ **72.** $\dfrac{-46}{0}$

73. $\dfrac{100}{-5}$ **74.** $\dfrac{-72}{8}$ **75.** $\dfrac{-38}{-1}$ **76.** $\dfrac{45}{-9}$

Solve.

77. A football team lost 5 yards on each of two consecutive plays. Represent the total loss by a product of integers, and find the product.

78. A horse race bettor lost $50 on each of four consecutive races. Represent the total loss by a product of integers, and find the product.

79. A person has a debt of $1024 and is ordered to pay it back in four equal payments. Represent the amount of each payment by a quotient of integers, and find the quotient.

80. Overnight, the temperature dropped 45 degrees Fahrenheit. If this took place over a time period of nine hours, represent the average temperature drop each hour by a quotient of integers. Then find the quotient.

(2.5) *Simplify.*

81. $(-7)^2$ **82.** -7^2 **83.** $5 - 8 + 3$ **84.** $-3 + 12 + (-7) - 10$

85. $-10 + 3 \cdot (-2)$ **86.** $5 - 10 \cdot (-3)$ **87.** $16 \div (-2) \cdot 4$ **88.** $-20 \div 5 \cdot 2$

89. $16 + (-3) \cdot 12 \div 4$ **90.** $-12 + 10 \div (-5)$

91. $4^3 - (8 - 3)^2$ **92.** $(-3)^3 - 90$

93. $\dfrac{(-4)(-3) - (-2)(-1)}{-10 + 5}$ **94.** $\dfrac{4(12 - 18)}{-10 \div (-2 - 3)}$

Find the average of each list of numbers.

95. $-18, 25, -30, 7, 0, -2$

96. $-45, -40, -30, -25$

Evaluate each expression for $x = -2$ and $y = 1$.

97. $2x - y$ **98.** $y^2 + x^2$ **99.** $\dfrac{3x}{6}$ **100.** $\dfrac{5y - x}{-y}$

(2.6) *For Exercises 101 and 102, answer "yes" or "no."*

101. Is -5 a solution of $2n - 6 = 16$?

102. Is -2 a solution of $2(c - 8) = -20$?

Solve.

103. $n - 7 = -20$

104. $-5 = n + 15$

105. $10x = -30$

106. $-8x = 72$

107. $-20 + 7 = y$

108. $x - 31 = -62$

109. $\dfrac{n}{-4} = -11$

110. $\dfrac{x}{-2} = 13$

111. $n + 12 = -7$

112. $n - 40 = -2$

113. $-36 = -6x$

114. $-40 = 8y$

Mixed Review

Perform the indicated operations.

115. $-6 + (-9)$

116. $-16 - 3$

117. $-4(-12)$

118. $\dfrac{84}{-4}$

119. $-76 - (-97)$

120. $-9 + 4$

Solve.

121. Wednesday's lowest temperature was $-18°C$. The cold weather continued and by Friday, it had dropped another $9°C$. What was the temperature on Friday?

122. The temperature at noon on a Monday in December was $-11°C$. By noon on Tuesday, it had warmed by $17°C$. What was the temperature at noon on Tuesday?

123. The top of a mountain has an altitude of 12,923 feet. The bottom of a valley is 195 feet below sea level. Find the difference between these two elevations.

124. Joe owed his mother $32. He gave her $23. Write his financial situation as a signed number.

Simplify.

125. $(3 - 7)^2 \div (6 - 4)^3$

126. $3(4 + 2) + (-6) - 3^2$

127. $2 - 4 \cdot 3 + 5$

128. $4 - 6 \cdot 5 + 1$

129. $\dfrac{-|-14| - 6}{7 + 2(-3)}$

130. $5(7 - 6)^3 - 4(2 - 3)^2 + 2^4$

Solve.

131. $n - 9 = -30$

132. $n + 18 = 1$

133. $-4x = -48$

134. $9x = -81$

135. $\dfrac{n}{-2} = 100$

136. $\dfrac{y}{-1} = -3$

Chapter 2 Getting Ready for the Test LC

MULTIPLE CHOICE All the exercises below are **Multiple Choice**. Choose the correct letter. Not all letters may be used and some may be used more than once.

For Exercises 1 through 4, identify each answer as

    A. 2   **B.** -2 **C.** 0

1. -2; choose the opposite
2. -2; choose the absolute value
3. 2; choose the absolute value
4. 2; choose the opposite

For Exercises 5 through 10, identify each answer as

    A. addition   **B.** subtraction   **C.** multiplication   **D.** division

5. For xy, the operation is _____.
6. For $-12(+3)$, the operation is _____.
7. For $-12 + 3$, the operation is _____.
8. For $\dfrac{-12}{+3}$, the operation is _____.
9. For $4 + 6 \cdot 2$, which operation is performed first?
10. For $4 + 6 \div 2$, which operation is performed first?

For Exercises 11 through 14, choose the solution of each equation as

    A. 2   **B.** -2 **C.** 18   **D.** -18

11. $-3x = 6$ 12. $x - 2 = -4$ 13. $\dfrac{x}{-3} = 6$ 14. $x + 2 = 4$

Chapter 2 Test MyLab Math or YouTube

For additional practice go to your study plan in MyLab Math.

Answers

1. ______
2. ______
3. ______
4. ______
5. ______
6. ______
7. ______
8. ______
9. ______
10. ______
11. ______
12. ______
13. ______
14. ______
15. ______
16. ______
17. ______
18. ______

Simplify each expression.

1. $-5 + 8$

2. $18 - 24$

3. $5 \cdot (-20)$

4. $-16 \div (-4)$

5. $-18 + (-12)$

6. $-7 - (-19)$

7. $-5 \cdot (-13)$

8. $\frac{-25}{-5}$

9. $|-25| + (-13)$

10. $14 - |-20|$

11. $|5| \cdot |-10|$

12. $\frac{|-10|}{-|-5|}$

13. $-8 + 9 \div (-3)$

14. $-7 + (-32) - 12 + 5$

15. $(-5)^3 - 24 \div (-3)$

16. $(5 - 9)^2 \cdot (8 - 2)^3$

17. $-(-7)^2 \div 7 \cdot (-4)$

18. $3 - (8 - 2)^3$

19. $\dfrac{4}{2} - \dfrac{8^2}{16}$

20. $\dfrac{-3(-2) + 12}{-1(-4 - 5)}$

21. $\dfrac{|25 - 30|^2}{2(-6) + 7}$

22. $5(-8) - [6 - (2 - 4)] + (12 - 16)^2$

Evaluate each expression for $x = 0$, $y = -3$, and $z = 2$.

23. $7x + 3y - 4z$

24. $10 - y^2$

25. $\dfrac{3z}{2y}$

26. Mary Dunstan, a diver, starts at sea level and then makes 4 successive descents of 22 feet. After the descents, what is her elevation?

27. Aaron Hawn has $129 in his checking account. He writes a check for $79, withdraws $40 from an ATM, and then deposits $35. Represent the new balance in his account by an integer.

28. Mt. Washington in New Hampshire has an elevation of 6288 feet above sea level. The Romanche Gap in the Atlantic Ocean has an elevation of 25,354 feet below sea level. Represent the difference in elevation between these two points by an integer. (*Source:* National Geographic Society and Defense Mapping Agency)

29. Lake Baykal in Siberian Russia is the deepest lake in the world, with a maximum depth of 5315 feet. The elevation of the lake's surface is 1495 feet above sea level. What is the elevation (with respect to sea level) of the deepest point in the lake? (*Source:* U.S. Geological Survey)

30. Find the average of $-12, -13, 0, 9$.

31. Translate the following phrases into mathematical expressions. Use x to represent "a number."
   a. The product of a number and 17
   b. A number subtracted from 20

Solve.

32. $-9n = -45$

33. $\dfrac{n}{-7} = 4$

34. $x - 16 = -36$

35. $-20 + 8 + 8 = x$

Chapters 1–2 Cumulative Review

Find the place value of the digit 3 in each whole number.

1. 396,418

2. 4308

3. 93,192

4. 693,298

5. 534,275,866

6. 267,301,818

7. Insert < or > to make a true statement.
   a. −7 7
   b. 0 −4
   c. −9 −11

8. Insert < or > to make a true statement
   a. 12 −4
   b. −13 −31
   c. −82 79

9. Add: 13 + 2 + 7 + 8 + 9

10. Add: 11 + 3 + 9 + 16

11. Subtract: 7826 − 505 Check by adding.

12. Subtract: 3285 − 272 Check by adding.

13. The radius of Jupiter is 43,441 miles. The radius of Saturn is 7257 miles less than the radius of Jupiter. Find the radius of Saturn. (*Source:* National Space Science Data Center)

14. C. J. Dufour wants to buy a digital camera. She has $762 in her savings account. If the camera costs $237, how much money will she have in her account after buying the camera?

15. Round 568 to the nearest ten.

16. Round 568 to the nearest hundred.

17. Round each number to the nearest hundred to find an estimated difference.
   4725
   −2879

18. Round each number to the nearest thousand to find an estimated difference.
   8394
   −2913

19. Rewrite each using the distributive property.
   a. 5(6 + 5)
   b. 20(4 + 7)
   c. 2(7 + 9)

20. Rewrite each using the distributive property.
   a. 5(2 + 12)
   b. 9(3 + 6)
   c. 4(8 + 1)

21. Multiply: 631 × 125

22. Multiply: 299 × 104

23. Find each quotient. Check by multiplying.
   a. 42 ÷ 7
   b. $\frac{64}{8}$
   c. $3\overline{)21}$

24. Find each quotient. Check by multiplying.
   a. $\frac{35}{5}$
   b. 64 ÷ 8
   c. $4\overline{)48}$

25. Divide: $3705 \div 5$. Check by multiplying.

26. Divide: $3648 \div 8$. Check by multiplying.

27. As part of a promotion, an executive receives 238 cards, each good for one free song download. If she wants to share them evenly with 19 friends, how many download cards will each friend receive? How many will be left over?

28. Mrs. Mallory's first-grade class is going to the zoo. She pays a total of $324 for 36 admission tickets. How much does each ticket cost?

Evaluate.

29. 9^2

30. 5^3

31. 6^1

32. 4^1

33. $5 \cdot 6^2$

34. $2^3 \cdot 7$

35. Simplify: $\dfrac{7 - 2 \cdot 3 + 3^2}{5(2 - 1)}$

36. Simplify: $\dfrac{6^2 + 4 \cdot 4 + 2^3}{37 - 5^2}$

37. Evaluate $x + 6$ if x is 8.

38. Evaluate $5 + x$ if x is 9.

39. Simplify:
   a. $|-9|$
   b. $|8|$
   c. $|0|$

40. Simplify:
   a. $|4|$
   b. $|-7|$

41. Add: $-2 + 25$

42. Add: $8 + (-3)$

43. Evaluate $2a - b$ for $a = 8$ and $b = -6$.

44. Evaluate $x - y$ for $x = -2$ and $y = -7$.

45. Multiply: $-7 \cdot 3$

46. Multiply: $5(-2)$

47. Multiply: $0 \cdot (-4)$

48. Multiply: $-6 \cdot 9$

49. Simplify: $3(4 - 7) + (-2) - 5$

50. Simplify: $4 - 8(7 - 3) - (-1)$

3 Solving Equations and Problem Solving

In this chapter, we continue making the transition from arithmetic to algebra. Recall that in algebra, letters (called variables) represent unknown quantities. Using variables is a very powerful method for solving problems that cannot be solved with arithmetic alone. This chapter introduces operations on algebraic expressions, and we continue solving variable equations.

Sections

3.1 Simplifying Algebraic Expressions

3.2 Solving Equations: Review of the Addition and Multiplication Properties

Integrated Review— Expressions and Equations

3.3 Solving Linear Equations in One Variable

3.4 Linear Equations in One Variable and Problem Solving

Check Your Progress

Vocabulary Check
Chapter Highlights
Chapter Review
Getting Ready for the Test
Chapter Test
Cumulative Review

Soccer, or Football as It Is Known Outside the United States, Is the Most Popular Sport in the World.

The 2016 FIFA World Cup was held in Japan and attracted an audience of 110 million viewers in Europe and more than 26 million in the United States watching the final.

FIFA, the international governing body of soccer, has strict regulations for the dimensions of the field. There is a minimum and maximum length allowed as well as a minimum and maximum width.

Soccer is a game that can be played by anyone who can walk, from toddler to old age. To accommodate the various skill levels as well as developmental levels, the standard size of the fields changes with skill and age. The bar graph below demonstrates how the maximum length, in yards, of a soccer field varies based on these criteria.

In Section 3.4, Exercises 45 and 46, we will solve applications that give us the unknown heights in the bar graph. (*Source:* FIFA and Trumark Athletics)

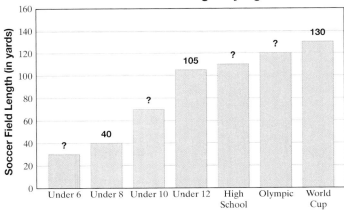

164

3.1 Simplifying Algebraic Expressions

Recall from Section 1.8 that a combination of numbers, letters (variables), and operation symbols is called an **algebraic expression** or simply an **expression**. Examples of expressions are below.

Algebraic Expressions

$4 \cdot x, \quad n + 7, \quad$ and $\quad 3y - 5 - x$

Recall that if two variables or a number and a variable are next to each other, with no operation sign between them, the indicated operation is multiplication. For example,

$3y$ means $3 \cdot y$

and

xy or $x(y)$ means $x \cdot y$

Also, the meaning of an exponent remains the same when the base is a variable. For example,

$y^2 = \underbrace{y \cdot y}_{\text{2 factors of } y}$ and $x^4 = \underbrace{x \cdot x \cdot x \cdot x}_{\text{4 factors of } x}$

Just as we can add, subtract, multiply, and divide numbers, we can add, subtract, multiply, and divide algebraic expressions. In previous sections we evaluated algebraic expressions like $x + 3$, $4x$, and $x + 2y$ for particular values of the variables. In this section, we explore working with variable expressions without evaluating them. We begin with a definition of a term.

Objectives

A Use Properties of Numbers to Combine Like Terms.

B Use Properties of Numbers to Multiply Expressions.

C Simplify Expressions by Multiplying and Then Combining Like Terms.

D Find the Perimeter and Area of Figures.

Objective A Combining Like Terms

The addends of an algebraic expression are called the **terms** of the expression.

$x + 3$ — 2 terms

$3y^2 + (-6y) + 4$ — 3 terms

A term that is only a number has a special name. It is called a **constant term**, or simply a **constant**. A term that contains a variable is called a **variable term**.

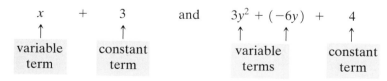

The number factor of a variable term is called the **numerical coefficient**. A numerical coefficient of 1 is usually not written.

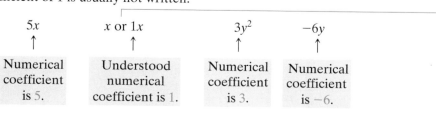

> **Helpful Hint**
> Recall that $1 \cdot$ any number = that number. This means that
> $1 \cdot x = x$ or that $1x = x$
> Thus x can always be replaced by $1x$ or $1 \cdot x$.

165

Chapter 3 | Solving Equations and Problem Solving

Terms with the same variable factors, except that they may have different numerical coefficients, are called **like terms**.

Like Terms	Unlike Terms
$3x, -4x$	$5x, x^2$
$-6y, 2y, y$	$7x, 7y$

✓**Concept Check** True or false? The terms $-7xz^2$ and $3z^2x$ are like terms. Explain.

A sum or difference of like terms can be simplified using the **distributive property**. Recall from Section 1.5 that the distributive property says that multiplication distributes over addition (and subtraction). Using variables, we can write the distributive property as follows:

$$(a + b)c = ac + bc.$$

If we write the right side of the equation first, then the left side, we have the following.

Distributive Property

If $a, b,$ and c are numbers, then

$$ac + bc = (a + b)c$$

Also,

$$ac - bc = (a - b)c$$

The distributive property guarantees that, no matter what number x is, $7x + 2x$ (for example) has the same value as $(7 + 2)x$, or $9x$. We then have that

$$7x + 2x = (7 + 2)x = 9x$$

This is an example of **combining like terms**. An algebraic expression is **simplified** when all like terms have been combined.

Example 1 Simplify each expression by combining like terms.

a. $4x + 6x$ **b.** $y - 5y$ **c.** $3x^2 + 5x^2 - 2$

Solution: Add or subtract like terms.

a. $4x + 6x = (4 + 6)x$
$= 10x$

b. $y - 5y = \overset{\text{Understood 1}}{1y} - 5y$
$= (1 - 5)y$
$= -4y$

c. $3x^2 + 5x^2 - 2 = (3 + 5)x^2 - 2$
$= 8x^2 - 2$

■ Work Practice 1

Practice 1

Simplify each expression by combining like terms.
a. $8m - 14m$
b. $6a + a$
c. $-y^2 + 3y^2 + 7$

Answers
1. a. $-6m$ **b.** $7a$ **c.** $2y^2 + 7$

✓**Concept Check Answer**
true

Section 3.1 | Simplifying Algebraic Expressions

Helpful Hint

In this section, we are simplifying expressions. Try not to confuse the two processes below.

Expression: $5y - 8y$

Simplify the **expression**:

$5y - 8y = (5 - 8)y$
$= -3y$

Equation: $8n = -40$

Solve the **equation**:

$8n = -40$

$\dfrac{8n}{8} = \dfrac{-40}{8}$ Divide both sides by 8.

$n = -5$ The solution is -5.

The commutative and associative properties of addition and multiplication can also help us simplify expressions. We presented these properties in Sections 1.3 and 1.5 and state them again using variables.

Properties of Addition and Multiplication

If a, b, and c are numbers, then

$a + b = b + a$ Commutative property of addition
$a \cdot b = b \cdot a$ Commutative property of multiplication

That is, the **order** of adding or multiplying two numbers can be changed without changing their sum or product.

$(a + b) + c = a + (b + c)$ Associative property of addition
$(a \cdot b) \cdot c = a \cdot (b \cdot c)$ Associative property of multiplication

That is, the **grouping** of numbers in addition or multiplication can be changed without changing their sum or product.

Helpful Hint

- Examples of these properties are

  $2 + 3 = 3 + 2$ Commutative property of addition
  $7 \cdot 9 = 9 \cdot 7$ Commutative property of multiplication
  $(1 + 8) + 10 = 1 + (8 + 10)$ Associative property of addition
  $(4 \cdot 2) \cdot 3 = 4 \cdot (2 \cdot 3)$ Associative property of multiplication

- These properties are not true for subtraction or division.

Example 2 Simplify: $2y - 6 + 4y + 8$

Solution: We begin by writing subtraction as the opposite of addition.

$2y - 6 + 4y + 8 = 2y + (-6) + 4y + 8$ Apply the commutative property of addition.
$= 2y + 4y + (-6) + 8$
$= (2 + 4)y + (-6) + 8$ Apply the distributive property.
$= 6y + 2$ Simplify.

Work Practice 2

Practice 2

Simplify: $6z + 5 + z - 4$

Answer
2. $7z + 1$

Practice 3–5

Simplify each expression by combining like terms.

3. $6y + 12y - 6$
4. $7y - 5 + y + 8$
5. $-7y + 2 - 2y - 9x + 12 - x$

Examples Simplify each expression by combining like terms.

3. $6x + 2x - 5$
$= 8x - 5$

4. $4x + 2 - 5x + 3$
$= 4x + 2 + (-5x) + 3$
$= 4x + (-5x) + 2 + 3$
$= -1x + 5 \quad \text{or} \quad -x + 5$

5. $2x - 5 + 3y + 4x - 10y + 11$
$= 2x + (-5) + 3y + 4x + (-10y) + 11$
$= 2x + 4x + 3y + (-10y) + (-5) + 11$
$= 6x - 7y + 6$

■ Work Practice 3–5

As we practice combining like terms, keep in mind that some of the steps may be performed mentally.

Objective B Multiplying Expressions

We can also use properties of numbers to multiply expressions such as $3(2x)$. By the associative property of multiplication, we can write the product $3(2x)$ as $(3 \cdot 2)x$, which simplifies to $6x$.

Practice 6–7

Multiply.
6. $6(4a)$
7. $-8(9x)$

Examples Multiply.

6. $5(3y) = (5 \cdot 3)y$ Apply the associative property of multiplication.
$= 15y$ Multiply.

7. $-2(4x) = (-2 \cdot 4)x$ Apply the associative property of multiplication.
$= -8x$ Multiply.

■ Work Practice 6–7

We can use the distributive property to combine like terms, which we have done, and also to multiply expressions such as $2(3 + x)$. By the distributive property, we have

$2(3 + x) = 2 \cdot 3 + 2 \cdot x$ Apply the distributive property.
$= 6 + 2x$ Multiply.

Practice 8

Use the distributive property to multiply: $8(y + 2)$

Example 8 Use the distributive property to multiply: $5(x + 4)$

Solution: By the distributive property,

$5(x + 4) = 5 \cdot x + 5 \cdot 4$ Apply the distributive property.
$= 5x + 20$ Multiply.

■ Work Practice 8

Answers
3. $18y - 6$ **4.** $8y + 3$
5. $-9y - 10x + 14$ **6.** $24a$
7. $-72x$ **8.** $8y + 16$

✓ **Concept Check** What's wrong with the following?
$8(a - b) = 8a - b$

✓ **Concept Check Answer**
did not distribute the 8;
$8(a - b) = 8a - 8b$

Section 3.1 | Simplifying Algebraic Expressions 169

Example 9 Multiply: $-3(5a + 2)$

Solution: By the distributive property,

$-3(5a + 2) = -3(5a) + (-3)(2)$ Apply the distributive property.
$= (-3 \cdot 5)a + (-6)$ Use the associative property and multiply.
$= -15a - 6$ Multiply.

Work Practice 9

Practice 9
Multiply: $3(7a - 5)$

Example 10 Multiply: $8(x - 4)$

Solution:

$8(x - 4) = 8 \cdot x - 8 \cdot 4$
$= 8x - 32$

Work Practice 10

Practice 10
Multiply: $6(5 - y)$

Objective C Simplifying Expressions

Next we will **simplify** expressions by first using the distributive property to multiply and then **combining** any like terms.

Example 11 Simplify: $2(3 + 7x) - 15$

Solution: First we use the distributive property to remove parentheses.

$2(3 + 7x) - 15 = 2(3) + 2(7x) - 15$ Apply the distributive property.
$= 6 + 14x - 15$ Multiply.
$= 14x + (-9)$ or $14x - 9$ Combine like terms.

Work Practice 11

Practice 11
Simplify: $5(2y - 3) - 8$

Helpful Hint
2 is *not* distributed to the -15 since it is not within the parentheses.

Example 12 Simplify: $-2(x - 5) + 4(2x + 2)$

Solution: First we use the distributive property to remove parentheses.

$-2(x - 5) + 4(2x + 2) = -2(x) - (-2)(5) + 4(2x) + 4(2)$ Apply the distributive property.
$= -2x + 10 + 8x + 8$ Multiply.
$= 6x + 18$ Combine like terms.

Work Practice 12

Practice 12
Simplify:
$-7(x - 1) + 5(2x + 3)$

Example 13 Simplify: $-(x + 4) + 5x + 16$

Solution: The expression $-(x + 4)$ means $-1(x + 4)$.

$-(x + 4) + 5x + 16 = -1(x + 4) + 5x + 16$
$= -1 \cdot x + (-1)(4) + 5x + 16$ Apply the distributive property.
$= -x + (-4) + 5x + 16$ Multiply.
$= 4x + 12$ Combine like terms.

Work Practice 13

Practice 13
Simplify: $-(y + 1) + 3y - 12$

Answers
9. $21a - 15$ **10.** $30 - 6y$
11. $10y - 23$ **12.** $3x + 22$
13. $2y - 13$

Objective D Finding Perimeter and Area

Practice 14

Find the perimeter of the square.

Example 14 Find the perimeter of the triangle.

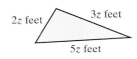

Solution: Recall that the perimeter of a figure is the distance around the figure. To find the perimeter, then, we find the sum of the lengths of the sides. We use the letter P to represent perimeter.

$$P = 2z + 3z + 5z$$
$$= 10z$$

Helpful Hint Don't forget to insert proper units.

The perimeter is $10z$ feet.

■ Work Practice 14

Practice 15

Find the area of the rectangular garden.

Example 15 Finding the Area of a Basketball Court

Find the area of this YMCA basketball court.

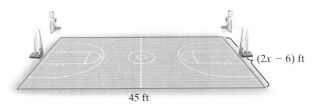

Solution: Recall how to find the area of a rectangle. **Area = Length · Width** or if A represents area, l represents length, and w represents width, we have $A = l \cdot w$.

$$A = l \cdot w$$
$$= 45(2x - 6) \quad \text{Let length} = 45 \text{ and width} = (2x - 6).$$
$$= 90x - 270 \quad \text{Multiply.}$$

The area is $(90x - 270)$ *square* feet.

■ Work Practice 15

Helpful Hint

Don't forget…

Area:	Perimeter:
• surface enclosed	• distance around
• measured in square units	• measured in units

Answers
14. $8x$ cm **15.** $(36y + 27)$ sq yd

Section 3.1 | Simplifying Algebraic Expressions

Vocabulary, Readiness & Video Check

Use the choices below to fill in each blank. Some choices may be used more than once.

| numerical coefficient | combine like terms | like | term | variable | associative |
| constant | expression | unlike | distributive | commutative | |

1. $14y^2 + 2x - 23$ is called a(n) _____ while $14y^2$, $2x$, and -23 are each called a(n) _____.
2. To multiply $3(-7x + 1)$, we use the _____ property.
3. To simplify an expression like $y + 7y$, we _____.
4. By the _____ properties, the *order* of adding or multiplying two numbers can be changed without changing their sum or product.
5. The term $5x$ is called a(n) _____ term while the term 7 is called a(n) _____ term.
6. The term z has an understood _____ of 1.
7. By the _____ properties, the *grouping* of adding or multiplying numbers can be changed without changing their sum or product.
8. The terms $-x$ and $5x$ are _____ terms and the terms $5x$ and $5y$ are _____ terms.
9. For the term $-3x^2y$, -3 is called the _____.

Martin-Gay Interactive Videos   *Watch the section lecture video and answer the following questions.*

Objective A 10. In Example 2, why can't the expression be simplified after the first step?
Objective B 11. In Example 5, what property is used to multiply?
Objective C 12. In Example 6, why is the 20 not multiplied by the 2?
Objective D 13. In Example 8, what operation is used to find P? What operation is used to find A? What do P and A stand for?

See Video 3.1

3.1 Exercise Set MyLab Math

Objective A *Simplify each expression by combining like terms. See Examples 1 through 5.*

1. $3x + 5x$
2. $8y + 3y$
3. $2n - 3n$
4. $7z - 10z$
5. $4c + c - 7c$
6. $5b - 8b - b$
7. $4x - 6x + x - 5x$
8. $8y + y - 2y - 8y$
9. $3a + 2a + 7a - 5$
10. $5b - 4b + b - 15$

Objective B *Multiply. See Examples 6 and 7.*

11. $6(7x)$
12. $4(4x)$
13. $-3(11y)$
14. $-3(21z)$
15. $12(6a)$
16. $13(5b)$

Multiply. See Examples 8 through 10.

17. $2(y + 3)$

18. $3(x + 1)$

19. $3(a - 6)$

20. $4(y - 6)$

21. $-4(3x + 7)$

22. $-8(8y + 10)$

Objective C *Simplify each expression. First use the distributive property to multiply and remove parentheses. See Examples 11 through 13.*

23. $2(x + 4) - 7$

24. $5(6 - y) - 2$

25. $8 + 5(3c - 1)$

26. $10 + 4(6d - 2)$

27. $-4(6n - 5) + 3n$

28. $-3(5 - 2b) - 4b$

29. $3 + 6(w + 2) + w$

30. $8z + 5(6 + z) + 20$

31. $2(3x + 1) + 5(x - 2)$

32. $3(5x - 2) + 2(3x + 1)$

33. $-(2y - 6) + 10$

34. $-(5x - 1) - 10$

Objectives A B C Mixed Practice *Simplify each expression. See Examples 1 through 13.*

35. $18y - 20y$

36. $x + 12x$

37. $z - 8z$

38. $-12x + 8x$

39. $9d - 3c - d$

40. $8r + s - 7s$

41. $2y - 6 + 4y - 8$

42. $a + 4 - 7a - 5$

43. $5q + p - 6q - p$

44. $m - 8n + m + 8n$

45. $2(x + 1) + 20$

46. $5(x - 1) + 18$

47. $5(x - 7) - 8x$

48. $3(x + 2) - 11x$

49. $-5(z + 3) + 2z$

50. $-8(1 + v) + 6v$

51. $8 - x + 4x - 2 - 9x$

52. $5y - 4 + 9y - y + 15$

53. $-7(x + 5) + 5(2x + 1)$

54. $-2(x + 4) + 8(3x - 1)$

55. $3r - 5r + 8 + r$

56. $6x - 4 + 2x - x + 3$

57. $-3(n - 1) - 4n$

58. $5(c + 2) + 7c$

59. $4(z - 3) + 5z - 2$

60. $8(m + 3) - 20 + m$

61. $6(2x - 1) - 12x$

62. $5(2a + 3) - 10a$

63. $-(4x - 5) + 5$

64. $-(7y - 2) + 6$

65. $-(4x - 10) + 2(3x + 5)$

66. $-(12b - 20) + 5(3b - 2)$

67. $3a + 4(a + 3)$

68. $b + 2(b - 5)$

69. $5y - 2(y - 1) + 3$

70. $3x - 4(x + 2) + 1$

Section 3.1 | Simplifying Algebraic Expressions

Objective D *Find the perimeter of each figure. See Example 14.*

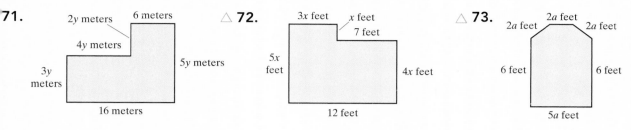

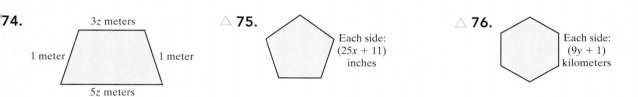

Find the area of each rectangle. See Example 15.

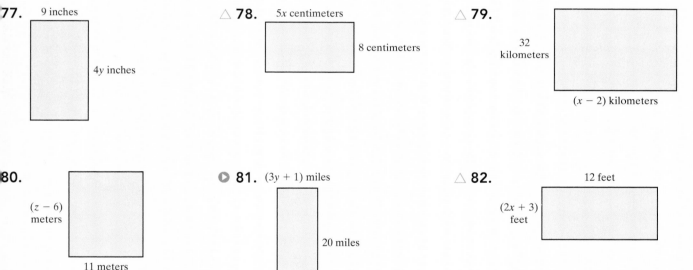

Objectives A B C D Mixed Practice *Solve. See Examples 1 through 15.*

83. Find the area of a regulation NCAA basketball court that is 94 feet long and 50 feet wide.

84. Find the area of a rectangular movie screen that is 50 feet long and 40 feet high.

85. A decorator wishes to put a wallpaper border around a rectangular room that measures 14 feet by 18 feet. Find the room's perimeter.

86. How much fencing will a rancher need for a rectangular cattle lot that measures 80 feet by 120 feet?

87. Find the perimeter of a triangular garden that measures 5 feet by x feet by $(2x + 1)$ feet.

88. Find the perimeter of a triangular picture frame that measures x inches by x inches by $(x - 14)$ inches.

Review

Perform each indicated operation. See Sections 2.2 and 2.3.

89. $-13 + 10$

90. $-15 + 23$

91. $-4 - (-12)$

92. $-7 - (-4)$

93. $-4 + 4$

94. $8 + (-8)$

Concept Extensions

If the expression on the left side of the equal sign is equivalent to the right side, write "correct." If not, write "incorrect" and then write an expression that is equivalent to the left side. See the second Concept Check in this section.

95. $5(3x - 2) \stackrel{?}{=} 15x - 2$

96. $-2(4x - 1) \stackrel{?}{=} -8x - 2$

97. $2(xy) \stackrel{?}{=} 2x \cdot 2y$

98. $-8(ab) \stackrel{?}{=} -8a \cdot (-8b)$

99. $7x - (x + 2) \stackrel{?}{=} 7x - x - 2$

100. $12y - (3y - 1) \stackrel{?}{=} 12y - 3y + 1$

101. $4(y - 3) + 11 \stackrel{?}{=} 4y - 7 + 11$

102. $6(x + 5) + 2 \stackrel{?}{=} 6x + 30 + 12$

Review commutative, associative, and distributive properties. Then identify which property allows us to write the equivalent expression on the right side of the equal sign.

103. $6(2x - 3) + 5 = 12x - 18 + 5$

104. $9 + 7x + (-2) = 7x + 9 + (-2)$

105. $-7 + (4 + y) = (-7 + 4) + y$

106. $(x + y) + 11 = 11 + (x + y)$

Write the expression that represents the area of each composite figure. Then simplify to find the total area.

△ **107.**

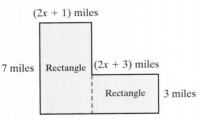

△ **108.**

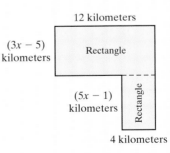

Simplify.

109. $9684q - 686 - 4860q + 12{,}960$

110. $76(268x + 592) - 2960$

111. If x is a whole number, which expression is greater: $2x$ or $5x$? Explain your answer.

112. If x is a whole number, which expression is greater: $-2x$ or $-5x$? Explain your answer.

113. Explain what makes two terms "like terms."

114. Explain how to combine like terms.

3.2 Solving Equations: Review of the Addition and Multiplication Properties

Objective A Using the Addition Property or the Multiplication Property

Objectives

A Use the Addition Property or the Multiplication Property to Solve Equations.

B Use Both Properties to Solve Equations.

C Translate Word Phrases to Mathematical Expressions.

In this section, we continue solving equations using the properties first introduced in Section 2.6.

First, let's recall the difference between an **expression** and an **equation.** Remember—an equation contains an equal sign and an expression does not.

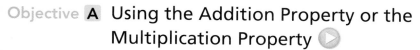

Equations	Expressions
$7x = 6x + 4$	$7x - 6x + 4$
$3(3y - 5) = 10y$	$y - 1 + 11y - 21$

equal signs ← → no equal signs

Thus far in this text, we have

Solved some equations (Section 2.6)

and

Simplified some expressions (Section 3.1)

As we will see in this section, the ability to simplify expressions will help us as we solve more equations.

The addition and multiplication properties are reviewed below.

Addition Property of Equality

Let a, b, and c represent numbers. Then

$a = b$
and $a + c = b + c$
are equivalent equations.

Also, $a = b$
and $a - c = b - c$
are equivalent equations.

Multiplication Property of Equality

Let a, b, and c represent numbers and let $c \neq 0$. Then

$a = b$
and $a \cdot c = b \cdot c$
are equivalent equations.

Also, $a = b$
and $\dfrac{a}{c} = \dfrac{b}{c}$
are equivalent equations.

In other words, the same number may be added to or subtracted from *both* sides of an equation without changing the solution of the equation. Also, *both* sides of an equation may be multiplied or divided by the same nonzero number without changing the solution of the equation.

Many equations in this section will contain expressions that can be simplified. If one or both sides of an equation can be simplified, do that first.

Example 1 Solve: $y - 5 = -2 - 6$

Solution: First we simplify the right side of the equation.

$y - 5 = -2 - 6$
$y - 5 = -8$ Combine like terms.

(Continued on next page)

Practice 1

Solve: $x + 6 = 1 - 3$

Answer

1. -8

175

Next we get y alone by using the addition property of equality. We add 5 to both sides of the equation.

$y - 5 + 5 = -8 + 5$ Add 5 to both sides.
$y = -3$ Simplify.

Check: To see that -3 is the solution, replace y with -3 in the original equation.

$y - 5 = -2 - 6$
$-3 - 5 \stackrel{?}{=} -2 - 6$ Replace y with -3.
$-8 = -8$ True

Since $-8 = -8$ is true, the solution is -3.

■ Work Practice 1

Practice 2

Solve: $10 = 2m - 4m$

Example 2   Solve: $3y - 7y = 12$

Solution: First, simplify the left side of the equation by combining like terms.

$3y - 7y = 12$
$-4y = 12$ Combine like terms.

Next, we get y alone by using the multiplication property of equality and dividing both sides by -4.

$\dfrac{-4y}{-4} = \dfrac{12}{-4}$ Divide both sides by -4.
$y = -3$ Simplify.

Check: Replace y with -3 in the original equation.

$3y - 7y = 12$
$3(-3) - 7(-3) \stackrel{?}{=} 12$
$-9 + 21 \stackrel{?}{=} 12$
$12 = 12$ True

The solution is -3.

■ Work Practice 2

✓**Concept Check**   What's wrong with the following solution?

$4x - 6x = 10$
$2x = 10$
$\dfrac{2x}{2} = \dfrac{10}{2}$
$x = 5$

Practice 3

Solve: $-8 + 6 = \dfrac{a}{3}$

Example 3   Solve: $\dfrac{z}{-4} = 11 - 5$

Solution: Simplify the right side of the equation first.

$\dfrac{z}{-4} = 11 - 5$
$\dfrac{z}{-4} = 6$

Answers

2. -5 **3.** -6

✓**Concept Check answer**

On the left side of the equation, $4x - 6x$ simplifies to $-2x$ not $2x$.

Section 3.2 | Solving Equations: Review of the Addition and Multiplication Properties 177

Next, to get z alone, multiply both sides by -4.

$$-4 \cdot \frac{z}{-4} = -4 \cdot 6 \qquad \text{Multiply both sides by } -4.$$

$$\frac{-4}{-4} \cdot z = -4 \cdot 6$$

$$1z = -24 \quad \text{or} \quad z = -24$$

Check to see that -24 is the solution.

■ Work Practice 3

Example 4 Solve: $5x + 2 - 4x = 7 - 19$

Solution: First we simplify each side of the equation separately.

$$5x + 2 - 4x = 7 - 19$$
$$\underbrace{5x - 4x}_{} + 2 = \underbrace{7 - 19}_{}$$
$$1x + 2 = -12 \quad \text{Combine like terms.}$$

To get x alone on the left side, we subtract 2 from both sides.

$$1x + 2 - 2 = -12 - 2 \qquad \text{Subtract 2 from both sides.}$$
$$1x = -14 \quad \text{or} \quad x = -14 \qquad \text{Simplify.}$$

Check to see that -14 is the solution.

■ Work Practice 4

Practice 4
Solve:
$-6y - 1 + 7y = 17 + 2$

Example 5 Solve: $8x - 9x = 12 - 17$

Solution: First combine like terms on each side of the equation.

$$8x - 9x = 12 - 17$$
$$-x = -5$$

Recall that $-x$ means $-1x$ and divide both sides by -1.

$$\frac{-1x}{-1} = \frac{-5}{-1} \qquad \text{Divide both sides by } -1.$$
$$x = 5 \qquad \text{Simplify.}$$

Check to see that the solution is 5.

■ Work Practice 5

Practice 5
Solve: $-4 - 10 = 4y - 5y$

Example 6 Solve: $3(3x - 5) = 10x$

Solution: First we multiply on the left side to remove the parentheses.

$$3(3x - 5) = 10x$$
$$3 \cdot 3x - 3 \cdot 5 = 10x \qquad \text{Use the distributive property.}$$
$$9x - 15 = 10x$$

Now we subtract $9x$ from both sides.

$$9x - 15 - 9x = 10x - 9x \qquad \text{Subtract } 9x \text{ from both sides.}$$
$$-15 = 1x \quad \text{or} \quad x = -15 \qquad \text{Simplify.}$$

■ Work Practice 6

Practice 6
Solve: $13x = 4(3x - 1)$

Answers
4. 20 **5.** 14 **6.** -4

Objective B Using Both Properties to Solve Equations

We now solve equations in one variable using more than one property of equality. To solve an equation such as $2x - 6 = 18$, we first get the variable term $2x$ alone on one side of the equation.

Practice 7

Solve: $5y + 2 = 17$

Example 7 Solve: $2x - 6 = 18$

Solution: We start by adding 6 to both sides to get the variable term $2x$ alone.

$$2x - 6 = 18$$
$$2x - 6 + 6 = 18 + 6 \quad \text{Add 6 to both sides.}$$
$$2x = 24 \quad \text{Simplify.}$$

To finish solving, we divide both sides by 2.

$$\frac{2x}{2} = \frac{24}{2} \quad \text{Divide both sides by 2.}$$
$$x = 12 \quad \text{Simplify.}$$

Check:

$$2x - 6 = 18$$
$$2(12) - 6 \stackrel{?}{=} 18 \quad \text{Replace } x \text{ with 12 and simplify.}$$
$$24 - 6 \stackrel{?}{=} 18$$
$$18 = 18 \quad \text{True}$$

The solution is 12.

Helpful Hint Don't forget to check the proposed solution in the *original* equation.

■ Work Practice 7

Don't forget, if one or both sides of an equation can be simplified, do that first.

Practice 8

Solve:
$-4(x + 2) - 60 = 2 - 10$

Example 8 Solve: $2 - 6 = -5(x + 4) - 39$

Solution: First, simplify each side of the equation.

$$2 - 6 = -5(x + 4) - 39$$
$$2 - 6 = -5x - 20 - 39 \quad \text{Use the distributive property.}$$
$$-4 = -5x - 59 \quad \text{Combine like terms on each side.}$$
$$-4 + 59 = -5x - 59 + 59 \quad \text{Add 59 to both sides to get the variable term alone.}$$
$$55 = -5x \quad \text{Simplify.}$$
$$\frac{55}{-5} = \frac{-5x}{-5} \quad \text{Divide both sides by } -5.$$
$$-11 = x \quad \text{or} \quad x = -11 \quad \text{Simplify.}$$

Check to see that -11 is the solution.

■ Work Practice 8

Answers

7. 3 **8.** -15

Section 3.2 | Solving Equations: Review of the Addition and Multiplication Properties

Objective C Translating Word Phrases into Expressions

Section 3.4 in this chapter contains a formal introduction to problem solving. To prepare for this section, let's once again review writing phrases as algebraic expressions using the following key words and phrases as a guide:

Addition	Subtraction	Multiplication	Division
sum	difference	product	quotient
plus	minus	times	divided
added to	subtracted from	multiply	shared equally among
more than	less than	twice	per
increased by	decreased by	of	divided by
total	less	double/triple	divided into

Helpful Hint Don't forget: Order matters for subtraction and division, so be careful when translating.

Example 9
Write each phrase as an algebraic expression. Use x to represent "a number."

a. a number increased by -5
b. the product of -7 and a number
c. a number less 20
d. the quotient of -18 and a number
e. a number subtracted from -2

Solution:

a. In words: a number / increased by / -5
   Translate: $x + (-5)$ or $x - 5$

b. In words: the product of / -7 / and / a number
   Translate: $-7 \cdot x$ or $-7x$

c. In words: a number / less / 20
   Translate: $x - 20$

d. In words: the quotient of / -18 / and / a number
   Translate: $-18 \div x$ or $\dfrac{-18}{x}$ or $-\dfrac{18}{x}$

e. In words: a number / subtracted from / -2
   Translate: $-2 - x$

Work Practice 9

Practice 9
Write each phrase as an algebraic expression. Use x to represent "a number."

a. the sum of -3 and a number
b. -5 decreased by a number
c. three times a number
d. a number subtracted from 83
e. the quotient of a number and -4

Answers
9. a. $-3 + x$ b. $-5 - x$ c. $3x$
   d. $83 - x$ e. $\dfrac{x}{-4}$ or $-\dfrac{x}{4}$

Helpful Hint

As we mentioned in Chapter 1, don't forget that order is important when subtracting. Notice the translation order of numbers and variables below.

Phrase	Translation
a number "less" 9	$x - 9$
a number "subtracted from" 9	$9 - x$

Practice 10

Translate each phrase into an algebraic expression. Let x be the unknown number.

a. The product of 5 and a number, decreased by 25

b. Twice the sum of a number and 3

c. The quotient of 39 and twice a number

Example 10

Write each phrase as an algebraic expression. Let x be the unknown number.

a. Twice a number, increased by -9
b. Three times the difference of a number and 11
c. The quotient of 5 times a number and 17

Solution:

a. In words: twice a number   increased by -9

Translate: $2x$ $+$ (-9) or $2x - 9$

b. In words: three times   the difference of   a number   and   11

Translate: 3   (x $-$ 11)   or $3(x - 11)$

c. In words: the quotient of   5 times a number   and   17

Translate: $5x$ $\div$ 17   or $\dfrac{5x}{17}$

■ Work Practice 10

Answers

10. a. $5x - 25$ **b.** $2(x + 3)$
c. $39 \div (2x)$ or $\dfrac{39}{2x}$

Vocabulary, Readiness & Video Check

Use the choices below to fill in each blank.

| equation | multiplication | equivalent | expression |
| solving | addition | simplifying | |

1. The equations $-3x = 51$ and $\dfrac{-3x}{-3} = \dfrac{51}{-3}$ are called _____ equations.

2. The difference between an equation and an expression is that a(n) _____ contains an equal sign, while a(n) _____ does not.

3. The process of writing $-3x + 10x$ as $7x$ is called _____ the expression.

4. For the equation $-5x - 1 = -21$, the process of finding that 4 is the solution is called _____ the equation.

Section 3.2 | Solving Equations: Review of the Addition and Multiplication Properties

5. By the _____ property of equality, $x = -2$ and $x + 7 = -2 + 7$ are equivalent equations.

6. By the _____ property of equality, $y = 8$ and $3 \cdot y = 3 \cdot 8$ are equivalent equations.

Martin-Gay Interactive Videos Watch the section lecture video and answer the following questions.

See Video 3.2

Objective A **7.** When solving Example 1, what must be done before applying the multiplication property of equality?

8. When solving Example 4, what must be done before applying the addition property of equality?

Objective B **9.** In Examples 5 and 6, both the addition and the multiplication properties of equality are used. Which property is used first both times?

Objective C **10.** In Example 7, why are we told to be careful when translating a subtraction phrase?

3.2 Exercise Set MyLab Math

Objective A *Solve each equation. First combine any like terms on each side of the equation. See Examples 1 through 5.*

1. $x - 3 = -1 + 4$
2. $x + 7 = 2 + 3$
3. $-7 + 10 = m - 5$

4. $1 - 8 = n + 2$
5. $2w - 12w = 40$
6. $10y - y = 45$

7. $24 = t + 3t$
8. $100 = 15y + 5y$
9. $2z = 12 - 14$

10. $-3x = 11 - 2$
11. $4 - 10 = \dfrac{z}{-3}$
12. $20 - 22 = \dfrac{z}{-4}$

13. $-3x - 3x = 50 - 2$
14. $5y - 9y = -14 + (-14)$
15. $\dfrac{x}{5} = -26 + 16$

16. $\dfrac{y}{3} = 32 - 52$
17. $7x + 7 - 6x = 10$
18. $-3 + 5x - 4x = 13$

19. $-8 - 9 = 3x + 5 - 2x$
20. $-7 + 10 = 4x - 6 - 3x$

Solve. First multiply to remove parentheses. See Example 6.

21. $2(5x - 3) = 11x$
22. $6(3x + 1) = 19x$
23. $3y = 2(y + 12)$

24. $17x = 4(4x - 6)$
25. $21y = 5(4y - 6)$
26. $28z = 9(3z - 2)$

27. $-3(-4 - 2z) = 7z$
28. $-2(-1 - 3y) = 7y$

Objective B *Solve each equation. See Examples 7 and 8.*

29. $2x - 8 = 0$
30. $3y - 12 = 0$
31. $7y + 3 = 24$
32. $5m + 1 = 46$

33. $-7 = 2x - 1$
34. $-11 = 3t - 2$
35. $6(6 - 4y) = 12y$
36. $4(3y - 5) = 14y$

37. $11(x - 6) = -4 - 7$
38. $5(x - 6) = -2 - 8$
39. $-3(x + 1) - 10 = 12 + 8$

40. $-2(x + 5) - 2 = -8 - 4$
41. $y - 20 = 6y$
42. $x - 63 = 10x$

43. $22 - 42 = 4(x - 1) - 4$
44. $35 - (-3) = 3(x - 2) + 17$

Objectives **A B** Mixed Practice *Solve each equation. See Examples 1 through 8.*

45. $-2 - 3 = -4 + x$
46. $7 - (-10) = x - 5$
47. $y + 1 = -3 + 4$
48. $y - 8 = -5 - 1$
49. $3w - 12w = -27$
50. $y - 6y = 20$
51. $-4x = 20 - (-4)$
52. $6x = 5 - 35$
53. $18 - 11 = \dfrac{x}{-5}$
54. $9 - 14 = \dfrac{x}{-12}$
55. $9x - 12 = 78$
56. $8x - 8 = 32$
57. $10 = 7t - 12t$
58. $-30 = t + 9t$
59. $5 - 5 = 3x + 2x$
60. $-42 + 20 = -2x + 13x$
61. $50y = 7(7y + 4)$
62. $65y = 8(8y - 9)$
63. $8x = 2(6x + 10)$
64. $10x = 6(2x - 3)$
65. $7x + 14 - 6x = -4 - 10$
66. $-10x + 11x + 5 = 9 - 5$
67. $\dfrac{x}{-4} = -1 - (-8)$
68. $\dfrac{y}{-6} = 6 - (-1)$
69. $23x + 8 - 25x = 7 - 9$
70. $8x - 4 - 6x = 12 - 22$
71. $-3(x + 9) - 41 = 4 - 60$
72. $-4(x + 7) - 30 = 3 - 37$

Objective **C** Translating *Write each phrase as a variable expression. Use x to represent "a number." See Examples 9 and 10.*

73. The sum of -7 and a number
74. Negative eight plus a number
75. Eleven subtracted from a number
76. A number subtracted from twelve
77. The product of -13 and a number
78. Twice a number
79. A number divided by -12
80. The quotient of negative six and a number
81. The product of -11 and a number, increased by 5
82. Negative four times a number, increased by 18
83. Negative ten decreased by 7 times a number
84. Twice a number, decreased by thirty
85. Seven added to the product of 4 and a number
86. The product of 7 and a number, added to 100
87. Twice a number, decreased by 17
88. The difference of -9 times a number, and 1
89. The product of -6 and the sum of a number and 15
90. Twice the sum of a number and -5
91. The quotient of 45 and the product of a number and -5
92. The quotient of ten times a number, and -4
93. The quotient of seventeen and a number, increased by -15
94. The quotient of -20 and a number, decreased by three

Review

This horizontal bar graph shows the top ten expensive cities for business travelers in 2016. Use this graph to answer Exercises 95 through 98. See Sections 1.2 and 1.3.

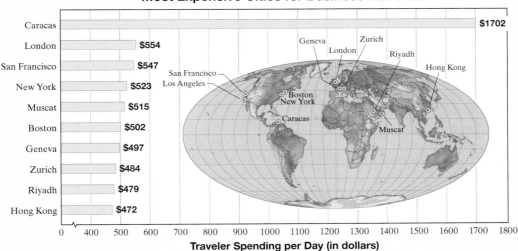

Most Expensive Cities for Business Travelers

City	Spending
Caracas	$1702
London	$554
San Francisco	$547
New York	$523
Muscat	$515
Boston	$502
Geneva	$497
Zurich	$484
Riyadh	$479
Hong Kong	$472

Traveler Spending per Day (in dollars)

Source: Business Travel News, 2016

95. For which city is the daily cost of business travel the highest?

96. For the cities shown, which cities have daily costs less than $510?

97. What is the combined daily cost for a business traveler going to London, Geneva, and Zurich?

98. What is the combined daily cost for a business traveler going to San Francisco, New York, and Boston?

Concept Extensions

99. In your own words, explain the addition property of equality.

100. Write an equation that can be solved using the addition property of equality.

For exercises 101 and 102, see the Concept Check in this section.

101. Are the equations below equivalent? Why or why not?
$$x + 7 = 4 + (-9)$$
$$x + 7 \stackrel{?}{=} 5$$

102. Are the equations below equivalent? Why or why not?
$$3x - 6x = 12$$
$$3x \stackrel{?}{=} 12$$

103. Why does the multiplication property of equality not allow us to divide both sides of an equation by zero?

104. Is the equation $-x = 6$ solved for the variable? Explain why or why not.

Solve.

105. $\dfrac{y}{72} = -86 - (-1029)$

106. $\dfrac{x}{-13} = 4^6 - 5^7$

107. $\dfrac{x}{-2} = 5^2 - |-10| - (-9)$

108. $\dfrac{y}{10} = (-8)^2 - |20| + (-2)^2$

109. $|-13| + 3^2 = 100y - |-20| - 99y$

110. $4(x - 11) + |90| - |-86| + 2^5 = 5x$

Integrated Review Sections 3.1–3.2

Expressions and Equations

Answers

For the table below, identify each as an expression or an equation.

Expression or Equation	
1. $7x - 5y + 14$	
2. $7x = 35 + 14$	
3. $3(x - 2) = 5(x + 1) - 17$	
4. $-9(2x + 1) - 4(x - 2) + 14$	

Fill in each blank with "simplify" or "solve."

5. To _____ an expression, we combine any like terms.

6. To _____ an equation, we use the properties of equality to find any value of the variable that makes the equation a true statement.

Simplify each expression by combining like terms.

7. $7x + x$

8. $6y - 10y$

9. $2a + 5a - 9a - 2$

10. $6a - 12 - a - 14$

Multiply and simplify if possible.

11. $-2(4x + 7)$

12. $-3(2x - 10)$

13. $5(y + 2) - 20$

14. $12x + 3(x - 6) - 13$

△ 15. Find the area of the rectangle.

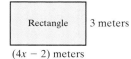

△ 16. Find the perimeter of the triangle.

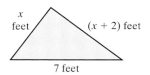

Solve and check.

17. $12 = 11x - 14x$

18. $8y + 7y = -45$

19. $x - 12 = -45 + 23$

20. $6 - (-5) = x + 5$

Solve and check.

21. $\dfrac{x}{3} = -14 + 9$

22. $\dfrac{z}{4} = -23 - 7$

23. $-6 + 2 = 4x + 1 - 3x$

24. $5 - 8 = 5x + 10 - 4x$

25. $6(3x - 4) = 19x$

26. $25x = 6(4x - 9)$

27. $-36x - 10 + 37x = -12 - (-14)$

28. $-8 + (-14) = -80y + 20 + 81y$

29. $3x - 16 = -10$

30. $4x - 21 = -13$

31. $-8z - 2z = 26 - (-4)$

32. $-12 + (-13) = 5x - 10x$

33. $-4(x + 8) - 11 = 3 - 26$

34. $-6(x - 2) + 10 = -4 - 10$

Translating *Write each phrase as an algebraic expression. Use x to represent "a number."*

35. The difference of a number and 10

36. The sum of −20 and a number

37. The product of 10 and a number

38. The quotient of 10 and a number

39. Five added to the product of −2 and a number

40. The product of −4 and the difference of a number and 1

21. _____

22. _____

23. _____

24. _____

25. _____

26. _____

27. _____

28. _____

29. _____

30. _____

31. _____

32. _____

33. _____

34. _____

35. _____

36. _____

37. _____

38. _____

39. _____

40. _____

3.3 Solving Linear Equations in One Variable

Objectives

A Solve Linear Equations Using the Addition and Multiplication Properties.

B Solve Linear Equations Containing Parentheses.

C Write Numerical Sentences as Equations.

In this chapter, the equations we are solving are called **linear equations in one variable** or **first-degree equations in one variable.** For example, an equation such as $5x - 2 = 6x$ is a linear equation in one variable. It is called linear or first degree because the exponent on each x is 1 and there is no variable below a fraction bar. It is an equation in one variable because it contains one variable, x.

Let's continue solving linear equations in one variable.

Objective A Solving Equations Using the Addition and Multiplication Properties

If an equation contains variable terms on both sides, we use the addition property of equality to get all the variable terms on one side and all the constants or numbers on the other side.

Practice 1

Solve: $7x + 12 = 3x - 4$

Example 1 Solve: $3a - 6 = a + 4$

Solution: Although it makes no difference which side we choose, let's move variable terms to the left side and constants to the right side.

$$3a - 6 = a + 4$$
$$3a - 6 + 6 = a + 4 + 6 \quad \text{Add 6 to both sides.}$$
$$3a = a + 10 \quad \text{Simplify.}$$
$$3a - a = a + 10 - a \quad \text{Subtract } a \text{ from both sides.}$$
$$2a = 10 \quad \text{Simplify.}$$
$$\frac{2a}{2} = \frac{10}{2} \quad \text{Divide both sides by 2.}$$
$$a = 5 \quad \text{Simplify.}$$

Check:
$$3a - 6 = a + 4 \quad \text{Original equation}$$
$$3 \cdot 5 - 6 \stackrel{?}{=} 5 + 4 \quad \text{Replace } a \text{ with 5.}$$
$$15 - 6 \stackrel{?}{=} 9 \quad \text{Simplify.}$$
$$9 = 9 \quad \text{True}$$

The solution is 5.

■ Work Practice 1

Helpful Hint

Make sure you understand which property to use to solve an equation.

Addition
$$x + 2 = 10$$

To undo addition of 2, we subtract 2 from both sides.
$$x + 2 - 2 = 10 - 2 \quad \text{Use addition property of equality.}$$
$$x = 8$$
Check: $x + 2 = 10$
$$8 + 2 \stackrel{?}{=} 10$$
$$10 = 10 \quad \text{True}$$

Understood multiplication
$$2x = 10$$

To undo multiplication of 2, we divide both sides by 2.
$$\frac{2x}{2} = \frac{10}{2} \quad \text{Use multiplication property of equality.}$$
$$x = 5$$
Check: $2x = 10$
$$2 \cdot 5 \stackrel{?}{=} 10$$
$$10 = 10 \quad \text{True}$$

Answer
1. -4

Section 3.3 | Solving Linear Equations in One Variable

Example 2 Solve: $17 - 7x + 3 = -3x + 21 - 3x$

Solution: First, simplify both sides of the equation.

$17 - 7x + 3 = -3x + 21 - 3x$
$20 - 7x = -6x + 21$ Simplify.

Next, move variable terms to one side of the equation and constants, or numbers, to the other side. To begin, let's add $6x$ to both sides.

$20 - 7x + 6x = -6x + 21 + 6x$ Add $6x$ to each side.
$20 - x = 21$ Simplify.
$20 - x - 20 = 21 - 20$ Subtract 20 from both sides.
$-1x = 1$ Simplify. Recall that $-x$ means $-1x$.
$\dfrac{-1x}{-1} = \dfrac{1}{-1}$ Divide both sides by -1.
$x = -1$ Simplify.

Check:
$17 - 7x + 3 = -3x + 21 - 3x$
$17 - 7(-1) + 3 \stackrel{?}{=} -3(-1) + 21 - 3(-1)$
$17 + 7 + 3 \stackrel{?}{=} 3 + 21 + 3$
$27 = 27$ True

The solution is -1.

Work Practice 2

Practice 2
Solve:
$40 - 5y + 5 = -2y - 10 - 4y$

Objective B Solving Equations Containing Parentheses

Recall from the previous section that if an equation contains parentheses, we will first use the distributive property to remove them.

Example 3 Solve: $7(x - 2) = 9x - 6$

Solution: First we apply the distributive property.

$7(x - 2) = 9x - 6$
$7x - 14 = 9x - 6$ Apply the distributive property.

Next we move variable terms to one side of the equation and constants to the other side.

$7x - 14 - 9x = 9x - 6 - 9x$ Subtract $9x$ from both sides.
$-2x - 14 = -6$ Simplify.
$-2x - 14 + 14 = -6 + 14$ Add 14 to both sides.
$-2x = 8$ Simplify.
$\dfrac{-2x}{-2} = \dfrac{8}{-2}$ Divide both sides by -2.
$x = -4$ Simplify.

Check to see that -4 is the solution.

Work Practice 3

Practice 3
Solve: $6(a - 5) = 4a + 4$

✓ **Concept Check** In Example 3, the solution is -4. To check this solution, what equation should we use?

Answers
2. -55 **3.** 17

✓ **Concept Check Answer**
$7(x - 2) = 9x - 6$

You may want to use the following steps to solve equations.

Steps for Solving an Equation

Step 1: If parentheses are present, use the distributive property.
Step 2: Combine any like terms on each side of the equation.
Step 3: Use the addition property of equality to rewrite the equation so that variable terms are on one side of the equation and constant terms are on the other side.
Step 4: Use the multiplication property of equality to divide both sides by the numerical coefficient of the variable to solve.
Step 5: Check the solution in the *original equation*.

Practice 4
Solve: $4(x + 3) + 1 = 13$

Example 4 Solve: $3(2x - 6) + 6 = 0$

Solution: $3(2x - 6) + 6 = 0$

Step 1:	$6x - 18 + 6 = 0$	Apply the distributive property.
Step 2:	$6x - 12 = 0$	Combine like terms on the left side of the equation.
Step 3:	$6x - 12 + 12 = 0 + 12$	Add 12 to both sides.
	$6x = 12$	Simplify.
Step 4:	$\dfrac{6x}{6} = \dfrac{12}{6}$	Divide both sides by 6.
	$x = 2$	Simplify.

Check:
Step 5:
$3(2x - 6) + 6 = 0$
$3(2 \cdot 2 - 6) + 6 \stackrel{?}{=} 0$
$3(4 - 6) + 6 \stackrel{?}{=} 0$
$3(-2) + 6 \stackrel{?}{=} 0$
$-6 + 6 \stackrel{?}{=} 0$
$0 = 0$ True

The solution is 2.

■ Work Practice 4

Objective C Writing Numerical Sentences as Equations

Next, we practice translating sentences into equations. Below are key words and phrases that translate to an equal sign. (*Note:* For a review of key words and phrases that translate to addition, subtraction, multiplication, and division, see Sections 1.8 and 3.2.)

Key Words or Phrases	Examples	Symbols
equals	3 equals 2 plus 1	$3 = 2 + 1$
gives	the quotient of 10 and -5 gives -2	$\dfrac{10}{-5} = -2$
is/was	17 minus 12 is 5	$17 - 12 = 5$
yields	11 plus 2 yields 13	$11 + 2 = 13$
amounts to	twice -15 amounts to -30	$2(-15) = -30$
is equal to	-24 is equal to 2 times -12	$-24 = 2(-12)$

Answer
4. 0

Section 3.3 | Solving Linear Equations in One Variable

Example 5
Translate each sentence into an equation.

a. The product of 7 and 6 is 42.
b. Twice the sum of 3 and 5 is equal to 16.
c. The quotient of −45 and 5 yields −9.

Solution:

a. In words: the product of 7 and 6   is   42

Translate: $7 \cdot 6$ $=$ 42

b. In words: twice   the sum of 3 and 5   is equal to   16

Translate: 2 $(3 + 5)$ $=$ 16

c. In words: the quotient of −45 and 5   yields   −9

Translate: $\dfrac{-45}{5}$ $=$ -9

■ Work Practice 5

Practice 5
Translate each sentence into an equation.

a. The difference of 110 and 80 is 30.
b. The product of 3 and the sum of −9 and 11 amounts to 6.
c. The quotient of 24 and −6 yields −4.

Answers

5 a. $110 - 80 = 30$
b. $3(-9 + 11) = 6$ **c.** $\dfrac{24}{-6} = -4$

Calculator Explorations — Checking Equations

A calculator can be used to check possible solutions of equations. To do this, replace the variable by the possible solution and evaluate each side of the equation separately. For example, to see whether 7 is a solution of the equation $52x = 15x + 259$, replace x with 7 and use your calculator to evaluate each side separately.

Equation: $52x = 15x + 259$

$52 \cdot 7 \stackrel{?}{=} 15 \cdot 7 + 259$ Replace x with 7.

Evaluate left side: $\boxed{52}$ $\boxed{\times}$ $\boxed{7}$ then $\boxed{=}$ or $\boxed{\text{ENTER}}$.
Display: $\boxed{364}$.
Evaluate right side: $\boxed{15}$ $\boxed{\times}$ $\boxed{7}$ $\boxed{+}$ $\boxed{259}$ then $\boxed{=}$ or $\boxed{\text{ENTER}}$.
Display: $\boxed{364}$.

Since the left side equals the right side, 7 is a solution of the equation $52x = 15x + 259$.

Use a calculator to determine whether the numbers given are solutions of each equation.

1. $76(x - 25) = -988$;   12
2. $-47x + 862 = -783$;   35
3. $x + 562 = 3x + 900$;   −170
4. $55(x + 10) = 75x + 910$;   −18
5. $29x - 1034 = 61x - 362$;   −21
6. $-38x + 205 = 25x + 120$;   25

Vocabulary, Readiness & Video Check

Use the choices below to fill in each blank. Some choices may be used more than once.

addition	multiplication	combine like terms
$5(2x + 6) - 1 = 39$	$3x - 9 + x - 16$	distributive

1. An example of an expression is _____ while an example of an equation is _____.
2. To solve $\dfrac{x}{-7} = -10$, we use the _____ property of equality.
3. To solve $x - 7 = -10$, we use the _____ property of equality.

Use the order of the Steps for Solving an Equation in this section to answer Exercises 4 through 6.

4. To solve $9x - 6x = 10 + 6$, first _____.
5. To solve $5(x - 1) = 25$, first use the _____ property.
6. To solve $4x + 3 = 19$, first use the _____ property of equality.

Martin-Gay Interactive Videos Watch the section lecture video and answer the following questions.

See Video 3.3

Objective A 7. In Example 1, the number 1 is subtracted from the left side of the equation. What property tells us we must also subtract 1 from the right side? Why is it important to do the same thing to both sides?

Objective B 8. From Example 3, what is the first step when solving an equation that contains parentheses? What property do we use to perform this step?

Objective C 9. What word or phrase translates to "equals" in Example 5? In Example 6?

3.3 Exercise Set MyLab Math

Objective A *Solve each equation. See Examples 1 and 2.*

1. $3x - 7 = 4x + 5$
2. $7x - 1 = 8x + 4$
3. $10x + 15 = 6x + 3$
4. $5x - 3 = 2x - 18$
5. $19 - 3x = 14 + 2x$
6. $4 - 7m = -3m + 4$
7. $-14x - 20 = -12x + 70$
8. $57y + 140 = 54y - 100$
9. $x + 20 + 2x = -10 - 2x - 15$
10. $2x + 10 + 3x = -12 - x - 20$
11. $40 + 4y - 16 = 13y - 12 - 3y$
12. $19x - 2 - 7x = 31 + 6x - 15$

Objective B *Solve each equation. See Examples 3 and 4.*

13. $35 - 17 = 3(x - 2)$
14. $22 - 42 = 4(x - 1)$
15. $3(x - 1) - 12 = 0$
16. $2(x + 5) + 8 = 0$
17. $2(y - 3) = y - 6$
18. $3(z + 2) = 5z + 6$
19. $-2(y + 4) = 2$
20. $-1(y + 3) = 10$
21. $2t - 1 = 3(t + 7)$
22. $-4 + 3c = 4(c + 2)$
23. $3(5c + 1) - 12 = 13c + 3$
24. $4(3t + 4) - 20 = 3 + 5t$

Mixed Practice (*Sections 2.6, 3.2, and 3.3*) *Solve each equation.*

25. $-4x = 44$
26. $-3x = 51$
27. $x + 9 = 2$
28. $y - 6 = -11$
29. $8 - b = 13$
30. $7 - z = 15$

31. $-20 - (-50) = \dfrac{x}{9}$

32. $-2 - 10 = \dfrac{z}{10}$

33. $3r + 4 = 19$

34. $7y + 3 = 38$

⬤ 35. $-7c + 1 = -20$

36. $-2b + 5 = -7$

37. $8y - 13y = -20 - 25$

38. $4x - 11x = -14 - 14$

39. $6(7x - 1) = 43x$

40. $5(3y - 2) = 16y$

41. $-4 + 12 = 16x - 3 - 15x$

42. $-9 + 20 = 19x - 4 - 18x$

43. $-10(x + 3) + 28 = -16 - 16$

44. $-9(x + 2) + 25 = -19 - 19$

45. $4x + 3 = 2x + 11$

46. $6y - 8 = 3y + 7$

47. $-2y - 10 = 5y + 18$

48. $7n + 5 = 12n - 10$

49. $-8n + 1 = -6n - 5$

50. $10w + 8 = w - 10$

51. $9 - 3x = 14 + 2x$

52. $4 - 7m = -3m$

53. $9a + 29 + 7 = 0$

54. $10 + 4v + 6 = 0$

55. $7(y - 2) = 4y - 29$

56. $2(z - 2) = 5z + 17$

57. $12 + 5t = 6(t + 2)$

58. $4 + 3c = 2(c + 2)$

⬤ 59. $3(5c - 1) - 2 = 13c + 3$

60. $4(2t + 5) - 21 = 7t - 6$

61. $10 + 5(z - 2) = 4z + 1$

62. $14 + 4(w - 5) = 6 - 2w$

63. $7(6 + w) = 6(2 + w)$

64. $6(5 + c) = 5(c - 4)$

Objective C Translating *Write each sentence as an equation. See Example 5.*

65. The sum of -42 and 16 is -26.

66. The difference of -30 and 10 equals -40.

67. The product of -5 and -29 gives 145.

68. The quotient of -16 and 2 yields -8.

69. Three times the difference of -14 and 2 amounts to -48.

70. Negative 2 times the sum of 3 and 12 is -30.

71. The quotient of 100 and twice 50 is equal to 1.

72. Seventeen subtracted from -12 equals -29.

Review

The following bar graph shows the number of U.S. federal individual income tax returns that are filed electronically during the years shown using efile. Use this graph to answer Exercises 73 through 76. Write number answers in standard form. See Sections 1.2 and 1.3.

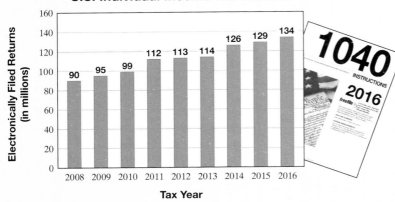

73. Determine the number of electronically filed returns predicted for 2016.

74. Determine the number of electronically filed returns for 2012.

75. By how much did the number of electronically filed returns increase from 2008 to 2015?

76. Describe any trends shown in this graph.

Evaluate each expression for $x = 3$, $y = -1$, and $z = 0$. See Section 2.5.

77. $x^3 - 2xy$ **78.** $y^3 + 3xyz$ **79.** $y^5 - 4x^2$ **80.** $(-y)^3 + 3xyz$

Concept Extensions

Using the Steps for Solving an Equation, choose the next operation for solving the given equation.

81. $2x - 5 = -7$
  a. Add 7 to both sides.
  b. Add 5 to both sides.
  c. Divide both sides by 2.

82. $3x + 2x = -x - 4$
  a. Add 4 to both sides.
  b. Subtract $2x$ from both sides.
  c. Add $3x$ and $2x$.

83. $-3x = -12$
  a. Divide both sides by -3.
  b. Add 12 to both sides.
  c. Add $3x$ to both sides.

84. $9 - 5x = 15$
  a. Divide both sides by -5.
  b. Subtract 15 from both sides.
  c. Subtract 9 from both sides.

A classmate shows you steps for solving an equation. The solution does not check, but the classmate is unable to find the error. For each set of steps, check the solution, find the error, and correct it.

85.
$$2(3x - 5) = 5x - 7$$
$$6x - 5 = 5x - 7$$
$$6x - 5 + 5 = 5x - 7 + 5$$
$$6x = 5x - 2$$
$$6x - 5x = 5x - 2 - 5x$$
$$x = -2$$

86.
$$37x + 1 = 9(4x - 7)$$
$$37x + 1 = 36x - 7$$
$$37x + 1 - 1 = 36x - 7 - 1$$
$$37x = 36x - 8$$
$$37x - 36x = 36x - 8 - 36x$$
$$x = -8$$

Solve.

87. $(-8)^2 + 3x = 5x + 4^3$ **88.** $3^2 \cdot x = (-9)^3$ **89.** $2^3(x + 4) = 3^2(x + 4)$ **90.** $x + 45^2 = 54^2$

91. A classmate tries to solve $3x = 39$ by subtracting 3 from both sides of the equation. Will this step solve the equation for x? Why or why not?

92. A classmate tries to solve $2 + x = 20$ by dividing both sides by 2. Will this step solve the equation for x? Why or why not?

3.4 Linear Equations in One Variable and Problem Solving

Objective A Writing Sentences as Equations

Now that we have practiced solving equations for a variable, we can extend our problem-solving skills considerably. We begin by writing sentences as equations using the following key words and phrases as a guide:

Addition	Subtraction	Multiplication	Division	Equal Sign
sum	difference	product	quotient	equals
plus	minus	times	divide	gives
added to	subtracted from	multiply	shared equally among	is/was
more than	less than	twice	per	yields
increased by	decreased by	of	divided by	amounts to
total	less	double/triple	divided into	is equal to

Objectives
A Write Sentences as Equations.
B Use Problem-Solving Steps to Solve Problems.

Example 1 Write each sentence as an equation. Use x to represent "a number."

a. Twenty increased by a number is 5.
b. Twice a number equals -10.
c. A number minus 11 amounts to 168.
d. Three times the sum of a number and 5 is -30.
e. The quotient of twice a number and 8 is equal to 2.

Solution:

a. In words: twenty | increased by | a number | is | 5
   Translate: $20 + x = 5$

b. In words: twice a number | equals | -10
   Translate: $2x = -10$

c. In words: a number | minus | 11 | amounts to | 168
   Translate: $x - 11 = 168$

d. In words: three times | the sum of a number and 5 | is | -30
   Translate: $3(x + 5) = -30$

(*Continued on next page*)

Practice 1
Write each sentence as an equation. Use x to represent "a number."

a. Four times a number is 20.
b. The sum of a number and -5 yields 32.
c. Fifteen subtracted from a number amounts to -23.
d. Five times the difference of a number and 7 is equal to -8.
e. The quotient of triple a number and 5 gives 1.

Answers
1. a. $4x = 20$ b. $x + (-5) = 32$
   c. $x - 15 = -23$
   d. $5(x - 7) = -8$
   e. $\dfrac{3x}{5} = 1$

e. In words: the quotient of twice a number and 8 is equal to 2

Translate: $2x \div 8 = 2$

or $\dfrac{2x}{8} = 2$

■ Work Practice 1

Objective B Using Problem-Solving Steps to Solve Problems

Our main purpose for studying arithmetic and algebra is to solve problems. In previous sections, we have prepared for problem solving by writing phrases as algebraic expressions and sentences as equations. We now draw upon this experience as we solve problems. The following problem-solving steps will be used throughout this text.

Problem-Solving Steps

1. UNDERSTAND the problem. During this step, become comfortable with the problem. Some ways of doing this are as follows:
   - Read and reread the problem.
   - Construct a drawing.
   - Propose a solution and check. Pay careful attention to how you check your proposed solution. This will help when writing an equation to model the problem.
   - Choose a variable to represent the unknown. Use this variable to represent any other unknowns.
2. TRANSLATE the problem into an equation.
3. SOLVE the equation.
4. INTERPRET the results: *Check* the proposed solution in the stated problem and *state* your conclusion.

The first problem that we solve consists of finding an unknown number.

Example 2 Finding an Unknown Number

Twice a number plus 3 is the same as the number minus 6. Find the unknown number.

Solution:

1. UNDERSTAND the problem. To do so, we read and reread the problem.
   Let's propose a solution to help us understand. Suppose the unknown number is 5. Twice this number plus 3 is $2 \cdot 5 + 3$ or 13. Is this the same as the number minus 6, or $5 - 6$, or -1? Since 13 is not the same as -1, we know that 5 is not the solution. However, remember that the purpose of proposing a solution is not to guess correctly, but to better understand the problem.
   Now let's choose a variable to represent the unknown. Let's let

   $x =$ unknown number

Practice 2

Translate "the sum of a number and 2 equals 6 added to three times the number" into an equation and solve.

Answer
2. -2

2. **TRANSLATE** the problem into an equation.

In words: twice a number | plus 3 | is the same as | the number minus 6
↓ | ↓ | ↓ | ↓
Translate: $2x$ | $+3$ | $=$ | $x - 6$

3. **SOLVE** the equation. To solve the equation, we first subtract x from both sides.

$$2x + 3 = x - 6$$
$$2x + 3 - x = x - 6 - x$$
$$x + 3 = -6 \quad \text{Simplify.}$$
$$x + 3 - 3 = -6 - 3 \quad \text{Subtract 3 from both sides.}$$
$$x = -9 \quad \text{Simplify.}$$

4. **INTERPRET** the results. First, *check* the proposed solution in the stated problem. Twice "-9" is -18 and $-18 + 3$ is -15. This is equal to the number minus 6, or "-9" $- 6$, or -15. Then *state* your conclusion: The unknown number is -9.

■ Work Practice 2

✓**Concept Check** Suppose you have solved an equation involving perimeter to find the length of a rectangular table. Explain why you would want to recheck your math if you obtain the result of -5.

Example 3 Determining Distances

The distance by road from Chicago, Illinois, to Los Angeles, California, is 1091 miles *more* than the distance from Chicago to Boston, Massachusetts. If the total of these two distances is 3017 miles, find the distance from Chicago to Boston. (*Source: World Almanac*)

Solution:

1. **UNDERSTAND** the problem. We read and reread the problem.

   Let's propose and check a solution to help us better understand the problem. Suppose the distance from Chicago to Boston is 600 miles. Since the distance from Chicago to Los Angeles is 1091 miles *more*, then this distance is $600 + 1091 = 1691$ miles. With these numbers, the total of the distances is $600 + 1691 = 2291$ miles. This is less than the given total of 3017 miles, so we are incorrect. But not only do we have a better understanding of this exercise, we also know that the distance from Boston to Chicago is greater than 600 miles since this proposed solution led to a total too small. Now let's choose a variable to represent an unknown. Then we'll use this variable to represent any other unknown quantities. Let

   x = distance from Chicago to Boston

   Then

   $x + 1091$ = distance from Chicago to Los Angeles

   Since that distance is 1091 miles more.

2. **TRANSLATE** the problem into an equation.

In words: Chicago to Boston distance | + | Chicago to Los Angeles distance | = | total miles
↓ | | ↓ | | ↓
Translate: x | $+$ | $x + 1091$ | $=$ | 3017

(*Continued on next page*)

Practice 3

The distance by road from Cincinnati, Ohio, to Denver, Colorado, is 71 miles *less* than the distance from Denver to San Francisco, California. If the total of these two distances is 2399 miles, find the distance from Denver to San Francisco.

Answer
3. 1235 miles

✓**Concept Check Answer**
Length cannot be negative.

3. SOLVE the equation:

$$x + x + 1091 = 3017$$
$$2x + 1091 = 3017 \quad \text{Combine like terms.}$$
$$2x + 1091 - 1091 = 3017 - 1091 \quad \text{Subtract 1091 from both sides.}$$
$$2x = 1926 \quad \text{Simplify.}$$
$$\frac{2x}{2} = \frac{1926}{2} \quad \text{Divide both sides by 2.}$$
$$x = 963 \quad \text{Simplify.}$$

4. INTERPRET the results. First *check* the proposed solution in the stated problem. Since x represents the distance from Chicago to Boston, this is 963 miles. The distance from Chicago to Los Angeles is $x + 1091 = 963 + 1091 = 2054$ miles. To check, notice that the total number of miles is $963 + 2054 = 3017$ miles, the given total of miles. Also, 2054 is 1091 more miles than 963, so the solution checks. Then, *state* your conclusion: The distance from Chicago to Boston is 963 miles.

■ Work Practice 3

Practice 4

A woman's $57,000 estate is to be divided so that her husband receives twice as much as her son. How much will each receive?

Example 4 Calculating Separate Costs

A salesperson at an electronics store sold a computer system and software for $2100, receiving four times as much money for the computer system as for the software. Find the price of each.

Solution:

1. UNDERSTAND the problem. We read and reread the problem. Then we choose a variable to represent an unknown. We use this variable to represent any other unknown quantities. We let

   x = the software price

   Then

   $4x$ = the computer system price

2. TRANSLATE the problem into an equation.

In words:	software price	and	computer price	is	2100
	↓	↓	↓	↓	↓
Translate:	x	$+$	$4x$	$=$	2100

3. SOLVE the equation:

$$x + 4x = 2100$$
$$5x = 2100 \quad \text{Combine like terms.}$$
$$\frac{5x}{5} = \frac{2100}{5} \quad \text{Divide both sides by 5.}$$
$$x = 420 \quad \text{Simplify.}$$

4. INTERPRET the results. *Check* the proposed solution in the stated problem. The software sold for $420. The computer system sold for $4x = 4(\$420) = \$1680$. Since $420 + 1680 = 2100$, the total price, and $1680 is four times $420, the solution checks. *State* your conclusion: The software sold for $420, and the computer system sold for $1680.

■ Work Practice 4

Answer
4. husband: $38,000; son: $19,000

Vocabulary, Readiness & Video Check

Martin-Gay Interactive Videos Watch the section lecture video and answer the following questions.

See Video 3.4

Objective A 1. In Example 2, why does the left side of the equation translate to $-20 - x$ and not $x - (-20)$?

Objective B 2. Why are parentheses used in the translation of the left side of the equation in Example 4?

3. In Example 5, the solution to the equation is $x = 37$. Why is this not the solution to the application?

3.4 Exercise Set MyLab Math

Objective A Translating *Write each sentence as an equation. Use x to represent "a number." See Example 1.*

1. A number added to -5 is -7.

2. Five subtracted from a number equals 10.

3. Three times a number yields 27.

4. The quotient of 8 and a number is -2.

5. A number subtracted from -20 amounts to 104.

6. Two added to twice a number gives -14.

7. Twice a number gives 108.

8. Five times a number is equal to -75.

9. The product of 5 and the sum of -3 and a number is -20.

10. Twice the sum of -17 and a number is -14.

Objective B Translating *Translate each sentence into an equation. Then solve the equation. See Example 2.*

11. Three times a number, added to 9, is 33. Find the number.

12. Twice a number, subtracted from 60, is 20. Find the number.

13. The sum of 3, 4, and a number amounts to 16. Find the number.

14. The sum of 7, 9, and a number is 40. Find the number.

15. The difference of a number and 3 is equal to the quotient of 10 and 5. Find the number.

16. Eight decreased by a number equals the quotient of 15 and 5. Find the number.

17. Thirty less a number is equal to the product of 3 and the sum of the number and 6. Find the number.

18. The product of a number and 3 is twice the sum of that number and 5. Find the number.

19. 40 subtracted from five times a number is 8 more than the number. Find the number.

20. Five times the sum of a number and 2 is 11 less than the number times 8. Find the number.

21. Three times the difference of some number and 5 amounts to the quotient of 108 and 12. Find the number.

22. Seven times the difference of some number and 1 gives the quotient of 70 and 10. Find the number.

23. The product of 4 and a number is the same as 30 less twice that same number. Find the number.

24. Twice a number equals 25 less triple that same number. Find the number.

Solve. For Exercises 25 and 26, the solutions have been started for you. See Examples 3 and 4.

25. Florida has 26 fewer electoral votes for president than California. If the total number of electoral votes for these two states is 84, find the number for each state. (*Source:* state.1Keydata.com)

    Start the solution:
    1. UNDERSTAND the problem. Reread it as many times as needed. Let's let
       x = number of electoral votes for California
       Then
       $x - 26$ = number of electoral votes for Florida
    2. TRANSLATE into an equation. (Fill in the blanks below.)

       votes for California + votes for Florida = 84

       _____ + _____ = 84

       Now, you finish with
    3. SOLVE the equation.
    4. INTERPRET the results.

26. Ohio has twice the number of electoral votes for president as Alabama. If the total number of electoral votes for these two states is 27, find the number for each state. (*Source:* state.1Keydata.com)

    Start the solution:
    1. UNDERSTAND the problem. Reread it as many times as needed. Let's let
       x = number of electoral votes for Alabama
       Then
       $2x$ = number of electoral votes for Ohio
    2. TRANSLATE into an equation. (Fill in the blanks below.)

       votes for Alabama + votes for Ohio = 27

       _____ + _____ = 27

       Now, you finish with
    3. SOLVE the equation.
    4. INTERPRET the results.

27. A falcon, when diving, can travel five times as fast as a pheasant's top speed. If the total speed for these two birds is 222 miles per hour, find the fastest speed of the falcon and the fastest speed of the pheasant.
    (*Source: Fantastic Book of Comparisons*)

28. Norway has had three times as many rulers as Liechtenstein. If the total number of rulers for both countries is 56, find the number of rulers for Norway and the number for Liechtenstein.

29. The U.S. newspaper with the greatest circulation is *USA Today,* followed by *The Wall Street Journal*. If the average daily circulation for *USA Today* is 226 thousand more than the average daily circulation of *The Wall Street Journal* and their combined circulation is 4340 thousand, find the circulation for each paper. (*Source: USA Today*)

30. The average life expectancy for an elephant is 24 years longer than the life expectancy for a chimpanzee. If the total of these life expectancies is 130 years, find the life expectancy of each.

31. Apple was recently named the Most Valuable Global Brand by *Forbes*. It has a recognition value worth $71 billion more than Google's. If the total value of these two global brands is $237 billion, find the value of each of the brands.

32. The global brand Microsoft has a recognition value worth $16 billion more than Coca-Cola's. If the total value of these two global brands is $134 billion, find the value of each of the brands.

33. An Xbox 360 game system and several games are sold for $440. The cost of the Xbox 360 is 3 times as much as the cost of the games. Find the cost of the Xbox 360 and the cost of the games.

34. The two top-selling Xbox 360 games are *Kinect Adventures* and *Grand Theft Auto V*. A price for *Grand Theft Auto V* is $24 more than a price for *Kinect Adventures*. If the total of these two prices is $36, find the price of each game. (*Source:* Gamestop.com)

35. By air, the distance from New York City to London is 2001 miles *less* than the distance from Los Angeles to Tokyo. If the total of these two distances is 8939 miles, find the distance from Los Angeles to Tokyo.

36. By air, the distance from Melbourne, Australia, to Cairo, Egypt, is 2338 miles *more* than the distance from Madrid, Spain, to Bangkok, Thailand. If the total of these distances is 15,012 miles, find the distance from Madrid to Bangkok.

37. The two NCAA stadiums with the largest capacities are Beaver Stadium (Penn State) and Michigan Stadium (Univ. of Michigan). Beaver Stadium has a capacity of 1081 more than Michigan Stadium. If the combined capacity for the two stadiums is 213,483, find the capacity for each stadium. (*Source:* National Collegiate Athletic Association)

38. A National Hot Rod Association (NHRA) top fuel dragster has a top speed of 95 mph faster than an Indy Racing League car. If the top speed for these two cars combined is 565 mph, find the top speed of each car. (*Source:* USA Today)

39. In 2020, China is projected to be the country with the greatest number of visiting tourists. This number is twice the number of tourists projected for Spain. If the total number of tourists for these two countries is projected to be 210 million, find the number projected for each. (*Source: The State of the World Atlas* by Dan Smith)

40. As of 2015, California contained the largest state population of Native Americans. This population was five times the Native American population of Washington State. If the total of these two populations is 792,000, find the Native American population in each of these two states. (*Source:* U.S. Census Bureau)

41. During 2015, twice as many motor vehicles were manufactured in the United States as in Germany. If the total number of vehicles manufactured in these two countries was approximately 18,300,000, find the number manufactured in the United States and the number manufactured in Germany. (*Source:* Organization of Motor Vehicle Manufacturers)

42. A Toyota Camry is traveling twice as fast as a Dodge truck. If their combined speed is 105 miles per hour, find the speed of the car and find the speed of the truck.

43. A biker sold his used mountain bike and accessories for $270. If he received five times as much money for the bike as he did for the accessories, find how much money he received for the bike.

44. A tractor and a plow attachment are worth $1200. The tractor is worth seven times as much money as the plow. Find the value of the tractor and the value of the plow.

45. The maximum length of a soccer field for Olympic play is 50 yards longer than the maximum length of a soccer field for play by under-10-year-olds. If the total length of these two categories of soccer fields is 190 yards, what is the maximum length for each field? (See the Chapter 3 Opener.)

46. The maximum length recommended for a high school soccer field is 80 yards longer than that for an under-6-year-old team. If the total length of these two categories of soccer fields is 140 yards, what is the maximum length for each field? (See the Chapter 3 Opener.)

47. During the 2016 men's NCAA Division 1 basketball championship game, the North Carolina Tar Heels scored 3 points fewer than the Villanova Wildcats. Together, both teams scored 151 points. How many points did the 2016 champion Villanova Wildcats score during the game? (*Source:* National Collegiate Athletic Association)

48. During the 2016 women's NCAA Division 1 basketball championship game, the Connecticut Huskies scored 31 points more than the Syracuse Orange. Together, both teams scored a total of 133 points. How many points did the 2016 champion Connecticut Huskies score during this game?

Advertisers in the United States use many media: newspapers, magazines, mailings, Internet, and TV. See Exercises 49 and 50.

49. By 2017, the money spent on Internet advertising in the United States is projected to exceed the amount spent on TV advertising. If it is estimated that 0.6 billion dollars more will be spent on Internet advertising in 2017 and the total amount spent on Internet and TV advertising is expected to be $150 billion, find the amount of money projected to be spent on each medium. (*Source:* USA Today)

50. By 2020, the money spent on Internet advertising in the United States is projected to continue to exceed the amount spent on TV advertising. If it is estimated that 11.8 billion dollars more will be spent on Internet advertising in 2020 and the total amount spent on Internet and TV advertising is expected to be about $175 billion, find the amount of money projected to be spent on each medium. (*Source:* USA Today)

Review

Round each number to the given place value. See Section 1.4.

51. 586 to the nearest ten

52. 82 to the nearest ten

53. 1026 to the nearest hundred

54. 52,333 to the nearest thousand

55. 2986 to the nearest thousand

56. 101,552 to the nearest hundred

Concept Extensions

57. Solve Example 3 again, but this time let x be the distance from Chicago to Los Angeles. Did you get the same results? Explain why or why not.

58. Solve Exercise 25 again, but this time let x be the number of electoral votes for Florida. Did you get the same results? Explain why or why not.

In real estate, a house's selling price P is found by adding the real estate agent's commission C to the amount A that the seller of the house receives: $P = A + C$.

59. A house sold for $230,000. The owner's real estate agent received a commission of $13,800. How much did the seller receive? (*Hint:* Substitute the known values into the equation, then solve the equation for the remaining unknown.)

60. A homeowner plans to use a real estate agent to sell his house. He hopes to sell the house for $165,000 and keep $156,750 of that. If everything goes as he has planned, how much will his real estate agent receive as a commission?

In retailing, the retail price P of an item can be computed using the equation $P = C + M$, where C is the wholesale cost of the item and M is the amount of markup.

61. The retail price of a computer system is $999 after a markup of $450. What is the wholesale cost of the computer system? (*Hint:* Substitute the known values into the equation, then solve the equation for the remaining unknown.)

62. Slidell Feed and Seed sells a bag of cat food for $12. If the store paid $7 for the cat food, what is the markup on the cat food?

Chapter 3 Group Activity

Modeling Equation Solving with Addition and Subtraction

Sections 3.1–3.4

We can use positive counters ● and negative counters ● to help us model the equation-solving process. We also need to use an object that represents a variable. We use small slips of paper with the variable name written on them.

Recall that taking a ● and ● together creates a neutral or zero pair. After a neutral pair has been formed, it can be removed from or added to an equation model without changing the overall value. We also need to remember that we can add or remove the same number of positive or negative counters from both sides of an equation without changing the overall value.

We can represent the equation $x + 5 = 2$ as follows:

To get the variable by itself, we must remove 5 black counters from both sides of the model. Because there are only 2 counters on the right side, we must add 5 negative counters to both sides of the model. Then we can remove neutral pairs: 5 from the left side and 2 from the right side.

We are left with the following model, which represents the solution, $x = -3$.

Similarly, we can represent the equation $x - 4 = -6$ as follows:

To get the variable by itself, we must remove 4 red counters from both sides of the model.

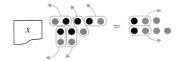

We are left with the following model, which represents the solution, $x = -2$.

Use the counter model to solve each equation.

1. $x - 3 = -7$
2. $x - 1 = -9$
3. $x + 2 = 8$
4. $x + 4 = 5$
5. $x + 8 = 3$
6. $x - 5 = -1$
7. $x - 2 = 1$
8. $x - 5 = 10$
9. $x + 3 = -7$
10. $x + 8 = -2$

Chapter 3 Vocabulary Check

Fill in each blank with one of the words or phrases listed below.

variable	addition	constant	algebraic expression	equation
terms	simplified	multiplication	evaluating the expression	solution
like	combined	numerical coefficient	distributive	

1. An algebraic expression is _____ when all like terms have been _____.
2. Terms that are exactly the same, except that they may have different numerical coefficients, are called _____ terms.
3. A letter used to represent a number is called a(n) _____.
4. A combination of operations on variables and numbers is called a(n) _____.
5. The addends of an algebraic expression are called the _____ of the expression.
6. The number factor of a variable term is called the _____.
7. Replacing a variable in an expression by a number and then finding the value of the expression is called _____ for the variable.
8. A term that is a number only is called a(n) _____.
9. A(n) _____ is of the form expression = expression.
10. A(n) _____ of an equation is a value for the variable that makes the equation a true statement.

11. To multiply $-3(2x + 1)$, we use the _____ property.

12. By the _____ property of equality, we may multiply or divide both sides of an equation by any nonzero number without changing the solution of the equation.

13. By the _____ property of equality, the same number may be added to or subtracted from both sides of an equation without changing the solution of the equation.

> **Helpful Hint**
> ▶ Are you preparing for your test? To help, don't forget to take these:
> - Chapter 3 Getting Ready for the Test on page 209
> - Chapter 3 Test on page 210
>
> Then check all of your answers at the back of this text. For further review, the step-by-step video solutions to any of these exercises are located in MyLab Math.

3 Chapter Highlights

Definitions and Concepts	Examples
Section 3.1 Simplifying Algebraic Expressions	
The addends of an algebraic expression are called the **terms** of the expression.	$5x^2 + (-4x) + (-2)$ — 3 terms
The number factor of a variable term is called the **numerical coefficient.**	Term Numerical Coefficient $7x$ 7 $-6y$ -6 x or $1x$ 1
Terms that are exactly the same, except that they may have different numerical coefficients, are called **like terms.**	$5x + 11x = (5 + 11)x = 16x$ like terms $y - 6y = (1 - 6)y = -5y$
An algebraic expression is **simplified** when all like terms have been **combined.**	
Use the **distributive property** to multiply an algebraic expression by a term.	Simplify: $-4(x + 2) + 3(5x - 7)$ $= -4 \cdot x + (-4) \cdot 2 + 3 \cdot 5x - 3 \cdot 7$ $= -4x + (-8) + 15x + (-21)$ $= 11x + (-29)$ or $11x - 29$
Section 3.2 Solving Equations: Review of the Addition and Multiplication Properties	
Addition Property of Equality Let a, b, and c represent numbers. Then $a = b$ and $a + c = b + c$ are equivalent equations. Also, $a = b$ and $a - c = b - c$ are equivalent equations. In other words, the same number may be added to or subtracted from both sides of an equation without changing the solution of the equation.	Solve for x: $x + 8 = 2 + (-1)$ $x + 8 = 1$ Combine like terms. $x + 8 - 8 = 1 - 8$ Subtract 8 from both sides. $x = -7$ Simplify. The solution is -7.

(continued)

Definitions and Concepts	Examples

Section 3.2 Solving Equations: Review of the Addition and Multiplication Properties *(continued)*

Multiplication Property of Equality Let a, b, and c represent numbers and let $c \neq 0$. Then $\quad a = b \qquad\qquad$ Also, $a = b$ and $a \cdot c = b \cdot c \qquad$ and $\dfrac{a}{c} = \dfrac{b}{c}$ are equivalent equations. $\quad$ are equivalent equations. In other words, both sides of an equation may be multiplied or divided by the same nonzero number without changing the solution of the equation.	Solve: $\;y - 7y = 30$ $\qquad\quad -6y = 30 \qquad$ Combine like terms. $\qquad\quad \dfrac{-6y}{-6} = \dfrac{30}{-6} \qquad$ Divide both sides by -6. $\qquad\qquad y = -5 \qquad$ Simplify. The solution is -5.

Section 3.3 Solving Linear Equations in One Variable

Steps for Solving an Equation **Step 1:** If parentheses are present, use the distributive property. **Step 2:** Combine any like terms on each side of the equation. **Step 3:** Use the addition property of equality to rewrite the equation so that variable terms are on one side of the equation and constant terms are on the other side. **Step 4:** Use the multiplication property of equality to divide both sides by the numerical coefficient of the variable to solve. **Step 5:** Check the solution in the *original equation*.	Solve for x: $5(3x - 1) + 15 = -5$ **Step 1:** $15x - 5 + 15 = -5 \qquad$ Apply the distributive property. **Step 2:** $\quad\;\; 15x + 10 = -5 \qquad$ Combine like terms. **Step 3:** $15x + 10 - 10 = -5 - 10 \quad$ Subtract 10 from both sides. $\qquad\qquad\quad 15x = -15$ **Step 4:** $\qquad\quad \dfrac{15x}{15} = \dfrac{-15}{15} \qquad$ Divide both sides by 15. $\qquad\qquad\qquad x = -1$ **Step 5:** Check to see that -1 is the solution.

Section 3.4 Linear Equations in One Variable and Problem Solving

Problem-Solving Steps 1. UNDERSTAND the problem. Some ways of doing this are $\quad$ Read and reread the problem. $\quad$ Construct a drawing. $\quad$ Choose a variable to represent an unknown in the problem. 2. TRANSLATE the problem into an equation.	The incubation period for a golden eagle is three times the incubation period for a hummingbird. If the total of their incubation periods is 60 days, find the incubation period for each bird. (*Source: Wildlife Fact File*, International Masters Publishers) 1. UNDERSTAND the problem. Then choose a variable to represent an unknown. Let $\quad x =$ incubation period of a hummingbird Then $\quad 3x =$ incubation period of a golden eagle 2. TRANSLATE. $\quad$ incubation of hummingbird $+$ incubation of golden eagle $\;$ is $\;$ 60 $\qquad\qquad\quad \downarrow \qquad\qquad\qquad\quad \downarrow \qquad\quad \downarrow \;\; \downarrow$ $\qquad\qquad\quad x \qquad + \qquad\qquad 3x \qquad = \;\; 60$

Definitions and Concepts	Examples
Section 3.4 Linear Equations in One Variable and Problem Solving *(continued)*	
3. SOLVE the equation.	3. SOLVE: $$x + 3x = 60$$ $$4x = 60$$ $$\frac{4x}{4} = \frac{60}{4}$$ $$x = 15$$
4. INTERPRET the results. *Check* the proposed solution in the stated problem and *state* your conclusion.	4. INTERPRET the solution in the stated problem. The incubation period for a hummingbird is 15 days. The incubation period for a golden eagle is $$3x = 3 \cdot 15 = 45 \text{ days.}$$ Since 15 days + 45 days = 60 days and 45 is 3(15), the solution checks. *State* your conclusion: The incubation period for a hummingbird is 15 days. The incubation period for a golden eagle is 45 days.

Chapter 3 Review

(3.1) *Simplify each expression by combining like terms.*

1. $3y + 7y - 15$

2. $2y - 10 - 8y$

3. $8a + a - 7 - 15a$

4. $y + 3 - 9y - 1$

Multiply.

5. $2(x + 5)$

6. $-3(y + 8)$

Simplify.

7. $7x + 3(x - 4) + x$

8. $-(3m + 2) - m - 10$

9. $3(5a - 2) - 20a + 10$

10. $6y + 3 + 2(3y - 6)$

11. $6y - 7 + 11y - y + 2$

12. $10 - x + 5x - 12 - 3x$

Find the perimeter of each figure.

△ **13.**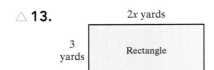

△ **14.** Square, 5y meters

Find the area of each figure.

△ **15.**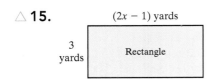

△ **16.** Rectangle $(x-2)$ centimeters, 10 centimeters; Rectangle $(5x+4)$ centimeters, 7 centimeters

(3.2) *Solve each equation.*

17. $z - 5 = -7$

18. $3x + 10 = 4x$

19. $3y = -21$

20. $-3a = -15$

21. $\dfrac{x}{-6} = 2$

22. $\dfrac{y}{-15} = -3$

23. $n + 18 = 10 - (-2)$

24. $c - 5 = -13 + 7$

25. $7x + 5 - 6x = -20$

26. $17x = 2(8x - 4)$

27. $5x + 7 = -3$

28. $-14 = 9y + 4$

29. $\dfrac{z}{4} = -8 - (-6)$

30. $-1 + (-8) = \dfrac{x}{5}$

31. $6y - 7y = 100 - 105$

32. $19x - 16x = 45 - 60$

33. $9(2x - 7) = 19x$

34. $-5(3x + 3) = -14x$

35. $3x - 4 = 11$

36. $6y + 1 = 73$

37. $2(x + 4) - 10 = -2(7)$

38. $-3(x - 6) + 13 = 20 - 1$

Translating *Translate each phrase into an algebraic expression. Let x represent "a number."*

39. The product of -5 and a number

40. Three subtracted from a number

41. The sum of -5 and a number

42. The quotient of -2 and a number

43. The product of -5 and a number, decreased by 50

44. Eleven added to twice a number

45. The quotient of 70 and the sum of a number and 6

46. Twice the difference of a number and 13

(3.3) *Solve each equation.*

47. $2x + 5 = 7x - 100$

48. $-6x - 4 = x + 66$

49. $2x + 7 = 6x - 1$

50. $5x - 18 = -4x$

51. $5(n - 3) = 7 + 3n$

52. $7(2 + x) = 4x - 1$

53. $6x + 3 - (-x) = -20 + 5x - 7$

54. $x - 25 + 2x = -5 + 2x - 10$

55. $3(x - 4) = 5x - 8$

56. $4(x - 3) = -2x - 48$

57. $6(2n - 1) + 18 = 0$

58. $7(3y - 2) - 7 = 0$

59. $95x - 14 = 20x - 10 + 10x - 4$

60. $32z + 11 - 28z = 50 + 2z - (-1)$

Translating *Write each sentence as an equation.*

61. The difference of 20 and -8 is 28.

62. Nineteen subtracted from -2 amounts to -21.

63. The quotient of -75 and the sum of 5 and 20 is equal to -3.

64. Five times the sum of 2 and -6 yields -20.

(3.4) Translating *Write each sentence as an equation using x as the variable.*

65. Twice a number minus 8 is 40.

66. The product of a number and 6 is equal to the sum of the number and 20.

67. Twelve subtracted from the quotient of a number and 2 is 10.

68. The difference of a number and 3 is the quotient of 8 and 4.

Solve.

69. Five times a number subtracted from 40 is the same as three times the number. Find the number.

70. The product of a number and 3 is twice the difference of that number and 8. Find the number.

71. In an election, the incumbent received 14,000 votes of the 18,500 votes cast. Of the remaining votes, the Democratic candidate received 272 more than the Independent candidate. Find how many votes the Democratic candidate received.

72. Rajiv Puri has twice as many movies on DVD as he has on Blu-ray Disc. Find the number of DVDs if he has a total of 126 movie recordings.

Mixed Review

Simplify.

73. $9x - 20x$

74. $-5(7x)$

75. $12x + 5(2x - 3) - 4$

76. $-7(x + 6) - 2(x - 5)$

Solve.

77. $c - 5 = -13 + 7$

78. $7x + 5 - 6x = -20$

79. $-7x + 3x = -50 - 2$

80. $-x + 8x = -38 - 4$

81. $9x + 12 - 8x = -6 + (-4)$

82. $-17x + 14 + 20x - 2x = 5 - (-3)$

83. $5(2x - 3) = 11x$

84. $\dfrac{y}{-3} = -1 - 5$

85. $12y - 10 = -70$

86. $4n - 8 = 2n + 14$

87. $-6(x - 3) = x + 4$

88. $9(3x - 4) + 63 = 0$

89. $-5z + 3z - 7 = 8z - 1 - 6$

90. $4x - 3 + 6x = 5x - 3 - 30$

91. Three times a number added to twelve is 27. Find the number.

92. Twice the sum of a number and four is ten. Find the number.

93. Out of 50 states and the District of Columbia, Vermont ranks 51st in the amount of energy consumption, followed by the District of Columbia. If the District of Columbia used 24.3 trillion BTU more than Vermont and the total number of BTU used by both entities was 240.5 trillion BTU, find the amount of energy consumption for each. (*Source:* Bureau of Transportation Statistics)

94. North and South Dakota both have over 80,000 roadway miles. North Dakota has 4489 more miles and the total number of roadway miles for both states is 169,197. Find the number of roadway miles for North Dakota and for South Dakota. (*Source:* U.S. Federal Highway Administration)

Chapter 3 Getting Ready for the Test LC

MULTIPLE CHOICE Simplify each expression in Exercises 1 through 6. Then choose the simplified form: A, B, C, D, or E. Choices may be used more than once or not at all.

A. $-6x$ **B.** $6x$ **C.** $4x + 2$ **D.** $4x + 1$ **E.** $4x + 6$

1. $x - 7x$
2. $-2x + 3 + 8x - 3$
3. $-3(2x)$
4. $2(2x + 1)$
5. $-(x + 2) + 5x + 4$
6. $3(1 + 2x) - 3$

MULTIPLE CHOICE

7. To solve $5x - 10 = 0$, we will first add 10 to each side of the equation. Once this is done, the equivalent equation is
   A. $5x = 0$ **B.** $5x = 10$ **C.** $5x = -10$

8. To solve $4(2x + 1) - 8 = 14x - 10$, we will first simplify the left side of the equation. Once this is done, the equivalent equation is
   A. $8x - 4 = 14x - 10$ **B.** $8x - 7 = 14x - 10$ **C.** $2x - 7 = 14x - 10$

MATCHING Let x represent "a number." Match each phrase in the first column with its translated variable expression in the second column.

9. the sum of -3 and a number
10. the product of -3 and a number
11. 5 subtracted from a number
12. 5 minus a number

A. $-3x$
B. $3 + x$
C. $-3 + x$
D. $x - 5$
E. $5 - x$
F. $5x$

MATCHING Let x represent "a number." Match each sentence in the first column with its translated equation in the second or third column.

13. Twice a number gives 14.
14. A number divided by 6 yields 14.
15. Two subtracted from a number equals 14.
16. Six times a number, added to 2 is 14.

A. $6x = 14$
B. $2x = 14$
C. $\dfrac{x}{6} = 14$
D. $2 - x = 14$

E. $x - 2 = 14$
F. $2 + 6x = 14$
G. $6(2 + x) = 14$

Chapter 3 Test

Answers

1. Simplify $7x - 5 - 12x + 10$ by combining like terms.

2. Multiply: $-2(3y + 7)$

3. Simplify: $-(3z + 2) - 5z - 18$

4. Write an expression that represents the perimeter of the equilateral triangle (a triangle with three sides of equal length). Simplify the expression.

$(5x + 5)$ inches

5. Write an expression that represents the area of the rectangle. Simplify the expression.

4 meters

Rectangle $(3x - 1)$ meters

Solve each equation.

6. $12 = y - 3y$

7. $\dfrac{x}{2} = -5 - (-2)$

8. $5x + 12 - 4x - 14 = 22$

9. $-4x + 7 = 15$

10. $2(x - 6) = 0$

11. $-4(x - 11) - 34 = 10 - 12$

12. $5x - 2 = x - 10$

13. $4(5x + 3) = 2(7x + 6)$

14. $6 + 2(3n - 1) = 28$

Translate the following phrases into mathematical expressions. If needed, use x to represent "a number."

15. The sum of -23 and a number

16. Three times a number, subtracted from -2

1. _____
2. _____
3. _____
4. _____
5. _____
6. _____
7. _____
8. _____
9. _____
10. _____
11. _____
12. _____
13. _____
14. _____
15. _____
16. _____

Translate each sentence into an equation. If needed, use x to represent "a number."

17. The sum of twice 5 and −15 is −5.

18. Six added to three times a number equals −30.

Solve.

19. The difference of three times a number and five times the same number is 4. Find the number.

20. In a championship basketball game, Paula made twice as many free throws as Maria. If the total number of free throws made by both women was 12, find how many free throws Paula made.

21. In a 10-kilometer race, there are 112 more men entered than women. Find the number of female runners if the total number of runners in the race is 600.

Chapters 1–3 Cumulative Review

1. Write 308,063,557 in words.

2. Write 276,004 in words.

3. Find the perimeter of the polygon shown.

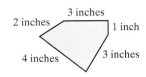

4. Find the perimeter of the rectangle shown.

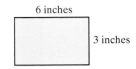

5. Subtract: $900 - 174$. Check by adding.

6. Subtract: $17{,}801 - 8216$. Check by adding.

7. Round 248,982 to the nearest hundred.

8. Round 844,497 to the nearest thousand.

9. Multiply: 25×8

10. Multiply: 395×74

11. Divide and check: $1872 \div 9$

12. Divide and check: $3956 \div 46$

13. Simplify: $2 \cdot 4 - 3 \div 3$

14. Simplify: $8 \cdot 4 + 9 \div 3$

15. Evaluate $x^2 + z - 3$ for $x = 5$ and $z = 4$.

16. Evaluate $2a^2 + 5 - c$ for $a = 2$ and $c = 3$.

17. Determine which numbers in the set $\{26, 40, 20\}$ are solutions of the equation $2n - 30 = 10$.

18. Insert $<$ or $>$ to make a true statement.
    a. $-14 \quad\quad 0$
    b. $-(-7) \quad\quad -8$

19. Add using a number line: $5 + (-2)$

20. Add using a number line: $-3 + (-4)$

212

Add.

21. $-15 + (-10)$

22. $3 + (-7)$

23. $-2 + 25$

24. $21 + 15 + (-19)$

Subtract.

25. $-4 - 10$

26. $-2 - 3$

27. $6 - (-5)$

28. $19 - (-10)$

29. $-11 - (-7)$

30. $-16 - (-13)$

Divide.

31. $\dfrac{-12}{6}$

32. $\dfrac{-30}{-5}$

33. $-20 \div (-4)$

34. $26 \div (-2)$

35. $\dfrac{48}{-3}$

36. $\dfrac{-120}{12}$

Find the value of each expression.

37. $(-3)^2$

38. -2^5

39. -3^2

40. $(-5)^2$

41. Simplify: $2y - 6 + 4y + 8$

42. Simplify: $6x + 2 - 3x + 7$

43. Determine whether -1 is a solution of the equation $3y + 1 = 3$.

44. Determine whether 2 is a solution of $5x - 3 = 7$.

45. Solve: $-12x = -36$

46. Solve: $-3y = 15$

47. Solve: $2x - 6 = 18$

48. Solve: $3a + 5 = -1$

49. A salesperson at an electronics store sold a computer system and software for $2100, receiving four times as much money for the computer system as for the software. Find the price of each.

50. Rose Daunis is thinking of a number. Two times the number plus four is the same amount as three times the number minus seven. Find Rose's number.

4 Fractions and Mixed Numbers

Fractions are numbers and, like whole numbers and integers, they can be added, subtracted, multiplied, and divided. Fractions are very useful and appear frequently in everyday language, in common phrases such as "half an hour," "quarter of a pound," and "third of a cup." This chapter reviews the concepts of fractions and mixed numbers and demonstrates how to add, subtract, multiply, and divide these numbers.

Sections

4.1 Introduction to Fractions and Mixed Numbers
4.2 Factors and Simplest Form
4.3 Multiplying and Dividing Fractions
4.4 Adding and Subtracting Like Fractions, Least Common Denominator, and Equivalent Fractions
4.5 Adding and Subtracting Unlike Fractions

Integrated Review— Summary on Fractions and Operations on Fractions

4.6 Complex Fractions and Review of Order of Operations
4.7 Operations on Mixed Numbers
4.8 Solving Equations Containing Fractions

Check Your Progress

Vocabulary Check
Chapter Highlights
Chapter Review
Getting Ready for the Test
Chapter Test
Cumulative Review

Pinnacles National Park in California was signed into law on January 10, 2013, as the United States' 59th and newest national park. (The former Pinnacles National Monument was established in 1908 by U.S. President Theodore Roosevelt.)

The National Park Service Turned 100 on August 25, 2016.

The National Park Service is well known for its stewardship of national parks and protecting natural resources for future generations. Many of us recognize the National Park ranger by the "Smokey Bear" hat, but the National Park Service also employs archeologists, architects, curators, historians, and other professionals to help manage and preserve the vastly different properties under its oversight.

The following graph is called a circle graph or pie chart. Each sector (shaped like a piece of a pie) shows the fraction of the areas managed by the National Park System by type of national designation.

In Section 4.2, Exercises 97–100 we show this same circle graph but in 3-D design. We simplify some of the fractions in it and also study sector size versus fraction value.

Areas Maintained by the National Park Service

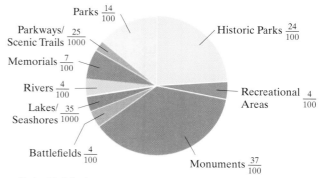

Source: National Park Service

4.1 Introduction to Fractions and Mixed Numbers

Objective A Identifying Numerators and Denominators

Whole numbers are used to count whole things or units, such as cars, horses, dollars, and people. To refer to a part of a whole, fractions can be used. Here are some examples of **fractions**. Study these examples for a moment.

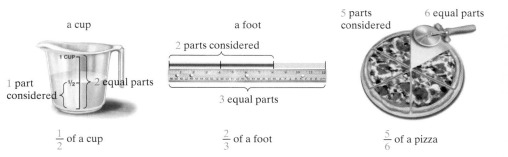

In a fraction, the top number is called the **numerator** and the bottom number is called the **denominator**. The bar between the numbers is called the **fraction bar.**

Name	Fraction	Meaning
numerator → denominator →	$\frac{5}{6}$	← number of parts being considered ← number of equal parts in the whole

Objectives

A Identify the Numerator and the Denominator of a Fraction.

B Write a Fraction to Represent Parts of Figures or Real-Life Data.

C Graph Fractions on a Number Line.

D Review Division Properties of 0 and 1.

E Write Mixed Numbers as Improper Fractions.

F Write Improper Fractions as Mixed Numbers or Whole Numbers.

Examples Identify the numerator and the denominator of each fraction.

1. $\frac{3}{7}$ ← numerator
         ← denominator

2. $\frac{13}{5x}$ ← numerator
         ← denominator

■ Work Practice 1–2

Practice 1–2

Identify the numerator and the denominator of each fraction.

1. $\frac{11}{2}$

2. $\frac{10y}{17}$

Helpful Hint

$\frac{3}{7}$ ← Remember that the bar in a fraction means division. Since division by 0 is undefined, a fraction with a denominator of 0 is undefined. For example, $\frac{3}{0}$ is undefined.

Objective B Writing Fractions to Represent Parts of Figures or Real-Life Data

One way to become familiar with the concept of fractions is to visualize fractions with shaded figures. We can then write a fraction to represent the shaded area of the figure (or diagram).

Answers

1. numerator = 11, denominator = 2
2. numerator = 10y, denominator = 17

Practice 3–4

Write a fraction to represent the shaded part of each figure.

3.

4.

Examples

Write a fraction to represent the shaded part of each figure.

3. In this figure, 2 of the 5 equal parts are shaded. Thus, the fraction is $\frac{2}{5}$.

$\frac{2}{5}$ ← number of parts shaded
  ← number of equal parts

4. In this figure, 3 of the 10 rectangles are shaded. Thus, the fraction is $\frac{3}{10}$.

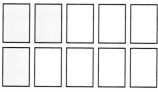

$\frac{3}{10}$ ← number of parts shaded
   ← number of equal parts

■ Work Practice 3–4

Practice 5–6

Write a fraction to represent the part of the whole shown.

5.

6.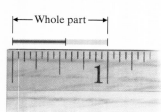

Examples

Write a fraction to represent the shaded part of the diagram.

5. The fraction is $\frac{3}{10}$.

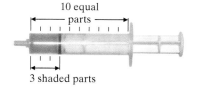

6. The fraction is $\frac{1}{3}$.

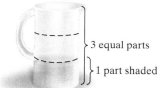

■ Work Practice 5–6

Practice 7

Draw and shade a part of a figure to represent the fraction.

7. $\frac{2}{3}$ of a figure

Examples

Draw a figure and then shade a part of it to represent each fraction.

7. $\frac{5}{6}$ of a figure

We will use a geometric figure such as a rectangle. Since the denominator is 6, we divide it into 6 equal parts. Then we shade 5 of the equal parts.

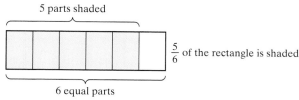

Answers

3. $\frac{3}{8}$ 4. $\frac{1}{6}$ 5. $\frac{7}{10}$ 6. $\frac{9}{16}$

7. answers may vary; for example,

Section 4.1 | Introduction to Fractions and Mixed Numbers

8. $\frac{3}{8}$ of a figure

If you'd like, our figure can consist of 8 triangles of the same size. We will shade 3 of the triangles.

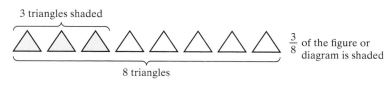

$\frac{3}{8}$ of the figure or diagram is shaded

Work Practice 7–8

Practice 8

Draw and shade a part of a figure to represent the fraction.

8. $\frac{7}{11}$ of a figure

✓ **Concept Check** If 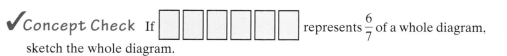 represents $\frac{6}{7}$ of a whole diagram, sketch the whole diagram.

Example 9 Writing Fractions from Real-Life Data

Of the eight planets in our solar system (Pluto is now a dwarf planet), three are closer to the Sun than Mars. What fraction of the planets are closer to the Sun than Mars?

Practice 9

Of the eight planets in our solar system, five are farther from the Sun than Earth is. What fraction of the planets are farther from the Sun than Earth is?

Solution: The fraction of planets closer to the Sun than Mars is:

$\frac{3}{8}$ ← number of planets closer
     ← number of planets in our solar system

Thus, $\frac{3}{8}$ of the planets in our solar system are closer to the Sun than Mars.

Work Practice 9

The definitions and statements below apply to positive fractions.

A **proper fraction** is a fraction whose numerator is less than its denominator. Proper fractions are less than 1. For example, the shaded portion of the triangle is represented by $\frac{2}{3}$.

An **improper fraction** is a fraction whose numerator is greater than or equal to its denominator. Improper fractions are greater than or equal to 1.

Answers
8. answers may vary; for example,

9. $\frac{5}{8}$

✓ **Concept Check Answer**

The shaded part of the group of circles below is $\frac{9}{4}$. The shaded part of the rectangle is $\frac{6}{6}$. Recall that $\frac{6}{6}$ simplifies to 1 and notice that the entire rectangle (1 whole figure) is shaded below.

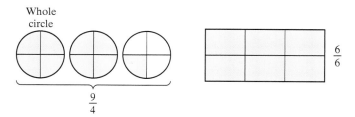

A **mixed number** contains a whole number and a fraction. Mixed numbers are greater than 1. Above, we wrote the shaded part of the group of circles as the improper fraction $\frac{9}{4}$. Now let's write the shaded part as a mixed number. The shaded part of the group of circles' area is $2\frac{1}{4}$. Read this as "two and one-fourth."

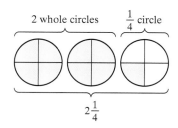

Note: The mixed number $2\frac{1}{4}$, diagrammed above, represents $2 + \frac{1}{4}$.

The mixed number $-3\frac{1}{5}$ represents $-\left(3 + \frac{1}{5}\right)$ or $-3 - \frac{1}{5}$. We review this later in this chapter.

Practice 10–11

Represent the shaded part of each figure group as both an improper fraction and a mixed number.

10.

11.

Examples Represent the shaded part of each figure group as both an improper fraction and a mixed number.

10.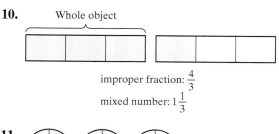

improper fraction: $\frac{4}{3}$

mixed number: $1\frac{1}{3}$

11.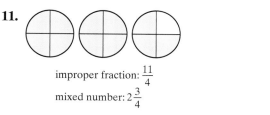

improper fraction: $\frac{11}{4}$

mixed number: $2\frac{3}{4}$

■ Work Practice 10–11

Answers

10. $\frac{8}{3}$, $2\frac{2}{3}$ **11.** $\frac{5}{4}$, $1\frac{1}{4}$

✓**Concept Check Answer**
3; answers may vary

✓**Concept Check** If you were to round $2\frac{3}{4}$, shown in Example 11 above, to the nearest whole number, would you choose 2 or 3? Why?

Objective C Graphing Fractions on a Number Line

Another way to visualize fractions is to graph them on a number line. To do this, think of 1 unit on the number line as a whole. To graph $\frac{2}{5}$, for example, divide the distance from 0 to 1 into 5 equal parts. Then start at 0 and count 2 parts to the right.

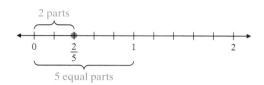

Notice that the graph of $\frac{2}{5}$ lies between 0 and 1. This means

$0 < \frac{2}{5} \left(\text{or } \frac{2}{5} > 0 \right)$ and also $\frac{2}{5} < 1$

Example 12 Graph each proper fraction on a number line.

a. $\frac{3}{4}$ b. $\frac{1}{2}$ c. $\frac{3}{6}$

Solution:

a. To graph $\frac{3}{4}$, divide the distance from 0 to 1 into 4 parts. Then start at 0 and count over 3 parts.

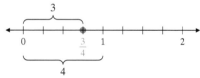

b.

c.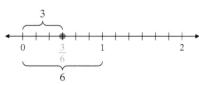

■ Work Practice 12

The statements below apply to positive fractions.
The fractions in Example 12 are all proper fractions. Notice that the value of each is less than 1. This is always true for proper fractions since the numerator of a proper fraction is less than the denominator.
On the next page, we graph improper fractions. Notice that improper fractions are greater than or equal to 1. This is always true since the numerator of an improper fraction is greater than or equal to the denominator.

Practice 12

Graph each proper fraction on a number line.

a. $\frac{5}{7}$ b. $\frac{2}{3}$ c. $\frac{4}{6}$

Answers

12.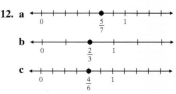

Practice 13

Graph each improper fraction on a number line.

a. $\dfrac{8}{3}$ b. $\dfrac{5}{4}$ c. $\dfrac{7}{7}$

Example 13 Graph each improper fraction on a number line.

a. $\dfrac{7}{6}$ b. $\dfrac{9}{5}$ c. $\dfrac{6}{6}$ d. $\dfrac{3}{1}$

Solution:

a.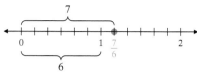

b.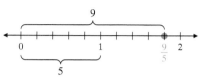

c.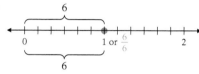

d. Each 1-unit distance has 1 equal part. Count over 3 parts.

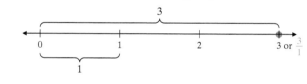

■ Work Practice 13

Note: We will graph mixed numbers at the end of this chapter.

Objective D Reviewing Division Properties of 0 and 1

Before we continue further, don't forget from Section 1.6 that a fraction bar indicates division. Let's review some division properties of 1 and 0.

$\dfrac{9}{9} = 1$ because $1 \cdot 9 = 9$ $\dfrac{-11}{1} = -11$ because $-11 \cdot 1 = -11$

$\dfrac{0}{6} = 0$ because $0 \cdot 6 = 0$ $\dfrac{6}{0}$ *is undefined* because there is no number that when multiplied by 0 gives 6.

In general, we can say the following.

Let n be any integer except 0.

$\dfrac{n}{n} = 1$ $\dfrac{0}{n} = 0$

$\dfrac{n}{1} = n$ $\dfrac{n}{0}$ is undefined.

Answers

13.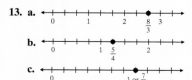

Examples Simplify.

14. $\dfrac{5}{5} = 1$ **15.** $\dfrac{-2}{-2} = 1$ **16.** $\dfrac{0}{-5} = 0$

17. $\dfrac{-5}{1} = -5$ **18.** $\dfrac{41}{1} = 41$ **19.** $\dfrac{19}{0}$ is undefined

Work Practice 14–19

Notice from Example 17 that we can have negative fractions. In fact,

$$\dfrac{-5}{1} = -5, \quad \dfrac{5}{-1} = -5, \quad \text{and} \quad -\dfrac{5}{1} = -5$$

Because all of the fractions equal -5, we have

$$\dfrac{-5}{1} = \dfrac{5}{-1} = -\dfrac{5}{1}$$

This means that the negative sign in a fraction can be written in the numerator, in the denominator, or in front of the fraction. Remember this as we work with negative fractions.

Practice 14–19
Simplify.

14. $\dfrac{9}{9}$ **15.** $\dfrac{-6}{-6}$

16. $\dfrac{0}{-1}$ **17.** $\dfrac{4}{1}$

18. $\dfrac{-13}{0}$ **19.** $\dfrac{-13}{1}$

Helpful Hint Remember, for example, that

$$-\dfrac{2}{3} = \dfrac{-2}{3} = \dfrac{2}{-3}$$

Objective E Writing Mixed Numbers as Improper Fractions

Earlier in this section, mixed numbers and improper fractions were both used to represent the shaded part of figure groups. For example,

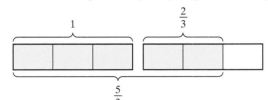

$1\dfrac{2}{3}$ or $\dfrac{5}{3}$ Thus, $1\dfrac{2}{3} = \dfrac{5}{3}$

The following steps may be used to write a mixed number as an improper fraction:

Writing a Mixed Number as an Improper Fraction

To write a mixed number as an improper fraction:
Step 1: Multiply the denominator of the fraction by the whole number.
Step 2: Add the numerator of the fraction to the product from Step 1.
Step 3: Write the sum from Step 2 as the numerator of the improper fraction over the original denominator.

For example,

$$1\dfrac{2}{3} = \dfrac{\overset{\text{Step 1}}{3 \cdot 1} + \overset{\text{Step 2}}{2}}{3} = \dfrac{3+2}{3} = \dfrac{5}{3} \quad \text{or} \quad 1\dfrac{2}{3} = \dfrac{5}{3}, \text{ as stated above.}$$

(Step 3)

Example 20 Write each as an improper fraction.

a. $4\dfrac{2}{9} = \dfrac{9 \cdot 4 + 2}{9} = \dfrac{36 + 2}{9} = \dfrac{38}{9}$ **b.** $1\dfrac{8}{11} = \dfrac{11 \cdot 1 + 8}{11} = \dfrac{11 + 8}{11} = \dfrac{19}{11}$

Work Practice 20

Practice 20
Write each as an improper fraction.

a. $5\dfrac{2}{7}$ **b.** $6\dfrac{2}{3}$

c. $10\dfrac{9}{10}$ **d.** $4\dfrac{1}{5}$

Answers
14. 1   **15.** 1   **16.** 0   **17.** 4
18. undefined   **19.** -13
20. a. $\dfrac{37}{7}$ **b.** $\dfrac{20}{3}$ **c.** $\dfrac{109}{10}$ **d.** $\dfrac{21}{5}$

Objective F Writing Improper Fractions as Mixed Numbers or Whole Numbers

Just as there are times when an improper fraction is preferred, sometimes a mixed or a whole number better suits a situation. To write improper fractions as mixed or whole numbers, we use division. Recall once again from Section 1.6 that the fraction bar means division. This means that the fraction

$$\frac{5}{3} \begin{array}{l}\text{numerator}\\ \text{denominator}\end{array} \quad \text{means} \quad 3\overline{)5} \begin{array}{l}\leftarrow \text{numerator}\\ \leftarrow \text{denominator}\end{array}$$

Writing an Improper Fraction as a Mixed Number or a Whole Number

To write an improper fraction as a mixed number or a whole number:

Step 1: Divide the denominator into the numerator.

Step 2: The whole number part of the mixed number is the quotient. The fraction part of the mixed number is the remainder over the original denominator.

$$\text{quotient} \, \frac{\text{remainder}}{\text{original denominator}}$$

For example,

$$\frac{5}{3} : 3\overline{)5} \quad \frac{5}{3} = 1\frac{2}{3} \begin{array}{l}\leftarrow \text{remainder}\\ \leftarrow \text{original denominator}\end{array}$$

Practice 21

Write each as a mixed number or a whole number.

a. $\dfrac{9}{5}$ b. $\dfrac{23}{9}$ c. $\dfrac{48}{4}$

d. $\dfrac{62}{13}$ e. $\dfrac{51}{7}$ f. $\dfrac{21}{20}$

Example 21 Write each as a mixed number or a whole number.

a. $\dfrac{30}{7}$ b. $\dfrac{16}{15}$ c. $\dfrac{84}{6}$

Solution:

a. $\dfrac{30}{7} : 7\overline{)30} \quad \dfrac{30}{7} = 4\dfrac{2}{7}$

b. $\dfrac{16}{15} : 15\overline{)16} \quad \dfrac{16}{15} = 1\dfrac{1}{15}$

c. $\dfrac{84}{6} : 6\overline{)84} \quad \dfrac{84}{6} = 14$ Since the remainder is 0, the result is the whole number 14.

Helpful Hint When the remainder is 0, the improper fraction is a whole number. For example, $\dfrac{92}{4} = 23$.

$4\overline{)92}$ with quotient 23.

■ Work Practice 21

Answers

21. a. $1\dfrac{4}{5}$ b. $2\dfrac{5}{9}$ c. 12 d. $4\dfrac{10}{13}$ e. $7\dfrac{2}{7}$ f. $1\dfrac{1}{20}$

Section 4.1 Introduction to Fractions and Mixed Numbers

Vocabulary, Readiness & Video Check

Use the choices below to fill in each blank.

| improper | fraction | proper | is undefined | mixed number | = 0 |
| ≥ 1 | denominator | = 1 | < 1 | numerator | |

1. The number $\frac{17}{31}$ is called a(n) _____. The number 31 is called its _____ and 17 is called its _____.

2. If we simplify each fraction, $\frac{-9}{-9}$ _____, $\frac{0}{-4}$ _____, and we say $\frac{-4}{0}$ _____.

3. The fraction $\frac{8}{3}$ is called a(n) _____ fraction, the fraction $\frac{3}{8}$ is called a(n) _____ fraction, and $10\frac{3}{8}$ is called a(n) _____.

4. The value of an improper fraction is always _____, and the value of a proper fraction is always _____.

Martin-Gay Interactive Videos

See Video 4.1

Watch the section lecture video and answer the following questions.

Objective A 5. Complete this statement based on Example 1: When the numerator is greater than or _____ to the denominator, you have a(n) _____ fraction.

Objective B 6. In Example 4, there are two shapes in the diagram, so why do the representative fractions have a denominator 3?

Objective C 7. From Examples 6 and 7, when graphing a positive fraction on a number line, how does the denominator help? What does the denominator tell you?

Objective D 8. From Example 10, what can you conclude about any fraction where the numerator and denominator are the same nonzero number?

Objective E 9. Complete this statement based on the lecture before Example 13: The operation of _____ is understood in a mixed number notation; for example, $1\frac{1}{3}$ means 1 _____ $\frac{1}{3}$.

Objective F 10. From the lecture before Example 16, what operation is used to write an improper fraction as a mixed number?

4.1 Exercise Set MyLab Math

Objectives A B *Identify the numerator and the denominator of each fraction and identify each fraction as proper or improper. See Examples 1, 2, 10, and 11.*

1. $\frac{1}{2}$

2. $\frac{1}{4}$

3. $\frac{10}{3}$

4. $\frac{53}{21}$

5. $\frac{15}{15}$

6. $\frac{26}{26}$

Objective B *Write a proper or improper fraction to represent the shaded part of each diagram. If an improper fraction is appropriate, write the shaded part of the diagram as (**a**) an improper fraction and (**b**) a mixed number. (Note to students: In case you know how to simplify fractions, none of the fractions in this section are simplified.) See Examples 3 through 6, and 10, and 11.*

7.

8.

9.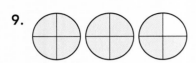

10.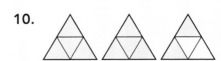

11.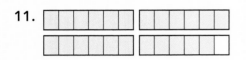

12.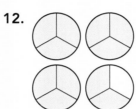

13.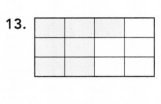

14.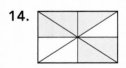

15.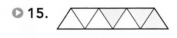

16.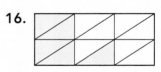

17.

18.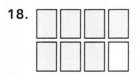

19.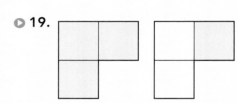

20.

21.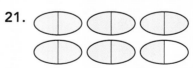

22.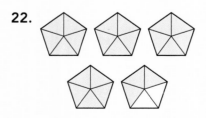

23.

24.

25.

26.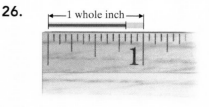

Objective B *Draw and shade a part of a diagram to represent each fraction. See Examples 7 and 8.*

27. $\frac{1}{5}$ of a diagram

28. $\frac{1}{16}$ of a diagram

29. $\frac{6}{7}$ of a diagram

30. $\frac{7}{9}$ of a diagram

31. $\frac{4}{4}$ of a diagram

32. $\frac{6}{6}$ of a diagram

Write each fraction. (Note to students: In case you know how to simplify fractions, none of the fractions in this section are simplified.) See Example 9.

33. Of the 131 students at a small private school, 42 are freshmen. What fraction of the students are freshmen?

34. Of the 63 employees at a new biomedical engineering firm, 22 are men. What fraction of the employees are men?

35. Use Exercise 33 to answer **a** and **b**.
    a. How many students are *not* freshmen?
    b. What fraction of the students are *not* freshmen?

36. Use Exercise 34 to answer **a** and **b**.
    a. How many of the employees are women?
    b. What fraction of the employees are women?

37. As of 2017, the United States has had 44 different men as presidents. A total of seven U.S. presidents were born in the state of Ohio, second only to the state of Virginia in producing U.S. presidents. What fraction of U.S. presidents were born in Ohio? (*Source:* World Almanac, 2017; *Note?* Grover Cleveland was our 22nd and 24th president, thus was counted once.)

38. Of the eight planets in our solar system, four have days that are longer than the 24-hour Earth day. What fraction of the planets have longer days than Earth has? (*Source:* National Space Science Data Center)

39. The 2016 Atlantic hurricane season was the deadliest since 2005. It was an above-average season, producing 15 named tropical storms, 10 of which turned into hurricanes. What fraction of the 2016 Atlantic tropical storms escalated into hurricanes?

40. There are 12 inches in a foot. What fractional part of a foot do 5 inches represent?

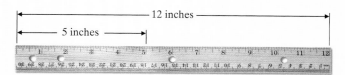

41. There are 31 days in the month of March. What fraction of the month do 11 days represent?

42. There are 60 minutes in an hour. What fraction of an hour do 37 minutes represent?

Mon.	Tue.	Wed.	Thu.	Fri.	Sat.	Sun.
					1	2
3	4	5	6	7	8	9
10	11	12	13	14	15	16
17	18	19	20	21	22	23
24	25	26	27	28	29	30
31						

43. In a prealgebra class containing 31 students, there are 18 freshmen, 10 sophomores, and 3 juniors. What fraction of the class is sophomores?

44. In a sports team with 20 children, there are 9 boys and 11 girls. What fraction of the team is boys?

45. Thirty-three out of the fifty total states in the United States contain federal Indian reservations.
   a. What fraction of the states contain federal Indian reservations?
   b. How many states do not contain federal Indian reservations?
   c. What fraction of the states do not contain federal Indian reservations? (*Source:* Tiller Research, Inc., Albuquerque, NM)

46. Consumer fireworks are legal in 44 out of the 50 total states in the United States.
   a. In what fraction of the states are consumer fireworks legal?
   b. In how many states are consumer fireworks illegal?
   c. In what fraction of the states are consumer fireworks illegal? (*Source:* United States Fireworks Safety Council)

47. A bag contains 50 red or blue marbles. If 21 marbles are blue, answer each question.
   a. What fraction of the marbles are blue?
   b. How many marbles are red?
   c. What fraction of the marbles are red?

48. An art dealer is taking inventory. His shop contains a total of 37 pieces, which are all sculptures, watercolor paintings, or oil paintings. If there are 15 watercolor paintings and 17 oil paintings, answer each question.
   a. What fraction of the inventory is watercolor paintings?
   b. What fraction of the inventory is oil paintings?
   c. How many sculptures are there?
   d. What fraction of the inventory is sculptures?

Objective C *Graph each fraction on a number line. See Examples 12 and 13.*

49. $\dfrac{1}{4}$

50. $\dfrac{1}{3}$

51. $\dfrac{4}{7}$

52. $\dfrac{5}{6}$

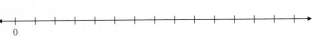

53. $\dfrac{8}{5}$

54. $\dfrac{9}{8}$

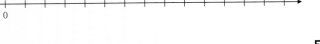

55. $\dfrac{7}{3}$

56. $\dfrac{13}{7}$

Objective D *Simplify by dividing. See Examples 14 through 19.*

57. $\dfrac{12}{12}$ **58.** $\dfrac{-3}{-3}$ **59.** $\dfrac{-5}{1}$ **60.** $\dfrac{-20}{1}$

Section 4.1 | Introduction to Fractions and Mixed Numbers

61. $\dfrac{0}{-2}$
62. $\dfrac{0}{-8}$
63. $\dfrac{-8}{-8}$
64. $\dfrac{-14}{-14}$

65. $\dfrac{-9}{0}$
66. $\dfrac{-7}{0}$
67. $\dfrac{3}{1}$
68. $\dfrac{5}{5}$

Objective E *Write each mixed number as an improper fraction. See Example 20.*

69. $2\dfrac{1}{3}$
70. $1\dfrac{13}{17}$
71. $3\dfrac{3}{5}$
72. $2\dfrac{5}{9}$

73. $6\dfrac{5}{8}$
74. $7\dfrac{3}{8}$
75. $11\dfrac{6}{7}$
76. $12\dfrac{2}{5}$

77. $9\dfrac{7}{20}$
78. $10\dfrac{14}{27}$
79. $166\dfrac{2}{3}$
80. $114\dfrac{2}{7}$

Objective F *Write each improper fraction as a mixed number or a whole number. See Example 21.*

81. $\dfrac{17}{5}$
82. $\dfrac{13}{7}$
83. $\dfrac{37}{8}$
84. $\dfrac{64}{9}$

85. $\dfrac{47}{15}$
86. $\dfrac{65}{12}$
87. $\dfrac{225}{15}$
88. $\dfrac{196}{14}$

89. $\dfrac{182}{175}$
90. $\dfrac{149}{143}$
91. $\dfrac{737}{112}$
92. $\dfrac{901}{123}$

Review

Simplify. See Section 1.7.

93. 3^2
94. 4^3
95. 5^3
96. 3^4

Concept Extensions

Write each fraction in two other equivalent ways by inserting the negative sign in different places.

97. $-\dfrac{11}{2} = = $

98. $-\dfrac{21}{4} = = $

99. $\dfrac{-13}{15} = = $

100. $\dfrac{45}{-57} = = $

101. In your own words, explain why $\dfrac{0}{10} = 0$ and $\dfrac{10}{0}$ is undefined.

102. In your own words, explain why $\dfrac{0}{-3} = 0$ and $\dfrac{-3}{0}$ is undefined.

Solve. See the Concept Checks in this section.

103. If ○○○○ represents $\frac{4}{9}$ of a whole diagram, sketch the whole diagram.

104. If △△ represents $\frac{1}{3}$ of a whole diagram, sketch the whole diagram.

105. Round the mixed number $7\frac{1}{8}$ to the nearest whole number.

106. Round the mixed number $5\frac{11}{12}$ to the nearest whole number.

107. The Gap Corporation owns stores in the United States with five different brand names, with the three most popular shown on the bar graph. What fraction of the total stores shown on the graph are named "Old Navy"?

108. There are currently 59 national parks in the United States. The bar graph below shows the four states with the greatest number of national parks. What fraction of the national parks found in these four states are located in California? (*Source:* The National Park Service)

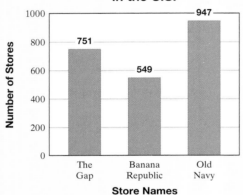

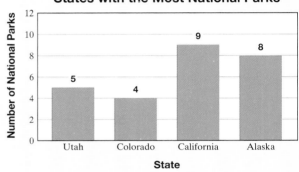

109. Habitat for Humanity is a nonprofit organization that helps provide affordable housing to families in need. Habitat for Humanity does its work of building and renovating houses through 1500 local affiliates in the United States and 80 international affiliates. What fraction of the total Habitat for Humanity affiliates are located in the United States? (*Hint:* First find the total number of affiliates.) (*Source:* Habitat for Humanity International)

110. The United States Marine Corps (USMC) has five principal training centers in California, three in North Carolina, two in South Carolina, one in Arizona, one in Hawaii, and one in Virginia. What fraction of the total USMC principal training centers are located in California? (*Hint:* First find the total number of USMC training centers.) (*Source:* U.S. Department of Defense)

4.2 Factors and Simplest Form

Objective A Writing a Number as a Product of Prime Numbers

Recall from Section 1.5 that since $12 = 2 \cdot 2 \cdot 3$, the numbers 2 and 3 are called *factors* of 12. A **factor** is any number that divides a number evenly (with a remainder of 0).

To perform operations on fractions, it is necessary to be able to factor a number. **Factoring** a number means writing a number as a product. We first practice writing a number as a product of prime numbers.

> A **prime number** is a natural number greater than 1 whose only factors are 1 and itself. The first few prime numbers are 2, 3, 5, 7, 11, 13, 17, 19, 23, 29, ….
> A **composite number** is a natural number greater than 1 that is not prime.

Objectives

A Write a Number as a Product of Prime Numbers.

B Write a Fraction in Simplest Form.

C Determine Whether Two Fractions Are Equivalent.

D Solve Problems by Writing Fractions in Simplest Form.

Helpful Hint

The natural number 1 is neither prime nor composite.

When a composite number is written as a product of prime numbers, this product is called the **prime factorization** of the number. For example, the prime factorization of 12 is $2 \cdot 2 \cdot 3$ because

$12 = \underline{2 \cdot 2 \cdot 3}$
  This product is 12 and each
  number is a prime number.

Because multiplication is commutative, the order of the factors is not important. We can write the factorization $2 \cdot 2 \cdot 3$ as $2 \cdot 3 \cdot 2$ or $3 \cdot 2 \cdot 2$. Any of these is called the prime factorization of 12.

> Every whole number greater than 1 has exactly one prime factorization.

One method for finding the prime factorization of a number is by using a factor tree, as shown in the next example.

Example 1 Write the prime factorization of 45.

Solution: We can begin by writing 45 as the product of two numbers, say, 5 and 9.

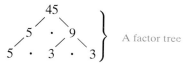

The number 5 is prime but 9 is not, so we write 9 as $3 \cdot 3$.

A factor tree

Each factor is now a prime number, so the prime factorization of 45 is $3 \cdot 3 \cdot 5$ or $3^2 \cdot 5$.

Work Practice 1

Practice 1

Use a factor tree to find the prime factorization of each number.

a. 30   **b.** 56   **c.** 72

Answers
1. a. $2 \cdot 3 \cdot 5$ **b.** $2^3 \cdot 7$ **c.** $2^3 \cdot 3^2$

✓ **Concept Check** True or false? Two different numbers can have exactly the same prime factorization. Explain your answer.

Practice 2
Write the prime factorization of 60.

Example 2 Write the prime factorization of 80.

Solution: Write 80 as a product of two numbers. Continue this process until all factors are prime.

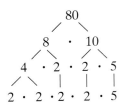

All factors are now prime, so the prime factorization of 80 is

$2 \cdot 2 \cdot 2 \cdot 2 \cdot 5$ or $2^4 \cdot 5$.

■ Work Practice 2

Helpful Hint

It makes no difference which factors you start with. The prime factorization of a number will be the same.

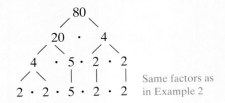

Same factors as in Example 2

There are a few quick **divisibility tests** to determine whether a number is divisible by the primes 2, 3, or 5. (A number is divisible by 2, for example, if 2 divides it evenly so that the remainder is 0.)

Divisibility Tests

A whole number is divisible by:
- 2 if the last digit is 0, 2, 4, 6, or 8.

  132 is divisible by 2 since the last digit is a 2.
- 3 if the sum of the digits is divisible by 3.

  144 is divisible by 3 since $1 + 4 + 4 = 9$ is divisible by 3.
- 5 if the last digit is 0 or 5.

  1115 is divisible by 5 since the last digit is a 5.

Answer
2. $2^2 \cdot 3 \cdot 5$

✓ **Concept Check Answer**
false; answers may vary

Helpful Hint

Here are a few other divisibility tests you may want to use. A whole number is divisible by:

- 4 if its last two digits are divisible by 4.

    17**12** is divisible by 4.

- 6 if it's divisible by 2 and 3.

    9858 is divisible by 6.

- 9 if the sum of its digits is divisible by 9.

    5238 is divisible by 9 since $5 + 2 + 3 + 8 = 18$ is divisible by 9.

When finding the prime factorization of larger numbers, you may want to use the procedure shown in Example 3.

Example 3 Write the prime factorization of 252.

Solution: For this method, we divide prime numbers into the given number. Since the ones digit of 252 is 2, we know that 252 is divisible by 2.

$$\begin{array}{r} 126 \\ 2\overline{)252} \end{array}$$

126 is divisible by 2 also.

$$\begin{array}{r} 63 \\ 2\overline{)126} \\ 2\overline{)252} \end{array}$$

63 is not divisible by 2 but is divisible by 3. Divide 63 by 3 and continue in this same manner until the quotient is a prime number.

$$\begin{array}{r} 7 \\ 3\overline{)21} \\ 3\overline{)63} \\ 2\overline{)126} \\ 2\overline{)252} \end{array}$$

Helpful Hint The order of choosing prime numbers does not matter. For consistency, we use the order 2, 3, 5, 7, ...

The prime factorization of 252 is $2 \cdot 2 \cdot 3 \cdot 3 \cdot 7$ or $2^2 \cdot 3^2 \cdot 7$.

■ Work Practice 3

In this text, we will write the factorization of a number from the smallest factor to the largest factor.

✓**Concept Check** True or false? The prime factorization of 117 is $9 \cdot 13$. Explain your reasoning.

Objective B Writing Fractions in Simplest Form

Fractions that represent the same portion of a whole or the same point on a number line are called **equivalent fractions**. Study the table on the next page to see two ways to visualize equivalent fractions.

Practice 3
Write the prime factorization of 297.

Answer
3. $3^3 \cdot 11$

✓**Concept Check Answer**
False; 9 is not prime.

Equivalent Fractions

Figures	Number Line
When we shade $\frac{1}{3}$ and $\frac{2}{6}$ on the same-sized figures, 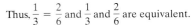 notice that both $\frac{1}{3}$ and $\frac{2}{6}$ represent the same portion of a whole. These fractions are called **equivalent fractions** and we write $\frac{1}{3} = \frac{2}{6}$.	When we graph $\frac{1}{3}$ and $\frac{2}{6}$ on a number line, notice that both $\frac{1}{3}$ and $\frac{2}{6}$ correspond to the same point. These fractions are called **equivalent fractions**, and we write $\frac{1}{3} = \frac{2}{6}$.

Thus, $\frac{1}{3} = \frac{2}{6}$ and $\frac{1}{3}$ and $\frac{2}{6}$ are equivalent.

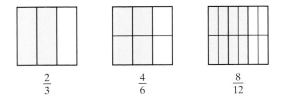

For example, $\frac{2}{3}, \frac{4}{6},$ and $\frac{8}{12}$ all represent the same shaded portion of the rectangle's area, so they are equivalent fractions. To show that these fractions are equivalent, we place an equal sign between them. In other words,

$$\frac{2}{3} = \frac{4}{6} = \frac{8}{12}$$

There are many equivalent forms of a fraction. A special equivalent form of a fraction is called **simplest form.**

Simplest Form of a Fraction

A fraction is written in **simplest form** or **lowest terms** when the numerator and the denominator have no common factors other than 1.

For example, the fraction $\frac{2}{3}$ is in simplest form because 2 and 3 have no common factor other than 1. The fraction $\frac{4}{6}$ *is not* in simplest form because 4 and 6 both have a factor of 2. That is, 2 is a common factor of 4 and 6. The process of writing a fraction in simplest form is called **simplifying** the fraction.

To simplify $\frac{4}{6}$ and write it as $\frac{2}{3}$, let's first study a few properties. Recall from Section 4.1 that any nonzero whole number n divided by itself is 1.

Any nonzero number n divided by itself is 1.

$$\frac{5}{5} = 1, \quad \frac{17}{17} = 1, \quad \frac{24}{24} = 1, \text{ or, in general, } \frac{n}{n} = 1$$

Section 4.2 | Factors and Simplest Form   233

Also, in general, if $\dfrac{a}{b}$ and $\dfrac{c}{d}$ are fractions (with b and d not 0), the following is true.

$$\dfrac{a \cdot c}{b \cdot d} = \dfrac{a}{b} \cdot \dfrac{c}{d}$$

Helpful Hint
These two properties together are called the Fundamental Property of Fractions
$$\dfrac{a \cdot c}{b \cdot c} = \dfrac{a}{b} \cdot \dfrac{c}{c} = \dfrac{a}{b}$$

These properties allow us to do the following:

$$\dfrac{4}{6} = \dfrac{2 \cdot 2}{2 \cdot 3} = \dfrac{2}{2} \cdot \dfrac{2}{3} = 1 \cdot \dfrac{2}{3} = \dfrac{2}{3}$$ When 1 is multiplied by a number, the result is the same number.
 ↑ This is 1.

Example 4 Write in simplest form: $\dfrac{12}{20}$

Solution: Notice that 12 and 20 have a common factor of 4.

$$\dfrac{12}{20} = \dfrac{4 \cdot 3}{4 \cdot 5} = \dfrac{4}{4} \cdot \dfrac{3}{5} = 1 \cdot \dfrac{3}{5} = \dfrac{3}{5}$$

Since 3 and 5 have no common factors (other than 1), $\dfrac{3}{5}$ is in simplest form.

■ Work Practice 4

Practice 4
Write in simplest form: $\dfrac{30}{45}$

If you have trouble finding common factors, write the prime factorization of the numerator and the denominator.

Example 5 Write in simplest form: $\dfrac{42x}{66}$

Solution: Let's write the prime factorizations of 42 and 66. Remember that $42x$ means $42 \cdot x$.

$$\dfrac{42x}{66} = \dfrac{2 \cdot 3 \cdot 7 \cdot x}{2 \cdot 3 \cdot 11} = \dfrac{2}{2} \cdot \dfrac{3}{3} \cdot \dfrac{7x}{11} = 1 \cdot 1 \cdot \dfrac{7x}{11} = \dfrac{7x}{11}$$

■ Work Practice 5

Practice 5
Write in simplest form: $\dfrac{39x}{51}$

In the example above, you may have saved time by noticing that 42 and 66 have a common factor of 6.

$$\dfrac{42x}{66} = \dfrac{6 \cdot 7x}{6 \cdot 11} = \dfrac{6}{6} \cdot \dfrac{7x}{11} = 1 \cdot \dfrac{7x}{11} = \dfrac{7x}{11}$$

Helpful Hint
Writing the prime factorizations of the numerator and the denominator is helpful in finding any common factors.

The method for simplifying negative fractions is the same as for positive fractions.

Example 6 Write in simplest form: $-\dfrac{10}{27}$

Solution:

$$-\dfrac{10}{27} = -\dfrac{2 \cdot 5}{3 \cdot 3 \cdot 3}$$ Prime factorizations of 10 and 27

Since 10 and 27 have no common factors, $-\dfrac{10}{27}$ is already in simplest form.

■ Work Practice 6

Note: We will study this concept further in the next section.

Practice 6
Write in simplest form: $-\dfrac{9}{50}$

Answers
4. $\dfrac{2}{3}$ **5.** $\dfrac{13x}{17}$ **6.** $-\dfrac{9}{50}$

Practice 7

Write in simplest form: $\dfrac{49}{112}$

Example 7 Write in simplest form: $\dfrac{30}{108}$

Solution:

$$\dfrac{30}{108} = \dfrac{2\cdot 3\cdot 5}{2\cdot 2\cdot 3\cdot 3\cdot 3} = \dfrac{2}{2}\cdot\dfrac{3}{3}\cdot\dfrac{5}{2\cdot 3\cdot 3} = 1\cdot 1\cdot\dfrac{5}{18} = \dfrac{5}{18}$$

■ Work Practice 7

We can use a shortcut procedure with common factors when simplifying.

$$\dfrac{4}{6} = \dfrac{\overset{1}{\cancel{2}}\cdot 2}{\underset{1}{\cancel{2}}\cdot 3} = \dfrac{1\cdot 2}{1\cdot 3} = \dfrac{2}{3}$$ Divide out the common factor of 2 in the numerator and denominator.

This procedure is possible because dividing out a common factor in the numerator and denominator is the same as removing a factor of 1 in the product.

Writing a Fraction in Simplest Form

To write a fraction in simplest form, write the prime factorization of the numerator and the denominator and then divide both by all common factors.

Practice 8

Write in simplest form: $-\dfrac{64}{20}$

Example 8 Write in simplest form: $-\dfrac{72}{26}$

Solution:

$$-\dfrac{72}{26} = -\dfrac{\overset{1}{\cancel{2}}\cdot 2\cdot 2\cdot 3\cdot 3}{\underset{1}{\cancel{2}}\cdot 13} = -\dfrac{1\cdot 2\cdot 2\cdot 3\cdot 3}{1\cdot 13} = -\dfrac{36}{13}$$

■ Work Practice 8

✓ **Concept Check** Which is the correct way to simplify the fraction $\dfrac{15}{25}$? Or are both correct? Explain.

a. $\dfrac{15}{25} = \dfrac{3\cdot\cancel{5}}{5\cdot\cancel{5}} = \dfrac{3}{5}$

b. $\dfrac{1\cancel{5}}{2\cancel{5}} = \dfrac{11}{21}$

In this chapter, we will simplify and perform operations on fractions containing variables. When the denominator of a fraction contains a variable, such as $\dfrac{6x^2}{60x^3}$, we will assume that the variable does not represent 0. Recall that the denominator of a fraction cannot be 0.

Practice 9

Write in simplest form: $\dfrac{7a^3}{56a^2}$

Example 9 Write in simplest form: $\dfrac{6x^2}{60x^3}$

Solution: Notice that 6 and 60 have a common factor of 6. Let's also use the definition of an exponent to factor x^2 and x^3.

$$\dfrac{6x^2}{60x^3} = \dfrac{\cancel{6}\cdot\cancel{x}\cdot\cancel{x}}{\cancel{6}\cdot 10\cdot\cancel{x}\cdot\cancel{x}\cdot x} = \dfrac{1\cdot 1\cdot 1}{1\cdot 10\cdot 1\cdot 1\cdot x} = \dfrac{1}{10x}$$

■ Work Practice 9

Answers

7. $\dfrac{7}{16}$ 8. $-\dfrac{16}{5}$ 9. $\dfrac{a}{8}$

✓ **Concept Check Answers**

a. correct   b. incorrect

Helpful Hint

Be careful when all factors of the numerator or denominator are divided out. In Example 9, the numerator was $1 \cdot 1 \cdot 1 = 1$, so the final result was $\dfrac{1}{10x}$.

Objective C Determining Whether Two Fractions Are Equivalent

Recall from Objective B that two fractions are equivalent if they represent the same part of a whole. One way to determine whether two fractions are equivalent is to see whether they simplify to the same fraction.

Example 10 Determine whether $\dfrac{16}{40}$ and $\dfrac{10}{25}$ are equivalent.

Solution: Simplify each fraction.

$$\dfrac{16}{40} = \dfrac{\overset{1}{\cancel{8}} \cdot 2}{\cancel{8} \cdot 5} = \dfrac{1 \cdot 2}{1 \cdot 5} = \dfrac{2}{5}$$

$$\dfrac{10}{25} = \dfrac{2 \cdot \overset{1}{\cancel{5}}}{5 \cdot \cancel{5}} = \dfrac{2 \cdot 1}{5 \cdot 1} = \dfrac{2}{5}$$

Since these fractions are the same, $\dfrac{16}{40} = \dfrac{10}{25}$.

Practice 10
Determine whether $\dfrac{7}{9}$ and $\dfrac{21}{27}$ are equivalent.

Work Practice 10

There is a shortcut method you may use to check or test whether two fractions are equivalent. In the example above, we learned that the fractions are equivalent, or

$$\dfrac{16}{40} = \dfrac{10}{25}$$

In this example above, we call $25 \cdot 16$ and $40 \cdot 10$ **cross products** because they are the products one obtains by multiplying diagonally across the equal sign, as shown below.

$$\underbrace{25 \cdot 16 \qquad\qquad 40 \cdot 10}_{\text{Cross Products}}$$

$$\dfrac{16}{40} = \dfrac{10}{25}$$

Notice that these cross products are equal:

$$25 \cdot 16 = 400, \quad 40 \cdot 10 = 400$$

In general, this is true for equivalent fractions.

Equality of Fractions

$$8 \cdot 6 \qquad\qquad 24 \cdot 2$$

$$\dfrac{6}{24} \overset{?}{=} \dfrac{2}{8}$$

Since the cross products ($8 \cdot 6 = 48$ and $24 \cdot 2 = 48$) are equal, the fractions are equal.

Note: If the cross products are not equal, the fractions are not equal.

Answer
10. equivalent

Practice 11
Determine whether $\frac{4}{13}$ and $\frac{5}{18}$ are equivalent.

Example 11
Determine whether $\frac{8}{11}$ and $\frac{19}{26}$ are equivalent.

Solution: Let's check cross products.

$26 \cdot 8 = 208$ $\frac{8}{11} \stackrel{?}{=} \frac{19}{26}$ $11 \cdot 19 = 209$

Since $208 \neq 209$, then $\frac{8}{11} \neq \frac{19}{26}$.

Helpful Hint: "Not equal to" symbol $\neq$

■ Work Practice 11

Objective D Solving Problems by Writing Fractions in Simplest Form

Many real-life problems can be solved by writing fractions. To make the answers clearer, these fractions should be written in simplest form.

Practice 12
There are only 14 mountains over 8000 meters in the world, all located in Asia. If 7 of these mountains are located in Nepal, determine what fraction of the mountains over 8000 meters are located in Nepal. Write the fraction in simplest form.

Example 12 Calculating Fraction of High Mountain Peaks

There are 49 mountain peaks over 7500 meters in the world, all located in Asia. Many of them form the borders between two countries and are thus counted twice on the bar graph below. Use the graph below to determine what fraction of the 49 mountain peaks over 7500 meters are located in India. Write the fraction in simplest form.

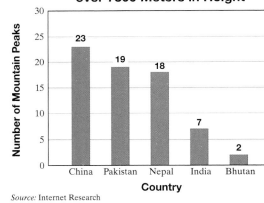

Location of Mountain Peaks over 7500 Meters in Height

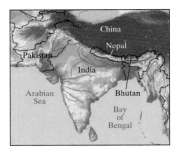

Source: Internet Research

Solution: First we determine the fraction of highest mountain peaks found in India.

$\frac{7}{49}$ ← mountain peaks over 7500 meters in India
      ← total mountain peaks over 7500 meters

Next, we simplify the fraction.

$\frac{7}{49} = \frac{\overset{1}{\cancel{7}}}{\underset{1}{\cancel{7}} \cdot 7} = \frac{1}{1 \cdot 7} = \frac{1}{7}$

Thus, $\frac{1}{7}$ of the world's mountain peaks over 7500 meters in height are located in India.

■ Work Practice 12

Answers
11. not equivalent **12.** $\frac{1}{2}$

Calculator Explorations Simplifying Fractions

Scientific Calculator

Many calculators have a fraction key, such as $\boxed{a^b/c}$, that allows you to simplify a fraction on the calculator. For example, to simplify $\frac{324}{612}$, enter

$\boxed{324}$ $\boxed{a^b/c}$ $\boxed{612}$ $\boxed{=}$

The display will read

$\boxed{9\ |\ 17}$

which represents $\frac{9}{17}$, the original fraction simplified.

Graphing Calculator

Graphing calculators also allow you to simplify fractions. The fraction option on a graphing calculator may be found under the $\boxed{\text{MATH}}$ menu.

To simplify $\frac{324}{612}$, enter

$\boxed{324}$ $\boxed{\div}$ $\boxed{612}$ $\boxed{\text{MATH}}$ $\boxed{\text{ENTER}}$ $\boxed{\text{ENTER}}$

The display will read

$\boxed{324/612 \blacktriangleright \text{Frac}\ 9/17}$

> **Helpful Hint**
> The Calculator Explorations boxes in this chapter provide only an introduction to fraction keys on calculators. Any time you use a calculator, there are both advantages and limitations to its use. Never rely solely on your calculator. It is very important that you understand how to perform all operations on fractions by hand in order to progress through later topics. For further information, talk to your instructor.

Use your calculator to simplify each fraction.

1. $\frac{128}{224}$
2. $\frac{231}{396}$
3. $\frac{340}{459}$
4. $\frac{999}{1350}$
5. $\frac{432}{810}$
6. $\frac{225}{315}$
7. $\frac{54}{243}$
8. $\frac{455}{689}$

Vocabulary, Readiness & Video Check

Use the choices below to fill in each blank.

| cross products | equivalent | composite |
| simplest form | prime factorization | prime |

1. The number 40 equals 2 · 2 · 2 · 5. Since each factor is prime, we call 2 · 2 · 2 · 5 the _____ of 40.
2. A natural number, other than 1, that is not prime is called a(n) _____ number.
3. A natural number that has exactly two different factors, 1 and itself, is called a(n) _____ number.
4. In $\frac{11}{48}$, since 11 and 48 have no common factors other than 1, $\frac{11}{48}$ is in _____.
5. Fractions that represent the same portion of a whole are called _____ fractions.
6. In the statement $\frac{5}{12} = \frac{15}{36}$, 5 · 36 and 12 · 15 are called _____.

Martin-Gay Interactive Videos Watch the section lecture video and answer the following questions.

See Video 4.2

Objective A 7. From Example 1, what two things should you check to make sure your prime factorization of a number is correct?

Objective B 8. From the lecture before Example 3, when you have a common factor in the numerator and denominator of a fraction, essentially you have what?

Objective C 9. Describe another way to solve Example 7 besides using cross products.

Objective D 10. Why isn't $\frac{10}{24}$ the final answer to Example 8? What is the final answer?

4.2 Exercise Set MyLab Math

Objective A Write the prime factorization of each number. See Examples 1 through 3.

1. 20
2. 12
3. 48
4. 75
5. 81
6. 64
7. 162
8. 128
9. 110
10. 130
11. 85
12. 93
13. 240
14. 836
15. 828
16. 504

Objective B Write each fraction in simplest form. See Examples 4 through 9.

17. $\frac{3}{12}$
18. $\frac{5}{30}$
19. $\frac{4x}{42}$
20. $\frac{9y}{48}$
21. $\frac{14}{16}$

22. $\frac{22}{34}$
23. $\frac{20}{30}$
24. $\frac{70}{80}$
25. $\frac{35a}{50a}$
26. $\frac{25z}{55z}$

27. $-\frac{63}{81}$
28. $-\frac{21}{49}$
29. $\frac{30x^2}{36x}$
30. $\frac{45b}{80b^2}$
31. $\frac{27}{64}$

32. $\frac{32}{63}$
33. $\frac{25xy}{40y}$
34. $\frac{36y}{42yz}$
35. $-\frac{40}{64}$
36. $-\frac{28}{60}$

37. $\frac{36x^3y^2}{24xy}$
38. $\frac{60a^2b}{36ab^3}$
39. $\frac{90}{120}$
40. $\frac{60}{150}$
41. $\frac{40xy}{64xyz}$

Section 4.2 | Factors and Simplest Form

42. $\dfrac{28abc}{60ac}$ **43.** $\dfrac{66}{308}$ **44.** $\dfrac{65}{234}$ **45.** $-\dfrac{55}{85y}$ **46.** $-\dfrac{78}{90x}$

47. $\dfrac{189z}{216z}$ **48.** $\dfrac{144y}{162y}$ **49.** $\dfrac{224a^3b^4c^2}{16ab^4c^2}$ **50.** $\dfrac{270x^4y^3z^3}{15x^3y^3z^3}$

Objective C *Determine whether each pair of fractions is equivalent. See Examples 10 and 11.*

51. $\dfrac{2}{6}$ and $\dfrac{4}{12}$ **52.** $\dfrac{3}{6}$ and $\dfrac{5}{10}$ ◐ **53.** $\dfrac{7}{11}$ and $\dfrac{5}{8}$

54. $\dfrac{2}{5}$ and $\dfrac{4}{11}$ **55.** $\dfrac{10}{15}$ and $\dfrac{6}{9}$ **56.** $\dfrac{4}{10}$ and $\dfrac{6}{15}$

57. $\dfrac{3}{9}$ and $\dfrac{6}{18}$ **58.** $\dfrac{2}{8}$ and $\dfrac{7}{28}$ **59.** $\dfrac{10}{13}$ and $\dfrac{13}{15}$

60. $\dfrac{16}{20}$ and $\dfrac{9}{16}$ **61.** $\dfrac{8}{18}$ and $\dfrac{12}{24}$ **62.** $\dfrac{6}{21}$ and $\dfrac{14}{35}$

Objective D *Solve. Write each fraction in simplest form. See Example 12.*

63. A work shift for an employee at Starbucks consists of 8 hours. What fraction of the employee's work shift is represented by 2 hours?

64. Two thousand baseball caps were sold one year at the U.S. Open Golf Tournament. What fractional part of this total do 200 caps represent?

65. There are 5280 feet in a mile. What fraction of a mile is represented by 2640 feet?

66. There are 100 centimeters in 1 meter. What fraction of a meter is 20 centimeters?

67. Sixteen out of the total fifty states in the United States have Ritz-Carlton hotels. (*Source:* Ritz-Carlton Hotel Company, LLC)
   a. What fraction of states can claim at least one Ritz-Carlton hotel?
   b. How many states do not have a Ritz-Carlton hotel?
   c. Write the fraction of states without a Ritz-Carlton hotel.

68. There are 152 national monuments in the United States. Twelve of these monuments are located in New Mexico. (*Source:* National Park Service)
   a. What fraction of the national monuments in the United States can be found in New Mexico?
   b. How many of the national monuments in the United States are found outside New Mexico?
   c. Write the fraction of national monuments found in states other than New Mexico.

69. The outer wall of the Pentagon is 24 inches thick. Ten inches is concrete, 8 inches is brick, and 6 inches is limestone. What fraction of the wall is concrete?

70. There are 35 students in a biology class. If 10 students made an A on the first test, what fraction of the students made an A?

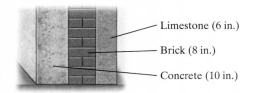
Limestone (6 in.)
Brick (8 in.)
Concrete (10 in.)

71. In the United States, Alaska has no income tax and Hawaii has state and federal income tax that can be filed at the same time. Thirty-six out of the 48 contiguous United States (the "lower" 48) allow residents to file their federal and state income tax electronically at the same time via efile.
   a. How many of the contiguous states do not have this type of service?
   b. What fraction of the contiguous states do not have this type of service? (*Source:* efile.com)

72. Katy Scarpulla bought a brand-new 2017 Toyota Camry Hybrid for $26,500. Her old car was traded in for $15,000.
   a. How much of her purchase price was not covered by her trade-in?
   b. What fraction of the purchase price was not covered by the trade-in?

73. As of this writing, a total of 366 individuals from the United States have been active astronauts. Of these, 45 are currently active and available for assignment in space. What fraction of U.S. astronauts are currently active? (*Source:* NASA)

74. Worldwide, Hallmark employs about 27,300 individuals. About 2700 employees work at the Hallmark headquarters in Kansas City, Missouri. What fraction of Hallmark employees work in Kansas City? (*Source:* Hallmark Card, Inc.)

Review

Evaluate each expression using the given replacement numbers. See Section 2.5.

75. $\dfrac{x^3}{9}$ when $x = -3$ **76.** $\dfrac{y^3}{5}$ when $y = -5$ **77.** $2y$ when $y = -7$ **78.** $-5a$ when $a = -4$

Concept Extensions

79. In your own words, define equivalent fractions.

80. Given a fraction, say $\dfrac{15}{40}$, how many fractions are there that are equivalent to it, but in simplest form or lowest terms? Explain your answer.

Write each fraction in simplest form.

81. $\dfrac{3975}{6625}$

82. $\dfrac{9506}{12,222}$

There are generally considered to be eight basic blood types. The table shows the number of people with the various blood types in a typical group of 100 blood donors. Use the table to answer Exercises 83 through 86. Write each answer in simplest form.

Distribution of Blood Types in Blood Donors	
Blood Type	**Number of People**
O Rh-positive	37
O Rh-negative	7
A Rh-positive	36
A Rh-negative	6
B Rh-positive	9
B Rh-negative	1
AB Rh-positive	3
AB Rh-negative	1
(*Source:* American Red Cross Biomedical Services)	

83. What fraction of blood donors have blood type A Rh-positive?

84. What fraction of blood donors have an O blood type?

85. What fraction of blood donors have an AB blood type?

86. What fraction of blood donors have a B blood type?

Section 4.2 | Factors and Simplest Form

Find the prime factorization of each number.

87. 34,020

88. 131,625

89. In your own words, define a prime number.

90. The number 2 is a prime number. All other even natural numbers are composite numbers. Explain why.

91. Two students have different prime factorizations for the same number. Is this possible? Explain.

92. Two students work to prime factor 120. One student starts by writing 120 as 12×10. The other student writes 120 as 24×5. Finish each prime factorization. Are they the same? Why or why not?

The following graph is called a circle graph or pie chart. Each sector (shaped like a piece of pie) shows the fraction of entering college freshmen who choose to major in each discipline shown. The whole circle represents the entire class of college freshmen. Use this graph to answer Exercises 93 through 96. Write each fraction answer in simplest form.

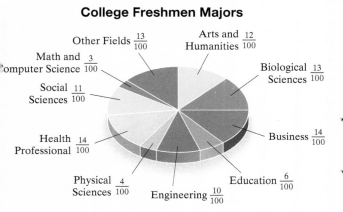

Source: The Higher Education Research Institute

93. What fraction of entering college freshmen plan to major in education?

94. What fraction of entering college freshmen plan to major in engineering?

95. Why is the business sector the same size as the health professional sector?

96. Why is the physical sciences sector smaller than the business sector?

Use this circle graph to answer Exercises 97 through 100. Write each fraction answer in simplest form.

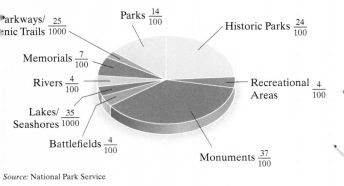

Source: National Park Service

97. What fraction of National Park Service areas are historic parks?

98. What fraction of National Park Service areas are parks?

99. Why is the national battlefields sector smaller than the national monuments sector?

100. Why is the recreation areas sector the same size as the battlefields sector?

Use the following numbers for Exercises 101 through 104.

8691   786   1235   2235   85   105   22   222   900   1470

101. List the numbers divisible by both 2 and 3.

102. List the numbers that are divisible by both 3 and 5.

103. The answers to Exercise 101 are also divisible by what number? Tell why.

104. The answers to Exercise 102 are also divisible by what number? Tell why.

4.3 Multiplying and Dividing Fractions

Objectives

A Multiply Fractions

B Evaluate Exponential Expressions with Fractional Bases.

C Divide Fractions.

D Multiply and Divide Given Fractional Replacement Values.

E Solve Applications That Require Multiplication of Fractions.

Objective A Multiplying Fractions

Let's use a diagram to discover how fractions are multiplied. For example, to multiply $\frac{1}{2}$ and $\frac{3}{4}$, we find $\frac{1}{2}$ of $\frac{3}{4}$. To do this, we begin with a diagram showing $\frac{3}{4}$ of a rectangle's area shaded.

$\frac{3}{4}$ of the rectangle's area is shaded.

To find $\frac{1}{2}$ of $\frac{3}{4}$, we heavily shade $\frac{1}{2}$ of the part that is already shaded.

By counting smaller rectangles, we see that $\frac{3}{8}$ of the larger rectangle is now heavily shaded, so that

$\frac{1}{2}$ of $\frac{3}{4}$ is $\frac{3}{8}$, or $\frac{1}{2} \cdot \frac{3}{4} = \frac{3}{8}$ Notice that $\frac{1}{2} \cdot \frac{3}{4} = \frac{1 \cdot 3}{2 \cdot 4} = \frac{3}{8}$.

Multiplying Fractions

To multiply two fractions, multiply the numerators and multiply the denominators. If $a, b, c,$ and d represent numbers, and b and d are not 0, we have

$$\frac{a}{b} \cdot \frac{c}{d} = \frac{a \cdot c}{b \cdot d}$$

Practice 1–2
Multiply.
1. $\frac{3}{7} \cdot \frac{5}{11}$ 2. $\frac{1}{3} \cdot \frac{1}{9}$

Examples Multiply.

1. $\frac{2}{3} \cdot \frac{5}{11} = \frac{2 \cdot 5}{3 \cdot 11} = \frac{10}{33}$ Multiply numerators. Multiply denominators.

This fraction is in simplest form since 10 and 33 have no common factors other than 1.

2. $\frac{1}{4} \cdot \frac{1}{2} = \frac{1 \cdot 1}{4 \cdot 2} = \frac{1}{8}$ This fraction is in simplest form.

■ Work Practice 1–2

Practice 3
Multiply and simplify: $\frac{6}{77} \cdot \frac{7}{8}$

Example 3 Multiply and simplify: $\frac{6}{7} \cdot \frac{14}{27}$

Solution:

$$\frac{6}{7} \cdot \frac{14}{27} = \frac{6 \cdot 14}{7 \cdot 27}$$

We can simplify by finding the prime factorizations and using our shortcut procedure of dividing out common factors in the numerator and denominator.

$$\frac{6 \cdot 14}{7 \cdot 27} = \frac{2 \cdot \overset{1}{\cancel{3}} \cdot 2 \cdot \overset{1}{\cancel{7}}}{\underset{1}{\cancel{7}} \cdot \underset{1}{\cancel{3}} \cdot 3 \cdot 3} = \frac{2 \cdot 2}{3 \cdot 3} = \frac{4}{9}$$

■ Work Practice 3

Answers
1. $\frac{15}{77}$ 2. $\frac{1}{27}$ 3. $\frac{3}{44}$

Section 4.3 | Multiplying and Dividing Fractions

Helpful Hint

Remember that the shortcut procedure in Example 3 is the same as removing factors of 1 in the product.

$$\frac{6 \cdot 14}{7 \cdot 27} = \frac{2 \cdot 3 \cdot 2 \cdot 7}{7 \cdot 3 \cdot 3 \cdot 3} = \frac{7}{7} \cdot \frac{3}{3} \cdot \frac{2 \cdot 2}{3 \cdot 3} = 1 \cdot 1 \cdot \frac{4}{9} = \frac{4}{9}$$

Example 4 Multiply and simplify: $\frac{23}{32} \cdot \frac{4}{7}$

Solution: Notice that 4 and 32 have a common factor of 4.

$$\frac{23}{32} \cdot \frac{4}{7} = \frac{23 \cdot 4}{32 \cdot 7} = \frac{23 \cdot \overset{1}{\cancel{4}}}{\underset{1}{\cancel{4}} \cdot 8 \cdot 7} = \frac{23}{8 \cdot 7} = \frac{23}{56}$$

■ Work Practice 4

Practice 4
Multiply and simplify: $\frac{4}{27} \cdot \frac{3}{8}$

Helpful Hint Don't forget that we may identify common factors that are not prime numbers.

After multiplying two fractions, always check to see whether the product can be simplified.

Example 5 Multiply: $-\frac{1}{4} \cdot \frac{1}{2}$

Solution: Recall that the product of a negative number and a positive number is a negative number.

$$-\frac{1}{4} \cdot \frac{1}{2} = -\frac{1 \cdot 1}{4 \cdot 2} = -\frac{1}{8}$$

■ Work Practice 5

Practice 5
Multiply.
$$\frac{1}{2} \cdot \left(-\frac{11}{28}\right)$$

Examples Multiply.

6. $\left(-\frac{6}{13}\right)\left(-\frac{26}{30}\right) = \frac{6 \cdot 26}{13 \cdot 30} = \frac{\overset{1}{\cancel{6}} \cdot \overset{1}{\cancel{13}} \cdot 2}{\underset{1}{\cancel{13}} \cdot \underset{1}{\cancel{6}} \cdot 5} = \frac{2}{5}$ The product of two negative numbers is a positive number.

7. $\frac{1}{3} \cdot \frac{2}{5} \cdot \frac{9}{16} = \frac{1 \cdot 2 \cdot 9}{3 \cdot 5 \cdot 16} = \frac{1 \cdot \overset{1}{\cancel{2}} \cdot \overset{1}{\cancel{3}} \cdot 3}{\underset{1}{\cancel{3}} \cdot 5 \cdot \underset{1}{\cancel{2}} \cdot 8} = \frac{3}{40}$

■ Work Practice 6–7

Practice 6–7
Multiply.
6. $\left(-\frac{4}{11}\right)\left(-\frac{33}{16}\right)$
7. $\frac{1}{6} \cdot \frac{3}{10} \cdot \frac{25}{16}$

We multiply fractions in the same way if variables are involved.

Example 8 Multiply: $\frac{3x}{4} \cdot \frac{8}{5x}$

Solution: Notice that 8 and 4 have a common factor of 4.

$$\frac{3x}{4} \cdot \frac{8}{5x} = \frac{3 \cdot \overset{1}{\cancel{x}} \cdot \overset{1}{\cancel{4}} \cdot 2}{\underset{1}{\cancel{4}} \cdot 5 \cdot \underset{1}{\cancel{x}}} = \frac{3 \cdot 1 \cdot 1 \cdot 2}{1 \cdot 5 \cdot 1} = \frac{6}{5}$$

■ Work Practice 8

Practice 8
Multiply: $\frac{2}{3} \cdot \frac{3y}{2}$

Helpful Hint

Recall that when the denominator of a fraction contains a variable, such as $\frac{8}{5x}$, we assume that the variable does not represent 0.

Answers

4. $\frac{1}{18}$ 5. $-\frac{11}{56}$ 6. $\frac{3}{4}$ 7. $\frac{5}{64}$ 8. y

Practice 9

Multiply: $\dfrac{a^3}{b^2} \cdot \dfrac{b}{a^2}$

Example 9 Multiply: $\dfrac{x^2}{y} \cdot \dfrac{y^3}{x}$

Solution:

$$\dfrac{x^2}{y} \cdot \dfrac{y^3}{x} = \dfrac{x^2 \cdot y^3}{y \cdot x} = \dfrac{\cancel{x} \cdot x \cdot \cancel{y} \cdot y \cdot y}{\cancel{y} \cdot \cancel{x}} = \dfrac{x \cdot y \cdot y}{1} = xy^2$$

■ Work Practice 9

Objective B Evaluating Expressions with Fractional Bases

The base of an exponential expression can also be a fraction.

$$\left(\dfrac{1}{3}\right)^4 = \underbrace{\dfrac{1}{3} \cdot \dfrac{1}{3} \cdot \dfrac{1}{3} \cdot \dfrac{1}{3}}_{\tfrac{1}{3} \text{ is a factor 4 times.}} = \dfrac{1 \cdot 1 \cdot 1 \cdot 1}{3 \cdot 3 \cdot 3 \cdot 3} = \dfrac{1}{81}$$

Practice 10

Evaluate.

a. $\left(\dfrac{3}{4}\right)^3$ b. $\left(-\dfrac{4}{5}\right)^2$

Example 10 Evaluate.

a. $\left(\dfrac{2}{5}\right)^4 = \dfrac{2}{5} \cdot \dfrac{2}{5} \cdot \dfrac{2}{5} \cdot \dfrac{2}{5} = \dfrac{2 \cdot 2 \cdot 2 \cdot 2}{5 \cdot 5 \cdot 5 \cdot 5} = \dfrac{16}{625}$

b. $\left(-\dfrac{1}{4}\right)^2 = \left(-\dfrac{1}{4}\right) \cdot \left(-\dfrac{1}{4}\right) = \dfrac{1 \cdot 1}{4 \cdot 4} = \dfrac{1}{16}$ The product of two negative numbers is a positive number.

■ Work Practice 10

Objective C Dividing Fractions

Before we can divide fractions, we need to know how to find the **reciprocal** of a fraction.

> **Reciprocal of a Fraction**
>
> Two numbers are **reciprocals** of each other if their product is 1. The reciprocal of the fraction $\dfrac{a}{b}$ is $\dfrac{b}{a}$ because $\dfrac{a}{b} \cdot \dfrac{b}{a} = \dfrac{a \cdot b}{b \cdot a} = 1.$

> **Helpful Hint**
>
> Every number has a reciprocal except 0. The number 0 has no reciprocal because there is no number a such that $0 \cdot a = 1.$

For example,

• The reciprocal of $\dfrac{7}{2}$ is $\dfrac{2}{7}$ because $\dfrac{7}{2} \cdot \dfrac{2}{7} = \dfrac{14}{14} = 1.$

• The reciprocal of 3 is $\dfrac{1}{3}$ because $3 \cdot \dfrac{1}{3} = \dfrac{3}{1} \cdot \dfrac{1}{3} = \dfrac{3}{3} = 1.$

• The reciprocal of $-\dfrac{5}{6}$ is $-\dfrac{6}{5}$ because $-\dfrac{5}{6} \cdot -\dfrac{6}{5} = \dfrac{30}{30} = 1.$

Answers

9. $\dfrac{a}{b}$ 10. a. $\dfrac{27}{64}$ b. $\dfrac{16}{25}$

Division of fractions has the same meaning as division of whole numbers. For example,

10 ÷ 5 means: How many 5s are there in 10?

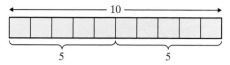

There are two 5s in 10, so 10 ÷ 5 = 2.

$\frac{3}{4} \div \frac{1}{8}$ means: How many $\frac{1}{8}$s are there in $\frac{3}{4}$?

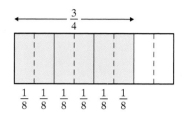

There are six $\frac{1}{8}$s in $\frac{3}{4}$, so $\frac{3}{4} \div \frac{1}{8} = 6$.

We use reciprocals to divide fractions.

Dividing Fractions

To divide two fractions, multiply the first fraction by the reciprocal of the second fraction.

If a, b, c, and d represent numbers, and b, c, and d are not 0, then

$$\frac{a}{b} \div \frac{c}{d} = \frac{a}{b} \cdot \frac{d}{c} = \frac{a \cdot d}{b \cdot c}$$

↑ reciprocal

For example,

$$\frac{3}{4} \div \frac{1}{8} = \frac{3}{4} \cdot \frac{8}{1} = \frac{3 \cdot 8}{4 \cdot 1} = \frac{3 \cdot 2 \cdot \cancel{4}}{\cancel{4} \cdot 1} = \frac{6}{1} \text{ or } 6$$

(multiply by reciprocal)

After dividing fractions, *always* check to see whether the result can be simplified.

Examples Divide and simplify.

11. $\frac{5}{16} \div \frac{3}{4} = \frac{5}{16} \cdot \frac{4}{3} = \frac{5 \cdot 4}{16 \cdot 3} = \frac{5 \cdot \cancel{4}}{\cancel{4} \cdot 4 \cdot 3} = \frac{5}{12}$

12. $\frac{2}{5} \div \frac{1}{2} = \frac{2}{5} \cdot \frac{2}{1} = \frac{2 \cdot 2}{5 \cdot 1} = \frac{4}{5}$

Work Practice 11–12

Practice 11–12

Divide and simplify.

11. $\frac{8}{7} \div \frac{2}{9}$ **12.** $\frac{4}{9} \div \frac{1}{2}$

Answers

11. $\frac{36}{7}$ **12.** $\frac{8}{9}$

Helpful Hint

When dividing by a fraction, do not look for common factors to divide out until you rewrite the division as multiplication.

Do not try to divide out these two 2s.

$$\frac{1}{2} \div \frac{2}{3} = \frac{1}{2} \cdot \frac{3}{2} = \frac{3}{4}$$

Practice 13

Divide: $-\dfrac{10}{4} \div \dfrac{2}{9}$

Example 13 Divide: $-\dfrac{7}{12} \div -\dfrac{5}{6}$

Solution: Recall that the quotient (or product) of two negative numbers is a positive number.

$$-\frac{7}{12} \div -\frac{5}{6} = -\frac{7}{12} \cdot -\frac{6}{5} = \frac{7 \cdot \cancel{6}}{2 \cdot \cancel{6} \cdot 5} = \frac{7}{10}$$

■ Work Practice 13

Practice 14

Divide: $\dfrac{3y}{4} \div 5y^3$

Example 14 Divide: $\dfrac{2x}{3} \div 3x^2$

Solution:

$$\frac{2x}{3} \div 3x^2 = \frac{2x}{3} \div \frac{3x^2}{1} = \frac{2x}{3} \cdot \frac{1}{3x^2} = \frac{2 \cdot \cancel{x} \cdot 1}{3 \cdot 3 \cdot \cancel{x} \cdot x} = \frac{2}{9x}$$

■ Work Practice 14

Practice 15

Simplify: $\left(-\dfrac{2}{3} \cdot \dfrac{9}{14}\right) \div \dfrac{7}{15}$

Example 15 Simplify: $\left(\dfrac{4}{7} \cdot \dfrac{3}{8}\right) \div -\dfrac{3}{4}$

Solution: Remember to perform the operations inside the ( ) first.

$$\left(\frac{4}{7} \cdot \frac{3}{8}\right) \div -\frac{3}{4} = \left(\frac{\cancel{4} \cdot 3}{7 \cdot 2 \cdot \cancel{4}}\right) \div -\frac{3}{4} = \frac{3}{14} \div -\frac{3}{4}$$

Now divide.

$$\frac{3}{14} \div -\frac{3}{4} = \frac{3}{14} \cdot -\frac{4}{3} = -\frac{\cancel{3} \cdot \cancel{2} \cdot 2}{\cancel{2} \cdot 7 \cdot \cancel{3}} = -\frac{2}{7}$$

■ Work Practice 15

Answers

13. $-\dfrac{45}{4}$ 14. $\dfrac{3}{20y^2}$ 15. $-\dfrac{45}{49}$

✓**Concept Check Answers**

a. incorrect   b. correct

✓**Concept Check** Which is the correct way to divide $\dfrac{3}{5}$ by $\dfrac{5}{12}$? Explain.

a. $\dfrac{3}{5} \div \dfrac{5}{12} = \dfrac{5}{3} \cdot \dfrac{5}{12}$ b. $\dfrac{3}{5} \div \dfrac{5}{12} = \dfrac{3}{5} \cdot \dfrac{12}{5}$

Objective D Multiplying and Dividing with Fractional Replacement Values

Recall the difference between an expression and an equation. For example, xy and $x \div y$ are expressions. They contain no equal signs. In Example 16, we practice *simplifying* expressions given fractional replacement values.

Example 16 If $x = \frac{7}{8}$ and $y = -\frac{1}{3}$, evaluate (a) xy and (b) $x \div y$.

Solution: Replace x with $\frac{7}{8}$ and y with $-\frac{1}{3}$.

a. $xy = \frac{7}{8} \cdot -\frac{1}{3}$
$= -\frac{7 \cdot 1}{8 \cdot 3}$
$= -\frac{7}{24}$

b. $x \div y = \frac{7}{8} \div -\frac{1}{3}$
$= \frac{7}{8} \cdot -\frac{3}{1}$
$= -\frac{7 \cdot 3}{8 \cdot 1}$
$= -\frac{21}{8}$

■ Work Practice 16

Practice 16

If $x = -\frac{3}{4}$ and $y = \frac{9}{2}$, evaluate (a) xy and (b) $x \div y$.

Example 17 Is $-\frac{2}{3}$ a solution of the equation $-\frac{1}{2}x = \frac{1}{3}$?

Solution: To check whether a number is a solution of an equation, recall that we replace the variable with the given number and see if a true statement results.

$-\frac{1}{2} \cdot x = \frac{1}{3}$ Recall that $-\frac{1}{2}x$ means $-\frac{1}{2} \cdot x$.

$-\frac{1}{2} \cdot -\frac{2}{3} \stackrel{?}{=} \frac{1}{3}$ Replace x with $-\frac{2}{3}$.

$\frac{1 \cdot \overset{1}{2}}{\underset{1}{2} \cdot 3} \stackrel{?}{=} \frac{1}{3}$ The product of two negative numbers is a positive number.

$\frac{1}{3} = \frac{1}{3}$ True

Since we have a true statement, $-\frac{2}{3}$ is a solution.

■ Work Practice 17

Practice 17

Is $-\frac{9}{8}$ a solution of the equation $2x = -\frac{9}{4}$?

Objective E Solving Problems by Multiplying Fractions

To solve real-life problems that involve multiplying fractions, we use our four problem-solving steps from Chapter 3. In Example 18, a new key word that implies multiplication is used. That key word is "**of.**"

Helpful Hint "of" usually translates to multiplication.

Answers

16. a. $-\frac{27}{8}$ **b.** $-\frac{1}{6}$ **17.** yes

Practice 18

Hershey Park is an amusement park in Hershey, Pennsylvania. Of its 66 rides, $\frac{1}{6}$ of these are roller coasters. How many roller coasters are in Hershey Park?

Example 18 — Finding the Number of Roller Coasters in an Amusement Park

Cedar Point is an amusement park located in Sandusky, Ohio. Its collection of 72 rides is the largest in the world. Of the rides, $\frac{2}{9}$ are roller coasters. How many roller coasters are in Cedar Point's collection of rides? (*Source:* Wikipedia)

Solution:

1. **UNDERSTAND** the problem. To do so, read and reread the problem. We are told that $\frac{2}{9}$ of Cedar Point's rides are roller coasters. The word "of" here means multiplication.

2. **TRANSLATE.**

    In words: number of roller coasters **is** $\frac{2}{9}$ **of** total rides at Cedar Point

    Translate: number of roller coasters $= \frac{2}{9} \cdot 72$

3. **SOLVE:** Before we solve, let's estimate a reasonable answer. The fraction $\frac{2}{9}$ is less than $\frac{1}{4}$ (draw a diagram, if needed), and $\frac{1}{4}$ of 72 rides is 18 rides, so the number of roller coasters should be less than 18.

$$\frac{2}{9} \cdot 72 = \frac{2}{9} \cdot \frac{72}{1} = \frac{2 \cdot 72}{9 \cdot 1} = \frac{2 \cdot \overset{1}{\cancel{9}} \cdot 8}{\cancel{9} \cdot 1} = \frac{16}{1} \text{ or } 16$$

4. **INTERPRET.** *Check* your work. From our estimate, our answer is reasonable. *State* your conclusion: The number of roller coasters at Cedar Point is 16.

■ Work Practice 18

Helpful Hint

To help visualize a fractional part of a whole number, look at the diagram below.

$\frac{1}{5}$ of 60 = ?

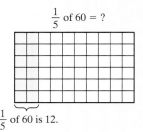

$\frac{1}{5}$ of 60 is 12.

Answer
18. 11 roller coasters

Section 4.3 | Multiplying and Dividing Fractions

Vocabulary, Readiness & Video Check

Use the choices below to fill in each blank. Not all choices will be used.

multiplication $\dfrac{a \cdot d}{b \cdot c}$ $\dfrac{a \cdot c}{b \cdot d}$ $\dfrac{2 \cdot 2 \cdot 2}{7}$ $\dfrac{2}{7} \cdot \dfrac{2}{7} \cdot \dfrac{2}{7}$

division   0   reciprocals

1. To multiply two fractions, we write $\dfrac{a}{b} \cdot \dfrac{c}{d} = $ _____ .
2. Two numbers are _____ of each other if their product is 1.
3. The expression $\dfrac{2^3}{7} = $ _____ while $\left(\dfrac{2}{7}\right)^3 = $ _____ .
4. Every number has a reciprocal except _____ .
5. To divide two fractions, we write $\dfrac{a}{b} \div \dfrac{c}{d} = $ _____ .
6. The word "of" indicates _____ .

See Video 4.3

Martin-Gay Interactive Videos   Watch the section lecture video and answer the following questions.

Objective A   7. In ▸ Example 2, how do we know that the answer is negative? ▸

Objective B   8. In ▸ Example 4, does the exponent apply to the negative sign? Why or why not? ▸

Objective C   9. Complete this statement based on ▸ Example 5: When writing the reciprocal of a fraction, the denominator becomes the _____ , and the numerator becomes the _____ . ▸

Objective D   10. In ▸ Example 11a, why don't we write out the prime factorizations of 4 and 9 in the numerator? ▸

Objective E   11. What formula is used to solve ▸ Example 12? ▸

4.3 Exercise Set   MyLab Math

Objective A   *Multiply. Write the product in simplest form. See Examples 1 through 9.*

1. $\dfrac{6}{11} \cdot \dfrac{3}{7}$
2. $\dfrac{5}{9} \cdot \dfrac{7}{4}$
3. $-\dfrac{2}{7} \cdot \dfrac{5}{8}$
4. $\dfrac{4}{15} \cdot -\dfrac{1}{20}$

5. $-\dfrac{1}{2} \cdot -\dfrac{2}{15}$
6. $-\dfrac{3}{11} \cdot -\dfrac{11}{12}$
7. $\dfrac{18x}{20} \cdot \dfrac{36}{99}$
8. $\dfrac{5}{32} \cdot \dfrac{64y}{100}$

9. $3a^2 \cdot \dfrac{1}{4}$
10. $-\dfrac{2}{3} \cdot 6y^3$
11. $\dfrac{x^3}{y^3} \cdot \dfrac{y^2}{x}$
12. $\dfrac{a}{b^3} \cdot \dfrac{b}{a^3}$

13. $0 \cdot \dfrac{8}{9}$
14. $\dfrac{11}{12} \cdot 0$
15. $-\dfrac{17y}{20} \cdot \dfrac{4}{5y}$
16. $-\dfrac{13x}{20} \cdot \dfrac{5}{6x}$

17. $\dfrac{11}{20} \cdot \dfrac{1}{7} \cdot \dfrac{5}{22}$
18. $\dfrac{27}{32} \cdot \dfrac{10}{13} \cdot \dfrac{16}{30}$

Objective **B** *Evaluate. See Example 10.*

19. $\left(\dfrac{1}{5}\right)^3$

20. $\left(\dfrac{8}{9}\right)^2$

21. $\left(-\dfrac{2}{3}\right)^2$

22. $\left(-\dfrac{1}{2}\right)^4$

23. $\left(-\dfrac{2}{3}\right)^3 \cdot \dfrac{1}{2}$

24. $\left(-\dfrac{3}{4}\right)^3 \cdot \dfrac{1}{3}$

Objective **C** *Divide. Write all quotients in simplest form. See Examples 11 through 14.*

25. $\dfrac{2}{3} \div \dfrac{5}{6}$

26. $\dfrac{5}{8} \div \dfrac{3}{4}$

27. $-\dfrac{6}{15} \div \dfrac{12}{5}$

28. $-\dfrac{4}{15} \div -\dfrac{8}{3}$

29. $-\dfrac{8}{9} \div \dfrac{x}{2}$

30. $\dfrac{10}{11} \div -\dfrac{4}{5x}$

31. $\dfrac{11y}{20} \div \dfrac{3}{11}$

32. $\dfrac{9z}{20} \div \dfrac{2}{9}$

33. $-\dfrac{2}{3} \div 4$

34. $-\dfrac{5}{6} \div 10$

35. $\dfrac{1}{5x} \div \dfrac{5}{x^2}$

36. $\dfrac{3}{y^2} \div \dfrac{9}{y^3}$

Objectives **A B C** **Mixed Practice** *Perform each indicated operation. See Examples 1 through 15.*

37. $\dfrac{2}{3} \cdot \dfrac{5}{9}$

38. $\dfrac{8}{15} \cdot \dfrac{5}{32}$

39. $\dfrac{3x}{7} \div \dfrac{5}{6x}$

40. $\dfrac{2}{5y} \div \dfrac{5y}{11}$

41. $\dfrac{16}{27y} \div \dfrac{8}{15y}$

42. $\dfrac{12y}{21} \div \dfrac{4y}{7}$

43. $-\dfrac{5}{28} \cdot \dfrac{35}{25}$

44. $\dfrac{24}{45} \cdot -\dfrac{5}{8}$

45. $\left(-\dfrac{3}{4}\right)^2$

46. $\left(-\dfrac{1}{2}\right)^5$

47. $\dfrac{x^2}{y} \cdot \dfrac{y^3}{x}$

48. $\dfrac{b}{a^2} \cdot \dfrac{a^3}{b^3}$

49. $7 \div \dfrac{2}{11}$

50. $-100 \div \dfrac{1}{2}$

51. $-3x \div \dfrac{x^2}{12}$

52. $-7x^2 \div \dfrac{14x}{3}$

53. $\left(\dfrac{2}{7} \div \dfrac{7}{2}\right) \cdot \dfrac{3}{4}$

54. $\dfrac{1}{2} \cdot \left(\dfrac{5}{6} \div \dfrac{1}{12}\right)$

55. $-\dfrac{19}{63y} \cdot 9y^2$

56. $16a^2 \cdot -\dfrac{31}{24a}$

57. $-\dfrac{2}{3} \cdot -\dfrac{6}{11}$

58. $-\dfrac{1}{5} \cdot -\dfrac{6}{7}$

59. $\dfrac{4}{8} \div \dfrac{3}{16}$

60. $\dfrac{9}{2} \div \dfrac{16}{15}$

61. $\dfrac{21x^2}{10y} \div \dfrac{14x}{25y}$

62. $\dfrac{17y^2}{24x} \div \dfrac{13y}{18x}$

63. $\left(1 \div \dfrac{3}{4}\right) \cdot \dfrac{2}{3}$

64. $\left(33 \div \dfrac{2}{11}\right) \cdot \dfrac{5}{9}$

65. $\dfrac{a^3}{2} \div 30a^3$

66. $15c^3 \div \dfrac{3c^2}{5}$

67. $\dfrac{ab^2}{c} \cdot \dfrac{c}{ab}$

68. $\dfrac{ac}{b} \cdot \dfrac{b^3}{a^2 c}$

69. $\left(\dfrac{1}{2} \cdot \dfrac{2}{3}\right) \div \dfrac{5}{6}$

70. $\left(\dfrac{3}{4} \cdot \dfrac{8}{9}\right) \div \dfrac{2}{5}$

71. $-\dfrac{4}{7} \div \left(\dfrac{4}{5} \cdot \dfrac{3}{7}\right)$

72. $\dfrac{5}{8} \div \left(\dfrac{4}{7} \cdot -\dfrac{5}{16}\right)$

Section 4.3 | Multiplying and Dividing Fractions

Objective D *Given the following replacement values, evaluate (**a**) xy and (**b**) $x \div y$. See Example 16.*

73. $x = \dfrac{2}{5}$ and $y = \dfrac{5}{6}$
74. $x = \dfrac{8}{9}$ and $y = \dfrac{1}{4}$
75. $x = -\dfrac{4}{5}$ and $y = \dfrac{9}{11}$
76. $x = \dfrac{7}{6}$ and $y = -\dfrac{1}{2}$

Determine whether the given replacement values are solutions of the given equations. See Example 17.

77. Is $-\dfrac{5}{18}$ a solution to $3x = -\dfrac{5}{6}$?

78. Is $\dfrac{9}{11}$ a solution to $\dfrac{2}{3}y = \dfrac{6}{11}$?

79. Is $\dfrac{2}{5}$ a solution to $-\dfrac{1}{2}z = \dfrac{1}{10}$?

80. Is $\dfrac{3}{5}$ a solution to $5x = \dfrac{1}{3}$?

Objective E Translating *Solve. Write each answer in simplest form. For Exercises 81 through 84, recall that "of" translates to multiplication. See Example 18.*

81. Find $\dfrac{1}{4}$ of 200.
82. Find $\dfrac{1}{5}$ of 200.
83. Find $\dfrac{5}{6}$ of 24.
84. Find $\dfrac{5}{8}$ of 24.

Solve. For Exercises 85 and 86, the solutions have been started for you. See Example 18.

85. In the United States, $\dfrac{7}{50}$ of college freshmen major in business. A community college in Pennsylvania has a freshman enrollment of approximately 800 students. How many of these freshmen might we project are majoring in business?

Start the solution:
1. UNDERSTAND the problem. Reread it as many times as needed.
2. TRANSLATE into an equation. (Fill in the blank below.)

freshmen majoring in business	is	$\dfrac{7}{50}$	of	community college freshmen enrollment
↓	↓	↓	↓	↓
freshmen majoring in business	=	$\dfrac{7}{50}$	·	_____

Finish with:
3. SOLVE
4. INTERPRET

86. A patient was told that, at most, $\dfrac{1}{5}$ of his calories should come from fat. If his diet consists of 3000 calories a day, find the maximum number of calories that can come from fat.

Start the solution:
1. UNDERSTAND the problem. Reread it as many times as needed.
2. TRANSLATE into an equation. (Fill in the blank below.)

patient's fat calories	is	$\dfrac{1}{5}$	of	his daily calories
↓	↓	↓	↓	↓
patient's fat calories	=	$\dfrac{1}{5}$	·	_____

Finish with:
3. SOLVE
4. INTERPRET

87. In 2015, there were approximately 240 million moviegoers in the United States and Canada. Of these, about $\dfrac{7}{20}$ viewed at least one 3-D movie. Find the approximate number of people who viewed at least one 3-D movie. (*Source:* Motion Picture Association of America)

88. In 2015, movie theater owners received a total of $11,100 million in movie admission tickets. About $\dfrac{3}{20}$ of this amount was for 3-D movies. Find the amount of money received from 3-D movies. (*Source:* Motion Picture Association of America)

89. The Oregon National Historic Trail is 2170 miles long. It begins in Independence, Missouri, and ends in Oregon City, Oregon. Manfred Coulon has hiked $\frac{2}{5}$ of the trail before. How many miles has he hiked? (*Source:* National Park Service)

90. Each turn of a screw sinks it $\frac{3}{16}$ of an inch deeper into a piece of wood. Find how deep the screw is after 8 turns.

91. The radius of a circle is one-half of its diameter, as shown. If the diameter of a circle is $\frac{3}{8}$ of an inch, what is its radius?

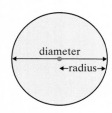

92. The diameter of a circle is twice its radius, as shown in the Exercise 91 illustration. If the radius of a circle is $\frac{7}{20}$ of a foot, what is its diameter?

93. A special on a cruise to the Bahamas is advertised to be $\frac{2}{3}$ of the regular price. If the regular price is $2757, what is the sale price?

94. A family recently sold their house for $102,000, but $\frac{3}{50}$ of this amount goes to the real estate companies that helped them sell their house. How much money does the family pay to the real estate companies?

95. The state of Mississippi houses $\frac{1}{184}$ of the total U.S. libraries. If there are about 9200 libraries in the United States, how many libraries are in Mississippi?

96. There have been about 595 contestants on the reality television show *Survivor* over 33 seasons. Some of these contestants have appeared in multiple seasons. If the number of repeat contestants is $\frac{19}{119}$ of the total number of participants in the first 33 seasons, how many contestants have participated more than once? (*Source:* Survivor.com)

Find the area of each rectangle. Recall that area = length · width.

△ 97.

△ 98.

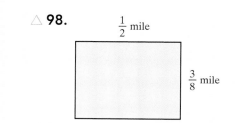

Recall from Section 4.2 that the following graph is called a **circle graph** or **pie chart.** Each sector (shaped like a piece of pie) shows the fractional part of a car's total mileage that falls into a particular category. The whole circle represents a car's total mileage.

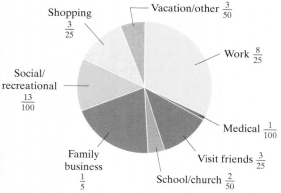

Source: The American Automobile Manufacturers Association and The National Automobile Dealers Association

In one year, a family drove 12,000 miles in the family car. Use the circle graph to determine how many of these miles might be expected to fall in the categories shown in Exercises 99 through 102.

99. Work

100. Shopping

101. Family business

102. Medical

Review

Perform each indicated operation. See Section 1.3.

103. $\begin{array}{r} 27 \\ 76 \\ + 98 \end{array}$

104. $\begin{array}{r} 811 \\ 42 \\ + 69 \end{array}$

105. $\begin{array}{r} 968 \\ - 772 \end{array}$

106. $\begin{array}{r} 882 \\ - 773 \end{array}$

Concept Extensions

107. In your own words, describe how to divide fractions.

108. In your own words, explain how to multiply fractions.

Simplify.

109. $\dfrac{42}{25} \cdot \dfrac{125}{36} \div \dfrac{7}{6}$

110. $\left(\dfrac{8}{13} \cdot \dfrac{39}{16} \cdot \dfrac{8}{9}\right)^2 \div \dfrac{1}{2}$

111. Approximately $\dfrac{1}{8}$ of the U.S. population lives in the state of California. If the U.S. population is approximately 325,040,000, find the approximate population of California. (*Source:* U.S. Census Bureau)

112. In 2016, there were approximately 4725 degree-granting postsecondary institutions in the United States. Of these, approximately $\dfrac{9}{25}$ were two-year colleges. How many two-year colleges were there in 2016? (Round to the nearest whole.) (*Source:* National Center for Educational Statistics)

113. The National Park Service is charged with maintaining 27,000 historic structures. Monuments and statues make up $\dfrac{63}{200}$ of these historic structures. How many monuments and statues is the National Park Service charged with maintaining? (*Source:* National Park Service)

114. If $\dfrac{3}{4}$ of 36 students on a first bus are girls and $\dfrac{2}{3}$ of the 30 students on a second bus are *boys*, how many students on the two buses are girls?

4.4 Adding and Subtracting Like Fractions, Least Common Denominator, and Equivalent Fractions

Objectives

A Add or Subtract Like Fractions.

B Add or Subtract Given Fractional Replacement Values.

C Solve Problems by Adding or Subtracting Like Fractions.

D Find the Least Common Denominator of a List of Fractions.

E Write Equivalent Fractions.

Fractions with the same denominator are called **like fractions**. Fractions that have different denominators are called **unlike fractions**.

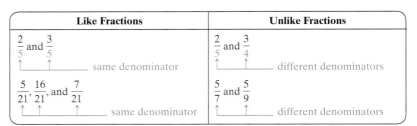

Objective A Adding or Subtracting Like Fractions

To see how we add like fractions (fractions with the same denominator), study one or both illustrations below.

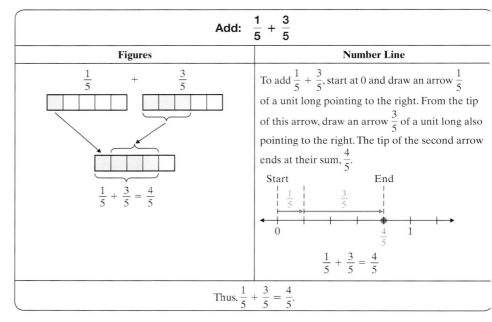

Notice that the numerator of the sum is the sum of the numerators. Also, the denominator of the sum is the **common denominator.** This is how we add fractions. Similar illustrations can be shown for subtracting fractions.

> **Adding or Subtracting Like Fractions (Fractions with the Same Denominator)**
>
> If a, b, and c are numbers and b is not 0, then
>
> $$\frac{a}{b} + \frac{c}{b} = \frac{a+c}{b} \qquad \text{and also} \qquad \frac{a}{b} - \frac{c}{b} = \frac{a-c}{b}$$

In other words, to add or subtract fractions with the same denominator, add or subtract their numerators and write the sum or difference over the **common** denominator.

Section 4.4 | Adding and Subtracting Like Fractions, Least Common Denominator

For example,

$$\frac{1}{4} + \frac{2}{4} = \frac{1+2}{4} = \frac{3}{4}$$ Add the numerators. Keep the denominator.

$$\frac{4}{5} - \frac{2}{5} = \frac{4-2}{5} = \frac{2}{5}$$ Subtract the numerators. Keep the denominator.

> **Helpful Hint**
> As usual, don't forget to write all answers in simplest form.

Examples Add and simplify.

1. $\frac{2}{7} + \frac{3}{7} = \frac{2+3}{7} = \frac{5}{7}$ ← Add the numerators. ← Keep the common denominator.

2. $\frac{3}{16x} + \frac{7}{16x} = \frac{3+7}{16x} = \frac{10}{16x} = \frac{\overset{1}{\cancel{2}} \cdot 5}{\underset{1}{\cancel{2}} \cdot 8 \cdot x} = \frac{5}{8x}$

3. $\frac{7}{8} + \frac{6}{8} + \frac{3}{8} = \frac{7+6+3}{8} = \frac{16}{8}$ or 2

Work Practice 1–3

Practice 1–3
Add and simplify.

1. $\frac{6}{13} + \frac{2}{13}$

2. $\frac{5}{8x} + \frac{1}{8x}$

3. $\frac{20}{11} + \frac{6}{11} + \frac{7}{11}$

✓ **Concept Check** Find and correct the error in the following:

$$\frac{1}{5} + \frac{1}{5} = \frac{2}{10}$$

Examples Subtract and simplify.

4. $\frac{8}{9} - \frac{1}{9} = \frac{8-1}{9} = \frac{7}{9}$ ← Subtract the numerators. ← Keep the common denominator.

5. $\frac{7}{8} - \frac{5}{8} = \frac{7-5}{8} = \frac{2}{8} = \frac{\overset{1}{\cancel{2}}}{\underset{1}{\cancel{2}} \cdot 4} = \frac{1}{4}$

Work Practice 4–5

Practice 4–5
Subtract and simplify.

4. $\frac{11}{12} - \frac{6}{12}$

5. $\frac{7}{15} - \frac{2}{15}$

From our earlier work, we know that

$$\frac{-12}{6} = \frac{12}{-6} = -\frac{12}{6}$$ since these all simplify to −2.

In general, the following is true:

$$\frac{-a}{b} = \frac{a}{-b} = -\frac{a}{b}$$ as long as b is not 0.

Example 6 Add: $-\frac{11}{8} + \frac{6}{8}$

Solution: $-\frac{11}{8} + \frac{6}{8} = \frac{-11+6}{8}$

$$= \frac{-5}{8} \text{ or } -\frac{5}{8}$$

Work Practice 6

Practice 6
Add: $-\frac{8}{17} + \frac{4}{17}$

Answers

1. $\frac{8}{13}$ 2. $\frac{3}{4x}$ 3. 3 4. $\frac{5}{12}$ 5. $\frac{1}{3}$
6. $-\frac{4}{17}$

✓ **Concept Check Answer**
We don't add denominators together;
correct solution: $\frac{1}{5} + \frac{1}{5} = \frac{2}{5}$.

Chapter 4 | Fractions and Mixed Numbers

Practice 7

Subtract: $\dfrac{2}{5} - \dfrac{7y}{5}$

Example 7 Subtract: $\dfrac{3x}{4} - \dfrac{7}{4}$

Solution: $\dfrac{3x}{4} - \dfrac{7}{4} = \dfrac{3x - 7}{4}$

Recall from Section 3.1 that the terms in the numerator are unlike terms and cannot be combined.

■ Work Practice 7

Practice 8

Subtract: $\dfrac{4}{11} - \dfrac{6}{11} - \dfrac{3}{11}$

Example 8 Subtract: $\dfrac{3}{7} - \dfrac{6}{7} - \dfrac{3}{7}$

Solution: $\dfrac{3}{7} - \dfrac{6}{7} - \dfrac{3}{7} = \dfrac{3 - 6 - 3}{7} = \dfrac{-6}{7}$ or $-\dfrac{6}{7}$

■ Work Practice 8

Helpful Hint

Recall that $\dfrac{-6}{7} = -\dfrac{6}{7}$ $\left(\text{Also, } \dfrac{6}{-7} = -\dfrac{6}{7}, \text{ if needed.}\right)$

Objective B Adding or Subtracting Given Fractional Replacement Values

Practice 9

Evaluate $x + y$ if $x = -\dfrac{10}{12}$ and $y = \dfrac{5}{12}$.

Example 9 Evaluate $y - x$ if $x = -\dfrac{3}{10}$ and $y = -\dfrac{8}{10}$.

Solution: Be very careful when replacing x and y with replacement values.

$y - x = -\dfrac{8}{10} - \left(-\dfrac{3}{10}\right)$ Replace x with $-\dfrac{3}{10}$ and y with $-\dfrac{8}{10}$.

$= \dfrac{-8 - (-3)}{10}$

$= \dfrac{-5}{10} = \dfrac{-1 \cdot 5}{2 \cdot 5} = \dfrac{-1}{2}$ or $-\dfrac{1}{2}$

■ Work Practice 9

✓**Concept Check** Fill in each blank with the best choice given.

expression   equation   simplified   solved

A(n) _____ contains an equal sign and may be _____ for the variable.

A(n) _____ does not contain an equal sign but may be _____.

Objective C Solving Problems by Adding or Subtracting Like Fractions

Many real-life problems involve finding the perimeters of square or rectangular-shaped figures such as pastures, swimming pools, and so on. We can use our knowledge of adding fractions to find perimeters.

Answers

7. $\dfrac{2 - 7y}{5}$ 8. $-\dfrac{5}{11}$ 9. $-\dfrac{5}{12}$

✓**Concept Check Answer**

equation; solved; expression; simplified

Example 10

Find the perimeter of the rectangle.

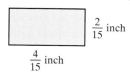

Solution: Recall that perimeter means distance around and that opposite sides of a rectangle are the same length.

Perimeter = $\dfrac{2}{15} + \dfrac{4}{15} + \dfrac{2}{15} + \dfrac{4}{15} = \dfrac{2+4+2+4}{15}$

$= \dfrac{12}{15} = \dfrac{\cancel{3} \cdot 4}{\cancel{3} \cdot 5} = \dfrac{4}{5}$

The perimeter of the rectangle is $\dfrac{4}{5}$ inch.

■ Work Practice 10

Practice 10

Find the perimeter of the square.

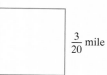

We can combine our skills in adding and subtracting fractions with our four problem-solving steps from Section 3.4 to solve many kinds of real-life problems.

Example 11 Calculating Distance

The distance from home to the Community Fitness center is $\dfrac{7}{8}$ of a mile and from home to the post office is $\dfrac{3}{8}$ of a mile. How much farther is it from home to the Community Fitness center than from home to the post office?

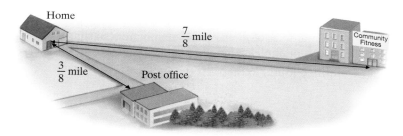

Practice 11

A jogger ran $\dfrac{13}{4}$ miles on Monday and $\dfrac{11}{4}$ miles on Wednesday. How much farther did he run on Monday than on Wednesday?

Solution:

1. **UNDERSTAND.** Read and reread the problem. The phrase "How much farther" tells us to subtract distances.
2. **TRANSLATE.**

   In words: distance farther **is** home to Community Fitness distance **minus** home to post office distance

   Translate: distance farther = $\dfrac{7}{8}$ − $\dfrac{3}{8}$

3. **SOLVE:** $\dfrac{7}{8} - \dfrac{3}{8} = \dfrac{7-3}{8} = \dfrac{4}{8} = \dfrac{\cancel{4}}{2 \cdot \cancel{4}} = \dfrac{1}{2}$

4. **INTERPRET.** *Check* your work. *State* your conclusion: The distance from home to Community Fitness is $\dfrac{1}{2}$ mile farther than from home to the post office.

■ Work Practice 11

Answers

10. $\dfrac{3}{5}$ mi   **11.** $\dfrac{1}{2}$ mi

Objective D Finding the Least Common Denominator

In the next section, we will add and subtract fractions that have different, or unlike, denominators. To do so, we first write them as equivalent fractions with a common denominator.

Although any common denominator can be used to add or subtract unlike fractions, we will use the **least common denominator (LCD)**. The LCD of a list of fractions is the same as the **least common multiple (LCM)** of the denominators. Why do we use this number as the common denominator? Since the LCD is the *smallest* of all common denominators, operations are usually less tedious with this number.

> The **least common denominator (LCD)** of a list of fractions is the smallest positive number divisible by all the denominators in the list. (The least common denominator is also the **least common multiple (LCM) of the denominators.**)

For example, the LCD of $\frac{1}{4}$ and $\frac{3}{10}$ is 20 because 20 is the smallest positive number divisible by both 4 and 10.

Finding the LCD: Method 1

One way to find the LCD is to see whether the larger denominator is divisible by the smaller denominator. If so, the larger number is the LCD. If not, then check consecutive multiples of the larger denominator until the LCD is found.

> **Method 1: Finding the LCD of a List of Fractions Using Multiples of the Largest Number**
>
> **Step 1:** Write the multiples of the largest denominator (starting with the number itself) until a multiple common to all denominators in the list is found.
>
> **Step 2:** The multiple found in Step 1 is the LCD.

Practice 12

Find the LCD of $\frac{7}{8}$ and $\frac{11}{16}$.

Example 12 Find the LCD of $\frac{3}{7}$ and $\frac{5}{14}$.

Solution: We write the multiples of 14 until we find one that is also a multiple of 7.

$14 \cdot 1 = 14$ A multiple of 7

The LCD is 14.

Work Practice 12

Practice 13

Find the LCD of $\frac{23}{25}$ and $\frac{1}{30}$.

Example 13 Find the LCD of $\frac{11}{12}$ and $\frac{7}{20}$.

Solution: We write the multiples of 20 until we find one that is also a multiple of 12.

$20 \cdot 1 = 20$ Not a multiple of 12
$20 \cdot 2 = 40$ Not a multiple of 12
$20 \cdot 3 = 60$ A multiple of 12

The LCD is 60.

Work Practice 13

Answers
12. 16   **13.** 150

Section 4.4 | Adding and Subtracting Like Fractions, Least Common Denominator

Method 1 for finding multiples works fine for smaller numbers, but may get tedious for larger numbers. For this reason, let's study a second method, which uses prime factorization.

Finding the LCD: Method 2

For example, to find the LCD of $\frac{11}{12}$ and $\frac{7}{20}$, such as in Example 13, let's look at the prime factorization of each denominator.

$$12 = 2 \cdot 2 \cdot 3$$
$$20 = 2 \cdot 2 \cdot 5$$

Recall that the LCD must be a multiple of both 12 and 20. Thus, to build the LCD, we will circle the greatest number of factors for each different prime number. The LCD is the product of the circled factors.

Prime Number Factors

12 = ⓶·② ③
20 = 2·2 ⑤   Circle either pair of 2s, but not both.
LCD = 2 · 2 · 3 · 5 = 60

The number 60 is the smallest number that both 12 and 20 divide into evenly. This method is summarized below:

Method 2: Finding the LCD of a List of Denominators Using Prime Factorization

Step 1: Write the prime factorization of each denominator.
Step 2: For each different prime factor in Step 1, circle the *greatest* number of times that factor occurs in any one factorization.
Step 3: The LCD is the product of the circled factors.

Example 14 Find the LCD of $-\frac{23}{72}$ and $\frac{17}{60}$.

Solution: First we write the prime factorization of each denominator.

$$72 = 2 \cdot 2 \cdot 2 \cdot 3 \cdot 3$$
$$60 = 2 \cdot 2 \cdot 3 \cdot 5$$

For the prime factors shown, we circle the greatest number of factors found in either factorization.

$$72 = \underbrace{2 \cdot 2 \cdot 2} \cdot \underbrace{3 \cdot 3}$$
$$60 = 2 \cdot 2 \cdot 3 \cdot \underbrace{5}$$

The LCD is the product of the circled factors.

LCD = 2 · 2 · 2 · 3 · 3 · 5 = 360

The LCD is 360.

■ Work Practice 14

Practice 14

Find the LCD of $-\frac{3}{40}$ and $\frac{11}{108}$.

Helpful Hint
If you prefer working with exponents, circle the factor with the greatest exponent.

Example 14:
$$72 = \underbrace{2^3} \cdot \underbrace{3^2}$$
$$60 = 2^2 \cdot 3 \cdot \underbrace{5}$$
$$LCD = 2^3 \cdot 3^2 \cdot 5 = 360$$

Answer
14. 1080

Chapter 4 | Fractions and Mixed Numbers

> **Helpful Hint**
> If the number of factors of a prime number is equal, circle either one, but not both. For example,
>
> $12 = \boxed{2 \cdot 2} \cdot 3$
> $15 = 3 \cdot \boxed{5}$ — Circle either 3 but not both.
>
> The LCD is $2 \cdot 2 \cdot 3 \cdot 5 = 60$.

Practice 15

Find the LCD of $\dfrac{7}{20}, \dfrac{1}{24}$, and $\dfrac{13}{45}$.

Example 15 Find the LCD of $\dfrac{1}{15}, \dfrac{5}{18}$, and $\dfrac{53}{54}$.

Solution: $15 = 3 \cdot \boxed{5}$
$18 = \boxed{2} \cdot 3 \cdot 3$
$54 = 2 \cdot \boxed{3 \cdot 3 \cdot 3}$

The LCD is $2 \cdot 3 \cdot 3 \cdot 3 \cdot 5$ or 270.

■ Work Practice 15

Practice 16

Find the LCD of $\dfrac{7}{y}$ and $\dfrac{6}{11}$.

Example 16 Find the LCD of $\dfrac{3}{5}, \dfrac{2}{x}$, and $\dfrac{7}{x^3}$.

Solution: $5 = \boxed{5}$
$x = x$
$x^3 = \boxed{x \cdot x \cdot x}$

LCD $= 5 \cdot x \cdot x \cdot x = 5x^3$

■ Work Practice 16

✓ **Concept Check** True or false? The LCD of the fractions $\dfrac{1}{6}$ and $\dfrac{1}{8}$ is 48.

Objective E Writing Equivalent Fractions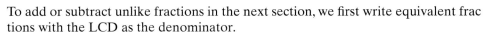

To add or subtract unlike fractions in the next section, we first write equivalent fractions with the LCD as the denominator.

To write $\dfrac{1}{3}$ as an equivalent fraction with a denominator of 6, we multiply by 1 in the form of $\dfrac{2}{2}$. Why? Because $3 \cdot 2 = 6$, so the new denominator will become 6 as shown below.

$$\dfrac{1}{3} = \dfrac{1}{3} \cdot 1 = \dfrac{1}{3} \cdot \dfrac{2}{2} = \dfrac{1 \cdot 2}{3 \cdot 2} = \dfrac{2}{6}$$

$\dfrac{2}{2} = 1$

So $\dfrac{1}{3} = \dfrac{2}{6}$.

> **Helpful Hint** Recall from the Helpful Hint on p. 233 that this is also called the Fundamental Property of Fractions.
>
> $\dfrac{a}{b} = \dfrac{a}{b} \cdot \dfrac{c}{c} = \dfrac{a \cdot c}{b \cdot c}$

To write an equivalent fraction,

$$\dfrac{a}{b} = \dfrac{a}{b} \cdot \dfrac{c}{c} = \dfrac{a \cdot c}{b \cdot c}$$

where a, b, and c are nonzero numbers.

Answers
15. 360 16. $11y$

✓ **Concept Check Answer**
false; it is 24

Section 4.4 | Adding and Subtracting Like Fractions, Least Common Denominator

✓**Concept Check** Which of the following is *not* equivalent to $\frac{3}{4}$?

a. $\frac{6}{8}$ b. $\frac{18}{24}$ c. $\frac{9}{14}$ d. $\frac{30}{40}$

Example 17 Write $\frac{3}{4}$ as an equivalent fraction with a denominator of 20.

$$\frac{3}{4} = \frac{}{20}$$

Solution: In the denominators, since $4 \cdot 5 = 20$, we will multiply by 1 in the form of $\frac{5}{5}$.

$$\frac{3}{4} = \frac{3}{4} \cdot \frac{5}{5} = \frac{3 \cdot 5}{4 \cdot 5} = \frac{15}{20}$$

Thus, $\frac{3}{4} = \frac{15}{20}$.

■ Work Practice 17

Practice 17

Write $\frac{7}{8}$ as an equivalent fraction with a denominator of 56.

$$\frac{7}{8} = \frac{}{56}$$

Helpful Hint

To check Example 17, write $\frac{15}{20}$ in simplest form.

$$\frac{15}{20} = \frac{3 \cdot \overset{1}{\cancel{5}}}{4 \cdot \underset{1}{\cancel{5}}} = \frac{3}{4}, \text{ the original fraction.}$$

If the original fraction is in lowest terms, we can check our work by writing the new, equivalent fraction in simplest form. This form should be the original fraction.

✓**Concept Check** True or false? When the fraction $\frac{2}{9}$ is rewritten as an equivalent fraction with 27 as the denominator, the result is $\frac{2}{27}$.

Example 18 Write an equivalent fraction with the given denominator.

$$\frac{2}{5} = \frac{}{15}$$

Solution: Since $5 \cdot 3 = 15$, we multiply by 1 in the form of $\frac{3}{3}$.

$$\frac{2}{5} = \frac{2}{5} \cdot \frac{3}{3} = \frac{2 \cdot 3}{5 \cdot 3} = \frac{6}{15}$$

Then $\frac{2}{5}$ is equivalent to $\frac{6}{15}$. They both represent the same part of a whole.

■ Work Practice 18

Practice 18

Write an equivalent fraction with the given denominator.

$$\frac{1}{4} = \frac{}{20}$$

Answers

17. $\frac{49}{56}$ 18. $\frac{5}{20}$

✓**Concept Check Answers**

c

false; the correct result would be $\frac{6}{27}$

Chapter 4 | Fractions and Mixed Numbers

Practice 19

Write an equivalent fraction with the given denominator.

$$\frac{3x}{7} = \frac{}{42}$$

Example 19 Write an equivalent fraction with the given denominator.

$$\frac{9x}{11} = \frac{}{44}$$

Solution: Since $11 \cdot 4 = 44$, we multiply by 1 in the form of $\frac{4}{4}$.

$$\frac{9x}{11} = \frac{9x}{11} \cdot \frac{4}{4} = \frac{9x \cdot 4}{11 \cdot 4} = \frac{36x}{44}$$

Then $\frac{9x}{11}$ is equivalent to $\frac{36x}{44}$.

■ Work Practice 19

Practice 20

Write an equivalent fraction with the given denominator.

$$4 = \frac{}{6}$$

Example 20 Write an equivalent fraction with the given denominator.

$$3 = \frac{}{7}$$

Solution: Recall that $3 = \frac{3}{1}$. Since $1 \cdot 7 = 7$, multiply by 1 in the form $\frac{7}{7}$.

$$\frac{3}{1} = \frac{3}{1} \cdot \frac{7}{7} = \frac{3 \cdot 7}{1 \cdot 7} = \frac{21}{7}$$

■ Work Practice 20

Don't forget that when the denominator of a fraction contains a variable, such as $\frac{8}{3x}$, we will assume that the variable does not represent 0. Recall that the denominator of a fraction cannot be 0.

Practice 21

Write an equivalent fraction with the given denominator.

$$\frac{9}{4x} = \frac{}{36x}$$

Example 21 Write an equivalent fraction with the given denominator.

$$\frac{8}{3x} = \frac{}{24x}$$

Solution: Since $3x \cdot 8 = 24x$, multiply by 1 in the form $\frac{8}{8}$.

$$\frac{8}{3x} = \frac{8}{3x} \cdot \frac{8}{8} = \frac{8 \cdot 8}{3x \cdot 8} = \frac{64}{24x}$$

■ Work Practice 21

Answers

19. $\frac{18x}{42}$ **20.** $\frac{24}{6}$ **21.** $\frac{81}{36x}$

✓ **Concept Check Answer**
answers may vary

✓ **Concept Check** What is the first step in writing $\frac{3}{10}$ as an equivalent fraction whose denominator is 100?

Section 4.4 Adding and Subtracting Like Fractions, Least Common Denominator

Vocabulary, Readiness & Video Check

Use the choices below to fill in each blank. Not all choices will be used.

least common denominator (LCD)   like $-\dfrac{a}{b}$ $\dfrac{a-c}{b}$ $\dfrac{a+c}{b}$ $-\dfrac{a}{-b}$

perimeter   unlike

equivalent

1. The fractions $\dfrac{9}{11}$ and $\dfrac{13}{11}$ are called _____ fractions while $\dfrac{3}{4}$ and $\dfrac{1}{3}$ are called _____ fractions.

2. $\dfrac{a}{b} + \dfrac{c}{b} =$ _____ and $\dfrac{a}{b} - \dfrac{c}{b} =$ _____.

3. As long as b is not 0, $\dfrac{-a}{b} = \dfrac{a}{-b} =$ _____.

4. The distance around a figure is called its _____.

5. The smallest positive number divisible by all the denominators of a list of fractions is called the _____.

6. Fractions that represent the same portion of a whole are called _____ fractions.

Martin-Gay Interactive Videos   Watch the section lecture video and answer the following questions.

See Video 4.4

Objective A 7. Complete this statement based on the lecture before Example 1: To add like fractions, we add the _____ and keep the same _____.

Objective B 8. In Example 6, why are we told to be careful when substituting the replacement value for y?

Objective C 9. What is the perimeter equation used to solve Example 7? What is the final answer?

Objective D 10. In Example 8, the LCD is found to be 45. What does this mean in terms of the specific fractions in the problem?

Objective E 11. From Example 10, why can we multiply a fraction by a form of 1 to get an equivalent fraction?

4.4 Exercise Set MyLab Math

Objective A *Add and simplify. See Examples 1 through 3 and 6.*

1. $\dfrac{5}{11} + \dfrac{2}{11}$

2. $\dfrac{9}{17} + \dfrac{2}{17}$

3. $\dfrac{2}{9} + \dfrac{4}{9}$

4. $\dfrac{3}{10} + \dfrac{2}{10}$

5. $-\dfrac{6}{20} + \dfrac{1}{20}$

6. $-\dfrac{3}{8} + \dfrac{1}{8}$

7. $-\dfrac{3}{14} + \left(-\dfrac{4}{14}\right)$

8. $-\dfrac{5}{24} + \left(-\dfrac{7}{24}\right)$

9. $\dfrac{2}{9x} + \dfrac{4}{9x}$

10. $\dfrac{3}{10y} + \dfrac{2}{10y}$

11. $-\dfrac{7x}{18} + \dfrac{3x}{18} + \dfrac{2x}{18}$

12. $-\dfrac{7z}{15} + \dfrac{3z}{15} + \dfrac{z}{15}$

Subtract and simplify. See Examples 4, 5, 7, and 8.

13. $\dfrac{10}{11} - \dfrac{4}{11}$
14. $\dfrac{9}{13} - \dfrac{5}{13}$
15. $\dfrac{7}{8} - \dfrac{1}{8}$
16. $\dfrac{5}{6} - \dfrac{1}{6}$

17. $\dfrac{1}{y} - \dfrac{4}{y}$
18. $\dfrac{4}{z} - \dfrac{7}{z}$
19. $-\dfrac{27}{33} - \left(-\dfrac{8}{33}\right)$
20. $-\dfrac{37}{45} - \left(-\dfrac{18}{45}\right)$

21. $\dfrac{20}{21} - \dfrac{10}{21} - \dfrac{17}{21}$
22. $\dfrac{27}{28} - \dfrac{5}{28} - \dfrac{28}{28}$
23. $\dfrac{7a}{4} - \dfrac{3}{4}$
24. $\dfrac{18b}{5} - \dfrac{3}{5}$

Mixed Practice *Perform the indicated operation. See Examples 1 through 8.*

25. $-\dfrac{9}{100} + \dfrac{99}{100}$
26. $-\dfrac{15}{200} + \dfrac{85}{200}$
27. $-\dfrac{13x}{28} - \dfrac{13x}{28}$
28. $-\dfrac{15}{26y} - \dfrac{15}{26y}$

29. $\dfrac{9x}{15} + \dfrac{1}{15}$
30. $\dfrac{2x}{15} + \dfrac{7}{15}$
31. $\dfrac{7x}{16} - \dfrac{15x}{16}$
32. $\dfrac{3}{16z} - \dfrac{15}{16z}$

33. $\dfrac{9}{12} - \dfrac{7}{12} - \dfrac{10}{12}$
34. $\dfrac{1}{8} - \dfrac{15}{8} + \dfrac{2}{8}$
35. $\dfrac{x}{4} + \dfrac{3x}{4} - \dfrac{2x}{4} + \dfrac{x}{4}$
36. $\dfrac{9y}{8} + \dfrac{2y}{8} + \dfrac{5y}{8} - \dfrac{4y}{8}$

Objective B *Evaluate each expression for the given replacement values. See Example 9.*

37. $x + y$; $x = \dfrac{3}{4}, y = \dfrac{2}{4}$
38. $x - y$; $x = \dfrac{7}{8}, y = \dfrac{9}{8}$

39. $x - y$; $x = -\dfrac{1}{5}, y = -\dfrac{3}{5}$
40. $x + y$; $x = -\dfrac{1}{6}, y = \dfrac{5}{6}$

Objective C *Find the perimeter of each figure. (Hint: Recall that perimeter means distance around.) See Example 10.*

41.
42.

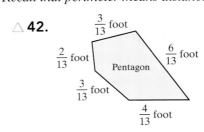

43.
44.

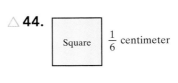

Section 4.4 | Adding and Subtracting Like Fractions, Least Common Denominator

Solve. For Exercises 45 and 46, the solutions have been started for you. Write each answer in simplest form. See Example 11.

45. A railroad inspector must inspect $\frac{19}{20}$ of a mile of railroad track. If she has already inspected $\frac{5}{20}$ of a mile, how much more does she need to inspect?

Start the solution:
1. UNDERSTAND the problem. Reread it as many times as needed.
2. TRANSLATE into an equation. (Fill in the blanks.)

distance left to inspect	is	distance needed to inspect	minus	distance already inspected
↓	↓	↓	↓	↓

distance left to inspect = _____ − _____

Finish with:
3. SOLVE. and 4. INTERPRET.

46. Scott Davis has run $\frac{10}{8}$ miles already and plans to complete $\frac{16}{8}$ miles. To do this, how much farther must he run?

Start the solution:
1. UNDERSTAND the problem. Reread it as many times as needed.
2. TRANSLATE into an equation. (Fill in the blanks.)

distance left to run	is	distance planned to run	minus	distance already run
↓	↓	↓	↓	↓

distance left to run = _____ − _____

Finish with:
3. SOLVE. and 4. INTERPRET.

47. As of early 2017, the fraction of states in the United States with maximum interstate highway speed limits 70 mph or greater was $\frac{41}{50}$. The fraction of states with 70 mph speed limits was $\frac{25}{50}$. What fraction of states had speed limits that were greater than 70 mph? (*Source:* Insurance Institute for Highway Safety)

48. When people take aspirin, $\frac{31}{50}$ of the time it is used to treat some type of pain. Approximately $\frac{7}{50}$ of all aspirin use is for treating headaches. What fraction of aspirin use is for treating pain other than headaches? (*Source:* Bayer Market Research)

The map of the world below shows the fraction of the world's surface land area taken up by each continent. In other words, the continent of Africa, for example, makes up $\frac{20}{100}$ of the land in the world. Use this map to solve Exercises 49 through 52. Write answers in simplest form.

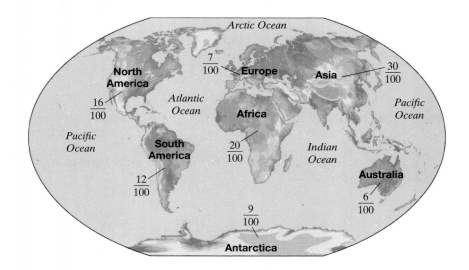

49. Find the fractional part of the world's land area within the continents of North America and South America.

50. Find the fractional part of the world's land area within the continents of Asia and Africa.

51. How much greater is the fractional part of the continent of Antarctica than the fractional part of the continent of Europe?

52. How much greater is the fractional part of the continent of Asia than the continent of Australia?

Objective **D** *Find the LCD of each list of fractions. See Examples 12 through 16.*

53. $\dfrac{2}{9}, \dfrac{6}{15}$

54. $\dfrac{7}{12}, \dfrac{3}{20}$

55. $-\dfrac{1}{36}, \dfrac{1}{24}$

56. $-\dfrac{1}{15}, \dfrac{1}{90}$

57. $\dfrac{2}{25}, \dfrac{3}{15}, \dfrac{5}{6}$

58. $\dfrac{3}{4}, \dfrac{1}{14}, \dfrac{13}{20}$

59. $-\dfrac{7}{24}, -\dfrac{5}{x}$

60. $-\dfrac{11}{y}, -\dfrac{13}{70}$

61. $\dfrac{23}{18}, \dfrac{1}{21}$

62. $\dfrac{45}{24}, \dfrac{2}{45}$

63. $\dfrac{4}{3}, \dfrac{8}{21}, \dfrac{3}{56}$

64. $\dfrac{12}{11}, \dfrac{20}{33}, \dfrac{12}{121}$

Objective **E** *Write each fraction as an equivalent fraction with the given denominator. See Examples 17 through 21.*

65. $\dfrac{2}{3} = \dfrac{}{21}$

66. $\dfrac{5}{6} = \dfrac{}{24}$

67. $\dfrac{4}{7} = \dfrac{}{35}$

68. $\dfrac{3}{5} = \dfrac{}{100}$

69. $\dfrac{1}{2} = \dfrac{}{50}$

70. $\dfrac{1}{5} = \dfrac{}{50}$

71. $\dfrac{14x}{17} = \dfrac{}{68}$

72. $\dfrac{19z}{21} = \dfrac{}{126}$

73. $\dfrac{2y}{3} = \dfrac{}{12}$

74. $\dfrac{3x}{2} = \dfrac{}{12}$

75. $\dfrac{5}{9} = \dfrac{}{36a}$

76. $\dfrac{7}{6} = \dfrac{}{36a}$

The table on the next page shows the fraction of goods sold online by type of goods in a particular year. Use this table to answer Exercises 77 through 80.

77. Complete the table by writing each fraction as an equivalent fraction with a denominator of 100.

78. Which of these types of goods has the largest fraction sold online?

Section 4.4 | Adding and Subtracting Like Fractions, Least Common Denominator   267

79. Which of these types of goods has the smallest fraction sold online?

80. Which of the types of goods has **more than** $\frac{3}{5}$ of the goods sold online? (*Hint:* Write $\frac{3}{5}$ as an equivalent fraction with a denominator of 100.)

Type of Goods	Fraction of All Goods That Are Sold Online	Equivalent Fraction with a Denominator of 100
books and magazines	$\frac{27}{50}$	
clothing and accessories	$\frac{1}{2}$	
computer hardware	$\frac{23}{50}$	
computer software	$\frac{1}{2}$	
drugs, health and beauty aids	$\frac{3}{20}$	
electronics and appliances	$\frac{13}{20}$	
food, beer, and wine	$\frac{9}{20}$	
home furnishings	$\frac{13}{25}$	
music and videos	$\frac{3}{5}$	
office equipment and supplies	$\frac{61}{100}$	
sporting goods	$\frac{12}{25}$	
toys, hobbies, and games	$\frac{1}{2}$	

(*Source:* Fedstats.gov)

Review

Simplify. See Section 1.7.

81. 3^2
82. 4^3
83. 5^3
84. 3^4

85. 7^2
86. 5^4
87. $2^3 \cdot 3$
88. $4^2 \cdot 5$

Concept Extensions

Find and correct the error. See the first Concept Check in this section.

89. $\frac{2}{7} + \frac{9}{7} = \frac{11}{14}$

90. $\frac{3}{4} - \frac{1}{4} = \frac{2}{8} = \frac{1}{4}$

Solve.

91. In your own words, explain how to add like fractions.

92. In your own words, explain how to subtract like fractions.

93. Use the map of the world for Exercises 49 through 52 and find the sum of all the continents' fractions. Explain your answer.

94. Mike Cannon jogged $\frac{3}{8}$ of a mile from home and then rested. Then he continued jogging farther from home for another $\frac{3}{8}$ of a mile until he discovered his watch had fallen off. He walked back along the same path for $\frac{4}{8}$ of a mile until he found his watch. Find how far he was from his home.

Write each fraction as an equivalent fraction with the indicated denominator.

95. $\frac{37x}{165} = \frac{}{3630}$

96. $\frac{108}{215y} = \frac{}{4085y}$

97. In your own words, explain how to find the LCD of two fractions.

98. In your own words, explain how to write a fraction as an equivalent fraction with a given denominator.

Solve. See the third and fourth Concept Checks in this section.

99. Which of the following are equivalent to $\frac{2}{3}$?

  a. $\frac{10}{15}$
  b. $\frac{40}{60}$
  c. $\frac{16}{20}$
  d. $\frac{200}{300}$

100. True or false? When the fraction $\frac{7}{12}$ is rewritten with a denominator of 48, the result is $\frac{11}{48}$. If false, give the correct fraction.

4.5 Adding and Subtracting Unlike Fractions

Objectives

A Add or Subtract Unlike Fractions.

B Write Fractions in Order.

C Evaluate Expressions Given Fractional Replacement Values.

D Solve Problems by Adding or Subtracting Unlike Fractions.

Objective A Adding and Subtracting Unlike Fractions

In this section we add and subtract fractions with unlike denominators. To add or subtract these unlike fractions, we first write the fractions as equivalent fractions with a common denominator and then add or subtract the like fractions. Recall from the previous section that the common denominator we use is called the **least common denominator (LCD).**

To begin, let's add the unlike fractions $\frac{3}{4} + \frac{1}{6}$.

The LCD of these fractions is 12. So we write each fraction as an equivalent fraction with a denominator of 12, then add as usual. This addition process is shown next and also illustrated by figures.

Add: $\frac{3}{4} + \frac{1}{6}$	The LCD is **12**.
Figures	**Algebra**
$\frac{3}{4} + \frac{1}{6}$	$\frac{3}{4} = \frac{3}{4} \cdot \frac{3}{3} = \frac{9}{12}$ and $\frac{1}{6} = \frac{1}{6} \cdot \frac{2}{2} = \frac{2}{12}$
$\frac{9}{12} + \frac{2}{12}$	Remember $\frac{3}{3} = 1$ and $\frac{2}{2} = 1$.
	Now we can add just as we did in Section 4.4.
$\frac{9}{12} + \frac{2}{12} = \frac{11}{12}$	$\frac{3}{4} + \frac{1}{6} = \frac{9}{12} + \frac{2}{12} = \frac{11}{12}$

Thus, the sum is $\frac{11}{12}$.

Adding or Subtracting Unlike Fractions

Step 1: Find the least common denominator (LCD) of the fractions.
Step 2: Write each fraction as an equivalent fraction whose denominator is the LCD.
Step 3: Add or subtract the like fractions.
Step 4: Write the sum or difference in simplest form.

Example 1 Add: $\frac{2}{5} + \frac{4}{15}$

Solution:

Step 1: The LCD of the fractions is 15. In later examples, we shall simply say, for example, that the LCD of the denominators 5 and 15 is 15.

Step 2: $\frac{2}{5} = \frac{2}{5} \cdot \frac{3}{3} = \frac{6}{15}$, $\frac{4}{15} = \frac{4}{15}$ ← This fraction already has a denominator of 15.

⎯Multiply by 1 in the form $\frac{3}{3}$.

Step 3: $\frac{2}{5} + \frac{4}{15} = \frac{6}{15} + \frac{4}{15} = \frac{10}{15}$

Step 4: Write in simplest form.

$\frac{10}{15} = \frac{2 \cdot \cancel{5}}{3 \cdot \cancel{5}} = \frac{2}{3}$

■ Work Practice 1

When the fractions contain variables, we add and subtract the same way.

Practice 1
Add: $\frac{2}{7} + \frac{8}{21}$

Answer
1. $\frac{2}{3}$

Practice 2

Add: $\dfrac{5y}{6} + \dfrac{2y}{9}$

Example 2 Add: $\dfrac{2x}{15} + \dfrac{3x}{10}$

Solution:

Step 1: The LCD of the denominators 15 and 10 is 30.

Step 2: $\dfrac{2x}{15} = \dfrac{2x}{15} \cdot \dfrac{2}{2} = \dfrac{4x}{30}$ $\dfrac{3x}{10} = \dfrac{3x}{10} \cdot \dfrac{3}{3} = \dfrac{9x}{30}$

Step 3: $\dfrac{2x}{15} + \dfrac{3x}{10} = \dfrac{4x}{30} + \dfrac{9x}{30} = \dfrac{13x}{30}$

Step 4: $\dfrac{13x}{30}$ is in simplest form.

■ Work Practice 2

Practice 3

Add: $-\dfrac{1}{5} + \dfrac{9}{20}$

Example 3 Add: $-\dfrac{1}{6} + \dfrac{1}{2}$

Solution: The LCD of the denominators 6 and 2 is 6.

$$-\dfrac{1}{6} + \dfrac{1}{2} = \dfrac{-1}{6} + \dfrac{1 \cdot 3}{2 \cdot 3}$$
$$= \dfrac{-1}{6} + \dfrac{3}{6}$$
$$= \dfrac{2}{6}$$

Helpful Hint Recall that $-\dfrac{1}{6} = \dfrac{-1}{6} = \dfrac{1}{-6}$

Next, simplify $\dfrac{2}{6}$.

$$\dfrac{2}{6} = \dfrac{\overset{1}{\cancel{2}}}{\underset{1}{\cancel{2}} \cdot 3} = \dfrac{1}{3}$$

■ Work Practice 3

✓ **Concept Check** Find and correct the error in the following:

$$\dfrac{2}{9} + \dfrac{4}{11} \ne \dfrac{6}{20} \ne \dfrac{3}{10}$$

Practice 4

Subtract: $\dfrac{5}{7} - \dfrac{9}{10}$

Answers

2. $\dfrac{19y}{18}$ **3.** $\dfrac{1}{4}$ **4.** $-\dfrac{13}{70}$

✓ **Concept Check Answer**

When adding fractions, we don't add the denominators. Correct solution:

$\dfrac{2}{9} + \dfrac{4}{11} = \dfrac{22}{99} + \dfrac{36}{99} = \dfrac{58}{99}$

Example 4 Subtract: $\dfrac{2}{3} - \dfrac{10}{11}$

Solution:

Step 1: The LCD of the denominators 3 and 11 is 33.

Step 2: $\dfrac{2}{3} = \dfrac{2}{3} \cdot \dfrac{11}{11} = \dfrac{22}{33}$ $\dfrac{10}{11} = \dfrac{10}{11} \cdot \dfrac{3}{3} = \dfrac{30}{33}$

Step 3: $\dfrac{2}{3} - \dfrac{10}{11} = \dfrac{22}{33} - \dfrac{30}{33} = \dfrac{-8}{33}$ or $-\dfrac{8}{33}$

Step 4: $-\dfrac{8}{33}$ is in simplest form.

■ Work Practice 4

Example 5 Find: $-\dfrac{3}{4} - \dfrac{1}{14} + \dfrac{6}{7}$

Solution: The LCD of the denominators 4, 14, and 7 is 28.

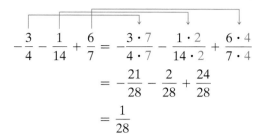

$$-\dfrac{3}{4} - \dfrac{1}{14} + \dfrac{6}{7} = -\dfrac{3 \cdot 7}{4 \cdot 7} - \dfrac{1 \cdot 2}{14 \cdot 2} + \dfrac{6 \cdot 4}{7 \cdot 4}$$

$$= -\dfrac{21}{28} - \dfrac{2}{28} + \dfrac{24}{28}$$

$$= \dfrac{1}{28}$$

Work Practice 5

Practice 5
Find: $\dfrac{5}{8} - \dfrac{1}{3} - \dfrac{1}{12}$

✓ **Concept Check** Find and correct the error in the following:

$$\dfrac{7}{12} - \dfrac{3}{4} = \dfrac{4}{8} = \dfrac{1}{2}$$

Example 6 Subtract: $2 - \dfrac{x}{3}$

Solution: Recall that $2 = \dfrac{2}{1}$. The LCD of the denominators 1 and 3 is 3.

$$\dfrac{2}{1} - \dfrac{x}{3} = \dfrac{2 \cdot 3}{1 \cdot 3} - \dfrac{x}{3}$$

$$= \dfrac{6}{3} - \dfrac{x}{3}$$

$$= \dfrac{6 - x}{3}$$

Helpful Hint
The expression $\dfrac{6-x}{3}$ from Example 6 *does not simplify* to $2 - x$. The number 3 must be a factor of both terms in the numerator (not just 6) in order to simplify.

The numerator $6 - x$ cannot be simplified further since 6 and $-x$ are unlike terms.

Work Practice 6

Practice 6
Subtract: $5 - \dfrac{y}{4}$

Objective B Writing Fractions in Order ▶

One important application of the least common denominator is to use the LCD to help order or compare fractions.

Example 7 Insert $<$ or $>$ to form a true sentence.

$$\dfrac{3}{4} \quad \dfrac{9}{11}$$

Solution: The LCD for these fractions is 44. Let's write each fraction as an equivalent fraction with a denominator of 44.

$$\dfrac{3}{4} = \dfrac{3 \cdot 11}{4 \cdot 11} = \dfrac{33}{44} \qquad \dfrac{9}{11} = \dfrac{9 \cdot 4}{11 \cdot 4} = \dfrac{36}{44}$$

(*Continued on next page*)

Practice 7
Insert $<$ or $>$ to form a true sentence.

$$\dfrac{5}{8} \quad \dfrac{11}{20}$$

Answers
5. $\dfrac{5}{24}$ **6.** $\dfrac{20 - y}{4}$ **7.** $>$

✓ **Concept Check Answers**
When adding fractions, we don't add the denominators. Correct solutions:
$\dfrac{7}{12} - \dfrac{3}{4} = \dfrac{7}{12} - \dfrac{9}{12} = -\dfrac{2}{12} = -\dfrac{1}{6}$

Since $33 < 36$, then

$$\frac{33}{44} < \frac{36}{44} \text{ or}$$

$$\frac{3}{4} < \frac{9}{11}$$

■ Work Practice 7

Practice 8

Insert $<$ or $>$ to form a true sentence.

$$-\frac{17}{20} \quad -\frac{4}{5}$$

Example 8 Insert $<$ or $>$ to form a true sentence.

$$-\frac{2}{7} \quad -\frac{1}{3}$$

Solution: The LCD for these fractions is 21.

$$-\frac{2}{7} = -\frac{2 \cdot 3}{7 \cdot 3} = -\frac{6}{21} \text{ or } \frac{-6}{21} \qquad -\frac{1}{3} = -\frac{1 \cdot 7}{3 \cdot 7} = -\frac{7}{21} \text{ or } \frac{-7}{21}$$

Since $-6 > -7$, then

$$-\frac{6}{21} > -\frac{7}{21} \text{ or}$$

$$-\frac{2}{7} > -\frac{1}{3}$$

■ Work Practice 8

Objective C Evaluating Expressions Given Fractional Replacement Values

Practice 9

Evaluate $x - y$ if $x = \frac{5}{11}$ and $y = \frac{4}{9}$.

Example 9 Evaluate $x - y$ if $x = \frac{7}{18}$ and $y = \frac{2}{9}$.

Solution: Replace x with $\frac{7}{18}$ and y with $\frac{2}{9}$ in the expression $x - y$.

$$x - y = \frac{7}{18} - \frac{2}{9}$$

The LCD of the denominators 18 and 9 is 18. Then

$$\frac{7}{18} - \frac{2}{9} = \frac{7}{18} - \frac{2 \cdot 2}{9 \cdot 2}$$

$$= \frac{7}{18} - \frac{4}{18}$$

$$= \frac{3}{18} = \frac{1}{6} \qquad \text{Simplified}$$

■ Work Practice 9

Objective D Solving Problems by Adding or Subtracting Unlike Fractions

Very often, real-world problems involve adding or subtracting unlike fractions.

Answers

8. $<$ **9.** $\frac{1}{99}$

Example 10 Finding Total Weight

A freight truck has $\frac{1}{4}$ of a ton of computers, $\frac{1}{3}$ of a ton of televisions, and $\frac{3}{8}$ of a ton of small appliances. Find the total weight of its load.

Practice 10
To repair her sidewalk, a homeowner must pour cement in three different locations. She needs $\frac{3}{5}$ of a cubic yard, $\frac{3}{10}$ of a cubic yard, and $\frac{1}{15}$ of a cubic yard for these locations. Find the total amount of cement the homeowner needs.

Solution:

1. **UNDERSTAND.** Read and reread the problem. The phrase "total weight" tells us to add.
2. **TRANSLATE.**

   In words: total weight is weight of computers plus weight of televisions plus weight of appliances

   Translate: total weight $= \frac{1}{4} + \frac{1}{3} + \frac{3}{8}$

3. **SOLVE:** The LCD for these fractions is 24.

$$\frac{1}{4} + \frac{1}{3} + \frac{3}{8} = \frac{1}{4} \cdot \frac{6}{6} + \frac{1}{3} \cdot \frac{8}{8} + \frac{3}{8} \cdot \frac{3}{3}$$

$$= \frac{6}{24} + \frac{8}{24} + \frac{9}{24}$$

$$= \frac{23}{24}$$

4. **INTERPRET.** *Check* the solution. *State* your conclusion: The total weight of the truck's load is $\frac{23}{24}$ ton.

■ Work Practice 10

Example 11 Calculating Flight Time

A flight from Tucson to Phoenix, Arizona, requires $\frac{5}{12}$ of an hour. If the plane has been flying $\frac{1}{4}$ of an hour, find how much time remains before landing.

Solution:

1. **UNDERSTAND.** Read and reread the problem. The phrase "how much time remains" tells us to subtract.

(*Continued on next page*)

Practice 11
Find the difference in length of two boards if one board is $\frac{3}{4}$ of a foot long and the other is $\frac{2}{3}$ of a foot long.

Answers

10. $\frac{29}{30}$ cu yd   11. $\frac{1}{12}$ ft

2. **TRANSLATE.**

In words:	time remaining	is	flight time from Tucson to Phoenix	minus	flight time already passed
	↓	↓	↓	↓	↓
Translate:	time remaining	=	$\dfrac{5}{12}$	−	$\dfrac{1}{4}$

3. **SOLVE:** The LCD for these fractions is 12.

$$\frac{5}{12} - \frac{1}{4} = \frac{5}{12} - \frac{1}{4} \cdot \frac{3}{3}$$

$$= \frac{5}{12} - \frac{3}{12}$$

$$= \frac{2}{12} = \frac{\cancel{2}}{\cancel{2} \cdot 6} = \frac{1}{6}$$

4. **INTERPRET.** *Check* the solution. *State* your conclusion: The remaining flight time is $\dfrac{1}{6}$ of an hour.

■ Work Practice 11

Calculator Explorations Performing Operations on Fractions

Scientific Calculator

Many calculators have a fraction key, such as $\boxed{a^b/c}$, that allows you to enter fractions and perform operations on fractions, and gives the result as a fraction. If your calculator has a fraction key, use it to calculate

$$\frac{3}{5} + \frac{4}{7}$$

Enter the keystrokes

$\boxed{3}\ \boxed{a^b/c}\ \boxed{5}\ \boxed{+}\ \boxed{4}\ \boxed{a^b/c}\ \boxed{7}\ \boxed{=}$

The display should read $\boxed{1\_6\ \ 35}$

which represents the mixed number $1\dfrac{6}{35}$. Let's write the result as a fraction. To convert from mixed number notation to fractional notation, press

$\boxed{2^{nd}}\ \boxed{d/c}$

The display now reads $\boxed{41\ |\ 35}$

which represents $\dfrac{41}{35}$, the sum in fractional notation.

Graphing Calculator

Graphing calculators also allow you to perform operations on fractions and will give exact fractional results. The fraction option on a graphing calculator may be found under the $\boxed{\text{MATH}}$ menu. To perform the addition in the first column, try the keystrokes

$\boxed{3}\ \boxed{\div}\ \boxed{5}\ \boxed{+}\ \boxed{4}\ \boxed{\div}\ \boxed{7}\ \boxed{\text{MATH}}\ \boxed{\text{ENTER}}$

$\boxed{\text{ENTER}}$

The display should read

$\boxed{3/5\ +\ 4/7 \blacktriangleright \text{Frac}\ 41/35}$

Use a calculator to add the following fractions. Give each sum as a fraction.

1. $\dfrac{1}{16} + \dfrac{2}{5}$
2. $\dfrac{3}{20} + \dfrac{2}{25}$
3. $\dfrac{4}{9} + \dfrac{7}{8}$
4. $\dfrac{9}{11} + \dfrac{5}{12}$
5. $\dfrac{10}{17} + \dfrac{12}{19}$
6. $\dfrac{14}{31} + \dfrac{15}{21}$

Section 4.5 Adding and Subtracting Unlike Fractions

Vocabulary, Readiness & Video Check

Use the choices below to fill in each blank. Any numerical answers are not listed. Not all choices may be used.

- expression
- equation
- least common denominator
- equivalent
- $>$
- $<$

1. To add or subtract unlike fractions, we first write the fractions as _____ fractions with a common denominator. The common denominator we use is called the _____.

2. The LCD for $\dfrac{1}{6}$ and $\dfrac{5}{8}$ is _____.

3. $\dfrac{1}{6} + \dfrac{5}{8} = \dfrac{1}{6} \cdot \dfrac{4}{4} + \dfrac{5}{8} \cdot \dfrac{3}{3} = \underline{} + \underline{} = \underline{}$.

4. $\dfrac{1}{6} - \dfrac{5}{8} = \dfrac{1}{6} \cdot \dfrac{4}{4} - \dfrac{5}{8} \cdot \dfrac{3}{3} = \underline{} - \underline{} = \underline{}$.

5. $x - y$ is a(n) _____ while $3x = \dfrac{1}{5}$ is a(n) _____.

6. Since $-10 < -1$, we know that $-\dfrac{10}{13}$ _____ $-\dfrac{1}{13}$.

Martin-Gay Interactive Videos Watch the section lecture video and answer the following questions.

See Video 4.5

Objective A 7. In Example 3, why can't we add the two terms in the numerator?

Objective B 8. In Example 5, when comparing two fractions, how does writing each fraction with the same denominator help?

Objective C 9. In Example 6, if we had chosen to simplify the first fraction before adding, what would our addition problem have become and what would our LCD have been?

Objective D 10. What are the two forms of the answer to Example 7?

4.5 Exercise Set MyLab Math

Objective A *Add or subtract as indicated. See Examples 1 through 6.*

1. $\dfrac{2}{3} + \dfrac{1}{6}$

2. $\dfrac{5}{6} + \dfrac{1}{12}$

3. $\dfrac{1}{2} - \dfrac{1}{3}$

4. $\dfrac{2}{3} - \dfrac{1}{4}$

5. $-\dfrac{2}{11} + \dfrac{2}{33}$

6. $-\dfrac{5}{9} + \dfrac{1}{3}$

7. $\dfrac{3}{14} - \dfrac{3}{7}$

8. $\dfrac{2}{15} - \dfrac{2}{5}$

9. $\dfrac{11x}{35} + \dfrac{2x}{7}$

10. $\dfrac{2y}{5} + \dfrac{3y}{25}$

11. $2 - \dfrac{y}{12}$

12. $5 - \dfrac{y}{20}$

13. $\dfrac{5}{12} - \dfrac{1}{9}$

14. $\dfrac{7}{12} - \dfrac{5}{18}$

15. $-7 + \dfrac{5}{7}$

16. $-10 + \dfrac{7}{10}$

17. $\dfrac{5a}{11} + \dfrac{4a}{9}$
18. $\dfrac{7x}{18} + \dfrac{2x}{9}$
19. $\dfrac{2y}{3} - \dfrac{1}{6}$
20. $\dfrac{5z}{6} - \dfrac{1}{12}$

21. $\dfrac{1}{2} + \dfrac{3}{x}$
22. $\dfrac{2}{5} + \dfrac{3}{x}$
23. $-\dfrac{2}{11} - \dfrac{2}{33}$
24. $-\dfrac{5}{9} - \dfrac{1}{3}$

25. $\dfrac{9}{14} - \dfrac{3}{7}$
26. $\dfrac{4}{5} - \dfrac{2}{15}$
27. $\dfrac{11y}{35} - \dfrac{2}{7}$
28. $\dfrac{2b}{5} - \dfrac{3}{25}$

29. $\dfrac{1}{9} - \dfrac{5}{12}$
30. $\dfrac{5}{18} - \dfrac{7}{12}$
◉ 31. $\dfrac{7}{15} - \dfrac{5}{12}$
32. $\dfrac{5}{8} - \dfrac{3}{20}$

33. $\dfrac{5}{7} - \dfrac{1}{8}$
34. $\dfrac{10}{13} - \dfrac{7}{10}$
◉ 35. $\dfrac{7}{8} + \dfrac{3}{16}$
36. $\dfrac{7}{18} + \dfrac{2}{9}$

37. $\dfrac{3}{9} - \dfrac{5}{9}$
38. $\dfrac{1}{13} - \dfrac{4}{13}$
39. $-\dfrac{2}{5} + \dfrac{1}{3} - \dfrac{3}{10}$
40. $-\dfrac{1}{3} - \dfrac{1}{4} + \dfrac{2}{5}$

41. $\dfrac{5}{11} + \dfrac{y}{3}$
42. $\dfrac{5z}{13} + \dfrac{3}{26}$
43. $-\dfrac{5}{6} - \dfrac{3}{7}$
44. $-\dfrac{1}{2} - \dfrac{3}{29}$

45. $\dfrac{x}{2} + \dfrac{x}{4} + \dfrac{2x}{16}$
46. $\dfrac{z}{4} + \dfrac{z}{8} + \dfrac{2z}{16}$
47. $\dfrac{7}{9} - \dfrac{1}{6}$
48. $\dfrac{9}{16} - \dfrac{3}{8}$

49. $\dfrac{2a}{3} + \dfrac{6a}{13}$
50. $\dfrac{3y}{4} + \dfrac{y}{7}$
51. $\dfrac{7}{30} - \dfrac{5}{12}$
52. $\dfrac{7}{30} - \dfrac{3}{20}$

◉ 53. $\dfrac{5}{9} + \dfrac{1}{y}$
54. $\dfrac{1}{12} - \dfrac{5}{x}$
55. $\dfrac{6}{5} - \dfrac{3}{4} + \dfrac{1}{2}$
56. $\dfrac{6}{5} + \dfrac{3}{4} - \dfrac{1}{2}$

57. $\dfrac{4}{5} + \dfrac{4}{9}$
58. $\dfrac{11}{12} - \dfrac{7}{24}$
59. $\dfrac{5}{9x} + \dfrac{1}{8}$
60. $\dfrac{3}{8} + \dfrac{5}{12x}$

◉ 61. $-\dfrac{9}{12} + \dfrac{17}{24} - \dfrac{1}{6}$
62. $-\dfrac{5}{14} + \dfrac{3}{7} - \dfrac{1}{2}$
63. $\dfrac{3x}{8} + \dfrac{2x}{7} - \dfrac{5}{14}$
64. $\dfrac{9x}{10} - \dfrac{1}{2} + \dfrac{x}{5}$

Objective B *Insert < or > to form a true sentence. See Examples 7 and 8.*

◉ 65. $\dfrac{2}{7}\ \ \dfrac{3}{10}$
66. $\dfrac{5}{9}\ \ \dfrac{6}{11}$
67. $-\dfrac{5}{6}\ \ -\dfrac{13}{15}$
68. $-\dfrac{7}{8}\ \ -\dfrac{5}{6}$
69. $-\dfrac{3}{4}\ \ -\dfrac{11}{14}$
70. $-\dfrac{2}{9}\ \ -\dfrac{3}{13}$

Objective C *Evaluate each expression if $x = \dfrac{1}{3}$ and $y = \dfrac{3}{4}$. See Example 9.*

71. $x + y$
72. $x - y$
73. xy
74. $x \div y$
◉ 75. $2y + x$
76. $2x + y$

Objective D Find the perimeter of each geometric figure. (Hint: Recall that perimeter means distance around.)

77.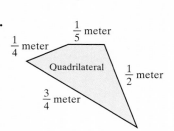

78.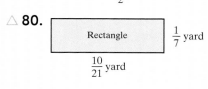

79.

Quadrilateral with sides $\frac{1}{4}$ meter, $\frac{1}{5}$ meter, $\frac{1}{2}$ meter, $\frac{3}{4}$ meter

80.

Rectangle with sides $\frac{1}{7}$ yard and $\frac{10}{21}$ yard

Translating Translate each phrase into an algebraic expression. Use "x" to represent "a number." See Examples 10 and 11.

81. The sum of a number and $\frac{1}{2}$

82. A number increased by $-\frac{2}{5}$

83. A number subtracted from $-\frac{3}{8}$

84. The difference of a number and $\frac{7}{20}$

Solve. For Exercises 85 and 86, the solutions have been started for you. See Examples 10 and 11.

85. The slowest mammal is the three-toed sloth from South America. The sloth has an average ground speed of $\frac{1}{10}$ mph. In trees, it can accelerate to $\frac{17}{100}$ mph. How much faster can a sloth travel in trees? (*Source: The Guinness Book of World Records*)

Start the solution:
1. UNDERSTAND the problem. Reread it as many times as needed.
2. TRANSLATE into an equation. (Fill in the blanks.)

how much faster sloth travels in trees	is	sloth speed in trees	minus	sloth speed on ground
↓	↓	↓	↓	↓
how much faster sloth travels in trees	=	____	−	____

Finish with:
3. SOLVE. and
4. INTERPRET.

86. Killer bees have been known to chase people for up to $\frac{1}{4}$ of a mile, while domestic European honeybees will normally chase a person for no more than 100 feet, or $\frac{5}{264}$ of a mile. How much farther will a killer bee chase a person than a domestic honeybee? (*Source:* Coachella Valley Mosquito & Vector Control District)

Start the solution:
1. UNDERSTAND the problem. Reread it as many times as needed.
2. TRANSLATE into an equation. (Fill in the blanks.)

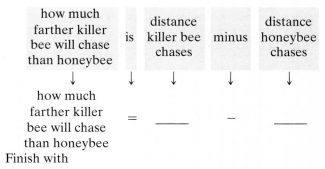

Finish with
3. SOLVE. and
4. INTERPRET.

87. Find the inner diameter of the washer.

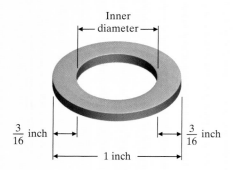

88. Find the inner diameter of the tubing.

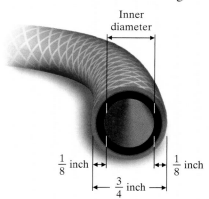

89. About $\frac{13}{20}$ of American students ages 10 to 17 name math, science, or art as their favorite subject in school. Art is the favorite subject for about $\frac{4}{25}$ of the American students ages 10 to 17. For what fraction of students this age is math or science their favorite subject? (*Source:* Peter D. Hart Research Associates for the National Science Foundation)

90. In the 2016 FIA Formula One World Championship, the Mercedes team wom $\frac{19}{21}$ of all races. If Mercedes driver Nico Rosberg won $\frac{3}{7}$ of the races, how many did Mercedes driver Lewis Hamilton win? (*Source:* formula1.com)

91. Given the following diagram, find its total length.

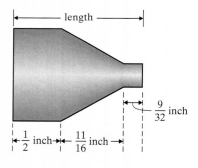

92. Given the following diagram, find its total width.

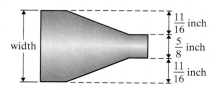

The table below shows the fraction of the Earth's water area taken up by each ocean. Use this table for Exercises 93 and 94.

Fraction of Earth's Water Area per Ocean	
Ocean	Fraction
Arctic	$\frac{1}{25}$
Atlantic	$\frac{13}{50}$
Pacific	$\frac{1}{2}$
Indian	$\frac{1}{5}$

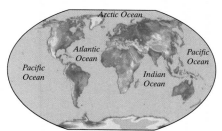

93. What fraction of the world's water surface area is accounted for by the Pacific and Atlantic Oceans?

94. What fraction of the world's water surface area is accounted for by the Arctic and Indian Oceans?

Section 4.5 | Adding and Subtracting Unlike Fractions

Use this circle graph to answer Exercises 95 through 98.

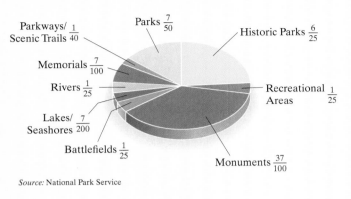

Areas Maintained by the National Park Service

Parkways/Scenic Trails $\frac{1}{40}$
Parks $\frac{7}{50}$
Historic Parks $\frac{6}{25}$
Memorials $\frac{7}{100}$
Rivers $\frac{1}{25}$
Lakes/Seashores $\frac{7}{200}$
Recreational Areas $\frac{1}{25}$
Battlefields $\frac{1}{25}$
Monuments $\frac{37}{100}$

Source: National Park Service

95. What fraction of areas maintained by the National Park Service are designated as rivers or lakes/seashores?

96. What fraction of areas maintained by the National Park Service are designated as recreation areas or memorials?

97. What fraction of areas maintained by the National Park Service are NOT national monuments?

98. What fraction of areas maintained by the National Park Service are NOT national parkways or scenic trails?

Review

Use order of operations to simplify. See Section 2.5.

99. $-50 \div 5 \cdot 2$

100. $8 - 6 \cdot 4 - 7$

101. $(8 - 6) \cdot (4 - 7)$

102. $50 \div (5 \cdot 2)$

Concept Extensions

For Exercises 103 and 104 below, do the following:

a) *Draw three rectangles of the same size and represent each fraction in the sum or difference, one fraction per rectangle, by shading.*

b) *Using these rectangles as estimates, determine whether there is an error in the sum or difference.*

c) *If there is an error, correctly calculate the sum or difference.*

See the Concept Checks in this section.

103. $\frac{3}{5} + \frac{4}{5} \stackrel{?}{=} \frac{7}{10}$

104. $\frac{5}{8} - \frac{3}{4} \stackrel{?}{=} \frac{2}{4}$

Subtract from left to right.

105. $\frac{2}{3} - \frac{1}{4} - \frac{2}{540}$

106. $\frac{9}{10} - \frac{7}{200} - \frac{1}{3}$

Perform each indicated operation.

107. $\frac{30}{55} + \frac{1000}{1760}$

108. $\frac{19}{26} - \frac{968}{1352}$

109. In your own words, describe how to add or subtract two fractions with different denominators.

110. Find the sum of the fractions in the circle graph above. Did the sum surprise you? Why or why not?

111. In 2015, about $\frac{2}{5}$ of the total number of pieces of mail delivered by the United States Postal Service was first-class mail. That same year, about $\frac{1}{2}$ of the total number of pieces of mail delivered by the United States Postal Service was standard mail. Which of these two categories account for a greater portion of the mail handled by volume? (*Source:* United States Postal Service)

Integrated Review Sections 4.1–4.5

Summary on Fractions and Operations on Fractions

Answers

Use a fraction to represent the shaded area of each figure. If the fraction is improper, also write the fraction as a mixed number.

1.
2.

Solve.

3. In a survey, 73 people out of 85 get fewer than 8 hours of sleep each night. What fraction of people in the survey get fewer than 8 hours of sleep?

4. Sketch a diagram to represent $\frac{9}{13}$.

Simplify.

5. $\frac{11}{-11}$ 6. $\frac{17}{1}$ 7. $\frac{0}{-3}$ 8. $\frac{7}{0}$

Write the prime factorization of each composite number. Write any repeated factors using exponents.

9. 65   10. 70   11. 315   12. 441

Write each fraction in simplest form.

13. $\frac{2}{14}$ 14. $\frac{24}{20}$ 15. $-\frac{56}{60}$ 16. $-\frac{72}{80}$

17. $\frac{54x}{135}$ 18. $\frac{90}{240y}$ 19. $\frac{165z^3}{210z}$ 20. $\frac{245ab}{385a^2b^3}$

Determine whether each pair of fractions is equivalent.

21. $\frac{7}{8}$ and $\frac{9}{10}$ 22. $\frac{10}{12}$ and $\frac{15}{18}$

23. Of the 50 states, 2 states are not adjacent to any other states.
   a. What fraction of the states are not adjacent to other states?
   b. How many states are adjacent to other states?
   c. What fraction of the states are adjacent to other states?

24. There are approximately 141,000 digital movie screens in the world. Of these, about 33,000 are in the U.S./Canada. (*Source:* Motion Picture Association of America)
   a. What fraction of digital movie screens are in the U.S./Canada?
   b. How many digital movie screens are not in the U.S./Canada?
   c. What fraction of digital movie screens are not in the U.S./Canada?

Integrated Review

Find the LCM of each list of numbers.

25. 5, 6   **26.** 2, 14   **27.** 6, 18, 30

Write each fraction as an equivalent fraction with the indicated denominator.

28. $\dfrac{7}{9} = \dfrac{}{36}$ **29.** $\dfrac{11}{15} = \dfrac{}{75}$ **30.** $\dfrac{5}{6} = \dfrac{}{48}$

The following summary will help you with the following review of operations on fractions.

Operations on Fractions

Let $a, b, c,$ and d be integers.

Addition: $\dfrac{a}{b} + \dfrac{c}{b} = \dfrac{a+c}{b}$
$b \neq 0$ ↑   ↑
common denominator

Subtraction: $\dfrac{a}{b} - \dfrac{c}{b} = \dfrac{a-c}{b}$
$(b \neq 0)$ ↑   ↑
common denominator

Multiplication: $\dfrac{a}{b} \cdot \dfrac{c}{d} = \dfrac{a \cdot c}{b \cdot d}$
$(b \neq 0, d \neq 0)$

Division: $\dfrac{a}{b} \div \dfrac{c}{d} = \dfrac{a}{b} \cdot \dfrac{d}{c} = \dfrac{a \cdot d}{b \cdot c}$
$(b \neq 0, d \neq 0, c \neq 0)$

Perform each indicated operation.

31. $\dfrac{1}{5} + \dfrac{3}{5}$ **32.** $\dfrac{1}{5} - \dfrac{3}{5}$ **33.** $\dfrac{1}{5} \cdot \dfrac{3}{5}$ **34.** $\dfrac{1}{5} \div \dfrac{3}{5}$

35. $\dfrac{2}{3} \div \dfrac{5}{6}$ **36.** $\dfrac{2a}{3} \cdot \dfrac{5}{6a}$ **37.** $\dfrac{2}{3y} - \dfrac{5}{6y}$ **38.** $\dfrac{2x}{3} + \dfrac{5x}{6}$

39. $-\dfrac{1}{7} \cdot -\dfrac{7}{18}$ **40.** $-\dfrac{4}{9} \cdot -\dfrac{3}{7}$ **41.** $-\dfrac{7z}{8} \div 6z^2$ **42.** $-\dfrac{9}{10} \div 5$

43. $\dfrac{7}{8} + \dfrac{1}{20}$ **44.** $\dfrac{5}{12} - \dfrac{1}{9}$ **45.** $\dfrac{2}{9} + \dfrac{1}{18} + \dfrac{1}{3}$ **46.** $\dfrac{3y}{10} + \dfrac{y}{5} + \dfrac{6}{25}$

Translating *Translate each to an expression. Use x to represent "a number."*

47. $\dfrac{2}{3}$ of a number   **48.** The quotient of a number and $-\dfrac{1}{5}$

49. A number subtracted from $-\dfrac{8}{9}$ **50.** $\dfrac{6}{11}$ increased by a number

Solve.

51. Find $\dfrac{2}{3}$ of 1530.

52. A contractor is using 18 acres of his land to sell $\dfrac{3}{4}$-acre lots. How many lots can he sell?

53. Suppose that the cross-section of a piece of pipe looks like the diagram shown. What is the inner diameter?

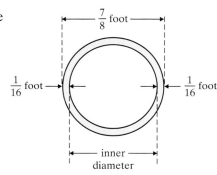

4.6 Complex Fractions and Review of Order of Operations

Objectives

A Simplify Complex Fractions.

B Review the Order of Operations.

C Evaluate Expressions Given Replacement Values.

Objective A Simplifying Complex Fractions

Thus far, we have studied operations on fractions. We now practice simplifying fractions whose numerators or denominators themselves contain fractions. These fractions are called **complex fractions**.

> **Complex Fraction**
>
> A fraction whose numerator or denominator or both numerator and denominator contain fractions is called a **complex fraction**.

Examples of complex fractions are

$$\frac{\dfrac{x}{4}}{\dfrac{3}{2}} \qquad \frac{\dfrac{1}{2}+\dfrac{3}{8}}{\dfrac{3}{4}-\dfrac{1}{6}} \qquad \frac{\dfrac{y}{5}-2}{\dfrac{3}{10}} \qquad \frac{-4z}{\dfrac{3}{5}}$$

Method 1 for Simplifying Complex Fractions

Two methods are presented to simplify complex fractions. The first method makes use of the fact that a fraction bar means division.

Practice 1

Simplify: $\dfrac{\dfrac{7y}{10}}{\dfrac{1}{5}}$

Example 1 Simplify: $\dfrac{\dfrac{x}{4}}{\dfrac{3}{2}}$

Solution: Since a fraction bar means division, the complex fraction $\dfrac{\dfrac{x}{4}}{\dfrac{3}{2}}$ can be written as $\dfrac{x}{4} \div \dfrac{3}{2}$. Then divide as usual to simplify.

$$\dfrac{x}{4} \div \dfrac{3}{2} = \dfrac{x}{4} \cdot \dfrac{2}{3} \qquad \text{Multiply by the reciprocal.}$$

$$= \dfrac{x \cdot \overset{1}{\cancel{2}}}{\underset{1}{\cancel{2}} \cdot 2 \cdot 3}$$

$$= \dfrac{x}{6}$$

■ Work Practice 1

Practice 2

Simplify: $\dfrac{\dfrac{1}{2}+\dfrac{1}{6}}{\dfrac{3}{4}-\dfrac{2}{3}}$

Example 2 Simplify: $\dfrac{\dfrac{1}{2}+\dfrac{3}{8}}{\dfrac{3}{4}-\dfrac{1}{6}}$

Solution: Recall the order of operations. Since the fraction bar is considered a grouping symbol, we simplify the numerator and the denominator of the complex fraction separately. Then we divide.

$$\dfrac{\dfrac{1}{2}+\dfrac{3}{8}}{\dfrac{3}{4}-\dfrac{1}{6}} = \dfrac{\dfrac{1 \cdot 4}{2 \cdot 4}+\dfrac{3}{8}}{\dfrac{3 \cdot 3}{4 \cdot 3}-\dfrac{1 \cdot 2}{6 \cdot 2}} = \dfrac{\dfrac{4}{8}+\dfrac{3}{8}}{\dfrac{9}{12}-\dfrac{2}{12}} = \dfrac{\dfrac{7}{8}}{\dfrac{7}{12}}$$

Answers

1. $\dfrac{7y}{2}$ **2.** $\dfrac{8}{1}$ or 8

Section 4.6 | Complex Fractions and Review of Order of Operations

Thus,

$$\frac{\frac{1}{2}+\frac{3}{8}}{\frac{3}{4}-\frac{1}{6}} = \frac{\frac{7}{8}}{\frac{7}{12}}$$

$$= \frac{7}{8} \div \frac{7}{12} \quad \text{Rewrite the quotient using the } \div \text{ sign.}$$

$$= \frac{7}{8} \cdot \frac{12}{7} \quad \text{Multiply by the reciprocal.}$$

$$= \frac{\overset{1}{\cancel{7}} \cdot 3 \cdot \overset{1}{\cancel{4}}}{2 \cdot \underset{1}{\cancel{4}} \cdot \underset{1}{\cancel{7}}} \quad \text{Multiply.}$$

$$= \frac{3}{2} \quad \text{Simplify.}$$

■ Work Practice 2

Method 2 for Simplifying Complex Fractions

The second method for simplifying complex fractions is to multiply the numerator and the denominator of the complex fraction by the LCD of all the fractions in its numerator and its denominator. This has the effect of leaving sums and differences of integers in the numerator and the denominator, as we shall see in the example below.

Let's use this second method to simplify the complex fraction in Example 2 again.

Example 3 Simplify: $\dfrac{\frac{1}{2}+\frac{3}{8}}{\frac{3}{4}-\frac{1}{6}}$

Solution: The complex fraction contains fractions with denominators 2, 8, 4, and 6. The LCD is 24. By the fundamental property of fractions (See the Helpful Hints on p. 233 and p. 260), we can multiply the numerator and the denominator of the complex fraction by 24. Notice below that by the distributive property, this means that we multiply each term in the numerator and denominator by 24.

$$\frac{\frac{1}{2}+\frac{3}{8}}{\frac{3}{4}-\frac{1}{6}} = \frac{24\left(\frac{1}{2}+\frac{3}{8}\right)}{24\left(\frac{3}{4}-\frac{1}{6}\right)}$$

$$= \frac{\left(\overset{12}{\cancel{24}} \cdot \frac{1}{\underset{1}{\cancel{2}}}\right) + \left(\overset{3}{\cancel{24}} \cdot \frac{3}{\underset{1}{\cancel{8}}}\right)}{\left(\overset{6}{\cancel{24}} \cdot \frac{3}{\underset{1}{\cancel{4}}}\right) - \left(\overset{4}{\cancel{24}} \cdot \frac{1}{\underset{1}{\cancel{6}}}\right)} \quad \text{Apply the distributive property. Then divide out common factors to aid in multiplying.}$$

$$= \frac{12 + 9}{18 - 4} \quad \text{Multiply.}$$

$$= \frac{21}{14}$$

$$= \frac{\overset{1}{\cancel{7}} \cdot 3}{\underset{1}{\cancel{7}} \cdot 2} = \frac{3}{2} \quad \text{Simplify.}$$

■ Work Practice 3

The simplified result is the same, of course, no matter which method is used.

Practice 3

Use Method 2 to simplify:

$$\dfrac{\frac{1}{2}+\frac{1}{6}}{\frac{3}{4}-\frac{2}{3}}$$

Answer
3. $\dfrac{8}{1}$ or 8

Practice 4

Simplify: $\dfrac{\dfrac{3}{4}}{\dfrac{x}{5} - 1}$

> **Helpful Hint**
> Don't forget to multiply the numerator and the denominator of the complex fraction by the same number—the LCD.

Example 4 Simplify: $\dfrac{\dfrac{y}{5} - 2}{\dfrac{3}{10}}$

Solution: Use the second method and multiply the numerator and the denominator of the complex fraction by the LCD of all fractions. Recall that $2 = \dfrac{2}{1}$. The LCD of the denominators 5, 1, and 10 is 10.

$$\dfrac{\dfrac{y}{5} - \dfrac{2}{1}}{\dfrac{3}{10}} = \dfrac{10\left(\dfrac{y}{5} - \dfrac{2}{1}\right)}{10\left(\dfrac{3}{10}\right)}$$ Multiply the numerator and denominator by 10.

$$= \dfrac{\left(\overset{2}{\cancel{10}} \cdot \dfrac{y}{\cancel{5}}\right) - \left(10 \cdot \dfrac{2}{1}\right)}{\cancel{10} \cdot \dfrac{3}{\cancel{10}}}$$ Apply the distributive property. Then divide out common factors to aid in multiplying.

$$= \dfrac{2y - 20}{3}$$ Multiply.

■ Work Practice 4

Objective B Reviewing the Order of Operations

At this time, it is probably a good idea to review the order of operations on expressions containing fractions. Before we do so, let's review how we perform operations on fractions.

Review of Operations on Fractions		
Operation	Procedure	Example
Multiply	Multiply the numerators and multiply the denominators.	$\dfrac{5}{9} \cdot \dfrac{1}{2} = \dfrac{5 \cdot 1}{9 \cdot 2} = \dfrac{5}{18}$
Divide	Multiply the first fraction by the reciprocal of the second fraction.	$\dfrac{2}{3} \div \dfrac{11}{13} = \dfrac{2}{3} \cdot \dfrac{13}{11} = \dfrac{2 \cdot 13}{3 \cdot 11} = \dfrac{26}{33}$
Add or Subtract	1. Write each fraction as an equivalent fraction whose denominator is the LCD. 2. Add or subtract numerators and write the result over the common denominator.	$\dfrac{3}{4} + \dfrac{1}{8} = \dfrac{3}{4} \cdot \dfrac{2}{2} + \dfrac{1}{8} = \dfrac{6}{8} + \dfrac{1}{8} = \dfrac{7}{8}$

Now let's review order of operations.

> **Order of Operations**
> 1. Perform all operations within parentheses ( ), brackets [ ], or other grouping symbols such as fraction bars, starting with the innermost set.
> 2. Evaluate any expressions with exponents.
> 3. Multiply or divide in order from left to right.
> 4. Add or subtract in order from left to right.

Practice 5

Simplify: $\left(\dfrac{2}{3}\right)^3 - 2$

Example 5 Simplify: $\left(\dfrac{4}{5}\right)^2 - 1$

Solution: According to the order of operations, first evaluate $\left(\dfrac{4}{5}\right)^2$.

$$\left(\dfrac{4}{5}\right)^2 - 1 = \dfrac{16}{25} - 1 \quad \text{Write } \left(\dfrac{4}{5}\right)^2 \text{ as } \dfrac{16}{25}.$$

Answers

4. $\dfrac{15}{4x - 20}$ 5. $-\dfrac{46}{27}$

Section 4.6 | Complex Fractions and Review of Order of Operations

Next, combine the fractions. The LCD of the denominators 25 and 1 is 25.

$$\frac{16}{25} - 1 = \frac{16}{25} - \frac{25}{25} \quad \text{Write 1 as } \frac{25}{25}.$$

$$= \frac{-9}{25} \text{ or } -\frac{9}{25} \quad \text{Subtract.}$$

■ Work Practice 5

Example 6 Simplify: $\left(\frac{1}{4} + \frac{2}{3}\right)\left(\frac{11}{12} + \frac{1}{4}\right)$

Solution: First perform operations inside parentheses. Then multiply.

$$\left(\frac{1}{4} + \frac{2}{3}\right)\left(\frac{11}{12} + \frac{1}{4}\right) = \left(\frac{1 \cdot 3}{4 \cdot 3} + \frac{2 \cdot 4}{3 \cdot 4}\right)\left(\frac{11}{12} + \frac{1 \cdot 3}{4 \cdot 3}\right) \quad \text{Each LCD is 12.}$$

$$= \left(\frac{3}{12} + \frac{8}{12}\right)\left(\frac{11}{12} + \frac{3}{12}\right)$$

$$= \left(\frac{11}{12}\right)\left(\frac{14}{12}\right) \quad \text{Add.}$$

$$= \frac{11 \cdot \overset{1}{\cancel{2}} \cdot 7}{\underset{1}{\cancel{2}} \cdot 6 \cdot 12} \quad \text{Multiply.}$$

$$= \frac{77}{72} \quad \text{Simplify.}$$

■ Work Practice 6

✓ **Concept Check** What should be done first to simplify the expression $\frac{1}{5} \cdot \frac{5}{2} - \left(\frac{2}{3} + \frac{4}{5}\right)^2$?

Practice 6

Simplify: $\left(-\frac{1}{2} + \frac{1}{5}\right)\left(\frac{7}{8} + \frac{1}{8}\right)$

Helpful Hint
If you find it difficult replacing a variable with a number, try the following. First, replace the variable with a set of parentheses, then place the replacement number between the parentheses.

If $x = \frac{4}{5}$, find $2x + x^2$.

$2x + x^2 = 2(\ ) + (\ )^2$

$= 2\left(\frac{4}{5}\right) + \left(\frac{4}{5}\right)^2 \ldots$

then continue simplifying.

Objective C Evaluating Algebraic Expressions

Example 7 Evaluate $2x + y^2$ if $x = -\frac{1}{2}$ and $y = \frac{1}{3}$.

Solution: Replace x and y with the given values and simplify.

$$2x + y^2 = 2\left(-\frac{1}{2}\right) + \left(\frac{1}{3}\right)^2 \quad \text{Replace } x \text{ with } -\frac{1}{2} \text{ and } y \text{ with } \frac{1}{3}.$$

$$= 2\left(-\frac{1}{2}\right) + \frac{1}{9} \quad \text{Write } \left(\frac{1}{3}\right)^2 \text{ as } \frac{1}{9}.$$

$$= -1 + \frac{1}{9} \quad \text{Multiply.}$$

$$= -\frac{9}{9} + \frac{1}{9} \quad \text{The LCD is 9.}$$

$$= -\frac{8}{9} \quad \text{Add.}$$

■ Work Practice 7

Practice 7

Evaluate $-\frac{3}{5} - xy$ if $x = \frac{3}{10}$ and $y = \frac{2}{3}$.

Answers

6. $-\frac{3}{10}$ **7.** $-\frac{4}{5}$

✓ **Concept Check Answer**
Add inside parentheses.

Vocabulary, Readiness & Video Check

Use the choices below to fill in each blank.

addition	multiplication	evaluate the exponential expression
subtraction	division	complex

1. A fraction whose numerator or denominator or both numerator and denominator contain fractions is called a(n) _____ fraction.

2. To simplify $-\dfrac{1}{2} + \dfrac{2}{3} \cdot \dfrac{7}{8}$, which operation do we perform first? _____

3. To simplify $-\dfrac{1}{2} \div \dfrac{2}{3} \cdot \dfrac{7}{8}$, which operation do we perform first? _____

4. To simplify $\dfrac{7}{8} \cdot \left(\dfrac{1}{2} - \dfrac{2}{3}\right)$, which operation do we perform first? _____

5. To simplify $\dfrac{1}{3} \div \dfrac{1}{4} \cdot \left(\dfrac{9}{11} + \dfrac{3}{8}\right)^3$, which operation do we perform first? _____

6. To simplify $9 - \left(-\dfrac{3}{4}\right)^2$, which operation do we perform first? _____

Martin-Gay Interactive Videos   Watch the section lecture video and answer the following questions.

See Video 4.6

Objective A   7. In Example 2, what property is used to simplify the denominator of the complex fraction?

Objective B   8. In Example 3, why can we add the fractions in the first set of parentheses right away?

Objective C   9. In Example 4, why do we use parentheses when substituting the replacement value for x? What would happen if we didn't use parentheses?

4.6 Exercise Set MyLab Math

Objective A   Simplify each complex fraction. See Examples 1 through 4.

1. $\dfrac{\frac{1}{8}}{\frac{3}{4}}$

2. $\dfrac{\frac{5}{12}}{\frac{15}{12}}$

3. $\dfrac{\frac{2}{3}}{\frac{2}{7}}$

4. $\dfrac{\frac{9}{25}}{\frac{6}{25}}$

5. $\dfrac{\frac{2x}{27}}{\frac{4}{9}}$

6. $\dfrac{\frac{3y}{11}}{\frac{1}{2}}$

7. $\dfrac{\frac{3}{4} + \frac{2}{5}}{\frac{1}{2} + \frac{3}{5}}$

8. $\dfrac{\frac{7}{6} + \frac{2}{3}}{\frac{3}{2} - \frac{8}{9}}$

9. $\dfrac{\frac{3x}{4}}{5 - \frac{1}{8}}$

10. $\dfrac{\frac{3}{10} + 2}{\frac{2}{5y}}$

Section 4.6 | Complex Fractions and Review of Order of Operations

Objective B *Use the order of operations to simplify each expression. See Examples 5 and 6.*

11. $\dfrac{1}{5} + \dfrac{1}{3} \cdot \dfrac{1}{4}$

12. $\dfrac{1}{2} + \dfrac{1}{6} \cdot \dfrac{1}{3}$

13. $\dfrac{5}{6} \div \dfrac{1}{3} \cdot \dfrac{1}{4}$

14. $\dfrac{7}{8} \div \dfrac{1}{4} \cdot \dfrac{1}{7}$

15. $2^2 - \left(\dfrac{1}{3}\right)^2$

16. $3^2 - \left(\dfrac{1}{2}\right)^2$

17. $\left(\dfrac{2}{9} + \dfrac{4}{9}\right)\left(\dfrac{1}{3} - \dfrac{9}{10}\right)$

18. $\left(\dfrac{1}{5} - \dfrac{1}{10}\right)\left(\dfrac{1}{5} + \dfrac{1}{10}\right)$

19. $\left(\dfrac{7}{8} - \dfrac{1}{2}\right) \div \dfrac{3}{11}$

20. $\left(-\dfrac{2}{3} - \dfrac{7}{3}\right) \div \dfrac{4}{9}$

21. $2 \cdot \left(\dfrac{1}{4} + \dfrac{1}{5}\right) + 2$

22. $\dfrac{2}{5} \cdot \left(5 - \dfrac{1}{2}\right) - 1$

23. $\left(\dfrac{3}{4}\right)^2 \div \left(\dfrac{3}{4} - \dfrac{1}{12}\right)$

24. $\left(\dfrac{8}{9}\right)^2 \div \left(2 - \dfrac{2}{3}\right)$

25. $\left(\dfrac{2}{5} - \dfrac{3}{10}\right)^2$

26. $\left(\dfrac{3}{2} - \dfrac{4}{3}\right)^3$

27. $\left(\dfrac{3}{4} + \dfrac{1}{8}\right)^2 - \left(\dfrac{1}{2} + \dfrac{1}{8}\right)$

28. $\left(\dfrac{1}{6} + \dfrac{1}{3}\right)^3 + \left(\dfrac{2}{5} \cdot \dfrac{3}{4}\right)^2$

Objective C *Evaluate each expression if $x = -\dfrac{1}{3}$, $y = \dfrac{2}{5}$, and $z = \dfrac{5}{6}$. See Example 7.*

29. $5y - z$

30. $2z - x$

31. $\dfrac{x}{z}$

32. $\dfrac{y + x}{z}$

33. $x^2 - yz$

34. $x^2 - z^2$

35. $(1 + x)(1 + z)$

36. $(1 - x)(1 - z)$

Objectives A B Mixed Practice *Simplify the following. See Examples 1 through 6.*

37. $\dfrac{\frac{5a}{24}}{\frac{1}{12}}$

38. $\dfrac{\frac{7}{10}}{\frac{14z}{25}}$

39. $\left(\dfrac{3}{2}\right)^3 + \left(\dfrac{1}{2}\right)^3$

40. $\left(\dfrac{5}{21} \div \dfrac{1}{2}\right) + \left(\dfrac{1}{7} \cdot \dfrac{1}{3}\right)$

41. $\left(-\dfrac{1}{2}\right)^2 + \dfrac{1}{5}$

42. $\left(-\dfrac{3}{4}\right)^2 + \dfrac{3}{8}$

43. $\dfrac{2+\dfrac{1}{6}}{1-\dfrac{4}{3}}$

44. $\dfrac{3-\dfrac{1}{2}}{4+\dfrac{1}{5}}$

45. $\left(1-\dfrac{2}{5}\right)^2$

46. $\left(-\dfrac{1}{2}\right)^2 - \left(\dfrac{3}{4}\right)^2$

47. $\left(\dfrac{3}{4}-1\right)\left(\dfrac{1}{8}+\dfrac{1}{2}\right)$

48. $\left(\dfrac{1}{10}+\dfrac{3}{20}\right)\left(\dfrac{1}{5}-1\right)$

49. $\left(-\dfrac{2}{9}-\dfrac{7}{9}\right)^4$

50. $\left(\dfrac{5}{9}-\dfrac{2}{3}\right)^2$

51. $\dfrac{\dfrac{1}{3}-\dfrac{5}{6}}{\dfrac{3}{4}+\dfrac{1}{2}}$

52. $\dfrac{\dfrac{7}{10}+\dfrac{1}{2}}{\dfrac{4}{5}+\dfrac{3}{4}}$

53. $\left(\dfrac{3}{4}\div\dfrac{6}{5}\right)-\left(\dfrac{3}{4}\cdot\dfrac{6}{5}\right)$

54. $\left(\dfrac{1}{2}\cdot\dfrac{2}{7}\right)-\left(\dfrac{1}{2}\div\dfrac{2}{7}\right)$

55. $\dfrac{\dfrac{x}{3}+2}{5+\dfrac{1}{3}}$

56. $\dfrac{1-\dfrac{x}{4}}{2+\dfrac{3}{8}}$

Review

Perform each indicated operation. If the result is an improper fraction, also write the improper fraction as a mixed number. See Sections 4.1 and 4.5.

57. $3+\dfrac{1}{2}$

58. $2+\dfrac{2}{3}$

59. $9-\dfrac{5}{6}$

60. $4-\dfrac{1}{5}$

Concept Extensions

61. Calculate $\dfrac{2^3}{3}$ and $\left(\dfrac{2}{3}\right)^3$. Do both of these expressions simplify to the same number? Explain why or why not.

62. Calculate $\left(\dfrac{1}{2}\right)^2 \cdot \left(\dfrac{3}{4}\right)^2$ and $\left(\dfrac{1}{2}\cdot\dfrac{3}{4}\right)^2$. Do both of these expressions simplify to the same number? Explain why or why not.

Recall that to find the average of two numbers, we find their sum and divide by 2. For example, the average of $\dfrac{1}{2}$ and $\dfrac{3}{4}$ is $\dfrac{\dfrac{1}{2}+\dfrac{3}{4}}{2}$. Find the average of each pair of numbers.

63. $\dfrac{1}{2}, \dfrac{3}{4}$

64. $\dfrac{3}{5}, \dfrac{9}{10}$

65. $\dfrac{1}{4}, \dfrac{2}{14}$

66. $\dfrac{5}{6}, \dfrac{7}{9}$

67. Two positive numbers, *a* and *b*, are graphed below. Where should the graph of their average lie?

68. Study Exercise 67. Without calculating, can $\frac{1}{3}$ be the average of $\frac{1}{2}$ and $\frac{8}{9}$? Explain why or why not.

Answer true or false for each statement.

69. It is possible for the average of two numbers to be greater than both numbers.

70. It is possible for the average of two numbers to be less than both numbers.

71. The sum of two negative fractions is always a negative number.

72. The sum of a negative fraction and a positive fraction is always a positive number.

73. It is possible for the sum of two fractions to be a whole number.

74. It is possible for the difference of two fractions to be a whole number.

75. What operation should be performed first to simplify

$$\frac{1}{5} \cdot \frac{5}{2} - \left(\frac{2}{3} + \frac{4}{5}\right)^2?$$

Explain your answer.

76. A student is to evaluate $x - y$ when $x = \frac{1}{5}$ and $y = -\frac{1}{7}$. This student is asking you if he should evaluate $\frac{1}{5} - \frac{1}{7}$. What do you tell this student and why?

Each expression contains one addition, one subtraction, one multiplication, and one division. Write the operations in the order that they should be performed. Do not actually simplify. See the Concept Check in this section.

77. $[9 + 3(4 - 2)] \div \frac{10}{21}$

78. $[30 - 4(3 + 2)] \div \frac{5}{2}$

79. $\frac{1}{3} \div \left(\frac{2}{3}\right)\left(\frac{4}{5}\right) - \frac{1}{4} + \frac{1}{2}$

80. $\left(\frac{5}{6} - \frac{1}{3}\right) \cdot \frac{1}{3} + \frac{1}{2} \div \frac{9}{8}$

Evaluate each expression if $x = \frac{3}{4}$ and $y = -\frac{4}{7}$.

81. $\frac{2 + x}{y}$

82. $4x + y$

83. $x^2 + 7y$

84. $\dfrac{\frac{9}{14}}{x + y}$

4.7 Operations on Mixed Numbers

Objectives

- **A** Graph Positive and Negative Fractions and Mixed Numbers
- **B** Multiply or Divide Mixed or Whole Numbers.
- **C** Add or Subtract Mixed Numbers.
- **D** Solve Problems Containing Mixed Numbers.
- **E** Perform Operations on Negative Mixed Numbers.

Objective A Graphing Fractions and Mixed Numbers

Let's review graphing fractions and practice graphing mixed numbers on a number line. This will help us visualize rounding and estimating operations with mixed numbers.

Recall that $5\frac{2}{3}$ means $5 + \frac{2}{3}$ and

$-4\frac{1}{6}$ means $-\left(4 + \frac{1}{6}\right)$ or $-4 - \frac{1}{6}$ or $-4 + \left(-\frac{1}{6}\right)$

Example 1 Graph the numbers on a number line:

$\frac{1}{2}, -\frac{3}{4}, 2\frac{2}{3}, -3, -3\frac{1}{8}$

Solution: Remember that $2\frac{2}{3}$ means $2 + \frac{2}{3}$.

Also, $-3\frac{1}{8}$ means $-3 - \frac{1}{8}$, so $-3\frac{1}{8}$ lies to the left of -3.

■ Work Practice 1

Practice 1

Graph the numbers on a number line.

$-5, -4\frac{1}{2}, 2\frac{3}{4}, \frac{1}{8}, -\frac{1}{2}$

✓ **Concept Check** Which of the following is/are equivalent to 9?

a. $7\frac{6}{3}$ b. $8\frac{4}{4}$ c. $8\frac{9}{9}$ d. $\frac{18}{2}$ e. all of these

Objective B Multiplying or Dividing with Mixed Numbers or Whole Numbers

When multiplying or dividing a fraction and a mixed or a whole number, remember that mixed and whole numbers can be written as improper fractions.

> **Multiplying or Dividing Fractions and Mixed Numbers or Whole Numbers**
>
> To multiply or divide with mixed numbers or whole numbers, first write any mixed or whole numbers as improper fractions and then multiply or divide as usual.

(*Note:* If an exercise contains a mixed number, we will write the answer as a mixed number, if possible.)

Answer

1.

✓ **Concept Check Answer**
e

Section 4.7 | Operations on Mixed Numbers

Example 2 Multiply: $3\dfrac{1}{3} \cdot \dfrac{7}{8}$

Solution: Recall from Section 4.1 that the mixed number $3\dfrac{1}{3}$ can be written as the fraction $\dfrac{10}{3}$. Then

$$3\dfrac{1}{3} \cdot \dfrac{7}{8} = \dfrac{10}{3} \cdot \dfrac{7}{8} = \dfrac{\overset{1}{2} \cdot 5 \cdot 7}{3 \cdot \underset{1}{2} \cdot 4} = \dfrac{35}{12} \text{ or } 2\dfrac{11}{12}$$

Practice 2
Multiply and simplify: $1\dfrac{2}{3} \cdot \dfrac{11}{15}$

■ Work Practice 2

Don't forget that a whole number can be written as a fraction by writing the whole number over 1. For example,

$$20 = \dfrac{20}{1} \quad \text{and} \quad 7 = \dfrac{7}{1}$$

Example 3 Multiply: $\dfrac{3}{4} \cdot 20$

Solution: $\dfrac{3}{4} \cdot 20 = \dfrac{3}{4} \cdot \dfrac{20}{1} = \dfrac{3 \cdot 20}{4 \cdot 1} = \dfrac{3 \cdot \overset{1}{\cancel{4}} \cdot 5}{\underset{1}{\cancel{4}} \cdot 1} = \dfrac{15}{1} \text{ or } 15$

Practice 3
Multiply: $\dfrac{5}{6} \cdot 18$

■ Work Practice 3

When both numbers to be multiplied are mixed or whole numbers, it is a good idea to estimate the product to see if your answer is reasonable. To do this, we first practice rounding mixed numbers to the nearest whole. If the fraction part of the mixed number is $\dfrac{1}{2}$ or greater, we round the whole number part up. If the fraction part of the mixed number is less than $\dfrac{1}{2}$, then we do not round the whole number part up. Study the table below for examples.

Mixed Number	Rounding
$5\dfrac{1}{4}$ $\dfrac{1}{4}$ is less than $\dfrac{1}{2}$	Thus, $5\dfrac{1}{4}$ rounds to 5.
$3\dfrac{9}{16}$ ← 9 is greater than 8. → Half of 16 is 8.	Thus, $3\dfrac{9}{16}$ rounds to 4.
$1\dfrac{3}{7}$ ← 3 is less than $3\dfrac{1}{2}$. → Half of 7 is $3\dfrac{1}{2}$.	Thus, $1\dfrac{3}{7}$ rounds to 1.

Example 4 Multiply $1\dfrac{2}{3} \cdot 2\dfrac{1}{4}$. Check by estimating.

Solution: $1\dfrac{2}{3} \cdot 2\dfrac{1}{4} = \dfrac{5}{3} \cdot \dfrac{9}{4} = \dfrac{5 \cdot 9}{3 \cdot 4} = \dfrac{5 \cdot \overset{1}{\cancel{3}} \cdot 3}{\underset{1}{\cancel{3}} \cdot 4} = \dfrac{15}{4} \text{ or } 3\dfrac{3}{4}$ Exact

Let's check by estimating.

$1\dfrac{2}{3}$ rounds to 2, $2\dfrac{1}{4}$ rounds to 2, and $2 \cdot 2 = 4$ Estimate

The estimate is close to the exact value, so our answer is reasonable.

Practice 4
Multiply. Check by estimating.
$3\dfrac{1}{5} \cdot 2\dfrac{3}{4}$

Answers
2. $1\dfrac{2}{9}$ **3.** 15   **4.** $8\dfrac{4}{5}$

■ Work Practice 4

Practice 5

Multiply. Check by estimating.

$3 \cdot 6\frac{7}{15}$

Example 5 Multiply: $7 \cdot 2\frac{11}{14}$. Check by estimating.

Solution: $7 \cdot 2\frac{11}{14} = \frac{7}{1} \cdot \frac{39}{14} = \frac{7 \cdot 39}{1 \cdot 14} = \frac{\overset{1}{7} \cdot 39}{1 \cdot 2 \cdot \underset{1}{7}} = \frac{39}{2}$ or $19\frac{1}{2}$ Exact

To estimate,

$2\frac{11}{14}$ rounds to 3 and $7 \cdot 3 = 21$.   Estimate

The estimate is close to the exact value, so our answer is reasonable.

■ Work Practice 5

✓ **Concept Check** Find the error.

$2\frac{1}{4} \cdot \frac{1}{2} = 2\frac{1 \cdot 1}{4 \cdot 2} = 2\frac{1}{8}$

Practice 6–8

Divide.

6. $\frac{4}{9} \div 7$ 7. $\frac{8}{15} \div 3\frac{4}{5}$

8. $3\frac{2}{7} \div 2\frac{3}{14}$

Examples Divide.

6. $\frac{3}{4} \div 5 = \frac{3}{4} \div \frac{5}{1} = \frac{3}{4} \cdot \frac{1}{5} = \frac{3 \cdot 1}{4 \cdot 5} = \frac{3}{20}$

7. $\frac{11}{18} \div 2\frac{5}{6} = \frac{11}{18} \div \frac{17}{6} = \frac{11}{18} \cdot \frac{6}{17} = \frac{11 \cdot 6}{18 \cdot 17} = \frac{11 \cdot \overset{1}{6}}{\underset{1}{6} \cdot 3 \cdot 17} = \frac{11}{51}$

8. $5\frac{2}{3} \div 2\frac{5}{9} = \frac{17}{3} \div \frac{23}{9} = \frac{17}{3} \cdot \frac{9}{23} = \frac{17 \cdot 9}{3 \cdot 23} = \frac{17 \cdot \overset{1}{3} \cdot 3}{\underset{1}{3} \cdot 23} = \frac{51}{23}$ or $2\frac{5}{23}$

■ Work Practice 6–8

Objective C Adding or Subtracting Mixed Numbers

We can add or subtract mixed numbers, too, by first writing each mixed number as an improper fraction. But it is often easier to add or subtract the whole-number parts and add or subtract the proper-fraction parts vertically.

> **Adding or Subtracting Mixed Numbers**
>
> To add or subtract mixed numbers, add or subtract the fraction parts and then add or subtract the whole number parts.

Practice 9

Add: $2\frac{1}{6} + 4\frac{2}{5}$.

Check by estimating.

Example 9 Add: $2\frac{1}{3} + 5\frac{3}{8}$. Check by estimating.

Solution: The LCD of the denominators 3 and 8 is 24.

$$2\frac{1 \cdot 8}{3 \cdot 8} = 2\frac{8}{24}$$
$$+5\frac{3 \cdot 3}{8 \cdot 3} = +5\frac{9}{24}$$
$$= 7\frac{17}{24} \quad \leftarrow \text{Add the fractions.}$$
$$\qquad\qquad\quad \leftarrow \text{Add the whole numbers.}$$

To check by estimating, we round as usual. The fraction $2\frac{1}{3}$ rounds to 2, $5\frac{3}{8}$ rounds to 5, and $2 + 5 = 7$, our estimate.

Our exact answer is close to 7, so our answer is reasonable.

Answers

5. $19\frac{2}{5}$ 6. $\frac{4}{63}$ 7. $\frac{8}{57}$ 8. $1\frac{15}{31}$

9. $6\frac{17}{30}$

✓ **Concept Check Answer**

forgot to change mixed number to improper fraction

■ Work Practice 9

Section 4.7 | Operations on Mixed Numbers

> **Helpful Hint**
> When adding or subtracting mixed numbers and whole numbers, it is a good idea to estimate to see if your answer is reasonable.

For the rest of this section, we leave most of the checking by estimating to you.

Example 10 Add: $3\frac{4}{5} + 1\frac{4}{15}$

Solution: The LCD of the denominators 5 and 15 is 15.

$$\begin{aligned} 3\frac{4}{5} &= 3\frac{12}{15} \\ +1\frac{4}{15} &= +1\frac{4}{15} \\ &4\frac{16}{15} \end{aligned}$$

Add the fractions; then add the whole numbers.
Notice that the fraction part is improper.

Since $\frac{16}{15}$ is $1\frac{1}{15}$, we can write the sum as

$$4\frac{16}{15} + 4 + 1\frac{1}{15} = 5\frac{1}{15}$$

Work Practice 10

Practice 10
Add: $3\frac{5}{14} + 2\frac{6}{7}$

✓ **Concept Check** Explain how you could estimate the following sum:
$5\frac{1}{9} + 14\frac{10}{11}$.

Example 11 Add: $2\frac{4}{5} + 5 + 1\frac{1}{2}$

Solution: The LCD of the denominators 5 and 2 is 10.

$$\begin{aligned} 2\frac{4}{5} &= 2\frac{8}{10} \\ 5 &= 5 \\ +1\frac{1}{2} &= +1\frac{5}{10} \\ &8\frac{13}{10} = 8 + 1\frac{3}{10} = 9\frac{3}{10} \end{aligned}$$

Work Practice 11

Practice 11
Add: $12 + 3\frac{6}{7} + 2\frac{1}{5}$

Example 12 Subtract: $8\frac{3}{7} - 5\frac{2}{21}$. Check by estimating.

Solution: The LCD of the denominators 7 and 21 is 21.

$$\begin{aligned} 8\frac{3}{7} &= 8\frac{9}{21} \quad \leftarrow \text{The LCD of 7 and 21 is 21.} \\ -5\frac{2}{21} &= -5\frac{2}{21} \\ &3\frac{7}{21} \quad \leftarrow \text{Subtract the fractions.} \end{aligned}$$

↑ Subtract the whole numbers.

Practice 12
Subtract: $32\frac{7}{9} - 16\frac{5}{18}$

Answers
10. $6\frac{3}{14}$ **11.** $18\frac{2}{35}$ **12.** $16\frac{1}{2}$

✓ **Concept Check Answer**
Round each mixed number to the nearest whole number and add. $5\frac{1}{9}$ rounds to 5 and $14\frac{10}{11}$ rounds to 15, and the estimated sum is $5 + 15 = 20$.

(*Continued on next page*)

Then $3\frac{7}{21}$ simplifies to $3\frac{1}{3}$. The difference is $3\frac{1}{3}$.

To check, $8\frac{3}{7}$ rounds to 8, $5\frac{2}{21}$ rounds to 5, and $8 - 5 = 3$, our estimate.

Our exact answer is close to 3, so our answer is reasonable.

■ Work Practice 12

When subtracting mixed numbers, borrowing may be needed, as shown in the next example.

Practice 13

Subtract: $9\frac{7}{15} - 4\frac{3}{5}$

Example 13 Subtract: $7\frac{3}{14} - 3\frac{6}{7}$

Solution: The LCD of the denominators 7 and 14 is 14.

$$7\frac{3}{14} = 7\frac{3}{14}$$
$$-3\frac{6}{7} = -3\frac{12}{14}$$

Notice that we cannot subtract $\frac{12}{14}$ from $\frac{3}{14}$, so we borrow from the whole number 7.

borrow 1 from 7

$$7\frac{3}{14} = 6 + 1\frac{3}{14} = 6 + \frac{17}{14} \text{ or } 6\frac{17}{14}$$

Now subtract.

$$7\frac{3}{14} = 7\frac{3}{14} = 6\frac{17}{14}$$
$$-3\frac{6}{7} = -3\frac{12}{14} = -3\frac{12}{14}$$
$$\phantom{-3\frac{6}{7} = -3\frac{12}{14} =\ } 3\frac{5}{14} \leftarrow \text{Subtract the fractions.}$$

↑ Subtract the whole numbers.

■ Work Practice 13

✓ **Concept Check** In the subtraction problem $5\frac{1}{4} - 3\frac{3}{4}$, $5\frac{1}{4}$ must be rewritten because $\frac{3}{4}$ cannot be subtracted from $\frac{1}{4}$. Why is it incorrect to rewrite $5\frac{1}{4}$ as $5\frac{5}{4}$?

Practice 14

Subtract: $25 - 10\frac{2}{9}$

Example 14 Subtract: $14 - 8\frac{3}{7}$

Solution:

$$14 = 13\frac{7}{7}$$
$$-8\frac{3}{7} = -8\frac{3}{7}$$
$$\phantom{-8\frac{3}{7} =\ } 5\frac{4}{7} \leftarrow \text{Subtract the fractions.}$$

Borrow 1 from 14 and write it as $\frac{7}{7}$.

↑ Subtract the whole numbers.

■ Work Practice 14

Answers

13. $4\frac{13}{15}$ 14. $14\frac{7}{9}$

✓ **Concept Check Answer**

Rewrite $5\frac{1}{4}$ as $4\frac{5}{4}$ by borrowing from the 5.

Objective D Solving Problems Containing Mixed Numbers

Now that we know how to perform operations on mixed numbers, we can solve real-life problems.

Example 15 Finding Legal Lobster Size

Lobster fishermen must measure the upper body shells of the lobsters they catch. Lobsters that are too small are thrown back into the ocean. Each state has its own size standard for lobsters to help control the breeding stock. Massachusetts divides its waters into four Lobster Conservation Management Areas, with a different minimum lobster size permitted in each area. In area three, the legal lobster size increased from $3\frac{13}{32}$ inches to $3\frac{1}{2}$ inches. How much of an increase was this? (*Source: Massachusetts Division of Marine Fisheries*)

Solution:

1. **UNDERSTAND.** Read and reread the problem carefully. The word "increase" found in the problem might make you think that we add to solve the problem. But the phrase "how much of an increase" tells us to subtract to find the increase.

2. **TRANSLATE.**

In words:	increase	is	new lobster size	minus	old lobster size
	↓	↓	↓	↓	↓
Translate:	increase	=	$3\frac{1}{2}$	−	$3\frac{13}{32}$

3. **SOLVE.** Before we solve, let's estimate by rounding to the nearest wholes. The fraction $3\frac{1}{2}$ can be rounded up to 4, $3\frac{13}{32}$ rounds to 3, and $4 - 3 = 1$. The increase is not 1, but will be smaller since we rounded $3\frac{1}{2}$ up and rounded $3\frac{13}{32}$ down.

$$\begin{aligned} 3\frac{1}{2} &= 3\frac{16}{32} \\ -3\frac{13}{32} &= -3\frac{13}{32} \\ \hline &\ \ \frac{3}{32} \end{aligned}$$

4. **INTERPRET.** *Check* your work. Our estimate tells us that the exact increase of $\frac{3}{32}$ is reasonable. *State* your conclusion: The increase in lobster size was $\frac{3}{32}$ of an inch.

Work Practice 15

Practice 15

The measurement around the trunk of a tree just below shoulder height is called its girth. The largest known American beech tree in the United States has a girth of $23\frac{1}{4}$ feet. The largest known sugar maple tree in the United States has a girth of $19\frac{5}{12}$ feet. How much larger is the girth of the largest known American beech tree than the girth of the largest known sugar maple tree? (*Source: American Forests*)

Answer

15. $3\frac{5}{6}$ ft

Practice 16

A designer of women's clothing designs a woman's dress that requires $3\frac{1}{7}$ yards of material. How many dresses can be made from a 44-yard bolt of material?

Example 16 — Calculating Manufacturing Materials Needed

In a manufacturing process, a metal-cutting machine cuts strips $1\frac{3}{5}$ inches long from a piece of metal stock. How many such strips can be cut from a 48-inch piece of stock?

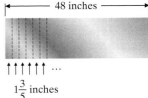

Solution:

1. **UNDERSTAND** the problem. To do so, read and reread the problem. Then draw a diagram:

   We want to know how many $1\frac{3}{5}$s there are in 48.

2. **TRANSLATE.**

   In words: number of strips **is** 48 **divided by** $1\frac{3}{5}$

   Translate: number of strips $= 48 \div 1\frac{3}{5}$

3. **SOLVE:** Let's estimate a reasonable answer. The mixed number $1\frac{3}{5}$ rounds to 2, and $48 \div 2 = 24$.

   $$48 \div 1\frac{3}{5} = 48 \div \frac{8}{5} = \frac{48}{1} \cdot \frac{5}{8} = \frac{48 \cdot 5}{1 \cdot 8} = \frac{\overset{1}{8} \cdot 6 \cdot 5}{1 \cdot \underset{1}{8}} = \frac{30}{1} \text{ or } 30$$

4. **INTERPRET.** *Check* your work. Since the exact answer of 30 is close to our estimate of 24, our answer is reasonable. *State* your conclusion: Thirty strips can be cut from the 48-inch piece of stock.

■ Work Practice 16

Objective E — Operating on Negative Mixed Numbers

To perform operations on negative mixed numbers, let's first practice writing these numbers as negative fractions and negative fractions as negative mixed numbers.

To understand negative mixed numbers, we simply need to know that, for example,

$$-3\frac{2}{5} \text{ means } -\left(3\frac{2}{5}\right)$$

Thus, to write a negative mixed number as a fraction, we do the following.

$$-3\frac{2}{5} = -\left(3\frac{2}{5}\right) = -\left(\frac{5 \cdot 3 + 2}{5}\right) = -\left(\frac{17}{5}\right) \text{ or } -\frac{17}{5}$$

Practice 17–18

Write each as a fraction.

17. $-9\frac{3}{7}$ **18.** $-5\frac{10}{11}$

Examples — Write each as a fraction.

17. $-1\frac{7}{8} = -\frac{8 \cdot 1 + 7}{8} = -\frac{15}{8}$ Write $1\frac{7}{8}$ as an improper fraction and keep the negative sign.

18. $-23\frac{1}{2} = -\frac{2 \cdot 23 + 1}{2} = -\frac{47}{2}$ Write $23\frac{1}{2}$ as an improper fraction and keep the negative sign.

■ Work Practice 17–18

Answers

16. 14 dresses   **17.** $-\frac{66}{7}$

18. $-\frac{65}{11}$

To write a negative fraction as a negative mixed number, we use a similar procedure. We simply disregard the negative sign, convert the improper fraction to a mixed number, and then reinsert the negative sign.

Examples Write each as a mixed number.

19. $-\dfrac{22}{5} = -4\dfrac{2}{5}$

$$5\overline{)22} \quad \dfrac{22}{5} = 4\dfrac{2}{5}$$
$$\underline{-20}$$
$$2$$

20. $-\dfrac{9}{4} = -2\dfrac{1}{4}$

$$4\overline{)9} \quad \dfrac{9}{4} = 2\dfrac{1}{4}$$
$$\underline{-8}$$
$$1$$

Practice 19–20

Write each as a mixed number.

19. $-\dfrac{37}{8}$ **20.** $-\dfrac{46}{5}$

Work Practice 19–20

We multiply or divide with negative mixed numbers the same way that we multiply or divide with positive mixed numbers. We first write each mixed number as a fraction.

Examples Perform the indicated operations.

21. $-4\dfrac{2}{5} \cdot 1\dfrac{3}{11} = -\dfrac{22}{5} \cdot \dfrac{14}{11} = -\dfrac{22 \cdot 14}{5 \cdot 11} = -\dfrac{2 \cdot \cancel{11} \cdot 14}{5 \cdot \cancel{11}} = -\dfrac{28}{5}$ or $-5\dfrac{3}{5}$

22. $-2\dfrac{1}{3} \div \left(-2\dfrac{1}{2}\right) = -\dfrac{7}{3} \div \left(-\dfrac{5}{2}\right) = -\dfrac{7}{3} \cdot \left(-\dfrac{2}{5}\right) = \dfrac{7 \cdot 2}{3 \cdot 5} = \dfrac{14}{15}$

Practice 21–22

21. $2\dfrac{3}{4} \cdot \left(-3\dfrac{3}{5}\right)$

22. $-4\dfrac{2}{7} \div 1\dfrac{1}{4}$

Work Practice 21–22

Helpful Hint Recall that $(-) \cdot (-) = +$

To add or subtract with negative mixed numbers, we must be very careful! Problems arise because recall that

$-3\dfrac{2}{5}$ means $-\left(3\dfrac{2}{5}\right)$

This means that

$-3\dfrac{2}{5} = -\left(3\dfrac{2}{5}\right) = -\left(3 + \dfrac{2}{5}\right) = -3 - \dfrac{2}{5}$ This can sometimes be easily overlooked.

To avoid problems, we will add or subtract negative mixed numbers by rewriting as addition and recalling how to add signed numbers.

Answers

19. $-4\dfrac{5}{8}$ **20.** $-9\dfrac{1}{5}$

21. $-9\dfrac{9}{10}$ **22.** $-3\dfrac{3}{7}$

Practice 23

Add: $6\frac{2}{3} + \left(-12\frac{3}{4}\right)$

Example 23 Add: $6\frac{3}{5} + \left(-9\frac{7}{10}\right)$

Solution: Here we are adding two numbers with different signs. Recall that we then subtract the absolute values and keep the sign of the larger absolute value.

Since $-9\frac{7}{10}$ has the larger absolute value, the answer is negative.

First, subtract absolute values:

$$9\frac{7}{10} = 9\frac{7}{10}$$
$$-6\frac{3 \cdot 2}{5 \cdot 2} = -6\frac{6}{10}$$
$$\overline{\phantom{-6\frac{3 \cdot 2}{5 \cdot 2}}} \quad \overline{3\frac{1}{10}}$$

Thus,

$$6\frac{3}{5} + \left(-9\frac{7}{10}\right) = -3\frac{1}{10} \quad \text{The result is negative since } -9\frac{7}{10} \text{ has the larger absolute value.}$$

■ Work Practice 23

Practice 24

Subtract: $-9\frac{2}{7} - 30\frac{11}{14}$

Example 24 Subtract: $-11\frac{5}{6} - 20\frac{4}{9}$

Solution: Let's write as an equivalent addition: $-11\frac{5}{6} + \left(-20\frac{4}{9}\right)$. Here, we are adding two numbers with like signs. Recall that we add their absolute values and keep the common negative sign.

First, add absolute values:

$$11\frac{5 \cdot 3}{6 \cdot 3} = 11\frac{15}{18}$$
$$+20\frac{4 \cdot 2}{9 \cdot 2} = +20\frac{8}{18}$$
$$\overline{\phantom{+20\frac{4 \cdot 2}{9 \cdot 2}}} \quad 31\frac{23}{18} \text{ or } 32\frac{5}{18}$$

Since $\frac{23}{18} = 1\frac{5}{18}$

Thus,

$$-11\frac{5}{6} - 20\frac{4}{9} = -32\frac{5}{18}$$

Keep the common sign.

■ Work Practice 24

Answers

23. $-6\frac{1}{12}$ **24.** $-40\frac{1}{14}$

Calculator Explorations — Converting Between Mixed Number and Fraction Notation

If your calculator has a fraction key, such as $\boxed{a^b/c}$, you can use it to convert between mixed number notation and fraction notation.

To write $13\frac{7}{16}$ as an improper fraction, press

$\boxed{13}$ $\boxed{a^b/c}$ $\boxed{7}$ $\boxed{a^b/c}$ $\boxed{16}$ $\boxed{2nd}$ $\boxed{d/c}$

The display will read

$\boxed{215 | 16}$

which represents $\frac{215}{16}$. Thus $13\frac{7}{16} = \frac{215}{16}$.

To convert $\frac{190}{13}$ to a mixed number, press

$\boxed{190}$ $\boxed{a^b/c}$ $\boxed{13}$ $=$

The display will read

$\boxed{14\_8/13}$

which represents $14\frac{8}{13}$. Thus $\frac{190}{13} = 14\frac{8}{13}$.

Write each mixed number as a fraction and each fraction as a mixed number.

1. $25\frac{5}{11}$
2. $67\frac{14}{15}$
3. $107\frac{31}{35}$
4. $186\frac{17}{21}$
5. $\frac{365}{14}$
6. $\frac{290}{13}$
7. $\frac{2769}{30}$
8. $\frac{3941}{17}$

Vocabulary, Readiness & Video Check

Use the choices below to fill in each blank.

round   fraction   whole number
improper   mixed number

The number $5\frac{3}{4}$ is called a(n) _____.

For $5\frac{3}{4}$, the 5 is called the _____ part and $\frac{3}{4}$ is called the _____ part.

To estimate operations on mixed numbers, we _____ mixed numbers to the nearest whole number.

The mixed number $2\frac{5}{8}$ written as a(n) _____ fraction is $\frac{21}{8}$.

Martin-Gay Interactive Videos Watch the section lecture video and answer the following questions.

See Video 4.7

Objective A 5. In Example 1, why is the unit distance between −4 and −3 on the number line split into 5 equal parts?

Objective B 6. Why do we need to know how to multiply fractions to solve Example 2?

Objective C 7. In Example 4, why is the first form of the answer not an appropriate form?

Objective D 8. Why do we need to know how to subtract fractions to solve Example 5?

Objective E 9. In Example 6, how is it determined whether the answer is positive or negative?

4.7 Exercise Set MyLab Math

Objective A *Graph each list of numbers on the given number line. See Example 1.*

1. $-2, -2\frac{2}{3}, 0, \frac{7}{8}, -\frac{1}{3}$

2. $-1, -1\frac{1}{4}, -\frac{1}{4}, 3\frac{1}{4}, 3$

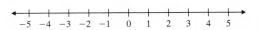

3. $4, \frac{1}{3}, -3, -3\frac{4}{5}, 1\frac{1}{3}$

4. $3, \frac{3}{8}, -4, -4\frac{1}{3}, -\frac{9}{10}$

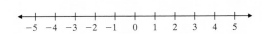

Objective B *Choose the best estimate for each product or quotient. See Examples 4 and 5.*

5. $2\frac{11}{12} \cdot 1\frac{1}{4}$
   a. 2   b. 3   c. 1   d. 12

6. $5\frac{1}{6} \cdot 3\frac{5}{7}$
   a. 9   b. 15   c. 8   d. 20

7. $12\frac{2}{11} \div 3\frac{9}{10}$
   a. 3   b. 4   c. 36   d. 9

8. $20\frac{3}{14} \div 4\frac{8}{11}$
   a. 5   b. 80   c. 4   d. 16

Multiply or divide. For Exercises 13 through 16, find the exact answer and an estimated answer. See Examples 2 through 8.

9. $2\frac{2}{3} \cdot \frac{1}{7}$

10. $\frac{5}{9} \cdot 4\frac{1}{5}$

11. $7 \div 1\frac{3}{5}$

12. $9 \div 1\frac{2}{3}$

13. $2\frac{1}{5} \cdot 3\frac{1}{2}$
    Exact:
    Estimate:

14. $2\frac{1}{4} \cdot 7\frac{1}{8}$
    Exact:
    Estimate:

15. $3\frac{4}{5} \cdot 6\frac{2}{7}$
    Exact:
    Estimate:

16. $5\frac{5}{6} \cdot 7\frac{3}{5}$
    Exact:
    Estimate:

17. $5 \cdot 2\frac{1}{2}$

18. $6 \cdot 3\frac{1}{3}$

19. $3\frac{2}{3} \cdot 1\frac{1}{2}$

20. $2\frac{4}{5} \cdot 2\frac{5}{8}$

21. $2\frac{2}{3} \div \frac{1}{7}$

22. $\frac{5}{9} \div 4\frac{1}{5}$

Objective C *Choose the best estimate for each sum or difference. See Examples 9 and 12.*

23. $3\frac{7}{8} + 2\frac{1}{5}$
   a. 6   b. 5   c. 1   d. 2

24. $3\frac{7}{8} - 2\frac{1}{5}$
   a. 6   b. 5   c. 1   d. 2

25. $8\frac{1}{3} + 1\frac{1}{2}$
   a. 4   b. 10   c. 6   d. 16

26. $8\frac{1}{3} - 1\frac{1}{2}$
   a. 4   b. 10   c. 6   d. 16

Add. For Exercises 27 through 30, find the exact sum and an estimated sum. See Examples 9 through 11.

27. $4\frac{7}{12}$
    $+ 2\frac{1}{12}$
    Exact:
    Estimate:

28. $7\frac{4}{11}$
    $+ 3\frac{2}{11}$
    Exact:
    Estimate:

29. $10\frac{3}{14}$
    $+ 3\frac{4}{7}$
    Exact:
    Estimate:

30. $12\frac{5}{12}$
    $+ 4\frac{1}{6}$
    Exact:
    Estimate:

31. $9\frac{1}{5}$
    $+ 8\frac{2}{25}$

32. $6\frac{2}{13}$
    $+ 8\frac{7}{26}$

33. $12\frac{3}{14}$
    10
    $+ 25\frac{5}{12}$

34. $8\frac{2}{9}$
    32
    $+ 9\frac{10}{21}$

35. $15\frac{4}{7}$
    $+ 9\frac{11}{14}$

36. $23\frac{3}{5}$
    $+ 8\frac{8}{15}$

37. $3\frac{5}{8}$
    $2\frac{1}{6}$
    $+ 7\frac{3}{4}$

38. $4\frac{1}{3}$
    $9\frac{2}{5}$
    $+ 3\frac{1}{6}$

Subtract. For Exercises 39 through 42, find the exact difference and an estimated difference. See Examples 12 through 14.

39. $4\frac{7}{10}$
    $- 2\frac{1}{10}$
    Exact:
    Estimate:

40. $7\frac{4}{9}$
    $- 3\frac{2}{9}$
    Exact:
    Estimate:

41. $10\frac{13}{14}$
    $- 3\frac{4}{7}$
    Exact:
    Estimate:

42. $12\frac{5}{12}$
    $- 4\frac{1}{6}$
    Exact:
    Estimate:

43. $9\dfrac{1}{5} - 8\dfrac{6}{25}$

44. $5\dfrac{2}{13} - 4\dfrac{7}{26}$

45. $6 - 2\dfrac{4}{9}$

46. $8 - 1\dfrac{7}{10}$

47. $63\dfrac{1}{6} - 47\dfrac{5}{12}$

48. $86\dfrac{2}{15} - 27\dfrac{3}{10}$

Objectives B C Mixed Practice *Perform each indicated operation. See Examples 2 through 14.*

49. $2\dfrac{3}{4} + 1\dfrac{1}{4}$

50. $5\dfrac{5}{8} + 2\dfrac{3}{8}$

51. $15\dfrac{4}{7} - 9\dfrac{11}{14}$

52. $23\dfrac{8}{15} - 8\dfrac{3}{5}$

53. $3\dfrac{1}{9} \cdot 2$

54. $4\dfrac{1}{2} \cdot 3$

55. $1\dfrac{2}{3} \div 2\dfrac{1}{5}$

56. $5\dfrac{1}{5} \div 3\dfrac{1}{4}$

57. $22\dfrac{4}{9} + 13\dfrac{5}{18}$

58. $15\dfrac{3}{25} + 5\dfrac{2}{5}$

59. $5\dfrac{2}{3} - 3\dfrac{1}{6}$

60. $5\dfrac{3}{8} - 2\dfrac{3}{16}$

61. $15\dfrac{1}{5} + 20\dfrac{3}{10} + 37\dfrac{2}{15}$

62. $3\dfrac{7}{16} + 6\dfrac{1}{2} + 9\dfrac{3}{8}$

63. $6\dfrac{4}{7} - 5\dfrac{11}{14}$

64. $47\dfrac{5}{12} - 23\dfrac{19}{24}$

65. $4\dfrac{2}{7} \cdot 1\dfrac{3}{10}$

66. $6\dfrac{2}{3} \cdot 2\dfrac{3}{4}$

67. $6\dfrac{2}{11} + 3 + 4\dfrac{10}{33}$

68. $7\dfrac{3}{7} + 15 + 20\dfrac{1}{2}$

Objective D Translating

Translate each phrase into an algebraic expression. Use x to represent "a number." See Examples 15 and 16.

69. $-5\frac{2}{7}$ decreased by a number

70. The sum of $8\frac{3}{4}$ and a number

71. Multiply $1\frac{9}{10}$ by a number.

72. Divide a number by $-6\frac{1}{11}$.

Solve. For Exercises 73 and 74, the solutions have been started for you. Write each answer in simplest form. See Examples 15 and 16.

73. A heart attack patient in rehabilitation walked on a treadmill $12\frac{3}{4}$ miles over 4 days. How many miles is this per day on average?

Start the solution:
1. UNDERSTAND the problem. Reread it as many times as needed.
2. TRANSLATE into an equation. (Fill in the blanks.)

miles per day	is	total miles	divided by	number of days
↓	↓	↓	↓	↓
miles per day	=	___	÷	___

Finish with:
3. SOLVE and 4. INTERPRET

74. A local restaurant is selling hamburgers from a booth on Memorial Day. A total of $27\frac{3}{4}$ pounds of hamburger have been ordered. How many quarter-pound hamburgers can this make?

Start the solution:
1. UNDERSTAND the problem. Reread it as many times as needed.
2. TRANSLATE into an equation. (Fill in the blanks.)

how many quarter-pound hamburgers	is	total pounds of hamburger	divided by	a quarter pound
↓	↓	↓	↓	↓
how many quarter-pound hamburgers	=	___	÷	___

Finish with:
3. SOLVE and 4. INTERPRET

75. The Gauge Act of 1846 set the standard gauge for U.S. railroads at $56\frac{1}{2}$ inches. (See figure.) If the standard gauge in Spain is $65\frac{9}{10}$ inches, how much wider is Spain's standard gauge than the U.S. standard gauge? (*Source:* San Diego Railroad Museum)

76. The standard railroad track gauge (see figure) in Spain is $65\frac{9}{10}$ inches, while in neighboring Portugal it is $65\frac{11}{20}$ inches. Which gauge is wider and by how much? (*Source:* San Diego Railroad Museum)

Track gauge (U.S. $56\frac{1}{2}$ inches)

$\frac{5}{8}$ inch

Point of measurement of gauge

77. If Tucson's average rainfall is $11\frac{1}{4}$ inches and Yuma's is $3\frac{3}{5}$ inches, how much more rain, on the average, does Tucson get than Yuma?

78. A pair of crutches needs adjustment. One crutch is 43 inches and the other is $41\frac{5}{8}$ inches. Find how much the shorter crutch should be lengthened to make both crutches the same length.

For Exercises 79 and 80, find the area of each figure.

79.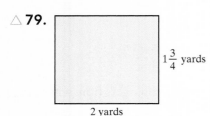

80. 5 inches, $3\frac{1}{2}$ inches

81. A model for a proposed computer chip measures $\frac{3}{4}$ inch by $1\frac{1}{4}$ inches. Find its area.

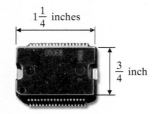

82. The Saltalamachios are planning to build a deck that measures $4\frac{1}{2}$ yards by $6\frac{1}{3}$ yards. Find the area of their proposed deck.

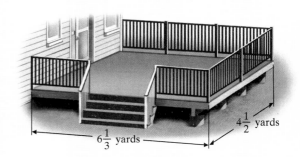

For Exercises 83 and 84, find the perimeter of each figure.

83.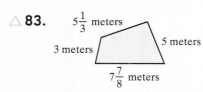

$5\frac{1}{3}$ meters, 5 meters, 3 meters, $7\frac{7}{8}$ meters

84. $3\frac{1}{4}$ yards (each of 5 sides)

85. A homeowner has $15\frac{2}{3}$ feet of plastic pipe. She cuts off a $2\frac{1}{2}$-foot length and then a $3\frac{1}{4}$-foot length. If she now needs a 10-foot piece of pipe, will the remaining piece do? If not, by how much will the piece be short?

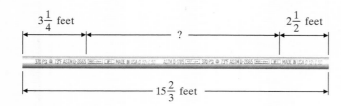

86. A trim carpenter cuts a board $3\frac{3}{8}$ feet long from one 6 feet long. How long is the remaining piece?

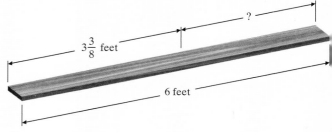

87. The area of the rectangle below is 12 square meters. If its width is $2\frac{4}{7}$ meters, find its length.

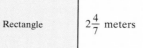

△ 88. The perimeter of the square below is $23\frac{1}{2}$ feet. Find the length of each side.

The following table lists three upcoming total eclipses of the Sun that will be visible in North America. The duration of each eclipse is listed in the table. Use the table to answer Exercises 89 through 92.

Total Solar Eclipses Visible from North America	
Date of Eclipse	**Duration (in minutes)**
August 21, 2017	$2\frac{2}{3}$
April 8, 2024	$4\frac{7}{15}$
March 30, 2033	$2\frac{37}{60}$
(*Source:* NASA/Goddard Space Flight Center)	

89. What is the total duration for the three eclipses?

90. What is the total duration for the two eclipses occurring in odd-numbered years?

91. How much longer will the April 8, 2024, eclipse be than the August 21, 2017, eclipse?

92. How much longer will the April 8, 2024, eclipse be than the March 30, 2033, eclipse?

Objective E *Perform the indicated operations. See Examples 17 through 24.*

93. $-4\frac{2}{5} \cdot 2\frac{3}{10}$

94. $-3\frac{5}{6} \div \left(-3\frac{2}{3}\right)$

95. $-5\frac{1}{8} - 19\frac{3}{4}$

96. $17\frac{5}{9} + \left(-14\frac{2}{3}\right)$

97. $-31\frac{2}{15} + 17\frac{3}{20}$

98. $-31\frac{7}{8} - \left(-26\frac{5}{12}\right)$

99. $-1\frac{5}{7} \cdot \left(-2\frac{1}{2}\right)$

100. $1\frac{3}{4} \div \left(-3\frac{1}{2}\right)$

101. $11\frac{7}{8} - 13\frac{5}{6}$

102. $-20\frac{2}{5} + \left(-30\frac{3}{10}\right)$

103. $-7\frac{3}{10} \div (-100)$

104. $-4\frac{1}{4} \div 2\frac{3}{8}$

Review

Multiply. See Section 4.3.

105. $\frac{1}{3}(3x)$ **106.** $\frac{1}{5}(5y)$ **107.** $\frac{2}{3}\left(\frac{3}{2}a\right)$ **108.** $-\frac{9}{10}\left(-\frac{10}{9}m\right)$

Concept Extensions

Solve. See the first Concept Check in this section.

109. Which of the following are equivalent to 10?
  a. $9\frac{5}{5}$ b. $9\frac{100}{100}$ c. $6\frac{44}{11}$ d. $8\frac{13}{13}$

110. Which of the following are equivalent to $7\frac{3}{4}$?
  a. $6\frac{7}{4}$ b. $5\frac{11}{4}$ c. $7\frac{12}{16}$ d. all of them

Solve. See the second Concept Check in this section.

111. A student asked you to check her work below. Is it correct? If not, where is the error?
$$20\frac{2}{3} \div 10\frac{1}{2} \stackrel{?}{=} 2\frac{1}{3}$$

112. A student asked you to check his work below. Is it correct? If not, where is the error?
$$3\frac{2}{3} \cdot 1\frac{1}{7} \stackrel{?}{=} 3\frac{2}{21}$$

113. In your own words, describe how to divide mixed numbers.

114. In your own words, explain how to multiply
  a. fractions
  b. mixed numbers

Solve. See the third Concept Check in this section.

115. In your own words, explain how to round a mixed number to the nearest whole number.

116. Use rounding to estimate the best sum for $11\frac{19}{20} + 9\frac{1}{10}$.
  a. 2   b. 3   c. 20   d. 21

Solve.

117. Explain in your own words why $9\frac{13}{9}$ is equal to $10\frac{4}{9}$.

118. In your own words, explain
  a. when to borrow when subtracting mixed numbers, and
  b. how to borrow when subtracting mixed numbers.

4.8 Solving Equations Containing Fractions

Objective A Solving Equations Containing Fractions

In Chapter 3, we solved linear equations in one variable. In this section, we practice this skill by solving linear equations containing fractions. To help us solve these equations, let's review the properties of equality.

Objectives

A Solve Equations Containing Fractions.

B Solve Equations by Multiplying by the LCD.

C Review Adding and Subtracting Fractions.

Addition Property of Equality

Let a, b, and c represent numbers. Then

$a = b$	Also, $a = b$
and $a + c = b + c$	and $a - c = b - c$
are equivalent equations.	are equivalent equations.

Multiplication Property of Equality

Let a, b, and c represent numbers and let $c \neq 0$. Then

$a = b$	Also, $a = b$
and $a \cdot c = b \cdot c$	and $\dfrac{a}{c} = \dfrac{b}{c}$
are equivalent equations.	are equivalent equations.

In other words, the same number may be added to or subtracted from both sides of an equation without changing the solution of the equation. Also, both sides of an equation may be multiplied or divided by the same nonzero number without changing the solution.

Also, don't forget that to solve an equation in x, our goal is to use properties of equality to write simpler equations, all equivalent to the original equation, until the final equation has the form

$x =$ **number**   or   **number** $= x$

Example 1 Solve: $x - \dfrac{3}{4} = \dfrac{1}{20}$

Solution: To get x by itself, add $\dfrac{3}{4}$ to both sides.

$$x - \dfrac{3}{4} = \dfrac{1}{20}$$

$$x - \dfrac{3}{4} + \dfrac{3}{4} = \dfrac{1}{20} + \dfrac{3}{4} \quad \text{Add } \dfrac{3}{4} \text{ to both sides.}$$

$$x = \dfrac{1}{20} + \dfrac{3 \cdot 5}{4 \cdot 5} \quad \text{The LCD of the denominators 20 and 4 is 20.}$$

$$x = \dfrac{1}{20} + \dfrac{15}{20}$$

$$x = \dfrac{16}{20}$$

$$x = \dfrac{\cancel{4} \cdot 4}{\cancel{4} \cdot 5} = \dfrac{4}{5} \quad \text{Write } \dfrac{16}{20} \text{ in simplest form.}$$

Practice 1

Solve: $y - \dfrac{2}{3} = \dfrac{5}{12}$

Answer

1. $\dfrac{13}{12}$

(Continued on next page)

Check: To check, replace x with $\frac{4}{5}$ in the original equation.

$$x - \frac{3}{4} = \frac{1}{20}$$

$$\frac{4}{5} - \frac{3}{4} \stackrel{?}{=} \frac{1}{20} \quad \text{Replace } x \text{ with } \frac{4}{5}.$$

$$\frac{4 \cdot 4}{5 \cdot 4} - \frac{3 \cdot 5}{4 \cdot 5} \stackrel{?}{=} \frac{1}{20} \quad \text{The LCD of the denominators 5 and 4 is 20.}$$

$$\frac{16}{20} - \frac{15}{20} \stackrel{?}{=} \frac{1}{20}$$

$$\frac{1}{20} = \frac{1}{20} \quad \text{True}$$

Thus $\frac{4}{5}$ is the solution of $x - \frac{3}{4} = \frac{1}{20}$.

■ Work Practice 1

Practice 2

Solve: $\frac{1}{5}y = 2$

Example 2 Solve: $\frac{1}{3}x = 7$

Solution: Recall that isolating x means that we want the coefficient of x to be 1. To do so, we use the multiplication property of equality and multiply both sides of the equation by the reciprocal of $\frac{1}{3}$, or 3. Since $\frac{1}{3} \cdot 3 = 1$, we will have isolated x.

$$\frac{1}{3}x = 7$$

$$3 \cdot \frac{1}{3}x = 3 \cdot 7 \quad \text{Multiply both sides by 3.}$$

$$1 \cdot x = 21 \text{ or } x = 21 \quad \text{Simplify.}$$

Check: To check, replace x with 21 in the original equation.

$$\frac{1}{3}x = 7 \quad \text{Original equation}$$

$$\frac{1}{3} \cdot 21 \stackrel{?}{=} 7 \quad \text{Replace } x \text{ with 21.}$$

$$7 = 7 \quad \text{True}$$

Since $7 = 7$ is a true statement, 21 is the solution of $\frac{1}{3}x = 7$.

■ Work Practice 2

Practice 3

Solve: $\frac{5}{7}b = 25$

Example 3 Solve: $\frac{3}{5}a = 9$

Solution: Multiply both sides by $\frac{5}{3}$, the reciprocal of $\frac{3}{5}$, so that the coefficient of a is 1.

$$\frac{3}{5}a = 9$$

$$\frac{5}{3} \cdot \frac{3}{5}a = \frac{5}{3} \cdot 9 \quad \text{Multiply both sides by } \frac{5}{3}.$$

$$1a = \frac{5 \cdot 9}{3} \quad \text{Multiply.}$$

$$a = 15 \quad \text{Simplify.}$$

Answers

2. 10 **3.** 35

Section 4.8 | Solving Equations Containing Fractions

Check: To check, replace a with 15 in the original equation.

$$\frac{3}{5}a = 9$$

$$\frac{3}{5} \cdot 15 \stackrel{?}{=} 9 \quad \text{Replace } a \text{ with 15.}$$

$$\frac{3 \cdot 15}{5} \stackrel{?}{=} 9 \quad \text{Multiply.}$$

$$9 = 9 \quad \text{True}$$

Since $9 = 9$ is true, 15 is the solution of $\frac{3}{5}a = 9$.

■ Work Practice 3

Example 4 Solve: $\frac{3}{4}x = -\frac{1}{8}$

Solution: Multiply both sides of the equation by $\frac{4}{3}$, the reciprocal of $\frac{3}{4}$.

$$\frac{3}{4}x = -\frac{1}{8}$$

$$\frac{4}{3} \cdot \frac{3}{4}x = \frac{4}{3} \cdot -\frac{1}{8} \quad \text{Multiply both sides by } \frac{4}{3}.$$

$$1x = -\frac{4 \cdot 1}{3 \cdot 8} \quad \text{Multiply.}$$

$$x = -\frac{1}{6} \quad \text{Simplify.}$$

Practice 4

Solve: $-\frac{7}{10}x = \frac{2}{5}$

Check: To check, replace x with $-\frac{1}{6}$ in the original equation.

$$\frac{3}{4}x = -\frac{1}{8} \quad \text{Original equation}$$

$$\frac{3}{4} \cdot -\frac{1}{6} \stackrel{?}{=} -\frac{1}{8} \quad \text{Replace } x \text{ with } -\frac{1}{6}.$$

$$-\frac{1}{8} = -\frac{1}{8} \quad \text{True}$$

Since we arrived at a true statement, $-\frac{1}{6}$ is the solution of $\frac{3}{4}x = -\frac{1}{8}$.

■ Work Practice 4

Example 5 Solve: $3y = -\frac{2}{11}$

Solution: We can either divide both sides by 3 or multiply both sides by the reciprocal of 3, which is $\frac{1}{3}$.

$$3y = -\frac{2}{11}$$

$$\frac{1}{3} \cdot 3y = \frac{1}{3} \cdot -\frac{2}{11} \quad \text{Multiply both sides by } \frac{1}{3}.$$

$$1y = -\frac{1 \cdot 2}{3 \cdot 11} \quad \text{Multiply.}$$

$$y = -\frac{2}{33} \quad \text{Simplify.}$$

Check to see that the solution is $-\frac{2}{33}$.

Practice 5

Solve: $5x = -\frac{3}{4}$

Answers

4. $-\frac{4}{7}$ **5.** $-\frac{3}{20}$

■ Work Practice 5

Chapter 4 | Fractions and Mixed Numbers

Objective B Solving Equations by Multiplying by the LCD

Solving equations with fractions can be tedious. If an equation contains fractions, it is often helpful to first multiply both sides of the equation by the LCD of the fractions. This has the effect of eliminating the fractions in the equation, as shown in the next example.

Let's solve the equation in Example 4 again. This time, we will multiply both sides by the LCD.

Practice 6

Solve: $\dfrac{11}{15}x = -\dfrac{3}{5}$

Example 6 Solve: $\dfrac{3}{4}x = -\dfrac{1}{8}$ (Example 4 solved an alternate way.)

Solution: First, multiply both sides of the equation by the LCD of the fractions $\dfrac{3}{4}$ and $-\dfrac{1}{8}$. The LCD of the denominators is 8.

$$\dfrac{3}{4}x = -\dfrac{1}{8}$$

$$8 \cdot \dfrac{3}{4}x = 8 \cdot -\dfrac{1}{8} \quad \text{Multiply both sides by 8.}$$

$$\dfrac{8 \cdot 3}{1 \cdot 4}x = -\dfrac{\overset{1}{\cancel{8}} \cdot 1}{1 \cdot \cancel{8}} \quad \text{Multiply the fractions.}$$

$$\dfrac{2 \cdot \overset{1}{\cancel{4}} \cdot 3}{1 \cdot \cancel{4}}x = -\dfrac{1 \cdot 1}{1 \cdot 1} \quad \text{Simplify.}$$

$$6x = -1$$

$$\dfrac{6x}{6} = \dfrac{-1}{6} \quad \text{Divide both sides by 6.}$$

$$x = \dfrac{-1}{6} \text{ or } -\dfrac{1}{6} \quad \text{Simplify.}$$

As seen in Example 4, the solution is $-\dfrac{1}{6}$.

■ Work Practice 6

Practice 7

Solve: $\dfrac{y}{8} + \dfrac{3}{4} = 2$

Example 7 Solve: $\dfrac{x}{6} + 1 = \dfrac{4}{3}$

Solution:

Solve by multiplying by the LCD:	Solve with fractions:
The LCD of the denominators 6 and 3 is 6. $\dfrac{x}{6} + 1 = \dfrac{4}{3}$ $6\left(\dfrac{x}{6} + 1\right) = 6\left(\dfrac{4}{3}\right)$ Multiply both sides by 6. $\overset{1}{\cancel{6}}\left(\dfrac{x}{\cancel{6}}\right) + 6(1) = \overset{2}{\cancel{6}}\left(\dfrac{4}{\cancel{3}}\right)$ Apply the distributive property. $x + 6 = 8$ Simplify. $x + 6 + (-6) = 8 + (-6)$ Add -6 to both sides. $x = 2$ Simplify.	$\dfrac{x}{6} + 1 = \dfrac{4}{3}$ $\dfrac{x}{6} + 1 - 1 = \dfrac{4}{3} - 1$ Subtract 1 from both sides. $\dfrac{x}{6} = \dfrac{4}{3} - \dfrac{3}{3}$ Write 1 as $\dfrac{3}{3}$. $\dfrac{x}{6} = \dfrac{1}{3}$ Subtract. $6 \cdot \dfrac{x}{6} = 6 \cdot \dfrac{1}{3}$ Multiply both sides by 6. $6 \cdot \dfrac{1}{6} \cdot x = \dfrac{\overset{2}{\cancel{6}} \cdot 1}{1 \cdot \cancel{3}}$ $1x = 2 \quad \text{or} \quad x = 2$

Answers

6. $-\dfrac{9}{11}$ **7.** 10

Check: To check, replace x with 2 in the original equation.

$$\frac{x}{6} + 1 = \frac{4}{3} \quad \text{Original equation}$$

$$\frac{2}{6} + 1 \stackrel{?}{=} \frac{4}{3} \quad \text{Replace } x \text{ with 2.}$$

$$\frac{1}{3} + \frac{3}{3} \stackrel{?}{=} \frac{4}{3} \quad \text{Simplify } \frac{2}{6}. \text{ The LCD of 3 and 1 is 3.}$$

$$\frac{4}{3} = \frac{4}{3} \quad \text{True}$$

Since we arrived at a true statement, 2 is the solution of $\frac{x}{6} + 1 = \frac{4}{3}$.

■ Work Practice 7

Let's review the steps for solving equations in x. An extra step is now included to handle equations containing fractions.

Solving an Equation in x

Step 1: If fractions are present, multiply both sides of the equation by the LCD of the fractions.
Step 2: If parentheses are present, use the distributive property.
Step 3: Combine any like terms on each side of the equation.
Step 4: Use the addition property of equality to rewrite the equation so that variable terms are on one side of the equation and constant terms are on the other side.
Step 5: Divide both sides of the equation by the numerical coefficient of x to solve.
Step 6: Check the answer in the **original equation.**

Example 8 Solve: $\frac{z}{5} - \frac{z}{3} = 6$

Solution:

$$\frac{z}{5} - \frac{z}{3} = 6$$

$$15\left(\frac{z}{5} - \frac{z}{3}\right) = 15(6) \quad \text{Multiply both sides by the LCD, 15.}$$

$$\overset{3}{\cancel{15}}\left(\frac{z}{\cancel{5}}\right) - \overset{5}{\cancel{15}}\left(\frac{z}{\cancel{3}}\right) = 15(6) \quad \text{Apply the distributive property.}$$

$$3z - 5z = 90 \quad \text{Simplify.}$$

$$-2z = 90 \quad \text{Combine like terms.}$$

$$\frac{-2z}{-2} = \frac{90}{-2} \quad \text{Divide both sides by } -2, \text{ the coefficient of } z.$$

$$z = -45 \quad \text{Simplify.}$$

To check, replace z with -45 in the **original equation** to see that a true statement results.

■ Work Practice 8

Practice 8

Solve: $\frac{x}{5} - x = \frac{1}{5}$

> **Helpful Hint** Don't forget to multiply *both* sides of the equation by the LCD.

Answer
8. $-\frac{1}{4}$

Practice 9

Solve: $\dfrac{y}{2} = \dfrac{y}{5} + \dfrac{3}{2}$

Example 9

Solve: $\dfrac{x}{2} = \dfrac{x}{3} + \dfrac{1}{2}$

Solution: First multiply both sides by the LCD, 6.

$$\dfrac{x}{2} = \dfrac{x}{3} + \dfrac{1}{2}$$

$$6\left(\dfrac{x}{2}\right) = 6\left(\dfrac{x}{3} + \dfrac{1}{2}\right) \quad \text{Multiply both sides by the LCD, 6.}$$

$$\overset{3}{\cancel{6}}\left(\dfrac{x}{2}\right) = \overset{2}{\cancel{6}}\left(\dfrac{x}{3}\right) + \overset{3}{\cancel{6}}\left(\dfrac{1}{2}\right) \quad \text{Apply the distributive property.}$$

$$3x = 2x + 3 \quad \text{Simplify.}$$

$$3x - 2x = 2x + 3 - 2x \quad \text{Subtract } 2x \text{ from both sides.}$$

$$x = 3 \quad \text{Simplify.}$$

To check, replace x with 3 in the original equation to see that a true statement results.

■ Work Practice 9

Objective C Review of Adding and Subtracting Fractions ▶

Make sure you understand the difference between **solving an equation** containing fractions and **adding or subtracting two fractions.** To solve an equation containing fractions, we use the multiplication property of equality and multiply both sides by the LCD of the fractions, thus eliminating the fractions. This method does not apply to adding or subtracting fractions. The multiplication property of equality applies only to equations. To add or subtract unlike fractions, we write each fraction as an equivalent fraction using the LCD of the fractions as the denominator. See the next example for a review.

Practice 10

Subtract: $\dfrac{9}{10} - \dfrac{y}{3}$

Example 10

Add: $\dfrac{x}{3} + \dfrac{2}{5}$

Solution: This is an expression, not an equation. Here, we are adding two unlike fractions. To add unlike fractions, we need to find the LCD. The LCD of the denominators 3 and 5 is 15. Write each fraction as an equivalent fraction with a denominator of 15.

$$\dfrac{x}{3} + \dfrac{2}{5} = \dfrac{x}{3} \cdot \dfrac{5}{5} + \dfrac{2}{5} \cdot \dfrac{3}{3} = \dfrac{x \cdot 5}{3 \cdot 5} + \dfrac{2 \cdot 3}{5 \cdot 3}$$

$$= \dfrac{5x}{15} + \dfrac{6}{15}$$

$$= \dfrac{5x + 6}{15}$$

■ Work Practice 10

✓ **Concept Check** Which of the following are equations and which are expressions?

a. $\dfrac{1}{2} + 3x = 5$

b. $\dfrac{2}{3}x - \dfrac{x}{5}$

c. $\dfrac{x}{12} + \dfrac{5x}{24}$

d. $\dfrac{x}{5} = \dfrac{1}{10}$

Answers

9. 5   10. $\dfrac{27 - 10y}{30}$

✓ **Concept Check Answers**

equations: a, d; expressions: b, c

Section 4.8 | Solving Equations Containing Fractions

Vocabulary, Readiness & Video Check

Fill in the blank with the least common denominator (LCD). Do not solve these equations.

1. Equation: $\frac{2}{3} + x = \frac{5}{6}$; LCD = ___
2. Equation: $\frac{x}{21} - 1 = \frac{1}{7}$; LCD = ___
3. Equation: $\frac{y}{5} + \frac{1}{3} = 2$; LCD = ___
4. Equation: $\frac{-2n}{11} + \frac{1}{2} = 5$; LCD = ___

See Video 4.8

Martin-Gay Interactive Videos Watch the section lecture video and answer the following questions.

Objective A
5. In Example 1, what property is used to get x by itself on one side of the equation?
6. Explain how a reciprocal is used to solve Example 2.

Objective B
7. Why are both sides of the equation multiplied by 12 in Example 3? What effect does this have on the fractions in the equation?

Objective C
8. In Example 5, why can't we multiply by the LCD of all fractions?

4.8 Exercise Set MyLab Math

Objective A *Solve each equation. Check your proposed solution. See Example 1.*

1. $x + \frac{1}{3} = -\frac{1}{3}$
2. $x + \frac{1}{9} = -\frac{7}{9}$
3. $y - \frac{3}{13} = -\frac{2}{13}$

4. $z - \frac{5}{14} = \frac{4}{14}$
5. $3x - \frac{1}{5} - 2x = \frac{1}{5} + \frac{2}{5}$
6. $5x + \frac{1}{11} - 4x = \frac{2}{11} - \frac{5}{11}$

7. $x - \frac{1}{12} = \frac{5}{6}$
8. $y - \frac{8}{9} = \frac{1}{3}$
9. $\frac{2}{5} + y = -\frac{3}{10}$
10. $\frac{1}{2} + a = -\frac{3}{8}$

11. $7z + \frac{1}{16} - 6z = \frac{3}{4}$
12. $9x - \frac{2}{7} - 8x = \frac{11}{14}$
13. $-\frac{2}{9} = x - \frac{5}{6}$
14. $-\frac{1}{4} = y - \frac{7}{10}$

Solve each equation. See Examples 2 through 5.

15. $7x = 2$

16. $-5x = 4$

17. $\frac{1}{4}x = 3$

18. $\frac{1}{3}x = 6$

19. $\frac{2}{9}y = -6$

20. $\frac{4}{7}x = -8$

21. $-\frac{4}{9}z = -\frac{3}{2}$

22. $-\frac{11}{10}x = -\frac{2}{7}$

23. $7a = \frac{1}{3}$

24. $2z = -\frac{5}{12}$

25. $-3x = -\frac{6}{11}$

26. $-4z = -\frac{12}{25}$

Objective B *Solve each equation. See Examples 6 through 9.*

27. $\frac{5}{9}x = -\frac{3}{18}$

28. $\frac{3}{5}y = -\frac{7}{20}$

29. $\frac{x}{3} + 2 = \frac{7}{3}$

30. $\frac{x}{5} - 1 = \frac{7}{5}$

31. $\frac{x}{5} - x = -8$

32. $\frac{x}{3} - x = -6$

33. $\frac{1}{2} - \frac{3}{5} = \frac{x}{10}$

34. $\frac{2}{3} - \frac{1}{4} = \frac{x}{12}$

35. $\frac{x}{3} = \frac{x}{5} - 2$

36. $\frac{a}{2} = \frac{a}{7} + \frac{5}{2}$

Objective C *Add or subtract as indicated. See Example 10.*

37. $\frac{x}{7} - \frac{4}{3}$

38. $-\frac{5}{9} + \frac{y}{8}$

39. $\frac{y}{2} + 5$

40. $2 + \frac{7x}{3}$

41. $\frac{3x}{10} + \frac{x}{6}$

42. $\frac{9x}{8} - \frac{5x}{6}$

Objectives A B C Mixed Practice *Solve. If no equation is given, perform the indicated operation. See Examples 1 through 10.*

43. $\frac{3}{8}x = \frac{1}{2}$

44. $\frac{2}{5}y = \frac{3}{10}$

45. $\frac{2}{3} - \frac{x}{5} = \frac{4}{15}$

46. $\frac{4}{5} + \frac{x}{4} = \frac{21}{20}$

47. $\frac{9}{14}z = \frac{27}{20}$

48. $\frac{5}{16}a = \frac{5}{6}$

49. $-3m - 5m = \frac{4}{7}$

50. $30n - 34n = \frac{3}{20}$

51. $\dfrac{x}{4} + 1 = \dfrac{1}{4}$

52. $\dfrac{y}{7} - 2 = \dfrac{1}{7}$

53. $\dfrac{5}{9} - \dfrac{2}{3}$

54. $\dfrac{8}{11} - \dfrac{1}{2}$

55. $\dfrac{1}{5}y = 10$

56. $\dfrac{1}{4}x = -2$

57. $\dfrac{5}{7}y = -\dfrac{15}{49}$

58. $-\dfrac{3}{4}x = \dfrac{9}{2}$

59. $\dfrac{x}{2} - x = -2$

60. $\dfrac{y}{3} = -4 + y$

61. $-\dfrac{5}{8}y = \dfrac{3}{16} - \dfrac{9}{16}$

62. $-\dfrac{7}{9}x = -\dfrac{5}{18} - \dfrac{4}{18}$

63. $17x - 25x = \dfrac{1}{3}$

64. $27x - 30x = \dfrac{4}{9}$

65. $\dfrac{7}{6}x = \dfrac{1}{4} - \dfrac{2}{3}$

66. $\dfrac{5}{4}y = \dfrac{1}{2} - \dfrac{7}{10}$

67. $\dfrac{b}{4} = \dfrac{b}{12} + \dfrac{2}{3}$

68. $\dfrac{a}{6} = \dfrac{a}{3} + \dfrac{1}{2}$

69. $\dfrac{x}{3} + 2 = \dfrac{x}{2} + 8$

70. $\dfrac{y}{5} - 2 = \dfrac{y}{3} - 4$

Review

Round each number to the given place value. See Section 1.4.

71. 57,236 to the nearest hundred

72. 576 to the nearest hundred

73. 327 to the nearest ten

74. 2333 to the nearest ten

Concept Extensions

75. Explain why the method for eliminating fractions in an **equation** does not apply to simplifying **expressions** containing fractions.

76. Think about which exercise (part **a** or part **b**) may be completed by multiplying by 6. Now complete each exercise.

    a. Solve: $\dfrac{x}{6} - \dfrac{5}{3} = 2$

    b. Subtract: $\dfrac{x}{6} - \dfrac{5}{3}$

Solve.

77. $\dfrac{14}{11} + \dfrac{3x}{8} = \dfrac{x}{2}$

78. $\dfrac{19}{53} = \dfrac{353x}{1431} + \dfrac{23}{27}$

79. Find the area and the perimeter of the rectangle. Remember to attach proper units.

    Rectangle: $\dfrac{1}{4}$ inch by $\dfrac{3}{4}$ inch

80. The area of the rectangle is $\dfrac{5}{12}$ square inch. Find its length, x.

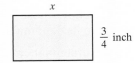

Chapter 4 Group Activity

Lobster Classification

Sections 4.1, 4.7, 4.8

This activity may be completed by working in groups or individually.

Lobsters are normally classified by weight. Use the weight classification table to answer the questions in this activity.

Classification of Lobsters	
Class	**Weight (in pounds)**
Chicken	1 to $1\frac{1}{8}$
Eighths	$1\frac{1}{8}$ to $1\frac{1}{4}$
Quarter	$1\frac{1}{4}$ to $1\frac{1}{2}$
Large (or select)	$1\frac{1}{2}$ to $2\frac{1}{2}$
Jumbo	Over $2\frac{1}{2}$

(*Source:* The Maine Lobster Marketing Collaborative)

1. A lobsterman has kept four lobsters from a lobster trap. Classify each lobster if they have the following weights:
   a. $1\frac{7}{8}$ pounds
   b. $1\frac{1}{16}$ pounds
   c. $2\frac{3}{4}$ pounds
   d. $1\frac{3}{8}$ pounds

2. A recipe requires 5 pounds of lobster. Using the minimum weight for each class, decide whether a chicken, a quarter, and a jumbo lobster will be enough for the recipe, and explain your reasoning. If not, suggest a better choice of lobsters to meet the recipe requirements.

3. A lobster market customer has selected two chickens, a large and a jumbo. The jumbo lobster weighs $3\frac{1}{4}$ pounds. What is the most that these four lobsters could weigh? What is the least that these four lobsters could weigh?

4. A lobster market customer wishes to buy three quarters, each weighing $1\frac{1}{4}$ pounds. If lobsters sell for $7 per pound, how much will the customer owe for her purchase?

5. Why do you think there is no classification for lobsters weighing under 1 pound?

Chapter 4 Vocabulary Check

Fill in each blank with one of the words or phrases listed below.

mixed number   complex fraction   like   numerator   prime factorization
composite number   equivalent   cross products   least common denominator   denominator
prime number   improper fraction   simplest form   undefined   0
reciprocals   proper fraction

1. Two numbers are _____ of each other if their product is 1.
2. A(n) _____ is a natural number greater than 1 that is not prime.
3. Fractions that represent the same portion of a whole are called _____ fractions.
4. A(n) _____ is a fraction whose numerator is greater than or equal to its denominator.
5. A(n) _____ is a natural number greater than 1 whose only factors are 1 and itself.
6. A fraction is in _____ when the numerator and the denominator have no factors in common other than 1.
7. A(n) _____ is one whose numerator is less than its denominator.
8. A(n) _____ contains a whole number part and a fraction part.
9. In the fraction $\frac{7}{9}$, the 7 is called the _____ and the 9 is called the _____.

10. The _____ of a number is the factorization in which all the factors are prime numbers.

11. The fraction $\frac{3}{0}$ is _____.

12. The fraction $\frac{0}{5}$ = _____.

13. Fractions that have the same denominator are called _____ fractions.

14. The LCM of the denominators in a list of fractions is called the _____.

15. A fraction whose numerator or denominator or both numerator and denominator contain fractions is called a(n) _____.

16. In $\frac{a}{b} = \frac{c}{d}$, $a \cdot d$ and $b \cdot c$ are called _____.

> **Helpful Hint**
> Are you preparing for your test? To help, don't forget to take these:
> - Chapter 4 Getting Ready for the Test on page 326
> - Chapter 4 Test on page 327
>
> Then check all of your answers at the back of this text. For further review, the step-by-step video solutions to any of these exercises are located in MyLab Math.

4 Chapter Highlights

Definitions and Concepts	Examples
Section 4.1 Introduction to Fractions and Mixed Numbers	
A **fraction** is of the form $\frac{\text{numerator}}{\text{denominator}}$ ← number of parts being considered ← number of equal parts in the whole	Write a fraction to represent the shaded part of the figure. $\frac{3}{8}$ ← number of parts shaded ← number of equal parts
A fraction is called a **proper fraction** if its numerator is less than its denominator.	Proper Fractions: $\frac{1}{3}, \frac{2}{5}, \frac{7}{8}, \frac{100}{101}$
A fraction is called an **improper fraction** if its numerator is greater than or equal to its denominator.	Improper Fractions: $\frac{5}{4}, \frac{2}{2}, \frac{9}{7}, \frac{101}{100}$
A **mixed number** contains a whole number and a fraction.	Mixed Numbers: $1\frac{1}{2}, 5\frac{7}{8}, 25\frac{9}{10}$
To Write a Mixed Number as an Improper Fraction 1. Multiply the denominator of the fraction by the whole number. 2. Add the numerator of the fraction to the product from Step 1. 3. Write the sum from Step 2 as the numerator of the improper fraction over the original denominator.	$5\frac{2}{7} = \frac{7 \cdot 5 + 2}{7} = \frac{35 + 2}{7} = \frac{37}{7}$
To Write an Improper Fraction as a Mixed Number or a Whole Number 1. Divide the denominator into the numerator. 2. The whole number part of the mixed number is the quotient. The fraction is the remainder over the original denominator. quotient $\frac{\text{remainder}}{\text{original denominator}}$	$\frac{17}{3} = 5\frac{2}{3}$ $\begin{array}{r}5\\3\overline{)17}\\-15\\\hline 2\end{array}$

Definitions and Concepts	Examples
Section 4.2 Factors and Simplest Form	
A **prime number** is a natural number that has exactly two different factors, 1 and itself.	2, 3, 5, 7, 11, 13, 17, ...
A **composite number** is any natural number other than 1 that is not prime.	4, 6, 8, 9, 10, 12, 14, 15, 16, ...
The **prime factorization** of a number is the factorization in which all the factors are prime numbers.	Write the prime factorization of 60. $60 = 6 \cdot 10$ $= 2 \cdot 3 \cdot 2 \cdot 5$ or $2^2 \cdot 3 \cdot 5$
Fractions that represent the same portion of a whole are called **equivalent fractions.**	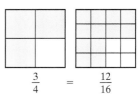 $\frac{3}{4} = \frac{12}{16}$
A fraction is in **simplest form** or **lowest terms** when the numerator and the denominator have no common factors other than 1.	The fraction $\frac{2}{3}$ is in simplest form.
To write a fraction in simplest form, write the prime factorizations of the numerator and the denominator and then divide both by all common factors.	Write in simplest form: $\frac{30}{36}$ $\frac{30}{36} = \frac{2 \cdot 3 \cdot 5}{2 \cdot 2 \cdot 3 \cdot 3} = \frac{2}{2} \cdot \frac{3}{3} \cdot \frac{5}{2 \cdot 3} = 1 \cdot 1 \cdot \frac{5}{6} = \frac{5}{6}$ or $\frac{30}{36} = \frac{\cancel{2} \cdot \cancel{3} \cdot 5}{\cancel{2} \cdot 2 \cdot \cancel{3} \cdot 3} = \frac{5}{6}$
Two fractions are equivalent if	Determine whether $\frac{7}{8}$ and $\frac{21}{24}$ are equivalent.
Method 1. They simplify to the same fraction.	$\frac{7}{8}$ is in simplest form.
Method 2. Their cross products are equal. $\begin{array}{c} 24 \cdot 7 \\ = 168 \end{array} \quad \frac{7}{8} = \frac{21}{24} \quad \begin{array}{c} 8 \cdot 21 \\ = 168 \end{array}$ Since $168 = 168$, $\frac{7}{8} = \frac{21}{24}$.	$\frac{21}{24} = \frac{\cancel{3} \cdot 7}{\cancel{3} \cdot 8} = \frac{1 \cdot 7}{1 \cdot 8} = \frac{7}{8}$ Since both simplify to $\frac{7}{8}$, then $\frac{7}{8} = \frac{21}{24}$.
Section 4.3 Multiplying and Dividing Fractions	
To multiply two fractions, multiply the numerators and multiply the denominators.	Multiply. $\frac{2x}{3} \cdot \frac{5}{7} = \frac{2x \cdot 5}{3 \cdot 7} = \frac{10x}{21}$ $\frac{3}{4} \cdot \frac{1}{6} = \frac{3 \cdot 1}{4 \cdot 6} = \frac{\cancel{3} \cdot 1}{4 \cdot \cancel{3} \cdot 2} = \frac{1}{8}$
To find the **reciprocal** of a fraction, interchange its numerator and denominator.	The reciprocal of $\frac{3}{5}$ is $\frac{5}{3}$.
To divide two fractions, multiply the first fraction by the reciprocal of the second fraction.	Divide. $-\frac{3}{10} \div \frac{7}{9} = -\frac{3}{10} \cdot \frac{9}{7} = -\frac{3 \cdot 9}{10 \cdot 7} = -\frac{27}{70}$

Definitions and Concepts	Examples

Section 4.4 Adding and Subtracting Like Fractions, Least Common Denominator, and Equivalent Fractions

Fractions that have the same denominator are called **like fractions**.	$-\dfrac{1}{3}$ and $\dfrac{2}{3}$; $\dfrac{5x}{7}$ and $\dfrac{6}{7}$
To add or subtract like fractions, combine the numerators and place the sum or difference over the common denominator.	$\dfrac{2}{7} + \dfrac{3}{7} = \dfrac{5}{7}$ ← Add the numerators. ← Keep the common denominator. $\dfrac{7}{8} - \dfrac{4}{8} = \dfrac{3}{8}$ ← Subtract the numerators. ← Keep the common denominator.
The **least common denominator (LCD)** of a list of fractions is the smallest positive number divisible by all the denominators in the list.	The LCD of $\dfrac{1}{2}$ and $\dfrac{5}{6}$ is 6 because 6 is the smallest positive number that is divisible by both 2 and 6.
Method 1 for finding the LCD of a list of fractions using multiples **Step 1:** Write the multiples of the largest denominator (starting with the number itself) until a multiple common to all denominators in the list is found. **Step 2:** The multiple found in Step 1 is the LCD.	Find the LCD of $\dfrac{1}{4}$ and $\dfrac{5}{6}$ using Method 1. $6 \cdot 1 = 6$ Not a multiple of 4 $6 \cdot 2 = 12$ A multiple of 4 The LCD is 12.
Method 2 for finding the LCD of a list of a fractions using prime factorization **Step 1:** Write the prime factorization of each denominator. **Step 2:** For each different prime factor in Step 1, circle the greatest number of times that factor occurs in any one factorization. **Step 3:** The LCD is the product of the circled factors.	Find the LCD of $\dfrac{5}{6}$ and $\dfrac{11}{20}$ using Method 2. $6 = 2 \cdot ③$ $20 = ②\cdot②\cdot⑤$ The LCD is $2 \cdot 2 \cdot 3 \cdot 5 = 60$
Equivalent fractions represent the same portion of a whole.	Write an equivalent fraction with the indicated denominator. $\dfrac{2}{8} = \dfrac{}{16}$ $\dfrac{2 \cdot 2}{8 \cdot 2} = \dfrac{4}{16}$

Section 4.5 Adding and Subtracting Unlike Fractions

To add or subtract fractions with unlike denominators	Add: $\dfrac{3}{20} + \dfrac{2}{5}$
Step 1: Find the LCD.	**Step 1:** The LCD of the denominators 20 and 5 is 20.
Step 2: Write each fraction as an equivalent fraction whose denominator is the LCD.	**Step 2:** $\dfrac{3}{20} = \dfrac{3}{20}$; $\dfrac{2}{5} = \dfrac{2}{5} \cdot \dfrac{4}{4} = \dfrac{8}{20}$
Step 3: Add or subtract the like fractions.	**Step 3:** $\dfrac{3}{20} + \dfrac{2}{5} = \dfrac{3}{20} + \dfrac{8}{20} = \dfrac{11}{20}$
Step 4: Write the sum or difference in simplest form.	**Step 4:** $\dfrac{11}{20}$ is in simplest form.

Definitions and Concepts	Examples
Section 4.6 Complex Fractions and Review of Order of Operations	
A fraction whose numerator or denominator or both contain fractions is called a **complex fraction**.	Complex Fractions: $$\dfrac{\dfrac{11}{4}}{\dfrac{7}{10}}, \quad \dfrac{\dfrac{y}{6}-11}{\dfrac{4}{3}}$$
One method for simplifying complex fractions is to multiply the numerator and the denominator of the complex fraction by the LCD of all fractions in its numerator and its denominator.	$$\dfrac{\dfrac{y}{6}-11}{\dfrac{4}{3}} = \dfrac{6\left(\dfrac{y}{6}-11\right)}{6\left(\dfrac{4}{3}\right)} = \dfrac{6\left(\dfrac{y}{6}\right)-6(11)}{6\left(\dfrac{4}{3}\right)}$$ $$= \dfrac{y-66}{8}$$
Section 4.7 Operations on Mixed Numbers	
To multiply with mixed numbers or whole numbers, first write any mixed or whole numbers as improper fractions and then multiply as usual.	$$2\dfrac{1}{3} \cdot \dfrac{1}{9} = \dfrac{7}{3} \cdot \dfrac{1}{9} = \dfrac{7 \cdot 1}{3 \cdot 9} = \dfrac{7}{27}$$
To divide with mixed numbers or whole numbers, first write any mixed or whole numbers as fractions and then divide as usual.	$$2\dfrac{5}{8} \div 3\dfrac{7}{16} = \dfrac{21}{8} \div \dfrac{55}{16} = \dfrac{21}{8} \cdot \dfrac{16}{55} = \dfrac{21 \cdot 16}{8 \cdot 55}$$ $$= \dfrac{21 \cdot 2 \cdot \cancel{8}}{\cancel{8} \cdot 55} = \dfrac{42}{55}$$
To add or subtract with mixed numbers, add or subtract the fractions and then add or subtract the whole numbers.	Add: $2\dfrac{1}{2} + 5\dfrac{7}{8}$ $$\begin{aligned} 2\dfrac{1}{2} &= 2\dfrac{4}{8} \\ +\,5\dfrac{7}{8} &= +\,5\dfrac{7}{8} \\ \hline &\;7\dfrac{11}{8} = 7 + 1\dfrac{3}{8} = 8\dfrac{3}{8} \end{aligned}$$
Section 4.8 Solving Equations Containing Fractions	
To Solve an Equation in x **Step 1:** If fractions are present, multiply both sides of the equation by the LCD of the fractions. **Step 2:** If parentheses are present, use the distributive property. **Step 3:** Combine any like terms on each side of the equation. **Step 4:** Use the addition property of equality to rewrite the equation so that variable terms are on one side of the equation and constant terms are on the other side. **Step 5:** Divide both sides by the numerical coefficient of x to solve. **Step 6:** Check the answer in the *original equation*.	Solve: $\dfrac{x}{15} + 2 = \dfrac{7}{3}$ $15\left(\dfrac{x}{15} + 2\right) = 15\left(\dfrac{7}{3}\right)$ Multiply by the LCD 15. $15\left(\dfrac{x}{15}\right) + 15 \cdot 2 = 15\left(\dfrac{7}{3}\right)$ $x + 30 = 35$ $x + 30 - 30 = 35 - 30$ $x = 5$ Check to see that 5 is the solution.

Chapter 4 Review

(4.1) *Write a fraction to represent the shaded area. If the fraction is improper, write the shaded area as a mixed number also. Do not simplify these answers.*

1.
2.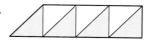
3.
4.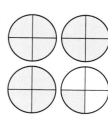

Solve.

5. A basketball player made 11 free throws out of 12 tries during a game. What fraction of free throws did the player make?

6. A new car lot contains 23 blue cars out of a total of 131 cars.
   a. How many cars on the lot are not blue?
   b. What fraction of cars on the lot are not blue?

Simplify by dividing.

7. $-\dfrac{3}{3}$
8. $\dfrac{-20}{-20}$
9. $\dfrac{0}{-1}$
10. $\dfrac{4}{0}$

Graph each fraction on a number line.

11. $\dfrac{7}{9}$

12. $\dfrac{4}{7}$

13. $\dfrac{5}{4}$

14. $\dfrac{7}{5}$

Write each improper fraction as a mixed number or a whole number.

15. $\dfrac{15}{4}$
16. $\dfrac{39}{13}$

Write each mixed number as an improper fraction.

17. $2\dfrac{1}{5}$
18. $3\dfrac{8}{9}$

321

(4.2) *Write each fraction in simplest form.*

19. $\dfrac{12}{28}$ **20.** $\dfrac{15}{27}$ **21.** $-\dfrac{25x}{75x^2}$ **22.** $-\dfrac{36y^3}{72y}$

23. $\dfrac{29ab}{32abc}$ **24.** $\dfrac{18xyz}{23xy}$ **25.** $\dfrac{45x^2y}{27xy^3}$ **26.** $\dfrac{42ab^2c}{30abc^3}$

27. There are 12 inches in a foot. What fractional part of a foot does 8 inches represent?

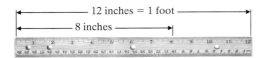

28. Six out of 15 cars are white. What fraction of the cars are *not* white?

Determine whether each two fractions are equivalent.

29. $\dfrac{10}{34}$ and $\dfrac{4}{14}$ **30.** $\dfrac{30}{50}$ and $\dfrac{9}{15}$

(4.3) *Multiply.*

31. $\dfrac{3}{5} \cdot \dfrac{1}{2}$ **32.** $-\dfrac{6}{7} \cdot \dfrac{5}{12}$ **33.** $-\dfrac{24x}{5} \cdot -\dfrac{15}{8x^3}$

34. $\dfrac{27y^3}{21} \cdot \dfrac{7}{18y^2}$ **35.** $\left(-\dfrac{1}{3}\right)^3$ **36.** $\left(-\dfrac{5}{12}\right)^2$

Divide.

37. $-\dfrac{3}{4} \div \dfrac{3}{8}$ **38.** $\dfrac{21a}{4} \div \dfrac{7a}{5}$ **39.** $-\dfrac{9}{2} \div -\dfrac{1}{3}$

40. $-\dfrac{5}{3} \div 2y$ **41.** Evaluate $x \div y$ if $x = \dfrac{9}{7}$ and $y = \dfrac{3}{4}$.   **42.** Evaluate ab if $a = -7$ and $b = \dfrac{9}{10}$.

Find the area of each figure.

 43.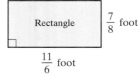

 44.

(4.4) *Add or subtract as indicated.*

45. $\dfrac{7}{11} + \dfrac{3}{11}$ **46.** $\dfrac{4}{9} + \dfrac{2}{9}$ **47.** $\dfrac{1}{12} - \dfrac{5}{12}$

48. $\dfrac{11x}{15} + \dfrac{x}{15}$ **49.** $\dfrac{4y}{21} - \dfrac{3}{21}$ **50.** $\dfrac{4}{15} - \dfrac{3}{15} - \dfrac{2}{15}$

Chapter 4 Review

Find the LCD of each list of fractions.

51. $\dfrac{2}{3}, \dfrac{5}{x}$

52. $\dfrac{3}{4}, \dfrac{3}{8}, \dfrac{7}{12}$

Write each fraction as an equivalent fraction with the given denominator.

53. $\dfrac{2}{3} = \dfrac{?}{30}$

54. $\dfrac{5}{8} = \dfrac{?}{56}$

55. $\dfrac{7a}{6} = \dfrac{?}{42}$

56. $\dfrac{9b}{4} = \dfrac{?}{20}$

57. $\dfrac{4}{5x} = \dfrac{?}{50x}$

58. $\dfrac{5}{9y} = \dfrac{?}{18y}$

Solve.

59. One evening Mark Alorenzo did $\dfrac{3}{8}$ of his homework before supper, another $\dfrac{2}{8}$ of it while his children did their homework, and $\dfrac{1}{8}$ after his children went to bed. What part of his homework did he do that evening?

△ **60.** The Simpsons will be fencing in their land, which is in the shape of a rectangle. In order to do this, they need to find its perimeter. Find the perimeter of their land.

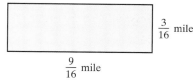

(4.5) *Add or subtract as indicated.*

61. $\dfrac{7}{18} + \dfrac{2}{9}$

62. $\dfrac{4}{13} - \dfrac{1}{26}$

63. $-\dfrac{1}{3} + \dfrac{1}{4}$

64. $-\dfrac{2}{3} + \dfrac{1}{4}$

65. $\dfrac{5x}{11} + \dfrac{2}{55}$

66. $\dfrac{4}{15} + \dfrac{b}{5}$

67. $\dfrac{5y}{12} - \dfrac{2y}{9}$

68. $\dfrac{7x}{18} + \dfrac{2x}{9}$

69. $\dfrac{4}{9} + \dfrac{5}{y}$

70. $\dfrac{9}{11} - \dfrac{3}{y}$

71. $\dfrac{4}{25} + \dfrac{23}{75} + \dfrac{7}{50}$

72. $\dfrac{2}{3} - \dfrac{2}{9} - \dfrac{1}{6}$

Find the perimeter of each figure.

73. Rectangle, $\dfrac{2}{9}$ meter, $\dfrac{5}{6}$ meter

△ **74.**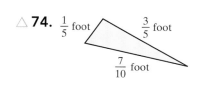

75. In a group of 100 blood donors, typically $\frac{9}{25}$ have type A Rh-positive blood and $\frac{3}{50}$ have type A Rh-negative blood. What fraction have type A blood?

76. Find the difference in length of two scarves if one scarf is $\frac{5}{12}$ of a yard long and the other is $\frac{2}{3}$ of a yard long.

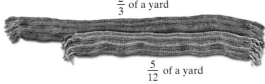

(4.6) *Simplify each complex fraction.*

77. $\dfrac{\frac{2x}{5}}{\frac{7}{10}}$

78. $\dfrac{\frac{3y}{7}}{\frac{11}{7}}$

79. $\dfrac{\frac{2}{5} - \frac{1}{2}}{\frac{3}{4} - \frac{7}{10}}$

80. $\dfrac{\frac{5}{6} - \frac{1}{4}}{-\frac{1}{12y}}$

Evaluate each expression if $x = \frac{1}{2}$, $y = -\frac{2}{3}$, *and* $z = \frac{4}{5}$.

81. $\dfrac{x}{y + z}$

82. $\dfrac{x + y}{z}$

Evaluate each expression. Use the order of operations to simplify.

83. $\frac{5}{13} \div \frac{1}{2} \cdot \frac{4}{5}$

84. $\frac{2}{27} - \left(\frac{1}{3}\right)^2$

85. $\frac{9}{10} \cdot \frac{1}{3} - \frac{2}{5} \cdot \frac{1}{11}$

86. $-\frac{2}{7} \cdot \left(\frac{1}{5} + \frac{3}{10}\right)$

(4.7) *Perform operations as indicated. Simplify your answers. Estimate where noted.*

87.
$7\frac{3}{8}$
$9\frac{5}{6}$
$+\ 3\frac{1}{12}$

88.
$8\frac{1}{5}$
$-\ 5\frac{3}{11}$
Exact:
Estimate:

89. $1\frac{5}{8} \cdot 3\frac{1}{5}$
Exact:
Estimate:

90. $6\frac{3}{4} \div 1\frac{2}{7}$

91. A truck traveled 341 miles on $15\frac{1}{2}$ gallons of gas. How many miles might we expect the truck to travel on 1 gallon of gas?

92. There are $7\frac{1}{3}$ grams of fat in each ounce of hamburger. How many grams of fat are in a 5-ounce hamburger patty?

Find the unknown measurements.

△ **93.**

△ **94.**

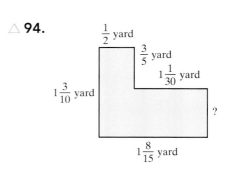

Chapter 4 Review

Perform the indicated operations.

95. $-12\frac{1}{7} + \left(-15\frac{3}{14}\right)$ **96.** $23\frac{7}{8} - 24\frac{7}{10}$ **97.** $-3\frac{1}{5} \div \left(-2\frac{7}{10}\right)$ **98.** $-2\frac{1}{4} \cdot 1\frac{3}{4}$

(4.8) *Solve each equation.*

99. $a - \frac{2}{3} = \frac{1}{6}$ **100.** $9x + \frac{1}{5} - 8x = -\frac{7}{10}$ **101.** $-\frac{3}{5}x = 6$ **102.** $\frac{2}{9}y = -\frac{4}{3}$

103. $\frac{x}{7} - 3 = -\frac{6}{7}$ **104.** $\frac{y}{5} + 2 = \frac{11}{5}$ **105.** $\frac{1}{6} + \frac{x}{4} = \frac{17}{12}$ **106.** $\frac{x}{5} - \frac{5}{4} = \frac{x}{2} - \frac{1}{20}$

Mixed Review

Perform the indicated operations. Write each answer in simplest form. Estimate where noted.

107. $\frac{6}{15} \cdot \frac{5}{8}$ **108.** $\frac{5x^2}{y} \div \frac{10x^3}{y^3}$ **109.** $\frac{3}{10} - \frac{1}{10}$ **110.** $\frac{7}{8x} \cdot -\frac{2}{3}$

111. $\frac{2x}{3} + \frac{x}{4}$ **112.** $-\frac{5}{11} + \frac{2}{55}$ **113.** $-1\frac{3}{5} \div \frac{1}{4}$ **114.** $\begin{array}{r} 2\frac{7}{8} \\ +9\frac{1}{2} \\ \hline \end{array}$ Exact: Estimate:

115. $\begin{array}{r} 12\frac{1}{7} \\ -9\frac{3}{5} \\ \hline \end{array}$ Exact: Estimate: **116.** Simplify: $\dfrac{2 + \frac{3}{4}}{1 - \frac{1}{8}}$ **117.** Evaluate: $-\frac{3}{8} \cdot \left(\frac{2}{3} - \frac{4}{9}\right)$

Solve.

118. $11x - \frac{2}{7} - 10x = -\frac{13}{14}$ **119.** $-\frac{3}{5}x = \frac{4}{15}$ **120.** $\frac{x}{12} + \frac{5}{6} = -\frac{3}{4}$

121. A ribbon $5\frac{1}{2}$ yards long is cut from a reel of ribbon with 50 yards on it. Find the length of the piece remaining on the reel.

△ **122.** A slab of natural granite is purchased and a rectangle with length $7\frac{4}{11}$ feet and width $5\frac{1}{2}$ feet is cut from it. Find the area of the rectangle.

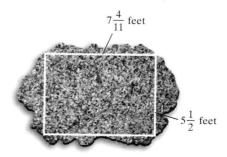

Chapter 4 — Getting Ready for the Test

MULTIPLE CHOICE Exercises 1–10 are **Multiple Choice**. Choose the correct answer.

For Exercises 1 through 4, choose whether the expression simplifies to

A. 1 **B.** −1 **C.** 0 **D.** undefined

1. $\dfrac{-2}{-2}$
2. $\dfrac{-2}{2}$
3. $\dfrac{2}{0}$
4. $\dfrac{0}{-2}$

5. The mixed number $4\dfrac{3}{5}$ written as a fraction is:

   A. $\dfrac{12}{5}$ **B.** $\dfrac{5}{23}$ **C.** $\dfrac{23}{5}$ **D.** $\dfrac{23}{3}$

6. The improper fraction $\dfrac{23}{8}$ written as a mixed number is:

   A. 2.3 **B.** $2\dfrac{7}{8}$ **C.** $8\dfrac{2}{3}$ **D.** $2\dfrac{8}{7}$

For Exercises 7 through 10, the exercise statement and correct answer are given. Choose whether the correct operation in the box should be:

A. +(addition) **B.** −(subtraction) **C.** ·(multiplication) **D.** ÷(division)

7. $\dfrac{8}{11} \square \dfrac{2}{11}$; Answer: $\dfrac{16}{121}$
8. $\dfrac{8}{11} \square \dfrac{2}{11}$; Answer: $\dfrac{8}{2}$ or 4
9. $\dfrac{8}{11} \square \dfrac{2}{11}$; Answer: $\dfrac{6}{11}$
10. $\dfrac{8}{11} \square \dfrac{2}{11}$; Answer: $\dfrac{10}{11}$

MATCHING For Exercises 11 through 14, **Match** each operation of fractions in the first column with the correct answer the second or third column.

11. $\dfrac{5}{7} + \dfrac{1}{7}$
12. $\dfrac{5}{7} \cdot \dfrac{1}{7}$
13. $\dfrac{5}{7} \div \dfrac{1}{7}$
14. $\dfrac{5}{7} - \dfrac{1}{7}$

A. $\dfrac{5}{7}$
B. $\dfrac{5}{1}$ or 5
C. $\dfrac{6}{14}$ or $\dfrac{3}{7}$
D. $\dfrac{4}{7}$
E. undefined
F. $\dfrac{6}{7}$
G. $\dfrac{6}{49}$
H. $\dfrac{5}{49}$

MULTIPLE CHOICE Exercises 15 through 21 are **Multiple Choice**. Choose the correct answer.

For each expression in Exercises 15 through 18, use order of operations and decide which operation below should performed first to simplify.

A. Addition **B.** Subtraction **C.** Multiplication **D.** Division

15. $\dfrac{1}{2} + \dfrac{1}{5} \cdot \dfrac{1}{4}$
16. $\left(\dfrac{1}{2} + \dfrac{1}{5}\right) \cdot \dfrac{1}{4}$
17. $\dfrac{1}{2} \div \dfrac{1}{5} \cdot \dfrac{1}{4}$
18. $\dfrac{1}{2} \div \left(\dfrac{1}{5} \cdot \dfrac{1}{4}\right)$

19. To solve $\dfrac{x}{10} + \dfrac{3}{5} = 2$, let's multiply both sides of the equation by 10. Once this is done, the equivalent equation is:

   A. $x + 6 = 2$ **B.** $x + 3 = 2$ **C.** $x + 6 = 12$ **D.** $x + 6 = 20$

Equations are solved and expressions are simplified. For Exercises 20 through 21, decide whether each exercise is an

A. expression *or an* **B.** equation

20. $\dfrac{2}{3} - \dfrac{1}{9}$
21. $\dfrac{2}{3} - \dfrac{1}{9} = \dfrac{x}{9}$

Chapter 4 Test

Write a fraction to represent the shaded area.

1.

Write the mixed number as an improper fraction.

2. $7\frac{2}{3}$

Write the improper fraction as a mixed number.

3. $\frac{75}{4}$

Write each fraction in simplest form.

4. $\frac{24}{210}$

5. $-\frac{42x}{70}$

Determine whether these fractions are equivalent.

6. $\frac{5}{7}$ and $\frac{8}{11}$

7. $\frac{6}{27}$ and $\frac{14}{63}$

Find the prime factorization of each number.

8. 84

9. 495

Perform each indicated operation and write the answers in simplest form.

10. $\frac{4}{4} \div \frac{3}{4}$

11. $-\frac{4}{3} \cdot \frac{4}{4}$

12. $\frac{7x}{9} + \frac{x}{9}$

13. $\frac{1}{7} - \frac{3}{x}$

14. $\frac{xy^3}{z} \cdot \frac{z}{xy}$

15. $-\frac{2}{3} \cdot -\frac{8}{15}$

16. $\frac{9a}{10} + \frac{2}{5}$

17. $-\frac{8}{15y} - \frac{2}{15y}$

18. $\frac{3a}{8} \cdot \frac{16}{6a^3}$

19. $\frac{11}{12} - \frac{3}{8} + \frac{5}{24}$

20. $3\frac{7}{8}$
    $7\frac{2}{5}$
    $+ 2\frac{3}{4}$

21. 19
    $-2\frac{3}{11}$

Answers

1. _____
2. _____
3. _____
4. _____
5. _____
6. _____
7. _____
8. _____
9. _____
10. _____
11. _____
12. _____
13. _____
14. _____
15. _____
16. _____
17. _____
18. _____
19. _____
20. _____
21. _____

22. $-\dfrac{16}{3} \div -\dfrac{3}{12}$ **23.** $3\dfrac{1}{3} \cdot 6\dfrac{3}{4}$ **24.** $-\dfrac{2}{7} \cdot \left(6 - \dfrac{1}{6}\right)$ **25.** $\dfrac{1}{2} \div \dfrac{2}{3} \cdot \dfrac{3}{4}$

26. $\left(-\dfrac{3}{4}\right)^2 \div \left(\dfrac{2}{3} + \dfrac{5}{6}\right)$ **27.** Find the average of $\dfrac{5}{6}, \dfrac{4}{3},$ and $\dfrac{7}{12}$.

Simplify each complex fraction.

28. $\dfrac{\dfrac{5x}{7}}{\dfrac{20x^2}{21}}$ **29.** $\dfrac{5 + \dfrac{3}{7}}{2 - \dfrac{1}{2}}$

Solve.

30. $-\dfrac{3}{8}x = \dfrac{3}{4}$ **31.** $\dfrac{x}{5} + x = -\dfrac{24}{5}$ **32.** $\dfrac{2}{3} + \dfrac{x}{4} = \dfrac{5}{12} + \dfrac{x}{2}$

Evaluate each expression for the given replacement values.

33. $-5x; x = -\dfrac{1}{2}$ **34.** $x \div y; x = \dfrac{1}{2}, y = 3\dfrac{7}{8}$

Solve.

35. A carpenter cuts a piece $2\dfrac{3}{4}$ feet long from a cedar plank that is $6\dfrac{1}{2}$ feet long. How long is the remaining piece?

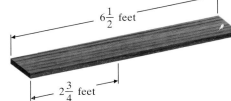

The circle graph below shows us how the average consumer spends money. For example, $\dfrac{7}{50}$ of spending goes for food. Use this information for Exercises 36 through 38.

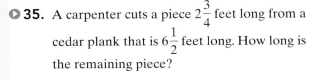

Consumer Spending

Other $\dfrac{2}{25}$; Education $\dfrac{1}{50}$; Insurance and pension $\dfrac{1}{10}$; Entertainment $\dfrac{1}{25}$; Health care $\dfrac{3}{50}$; Transportation $\dfrac{1}{5}$; Clothing $\dfrac{1}{25}$; Food $\dfrac{7}{50}$; Housing $\dfrac{8}{25}$

Source: U.S. Bureau of Labor Statistics; based on survey

36. What fraction of spending goes for housing and food combined?

37. What fraction of spending goes for education, transportation, and clothing?

38. Suppose your family spent $47,000 on the items in the graph. How much might we expect was spent on health care?

Find the perimeter and area of the figure.

39.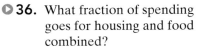

Rectangle, $\dfrac{2}{3}$ foot by 1 foot

40. During a 258-mile trip, a car used $10\dfrac{3}{4}$ gallons of gas. How many miles would we expect the car to travel on 1 gallon of gas

Cumulative Review — Chapters 1–4

Write each number in words.

1. 546

2. 115

3. 27,034

4. 6573

5. Add: $46 + 713$

6. Add: $587 + 44$

7. Subtract: $543 - 29$. Check by adding.

8. Subtract: $995 - 62$. Check by adding.

9. Round 278,362 to the nearest thousand.

10. Round 1436 to the nearest ten.

11. A digital video disc (DVD) can hold about 4800 megabytes (MB) of information. How many megabytes can 12 DVDs hold?

12. On a trip across the country, Daniel Daunis travels 435 miles per day. How many total miles does he travel in 3 days?

13. Divide and check: $56{,}717 \div 8$

14. Divide and check: $4558 \div 12$

Write using exponential notation.

15. $7 \cdot 7 \cdot 7$

16. $7 \cdot 7$

17. $3 \cdot 3 \cdot 3 \cdot 3 \cdot 9 \cdot 9 \cdot 9$

18. $9 \cdot 9 \cdot 9 \cdot 9 \cdot 5 \cdot 5$

19. Evaluate $2(x - y)$ for $x = 6$ and $y = 3$.

20. Evaluate $8a + 3(b - 5)$ for $a = 5$ and $b = 9$.

21. The world's deepest cave is Krubera (or Voronja), in the country of Georgia, located by the Black Sea in Asia. It has been explored to a depth of 7188 feet below the surface of Earth. Represent this position using an integer. (*Source:* messagetoeagle.com and Wikipedia)

22. The temperature on a cold day in Minneapolis, MN, was 21°F below zero. Represent this temperature using an integer.

Answers

1. _____
2. _____
3. _____
4. _____
5. _____
6. _____
7. _____
8. _____
9. _____
10. _____
11. _____
12. _____
13. _____
14. _____
15. _____
16. _____
17. _____
18. _____
19. _____
20. _____
21. _____
22. _____

23. Add using a number line: $-7 + 3$

24. Add using a number line: $-3 + 8$

25. Simplify: $7 - 8 - (-5) - 1$

26. Simplify: $6 + (-8) - (-9) + 3$

27. Evaluate: $(-5)^2$

28. Evaluate: -2^4

29. Simplify: $3(4 - 7) + (-2) - 5$

30. Simplify: $(20 - 5^2)^2$

31. Simplify: $2y - 6 + 4y + 8$

32. Simplify: $5x - 1 + x + 10$

Solve.

33. $5x + 2 - 4x = 7 - 19$

34. $9y + 1 - 8y = 3 - 20$

35. $17 - 7x + 3 = -3x + 21 - 3x$

36. $9x - 2 = 7x - 24$

37. Write a fraction to represent the shaded part of the figure.

38. Write the prime factorization of 156.

39. Write each as an improper fraction.

a. $4\frac{2}{9}$ b. $1\frac{8}{11}$

40. Write $\frac{39}{5}$ as a mixed number.

41. Write in simplest form: $\frac{42x}{66}$

42. Write in simplest form: $\frac{70}{105y}$

43. Multiply: $3\frac{1}{3} \cdot \frac{7}{8}$

44. Multiply: $\frac{2}{3} \cdot 4$

45. Divide and simplify: $\frac{5}{16} \div \frac{3}{4}$

46. Divide: $1\frac{1}{10} \div 5\frac{3}{5}$

Decimals

5

Decimal numbers represent parts of a whole, just like fractions. For example, one penny is 0.01 or $\frac{1}{100}$ of a dollar. In this chapter, we learn to perform arithmetic operations on decimals and to analyze the relationship between fractions and decimals. We also learn how decimals are used in the real world.

Sections

5.1 Introduction to Decimals

5.2 Adding and Subtracting Decimals

5.3 Multiplying Decimals and Circumference of a Circle

5.4 Dividing Decimals

Integrated Review—Operations on Decimals

5.5 Fractions, Decimals, and Order of Operations

5.6 Solving Equations Containing Decimals

5.7 Decimal Applications: Mean, Median, and Mode

Check Your Progress

Vocabulary Check

Chapter Highlights

Chapter Review

Getting Ready for the Test

Chapter Test

Cumulative Review

Attendence to U.S. Movie Theaters Is Increasing

Admissions to U.S. movie theaters recently increased by 4%. More than 2/3 of the U.S. and Canadian population went to the movies in 2016. There were 681 films released in 2016, which is consistent with previous years and appears likely to continue in the future. Not all films are successes, and even fewer are true blockbusters. The chart below lists the top all-time highest-money-grossing films in U.S. movie history.

While we have practiced calculating averages before in this text, an average does not always simplify to a whole number. While fractions are useful, decimals are an important system of numbers that can also be used to show values between whole numbers. Data are easy to round when they are in the form of a decimal. In Section 5.1. Exercise 109, and Section 5.2, Exercise 85, we will continue to study the all-time highest-grossing American movies.

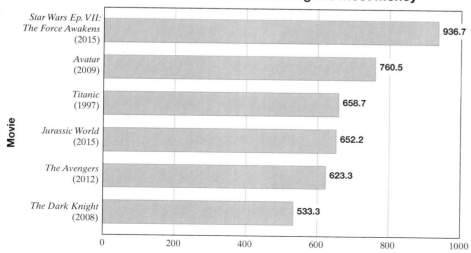

Source: the-numbers.com; Yearly Box Office

5.1 Introduction to Decimals

Objectives

A Know the Meaning of Place Value for a Decimal Number and Write Decimals in Words.

B Write Decimals in Standard Form.

C Write Decimals as Fractions.

D Compare Decimals.

E Round Decimals to Given Place Values.

Objective A Decimal Notation and Writing Decimals in Words

Like fractional notation, decimal notation is used to denote a part of a whole. Numbers written in decimal notation are called **decimal numbers,** or simply **decimals**. The decimal 17.758 has three parts.

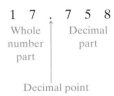

In Section 1.2, we introduced place value for whole numbers. Place names and place values for the whole number part of a decimal number are exactly the same. Place names and place values for the decimal part are shown below.

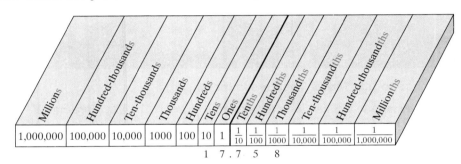

Notice that the value of each place is $\frac{1}{10}$ of the value of the place to its left. For example,

$$1 \cdot \frac{1}{10} = \frac{1}{10} \quad \text{and} \quad \frac{1}{10} \cdot \frac{1}{10} = \frac{1}{100}$$

↑ ones ↑ tenths   ↑ tenths ↑ hundredths

> **Helpful Hint** Notice that place values to the left of the decimal point end in "s." Place values to the right of the decimal point end in "ths."

The decimal number 17.758 means

1 ten + 7 ones + 7 tenths + 5 hundredths + 8 thousandths

or $1 \cdot 10 + 7 \cdot 1 + 7 \cdot \frac{1}{10} + 5 \cdot \frac{1}{100} + 8 \cdot \frac{1}{1000}$

or $10 + 7 + \frac{7}{10} + \frac{5}{100} + \frac{8}{1000}$

Writing (or Reading) a Decimal in Words

Step 1: Write the whole number part in words.

Step 2: Write "and" for the decimal point.

Step 3: Write the decimal part in words as though it were a whole number, followed by the place value of the last digit.

Section 5.1 | Introduction to Decimals 333

Example 1 Write each decimal in words.

a. 0.7   b. −50.82   c. 21.093

Solution:
a. seven tenths
b. negative fifty and eighty-two hundredths
c. twenty-one and ninety-three thousandths

Work Practice 1

Practice 1
Write each decimal in words.
a. 0.06
b. −200.073
c. 0.0829

Example 2 Write the decimal in the following sentence in words: The Golden Jubilee Diamond is a 545.67-carat cut diamond. (*Source: The Guinness Book of Records*)

Solution: five hundred forty-five and sixty-seven hundredths

Work Practice 2

Practice 2
Write the decimal 87.31 in words.

Example 3 Write the decimal in the following sentence in words: The oldest known fragments of the Earth's crust are Zircon crystals; they were discovered in Australia and are thought to be 4.276 billion years old. (*Source: The Guinness Book of Records*)

Solution: four and two hundred seventy-six thousandths

Work Practice 3

Practice 3
Write the decimal 52.1085 in words.

Suppose that you are paying for a purchase of $368.42 at Circuit City by writing check. Checks are usually written using the following format.

Helpful Hint Note: Although check-writing has decreased, checks are still used for larger sums such as payroll and rent.

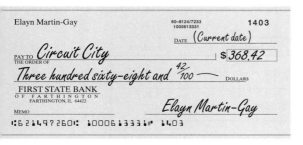

Average Daily Volume of Commercial Checks

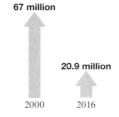

67 million (2000)
20.9 million (2016)

Source: Federal Reserve System

Answers

1. a. six hundredths **b.** negative two hundred and seventy-three thousandths **c.** eight hundred twenty-nine ten-thousandths

2. eighty-seven and thirty-one hundredths

3. fifty-two and one thousand eighty-five ten-thousandths

Practice 4

Fill in the check to CLECO (Central Louisiana Electric Company) to pay for your monthly electric bill of $207.40.

Example 4

Fill in the check to Camelot Music to pay for your purchase of $92.98.

Solution:

■ Work Practice 4

Objective B Writing Decimals in Standard Form

A decimal written in words can be written in standard form by reversing the procedure in Objective A.

Examples Write each decimal in standard form.

5. Forty-eight and twenty-six hundredths is

6. Six and ninety-five thousandths is

■ Work Practice 5–6

Helpful Hint

When converting a decimal from words to decimal notation, make sure the last digit is in the correct place by inserting 0s if necessary. For example,

Two and thirty-eight thousandths is 2.038
                                        ↑
                                  thousandths place

Practice 5–6

Write each decimal in standard form.
5. Five hundred and ninety-six hundredths
6. Thirty-nine and forty-two thousandths

Objective C Writing Decimals as Fractions

Once you master reading and writing decimals, writing a decimal as a fraction follows naturally.

Decimal	In Words	Fraction
0.7	seven tenths	$\dfrac{7}{10}$
0.51	fifty-one hundredths	$\dfrac{51}{100}$
0.009	nine thousandths	$\dfrac{9}{1000}$
0.05	five hundredths	$\dfrac{5}{100} = \dfrac{1}{20}$

Answers
4. [check filled out to CLECO for $207.40]
5. 500.96
6. 39.042

Notice that the number of decimal places in a decimal number is the same as the number of zeros in the denominator of the equivalent fraction. We can use this fact to write decimals as fractions.

$$0.31 = \frac{31}{100}$$

2 decimal places — 2 zeros

$$0.007 = \frac{7}{1000}$$

3 decimal places — 3 zeros

Example 7
Write 0.47 as a fraction.

Solution: $0.47 = \frac{47}{100}$

2 decimal places — 2 zeros

■ Work Practice 7

Practice 7
Write 0.051 as a fraction.

Example 8
Write 5.9 as a mixed number.

Solution: $5.9 = 5\frac{9}{10}$

1 decimal place — 1 zero

■ Work Practice 8

Practice 8
Write 29.97 as a mixed number.

Examples
Write each decimal as a fraction or a mixed number. Write your answer in simplest form.

9. $0.125 = \frac{125}{1000} = \frac{\overset{1}{\cancel{125}}}{8 \cdot \cancel{125}} = \frac{1}{8}$

10. $43.5 = 43\frac{5}{10} = 43\frac{\overset{1}{\cancel{5}}}{2 \cdot \cancel{5}} = 43\frac{1}{2 \cdot 1} = 43\frac{1}{2}$

11. $-105.083 = -105\frac{83}{1000}$

■ Work Practice 9–11

Practice 9–11
Write each decimal as a fraction or mixed number. Write your answer in simplest form.

9. 0.12
10. 64.8
11. −209.986

Later in the chapter, we write fractions as decimals. If you study Examples 7–11, you already know how to write fractions with denominators of 10, 100, 1000, and so on, as decimals.

Objective D Comparing Decimals

One way to compare positive decimals is by comparing digits in corresponding places. To see why this works, let's compare 0.5 or $\frac{5}{10}$ and 0.8 or $\frac{8}{10}$. We know

$\frac{5}{10} < \frac{8}{10}$ since $5 < 8$, so

$0.5 < 0.8$ since $5 < 8$

This leads to the following.

Answers

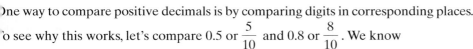

7. $\frac{51}{1000}$ **8.** $29\frac{97}{100}$ **9.** $\frac{3}{25}$
10. $64\frac{4}{5}$ **11.** $-209\frac{493}{500}$

Comparing Two Positive Decimals

Compare digits in the same places from left to right. When two digits are not equal, the number with the larger digit is the larger decimal. If necessary, insert 0s after the last digit to the right of the decimal point to continue comparing.

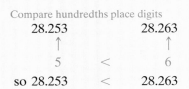

Helpful Hint

For any decimal, writing 0s after the last digit to the right of the decimal point does not change the value of the number.

$7.6 = 7.60 = 7.600$, and so on

When a whole number is written as a decimal, the decimal point is placed to the right of the ones digit.

$25 = 25.0 = 25.00$, and so on

Practice 12

Insert $<, >,$ or $=$ to form a true statement.

26.208   26.28

Example 12 Insert $<, >,$ or $=$ to form a true statement.

0.378   0.368

Solution:

0.3 7 8   0.3 6 8   The tenths places are the same.

0.3 7 8   0.3 6 8   The hundredths places are different.

Since $7 > 6$, then $0.378 > 0.368$.

■ Work Practice 12

Practice 13

Insert $<, >,$ or $=$ to form a true statement.

0.12   0.026

Example 13 Insert $<, >,$ or $=$ to form a true statement.

0.052   0.236

Solution: $0.\,0\,52 < 0.\,2\,36$ 0 is smaller than 2 in the tenths place.

■ Work Practice 13

We can also use a number line to compare decimals. This is especially helpful when comparing negative decimals. Remember, the number whose graph is to the left is smaller, and the number whose graph is to the right is larger.

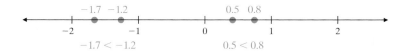

Answers
12. $<$ 13. $>$

Section 5.1 | Introduction to Decimals

Helpful Hint

If you have trouble comparing two negative decimals, try the following: Compare their absolute values. Then to correctly compare the negative decimals, *reverse* the direction of the inequality symbol.

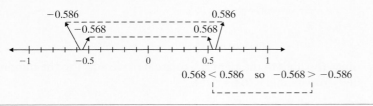

$0.568 < 0.586$ so $-0.568 > -0.586$

Example 14 Insert $<$, $>$, or $=$ to form a true statement.

$-0.0101 \quad -0.00109$

Solution: Since $0.0101 > 0.00109$, then $-0.0101 < -0.00109$.

Work Practice 14

Practice 14

Insert $<$, $>$, or $=$ to form a true statement.

$-0.039 \quad -0.0309$

Objective E Rounding Decimals

We **round the decimal part** of a decimal number in nearly the same way as we round whole numbers. The only difference is that we drop digits to the right of the rounding place, instead of replacing these digits with 0s. For example,

36.954 rounded to the nearest hundredth is 36.95.

Rounding Decimals to a Place Value to the Right of the Decimal Point

Step 1: Locate the digit to the right of the given place value.

Step 2: If this digit is 5 or greater, add 1 to the digit in the given place value and drop all digits to its right. If this digit is less than 5, drop all digits to the right of the given place.

Example 15 Round 736.2359 to the nearest tenth.

Solution:

Step 1: We locate the digit to the right of the tenths place.

736.2 3 59 (tenths place; digit to the right)

Step 2: Since the digit to the right is less than 5, we drop it and all digits to its right.

Thus, 736.2359 rounded to the nearest tenth is 736.2.

Work Practice 15

The same steps for rounding can be used when the decimal is negative.

Practice 15

Round 482.7817 to the nearest thousandth.

Answers
14. $<$ **15.** 482.782

Practice 16

Round −0.032 to the nearest hundredth.

Example 16 Round −0.027 to the nearest hundredth.

Solution:

Step 1: Locate the digit to the right of the hundredths place.

$$-0.02\,7$$

with hundredths place indicated above the 2 and digit to the right indicated below the 7.

Step 2: Since the digit to the right is 5 or greater, we add 1 to the hundredths digit and drop all digits to its right.

Thus, −0.027 is −0.03 rounded to the nearest hundredth.

■ Work Practice 16

The following number line illustrates the rounding of negative decimals.

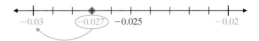

In Section 5.3, we will introduce a formula for the distance around a circle. The distance around a circle is given the special name **circumference.**

The symbol π is the Greek letter pi, pronounced "pie." We use π to denote the following constant:

$$\pi = \frac{\text{circumference of a circle}}{\text{diameter of a circle}}$$

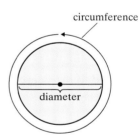

The value π is an **irrational number.** This means if we try to write it as a decimal, it neither ends nor repeats in a pattern.

Practice 17

$\pi \approx 3.14159265$. Round π to the nearest ten-thousandth.

Example 17 $\pi \approx 3.14159265$. Round π to the nearest hundredth.

Solution:

hundredths place — 1 is less than 5.

$$3.14159265$$

Delete these digits.

Thus, 3.14159265 rounded to the nearest hundredth is 3.14. In other words, $\pi \approx 3.14$.

■ Work Practice 17

Rounding often occurs with money amounts. Since there are 100 cents in a dollar, each cent is $\frac{1}{100}$ of a dollar. This means that if we want to round to the nearest cent, we round to the nearest hundredth of a dollar.

Answers
16. −0.03 **17.** $\pi \approx 3.1416$

✓**Concept Check Answer**
c

✓**Concept Check** 1756.0894 rounded to the nearest *ten* is
a. 1756.1 **b.** 1760.0894 **c.** 1760 **d.** 1750

Example 18 Determining State Taxable Income

A high school teacher's taxable income is $41,567.72. The tax tables in the teacher's state use amounts rounded to the nearest dollar. Round the teacher's income to the nearest whole dollar.

Solution: Rounding to the nearest whole dollar means rounding to the ones place.

ones place ⟶ 7 is greater than 5.
$41,567.72
↑ Add 1. ⟵ Delete these digits.

Thus, the teacher's income rounded to the nearest dollar is $41,568.

■ Work Practice 18

Practice 18
Water bills in Mexia are always rounded to the nearest dollar. Round a water bill of $24.62 to the nearest dollar.

Answer
18. $25

Vocabulary, Readiness & Video Check

Use the choices below to fill in each blank.

words	decimals	tenths	after
tens	circumference	and	standard form

1. The number "twenty and eight hundredths" is written in _____ and "20.08" is written in _____.
2. Another name for the distance around a circle is its _____.
3. Like fractions, _____ are used to denote part of a whole.
4. When writing a decimal number in words, the decimal point is written as _____.
5. The place value _____ is to the right of the decimal point while _____ is to the left of the decimal point.
6. The decimal point in a whole number is _____ the last digit.

Martin-Gay Interactive Videos   Watch the section lecture video and answer the following questions.

See Video 5.1

Objective A 7. In Example 1, how is the decimal point written?
Objective B 8. Why is 9.8 not the correct answer to Example 3? What is the correct answer?
Objective C 9. From Example 5, why does reading a decimal number correctly help you write it as an equivalent fraction?
Objective D 10. In Example 7, we compare place value by place value in which direction?
Objective E 11. Example 8 is being rounded to the nearest tenth, so why is the digit 7, which is not in the tenths place, looked at?

5.1 Exercise Set MyLab Math

Objective A *Write each decimal number in words. See Examples 1 through 3.*

1. 5.62
2. 9.57
3. 16.23
4. 47.65

5. −0.205

6. −0.495

7. 167.009

8. 233.056

9. 3000.04

10. 5000.02

11. 105.6

12. 410.3

13. The Akashi Kaikyo Bridge, between Kobe and Awaji-Shima, Japan, is approximately 2.43 miles long.

14. The English Channel Tunnel is a 31.04-mile long undersea rail tunnel connecting England and France. (*Source: Railway Directory & Year Book*)

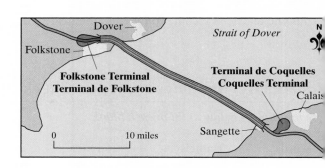

Fill in each check for the described purchase. See Example 4.

15. Your monthly car loan of $321.42 to R. W. Financial.

16. Your part of the monthly apartment rent, which is $213.70. You pay this to Amanda Dupre.

17. Your bill of $91.68 to Verizon wireless.

18. Your grocery bill of $387.49 at Kroger.

Objective B *Write each decimal number in standard form. See Examples 5 and 6.*

19. Two and eight tenths

20. Five and one tenth

21. Nine and eight hundredths

22. Twelve and six hundredths

23. Negative seven hundred five and six hundred twenty-five thousandths

24. Negative eight hundred four and three hundred ninety-nine thousandths

25. Forty-six ten-thousandths

26. Eighty-three ten-thousandths

Section 5.1 | Introduction to Decimals

Objective C *Write each decimal as a fraction or a mixed number. Write your answer in simplest form. See Examples 7 through 11.*

27. 0.7　　　　　**28.** 0.9　　　　　**29.** 0.27　　　　　**30.** 0.39

31. 0.4　　　　　**32.** 0.8　　　　　**33.** 5.4　　　　　**34.** 6.8

35. −0.058　　　**36.** −0.024　　　**37.** 7.008　　　　**38.** 9.005

39. 15.802　　　**40.** 11.406　　　**41.** 0.3005　　　**42.** 0.2006

Objectives A B C Mixed Practice *Fill in the chart. The first row is completed for you. See Examples 1 through 11.*

Decimal Number in Standard Form	In Words	Fraction
0.37	thirty-seven hundredths	$\frac{37}{100}$
43.	eight tenths	
44.	five tenths	
45. 0.077		
46. 0.019		

Objective D *Insert <, >, or = between each pair of numbers to form a true statement. See Examples 12 through 14.*

47. 0.15　0.16　　**48.** 0.12　0.15　　**49.** −0.57　−0.54　　**50.** −0.59　−0.52

51. 0.098　0.1　　**52.** 0.0756　0.2　　**53.** 0.54900　0.549　　**54.** 0.98400　0.984

55. 167.908　167.980　　**56.** 519.3405　519.3054　　**57.** −1.062　−1.07　　**58.** −18.1　−18.01

59. −7.052　7.0052　　**60.** 0.01　−0.1　　**61.** −0.023　−0.024　　**62.** −0.562　−0.652

Objective E *Round each decimal to the given place value. See Examples 15 through 18.*

63. 0.57, nearest tenth　　**64.** 0.64, nearest tenth　　**65.** 98,207.23, nearest ten

66. 68,934.543, nearest ten　　**67.** −0.234, nearest hundredth　　**68.** −0.892, nearest hundredth

69. 0.5942, nearest thousandth　　**70.** 63.4523, nearest thousandth

Recall that the number π, written as a decimal, neither ends nor repeats in a pattern. Given that π ≈ 3.14159265, round π to the given place values below. (We study π further in Section 5.3.) See Example 17.

71. tenth **72.** ones **73.** thousandth **74.** hundred-thousandth

Round each monetary amount to the nearest cent or dollar as indicated. See Example 18.

75. $26.95, to the nearest dollar

76. $14,769.52, to the nearest dollar

77. $0.1992, to the nearest cent

78. $0.7633, to the nearest cent

Round each number to the given place value. See Example 18.

79. At the time of this writing, the Apple MacBook Air is the thinnest Mac in production. At its thickest point, it measures 0.68 in. Round this number to the nearest tenth. (*Source:* Apple, Inc.)

13-inch MacBook Air

80. A large tropical cockroach of the family Dictyoptera is the fastest-moving insect. This insect was clocked at a speed of 3.36 miles per hour. Round this number to the nearest tenth. (*Source:* University of California, Berkeley)

81. The most decorated athlete in Olympic history, Michael Phelps of the United States, won his fourth gold medal of the 2016 Rio Summer Olympics in the 200-meter individual medley with a time of 1.918 minutes. Round this time to the nearest hundredth of a minute. (*Source:* ESPN)

82. The population density of the state of Oklahoma is 56.411 people per square mile. Round this population density to the nearest tenth. (*Source:* World Population Review)

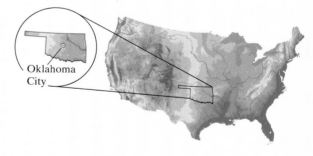

83. A used biology textbook is priced at $67.89. Round this price to the nearest dollar.

84. A used office desk is advertised at $19.95 by Drawley's Office Furniture. Round this price to the nearest dollar.

85. Venus makes a complete orbit around the Sun every 224.695 days. Round this figure to the nearest whole day. (*Source:* National Space Science Data Center)

86. The length of a day on Mars, a full rotation about its axis, is 24.6229 hours. Round this figure to the nearest thousandth. (*Source:* National Space Science Data Center)

Review

Perform each indicated operation. See Section 1.3.

87. 3452 + 2314 **88.** 8945 + 4536 **89.** 82 − 47 **90.** 4002 − 3897

Concept Extensions

Solve. See the Concept Check in this section.

91. 2849.1738 rounded to the nearest hundred is
a. 2849.17 b. 2800 c. 2850 d. 2849.174

92. 146.059 rounded to the nearest ten is
a. 146.0 b. 146.1 c. 140 d. 150

93. 2849.1738 rounded to the nearest hundredth is
a. 2849.17 b. 2800 c. 2850 d. 2849.18

94. 146.059 rounded to the nearest tenth is
a. 146.0 b. 146.1 c. 140 d. 150

95. In your own words, describe how to write a decimal as a fraction or a mixed number.

96. Explain how to identify the value of the 9 in the decimal 486.3297.

97. Write $7\frac{12}{100}$ as a decimal.

98. Write $17\frac{268}{1000}$ as a decimal.

99. Write 0.00026849577 as a fraction.

100. Write 0.00026849576 in words.

101. Write a 5-digit number that rounds to 1.7.

102. Write a 4-digit number that rounds to 26.3.

103. Write a decimal number that is greater than 8 but less than 9.

104. Write a decimal number that is greater than 48.1, but less than 48.2.

105. Which number(s) rounds to 0.26?
0.26559   0.26499   0.25786   0.25186

106. Which number(s) rounds to 0.06?
0.0612   0.066   0.0586   0.0506

Write these numbers from smallest to largest.

107. 0.9
0.1038
0.10299
0.1037

108. 0.01
0.0839
0.09
0.1

109. The all-time top six movies (those that have earned the most money in the United States) along with the approximate amount of money they have earned are listed in the table. Estimate the total amount of money that these movies have earned by first rounding each earning to the nearest hundred-million. (*Source:* the-numbers.com)

Top All-Time American Movies	
Movie	**Gross Domestic Earnings**
Star Wars Ep. VII: The Force Awakens (2015)	$936.7 million
Avatar (2009)	$760.5 million
Titanic (1997)	$658.7 million
Jurassic World (2015)	$652.2 million
The Avengers (2012)	$623.3 million
The Dark Knight (2008)	$533.3 million

110. In 2015, there were 1199.1 million singles downloaded at an average price of $1.20 each. Find an estimate of the total revenue from downloaded singles by answering parts **a–c**. (*Source:* Recording Industry Association of America)

a. Round 1199.1 million to the nearest ten million.

b. Multiply the rounded value in part **a** by 12.

c. Move the decimal point in the product from part **b** one place to the left. This number is the total revenue in millions of dollars.

5.2 Adding and Subtracting Decimals

Objectives

A Add or Subtract Decimals.

B Estimate When Adding or Subtracting Decimals.

C Evaluate Expressions with Decimal Replacement Values.

D Simplify Expressions Containing Decimals.

E Solve Problems That Involve Adding or Subtracting Decimals.

Objective A Adding or Subtracting Decimals

Adding or subtracting decimals is similar to adding or subtracting whole numbers. We add or subtract digits in corresponding place values from right to left, carrying or borrowing if necessary. To make sure that digits in corresponding place values are added or subtracted, we line up the decimal points vertically.

Adding or Subtracting Decimals

Step 1: Write the decimals so that the decimal points line up vertically.

Step 2: Add or subtract as with whole numbers.

Step 3: Place the decimal point in the sum or difference so that it lines up vertically with the decimal points in the problem.

In this section, we will insert zeros in decimal numbers so that place-value digits line up neatly. This is shown in Example 1.

Example 1 Add: $23.85 + 1.604$

Solution: First we line up the decimal points vertically.

$$\begin{array}{r} 23.850 \\ +\ 1.604 \\ \hline \end{array}$$ Insert one 0 so that digits line up neatly.

↑ Line up decimal points.

Then we add the digits from right to left as for whole numbers.

$$\begin{array}{r} \overset{1}{2}3.850 \\ +\ 1.604 \\ \hline 25.454 \end{array}$$

Place the decimal point in the sum so that all decimal points line up.

■ Work Practice 1

Practice 1

Add.
a. $19.52 + 5.371$
b. $40.08 + 17.612$
c. $0.125 + 422.8$

Helpful Hint

Recall that 0s may be placed after the last digit to the right of the decimal point without changing the value of the decimal. This may be used to help line up place values when adding decimals.

$$\begin{array}{r} 3.2 \\ 15.567 \\ +\ 0.11 \\ \hline \end{array}$$ becomes $$\begin{array}{r} 3.200 \\ 15.567 \\ +\ 0.110 \\ \hline 18.877 \end{array}$$ Insert two 0s.

Insert one 0.

Add.

Answers
1. a. 24.891 **b.** 57.692 **c.** 422.925

Section 5.2 | Adding and Subtracting Decimals

Example 2 Add: 763.7651 + 22.001 + 43.89

Solution: First we line up the decimal points.

$$\begin{array}{r} \overset{111}{763.7651} \\ 22.0010 \\ + 43.8900 \\ \hline 829.6561 \end{array}$$

Insert one 0.
Insert two 0s.
Add.

■ Work Practice 2

Practice 2
Add.
a. 34.567 + 129.43 + 2.8903
b. 11.21 + 46.013 + 362.526

Helpful Hint
Don't forget that the decimal point in a whole number is positioned after the last digit.

Example 3 Add: 45 + 2.06

Solution:

$$\begin{array}{r} 45.00 \\ + 2.06 \\ \hline 47.06 \end{array}$$

Insert a decimal point and two 0s.
Line up decimal points.
Add.

■ Work Practice 3

Practice 3
Add: 19 + 26.072

✓ **Concept Check** What is wrong with the following calculation of the sum of 7.03, 2.008, 19.16, and 3.1415?

$$\begin{array}{r} 7.03 \\ 2.008 \\ 19.16 \\ + 3.1415 \\ \hline 3.6042 \end{array}$$ ✗

Example 4 Add: 3.62 + (−4.78)

Solution: Recall from Chapter 2 that to add two numbers with different signs, we find the difference of the larger absolute value and the smaller absolute value. The sign of the answer is the same as the sign of the number with the larger absolute value.

$$\begin{array}{r} 4.78 \\ - 3.62 \\ \hline 1.16 \end{array}$$ Subtract the absolute values.

Thus, 3.62 + (−4.78) = −1.16

The sign of the number with the larger absolute value; −4.78 has the larger absolute value.

■ Work Practice 4

Practice 4
Add: 7.12 + (−9.92)

Answers
2. a. 166.8873 b. 419.749
3. 45.072 4. −2.8

✓ **Concept Check Answer**
The decimal places are not lined up properly.

Subtracting decimals is similar to subtracting whole numbers. We line up digits and subtract from right to left, borrowing when needed.

Practice 5
Subtract. Check your answers.
a. $6.7 - 3.92$
b. $9.72 - 4.068$

Example 5 Subtract: $3.5 - 0.068$. Check your answer.

Solution:
$$\begin{array}{r} 3.\overset{4}{\cancel{5}}\overset{9}{\cancel{0}}\overset{10}{\cancel{0}} \\ -0.0\,6\,8 \\ \hline 3.4\,3\,2 \end{array}$$
Insert two 0s.
Line up decimal points.
Subtract.

Check: Recall that we can check a subtraction problem by adding.

$$\begin{array}{r} 3.432 \quad \text{Difference} \\ +0.068 \quad \text{Subtrahend} \\ \hline 3.500 \quad \text{Minuend} \end{array}$$

■ Work Practice 5

Practice 6
Subtract. Check your answers.
a. $73 - 29.31$
b. $210 - 68.22$

Example 6 Subtract: $85 - 17.31$. Check your answer.

Solution:
$$\begin{array}{r} 7\,\overset{14}{\cancel{4}}\,\overset{9}{\cancel{.}}\,\overset{}{\cancel{0}}\,10 \\ 8\,\overset{}{\cancel{5}}.\,\overset{}{\cancel{0}}\,\overset{}{\cancel{0}} \\ -1\,7\,.3\,1 \\ \hline 6\,7\,.6\,9 \end{array}$$

Check:
$$\begin{array}{r} 67.69 \quad \text{Difference} \\ +17.31 \quad \text{Subtrahend} \\ \hline 85.00 \quad \text{Minuend} \end{array}$$

■ Work Practice 6

Practice 7
Subtract 19 from 25.91

Example 7 Subtract 3 from 6.98.

Solution:
$$\begin{array}{r} 6.98 \\ -3.00 \\ \hline 3.98 \end{array}$$ Insert two 0s.

Check:
$$\begin{array}{r} 3.98 \quad \text{Difference} \\ +3.00 \quad \text{Subtrahend} \\ \hline 6.98 \quad \text{Minuend} \end{array}$$

■ Work Practice 7

Practice 8
Subtract: $-5.4 - 9.6$

Example 8 Subtract: $-5.8 - 1.7$

Solution: Recall from Chapter 2 that to subtract 1.7, we add the opposite of 1.7, or -1.7. Thus

$-5.8 - 1.7 = -5.8 + (-1.7)$ To subtract, add the opposite of 1.7, which is -1.7.

Add the absolute values.

$= -7.5.$

Use the common negative sign.

■ Work Practice 8

Practice 9
Subtract: $-1.05 - (-7.23)$

Example 9 Subtract: $-2.56 - (-4.01)$

Solution: $-2.56 - (-4.01) = -2.56 + 4.01$ To subtract, add the opposite of -4.01, which is 4.01.

Subtract the absolute values.

$= 1.45$

The answer is positive since 4.01 has the larger absolute value.

■ Work Practice 9

Answers
5. a. 2.78 b. 5.652
6. a. 43.69 b. 141.78
7. 6.91 8. -15 9. 6.18

Objective B Estimating When Adding or Subtracting Decimals

To help avoid errors, we can also estimate to see if our answer is reasonable when adding or subtracting decimals. Although only one estimate is needed per operation, we show two for variety.

Example 10 Add or subtract as indicated. Then estimate to see if the answer is reasonable by rounding the given numbers and adding or subtracting the rounded numbers.

a. $27.6 + 519.25$

Exact		Estimate 1		Estimate 2
27.60	rounds to	30		30
$+519.25$	rounds to	$+500$	or	$+520$
546.85		530		550

Since the exact answer is close to either estimate, it is reasonable. (In the first estimate, each number is rounded to the place value of the leftmost digit. In the second estimate, each number is rounded to the nearest ten.)

b. $11.01 - 0.862$

Exact		Estimate 1		Estimate 2
11.010	rounds to	10		11
-0.862	rounds to	-1	or	-1
10.148		9		10

In the first estimate, we rounded the first number to the nearest ten and the second number to the nearest one. In the second estimate, we rounded both numbers to the nearest one. Both estimates show us that our answer is reasonable.

■ Work Practice 10

Practice 10

Add or subtract as indicated. Then estimate to see if the answer is reasonable by rounding the given numbers and adding or subtracting the rounded numbers.

a. $58.1 + 326.97$

b. $16.08 - 0.925$

Helpful Hint Remember that estimates are used for our convenience to quickly check the reasonableness of an answer.

✓**Concept Check** Why shouldn't the sum $21.98 + 42.36$ be estimated as $30 + 50 = 80$?

Objective C Using Decimals as Replacement Values

Let's review evaluating expressions with given replacement values. This time the replacement values are decimals.

Example 11 Evaluate $x - y$ for $x = 2.8$ and $y = 0.92$.

Solution: Replace x with 2.8 and y with 0.92 and simplify.

$$x - y = 2.8 - 0.92$$
$$= 1.88$$

$$\begin{array}{r} 2.80 \\ -0.92 \\ \hline 1.88 \end{array}$$

■ Work Practice 11

Practice 11

Evaluate $y - z$ for $y = 11.6$ and $z = 10.8$.

Answers
10. a. Exact: 385.07; an estimate: 390 **b.** Exact: 15.155; an estimate: 15 **11.** 0.8

✓**Concept Check Answer**
Each number is rounded incorrectly. The estimate is too high.

Practice 12
Is 12.1 a solution of the equation $y - 4.3 = 7.8$?

Example 12 Is 2.3 a solution of the equation $6.3 = x + 4$?

Solution: Replace x with 2.3 in the equation $6.3 = x + 4$ to see if the result is a true statement.

$6.3 = x + 4$
$6.3 \stackrel{?}{=} 2.3 + 4$ Replace x with 2.3.
$6.3 = 6.3$ True

Since $6.3 = 6.3$ is a true statement, 2.3 is a solution of $6.3 = x + 4$.

■ Work Practice 12

Objective D Simplifying Expressions Containing Decimals

Practice 13
Simplify by combining like terms:

$-4.3y + 7.8 - 20.1y + 14.6$

Example 13 Simplify by combining like terms:

$11.1x - 6.3 + 8.9x - 4.6$

Solution:

$11.1x - 6.3 + 8.9x - 4.6 = 11.1x + 8.9x + (-6.3) + (-4.6)$
$= 20x + (-10.9)$
$= 20x - 10.9$

■ Work Practice 13

Objective E Solving Problems by Adding or Subtracting Decimals

Decimals are very common in real-life problems.

Practice 14
Find the total monthly cost of owning and operating a certain automobile given the expenses shown.

Monthly car payment:	$563.52
Monthly insurance cost:	$52.68
Average gasoline bill per month:	$127.50

Example 14 Calculating the Cost of Owning an Automobile

Find the total monthly cost of owning and operating a certain automobile given the expenses shown.

Monthly car payment:	$256.63
Monthly insurance cost:	$47.52
Average gasoline bill per month:	$195.33

Solution:
1. UNDERSTAND. Read and reread the problem. The phrase "total monthly cost" tells us to add.
2. TRANSLATE.

In words:	total monthly cost	is	car payment	plus	insurance cost	plus	gasoline bill
	↓	↓	↓	↓	↓	↓	↓
Translate:	total monthly cost	=	$256.63	+	$47.52	+	$195.33

Answers
12. yes 13. $-24.4y + 22.4$
14. $743.70

3. SOLVE: Let's also estimate by rounding each number to the nearest ten.

$$
\begin{array}{r}
\overset{1\;1\;1}{256.63} \\
47.52 \\
+\;195.33 \\
\hline
499.48
\end{array}
\quad
\begin{array}{l}
\text{rounds to} \quad 260 \\
\text{rounds to} \quad\;\; 50 \\
\text{rounds to} \quad 200 \\
\hline
\text{Exact} \quad\quad 510 \quad \text{Estimate}
\end{array}
$$

4. INTERPRET. *Check* your work. Since our estimate is close to our exact answer, our answer is reasonable. *State* your conclusion: The total monthly cost is $499.48.

▰ Work Practice 14

The next bar graph has horizontal bars. To visualize the value represented by a bar, see how far it extends to the right. The value of each bar is labeled. and we will study bar graphs further in a later chapter.

Example 15 — Comparing Average Heights

The bar graph shows the current average heights for adults in various countries. How much greater is the average height in Denmark than the average height in the United States?

Average Adult Height

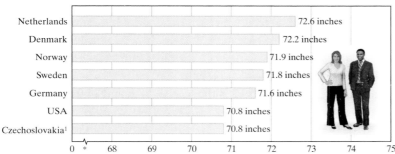

Netherlands	72.6 inches
Denmark	72.2 inches
Norway	71.9 inches
Sweden	71.8 inches
Germany	71.6 inches
USA	70.8 inches
Czechoslovakia[1]	70.8 inches

[1]Average for Czech Republic, Slovakia
Source: USA Today

* The ⌇ means that some numbers are purposefully missing on the axis.

Practice 15

Use the bar graph in Example 15. How much greater is the average height in the Netherlands than the average height in Czechoslovakia?

Solution:

1. UNDERSTAND. Read and reread the problem. Since we want to know "how much greater," we subtract.

2. TRANSLATE.

In words:	How much greater	is	Denmark's average height	minus	U.S. average height
	↓	↓	↓	↓	↓
Translate:	How much greater	=	72.2	−	70.8

3. SOLVE: We estimate by rounding each number to the nearest whole.

$$
\begin{array}{r}
\overset{\;\;1\;\;12}{7\,2.\,\cancel{2}} \\
-\;7\,0.\,8 \\
\hline
1.\,4
\end{array}
\quad
\begin{array}{l}
\text{rounds to} \quad\;\; 72 \\
\text{rounds to} \quad -71 \\
\hline
\text{Exact} \quad\quad\;\; 1 \quad \text{Estimate}
\end{array}
$$

4. INTERPRET. *Check* your work. Since our estimate is close to our exact answer, 1.4 inches is reasonable. *State* your conclusion: The average height in Denmark is 1.4 inches greater than the average U.S. height.

▰ Work Practice 15

Answer
15. 1.8 in.

Calculator Explorations Decimals

Entering Decimal Numbers

To enter a decimal number, find the key marked [.]. To enter the number 2.56, for example, press the keys [2] [.] [56].
The display will read [2.56].

Operations on Decimal Numbers

Operations on decimal numbers are performed in the same way as operations on whole or signed numbers. For example, to find 8.625 − 4.29, press the keys [8.625] [−] [4.29] then [=] or [ENTER].

The display will read [4.335]. (Although entering 8.625, for example, requires pressing more than one key, we group numbers together here for easier reading.)

Use a calculator to perform each indicated operation.

1. 315.782 + 12.96
2. 29.68 + 85.902
3. 6.249 − 1.0076
4. 5.238 − 0.682
5. 12.555
   224.987
   5.2
   +622.65
6. 47.006
   0.17
   313.259
   +139.088

Vocabulary, Readiness & Video Check

Use the choices below to fill in each blank. Not all choices will be used.

| minuend | vertically | like | true |
| difference | subtrahend | last | false |

1. The decimal point in a whole number is positioned after the _____ digit.
2. In 89.2 − 14.9 = 74.3, the number 74.3 is called the _____, 89.2 is the _____, and 14.9 is the _____.
3. To simplify an expression, we combine any _____ terms.
4. To add or subtract decimals, we line up the decimal points _____.
5. True or false: If we replace x with 11.2 and y with −8.6 in the expression $x - y$, we have 11.2 − 8.6. _____
6. True or false: If we replace x with −9.8 and y with −3.7 in the expression $x + y$, we have −9.8 + 3.7. _____

Martin-Gay Interactive Videos

See Video 5.2

Watch the section lecture video and answer the following questions.

Objective A 7. From Examples 1–3, why do you think we line up decimal points?

Objective B 8. In Example 4, estimating is used to check whether the answer to the subtraction problem is reasonable, but what is the best way to fully check?

Objective C 9. In Example 5, why is the actual subtraction performed to the side?

Objective D 10. How many sets of like terms are there in Example 6?

Objective E 11. In Example 7, to calculate the amount of border material needed, we are actually calculating the _____ of the triangle.

5.2 Exercise Set MyLab Math

Objectives A B Mixed Practice *Add. See Examples 1 through 4 and 10. For those exercises marked, also estimate to see if the answer is reasonable.*

1. $5.6 + 2.1$
2. $3.6 + 4.1$
3. $8.2 + 2.15$
4. $5.17 + 3.7$
5. $24.6 + 2.39 + 0.0678$
6. $32.4 + 1.58 + 0.0934$
7. $-2.6 + (-5.97)$
8. $-18.2 + (-10.8)$
9. $18.56 + (-8.23)$
10. $4.38 + (-6.05)$
11. $234.89 + 230.67$ Exact: _____ Estimate: _____
12. $734.89 + 640.56$ Exact: _____ Estimate: _____
13. $100.009 + 6.08 + 9.034$ Exact: _____ Estimate: _____
14. $200.89 + 7.49 + 62.83$ Exact: _____ Estimate: _____
15. Find the sum of 39, 3.006, and 8.403
16. Find the sum of 65, 5.0903, and 6.9003

Subtract and check. See Examples 5 through 10. For those exercises marked, also estimate to see if the answer is reasonable.

17. $12.6 - 8.2$
18. $8.9 - 3.1$
19. $18 - 2.7$
20. $28 - 3.3$
21. $654.9 - 56.67$
22. $863.23 - 39.453$
23. $5.9 - 4.07$ Exact: _____ Estimate: _____
24. $6.4 - 3.04$ Exact: _____ Estimate: _____
25. $1000 - 123.4$ Exact: _____ Estimate: _____
26. $2000 - 327.47$ Exact: _____ Estimate: _____
27. $200 - 5.6$
28. $800 - 8.9$
29. $-1.12 - 5.2$
30. $-8.63 - 5.6$
31. $5.21 - 11.36$
32. $8.53 - 17.84$
33. $-2.6 - (-5.7)$
34. $-9.4 - (-10.4)$
35. $3 - 0.0012$
36. $7 - 0.097$
37. Subtract 6.7 from 23.
38. Subtract 9.2 from 45.

Objective A *Perform the indicated operation. See Examples 1 through 9.*

39. $0.9 + 2.2$ **40.** $0.7 + 3.4$ **41.** $-6.06 + 0.44$ **42.** $-5.05 + 0.88$

43. $500.21 - 136.85$ **44.** $600.47 - 254.68$ **45.** $50.2 - 600$ **46.** $40.3 - 700$

47. Subtract 61.9 from 923.5.   **48.** Subtract 45.8 from 845.9.   **49.** Add 200.008 and 7.06 and 9.046.

50. Add 100.007 and 2.08 and 5.014.   **51.** $-0.003 + 0.091$ **52.** $-0.004 + 0.085$

53. $-102.4 - 78.04$ **54.** $-36.2 - 10.02$ **55.** $-2.9 - (-1.8)$ **56.** $-6.5 - (-3.3)$

Objective C *Evaluate each expression for $x = 3.6$, $y = 5$, and $z = 0.21$. See Example 11.*

57. $x + z$ **58.** $y + x$ **59.** $x - z$

60. $y - z$ **61.** $y - x + z$ **62.** $x + y + z$

Determine whether the given values are solutions to the given equations. See Example 12.

63. Is 7 a solution to $x + 2.7 = 9.3$?   **64.** Is 3.7 a solution to $x + 5.9 = 8.6$?

65. Is -11.4 a solution to $27.4 + y = 16$?   **66.** Is -22.9 a solution to $45.9 + z = 23$?

67. Is 1 a solution to $2.3 + x = 5.3 - x$?   **68.** Is 0.9 a solution to $1.9 - x = x + 0.1$?

Objective D *Simplify by combining like terms. See Example 13.*

69. $30.7x + 17.6 - 23.8x - 10.7$ **70.** $14.2z + 11.9 - 9.6z - 15.2$

71. $-8.61 + 4.23y - 2.36 - 0.76y$ **72.** $-8.96x - 2.31 - 4.08x + 9.68$

Objective E *Solve. For Exercises 73 and 74, the solutions have been started for you. See Examples 14 and 15.*

73. Ann-Margaret Tober bought a book for $32.48. If she paid with two $20 bills, what was her change?

Start the solution:
1. UNDERSTAND the problem. Reread it as many times as needed.
2. TRANSLATE into an equation. (Fill in the blank.)

change	is	two $20 bills	minus	cost of book
↓	↓	↓	↓	↓
change	=	40	−	_____

Finish with
3. SOLVE and 4. INTERPRET

74. Phillip Guillot bought a car part for $18.26. If he paid with two $10 bills, what was his change?

Start the solution:
1. UNDERSTAND the problem. Reread it as many times as needed.
2. TRANSLATE into an equation. (Fill in the blank.)

change	is	two $10 bills	minus	cost of car part
↓	↓	↓	↓	↓
change	=	20	−	_____

Finish with
3. SOLVE and 4. INTERPRET

75. Microsoft stock opened the day at $35.17 per share, and the closing price the same day was $34.75. By how much did the price of each share change?

76. A pair of eyeglasses costs a total of $347.89. The frames of the glasses are $97.23. How much do the lenses of the eyeglasses cost?

△ 77. Find the perimeter.

Square 7.14 meters

△ 78. Find the perimeter.

4.2 in.   5.78 in.   7.8 in.

79. The Apple iPhone 7 was released in 2016. It measures 5.44 inches by 2.64 inches. Find the perimeter of this phone. (*Source:* Apple.com)

80. The Samsung Galaxy 57 Edge was released in 2016. It measures 5.91 inches by 2.86 inches. Find the perimeter of the phone. (*Source:* Google.com)

81. The average wind speed at the weather station on Mt. Washington in New Hampshire is 35.2 miles per hour. The highest speed ever recorded at the station is 231.0 miles per hour. How much faster is the highest speed than the average wind speed? (*Source:* National Centers for Environmental Information)

82. The average annual rainfall in Omaha, Nebraska, is 30.08 inches. The average annual rainfall in New Orleans, Louisiana, is 64.16 inches. On average, how much more rain does New Orleans receive annually than Omaha? (*Source:* National Centers for Environmental Information)

This bar graph shows the increase in the total number of text messages per person per day in the United States. Use this graph for Exercises 83 and 84. (Source: Pew Research Center.)

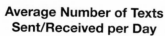
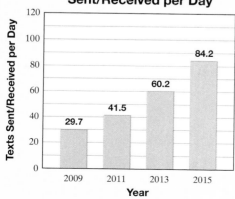

Average Number of Texts Sent/Received per Day

83. Find the increase in the number of texts sent or received per day from 2011 to 2015.

84. Find the increase in the number of texts sent or received per day from 2013 to 2015.

85. As of this writing, the top three U.S. movies that made the most money through movie ticket sales are *Star Wars Episode VII: The Force Awakens* (2015), $936.7 million; *Avatar* (2009), $760.5 million; and *Titanic* (1997), $658.7 million. What was the total amount of ticket sales for these three movies? Round your answer to the nearest million. (*Source:* the-numbers.com)

86. In 2011, the average credit card late fee was $23.15. By 2015, the average credit card late fee had increased by $3.85. Find the average credit card late fee in 2015. (*Source:* Consumer Financial Protection Bureau)

87. The snowiest city in the United States is Valdez, AK, which receives an average of 110.5 more inches of snow than the second snowiest city. The second snowiest city in the United States is Crested Butte, CO. Crested Butte receives an average of 215.8 inches annually. How much snow does Valdez receive on average each year? (*Source:* The Weather Channel)

88. The driest place in the world is the Atacama Desert in Chile, which receives an average of only 0.004 inch of rain per year. Yuma, Arizona, is the driest city in the United States. Yuma receives an average of 3.006 more inches of rain each year than the Atacama Desert. What is the average annual rainfall in Yuma? (*Source:* National Climatic Data Center)

89. A landscape architect is planning a border for a flower garden shaped like a triangle. The sides of the garden measure 12.4 feet, 29.34 feet, and 25.7 feet. Find the amount of border material needed.

90. A contractor purchased enough railing to completely enclose the newly built deck shown below. Find the amount of railing purchased.

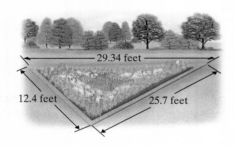

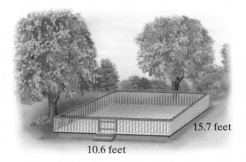

The table shows the average speeds for the Daytona 500 winners for the years shown. Use this table to answer Exercises 91 and 92. (Source: Daytona International Speedway)

Daytona 500 Winners		
Year	Winner	Average Speed
2001	Michael Waltrip	161.783
2004	Dale Earnhardt Jr.	156.341
2007	Kevin Harvick	149.333
2010	Jamie McMurray	137.284
2013	Jimmie Johnson	159.250
2016	Denny Hamlin	157.549

91. How much slower was the average Daytona 500 winning speed in 2016 than in 2001?

92. How much faster was Dale Earnhardt Jr.'s average Daytona 500 winning speed in 2004 than Jamie McMurray's Daytona 500 winning speed in 2010?

The bar graph shows the top five chocolate-consuming nations in the world. Use this table to answer Exercises 93 through 97.

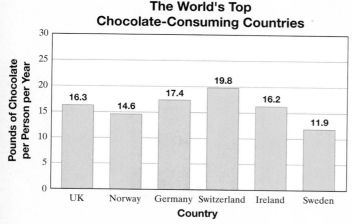

93. Which country in the table has the greatest chocolate consumption per person?

94. Which country in the table has the least chocolate consumption per person?

95. How much more is the greatest chocolate consumption than the least chocolate consumption shown in the table?

96. How much more chocolate does the average German consume per year than the average Norwegian?

97. Make a chart listing the countries and their corresponding chocolate consumptions in order from greatest to least.

Review

Multiply. See Sections 1.5 and 4.3.

98. $23 \cdot 2$ **99.** $46 \cdot 3$ **100.** $39 \cdot 3$ **101.** $\left(\dfrac{2}{3}\right)^2$ **102.** $\left(\dfrac{1}{5}\right)^3$

Concept Extensions

A friend asks you to check his calculations for Exercises 103 and 104. Are they correct? If not, explain your friend's errors and correct the calculations. See the first Concept Check in this section.

103.
$$\begin{array}{r} \overset{1}{9}.2 \\ \overset{1}{8}.6\,3 \\ +\ 4.0\,0\,5 \\ \hline 4.9\,6\,0 \end{array}$$

104.
$$\begin{array}{r} \overset{8\ 9\ 9\ 9}{9\cancel{0}\cancel{0}.\cancel{0}} \\ -\ \ \ 96.4 \\ \hline 8\,03.5 \end{array}$$

Find the unknown length in each figure.

105.

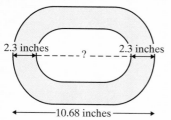

△ **106.**

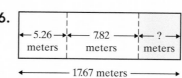

Let's review the values of these common U.S. coins in order to answer the following exercises.

Penny   Nickel   Dime   Quarter

$0.01 $0.05 $0.10 $0.25

For Exercises 107 and 108, write the value of each group of coins. To do so, it is usually easiest to start with the coin(s) of greatest value and end with the coin(s) of least value.

107.

108.

109. Name the different ways that coins can have a value of $0.17 given that you may use no more than 10 coins.

110. Name the different ways that coin(s) can have a value of $0.25 given that there are no pennies.

111. Why shouldn't the sum
$$82.95 + 51.26$$
be estimated as $90 + 60 = 150$?
See the second Concept Check in this section.

112. Laser beams can be used to measure the distance to the moon. One measurement showed the distance to the moon to be 256,435.235 miles. A later measurement showed that the distance is 256,436.012 miles. Find how much farther away the moon is in the second measurement as compared to the first.

113. Explain how adding or subtracting decimals is similar to adding or subtracting whole numbers.

114. Can the sum of two negative decimals ever be a positive decimal? Why or why not?

Combine like terms and simplify.

115. $-8.689 + 4.286x - 14.295 - 12.966x + 30.861x$

116. $14.271 - 8.968x + 1.333 - 201.815x + 101.239x$

5.3 Multiplying Decimals and Circumference of a Circle

Objective A Multiplying Decimals

Multiplying decimals is similar to multiplying whole numbers. The only difference is that we place a decimal point in the product. To discover where a decimal point is placed in a product, let's multiply 0.6×0.03. We first write each decimal as an equivalent fraction and then multiply.

$$\underset{\text{1 decimal place}}{0.6} \times \underset{\text{2 decimal places}}{0.03} = \frac{6}{10} \times \frac{3}{100} = \frac{18}{1000} = \underset{\text{3 decimal places}}{0.018}$$

Notice that $1 + 2 = 3$, the number of decimal places in the product. Now let's multiply 0.03×0.002.

$$\underset{\text{2 decimal places}}{0.03} \times \underset{\text{3 decimal places}}{0.002} = \frac{3}{100} \times \frac{2}{1000} = \frac{6}{100,000} = \underset{\text{5 decimal places}}{0.00006}$$

Again, we see that $2 + 3 = 5$, the number of decimal places in the product.

Instead of writing decimals as fractions each time we want to multiply, we notice a pattern from these examples and state a rule that we can use:

Multiplying Decimals

Step 1: Multiply the decimals as though they are whole numbers.

Step 2: The decimal point in the product is placed so that the number of decimal places in the product is equal to the *sum* of the number of decimal places in the factors.

Objectives

A Multiply Decimals.
B Estimate When Multiplying Decimals.
C Multiply Decimals by Powers of 10.
D Evaluate Expressions with Decimal Replacement Values.
E Find the Circumference of Circles.
F Solve Problems by Multiplying Decimals.

Example 1   Multiply: 23.6×0.78

Solution:
```
    23.6    1 decimal place
  × 0.78    2 decimal places
  ------
    1888
   16520
  ------
  18.408
```
Since $1 + 2 = 3$, insert the decimal point in the product so that there are 3 decimal places.

■ Work Practice 1

Practice 1
Multiply: 34.8×0.62

Example 2   Multiply: 0.0531×16

Solution:
```
   0.0531    4 decimal places
  ×    16    0 decimal places
  -------
     3186
     5310
  -------
   0.8496
```
4 decimal places $(4 + 0 = 4)$

■ Work Practice 2

Practice 2
Multiply: 0.0641×27

Answers
1. 21.576 **2.** 1.7307

Practice 3
Multiply: $(7.3)(-0.9)$

✓ **Concept Check** True or false? The number of decimal places in the product of 0.261 and 0.78 is 6. Explain.

Example 3 Multiply: $(-2.6)(0.8)$

Solution: Recall that the product of a negative number and a positive number is a negative number.

$(-2.6)(0.8) = -2.08$

■ Work Practice 3

Objective B Estimating When Multiplying Decimals

Just as for addition and subtraction, we can estimate when multiplying decimals to check the reasonableness of our answer.

Practice 4
Multiply: 30.26×2.89. Then estimate to see whether the answer is reasonable.

Example 4 Multiply: 28.06×1.95. Then estimate to see whether the answer is reasonable by rounding each factor, then multiplying the rounded numbers.

Solution:

Exact	Estimate 1		or	Estimate 2	
28.06	28			30	
× 1.95	× 2	Rounded to ones		× 2	Rounded to one nonzero digit
14030	56			60	
252540					
280600					
54.7170					

The answer 54.7170 or 54.717 is reasonable.

■ Work Practice 4

As shown in Example 4, estimated results will vary depending on what estimates are used. Notice that estimating results is a good way to see whether the decimal point has been correctly placed.

Objective C Multiplying Decimals by Powers of 10

There are some patterns that occur when we multiply a number by a power of 10, such as 10, 100, 1000, 10,000, and so on.

$23.6951 \times 10 = 236.951$ Move the decimal point *1 place* to the *right*.
          ↑
       1 zero

$23.6951 \times 100 = 2369.51$ Move the decimal point *2 places* to the *right*.
          ↑
       2 zeros

$23.6951 \times 100{,}000 = 2{,}369{,}510.$ Move the decimal point *5 places* to the *right* (insert a zero)
          ↑
       5 zeros

Notice that we move the decimal point the same number of places as there are zeros in the power of 10.

Answers
3. -6.57
4. Exact: 87.4514; Estimate: $30 \cdot 3 = 90$

✓ **Concept Check Answer**
false: 3 decimal places and 2 decimal places means 5 decimal places in the product

Section 5.3 | Multiplying Decimals and Circumference of a Circle

Multiplying Decimals by Powers of 10 Such as 10, 100, 1000, 10,000

Move the decimal point to the *right* the same number of places as there are *zeros* in the power of 10.

Examples Multiply.

5. $7.68 \times 10 = 76.8$ 7.68
6. $23.702 \times 100 = 2370.2$ 23.702
7. $(-76.3)(1000) = -76{,}300$ 76.300

Work Practice 5–7

Practice 5–7

Multiply.
5. 46.8×10
6. 203.004×100
7. $(-2.33)(1000)$

There are also powers of 10 that are less than 1. The decimals 0.1, 0.01, 0.001, 0.0001, and so on, are examples of powers of 10 less than 1. Notice the pattern when we multiply by these powers of 10:

$569.2 \times 0.1 = 56.92$ Move the decimal point *1 place* to the *left*.
    ↑
 1 decimal place

$569.2 \times 0.01 = 5.692$ Move the decimal point *2 places* to the *left*.
    ↑
 2 decimal places

$569.2 \times 0.0001 = 0.05692$ Move the decimal point *4 places* to the *left* (insert one 0).
    ↑
 4 decimal places

Multiplying Decimals by Powers of 10 Such as 0.1, 0.01, 0.001, 0.0001

Move the decimal point to the *left* the same number of places as there are *decimal places* in the power of 10.

Examples Multiply.

8. $42.1 \times 0.1 = 4.21$ 42.1
9. $76{,}805 \times 0.01 = 768.05$ 76,805.
10. $(-9.2)(-0.001) = 0.0092$ 0009.2

Work Practice 8–10

Practice 8–10

Multiply.
8. 6.94×0.1
9. 3.9×0.01
10. $(-7682)(-0.001)$

Many times we see large numbers written, for example, in the form 297.9 million rather than in the longer standard notation. The next example shows us how to interpret these numbers.

Answers
5. 468 **6.** 20,300.4 **7.** −2330
8. 0.694 **9.** 0.039 **10.** 7.682

Practice 11

In 2016, there were 60.25 million married couples in the United States. Write this number in standard notation. (*Source:* U.S. Census Bureau)

Example 11 In 2050, the population of the United States is projected to be 438.2 million. Write this number in standard notation. (*Source:* Pew Research Center)

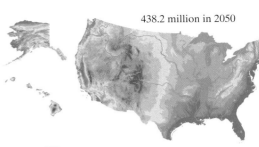

438.2 million in 2050

Solution: 438.2 million = 438.2 × 1 million
= 438.2 × 1,000,000 = 438,200,000

■ Work Practice 11

Objective D Using Decimals as Replacement Values

Now let's practice working with variables.

Practice 12

Evaluate $7y$ for $y = -0.028$.

Example 12 Evaluate xy for $x = 2.3$ and $y = 0.44$.

Solution: Recall that xy means $x \cdot y$.

$xy = (2.3)(0.44)$

$ = 1.012$

$$\begin{array}{r} 2.3 \\ \times\, 0.44 \\ \hline 92 \\ 920 \\ \hline 1.012 \end{array}$$

■ Work Practice 12

Practice 13

Is -5.5 a solution of the equation $-6x = 33$?

Example 13 Is -9 a solution of the equation $3.7y = -3.33$?

Solution: Replace y with -9 in the equation $3.7y = -3.33$ to see if a tr equation results.

$3.7y = -3.33$

$3.7(-9) \stackrel{?}{=} -3.33$ Replace y with -9.

$-33.3 = -3.33$ False

Since $-33.3 = -3.33$ is a false statement, -9 is **not** a solution of $3.7y = -3.33$.

■ Work Practice 13

Objective E Finding the Circumference of a Circle

Recall from Section 1.3 that the distance around a polygon is called its perimet The distance around a circle is given the special name **circumference,** and this d tance depends on the radius (or the diameter) of the circle.

Circumference of a Circle

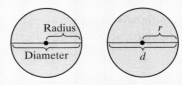

Circumference = $2 \cdot \pi \cdot$ **r**adius or Circumference = $\pi \cdot$ **d**iameter

$C = 2\pi r$ or $C = \pi d$

Answers
11. 60,250,000 **12.** -0.196 **13.** yes

Section 5.3 | Multiplying Decimals and Circumference of a Circle

In Section 5.1, we learned about the symbol π as the Greek letter pi, pronounced "pie." It is a constant between 3 and 4. A decimal approximation for π is 3.14. Also, a fraction approximation for π is $\frac{22}{7}$.

Example 14 Circumference of a Circle

Find the circumference of a circle whose radius is 5 inches. Then use the approximation 3.14 for π to approximate the circumference.

Solution: Let $r = 5$ in the formula $C = 2\pi r$.

$$C = 2\pi r$$
$$= 2\pi \cdot 5$$
$$= 10\pi$$

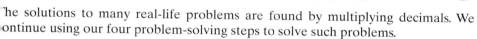

Next, replace π with the approximation 3.14.

$$C = 10\pi$$
(is approximately) $\longrightarrow \approx 10(3.14)$
$$\approx 31.4$$

The **exact** circumference or distance around the circle is 10π inches, which is **approximately** 31.4 inches.

■ Work Practice 14

Practice 14
Find the circumference of a circle whose radius is 11 meters. Then use the approximation 3.14 for π to approximate this circumference.

Objective F Solving Problems by Multiplying Decimals ▶

The solutions to many real-life problems are found by multiplying decimals. We continue using our four problem-solving steps to solve such problems.

Example 15 Finding the Total Cost of Materials for a Job

A college student is hired to paint a billboard with paint costing $2.49 per quart. If the job requires 3 quarts of paint, what is the total cost of the paint?

Solution:

1. UNDERSTAND. Read and reread the problem. The phrase "total cost" might make us think addition, but since this problem requires repeated addition, let's multiply.

2. TRANSLATE.

In words: total cost   is   cost per quart of paint   times   number of quarts

Translate: total cost   =   2.49   ×   3

3. SOLVE. We can estimate to check our calculations. The number 2.49 rounds to 2 and $2 \times 3 = 6$.

$$\begin{array}{r} 2.49 \\ \times 3 \\ \hline 7.47 \end{array}$$

4. INTERPRET. *Check* your work. Since 7.47 is close to our estimate of 6, our answer is reasonable. *State* your conclusion: The total cost of the paint is $7.47.

■ Work Practice 15

Practice 15
A biology major is fertilizing her personal garden. She uses 5.6 ounces of fertilizer per square yard. The garden measures 60.5 square yards. How much fertilizer does she need?

Answers
14. 22π m $\approx$ 69.08 m   **15.** 338.8 oz

Vocabulary, Readiness & Video Check

Use the choices below to fill in each blank.

circumference   left   sum   zeros
decimal places   right   product   factor

1. When multiplying decimals, the number of decimal places in the product is equal to the _____ of the number of decimal places in the factors.
2. In $8.6 \times 5 = 43$, the number 43 is called the _____, while 8.6 and 5 are each called a _____.
3. When multiplying a decimal number by powers of 10 such as 10, 100, 1000, and so on, we move the decimal point in the number to the _____ the same number of places as there are _____ in the power of 10.
4. When multiplying a decimal number by powers of 10 such as 0.1, 0.01, and so on, we move the decimal point in the number to the _____ the same number of places as there are _____ in the power of 10.
5. The distance around a circle is called its _____.

Martin-Gay Interactive Videos

See Video 5.3

Watch the section lecture video and answer the following questions.

Objective A   6. From the lecture before Example 1, what's the main difference between multiplying whole numbers and multiplying decimal numbers?
Objective B   7. From Example 3, what does estimating especially help us with?
Objective C   8. Why don't we do any actual multiplying in Example 5?
Objective D   9. In Example 8, once all replacement values are inserted in the variable expression, what is the resulting expression to evaluate?
Objective E   10. Why is 31.4 cm not the exact answer to Example 9?
Objective F   11. In Example 10, why is 24.8 not the complete answer? What is the complete answer?

5.3 Exercise Set MyLab Math

Objectives A B Mixed Practice *Multiply. See Examples 1 through 4. For those exercises marked, also estimate to see if the answer is reasonable.*

1. 0.17×8
2. 0.23×9
3. 1.2
   $\underline{\times 0.5}$
4. 6.8
   $\underline{\times 0.3}$

5. $(-2.3)(7.65)$
6. $(4.7)(-9.02)$
7. $(-5.73)(-9.6)$
8. $(-7.84)(-3.5)$

9. 6.8×4.2
   Exact:
   Estimate:
10. 8.3×2.7
    Exact:
    Estimate:
11. 0.347
    $\underline{\times 0.3}$
12. 0.864
    $\underline{\times 0.4}$

Section 5.3 | Multiplying Decimals and Circumference of a Circle

13. 1.0047 × 8.2 Exact: _____ Estimate: _____
14. 2.0005 × 5.5 Exact: _____ Estimate: _____
15. 490.2 × 0.023
16. 300.9 × 0.032

Objective C *Multiply. See Examples 5 through 10.*

17. 6.5×10
18. 7.2×100
19. 8.3×0.1
20. 23.4×0.1

21. $(-7.093)(1000)$
22. $(-1.123)(1000)$
23. 0.7×100
24. 0.5×100

25. $(-9.83)(-0.01)$
26. $(-4.72)(-0.01)$
27. 25.23×0.001
28. 36.41×0.001

Objectives A B C Mixed Practice *Multiply. See Examples 1 through 10.*

29. 0.123×0.4
30. 0.216×0.3
31. $(147.9)(100)$
32. $(345.2)(100)$

33. 8.6×0.15
34. 0.42×5.7
35. $(937.62)(-0.01)$
36. $(-0.001)(562.01)$

37. 562.3×0.001
38. 993.5×0.001
39. 6.32 × 5.7
40. 9.21 × 3.8

Write each number in standard notation. See Example 11.

41. The cost of the Hubble Space Telescope at launch was $1.5 billion. (*Source:* NASA)
42. About 56.7 million American households own at least one dog. (*Source:* American Pet Products Manufacturers Association)
43. The Blue Streak is the oldest roller coaster at Cedar Point, an amusement park in Sandusky, Ohio. Since 1964, it has given more than 49.8 million rides. (*Source:* Cedar Fair, L.P.)
44. In 2017, the restaurant industry had projected sales of $798.7 billion. (*Source:* National Restaurant Association)

Objective D *Evaluate each expression for $x = 3$, $y = -0.2$, and $z = 5.7$. See Example 12.*

45. xy
46. yz
47. $xz - y$
48. $-5y + z$

Determine whether the given value is a solution of each given equation. See Example 13.

49. Is 14.2 a solution of $0.6x = 4.92$?
50. Is 1414 a solution of $100z = 14.14$?
51. Is -4 a solution of $3.5y = -14$?
52. Is -3.6 a solution of $0.7x = -2.52$?

Objective E *Find the circumference of each circle. Then use the approximation 3.14 for π and approximate each circumference. See Example 14.*

53.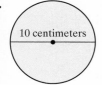
54.
55.
56.

Objectives **E F** Mixed Practice *Solve. For Exercises 57 and 58, the solutions have been started for you. For circumference applications find the exact circumference and then use 3.14 for π to approximate the circumference. See Examples 14 and 15.*

57. An electrician for Central Power and Light worked 40 hours last week. Calculate his pay before taxes for last week if his hourly wage is $17.88.

Start the solution:

1. UNDERSTAND the problem. Reread it as many times as needed.
2. TRANSLATE into an equation. (Fill in the blanks.)

pay before taxes	is	hourly wage	times	hours worked
↓	↓	↓	↓	↓
pay before taxes	=	____	×	____

Finish with:
3. SOLVE and 4. INTERPRET.

58. An assembly line worker worked 20 hours last week. Her hourly rate is $19.52 per hour. Calculate her pay before taxes.

Start the solution:

1. UNDERSTAND the problem. Reread it as many times as needed.
2. TRANSLATE into an equation. (Fill in the blanks.)

pay before taxes	is	hourly rate	times	hours worked
↓	↓	↓	↓	↓
pay before taxes	=	____	×	____

Finish with:
3. SOLVE and 4. INTERPRET.

59. A 1-ounce serving of cream cheese contains 6.2 grams of saturated fat. How much saturated fat is in 4 ounces of cream cheese? (*Source: Home and Garden Bulletin No. 72;* U.S. Department of Agriculture)

60. A 3.5-ounce serving of lobster meat contains 0.1 gram of saturated fat. How much saturated fat do 3 servings of lobster meat contain? (*Source:* The National Institutes of Health)

△ **61.** Recall that the face of the Apple iPhone 7 (see Section 5.2) measures approximately 5.44 inches by 2.6 inches. Find the area of the face of the Apple iPhone 7.

△ **62.** Recall that the face of the Samsung Galaxy 57 Edge (see Section 5.2) measures approximately 5.91 inches by 2.9 inches. Find the area of the face of the Samsung Galaxy 57 Edge.

△ **63.** In 1893, the first ride called a Ferris wheel was constructed by Washington Gale Ferris. Its diameter was 250 feet. Find its circumference. Give the exact answer and an approximation using 3.14 for π. (*Source: The Handy Science Answer Book,* Visible Ink Press, 1994)

△ **64.** The radius of Earth is approximately 3950 miles. Find the distance around Earth at the equator. Give the exact answer and an approximation using 3.14 for π. (*Hint:* Find the circumference of a circle with radius 3950 miles.)

△ **65.** The London Eye, built for the millennium celebration in London, resembles a gigantic Ferris wheel with a diameter of 135 meters. If Adam Hawn rides the Eye for one revolution, find how far he travels. Give the exact answer and an approximation using 3.14 for π. (*Source:* Londoneye.com)

△ **66.** The world's longest suspension bridge is the Akashi Kaikyo Bridge in Japan. This bridge has two circular caissons, which are underwater foundations. If the diameter of a caisson is 80 meters, find its circumference. Give the exact answer and an approximation using 3.14 for π. (*Source: Scientific American; How Things Work Today*)

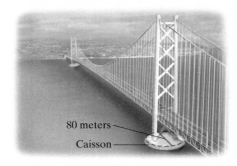

67. A meter is a unit of length in the metric system that is approximately equal to 39.37 inches. Sophia Wagner is 1.65 meters tall. Find her approximate height in inches.

68. The doorway to a room is 2.15 meters tall. Approximate this height in inches. (*Hint:* See Exercise 67.)

69. a. Approximate the circumference of each circle.

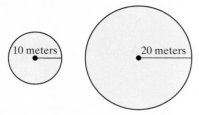

b. If the radius of a circle is doubled, is its corresponding circumference doubled?

△ **70. a.** Approximate the circumference of each circle.

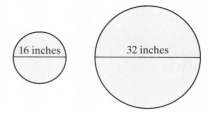

b. If the diameter of a circle is doubled, is its corresponding circumference doubled?

71. In December 2016, the price of wheat was $4.0425 per bushel. How much would 100 bushels of wheat cost at this price? (*Source:* ndwheat.com)

72. In December 2016, the price of corn was $3.47 per bushel. How much would a company pay for 10,000 bushels of corn? (*Source:* quotecorn.com)

The table shows currency exchange rates for various countries on February 1, 2017. To find the amount of foreign currency equivalent to an amount of U.S. dollars, multiply the U.S. dollar amount by the exchange rate listed in the table. Use this table to answer Exercises 73 through 76. If needed, round answers to the nearest hundredth.

Foreign Currency Exchange Rates	
Country	**Exchange Rate**
Canadian dollar	1.3032
European Union euro	0.9292
Mexican peso	20.5077
Chinese yuan	6.88
Japanese yen	112.73
Swiss franc	0.9927

73. How many Canadian dollars are equivalent to $750 U.S.?

74. Suppose you wish to exchange 300 American dollars for Chinese yuan. How much money, in Chinese yuan, would you receive?

75. The Scarpulla family is traveling to Mexico. How many Mexican pesos can they "buy" with 800 U.S. dollars?

76. A French tourist to the United States spent $130 for souvenirs at the *Head of the Charles Regatta* in Boston. How much money did he spend in euros? Round to the nearest hundredth.

Review

Divide. See Sections 1.6 and 4.3.

77. $2916 \div 6$

78. $2920 \div 365$

79. $-\dfrac{24}{7} \div \dfrac{8}{21}$

80. $\dfrac{162}{25} \div -\dfrac{9}{75}$

Concept Extensions

Mixed Practice (*Sections 5.2, 5.3*) *Perform the indicated operations.*

81. $3.6 + 0.04$

82. $7.2 + 0.14 + 98.6$

83. $3.6 - 0.04$

84. $100 - 48.6$

85. -0.221×0.5

86. -3.6×0.04

87. Find how far radio waves travel in 20.6 seconds. (Radio waves travel at a speed of 186,000 miles per second.)

88. If it takes radio waves approximately 8.3 minutes to travel from the sun to the earth, find approximately how far it is from the sun to the earth. (*Hint:* See Exercise **87**.)

89. In your own words, explain how to find the number of decimal places in a product of decimal numbers.

90. In your own words, explain how to multiply by a power of 10.

91. Write down two decimal numbers whose product will contain 5 decimal places. Without multiplying, explain how you know your answer is correct.

5.4 Dividing Decimals

Objectives

A Divide Decimals.

B Estimate When Dividing Decimals.

C Divide Decimals by Powers of 10.

D Evaluate Expressions with Decimal Replacement Values.

E Solve Problems by Dividing Decimals.

Objective A Dividing Decimals

Dividing decimal numbers is similar to dividing whole numbers. The only difference is that we place a decimal point in the quotient. If the divisor is a whole number, we place the decimal point in the quotient directly above the decimal point in the dividend, and then divide as with whole numbers. Recall that division can be checked by multiplication.

Example 1 Divide: $270.2 \div 7$. Check your answer.

Solution: We divide as usual. The decimal point in the quotient is directly above the decimal point in the dividend.

$$\begin{array}{r} 38.6 \\ 7\overline{)270.2} \\ -21 \\ \hline 60 \\ -56 \\ \hline 4\,2 \\ -4\,2 \\ \hline 0 \end{array}$$

divisor → 7)270.2 ← dividend, 38.6 ← quotient — Write the decimal point

Check:
$$\begin{array}{r} 38.6 \\ \times\ \ \ 7 \\ \hline 270.2 \end{array}$$

The quotient is 38.6.

Practice 1
Divide: $370.4 \div 8$. Check your answer.

Work Practice 1

Example 2 Divide: $32\overline{)8.32}$

Solution: We divide as usual. The decimal point in the quotient is directly above the decimal point in the dividend.

dividend → 32)8.32 ← divisor, 0.26 ← quotient

$$\begin{array}{r} 0.26 \\ 32\overline{)8.32} \\ -64 \\ \hline 192 \\ -192 \\ \hline 0 \end{array}$$

Check:
$$\begin{array}{r} 0.26 \text{ quotient}\\ \times\ \ 32 \text{ divisor}\\ \hline 52 \\ 7\,80 \\ \hline 8.32 \text{ dividend} \end{array}$$

Practice 2
Divide: $48\overline{)34.08}$. Check your answer.

Work Practice 2

Answers
1. 46.3   **2.** 0.71

Sometimes to continue dividing we need to insert zeros after the last digit in the dividend.

Example 3   Divide: $-5.98 \div 115$

Solution: Recall that a negative number divided by a positive number gives a negative quotient.

$$
\begin{array}{r}
0.052 \\
115{\overline{\smash{)}5.980}} \quad \leftarrow \text{Insert one 0.}\\
\underline{-5\,75} \\
230 \\
\underline{-230} \\
0
\end{array}
$$

Thus $-5.98 \div 115 = -0.052$.

■ Work Practice 3

Practice 3

Divide and check.
a. $-15.89 \div 14$
b. $-2.808 \div (-104)$

If the divisor is not a whole number, before we divide we need to move the decimal point to the right until the divisor is a whole number.

$$1.5{\overline{\smash{)}64.85}}$$

divisor ⎯⏐   ⏐⎯ dividend

To understand how this works, let's rewrite

$1.5{\overline{\smash{)}64.85}}$ as $\dfrac{64.85}{1.5}$

and then multiply by 1 in the form of $\dfrac{10}{10}$. We use the form $\dfrac{10}{10}$ so that the denominator (divisor) becomes a whole number.

$$\dfrac{64.85}{1.5} = \dfrac{64.85}{1.5} \cdot 1 = \dfrac{64.85}{1.5} \cdot \dfrac{10}{10} = \dfrac{64.85 \cdot 10}{1.5 \cdot 10} = \dfrac{648.5}{15},$$

which can be written as $15.{\overline{\smash{)}648.5}}$. Notice that

$1.5{\overline{\smash{)}64.85}}$ is equivalent to $15.{\overline{\smash{)}648.5}}$

The decimal points in the dividend and the divisor were both moved one place to the right, and the divisor is now a whole number. This procedure is summarized next:

Dividing by a Decimal

Step 1: Move the decimal point in the divisor to the right until the divisor is a whole number.

Step 2: Move the decimal point in the dividend to the right the *same number of places* as the decimal point was moved in Step 1.

Step 3: Divide. Place the decimal point in the quotient directly over the moved decimal point in the dividend.

Answers

3. a. -1.135 b. 0.027

Practice 4

Divide: $166.88 \div 5.6$

Example 4 Divide: $10.764 \div 2.3$

Solution: We move the decimal points in the divisor and the dividend one place to the right so that the divisor is a whole number.

$2.3\overline{)10.764}$ becomes
$$\begin{array}{r} 4.68 \\ 23.\overline{)107.64} \\ -92\downarrow \\ \hline 15\,6 \\ -13\,8\downarrow \\ \hline 1\,84 \\ -1\,84 \\ \hline 0 \end{array}$$

■ Work Practice 4

Practice 5

Divide: $1.976 \div 0.16$

Example 5 Divide: $5.264 \div 0.32$

Solution:

$0.32\overline{)5.264}$ becomes
$$\begin{array}{r} 16.45 \\ 32\overline{)526.40} \\ -32\downarrow \\ \hline 206 \\ -192\downarrow \\ \hline 14\,4 \\ -12\,8\downarrow \\ \hline 1\,60 \\ -1\,60 \\ \hline 0 \end{array}$$ Insert one 0.

■ Work Practice 5

✓**Concept Check** Is it always true that the number of decimal places in a quotient equals the sum of the decimal places in the dividend and divisor?

Practice 6

Divide: $23.4 \div 0.57$. Round the quotient to the nearest hundredth.

Example 6 Divide: $17.5 \div 0.48$. Round the quotient to the nearest hundredth.

Solution: First we move the decimal points in the divisor and the dividend two places. Then we divide and round the quotient to the nearest hundredth.

hundredths place
$$\begin{array}{r} 36.458 \approx 36.46 \\ 48.\overline{)1750.000} \\ -144\downarrow \\ \hline 310 \\ -288\downarrow \\ \hline 22\,0 \\ -19\,2\downarrow \\ \hline 2\,80 \\ -2\,40\downarrow \\ \hline 400 \\ -384 \\ \hline 16 \end{array}$$

"is approximately"

When rounding to the nearest hundredth, carry the division process out to one more decimal place, the thousandths place.

■ Work Practice 6

Answers

4. 29.8 **5.** 12.35 **6.** 41.05

✓**Concept Check Answer**

no

✓ **Concept Check** If a quotient is to be rounded to the nearest thousandth, to what place should the division be carried out? (Assume that the division carries out to your answer.)

Objective B Estimating When Dividing Decimals

Just as for addition, subtraction, and multiplication of decimals, we can estimate when dividing decimals to check the reasonableness of our answer.

Example 7 Divide: $272.356 \div 28.4$. Then estimate to see whether the proposed result is reasonable.

Solution:

Exact:	Estimate 1		Estimate 2
9.59	9		10
284.)2723.56	30)270	or	30)300
−2556			
167 5			
−142 0			
25 56			
−25 56			
0			

The estimate is 9 or 10, so 9.59 is reasonable.

Work Practice 7

Practice 7
Divide: $722.85 \div 91.5$. Then estimate to see whether the proposed answer is reasonable.

Objective C Dividing Decimals by Powers of 10

As with multiplication, there are patterns that occur when we divide decimals by powers of 10 such as 10, 100, 1000, and so on.

$\dfrac{569.2}{10} = 56.92$ Move the decimal point *1 place* to the *left*.
     ↑———— 1 zero

$\dfrac{569.2}{10,000} = 0.05692$ Move the decimal point *4 places* to the *left*.
     ↑———— 4 zeros

This pattern suggests the following rule:

Dividing Decimals by Powers of 10 Such as 10, 100, or 1000

Move the decimal point of the dividend to the *left* the same number of places as there are *zeros* in the power of 10.

Examples Divide.

8. $\dfrac{786.1}{1000} = 0.7861$ Move the decimal point *3 places* to the *left*.
         ↑———— 3 zeros

9. $\dfrac{0.12}{10} = 0.012$ Move the decimal point *1 place* to the *left*.
         ↑———— 1 zero

Work Practice 8–9

Practice 8–9
Divide.

8. $\dfrac{362.1}{1000}$ **9.** $-\dfrac{0.49}{10}$

Answers
7. Exact: 7.9; an estimate: 8
8. 0.3621 **9.** −0.049

✓ **Concept Check Answer**
ten-thousandths place

Objective D Using Decimals as Replacement Values

Practice 10
Evaluate $x \div y$ for $x = 0.035$ and $y = 0.02$.

Example 10 Evaluate $x \div y$ for $x = 2.5$ and $y = 0.05$.

Solution: Replace x with 2.5 and y with 0.05.

$$x \div y = 2.5 \div 0.05 \quad 0.05\overline{)2.5} \text{ becomes } 5\overline{)250}$$
$$= 50$$

Work Practice 10

Practice 11
Is 39 a solution of the equation $\dfrac{x}{100} = 3.9$?

Example 11 Is 720 a solution of the equation $\dfrac{y}{100} = 7.2$?

Solution: Replace y with 720 to see if a true statement results.

$$\dfrac{y}{100} = 7.2 \quad \text{Original equation}$$
$$\dfrac{720}{100} \stackrel{?}{=} 7.2 \quad \text{Replace } y \text{ with 720.}$$
$$7.2 = 7.2 \quad \text{True}$$

Since $7.2 = 7.2$ is a true statement, 720 is a solution of the equation.

Work Practice 11

Objective E Solving Problems by Dividing Decimals

Many real-life problems involve dividing decimals.

Practice 12
A bag of fertilizer covers 1250 square feet of lawn. Tim Parker's lawn measures 14,800 square feet. How many bags of fertilizer does he need? If he can buy only whole bags of fertilizer, how many whole bags does he need?

Example 12 Calculating Materials Needed for a Job

A gallon of paint covers a 250-square-foot area. If Betty Adkins wishes to paint wall that measures 1450 square feet, how many gallons of paint does she need? If sh can buy only gallon containers of paint, how many gallon containers does she need

Solution:

1. UNDERSTAND. Read and reread the problem. We need to know how many 250s are in 1450, so we divide.

2. TRANSLATE.

   In words: number of gallons   is   square feet   divided by   square feet per gallon

   Translate: number of gallons = 1450 ÷ 250

3. SOLVE. Let's see if our answer is reasonable by estimating. The dividend 145(rounds to 1500 and the divisor 250 rounds to 300. Then $1500 \div 300 = 5$.

$$\begin{array}{r} 5.8 \\ 250\overline{)1450.0} \\ -1250 \\ \hline 200\ 0 \\ -200\ 0 \\ \hline 0 \end{array}$$

Answers
10. 1.75 **11.** no **12.** 11.84 bags; 12 bags

4. INTERPRET. *Check* your work. Since our estimate is close to our answer of 5, our answer is reasonable. *State* your conclusion: Betty needs 5.8 gallons of paint. If she can buy only gallon containers of paint, she needs 6 gallon containers of paint to complete the job.

Work Practice 12

 Calculator Explorations   Estimation

Calculator errors can easily be made by pressing an incorrect key, holding a key too long, or by not pressing a correct key hard enough. Estimation is a valuable tool that can be used to check calculator results.

Example Use estimation to determine whether the calculator result is reasonable or not. (For example, a result that is not reasonable can occur if proper keys are not pressed.)

Divide: $82.064 \div 23$
Calculator display: 35.68

Solution: Round each number to the nearest 10. Since $80 \div 20 = 4$, the calculator display 35.68 is not reasonable.

Use estimation to determine whether each result is reasonable or not.

1. 102.62×41.8; Result: 428.9516
2. $174.835 \div 47.9$; Result: 3.65
3. $1025.68 - 125.42$; Result: 900.26
4. $562.781 + 2.96$; Result: 858.781

Vocabulary, Readiness & Video Check

Use the choices below to fill in each blank. Some choices may be used more than once, and some not used at all.

| dividend | divisor | quotient | true |
| zeros | left | right | false |

1. In $6.5 \div 5 = 1.3$, the number 1.3 is called the _____, 5 is the _____, and 6.5 is the _____.
2. To check a division exercise, we can perform the following multiplication: quotient · _____ = _____.
3. To divide a decimal number by a power of 10 such as 10, 100, 1000, and so on, we move the decimal point in the number to the _____ the same number of places as there are _____ in the power of 10.
4. True or false: If we replace x with -12.6 and y with 0.3 in the expression $y \div x$, we have $0.3 \div (-12.6)$. _____

Martin-Gay Interactive Videos   Watch the section lecture video and answer the following questions.

See Video 5.4

Objective A 5. From the lecture before ▣ Example 1, what must we make sure the divisor is before dividing decimals? ▶

Objective B 6. From ▣ Example 4, what does estimating especially help us with? ▶

Objective C 7. Why don't we do any actual dividing in ▣ Example 5? ▶

Objective D 8. In ▣ Example 6, 4 does not divide into 1. How does this affect the quotient? ▶

Objective E 9. In ▣ Example 7, why is the division carried to the hundredths place? ▶

5.4 Exercise Set MyLab Math

Objectives A B Mixed Practice *Divide. See Examples 1 through 5 and 7. For those exercises marked, also estimate to see if the answer is reasonable.*

1. $6\overline{)27.6}$
2. $4\overline{)23.6}$
3. $5\overline{)0.47}$
4. $6\overline{)0.51}$
5. $0.06\overline{)18}$
6. $0.04\overline{)20}$
7. $0.42\overline{)3.066}$
8. $0.36\overline{)1.764}$
9. $5.5\overline{)36.3}$
   Exact:
   Estimate:
10. $2.2\overline{)21.78}$
    Exact:
    Estimate:
11. $7.434 \div 18$
12. $8.304 \div 16$
13. $36 \div (-0.06)$
14. $36 \div (-0.04)$
15. Divide -4.2 by -0.6.
16. Divide -3.6 by -0.9.
17. $0.27\overline{)1.296}$
18. $0.34\overline{)2.176}$
19. $0.02\overline{)42}$
20. $0.03\overline{)24}$
21. $4.756 \div 0.82$
22. $3.312 \div 0.92$
23. $-36.3 \div (-6.6)$
24. $-21.78 \div (-9.9)$
25. $7.2\overline{)70.56}$
    Exact:
    Estimate:
26. $6.3\overline{)54.18}$
    Exact:
    Estimate:
27. $5.4\overline{)51.84}$
28. $7.7\overline{)33.88}$
29. $\dfrac{1.215}{0.027}$
30. $\dfrac{3.213}{0.051}$
31. $0.25\overline{)13.648}$
32. $0.75\overline{)49.866}$
33. $3.78\overline{)0.02079}$
34. $2.96\overline{)0.01332}$

Divide. Round the quotients as indicated. See Example 6.

35. Divide: $0.549 \div 0.023$. Round the quotient to the nearest hundredth.
36. Divide: $0.0453 \div 0.98$. Round the quotient to the nearest thousandth.
37. Divide: $68.39 \div 0.6$. Round the quotient to the nearest tenth.
38. Divide: $98.83 \div 3.5$. Round the quotient to the nearest tenth.

Objective C *Divide. See Examples 8 and 9.*

39. $\dfrac{83.397}{100}$
40. $\dfrac{64.423}{100}$
41. $\dfrac{26.87}{10}$
42. $\dfrac{13.49}{10}$
43. $12.9 \div (-1000)$
44. $13.49 \div (-10,000)$

Objectives A C Mixed Practice *Divide. See Examples 1 through 5, 8, and 9.*

45. $7\overline{)88.2}$
46. $9\overline{)130.5}$
47. $\dfrac{13.1}{10}$
48. $\dfrac{17.7}{10}$
49. $\dfrac{456.25}{10,000}$
50. $\dfrac{986.11}{10,000}$
51. $1.239 \div 3$
52. $0.54 \div 12$

53. Divide 4.8 by −0.6. **54.** Divide 4.9 by −0.7. **55.** −1.224 ÷ 0.17 **56.** −1.344 ÷ 0.42

57. Divide 42 by 0.03. **58.** Divide 27 by 0.03. **59.** Divide −18 by −0.6. **60.** Divide 20 by 0.4.

61. Divide 87 by −0.0015. **62.** Divide 35 by −0.0007. **63.** −1.104 ÷ 1.6 **64.** −2.156 ÷ 0.98

65. −2.4 ÷ (−100) **66.** −86.79 ÷ (−1000) **67.** $\dfrac{4.615}{0.071}$ **68.** $\dfrac{23.8}{0.035}$

Objective D *Evaluate each expression for $x = 5.65$, $y = -0.8$, and $z = 4.52$. See Example 10.*

69. $z \div y$ **70.** $z \div x$ **71.** $x \div y$ **72.** $y \div 2$

Determine whether the given values are solutions of the given equations. See Example 11.

73. $\dfrac{x}{4} = 3.04$; $x = 12.16$ **74.** $\dfrac{y}{8} = 0.89$; $y = 7.12$ **75.** $\dfrac{z}{100} = 0.8$; $z = 8$ **76.** $\dfrac{x}{10} = 0.23$; $x = 23$

Objective E *Solve. For Exercises 77 and 78, the solutions have been started for you. See Example 12.*

77. A new homeowner is painting the walls of a room. The walls have a total area of 546 square feet. A quart of paint covers 52 square feet. If the paint is sold in whole quarts only, how many quarts are needed?

Start the solution:

1. UNDERSTAND the problem. Reread it as many times as needed.
2. TRANSLATE into an equation. (Fill in the blanks.)

   number of quarts is square feet divided by square feet per quart
   ↓ ↓ ↓ ↓ ↓
   number of quarts = _____ ÷ _____

3. SOLVE. Don't forget to round up your quotient.
4. INTERPRET.

78. A shipping box can hold 36 books. If 486 books must be shipped, how many boxes are needed?

Start the solution:

1. UNDERSTAND the problem. Reread it as many times as needed.
2. TRANSLATE into an equation. (Fill in the blanks.)

   number of boxes is number of books divided by books per box
   ↓ ↓ ↓ ↓ ↓
   number of boxes = _____ ÷ _____

3. SOLVE. Don't forget to round up your quotient.
4. INTERPRET.

79. There are approximately 39.37 inches in 1 meter. How many meters, to the nearest tenth of a meter, are there in 200 inches?

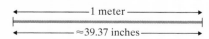

80. There are 2.54 centimeters in 1 inch. How many inches are there in 50 centimeters? Round to the nearest tenth.

81. In the United States, an average child will wear down 730 crayons by his or her tenth birthday. Find the number of boxes of 64 crayons this is equivalent to. Round to the nearest tenth. (*Source:* Binney & Smith Inc.)

82. In 2016, American farmers received an average of $84.20 per hundred pounds of chicken. What was the average price per pound for chickens? Round to the nearest cent. (*Source:* National Agricultural Statistics Service)

A child is to receive a dose of 0.5 teaspoon of cough medicine every 4 to 6 hours. If the bottle contains 4 fluid ounces, answer Exercises 83 through 86.

83. A fluid ounce equals 6 teaspoons. How many teaspoons are in 4 fluid ounces?

84. The bottle of medicine contains how many doses for the child? (*Hint:* See Exercise **83**.)

85. If the child takes a dose every four hours, how many days will the medicine last?

86. If the child takes a dose every six hours, how many days will the medicine last?

87. Americans ages 16–19 drive, on average, 7624 miles per year. About how many miles each week is that? Round to the nearest tenth. (*Note:* There are 52 weeks in a year.) (*Source:* Federal Highway Administration)

88. Drake Saucier was interested in the gas mileage on his "new" used car. He filled the tank, drove 423.8 miles, and filled the tank again. When he refilled the tank, it took 19.35 gallons of gas. Calculate the miles per gallon for Drake's car. Round to the nearest tenth.

89. The book *Harry Potter and the Deathly Hallows* was released to the public on July 21, 2007. Booksellers in the United States sold approximately 8292 thousand copies in the first 24 hours after release. If the same number of books was sold each hour, calculate the number of books sold each hour in the United States for that first day.

90. The greatest money winner in men's professional golf from January 2016 through November 2016 was Hideki Matsuyama. He earned approximately $2,524,750. Suppose he had earned this working 40 hours each week for a year. Determine his hourly wage to the nearest cent. (*Note:* There are 52 weeks in a year) (*Source:* Professional Golfers' Association)

Review

Perform the indicated operation. See Sections 4.3 and 4.5.

91. $\dfrac{3}{5} \cdot \dfrac{7}{10}$

92. $\dfrac{3}{5} \div \dfrac{7}{10}$

93. $\dfrac{3}{5} - \dfrac{7}{10}$

94. $-\dfrac{3}{4} - \dfrac{1}{14}$

Concept Extensions

Mixed Practice (Sections 5.2, 5.3, 5.4) *Perform the indicated operation.*

95. $1.278 \div 0.3$

96. 1.278×0.3

97. $1.278 + 0.3$

98. $1.278 - 0.3$

99. $(-8.6)(3.1)$

100. $7.2 + 0.05 + 49.1$

101. $\begin{array}{r} 1000 \\ -95.71 \\ \hline \end{array}$

102. $\dfrac{87.2}{-10{,}000}$

Choose the best estimate.

103. 8.62×41.7
a. 36
b. 32
c. 360
d. 3.6

104. $1.437 + 20.69$
a. 34
b. 22
c. 3.4
d. 2.2

105. $78.6 \div 97$
a. 7.86
b. 0.786
c. 786
d. 7860

106. $302.729 - 28.697$
a. 270
b. 20
c. 27
d. 300

Recall from Section 1.6 that the average of a list of numbers is their total divided by how many numbers there are in the list. Use this procedure to find the average of the test scores listed in Exercises 107 and 108. If necessary, round to the nearest tenth.

107. 86, 78, 91, 87

108. 56, 75, 80

109. The area of a rectangle is 38.7 square feet. If its width is 4.5 feet, find its length.

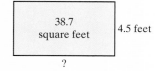

110. The perimeter of a square is 180.8 centimeters. Find the length of a side.

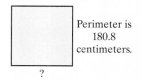

111. When dividing decimals, describe the process you use to place the decimal point in the quotient.

112. In your own words, describe how to quickly divide a number by a power of 10 such as 10, 100, 1000, etc.

To convert wind speeds in miles per hour to knots, divide by 1.15. Use this information and the Saffir-Simpson Hurricane Intensity chart below to answer Exercises 113 and 114. Round to the nearest tenth.

Saffir-Simpson Hurricane Intensity Scale				
Category	Wind Speed	Barometric Pressure [inches of mercury (Hg)]	Storm Surge	Damage Potential
1 (Weak)	75–95 mph	≥28.94 in.	4–5 ft	Minimal damage to vegetation
2 (Moderate)	96–110 mph	28.50–28.93 in.	6–8 ft	Moderate damage to houses
3 (Strong)	111–130 mph	27.91–28.49 in.	9–12 ft	Extensive damage to small buildings
4 (Very Strong)	131–155 mph	27.17–27.90 in.	13–18 ft	Extreme structural damage
5 (Devastating)	>155 mph	<27.17 in.	>18 ft	Catastrophic building failures possible

113. The chart gives wind speeds in miles per hour. What is the range of wind speeds for a Category 1 hurricane in knots?

114. What is the range of wind speeds for a Category 4 hurricane in knots?

115. A rancher is building a horse corral that's shaped like a rectangle with a width of 24.3 meters. He plans to make a four-wire fence; that is, he will string four wires around the corral. If he plans to use all of his 412.8 meters of wire, find the length of the corral he can construct.

116. A college student signed up for a new credit card that guarantees her no interest charges on transferred balances for a year. She transferred over a $2523.86 balance from her old credit card. Her minimum payment is $185.35 per month. If she pays only the minimum, will she pay off her balance before interest charges start again?

Integrated Review

Sections 5.1–5.4

Operations on Decimals

Answers

1. ______
2. ______
3. ______
4. ______
5. ______
6. ______
7. ______
8. ______
9. ______
10. ______
11. ______
12. ______
13. ______
14. ______
15. ______
16. ______
17. ______
18. ______
19. ______
20. ______
21. ______
22. ______
23. ______
24. ______

Perform the indicated operation.

1. $1.6 + 0.97$

2. $3.2 + 0.85$

3. $9.8 - 0.9$

4. $10.2 - 6.7$

5. $\begin{array}{r} 0.8 \\ \times 0.2 \\ \hline \end{array}$

6. $\begin{array}{r} 0.6 \\ \times 0.4 \\ \hline \end{array}$

7. $8\overline{)2.16}$

8. $6\overline{)3.12}$

9. $(9.6)(-0.5)$

10. $(-8.7)(-0.7)$

11. $\begin{array}{r} 123.6 \\ -\ 48.04 \\ \hline \end{array}$

12. $\begin{array}{r} 325.2 \\ -\ 36.08 \\ \hline \end{array}$

13. $-25 + 0.026$

14. $0.125 + (-44)$

15. $29.24 \div (-3.4)$

16. $-10.26 \div (-1.9$

17. -2.8×100

18. 1.6×1000

19. $\begin{array}{r} 96.21 \\ 7.028 \\ +\ 121.7 \\ \hline \end{array}$

20. $\begin{array}{r} 0.268 \\ 1.93 \\ +\ 142.881 \\ \hline \end{array}$

21. $-25.76 \div (-46)$

22. $-27.09 \div 43$

23. $\begin{array}{r} 12.004 \\ \times \quad 2.3 \\ \hline \end{array}$

24. $\begin{array}{r} 28.006 \\ \times \quad 5.2 \\ \hline \end{array}$

25. Subtract 4.6 from 10. **26.** Subtract 18 from 0.26. **27.** $-268.19 - 146.25$

28. $-860.18 - 434.85$ **29.** $\dfrac{2.958}{-0.087}$ **30.** $\dfrac{-1.708}{0.061}$

31. $160 - 43.19$ **32.** $120 - 101.21$ **33.** 15.62×10

34. $15.62 \div 10$ **35.** $15.62 + 10$ **36.** $15.62 - 10$

37. Find the distance in miles between Garden City, Kansas, and Wichita, Kansas. Next, estimate the distance by rounding each given distance to the nearest ten.

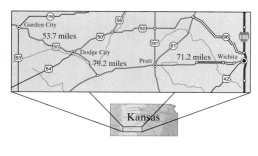

38. It costs $7.96 to send a 5-pound package locally via parcel post at a U.S. Post Office. To send the same package Priority Mail costs $9.85. How much more does it cost to send a package as Priority Mail? (*Source:* United States Postal Service)

39. In 2011, sales of Blu-ray Discs were $2 billion, but DVD sales dropped to $6.8 billion. Find the total spent to buy Blu-ray Discs or DVDs in 2011. Write the total in billions of dollars and also in standard notation. (*Source: USA Today*)

5.5 Fractions, Decimals, and Order of Operations

Objectives

A Write Fractions as Decimals.

B Compare Decimals and Fractions.

C Simplify Expressions Containing Decimals and Fractions Using Order of Operations.

D Solve Area Problems Containing Fractions and Decimals.

E Evaluate Expressions Given Decimal Replacement Values.

Objective A Writing Fractions as Decimals

To write a fraction as a decimal, we interpret the fraction bar to mean division and find the quotient.

Writing Fractions as Decimals

To write a fraction as a decimal, divide the numerator by the denominator.

Example 1 Write $\frac{1}{4}$ as a decimal.

Solution: $\frac{1}{4} = 1 \div 4$

$$\begin{array}{r} 0.25 \\ 4\overline{)1.00} \\ -8 \\ \hline 20 \\ -20 \\ \hline 0 \end{array}$$

Thus, $\frac{1}{4}$ written as a decimal is 0.25.

■ Work Practice 1

Practice 1

a. Write $\frac{2}{5}$ as a decimal.

b. Write $\frac{9}{40}$ as a decimal.

Example 2 Write $-\frac{5}{8}$ as a decimal.

Solution: $-\frac{5}{8} = -(5 \div 8) = -0.625$

$$\begin{array}{r} 0.625 \\ 8\overline{)5.000} \\ -4\,8 \\ \hline 20 \\ -16 \\ \hline 40 \\ -40 \\ \hline 0 \end{array}$$

■ Work Practice 2

Practice 2

Write $-\frac{3}{8}$ as a decimal.

Example 3 Write $\frac{2}{3}$ as a decimal.

Solution:
$$\begin{array}{r} 0.666\ldots \\ 3\overline{)2.000} \\ -1\,8 \\ \hline 20 \\ -18 \\ \hline 20 \\ -18 \\ \hline 2 \end{array}$$

Remainder is 2, then 0 is brought down.

Remainder is 2, then 0 is brought down.

Remainder is 2.

Practice 3

a. Write $\frac{5}{6}$ as a decimal.

b. Write $\frac{2}{9}$ as a decimal.

Answers

1. a. 0.4 **b.** 0.225 **2.** −0.375
3. a. $0.8\overline{3}$ **b.** $0.\overline{2}$

Section 5.5 | Fractions, Decimals, and Order of Operations

Notice that the digit 2 keeps occurring as the remainder. This will continue so that the digit 6 will keep repeating in the quotient. We place a bar over the digit 6 to indicate that it repeats.

$$\frac{2}{3} = 0.666\ldots = 0.\overline{6}$$

We can also write a decimal approximation for $\frac{2}{3}$. For example, $\frac{2}{3}$ rounded to the nearest hundredth is 0.67. This can be written as $\frac{2}{3} \approx 0.67$.

■ Work Practice 3

Example 4 Write $\frac{22}{7}$ as a decimal. (Recall that the fraction $\frac{22}{7}$ is an approximation for π.) Round to the nearest hundredth.

Solution:
$$\begin{array}{r} 3.142 \approx 3.14 \\ 7\overline{)22.000} \\ -21 \\ \hline 10 \\ -7 \\ \hline 30 \\ -28 \\ \hline 20 \\ -14 \\ \hline 6 \end{array}$$

Carry the division out to the thousandths place.

The fraction $\frac{22}{7}$ in decimal form is approximately 3.14.

■ Work Practice 4

Practice 4

Write $\frac{28}{13}$ as a decimal. Round to the nearest thousandth.

Example 5 Write $2\frac{3}{16}$ as a decimal.

Solution:

Option 1. Write the fractional part only as a decimal.

$$\frac{3}{16} \longrightarrow \begin{array}{r} 0.1875 \\ 16\overline{)3.0000} \\ -1\,6 \\ \hline 1\,40 \\ -1\,28 \\ \hline 120 \\ -112 \\ \hline 80 \\ -80 \\ \hline 0 \end{array}$$

Thus $2\frac{3}{16} = 2.1875$

Option 2. Write $2\frac{3}{16}$ as an improper fraction, and divide.

$$2\frac{3}{16} = \frac{35}{16} \longrightarrow \begin{array}{r} 2.1875 \\ 16\overline{)35.0000} \\ -32 \\ \hline 3\,0 \\ -1\,6 \\ \hline 1\,40 \\ -1\,28 \\ \hline 120 \\ -112 \\ \hline 80 \\ -80 \\ \hline 0 \end{array}$$

Practice 5

Write $3\frac{5}{16}$ as a decimal.

■ Work Practice 5

Answers
4. 2.154 **5.** 3.3125

Some fractions may be written as decimals using our knowledge of decimals. From Section 5.1, we know that if the denominator of a fraction is 10, 100, 1000, or so on, we can immediately write the fraction as a decimal. For example,

$$\frac{4}{10} = 0.4, \quad \frac{12}{100} = 0.12, \text{ and so on}$$

Practice 6

Write $\frac{3}{5}$ as a decimal.

Example 6 Write $\frac{4}{5}$ as a decimal.

Solution: Let's write $\frac{4}{5}$ as an equivalent fraction with a denominator of 10.

$$\frac{4}{5} = \frac{4}{5} \cdot \frac{2}{2} = \frac{8}{10} = 0.8$$

Work Practice 6

Practice 7

Write $\frac{3}{50}$ as a decimal.

Example 7 Write $\frac{1}{25}$ as a decimal.

Solution: $\frac{1}{25} = \frac{1}{25} \cdot \frac{4}{4} = \frac{4}{100} = 0.04$

Work Practice 7

✓**Concept Check** Suppose you are writing the fraction $\frac{9}{16}$ as a decimal. How do you know you have made a mistake if your answer is 1.735?

Objective B Comparing Decimals and Fractions

Now we can compare decimals and fractions by writing fractions as equivalent decimals.

Practice 8

Insert <, >, or = to form a true statement.

$$\frac{1}{5} \quad 0.25$$

Example 8 Insert <, >, or = to form a true statement.

$$\frac{1}{8} \quad 0.12$$

Solution: First we write $\frac{1}{8}$ as an equivalent decimal. Then we compare decimal places.

```
   0.125
8)1.000
  −8
   20
  −16
   40
  −40
    0
```

Original numbers	$\frac{1}{8}$	0.12
Decimals	0.125	0.120
Compare	0.125 > 0.12	

Thus, $\frac{1}{8} > 0.12$

Work Practice 8

Answers

6. 0.6 **7.** 0.06 **8.** <

✓**Concept Check Answer**

$\frac{9}{16}$ is less than 1 while 1.735 is greater than 1.

Section 5.5 | Fractions, Decimals, and Order of Operations

Example 9
Insert <, >, or = to form a true statement.

$0.\overline{7}$ $\dfrac{7}{9}$

Solution: We write $\dfrac{7}{9}$ as a decimal and then compare.

$$\begin{array}{r} 0.77\ldots \\ 9\overline{)7.00} \\ -6.3 \\ \hline 70 \\ -63 \\ \hline 7 \end{array} = 0.\overline{7}$$

Original numbers	$0.\overline{7}$	$\dfrac{7}{9}$
Decimals	$0.\overline{7}$	$0.\overline{7}$
Compare	$0.\overline{7} = 0.\overline{7}$	

Thus, $0.\overline{7} = \dfrac{7}{9}$

Work Practice 9

Practice 9
Insert <, >, or = to form a true statement.

a. $\dfrac{1}{2}$ 0.54   b. $0.\overline{5}$ $\dfrac{5}{9}$

c. $\dfrac{5}{7}$ 0.72

Example 10
Write the numbers in order from smallest to largest.

$\dfrac{9}{20}, \dfrac{4}{9}, 0.456$

Solution:

Original numbers	$\dfrac{9}{20}$	$\dfrac{4}{9}$	0.456
Decimals	0.450	0.444…	0.456
Compare in order	2nd	1st	3rd

Written in order, we have

1st 2nd 3rd
↓ ↓ ↓
$\dfrac{4}{9}, \dfrac{9}{20}, 0.456$

Work Practice 10

Practice 10
Write the numbers in order from smallest to largest.

a. $\dfrac{1}{3}, 0.302, \dfrac{3}{8}$

b. $1.26, 1\dfrac{1}{4}, 1\dfrac{2}{5}$

c. $0.4, 0.41, \dfrac{3}{7}$

Objective C   Simplifying Expressions with Decimals and Fractions

In the remaining examples, we will review the order of operations by simplifying expressions that contain decimals.

Order of Operations

1. Perform all operations within parentheses ( ), brackets [ ], or other grouping symbols such as fraction bars.
2. Evaluate any expressions with exponents.
3. Multiply or divide in order from left to right.
4. Add or subtract in order from left to right.

Answers
9. a. <   b. =   c. <
10. a. $0.302, \dfrac{1}{3}, \dfrac{3}{8}$ b. $1\dfrac{1}{4}, 1.26, 1\dfrac{2}{5}$
c. $0.4, 0.41, \dfrac{3}{7}$

Practice 11
Simplify: $897.8 \div 100 \times 10$

Example 11 Simplify: $723.6 \div 1000 \times 10$

Solution: Multiply or divide in order from left to right.

$723.6 \div 1000 \times 10 = 0.7236 \times 10$ Divide.
$ = 7.236$ Multiply.

■ Work Practice 11

Practice 12
Simplify: $-8.69(3.2 - 1.8)$

Example 12 Simplify: $-0.5(8.6 - 1.2)$

Solution: According to the order of operations, we simplify inside the parentheses first.

$-0.5(8.6 - 1.2) = -0.5(7.4)$ Subtract.
$ = -3.7$ Multiply.

■ Work Practice 12

Practice 13
Simplify: $(-0.7)^2 + 2.1$

Example 13 Simplify: $(-1.3)^2 + 2.4$

Solution: Recall the meaning of an exponent.

$(-1.3)^2 + 2.4 = (-1.3)(-1.3) + 2.4$ Use the definition of an exponent.
$ = 1.69 + 2.4$ Multiply. The product of two negative numbers is a positive number.
$ = 4.09$ Add.

■ Work Practice 13

Practice 14
Simplify: $\dfrac{20.06 - (1.2)^2 \div 10}{0.02}$

Example 14 Simplify: $\dfrac{5.68 + (0.9)^2 \div 100}{0.2}$

Solution: First we simplify the numerator of the fraction. Then we divide.

$\dfrac{5.68 + (0.9)^2 \div 100}{0.2} = \dfrac{5.68 + 0.81 \div 100}{0.2}$ Simplify $(0.9)^2$.

$\phantom{\dfrac{5.68 + (0.9)^2 \div 100}{0.2}} = \dfrac{5.68 + 0.0081}{0.2}$ Divide.

$\phantom{\dfrac{5.68 + (0.9)^2 \div 100}{0.2}} = \dfrac{5.6881}{0.2}$ Add.

$\phantom{\dfrac{5.68 + (0.9)^2 \div 100}{0.2}} = 28.4405$ Divide.

■ Work Practice 14

Objective D Solving Area Problems Containing Fractions and Decimals

Sometimes real-life problems contain both fractions and decimals. In the next example, we review the area of a triangle, and when values are substituted, the result may be an expression containing both fractions and decimals.

Answers
11. 89.78   **12.** −12.166
13. 2.59   **14.** 995.8

Section 5.5 | Fractions, Decimals, and Order of Operations

Example 15 The area of a triangle is Area = $\frac{1}{2}$ · base · height. Find the area of the triangle shown.

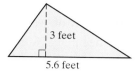

Practice 15
Find the area of the triangle.

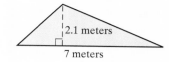

Solution:

Area = $\frac{1}{2}$ · base · height

= $\frac{1}{2}$ · 5.6 · 3

= 0.5 · 5.6 · 3   Write $\frac{1}{2}$ as the decimal 0.5.

= 8.4

The area of the triangle is 8.4 square feet.

Work Practice 15

Objective E Using Decimals as Replacement Values

Example 16 Evaluate $-2x + 5$ for $x = 3.8$.

Solution: Replace x with 3.8 in the expression $-2x + 5$ and simplify.

$-2x + 5 = -2(3.8) + 5$ Replace x with 3.8.

$= -7.6 + 5$ Multiply.

$= -2.6$ Add.

Work Practice 16

Practice 16
Evaluate $1.7y - 2$ for $y = 2.3$.

Answers
15. 7.35 sq m   16. 1.91

Vocabulary, Readiness & Video Check

Answer each exercise "true" or "false."

1. The number $0.\overline{5}$ means 0.555. _____

2. To write $\frac{9}{19}$ as a decimal, perform the division $19\overline{)9}$. _____

3. $(-1.2)^2$ means $(-1.2)(-1.2)$ or -1.44. _____

4. To simplify $8.6(4.8 - 9.6)$, we first subtract. _____

Martin-Gay Interactive Videos

See Video 5.5

Watch the section lecture video and answer the following questions.

Objective A 5. In Example 2, why is the bar placed over just the 6?

Objective B 6. In Example 3, why do we write the fraction as a decimal rather than the decimal as a fraction?

Objective C 7. In Example 4, besides meaning division, what other purpose does the fraction bar serve?

Objective D 8. What formula is used to solve Example 5? What is the final answer?

Objective E 9. In Example 6, once all replacement values are put into the variable expression, what is the resulting expression to evaluate?

5.5 Exercise Set MyLab Math

Objective A *Write each number as a decimal. See Examples 1 through 7.*

1. $\dfrac{1}{5}$
2. $\dfrac{1}{20}$
3. $\dfrac{17}{25}$
4. $\dfrac{13}{25}$
5. $\dfrac{3}{4}$

6. $\dfrac{1}{8}$
7. $-\dfrac{2}{25}$
8. $-\dfrac{3}{25}$
9. $\dfrac{9}{4}$
10. $\dfrac{8}{5}$

11. $\dfrac{11}{12}$
12. $\dfrac{5}{12}$
13. $\dfrac{17}{40}$
14. $\dfrac{19}{25}$
15. $\dfrac{18}{40}$

16. $\dfrac{31}{40}$
17. $-\dfrac{1}{3}$
18. $-\dfrac{7}{9}$
19. $\dfrac{7}{16}$
20. $\dfrac{9}{16}$

21. $\dfrac{7}{11}$
22. $\dfrac{9}{11}$
23. $5\dfrac{17}{20}$
24. $4\dfrac{7}{8}$
25. $\dfrac{78}{125}$
26. $\dfrac{159}{375}$

Round each number as indicated. See Example 4.

27. Round your decimal answer to Exercise **17** to the nearest hundredth.

28. Round your decimal answer to Exercise **18** to the nearest hundredth.

29. Round your decimal answer to Exercise **19** to the nearest hundredth.

30. Round your decimal answer to Exercise **20** to the nearest hundredth.

31. Round your decimal answer to Exercise **21** to the nearest tenth.

32. Round your decimal answer to Exercise **22** to the nearest tenth.

Write each fraction as a decimal. If necessary, round to the nearest hundredth. See Examples 1 through 7.

33. Of the U.S. mountains that are over 14,000 feet in elevation, $\dfrac{56}{91}$ are located in Colorado. (*Source:* U.S. Geological Survey)

34. About $\dfrac{9}{20}$ of all U.S. citizens have type O blood. (*Source:* American Red Cross Biomedical Services)

35. About $\frac{43}{50}$ of Americans are Internet users. (*Source:* Digitalcenter)

36. About $\frac{14}{25}$ of Americans use the Internet through a wireless device. (*Source:* Digitalcenter)

37. When first launched, the Hubble Space Telescope's primary mirror was out of shape on the edges by $\frac{1}{50}$ of a human hair. This very small defect made it difficult to focus faint objects being viewed. Because the HST was in low Earth orbit, it was serviced by a shuttle and the defect was corrected.

38. The two mirrors currently in use in the Hubble Space Telescope were ground so that they do not deviate from a perfect curve by more than $\frac{1}{800,000}$ of an inch. Do not round this number.

Objective B *Insert* $<, >$, *or* $=$ *to form a true statement. See Examples 8 and 9.*

39. 0.562   0.569

40. 0.983   0.988

41. 0.215 $\frac{43}{200}$

42. $\frac{29}{40}$ 0.725

43. -0.0932 -0.0923

44. -0.00563 -0.00536

45. $0.\overline{6}$ $\frac{5}{6}$

46. $0.\overline{1}$ $\frac{2}{17}$

47. $\frac{51}{91}$ $0.56\overline{4}$

48. $0.58\overline{3}$ $\frac{6}{11}$

49. $\frac{4}{7}$ 0.14

50. $\frac{5}{9}$ 0.557

51. 1.38 $\frac{18}{13}$

52. 0.372 $\frac{22}{59}$

53. 7.123 $\frac{456}{64}$

54. 12.713 $\frac{89}{7}$

Write the numbers in order from smallest to largest. See Example 10.

55. 0.34, 0.35, 0.32

56. 0.47, 0.42, 0.40

57. 0.49, 0.491, 0.498

58. 0.72, 0.727, 0.728

59. 5.23, $\frac{42}{8}$, 5.34

60. 7.56, $\frac{67}{9}$, 7.562

61. $\frac{5}{8}$, 0.612, 0.649

62. $\frac{5}{6}$, 0.821, 0.849

Objective C *Simplify each expression. See Examples 11 through 14.*

63. $(0.3)^2 + 0.5$

64. $(-2.5)(3) - 4.7$

65. $\dfrac{1 + 0.8}{-0.6}$

66. $(-0.05)^2 + 3.13$

67. $(-2.3)^2(0.3 + 0.7)$

68. $(8.2)(100) - (8.2)(10)$

69. $(5.6 - 2.3)(2.4 + 0.4)$

70. $\dfrac{0.222 - 2.13}{12}$

71. $\dfrac{(4.5)^2}{100}$

72. $0.9(5.6 - 6.5)$

73. $\dfrac{7 + 0.74}{-6}$

74. $(1.5)^2 + 0.5$

Find the value of each expression. Give the result as a decimal. See Examples 11 through 14.

75. $\dfrac{1}{5} - 2(7.8)$

76. $\dfrac{3}{4} - (9.6)(5)$

77. $\dfrac{1}{4}(-9.6 - 5.2)$

78. $\dfrac{3}{8}(4.7 - 5.9)$

Objective D *Find the area of each triangle or rectangle. See Example 15.*

△ 79.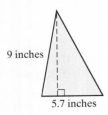

△ 80.

81.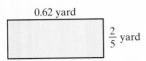

△ 82.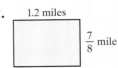

Objective E *Evaluate each expression for $x = 6$, $y = 0.3$, and $z = -2.4$. See Example 16.*

83. z^2 84. y^2 85. $x - y$ 86. $x - z$ 87. $4y - z$ 88. $\dfrac{x}{y} + 2z$

Review

Simplify. See Sections 4.3 and 4.5.

89. $\dfrac{9}{10} + \dfrac{16}{25}$ 90. $\dfrac{4}{11} - \dfrac{19}{22}$ 91. $\left(\dfrac{2}{5}\right)\left(\dfrac{5}{2}\right)^2$ 92. $\left(\dfrac{2}{3}\right)^2 \left(\dfrac{3}{2}\right)^3$

Concept Extensions

Without calculating, describe each number as < 1, $= 1$, or > 1. See the Concept Check in this section.

93. 1.0 94. 1.0000 95. 1.00001 96. $\dfrac{101}{99}$ 97. $\dfrac{99}{100}$ 98. $\dfrac{99}{99}$

In 2016, there were 10,181 commercial radio stations in the United States. The most popular formats are listed in the table along with their counts. Use this graph to answer Exercises 99 through 102.

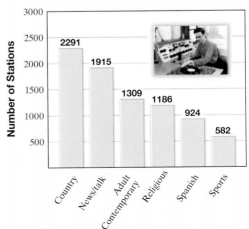

Top Commercial Radio Station Formats in 2016
Format (Total stations: 10,181)

99. Write as a decimal the fraction of radio stations that are sports. Round to the nearest thousandth.

100. Write as a decimal the fraction of radio stations with a Spanish format. Round to the nearest hundredth.

101. Estimate, by rounding each number in the table to the nearest hundred, the total number of stations with the top six formats in 2016.

102. Use your estimate from Exercise **101** to write the fraction of radio stations accounted for by the top six formats as a decimal. Round to the nearest hundredth.

103. Describe two ways to write fractions as decimals.

104. Describe two ways to write mixed numbers as decimals.

5.6 Solving Equations Containing Decimals

Objective A Solving Equations Containing Decimals

In this section, we continue our work with decimals and algebra by solving equations containing decimals. First, we review the steps given earlier for solving an equation.

Objective

A Solve Equations Containing Decimals.

Steps for Solving an Equation in x

Step 1: If fractions are present, multiply both sides of the equation by the LCD of the fractions.
Step 2: If parentheses are present, use the distributive property.
Step 3: Combine any like terms on each side of the equation.
Step 4: Use the addition property of equality to rewrite the equation so that variable terms are on one side of the equation and constant terms are on the other side.
Step 5: Divide both sides by the numerical coefficient of x to solve.
Step 6: Check your answer in the **original equation**.

Example 1 Solve: $x - 1.5 = 8$

Solution: Steps 1 through 3 are not needed for this equation, so we begin with Step 4. To get x alone on one side of the equation, add 1.5 to both sides.

$$x - 1.5 = 8 \quad \text{Original equation}$$
$$x - 1.5 + 1.5 = 8 + 1.5 \quad \text{Add 1.5 to both sides.}$$
$$x = 9.5 \quad \text{Simplify.}$$

Check: To check, replace x with 9.5 in the *original equation*.

$$x - 1.5 = 8 \quad \text{Original equation}$$
$$9.5 - 1.5 \stackrel{?}{=} 8 \quad \text{Replace } x \text{ with 9.5.}$$
$$8 = 8 \quad \text{True}$$

Since $8 = 8$ is a true statement, 9.5 is the solution of the equation.

Practice 1
Solve: $z + 0.9 = 1.3$

Work Practice 1

Example 2 Solve: $-2y = 6.7$

Solution: Steps 1 through 4 are not needed for this equation, so we begin with Step 5. To solve for y, divide both sides by the coefficient of y, which is -2.

$$-2y = 6.7 \quad \text{Original equation}$$
$$\frac{-2y}{-2} = \frac{6.7}{-2} \quad \text{Divide both sides by } -2.$$
$$y = -3.35 \quad \text{Simplify.}$$

Check: To check, replace y with -3.35 in the original equation.

$$-2y = 6.7 \quad \text{Original equation}$$
$$-2(-3.35) \stackrel{?}{=} 6.7 \quad \text{Replace } y \text{ with } -3.35.$$
$$6.7 = 6.7 \quad \text{True}$$

Thus -3.35 is the solution of the equation $-2y = 6.7$.

Practice 2
Solve: $0.17x = -0.34$

Work Practice 2

Answers
1. 0.4 **2.** -2

Practice 3

Solve: $2.9 = 1.7 + 0.3x$

Example 3 Solve: $1.2x + 5.8 = 8.2$

Solution: We begin with Step 4 and get the variable term alone by subtracting 5.8 from both sides.

$$1.2x + 5.8 = 8.2$$
$$1.2x + 5.8 - 5.8 = 8.2 - 5.8 \quad \text{Subtract 5.8 from both sides.}$$
$$1.2x = 2.4 \quad \text{Simplify.}$$
$$\frac{1.2x}{1.2} = \frac{2.4}{1.2} \quad \text{Divide both sides by 1.2.}$$
$$x = 2 \quad \text{Simplify.}$$

To check, replace x with 2 in the original equation. The solution is 2.

■ Work Practice 3

Practice 4

Solve: $8x + 4.2 = 10x + 11.6$

Example 4 Solve: $7x + 3.2 = 4x - 1.6$

Solution: We start with Step 4 to get variable terms on one side and constant terms on the other.

$$7x + 3.2 = 4x - 1.6$$
$$7x + 3.2 - 3.2 = 4x - 1.6 - 3.2 \quad \text{Subtract 3.2 from both sides.}$$
$$7x = 4x - 4.8 \quad \text{Simplify.}$$
$$7x - 4x = 4x - 4.8 - 4x \quad \text{Subtract } 4x \text{ from both sides.}$$
$$3x = -4.8 \quad \text{Simplify.}$$
$$\frac{3x}{3} = -\frac{4.8}{3} \quad \text{Divide both sides by 3.}$$
$$x = -1.6 \quad \text{Simplify.}$$

Check to see that -1.6 is the solution.

■ Work Practice 4

Practice 5

Solve: $6.3 - 5x = 3(x + 2.9)$

Example 5 Solve: $5(x - 0.36) = -x + 2.4$

Solution: First use the distributive property to distribute the factor 5.

$$5(x - 0.36) = -x + 2.4 \quad \text{Original equation}$$
$$5x - 1.8 = -x + 2.4 \quad \text{Apply the distributive property.}$$

Next, get the variable terms alone on the left side of the equation by adding 1.8 to both sides of the equation and then adding x to both sides of the equation.

$$5x - 1.8 + 1.8 = -x + 2.4 + 1.8 \quad \text{Add 1.8 to both sides.}$$
$$5x = -x + 4.2 \quad \text{Simplify.}$$
$$5x + x = -x + 4.2 + x \quad \text{Add } x \text{ to both sides.}$$
$$6x = 4.2 \quad \text{Simplify.}$$
$$\frac{6x}{6} = \frac{4.2}{6} \quad \text{Divide both sides by 6.}$$
$$x = 0.7 \quad \text{Simplify.}$$

To verify that 0.7 is the solution, replace x with 0.7 in the original equation.

■ Work Practice 5

Answers

3. 4   **4.** -3.7 **5.** -0.3

Instead of solving equations with decimals, sometimes it may be easier to first rewrite the equation so that it contains integers only. Recall that multiplying a decimal by a power of 10 such as 10, 100, or 1000 has the effect of moving the decimal point to the right. We can use the multiplication property of equality to multiply both sides of an equation by an appropriate power of 10. The resulting equivalent equation will contain integers only.

Example 6 Solve: $0.5y + 2.3 = 1.65$

Solution: Multiply both sides of the equation by 100. This will move the decimal point in each term two places to the right.

$$0.5y + 2.3 = 1.65 \quad \text{Original equation}$$
$$100(0.5y + 2.3) = 100(1.65) \quad \text{Multiply both sides by 100.}$$
$$100(0.5y) + 100(2.3) = 100(1.65) \quad \text{Apply the distributive property.}$$
$$50y + 230 = 165 \quad \text{Simplify.}$$

Now the equation contains integers only. Continue solving by subtracting 230 from both sides.

$$50y + 230 = 165$$
$$50y + 230 - 230 = 165 - 230 \quad \text{Subtract 230 from both sides.}$$
$$50y = -65 \quad \text{Simplify.}$$
$$\frac{50y}{50} = \frac{-65}{50} \quad \text{Divide both sides by 50.}$$
$$y = -1.3 \quad \text{Simplify.}$$

Check to see that -1.3 is the solution by replacing y with -1.3 in the original equation.

■ Work Practice 6

Practice 6

Solve: $0.2y + 2.6 = 4$

Answer
6. 7

✓ **Concept Check** By what number would you multiply both sides of the following equation to make calculations easier? Explain your choice.

$1.7x + 3.655 = -14.2$

✓ **Concept Check Answer**
Multiply by 1000.

Martin-Gay Interactive Videos

See Video 5.6

Watch the section lecture video and answer the following questions.

Objective A
1. In ▶ Example 3, why did we multiply both sides of the equation by 10? ▶
2. In ▶ Example 4, we subtracted $3x$ from both sides, but what would have been a potential benefit of subtracting $2x$ instead? ▶

5.6 Exercise Set MyLab Math

Objective A *Solve each equation. See Examples 1 and 2.*

1. $x + 1.2 = 7.1$
2. $y - 0.5 = 9$
3. $-5y = 2.15$
4. $-0.4x = 50$

5. $6.2 = y - 4$
6. $9.7 = x + 11.6$
7. $3.1x = -13.95$
8. $3y = -25.8$

Solve each equation. See Examples 3 through 5.

9. $-3.5x + 2.8 = -11.2$
10. $7.1 - 0.2x = 6.1$
11. $6x + 8.65 = 3x + 10$

12. $7x - 9.64 = 5x + 2.32$
13. $2(x - 1.3) = 5.8$
14. $5(x + 2.3) = 19.5$

Solve each equation by first multiplying both sides by an appropriate power of 10 so that the equation contains integers only. See Example 6.

15. $0.4x + 0.7 = -0.9$
16. $0.7x + 0.1 = 1.5$
17. $7x - 10.8 = x$

18. $3y = 7y + 24.4$
19. $2.1x + 5 - 1.6x = 10$
20. $1.5x + 2 - 1.2x = 12.2$

Solve. See Examples 1 through 6.

21. $y - 3.6 = 4$
22. $x + 5.7 = 8.4$
23. $-0.02x = -1.2$

24. $-9y = -0.162$
25. $6.5 = 10x + 7.2$
26. $2x - 4.2 = 8.6$

27. $2.7x - 25 = 1.2x + 5$
28. $9y - 6.9 = 6y - 11.1$
29. $200x - 0.67 = 100x + 0.81$

30. $2.3 + 500x = 600x - 0.2$
31. $3(x + 2.71) = 2x$
32. $7(x + 8.6) = 6x$

33. $8x - 5 = 10x - 8$
34. $24y - 10 = 20y - 17$
35. $1.2 + 0.3x = 0.9$

36. $1.5 = 0.4x + 0.5$
37. $-0.9x + 2.65 = -0.5x + 5.45$
38. $-50x + 0.81 = -40x - 0.48$

39. $4x + 7.6 = 2(3x - 3.2)$
40. $4(2x - 1.6) = 5x - 6.4$
41. $0.7x + 13.8 = x - 2.16$

42. $y - 5 = 0.3y + 4.1$

Review

Simplify each expression by combining like terms. See Section 3.1.

43. $2x - 7 + x - 9$

44. $x + 14 - 5x - 17$

Perform the indicated operation. See Sections 4.3, 4.5, and 4.7.

45. $\dfrac{6x}{5} \cdot \dfrac{1}{2x^2}$

46. $5\dfrac{1}{3} \div 9\dfrac{1}{6}$

47. $\dfrac{x}{3} + \dfrac{2x}{7}$

48. $50 - 14\dfrac{9}{13}$

Concept Extensions

Mixed Practice (Sections 5.2 and 5.6) *This section of exercises contains equations and expressions. If the exercise contains an equation, solve it for the variable. If the exercise contains an expression, simplify it by combining any like terms.*

49. $b + 4.6 = 8.3$

50. $y - 15.9 = -3.8$

51. $2x - 0.6 + 4x - 0.01$

52. $-x - 4.1 - x - 4.02$

53. $5y - 1.2 - 7y + 8$

54. $9a - 5.6 - 3a + 6$

55. $2.8 = z - 6.3$

56. $9.7 = x + 4.3$

57. $4.7x + 8.3 = -5.8$

58. $2.8x + 3.4 = -13.4$

59. $7.76 + 8z - 12z + 8.91$

60. $9.21 + x - 4x + 11.33$

61. $5(x - 3.14) = 4x$

62. $6(x + 1.43) = 5x$

63. $2.6y + 8.3 = 4.6y - 3.4$

64. $8.4z - 2.6 = 5.4z + 10.3$

65. $9.6z - 3.2 - 11.7z - 6.9$

66. $-3.2x + 12.6 - 8.9x - 15.2$

67. Explain in your own words the property of equality that allows us to multiply both sides of an equation by a power of 10.

68. By what number would you multiply both sides of $8x - 7.6 = 4.23$ to make calculations easier? Explain your choice.

69. Construct an equation whose solution is 1.4.

70. Construct an equation whose solution is -8.6.

Solve.

71. $-5.25x = -40.33575$

72. $7.68y = -114.98496$

73. $1.95y + 6.834 = 7.65y - 19.8591$

74. $6.11x + 4.683 = 7.51x + 18.235$

5.7 Decimal Applications: Mean, Median, and Mode

Objectives

A Find the Mean (or Average) of a List of Numbers.

B Find the Median of a List of Numbers.

C Find the Mode of a List of Numbers.

We are certainly familiar with the word "average." The next four examples sho real-life averages. For example,

- Adults employed in the United States report working an **average** of 47 hours pe week, according to a Gallup poll.
- The U.S. miles per gallon **average** for light vehicles is 25.5, according to autonew com (*Automotive News*).

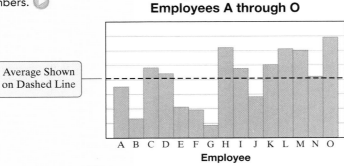

Average Annual Sales for Employees A through O

Average Shown on Dashed Line

The Average Retirement Age in the U.S

Based on U.S. Census Bureau labor force participation data.

As our accumulation of data increases, our ability to gather, store, and prese these tremendous amounts of data increases. Sometimes it is desirable to be able describe a set of data by a single "middle" number or a measure of central tendenc Three of the most common **measures of central tendency** are the **mean** (or average the **median**, and the **mode**.

Objective A Finding the Mean

The most common measure of central tendency is the mean (sometimes called t "arithmetic mean" or the "average"). Recall that we first introduced finding the ave age of a list of numbers in Section 1.6.

> The **mean (average)** of a set of number items is the sum of the items divided by the number of items.
>
> $$\text{mean} = \frac{\text{sum of items}}{\text{number of items}}$$

For example: To find the mean of four test scores—86, 82, 93, and 75—we fir the sum of the scores and then divide by the number of scores, 4.

$$\text{mean} = \frac{86 + 82 + 93 + 75}{4} = \frac{336}{4} = 84$$

Notice that by looking at a bar graph of the scores with a dashed line at the mean, it is reasonable that 84 is one *measure of central tendency.*

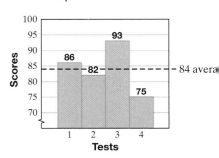

Section 5.7 | Decimal Applications: Mean, Median, and Mode

Example 1 Finding the Mean Time in an Experiment

Seven students in a psychology class conducted an experiment on mazes. Each student was given a pencil and asked to successfully complete the same maze. The timed results are below:

Student	Ann	Thanh	Carlos	Jesse	Melinda	Ramzi	Dayni
Time (Seconds)	13.2	11.8	10.7	16.2	15.9	13.8	18.5

a. Who completed the maze in the shortest time? Who completed the maze in the longest time?

b. Find the mean time.

c. How many students took longer than the mean time? How many students took shorter than the mean time?

Solution:

a. Carlos completed the maze in 10.7 seconds, the shortest time. Dayni completed the maze in 18.5 seconds, the longest time.

b. To find the mean (or average), we find the sum of the items and divide by 7, the number of items.

$$\text{mean} = \frac{\text{sum of items}}{\text{number of items}} \quad \frac{13.2 + 11.8 + 10.7 + 16.2 + 15.9 + 13.8 + 18.5}{7}$$

$$= \frac{100.1}{7} = 14.3$$

c. Three students, Jesse, Melinda, and Dayni, had times longer than the mean time. Four students, Ann, Thanh, Carlos, and Ramzi, had times shorter than the mean time.

Practice 1
Find the mean of the following test scores: 87, 75, 96, 91, and 78.

■ Work Practice 1

✓**Concept Check** Estimate the mean of the following set of data:

5, 10, 10, 10, 10, 15

The mean has one main disadvantage. This measure of central tendency can be greatly affected by *outliers*. (**Outliers** are values that are especially large or small when compared with the rest of the data set.) Let's see an example of this next.

Example 2 The table lists the rounded salary of 10 staff numbers.

Staff	A	B	C	D	E	F	G	H	I	J
Salary (in thousands)	$32	$34	$46	$38	$42	$95	$102	$50	$42	$41

a. Find the mean of all 10 staff members.
b. Find the mean of all staff members except F and G.

Solution:

a. $\text{mean} = \dfrac{32 + 34 + 46 + 38 + 42 + 95 + 102 + 50 + 42 + 41}{10} = \dfrac{522}{10}$

$= 52.2$

The mean salary is $52.2 thousand or $52,200.

(Continued on next page)

Practice 2
Use the table in Example 2 and find the mean salary of all staff members except G. Round thousands of dollars to 2 decimal places.

Answers
1. 85.4 **2.** $46.67 thousand or $46,670

✓**Concept Check Answer**
10

b. $\text{mean} = \dfrac{32 + 34 + 46 + 38 + 42 + 50 + 42 + 41}{8} = \dfrac{325}{8}$

$= 40.625$

Now, the mean salary is $40.625 thousand or $40,625.

■ Work Practice 2

The mean in part **a.** does not appear to be a measure of central tendency because this mean, $52.2 thousand, is greater than all salaries except 2 of the 10. Also notice the difference in the means for parts **a.** and **b.** By removing the 2 outliers, the mean was greatly reduced.

Although the mean was calculated correctly each time, parts **a.** and **b.** of Example 2 show one disadvantage of the mean. That is, a few numerical outliers can greatly affect the mean.

In Section 8.1 we will discuss the range of a data set as well as calculate measures of central tendency from frequency distribution tables and graphs.

Helpful Hint

Remember an important disadvantage of the mean.
If our data set has a few outliers, the mean may not be the best measure of central tendency.

Often in college, the calculation of a **grade point average** (GPA) is a **weighted mean** and is calculated as shown in Example 3.

Example 3 Calculating Grade Point Average (GPA)

The following grades were earned by a student during one semester. Find the student's grade point average.

Course	Grade	Credit Hours
College mathematics	A	3
Biology	B	3
English	A	3
PE	C	1
Social studies	D	2

Solution: To calculate the grade point average, we need to know the point values for the different possible grades. The point values of grades commonly used in colleges and universities are given below:

A: 4, B: 3, C: 2, D: 1, F: 0

Now, to find the grade point average, we multiply the number of credit hours for each course by the point value of each grade. The grade point average is the sum of these products divided by the sum of the credit hours.

Course	Grade	Point Value of Grade	Credit Hours	Point Value of Credit Hours
College mathematics	A	4	3	12
Biology	B	3	3	9
English	A	4	3	12
PE	C	2	1	2
Social studies	D	1	2	2
		Totals:	12	37

Practice 3

Find the grade point average if the following grades were earned in one semester. Round to 2 decimal places.

Grade	Credit Hours
A	2
B	4
C	5
D	2
A	2

Answer
3. 2.67

grade point average = $\frac{37}{12}$ ≈ 3.08 rounded to two decimal places

The student earned a grade point average of 3.08.

Work Practice 3

Objective B Finding the Median

You may have noticed that a very low number or a very high number can affect the mean of a list of numbers. Because of this, you may sometimes want to use another measure of central tendency. A second measure of central tendency is called the **median**. The median of a list of numbers is not affected by a low or high number in the list.

> The **median** of a set of numbers in numerical order is the middle number. If the number of items is odd, the median is the middle number. If the number of items is even, the median is the mean of the two middle numbers.

Example 4 Find the median of the following list of numbers:

25, 54, 56, 57, 60, 71, 98

Solution: Because this list is in numerical order, the median is the middle number, 57.

Work Practice 4

Practice 4
Find the median of the list of numbers: 5, 11, 14, 23, 24, 35, 38, 41, 43

Example 5 Find the median of the following list of scores: 67, 91, 75, 86, 55, 91

Solution: First we list the scores in numerical order and then we find the middle number.

55, 67, 75, 86, 91, 91

Since there is an even number of scores, there are two middle numbers, 75 and 86. The median is the mean of the two middle numbers.

median = $\frac{75 + 86}{2}$ = 80.5

The median is 80.5.

Helpful Hint Don't forget to write the numbers in order from smallest to largest before finding the median.

Work Practice 5

Practice 5
Find the median of the list of scores:
36, 91, 78, 65, 95, 95, 88, 71

Objective C Finding the Mode

The last common measure of central tendency is called the **mode**.

> The **mode** of a set of numbers is the number that occurs most often. (It is possible for a set of numbers to have more than one mode or to have no mode.)

Example 6 Find the mode of the list of numbers:

11, 14, 14, 16, 31, 56, 65, 77, 77, 78, 79

Solution: There are two numbers that occur the most often. They are 14 and 77. This list of numbers has two modes, 14 and 77.

Work Practice 6

Practice 6
Find the mode of the list of numbers:
14, 10, 10, 13, 15, 15, 15, 17, 18, 18, 20

Answers
4. 24   5. 83   6. 15

Practice 7

Find the median and the mode of the list of numbers:
26, 31, 15, 15, 26, 30, 16, 18, 15, 35

Example 7 Find the median and the mode of the following set of numbers. These numbers were high temperatures for 14 consecutive days in a city in Montana.

76, 80, 85, 86, 89, 87, 82, 77, 76, 79, 82, 89, 89, 92

Solution: First we write the numbers in numerical order.

76, 76, 77, 79, 80, 82, 82, 85, 86, 87, 89, 89, 89, 92

Since there is an even number of items, the median is the mean of the two middle numbers, 82 and 85.

$$\text{median} = \frac{82 + 85}{2} = 83.5$$

The mode is 89, since 89 occurs most often.

■ Work Practice 7

✓ **Concept Check** True or false? Every set of numbers *must* have a mean, median, and mode. Explain your answer.

Helpful Hint

Don't forget that it is possible for a list of numbers to have no mode. For example, the list

2, 4, 5, 6, 8, 9

has no mode. There is no number or numbers that occur more often than the others.

Answer
7. median: 22; mode: 15

✓**Concept Check Answer**
false; a set of numbers may have no mode

Remember, in Section 8.1, we will calculate mean, median and mode from frequency distribution tables and graphs.

Vocabulary, Readiness & Video Check

Use the choices below to fill in each blank. Some choices may be used more than once.

| mean | mode | grade point average |
| median | average | |

1. Another word for "mean" is _____.
2. The number that occurs most often in a set of numbers is called the _____.
3. The _____ of a set of number items is $\dfrac{\text{sum of items}}{\text{number of items}}$.
4. The _____ of a set of numbers is the middle number. If the number of numbers is even, it is the _____ of the two middle numbers.
5. An example of weighted mean is a calculation of _____.

Martin-Gay Interactive Videos

Watch the section lecture video and answer the following questions.

Objective A   **6.** Why is the ≈ symbol used in Example 1?

Objective B   **7.** From Example 3, what is always the first step when finding the median of a set of data numbers?

Objective C   **8.** From Example 4, why do you think it is helpful to have data numbers in numerical order when finding the mode?

See Video 5.7

5.7 Exercise Set MyLab Math

Objectives A B C Mixed Practice *For each set of numbers, find the mean, median, and mode. If necessary, round the mean to one decimal place. See Examples 1, 2, and 4 through 6.*

1. 15, 23, 24, 18, 25

2. 45, 36, 28, 46, 52

3. 7.6, 8.2, 8.2, 9.6, 5.7, 9.1

4. 4.9, 7.1, 6.8, 6.8, 5.3, 4.9

5. 0.5, 0.2, 0.2, 0.6, 0.3, 1.3, 0.8, 0.1, 0.5

6. 0.6, 0.6, 0.8, 0.4, 0.5, 0.3, 0.7, 0.8, 0.1

7. 231, 543, 601, 293, 588, 109, 334, 268

8. 451, 356, 478, 776, 892, 500, 467, 780

The 10 tallest buildings in the world, completed as of the end of 2016, are listed in the following table. Use this table to answer Exercises 9 through 14. If necessary, round results to one decimal place. See Examples 1, 2, and 4 through 6.

9. Find the mean height of the five tallest buildings.

10. Find the median height of the five tallest buildings.

11. Find the median height of the six tallest buildings.

12. Find the mean height of the six tallest buildings.

Building	Height in Feet
Burj Khalifa, Dubai	2717
Shanghai Tower, Shanghai	2073
Makkah Royal Clock Tower, Mecca	1972
Ping An Finance Center	1965
Lotte World Tower	1819
One World Trade Center, New York City	1776
Guangzhou CTF Finance Center, Guangzhou	1739
Taipei 101, Taipei	1667
Shanghai World Financial Center, Shanghai	1614
International Commerce Center, Hong Kong	1588
(*Source:* Council on Tall Buildings and Urban Habitat)	

13. Given the building heights, explain how you know, without calculating, that the answer to Exercise **10** is greater than the answer to Exercise **11**.

14. Given the building heights, explain how you know, without calculating, that the answer to Exercise **12** is less than the answer to Exercise **9**.

For Exercises 15 through 18, the grades are given for a student for a particular semester. Find the grade point average. If necessary, round the grade point average to the nearest hundredth. See Example 3.

15.

Grade	Credit Hours
B	3
C	3
A	4
C	4

16.

Grade	Credit Hours
D	1
F	1
C	4
B	5

17.

Grade	Credit Hours
A	3
A	3
A	4
B	3
C	1

18.

Grade	Credit Hours
B	2
B	2
C	3
A	3
B	3

During an experiment, the following times (in seconds) were recorded:
   7.8, 6.9, 7.5, 4.7, 6.9, 7.0.

19. Find the mean.

20. Find the median.

21. Find the mode.

In a mathematics class, the following test scores were recorded for a student: 93, 85, 89, 79, 88, 91.

22. Find the mean.

23. Find the median.

24. Find the mode.

The following pulse rates were recorded for a group of 15 students:
   78, 80, 66, 68, 71, 64, 82, 71, 70, 65, 70, 75, 77, 86, 72.

25. Find the mean.

26. Find the median.

27. Find the mode.

28. How many pulse rates were higher than the mean?

29. How many pulse rates were lower than the mean?

30. Explain how to find the position of the median.

Below are lengths for the six longest rivers in the world.

Name	Length (miles)
Nile	4160
Amazon	4000
Yangtze	3915
Mississippi-Missouri	3709
Ob-Irtysh	3459
Huang Ho	3395

Find the mean and the median for each of the following.

31. the six longest rivers

32. the three longest rivers

Review

Write each fraction in simplest form. See Section 4.2.

33. $\dfrac{18}{30y}$

34. $\dfrac{4x}{36}$

35. $\dfrac{55y^2}{75y^2}$

36. $\dfrac{35a^3}{100a^2}$

Concept Extensions

Find the missing numbers in each set of numbers.

37. ____, ____, ____, 40, ____. The mode is 35. The median is 37. The mean is 38.

38. 16, 18, ____, ____, ____. The mode is 21. The median is 20.

39. Without making any computations, decide whether the median of the following list of numbers will be a whole number. Explain your reasoning.

36, 77, 29, 58, 43

40. Write a list of numbers for which you feel the median would be a better measure of central tendency than the mean.

Chapter 5 Group Activity

Maintaining a Checking Account

(Sections 5.1, 5.2, 5.3)

This activity may be completed by working in groups or individually.

A checking account is a convenient way of handling money and paying bills. To open a checking account, the bank or savings and loan association requires a customer to make a deposit. Then the customer receives a checkbook that contains checks, deposit slips, and a register for recording checks written and deposits made. It is important to record all payments and deposits that affect the account. It is also important to keep the checkbook balance current by subtracting checks written and adding deposits made.

About once a month, checking customers receive a statement from the bank listing all activity that the account has had in the last month. The statement lists a beginning balance, all checks and deposits, any service charges made against the account, and an ending balance. Because it may take several days for checks that a customer has written to clear the banking system, the check register may list checks that do not appear on the monthly bank statement. These checks are called **outstanding checks.** Deposits that are recorded in the check register but do not appear on the statement are called **deposits in transit.** Because of these differences, it is important to balance, or reconcile, the checkbook against the monthly statement. The steps for doing so are listed below.

Balancing or Reconciling a Checkbook

Step 1: Place a check mark in the checkbook register next to each check and deposit listed on the monthly bank statement. Any entries in the register without a check mark are outstanding checks or deposits in transit.

Step 2: Find the ending checkbook register balance and add to it any outstanding checks and any interest paid on the account.

Step 3: From the total in Step 2, subtract any deposits in transit and any service charges.

Step 4: Compare the amount found in Step 3 with the ending balance listed on the bank statement. If they are the same, the checkbook balances with the bank statement. Be sure to update the check register with service charges and interest.

Step 5: If the checkbook does not balance, recheck the balancing process. Next, make sure that the running checkbook register balance was calculated correctly. Finally, compare the checkbook register with the statement to make sure that each check was recorded for the correct amount.

For the checkbook register and monthly bank statement given:
a. update the checkbook register **b.** list the outstanding checks and their total, and deposits in transit
c. balance the checkbook—be sure to update the register with any interest or service fees

Checkbook Register

#	Date	Description	Payment	✓	Deposit	Balance
						425.86
114	4/1	Market Basket	30.27			
115	4/3	May's Texaco	8.50			
	4/4	Cash at ATM	50.00			
116	4/6	UNO Bookstore	121.38			
	4/7	Deposit			100.00	
117	4/9	MasterCard	84.16			
118	4/10	Redbox	6.12			
119	4/12	Kroger	18.72			
120	4/14	Parking sticker	18.50			
	4/15	Direct deposit			294.36	
121	4/20	Rent	395.00			
122	4/25	Student fees	20.00			
	4/28	Deposit			75.00	

First National Bank Monthly Statement 4/30

BEGINNING BALANCE:		425.86
Date	Number	Amount
CHECKS AND ATM WITHDRAWALS		
4/3	114	30.27
4/4	ATM	50.00
4/11	117	84.16
4/13	115	8.50
4/15	119	18.72
4/22	121	395.00
DEPOSITS		
4/7		100.00
4/15	Direct deposit	294.36
SERVICE CHARGES		
Low balance fee		7.50
INTEREST		
Credited 4/30		1.15
ENDING BALANCE:		227.22

Chapter 5 Vocabulary Check

Fill in each blank with one of the choices listed below. Some choices may be used more than once and some not at all.

vertically	decimal	and	right triangle
standard form	mean	median	circumference
sum	denominator	numerator	mode

1. Like fractional notation, _____ notation is used to denote a part of a whole.
2. To write fractions as decimals, divide the _____ by the _____.
3. To add or subtract decimals, write the decimals so that the decimal points line up _____.
4. When writing decimals in words, write "_____" for the decimal point.
5. When multiplying decimals, the decimal point in the product is placed so that the number of decimal places in the product is equal to the _____ of the number of decimal places in the factors.
6. The _____ of a set of numbers is the number that occurs most often.
7. The distance around a circle is called the _____.
8. The _____ of a set of numbers in numerical order is the middle number. If there is an even number of numbers, the median is the _____ of the two middle numbers.
9. The _____ of a list of items with number values is $\frac{\text{sum of items}}{\text{number of items}}$.
10. When 2 million is written as 2,000,000, we say it is written in _____.

> **Helpful Hint**
> ● Are you preparing for your test? To help, don't forget to take these:
> - Chapter 5 Getting Ready for the Test on page 410
> - Chapter 5 Test on page 411
>
> Then check all of your answers at the back of this text. For further review, the step-by-step video solutions to any of these exercises are located in MyLab Math.

5 Chapter Highlights

Definitions and Concepts	Examples
Section 5.1 Introduction to Decimals	

Place-Value Chart

hundreds	tens	ones	.	tenths	hundredths	thousandths	ten-thousandths	hundred-thousandths
		4		2	6	5		
100	10	1	decimal point	$\frac{1}{10}$	$\frac{1}{100}$	$\frac{1}{1000}$	$\frac{1}{10,000}$	$\frac{1}{100,000}$

4.265 means

$$4 \cdot 1 + 2 \cdot \frac{1}{10} + 6 \cdot \frac{1}{100} + 5 \cdot \frac{1}{1000}$$

or

$$4 + \frac{2}{10} + \frac{6}{100} + \frac{5}{1000}$$

(continued)

Definitions and Concepts	Examples
Section 5.1 Introduction to Decimals (*continued*)	
Writing (or Reading) a Decimal in Words **Step 1:** Write the whole number part in words. **Step 2:** Write "and" for the decimal point. **Step 3:** Write the decimal part in words as though it were a whole number, followed by the place value of the last digit.	Write 3.08 in words. Three and eight hundredths
A decimal written in words can be written in standard form by reversing the above procedure.	Write "negative four and twenty-one thousandths" in standard form. -4.021
To Round a Decimal to a Place Value to the Right of the Decimal Point **Step 1:** Locate the digit to the right of the given place value. **Step 2:** If this digit is 5 or greater, add 1 to the digit in the given place value and drop all digits to its right. If this digit is less than 5, drop all digits to the right of the given place value.	Round 86.1256 to the nearest hundredth. **Step 1:** 86.12 5 6 (hundredths place; digit to the right) **Step 2:** Since the digit to the right is 5 or greater, we add 1 to the digit in the hundredths place and drop all digits to its right. 86.1256 rounded to the nearest hundredth is 86.13.
Section 5.2 Adding and Subtracting Decimals	
To Add or Subtract Decimals **Step 1:** Write the decimals so that the decimal points line up vertically. **Step 2:** Add or subtract as with whole numbers. **Step 3:** Place the decimal point in the sum or difference so that it lines up vertically with the decimal points in the problem.	Add: $4.6 + 0.28$ Subtract: $2.8 - 1.04$ 4.60 $2.8\overset{7\;10}{\cancel{0}}$ $+0.28$ -1.04 4.88 1.76
Section 5.3 Multiplying Decimals and Circumference of a Circle	
To Multiply Decimals **Step 1:** Multiply the decimals as though they are whole numbers. **Step 2:** The decimal point in the product is placed so that the number of decimal places in the product is equal to the *sum* of the number of decimal places in the factors.	Multiply: 1.48×5.9 1.48 ← 2 decimal places $\times\;5.9$ ← 1 decimal place 1332 7400 8.732 ← 3 decimal places

Definitions and Concepts	Examples

Section 5.3 Multiplying Decimals and Circumference of a Circle (*continued*)

The **circumference** of a circle is the distance around the circle. or $C = \pi d$ or $C = 2\pi r$ where $\pi \approx 3.14$ or $\pi \approx \dfrac{22}{7}$.	Find the exact circumference of a circle with radius 5 miles and an approximation by using 3.14 for π. 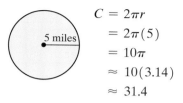 $\begin{aligned} C &= 2\pi r \\ &= 2\pi(5) \\ &= 10\pi \\ &\approx 10(3.14) \\ &\approx 31.4 \end{aligned}$ The circumference is exactly 10π miles and approximately 31.4 miles.

Section 5.4 Dividing Decimals

To Divide Decimals

Step 1: If the divisor is not a whole number, move the decimal point in the divisor to the right until the divisor is a whole number.

Step 2: Move the decimal point in the dividend to the right the *same number of places* as the decimal point was moved in Step 1.

Step 3: Divide. The decimal point in the quotient is directly over the moved decimal point in the dividend.

Divide: $1.118 \div 2.6$

$$\begin{array}{r} 0.43 \\ 2.6\overline{)1.118} \\ -104 \\ \hline 78 \\ -78 \\ \hline 0 \end{array}$$

Section 5.5 Fractions, Decimals, and Order of Operations

To **write fractions as decimals,** divide the numerator by the denominator.

Write $\dfrac{3}{8}$ as a decimal.

$$\begin{array}{r} 0.375 \\ 8\overline{)3.000} \\ -2\,4 \\ \hline 60 \\ -56 \\ \hline 40 \\ -40 \\ \hline 0 \end{array}$$

Order of Operations

1. Perform all operations within parentheses ( ), brackets [ ], or other grouping symbols such as fraction bars.
2. Evaluate any expressions with exponents.
3. Multiply or divide in order from left to right.
4. Add or subtract in order from left to right.

Simplify.

$\begin{aligned} -1.9(12.8 - 4.1) &= -1.9(8.7) &&\text{Subtract.} \\ &= -16.53 &&\text{Multiply.} \end{aligned}$

Definitions and Concepts	Examples
Section 5.6 Solving Equations Containing Decimals	

Steps for Solving an Equation in x

Step 1: If fractions are present, multiply both sides of the equation by the LCD of the fractions.

Step 2: If parentheses are present, use the distributive property.

Step 3: Combine any like terms on each side of the equation.

Step 4: Use the addition property of equality to rewrite the equation so that variable terms are on one side of the equation and constant terms are on the other side.

Step 5: Divide both sides by the numerical coefficient of x to solve.

Step 6: Check your answer in the *original equation*.

Solve:

$$3(x + 2.6) = 10.92$$
$$3x + 7.8 = 10.92 \quad \text{Apply the distributive property.}$$
$$3x + 7.8 - 7.8 = 10.92 - 7.8 \quad \text{Subtract 7.8 from both sides.}$$
$$3x = 3.12 \quad \text{Simplify.}$$
$$\frac{3x}{3} = \frac{3.12}{3} \quad \text{Divide both sides by 3.}$$
$$x = 1.04 \quad \text{Simplify.}$$

Check 1.04 in the original equation.

Section 5.7 Decimal Applications: Mean, Median, and Mode

The **mean** (or **average**) of a set of number items is

$$\text{mean} = \frac{\text{sum of items}}{\text{number of items}}$$

Find the mean, median, and mode of the following set of numbers: 33, 35, 35, 43, 68, 68

$$\text{mean} = \frac{33 + 35 + 35 + 43 + 68 + 68}{6} = 47$$

The **median** of a set of numbers in numerical order is the middle number. If the number of items is even, the median is the mean of the two middle numbers.

The median is the mean of the two middle numbers, 35 and 43

$$\text{median} = \frac{35 + 43}{2} = 39$$

The **mode** of a set of numbers is the number that occurs most often. (A set of numbers may have no mode or more than one mode.)

There are two modes because there are two numbers that occur twice:

35 and 68

Chapter 5 Review

(5.1) *Determine the place value of the number 4 in each decimal.*

1. 23.45

2. 0.000345

Write each decimal in words.

3. −23.45

4. 0.00345

5. 109.23

6. 200.000032

Write each decimal in standard form.

7. Eight and six hundredths

8. Negative five hundred three and one hundred two thousandths

9. Sixteen thousand twenty-five and fourteen ten-thousandths

10. Fourteen and eleven thousandths

Write each decimal as a fraction or a mixed number.

11. 0.16

12. -12.023

Write each fraction or mixed number as a decimal.

13. $\dfrac{231}{100,000}$

14. $25\dfrac{1}{4}$

Insert $<$, $>$, or $=$ between each pair of numbers to make a true statement.

15. 0.49   0.43

16. 0.973   0.9730

17. -38.0027 -38.00056

18. -0.230505 -0.23505

Round each decimal to the given place value.

19. 0.623, nearest tenth

20. 0.9384, nearest hundredth

21. -42.895, nearest hundredth

22. 16.34925, nearest thousandth

Write each number in standard notation.

23. Saturn is a distance of about 890.5 million miles from the Sun.

24. The tail of a comet can be over 600 thousand miles long.

(5.2) *Add.*

25. $8.6 + 9.5$

26. $3.9 + 1.2$

27. $-6.4 + (-0.88)$

28. $-19.02 + 6.98$

29. $200.49 + 16.82 + 103.002$

30. $0.00236 + 100.45 + 48.29$

Subtract.

31. $4.9 - 3.2$

32. $5.23 - 2.74$

33. $-892.1 - 432.4$

34. $0.064 - 10.2$

35. $100 - 34.98$

36. $200 - 0.00198$

Solve.

37. Find the total distance between Grove City and Jerome.

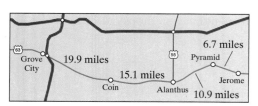

38. Evaluate $x - y$ for $x = 1.2$ and $y = 6.9$.

△ **39.** Find the perimeter.

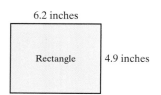

△ **40.** Find the perimeter.

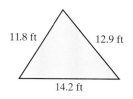

(5.3) *Multiply.*

41. 7.2×10

42. 9.345×1000

43. -34.02×2.3

44. $-839.02 \times (-87.3)$

Find the exact circumference of each circle. Then use the approximation 3.14 for π and approximate the circumference.

△ **45.**

△ **46.**

(5.4) *Divide. Round the quotient to the nearest thousandth if necessary.*

47. $3\overline{)0.2631}$

48. $20\overline{)316.5}$

49. $-21 \div (-0.3)$

50. $-0.0063 \div 0.03$

51. $0.34\overline{)2.74}$

52. $19.8\overline{)601.92}$

53. $\dfrac{23.65}{1000}$

54. $\dfrac{93}{-10}$

55. There are approximately 3.28 feet in 1 meter. Find how many meters are in 24 feet to the nearest tenth of a meter.

56. George Strait pays $69.71 per month to pay back a loan of $3136.95. In how many months will the loan be paid off?

(5.5) *Write each fraction or mixed number as a decimal. Round to the nearest thousandth if necessary.*

57. $\dfrac{4}{5}$

58. $-\dfrac{12}{13}$

59. $2\dfrac{1}{3}$

60. $\dfrac{13}{60}$

Chapter 5 Review

Insert <, >, or = to make a true statement.

61. 0.392   0.39200

62. −0.0231   −0.0221

63. $\dfrac{4}{7}$ 0.625

64. 0.293 $\dfrac{5}{17}$

Write the numbers in order from smallest to largest.

65. 0.837, 0.839, 0.832

66. 0.685, 0.626, $\dfrac{5}{8}$

67. $\dfrac{3}{7}$, 0.42, 0.43

68. $\dfrac{18}{11}$, 1.63, $\dfrac{19}{12}$

Simplify each expression.

69. $-7.6 \times 1.9 + 2.5$

70. $(-2.3)^2 - 1.4$

71. $0.0726 \div 10 \times 1000$

72. $0.9(6.5 - 5.6)$

73. $\dfrac{(1.5)^2 + 0.5}{0.05}$

74. $\dfrac{7 + 0.74}{-0.06}$

Find each area.

75.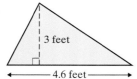

△ **76.**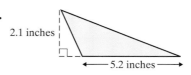

(5.6) *Solve.*

77. $x + 3.9 = 4.2$

78. $70 = y - 22.81$

79. $2x = 17.2$

80. $-1.1y = 88$

81. $3x - 0.78 = 1.2 + 2x$

82. $-x + 0.6 - 2x = -4x - 0.9$

83. $-1.3x - 9.4 = -0.4x + 8.6$

84. $3(x - 1.1) = 5x - 5.3$

(5.7) *Find the mean, median, and any mode(s) for each list of numbers. If necessary, round to the nearest tenth.*

85. 13, 23, 33, 14, 6

86. 45, 86, 21, 60, 86, 64, 45

87. 14,000, 20,000, 12,000, 20,000, 36,000, 45,000

88. 560, 620, 123, 400, 410, 300, 400, 780, 430, 450

For Exercises 89 and 90, the grades are given for a student for a particular semester. Find each grade point average. If necessary, round the grade point average to the nearest hundredth.

89.

Grade	Credit Hours
A	3
A	3
C	2
B	3
C	1

90.

Grade	Credit Hours
B	3
B	4
C	2
D	2
B	3

Mixed Review

91. Write 200.0032 in words.

92. Write negative sixteen and nine hundredths in standard form.

93. Write 0.0847 as a fraction or a mixed number.

94. Write the numbers $\frac{6}{7}, \frac{8}{9}, 0.75$ in order from smallest to largest.

Write each fraction as a decimal. Round to the nearest thousandth, if necessary.

95. $-\frac{7}{100}$

96. $\frac{9}{80}$ (Do not round.)

97. $\frac{8935}{175}$

Insert $<$, $>$, or $=$ to make a true statement.

98. -402.000032 ____ -402.00032

99. $\frac{6}{11}$ ____ 0.55

Round each decimal to the given place value.

100. 86.905, nearest hundredth

101. 3.11526, nearest thousandth

Round each money amount to the nearest dollar.

102. $123.46

103. $3645.52

Add or subtract as indicated.

104. $3.2 - 4.9$

105. $9.12 - 3.86$

106. $-102.06 + 89.3$

107. $-4.021 + (-10.83) + (-0.056)$

Multiply or divide as indicated. Round to the nearest thousandth, if necessary.

108. $\begin{array}{r}2.54\\ \times\ 3.2\end{array}$

109. $(-3.45)(2.1)$

110. $0.005\overline{)24.5}$

111. $2.3\overline{)54.98}$

Solve.

112. Tomaso is going to fertilize his lawn, a rectangle that measures 77.3 feet by 115.9 feet. Approximate the area of the lawn by rounding each measurement to the nearest ten feet.

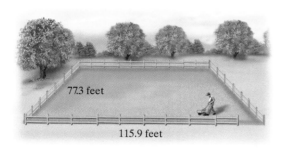

113. Estimate the cost of the items to see whether the groceries can be purchased with a $10 bill.

Simplify each expression.

114. $\dfrac{(3.2)^2}{100}$

115. $(2.6 + 1.4)(4.5 - 3.6)$

Find the mean, median, and any mode(s) for each list of numbers. If needed, round answers to the nearest hundredth.

116. 73, 82, 95, 68, 54

117. 952, 327, 566, 814, 327, 729

Chapter 5 — Getting Ready for the Test

MULTIPLE CHOICE All the exercises below are **Multiple Choice**. Choose the correct letter.

For Exercises 1 through 4, name the place value for the given digit. The number is 1026.89704 and choices are:

A. tens   **B.** hundreds   **C.** tenths   **D.** thousandths   **E.** ten-thousandths

1. the digit 8
2. the digit 2
3. the digit 0 to the left of decimal point
4. the digit 0 to the right of the decimal point

For Exercises 5 through 8, let $x = 3$ and $y = 0.2$. Choose the expression that gives each answer.

A. $x + y$ **B.** $x - y$ **C.** xy **D.** $x \div y$

5. answer: 0.6
6. answer: 3.2
7. answer: 2.8
8. answer: 15

For Exercises 9 through 11, choose the correct letter.

9. The expression $2(0.5)^2 + 0.3$ simplifies to
    A. 1.3   **B.** 0.8   **C.** 2.3   **D.** 1.28

10. One way to solve $5(x - 3.2) = 9.5$ is to first use the distributive property to multiply on the left side of the equation. Once this is done, the equivalent equation is
    A. $5x - 3.2 = 9.5$ **B.** $5x - 16 = 9.5$ **C.** $5x - 16 = 95$

11. One way to solve $0.3x + 1.8 = 2.16$ is to first multiply both sides of the equation by 100. Once this is done, the equivalent equation is
    A. $3x + 18 = 216$ **B.** $3x + 1.8 = 216$ **C.** $30x + 180 = 216$

For Exercises 12 through 14, choose the correct directions that lead to the given correct answer for the data set: 7, 9, 10, 13, 13.

A. Find the mean.   **B.** Find the median.   **C.** Find the mode.

12. answer: 10
13. answer: 13
14. answer: 10.4

Chapter 5 Test

Write each decimal as indicated.

1. 45.092, in words

2. Three thousand and fifty-nine thousandths, in standard form

Perform each indicated operation. Round the result to the nearest thousandth if necessary.

3. $2.893 + 4.21 + 10.492$

4. $-47.92 - 3.28$

5. $9.83 - 30.25$

6. 10.2×4.01

7. $-0.00843 \div (-0.23)$

Round each decimal to the indicated place value.

8. 34.8923, nearest tenth

9. 0.8623, nearest thousandth

Insert $<$, $>$, or $=$ between each pair of numbers to form a true statement.

10. 25.0909   25.9090

11. $\dfrac{4}{9}$ 0.445

Write each decimal as a fraction or a mixed number.

12. 0.345

13. -24.73

Write each fraction as a decimal. If necessary, round to the nearest thousandth.

14. $-\dfrac{13}{26}$

15. $\dfrac{16}{17}$

Simplify.

16. $(-0.6)^2 + 1.57$

17. $\dfrac{0.23 + 1.63}{-0.3}$

18. $2.4x - 3.6 - 1.9x - 9.8$

Answers

1. _____
2. _____
3. _____
4. _____
5. _____
6. _____
7. _____
8. _____
9. _____
10. _____
11. _____
12. _____
13. _____
14. _____
15. _____
16. _____
17. _____
18. _____

Solve.

19. $0.2x + 1.3 = 0.7$

20. $2(x + 5.7) = 6x - 3.4$

Find the mean, median, and mode of each list of numbers.

21. 26, 32, 42, 43, 49

22. 8, 10, 16, 16, 14, 12, 12, 13

Find the grade point average. If necessary, round to the nearest hundredth.

23.

Grade	Credit Hours
A	3
B	3
C	3
B	4
A	1

Solve.

24. At its farthest, Pluto is 4583 million miles from the Sun. Write this number using standard notation.

25. Find the area.

26. Find the exact circumference of the circle. Then use the approximation 3.14 for π and approximate the circumference.

27. Vivian Thomas is going to put insecticide on her lawn to control grubworms. The lawn is a rectangle that measures 123.8 feet by 80 feet. The amount of insecticide required is 0.02 ounce per square foot.

    a. Find the area of her lawn.

    b. Find how much insecticide Vivian needs to purchase.

28. Find the total distance from Bayette to Center City.

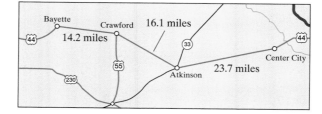

Cumulative Review — Chapters 1–5

Write each number in words.

1. 72

2. 107

3. 546

4. 5026

5. Add: 46 + 713

6. Find the perimeter.

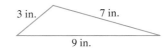

7. Subtract: 543 − 29. Then check by adding.

8. Divide: 3268 ÷ 27

9. Round 278,362 to the nearest thousand.

10. Write the prime factorization of 30.

11. Multiply: 236 × 86

12. Multiply: 236 × 86 × 0

13. Find each quotient and then check the answer by multiplying.
  a. $1\overline{)7}$
  b. 12 ÷ 1
  c. $\dfrac{6}{6}$
  d. 9 ÷ 9
  e. $\dfrac{20}{1}$
  f. $18\overline{)18}$

14. Find the average of 25, 17, 19, and 39.

15. Simplify: 2 · 4 − 3 ÷ 3

16. Simplify: 77 ÷ 11 · 7

Evaluate.

17. 9^2

18. 5^3

19. 3^4

20. 10^3

21. Evaluate $\dfrac{x - 5y}{y}$ for $x = 35$ and $y = 5$.

22. Evaluate $\dfrac{2a + 4}{c}$ for $a = 7$ and $c = 3$.

23. Find the opposite of each number.
   a. 13   b. -2 c. 0

24. Find the opposite of each number.
   a. -7 b. 4   c. -1

25. Add: $-2 + (-21)$

26. Add: $-7 + (-15)$

Find the value of each expression.

27. $5 \cdot 6^2$

28. $4 \cdot 2^3$

29. -7^2

30. $(-2)^5$

31. $(-5)^2$

32. -3^2

Represent the shaded part as an improper fraction and a mixed number.

33.

34.

35.

36.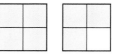

37. Write the prime factorization of 252.

38. Find the difference of 87 and 25.

Cumulative Review

39. Write $-\dfrac{72}{26}$ in simplest form.

40. Write $9\dfrac{7}{8}$ as an improper fraction.

41. Determine whether $\dfrac{16}{40}$ and $\dfrac{10}{25}$ are equivalent.

42. Insert < or > to form a true statement. $\quad \dfrac{4}{7} \quad \dfrac{5}{9}$

Multiply.

43. $\dfrac{2}{3} \cdot \dfrac{5}{11}$

44. $2\dfrac{5}{8} \cdot \dfrac{4}{7}$

45. $\dfrac{1}{4} \cdot \dfrac{1}{2}$

46. $7 \cdot 5\dfrac{2}{7}$

Solve.

47. $\dfrac{z}{-4} = 11 - 5$

48. $6x - 12 - 5x = -20$

49. Add: $763.7651 + 22.001 + 43.89$

50. Add: $89.27 + 14.361 + 127.2318$

51. Multiply: 23.6×0.78

52. Multiply: 43.8×0.645

39. _____

40. _____

41. _____

42. _____

43. _____

44. _____

45. _____

46. _____

47. _____

48. _____

49. _____

50. _____

51. _____

52. _____

6 Ratio, Proportion, and Percent

This chapter is devoted to ratio, proportion, and percent, concepts used virtually every day in ordinary and business life. Understanding percent and using it efficiently depend on understanding ratios, because a percent is a ratio whose denominator is 100. We present techniques to write percents as fractions and as decimals. We then solve percent problems by using equations and also with proportions. A proportion is an equation with two equal ratios. The rest of this chapter is devoted to select applications of percent—including interest rates, sales tax, discounts, and other real-life situations.

Sections

6.1 Ratio and Proportion
6.2 Percents, Decimals, and Fractions
6.3 Solving Percent Problems with Equations
6.4 Solving Percent Problems with Proportions
   Integrated Review—Ratio, Proportion, and Percent
6.5 Applications of Percent
6.6 Percent and Problem Solving: Sales Tax, Commission, and Discount
6.7 Percent and Problem Solving: Interest

Check Your Progress

Vocabulary Check
Chapter Highlights
Chapter Review
Getting Ready for the Test
Chapter Test
Cumulative Review

How Many Internet Users Are There in the World?

We have a world population that is approaching 7.5 billion. As our population grows, so does the access to the Internet. Below is an attempt to show the increase in the world use of the Internet via what we will call an "Internet user." For the use of our graph below, an Internet user is defined to be someone who can access the Internet, through either a computer or a mobile device in the location where he or she calls home. The graph on the left illustrates the percent of world Internet users increasing, and the circle graph on the right shows these Internet users by region.

In Section 6.5, Exercises 71 through 76, we study the number of Internet users in selected countries.

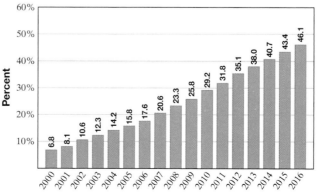
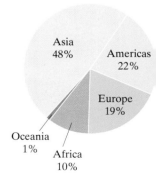

Source: Internet Live Stats (www.InternetLiveStats.com)

6.1 Ratio and Proportion

Objective A Writing Ratios as Fractions

A **ratio** is the quotient of two numbers or two quantities. A ratio, in fact, is no different from a fraction, except that a ratio is sometimes written using notation other than fractional notation. For example, the ratio of 1 to 2 can be written as

1 to 2   or $\dfrac{1}{2}$ or   1 : 2

fractional notation   colon notation

These ratios are all read as, "the ratio of 1 to 2."

Objectives

A Write Ratios as Fractions.
B Solve Proportions.
C Solve Problems Modeled by Proportions.

> **Ratio**
>
> The ratio of a number a to a number b is their quotient. Ways of writing ratios are
>
> a to b, $a : b$,   and $\dfrac{a}{b}$

Whenever possible, we will convert quantities in a ratio to the same unit of measurement.

Example 1 Write a ratio for each phrase. Use fractional notation.

a. The ratio of 2 parts salt to 5 parts water
b. The ratio of 18 inches to 2 feet

Solution:

a. The ratio of 2 parts salt to 5 parts water is $\dfrac{2}{5}$.

b. First we convert to the same unit of measurement. For example,

2 feet = 2 · 12 inches = 24 inches

The ratio of 18 inches to 2 feet is then $\dfrac{18}{24}$, or $\dfrac{3}{4}$ in lowest terms.

■ Work Practice 1

Practice 1

Write a ratio for each phrase. Use fractional notation.

a. The ratio of 3 parts oil to 7 parts gasoline
b. The ratio of 40 minutes to 3 hours

If a ratio compares two decimal numbers, we will write the simplified ratio as a ratio of whole numbers.

Example 2 Write the ratio of 2.5 to 3.15 as a fraction in simplest form.

Solution: The ratio is

$\dfrac{2.5}{3.15}$

Now let's clear the ratio of decimals.

$\dfrac{2.5}{3.15} = \dfrac{2.5 \cdot 100}{3.15 \cdot 100} = \dfrac{250}{315} = \dfrac{50}{63}$ Simplest form

■ Work Practice 2

Practice 2

Write the ratio of 1.68 to 4.8 as a fraction in simplest form.

Answers

1. a. $\dfrac{3}{7}$ b. $\dfrac{2}{9}$ 2. $\dfrac{7}{20}$

417

Practice 3

Given the triangle shown:

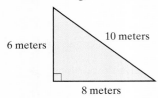

a. Find the ratio of the length of the shortest side to the length of the longest side in simplest form.

b. Find the ratio of the length of the longest side to the perimeter of the triangle in simplest form.

Example 3

Given the rectangle shown:

a. Find the ratio of its width (shorter side) to its length (longer side).

b. Find the ratio of its length to its perimeter.

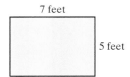

Solution:

a. The ratio of its width to its length is

$$\frac{\text{width}}{\text{length}} = \frac{5 \text{ feet}}{7 \text{ feet}} = \frac{5}{7}$$

b. Recall that the perimeter of a rectangle is the distance around the rectangle: $7 + 5 + 7 + 5 = 24$ feet. The ratio of its length to its perimeter is

$$\frac{\text{length}}{\text{perimeter}} = \frac{7 \text{ feet}}{24 \text{ feet}} = \frac{7}{24}$$

■ Work Practice 3

✓ **Concept Check** Explain why the answer $\frac{7}{5}$ would be incorrect for part (a) of Example 3.

Objective B Solving Proportions

Ratios can be used to form proportions. A **proportion** is a mathematical statement that two ratios are equal.

For example, the equation

$$\frac{1}{2} = \frac{4}{8}$$

is a proportion that says that the ratios $\frac{1}{2}$ and $\frac{4}{8}$ are equal.

Notice that a proportion contains four numbers. If any three numbers are known, we can solve and find the fourth number. One way to do so is to use cross products. To understand cross products, which were introduced in Section 4.2, let's start with the proportion

$$\frac{a}{b} = \frac{c}{d}$$

and multiply both sides by the LCD, bd.

$$\frac{a}{b} = \frac{c}{d}$$

$$bd\left(\frac{a}{b}\right) = bd\left(\frac{c}{d}\right) \quad \text{Multiply both sides by the LCD, } bd.$$

$$\underline{ad} = \underline{bc} \quad \text{Simplify.}$$

cross product cross product

Answers

3. **a.** $\frac{3}{5}$ **b.** $\frac{5}{12}$

✓ **Concept Check Answer**

$\frac{7}{5}$ is the ratio of the rectangle's length to its width.

Notice why *ad* and *bc* are called cross products.

$$ad \nwarrow \frac{a}{b} = \frac{c}{d} \nearrow bc$$

Cross Products

If $\frac{a}{b} = \frac{c}{d}$, then $ad = bc$.

Example 4 Solve for x: $\frac{45}{x} = \frac{5}{7}$

Solution: To solve, we set cross products equal.

$$\frac{45}{x} = \frac{5}{7}$$

$45 \cdot 7 = x \cdot 5$ Set cross products equal.
$315 = 5x$ Multiply.
$\frac{315}{5} = \frac{5x}{5}$ Divide both sides by 5.
$63 = x$ Simplify.

Check: To check, substitute 63 for x in the original proportion. The solution is 63.

Work Practice 4

Practice 4

Solve for x: $\frac{3}{8} = \frac{63}{x}$

Example 5 Solve for x: $\frac{x-5}{3} = \frac{x+2}{5}$

Solution:

$$\frac{x-5}{3} = \frac{x+2}{5}$$

$5(x - 5) = 3(x + 2)$ Set cross products equal.
$5x - 25 = 3x + 6$ Multiply.
$5x = 3x + 31$ Add 25 to both sides.
$2x = 31$ Subtract $3x$ from both sides.
$\frac{2x}{2} = \frac{31}{2}$ Divide both sides by 2.
$x = \frac{31}{2}$

Check: Verify that $\frac{31}{2}$ is the solution.

Work Practice 5

Practice 5

Solve for x: $\frac{2x+1}{7} = \frac{x-3}{5}$

✓ **Concept Check** For which of the following equations can we immediately use cross products to solve for x?

a. $\frac{2-x}{5} = \frac{1+x}{3}$ **b.** $\frac{2}{5} - x = \frac{1+x}{3}$

Answers

4. $x = 168$ **5.** $x = -\frac{26}{3}$

✓ **Concept Check Answer**

a

Objective C Solving Problems Modeled by Proportions

Writing proportions is a powerful tool for solving problems in almost every field including business, chemistry, biology, health sciences, and engineering, as well as in daily life. Given a specified ratio (or rate) of two quantities, a proportion can be used to determine an unknown quantity.

Example 6 Determining Distances from a Map

On a Chamber of Commerce map of Abita Springs, 5 miles corresponds to 2 inches. How many miles correspond to 7 inches?

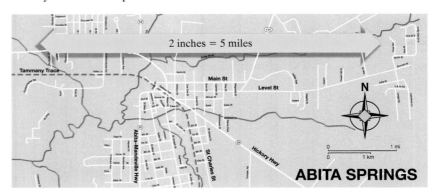

Solution:

1. UNDERSTAND. Read and reread the problem. You may want to draw a diagram.

   From the diagram we can see that a reasonable solution should be between 15 and 20 miles.

2. TRANSLATE. We will let x be our unknown number. Since 5 miles corresponds to 2 inches as x miles corresponds to 7 inches, we have the proportion

   $$\text{miles} \rightarrow \frac{5}{2} = \frac{x}{7} \leftarrow \text{miles}$$
   $$\text{inches} \rightarrow \qquad \qquad \leftarrow \text{inches}$$

3. SOLVE: In earlier sections, we estimated to obtain a reasonable answer. Notice we did this in Step 1 above.

   $$\frac{5}{2} = \frac{x}{7}$$
   $$5 \cdot 7 = 2 \cdot x \qquad \text{Set the cross products equal to each other.}$$
   $$35 = 2x \qquad \text{Multiply.}$$
   $$\frac{35}{2} = \frac{2x}{2} \qquad \text{Divide both sides by 2.}$$
   $$17\frac{1}{2} = x \text{ or } x = 17.5 \qquad \text{Simplify.}$$

4. INTERPRET. *Check* your work. This result is reasonable since it is between 15 and 20 miles. *State* your conclusion: 7 inches corresponds to 17.5 miles.

■ Work Practice 6

Practice 6

On an architect's blueprint, 1 inch corresponds to 4 feet. How long is a wall represented by a $4\frac{1}{4}$-inch line on the blueprint?

Answer
6. 17 ft

Helpful Hint

We can also solve Example 6 by writing the proportion

$$\frac{2 \text{ inches}}{5 \text{ miles}} = \frac{7 \text{ inches}}{x \text{ miles}}$$

Although other proportions may be used to solve Example 6, we will solve by writing proportions so that the numerators have the same unit measures and the denominators have the same unit measures.

Example 7 — Finding Medicine Dosage

The standard dose of an antibiotic is 4 cc (cubic centimeters) for every 25 pounds (lb) of body weight. At this rate, find the standard dose for a 140-lb woman.

Solution:

1. UNDERSTAND. Read and reread the problem. You may want to draw a diagram to estimate a reasonable solution.

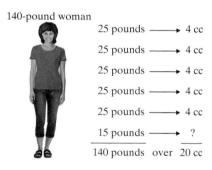

From the diagram, we can see that a reasonable solution is a little over 20 cc.

2. TRANSLATE. We will let x be the unknown number. From the problem, we know that 4 cc is to 25 pounds as x cc is to 140 pounds, or

cubic centimeters $\rightarrow$ $\quad\dfrac{4}{25} = \dfrac{x}{140}\quad$ $\leftarrow$ cubic centimeters
pounds $\rightarrow$ $\qquad\qquad\qquad$ $\leftarrow$ pounds

3. SOLVE:

$$\frac{4}{25} = \frac{x}{140}$$

$4 \cdot 140 = 25 \cdot x$ Set the cross products equal to each other.

$560 = 25x$ Multiply.

$\dfrac{560}{25} = \dfrac{25x}{25}$ Divide both sides by 25.

$22\dfrac{2}{5} = x$ or $x = 22.4$ Simplify.

4. INTERPRET. *Check* your work. This result is reasonable since it is a little over 20 cc. *State* your conclusion: The standard dose for a 140-lb woman is 22.4 cc.

Work Practice 7

Practice 7

An auto mechanic recommends that 5 ounces of isopropyl alcohol be mixed with a tankful of gas (16 gallons) to increase the octane of the gasoline for better engine performance. At this rate, how many gallons of gas can be treated with an 8-ounce bottle of alcohol?

Answer

7. $25\dfrac{3}{5}$ gal or 25.6 gal

Vocabulary, Readiness & Video Check

Use the words and phrases below to fill in each blank.

ratio   cross products   proportion

1. $\frac{4.2}{8.4} = \frac{1}{2}$ is called a _____ while the fraction $\frac{7}{8}$ may also be called a _____.

2. In $\frac{a}{b} = \frac{c}{d}$, $a \cdot d$ and $b \cdot c$ are called _____.

Martin-Gay Interactive Videos Watch the section lecture video and answer the following questions.

Objective A 3. In Example 2, why are dollars converted to nickels?
Objective B 4. Based on Examples 3 and 4, can proportions only be solved by using cross products? Explain.
Objective C 5. In Example 5 we are told there are many ways to set up a correct proportion. Why does this fact make it even more important to check that your solution is reasonable?

See Video 6.1

6.1 Exercise Set MyLab Math

Objective A *Write each ratio in fractional notation in lowest terms. See Examples 1 through 3.*

1. 2 megabytes to 15 megabytes
2. 18 disks to 41 disks
3. 10 inches to 12 inches
4. 15 miles to 40 miles

5. 4 nickels to 2 dollars
6. 12 quarters to 2 dollars
7. 190 minutes to 3 hours
8. 60 hours to 2 days

9. 7.7 to 10
10. 8.1 to 10
11. 3.5 to 12.25
12. 1.5 to 9.75

△ 13. Find the ratio of the length to the width of a regulation size basketball court.

△ 14. Find the ratio of the base to the height of the triangular mainsail.

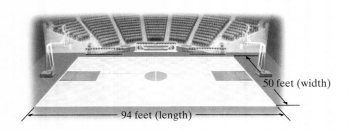

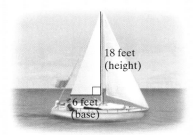

Section 6.1 | Ratio and Proportion

15. Find the ratio of the longest side to the perimeter of the right-triangular-shaped billboard.

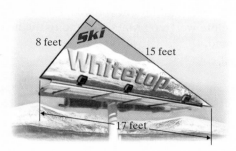

16. Find the ratio of the width to the perimeter of the rectangular vegetable garden.

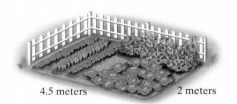

Blood contains three types of cells: red blood cells, white blood cells, and platelets. For approximately every 600 red blood cells in healthy humans, there are 40 platelets and 1 white blood cell. Use this for Exercises 17 and 18. (Source: American Red Cross Biomedical Services)

17. Write the ratio of red blood cells to platelet cells.

18. Write the ratio of white blood cells to red blood cells.

Exercises 19 through 21 have to do with disk storage.

19. A standard DVD has a diameter of 12 centimeters while a miniDVD has a diameter of 8 centimeters. Write the ratio of the miniDVD diameter to the standard DVD diameter.

12 cm Digital Versatile Disc (DVD)

8 cm mini DVD

20. LaserDiscs have a diameter of 12 centimeters and 20 centimeters. Write the ratio of the smaller diameter to the larger diameter.

20 cm LaserDisc

12 cm LaserDisc

21. Hard drives come in many diameters. Find the ratio of a $2\frac{1}{2}''$ diameter to a $5\frac{1}{4}''$ diameter hard drive. (Note: If you choose to work with mixed numbers, find $2\frac{1}{2} \div 5\frac{1}{4}$, or write each mixed number as an equivalent decimal number.)

22. The actual size of Olympic Gold, Silver, and Bronze medals varies for each Olympic Games. The diameter of the medals for the 2008 Beijing games was approximately $2\frac{3}{4}$ inches, while the diameter of the medals for the 2016 Rio games was approximately $3\frac{1}{3}$ inches. Find the ratio of the medal diameter for the Beijing games to the medal diameter for the Rio games.

Objective B *Solve each proportion. See Examples 4 and 5.*

23. $\dfrac{2}{3} = \dfrac{x}{6}$

24. $\dfrac{x}{2} = \dfrac{16}{6}$

25. $\dfrac{x}{10} = \dfrac{5}{9}$

26. $\dfrac{9}{4x} = \dfrac{6}{2}$

27. $\dfrac{4x}{6} = \dfrac{7}{2}$

28. $\dfrac{a}{5} = \dfrac{3}{2}$

29. $\dfrac{x-3}{x} = \dfrac{4}{7}$

30. $\dfrac{y}{y-16} = \dfrac{5}{3}$

31. $\dfrac{x+1}{2x+3} = \dfrac{2}{3}$

32. $\dfrac{x+1}{x+2} = \dfrac{5}{3}$

33. $\dfrac{9}{5} = \dfrac{12}{3x+2}$

34. $\dfrac{6}{11} = \dfrac{27}{3x-2}$

35. $\dfrac{3}{x+1} = \dfrac{5}{2x}$

36. $\dfrac{7}{x-3} = \dfrac{8}{2x}$

37. $\dfrac{16}{3x-4} = \dfrac{5}{x}$

38. $\dfrac{x}{3} = \dfrac{2x-5}{7}$

Objective C *Solve. See Examples 6 and 7.*

39. The ratio of the weight of an object on Earth to the weight of the same object on Jupiter is 2 to 5. If an elephant weighs 4100 pounds on Earth, find the elephant's weight on Jupiter.

40. If a 170-pound person weighs approximately 65 pounds on Mars, how much does a 9000-pound satellite weigh on Mars? Round to the nearest pound.

41. There are 110 calories per 28.4 grams of Crispy Rice cereal. Find how many calories are in 42.6 grams of this cereal.

42. On an architect's blueprint, 1 inch corresponds to 4 feet. Find the length of a wall represented by a line that is $3\frac{7}{8}$ inches long on the blueprint.

43. A 16-oz grande Shaken Sweet Tea at Starbucks has 100 calories. How many calories are there in a 24-oz venti Shaken Sweet Tea? (*Source:* Starbucks Coffee Company)

44. A 16-oz nonfat Skinny Mocha at Starbucks has 140 calories. How many calories are there in a 12-oz nonfat Skinny Mocha? (*Source:* Starbucks Coffee Company)

45. Mosquitos are annoying insects. To eliminate mosquito larvae, a certain granular substance can be applied to standing water in a ratio of 1 tsp per 25 sq ft of standing water.
  a. At this rate, find how many teaspoons of granules must be used for 450 square feet.
  b. If 3 tsp = 1 tbsp, how many tablespoons of granules must be used?

46. Another type of mosquito control is liquid, where 3 oz of pesticide is mixed with 100 oz of water. This mixture is sprayed on roadsides to control mosquito breeding grounds hidden by tall grass.
  a. If one mixture of water with this pesticide can treat 150 feet of roadway, how many ounces of pesticide are needed to treat one mile? (*Hint:* 1 mile = 5280 feet)
  b. If 8 liquid ounces equals one cup, write your answer to part **a** in cups. Round to the nearest cup.

47. The daily supply of oxygen for one person is provided by 625 square feet of lawn. A total of 3750 square feet of lawn would provide the daily supply of oxygen for how many people? (*Source:* Professional Lawn Care Association of America)

48. The Bureau of Labor Statistics predicts that by 2024, 130 million of the 160 million U.S. employees will work in the service industry. In a town of 6800 workers, how many would be expected to work in service-industry jobs? (*Source:* U.S. Bureau of Labor Statistics)

49. A student would like to estimate the height of the Statue of Liberty in New York City's harbor. The length of the Statue of Liberty's right arm is 42 feet. The student's right arm is 2 feet long and her height is $5\frac{1}{3}$ feet. Use this information to estimate the height of the Statue of Liberty. How close is your estimate to the statue's actual height of 111 feet, 1 inch from heel to top of head? (*Source:* National Park Service)

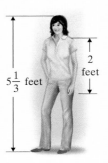

50. The length of the Statue of Liberty's index finger is 8 feet while the height to the top of the head is about 111 feet. Suppose your measurements are proportionaly the same as this statue and that your height is 5 feet.

   a. Use this information to find the proposed length of your index finger. Give an exact measurement and then a decimal rounded to the nearest hundredth.

   b. Measure your index finger and write it as decimal in feet rounded to the nearest hundredth. How close is the length of your index finger to the answer to **a**? Explain why.

51. There are 72 milligrams of cholesterol in a 3.5-ounce serving of lobster. How much cholesterol is in 5 ounces of lobster? Round to the nearest tenth of a milligram. (*Source:* The National Institutes of Health)

52. There are 76 milligrams of cholesterol in a 3-ounce serving of skinless chicken. How much cholesterol is in 8 ounces of chicken? (*Source:* USDA)

53. The Comcast Building in New York City is 850 feet tall and contains 69 stories. The Empire State Building contains 102 stories. If the Empire State Building has the same number of feet per floor as the Comcast Building, approximate its height rounded to the nearest foot. (*Source:* Skyscrapers.com)

54. In 2015, 9 out of 10 American preschool-aged children had received a measles, mumps, and rubella (MMR) vaccine. In a preschool with 130 students, how many students would be expected to have received an MMR vaccine? (*Source:* Centers for Disease Control and Prevention)

55. In 2015, nearly three out of every ten autos sold in the United States was a crossover utility vehicle (CUV). If an auto dealership sold 4500 vehicles in total that year, how many of these sales would you expect involved a CUV? (*Source:* Alliance of Automobile Manufacturers)

56. In the United States in 2016, 1 in every 26 private-sector jobs depended on the auto industry. In a group of 595 private-sector U.S. workers, how many would you expect to have worked in a job that depended on the auto industry? (*Source:* Alliance of Automobile Manufacturers)

57. In 2016, 1 out of every 5 Americans under the age of 50 used a cell phone to read e-books. If there were approximately 275 million Americans under the age of 50 in 2016, how many of them were likely to have read an e-book on a cell phone? (*Source:* Pew Research Center)

58. One pound of firmly packed brown sugar yields $2\frac{1}{4}$ cups. How many pounds of brown sugar will be required in a recipe that calls for 6 cups of firmly packed brown sugar? (*Source:* Based on data from *Family Circle* magazine)

When making homemade ice cream in an electric or a hand-cranked freezer, the tub containing the ice cream mix is surrounded by a brine (water/salt) solution. To freeze the ice cream mix rapidly so that smooth and creamy ice cream results, the brine solution should combine crushed ice and rock salt in a ratio of 5 to 1. Use this for Exercises 59 and 60. (Source: White Mountain Freezers, The Rival Company)

59. A small ice cream freezer requires 12 cups of crushed ice. How much rock salt should be mixed with the ice to create the necessary brine solution?

60. A large ice cream freezer requires $18\frac{3}{4}$ cups of crushed ice. How much rock salt will be needed?

61. The gas/oil ratio for a certain chainsaw is 50 to 1.
    a. How much oil (in gallons) should be mixed with 5 gallons of gasoline?
    b. If 1 gallon equals 128 fluid ounces, write the answer to part **a** in fluid ounces. Round to the nearest whole ounce.

62. The gas/oil ratio for a certain tractor mower is 20 to 1
    a. How much oil (in gallons) should be mixed with 10 gallons of gas?
    b. If 1 gallon equals 4 quarts, write the answer to part **a** in quarts.

63. The adult daily dosage for a certain medicine is 150 mg (milligrams) of medicine for every 20 pounds of body weight.
    a. At this rate, find the daily dose for a man who weighs 275 pounds.
    b. If the man is to receive 500 mg of this medicine every 8 hours, is he receiving the proper dosage?

64. The adult daily dosage for a certain medicine is 80 mg (milligrams) for every 25 pounds of body weight.
    a. At this rate, find the daily dose for a woman who weighs 190 pounds.
    b. If she is to receive this medicine every 6 hours, find the amount to be given every 6 hours.

Review

Find the prime factorization of each number. See Section 4.2.

65. 200   **66.** 300   **67.** 32   **68.** 81

Concept Extensions

69. Suppose someone tells you that the ratio of 11 inches to 2 feet is $\frac{11}{2}$. How would you correct that person and explain the error?

70. Write a real-life example of a ratio that can be written in fractional notation as $\frac{3}{2}$.

As we have seen earlier, proportions are often used in medicine dosage calculations. The exercises below have to do with liquid drug preparations, where the weight of the drug is contained in a volume of solution. The description of mg and ml below will help. We will study metric units further in Chapter 8.

mg means milligrams (A paper clip weighs about a gram. A milligram is about the weight of $\frac{1}{1000}$ of a paper clip.)

ml means milliliters (A liter is about a quart. A milliliter is about the amount of liquid in $\frac{1}{1000}$ of a quart.)

One way to solve the applications below is to set up the proportion $\frac{mg}{ml} = \frac{mg}{ml}$.

A solution strength of 15 mg of medicine in 1 ml of solution is available.

71. If a patient needs 12 mg of medicine, how many ml do you administer?

72. If a patient needs 33 mg of medicine, how many ml do you administer?

A solution strength of 8 mg of medicine in 1 ml of solution is available.

73. If a patient needs 10 mg of medicine, how many ml do you administer?

74. If a patient needs 6 mg of medicine, how many ml do you administer?

6.2 Percents, Decimals, and Fractions

Objective A Understanding Percent

The word **percent** comes from the Latin phrase *per centum*, which means **"per 100."** For example, 53% (53 percent) means 53 per 100. In the square below, 53 of the 100 squares are shaded. Thus, 53% of the figure is shaded.

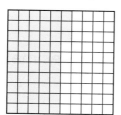

53 of 100 squares are shaded
or
53% is shaded.

Objectives

A Understand Percent.

B Write Percents as Decimals or Fractions.

C Write Decimals or Fractions as Percents.

D Solve Applications with Percents, Decimals, and Fractions.

Since 53% means 53 per 100, 53% is the ratio of 53 to 100, or $\frac{53}{100}$.

$$53\% = \frac{53}{100}$$

Also,

$$7\% = \frac{7}{100} \quad \text{7 parts per 100 parts}$$

$$73\% = \frac{73}{100} \quad \text{73 parts per 100 parts}$$

$$109\% = \frac{109}{100} \quad \text{109 parts per 100 parts}$$

Percent

Percent means **per one hundred.** The "%" symbol is used to denote percent.

Percent is used in a variety of everyday situations. For example,

- 88.5% of the U.S. population are Internet users.
- The store is having a 20%-off sale.
- For the past two years, the enrollment in community colleges increased 15%.
- The South is the home of 49% of all frequent paintball participants.
- 50% of the cinema screens in the U.S. are digital non-3-D.

Example 1 In a survey of 100 people, 17 people drive blue cars. What percent of the people drive blue cars?

Solution: Since 17 people out of 100 drive blue cars, the fraction is $\frac{17}{100}$. Then

$$\frac{17}{100} = 17\%$$

Work Practice 1

Practice 1

Of 100 students in a club, 27 are freshmen. What percent of the students are freshmen?

Answer
1. 27%

Practice 2

31 out of 100 college instructors are in their forties. What percent of these instructors are in their forties?

Example 2 45 out of every 100 young college graduates return home to live (at least temporarily). What percent of college graduates is this? (*Source:* Independent Insurance Agents of America)

Solution:

$$\frac{45}{100} = 45\%$$

■ Work Practice 2

Objective B Writing Percents as Decimals or Fractions

Since percent means "per hundred," we have that

$$1\% = \frac{1}{100} = 0.01$$

In other words, the percent symbol means "per hundred" or, equivalently, "$\frac{1}{100}$" or "0.01." Thus

Write 87% as a fraction: $87\% = 87 \times \frac{1}{100} = \frac{87}{100}$

or

Write 87% as a decimal: $87\% = 87 \times (0.01) = 0.87$

Results are the same.

Of course, we know that the end results are the same; that is,

$$\frac{87}{100} = 0.87$$

The above gives us two options for converting percents. We can replace the percent symbol, %, by $\frac{1}{100}$ or by 0.01 and then multiply.

> For consistency, when we
> - convert from a percent to a *decimal*, we will drop the % symbol and multiply by 0.01.
> - convert from a percent to a *fraction*, we will drop the % symbol and multiply by $\frac{1}{100}$.

Let's practice writing percents as decimals, then writing percents as fractions.

> **Writing a Percent as a Decimal**
>
> Replace the percent symbol with its decimal equivalent, 0.01; then multiply.
>
> $43\% = 43(0.01) = 0.43$

Helpful Hint

If it helps, think of writing a percent as a decimal by

Percent → | Remove the % symbol and move the decimal point 2 places to the left. | → Decimal

Answer
2. 31%

Section 6.2 | Percents, Decimals, and Fractions

Examples Write each percent as a decimal.

3. $23\% = 23(0.01) = 0.23$ Replace the percent symbol with 0.01. Then multiply.
4. $4.6\% = 4.6(0.01) = 0.046$ Replace the percent symbol with 0.01. Then multiply.
5. $190\% = 190(0.01) = 1.90$ or 1.9
6. $0.74\% = 0.74(0.01) = 0.0074$
7. $100\% = 100(0.01) = 1.00$ or 1

> **Helpful Hint**
> We just learned that $100\% = 1$.

Work Practice 3–7

Practice 3–7
Write each percent as a decimal.
3. 49%
4. 3.1%
5. 175%
6. 0.46%
7. 600%

✓ **Concept Check** Why is it incorrect to write the percent 0.033% as 3.3 in decimal form?

Now let's write percents as fractions.

Writing a Percent as a Fraction

Replace the percent symbol with its fraction equivalent, $\frac{1}{100}$; then multiply. Don't forget to simplify the fraction if possible.

$$43\% = 43 \cdot \frac{1}{100} = \frac{43}{100}$$

Examples Write each percent as a fraction or mixed number in simplest form.

8. $40\% = 40 \cdot \frac{1}{100} = \frac{40}{100} = \frac{2 \cdot \cancel{20}}{5 \cdot \cancel{20}} = \frac{2}{5}$

9. $1.9\% = 1.9 \cdot \frac{1}{100} = \frac{1.9}{100}$. We don't want the numerator of the fraction to contain a decimal, so we multiply by 1 in the form of $\frac{10}{10}$.

   $= \frac{1.9}{100} \cdot \frac{10}{10} = \frac{1.9 \cdot 10}{100 \cdot 10} = \frac{19}{1000}$

10. $125\% = 125 \cdot \frac{1}{100} = \frac{125}{100} = \frac{5 \cdot \cancel{25}}{4 \cdot \cancel{25}} = \frac{5}{4}$ or $1\frac{1}{4}$

11. $33\frac{1}{3}\% = 33\frac{1}{3} \cdot \frac{1}{100} = \frac{100}{3} \cdot \frac{1}{100} = \frac{\cancel{100} \cdot 1}{3 \cdot \cancel{100}} = \frac{1}{3}$
    ↳ Write as an improper fraction.

12. $100\% = 100 \cdot \frac{1}{100} = \frac{100}{100} = 1$

> **Helpful Hint**
> Just as in Example 7, we confirm that $100\% = 1$.

Work Practice 8–12

Practice 8–12
Write each percent as a fraction or mixed number in simplest form.
8. 50%
9. 2.3%
10. 150%
11. $66\frac{2}{3}\%$
12. 12%

Objective C Writing Decimals or Fractions as Percents ▶

To write a decimal or fraction as a percent, we use the result of Examples 7 and 12. In these examples, we found that $1 = 100\%$.

Write 0.38 as a percent: $0.38 = 0.38(1) = 0.38(100\%) = 38.\%$

Write $\frac{1}{5}$ as a percent: $\frac{1}{5} = \frac{1}{5}(1) = \frac{1}{5} \cdot 100\% = \frac{100}{5}\% = 20\%$

First, let's practice writing decimals as percents.

Answers
3. 0.49   4. 0.031   5. 1.75   6. 0.0046
7. 6   8. $\frac{1}{2}$ 9. $\frac{23}{1000}$ 10. $\frac{3}{2}$ or $1\frac{1}{2}$
11. $\frac{2}{3}$ 12. $\frac{3}{25}$

✓ **Concept Check Answer**
To write a percent as a decimal, the decimal point should be moved two places to the left, not to the right. So the correct answer is 0.00033.

Writing a Decimal as a Percent

Multiply by 1 in the form of 100%.

$$0.27 = 0.27(100\%) = 27.\%$$

Helpful Hint

If it helps, think of writing a decimal as a percent by reversing the steps in the Helpful Hint on page 428.

Percent ← | Move the decimal point 2 places to the right and attach a % symbol. | ← Decimal

Practice 13–16

Write each decimal as a percent.
13. 0.14   14. 1.75
15. 0.057   16. 0.5

Examples Write each decimal as a percent.

13. $0.65 = 0.65(100\%) = 65.\%$ or 65% Multiply by 100%.
14. $1.25 = 1.25(100\%) = 125.\%$ or 125%
15. $0.012 = 0.012(100\%) = 001.2\%$ or 1.2%
16. $0.6 = 0.6(100\%) = 060.\%$ or 60%

Helpful Hint A zero was inserted as a placeholder.

■ Work Practice 13–16

✓**Concept Check** Why is it incorrect to write the decimal 0.0345 as 34.5% percent form?

Now let's write fractions as percents.

Writing a Fraction as a Percent

Multiply by 1 in the form of 100%.

$$\frac{1}{8} = \frac{1}{8} \cdot 100\% = \frac{1}{8} \cdot \frac{100}{1}\% = \frac{100}{8}\% = 12\frac{1}{2}\% \text{ or } 12.5\%$$

Helpful Hint

From Examples 7 and 12, we know that

$$100\% = 1$$

Recall that when we multiply a number by 1, we are not changing the value of that number. This means that when we multiply a number by 100%, we are not changing its value but rather writing the number as an equivalent percent.

Answers

13. 14%   14. 175%   15. 5.7%
16. 50%

✓**Concept Check Answer**

To change a decimal to a percent, multiply by 100%, or move the decimal point *only* two places to the right. So the correct answer is 3.45%.

Section 6.2 | Percents, Decimals, and Fractions

Examples Write each fraction or mixed number as a percent.

17. $\dfrac{7}{20} = \dfrac{7}{20} \cdot 100\% = \dfrac{7}{20} \cdot \dfrac{100}{1}\% = \dfrac{700}{20}\% = 35\%$

18. $\dfrac{2}{3} = \dfrac{2}{3} \cdot 100\% = \dfrac{2}{3} \cdot \dfrac{100}{1}\% = \dfrac{200}{3}\% = 66\dfrac{2}{3}\%$

19. $2\dfrac{1}{4} = \dfrac{9}{4} \cdot 100\% = \dfrac{9}{4} \cdot \dfrac{100}{1}\% = \dfrac{900}{4}\% = 225\%$

Helpful Hint: $\dfrac{200}{3} = 66.\overline{6}$. Thus, another way to write $\dfrac{200}{3}\%$ is $66.\overline{6}\%$.

■ Work Practice 17–19

Practice 17–19
Write each fraction or mixed number as a percent.

17. $\dfrac{3}{25}$ 18. $\dfrac{9}{40}$ 19. $5\dfrac{1}{2}$

✓ **Concept Check** Which digit in the percent 76.4582% represents
a. A tenth percent?
b. A hundredth percent?
c. A thousandth percent?
d. A ten percent?

Example 20 Write $\dfrac{1}{12}$ as a percent. Round to the nearest hundredth percent.

Solution:

$\dfrac{1}{12} = \dfrac{1}{12} \cdot 100\% = \dfrac{1}{12} \cdot \dfrac{100\%}{1} = \dfrac{100}{12}\% \approx 8.33\%$ ← "approximately"

$\begin{array}{r} 8.333 \approx 8.33 \\ 12\overline{)100.000} \\ -96 \\ \hline 40 \\ -36 \\ \hline 40 \\ -36 \\ \hline 40 \\ -36 \\ \hline 4 \end{array}$

Thus, $\dfrac{1}{12}$ is approximately 8.33%.

■ Work Practice 20

Practice 20
Write $\dfrac{3}{17}$ as a percent. Round to the nearest hundredth percent.

Objective D Solving Applications with Percents, Decimals, and Fractions ▶

Let's summarize what we have learned so far about percents, decimals, and fractions:

Summary of Converting Percents, Decimals, and Fractions

- To write *a percent as a decimal*, replace the % symbol with its decimal equivalent, 0.01; then multiply.
- To write *a percent as a fraction*, replace the % symbol with its fraction equivalent, $\dfrac{1}{100}$; then multiply.
- To write *a decimal or fraction as a percent*, multiply by 100%.

Answers
17. 12%   18. $22\dfrac{1}{2}$%   19. 550%
20. 17.65%

✓ **Concept Check Answers**
a. 4   b. 5   c. 8   d. 7

If we let x represent a number, below we summarize using symbols.

Write a percent as a decimal:	Write a percent as a fraction:	Write a number as a percent:
$x\% = x(0.01)$	$x\% = x \cdot \dfrac{1}{100}$	$x = x \cdot 100\%$

Practice 21

A family decides to spend no more than 27.5% of its monthly income on rent. Write 27.5% as a decimal and as a fraction.

Example 21 In the last ten years, automobile thefts in the continental United States have decreased 40.8%. Write this percent as a decimal and as a fraction. (*Source:* Insurance Information Institute)

Solution:

As a decimal: $40.8\% = 40.8(0.01) = 0.408$

As a fraction: $40.8\% = 40.8 \cdot \dfrac{1}{100} = \dfrac{40.8}{100} = \dfrac{40.8}{100} \cdot \dfrac{10}{10} = \dfrac{408}{1000} = \dfrac{\cancel{8} \cdot 51}{\cancel{8} \cdot 125} = \dfrac{51}{125}$

Thus, 40.8% written as a decimal is 0.408, and written as a fraction is $\dfrac{51}{125}$.

■ Work Practice 21

Practice 22

Provincetown's budget for waste disposal increased by $1\dfrac{3}{4}$ times over the budget from last year. What percent of increase is this?

Example 22 An advertisement for a stereo system reads "$\dfrac{1}{4}$ off." What percent off is this?

Solution: Write $\dfrac{1}{4}$ as a percent.

$$\dfrac{1}{4} = \dfrac{1}{4} \cdot 100\% = \dfrac{1}{4} \cdot \dfrac{100\%}{1} = \dfrac{100}{4}\% = 25\%$$

Thus, "$\dfrac{1}{4}$ off" is the same as "25% off."

■ Work Practice 22

Answers

21. $0.275, \dfrac{11}{40}$ **22.** 175%

Note: It is helpful to know a few basic percent conversions. Appendix A.2 contains a handy reference of percent, decimal, and fraction equivalencies.

Also, Appendix A.3 shows how to find common percents of a number.

Section 6.2 | Percents, Decimals, and Fractions

Vocabulary, Readiness & Video Check

Use the choices below to fill in each blank. Some choices may be used more than once.

$\frac{1}{100}$ 0.01   100%   percent

_____ means "per hundred."

_____ = 1.

The % symbol is read as _____.

To write a decimal or a fraction as a percent, multiply by 1 in the form of _____.

To write a percent as a *decimal*, drop the % symbol and multiply by _____.

To write a percent as a *fraction*, drop the % symbol and multiply by _____.

Martin-Gay Interactive Videos   Watch the section lecture video and answer the following questions.

See Video 6.2

Objective A   7. From the lecture before Example 1, what is the most important thing to remember about percent?

Objective B   8. From Example 4, what is the percent equivalent of 1?

Objective C   9. Complete this statement based on Example 9: Multiplying by 100% is the same as multiplying by _____.

Objective D   10. From Example 15, what is the main difference between writing a percent as a decimal and writing a percent as a fraction?

6.2 Exercise Set   MyLab Math

Objective A   *Solve. See Examples 1 and 2.*

1. In a survey of 100 college students, 96 use the Internet. What percent use the Internet?

2. A basketball player makes 81 out of 100 attempted free throws. What percent of free throws are made?

One hundred adults were asked to name their favorite sport, and the results are shown in the circle graph.

3. What sport was preferred by most adults? What percent preferred this sport?

4. What sport was preferred by the least number of adults? What percent preferred this sport?

5. What percent of adults preferred football or soccer?

6. What percent of adults preferred basketball or baseball?

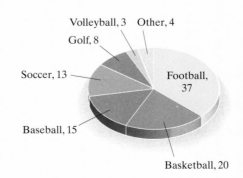

Objective B *Write each percent as a decimal. See Examples 3 through 7.*

7. 41% **8.** 62% **9.** 6% **10.** 3%

11. 100% **12.** 136% **13.** 73.6% **14.** 45.7%

15. 2.8% **16.** 1.4% **17.** 0.6% **18.** 0.9%

19. 300% **20.** 500% **21.** 32.58% **22.** 72.18%

Write each percent as a fraction or mixed number in simplest form. See Examples 8 through 12.

23. 8% **24.** 22% **25.** 4% **26.** 2%

27. 4.5% **28.** 7.5% **29.** 175% **30.** 275%

31. 6.25% **32.** 8.75% **33.** $10\frac{1}{3}$% **34.** $7\frac{3}{4}$%

35. $22\frac{3}{8}$% **36.** $21\frac{7}{8}$%

Objective C *Write each decimal as a percent. See Examples 13 through 16.*

37. 0.22 **38.** 0.44 **39.** 0.006 **40.** 0.008

41. 5.3 **42.** 2.7 **43.** 0.056 **44.** 0.019

45. 0.2228 **46.** 0.1115 **47.** 3.00 **48.** 9.00

49. 0.7 **50.** 0.8

Write each fraction or mixed number as a percent. See Examples 17 through 19.

51. $\frac{7}{10}$ **52.** $\frac{9}{10}$ **53.** $\frac{4}{5}$ **54.** $\frac{2}{5}$

55. $\frac{34}{50}$ **56.** $\frac{41}{50}$ **57.** $\frac{3}{8}$ **58.** $\frac{5}{16}$

59. $\frac{1}{3}$ **60.** $\frac{5}{6}$ **61.** $4\frac{1}{2}$ **62.** $6\frac{1}{5}$

63. $1\frac{9}{10}$ **64.** $2\frac{7}{10}$

Write each fraction as a percent. Round to the nearest hundredth percent. See Example 20.

65. $\frac{9}{11}$ **66.** $\frac{11}{12}$ **67.** $\frac{4}{15}$ **68.** $\frac{10}{11}$

Section 6.2 | Percents, Decimals, and Fractions

Objectives **A B C** Mixed Practice *Complete each table. See Examples 1 through 20.*

69.

Percent	Decimal	Fraction
60%		
	0.235	
		$\frac{4}{5}$
$33\frac{1}{3}\%$		
		$\frac{7}{8}$
7.5%		

70.

Percent	Decimal	Fraction
	0.525	
		$\frac{3}{4}$
$66\frac{1}{2}\%$		
		$\frac{5}{6}$
100%		
		$\frac{7}{50}$

71.

Percent	Decimal	Fraction
200%		
	2.8	
705%		
		$4\frac{27}{50}$

72.

Percent	Decimal	Fraction
800%		
	3.2	
608%		
		$9\frac{13}{50}$

Objective **D** *Write each percent as a decimal and a fraction. See Examples 21 and 22.*

73. People take aspirin for a variety of reasons. The most common use of aspirin is to prevent heart disease, accounting for 38% of all aspirin use. (*Source:* Bayer Market Research)

74. Spending for advertisements on the Internet in 2016 increased by 14.2% over spending in 2015. (*Source:* USA Today)

75. Together, the Greenland and Antarctic Ice sheets contain more than 99% of the freshwater ice on earth. (*Source:* National Snow and Ice Data Center)

76. In 2016, geothermal sources accounted for 66% of Iceland's energy use. (*Source:* National Energy Authority of Iceland)

77. In the United States recently, 48% of households had no landlines, just cell phones. (*Source:* CTIA— The Wireless Association)

78. From 2005 to 2015, use of smoked tobacco products in the United States decreased by 28.6%. (*Source:* Centers for Disease Control)

In Exercises 79 through 82, write the percent from the circle graph as a decimal and a fraction. See Examples 21 and 22.

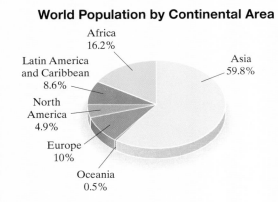

World Population by Continental Area
Africa 16.2%
Latin America and Caribbean 8.6%
North America 4.9%
Europe 10%
Oceania 0.5%
Asia 59.8%

79. Oceania: 0.5%

80. Europe: 10%

81. Africa: 16.2%

82. Asia: 59.8%

Solve. See Examples 21 and 22.

83. In a particular year, 0.781 of all electricity produced in France was nuclear generated. Write this decimal as a percent.

84. In 2015, the United States' share of global passenger vehicle production was 0.1496. Write this decimal as a percent. (*Source:* OICA)

85. The mirrors on the Hubble Space Telescope are able to lock onto a target without deviating more than $\frac{7}{1000}$ of an arc-second. Write this fraction as a percent. (*Source:* NASA)

86. In a particular year, $\frac{1}{4}$ of light trucks sold in the United States were white. Write this fraction as a percent.

87. In 2016, the U.S. Postal Service processed and delivered approximately 47% of the world's card and letter mail volume. Write this percent as a decimal. (*Source:* U.S. Postal Service)

88. In 2015, 81% of U.S. households had at least one high-definition television. Write this percent as a fraction. (*Source:* Leichtman Research Group)

Review

Perform the indicated operations. See Sections 4.6 and 4.7.

89. $\frac{3}{4} - \frac{1}{2} \cdot \frac{8}{9}$

90. $\left(\frac{2}{11} + \frac{5}{11}\right)\left(\frac{2}{11} - \frac{5}{11}\right)$

91. $6\frac{2}{3} - 4\frac{5}{6}$

92. $6\frac{2}{3} \div 4\frac{5}{6}$

Concept Extensions

Solve. See the Concept Checks in this section.

93. Given the percent 52.8647%, round as indicated.
  a. Round to the nearest tenth percent.
  b. Round to the nearest hundredth percent.

94. Given the percent 0.5269%, round as indicated.
  a. Round to the nearest tenth percent.
  b. Round to the nearest hundredth percent.

95. Which of the following are correct?
  a. 6.5% = 0.65
  b. 7.8% = 0.078
  c. 120% = 0.12
  d. 0.35% = 0.0035

96. Which of the following are correct?
  a. 0.231 = 23.1%
  b. 5.12 = 0.0512%
  c. 3.2 = 320%
  d. 0.0175 = 0.175%

Recall that 1 = 100%. This means that 1 whole is 100%. Use this for Exercises 97 and 98. (Source: Some Body, *by Dr. Pete Rowen)*

97. The four blood types are A, B, O, and AB. (Each blood type can also be further classified as Rh-positive or Rh-negative depending upon whether your blood contains protein or not.) Given the percent blood types for people in the United States below, calculate the percent of the U.S. population with AB blood type.

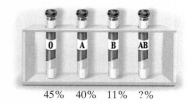

45%   40%   11%   ?%

98. The top four components of bone are below. Find the missing percent.
  1. Minerals—45%
  2. Living tissue—30%
  3. Water—20%
  4. Other—?

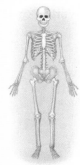

What percent of the figure is shaded?

99. ⬜⬜⬜⬜

100. (circle divided into 8 equal sectors)

Fill in the blanks.

101. A fraction written as a percent is greater than 100% when the numerator is _____ than the denominator. (greater/less)

102. A decimal written as a percent is less than 100% when the decimal is _____ than 1. (greater/less)

Write each fraction as a decimal and then write each decimal as a percent. Round the decimal to three decimal places (nearest thousandth) and the percent to the nearest tenth percent.

103. $\dfrac{21}{79}$

104. $\dfrac{56}{102}$

The bar graph shows the predicted fastest-growing occupations by percent that require an associate degree or more education. Use this graph for Exercises 105 through 108.

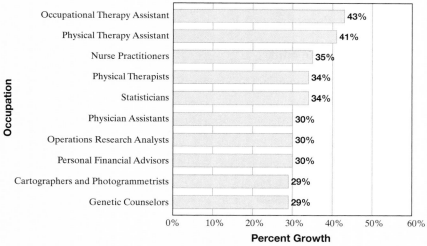

Fastest-Growing Occupations 2014–2024 (projected)

Occupation	Percent Growth
Occupational Therapy Assistant	43%
Physical Therapy Assistant	41%
Nurse Practitioners	35%
Physical Therapists	34%
Statisticians	34%
Physician Assistants	30%
Operations Research Analysts	30%
Personal Financial Advisors	30%
Cartographers and Photogrammetrists	29%
Genetic Counselors	29%

Source: Bureau of Labor Statistics

105. What occupation is predicted to be the fastest growing?

106. What occupation is predicted to be the second fastest growing?

107. Write the percent change for physician assistants as a decimal.

108. Write the percent change for statisticians as a decimal.

109. In your own words, explain how to write a percent as a decimal.

110. In your own words, explain how to write a decimal as a percent.

6.3 Solving Percent Problems with Equations

Objectives

A Write Percent Problems as Equations.

B Solve Percent Problems.

Sections 6.3 and 6.4 introduce two methods for solving percent problems. It may not be necessary for you to study both sections. You may want to check with your instructor for further advice.

To solve percent problems in this section, we will translate the problems into mathematical statements, or equations.

Objective A Writing Percent Problems as Equations

Recognizing key words in a percent problem is helpful in writing the problem as an equation. Three key words in the statement of a percent problem and their meanings are as follows:

of means **multiplication** ($\cdot$)
is means **equal** ($=$)
what (or some equivalent) means **the unknown number**

In our examples, we will let the letter x stand for the unknown number.

Practice 1

Translate: 8 is what percent of 48?

Example 1 Translate to an equation:

5 is what percent of 20?

Solution: 5   is   what percent of   20?
            ↓   ↓   ↓   ↓ ↓
            5   = x $\cdot$ 20

■ Work Practice 1

Helpful Hint

Remember that an equation is simply a mathematical statement that contains an equal sign ($=$).

$5 = 20x$
   ↑
equal sign

Practice 2

Translate: 2.6 is 40% of what number?

Example 2 Translate to an equation:

1.2 is 30% of what number?

Solution: 1.2   is   30%   of   what number?
            ↓   ↓   ↓   ↓   ↓
            1.2 =   30% $\cdot$ x

■ Work Practice 2

Practice 3

Translate: What number is 90% of 0.045?

Example 3 Translate to an equation:

What number is 25% of 0.008?

Solution: What number   is   25%   of   0.008?
              ↓   ↓   ↓   ↓   ↓
              x =   25% $\cdot$ 0.008

■ Work Practice 3

Answers
1. $8 = x \cdot 48$ **2.** $2.6 = 40\% \cdot x$
3. $x = 90\% \cdot 0.045$

Examples
Translate each question to an equation.

4. 38% of 200 is what number?
   38% · 200 = x

5. 40% of what number is 80?
   40% · x = 80

6. What percent of 85 is 34?
   x · 85 = 34

Work Practice 4–6

Practice 4–6
Translate each question to an equation.
4. 56% of 180 is what number?
5. 12% of what number is 21?
6. What percent of 95 is 76?

✓**Concept Check** In the equation $2x = 10$, what step is taken to solve the equation?

Objective B Solving Percent Problems

You may have noticed by now that each percent problem has contained three numbers—in our examples, two are known and one is unknown. Each of these numbers is given a special name.

15% of 60 is 9

15% (percent) · 60 (base) = 9 (amount)

We call this equation the **percent equation.**

> **Percent Equation**
> percent · base = amount

Once a percent problem has been written as a percent equation, we can use the equation to find the unknown number, whether it is the percent, the base, or the amount.

Example 7 Solving Percent Equations for the Amount

What number is 35% of 60?

Solution:
$x = 35\% \cdot 60$ Translate to an equation.
$x = 0.35 \cdot 60$ Write 35% as 0.35.
$x = 21$ Multiply:
  60
  × 0.35
  300
  1800
  21.00

Then 21 is 35% of 60. Is this reasonable? To see, round 35% to 40%. Then 40% of 60 or 0.40(60) is 24. Our result is reasonable since 21 is close to 24.

Work Practice 7

Practice 7
What number is 25% of 90?

Answers
4. $56\% \cdot 180 = x$ 5. $12\% \cdot x = 21$
6. $x \cdot 95 = 76$ 7. 22.5

✓**Concept Check Answer**
Divide both sides of the equation by 2.

Chapter 6 | Ratio, Proportion, and Percent

> **Helpful Hint**
> When solving a percent equation, write the percent as a decimal or fraction.

Practice 8
95% of 400 is what number?

Example 8 85% of 300 is what number?

Solution:
$85\% \cdot 300 = x$ Translate to an equation.
$0.85 \cdot 300 = x$ Write 85% as 0.85.
$255 = x$ Multiply: $0.85 \cdot 300 = 255$.

Then 85% of 300 is 255. Is this result reasonable? To see, round 85% to 90%. Then 90% of 300 or $0.90(300) = 270$, which is close to 255.

■ Work Practice 8

Practice 9
15% of what number is 2.4?

Example 9 Solving Percent Equations for the Base

12% of what number is 0.6?

Solution:
$12\% \cdot x = 0.6$ Translate to an equation.
$0.12 \cdot x = 0.6$ Write 12% as 0.12.
$\dfrac{0.12 \cdot x}{0.12} = \dfrac{0.6}{0.12}$ Divide both sides by 0.12.
$x = 5$

$$0.12\overline{)0.60}$$
$$\underline{-60}$$
$$0$$

Then 12% of 5 is 0.6. Is this reasonable? To see, round 12% to 10%. Then 10% of 5 or $0.10(5) = 0.5$, which is close to 0.6.

■ Work Practice 9

Practice 10
18 is $4\dfrac{1}{2}\%$ of what number?

Example 10 13 is $6\dfrac{1}{2}\%$ of what number?

Solution:
$13 = 6\dfrac{1}{2}\% \cdot x$ Translate to an equation.
$13 = 0.065 \cdot x$ $6\dfrac{1}{2}\% = 6.5\% = 0.065$
$\dfrac{13}{0.065} = \dfrac{0.065 \cdot x}{0.065}$ Divide both sides by 0.065.
$200 = x$

$$0.065\overline{)13.000}$$
$$\underline{-130}$$
$$0$$

Then 13 is $6\dfrac{1}{2}\%$ of 200. Check to see if this result is reasonable.

■ Work Practice 10

Answers
8. 380   **9.** 16   **10.** 400

Section 6.3 | Solving Percent Problems with Equations

Example 11 — Solving Percent Equations for the Percent

What percent of 12 is 9?

Solution:

$x \cdot 12 = 9$ Translate to an equation.

$\dfrac{x \cdot 12}{12} = \dfrac{9}{12}$ Divide both sides by 12.

or $x = 0.75$

Next, since we are looking for percent, we can write $\dfrac{9}{12}$ or 0.75 as a percent.

$x = 75\%$

Then 75% of 12 is 9. To check, see that $75\% \cdot 12 = 9$.

■ Work Practice 11

Practice 11

What percent of 90 is 27?

Helpful Hint

If your unknown in a percent equation is the percent, don't forget to convert your answer to a percent.

Example 12

78 is what percent of 65?

Solution:

$78 = x \cdot 65$ Translate to an equation.

$\dfrac{78}{65} = \dfrac{x \cdot 65}{65}$ Divide both sides by 65.

$1.2 = x$

$120\% = x$ Write 1.2 as a percent.

Then 78 is 120% of 65. Check this result.

■ Work Practice 12

Practice 12

63 is what percent of 45?

✓ Concept Check Consider these problems.

1. 75% of $50 =$
   a. 50
   b. a number greater than 50
   c. a number less than 50

2. 40% of a number is 10. Is the number
   a. 10?
   b. less than 10?
   c. greater than 10?

3. 800 is 120% of what number? Is the number
   a. 800?
   b. less than 800?
   c. greater than 800?

Helpful Hint

Use the following to see if your answers are reasonable.

(100%) of a number $=$ the number

$\begin{pmatrix} \text{a percent} \\ \text{greater than} \\ 100\% \end{pmatrix}$ of a number $=$ a number greater than the original number

$\begin{pmatrix} \text{a percent} \\ \text{less than } 100\% \end{pmatrix}$ of a number $=$ a number less than the original number

Answers
11. 30% 12. 140%

✓ **Concept Check Answers**
1. c   2. c   3. b

Vocabulary, Readiness & Video Check

Use the choices below to fill in each blank.

percent	amount	of	less
base	the number	is	greater

1. The word _____ translates to "=".
2. The word _____ usually translates to "multiplication."
3. In the statement "10% of 90 is 9," the number 9 is called the _____, 90 is called the _____, and 10 is called the _____.
4. 100% of a number = _____.
5. Any "percent greater than 100%" of "a number" = "a number _____ than the original number."
6. Any "percent less than 100%" of "a number" = "a number _____ than the original number."

Martin-Gay Interactive Videos Watch the section lecture video and answer the following questions.

See Video 6.3

Objective A 7. What are the three translations we need to remember from the lecture before Example 1?

Objective B 8. What is different about the translated equation in Example 5?

6.3 Exercise Set MyLab Math

Objective A Translating *Translate each to an equation. Do not solve. See Examples 1 through 6.*

1. 18% of 81 is what number?
2. 36% of 72 is what number?
3. 20% of what number is 105?
4. 40% of what number is 6?
5. 0.6 is 40% of what number?
6. 0.7 is 20% of what number?
7. What percent of 80 is 3.8?
8. 9.2 is what percent of 92?
9. What number is 9% of 43?
10. What number is 25% of 55?
11. What percent of 250 is 150?
12. What percent of 375 is 300?

Section 6.3 | Solving Percent Problems with Equations

Objective B *Solve. See Examples 7 and 8.*

13. 10% of 35 is what number?

14. 25% of 68 is what number?

15. What number is 14% of 205?

16. What number is 18% of 425?

Solve. See Examples 9 and 10.

17. 1.2 is 12% of what number?

18. 0.22 is 44% of what number?

19. $8\frac{1}{2}$% of what number is 51?

20. $4\frac{1}{2}$% of what number is 45?

Solve. See Examples 11 and 12.

21. What percent of 80 is 88?

22. What percent of 40 is 60?

23. 17 is what percent of 50?

24. 48 is what percent of 50?

Objectives A B Mixed Practice *Solve. See Examples 1 through 12.*

25. 0.1 is 10% of what number?

26. 0.5 is 5% of what number?

27. 150% of 430 is what number?

28. 300% of 56 is what number?

29. 82.5 is $16\frac{1}{2}$% of what number?

30. 7.2 is $6\frac{1}{4}$% of what number?

31. 2.58 is what percent of 50?

32. 2.64 is what percent of 25?

33. What number is 42% of 60?

34. What number is 36% of 80?

35. What percent of 184 is 64.4?

36. What percent of 120 is 76.8?

37. 120% of what number is 42?

38. 160% of what number is 40?

39. 2.4% of 26 is what number?

40. 4.8% of 32 is what number?

41. What percent of 600 is 3?

42. What percent of 500 is 2?

43. 6.67 is 4.6% of what number?

44. 9.75 is 7.5% of what number?

45. 1575 is what percent of 2500?

46. 2520 is what percent of 3500?

47. 2 is what percent of 50?

48. 2 is what percent of 40?

Review

Find the value of x in each proportion. See Section 6.1.

49. $\frac{27}{x} = \frac{9}{10}$

50. $\frac{35}{x} = \frac{7}{5}$

51. $\frac{x}{5} = \frac{8}{11}$

52. $\frac{x}{3} = \frac{6}{13}$

Write each sentence as a proportion. See Section 6.1.

53. 17 is to 12 as x is to 20. **54.** 20 is to 25 as x is to 10. **55.** 8 is to 9 as 14 is to x. **56.** 5 is to 6 as 15 is to x.

Concept Extensions

For each equation, determine the next step taken to find the value of n. See the first Concept Check in this section.

57. $5 \cdot n = 32$
   a. $n = 5 \cdot 32$ **b.** $n = \dfrac{5}{32}$ **c.** $n = \dfrac{32}{5}$ **d.** none of these

58. $68 = 8 \cdot n$
   a. $n = 8 \cdot 68$ **b.** $n = \dfrac{68}{8}$ **c.** $n = \dfrac{8}{68}$ **d.** none of these

59. $0.06 = n \cdot 7$
   a. $n = 0.06 \cdot 7$ **b.** $n = \dfrac{0.06}{7}$ **c.** $n = \dfrac{7}{0.06}$ **d.** none of these

60. $n = 0.7 \cdot 12$
   a. $n = 8.4$ **b.** $n = \dfrac{12}{0.7}$ **c.** $n = \dfrac{0.7}{12}$ **d.** none of these

61. Write a word statement for the equation $20\% \cdot x = 18.6$. Use the phrase "what number" for "x."

62. Write a word statement for the equation $x = 33\dfrac{1}{3}\% \cdot 24$. Use the phrase "what number" for "x."

For each exercise, determine whether the percent, x, is (a) 100%, (b) greater than 100%, or (c) less than 100%. See the second Concept Check in this section.

63. $x\%$ of 20 is 30   **64.** $x\%$ of 98 is 98   **65.** $x\%$ of 120 is 85   **66.** $x\%$ of 35 is 50

For each exercise, determine whether the number, y, is (a) equal to 45, (b) greater than 45, or (c) less than 45.

67. 55% of 45 is y **68.** 230% of 45 is y **69.** 100% of 45 is y

70. 30% of y is 45   **71.** 100% of y is 45   **72.** 180% of y is 45

73. In your own words, explain how to solve a percent equation.

74. Write a percent problem that uses the percent 50%.

Solve.

75. 1.5% of 45,775 is what number?

76. What percent of 75,528 is 27,945.36?

77. 22,113 is 180% of what number?

6.4 Solving Percent Problems with Proportions

There is more than one method that can be used to solve percent problems. (See the note at the beginning of Section 6.3.) In the last section, we used the percent equation. In this section, we will use proportions.

Objectives

A Write Percent Problems as Proportions.

B Solve Percent Problems.

Objective A Writing Percent Problems as Proportions

To understand the proportion method, recall that 70% means the ratio of 70 to 100, or $\frac{70}{100}$.

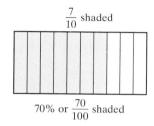

$$70\% = \frac{70}{100} = \frac{7}{10}$$

70% or $\frac{70}{100}$ shaded

Since the ratio $\frac{70}{100}$ is equal to the ratio $\frac{7}{10}$, we have the proportion

$$\frac{7}{10} = \frac{70}{100}$$

We call this proportion the **percent proportion**. In general, we can name the parts of this proportion as follows:

Percent Proportion

$$\frac{\text{amount}}{\text{base}} = \frac{\text{percent}}{100} \quad \leftarrow \text{always } 100$$

or

$$\begin{array}{l}\text{amount} \rightarrow \\ \text{base} \rightarrow \end{array} \frac{a}{b} = \frac{p}{100} \begin{array}{l}\leftarrow \text{percent} \\ \leftarrow \text{always } 100\end{array}$$

When we translate percent problems to proportions, the **percent**, p, can be identified by looking for the symbol % or the word *percent*. The **base**, b, usually follows the word *of*. The **amount**, a, is the part compared to the whole.

Helpful Hint

This table may be useful when identifying the parts of a proportion.

Part of Proportion	How It's Identified
Percent	% or *percent*
Base	Appears after *of*
Amount	Part compared to whole

Practice 1
Translate to a proportion.
27% of what number is 54?

Example 1 Translate to a proportion.

12% of what number is 47?

Solution:
- percent
- base — It appears after the word *of*.
- amount — It is the part compared to the whole.

$$\text{amount} \to \frac{47}{b} = \frac{12}{100} \leftarrow \text{percent}$$
$$\text{base} \to$$

■ Work Practice 1

Practice 2
Translate to a proportion.
30 is what percent of 90?

Example 2 Translate to a proportion.

101 is what percent of 200?

Solution:
- amount — It is the part compared to the whole.
- percent
- base — It appears after the word *of*.

$$\text{amount} \to \frac{101}{200} = \frac{p}{100} \leftarrow \text{percent}$$
$$\text{base} \to$$

■ Work Practice 2

Practice 3
Translate to a proportion.
What number is 25% of 116?

Example 3 Translate to a proportion.

What number is 90% of 45?

Solution:
- amount — It is the part compared to the whole.
- percent
- base — It appears after the word *of*.

$$\text{amount} \to \frac{a}{45} = \frac{90}{100} \leftarrow \text{percent}$$
$$\text{base} \to$$

■ Work Practice 3

Practice 4
Translate to a proportion.
680 is 65% of what number?

Example 4 Translate to a proportion.

238 is 40% of what number?

Solution: amount   percent   base

$$\frac{238}{b} = \frac{40}{100}$$

■ Work Practice 4

Answers

1. $\dfrac{54}{b} = \dfrac{27}{100}$ 2. $\dfrac{30}{90} = \dfrac{p}{100}$
3. $\dfrac{a}{116} = \dfrac{25}{100}$ 4. $\dfrac{680}{b} = \dfrac{65}{100}$

Section 6.4 | Solving Percent Problems with Proportions

Example 5 Translate to a proportion.

What percent of 30 is 75?
↓ ↓ ↓
percent base amount

Solution:

$$\frac{75}{30} = \frac{p}{100}$$

Work Practice 5

Practice 5

Translate to a proportion. What percent of 40 is 75?

Example 6 Translate to a proportion.

45% of 105 is what number?
↓ ↓ ↓
percent base amount

Solution:

$$\frac{a}{105} = \frac{45}{100}$$

Work Practice 6

Practice 6

Translate to a proportion. 46% of 80 is what number?

✓ **Concept Check** Consider the statement "78 is what percent of 350?" Which part of the percent proportion is unknown?
a. the amount
b. the base
c. the percent

Consider another statement: "14 is 10% of some number." Which part of the percent proportion is unknown?
a. the amount
b. the base
c. the percent

Objective B Solving Percent Problems

The proportions that we have written in this section contain three values that can change: the percent, the base, and the amount. If any two of these values are known, we can find the third (the unknown) value. To do this, we write a percent proportion and find the unknown value as we did in Section 6.1.

Example 7 Solving Percent Proportions for the Amount

What number is 30% of 9?
↓ ↓ ↓
amount percent base

Solution:

$$\frac{a}{9} = \frac{30}{100}$$

Practice 7

What number is 8% of 120?

(Continued on next page)

Answers

5. $\frac{75}{40} = \frac{p}{100}$ 6. $\frac{a}{80} = \frac{46}{100}$ 7. 9.6

✓ **Concept Check Answers**
c, b

Chapter 6 | Ratio, Proportion, and Percent

Helpful Hint The proportion in Example 7 contains the ratio $\frac{30}{100}$. A ratio in a proportion may be simplified before solving the proportion. The unknown number in both $\frac{a}{9} = \frac{30}{100}$ and $\frac{a}{9} = \frac{3}{10}$ is 2.7.

To solve, we set cross products equal to each other.

$$\frac{a}{9} = \frac{30}{100}$$

$a \cdot 100 = 9 \cdot 30$	Set cross products equal.
$100a = 270$	Multiply.
$\frac{100a}{100} = \frac{270}{100}$	Divide both sides by 100, the coefficient of a.
$a = 2.7$	Simplify.

Thus, 2.7 is 30% of 9.

■ Work Practice 7

Practice 8
65% of what number is 52?

Example 8 Solving Percent Problems for the Base

150% of what number is 30?
↓   ↓   ↓

Solution: percent   base   amount

$\frac{30}{b} = \frac{150}{100}$	Write the proportion.
$\frac{30}{b} = \frac{3}{2}$	Simplify $\frac{150}{100}$ and write as $\frac{3}{2}$.
$30 \cdot 2 = b \cdot 3$	Set cross products equal.
$60 = 3b$	Multiply.
$\frac{60}{3} = \frac{3b}{3}$	Divide both sides by 3.
$20 = b$	Simplify.

Thus, 150% of 20 is 30.

■ Work Practice 8

✓ **Concept Check** When solving a percent problem by using a proportion, describe how you can check the result.

Practice 9
15.4 is 5% of what number?

Example 9

20.8   is   40%   of   what number?
↓   ↓   ↓

Solution: amount   percent   base

$$\frac{20.8}{b} = \frac{40}{100} \quad \text{or} \quad \frac{20.8}{b} = \frac{2}{5} \quad \text{Write the proportion and simplify } \frac{40}{100}.$$

$20.8 \cdot 5 = b \cdot 2$	Set cross products equal.
$104 = 2b$	Multiply.
$\frac{104}{2} = \frac{2b}{2}$	Divide both sides by 2.
$52 = b$	Simplify.

So, 20.8 is 40% of 52.

■ Work Practice 9

Answers
8. 80   9. 308

✓ **Concept Check Answer**
by putting the result into the proportion and checking that the proportion is true

Section 6.4 | Solving Percent Problems with Proportions

Example 10 — Solving Percent Problems for the Percent

What percent of 50 is 8?
↓ ↓ ↓
percent   base   amount

Solution:

$$\frac{8}{50} = \frac{p}{100} \quad \text{or} \quad \frac{4}{25} = \frac{p}{100}$$ Write the proportion and simplify $\frac{8}{50}$.

$4 \cdot 100 = 25 \cdot p$ Set cross products equal.

$400 = 25p$ Multiply.

$\frac{400}{25} = \frac{25p}{25}$ Divide both sides by 25.

$16 = p$ Simplify.

So, 16% of 50 is 8.

■ Work Practice 10

Practice 10
What percent of 40 is 8?

Helpful Hint: Recall from our percent proportion that this number already is a percent. Just keep the number as is and attach a % symbol.

Example 11

504 is what percent of 360?
↓ ↓ ↓
Solution: amount   percent   base

$$\frac{504}{360} = \frac{p}{100}$$

Let's choose not to simplify the ratio $\frac{504}{360}$.

$504 \cdot 100 = 360 \cdot p$ Set cross products equal.

$50{,}400 = 360p$ Multiply.

$\frac{50{,}400}{360} = \frac{360p}{360}$ Divide both sides by 360.

$140 = p$ Simplify.

Notice that by choosing not to simplify $\frac{504}{360}$, we had larger numbers in our equation. Either way, we find that 504 is 140% of 360.

■ Work Practice 11

Practice 11
414 is what percent of 180?

Helpful Hint

Use the following to see whether your answers to the above examples and practice problems are reasonable.

100% of a number = the number

$\begin{pmatrix} \text{a percent} \\ \text{greater than} \\ 100\% \end{pmatrix}$ of a number = a number larger than the original number

$\begin{pmatrix} \text{a percent} \\ \text{less than } 100\% \end{pmatrix}$ of a number = a number less than the original number

Answers
10. 20% **11.** 230%

Vocabulary, Readiness & Video Check

Use the choices below to fill in each blank. These choices will be used more than once.

amount   base   percent

1. When translating the statement "20% of 15 is 3" to a proportion, the number 3 is called the _____, 15 is the _____, and 20 is the _____.
2. In the question "50% of what number is 28?", which part of the percent proportion is unknown? _____
3. In the question "What number is 25% of 200?", which part of the percent proportion is unknown? _____
4. In the question "38 is what percent of 380?", which part of the percent proportion is unknown? _____

Martin-Gay Interactive Videos Watch the section lecture video and answer the following questions.

Objective A 5. In Example 1, how did we identify what part of the percent proportion 45 is?

Objective B 6. From Examples 4–6, what number is *always* part of the cross product equation of a percent proportion?

See Video 6.4

6.4 Exercise Set MyLab Math

Objective A Translating *Translate each to a proportion. Do not solve. See Examples 1 through 6.*

1. 98% of 45 is what number?
2. 92% of 30 is what number?
3. What number is 4% of 150?
4. What number is 7% of 175?
5. 14.3 is 26% of what number?
6. 1.2 is 47% of what number?
7. 35% of what number is 84?
8. 85% of what number is 520?
9. What percent of 400 is 70?
10. What percent of 900 is 216?
11. 8.2 is what percent of 82?
12. 9.6 is what percent of 96?

Objective B *Solve. See Example 7.*

13. 40% of 65 is what number?
14. 25% of 84 is what number?
15. What number is 18% of 105?
16. What number is 60% of 29?

Solve. See Examples 8 and 9.

17. 15% of what number is 90?

18. 55% of what number is 55?

19. 7.8 is 78% of what number?

20. 1.1 is 44% of what number?

Solve. See Examples 10 and 11.

21. What percent of 35 is 42?

22. What percent of 98 is 147?

23. 14 is what percent of 50?

24. 24 is what percent of 50?

Objectives A B Mixed Practice Solve. See Examples 1 through 11.

25. 3.7 is 10% of what number?

26. 7.4 is 5% of what number?

27. 2.4% of 70 is what number?

28. 2.5% of 90 is what number?

29. 160 is 16% of what number?

30. 30 is 6% of what number?

31. 394.8 is what percent of 188?

32. 550.4 is what percent of 172?

33. What number is 89% of 62?

34. What number is 53% of 130?

35. What percent of 6 is 2.7?

36. What percent of 5 is 1.6?

37. 140% of what number is 105?

38. 170% of what number is 221?

39. 1.8% of 48 is what number?

40. 7.8% of 24 is what number?

41. What percent of 800 is 4?

42. What percent of 500 is 3?

43. 3.5 is 2.5% of what number?

44. 9.18 is 6.8% of what number?

45. 20% of 48 is what number?

46. 75% of 14 is what number?

47. 2486 is what percent of 2200?

48. 9310 is what percent of 3800?

Review

Add or subtract the fractions. See Sections 4.4, 4.5, and 4.7

49. $-\dfrac{11}{16} + \left(-\dfrac{3}{16}\right)$

50. $\dfrac{7}{12} - \dfrac{5}{8}$

51. $3\dfrac{1}{2} - \dfrac{11}{30}$

52. $2\dfrac{2}{3} + 4\dfrac{1}{2}$

Add or subtract the decimals. See Section 5.2.

53. 0.41
$\underline{+0.29}$

54. 10.78
4.3
$\underline{+0.21}$

55. 2.38
$\underline{-0.19}$

56. 16.37
$\underline{-2.61}$

Concept Extensions

57. Write a word statement for the proportion $\dfrac{x}{28} = \dfrac{25}{100}$. Use the phrase "what number" for "x."

58. Write a percent statement that translates to $\dfrac{16}{80} = \dfrac{20}{100}$.

Suppose you have finished solving four percent problems using proportions that you set up correctly. Check each answer to see if each makes the proportion a true proportion. If any proportion is not true, solve it to find the correct solution. See the Concept Checks in this section.

59. $\dfrac{a}{64} = \dfrac{25}{100}$
Is the amount equal to 17?

60. $\dfrac{520}{b} = \dfrac{65}{100}$
Is the base equal to 800?

61. $\dfrac{p}{100} = \dfrac{13}{52}$
Is the percent equal to 25 (25%)?

62. $\dfrac{36}{12} = \dfrac{p}{100}$
Is the percent equal to 50 (50%)?

63. In your own words, describe how to identify the percent, the base, and the amount in a percent problem.

64. In your own words, explain how to use a proportion to solve a percent problem.

Solve. Round to the nearest tenth, if necessary.

65. What number is 22.3% of 53,862?

66. What percent of 110,736 is 88,542?

67. 8652 is 119% of what number?

Sections 6.1–6.4 — Integrated Review

Ratio, Proportion, and Percent

Write each ratio as a ratio of whole numbers using fractional notation. Write the fraction in simplest form.

1. 18 to 20
2. 36 to 100
3. 8.6 to 10
4. 1.6 to 4.6

Find the ratio described in each problem.

5. Find the ratio of the width to the length of the sign below.

6. At the end of fiscal year 2016, Apple Inc. had approximately $320 billion in assets and approximately $192 billion in debts. Find the ratio of assets to debts. (*Source:* Apple Inc.)

Solve.

7. $\dfrac{3.5}{12.5} = \dfrac{7}{z}$

8. $\dfrac{x+7}{3} = \dfrac{2x}{5}$

An office uses 5 boxes of envelopes every 3 weeks.

9. Find how long a gross of envelope boxes is likely to last. (A gross of boxes is 144 boxes.) Round to the nearest week.

10. Find how many boxes should be purchased to last a month. Round to the nearest box.

Write each number as a percent.

11. 0.94
12. 0.17
13. $\dfrac{3}{8}$
14. $\dfrac{7}{2}$

15. 4.7
16. 8
17. $\dfrac{9}{20}$
18. $\dfrac{53}{50}$

19. $6\dfrac{3}{4}$
20. $3\dfrac{1}{4}$
21. 0.02
22. 0.06

Answers

1. _____
2. _____
3. _____
4. _____
5. _____
6. _____
7. _____
8. _____
9. _____
10. _____
11. _____
12. _____
13. _____
14. _____
15. _____
16. _____
17. _____
18. _____
19. _____
20. _____
21. _____
22. _____

Write each percent as a decimal.

23. 71% **24.** 31% **25.** 3% **26.** 4%

27. 224% **28.** 700% **29.** 2.9% **30.** 6.6%

Write each percent as a decimal and as a fraction or mixed number in simplest form. (If necessary when writing as a decimal, round to the nearest thousandth.)

31. 7% **32.** 5% **33.** 6.8% **34.** 11.25%

35. 74% **36.** 45% **37.** $16\frac{1}{3}$% **38.** $12\frac{2}{3}$%

Solve each percent problem.

39. 15% of 90 is what number? **40.** 78 is 78% of what number?

41. 297.5 is 85% of what number? **42.** 78 is what percent of 65?

43. 23.8 is what percent of 85? **44.** 38% of 200 is what number?

45. What number is 40% of 85? **46.** What percent of 99 is 128.7?

47. What percent of 250 is 115? **48.** What number is 45% of 84?

49. 42% of what number is 63? **50.** 95% of what number is 58.9?

6.5 Applications of Percent

Objective A Solving Applications Involving Percent

Percent is used in a variety of everyday situations. The next examples show just a few ways that percent occurs in real-life settings. (Each of these examples shows two ways of solving these problems. If you studied Section 6.3 only, see *Method 1*. If you studied Section 6.4 only, see *Method 2*.)

The next example has to do with the Appalachian Trail, a hiking trail conceived by a forester in 1921 and diagrammed to the right. (*Note:* The trail mileage changes from year to year as maintenance groups reroute the trail as needed.)

Objectives

A Solve Applications Involving Percent.

B Find Percent of Increase and Percent of Decrease.

Example 1 The circle graph in the margin shows the Appalachian Trail mileage by state. If the total mileage of the trail is 2174, use the circle graph to determine the number of miles in the state of New York. Round to the nearest whole mile.

Solution: *Method 1.* First, we state the problem in words.

In words: What number is 4% of 2174?

Translate: x $=$ 4% $\cdot$ 2174

To solve for x, we find $4\% \cdot 2174$.

$x = 0.04 \cdot 2174$ Write 4% as a decimal.
$x = 86.96$ Multiply.
$x \approx 87$ Round to the nearest whole.

Rounded to the nearest whole mile, we have that approximately 87 miles of the Appalachian Trail are in New York state.

Method 2. State the problem in words; then translate.

In words: What number is 4% of 2174?
              amount   percent   base

Translate: $\dfrac{a}{2174} = \dfrac{4}{100}$ amount → a, base → 2174, percent → 4

Next, we solve for a.

$a \cdot 100 = 2174 \cdot 4$ Set cross products equal.
$100a = 8696$ Multiply.
$\dfrac{100a}{100} = \dfrac{8696}{100}$ Divide both sides by 100.
$a = 86.96$ Simplify.
$a \approx 87$ Round to the nearest whole.

Rounded to the nearest whole mile, we have that approximately 87 miles of the Appalachian Trail are in New York state.

Work Practice 1

Practice 1

If the total mileage of the Appalachian Trail is 2174, use the circle graph to determine the number of miles in the state of Virginia.

Appalachian Trail Mileage by State Percent

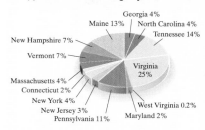

Total miles: 2174
(Due to rounding, these percents have a sum greater than 100%.)
Source: purebound.com

Answer
1. 543.5 mi

455

Practice 2

From 2014 to 2024, it is projected that the number of employed nurses will grow by 439,300. If the number of nurses employed in 2014 was 2,751,000, find the percent of increase in nurses employed from 2014 to 2024. Round to the nearest percent. (*Source:* Bureau of Labor Statistics)

Practice 3

The freshmen class of 864 students is 32% of all students at Euclid University. How many students go to Euclid University?

Answers
2. 16%   **3.** 2700

Example 2 Finding Percent of Nursing School Applications Accepted

There continues to be a shortage of nursing school facilities. Recently, of the 266 thousand applications to bachelor degree nursing schools, 120 thousand of these were accepted. What percent of these applications were accepted? Round to the nearest percent. (*Source:* Bureau of Labor Statistics)

Solution: *Method 1.* First, we state the problem in words.

In words: 120 is what percent of 266?

Translate: $120 = x \cdot 266$
or $120 = 266x$

Next, solve for x.

$$\frac{120}{266} = \frac{266x}{266}$$ Divide both sides by 266.

$0.45 \approx x$ Divide and round to the nearest hundredth.

$45\% \approx x$ Write as a percent.

About 45% of nursing school applications were accepted.

Method 2.

In words: 120 is what percent of 266?
                    amount   percent   base

Translate: amount → $\dfrac{120}{266} = \dfrac{p}{100}$ ← percent
           base →

Next, solve for p.

$120 \cdot 100 = 266 \cdot p$ Set cross products equal.

$12{,}000 = 266p$ Multiply.

$\dfrac{12{,}000}{266} = \dfrac{266p}{266}$ Divide both sides by 266.

$45 \approx p$

About 45% of nursing school applications were accepted.

■ Work Practice 2

Example 3 Finding the Base Number of Absences

Mr. Percy, the principal at Slidell High School, counted 31 freshmen absent during a particular day. If this is 4% of the total number of freshmen, how many freshmen are there at Slidell High School?

Solution: *Method 1.* First we state the problem in words; then we translate.

In words: 31 is 4% of what number?

Translate: $31 = 4\% \cdot x$

Next, we solve for x.

$31 = 0.04 \cdot x$ Write 4% as a decimal.

$\dfrac{31}{0.04} = \dfrac{0.04x}{0.04}$ Divide both sides by 0.04.

$775 = x$ Simplify.

There are 775 freshmen at Slidell High School.

Method 2. First we state the problem in words; then we translate.

In words: 31 is 4% of what number?
           ↓   ↓   ↓
        amount percent   base

Translate: amount → $\dfrac{31}{b} = \dfrac{4}{100}$ ← percent
            base →

Next, we solve for b.

$31 \cdot 100 = b \cdot 4$ Set cross products equal.
$3100 = 4b$ Multiply.
$\dfrac{3100}{4} = \dfrac{4b}{4}$ Divide both sides by 4.
$775 = b$ Simplify.

There are 775 freshmen at Slidell High School.

■ Work Practice 3

Example 4 Finding the Base Increase in Licensed Drivers

From 2015 to 2016, the number of licensed drivers on the road in the United States increased by 5.8%. In 2015, there were about 210 million licensed drivers on the road.
a. Find the increase in licensed drivers from 2015 to 2016.
b. Find the number of licensed drivers on the road in 2016. (*Source:* Federal Highway Administration)

Practice 4

From 2015 to 2016, the number of registered cars and light trucks on the road in the United States increased by 1.5%. In 2015, the number of registered cars and light trucks on the road was 260 million.
a. Find the increase in the number of registered cars and light trucks on the road in 2016.
b. Find the total number of registered cars and light trucks on the road in 2016.

(*Source:* Hedges Company and Federal Highway Administration)

Solution: *Method 1.* First we find the increase in licensed drivers.

In words: What number is 5.8% of 210?
               ↓   ↓   ↓   ↓
Translate: x =   5.8% · 210

Next, we solve for x.

$x = 0.058 \cdot 210$ Write 5.8% as a decimal.
$x = 12.18$ Multiply.

a. The increase in licensed drivers was 12.18 million.
b. This means that the number of licensed drivers in 2016 was

Number of licensed drivers in 2016 = Number of licensed drivers in 2015 + Increase in number of licensed drivers

= 210 million + 12.18 million
= 222.18 million

Answers
4. a. 3.9 million **b.** 263.9 million

(*Continued on next page*)

Method 2. First we find the increase in licensed drivers.

In words: What number is 5.8% of 210?

 amount percent base

Translate: $\dfrac{\text{amount} \to a}{\text{base} \to 210} = \dfrac{5.8}{100} \leftarrow \text{percent}$

Next, we solve for a.

$a \cdot 100 = 210 \cdot 5.8$ $\quad$ Set cross products equal.

$100a = 1218$ $\quad$ Multiply.

$\dfrac{100a}{100} = \dfrac{1218}{100}$ $\quad$ Divide both sides by 100.

$a = 12.18$ $\quad$ Simplify.

a. The increase in licensed drivers was 12.18 million.

b. This means that the number of licensed drivers in 2016 was

Number of licensed drivers in 2016 = Number of licensed drivers in 2015 + Increase in number of licensed drivers

$ = 210 \text{ million} + 12.18 \text{ million}$

$ = 222.18 \text{ million}$

■ Work Practice 4

Objective B Finding Percent of Increase and Percent of Decrease

We often use percents to show how much an amount has increased or decreased.

Suppose that the population of a town is 10,000 people and then it increases by 2000 people. The **percent of increase** is

$\dfrac{\text{amount of increase} \to 2000}{\text{original amount} \to 10{,}000} = 0.2 = 20\%$

In general, we have the following.

Percent of Increase

$\text{percent of increase} = \dfrac{\text{amount of increase}}{\text{original amount}}$

Then write the quotient as a percent.

Practice 5

The number of people attending the local play, *Peter Pan*, increased from 285 on Friday to 333 on Saturday. Find the percent of increase in attendance. Round to the nearest tenth percent.

Answer
5. 16.8%

Example 5 Finding Percent of Increase

The number of applications for a mathematics scholarship at one university increased from 34 to 45 in one year. What is the percent of increase? Round to the nearest whole percent.

Solution: First we find the amount of increase by subtracting the original number of applicants from the new number of applicants.

amount of increase = 45 − 34 = 11

The amount of increase is 11 applicants. To find the percent of increase,

Section 6.5 | Applications of Percent   459

$$\text{percent of increase} = \frac{\text{amount of increase}}{\text{original amount}} = \frac{11}{34} \approx 0.32 = 32\%$$

The number of applications increased by about 32%.

Work Practice 5

Helpful Hint Make sure that this number is the original number and not the new number.

✓ **Concept Check** A student is calculating the percent increase in enrollment from 180 students one year to 200 students the next year. Explain what is wrong with the following calculations:

$$\cancel{\frac{\text{Amount}}{\text{of increase}} = 200 - 180 = 20}$$

$$\cancel{\frac{\text{Percent of}}{\text{increase}} = \frac{20}{200} = 0.1 = 10\%}$$

Suppose that your income was $300 a week and then it decreased by $30. The **percent of decrease** is

$$\begin{array}{l}\text{amount of decrease} \rightarrow \\ \text{original amount} \rightarrow \end{array} \frac{\$30}{\$300} = 0.1 = 10\%$$

Percent of Decrease

$$\text{percent of decrease} = \frac{\text{amount of decrease}}{\text{original amount}}$$

Then write the quotient as a percent.

Example 6 Finding Percent of Decrease

In response to a decrease in sales, a company with 1500 employees reduces the number of employees to 1230. What is the percent of decrease?

Solution: First we find the amount of decrease by subtracting 1230 from 1500.

amount of decrease = 1500 − 1230 = 270

The amount of decrease is 270. To find the percent of decrease,

$$\frac{\text{percent of}}{\text{decrease}} = \frac{\text{amount of decrease}}{\text{original amount}} = \frac{270}{1500} = 0.18 = 18\%$$

The number of employees decreased by 18%.

Work Practice 6

Practice 6
A town with a population of 20,200 in 2003 decreased to 18,483 in 2013. What was the percent of decrease?

Answer
6. 8.5%

✓ **Concept Check Answers**
To find the percent of increase, you have to divide the amount of increase (20) by the original amount (180); 10% decrease.

✓ **Concept Check** An ice cream stand sold 6000 ice cream cones last summer. This year the same stand sold 5400 cones. Was there a 10% increase, a 10% decrease, or neither? Explain.

Martin-Gay Interactive Videos Watch the section lecture video and answer the following questions.

Objective A 1. How do we interpret the answer 175,000 in Example 1? ▶

Objective B 2. In Example 3, what does the improper fraction tell us? ▶

See Video 6.5

6.5 Exercise Set MyLab Math

Objective A *Solve. For Exercises 1 and 2, the solutions have been started for you. See Examples 1 through 4. If necessary, round percents to the nearest tenth and all other answers to the nearest whole.*

1. An inspector found 24 defective bolts during an inspection. If this is 1.5% of the total number of bolts inspected, how many bolts were inspected?

   Start the solution:
   1. UNDERSTAND the problem. Reread it as many times as needed.
   Go to *Method 1* or *Method 2*.
   Method 1.
   2. TRANSLATE into an equation. (Fill in the boxes.)

      24   is   1.5%   of   what number
      ↓   ↓   ↓   ↓   ↓
      24   □   1.5%   □   x

   3. SOLVE for *x*. (See Example 3, Method 1, for help.)
   4. INTERPRET. The total number of bolts inspected was _____.

   Method 2.
   2. TRANSLATE into a proportion. (Fill in the first two blanks with "amount" or "base.")

      24   is   1.5%   of   what number
      ↓   ↓   ↓
      _____   percent   _____

      amount → _____ = 1.5 ← percent
      base →   _____   100

   3. SOLVE the proportion. (See Example 3, Method 2, for help.)
   4. INTERPRET. The total number of bolts inspected was _____.

2. A day care worker found 28 children absent one day during an epidemic of chicken pox. If this was 35% of the total number of children attending the day care center, how many children attend this day care center?

   Start the solution:
   1. UNDERSTAND the problem. Reread it as many times as needed.
   Go to *Method 1* or *Method 2*.
   Method 1.
   2. TRANSLATE into an equation. (Fill in the boxes.

      28   is   35%   of   what number
      ↓   ↓   ↓   ↓   ↓
      28   □   35%   □   x

   3. SOLVE for *x*. (See Example 3, Method 1, for help.)
   4. INTERPRET. The total number of children attending the day care center is _____.

   Method 2.
   2. TRANSLATE into a proportion. (Fill in the firs two blanks with "amount" or "base.")

      28   is   35%   of   what number
      ↓   ↓   ↓
      _____   percent   _____

      amount → _____ = 35 ← percent
      base →   _____   100

   3. SOLVE the proportion. (See Example 3, Method 2, for help.)
   4. INTERPRET. The total number of children attending the day care center is _____.

3. One model of Total Gym provides weight resistance through adjustments of incline. The minimum weight resistance is 4% of the weight of the person using the Total Gym. Find the minimum weight resistance possible for a 220-pound man. (*Source:* Total Gym)

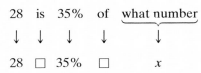

4. The maximum weight resistance for one model of Total Gym is 60% of the weight of the person using it. Find the maximum weight resistance possible for a 220-pound man. (See Exercise **3** if needed.)

5. A student's cost for last semester at her community college was $2700. She spent $378 of that on books. What percent of last semester's college costs was spent on books?

6. Pierre Sampeau belongs to his local food cooperative, where he receives a percentage of what he spends each year as a dividend. He spent $3850 last year at the food cooperative store and received a dividend of $154. What percent of his total spending at the food cooperative did he receive as a dividend?

Section 6.5 | Applications of Percent   461

2015, there were about 43,700 cinemas screens in the United States and Canada. Use this information for Exercises 7 and 8. *(Source: MPAA)*

7. If about 37.5% of the total screens in the United States and Canada were digital 3-D screens, find the approximate number of digital 3-D screens. Round to the nearest whole.

8. If about 2.5% of the total screens in the United States and Canada were analog screens, find the approximate number of analog screens. Round to the nearest whole.

Solve.

9. In 2016, there were 36,899 McDonald's restaurants worldwide, with 14,155 of them located in the United States. Determine the percent of McDonald's restaurants in the United States in 2016. Round to the nearest percent. *(Source: McDonald's Corporation)*

10. Of the 66,800 veterinarians in private practice in the United States in 2015, approximately 36,900 are women. Determine the percent of female veterinarians in private practice in the United States in 2015. Round to the nearest percent. *(Source: American Veterinary Medical Association)*

11. A furniture company currently produces 6200 chairs per month. If production decreases by 8%, find the decrease and the new number of chairs produced each month.

12. The enrollment at a local college decreased by 5% over last year's enrollment of 7640. Find the decrease in enrollment and the current enrollment.

13. From 2014 to 2024, the number of people employed as occupational therapy assistants in the United States is expected to increase by 43%. The number of people employed as occupational therapy assistants in 2014 was 41,900. Find the predicted number of occupational therapy assistants in 2024. *(Source: Bureau of Labor Statistics)*

14. From 2012 to 2016, the number of households owning reptiles increased by 63%. The number of households owning reptiles in 2012 was 3 million. Find the number of households owning reptiles in 2016. *(Source: American Veterinary Medical Association)*

Two states, West Virginia and Connecticut, decreased in population from 2014 to 2016. Their locations are shown on the U.S. map below. (Source: United States Census Bureau)

15. In 2014, the population of West Virginia was approximately 1,853,000. If the population decrease was about 0.49%, find the population of West Virginia in 2016. Round to the nearest whole number.

16. In 2014, the population of Connecticut was approximately 3,595,000. If the population decrease was about 0.001%, find the population of Connecticut in 2016.

A popular extreme sport is snowboarding. Ski trails are marked with difficulty levels of easy ●, intermediate ■, difficult ◆, expert ◆◆, and other variations. Use this information for Exercises 17 and 18. Round each percent to the nearest whole.

17. At Keystone ski area in Colorado, approximately 41 of the 135 total ski runs are rated intermediate. What percent of the runs are intermediate?

18. At Telluride ski area in Colorado, about 28 of the 115 total ski runs are rated easy. What percent of the runs are easy?

For each food described, find the percent of total calories from fat. If necessary, round to the nearest tenth percent. See Example 2.

19. Ranch dressing serving size of 2 tablespoons

	Calories
Total	40
From fat	20

20. Unsweetened cocoa powder serving size of 1 tablespoon

	Calories
Total	20
From fat	5

21.

Nutrition Facts
Serving Size 1 pouch (20g)
Servings Per Container 6

Amount Per Serving

Calories	80
Calories from fat	10

	% Daily Value*
Total Fat 1g	2%
Sodium 45mg	2%
Total Carbohydrate 17g	6%
Sugars 9g	
Protein 0g	
Vitamin C	25%

Not a significant source of saturated fat, cholesterol, dietary fiber, vitamin A, calcium and iron.

*Percent Daily Values are based on a 2,000 calorie diet.

Artificial Fruit Snacks

22.

Nutrition Facts
Serving Size $\frac{1}{4}$ cup (33g)
Servings Per Container About 9

Amount Per Serving

Calories 190   Calories from Fat 130

	% Daily Value
Total Fat 16g	24%
Saturated Fat 3g	16%
Cholesterol 0mg	0%
Sodium 135mg	6%
Total Carbohydrate 9g	3%
Dietary Fiber 1g	5%
Sugars 2g	
Protein 5g	

Vitamin A 0% • Vitamin C 0%
Calcium 0% • Iron 8%

Peanut Mixture

23.

Nutrition Facts
Serving Size 18 crackers (29g)
Servings Per Container About 9

Amount Per Serving

Calories 120   Calories from Fat 35

	% Daily Value*
Total Fat 4g	6%
Saturated Fat 0.5g	3%
Polyunsaturated Fat 0g	
Monounsaturated Fat 1.5g	
Cholesterol 0mg	0%
Sodium 220mg	9%
Total Carbohydrate 21g	7%
Dietary Fiber 2g	7%
Sugars 3g	
Protein 2g	

Vitamin A 0% • Vitamin C 0%
Calcium 2% • Iron 4%
Phosphorus 10%

Snack Crackers

24.

Nutrition Facts
Serving Size 28 crackers (31g)
Servings Per Container About 6

Amount Per Serving

Calories 130   Calories from Fat 35

	% Daily Value*
Total Fat 4g	6%
Saturated Fat 2g	10%
Polyunsaturated Fat 1g	
Monounsaturated Fat 1g	
Cholesterol 0mg	0%
Sodium 470mg	20%
Total Carbohydrate 23g	8%
Dietary Fiber 1g	4%
Sugars 4g	
Protein 2g	

Vitamin A 0% • Vitamin C 0%
Calcium 0% • Iron 2%

Snack Crackers

Solve. If necessary, round money amounts to the nearest cent and all other amounts to the nearest tenth. See Examples 1 through 4.

25. A family paid $26,250 as a down payment for a home. If this represents 15% of the price of the home, find the price of the home.

26. A banker learned that $842.40 is withheld from his monthly check for taxes and insurance. If this represents 18% of his total pay, find the total pay.

27. An owner of a repair service company estimates that for every 40 hours a repairperson is on the job, he can bill for only 78% of the hours. The remaining hours, the repairperson is idle or driving to or from a job. Determine the number of hours per 40-hour week the owner can bill for a repairperson.

28. A manufacturer of electronic components expects 1.04% of its products to be defective. Determine the number of defective components expected in a batch of 28,350 components. Round to the nearest whole component.

29. A car manufacturer announced that next year, the price of a certain model of car will increase by 4.5%. This year the price is $19,286. Find the increase in price and the new price.

30. A union contract calls for a 6.5% salary increase for all employees. Determine the increase and the new salary that a worker currently making $58,500 under this contract can expect.

A popular extreme sport is artificial wall climbing. The photo shown is an artificial climbing wall. Exercises 31 and 32 are about the Footsloggers Climbing Tower in Boone, North Carolina.

31. A climber is resting at a height of 21 feet while on the Footsloggers Climbing Tower. If this is 60% of the tower's total height, find the height of the tower.

32. A group plans to climb the Footsloggers Climbing Tower at the group rate, once they save enough money. Thus far, $126 has been saved. If this is 70% of the total amount needed for the group, find the total price.

33. Tuition for an Ohio resident at the Columbus campus of Ohio State University was $8994 in 2010. The tuition increased by 11.6% during the period from 2010 to 2016. Find the increase and the tuition for the 2016–2017 school year. Round the increase to the nearest whole dollar. (*Source:* Ohio State University)

34. The population of Americans aged 65 and older was 40 million in 2010. That population is projected to increase by 80.5% by 2030. Find the increase and the projected 2030 population. (*Source:* Bureau of the Census)

35. From 2014–2015 to 2024–2025, the number of students enrolled in associate degree programs is projected to increase by 22.3%. If the enrollment in associate degree programs in 2014–2015 was 6,700,000, find the increase and the projected number of students enrolled in an associate degree program in 2024–2025. (*Source:* National Center for Educational Statistics)

36. From 2010–2011 to 2021–2022, the number of bachelor degrees awarded is projected to increase by 17.4%. If the number of bachelor degrees awarded in 2010–2011 was 1,703,000, find the increase and the projected number of bachelor degrees awarded in the 2021–2022 school year. (*Source:* National Center for Educational Statistics)

Objective **B** *Find the amount of increase and the percent of increase. See Example 5.*

	Original Amount	New Amount	Amount of Increase	Percent of Increase
37.	50	80		
38.	8	12		
39.	65	117		
40.	68	170		

Find the amount of decrease and the percent of decrease. See Example 6.

	Original Amount	New Amount	Amount of Decrease	Percent of Decrease
41.	8	6		
42.	25	20		
43.	160	40		
44.	200	162		

Solve. Round percents to the nearest tenth, if necessary. See Examples 5 and 6.

45. There are 150 calories in a cup of whole milk and only 84 in a cup of skim milk. In switching to skim milk, find the percent of decrease in number of calories per cup.

46. In reaction to a slow economy, the number of employees at a soup company decreased from 530 to 477. What was the percent of decrease in the number of employees?

47. Before taking a typing course, Geoffry Landers could type 32 words per minute. By the end of the course, he was able to type 76 words per minute. Find the percent of increase.

48. The number of cable TV systems recently decreased from 10,845 to 10,700. Find the percent of decrease.

49. In 2016, there were approximately 71 million virtual reality devices in use worldwide. This is expected to grow to 337 million in 2020. What is the projected percent of increase? (*Source:* CTIA—The Wireless Association)

50. The population of Japan is expected to decrease from 127,799 thousand in 2011 to 97,076 thousand in 2050. Find the percent of decrease. (*Source:* International Programs Center, Bureau of the Census, U.S. Dept. of Commerce)

51. In 2014, there were 3782 thousand elementary and secondary teachers employed in the United States. This number is expected to increase to 4151 thousand teachers in 2021. What is the percent of increase? (*Source:* National Center for Educational Statistics)

52. In 2014, approximately 475 thousand correctional officers were employed in the United States. By 2024 this number is expected to increase to 493 thousand correctional officers. What is the percent of increase? (*Source:* Bureau of Labor Statistics)

53. Between 2014 and 2015, permanent digital downloads of singles decreased from approximately 1199 million to approximately 1021 million. What is the percent of decrease? (*Source:* Recording Industry Association of America)

54. As the largest health care occupation, registered nurses held about 3.1 million jobs in 2015. The number of registered nurses is expected to be 3.9 million by 2025. What is the percent of increase? (*Source:* American Association of Colleges of Nursing)

rd is a U.S. automobile brand, named after its founder, Henry Ford. Use the graph below to answer Exercises 55 and 56.

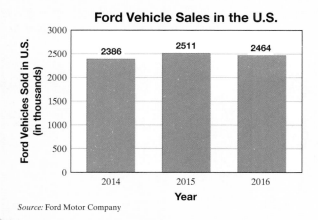

Ford Vehicle Sales in the U.S.

Source: Ford Motor Company

55. Find the percent of decrease in sales of Ford vehicles in the United States from 2015 to 2016.

56. Find the percent of increase in sales of Ford vehicles in the United States from 2014 to 2015.

7. In 1940, the average size of a farm in the United States was 174 acres. In a recent year, the average size of a farm in the United States had increased to 420 acres. What is this percent of increase? (*Source:* National Agricultural Statistics Service)

58. In 2012, there were 651 deaths from boating accidents in the United States. By 2015, the number of deaths from boating accidents had decreased to 626. What was the percent of decrease? (*Source:* U.S. Coast Guard)

1999, Napster, a free online file-sharing service, debuted. iTunes, which debuted in 2003, is given credit for getting people to rt paying for digital music. In 2015, for the first time, streaming music was the largest revenue producer in the music industry.

9. In 1999, total revenue from U.S. music sales and licensing was $14.6 billion. By 2015, this revenue had dropped to $7.1 billion. Find this percent of decrease in music revenue. (*Source:* Recording Industry Association of America)

60. By comparing prices, a particular music album downloads from a low of $2.99 to a high of $7.99. Find the percent of increase from $2.99 to $7.99.

eview

rform each indicated operation. See Sections 4.3 through 4.5, 4.7, and 5.2 through 5.4.

1. 0.12×38

62. $29.4 \div 0.7$

63. $9.20 + 1.98$

64. $78 - 19.46$

5. $-\dfrac{3}{8} + \dfrac{5}{12}$

66. $\left(-\dfrac{3}{8}\right)\left(-\dfrac{5}{12}\right)$

67. $2\dfrac{4}{5} \div 3\dfrac{9}{10}$

68. $2\dfrac{4}{5} - 3\dfrac{9}{10}$

oncept Extensions

9. If a number is increased by 100%, how does the increased number compare with the original number? Explain your answer.

70. In your own words, explain what is wrong with the following statement: "Last year we had 80 students attend. This year we have a 50% increase or a total of 160 students attend."

Check the Chapter Opener bar graph. If the number of Internet users in 2016 was 46.1% of the world population, it does not necessarily mean that each country of the world has 46.1% of its population identified as Internet users. In fact, there is a country that claims to have 100% of their population as Internet users. In the table below, we have listed a few selected countries along with Internet user data. Use your knowledge of percent and fill in the table. If needed, round percents to the nearest tenth and populations to the nearest hundred. (Source: Internet Live Stats)

	Country	Population	Internet Users % of Population	Number of Internet Users
71.	Iceland	331,778	100%	
72.	Faeroe Islands	48,239	98.5%	
73.	Monaco	37,863		35,196
74.	Samoa	194,523		56,373
75.	U.S.		88.5%	286,942,400
76.	Ecuador		43.1%	7,055,575

77. Check the bar graph for Exercises **55** and **56**. Are your answers for Exercises **55** and **56** reasonable? Explain why or why not.

78. Check the circle graph in Section 6.2, Exercises **79** through **82**. If the total world population of approximately 7417 million people represents 100% of this circle graph, do the percents calculated seem reasonable? Why or why not?

Explain what errors were made by each student when solving percent of increase or decrease problems and then correct the errors. See the Concept Checks in this section.

"The population of a certain rural town was 150 in 1990, 180 in 2000, and 150 in 2010."

79. Find the percent of increase in population from 1990 to 2000.

Miranda's solution: Percent of increase $= \dfrac{30}{180} = 0.1\overline{6} \approx 16.7\%$

80. Find the percent of decrease in population from 2000 to 2010.

Jeremy's solution: Percent of decrease $= \dfrac{30}{150} = 0.20 = 20\%$

81. The percent of increase from 1990 to 2000 is the same as the percent decrease from 2000 to 2010. True or false?

Chris's answer: True because they had the same amount of increase as the amount of decrease.

6.6 Percent and Problem Solving: Sales Tax, Commission, and Discount

Objective A Calculating Sales Tax and Total Price

Percents are frequently used in the retail trade. For example, most states charge a tax on certain items when purchased. This tax is called a **sales tax,** and retail stores collect it for the state. Sales tax is almost always stated as a percent of the purchase price.

A 9% sales tax rate on a purchase of a $10 calculator gives a sales tax of

sales tax = 9% of $10 = 0.09 · $10.00 = $0.90

The total price to the customer would be

purchase price plus sales tax
↓   ↓   ↓
$10.00   + $0.90 = $10.90

This example suggests the following equations:

Sales Tax and Total Price

sales tax = tax rate · purchase price
total price = purchase price + sales tax

In this section we round dollar amounts to the nearest cent.

Objectives

A Calculate Sales Tax and Total Price.

B Calculate Commissions.

C Calculate Discount and Sale Price.

Example 1 Finding Sales Tax and Purchase Price

Find the sales tax and the total price on the purchase of an $85.50 atlas in a city where the sales tax rate is 7.5%.

Solution: The purchase price is $85.50 and the tax rate is 7.5%.

sales tax = tax rate · purchase price
↓   ↓   ↓
sales tax = 7.5% · $85.50
         = 0.075 · $85.5   Write 7.5% as a decimal.
         ≈ $6.41   Rounded to the nearest cent

Thus, the sales tax is $6.41. Next find the total price.

total price = purchase price + sales tax
↓   ↓   ↓
total price = $85.50 + $6.41
           = $91.91

The sales tax on $85.50 is $6.41, and the total price is $91.91.

Work Practice 1

Practice 1

If the sales tax rate is 8.5%, what is the sales tax and the total amount due on a $59.90 Goodgrip tire? (Round the sales tax to the nearest cent.)

Answer
1. tax: $5.09; total: $64.99

468 Chapter 6 | Ratio, Proportion, and Percent

✓ **Concept Check** The purchase price of a textbook is $50 and sales tax is 10%. If you are told by the cashier that the total price is $75, how can you tell that a mistake has been made?

Practice 2

The sales tax on an $18,500 automobile is $1665. Find the sales tax rate.

Example 2 Finding a Sales Tax Rate

The sales tax on a $310 Sony flat-screen digital 32-inch television is $26.35. Find the sales tax rate.

Solution: Let r represent the unknown sales tax rate. Then

sales tax = tax rate · purchase price

$$\$26.35 = r \cdot \$310$$

$$\frac{26.35}{310} = \frac{r \cdot 310}{310} \quad \text{Divide both sides by 310.}$$

$$0.085 = r \quad \text{Simplify.}$$

$$8.5\% = r \quad \text{Write 0.085 as a percent.}$$

The sales tax rate is 8.5%.

■ Work Practice 2

Objective B Calculating Commissions

A **wage** is payment for performing work. Hourly wage, commissions, and salary are some of the ways wages can be paid. Many people who work in sales are paid a commission. An employee who is paid a **commission** is paid a percent of his or her total sales.

Commission

commission = commission rate · sales

Practice 3

A sales representative for Office Product Copiers sold $47,632 worth of copy equipment and supplies last month. What is his commission for the month if he is paid a commission of 6.6% of his total sales for the month?

Example 3 Finding the Amount of Commission

Sherry and Cindy Souter, real estate broker partners for Wealth Investments, sold a house for $214,000 last week. If their commission is 1.5% of the selling price of the home, find the amount of their commission.

Solution:

commission = commission rate · sales

commission = 1.5% · $214,000

= 0.015 · $214,000   Write 1.5% as 0.015.

= $3210   Multiply.

Their commission on the house is $3210.

■ Work Practice 3

Answers

2. 9%   3. $3143.71

✓ **Concept Check Answer**

Since $10\% = \frac{1}{10}$, the sales tax is $\frac{\$50}{10} = \$5$. The total price should have been $55.

Section 6.6 | Percent and Problem Solving: Sales Tax, Commission, and Discount

Example 4 — Finding a Commission Rate

A salesperson earned $1560 for selling $13,000 worth of electronics equipment. Find the commission rate.

Solution: Let r stand for the unknown commission rate. Then

commission = commission rate · sales

$$\$1560 = r \cdot \$13{,}000$$

$$\frac{1560}{13{,}000} = r \quad \text{Divide 1560 by 13,000, the number multiplied by } r.$$

$$0.12 = r \quad \text{Simplify.}$$

$$12\% = r \quad \text{Write 0.12 as a percent.}$$

The commission rate is 12%.

Work Practice 4

Practice 4
A salesperson earns $645 for selling $4300 worth of appliances. Find the commission rate.

Objective C — Calculating Discount and Sale Price

Suppose that an item that normally sells for $40 is on sale for 25% off. This means that the **original price** of $40 is reduced, or **discounted,** by 25% of $40, or $10. The **discount rate** is 25%, the **amount of discount** is $10, and the **sale price** is $40 − $10, or $30. Study the diagram below to visualize these terms.

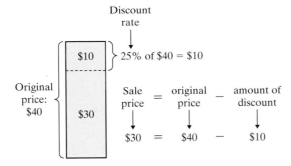

To calculate discounts and sale prices, we can use the following equations:

Discount and Sale Price

amount of discount = discount rate · original price

sale price = original price − amount of discount

Example 5 — Finding a Discount and a Sale Price

An electric rice cooker that normally sells for $65 is on sale for 25% off. What is the amount of discount and what is the sale price?

Solution: First we find the amount of discount, or simply the discount.

amount of discount = discount rate · original price

amount of discount = 25% · $65

= 0.25 · $65 Write 25% as 0.25.

= $16.25 Multiply.

(Continued on next page)

Practice 5
A discontinued washer and dryer combo is advertised on sale for 35% off the regular price of $700. Find the amount of discount and the sale price.

Answers
4. 15%   5. $245; $455

The discount is $16.25. Next, find the sale price.

sale price = original price − discount
sale price = $65 − $16.25
           = $48.75   Subtract.

The sale price is $48.75.

Work Practice 5

Vocabulary, Readiness & Video Check

Use the choices below to fill in each blank.

amount of discount   sale price   sales tax
commission   total price

1. _____ = tax rate · purchase price
2. _____ = purchase price + sales tax
3. _____ = commission rate · sales
4. _____ = discount rate · original price
5. _____ = original price − amount of discount

Martin-Gay Interactive Videos Watch the section lecture video and answer the following questions.

See Video 6.6

Objective A 6. In Example 1, what is our first step after translating the problem into an equation?

Objective B 7. What is our final step in solving Example 2?

Objective C 8. In the lecture before Example 3, since both equations shown involve the "amount of discount," how can the two equations be combined into one equation?

6.6 Exercise Set MyLab Math

Objective A Solve. See Examples 1 and 2.

1. What is the sales tax on a jacket priced at $150 if the sales tax rate is 5%?

2. If the sales tax rate is 6%, find the sales tax on a microwave oven priced at $188.

3. The purchase price of a camcorder is $799. What is the total price if the sales tax rate is 7.5%? Round to the nearest hundredth.

4. A stereo system has a purchase price of $426. What is the total price if the sales tax rate is 8%?

5. A new large-screen television has a purchase price of $4790. If the sales tax on this purchase is $335.30, find the sales tax rate.

6. The sales tax on the purchase of a $6800 used car is $374. Find the sales tax rate.

Section 6.6 | Percent and Problem Solving: Sales Tax, Commission, and Discount

7. The sales tax on a table saw is $10.20.
   a. What is the purchase price of the table saw (before tax) if the sales tax rate is 8.5%? (*Hint:* Use the sales tax equation and insert the replacement values.)
   b. Find the total price of the table saw.

8. The sales tax on a one-half-carat diamond ring is $76.
   a. Find the purchase price of the ring (before tax) if the sales tax rate is 9.5%. (See the hint for Exercise **7.a**.)
   b. Find the total price of the ring.

9. A gold and diamond bracelet sells for $1800. Find the sales tax and the total price if the sales tax rate is 6.5%.

10. The purchase price of a personal computer is $1890. If the sales tax rate is 8%, what is the sales tax and the total price?

11. The sales tax on the purchase of a futon is $24.25. If the tax rate is 5%, find the purchase price of the futon.

12. The sales tax on the purchase of a TV-DVD combination is $32.85. If the tax rate is 9%, find the purchase price of the TV-DVD.

13. The sales tax is $98.70 on a stereo sound system purchase of $1645. Find the sales tax rate.

14. The sales tax is $103.50 on a necklace purchase of $1150. Find the sales tax rate.

15. A cell phone costs $210, a battery recharger costs $15, and batteries cost $5. What is the sales tax and total price for purchasing these items if the sales tax rate is 7%?

16. Ms. Warner bought a blouse for $35, a skirt for $55, and a blazer for $95. Find the sales tax and the total price she paid, given a sales tax rate of 6.5%.

Objective B *Solve. See Examples 3 and 4.*

17. A sales representative for a large furniture warehouse is paid a commission rate of 4%. Find her commission if she sold $1,329,401 worth of furniture last year.

18. Rosie Davis-Smith is a beauty consultant for a home cosmetic business. She is paid a commission rate of 12.8%. Find her commission if she sold $1638 in cosmetics last month.

19. A salesperson earned a commission of $1380.40 for selling $9860 worth of paper products. Find the commission rate.

20. A salesperson earned a commission of $3575 for selling $32,500 worth of books to various bookstores. Find the commission rate.

21. How much commission will Jack Pruet make on the sale of a $325,900 house if he receives 1.5% of the selling price?

22. Frankie Lopez sold $9638 of jewelry this week. Find her commission for the week if she receives a commission rate of 5.6%.

23. A real estate agent earned a commission of $5565 for selling a house. If his rate is 3%, find the selling price of the house. (*Hint:* Use the commission equation and insert the replacement values.)

24. A salesperson earned $1750 for selling fertilizer. If her commission rate is 7%, find the selling price of the fertilizer. (See the hint for Exercise **23**.)

Objective **C** *Find the amount of discount and the sale price. See Example 5.*

	Original Price	Discount Rate	Amount of Discount	Sale Price
25.	$89	10%		
26.	$74	20%		
27.	$196.50	50%		
28.	$110.60	40%		
29.	$410	35%		
30.	$370	25%		
31.	$21,700	15%		
32.	$17,800	12%		

33. A $300 fax machine is on sale for 15% off. Find the amount of discount and the sale price.

34. A $4295 designer dress is on sale for 30% off. Find the amount of discount and the sale price.

Objectives **A B** Mixed Practice *Complete each table.*

	Purchase Price	Tax Rate	Sales Tax	Total Price
35.	$305	9%		
36.	$243	8%		
37.	$56	5.5%		
38.	$65	8.4%		

	Sale	Commission Rate	Commission
39.	$235,800	3%	
40.	$195,450	5%	
41.	$17,900		$1432
42.	$25,600		$2304

Review

Multiply. See Sections 4.3, 5.3, and 5.5.

43. $2000 \cdot \dfrac{3}{10} \cdot 2$

44. $500 \cdot \dfrac{2}{25} \cdot 3$

45. $400 \cdot \dfrac{3}{100} \cdot 11$

46. $1000 \cdot \dfrac{1}{20} \cdot 5$

47. $600 \cdot 0.04 \cdot \dfrac{2}{3}$

48. $6000 \cdot 0.06 \cdot \dfrac{3}{4}$

Concept Extensions

Solve. See the Concept Check in this section.

49. Your purchase price is $68 and the sales tax rate is 9.5%. Round each amount and use the rounded amounts to estimate the total price. Choose the best estimate.
    a. $105 b. $58 c. $93 d. $77

50. Your purchase price is $200 and the tax rate is 10%. Choose the best estimate of the total price.
    a. $190 b. $210 c. $220 d. $300

Tipping *One very useful application of percent is mentally calculating a tip. Recall that to find 10% of a number, simply move the decimal point one place to the left. To find 20% of a number, just double 10% of the number. To find 15% of a number, find 10% and then add to that number half of the 10% amount. Mentally fill in the chart below. To do so, start by rounding the bill amount to the nearest dollar.*

Tipping Chart

	Bill Amount	10%	15%	20%
51.	$40.21			
52.	$15.89			
53.	$72.17			
54.	$9.33			

55. Suppose that the original price of a shirt is $50. Which is better, a 60% discount or a discount of 30% followed by a discount of 35% of the reduced price? Explain your answer.

56. Which is better, a 30% discount followed by an additional 25% off or a 20% discount followed by an additional 40% off? To see, suppose an item costs $100 and calculate each discounted price. Explain your answer.

57. A diamond necklace sells for $24,966. If the tax rate is 7.5%, find the total price.

58. A house recently sold for $562,560. The commission rate on the sale is 5.5%. If the real estate agent is to receive 60% of the commission, find the amount received by the agent.

6.7 Percent and Problem Solving: Interest

Objective A Calculating Simple Interest

Interest is money charged for using other people's money. When you borrow money, you pay interest. When you loan or invest money, you earn interest. The money borrowed, loaned, or invested is called the **principal amount,** or simply **principal.** Interest is normally stated in terms of a percent of the principal for a given period of time. The **interest rate** is the percent used in computing the interest. Unless stated otherwise, *the rate is understood to be per year.* When the interest is computed on the original principal, it is called **simple interest.** Simple interest is calculated using the following equation:

Objectives

A Calculate Simple Interest.

B Calculate Compound Interest.

Simple Interest

Simple Interest = Principal · Rate · Time

$$I = P \cdot R \cdot T$$

where the rate is understood to be per year and time is in years.

Practice 1

Find the simple interest after 5 years on $875 at an interest rate of 7%.

Example 1 Finding Simple Interest

Find the simple interest after 2 years on $500 at an interest rate of 12%.

Solution: In this example, $P = \$500$, $R = 12\%$, and $T = 2$ years. Replace the variables with values in the formula $I = PRT$.

$$I = P \cdot R \cdot T$$
$$I = \$500 \cdot 12\% \cdot 2 \quad \text{Let } P = \$500, R = 12\%, \text{ and } T = 2.$$
$$= \$500 \cdot (0.12) \cdot 2 \quad \text{Write 12\% as a decimal.}$$
$$= \$120 \quad \text{Multiply.}$$

The simple interest is $120.

■ Work Practice 1

If time is not given in years, we need to convert the given time to years.

Practice 2

A student borrowed $1500 for 9 months on her credit card at a simple interest rate of 20%. How much interest did she pay?

Example 2 Finding Simple Interest

A recent college graduate borrowed $2400 at 10% simple interest for 8 months to buy a used Toyota Corolla. Find the simple interest he paid.

Solution: Since there are 12 months in a year, we first find what part of a year 8 months is.

$$8 \text{ months} = \frac{8}{12} \text{ year} = \frac{2}{3} \text{ year}$$

Now we find the simple interest.

$$I = P \cdot R \cdot T$$
$$= \$2400 \cdot (0.10) \cdot \frac{2}{3} \quad \text{Let } P = \$2400, R = 10\% \text{ or } 0.10, \text{ and } T = \frac{2}{3}.$$
$$= \$160$$

The interest on his loan is $160.

■ Work Practice 2

✓**Concept Check** Suppose in Example 2 you had obtained an answer of $16,000. How would you know that you had made a mistake in this problem?

When money is borrowed, the borrower pays the original amount borrowed, or the principal, as well as the interest. When money is invested, the investor receives the original amount invested, or the principal, as well as the interest. In either case, the **total amount** is the sum of the principal and the interest.

Finding the Total Amount of a Loan or Investment

total amount (paid or received) = principal + interest

Answers
1. $306.25 **2.** $225

✓**Concept Check Answer**
$16,000 is too much interest.

Example 3 Finding the Total Amount of an Investment

An accountant invested $2000 at a simple interest rate of 10% for 2 years. What total amount of money will she have from her investment in 2 years?

Solution: First we find her interest.

$$I = P \cdot R \cdot T$$
$$= \$2000 \cdot (0.10) \cdot 2 \quad \text{Let } P = \$2000, R = 10\% \text{ or } 0.10, \text{ and } T = 2.$$
$$= \$400$$

The interest is $400.

Next, we add the interest to the principal.

total amount = principal + interest

total amount = $2000 + $400
            = $2400

After 2 years, she will have a total amount of $2400.

■ Work Practice 3

Practice 3
If $2100 is borrowed at a simple interest rate of 13% for 6 months, find the total amount paid.

✓ **Concept Check** Which investment would earn more interest: an amount of money invested at 8% interest for 2 years, or the same amount of money invested at 8% for 3 years? Explain.

Objective B Calculating Compound Interest

Recall that simple interest depends on the original principal only. Another type of interest is compound interest. **Compound interest** is computed not only on the principal, but also on the interest already earned in previous compounding periods. Compound interest is used more often than simple interest.

Let's see how compound interest differs from simple interest. Suppose that $2000 is invested at 7% interest **compounded annually** for 3 years. This means that interest is added to the principal at the end of each year and that next year's interest is computed on this new amount. In this section, we round dollar amounts to the nearest cent.

	Amount at Beginning of Year	Principal	·	Rate	·	Time	= Interest	Amount at End of Year
1st year	$2000	$2000	·	0.07	·	1	= $140	$2000 + 140 = $2140
2nd year	$2140	$2140	·	0.07	·	1	= $149.80	$2140 + 149.80 = $2289.80
3rd year	$2289.80	$2289.80	·	0.07	·	1	≈ $160.29	$2289.80 + 160.29 = $2450.09

The compound interest earned can be found by

total amount − original principal = compound interest

$2450.09 − $2000 = $450.09

The simple interest earned would have been

principal · rate · time = interest

$2000 · 0.07 · 3 = $420

Since compound interest earns "interest on interest," compound interest earns more than simple interest.

Computing compound interest using the method above can be tedious. We can use a calculator and the compound interest formula on the next page to compute compound interest more quickly.

Answer
3. $2236.50

✓ **Concept Check Answers**
8% for 3 years. Since the interest rate is the same, the longer you keep the money invested, the more interest you earn.

Compound Interest Formula

The total amount A in an account is given by

$$A = P\left(1 + \frac{r}{n}\right)^{n \cdot t}$$

where P is the principal, r is the interest rate written as a decimal, t is the length of time in years, and n is the number of times compounded per year.

Practice 4
$3000 is invested at 4% interest compounded annually. Find the total amount after 6 years.

Example 4
$1800 is invested at 2% interest compounded annually. Find the total amount after 3 years.

Solution: "Compounded annually" means 1 time a year, so $n = 1$. Also, $P = \$1800$, $r = 2\% = 0.02$, and $t = 3$ years.

$$A = P\left(1 + \frac{r}{n}\right)^{n \cdot t}$$
$$= 1800\left(1 + \frac{0.02}{1}\right)^{1 \cdot 3}$$
$$= 1800(1.02)^3$$
$$\approx 1910.17 \quad \text{Round to 2 decimal places.}$$

Helpful Hint: Remember order of operations. **First** evaluate $(1.02)^3$, then multiply by 1800.

The total amount at the end of 3 years is $1910.17.

▪ Work Practice 4

Practice 5
$5500 is invested at $6\frac{1}{4}\%$ compounded *daily* for 5 years. Find the total amount at the end of 5 years. (Use 1 year = 365 days.)

Example 5 Finding Total Amount Received from an Investment
$4000 is invested at 5.3% compounded quarterly for 10 years. Find the total amount at the end of 10 years.

Solution: "Compounded quarterly" means 4 times a year, so $n = 4$. Also, $P = \$4000$, $r = 5.3\% = 0.053$, and $t = 10$ years.

$$A = P\left(1 + \frac{r}{n}\right)^{n \cdot t}$$
$$= 4000\left(1 + \frac{0.053}{4}\right)^{4 \cdot 10}$$
$$= 4000(1.01325)^{40}$$
$$\approx 6772.12$$

The total amount after 10 years is $6772.12.

▪ Work Practice 5

Answers
4. $3795.96   5. $7517.41

Calculator Explorations — Compound Interest Formula

For a review of using your calculator to evaluate compound interest, see this box.

Let's review the calculator keys pressed to evaluate the expression in Example 5,

$$4000\left(1 + \frac{0.053}{4}\right)^{40}$$

To evaluate, press the keys

[4000] [×] [(] [1] [+] [0.053] [÷] [4] [)] [y^x] or [∧] [40] then [=] or [ENTER]. The display will read [6772.117549]. Rounded to 2 decimal places, this is 6772.12.

Find the compound interest.

1. $600, 5 years, 9%, compounded quarterly
2. $10,000, 15 years, 4%, compounded daily
3. $1200, 20 years, 11%, compounded annually
4. $5800, 1 year, 7%, compounded semiannually
5. $500, 4 years, 6%, compounded quarterly
6. $2500, 19 years, 5%, compounded daily

Section 6.7 | Percent and Problem Solving: Interest

Vocabulary, Readiness & Video Check

Use the choices below to fill in each blank. Choices may be used more than once.

total amount   simple   principal amount   compound

1. To calculate _____ interest, use $I = P \cdot R \cdot T$.
2. To calculate _____ interest, use $A = P\left(1 + \dfrac{r}{n}\right)^{n \cdot t}$.
3. _____ interest is computed not only on the original principal, but also on interest already earned in previous compounding periods.
4. When interest is computed on the original principal only, it is called _____ interest.
5. _____ (paid or received) = principal + interest
6. The _____ is the money borrowed, loaned, or invested.

Martin-Gay Interactive Videos   Watch the section lecture video and answer the following questions.

See Video 6.7

Objective A   7. Complete this statement based on the lecture before Example 1: Simple interest is charged on the _____ only.

Objective B   8. In Example 2, how often is the interest compounded and what number does this translate to in the formula?

6.7 Exercise Set MyLab Math

Objective A *Find the simple interest. See Examples 1 and 2.*

	Principal	Rate	Time
1.	$200	8%	2 years
2.	$160	11.5%	4 years
3.	$5000	10%	$1\frac{1}{2}$ years
4.	$375	18%	6 months
5.	$2500	16%	21 months

	Principal	Rate	Time
6.	$800	9%	3 years
7.	$950	12.5%	5 years
8.	$1500	14%	$2\frac{1}{4}$ years
9.	$775	15%	8 months
10.	$1000	10%	18 months

Solve. See Examples 1 through 3.

11. A company borrows $162,500 for 5 years at a simple interest rate of 12.5%. Find the interest paid on the loan and the total amount paid back.

12. $265,000 is borrowed to buy a house. If the simple interest rate on the 30-year loan is 8.25%, find the interest paid on the loan and the total amount paid back.

13. A money market fund advertises a simple interest rate of 9%. Find the total amount received on an investment of $5000 for 15 months.

14. The Real Service Company takes out a 270-day (9-month) short-term, simple interest loan of $4500 to finance the purchase of some new equipment. If the interest rate is 14%, find the total amount that the company pays back.

15. Marsha borrows $8500 and agrees to pay it back in 4 years. If the simple interest rate is 17%, find the total amount she pays back.

16. An 18-year-old is given a high school graduation gift of $2000. If this money is invested at 8% simple interest for 5 years, find the total amount.

Objective B *Find the total amount in each compound interest account. See Examples 4 and 5.*

17. $6150 is compounded semiannually at a rate of 14% for 15 years.

18. $2060 is compounded annually at a rate of 15% for 10 years.

19. $1560 is compounded daily at a rate of 8% for 5 years.

20. $1450 is compounded quarterly at a rate of 10% for 15 years.

21. $10,000 is compounded semiannually at a rate of 9% for 20 years.

22. $3500 is compounded daily at a rate of 8% for 10 years.

23. $2675 is compounded annually at a rate of 9% for 1 year.

24. $6375 is compounded semiannually at a rate of 10% for 1 year.

25. $2000 is compounded annually at a rate of 8% for 5 years.

26. $2000 is compounded semiannually at a rate of 8% for 5 years.

27. $2000 is compounded quarterly at a rate of 8% for 5 years.

28. $2000 is compounded daily at a rate of 8% for 5 years.

Review

Find the perimeter of each figure. See Section 1.3.

29.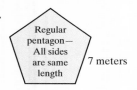

30.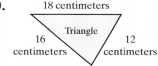

31.

32. Square, 21 miles

Perform the indicated operations. See Section 4.3 through 4.6.

33. $\dfrac{x}{4} + \dfrac{x}{5}$

34. $-\dfrac{x}{4} \div \left(-\dfrac{x}{5}\right)$

35. $\left(\dfrac{2}{3}\right)\left(-\dfrac{1}{3}\right) - \left(\dfrac{9}{10}\right)\left(\dfrac{2}{5}\right)$

36. $\dfrac{3}{11} \div \dfrac{9}{22} \cdot \dfrac{1}{3}$

Concept Extensions

37. Review Exercises 25 through 28. As the number of compoundings increases per year, how is the total amount affected?

38. Explain how to find the amount of interest in a compounded account.

39. Compare the following accounts: Account 1: $1000 is invested for 10 years at a simple interest rate of 6%. Account 2: $1000 is compounded semiannually at a rate of 6% for 10 years. Discuss how the interest is computed for each account. Determine which account earns more interest. Why?

Chapter 6 Group Activity

Investigating Scale Drawings

Section 6.1

Materials:

- ruler
- tape measure
- grid paper (optional)

This activity may be completed by working in groups or individually.

Scale drawings are used by architects, engineers, interior designers, shipbuilders, and others. In a scale drawing, each unit measurement on the drawing represents a fixed length on the object being drawn. For instance, in an architect's scale drawing, 1 inch on the drawing may represent 10 feet on a building. The scale describes the relationship between the measurements. If the measurements have the same units, the scale can be expressed as a ratio. In this case, the ratio would be 1:120, representing 1 inch to 120 inches (or 10 feet).

Use a ruler and the scale drawing of a college building below to answer the following questions.

1. How wide are each of the front doors of the college building?
2. How long is the front of the college building?
3. How tall is the front of the college building?

Now you will draw your own scale floor plan. First choose a room to draw—it can be your math classroom, your living room, your dormitory room, or any room that can be easily measured. Start by using a tape measure to measure the distances around the base of the walls in the room you are drawing.

4. Choose a scale for your floor plan.
5. Convert each measurement in the room you are drawing to the corresponding lengths needed for the scale drawing.
6. Complete your floor plan (you may find it helpful to use grid paper). Mark the locations of doors and windows on your floor plan. Be sure to indicate on the drawing the scale used in your floor plan.

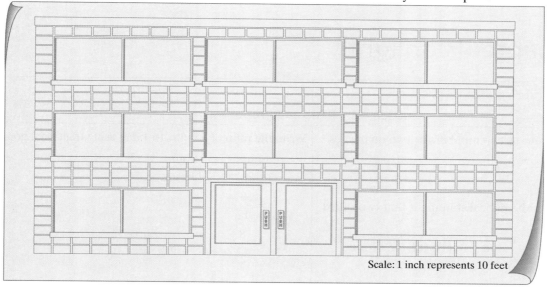

Scale: 1 inch represents 10 feet

Chapter 6 Vocabulary Check

Fill in each blank with one of the words or phrases listed below. Some words may be used more than once.

percent	sales tax	is	0.01	$\frac{1}{100}$	amount of discount
percent of decrease	total price	ratio	proportion	base	of
amount	100%	compound interest	percent of increase	sale price	commission

1. In a mathematical statement, _____ usually means "multiplication."
2. In a mathematical statement, _____ means "equal."
3. _____ means "per hundred."

4. _____ is computed not only on the principal, but also on interest already earned in previous compounding periods.

5. In the percent proportion, $\dfrac{\rule{1cm}{0.4pt}}{\rule{0pt}{0pt}} = \dfrac{\text{percent}}{100}$.

6. To write a decimal or fraction as a percent, multiply by _____.

7. The decimal equivalent of the % symbol is _____.

8. The fraction equivalent of the % symbol is _____.

9. The percent equation is _____ · percent = _____.

10. _____ = $\dfrac{\text{amount of decrease}}{\text{original amount}}$

11. _____ = $\dfrac{\text{amount of increase}}{\text{original amount}}$

12. _____ = tax rate · purchase price
13. _____ = purchase price + sales tax
14. _____ = commission rate · sales
15. _____ = discount rate · original price
16. _____ = original price − amount of discount
17. A(n) _____ is a mathematical statement that two ratios are equal.
18. A (n) _____ is the quotient of two numbers or two quantities

> **Helpful Hint**
>
> ▶ Are you preparing for your test? To help, don't forget to take these:
> - Chapter 6 Getting Ready for the Test on page 487
> - Chapter 6 Test on page 488
>
> Then check all of your answers at the back of this text. For further review, the step-by-step video solutions to any of these exercises are located in MyLab Math.

6 Chapter Highlights

Definitions and Concepts	Examples
Section 6.1 Ratio and Proportion	
A **ratio** is the quotient of two numbers or two quantities.	Write the ratio of 5 hours to 1 day using fractional notation. $$\dfrac{5 \text{ hours}}{1 \text{ day}} = \dfrac{5 \text{ hours}}{24 \text{ hours}} = \dfrac{5}{24}$$
A **proportion** is a mathematical statement that two ratios are equal. In the proportion $\dfrac{a}{b} = \dfrac{c}{d}$, the products ad and bc are called **cross products.** If $\dfrac{a}{b} = \dfrac{c}{d}$, then $ad = bc$.	$\dfrac{2}{3} = \dfrac{8}{12}$ $\dfrac{x}{7} = \dfrac{15}{35}$ $\dfrac{2}{3} \times \dfrac{8}{12}$ ⟶ 3 · 8 or 24, 2 · 12 or 24 Solve: $\dfrac{3}{4} = \dfrac{x}{x-1}$ $\dfrac{3}{4} = \dfrac{x}{x-1}$ $3(x - 1) = 4x$ Set cross products equal. $3x - 3 = 4x$ $-3 = x$
Section 6.2 Percents, Decimals, and Fractions	
Percent means "per hundred." The % symbol denotes percent. **To write a percent as a decimal,** replace the % symbol with its decimal equivalent, 0.01, and multiply.	$51\% = \dfrac{51}{100}$ 51 per 100 $7\% = \dfrac{7}{100}$ 7 per 100 $32\% = 32(0.01) = 0.32$

Definitions and Concepts	Examples
Section 6.2 Percents, Decimals, and Fractions (*continued*)	
To write a decimal as a percent, multiply by 100%.	$0.08 = 0.08(100\%) = 08.\% = 8\%$
To write a percent as a fraction, replace the % symbol with its fraction equivalent, $\frac{1}{100}$, and multiply.	$25\% = 25 \cdot \frac{1}{100} = \frac{25}{100} = \frac{25}{4 \cdot 25} = \frac{1}{4}$
To write a fraction as a percent, multiply by 100%.	$\frac{1}{6} = \frac{1}{6} \cdot 100\% = \frac{1}{6} \cdot \frac{100}{1}\% = \frac{100}{6}\% = 16\frac{2}{3}\%$
Section 6.3 Solving Percent Problems with Equations	
Three key words in the statement of a percent problem are **of,** which means multiplication ($\cdot$) **is,** which means equal ($=$) **what** (or some equivalent word or phrase), which stands for the unknown (x)	Solve: 6 is 12% of what number? ↓ ↓ ↓ ↓ ↓ 6 = 12% $\cdot$ x $6 = 0.12 \cdot x$ Write 12% as a decimal. $\frac{6}{0.12} = \frac{0.12 \cdot x}{0.12}$ Divide both sides by 0.12. $50 = x$ Thus, 6 is 12% of 50.
Section 6.4 Solving Percent Problems with Proportions	
Percent Proportion $\frac{\text{amount}}{\text{base}} = \frac{\text{percent}}{100}$ ← always 100 or amount → $\frac{a}{b} = \frac{p}{100}$ ← percent base →	Solve: 20.4 is what percent of 85? ↓ ↓ ↓ amount percent base amount → $\frac{20.4}{85} = \frac{p}{100}$ ← percent base → $20.4 \cdot 100 = 85 \cdot p$ Set cross products equal. $2040 = 85 \cdot p$ Multiply. $\frac{2040}{85} = \frac{85 \cdot p}{85}$ Divide both sides by 85. $24 = p$ Simplify. Thus, 20.4 is 24% of 85.
Section 6.5 Applications of Percent	
Percent of Increase percent of increase $= \frac{\text{amount of increase}}{\text{original amount}}$ **Percent of Decrease** percent of decrease $= \frac{\text{amount of decrease}}{\text{original amount}}$	A town with a population of 16,480 decreased to 13,870 over a 12-year period. Find the percent of decrease. Round to the nearest whole percent. amount of decrease $= 16{,}480 - 13{,}870$ $= 2610$ percent of decrease $= \frac{\text{amount of decrease}}{\text{original amount}}$ $= \frac{2610}{16{,}480} \approx 0.16$ $\approx 16\%$ The town's population decreased by about 16%.

Definitions and Concepts	Examples
Section 6.6 Percent and Problem Solving: Sales Tax, Commission, and Discount	

Sales Tax
   sales tax = sales tax rate · purchase price
   total price = purchase price + sales tax
	Find the sales tax and the total price of a purchase of $42.00 if the sales tax rate is 9%.
	
	sales tax = sales tax rate · purchase price
	sales tax = 9% · $42
	        = 0.09 · $42
	        = $3.78
	
	The total price is
	
	total price = purchase price + sales tax
	total price = $42.00 + $3.78
	           = $45.78
	
	The total price is $45.78.

Commission
   commission = commission rate · sales
	A salesperson earns a commission of 3%. Find the commission from sales of $12,500 worth of appliances.
	
	commission = commission rate · sales
	commission = 3% · $12,500
	          = 0.03 · 12,500
	          = $375
	
	The commission is $375.

Discount and Sale Price
   amount of discount = discount rate · original price
   sale price = original price − amount of discount
	A suit is priced at $320 and is on sale today for 25% off. What is the sale price?
	
	amount of discount = discount rate · original price
	amount of discount = 25% · $320
	                 = 0.25 · 320
	                 = $80
	
	sale price = original price − amount of discount
	sale price = $320 − $80
	          = $240
	
	The sale price is $240.

Definitions and Concepts	Examples
Section 6.7 Percent and Problem Solving: Interest	
Simple Interest interest = principal · rate · time where the rate is understood to be per year, unless told otherwise.	Find the simple interest after 3 years on $800 at an interest rate of 5%. interest = principal · rate · time interest = $800 · 5% · 3 = $800 · 0.05 · 3 Write 5% as 0.05. = $120 Multiply. The interest is $120.
Compound Interest Formula The total amount A in an account is $$A = P\left(1 + \frac{r}{n}\right)^{n \cdot t}$$ where P is the principal, r is the interest rate written as a decimal, t is time in years, and n is the number of times compounded per year.	$800 is invested at 5% compounded quarterly for 10 years. Find the total amount at the end of 10 years. "Compounded quarterly" means $n = 4$. Also, $P = \$800$, $r = 5\% = 0.05$, and $t = 10$ years. $$A = P\left(1 + \frac{r}{n}\right)^{n \cdot t}$$ $$= 800\left(1 + \frac{0.05}{4}\right)^{4 \cdot 10}$$ $$= 800(1.0125)^{40}$$ $$\approx 1314.90$$ The total amount is $1314.90.

Chapter 6   Review

(6.1) *Write each phrase as a ratio in fractional notation.*

1. 20 cents to 1 dollar

2. four parts red to six parts white

Solve each proportion.

3. $\dfrac{x}{2} = \dfrac{12}{4}$

4. $\dfrac{20}{1} = \dfrac{x}{25}$

5. $\dfrac{32}{100} = \dfrac{100}{x}$

6. $\dfrac{20}{2} = \dfrac{c}{5}$

7. $\dfrac{2}{x-1} = \dfrac{3}{x+3}$

8. $\dfrac{4}{y-3} = \dfrac{3}{y+2}$

9. $\dfrac{y+2}{y} = \dfrac{5}{3}$

10. $\dfrac{x-3}{3x+2} = \dfrac{2}{5}$

11. A machine can process 300 parts in 20 minutes. Find how many parts can be processed in 45 minutes.

12. As his consulting fee, Mr. Visconti charges $90.00 per day. Find how much he charges for 3 hours of consulting. Assume an 8-hour workday.

(6.2) *Solve.*

13. In a survey of 100 adults, 37 preferred pepperoni on their pizzas. What percent preferred pepperoni?

14. A basketball player made 77 out 100 attempted free throws. What percent of free throws was made?

Write each percent as a decimal.

15. 26%
16. 75%
17. 3.5%
18. 1.5%
19. 275%
20. 400%
21. 47.85%
22. 85.34%

Write each decimal as a percent.

23. 1.6
24. 0.055
25. 0.076
26. 0.085
27. 0.71
28. 0.65
29. 6
30. 9

Write each percent as a fraction or mixed number in simplest form.

31. 7%
32. 15%
33. 25%
34. 8.5%
35. 10.2%
36. $16\frac{2}{3}$%
37. $33\frac{1}{3}$%
38. 110%

Write each fraction or mixed number as a percent.

39. $\frac{2}{5}$
40. $\frac{7}{10}$
41. $\frac{7}{12}$
42. $1\frac{2}{3}$
43. $1\frac{1}{4}$
44. $\frac{3}{5}$
45. $\frac{1}{16}$
46. $\frac{5}{8}$

(6.3) Translating *Translate each to an equation and solve.*

47. 1250 is 1.25% of what number?
48. What number is $33\frac{1}{3}$% of 24,000?
49. 124.2 is what percent of 540?
50. 22.9 is 20% of what number?
51. What number is 17% of 640?
52. 693 is what percent of 462?

(6.4) Translating *Translate each to a proportion and solve.*

53. 104.5 is 25% of what number?
54. 16.5 is 5.5% of what number?
55. What number is 30% of 532?
56. 63 is what percent of 35?
57. 93.5 is what percent of 85?
58. What number is 33% of 500?

Chapter 6 Review **485**

6.5) *Solve.*

59. In a survey of 2000 people, it was found that 1320 have a microwave oven. Find the percent of people who own microwaves.

60. Of the 12,360 freshmen entering County College, 2000 are enrolled in prealgebra. Find the percent of entering freshmen who are enrolled in prealgebra. Round to the nearest whole percent.

61. The number of violent crimes in a city decreased from 675 to 534. Find the percent of decrease. Round to the nearest tenth percent.

62. The current charge for dumping waste in a local landfill is $16 per cubic foot. To cover new environmental costs, the charge will increase to $33 per cubic foot. Find the percent of increase.

63. This year the fund drive for a charity collected $215,000. Next year, a 4% decrease is expected. Find how much is expected to be collected in next year's drive.

64. A local union negotiated a new contract that increases the hourly pay 15% over last year's pay. The old hourly rate was $11.50. Find the new hourly rate rounded to the nearest cent.

6.6) *Solve.*

65. If the sales tax rate is 9.5%, what is the total amount charged for a $250 coat?

66. Find the sales tax paid on a $25.50 purchase if the sales tax rate is 8.5%. Round to the nearest cent.

67. Russ James is a sales representative for a chemical company and is paid a commission rate of 5% on all sales. Find his commission if he sold $100,000 worth of chemicals last month.

68. Carol Sell is a sales clerk in a clothing store. She receives a commission of 7.5% on all sales. Find her commission for the week if her sales for the week were $4005. Round to the nearest cent.

69. A $3000 mink coat is on sale for 30% off. Find the discount and the sale price.

70. A $90 calculator is on sale for 10% off. Find the discount and the sale price.

6.7) *Solve.*

71. Find the simple interest due on $4000 loaned for 4 months at 12% interest.

72. Find the simple interest due on $6500 loaned for 3 months at 20%.

73. Find the total amount in an account if $5500 is compounded annually at 12% for 15 years.

74. Find the total amount in an account if $6000 is compounded semiannually at 11% for 10 years.

75. Find the total amount in an account if $100 is compounded quarterly at 12% for 5 years.

76. Find the total amount in an account if $1000 is compounded quarterly at 18% for 20 years.

Mixed Review

Write each percent as a decimal.

77. 3.8%

78. 124.5%

Write each decimal as a percent.

79. 0.54

80. 95.2

Write each percent as a fraction or mixed number in simplest form.

81. 47%

82. 5.6%

Write each fraction or mixed number as a percent.

83. $\frac{3}{8}$

84. $\frac{6}{5}$

Translating *Translate each into an equation and solve.*

85. 43 is 16% of what number?

86. 27.5 is what percent of 25?

87. What number is 36% of 1968?

88. 67 is what percent of 50?

Translating *Translate each into a proportion and solve.*

89. 75 is what percent of 25?

90. What number is 16% of 240?

91. 28 is 5% of what number?

92. 52 is what percent of 16?

Solve.

93. The total number of cans in a soft drink machine is 300. If 78 soft drinks have been sold, find the percent of soft drink cans that have been sold.

94. A home valued at $96,950 last year has lost 7% of it value this year. Find the loss in value.

95. A dinette set sells for $568.00. If the sales tax rate is 8.75%, find the total price of the dinette set.

96. The original price of a video game is $23.00. It is on sale for 15% off. What is the amount of the discount?

97. A candy salesman makes a commission of $1.60 from each case of candy he sells. If a case of candy costs $12.80, what is his rate of commission?

98. Find the total amount due on a 6-month loan of $1400 at a simple interest rate of 13%.

99. Find the total amount due on a loan of $5500 for 9 years at 12.5% simple interest.

Chapter 6 — Getting Ready for the Test LC

MULTIPLE CHOICE All of the exercises are **Multiple Choice.** Choose the correct letter for each exercise.

1. The phrase "6 is to 7 as 18 is to 21" translates to
   A. $\dfrac{6}{7} = \dfrac{18}{21}$ **B.** $\dfrac{6}{18} = \dfrac{21}{7}$ **C.** $\dfrac{6}{21} = \dfrac{7}{18}$

2. Use cross products to determine which proportion is *not* equivalent to the proportion $\dfrac{3}{x} = \dfrac{5}{11}$.
   A. $\dfrac{3}{11} = \dfrac{x}{5}$ **B.** $\dfrac{x}{3} = \dfrac{11}{5}$ **C.** $\dfrac{3}{5} = \dfrac{x}{11}$ **D.** $\dfrac{11}{x} = \dfrac{5}{3}$

3. Since "percent" means "per hundred," choose the number that does *not* equal 12%.
   A. $\dfrac{12}{100}$ **B.** 0.12 **C.** $\dfrac{3}{25}$ **D.** 1.2

4. Choose the number that does *not* equal 100%.
   A. 10 **B.** 1 **C.** 1.00 **D.** $\dfrac{100}{100}$

5. Choose the number that does *not* equal 50%.
   A. 0.5 **B.** 5 **C.** $\dfrac{1}{2}$ **D.** $\dfrac{50}{100}$

6. Choose the number that does *not* equal 80%.
   A. 0.80 **B.** 0.8 **C.** 8 **D.** $\dfrac{4}{5}$

Use the information below for Exercises 7 through 10.
- 100% of a number is that original number.
- 50% of a number is half that number.
- 25% of a number is $\dfrac{1}{4}$ of that number.
- 10% of a number is $\dfrac{1}{10}$ of that number.

For Exercises 7 through 10, choose the letter that correctly fills in each blank.

   A. 100% **B.** 50% **C.** 25% **D.** 10%

7. _____ of 70 is 35.
8. _____ of 88 is 8.8.
9. _____ of 47 is 47.
10. _____ of 28 is 7.

For Exercise 11, choose the correct letter.

11. Your bill at a restaurant is $24.86. If you want to leave a 20% tip, which tip amount is closest to 20%?
    A. $2.48 **B.** $20 **C.** $5 **D.** $10

For Exercises 12 and 13, an amount of $150 is to be discounted by 10%.

12. Find the amount of discount.
    A. $15 **B.** $45 **C.** $30 **D.** $10

13. Find the new discounted price (original price − discount).
    A. $120 **B.** $10 **C.** $140 **D.** $135

For Exercise 14, choose the correct letter.

14. If the original price of a pair of shoes is $40 and the shoe price is to be discounted by 25% at the register, choose the closest amount for your shoes before tax.
    A. $10 **B.** $20 **C.** $30 **D.** $40

Chapter 6 Test MyLab Math or YouTube

For additional practice go to your study plan in MyLab Math.

Answers

Write each percent as a decimal.

1. 85%
2. 500%
3. 0.6%

Write each decimal as a percent.

4. 0.056
5. 6.1
6. 0.35

Write each percent as a fraction or a mixed number in simplest form.

7. 120%
8. 38.5%
9. 0.2%

Write each fraction or mixed number as a percent.

10. $\dfrac{11}{20}$
11. $\dfrac{3}{8}$
12. $1\dfrac{3}{4}$

13. In the past decade, Americans have increased their consumption of bottled water by $\dfrac{1}{4}$. Write $\dfrac{1}{4}$ as a percent. (*Source:* Bottled Water Organization)

14. As of 2016, 45% of all working-age households have no retirement plan in place. Write 45% as a fraction. (*Source:* National Institute on Retirement Security)

Solve.

15. What number is 42% of 80?
16. 0.6% of what number is 7.5?
17. 567 is what percent of 756?

Solve. If necessary, round percents to the nearest tenth, dollar amounts to the nearest cent, and all other numbers to the nearest whole.

18. An alloy is 12% copper. How much copper is contained in 320 pounds of this alloy?

19. A farmer in Nebraska estimates that 20% of his potential crop, or $11,350, has been lost to a hard freeze. Find the total value of his potential crop.

20. If the local sales tax rate is 8.25%, find the total amount charged for a stereo system priced at $354.

21. A town's population increased from 25,200 to 26,460. Find the percent of increase.

22. A $120 framed picture is on sale for 15% off. Find the discount and the sale price.

23. Randy Nguyen is paid a commission rate of 4% on all sales. Find Randy's commission if his sales were $9875.

24. A sales tax of $13.77 is added to an item's price of $152.99. Find the sales tax rate. Round to the nearest whole percent.

25. Find the simple interest earned on $2000 saved for $3\frac{1}{2}$ years at an interest rate of 9.25%.

26. $1365 is compounded annually at 8%. Find the total amount in the account after 5 years.

27. A couple borrowed $400 from a bank at 13.5% simple interest for 6 months for car repairs. Find the total amount due the bank at the end of the 6-month period.

28. In a recent 5-year period, the number of a category of crimes reported in New York City decreased from 51,209 to 50,008. Find the percent of decrease. (*Source:* New York State Division of Criminal Justice Services)

29. Write the ratio $75 to $10 as a fraction in simplest form.

30. Solve: $\dfrac{5}{y+1} = \dfrac{4}{y+2}$

31. In a sample of 85 fluorescent bulbs, 3 were found to be defective. At this rate, how many defective bulbs should be found in 510 bulbs?

18. _____

19. _____

20. _____

21. _____

22. _____

23. _____

24. _____

25. _____

26. _____

27. _____

28. _____

29. _____

30. _____

31. _____

Chapters 1–6 Cumulative Review

Answers

1. _____
2. _____
3. _____
4. _____
5. _____
6. _____
7. _____
8. _____
9. _____
10. _____
11. _____
12. _____
13. _____
14. _____
15. _____
16. _____
17. _____
18. _____
19. _____
20. _____
21. _____
22. _____

1. Multiply: 236×86

2. Multiply: 409×76

3. Subtract 7 from -3.

4. Subtract -2 from 8.

5. Solve: $x - 2 = -1$

6. Solve: $x + 4 = 3$

7. Solve: $3(2x - 6) + 6 = 0$

8. Solve: $5(x - 2) = 3x$

9. Write an equivalent fraction with the given denominator.

$$3 = \frac{}{7}$$

10. Write an equivalent fraction with the given denominator.

$$8 = \frac{}{5}$$

11. Write in simplest form: $-\dfrac{10}{27}$

12. Write in simplest form: $\dfrac{10y}{32}$

13. Divide: $-\dfrac{7}{12} \div -\dfrac{5}{6}$

14. Divide: $-\dfrac{2}{5} \div \dfrac{7}{10}$

15. Evaluate $y - x$ if $x = -\dfrac{3}{10}$ and $y = -\dfrac{8}{10}$.

16. Evaluate $2x + 3y$ if $x = \dfrac{2}{5}$ and $y = -$

17. Find: $-\dfrac{3}{4} - \dfrac{1}{14} + \dfrac{6}{7}$

18. Find: $\dfrac{2}{9} + \dfrac{7}{15} - \dfrac{1}{3}$

19. Simplify: $\dfrac{\frac{1}{2} + \frac{3}{8}}{\frac{3}{4} - \frac{1}{6}}$

20. Simplify: $\dfrac{\frac{2}{3} + \frac{1}{6}}{\frac{3}{4} - \frac{3}{5}}$

21. Solve: $\dfrac{x}{2} = \dfrac{x}{3} + \dfrac{1}{2}$

22. Solve: $\dfrac{x}{2} + \dfrac{1}{5} = 3 - \dfrac{x}{5}$

Cumulative Review

23. Write each mixed number as an improper fraction.
  a. $4\frac{2}{9}$
  b. $1\frac{8}{11}$

24. Write each mixed number as an improper fraction.
  a. $3\frac{2}{5}$
  b. $6\frac{2}{7}$

Write each decimal as a fraction or mixed number. Write your answer in simplest form.

25. 0.125

26. 0.85

27. −105.083

28. 17.015

29. Subtract: 85 − 17.31

30. Subtract: 38 − 10.06

Multiply.

31. 7.68 × 10

32. 12.483 × 100

33. (−76.3)(1000)

34. −853.75 × 10

35. Evaluate $x \div y$ for $x = 2.5$ and $y = 0.05$.

36. Is 470 a solution of the equation $\frac{x}{100} = 4.75$?

37. Find the median of the scores: 67, 91, 75, 86, 55, 91

38. Find the mean of 36, 40, 86, and 30.

39. Write the ratio of 2.5 to 3.15 as a fraction in simplest form.

40. Write the ratio of 5.8 to 7.6 as a fraction in simplest form.

41. Solve for x: $\frac{45}{x} = \frac{5}{7}$

42. Solve for x: $\frac{x-1}{3x+1} = \frac{3}{8}$

43. On a chamber of commerce map of Abita Springs, 5 miles corresponds to 2 inches. How many miles correspond to 7 inches?

44. A student doing math homework can complete 7 problems in about 6 minutes. At this rate, how many problems can be completed in 30 minutes?

Write each percent as a fraction or mixed number in simplest form.

45. 1.9%

46. 2.3%

47. $33\frac{1}{3}$%

48. 108%

49. What number is 35% of 60?

50. What number is 42% of 85?

7 Graphs, Triangle Applications, and Introduction to Statistics and Probability

In this chapter, we present data in a usable form on a graph and the basic ideas of probability. Also in this chapter, we introduce important triangle applications, such as the Pythagorean theorem and the concept of congruent and similar triangles.

BEV, or Battery Electric Vehicle

HEV, or Hybrid Electric Vehicle

PHEV, or Plug-in Hybrid Electric Vehicle

Sections

7.1 Pictographs, Bar Graphs, Histograms, Line Graphs, and Introduction to Statistics

7.2 Circle Graphs

Integrated Review—Reading Graphs

7.3 Square Roots and the Pythagorean Theorem

7.4 Congruent and Similar Triangles

7.5 Counting and Introduction to Probability

Check Your Progress

Vocabulary Check
Chapter Highlights
Chapter Review
Getting Ready for the Test
Chapter Test
Cumulative Review

What Are a BEV, HEV, and PHEV?

The car industry is getting complicated. In this Chapter Opener, we study the popularity of electric vehicles (EV). In general, all of the above are versions of electric vehicles (EV). Let's simplify the differences below:

BEV, or Battery Electric Vehicle: These are vehicles that operate solely on battery power that can be charged by plugging into an outlet or charging station. They are sometimes called "all electric vehicles."

HEV, or Hybrid Electric Vehicle: These vehicles run on a battery (electricity) and an engine (usually gasoline-powered). We can call these non-plug-in hybrids.

PHEV, or Plug-in Hybrid Electric Vehicle: These are also hybrid vehicles, but their batteries can only be charged by plugging in to an outlet or charging station.

The graph below is a double broken line graph (or double line graph) of HEV and BEV cars sold in 2016. A graph like this helps us visually see trends and make predictions. In Section 7.2, Exercise 28, we use percent data on electric vehicle sales to draw a circle graph.

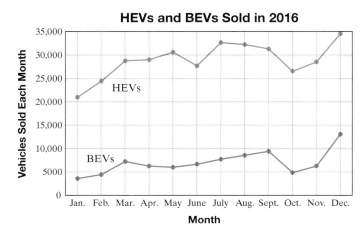

Source: Electric Drive Transportation Association

7.1 Pictographs, Bar Graphs, Histograms, Line Graphs, and Introduction to Statistics

Often data are presented visually in a graph. In this section, we practice reading several kinds of graphs including pictographs, bar graphs, and line graphs.

Objectives

A Read Pictographs.

B Read and Construct Bar Graphs.

C Read and Construct Histograms (or Frequency Distribution Graphs).

D Read Line Graphs.

E Calculate Range, Mean, Median, and Mode from a Frequency Distribution Table or Graph.

Objective A Reading Pictographs

A **pictograph** such as the one below is a graph in which pictures or symbols are used. This type of graph contains a key that explains the meaning of the symbol used. An advantage of using a pictograph to display information is that comparisons can easily be made. A disadvantage of using a pictograph is that it is often hard to tell what fractional part of a symbol is shown. For example, in the pictograph below, Arabic shows a part of a symbol, but it's hard to read with any accuracy what fractional part of a symbol is shown.

Example 1 Calculating Languages Spoken

The following pictograph shows the top eight most-spoken (primary) languages. Use this pictograph to answer the questions.

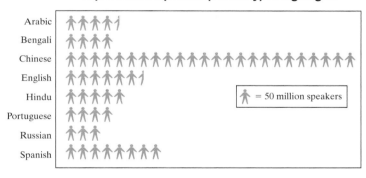

Source: www.ethnologue.com

a. Approximate the number of people who primarily speak Russian.
b. Approximate how many more people primarily speak English than Russian.

Solution:

a. Russian corresponds to 3 symbols, and each symbol represents 50 million speakers. This means that the number of people who primarily speak Russian is approximately 3 · (50 million) or 150 million people.

b. English shows $3\frac{1}{2}$ more symbols than Russian. This means that $3\frac{1}{2}$ · (50 million) or 175 million more people primarily speak English than Russian.

Work Practice 1

Practice 1

Use the pictograph shown in Example 1 to answer the following questions:

a. Approximate the number of people who primarily speak Spanish.

b. Approximate how many more people primarily speak Spanish than Arabic.

Answers
1. a. 400 million people
b. 175 million people

Objective B Reading and Constructing Bar Graphs

Another way to visually present data is with a **bar graph.** Bar graphs can appear with vertical bars or horizontal bars. Although we have studied bar graphs in previous sections, we now practice reading the height or length of the bars contained in a bar graph. An advantage to using bar graphs is that a scale is usually included for greater accuracy. Care must be taken when reading bar graphs, as well as other types of graphs—they may be misleading, as shown later in this section.

Example 2   Finding the Number of Endangered Species

The following bar graph shows the number of endangered species in the United States in 2016. Use this graph to answer the questions.

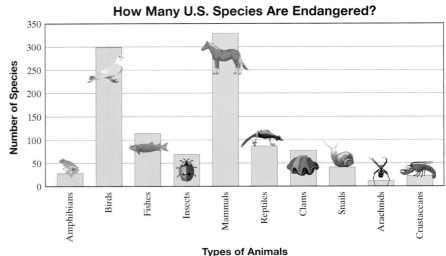

Source: U.S. Fish and Wildlife Service

Practice 2

Use the bar graph in Example 2 to answer the following questions:

a. Approximate the number of endangered species that are fishes.

b. Which category shows the fewest endangered species?

a. Approximate the number of endangered species that are clams.

b. Which category has the most endangered species?

Solution:

a. To approximate the number of endangered species that are clams, we go to the top of the bar that represents clams. From the top of this bar, we move horizontally to the left until the scale is reached. We read the height of the bar on the scale as approximately 75. There are approximately 75 clam species that are endangered, as shown.

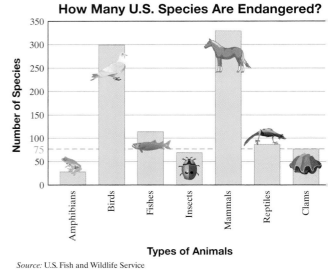

Source: U.S. Fish and Wildlife Service

b. The most endangered species is represented by the tallest (longest) bar. The tallest bar corresponds to mammals.

■ Work Practice 2

Answers

2. a. 115 **b.** arachnids

Section 7.1 | Pictographs, Bar Graphs, Histograms, Line Graphs, and Introduction to Statistics

Next, we practice constructing a bar graph.

Example 3 Draw a vertical bar graph using the information in the table below, which gives the caffeine content of selected foods.

Average Caffeine Content of Selected Foods			
Food	Milligrams	Food	Milligrams
Brewed coffee (percolator, 8 ounces)	124	Instant coffee (8 ounces)	104
Brewed decaffeinated coffee (8 ounces)	3	Brewed tea (U.S. brands, 8 ounces)	64
Coca-Cola Classic (8 ounces)	31	Mr. Pibb (8 ounces)	27
Dark chocolate (semisweet, $1\frac{1}{2}$ ounces)	30	Milk chocolate (8 ounces)	9

(*Sources:* International Food Information Council and the Coca-Cola Company)

Solution: We draw and label a vertical line and a horizontal line as shown below on the left. These lines are also called axes. We place the different food categories along the horizontal axis. Along the vertical axis, we place a scale.

There are many choices of scales that would be appropriate. Notice that the milligrams range from a low of 3 to a high of 124. From this information, we use a scale that starts at 0 and then shows multiples of 20 so that the scale is not too cluttered. The scale stops at 140, the smallest multiple of 20 that will allow all milligrams to be graphed. It may also be helpful to draw horizontal lines along the scale markings to help draw the vertical bars at the correct height. The finished bar graph is shown below on the right.

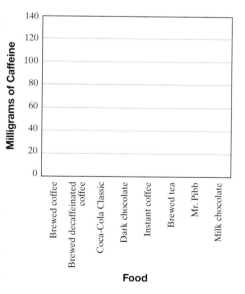
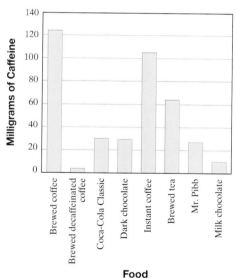

■ Work Practice 3

As mentioned previously, graphs can be misleading. Both graphs on the next page show the same information, but with different scales. Special care should be taken when forming conclusions from the appearance of a graph.

Practice 3

Draw a vertical bar graph using the information in the table about electoral votes for selected states.

Total Electoral Votes by Selected States	
State	Electoral Votes
Texas	38
California	55
Florida	29
Kansas	6
Indiana	20
Georgia	15

(*Source:* World Almanac)

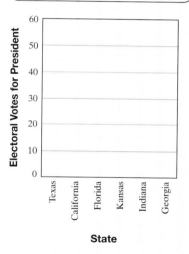

Answer

3.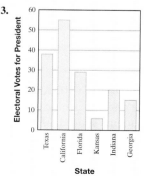

Notice the ⤵ symbol on each vertical scale on the graphs below. This symbol alerts us that numbers are missing from that scale

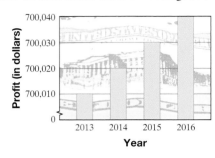
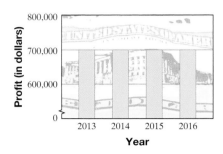

Are profits shown in the graphs above greatly increasing, or are they remaining about the same?

Objective C Reading and Constructing Histograms

Suppose that the test scores of 36 students are summarized in the table below. We call this table a **frequency distribution table** since one column gives the frequency or number of times the event in the other column occurred.

Student Scores	Frequency (Number of Students)
40–49	1
50–59	3
60–69	2
70–79	10
80–89	12
90–99	8

The results in this frequency distribution table can be displayed in a histogram. A **histogram** is a special bar graph. The width of each bar represents a range of numbers called a **class interval.** The height of each bar corresponds to how many times a number in the class interval occurs and is called the **class frequency.** The bars in a histogram lie side by side with no space between them. *Note:* Another name for this histogram is a **frequency distribution graph**.

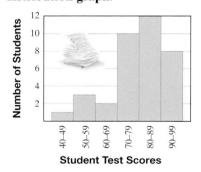

Practice 4

Use the histogram above Example 4 to determine how many students scored 80–89 on the test.

Example 4 Reading a Histogram on Student Test Scores

Use the preceding histogram to determine how many students scored 50–59 on the test.

Solution: We find the bar representing 50–59. The height of this bar is 3, which means 3 students scored 50–59 on the test.

■ Work Practice 4

Answer
4. 12

Section 7.1 | Pictographs, Bar Graphs, Histograms, Line Graphs, and Introduction to Statistics

Example 5 Reading a Histogram on Student Test Scores

Use the histogram above Example 4 to determine how many students scored 80 or above on the test.

Solution: We see that two different bars fit this description. There are 12 students who scored 80–89 and 8 students who scored 90–99. The sum of these two categories is 12 + 8 or 20 students. Thus, 20 students scored 80 or above on the test.

■ Work Practice 5

Now we will look at a way to construct histograms.

The daily high temperatures for 1 month in New Orleans, Louisiana, are recorded in the following list:

85°	90°	95°	89°	88°	94°
87°	90°	95°	92°	95°	94°
82°	92°	96°	91°	94°	92°
89°	89°	90°	93°	95°	91°
88°	90°	88°	86°	93°	89°

The data in this list have not been organized and can be hard to interpret. One way to organize the data is to place them in a **frequency distribution table.** We will do this in Example 6.

Example 6 Completing a Frequency Distribution on Temperature

Complete the frequency distribution table for the preceding temperature data.

Solution: Go through the data and place a tally mark in the second column of the table next to the class interval. Then count the tally marks and write each total in the third column of the table.

Class Intervals (Temperatures)	Tally	Class Frequency (Number of Days)
82°–84°			1
85°–87°	|||	3
88°–90°	||||| |||||		11
91°–93°	||||| ||	7
94°–96°	||||| |||	8

■ Work Practice 6

Example 7 Constructing a Histogram

Construct a histogram from the frequency distribution table in Example 6.

Solution:

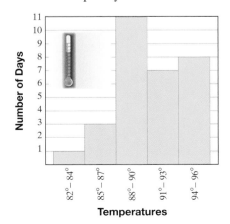

Practice 5
Use the histogram above Example 4 to determine how many students scored less than 80 on the test.

Practice 6
Complete the frequency distribution table for the data below. Each number represents a credit card owner's unpaid balance for one month.

0	53	89	125
265	161	37	76
62	201	136	42

Class Intervals (Credit Card Balances)	Tally	Class Frequency (Number of Months)
$0–$49		
$50–$99		
$100–$149		
$150–$199		
$200–$249		
$250–$299		

Practice 7
Construct a histogram from the frequency distribution table for Practice 6.

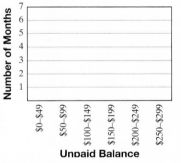

Answers

5. 16

6.
Tally	Class Frequency (Number Months)	Tally	Class Frequency (Number Months)
				3			1
					4			1
			2			1

7.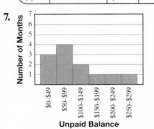

■ Work Practice 7

✓ **Concept Check** Which of the following sets of data is better suited to representation by a histogram? Explain.

Set 1		Set 2	
Grade on Final	# of Students	Section Number	Avg. Grade on Final
51–60	12	150	78
61–70	18	151	83
71–80	29	152	87
81–90	23	153	73
91–100	25		

Objective D Reading Line Graphs

Another common way to display information with a graph is by using a **line graph**. An advantage of a line graph is that it can be used to visualize relationships between two quantities. A line graph can also be very useful in showing changes over time.

Practice 8

Use the temperature graph in Example 8 to answer the following questions:

a. During what month is the average daily temperature the lowest?

b. During what month is the average daily temperature 25°F?

c. During what months is the average daily temperature greater than 70°F?

Example 8 Reading Temperatures from a Line Graph

The following line graph shows the average daily temperature for each month in Omaha, Nebraska. Use this graph to answer the questions below.

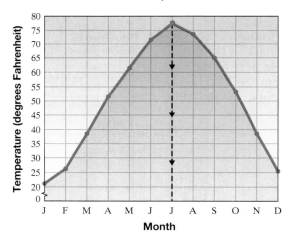

Source: National Climatic Data Center

a. During what month is the average daily temperature the highest?

b. During what month, from July through December, is the average daily temperature 65°F?

c. During what months is the average daily temperature less than 30°F?

Solution:

a. The month with the highest temperature corresponds to the highest point. This is the red point shown on the graph above. We follow this highest point downward to the horizontal month scale and see that this point corresponds to July.

Answers

8. a. January b. December
c. June, July, and August

✓ **Concept Check Answer**

Set 1; the grades are arranged in ranges of scores.

Section 7.1 | Pictographs, Bar Graphs, Histograms, Line Graphs, and Introduction to Statistics

b. The months July through December correspond to the right side of the graph. We find the 65°F mark on the vertical temperature scale and move to the right until a point on the right side of the graph is reached. From that point, we move downward to the horizontal month scale and read the corresponding month. During the month of September, the average daily temperature is 65°F.

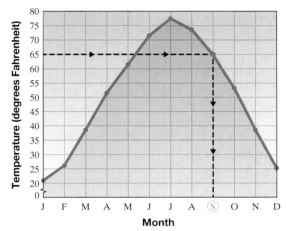

Source: National Climatic Data Center

c. To see what months the temperature is less than 30°F, we find what months correspond to points that fall below the 30°F mark on the vertical scale. These months are January, February, and December.

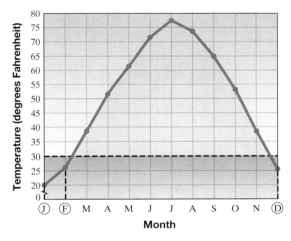

Source: National Climatic Data Center

Work Practice 8

Objective E Finding the Range of a Data Set and Reviewing Mean, Median, and Mode

In Section 5.7, we studied three measures of central tendency—the mean, the median, and the mode. All three of these can be used to calculate a number that might best describe a data set.

In this objective, we study one way to describe the dispersion of a data set, and we review mean, median, and mode. What is dispersion? In statistics, **dispersion** is a way to describe the degree to which the data values are scattered.

Range

The range of a data set is the difference between the largest data value and the smallest data value.

$$\text{range} = \text{largest data value} - \text{smallest data value}$$

Practice 9

The table lists the rounded salary of 10 staff members. Find the range.

Staff	Salary (in thousands)
A	$32
B	$34
C	$46
D	$38
E	$42
F	$95
G	$102
H	$50
J	$42

Example 9 The following pulse rates (for 1 minute) were recorded for a group of 15 students. Find the range.

78, 80, 66, 68, 71, 64, 82, 71, 70, 65, 70, 75, 77, 86, 72.

Solution:
$$\begin{aligned}\text{range} &= \text{largest data value} - \text{smallest data value}\\ &= 86 - 64\\ &= 22\end{aligned}$$

The range of this data set is 22.

■ Work Practice 9

Before we review mean, median, and mode, let's introduce a formula for finding the position of the median.

- Recall that the **median** of a set of numbers in numerical order is the middle number.
- If the number of items is odd, the median is the middle number.
- If the number of items is even, the median is the *mean* (average) of the two middle numbers.

For a long list of data items, this formula gives us the position of the median.

Position of the Median

For n data items in order from smallest to largest, the median is the item in the $\frac{n+1}{2}$ position.

Note:

If n is an even number, then the position formula $\frac{n+1}{2}$, will not be a whole number.

In this case, simply find the average of the two data items whose positions are closest to but before and after $\frac{n+1}{2}$.

Helpful Hint

The formula above, $\frac{n+1}{2}$, does not give the *value* of the median, just the **position** of the median.

Answer
9. $70 thousand

Example 10

Find the (a) range, (b) mean, (c) median, and (d) mode from this frequency distribution table of retirement ages. If needed, round answers to 1 decimal place.

Age	Frequency
60	3
61	1
62	1
63	2
64	2
65	2

Solution: Study the table for a moment. From the frequency column, we see that there are 11 data items $(3 + 1 + 1 + 2 + 2 + 2)$.

a. range = largest data value − smallest data value
= 65 − 60
= 5

The range of this data set is 5.

b. To find the mean, we use our mean formula:

$$\text{mean} = \frac{\text{sum of items}}{\text{number of items}} = \frac{60 \cdot 3 + 61 + 62 + 63 \cdot 2 + 64 \cdot 2 + 65 \cdot 2}{11}$$

$$= \frac{687}{11} \approx 62.5$$

The mean of the data set is approximately 62.5.

c. Since there are 11 data items and the items are arranged in numerical order in the table, we find $\frac{n+1}{2}$ to locate the middle item. This is $\frac{11+1}{2} = \frac{12}{2} = 6$, or the sixth item.

The median is the sixth number, or 63.

d. The mode has the greatest frequency, so the mode is 60.

■ Work Practice 10

Practice 10

One state with a young retirement age is Michigan. The table below is from a poll of retirement ages from that state.

Age	Frequency
50	1
59	3
60	3
62	5
63	2
67	1

Find the (a) range, (b) mean, (c) median, and (d) mode of these data. If needed, round answers to 1 decimal place.

Helpful Hint
Since there are three 60's, for example, we can either use:
$60 + 60 + 60$ or $3 \cdot 60$
(or $60 \cdot 3$).

Answers
10. a. range: 17 **b.** mean: 60.7,
c. median: 62, **d.** mode: 62

Vocabulary, Readiness & Video Check

Fill in each blank with one of the choices below. Not all choices will be used.

| pictograph | bar | class frequency | range | mean |
| histogram | line | class interval | median | mode |

1. A _____ graph presents data using vertical or horizontal bars.
2. A _____ is a graph in which pictures or symbols are used to visually present data.
3. A _____ graph displays information with a line that connects data points.
4. A _____ is a special bar graph in which the width of each bar represents a _____ and the height of each bar represents the _____.
5. _____ = greatest data value − smallest data value
6. The formula $\frac{n+1}{2}$ can be used to find the position of the _____.

Martin-Gay Interactive Videos

See Video 7.1

Watch the section lecture video and answer the following questions.

Objective A 7. From the pictograph in Example 1, how would you approximate the number of wildfires for any given year?

Objective B 8. What is one advantage of displaying data in a bar graph?

Objective C 9. Complete this statement based on the lecture before Example 6: A histogram is a special kind of _____.

Objective D 10. From the line graph in Examples 10–13, what year averaged the greatest number of goals per game and what was this average?

Objective E 11. Based on the results of Example 15, give an example of a data set of four data items whose range is 0.

7.1 Exercise Set MyLab Math

Objective A *The following pictograph shows the number of acres devoted to wheat production in 2016 in selected states. Use this graph to answer Exercises 1 through 8. See Example 1. (Source: U.S. Department of Agriculture)*

1. Which state plants the greatest acreage in wheat?

2. Which of the states shown plant the least amount of wheat acreage?

3. Approximate the number of acres of wheat planted in Montana.

4. Approximate the number of acres of wheat planted in Kansas.

5. Which state(s) plant less than 3,000,000 acres of wheat?

6. Which state(s) plant more than 7,000,000 acres of wheat?

7. Which state plants more wheat: Montana or Oklahoma?

8. Which states plant about the same amount of acreage?

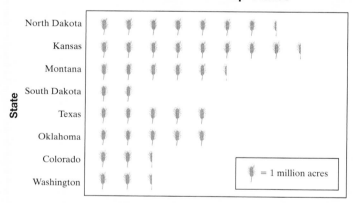

The following pictograph shows the average number of wildfires in the United States between 2010 and 2016. Use this graph to answer Exercises 9 through 16. See Example 1. (Source: National Interagency Fire Center)

9. Approximate the number of wildfires in 2013.

10. Approximately how many wildfires were there in 2012?

11. Which year, of the years shown, had the most wildfires?

12. In what years were the number of wildfires greater than 72,000?

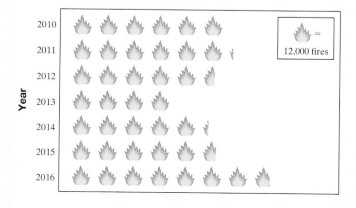

13. What was the amount of decrease in wildfires from 2012 to 2013?

14. What was the amount of increase in wildfires from 2015 to 2016?

15. What was the average annual number of wildfires from 2013 to 2015? (*Hint:* How do you calculate the average?)

16. Give a possible explanation for the sharp increase in the number of wildfires in 2016.

Objective **B** *The National Weather Service has exacting definitions for hurricanes; they are tropical storms with winds in excess of 74 mph. The following bar graph shows the number of hurricanes, by month, that have made landfall on the mainland United States between 1851 and 2016. Use this graph to answer Exercises 17 through 22. See Example 2. (Source: National Weather Service: National Hurricane Center)*

17. In which month did the most hurricanes make landfall in the United States?

18. In which month did the fewest hurricanes make landfall in the United States?

19. Approximate the number of hurricanes that made landfall in the United States during the month of August.

20. Approximate the number of hurricanes that made landfall in the United States in September.

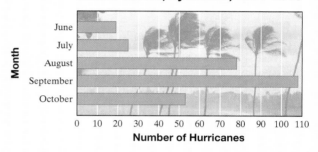

21. In 2008 alone, two hurricanes made landfall during the month of August. What fraction of all the 78 hurricanes that made landfall during August is this?

22. In 2007, only one hurricane made landfall on the United States during the entire season, in the month of September. If there have been 108 hurricanes to make landfall in the month of September since 1851, approximately what percent of these arrived in 2007?

The following horizontal bar graph shows the approximate 2016 population of the world's largest cities (including their suburbs). Use this graph to answer Exercises 23 through 28. See Example 2. (Source: CityPopulation)

23. Name the city with the largest population, and estimate its population.

24. Name the city whose population is between 23 million and 24 million.

25. Name the city in the United States with the largest population, and estimate its population.

26. Name the two cities that have approximately the same population.

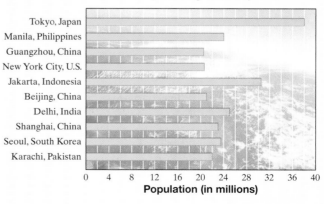

27. How much larger (in terms of population) is Manila, Philippines, than Beijing, China?

28. How much larger (in terms of population) is Jakarta, Indonesia, than Seoul, South Korea?

Use the information given to draw a vertical bar graph. Clearly label the bars. See Example 3.

29.

Fiber Content of Selected Foods	
Food	Grams of Total Fiber
Kidney beans $\left(\frac{1}{2} c\right)$	4.5
Oatmeal, cooked $\left(\frac{3}{4} c\right)$	3.0
Peanut butter, chunky (2 tbsp)	1.5
Popcorn (1 c)	1.0
Potato, baked with skin (1 med)	4.0
Whole wheat bread (1 slice)	2.5
(*Sources:* American Dietetic Association and National Center for Nutrition and Dietetics)	

30.

U.S. Annual Food Sales	
Year	Sales in Billions of Dollars
2013	1410
2014	1462
2015	1512
2016	1584
(*Source:* U.S. Department of Agriculture)	

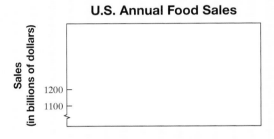

31.

Best-Selling Albums of All Time (U.S. Sales)	
Album	Estimated Sales (in millions)
Pink Floyd: *The Wall* (1979)	23
Michael Jackson: *Thriller* (1982)	29
Billy Joel: *Greatest Hits Volumes I–II* (1985)	23
Eagles: *Their Greatest Hits* (1976)	29
Led Zeppelin: *Led Zeppelin IV* (1971)	23
(*Source:* Recording Industry Association of America)	

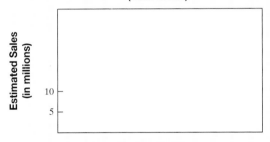

32.

Selected Worldwide Commercial Space Launches	
Location or Name	Total Commercial Space Launches 1990–2015
United States	181
Europe	169
Russia	162
China	23
Sea Launch*	41
*Sea Launch is an international venture involving four countries that uses its own launch facility outside national borders.	
(*Source:* Bureau of Transportation Statistics)	

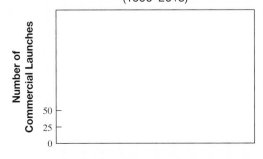

Objective C *The following histogram shows the number of miles that each adult, from a survey of 100 adults, drives per week. Use this histogram to answer Exercises 33 through 42. See Examples 4 and 5.*

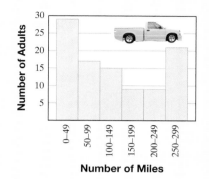

33. How many adults drive 100–149 miles per week?

34. How many adults drive 200–249 miles per week?

35. How many adults drive fewer than 150 miles per week?

36. How many adults drive 200 miles or more per week?

37. How many adults drive 100–199 miles per week?

38. How many adults drive 150–249 miles per week?

39. How many more adults drive 250–299 miles per week than 200–249 miles per week?

40. How many more adults drive 0–49 miles per week than 50–99 miles per week?

41. What is the ratio of adults who drive 150–199 miles per week to the total number of adults surveyed?

42. What is the ratio of adults who drive 50–99 miles per week to the total number of adults surveyed?

The following histogram shows the ages of householders for the year 2015. Use this histogram to answer Exercises 43 through 50. For Exercises 45 through 50, estimate to the nearest whole million. See Examples 4 and 5.

43. The most householders were in what age range?

44. The least number of householders were in what age range?

45. How many householders were 55–64 years old?

46. How many householders were 35–44 years old?

Source: U.S. Bureau of the Census, *Current Population Reports*

47. How many householders were 44 years old or younger?

48. How many householders were 55 years old or older?

49. How many more householders were 45–54 years old than 35–44 years old?

50. How many more householders were 45–54 years old than 75 and over?

The following list shows the golf scores for an amateur golfer. Use this list to complete the frequency distribution table to the right. See Example 6.

78	84	91	93	97
97	95	85	95	96
101	89	92	89	100

	Class Intervals (Scores)	Tally	Class Frequency (Number of Games)
▶ 51.	70–79		
▶ 52.	80–89		
▶ 53.	90–99		
▶ 54.	100–109		

Twenty-five people in a survey were asked to give their current checking account balances. Use the balances shown in the following list to complete the frequency distribution table to the right. See Example 6.

$53	$105	$162	$443	$109
$468	$47	$259	$316	$228
$207	$357	$15	$301	$75
$86	$77	$512	$219	$100
$192	$288	$352	$166	$292

	Class Intervals (Account Balances)	Tally	Class Frequency (Number of People)
55.	$0–$99		
56.	$100–$199		
57.	$200–$299		
58.	$300–$399		
59.	$400–$499		
60.	$500–$599		

▶ **61.** Use the frequency distribution table from Exercises **51** through **54** to construct a histogram. See Example 7.

62. Use the frequency distribution table from Exercises **55** through **60** to construct a histogram. See Example 7.

Golf Scores

Account Balances

Objective D *Beach Soccer World Cup is now held every two years. The following line graph shows the World Cup goals per game average for beach soccer during the years shown. Use this graph to answer Exercises 63 through 70. See Example 8.*

▶ **63.** Find the average number of goals per game in 2015.

64. Find the average number of goals per game in 2013.

▶ **65.** During what year shown was the average number of goals per game the highest?

66. During what year shown was the average number of goals per game the lowest?

Beach Soccer World Cup Goals per Game Average

Source: Wikipedia

Section 7.1 | Pictographs, Bar Graphs, Histograms, Line Graphs, and Introduction to Statistics

67. From 2013 to 2015, did the average number of goals per game increase or decrease?

68. From 2011 to 2013, did the average number of goals per game increase or decrease?

69. During what year(s) shown were the average goals per game less than 8?

70. During what year(s) shown were the average goals per game greater than 8?

Objective E *Find the range for each data set. See Example 9.*

71. 14, 16, 8, 10, 20

72. 25, 15, 11, 40, 37

73. 206, 206, 555, 556

74. 129, 188, 188, 276

75. 9, 9, 9, 9, 11

76. 7, 7, 7, 7, 10

77.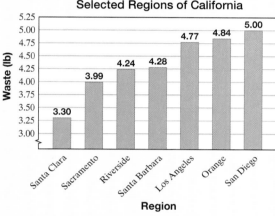

78. Scores on a Basic Math Test

*Use each frequency distribution table to find the **a.** mean, **b.** median, and **c.** mode. If needed, round the mean to 1 decimal place. See Example 10.*

79.

Data Item	Frequency
5	1
6	1
7	2
8	5
9	6
10	2

80.

Data Item	Frequency
3	2
4	1
5	4
6	7
7	2
8	1

81.

Data Item	Frequency
2	5
3	7
4	4
5	7
6	8
7	8
8	8
9	6
10	5

82.

Data Item	Frequency
4	3
5	8
6	5
7	8
8	2

Use each graph of data items to find the **a.** mean, **b.** median, and **c.** mode. If needed round the mean to one decimal place.

83.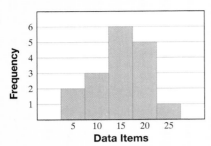

84.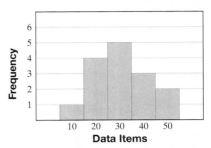

85.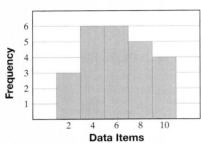

86.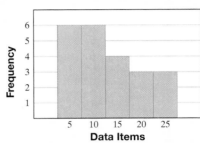

Review

Find each percent. See Section 6.3 or 6.4.

87. 30% of 12

88. 45% of 120

89. 10% of 62

90. 95% of 50

Write each fraction as a percent. See Section 6.2.

91. $\dfrac{1}{4}$

92. $\dfrac{2}{5}$

93. $\dfrac{17}{50}$

94. $\dfrac{9}{10}$

Concept Extensions

The following double line graph shows temperature highs and lows for a week. Use this graph to answer Exercises 95 through 100.

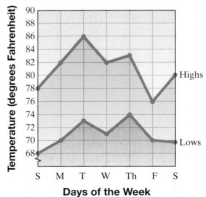

95. What was the high temperature reading on Thursday?

96. What was the low temperature reading on Thursday?

97. What day was the temperature the lowest? What was this low temperature?

98. What day of the week was the temperature the highest? What was this high temperature?

99. On what day of the week was the difference between the high temperature and the low temperature the greatest? What was this difference in temperature?

100. On what day of the week was the difference between the high temperature and the low temperature the least? What was this difference in temperature?

101. True or false? With a bar graph, the width of the bar is just as important as the height of the bar. Explain your answer.

102. Kansas plants about 17% of the wheat acreage in the United States. About how many acres of wheat are planted in the United States, according to the pictograph for Exercises **1** through **8**? Round to the nearest million acres.

7.2 Circle Graphs

Objective A Reading Circle Graphs

In Exercise Set 6.2, the following **circle graph** was shown. This particular graph shows the favorite sport for 100 adults.

Objectives

A Read Circle Graphs.

B Draw Circle Graphs.

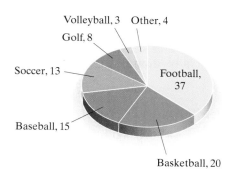

Each sector of the graph (shaped like a piece of pie) shows a category and the relative size of the category. In other words, the most popular sport is football, and it is represented by the largest sector.

Example 1
Find the ratio of adults preferring basketball to total adults. Write the ratio as a fraction in simplest form.

Solution: The ratio is

$$\frac{\text{people preferring basketball}}{\text{total adults}} = \frac{20}{100} = \frac{1}{5}$$

Work Practice 1

Practice 1

Find the ratio of adults preferring golf to total adults. Write the ratio as a fraction in simplest form.

A circle graph is often used to show percents in different categories, with the whole circle representing 100%.

Example 2 Using a Circle Graph

The following graph shows the percent of visitors to the United States in a recent year by various regions. Using the circle graph shown, determine the percent of visitors who came to the United States from Mexico or Canada.

Solution: To find this percent, we add the percents corresponding to Mexico and Canada. The percent of visitors to the United States that came from Mexico or Canada is

$$34\% + 21\% = 55\%$$

Work Practice 2

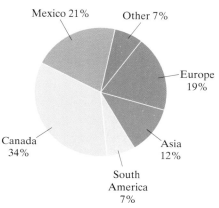

Source: Office of Travel and Tourism Industries, 2012

Practice 2

Using the circle graph shown in Example 2, determine the percent of visitors to the United States that came from Europe, Asia, or South America.

Answers

1. $\frac{2}{25}$ **2.** 38%

Helpful Hint

Since a circle graph represents a whole, the percents should add to 100% or 1. Notice this is true for Example 2.

Practice 3

Use the information in Example 3 and the circle graph from Example 2 to predict the number of tourists from Mexico in 2017.

Example 3 Finding Percent of Population

The U.S. Department of Commerce forecasts 81 million international visitors to the United States in 2017. Use the circle graph from Example 2 and predict the number of tourists that might be from Europe.

Solution: We use the percent equation.

amount = percent · base

amount = 0.19 · 81,000,000

= 0.19(81,000,000)

= 15,390,000

Thus, 15,390,000 tourists might come from Europe in 2017.

■ Work Practice 3

✓ **Concept Check** Can the following data be represented by a circle graph Why or why not?

Responses to the Question "In Which Activities Are You Involved?"	
Intramural sports	60%
On-campus job	42%
Fraternity/sorority	27%
Academic clubs	21%
Music programs	14%

Objective B Drawing Circle Graphs

To draw a circle graph, we use the fact that a whole circle contains 360° (degrees).

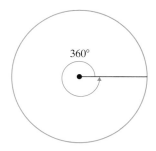

Answer

3. 17,010,000 tourists from Mexico

✓ **Concept Check Answer**

no; the percents add up to more than 100%

Example 4 — Drawing a Circle Graph for U.S. Armed Forces Personnel

The following table shows the percent of U.S. armed forces personnel that were in each branch of service in 2016. (*Source:* U.S. Department of Defense)

Branch of Service	Percent
Army	37
Navy	25
Marine Corps	14
Air Force	24

(Note: The Coast Guard is now under the Department of Homeland Security.)

Draw a circle graph showing this data.

Solution: First we find the number of degrees in each sector representing each branch of service. Remember that the whole circle contains 360°. (We will round degrees to the nearest whole degree.)

Sector	Degrees in Each Sector
Army	37% × 360° = 0.37 × 360° = 133.2° ≈ 133°
Navy	25% × 360° = 0.25 × 360° = 90°
Marine Corps	14% × 360° = 0.14 × 360° = 50.4° ≈ 50°
Air Force	24% × 360° = 0.24 × 360° = 86.4° ≈ 86°

Helpful Hint

Check your calculations by finding the sum of the degrees.

133° + 90° + 50° + 86° = 359°

The sum should be 360°. (Our sum varies slightly because of rounding.)

Next we draw a circle and mark its center. Then we draw a line from the center of the circle to the circle itself.

To construct the sectors, we will use a **protractor.** A protractor measures the number of degrees in an angle. We place the hole in the protractor over the center of the circle. Then we adjust the protractor so that 0° on the protractor is aligned with the line that we drew.

It makes no difference which sector we draw first. To construct the "Army" sector, we find 133° on the protractor and mark our circle. Then we remove the protractor and use this mark to draw a second line from the center to the circle itself.

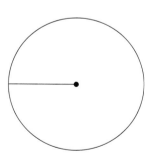

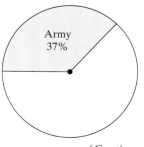

(*Continued on next page*)

Practice 4

Use the data shown to draw a circle graph.

Freshmen	30%
Sophomores	27%
Juniors	25%
Seniors	18%

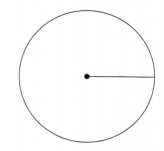

Answer

4.
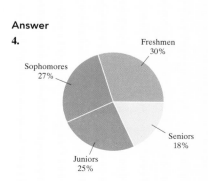

To construct the "Navy" sector, we follow the same procedure as above, except that we line up 0° with the second line we drew and mark the protractor at 90°.

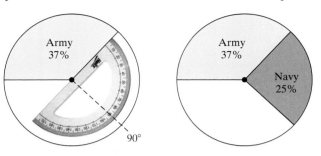

We continue in this manner until the circle graph is complete.

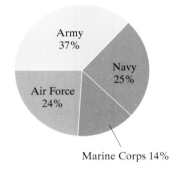

Work Practice 4

✓ **Concept Check** True or false? The larger a sector in a circle graph, the larger the percent of the total it represents. Explain your answer.

✓ Concept Check Answer
true

Vocabulary, Readiness & Video Check

Use the choices below to fill in each blank.

    sector   circle   100   360

1. In a _____ graph, each section (shaped like a piece of pie) shows a category and the relative size of the category.
2. A circle graph contains pie-shaped sections, each called a _____.
3. The number of degrees in a whole circle is _____.
4. If a circle graph has percent labels, the percents should add up to _____.

Martin-Gay Interactive Videos

See Video 7.2

Watch the section lecture video and answer the following questions.

Objective A 5. From Example 3, when a circle graph shows different parts or percents of some whole category, what is the sum of the percents in the whole circle graph?

Objective B 6. From Example 6, when looking at the sector degree measures of a circle graph, the whole circle graph corresponds to what degree measure?

7.2 Exercise Set MyLab Math

Objective A *The following circle graph is a result of surveying 700 college students. They were asked where they live while attending college. Use this graph to answer Exercises 1 through 6. Write all ratios as fractions in simplest form. See Example 1.*

1. Where do most of these college students live?

2. Besides the category "Other arrangements," where do the fewest of these college students live?

3. Find the ratio of students living in campus housing to total students.

4. Find the ratio of students living in off-campus rentals to total students.

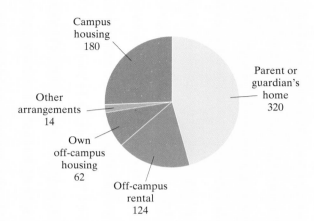

5. Find the ratio of students living in campus housing to students living in a parent or guardian's home.

6. Find the ratio of students living in off-campus rentals to students living in a parent or guardian's home.

The following circle graph shows the percent of the land area of the continents of Earth. Use this graph for Exercises 7 through 14. See Example 2.

7. Which continent is the largest?

8. Which continent is the smallest?

9. What percent of the land on Earth is accounted for by Asia and Europe together?

10. What percent of the land on Earth is accounted for by North and South America?

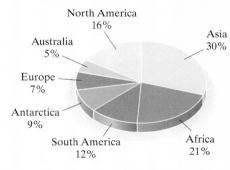

Source: National Geographic Society

The total amount of land from the continents is approximately 57,000,000 square miles. Use the graph to find the area of the continents given in Exercises 11 through 14. See Example 3.

11. Asia
12. South America
13. Australia
14. Europe

The following circle graph shows the percent of the types of books available at Midway Memorial Library. Use this graph for Exercises 15 through 24. See Example 2.

15. What percent of books are classified as some type of fiction?

16. What percent of books are nonfiction or reference?

17. What is the second-largest category of books?

18. What is the third-largest category of books?

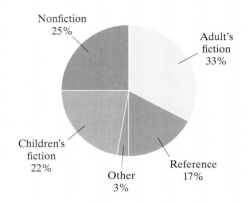

If this library has 125,600 books, find how many books are in each category given in Exercises 19 through 24. See Example 3.

19. Nonfiction

20. Reference

21. Children's fiction

22. Adult's fiction

23. Reference or other

24. Nonfiction or other

Objective B *Fill in each table. Round to the nearest degree. Then draw a circle graph to represent the information given in each table. (Remember: The total of "Degrees in Sector" column should equal 360° or very close to 360° because of rounding.) See Example 4.*

25.

Types of Apples Grown in Washington State		
Type of Apple	**Percent**	**Degrees in Sector**
Red Delicious	37%	
Golden Delicious	13%	
Fuji	14%	
Gala	15%	
Granny Smith	12%	
Other varieties	6%	
Braeburn	3%	
(*Source:* U.S. Apple Association)		

26.

Color Distribution of M&M's Milk Chocolate		
Color	**Percent**	**Degrees in Sector**
Blue	24%	
Orange	20%	
Green	16%	
Yellow	14%	
Red	13%	
Brown	13%	
(*Source:* M&M Mars)		

27.

Distribution of Large Dams by Continent		
Continent	Percent	Degrees in Sector
Europe	19%	
North America	32%	
South America	3%	
Asia	39%	
Africa	5%	
Australia	2%	

(*Source:* International Commission on Large Dams)

28.

2015 Electric Vehicle Sales by make of Car		
Company	Percent	Degrees
Nissan	15%	
Tesla	23%	
Chevrolet	16%	
Toyota	4%	
Ford	17%	
BMW	12%	
VW	4%	
Other	9%	

(*Source:* HybridCars.com)

Review

Write the prime factorization of each number. See Section 4.2.

29. 20

30. 25

31. 40

32. 16

33. 85

34. 105

Concept Extensions

The following circle graph shows the relative sizes of the great oceans.

35. Without calculating, determine which ocean is the largest. How can you answer this question by looking at the circle graph?

36. Without calculating, determine which ocean is the smallest. How can you answer this question by looking at the circle graph?

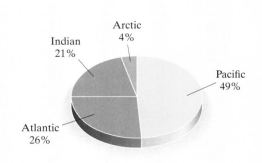

Source: Philip's World Atlas

These oceans together make up 264,489,800 square kilometers of Earth's surface. Find the square kilometers for each ocean.

37. Pacific Ocean **38.** Atlantic Ocean **39.** Indian Ocean **40.** Arctic Ocean

The following circle graph summarizes the results of online spending in America. Let's use these results to make predictions about the online spending behavior of a community of 2800 Internet users age 18 and over. Use this graph for Exercises 41 through 46. Round to the nearest whole. (Note: Because of rounding, these percents do not have a sum of 100%.)

41. How many of the survey respondents said that they spend $0 online each month?

42. How many of the survey repondents said that they spend $1–$100 online each month?

43. How many of the survey respondents said that they spend $0 to $100 online each month?

44. How many of the survey respondents said that they spend $1 to $1000 online each month?

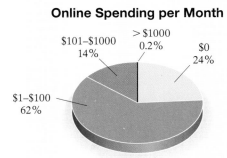

Source: The Digital Future Report, 2013

45. Find the ratio of *number* of respondents who spend $0 online to *number* of respondents who spend $1–$100 online. Write the ratio as a fraction. Simplify the fraction if possible.

46. Find the ratio of *percent* of respondents who spend $101–$1000 online to *percent* of those who spend $1–$100. Write the ratio as a fraction with integers in the numerator and denominator. Simplify the fraction if possible.

See the Concept Checks in this section.

47. Can the data below be represented by a circle graph? Why or why not?

Responses to the Question "What Classes Are You Taking?"	
Math	80%
English	72%
History	37%
Biology	21%
Chemistry	14%

48. True or false? The smaller a sector in a circle graph, the smaller the percent of the total it represents. Explain why.

Sections 7.1–7.2 **Integrated Review**

Reading Graphs

Answers

The following pictograph shows the six occupations with the largest estimated numerical increase (not salary increase) in employment in the United States between 2014 and 2024. Use this graph to answer Exercises 1 through 4.

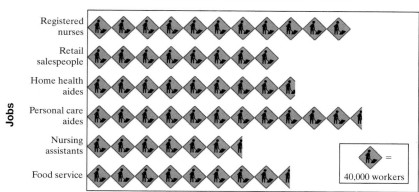

Source: Bureau of Labor Statistics

1. _____
2. _____
3. _____
4. _____

1. Approximate the increase in the number of retail salespeople from 2014 to 2024.

2. Approximate the increase in the number of registered nurses from 2014 to 2024.

3. Which occupation is expected to show the greatest increase in number of employees between the years shown?

4. Which of the listed occupations is expected to show the smallest increase in number of employees between the years shown?

The following bar graph shows the tallest U.S. dams. Use this graph to answer Exercises 5 through 8.

5. _____
6. _____
7. _____
8. _____

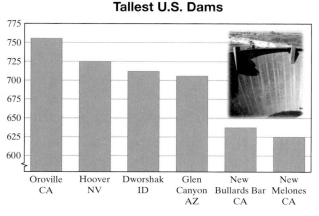

Source: Committee on Register of Dams

5. Name the U.S. dam with the greatest height and estimate its height.

6. Name the U.S. dam whose height is between 625 and 650 feet and estimate its height.

7. Estimate how much taller the Hoover Dam is than the Glen Canyon Dam.

8. How many U.S. dams have heights over 700 feet?

Chapter 7 | Graphs, Triangle Applications, and Introduction to Statistics and Probability

The following line graph shows the daily high temperatures for one week in Annapolis, Maryland. Use this graph to answer Exercises 9 through 12.

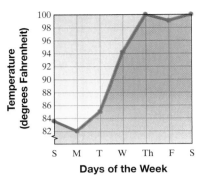

9. Name the day(s) of the week with the highest temperature and give that high temperature.

10. Name the day(s) of the week with the lowest temperature and give that low temperature.

11. On what days of the week was the temperature less than 90° Fahrenheit?

12. On what days of the week was the temperature greater than 90° Fahrenheit?

The following circle graph shows the types of milk beverages consumed in the United States. Use this graph for Exercises 13 through 16. If a store in Kerrville, Texas, sells 200 quart containers of milk per week, estimate how many quart containers are sold in each category below.

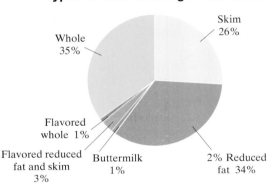

Source: U.S. Department of Agriculture

13. Whole milk

14. Skim milk

15. Buttermilk

16. Flavored reduced fat and skim milk

The following list shows weekly quiz scores for a student in prealgebra. Use this list to complete the frequency distribution table below.

| 50 | 80 | 80 | 90 | 88 | 97 | 67 | 89 | 71 |
| 78 | 93 | 99 | 75 | 53 | 93 | 83 | 86 | |

	Class Intervals (Scores)	Tally	Class Frequency (Number of Quizzes)
17.	50–59		
18.	60–69		
19.	70–79		
20.	80–89		
21.	90–99		

22. Use the frequency distribution table from Exercises 17 through 21 to construct a histogram.

7.3 Square Roots and the Pythagorean Theorem

Now that we know how to write ratios and solve proportions, in Section 7.4 we will use proportions to help us find unknown sides of similar triangles. In this section, we prepare for work on triangles by studying right triangles and their applications.

First, let's practice finding square roots.

Objectives
A Find the Square Root of a Number.
B Approximate Square Roots.
C Use the Pythagorean Theorem.

Objective A Finding Square Roots

The **square** of a number is the number times itself. For example,

The square of 5 is 25 because 5^2 or $5 \cdot 5 = 25$.
The square of -5 is also 25 because $(-5)^2$ or $(-5)(-5) = 25$.

The reverse process of squaring is finding a **square root.** For example,

A square root of 25 is 5 because $5 \cdot 5$ or $5^2 = 25$.
A square root of 25 is also -5 because $(-5)(-5)$ or $(-5)^2 = 25$.

Every positive number has two square roots. We see that the square roots of 25 are 5 and -5.

We use the symbol $\sqrt{}$, called a **radical sign,** to indicate the positive square root of a nonnegative number. For example,

$\sqrt{25} = 5$ because $5^2 = 25$ and 5 is positive.
$\sqrt{9} = 3$ because $3^2 = 9$ and 3 is positive.

Square Root of a Number

The square root, $\sqrt{}$, of a positive number a is the positive number b whose square is a. In symbols,

$\sqrt{a} = b$, if $b^2 = a$

Also, $\sqrt{0} = 0$.

Helpful Hint

Remember that the radical sign $\sqrt{}$ is used to indicate the **positive** (or principal) **square root** of a nonnegative number.

Examples Find each square root.

1. $\sqrt{49} = 7$ because $7^2 = 49$.
2. $\sqrt{36} = 6$ because $6^2 = 36$.
3. $\sqrt{1} = 1$ because $1^2 = 1$.
4. $\sqrt{81} = 9$ because $9^2 = 81$.
5. $\sqrt{\dfrac{1}{36}} = \dfrac{1}{6}$ because $\left(\dfrac{1}{6}\right)^2$ or $\dfrac{1}{6} \cdot \dfrac{1}{6} = \dfrac{1}{36}$.
6. $\sqrt{\dfrac{4}{25}} = \dfrac{2}{5}$ because $\left(\dfrac{2}{5}\right)^2$ or $\dfrac{2}{5} \cdot \dfrac{2}{5} = \dfrac{4}{25}$.

Work Practice 1–6

Practice 1–6

Find each square root.
1. $\sqrt{100}$ 2. $\sqrt{64}$
3. $\sqrt{169}$ 4. $\sqrt{0}$
5. $\sqrt{\dfrac{1}{4}}$ 6. $\sqrt{\dfrac{9}{16}}$

Answers
1. 10   2. 8   3. 13   4. 0   5. $\dfrac{1}{2}$ 6. $\dfrac{3}{4}$

Objective B Approximating Square Roots

Thus far, we have found square roots of perfect squares. Numbers like $\frac{1}{4}$, 36, $\frac{1}{2}$, and 1 are called **perfect squares** because their square root is a whole number or fraction. A square root such as $\sqrt{5}$ cannot be written as a whole number or a fraction since 5 is not a perfect square.

Although $\sqrt{5}$ cannot be written as a whole number or a fraction, it can be approximated by estimating, by using a table (as in the appendix), or by using a calculator.

Practice 7

Use Appendix A.4 or a calculator to approximate each square root to the nearest thousandth.

a. $\sqrt{10}$ b. $\sqrt{62}$

Example 7 Use Appendix A.4 or a calculator to approximate each square root to the nearest thousandth.

a. $\sqrt{43} \approx 6.557$ is approximately
b. $\sqrt{80} \approx 8.944$

■ Work Practice 7

Helpful Hint

$\sqrt{80}$ is *approximately* 8.944. This means that if we multiply 8.944 by 8.944, the product is *close* to 80.

$8.944 \times 8.944 \approx 79.995$

It is possible to approximate a square root to the nearest whole number without the use of a calculator or table. To do so, study the number line below and look for patterns.

Above the number line, notice that as the numbers under the radical signs increase, their value, and thus their placement on the number line, increase also.

Practice 8

Without a calculator or table, approximate $\sqrt{62}$ to the nearest whole.

Example 8 Without a calculator or table:

a. Determine which two whole numbers $\sqrt{78}$ is between.
b. Use part **a** to approximate $\sqrt{78}$ to the nearest whole.

Solution:

a. Review perfect squares and recall that $\sqrt{64} = 8$ and $\sqrt{81} = 9$. Since 78 is between 64 and 81, $\sqrt{78}$ is between $\sqrt{64}$ (or 8) and $\sqrt{81}$ (or 9).

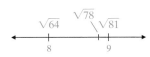

Thus, $\sqrt{78}$ is between 8 and 9.

b. Since $\sqrt{78}$ is closer to $\sqrt{81}$ (or 9) than $\sqrt{64}$ (or 8), then (as our number line shows), $\sqrt{78}$ approximated to the nearest whole is 9.

■ Work Practice 8

Answers
7. a. 3.162   **b.** 7.874   **8.** 8

Section 7.3 | Square Roots and the Pythagorean Theorem

Objective C Using the Pythagorean Theorem

One important application of square roots has to do with right triangles. Recall that a **right triangle** is a triangle in which one of the angles is a right angle, or measures 90° (degrees). The **hypotenuse** of a right triangle is the side opposite the right angle. The **legs** of a right triangle are the other two sides. These are shown in the following figure. The right angle in the triangle is indicated by the small square drawn in that angle.

The following theorem is true for all right triangles.

Pythagorean Theorem

If a and b are the lengths of the legs of a right triangle and c is the length of the hypotenuse, then

$a^2 + b^2 = c^2$

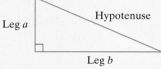

In other words, $(\text{leg})^2 + (\text{other leg})^2 = (\text{hypotenuse})^2$.

Example 9 Find the length of the hypotenuse of the given right triangle.

Solution: Let $a = 6$ and $b = 8$. According to the Pythagorean theorem,

$a^2 + b^2 = c^2$
$6^2 + 8^2 = c^2$ Let $a = 6$ and $b = 8$.
$36 + 64 = c^2$ Evaluate 6^2 and 8^2.
$100 = c^2$ Add.

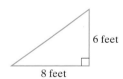

Practice 9

Find the length of the hypotenuse of the given right triangle.

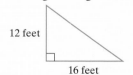

In the equation $c^2 = 100$, the solutions of c are the square roots of 100. Since $10 \cdot 10 = 100$ and $(-10)(-10) = 100$, both 10 and -10 are square roots of 100. Since c represents a length, we are only interested in the positive square root of c^2.

$c = \sqrt{100}$
$ = 10$

The hypotenuse is 10 feet long.

Work Practice 9

Example 10 Approximate the length of the hypotenuse of the given right triangle. Round the length to the nearest whole unit.

Solution: Let $a = 17$ and $b = 10$.

$a^2 + b^2 = c^2$
$17^2 + 10^2 = c^2$
$289 + 100 = c^2$
$389 = c^2$
$\sqrt{389} = c$ or $c \approx 20$ From Appendix A.4 or a calculator

The hypotenuse is exactly $\sqrt{389}$ meters, which is approximately 20 meters.

Work Practice 10

Practice 10

Approximate the length of the hypotenuse of the given right triangle. Round to the nearest whole unit.

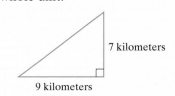

Answers

9. 20 feet   **10.** 11 kilometers

Practice 11

Find the length of the leg in the given right triangle. Give the exact length and a two-decimal-place approximation.

Example 11 Find the length of the leg in the given right triangle. Give the exact length and a two-decimal-place approximation.

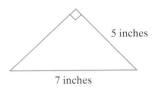

Solution: Notice that the hypotenuse measures 7 inches and that the length of one leg measures 5 inches. Thus, let $c = 7$ and a or b be 5. We will let $a = 5$.

$$a^2 + b^2 = c^2$$
$$5^2 + b^2 = 7^2 \qquad \text{Let } a = 5 \text{ and } c = 7.$$
$$25 + b^2 = 49 \qquad \text{Evaluate } 5^2 \text{ and } 7^2.$$
$$b^2 = 24 \qquad \text{Subtract 25 from both sides.}$$
$$b = \sqrt{24} \approx 4.90$$

The length of the leg is exactly $\sqrt{24}$ inches and approximately 4.90 inches.

■ Work Practice 11

✓ **Concept Check** The following lists are the lengths of the sides of two triangles. Which set forms a right triangle?
a. 8, 15, 17   **b.** 24, 30, 40

Practice 12

A football field is a rectangle measuring 100 yards by 53 yards. Draw a diagram and find the length of the diagonal of a football field to the nearest yard.

Example 12 Finding the Dimensions of a Park

An inner-city park is in the shape of a square that measures 300 feet on a side. A sidewalk is to be constructed along the diagonal of the park. Find the length of the sidewalk rounded to the nearest whole foot.

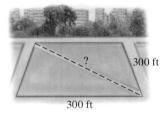

Solution: The diagonal is the hypotenuse of a right triangle, which we label c.

$$a^2 + b^2 = c^2$$
$$300^2 + 300^2 = c^2 \qquad \text{Let } a = 300 \text{ and } b = 300.$$
$$90{,}000 + 90{,}000 = c^2 \qquad \text{Evaluate } (300)^2.$$
$$180{,}000 = c^2 \qquad \text{Add.}$$
$$\sqrt{180{,}000} = c \text{ or } c \approx 424$$

The length of the sidewalk is approximately 424 feet.

■ Work Practice 12

Answers
11. $\sqrt{120}$ feet ≈ 10.95 feet
12. 113 yards

✓ **Concept Check Answer**
a

Calculator Explorations Finding and Approximating Square Roots

To simplify or approximate square roots using a calculator, locate the key marked $\boxed{\sqrt{}}$.

To simplify $\sqrt{64}$, for example, press the keys

$\boxed{64}$ $\boxed{\sqrt{}}$ or $\boxed{\sqrt{}}$ $\boxed{64}$

The display should read $\boxed{8}$. Then $\sqrt{64} = 8$

To *approximate* $\sqrt{10}$, press the keys

$\boxed{10}$ $\boxed{\sqrt{}}$ or $\boxed{\sqrt{}}$ $\boxed{10}$

The display should read $\boxed{3.16227766}$. This is an *approximation* for $\sqrt{10}$. A three-decimal-place approximation is

$\sqrt{10} \approx 3.162$

Is this answer reasonable? Since 10 is between the perfect squares 9 and 16, $\sqrt{10}$ is between $\sqrt{9} = 3$ and $\sqrt{16} = 4$. Our answer is reasonable since 3.162 is between 3 and 4.

Simplify.
1. $\sqrt{1024}$
2. $\sqrt{676}$

Approximate each square root. Round each answer to the nearest thousandth.
3. $\sqrt{15}$
4. $\sqrt{19}$
5. $\sqrt{97}$
6. $\sqrt{56}$

Vocabulary, Readiness & Video Check

Use the choices below to fill in each blank. Some choices will be used more than once.

| squaring | Pythagorean theorem | radical | −10 | leg |
| hypotenuse | perfect squares | 10 | c^2 | b^2 |

1. The square roots of 100 are _____ and _____ because $10 \cdot 10 = 100$ and $(-10)(-10) = 100$.
2. $\sqrt{100} = $ _____ because $10 \cdot 10 = 100$ and 10 is positive.
3. The _____ sign is used to denote the positive square root of a nonnegative number.
4. The reverse process of _____ a number is finding a square root of a number.
5. The numbers 9, 1, and $\frac{1}{25}$ are called _____.
6. Label the parts of the right triangle.
7. In the given triangle, $a^2 + $ _____ $ = $ _____.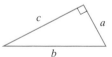
8. The _____ can be used for right triangles.

Martin-Gay Interactive Videos

Watch the section lecture video and answer the following questions.

See Video 7.3

Objective A 9. From the lecture before ▣ Example 1, list the square root(s) of 49. How is this different from finding $\sqrt{49}$?

Objective B 10. In ▣ Example 5, how do we know $\sqrt{15}$ is closer to 4 than to 3?

Objective C 11. From ▣ Example 6, how do we know which side of a right triangle is the hypotenuse?

524 Chapter 7 | Graphs, Triangle Applications, and Introduction to Statistics and Probability

7.3 Exercise Set MyLab Math

Objective A Find each square root. See Examples 1 through 6.

1. $\sqrt{4}$
2. $\sqrt{9}$
3. $\sqrt{121}$
4. $\sqrt{144}$
5. $\sqrt{\dfrac{1}{81}}$
6. $\sqrt{\dfrac{1}{64}}$
7. $\sqrt{\dfrac{16}{64}}$
8. $\sqrt{\dfrac{36}{81}}$

Objective B Use Appendix A.4 or a calculator to approximate each square root. Round the square root to the nearest thousandth. See Example 7.

9. $\sqrt{3}$
10. $\sqrt{5}$
11. $\sqrt{15}$
12. $\sqrt{17}$
13. $\sqrt{31}$
14. $\sqrt{85}$
15. $\sqrt{26}$
16. $\sqrt{35}$

Determine what two whole numbers each square root is between without using a calculator or table. Then use a calculator or Appendix A.4 to check. See Example 8.

17. $\sqrt{38}$
18. $\sqrt{27}$
19. $\sqrt{101}$
20. $\sqrt{85}$

Objectives A B Mixed Practice Find each square root. If necessary, round the square root to the nearest thousandth. See Examples 1 through 8.

21. $\sqrt{256}$
22. $\sqrt{625}$
23. $\sqrt{92}$
24. $\sqrt{18}$
25. $\sqrt{\dfrac{49}{144}}$
26. $\sqrt{\dfrac{121}{169}}$
27. $\sqrt{71}$
28. $\sqrt{62}$

Objective C Find the unknown length in each right triangle. If necessary, approximate the length to the nearest thousandth. See Examples 9 through 12.

29.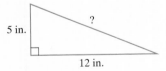

30.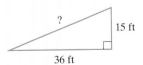

31.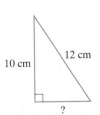

32.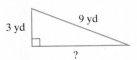

33.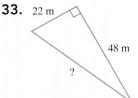

34.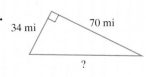

35.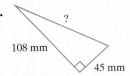

36.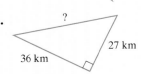

Sketch each right triangle and find the length of the side not given. If necessary, approximate the length to the nearest thousandth. (Each length is in units.) See Examples 9 through 12.

37. leg = 3, leg = 4   **38.** leg = 9, leg = 12   **39.** leg = 5, hypotenuse = 13

40. leg = 6, hypotenuse = 10   **41.** leg = 10, leg = 14   **42.** leg = 2, leg = 16

43. leg = 35, leg = 28   **44.** leg = 30, leg = 15   **45.** leg = 30, leg = 30

46. leg = 21, leg = 21   **47.** hypotenuse = 2, leg = 1   **48.** hypotenuse = 9, leg = 8

49. leg = 7.5, leg = 4   **50.** leg = 12, leg = 22.5

Solve. See Example 12.

51. A standard city block is a square with each side measuring 100 yards. Find the length of the diagonal of a city block to the nearest hundredth yard.

52. A section of land is a square with each side measuring 1 mile. Find the length of the diagonal of the section of land to the nearest thousandth mile.

53. Find the height of the tree. Round the height to one decimal place.

54. Find the height of the antenna. Round the height to one decimal place.

55. The playing field for football is a rectangle that is 300 feet long by 160 feet wide. Find the length of a straight-line run that started at one corner and went diagonally to end at the opposite corner. Round to the nearest foot, if necessary.

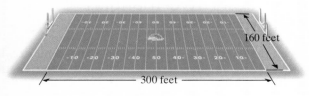

56. A soccer field is in the shape of a rectangle and its dimensions depend on the age of the players. The dimensions of the soccer field below are the minimum dimensions for international play. Find the length of the diagonal of this rectangle. Round the answer to the nearest tenth of a yard.

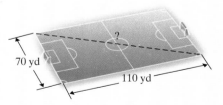

Review

Write each fraction in simplest form. See Section 4.2.

57. $\dfrac{10}{12}$ **58.** $\dfrac{10}{15}$ **59.** $\dfrac{2x}{60}$ **60.** $\dfrac{35}{75y}$

526   Chapter 7 | Graphs, Triangle Applications, and Introduction to Statistics and Probability

Perform the indicated operations. See Sections 4.3 and 4.4.

61. $\dfrac{9}{13y} + \dfrac{12}{13y}$ **62.** $\dfrac{3x}{9} - \dfrac{5}{9}$ **63.** $\dfrac{9}{8} \cdot \dfrac{x}{8}$ **64.** $\dfrac{7x}{11} \div \dfrac{8x}{11}$

Concept Extensions

Use the results of Exercises 17 through 20 and approximate each square root to the nearest whole without using a calculator or table. Then use a calculator or Appendix A.4 to check. See Example 8.

65. $\sqrt{38}$ **66.** $\sqrt{27}$ **67.** $\sqrt{101}$ **68.** $\sqrt{85}$

69. Without using a calculator, explain how you know that $\sqrt{105}$ is *not* approximately 9.875.

70. Without using a calculator, explain how you know that $\sqrt{27}$ is *not* approximately 3.296.

Does the set form the lengths of the sides of a right triangle? See the Concept Check in this section.

71. 25, 60, 65

72. 20, 45, 50

△ **73.** Find the exact length of x. Then give a two-decimal-place approximation.

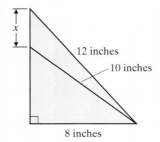

7.4 Congruent and Similar Triangles

Objectives

A Decide Whether Two Triangles Are Congruent.

B Find the Ratio of Corresponding Sides in Similar Triangles.

C Find Unknown Lengths of Sides in Similar Triangles.

Objective A Deciding Whether Two Triangles Are Congruent

Two triangles are **congruent** when they have the same shape and the same size. I congruent triangles, the measures of corresponding angles are equal and the length of corresponding sides are equal. The following triangles are congruent:

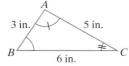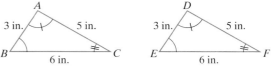

Since these triangles are congruent, the measures of corresponding angles are equa
   Angles with equal measure: $\angle A$ and $\angle D$, $\angle B$ and $\angle E$, $\angle C$ and $\angle F$. Also, th lengths of corresponding sides are equal.
   Equal corresponding sides: $\overline{AB}$ and $\overline{DE}$, $\overline{BC}$ and $\overline{EF}$, $\overline{CA}$ and $\overline{FD}$

Any one of the following may be used to determine whether two triangles are congruent:

Congruent Triangles

Angle-Side-Angle (ASA)

If the measures of two angles of a triangle equal the measures of two angles of another triangle, and the lengths of the sides between each pair of angles are equal, the triangles are congruent.

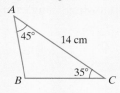 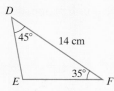

For example, these two triangles are congruent by Angle-Side-Angle.

Side-Side-Side (SSS)

If the lengths of the three sides of a triangle equal the lengths of the corresponding sides of another triangle, the triangles are congruent.

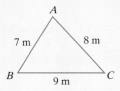 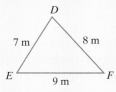

For example, these two triangles are congruent by Side-Side-Side.

Side-Angle-Side (SAS)

If the lengths of two sides of a triangle equal the lengths of corresponding sides of another triangle, and the measures of the angles between each pair of sides are equal, the triangles are congruent.

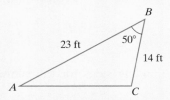 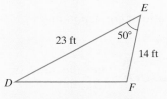

For example, these two triangles are congruent by Side-Angle-Side.

Example 1 Determine whether triangle ABC is congruent to triangle DEF.

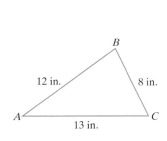 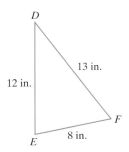

Solution: Since the lengths of all three sides of triangle ABC equal the lengths of all three sides of triangle DEF, the triangles are congruent.

■ Work Practice 1

Practice 1

a. Determine whether triangle MNO is congruent to triangle RQS.

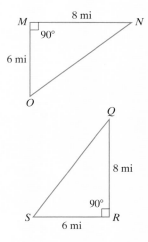

b. Determine whether triangle GHI is congruent to triangle JKL.

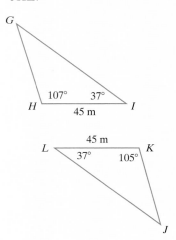

Answers

1. a. congruent **b.** not congruent

In Example 1, notice that as soon as we know that the two triangles are congruent, we know that all three corresponding angles are congruent.

Objective B Finding the Ratios of Corresponding Sides in Similar Triangles

Two triangles are **similar** when they have the same shape but not necessarily the same size. In similar triangles, the measures of corresponding angles are equal and corresponding sides are in proportion. The following triangles are similar:

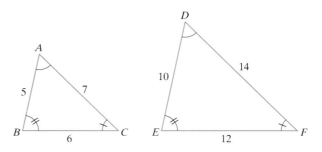

Since these triangles are similar, the measures of corresponding angles are equal. (Note: The triangles above are not drawn to scale.)

Angles with equal measure: $\angle A$ and $\angle D$, $\angle B$ and $\angle E$, $\angle C$ and $\angle F$. Also, the lengths of corresponding sides are in proportion.

Sides in proportion: $\dfrac{AB}{DE} = \dfrac{BC}{EF} = \dfrac{CA}{FD}$ or, in this particular case,

$$\dfrac{AB}{DE} = \dfrac{5}{10} = \dfrac{1}{2}, \quad \dfrac{BC}{EF} = \dfrac{6}{12} = \dfrac{1}{2}, \quad \dfrac{CA}{FD} = \dfrac{7}{14} = \dfrac{1}{2}$$

The ratio of corresponding sides is $\dfrac{1}{2}$.

Practice 2

Find the ratio of corresponding sides for the similar triangles QRS and XYZ.

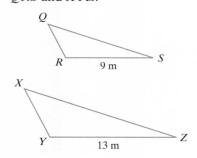

Example 2 Find the ratio of corresponding sides for the similar triangles ABC and DEF.

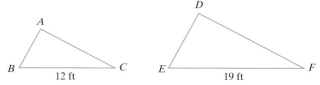

Solution: We are given the lengths of two corresponding sides. Their ratio is

$$\dfrac{12 \text{ feet}}{19 \text{ feet}} = \dfrac{12}{19}$$

■ Work Practice 2

Objective C Finding Unknown Lengths of Sides in Similar Triangles

Because the ratios of lengths of corresponding sides are equal, we can use proportions to find unknown lengths in similar triangles.

Answer

2. $\dfrac{9}{13}$

Example 3
Given that the triangles are similar, find the missing length y.

Solution: Since the triangles are similar, corresponding sides are in proportion. Thus, the ratio of 2 to 3 is the same as the ratio of 10 to y, or

$$\frac{2}{3} = \frac{10}{y}$$

To find the unknown length y, we set cross products equal.

$$\frac{2}{3} = \frac{10}{y}$$

$2 \cdot y = 3 \cdot 10$ Set cross products equal.
$2y = 30$ Multiply.
$\frac{2y}{2} = \frac{30}{2}$ Divide both sides by 2.
$y = 15$ Simplify.

The missing length is 15 units.

Work Practice 3

Practice 3
Given that the triangles are similar, find the missing length x.

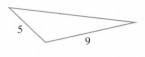

✓ **Concept Check** The following two triangles are similar. Which vertices of the first triangle appear to correspond to which vertices of the second triangle?

 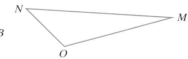

Many applications involve a diagram containing similar triangles. Surveyors, astronomers, and many other professionals continually use similar triangles in their work.

Practice 4
Tammy Shultz, a firefighter, needs to estimate the height of a burning building. She estimates the length of her shadow to be 8 feet long and the length of the building's shadow to be 60 feet long. Find the approximate height of the building if she is 5 feet tall.

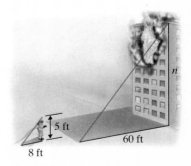

Example 4 Finding the Height of a Tree

Mel Wagstaff is a 6-foot-tall park ranger who needs to know the height of a particular tree. He measures the shadow of the tree to be 69 feet long when his own shadow is 9 feet long. Find the height of the tree.

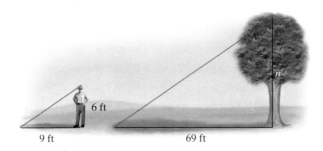

Answers
3. $x = \frac{10}{3}$ or $3\frac{1}{3}$ units
4. approximately 37.5 ft

✓ **Concept Check Answer**
A corresponds to O; B corresponds to N; C corresponds to M

(*Continued on next page*)

Solution:

1. **UNDERSTAND.** Read and reread the problem. Notice that the triangle formed by the sun's rays, Mel, and his shadow is similar to the triangle formed by the sun's rays, the tree, and its shadow.

2. **TRANSLATE.** Write a proportion from the similar triangles formed.

$$\frac{\text{Mel's height}}{\text{height of tree}} \;\rightarrow\; \frac{6}{n} = \frac{9}{69} \;\leftarrow\; \frac{\text{length of Mel's shadow}}{\text{length of tree's shadow}}$$

or $\dfrac{6}{n} = \dfrac{3}{23}$ Simplify $\dfrac{9}{69}$ (ratio in lowest terms).

3. **SOLVE** for n:

$$\frac{6}{n} = \frac{3}{23}$$

$6 \cdot 23 = n \cdot 3$ Set cross products equal.
$138 = 3n$ Multiply.
$\dfrac{138}{3} = \dfrac{3n}{3}$ Divide both sides by 3.
$46 = n$

4. **INTERPRET.** *Check* to see that replacing n with 46 in the proportion makes the proportion true. *State* your conclusion: The height of the tree is 46 feet.

■ Work Practice 4

Vocabulary, Readiness & Video Check

Answer each question true or false.

1. Two triangles that have the same shape but not necessarily the same size are congruent. _____
2. Two triangles are congruent if they have the same shape and size. _____
3. Congruent triangles are also similar. _____
4. Similar triangles are also congruent. _____
5. For the two similar triangles, the ratio of corresponding sides is $\dfrac{5}{6}$. _____

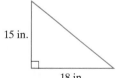

Martin-Gay Interactive Videos

See Video 7.4

Watch the section lecture video and answer the following questions.

Objective A 6. How did we decide which congruency rule to use to determine if the two triangles in Example 1 are congruent? ▶

Objective B 7. From Example 2, what does "corresponding sides are in proportion" mean? ▶

Objective C 8. In Example 4, what is another proportion we could have used to solve the application problem? ▶

Section 7.4 | Congruent and Similar Triangles   531

7.4 Exercise Set MyLab Math

Objective A *Determine whether each pair of triangles is congruent. If congruent, state the reason why, such as SSS, SAS, or ASA. See Example 1.*

1.
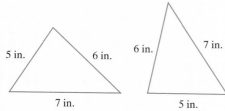

2.
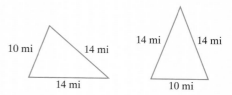

3.
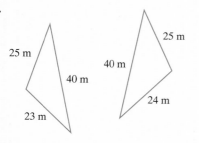

4.
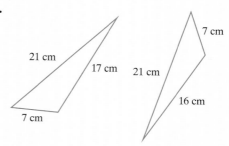

5.
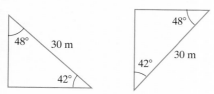

6.
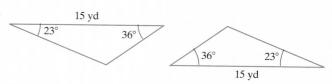

7.
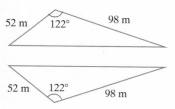

8.
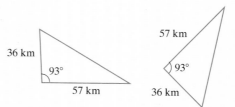

Objective B *Find each ratio of the corresponding sides of the given similar triangles. See Example 2.*

9.
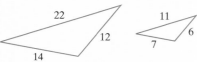

10.
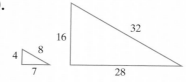

11.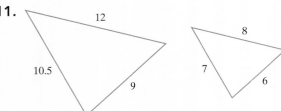

12.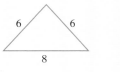

Objective C *Given that the pairs of triangles are similar, find the unknown length of the side labeled with a variable. See Example 3.*

13.

14.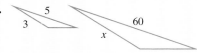

15.

16.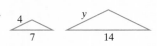

17.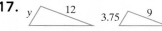

18.

19.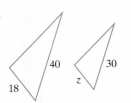

20.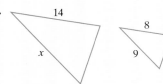

21.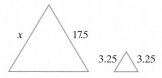

22.

23.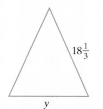

24.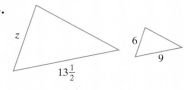

25.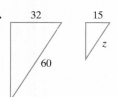

26.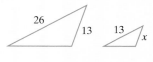

27.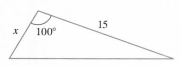

28.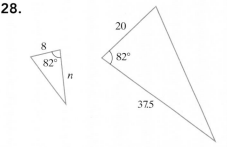

Solve. For Exercises 29 and 30, the solutions have been started for you. See Example 4.

29. Given the following diagram, approximate the height of the observation deck in the Seattle Space Needle in Seattle, Washington. (*Source:* Seattle Space Needle)

Start the solution:

1. UNDERSTAND the problem. Reread it as many times as needed.
2. TRANSLATE into a proportion using the similar triangles formed. (Fill in the blanks.)

   height of observation deck → $\dfrac{x}{13}$ = $\dfrac{}{}$ ← length of Space Needle shadow
   height of pole → ← length of pole shadow

3. SOLVE by setting cross products equal.
4. INTERPRET.

30. Fountain Hills, Arizona, boasts the tallest fountain in the United States. The fountain sits in a 28-acre lake and shoots up a column of water every hour. Based on the diagram below, what is the approximate height of the fountain?

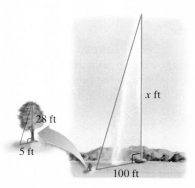

Start the solution:

1. UNDERSTAND the problem. Reread it as many times as needed.
2. TRANSLATE into a proportion using the similar triangles formed. (Fill in the blanks.)

   height of tree → $\dfrac{28}{x}$ = $\dfrac{}{}$ ← length of tree shadow
   height of fountain → ← length of fountain shadow

3. SOLVE by setting cross products equal.
4. INTERPRET.

31. Given the following diagram, approximate the height of the Bank One Tower in Oklahoma City, Oklahoma. (*Source: The World Almanac*)

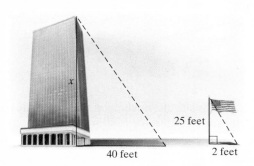

32. The tallest tree currently growing is Hyperion, a redwood located in the Redwood National Park in California. Given the following diagram, approximate its height. (*Source: Guinness World Records*) (*Note:* The tree's current recorded height is 379.1 ft.)

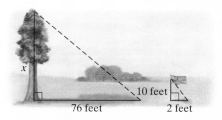

33. If a 30-foot tree casts an 18-foot shadow, find the length of the shadow cast by a 24-foot tree.

34. If a 24-foot flagpole casts a 32-foot shadow, find the length of the shadow cast by a 44-foot antenna. Round to the nearest tenth.

Review

Solve. See Section 6.1.

35. For the health of his fish, Pete's Sea World uses the standard that a 20-gallon tank should house only 19 neon tetras. Find the number of neon tetras that Pete would place into a 55-gallon tank.

36. A local package express deliveryman is traveling the city expressway at 45 mph when he is forced to slow down due to traffic ahead. His truck slows at the rate of 3 mph every 5 seconds. Find his speed 8 seconds after braking.

Solve. See Section 7.3.

37. Launch Umbilical Tower 1 is the name of the gantry used for the *Apollo* launch that took Neil Armstrong and Buzz Aldrin to the moon. Find the height of the gantry to the nearest whole foot.

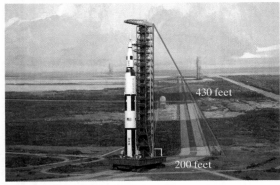

38. Arena polo, popular in the United States and England, is played on a field that is 100 yards long and usually 50 yards wide. Find the length, to the nearest yard, of the diagonal of this field.

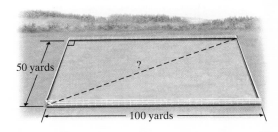

Perform the indicated operation. See Sections 5.2 through 5.4.

39. $3.6 + 0.41$

40. $0.41 - 3.6$

41. $(0.41)(-3)$

42. $-0.48 \div 3$

Concept Extensions

43. The print area on a particular page measures 7 inches by 9 inches. A printing shop is to copy the page and reduce the print area so that its length is 5 inches. What will its width be? Will the print now fit on a 3-by-5-inch index card?

44. The art sample for a banner measures $\frac{1}{3}$ foot in width by $1\frac{1}{2}$ feet in length. If the completed banner is to have a length of 9 feet, find its width.

Given that the pairs of triangles are similar, find the length of the side labeled n. Round your results to 1 decimal place.

45.

46.

47. In your own words, describe any differences in similar triangles and congruent triangles.

48. Describe a situation where similar triangles would be useful for a contractor building a house.

49. A triangular park is planned and waiting to be approved by the city zoning commission. A drawing of the park shows sides of length 5 inches, $7\frac{1}{2}$ inches, and $10\frac{5}{8}$ inches. If the scale on the drawing is $\frac{1}{4}$ in. = 10 ft, find the actual proposed dimensions of the park.

50. John and Robyn Costello draw a triangular deck on their house plans. Robyn measures sides of the deck drawing on the plans to be 3 inches, $4\frac{1}{2}$ inches, and 6 inches. If the scale on the drawing is $\frac{1}{4}$ in. = 1 foot, find the lengths of the sides of the deck they want built.

7.5 Counting and Introduction to Probability

Objective A Using a Tree Diagram

In our daily conversations, we often talk about the likelihood or **probability** of a given result occurring. For example:

The *chance* of thundershowers is 70 percent.
What are the *odds* that the New Orleans Saints will go to the Super Bowl?
What is the *probability* that you will finish cleaning your room today?

Each of these chance happenings—thundershowers, the New Orleans Saints playing in the Super Bowl, and finishing cleaning your room today—is called an **experiment.** The possible results of an experiment are called **outcomes.** For example, flipping a coin is an experiment, and the possible outcomes are heads (H) or tails (T).

One way to picture the outcomes of an experiment is to draw a **tree diagram.** Each outcome is shown on a separate branch. For example, the outcomes of flipping a coin are

Heads   Tails

Objectives

A Use a Tree Diagram to Count Outcomes.

B Find the Probability of an Event.

Example 1 Draw a tree diagram for tossing a coin twice. Then use the diagram to find the number of possible outcomes.

Solution:

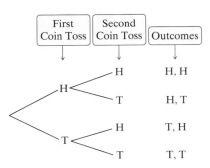

There are 4 possible outcomes when tossing a coin twice.

■ Work Practice 1

Example 2 Draw a tree diagram for an experiment consisting of rolling a die and then tossing a coin. Then use the diagram to find the number of possible outcomes.

Die

(Continued on next page)

Practice 1

Draw a tree diagram for tossing a coin three times. Then use the diagram to find the number of possible outcomes.

Practice 2

Draw a tree diagram for an experiment consisting of tossing a coin and then rolling a die. Then use the diagram to find the number of possible outcomes. (Answer appears on the next page.)

Answer

1.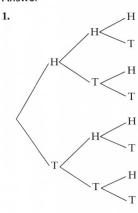

8 outcomes

Solution: Recall that a die has six sides and that each side represents a number, 1 through 6.

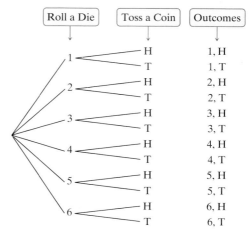

There are 12 possible outcomes for rolling a die and then tossing a coin.

■ Work Practice 2

Any number of outcomes considered together is called an **event.** For example, when tossing a coin twice, H, H is an event. The event is tossing heads first and tossing heads second. Another event would be tossing tails first and then heads (T, H), and so on.

Objective B Finding the Probability of an Event

As we mentioned earlier, the **probability of an event is a measure of the chance or likelihood of it occurring.** For example, if a coin is tossed, what is the probability that heads occurs? Since one of two equally likely possible outcomes is heads, the probability is $\frac{1}{2}$.

The Probability of an Event

$$\text{probability of an event} = \frac{\text{number of ways that the event can occur}}{\text{number of possible outcomes}}$$

Note from the definition of probability that the probability of an event is always between 0 and 1, inclusive (i.e., including 0 and 1). A probability of 0 means that an event won't occur, and a probability of 1 means that an event is certain to occur.

Example 3 If a coin is tossed twice, find the probability of tossing heads on the first toss and then heads again on the second toss (H, H).

Solution:

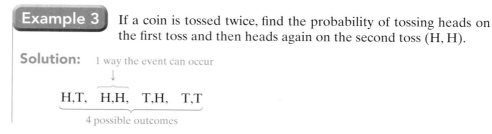

Practice 3
If a coin is tossed three times, find the probability of tossing tails, then heads, then tails (T, H, T).

Answers

2.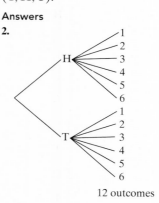
12 outcomes

3. $\frac{1}{8}$

$$\text{probability} = \frac{1}{4} \begin{array}{l} \text{Number of ways the event can occur} \\ \text{Number of possible outcomes} \end{array}$$

The probability of tossing heads and then heads is $\frac{1}{4}$.

■ Work Practice 3

Example 4 If a die is rolled one time, find the probability of rolling a 3 or a 4.

Solution: Recall that there are 6 possible outcomes when rolling a die.

possible outcomes: 1, 2, 3, 4, 5, 6 (2 ways that the event can occur; 6 possible outcomes)

$$\text{probability of a 3 or a 4} = \frac{2}{6} \begin{array}{l} \text{Number of ways the event can occur} \\ \text{Number of possible outcomes} \end{array}$$

$$= \frac{1}{3} \quad \text{Simplest form}$$

Practice 4

If a die is rolled one time, find the probability of rolling a 2 or a 5.

■ Work Practice 4

✓ **Concept Check** Suppose you have calculated a probability of $\frac{11}{9}$. How do you know that you have made an error in your calculation?

Example 5 Find the probability of choosing a red marble from a box containing 1 red, 1 yellow, and 2 blue marbles.

Solution:

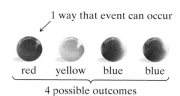

1 way that event can occur

red yellow blue blue

4 possible outcomes

$$\text{probability} = \frac{1}{4}$$

Practice 5

Use the diagram and information in Example 5 and find the probability of choosing a blue marble from the box.

Answers

4. $\frac{1}{3}$ 5. $\frac{1}{2}$

✓ **Concept Check Answer**

The number of ways an event can occur can't be larger than the number of possible outcomes.

■ Work Practice 5

Vocabulary, Readiness & Video Check

Use the choices below to fill in each blank. Choices may be used more than once.

0   probability   tree diagram
1   outcome

1. A possible result of an experiment is called a(n) _____.

2. A(n) _____ shows each outcome of an experiment as a separate branch.

3. The _____ of an event is a measure of the likelihood of it occurring.

4. _____ is calculated by the number of ways that an event can occur divided by the number of possible outcomes.
5. A probability of _____ means that an event won't occur.
6. A probability of _____ means that an event is certain to occur.

Martin-Gay Interactive Videos

See Video 7.5

Watch the section lecture video and answer the following questions.

Objective A 7. In Example 1, how was the possible number of outcomes for the experiment determined from the tree diagram?

Objective B 8. In Example 2, what is the probability of getting a 7? Explain this result.

7.5 Exercise Set MyLab Math

Objective A *Draw a tree diagram for each experiment. Then use the diagram to find the number of possible outcomes. See Examples 1 and 2.*

1. Choosing a letter in the word MATH and then a number (1, 2, or 3)

2. Choosing a number (1 or 2) and then a vowel (a, e, i, o, or u)

3. Spinning Spinner A once

4. Spinning Spinner B once

Spinner A

Spinner B

5. Spinning Spinner B twice

6. Spinning Spinner A twice

7. Spinning Spinner A and then Spinner B

8. Spinning Spinner B and then Spinner A

9. Tossing a coin and then spinning Spinner B

10. Spinning Spinner A and then tossing a coin

Objective **B** *If a single die is tossed once, find the probability of each event. See Examples 3 through 5.*

11. A 5

12. A 9

13. A 1 or a 6

14. A 2 or a 3

15. An even number

16. An odd number

17. A number greater than 2

18. A number less than 6

Suppose the spinner shown is spun once. Find the probability of each event. See Examples 3 through 5.

19. The result of the spin is 2.

20. The result of the spin is 3.

21. The result of the spin is 1, 2, or 3.

22. The result of the spin is not 3.

23. The result of the spin is an odd number.

24. The result of the spin is an even number.

If a single choice is made from the bag of marbles shown, find the probability of each event. See Examples 3 through 5.

25. A red marble is chosen.

26. A blue marble is chosen.

27. A yellow marble is chosen.

28. A green marble is chosen.

29. A green or red marble is chosen.

30. A blue or yellow marble is chosen.

A new drug is being tested that is supposed to lower blood pressure. This drug was given to 200 people, and the results are shown below.

Lower Blood Pressure	Higher Blood Pressure	Blood Pressure Not Changed
152	38	10

31. If a person is testing this drug, what is the probability that his or her blood pressure will be higher?

32. If a person is testing this drug, what is the probability that his or her blood pressure will be lower?

33. If a person is testing this drug, what is the probability that his or her blood pressure will not change?

34. What is the sum of the answers to Exercises **31**, **32**, and **33**? In your own words, explain why.

Review

Perform each indicated operation. See Sections 4.3 and 4.5.

35. $\dfrac{1}{2} + \dfrac{1}{3}$ **36.** $\dfrac{7}{10} - \dfrac{2}{5}$ **37.** $\dfrac{1}{2} \cdot \dfrac{1}{3}$ **38.** $\dfrac{7}{10} \div \dfrac{2}{5}$ **39.** $5 \div \dfrac{3}{4}$ **40.** $\dfrac{3}{5} \cdot 10$

Concept Extensions

Recall that a deck of cards contains 52 cards. These cards consist of four suits (hearts, spades, clubs, and diamonds) of each of the following: 2, 3, 4, 5, 6, 7, 8, 9, 10, jack, queen, king, and ace. If a card is chosen from a deck of cards, find the probability of each event.

41. The king of hearts **42.** The 10 of spades

43. A king **44.** A 10

45. A heart **46.** A club

47. A card in black ink **48.** A queen or ace

Two dice are tossed. Find the probability of each sum of the dice. (Hint: Draw a tree diagram of the possibilities of two tosses of a die, and then find the sum of the numbers on each branch.)

49. A sum of 6 **50.** A sum of 10 **51.** A sum of 13 **52.** A sum of 2

Solve. See the Concept Check in this section.

53. In your own words, explain why the probability of an event cannot be greater than 1.

54. In your own words, explain when the probability of an event is 0.

Chapter 7 Group Activity

Stem-and-Leaf Displays

Sections 7.1 and 7.2

Stem-and-leaf displays are another way to organize data. After data are logically organized, it can be much easier to draw conclusions from them.

Suppose we have collected the following set of data. It could represent the test scores for an algebra class or the pulse rates of a group of small children.

90	73	93	99	79	95	69	78	93	80
89	85	97	78	75	79	72	76	97	88
83	98	72	94	92	79	70	98	85	99

In a stem-and-leaf display, the last digit of each number forms the *leaf,* and the remaining digits to the left form the *stem.* For the first number in the list, 90, 9 is the stem and 0 is the leaf. To make the stem-and-leaf display, we write all of the stems in numerical order in a column. Then we write each leaf on the horizontal line next to its stem, aligning leaves in vertical columns. In this case, because the data range from 69 to 99, we use the stems 6, 7, 8, and 9. Each line of the display represents an interval of data; for instance, the line corresponding to the stem **7** represents all data that fall in the interval **70** to **79,** inclusive. After the data have been divided into stems and leaves on the display as shown in the table below, we simply rearrange the leaves on each line to appear in numerical order, as shown in the table in the rightmost column.

Stem	Leaf		Stem	Leaf
6	9	→	6	9
7	39885926290	→	7	02235688999
8	095835	→	8	035589
9	039537784289	→	9	023345778899

Now that the data have been organized into a stem-and-leaf display, it is easy to answer questions about the data such as: What are the least and greatest values in the set of data? Which data interval contains the most items from the data set? How many values fall between 74 and 84? Which data value occurs most frequently in the data set? What patterns or trends do you see in the data?

Make a stem-and-leaf display of the weekend emergency room admission data shown below. Then answer the following questions:

1. What is the difference between the least number and the greatest number of weekend ER admissions?

2. How many weekends had between 125 and 165 ER admissions?

3. What number of weekend ER admissions occurred most frequently?

4. Which interval contains the most weekend ER admissions?

Number of Emergency Room Admissions on Weekends				
198	168	117	185	159
160	177	169	112	175
170	188	137	117	145
198	169	154	163	192
167	179	155	133	121
162	188	124	145	146
128	181	198	149	140
122	162	161	180	177

Chapter 7 Vocabulary Check

Fill in each blank with one of the words or phrases listed below. Some items may be used more than once.

outcomes	class interval	experiment	probability	Pythagorean	leg
pictograph	bar	class frequency	hypotenuse	similar	
histogram	circle	tree diagram	right	congruent	

1. _____ triangles have the same shape and the same size.
2. _____ triangles have exactly the same shape but not necessarily the same size.
3–5. Label the sides of the right triangle.
6. A triangle with one right angle is called a(n) _____ triangle.

7. In the right triangle, $a^2 + b^2 = c^2$ is called the _____ theorem.
8. A(n) _____ graph presents data using vertical or horizontal bars.
9. The possible results of an experiment are the _____.
10. A(n) _____ is a graph in which pictures or symbols are used to visually present data.
11. A(n) _____ is one way to picture and count outcomes.
12. A(n) _____ is an activity being considered, such as tossing a coin or rolling a die.
13. In a(n) _____ graph, each section (shaped like a piece of pie) shows a category and the relative size of the category.
14. The _____ of an event is $\dfrac{\text{number of ways that the event can occur}}{\text{number of possible outcomes}}$.
15. A(n) _____ is a special bar graph in which the width of each bar represents a(n) _____ and the height of each bar represents the _____.

Helpful Hint
- Are you preparing for your test? To help, don't forget to take these:
  - Chapter 7 Getting Ready for the Test on page 553
  - Chapter 7 Test on page 555

Then check all of your answers at the back of this text. For further review, the step-by-step video solutions to any of these exercises are located in MyLab Math.

7 Chapter Highlights

Definitions and Concepts	Examples

Section 7.1 Pictographs, Bar Graphs, Histograms, Line Graphs, and Introduction to Statistics

A **pictograph** is a graph in which pictures or symbols are used to visually present data.

A **line graph** displays information with a line that connects data points.

A **bar graph** presents data using vertical or horizontal bars.

The bar graph on the right shows the number of acres of corn harvested in a recent year for selected states.

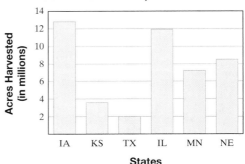

Source: U.S. Department of Agriculture

(Continued)

Definitions and Concepts	Examples
Section 7.1 Pictographs, Bar Graphs, Histograms, Line Graphs, and Introduction to Statistics *(continued)*	

	1. Approximately how many acres of corn were harvested in Iowa? 12,800,000 acres 2. About how many more acres of corn were harvested in Illinois than Nebraska? $$\begin{array}{r} 12 \text{ million} \\ -\ 8.5 \text{ million} \\ \hline 3.5 \text{ million} \end{array}$$ or 3,500,000 acres
A **histogram** is a special bar graph in which the width of each bar represents a **class interval** and the height of each bar represents the **class frequency.** The histogram on the right shows student quiz scores.	1. How many students received a score of 6–10? 4 students 2. How many students received a score of 11–20? $9 + 13 = 22$ students
The **range** of a data set is the difference between the largest data value and the smallest data value. range = largest data value − smallest data value	The range of data set 5, 7, 9, 11, 21, 21 is: range = 21 − 5 = 16

Section 7.2 Circle Graphs	

In a **circle graph,** each section (shaped like a piece of pie) shows a category and the relative size of the category. The circle graph on the right classifies tornadoes by wind speed.	**Tornado Wind Speeds** 110–205 mph 28% Greater than 205 mph 2% Less than 110 mph 70% *Source:* National Oceanic and Atmospheric Administration 1. What percent of tornadoes have wind speeds of 110 mph or greater? $28\% + 2\% = 30\%$ 2. If there were 939 tornadoes in the United States in 2012, how many of these might we expect to have had wind speeds less than 110 mph? Find 70% of 939. $70\%(939) = 0.70(939) = 657.3 \approx 657$ Around 657 tornadoes would be expected to have had wind speeds of less than 110 mph.

Chapter 7 Highlights

Definitions and Concepts	Examples
Section 7.3	**Square Roots and the Pythagorean Theorem**

Square Root of a Number

The square root of a positive number a is the positive number b whose square is a. In symbols,

$\sqrt{a} = b$, if $b^2 = a$

Also, $\sqrt{0} = 0$.

$\sqrt{9} = 3$, $\sqrt{100} = 10$, $\sqrt{1} = 1$

Pythagorean Theorem

If a and b are the lengths of the legs of a right triangle and c is the length of the hypotenuse, then

$a^2 + b^2 = c^2$

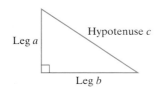

Find c.

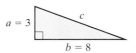

$a^2 + b^2 = c^2$
$3^2 + 8^2 = c^2$ Let $a = 3$ and $b = 8$.
$9 + 64 = c^2$ Multiply.
$73 = c^2$ Simplify.
$\sqrt{73} = c$ or $c \approx 8.5$

| **Section 7.4** | **Congruent and Similar Triangles** |

Congruent triangles have the same shape and the same size. Corresponding angles are equal, and corresponding sides are equal.

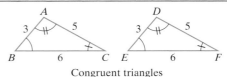

Congruent triangles

Similar triangles have exactly the same shape but not necessarily the same size. Corresponding angles are equal, and the ratios of the lengths of corresponding sides are equal.

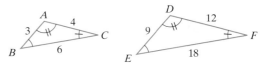

Similar triangles

$\dfrac{AB}{DE} = \dfrac{3}{9} = \dfrac{1}{3}$, $\dfrac{BC}{EF} = \dfrac{6}{18} = \dfrac{1}{3}$,

$\dfrac{CA}{FD} = \dfrac{4}{12} = \dfrac{1}{3}$

Definitions and Concepts	Examples
Section 7.5 Counting and Introduction to Probability	
An **experiment** is an activity being considered, such as tossing a coin or rolling a die. The possible results of an experiment are the **outcomes**. A **tree diagram** is one way to picture and count outcomes.	Draw a tree diagram for tossing a coin and then choosing a number from 1 to 4.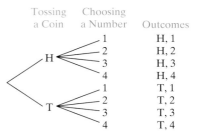
Any number of outcomes considered together is called an **event**. The **probability** of an event is a measure of the chance or likelihood of it occurring. probability of an event = $\dfrac{\text{number of ways that the event can occur}}{\text{number of possible outcomes}}$	Find the probability of tossing a coin twice and tails occurring each time. $(H, H), (H, T), (T, H), (T, T)$ 1 way the event can occur / 4 possible outcomes probability = $\dfrac{1}{4}$

Chapter 7   Review

(7.1) *The following pictograph shows the number of new homes constructed from October 2015 to October 2016, by region. Use this graph to answer Exercises 1 through 6.*

**New Home Construction
October 2015–October 2016**

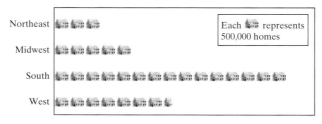

Source: U.S. Census Bureau

1. How many new homes were constructed in the Midwest during the given year?

2. How many new homes were constructed in the South during the given year?

3. Which region had the most new homes constructed?

4. Which region had the fewest new homes constructed?

5. Which region(s) had 3,000,000 or more new homes constructed?

6. Which region(s) had fewer than 3,000,000 new homes constructed?

The following bar graph shows the percent of persons age 25 or over who completed four or more years of college. Use this graph to answer Exercises 7 through 10.

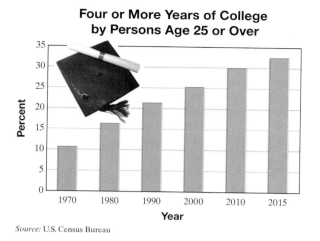

Source: U.S. Census Bureau

7. Approximate the percent of persons who had completed four or more years of college by 2010.

8. What year shown had the greatest percent of persons completing four or more years of college?

9. What years shown had 20% or more of persons completing four or more years of college?

10. Describe any patterns you notice in this graph.

The following line graph shows the total number of Olympic medals awarded during the Summer Olympics since 1996. Use this graph to answer Exercises 11 through 16.

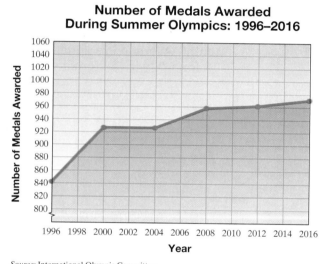

Source: International Olympic Committee

11. Approximate the number of medals awarded during the Summer Olympics of 2012.

12. Approximate the number of medals awarded during the Summer Olympics of 2000.

13. Approximate the number of medals awarded during the Summer Olympics of 2016.

14. Approximate the number of medals awarded during the Summer Olympics of 1996.

15. How many more medals were awarded at the Summer Olympics of 2008 than at the Summer Olympics of 2004?

16. The number of medals awarded at the Summer Olympics of 2016 is the greatest number of medals awarded at previous Summer Olympics. Why do you think this is so? Give your explanation in complete sentences.

The following histogram shows the hours worked per week by the employees of Southern Star Furniture. Use this histogram to answer Exercises 17 through 20.

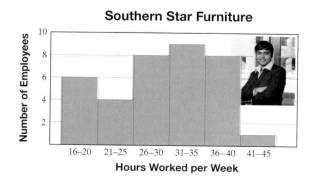

17. How many employees work 41–45 hours per week?

18. How many employees work 21–25 hours per week?

19. How many employees work 30 hours or less per week?

20. How many employees work 36 hours or more per week?

Following is a list of monthly record high temperatures for New Orleans, Louisiana. Use this list to complete the frequency distribution table below.

83	96	101	92
85	100	92	102
89	101	87	84

	Class Intervals (Temperatures)	Tally	Class Frequency (Number of Months)
21.	80°-89°		
22.	90°-99°		
23.	100°-109°		

24. Use the table from Exercises **21–23** to draw a histogram.

Find the range for each data set.

25. 4, 4, 4, 3, 3, 5, 6   **26.** 1, 1, 1, 8, 8   **27.** 70, 75, 95, 60, 88   **28.** 80, 87, 97, 99, 85

29.
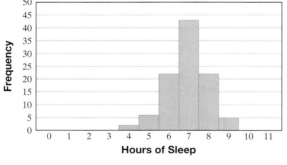

30.
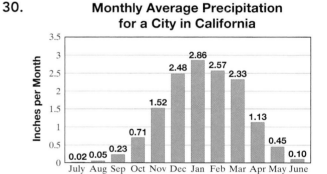

*Use each frequency distribution table to find the **a.** mean, **b.** median, and **c.** mode. If needed, round the mean to 1 decimal place*

31.

Data Item	Frequency
60	2
61	10
62	5
63	11
64	3

32.

Data Item	Frequency
60	5
61	7
62	8
63	10
64	15

33.

Data Item	Frequency
11	5
12	5
13	3
14	1
16	3
17	1

34.

Data Item	Frequency
15	3
16	2
17	1
18	6
19	4
20	6

Use each graph of data items to find the **a.** mean, **b.** median, and **c.** mode. If needed round the mean to 1 decimal place.

35.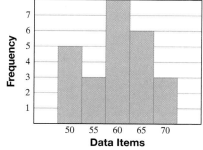

36.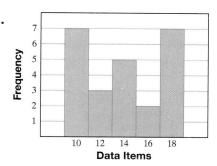

(7.2) The following circle graph shows a family's $4000 monthly budget. Use this graph to answer Exercises 37 through 42. Write all ratios as fractions in simplest form.

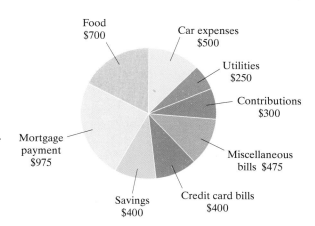

37. What is the largest budget item?

38. What is the smallest budget item?

39. How much money is budgeted for the mortgage payment and utilities?

40. How much money is budgeted for savings and contributions?

41. Find the ratio of the mortage payment to the total monthly budget.

42. Find the ratio of food to the total monthly budget.

In 2016, there were approximately 117 buildings 200 meters or taller completed in the world. The following circle graph shows the percent of these buildings by region. Use this graph to answer Exercises 43 through 46. Round each answer to the nearest whole.

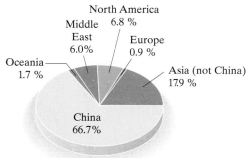

Percent of Tall Buildings Completed in 2016 200 Meters or Taller by Region

Source: Council on Tall Buildings and Urban Habitats

43. How many completed tall buildings were located in China?

44. How many completed tall buildings were located in the rest of Asia?

45. How many completed tall buildings were located in Oceania?

46. How many completed tall buildings were located in the Middle East?

(7.3) *Find each square root. If necessary, round the square root to the nearest thousandth.*

47. $\sqrt{64}$ **48.** $\sqrt{144}$ **49.** $\sqrt{12}$ **50.** $\sqrt{15}$ **51.** $\sqrt{0}$

52. $\sqrt{1}$ **53.** $\sqrt{50}$ **54.** $\sqrt{65}$ **55.** $\sqrt{\dfrac{4}{25}}$ **56.** $\sqrt{\dfrac{1}{100}}$

Find the unknown length in each given right triangle. If necessary, round to the nearest tenth.

△ **57.** leg = 12, leg = 5

△ **58.** leg = 20, leg = 21

△ **59.** leg = 9, hypotenuse = 14

△ **60.** leg = 66, hypotenuse = 86

△ **61.** Find the length, to the nearest hundredth, of the diagonal of a square that has a side of length 20 centimeters.

△ **62.** Find the height of the building rounded to the nearest tenth.

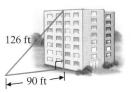

(7.4) *Determine whether each pair of triangles is congruent. If congruent, state the reason why, such as SSS, SAS, or ASA*

63.

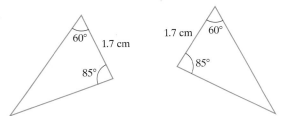

64.

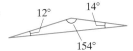

Given that the pairs of triangles are similar, find the unknown length x.

65.

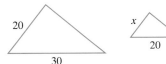

66.

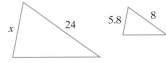

Solve.

67. A housepainter needs to estimate the height of a condominium. He estimates the length of his shadow to be 7 feet long and the length of the building's shadow to be 42 feet long. Find the approximate height of the building if the housepainter is $5\dfrac{1}{2}$ feet tall.

68. A design company is making a triangular sail for a model sailboat. The model sail is to be the same shape as a life-size sailboat's sail. Use the following diagram to find the unknown lengths x and y.

(7.5) *Draw a tree diagram for each experiment. Then use the diagram to determine the number of outcomes.*

Spinner 1

Spinner 2

69. Tossing a coin and then spinning Spinner 1

70. Spinning Spinner 2 and then tossing a coin

71. Spinning Spinner 1 twice

72. Spinning Spinner 1 and then Spinner 2

Find the probability of each event.

Die

73. Rolling a 4 on a die

74. Rolling a 3 on a die

75. Spinning a 4 on Spinner 1

76. Spinning a 3 on Spinner 1

77. Spinning either a 1, 3, or 5 on Spinner 1

78. Spinning either a 2 or a 4 on Spinner 1

Mixed Review

Given a bag containing 2 red marbles, 2 blue marbles, 3 yellow marbles, and 1 green marble, find the following:

79. The probability of choosing a blue marble from the bag

80. The probability of choosing a yellow marble from the bag

81. The probability of choosing a red marble from the bag

82. The probability of choosing a green marble from the bag

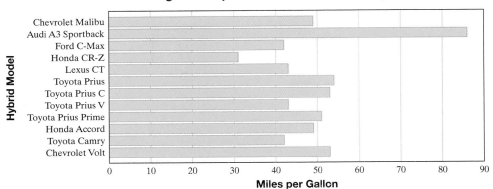

Source: http://www.greenhybrid.com

83. Name the hybrid model with the greatest number of miles per gallon, and estimate the miles per gallon.

84. Name the hybrid model with the least number of miles per gallon, and estimate the miles per gallon.

85. Name the hybrid model(s) whose miles per gallon is the same, and estimate the miles per gallon.

86. Name the hybrid model whose miles per gallon is the closest to 55, and estimate the miles per gallon.

Find each square root. If necessary, approximate and round to the nearest thousandth.

87. $\sqrt{36}$ **88.** $\sqrt{\dfrac{16}{81}}$ **89.** $\sqrt{105}$ **90.** $\sqrt{32}$

Find the unknown length of each given right triangle. If necessary, round to the nearest tenth.

91. leg = 66, leg = 56

92. leg = 12, hypotenuse = 24

Given that the pairs of triangles are similar, find the unknown length, n.

93.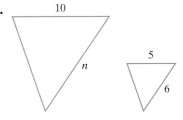

94.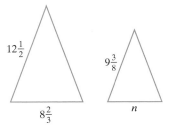

*For each graph, calculate parts **a.** through **d.** Round the mean to two decimal places.*

a. mean   **b.** median   **c.** mode   **d.** range

95.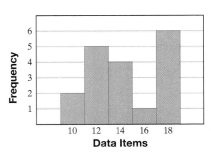

96.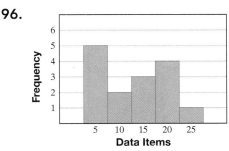

Chapter 7 Getting Ready for the Test

MULTIPLE CHOICE Exercises 1 through 6 are **Multiple Choice**. Choose the correct letter.

For Exercises 1 through 4, use the graph below.

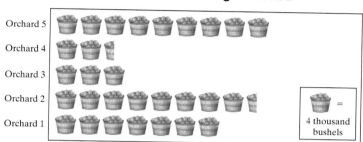

1. How many bushels of oranges did Orchard 1 produce?
   A. 7 bushels **B.** 28 bushels **C.** 28,000 bushels

2. Which orchard above produced the most bushels?
   A. Orchard 1 **B.** Orchard 2 **C.** Orchard 5 **D.** Orchards 2 and 5 produced the same.

3. How many bushels of oranges did Orchard 4 produce?
   A. 3 bushels **B.** $2\frac{1}{2}$ bushels **C.** 10 bushels **D.** 10 thousand bushels

4. How many more bushels did Orchard 2 produce than Orchard 4?
   A. 24 bushels **B.** $6\frac{1}{2}$ bushels **C.** 24,000 bushels **D.** 6000 bushels

5. Which square root is between the numbers 8 and 10?
   A. $\sqrt{36}$ **B.** $\sqrt{25}$ **C.** $\sqrt{49}$ **D.** $\sqrt{81}$

6. Which square root is between the numbers 6 and 7?
   A. $\sqrt{10}$ **B.** $\sqrt{13}$ **C.** $\sqrt{50}$ **D.** $\sqrt{40}$

TRUE OR FALSE *Choose the correct letter for Exercises 7 through 10.*
   A. True   or   **B.** False

7. The Pythagorean theorem applies to right triangles only.

8. A right triangle can have two 90° angles.

9. A right triangle has 3 sides, 1 side is called a leg and the other 2 sides are each called a hypotenuse.

10. The hypotenuse of a right triangle is the longest side.

MULTIPLE CHOICE Exercises 11 through 16 are **Multiple Choice**. Choose the correct letter.

11. Choose the degrees in the unknown sector of this circle graph.

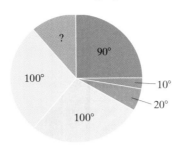

A. 60°   B. 50°   C. 40°   D. 80°

12. The five colored marbles are placed in a bag. What is the probability of choosing a red marble?

A. $\frac{1}{5}$ B. $\frac{2}{5}$ C. $\frac{3}{5}$ D. $\frac{4}{5}$

13. For the marbles in Exercise 12, What is the probability of choosing a green marble?

A. 0   B. 1   C. $\frac{2}{5}$ D. $\frac{3}{5}$

For Exercises 14 through 16, use the data sets and choices below to answer.
A. data set: 10, 10, 10, 10, 10
B. data set: 6, 8, 10, 12, 14
C. data set: 8, 9, 10, 11, 12
D. they are the same

14. Which data set has the greatest range?

15. Which data set has the greatest median?

16. Which data set has the greatest mean?

MATCHING *Choose* **two** *data sets (two letters) in the right column that make each exercise statement true. Data sets ma be used more than once or not at all.*

17. equal means

18. equal number of modes

19. equal ranges

20. equal medians

A. 10, 20, 30, 30, 40, 50
B. 9, 11, 11, 30, 48
C. 20, 25, 30, 35, 40
D. 50, 55, 60, 65, 70

Chapter 7 Test

The following pictograph shows the money collected each week from a wrapping paper fundraiser. Use this graph to answer Exercises 1 through 3.

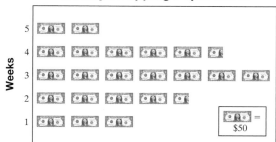

Weekly Wrapping Paper Sales

1. How much money was collected during the second week?

2. During which week was the most money collected? How much money was collected during that week?

3. What was the total amount of money collected for the fundraiser?

The bar graph shows the normal monthly precipitation in centimeters for Chicago, Illinois. Use this graph to answer Exercises 4 through 6.

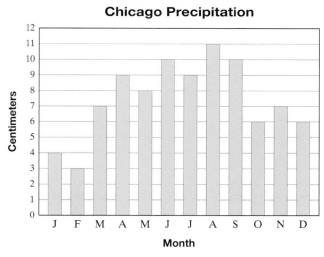

Source: U.S. National Oceanic and Atmospheric Administration, *Climatography of the United States*, No. 81

4. During which month(s) does Chicago normally have more than 9 centimeters of precipitation?

5. During which month does Chicago normally have the least amount of precipitation? How much precipitation occurs during that month?

Answers

1. _____

2. _____

3. _____

4. _____

5. _____

555

6. During which month(s) does 7 centimeters of precipitation normally occur?

7. Use the information in the table to draw a bar graph. Clearly label each bar.

Most Common Blood Types	
Blood Type	% of Population with This Blood Type
O+	38%
A+	34%
B+	9%
O−	7%
A−	6%
AB+	3%
B−	2%
AB−	1%

Most Common Blood Types by Percent in the Population

The following line graph shows the annual inflation rate in the United States for the years 2008–2016. Use this graph to answer Exercises 8 through 10.

U.S. Annual Inflation Rate

Source: Bureau of Labor Statistics

8. Approximate the annual inflation rate in 2014.

9. During which of the years shown was the inflation rate greater than 3%?

10. During which sets of years was the inflation rate decreasing?

The result of a survey of 200 people is shown in the following circle graph. Each person was asked to tell his or her favorite type of music. Use this graph to answer Exercises 11 and 12.

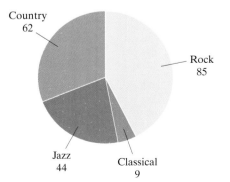

Country 62
Rock 85
Classical 9
Jazz 44

11. Find the ratio of those who prefer rock music to the total number surveyed.

12. Find the ratio of those who prefer country music to those who prefer jazz.

Chapter 7 Test 557

The following circle graph shows the projected age distribution of the population of the United States in 2020. There are projected to be 335 million people in the United States in 2020. Use the graph to find how many people are expected to be in the age groups given.

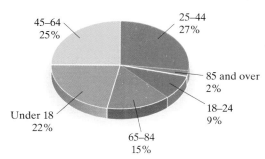

U.S. Population in 2020 by Age Groups

- 45–64: 25%
- 25–44: 27%
- 85 and over: 2%
- 18–24: 9%
- 65–84: 15%
- Under 18: 22%

Source: U.S. Census Bureau

13. Under 18 (Round to nearest whole million.)

14. 25–44 (Round to nearest whole million.)

A professor measures the heights of the students in her class. The results are shown in the following histogram. Use this histogram to answer Exercises 15 and 16.

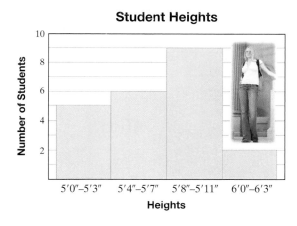

Student Heights

15. How many students are 5'8"–5'11" tall?

16. How many students are 5'7" or shorter?

17. The history test scores of 25 students are shown below. Use these scores to complete the frequency distribution table.

70	86	81	65	92
43	72	85	69	97
82	51	75	50	68
88	83	85	77	99
77	63	59	84	90

Class Intervals (Scores)	Tally	Class Frequency (Number of Students)
40–49		
50–59		
60–69		
70–79		
80–89		
90–99		

13. _____

14. _____

15. _____

16. _____

17. _____

18. Use the results of Exercise **17** to draw a histogram.

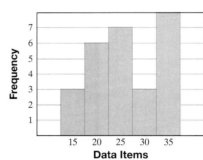
Scores

19. Find the range of the data items:
10, 18, 13, 16, 13

20. Use the data for Exercise **17** and find the range.

21. Use the frequency distribution table to find the following. Round the mean to 1 decimal place.

a. mean
c. mode
b. median
d. range

Data Item	Frequency
90	3
91	1
92	2
93	8
94	8

22. Use the graph of data items to find the following. Round the mean to 1 decimal place.

a. mean
c. mode
b. median
d. range

Find each square root and simplify. Round to the nearest thousandth if necessary.

23. $\sqrt{49}$

24. $\sqrt{157}$

25. $\sqrt{\dfrac{64}{100}}$

Solve.

26. Approximate, to the nearest hundredth of a centimeter the length of the missing side of a right triangle with legs of 4 centimeters each.

27. Given that the following triangles are similar, find the unknown length, n.

28. Tamara Watford, surveyor, needs to estimate the height of a tower. She estimates the length of her shadow to be 4 feet long and the length of the tower's shadow to be 48 feet long. Find the height of the tower if she is $5\frac{3}{4}$ feet tall.

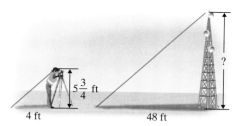

28. _____

29. Draw a tree diagram for the experiment of spinning the spinner twice. Then use the diagram to determine the number of outcomes.

29. _____

30. Draw a tree diagram for the experiment of tossing a coin twice. Then use the diagram to determine the number of outcomes.

Suppose that the numbers 1 through 10 are each written on same-size pieces of paper and placed in a bag. You then select one piece of paper from the bag.

30. _____

31. What is the probability of choosing a 6 from the bag?

31. _____

32. What is the probability of choosing a 3 or a 4 from the bag?

32. _____

Chapters 1–7 Cumulative Review

Answers

1. Simplify: $4^3 + [3^2 - (10 \div 2)] - 7 \cdot 3$

2. $7^2 + [5^3 - (6 \div 3)] + 4 \cdot 2$

3. Evaluate $x - y$ for $x = -3$ and $y = 9$.

4. Evaluate $x - y$ for $x = 7$ and $y = -2$.

5. Solve: $3y - 7y = 12$

6. Solve: $2x - 6x = 24$

7. Solve: $\dfrac{x}{6} + 1 = \dfrac{4}{3}$

8. Solve: $\dfrac{7}{2} + \dfrac{a}{4} = 1$

9. Add: $2\dfrac{1}{3} + 5\dfrac{3}{8}$

10. Add: $3\dfrac{2}{5} + 4\dfrac{3}{4}$

11. Write 5.9 as a mixed number.

12. Write 2.8 as a mixed number.

13. Subtract: $3.5 - 0.068$.

14. Subtract: $7.4 - 0.073$

15. Multiply: 0.0531×16

16. Multiply: 0.147×0.2

17. Divide: $-5.98 \div 115$

18. Divide: $27.88 \div 205$

19. Simplify: $(-1.3)^2 + 2.4$

20. Simplify: $(-2.7)^2$

21. Write $\dfrac{1}{4}$ as a decimal.

22. Write $\dfrac{3}{8}$ as a decimal.

23. Solve: $5(x - 0.36) = -x + 2.4$

24. Solve: $4(0.35 - x) = x - 7$

25. Use the appendix or a calculator to approximate $\sqrt{80}$ to the nearest thousandth.

26. Use the appendix or a calculator to approximate $\sqrt{60}$ to the nearest thousandth.

Cumulative Review 561

27. Find $\sqrt{\dfrac{1}{36}}$.

28. Find $\sqrt{\dfrac{16}{49}}$.

Sketch each right triangle and find the length of the side not given. Round to two decimal places.

29. leg = 5 in., hypotenuse = 7 in.

30. leg = 2, leg = 16

31. Solve $\dfrac{x-5}{3} = \dfrac{x+2}{5}$ for x and then check.

32. Solve $\dfrac{8}{5} = \dfrac{x}{10}$ for x and then check.

33. Find the ratio of corresponding sides for the similar triangles ABC and DEF.

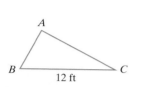
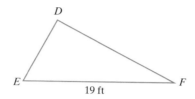

34. Find the ratio of corresponding sides for the similar triangles GHJ and KLM.

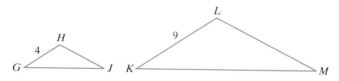

Write each percent as a decimal.

35. 4.6%

36. 32%

37. 0.74%

38. 2.7%

39. What number is 35% of 60?

40. What number is 40% of 36?

41. 20.8 is 40% of what number?

42. 9.5 is 25% of what number?

43. The sales tax on a $310 Sony flat-screen digital 32-inch television is $26.35. Find the sales tax rate.

44. The sales tax on a $2.00 yo-yo is $0.13. Find the sales tax rate.

45. $4000 is invested at 5.3% compounded quarterly for 10 years. Find the total amount at the end of 10 years.

46. Linda Bonnett borrows $1600 for 1 year. If the interest is $128.60, find the monthly payment. (Hint: Find the total amount due and divide by the number of months in a year.)

47. Find the median of the list of numbers: 25, 54, 56, 57, 60, 71, 98

48. Find the median of the list of numbers: 43, 46, 47, 50, 52, 83

49. If a die is rolled one time, find the probability of rolling a 3 or a 4.

50. If a die is rolled one time, find the probability of rolling an even number.

8 Geometry and Measurement

The word *geometry* is formed from the Greek words *geo*, meaning "Earth," and *metron*, meaning "measure." Geometry literally means to measure the Earth. In this chapter we learn about various geometric figures and their properties, such as perimeter, area, and volume. Knowledge of geometry can help us solve practical problems in all types of real-life situations.

Sections

8.1 Lines and Angles

8.2 Perimeter

8.3 Area, Volume, and Surface Area

   Integrated Review—Geometry Concepts

8.4 Linear Measurement

8.5 Weight and Mass

8.6 Capacity

8.7 Temperature and Conversions Between the U.S. and Metric Systems

Check Your Progress

Vocabulary Check

Chapter Highlights

Chapter Review

Getting Ready for the Test

Chapter Test

Cumulative Review

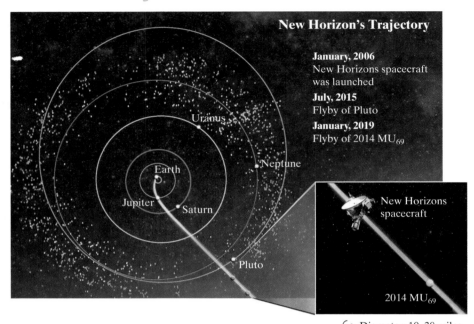

2014 MU_{69}
- Diameter: 10–30 miles
- 4 billion miles from earth
- An object in the Kuiper belt

Source: Minor Planet Center, www.cfeps.net and others

Where Is New Horizons Spacecraft Now, and How Do We Receive Data Collected?

New Horizons is NASA's robotic spacecraft mission. This spacecraft is about the size and shape of a grand piano with a satellite dish attached. It was launched in January 2006 and is now headed for a January 1, 2019, flyby past 2014 MU_{69}, a small object in the Kuiper asteroid belt. How do we receive the images of this spacecraft when it is so far into deep space?

The Deep Space Network (DSN) is a worldwide network of large antennas and communication facilities. When a mission is in deep space, fewer sites are needed for sending and receiving transmissions; thus, the DSN uses only three sites, shown below. The diagram below shows an overview of Earth from the vantage point of the North Pole and the location of these three sites.

We study some geometry of the DSN in Section 8.1, Exercises 65 and 66, and Section 8.7, Exercises 23 through 26.

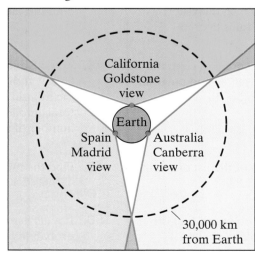

8.1 Lines and Angles

Objective A Identifying Lines, Line Segments, Rays, and Angles

Let's begin with a review of two important concepts—space and plane.

Space extends in all directions indefinitely. Examples of objects in space are houses, grains of salt, bushes, your *Prealgebra and Introductory Algebra* textbook, and you.

A **plane** is a flat surface that extends indefinitely. Surfaces like a plane are a classroom floor or a blackboard or whiteboard.

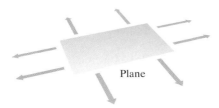
Plane

Objectives

A Identify Lines, Line Segments, Rays, and Angles.

B Classify Angles as Acute, Right, Obtuse, or Straight.

C Identify Complementary and Supplementary Angles.

D Find Measures of Angles.

The most basic concept of geometry is the idea of a point in space. A **point** has no length, no width, and no height, but it does have location. We represent a point by a dot, and we usually label points with capital letters.

P
•
Point P

A **line** is a set of points extending indefinitely in two directions. A line has no width or height, but it does have length. We can name a line by any two of its points or by a single lowercase letter. A **line segment** is a piece of a line with two endpoints.*

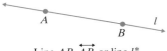
Line AB, $\overleftrightarrow{AB}$, or line l*

Line segment AB or $\overline{AB}$*

A **ray** is a part of a line with one endpoint. A ray extends indefinitely in one direction. An **angle** is made up of two rays that share the same endpoint. The common endpoint is called the **vertex.**

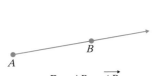
Ray AB or $\overrightarrow{AB}$

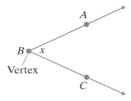
Vertex

The angle in the figure above can be named

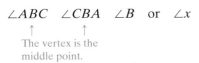

The vertex is the middle point.

Rays BA and BC are **sides** of the angle.

---

*Although line l is also line BA or $\overleftrightarrow{BA}$, we will use only one order of points to name lines or line segments.

564 Chapter 8 | Geometry and Measurement

> **Helpful Hint**
>
> **Naming an Angle**
> When there is no confusion as to what angle is being named, you may use the vertex alone.
>
>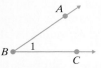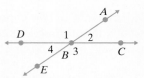
>
> Name of ∠B is all right. Name of ∠B is *not* all right.
> There is no confusion. ∠B means ∠1. There is confusion. Does ∠B mean ∠1, ∠2, ∠3, or ∠4?

Practice 1

Identify each figure as a line, a ray, a line segment, or an angle. Then name the figure using the given points.

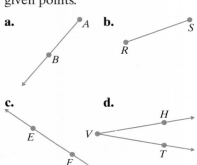

Example 1 Identify each figure as a line, a ray, a line segment, or an angle. Then name the figure using the given points.

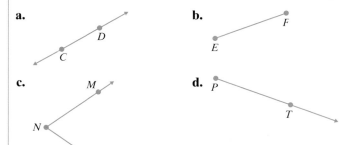

Solution:

Figure (**a**) extends indefinitely in two directions. It is line CD or $\overleftrightarrow{CD}$.

Figure (**b**) has two endpoints. It is line segment EF or $\overline{EF}$.

Figure (**c**) has two rays with a common endpoint. It is ∠MNO, ∠ONM, or ∠N.

Figure (**d**) is part of a line with one endpoint. It is ray PT or $\overrightarrow{PT}$.

■ Work Practice 1

Practice 2

Use the figure in Example 2 to list other ways to name ∠z.

Example 2 List other ways to name ∠y.

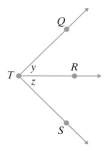

Solution: Two other ways to name ∠y are ∠QTR and ∠RTQ. We may *not* use the vertex alone to name this angle because three different angles have T as their vertex.

■ Work Practice 2

Answers

1. **a.** ray; ray AB or $\overrightarrow{AB}$
   b. line segment; line segment RS or $\overline{RS}$
   c. line; line EF or $\overleftrightarrow{EF}$
   d. angle; ∠TVH or ∠HVT or ∠V
2. ∠RTS, ∠STR

Section 8.1 | Lines and Angles   565

Objective B Classifying Angles as Acute, Right, Obtuse, or Straight

An angle can be measured in **degrees.** The symbol for degrees is a small, raised circle, °. There are 360° in a full revolution, or a full circle.

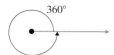

$\frac{1}{2}$ of a revolution measures $\frac{1}{2}(360°) = 180°$. An angle that measures 180° is called a **straight angle.**

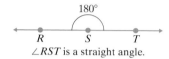

∠RST is a straight angle.

$\frac{1}{4}$ of a revolution measures $\frac{1}{4}(360°) = 90°$. An angle that measures 90° is called a **right angle.** The symbol ∟ is used to denote a right angle.

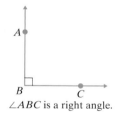

∠ABC is a right angle.

An angle whose measure is between 0° and 90° is called an **acute angle.**

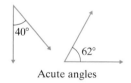

Acute angles

An angle whose measure is between 90° and 180° is called an **obtuse angle.**

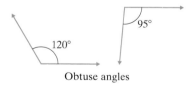

Obtuse angles

Example 3 Classify each angle as acute, right, obtuse, or straight.

a.

b.
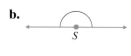

c.

d.
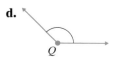

(*Continued on next page*)

Practice 3
Classify each angle as acute, right, obtuse, or straight.

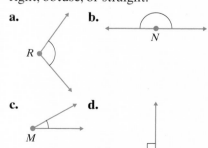

Answers
3. **a.** obtuse **b.** straight **c.** acute **d.** right

Solution:
a. ∠R is a right angle, denoted by ∟. It measures 90°.
b. ∠S is a straight angle. It measures 180°.
c. ∠T is an acute angle. It measures between 0° and 90°.
d. ∠Q is an obtuse angle. It measures between 90° and 180°.

■ Work Practice 3

Let's look at ∠B below, whose measure is 62°.

There is a shorthand notation for writing the measure of this angle. To write "The measure of ∠B is 62°," we can write,

$$m\angle B = 62°.$$

By the way, note that ∠B is an acute angle because $m\angle B$ is between 0° and 90°.

Objective C Identifying Complementary and Supplementary Angles

Two angles that have a sum of 90° are called **complementary angles**. We say that each angle is the **complement** of the other.

∠R and ∠S are complementary angles because

$$m\angle R + m\angle S = 60° + 30° = 90°$$

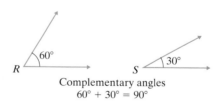

Complementary angles
60° + 30° = 90°

Two angles that have a sum of 180° are called **supplementary angles**. We say that each angle is the **supplement** of the other.

∠M and ∠N are supplementary angles because

$$m\angle M + m\angle N = 125° + 55° = 180°$$

Supplementary angles
125° + 55° = 180°

Practice 4
Find the complement of a 29° angle.

Example 4 Find the complement of a 48° angle.

Solution: Two angles that have a sum of 90° are complementary. This means that the complement of an angle that measures 48° is an angle that measures 90° − 48° = 42°.

■ Work Practice 4

Answer
4. 61°

Section 8.1 | Lines and Angles 567

Example 5 Find the supplement of a 107° angle.

Solution: Two angles that have a sum of 180° are supplementary. This means that the supplement of an angle that measures 107° is an angle that measures $180° - 107° = 73°$.

Work Practice 5

Practice 5
Find the supplement of a 67° angle.

✓ **Concept Check** True or false? The supplement of a 48° angle is 42°. Explain.

Objective D Finding Measures of Angles

Measures of angles can be added or subtracted to find measures of related angles.

Example 6 Find the measure of ∠x. Then classify ∠x as an acute, obtuse, or right angle.

Solution:
$m∠x = m∠QTS - m∠RTS$
$= 87° - 52°$
$= 35°$

Thus, the measure of ∠x (m∠x) is 35°.
Since ∠x measures between 0° and 90°, it is an acute angle.

Work Practice 6

Practice 6

a. Find the measure of ∠y.

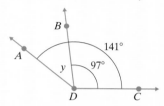

b. Find the measure of ∠x.

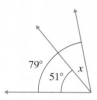

c. Classify ∠x and ∠y as acute, obtuse, or right angles.

Two lines in a plane can be either parallel or intersecting. **Parallel lines** never meet. **Intersecting lines** meet at a point. The symbol ∥ is used to indicate "is parallel to." For example, in the figure, $p \parallel q$.

Parallel lines   Intersecting lines

Some intersecting lines are perpendicular. Two lines are **perpendicular** if they form right angles when they intersect. The symbol ⊥ is used to denote "is perpendicular to." For example, in the figure below, $n \perp m$.

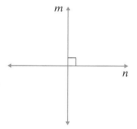

Perpendicular lines

When two lines intersect, four angles are formed. Two angles that are opposite each other are called **vertical angles.** Vertical angles have the same measure.

Two angles that share a common side are called **adjacent angles.** Adjacent angles formed by intersecting lines are supplementary. That is, they have a sum of 180°.

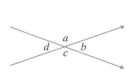

Vertical angles:
∠a and ∠c
∠d and ∠b

Adjacent angles:
∠a and ∠b
∠b and ∠c
∠c and ∠d
∠d and ∠a

Answers
5. 113° **6. a.** 44° **b.** 28° **c.** acute

✓ **Concept Check Answer**
false; the *complement* of a 48° angle is 42°; the *supplement* of a 48° angle is 132°

Here are a few real-life examples of the lines we just discussed.

Parallel lines   Vertical angles   Perpendicular lines

Practice 7
Find the measure of $\angle a$, $\angle b$, and $\angle c$.

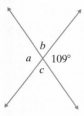

Example 7   Find the measure of $\angle x$, $\angle y$, and $\angle z$ if the measure of $\angle t$ is 42°.

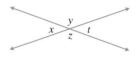

Solution: Since $\angle t$ and $\angle x$ are vertical angles, they have the same measure, so $\angle x$ measures 42°.

Since $\angle t$ and $\angle y$ are adjacent angles, their measures have a sum of 180°. So $\angle y$ measures $180° - 42° = 138°$.

Since $\angle y$ and $\angle z$ are vertical angles, they have the same measure. So $\angle z$ measures 138°.

■ Work Practice 7

A line that intersects two or more lines at different points is called a **transversal.** Line l is a transversal that intersects lines m and n. The eight angles formed have special names. Some of these names are:

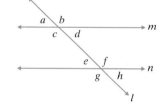

Corresponding angles: $\angle a$ and $\angle e$, $\angle c$ and $\angle g$, $\angle b$ and $\angle f$, $\angle d$ and $\angle h$

Alternate interior angles: $\angle c$ and $\angle f$, $\angle d$ and $\angle e$

When two lines cut by a transversal are *parallel*, the following statement is true:

Practice 8
Given that $m \parallel n$ and that the measure of $\angle w = 45°$, find the measures of all the angles shown.

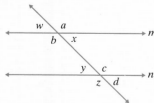

Parallel Lines Cut by a Transversal
If two parallel lines are cut by a transversal, then the measures of **corresponding angles are equal** and the measures of the **alternate interior angles are equal.**

Example 8   Given that $m \parallel n$ and that the measure of $\angle w$ is 100°, find the measures of $\angle x$, $\angle y$, and $\angle z$.

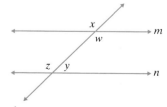

Solution:

$m\angle x = 100°$ $\angle x$ and $\angle w$ are vertical angles.

$m\angle z = 100°$ $\angle x$ and $\angle z$ are corresponding angles.

$m\angle y = 180° - 100° = 80°$ $\angle z$ and $\angle y$ are supplementary angles.

■ Work Practice 8

Answers
7. $m\angle a = 109°$; $m\angle b = 71°$; $m\angle c = 71°$
8. $m\angle x = 45°$; $m\angle y = 45°$; $m\angle z = 135°$; $m\angle a = 135°$; $m\angle b = 135°$; $m\angle c = 135°$; $m\angle d = 45°$

Section 8.1 Lines and Angles

Vocabulary, Readiness & Video Check

Use the choices below to fill in each blank.

acute	straight	degrees	adjacent	parallel	intersecting
obtuse	space	plane	point	vertical	vertex
right	angle	ray	line	perpendicular	transversal

1. A(n) _____ is a flat surface that extends indefinitely.
2. A(n) _____ has no length, no width, and no height.
3. _____ extends in all directions indefinitely.
4. A(n) _____ is a set of points extending indefinitely in two directions.
5. A(n) _____ is part of a line with one endpoint.
6. A(n) _____ is made up of two rays that share a common endpoint. The common endpoint is called the _____.
7. A(n) _____ angle measures 180°.
8. A(n) _____ angle measures 90°.
9. A(n) _____ angle measures between 0° and 90°.
10. A(n) _____ angle measures between 90° and 180°.
11. _____ lines never meet and _____ lines meet at a point.
12. Two intersecting lines are _____ if they form right angles when they intersect.
13. An angle can be measured in _____.
14. A line that intersects two or more lines at different points is called a(n) _____.
15. When two lines intersect, four angles are formed, called _____ angles.
16. Two angles that share a common side are called _____ angles.

Martin-Gay Interactive Videos

See Video 8.1

Watch the section lecture video and answer the following questions.

Objective A 17. In the lecture after Example 2, what are the four ways we can name the angle shown?

Objective B 18. In the lecture before Example 3, what type of angle forms a line? What is its measure?

Objective C 19. What calculation is used to find the answer to Example 6?

Objective D 20. In the lecture before Example 7, two lines in a plane that aren't parallel must what?

8.1 Exercise Set   MyLab Math

Objective A *Identify each figure as a line, a ray, a line segment, or an angle. Then name the figure using the given points. See Examples 1 and 2.*

1.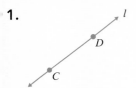

2.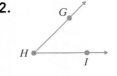

3.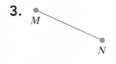

4.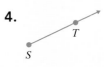

5.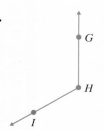
6.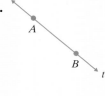
7.
8.

List two other ways to name each angle. See Example 2.

9. ∠x

10. ∠w

11. ∠z

12. ∠y

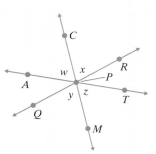

Objective B *Classify each angle as acute, right, obtuse, or straight. See Example 3.*

13.
14.
15.
16.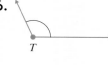

17.
18.
19.
20.

Objective C *Find each complementary or supplementary angle as indicated. See Examples 4 and 5.*

21. Find the complement of a 23° angle.

22. Find the complement of a 77° angle.

23. Find the supplement of a 17° angle.

24. Find the supplement of a 77° angle.

25. Find the complement of a 58° angle.

26. Find the complement of a 22° angle.

27. Find the supplement of a 150° angle.

28. Find the supplement of a 130° angle.

29. Identify the pairs of complementary angles.

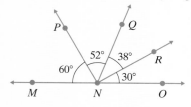

30. Identify the pairs of complementary angles.

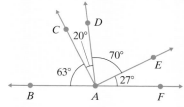

31. Identify the pairs of supplementary angles.

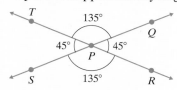

32. Identify the pairs of supplementary angles.

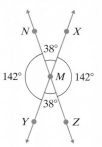

Objective D Find the measure of ∠x in each figure. See Example 6.

33.

34.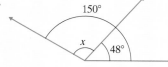

35.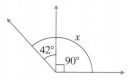

36.

Find the measures of angles x, y, and z in each figure. See Examples 7 and 8.

37.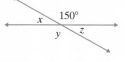

38.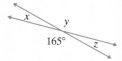

39.

40.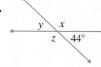

41. $m \parallel n$

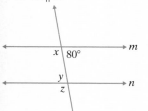

42. $m \parallel n$

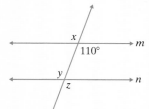

43. $m \parallel n$

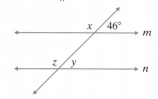

44. $m \parallel n$

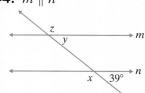

Objectives A D Mixed Practice *Find two other ways of naming each angle. See Example 2.*

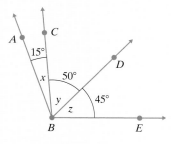

45. ∠x

46. ∠y

47. ∠z

48. ∠ABE (just name one other way)

Find the measure of each angle in the figure above. See Example 6.

49. ∠ABC   **50.** ∠EBD   **51.** ∠CBD   **52.** ∠CBA

53. ∠DBA   **54.** ∠EBC   **55.** ∠CBE   **56.** ∠ABE

Review

Perform each indicated operation. See Sections 4.3, 4.5, and 4.7.

57. $\frac{7}{8} + \frac{1}{4}$ **58.** $\frac{7}{8} - \frac{1}{4}$ **59.** $\frac{7}{8} \cdot \frac{1}{4}$ **60.** $\frac{7}{8} \div \frac{1}{4}$

61. $3\frac{1}{3} - 2\frac{1}{2}$ **62.** $3\frac{1}{3} + 2\frac{1}{2}$ **63.** $3\frac{1}{3} \div 2\frac{1}{2}$ **64.** $3\frac{1}{3} \cdot 2\frac{1}{2}$

Concept Extensions

Use this North Pole overhead view of the three sites of the Deep Space Network to answer Exercises 65 and 66. (See the Chapter Opener.)

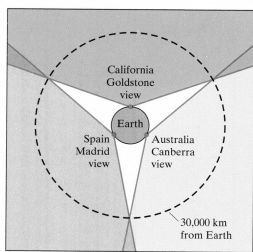

65. How many degrees are there around the Earth at the equator?

66. If the three sites of the Deep Space Network (red dots shown) are about the same number of degrees apart, how many degrees apart are they?

67. One great pyramid at Chichen Itza, Mexico, was the Temple of Kukulkan. The four faces of this pyramid have protruding stairways that rise at a 45° angle. Find the complement of this angle.

68. The UNESCO World Heritage Site of the Cahokia Mounds is located in Cahokia, Illinois. This site was once populated by the Mississippian People of North America and was built up on a series of mounds. These mounds were completely constructed of silt and dirt. The largest and best known of these is the Monk's Mound, which is considered a truncated pyramid, with the top being a flat base rather than extending to a point. On one side of this mound, modern archeologists have measured a 35° angle. Find the complement of this angle.

69. The angle between the two walls of the Vietnam Veterans Memorial in Washington, D.C., is 125.2°. Find the supplement of this angle. (*Source:* National Park Service)

70. The faces of Khafre's Pyramid at Giza, Egypt, are inclined at an angle of 53.13°. Find the complement of this angle. (*Source:* PBS *NOVA* Online)

Answer true or false for Exercises 71 through 74. See the Concept Check in this section. If false, explain why.

71. The complement of a 100° angle is an 80° angle.

72. It is possible to find the complement of a 120° angle.

73. It is possible to find the supplement of a 120° angle.

74. The supplement of a 5° angle is a 175° angle.

75. If lines m and n are parallel, find the measures of angles a through e.

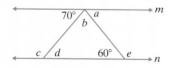

76. Below is a rectangle. List which segments, if extended, would be parallel lines.

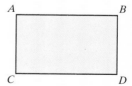

77. Can two supplementary angles both be acute? Explain why or why not.

78. In your own words, describe how to find the complement and the supplement of a given angle.

79. Find two complementary angles with the same measure.

80. Is the figure below possible? Why or why not?

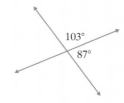

8.2 Perimeter

Objectives

A Use Formulas to Find Perimeters.

B Use Formulas to Find Circumferences.

Objective A Using Formulas to Find Perimeters

Recall from Section 1.3 that the perimeter of a polygon is the distance around the polygon. This means that the perimeter of a polygon is the sum of the lengths of its sides.

Example 1 Find the perimeter of the rectangle below.

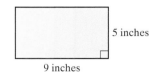

Solution:

$$\text{perimeter} = 9 \text{ inches} + 9 \text{ inches} + 5 \text{ inches} + 5 \text{ inches}$$
$$= 28 \text{ inches}$$

■ Work Practice 1

Notice that the perimeter of the rectangle in Example 1 can be written as
$2 \cdot (9 \text{ inches}) + 2 \cdot (5 \text{ inches})$.
   ↑   ↑
length   width

In general, we can say that the perimeter of a rectangle is always

$$2 \cdot \text{length} + 2 \cdot \text{width}$$

As we have just seen, the perimeters of some special figures such as rectangles form patterns. These patterns are given as **formulas.** The formula for the perimeter of a rectangle is shown next:

Perimeter of a Rectangle

Perimeter $= 2 \cdot \textbf{length} + 2 \cdot \textbf{width}$

In symbols, this can be written as

$P = 2l + 2w$

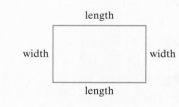

Example 2 Find the perimeter of a rectangle with a length of 11 inches and a width of 3 inches.

Solution: We use the formula for perimeter and replace the letters by their known lengths.

$$P = 2l + 2w$$
$$= 2 \cdot 11 \text{ in.} + 2 \cdot 3 \text{ in.} \quad \text{Replace } l \text{ with 11 in. and } w \text{ with 3 in.}$$
$$= 22 \text{ in.} + 6 \text{ in.}$$
$$= 28 \text{ in.}$$

The perimeter is 28 inches.

■ Work Practice 2

Practice 1

a. Find the perimeter of the rectangle.

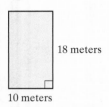

b. Find the perimeter of the rectangular lot shown below:

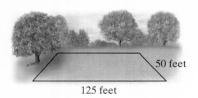

Practice 2

Find the perimeter of a rectangle with a length of 32 centimeters and a width of 15 centimeters.

Answers
1. a. 56 m **b.** 350 ft   **2.** 94 cm

Perimeter of a Square

Perimeter = side + side + side + side
         = 4 · side

In symbols,
$P = 4s$

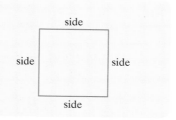

Example 3 Finding the Perimeter of a Field

How much fencing is needed to enclose a square field 50 yards on a side?

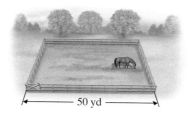

Solution: To find the amount of fencing needed, we find the distance around, or perimeter. The formula for the perimeter of a square is $P = 4s$. We use this formula and replace s by 50 yards.

$P = 4s$
$= 4 \cdot 50$ yd
$= 200$ yd

The amount of fencing needed is 200 yards.

■ Work Practice 3

The formula for the perimeter of a triangle with sides of lengths a, b, and c is given next:

Perimeter of a Triangle

Perimeter = side **a** + side **b** + side **c**

In symbols,
$P = a + b + c$

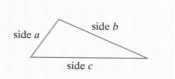

Practice 3
Find the perimeter of a square tabletop if each side is 4 feet long.

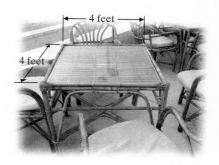

Example 4
Find the perimeter of a triangle if the sides are 3 inches, 7 inches, and 6 inches.

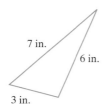

Practice 4
Find the perimeter of a triangle if the sides are 6 centimeters, 10 centimeters, and 8 centimeters in length.

Answers
3. 16 ft **4.** 24 cm

(*Continued on next page*)

Solution: The formula for the perimeter is $P = a + b + c$, where a, b, and c are the lengths of the sides. Thus,

$$P = a + b + c$$
$$= 3 \text{ in.} + 7 \text{ in.} + 6 \text{ in.}$$
$$= 16 \text{ in.}$$

The perimeter of the triangle is 16 inches.

■ Work Practice 4

The method for finding the perimeter of any polygon is given next:

Perimeter of a Polygon

The perimeter of a polygon is the sum of the lengths of its sides.

Practice 5

Find the perimeter of the trapezoid shown.

Example 5 Find the perimeter of the trapezoid shown below:

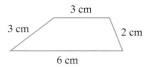

Solution: To find the perimeter, we find the sum of the lengths of its sides.

$$\text{perimeter} = 3 \text{ cm} + 2 \text{ cm} + 6 \text{ cm} + 3 \text{ cm} = 14 \text{ cm}$$

The perimeter is 14 centimeters.

■ Work Practice 5

Practice 6

Find the perimeter of the room shown.

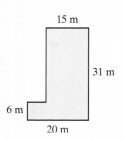

Example 6 Finding the Perimeter of a Room

Find the perimeter of the room shown below:

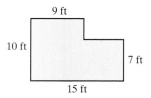

Solution: To find the perimeter of the room, we first need to find the lengths of all sides of the room.

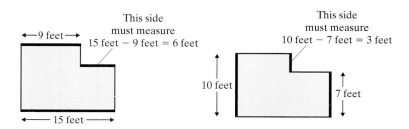

Answers
5. 23 km **6.** 102 m

Now that we know the measures of all sides of the room, we can add the measures to find the perimeter.

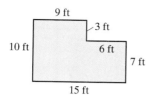

perimeter = 10 ft + 9 ft + 3 ft + 6 ft + 7 ft + 15 ft
= 50 ft

The perimeter of the room is 50 feet.

Work Practice 6

Example 7 Calculating the Cost of Wallpaper Border

A rectangular room measures 10 feet by 12 feet. Find the cost to hang a wallpaper border on the walls close to the ceiling if the cost of the wallpaper border is $1.09 per foot.

Solution: First we find the perimeter of the room.

$P = 2l + 2w$
$= 2 \cdot 12 \text{ ft} + 2 \cdot 10 \text{ ft}$ Replace *l* with 12 feet and *w* with 10 feet.
$= 24 \text{ ft} + 20 \text{ ft}$
$= 44 \text{ ft}$

The cost of the wallpaper is

cost = $1.09 · 44 ft = $47.96

The cost of the wallpaper is $47.96.

Work Practice 7

Practice 7

A rectangular lot measures 60 feet by 120 feet. Find the cost to install fencing around the lot if the cost of fencing is $1.90 per foot.

Objective B Using Formulas to Find Circumferences

Recall from Section 5.3 that the distance around a circle is called the **circumference.** This distance depends on the radius or the diameter of the circle.
The formulas for circumference are shown next:

Circumference of a Circle

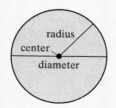

Circumference = $2 \cdot \pi \cdot$ **radius** or Circumference = $\pi \cdot$ **diameter**
In symbols,

$C = 2\pi r$ or $C = \pi d$

where $\pi \approx 3.14$ or $\pi \approx \frac{22}{7}$.

Answer
7. $684

To better understand circumference and π (pi), try the following experiment. Take any can and measure its circumference and its diameter.

The can in the figure above has a circumference of 23.5 centimeters and a diameter of 7.5 centimeters. Now divide the circumference by the diameter.

$$\frac{\text{circumference}}{\text{diameter}} = \frac{23.5 \text{ cm}}{7.5 \text{ cm}} \approx 3.13$$

Try this with other sizes of cylinders and circles—you should always get a number close to 3.1. The exact ratio of circumference to diameter is π. (Recall that $\pi \approx 3.14$ or $\approx \frac{22}{7}$.)

Practice 8

An irrigation device waters a circular region with a diameter of 20 yards. Find the exact circumference of the watered region, then use $\pi \approx 3.14$ to give an approximation.

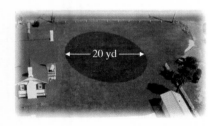

Example 8 Finding Circumference of a Circular Spa

A homeowner plans to install a border of new tiling around the circumference of her circular spa. If her spa has a diameter of 14 feet, find its exact circumference. Then use the approximation 3.14 for π to approximate the circumference.

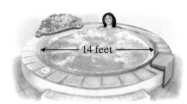

Solution: Because we are given the diameter, we use the formula $C = \pi d$.

$C = \pi d$
$ = \pi \cdot 14 \text{ ft}$ Replace d with 14 feet.
$ = 14\pi \text{ ft}$

The circumference of the spa is *exactly* 14π feet. By replacing π with the *approximation* 3.14, we find that the circumference is *approximately* 14 feet $\cdot$ 3.14 = 43.96 feet.

■ Work Practice 8

✓**Concept Check** The distance around which figure is greater: a square with side length 5 inches or a circle with radius 3 inches?

Answer
8. exactly 20π yd $\approx$ 62.8 yd

✓**Concept Check Answer**
a square with side length 5 in.

Section 8.2 | Perimeter   579

Vocabulary, Readiness & Video Check

Use the choices below to fill in each blank.

circumference	radius	π	$\dfrac{22}{7}$
diameter	perimeter	3.14	

1. The _____ of a polygon is the sum of the lengths of its sides.
2. The distance around a circle is called the _____.
3. The exact ratio of circumference to diameter is _____.
4. The diameter of a circle is double its _____.
5. Both _____ and _____ are approximations for π.
6. The radius of a circle is half its _____.

Martin-Gay Interactive Videos Watch the section lecture video and answer the following questions.

Objective A 7. In Example 1, how can the perimeter be found if we forget the formula?

Objective B 8. From the lecture before Example 6, circumference is a special name for what?

See Video 8.2

8.2 Exercise Set MyLab Math

Objective A *Find the perimeter of each figure. See Examples 1 through 6.*

1. Rectangle, 15 ft by 17 ft

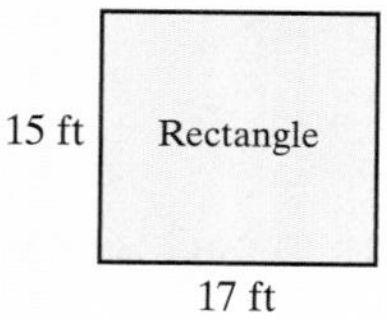

2. Rectangle, 14 m by 5 m

3. Parallelogram, 35 cm and 25 cm

4. Parallelogram, 3 yd and 2 yd

5. Triangle with sides 5 in., 7 in., 9 in.

6. Triangle with sides 5 units, 11 units, 10 units

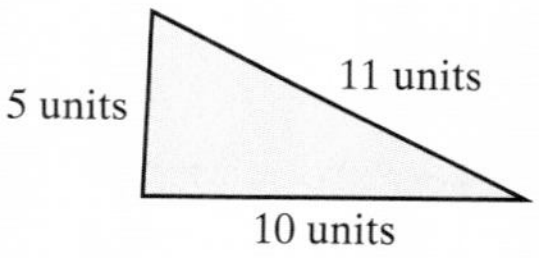

7. Figure with sides 10 ft, 8 ft, 8 ft, 15 ft, 7 ft

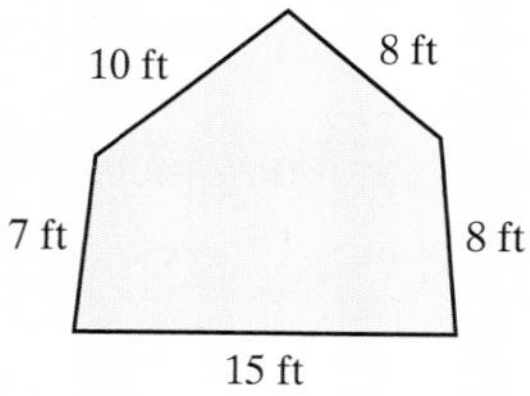

8. Figure with sides 10 m, 4 m, 10 m, 9 m, 20 m, 13 m

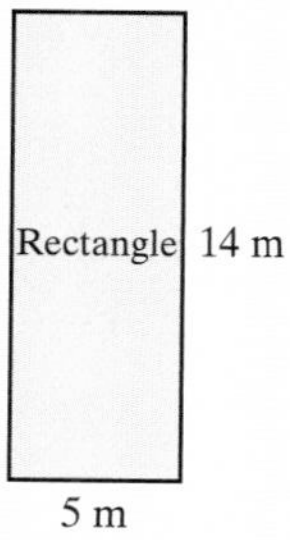

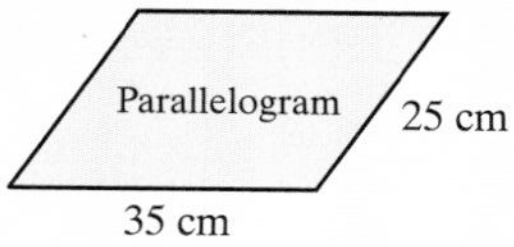

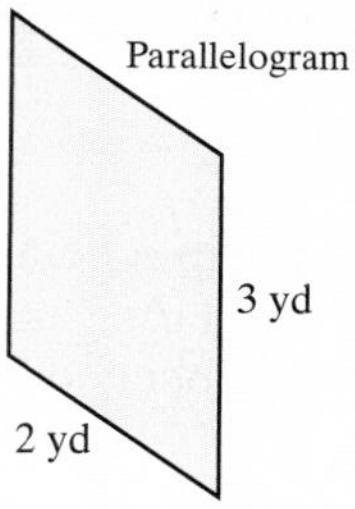

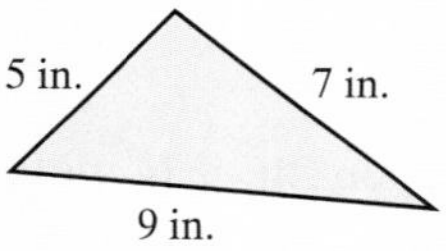

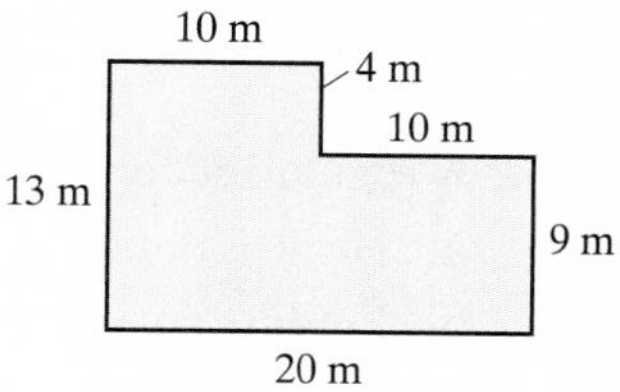

Find the perimeter of each regular polygon. (The sides of a regular polygon have the same length.)

9.  14 inches

10. 50 m

11. 31 cm

12. 15 yd

Solve. See Examples 1 through 7.

13. A polygon has sides of length 5 feet, 3 feet, 2 feet, 7 feet, and 4 feet. Find its perimeter.

14. A triangle has sides of length 8 inches, 12 inches, and 10 inches. Find its perimeter.

15. A line-marking machine lays down lime powder to mark both foul lines on a baseball field. If each foul line for this field measures 312 feet, how many feet of lime powder will be deposited?

16. A baseball diamond has 4 sides, with each side length 90 feet. If a baseball player hits a home run, how far does the player run (home plate, around the bases, then back to home plate)?

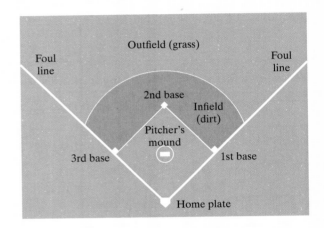

17. If a football field is 53 yards wide and 120 yards long, what is the perimeter?

18. A stop sign has eight equal sides of length 12 inches. Find its perimeter.

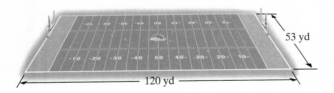

19. A metal strip is being installed around a workbench that is 8 feet long and 3 feet wide. Find how much stripping is needed for this project.

20. Find how much fencing is needed to enclose a rectangular garden 70 feet by 21 feet.

21. If the stripping in Exercise **19** costs $2.50 per foot, find the total cost of the stripping.

22. If the fencing in Exercise **20** costs $2 per foot, find the total cost of the fencing.

23. A regular octagon has a side length of 9 inches.
    a. How many sides does an octagon have?
    b. Find its perimeter.

24. A regular pentagon has a side length of 14 meters.
    a. How many sides does a pentagon have?
    b. Find its perimeter.

25. Find the perimeter of the top of a square compact disc case if the length of one side is 7 inches.

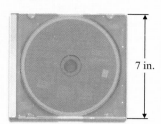

26. Find the perimeter of a square ceramic tile with a side of length 3 inches.

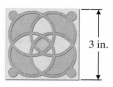

27. A rectangular room measures 10 feet by 11 feet. Find the cost of installing a strip of wallpaper around the room if the wallpaper costs $0.86 per foot.

28. A rectangular house measures 85 feet by 70 feet. Find the cost of installing gutters around the house if the cost is $2.36 per foot.

Find the perimeter of each figure. See Example 6.

29.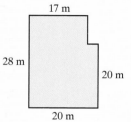

30.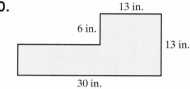

31.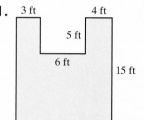

32.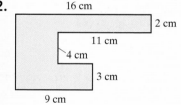

33.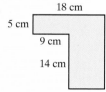

34.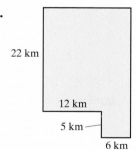

Objective B *Find the circumference of each circle. Give the exact circumference and then an approximation. Use π ≈ 3.14. See Example 8.*

35.

36.

37.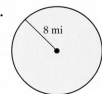

38. 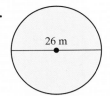 50 ft

39. 26 m

40. 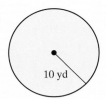 10 yd

Solve.

41. Wyley Robinson just bought a trampoline for his children to use. The trampoline has a diameter of 15 feet. If Wyley wishes to buy netting to go around the outside of the trampoline, how many feet of netting does he need? Use 3.14 for π.

42. The largest round barn in the world is located at the Marshfield Fairgrounds in Wisconsin. The barn has a diameter of 150 ft. What is the circumference of the barn? Use 3.14 for π. (*Source: The Milwaukee Journal Sentinel*)

43. Meteor Crater, near Winslow, Arizona, is 4000 feet in diameter. Approximate the distance around the crater. Use 3.14 for π. (*Source: The Handy Science Answer Book*)

44. The *Pearl of Lao-tze* has a diameter of $5\frac{1}{2}$ inches. Approximate the distance around the pearl. Use $\frac{22}{7}$ for π. (*Source: The Guinness World Records*)

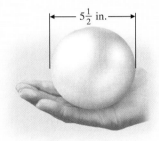

Objectives A B Mixed Practice *Find the distance around each figure. For circles, give the exact circumference and then an approximation. Use $\pi \approx 3.14$. See Examples 1 through 8.*

45. 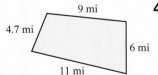 9 mi, 4.7 mi, 6 mi, 11 mi

46. 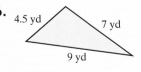 4.5 yd, 7 yd, 9 yd

47. 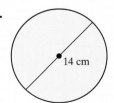 14 cm

48. 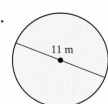 11 m

49.  Regular Pentagon 8 mm

50.  Regular Parallelogram 19 km

51. 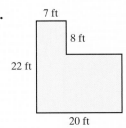 7 ft, 8 ft, 22 ft, 20 ft

52. 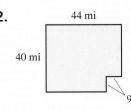 44 mi, 40 mi, 9

Review

Simplify. See Section 1.7.

53. $5 + 6 \cdot 3$ **54.** $25 - 3 \cdot 7$ **55.** $(20 - 16) \div 4$ **56.** $6 \cdot (8 + 2)$

57. $72 \div (2 \cdot 6)$ **58.** $(72 \div 2) \cdot 6$ **59.** $(18 + 8) - (12 + 4)$ **60.** $4^1 \cdot (2^3 - 8)$

Concept Extensions

There are a number of factors that determine the dimensions of a rectangular soccer field. Use the table below to answer Exercises 61 and 62.

Soccer Field Width and Length		
Age	Width Min–Max	Length Min–Max
Under 6/7:	15–20 yards	25–30 yards
Under 8:	20–25 yards	30–40 yards
Under 9:	30–35 yards	40–50 yards
Under 10:	40–50 yards	60–70 yards
Under 11:	40–50 yards	70–80 yards
Under 12:	40–55 yards	100–105 yards
Under 13:	50–60 yards	100–110 yards
International:	70–80 yards	110–120 yards

61. a. Find the minimum length and width of a soccer field for 8-year-old children. (Carefully consider the age.)
 b. Find the perimeter of this field.

62. a. Find the maximum length and width of a soccer field for 12-year-old children.
 b. Find the perimeter of this field.

Solve. See the Concept Check in this section. Choose the figure that has the greater distance around.

63. a. A square with side length 3 inches
 b. A circle with diameter 4 inches

64. a. A circle with diameter 7 inches
 b. A square with side length 7 inches

65. a. Find the circumference of each circle. Approximate the circumference by using 3.14 for π.

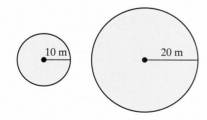

 b. If the radius of a circle is doubled, is its corresponding circumference doubled?

66. a. Find the circumference of each circle. Approximate the circumference by using 3.14 for π.

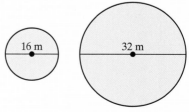

 b. If the diameter of a circle is doubled, is its corresponding circumference doubled?

67. In your own words, explain how to find the perimeter of any polygon.

68. In your own words, explain how perimeter and circumference are the same and how they are different.

Find the perimeter. Round your results to the nearest tenth.

69.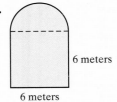
6 meters
6 meters

70.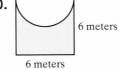
6 meters
6 meters

71.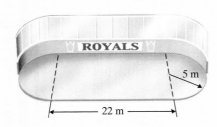
5 m
22 m

72.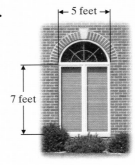
5 feet
7 feet

8.3 Area, Volume, and Surface Area

Objectives

A Find the Area of Plane Regions.

B Find the Volume and Surface Area of Solids.

Objective A Finding Area of Plane Regions

Recall that area measures the number of square units that cover the surface of a plane region; that is, a region that lies in a plane. Thus far, we know how to find the areas of a rectangle and a square. These formulas, as well as formulas for finding the areas of other common geometric figures, are given next.

Area Formulas of Common Geometric Figures

Geometric Figure	Area Formula
RECTANGLE 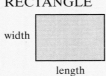 width, length	Area of a rectangle: **A**rea = **l**ength · **w**idth $A = lw$
SQUARE side, side	Area of a square: **A**rea = **s**ide · **s**ide $A = s \cdot s = s^2$

Section 8.3 | Area, Volume, and Surface Area

Geometric Figure	Area Formula
TRIANGLE 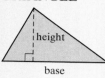	Area of a triangle: $$\text{Area} = \frac{1}{2} \cdot \text{base} \cdot \text{height}$$ $$A = \frac{1}{2}bh$$
PARALLELOGRAM 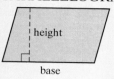	Area of a parallelogram: $$\text{Area} = \text{base} \cdot \text{height}$$ $$A = bh$$
TRAPEZOID 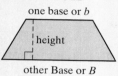 one base or b / other Base or B	Area of a trapezoid: $$\text{Area} = \frac{1}{2} \cdot (\text{one base} + \text{other Base}) \cdot \text{height}$$ $$A = \frac{1}{2}(b + B)h$$

Use these formulas for the following examples.

Helpful Hint
Area is always measured in square units.

Example 1 Find the area of the triangle.

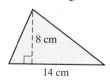
8 cm
14 cm

Solution:

$$A = \frac{1}{2}bh$$

$$= \frac{1}{2} \cdot 14 \text{ cm} \cdot 8 \text{ cm}$$

$$= \frac{\overset{1}{\cancel{2}} \cdot 7 \cdot 8}{\underset{1}{\cancel{2}}} \text{ square centimeters} \qquad \text{Write 14 as } 2 \cdot 7.$$

$$= 56 \text{ square centimeters}$$

The area is 56 square centimeters.

Work Practice 1

Practice 1
Find the area of the triangle.

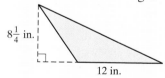
$8\frac{1}{4}$ in.
12 in.

Answer
1. $49\frac{1}{2}$ sq in.

Practice 2

Find the area of the trapezoid.

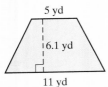

Example 2 Find the area of the parallelogram.

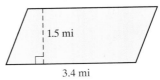

Solution:

$A = bh$
$= 3.4 \text{ miles} \cdot 1.5 \text{ miles}$ Replace b, base, with 3.4 miles and h, height, with 1.5 miles.
$= 5.1 \text{ square miles}$

The area is 5.1 square miles.

■ Work Practice 2

Helpful Hint

When finding the area of figures, check to make sure that all measurements are in the same units before calculations are made.

Practice 3

Find the area of the figure.

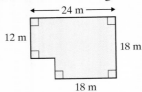

Example 3 Find the area of the figure.

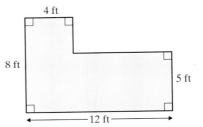

Solution: Split the figure into two rectangles. To find the area of the figure, we find the sum of the areas of the two rectangles.

area of Rectangle 1 $= lw$
$= 8 \text{ feet} \cdot 4 \text{ feet}$
$= 32 \text{ square feet}$

Notice that the length of Rectangle 2 is 12 feet − 4 feet or 8 feet.

area of Rectangle 2 $= lw$
$= 8 \text{ feet} \cdot 5 \text{ feet}$
$= 40 \text{ square feet}$

area of the figure = area of Rectangle 1 + area of Rectangle 2
$= 32 \text{ square feet} + 40 \text{ square feet}$
$= 72 \text{ square feet}$

■ Work Practice 3

Answers
2. 48.8 sq yd **3.** 396 sq m

Helpful Hint

The figure in Example 3 could also be split into two rectangles as shown.

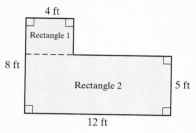

To better understand the formula for area of a circle, try the following. Cut a circle into many pieces, as shown.

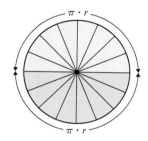

The circumference of a circle is $2\pi r$. This means that the circumference of half a circle is half of $2\pi r$, or πr.

Then unfold the two halves of the circle and place them together, as shown.

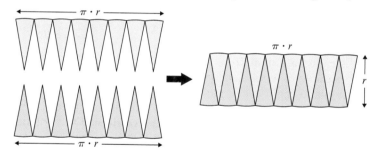

The figure on the right is almost a parallelogram with a base of πr and a height of r. The area is

$A = $ base $\cdot$ height
$ = (\pi r) \cdot r$
$ = \pi r^2$

This is the formula for the area of a circle.

Area Formula of a Circle

Circle

Area of a circle

Area $= \pi \cdot (\text{radius})^2$

$A = \pi r^2$

(A fraction approximation for π is $\dfrac{22}{7}$.)

(A decimal approximation for π is 3.14.)

Practice 4

Find the area of the given circle. Find the exact area and an approximation. Use 3.14 as an approximation for π.

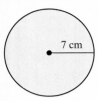

Example 4

Find the area of a circle with a radius of 3 feet. Find the exact area and an approximation. Use 3.14 as an approximation for π.

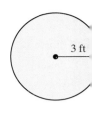

Solution: We let $r = 3$ feet and use the formula

$A = \pi r^2$
$= \pi \cdot (3 \text{ feet})^2$ Replace r with 3 feet.
$= 9 \cdot \pi$ square feet   Replace $(3 \text{ feet})^2$ with 9 sq ft.

To approximate this area, we substitute 3.14 for π.

$9 \cdot \pi$ square feet $\approx 9 \cdot 3.14$ square feet
$= 28.26$ square feet

The *exact* area of the circle is 9π square feet, which is *approximately* 28.26 square feet.

■ Work Practice 4

✓ **Concept Check** Use estimation to decide which figure would have a larger area: a circle of diameter 10 in. or a square 10 in. long on each side.

Objective B   Finding Volume and Surface Area of Solids

A **convex solid** is a set of points, S, not all in one plane, such that for any two points A and B in S, all points between A and B are also in S. In this section, we will find the volume and surface area of special types of solids called polyhedrons. A solid formed by the intersection of a finite number of planes is called a **polyhedron**. The box to the right is an example of a polyhedron.

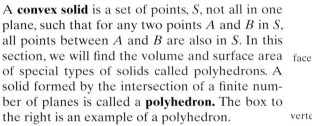

Each of the plane regions of a polyhedron is called a **face** of the polyhedron. If the intersection of two faces is a line segment, this line segment is an **edge** of the polyhedron. The intersections of the edges are the **vertices** of the polyhedron.

Volume is a measure of the space of a solid. The volume of a box or can, for example, is the amount of space inside. Volume can be used to describe the amount of juice in a pitcher or the amount of concrete needed to pour a foundation for a house.

The volume of a solid is the number of **cubic units** in the solid. A cubic centimeter and a cubic inch are illustrated.

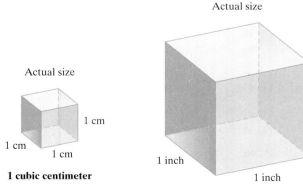

Answer
4. 49π sq cm ≈ 153.86 sq cm

✓ **Concept Check Answer**
A square 10 in. long on each side would have a larger area.

The **surface area** of a polyhedron is the sum of the areas of the faces of the polyhedron. For example, each face of the cube to the left on the previous page has an area of 1 square centimeter. Since there are 6 faces of the cube, the sum of the areas of the faces is 6 square centimeters. Surface area can be used to describe the amount of material needed to cover a solid. Surface area is measured in square units.

Formulas for finding the volumes, V, and surface areas, SA, of some common solids are given next. (Note: Spheres, circular cylinders, and cones are not polyhedrons, but they are solids and we will calculate surface areas and volumes of these solids.)

Volume and Surface Area Formulas of Common Solids

Solid	Formulas
RECTANGULAR SOLID	$V = lwh$ $SA = 2lh + 2wh + 2lw$ where h = height, w = width, l = length
CUBE	$V = s^3$ $SA = 6s^2$ where s = side
SPHERE	$V = \frac{4}{3}\pi r^3$ $SA = 4\pi r^2$ where r = radius
CIRCULAR CYLINDER	$V = \pi r^2 h$ $SA = 2\pi rh + 2\pi r^2$ where h = height, r = radius
CONE	$V = \frac{1}{3}\pi r^2 h$ $SA = \pi r\sqrt{r^2 + h^2} + \pi r^2$ where h = height, r = radius
SQUARE-BASED PYRAMID	$V = \frac{1}{3}s^2 h$ $SA = B + \frac{1}{2}pl$ where B = area of base, p = perimeter of base, h = height, s = side, l = slant height

Practice 5

Find the volume and surface area of a rectangular box that is 7 feet long, 3 feet wide, and 4 feet high.

Example 5 Find the volume and surface area of a rectangular box that is 12 inches long, 6 inches wide, and 3 inches high.

Solution: Let $h = 3$ in., $l = 12$ in., and $w = 6$ in.

$$V = lwh$$
$$V = 12 \text{ inches} \cdot 6 \text{ inches} \cdot 3 \text{ inches} = 216 \text{ cubic inches}$$

The volume of the rectangular box is 216 cubic inches.

$$SA = 2lh + 2wh + 2lw$$
$$= 2(12 \text{ in.})(3 \text{ in.}) + 2(6 \text{ in.})(3 \text{ in.}) + 2(12 \text{ in.})(6 \text{ in.})$$
$$= 72 \text{ sq in.} + 36 \text{ sq in.} + 144 \text{ sq in.}$$
$$= 252 \text{ sq in.}$$

The surface area of the rectangular box is 252 square inches.

■ Work Practice 5

✓**Concept Check** Juan is calculating the volume of the following rectangular solid. Find the error in his calculation.

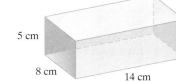

$$\text{Volume} = l + w + h$$
$$= 14 \text{ cm} + 8 \text{ cm} + 5 \text{ cm}$$
$$= 27 \text{ cu cm}$$

Practice 6

Find the volume and surface area of a ball of radius $\frac{1}{2}$ centimeter. Give the exact volume and surface area. Then use $\frac{22}{7}$ for π and approximate the values.

Example 6 Find the volume and surface area of a ball of radius 2 inches. Give the exact volume and surface area. Then use the approximation $\frac{22}{7}$ for π.

Solution:

$$V = \frac{4}{3}\pi r^3 \qquad \text{Formula for volume of a sphere}$$

$$V = \frac{4}{3} \cdot \pi (2 \text{ in.})^3 \qquad \text{Let } r = 2 \text{ inches.}$$

$$= \frac{32}{3}\pi \text{ cu in.} \qquad \text{Exact volume}$$

$$\approx \frac{32}{3} \cdot \frac{22}{7} \text{ cu in.} \qquad \text{Approximate } \pi \text{ with } \frac{22}{7}.$$

$$= \frac{704}{21} \text{ or } 33\frac{11}{21} \text{ cu in.} \qquad \text{Approximate volume}$$

Answers

5. $V = 84$ cu ft; $SA = 122$ sq ft
6. $V = \frac{1}{6}\pi$ cu cm $\approx \frac{11}{21}$ cu cm;

$SA = \pi$ sq cm $\approx 3\frac{1}{7}$ sq cm

✓**Concept Check Answer**
Volume = lwh
$= 14$ cm $\cdot 8$ cm $\cdot 5$ cm
$= 560$ cu cm

The volume of the sphere is exactly $\frac{32}{3}\pi$ cubic inches or approximately $33\frac{11}{21}$ cubic inches.

$$SA = 4\pi r^2 \qquad \text{Formula for surface area}$$
$$SA = 4 \cdot \pi (2\text{ in.})^2 \qquad \text{Let } r = 2 \text{ inches.}$$
$$= 16\pi \text{ sq in.} \qquad \text{Exact surface area}$$
$$\approx 16 \cdot \frac{22}{7} \text{ sq in.} \qquad \text{Approximate } \pi \text{ with } \frac{22}{7}.$$
$$= \frac{352}{7} \text{ or } 50\frac{2}{7} \text{ sq in.} \qquad \text{Approximate surface area}$$

The surface area of the sphere is exactly 16π square inches or approximately $50\frac{2}{7}$ square inches.

■ Work Practice 6

Example 7 Find the volume of a can that has a $3\frac{1}{2}$-inch radius and a height of 6 inches. Give the exact volume and an approximate volume. Use $\frac{22}{7}$ for π.

Practice 7

Find the volume of a cylinder of radius 5 inches and height 9 inches. Give the exact answer and an approximate answer. Use 3.14 for π.

Solution: Using the formula for a circular cylinder, we have

$$V = \pi \cdot r^2 \cdot h \qquad 3\frac{1}{2} = \frac{7}{2}$$
$$= \pi \cdot \left(\frac{7}{2} \text{ in.}\right)^2 \cdot 6 \text{ in.}$$
$$= \pi \cdot \frac{49}{4} \text{ sq in.} \cdot 6 \text{ in.}$$
$$= \frac{\pi \cdot 49 \cdot \overset{1}{\cancel{2}} \cdot 3}{\underset{1}{\cancel{2}} \cdot 2} \text{ cu in.}$$
$$= 73\frac{1}{2}\pi \text{ cu in. or } 73.5\pi \text{ cu in.}$$

This is the exact volume. To approximate the volume, use the approximation $\frac{22}{7}$ for π.

$$V = 73\frac{1}{2}\pi \approx \frac{147}{2} \cdot \frac{22}{7} \text{ cu in.} \qquad \text{Replace } \pi \text{ with } \frac{22}{7}.$$
$$= \frac{21 \cdot \overset{1}{\cancel{7}} \cdot \overset{1}{\cancel{2}} \cdot 11}{\underset{1}{\cancel{2}} \cdot \underset{1}{\cancel{7}}} \text{ cu in.}$$
$$= 231 \text{ cu in.}$$

The volume is approximately 231 cubic inches.

■ Work Practice 7

Answer
7. 225π cu in. ≈ 706.5 cu in.

Practice 8

Find the volume of a square-based pyramid that has a 3-meter side and a height of 5.1 meters.

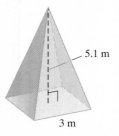

Example 8 Find the volume of a cone that has a height of 14 centimeters and a radius of 3 centimeters. Give the exact answer and an approximate answer. Use 3.14 for π.

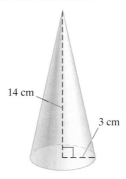

Solution: Using the formula for volume of a cone, we have

$$V = \frac{1}{3} \cdot \pi \cdot r^2 \cdot h$$

$$= \frac{1}{3} \cdot \pi \cdot (3 \text{ cm})^2 \cdot 14 \text{ cm} \quad \text{Replace } r \text{ with 3 cm and } h \text{ with 14 cm.}$$

$$= 42\pi \text{ cu cm}$$

Thus, 42π cubic centimeters is the exact volume. To approximate the volume, use the approximation 3.14 for π.

$$V \approx 42 \cdot 3.14 \text{ cu cm} \quad \text{Replace } \pi \text{ with 3.14.}$$

$$= 131.88 \text{ cu cm}$$

The volume is approximately 131.88 cubic centimeters.

■ Work Practice 8

Answer
8. 15.3 cu m

Vocabulary, Readiness & Video Check

Use the choices below to fill in each blank. Some choices may be used more than once.

area   surface area   cubic
volume   square

1. The _____ of a polyhedron is the sum of the areas of its faces.
2. The measure of the amount of space inside a solid is its _____.
3. _____ measures the amount of surface enclosed by a region.
4. Volume is measured in _____ units.
5. Area is measured in _____ units.
6. Surface area is measured in _____ units.

Section 8.3 | Area, Volume, and Surface Area

Martin-Gay Interactive Videos Watch the section lecture video and answer the following questions.

See Video 8.3

Objective A 7. Explain why we need to use a formula twice to solve Example 3.

Objective B 8. In Examples 7 and 8, explain the difference in the two volume answers found for each.

8.3 Exercise Set MyLab Math

Objective A Find the area of each geometric figure. If the figure is a circle, give the exact area and then use the given approximation for π to approximate the area. See Examples 1 through 4.

1. Rectangle, 2 m by 3.5 m

2. Rectangle, 1.2 ft by 3.5 ft

3. Triangle, height 3 yd, base $6\frac{1}{2}$ yd

4. Triangle, height 5 ft, base $4\frac{1}{2}$ ft

5. Right triangle, legs 5 yd and 6 yd

6. Right triangle, legs 5 ft and 7 ft

7. Use 3.14 for π. 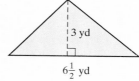 $d = 3$ in.

8. Use $\frac{22}{7}$ for π. 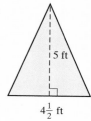 $r = 5$ cm

9. Parallelogram, height 5.25 ft, base 7 ft

10. Parallelogram, height 4.25 cm, base 3 cm

11. 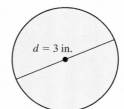 Trapezoid, parallel sides 5 m and 9 m, height 4 m

12. 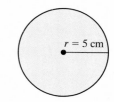 Trapezoid, parallel sides 6 in. and $8\frac{1}{2}$ in., height 5 in.

13. 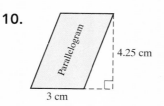 Trapezoid, parallel sides 4 yd and 7 yd, height 4 yd

14. 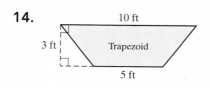 Trapezoid, parallel sides 10 ft and 5 ft, height 3 ft

15. 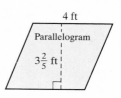 Parallelogram, base 4 ft, height $3\frac{2}{5}$ ft

16.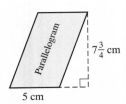

17.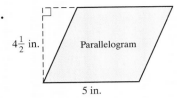

18.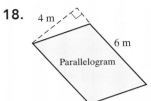

19.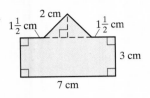

20.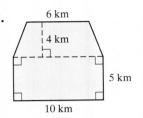

21.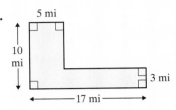

22.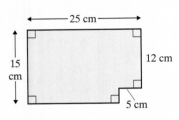

23.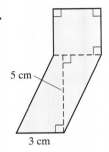

24.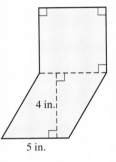

25. Use $\frac{22}{7}$ for π.

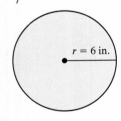

26. Use 3.14 for π.

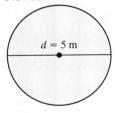

Objective B *Find the volume and surface area of each solid. See Examples 5 through 8. For formulas containing π, give the exact answer and then approximate using $\frac{22}{7}$ for π.*

27.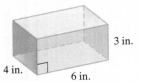

28.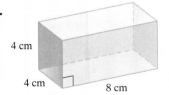

29.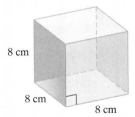

30.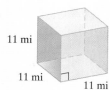

31. For surface area, use $\pi = 3.14$ and round to the nearest hundredth.

32. For surface area, use $\pi = 3.14$ and round to the nearest hundredth.

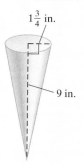

33.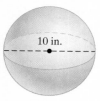

34.

35. Find the volume only.

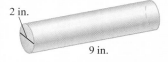

36. Find the volume only.

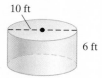

37. Find the volume only.

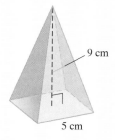

38. Find the volume only.

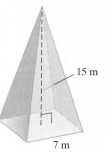

Objectives A B Mixed Practice *Solve. See Examples 1 through 8.*

39. Find the volume of a cube with edges of $1\frac{1}{3}$ inches.

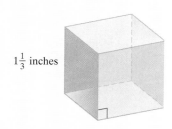

40. A water storage tank is in the shape of a cone with the pointed end down. If the radius is 14 ft and the depth of the tank is 15 ft, approximate the volume of the tank in cubic feet. Use $\frac{22}{7}$ for π.

41. Find the volume and surface area of a rectangular box 2 ft by 1.4 ft by 3 ft.

42. Find the volume and surface area of a box in the shape of a cube that is 5 ft on each side.

43. The largest American flag measures 505 feet by 225 feet. It's the U.S. "Super flag" owned by "Ski" Demski of Long Beach, California. Find its area. (*Source: Guinness World Records*)

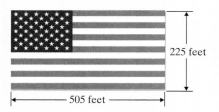

44. The largest indoor illuminated sign is a billboard at Dubai Airport. It measures 28 meters by 6.2 meters. Find its area. (*Source: The Guinness Book of World Records*)

45. A drapery panel measures 6 ft by 7 ft. Find how many square feet of material are needed for *four* panels.

46. A page in this book measures 27.6 cm by 21.5 cm. Find its area.

47. A paperweight is in the shape of a square-based pyramid 20 centimeters tall. If an edge of the base is 12 centimeters, find the volume of the paperweight.

48. A birdbath is made in the shape of a hemisphere (half-sphere). If its radius is 10 inches, approximate its volume. Use $\frac{22}{7}$ for π.

49. Find how many square feet of land are in the following plot:

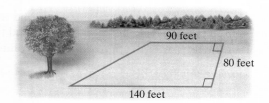

50. For Gerald Gomez to determine how much grass seed he needs to buy, he must know the size of his yard. Use the drawing to determine how many square feet are in his yard.

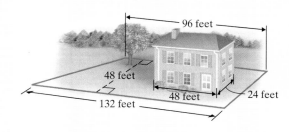

51. Find the exact volume and surface area of a sphere with a radius of 7 inches.

52. A tank is in the shape of a cylinder 8 feet tall and 3 feet in radius. Find the exact volume and surface area of the tank.

53. The outlined part of the roof shown is in the shape of a trapezoid and needs to be shingled. The number of shingles to buy depends on the area.
   a. Use the dimensions given to find the area of the outlined part of the roof to the nearest whole square foot.

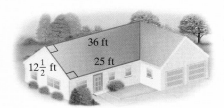

   b. Shingles are packaged in a unit called a "square." If a "square" covers 100 square feet, how many whole squares need to be purchased to shingle this part of the roof?

54. The entire side of the building shaded in the drawing is to be bricked. The number of bricks to buy depends on the area.
   a. Find the area.

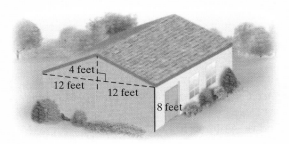

   b. If the side area of each brick (including mortar room) is $\frac{1}{6}$ square ft, find the number of bricks that are needed to brick the end of the building.

55. Find the exact volume of a waffle ice cream cone with a 3-in. diameter and a height of 7 inches.

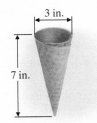

56. A snow globe has a diameter of 6 inches. Find its exact volume. Then approximate its volume using 3.14 for π.

57. Paul Revere's Pizza in the USA will bake and deliver a round pizza with a 4-foot diameter. This pizza is called the "Ultimate Party Pizza" and its current price is $99.99. Find the exact area of the top of the pizza and an approximation. Use 3.14 as an approximation for π.

58. The face of a circular watch has a diameter of 2 centimeters. What is its area? Find the exact area and an approximation. Use 3.14 as an approximation for π.

59. Zorbing is an extreme sport invented by two New Zealanders who joke that they were looking for a way to walk on water. A Zorb is a large sphere inside a second sphere with the space between the spheres pumped full of air. There is a tunnel-like opening so a person can crawl into the inner sphere. You are strapped in and sent down a Zorbing hill. A standard Zorb is approximately 3 m in diameter. Find the exact volume of a Zorb, and approximate the volume using 3.14 for π.

60. Mount Fuji, in Japan, is considered the most beautiful composite volcano in the world. The mountain is in the shape of a cone whose height is about 3.5 kilometers and whose base radius is about 3 kilometers. Approximate the volume of Mt. Fuji in cubic kilometers. Use $\frac{22}{7}$ for π.

61. A $10\frac{1}{2}$-foot by 16-foot concrete wall is to be built using concrete blocks. Find the area of the wall.

62. The floor of Terry's attic is 24 feet by 35 feet. Find how many square feet of insulation are needed to cover the attic floor.

63. Find the volume of a pyramid with a square base 5 inches on a side and a height of $1\frac{3}{10}$ inches.

64. Approximate to the nearest hundredth the volume of a sphere with a radius of 2 centimeters. Use 3.14 for π.

The Space Cube is supposed to be the world's smallest computer, with dimensions of 2 inches by 2 inches by 2.2 inches.

65. Find the volume of the Space Cube.

66. Find the surface area of the Space Cube.

Review

Evaluate. See Section 1.7.

67. 5^2
68. 7^2
69. 3^2
70. 20^2
71. $1^2 + 2^2$
72. $5^2 + 3^2$
73. $4^2 + 2^2$
74. $1^2 + 6^2$

Concept Extensions

Given the following situations, tell whether you are more likely to be concerned with area or perimeter.

75. ordering fencing to fence a yard

76. ordering grass seed to plant in a yard

77. buying carpet to install in a room

78. buying gutters to install on a house

79. ordering paint to paint a wall

80. ordering baseboards to install in a room

81. buying a wallpaper border to go on the walls around a room

82. buying fertilizer for your yard

Solve.

83. A pizza restaurant recently advertised two specials. The first special was a 12-inch pizza for $10. The second special was two 8-inch pizzas for $9. Determine the better buy. (*Hint:* First compare the areas of the two specials and then find a price per square inch for both specials.)

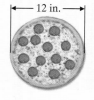
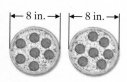

84. Find the approximate area of the state of Utah.

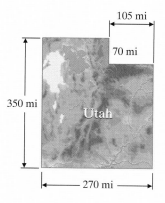

85. The Hayden Planetarium, at the Museum of Natural History in New York City, boasts a dome that has a diameter of 20 m. The dome is a hemisphere, or half a sphere. What is the volume enclosed by the dome at the Hayden Planetarium? Use 3.14 for π and round to the nearest hundredth. (*Source:* Hayden Planetarium)

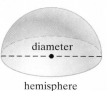

86. The Adler Museum in Chicago has a planetarium, its StarRider Theater, that has a diameter of 55 feet. Find the surface area of its hemispheric (half a sphere) dome. Use 3.14 for π. (*Source:* The Adler Museum)

87. Can you compute the volume of a rectangle? Why or why not?

88. In your own words, explain why perimeter is measured in units and area is measured in square units. (*Hint:* See Section 1.5 for an introduction to the meaning of area.)

89. Find the area of the shaded region. Use the approximation 3.14 for π.

90. Michigan is the home of the Uniroyal Giant Tyre sculpture, which was built originally as a Ferris wheel for the 1965 New York World's Fair. The diameter of this eight-story sculpture is 40 ft. Find the exact area of the face of the tyre and an approximation. Use $\pi \approx 3.14$. (*Source:* Guinness Book of World Records)

Find the area of each figure. If needed, use $\pi \approx 3.14$ and round results to the nearest tenth.

91. Find the skating area.

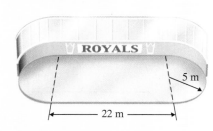

92.

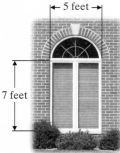

93. Do two rectangles with the same perimeter have the same area? To see, find the perimeter and the area of each rectangle.

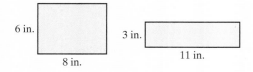

94. Do two rectangular solids with the same volume have the same surface area? To see, find the volume and surface area of each rectangular solid.

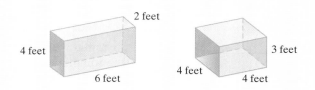

Integrated Review Sections 8.1–8.3

Geometry Concepts

1. Find the supplement and the complement of a 27° angle.

Find the measures of angles x, y, and z in each figure in Exercises 2 and 3.

2.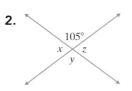

3. $m \parallel n$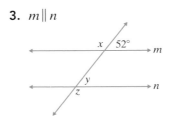

4. Find the measure of ∠x. (*Hint:* The sum of the angle measures of a triangle is 180°.)

5. Find the diameter.

6. Find the radius.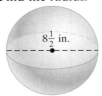

For Exercises 7 through 11, find the perimeter (or circumference) and area of each figure. For the circle, give the exact circumference and area. Then use $\pi \approx 3.14$ to approximate each. Don't forget to attach correct units.

7.

8.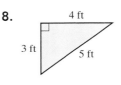

9.

10.

11.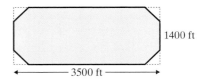

12. The smallest cathedral is in Highlandville, Missouri. The rectangular floor of the cathedral measures 14 feet by 17 feet. Find its perimeter and its area. (*Source: The Guinness Book of World Records*)

13. A new Tesla Giga factory is being constructed in Nevada. Although the factory will be in the shape of an emerald, find the perimeter and area of a rectangle with length 3500 feet and width 1400 feet.

Find the volume of each solid. Don't forget to attach correct units. For Exercises 14 and 15, find the surface area also.

14. A cube with edges of 4 inches each

15. A rectangular box 2 feet by 3 feet by 5.1 feet

16. A pyramid with a square base 10 centimeters on a side and a height of 12 centimeters

17. A sphere with a diameter of 3 miles (Give the exact volume and then use $\pi \approx \frac{22}{7}$ to approximate.)

8.4 Linear Measurement

Objective A Defining and Converting U.S. System Units of Length

In the United States, two systems of measurement are commonly used. They are the **United States (U.S.), or English, measurement system** and the **metric system.** The U.S. measurement system is familiar to most Americans. Units such as feet, miles, ounces, and gallons are used. However, the metric system is also commonly used in fields such as medicine, sports, international marketing, and certain physical sciences. We are accustomed to buying 2-liter bottles of soft drinks, watching televised coverage of the 100-meter dash at the Olympic Games, or taking a 200-milligram dose of pain reliever.

The U.S. system of measurement uses the **inch, foot, yard,** and **mile** to measure length. The following is a summary of equivalencies between units of length:

U.S. Units of Length

12 inches (in.) = 1 foot (ft)
3 feet = 1 yard (yd)
36 inches = 1 yard
5280 feet = 1 mile (mi)

Objectives

A Define U.S. Units of Length and Convert from One Unit to Another.

B Use Mixed U.S. Units of Length.

C Perform Arithmetic Operations on U.S. Units of Length.

D Define Metric Units of Length and Convert from One Unit to Another.

E Perform Arithmetic Operations on Metric Units of Length.

To convert from one unit of length to another, we will use **unit fractions.** We define a unit fraction to be a fraction that is equivalent to 1. Examples of unit fractions are as follows:

Unit Fractions

$\dfrac{12 \text{ in.}}{1 \text{ ft}} = 1$ or $\dfrac{1 \text{ ft}}{12 \text{ in.}} = 1$ (since 12 in. = 1 ft)

$\dfrac{3 \text{ ft}}{1 \text{ yd}} = 1$ or $\dfrac{1 \text{ yd}}{3 \text{ ft}} = 1$ (since 3 ft = 1 yd)

$\dfrac{5280 \text{ ft}}{1 \text{ mi}} = 1$ or $\dfrac{1 \text{ mi}}{5280 \text{ ft}} = 1$ (since 5280 ft = 1 mi)

Remember that multiplying a number by 1 does not change the value of the number.

Example 1 Convert 8 feet to inches.

Solution: We multiply 8 feet by a unit fraction that uses the equality 12 inches = 1 foot. The unit fraction should be in the form $\dfrac{\text{units to convert to}}{\text{original units}}$ or, in this case, $\dfrac{12 \text{ inches}}{1 \text{ foot}}$. We do this so that like units will divide out to 1, as shown.

$8 \text{ ft} = \dfrac{8 \text{ ft}}{1} \cdot 1$

$= \dfrac{8 \text{ ft}}{1} \cdot \dfrac{12 \text{ in.}}{1 \text{ ft}}$ Multiply by 1 in the form of $\dfrac{12 \text{ in.}}{1 \text{ ft}}$.

$= 8 \cdot 12 \text{ in.}$

$= 96 \text{ in.}$ Multiply.

Practice 1

Convert 6 feet to inches.

Answer
1. 72 in.

(Continued on next page)

Thus, 8 ft = 96 in., as shown in the diagram:

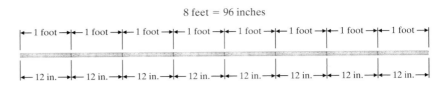

8 feet = 96 inches

■ Work Practice 1

Practice 2
Convert 8 yards to feet.

Example 2
Convert 7 feet to yards.

Solution: We multiply by a unit fraction that compares 1 yard to 3 feet.

$$7 \text{ ft} = \frac{7 \text{ ft}}{1} \cdot 1$$
$$= \frac{7 \text{ ft}}{1} \cdot \frac{1 \text{ yd}}{3 \text{ ft}} \quad \leftarrow \text{Units to convert to}$$
$$\phantom{= \frac{7 \text{ ft}}{1} \cdot \frac{1 \text{ yd}}{3 \text{ ft}}} \quad \leftarrow \text{Original units}$$
$$= \frac{7}{3} \text{ yd}$$
$$= 2\frac{1}{3} \text{ yd} \quad \text{Divide.}$$

Thus, 7 ft = $2\frac{1}{3}$ yd, as shown in the diagram.

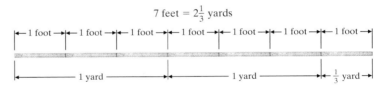

7 feet = $2\frac{1}{3}$ yards

Helpful Hint When converting from one unit to another, select a unit fraction with the properties below:

$$\frac{\text{units you are converting to}}{\text{original units}}$$

By using this unit fraction, the original units will divide out, as wanted.

■ Work Practice 2

Practice 3
Suppose the pelican's bill in the photo measures 18 inches. Convert 18 inches to feet, using decimals.

Example 3 Finding the Length of a Pelican's Bill

The Australian pelican has the longest bill, measuring from 13 to 18.5 inches long. The pelican in the photo has a 15-inch bill. Convert 15 inches to feet, using decimals in your final answer.

Solution:

$$15 \text{ in.} = \frac{15 \text{ in.}}{1} \cdot \frac{1 \text{ ft}}{12 \text{ in.}} \quad \begin{array}{l} \leftarrow \text{Units to convert to} \\ \leftarrow \text{Original units} \end{array}$$
$$= \frac{15}{12} \text{ ft}$$
$$= \frac{5}{4} \text{ ft} \quad \text{Simplify } \frac{15}{12}.$$
$$= 1.25 \text{ ft} \quad \text{Divide.}$$

Thus, 15 in. = 1.25 ft, as shown in the diagram.

15 inches = 1.25 ft

Answers
2. 24 ft 3. 1.5 ft

■ Work Practice 3

Objective B Using Mixed U.S. System Units of Length

Sometimes it is more meaningful to express a measurement of length with mixed units such as 1 ft and 5 in. We usually condense this and write 1 ft 5 in.

In Example 2, we found that 7 feet is the same as $2\frac{1}{3}$ yards. The measurement can also be written as a mixture of yards and feet. That is,

7 ft = ____ yd ____ ft

Because 3 ft = 1 yd, we divide 3 into 7 to see how many whole yards are in 7 feet. The quotient is the number of yards, and the remainder is the number of feet.

$$\begin{array}{r} 2 \text{ yd } 1 \text{ ft} \\ 3\overline{)7} \\ \underline{-6} \\ 1 \end{array}$$

Thus, 7 ft = 2 yd 1 ft, as seen in the diagram:

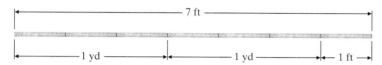

Example 4 Convert: 134 in. = ____ ft ____ in.

Solution: Because 12 in. = 1 ft, we divide 12 into 134. The quotient is the number of feet. The remainder is the number of inches. To see why we divide 12 into 134, notice that

$$134 \text{ in.} = \frac{134 \text{ in.}}{1} \cdot \frac{1 \text{ ft}}{12 \text{ in.}} = \frac{134}{12} \text{ ft}$$

$$\begin{array}{r} 11 \text{ ft } 2 \text{ in.} \\ 12\overline{)134} \\ \underline{-12} \\ 14 \\ \underline{-12} \\ 2 \end{array}$$

Thus, 134 in. = 11 ft 2 in.

Work Practice 4

Practice 4

Convert:
68 in. = ____ ft ____ in.

Example 5 Convert 3 feet 7 inches to inches.

Solution: First, we convert 3 feet to inches. Then we add 7 inches.

$$3 \text{ ft} = \frac{3 \text{ ft}}{1} \cdot \frac{12 \text{ in.}}{1 \text{ ft}} = 36 \text{ in.}$$

Then

$$3 \text{ ft } 7 \text{ in.} = 36 \text{ in.} + 7 \text{ in.} = 43 \text{ in.}$$

Work Practice 5

Practice 5

Convert 5 yards 2 feet to feet.

Answers

4. 5 ft 8 in. **5.** 17 ft

Objective C Performing Operations on U.S. System Units of Length

Finding sums or differences of measurements often involves converting units, as shown in the next example. Just remember that, as usual, only like units can be added or subtracted.

Practice 6
Add 4 ft 8 in. to 8 ft 11 in.

Example 6 Add 3 ft 2 in. and 5 ft 11 in.

Solution: To add, we line up the similar units.

$$\begin{array}{r} 3 \text{ ft } \;\, 2 \text{ in.} \\ + \; 5 \text{ ft } 11 \text{ in.} \\ \hline 8 \text{ ft } 13 \text{ in.} \end{array}$$

Since 13 inches is the same as 1 ft 1 in., we have

$$8 \text{ ft } 13 \text{ in.} = 8 \text{ ft } + 1 \text{ ft } 1 \text{ in.}$$
$$= 9 \text{ ft } 1 \text{ in.}$$

■ Work Practice 6

✓**Concept Check** How could you estimate the following sum?

$$\begin{array}{r} 7 \text{ yd } \;\, 4 \text{ in.} \\ + \; 3 \text{ yd } 27 \text{ in.} \end{array}$$

Practice 7
Multiply 4 ft 7 in. by 4.

Example 7 Multiply 8 ft 9 in. by 3.

Solution: By the distributive property, we multiply 8 ft by 3 and 9 in. by 3.

$$\begin{array}{r} 8 \text{ ft } \;\, 9 \text{ in.} \\ \times \;\;\;\;\;\;\;\; 3 \\ \hline 24 \text{ ft } 27 \text{ in.} \end{array}$$

Since 27 in. is the same as 2 ft 3 in., we simplify the product as

$$24 \text{ ft } 27 \text{ in.} = 24 \text{ ft } + 2 \text{ ft } 3 \text{ in.}$$
$$= 26 \text{ ft } 3 \text{ in.}$$

■ Work Practice 7

We divide in a similar manner as above.

Practice 8
A carpenter cuts 1 ft 9 in. from a board of length 5 ft 8 in. Find the remaining length of the board.

Example 8 Finding the Length of a Piece of Rope

A rope of length 6 yd 1 ft has 2 yd 2 ft cut from one end. Find the length of the remaining rope.

Solution: Subtract 2 yd 2 ft from 6 yd 1 ft.

$$\begin{array}{rl} \text{beginning length} \rightarrow & 6 \text{ yd } 1 \text{ ft} \\ - \;\;\;\; \text{amount cut} \rightarrow & -2 \text{ yd } 2 \text{ ft} \\ \hline \text{remaining length} & \end{array}$$

We cannot subtract 2 ft from 1 ft, so we borrow 1 yd from the 6 yd. One yard is converted to 3 ft and combined with the 1 ft already there.

Answers
6. 13 ft 7 in. **7.** 18 ft 4 in.
8. 3 ft 11 in.

✓**Concept Check Answer**
round each to the nearest yard:
7 yd + 4 yd = 11 yd

Borrow 1 yd = 3 ft

5 yd + (1 yd)(3 ft)

$$\begin{array}{r} \cancel{6\text{ yd}}\ 1\text{ ft} = 5\text{ yd }4\text{ ft} \\ -2\text{ yd }2\text{ ft} = -2\text{ yd }2\text{ ft} \\ \hline 3\text{ yd }2\text{ ft} \end{array}$$

The remaining rope is 3 yd 2 ft long.

Work Practice 8

Objective D Defining and Converting Metric System Units of Length ▶

The basic unit of length in the metric system is the **meter**. A meter is slightly longer than a yard. It is approximately 39.37 inches long. Recall that a yard is 36 inches long.

1 yard = 36 inches

1 meter ≈ 39.37 inches

All units of length in the metric system are based on the meter. The following is a summary of the prefixes used in the metric system. Also shown are equivalencies between units of length. Like the decimal system, the metric system uses powers of 10 to define units.

Metric Units of Length
1 **kilo**meter (km) = 1000 meters (m)
1 **hecto**meter (hm) = 100 m
1 **deka**meter (dam) = 10 m
1 **meter** (m) = 1 m
1 **deci**meter (dm) = 1/10 m or 0.1 m
1 **centi**meter (cm) = 1/100 m or 0.01 m
1 **milli**meter (mm) = 1/1000 m or 0.001 m

The figure below will help you with decimeters, centimeters, and millimeters.

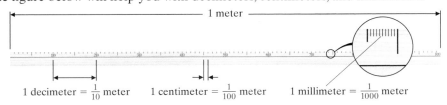

1 decimeter = $\frac{1}{10}$ meter   1 centimeter = $\frac{1}{100}$ meter   1 millimeter = $\frac{1}{1000}$ meter

Helpful Hint

Study the figure above for other equivalencies between metric units of length.

10 decimeters = 1 meter   10 millimeters = 1 centimeter
100 centimeters = 1 meter   10 centimeters = 1 decimeter
1000 millimeters = 1 meter

These same prefixes are used in the metric system for mass and capacity. The most commonly used measurements of length in the metric system are the **meter, millimeter, centimeter,** and **kilometer.**

✓ **Concept Check** Is this statement reasonable? "The screen of a home television set has a 30-meter diagonal." Why or why not?

✓ **Concept Check Answer**
no; answers may vary

Being comfortable with the metric units of length means gaining a "feeling" for metric lengths, just as you have a "feeling" for the lengths of an inch, a foot, and a mile. To help you accomplish this, study the following examples:

Examples of Metric System – Length

- A millimeter ia about the thickness of a large paper clip.
- A centimeter is about the width of a large paper clip.

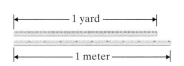

- A meter is slightly longer than a yard.

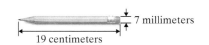

- Pencil dimensions

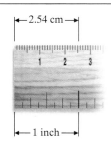

- 1 in. = 2.54 cm

- The width of this book is approximately 21.5 centimeters.

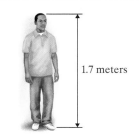

- Man's height

- A kilometer is about two-thirds of a mile.

- The distance between New York City and Philadelphia is about 130 kilometers by air.

As with the U.S. system of measurement, unit fractions may be used to convert from one unit of length to another. For example, let's convert 1200 meters to kilometers. To do so, we will multiply by 1 in the form of the unit fraction

$$\frac{1 \text{ km}}{1000 \text{ m}} \quad \leftarrow \text{Units to convert to} \\ \leftarrow \text{Original units}$$

$$1200 \text{ m} = \frac{1200 \text{ m}}{1} \cdot 1 = \frac{1200 \text{ m}}{1} \cdot \overbrace{\frac{1 \text{ km}}{1000 \text{ m}}}^{\text{Unit fraction}} = \frac{1200 \text{ km}}{1000} = 1.2 \text{ km}$$

The metric system does, however, have a distinct advantage over the U.S. system of measurement: the ease of converting from one unit of length to another. Since all units of length are powers of 10 of the meter, converting from one unit of length to another is as simple as moving the decimal point. Listing units of length in order from largest to smallest helps to keep track of how many places to move the decimal point when converting.

Let's again convert 1200 meters to kilometers. This time, to convert from meters to kilometers, we move along the chart shown, 3 units to the left, from meters to kilometers. This means that we move the decimal point 3 places to the left.

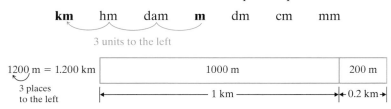

Thus, 1200 m = 1.2 km, as shown in the diagram.

Example 9 Convert 2.3 m to centimeters.

Solution: First we will convert by using a unit fraction.

$$2.3 \text{ m} = \frac{2.3 \text{ m}}{1} \cdot \frac{\overset{\text{Unit fraction}}{100 \text{ cm}}}{1 \text{ m}} = 230 \text{ cm}$$

Now we will convert by listing the units of length in order from left to right and moving from meters to centimeters.

km   hm   dam   m   dm   cm   mm
                        2 units to the right

2.30 m = 230. cm
2 places to the right

With either method, we get 230 cm.

■ Work Practice 9

Practice 9
Convert 2.5 m to millimeters.

Example 10 Convert 450,000 mm to meters.

Solution: We list the units of length in order from left to right and move from millimeters to meters.

km   hm   dam   m   dm   cm   mm
                        3 units to the left

Thus, move the decimal point 3 places to the left.

450,000 mm = 450.000 m or 450 m

■ Work Practice 10

Practice 10
Convert 3500 m to kilometers.

✓ **Concept Check** What is wrong with the following conversion of 150 cm to meters?

150.00 cm = 15,000 m

Objective E Performing Operations on Metric System Units of Length ▶

To add, subtract, multiply, or divide with metric measurements of length, we write all numbers using the same unit of length and then add, subtract, multiply, or divide as with decimals.

Answers
9. 2500 mm   **10.** 3.5 km

✓ **Concept Check Answer**
decimal point should be moved two places to the left: 1.5 m

Practice 11
Subtract 640 m from 2.1 km.

Example 11 Subtract 430 m from 1.3 km.

Solution: First we convert both measurements to kilometers or both to meters

$$430 \text{ m} = 0.43 \text{ km}$$

$$\begin{array}{r} 1.30 \text{ km} \\ -0.43 \text{ km} \\ \hline 0.87 \text{ km} \end{array}$$

or

$$1.3 \text{ km} = 1300 \text{ m}$$

$$\begin{array}{r} 1300 \text{ m} \\ -430 \text{ m} \\ \hline 870 \text{ m} \end{array}$$

The difference is 0.87 km or 870 m.

■ Work Practice 11

Practice 12
Multiply 18.3 hm by 5.

Example 12 Multiply 5.7 mm by 4.

Solution: Here we simply multiply the two numbers. Note that the unit of measurement remains the same.

$$\begin{array}{r} 5.7 \text{ mm} \\ \times \quad 4 \\ \hline 22.8 \text{ mm} \end{array}$$

■ Work Practice 12

Practice 13
Doris Blackwell is knitting a scarf that is currently 0.8 meter long. If she knits an additional 45 centimeters, how long will the scarf be?

Example 13 Finding a Person's Height

Fritz Martinson was 1.2 meters tall on his last birthday. Since then, he has grown 14 centimeters. Find his current height in meters.

Solution:

$$\begin{array}{rcl} \text{original height} & \rightarrow & 1.20 \text{ m} \\ + \text{ height grown} & \rightarrow & + \; 0.14 \text{ m} \\ \hline \text{current height} & & 1.34 \text{ m} \end{array}$$ (Since 14 cm = 0.14 m)

Fritz is now 1.34 meters tall.

■ Work Practice 13

Answers
11. 1.46 km or 1460 m   12. 91.5 hm
13. 125 cm or 1.25 m

Vocabulary, Readiness & Video Check

Use the choices below to fill in each blank. Some choices may be used more than once.

| inches | yard | unit fraction |
| feet | meter | |

1. The basic unit of length in the metric system is the _____.
2. The expression $\dfrac{1 \text{ foot}}{12 \text{ inches}}$ is an example of a(n) _____.
3. A meter is slightly longer than a(n) _____.
4. One foot equals 12 _____.
5. One yard equals 3 _____.
6. One yard equals 36 _____.
7. One mile equals 5280 _____.

Section 8.4 | Linear Measurement 609

Martin-Gay Interactive Videos   Watch the section lecture video and answer the following questions.

See Video 8.4

Objective A 8. In Example 3, what units are used in the denominator of the unit fraction and why was this decided?

Objective B 9. In Example 4, how is a mixed unit similar to a mixed number? Use examples in your answer.

Objective C 10. In Example 5, why is the sum of the addition problem not the final answer? Reference the sum in your answer.

Objective D 11. In the lecture before Example 6, why is it easier to convert metric units than U.S. units?

Objective E 12. What two answers did we get for Example 8? Explain why both answers are correct.

8.4 Exercise Set MyLab Math

Objective A Convert each measurement as indicated. See Examples 1 through 3.

1. 60 in. to feet
2. 84 in. to feet
3. 12 yd to feet
4. 18 yd to feet

5. 42,240 ft to miles
6. 36,960 ft to miles
7. $8\frac{1}{2}$ ft to inches
8. $12\frac{1}{2}$ ft to inches

9. 10 ft to yards
10. 25 ft to yards
11. 6.4 mi to feet
12. 3.8 mi to feet

13. 162 in. to yd (Write answer as a decimal.)
14. 7216 yd to mi (Write answer as a decimal.)
15. 3 in. to ft (Write answer as a decimal.)
16. 129 in. to ft (Write answer as a decimal.)

Objective B Convert each measurement as indicated. See Examples 4 and 5.

17. 40 ft = _____ yd _____ ft
18. 100 ft = _____ yd _____ ft
19. 85 in. = _____ ft _____ in.

20. 59 in. = _____ ft _____ in.
21. 10,000 ft = _____ mi _____ ft
22. 25,000 ft = _____ mi _____ ft

23. 5 ft 2 in. = _____ in.
24. 6 ft 10 in. = _____ in.
25. 8 yd 2 ft = _____ ft

26. 4 yd 1 ft = _____ ft
27. 2 yd 1 ft = _____ in.
28. 1 yd 2 ft = _____ in.

Objective C *Perform each indicated operation. Simplify the result if possible. See Examples 6 through 8.*

29. 3 ft 10 in. + 7 ft 4 in.

30. 12 ft 7 in. + 9 ft 11 in.

▶ **31.** 12 yd 2 ft + 9 yd 2 ft

32. 16 yd 2 ft + 8 yd 2 ft

33. 22 ft 8 in. − 16 ft 3 in.

34. 15 ft 5 in. − 8 ft 2 in.

35. 18 ft 3 in. − 10 ft 9 in.

36. 14 ft 8 in. − 3 ft 11 in.

37. 28 ft 8 in. ÷ 2

38. 34 ft 6 in. ÷ 2

39. 16 yd 2 ft × 5

40. 15 yd 1 ft × 8

Objective D *Convert as indicated. See Examples 9 and 10.*

41. 60 m to centimeters

42. 46 m to centimeters

43. 40 mm to centimeters

44. 14 mm to centimeters

45. 500 m to kilometers

46. 400 m to kilometers

47. 1700 mm to meters

48. 6400 mm to meters

▶ **49.** 1500 cm to meters

50. 6400 cm to meters

51. 0.42 km to centimeters

52. 0.95 km to centimeters

53. 7 km to meters

54. 5 km to meters

55. 8.3 cm to millimeters

56. 4.6 cm to millimeters

57. 20.1 mm to decimeters

58. 140.2 mm to decimeters

▶ **59.** 0.04 m to millimeters

60. 0.2 m to millimeters

Objective E *Perform each indicated operation. Remember to insert units when writing your answers. See Examples 11 through 13.*

61. 8.6 m + 0.34 m

62. 14.1 cm + 3.96 cm

63. 2.9 m + 40 mm

64. 30 cm + 8.9 m

▶ **65.** 24.8 mm − 1.19 cm

66. 45.3 m − 2.16 dam

67. 15 km − 2360 m

68. 14 cm − 15 mm

69. 18.3 m × 3

70. 14.1 m × 4

71. 6.2 km ÷ 4

72. 9.6 m ÷ 5

Objectives A C D E Mixed Practice *Solve. Remember to insert units when writing your answers. For Exercises 73 through 82, complete the charts. See Examples 1 through 13.*

		Yards	Feet	Inches
73.	Chrysler Building in New York City		1046	
74.	4-story building			792
75.	Python length		35	
76.	Ostrich height			108

		Meters	Millimeters	Kilometers	Centimeters
77.	Length of elephant	5			
78.	Height of grizzly bear	3			
79.	Tennis ball diameter				6.5
80.	Golf ball diameter				4.6
81.	Distance from London to Paris			342	
82.	Distance from Houston to Dallas			396	

83. The National Zoo maintains a small patch of bamboo, which it grows as a food supply for its pandas. Two weeks ago, the bamboo was 6 ft 10 in. tall. Since then, the bamboo has grown 3 ft 8 in. How tall is the bamboo now?

84. While exploring in the Marianas Trench, a submarine probe was lowered to a point 1 mile 1400 feet below the ocean's surface. Later it was lowered an additional 1 mile 4000 feet below this point. How far was the probe below the surface of the Pacific?

85. At its deepest point, the Grand Canyon of the Colorado River in Arizona is about 6000 ft. The Grand Canyon of the Yellowstone River, which is in Yellowstone National Park in Wyoming, is at most 900 feet deep. How much deeper is the Grand Canyon of the Colorado River than the Grand Canyon of the Yellowstone River? (*Source:* National Park Service)

86. The Grand Canyon of the Gunnison River, in Colorado, is often called the Black Canyon of the Gunnison because it is so steep that light rarely penetrates the depth of the canyon. The Black Canyon of the Gunnison is only 1150 ft wide at its narrowest point. At its narrowest, the Grand Canyon of the Yellowstone is $\frac{1}{2}$ mile wide. Find the difference in width between the Grand Canyon of the Yellowstone and the Black Canyon of the Gunnison. (*Note:* Notice that the dimensions are different.) (*Source:* National Park Service)

87. The tallest man in the world is recorded as Robert Pershing Wadlow of Alton, Illinois. Born in 1918, he measured 8 ft 11 in. at his tallest. The shortest man in the world is Chandra Bahadur Dangi of Nepal, who measures 21.5 in. How many times taller than Chandra is Robert? Round to one decimal place. (*Source: Guinness World Records*)

88. A 3.4-m rope is attached to a 5.8-m rope. However, when the ropes are tied, 8 cm of length is lost to form the knot. What is the length of the tied ropes?

89. The ice on a pond is 5.33 cm thick. For safe skating, the owner of the pond insists that it be 80 mm thick. How much thicker must the ice be before skating is allowed?

90. The sediment on the bottom of the Towamencin Creek is normally 14 cm thick, but a recent flood washed away 22 mm of sediment. How thick is it now?

91. The Amana Corporation stacks up its microwave ovens in a distribution warehouse. Each stack is 1 ft 9 in. wide. How far from the wall would 9 of these stacks extend?

92. The highway commission is installing concrete sound barriers along a highway. Each barrier is 1 yd 2 ft long. Find the total length of 25 barriers placed end to end.

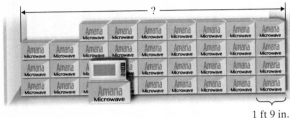

93. A logging firm needs to cut a 67-m-long redwood log into 20 equal pieces before loading it onto a truck for shipment. How long will each piece be?

94. A 112.5-foot-tall dead pinewood tree is removed by starting at the top and cutting off 9-foot-long sections. How many whole sections are removed?

95. The giant Coca-Cola billboard in Times Square is lit by more than 2.6 million LEDs, which are completely powered by wind energy. This huge advertisement measures approximately 44 feet by 65 feet. Find the perimeter of the sign in yards. (*Source:* Coca-Cola Company)

96. The longest truck in the world is operated by Gould Transport in Australia, and is the 182-ft Road Train. How many *yards* long are 2 of these trucks? (*Source: Guinness World Records*)

Review

Write each decimal as a fraction and each fraction as a decimal. See Sections 5.1 and 5.5.

97. 0.21

98. 0.86

99. $\dfrac{13}{100}$

100. $\frac{47}{100}$ 101. $\frac{1}{4}$ 102. $\frac{3}{20}$

Concept Extensions

Determine whether the measurement in each statement is reasonable.

103. The width of a twin-size bed is 20 meters.

104. A window measures 1 meter by 0.5 meter.

105. A drinking glass is made of glass 2 millimeters thick.

106. A paper clip is 4 kilometers long.

107. The distance across the Colorado River is 50 kilometers.

108. A model's hair is 30 centimeters long.

Estimate each sum or difference. See the first Concept Check in this section.

109. $\begin{aligned}&5 \text{ yd }2 \text{ in.}\\&+7 \text{ yd } 30 \text{ in.}\end{aligned}$

110. $\begin{aligned}&45 \text{ ft }1 \text{ in.}\\&-10 \text{ ft } 11 \text{ in.}\end{aligned}$

111. Using a unit other than the foot, write a length that is equivalent to 4 feet. (*Hint:* There are many possibilities.)

112. Using a unit other than the meter, write a length that is equivalent to 7 meters. (*Hint:* There are many possibilities.)

113. To convert from meters to centimeters, the decimal point is moved two places to the right. Explain how this relates to the fact that the prefix *centi* means $\frac{1}{100}$.

114. Explain why conversions in the metric system are easier to make than conversions in the U.S. system of measurement.

115. An advertisement sign outside Fenway Park in Boston measures 18.3 m by 18.3 m. What is the area of this sign?

8.5 Weight and Mass

Objectives

A Define U.S. Units of Weight and Convert from One Unit to Another.

B Perform Arithmetic Operations on U.S. Units of Weight.

C Define Metric Units of Mass and Convert from One Unit to Another.

D Perform Arithmetic Operations on Metric Units of Mass.

Objective A Defining and Converting U.S. System Units of Weight

Whenever we talk about how heavy an object is, we are concerned with the object **weight**. We discuss weight when we refer to a 12-ounce box of Rice Krispies, 15-pound tabby cat, or a barge hauling 24 tons of garbage.

12 ounces

15 pounds

24 tons of garbage

The most common units of weight in the U.S. measurement system are the **ounce**, the **pound**, and the **ton**. The following is a summary of equivalencies between units of weight:

U.S. Units of Weight	Unit Fractions
16 ounces (oz) = 1 pound (lb)	$\dfrac{16 \text{ oz}}{1 \text{ lb}} = \dfrac{1 \text{ lb}}{16 \text{ oz}} = 1$
2000 pounds = 1 ton	$\dfrac{2000 \text{ lb}}{1 \text{ ton}} = \dfrac{1 \text{ ton}}{2000 \text{ lb}} = 1$

✓**Concept Check** If you were describing the weight of a fully loaded semi trailer, which type of unit would you use: ounce, pound, or ton? Why?

Unit fractions that equal 1 are used to convert between units of weight in the U.S. system. When converting using unit fractions, recall that the numerator of a unit fraction should contain the units we are converting to and the denominator should contain the original units.

Practice 1
Convert 6500 pounds to tons.

Answer
1. $3\dfrac{1}{4}$ tons

✓**Concept Check Answer**
ton

Example 1 Convert 9000 pounds to tons.

Solution: We multiply 9000 lb by a unit fraction that uses the equality 2000 pounds = 1 ton.

Remember, the unit fraction should be $\dfrac{\text{units to convert to}}{\text{original units}}$ or $\dfrac{1 \text{ ton}}{2000 \text{ lb}}$.

$$9000 \text{ lb} = \dfrac{9000 \text{ lb}}{1} \cdot 1 = \dfrac{9000 \text{ lb}}{1} \cdot \dfrac{1 \text{ ton}}{2000 \text{ lb}} = \dfrac{9000 \text{ tons}}{2000} = \dfrac{9}{2} \text{ tons or } 4\dfrac{1}{2} \text{ tons}$$

Section 8.5 | Weight and Mass 615

Work Practice 1

Example 2 Convert 3 pounds to ounces.

Solution: We multiply by the unit fraction $\frac{16 \text{ oz}}{1 \text{ lb}}$ to convert from pounds to ounces.

$$3 \text{ lb} = \frac{3 \text{ lb}}{1} \cdot 1 = \frac{3 \text{ lb}}{1} \cdot \frac{16 \text{ oz}}{1 \text{ lb}} = 3 \cdot 16 \text{ oz} = 48 \text{ oz}$$

Practice 2

Convert 72 ounces to pounds.

3 lb = 48 oz

Work Practice 2

As with length, it is sometimes useful to simplify a measurement of weight by writing it in terms of mixed units.

Example 3 Convert: 33 ounces = _____ lb _____ oz

Solution: Because 16 oz = 1 lb, divide 16 into 33 to see how many pounds are in 33 ounces. The quotient is the number of pounds, and the remainder is the number of ounces. To see why we divide 16 into 33, notice that

$$33 \text{ oz} = 33 \text{ oz} \cdot \frac{1 \text{ lb}}{16 \text{ oz}} = \frac{33}{16} \text{ lb}$$

Thus, 33 ounces is the same as 2 lb 1 oz.

Practice 3

Convert:
47 ounces = _____ lb _____ oz

33 oz = 2 lb 1 oz

Work Practice 3

Answers

2. $4\frac{1}{2}$ lb   **3.** 2 lb 15 oz

Objective B Performing Operations on U.S. System Units of Weight

Performing arithmetic operations on units of weight works the same way as performing arithmetic operations on units of length.

Practice 4

Subtract 5 tons 1200 lb from 8 tons 100 lb.

Example 4 Subtract 3 tons 1350 lb from 8 tons 1000 lb.

Solution: To subtract, we line up similar units.

$$\begin{array}{r} 8 \text{ tons } 1000 \text{ lb} \\ -\ 3 \text{ tons } 1350 \text{ lb} \end{array}$$

Since we cannot subtract 1350 lb from 1000 lb, we borrow 1 ton from the 8 tons. To do so, we write 1 ton as 2000 lb and combine it with the 1000 lb.

7 tons + 1 ton 2000 lb

$$\begin{array}{r} 8 \text{ tons } 1000 \text{ lb} \\ -\ 3 \text{ tons } 1350 \text{ lb} \end{array} = \begin{array}{r} 7 \text{ tons } 3000 \text{ lb} \\ -\ 3 \text{ tons } 1350 \text{ lb} \\ \hline 4 \text{ tons } 1650 \text{ lb} \end{array}$$

To check, see that the sum of 4 tons 1650 lb and 3 tons 1350 lb is 8 tons 1000 lb.

■ Work Practice 4

Practice 5

Divide 5 lb 8 oz by 4.

Example 5 Divide 9 lb 6 oz by 2.

Solution: We divide each of the units by 2.

$$\begin{array}{r} 4 \text{ lb} \qquad 11 \text{ oz} \\ 2\overline{)9 \text{ lb} \qquad 6 \text{ oz}} \\ \underline{-8} \\ 1 \text{ lb} = \underline{16 \text{ oz}} \\ 22 \text{ oz} \end{array}$$ Divide 2 into 22 oz to get 11 oz.

To check, multiply 4 pounds 11 ounces by 2. The result is 9 pounds 6 ounces.

■ Work Practice 5

Practice 6

A 5-lb 14-oz batch of cookies is packed into a 6-oz container before it is mailed. Find the total weight.

Example 6 Finding the Weight of a Child

Bryan weighed 8 lb 8 oz at birth. By the time he was 1 year old, he had gained 11 lb 14 oz. Find his weight at age 1 year.

Solution:

$$\begin{array}{rcl} \text{birth weight} & \rightarrow & 8 \text{ lb } \ 8 \text{ oz} \\ +\ \text{weight gained} & \rightarrow & +\ 11 \text{ lb } 14 \text{ oz} \\ \hline \text{total weight} & \rightarrow & 19 \text{ lb } 22 \text{ oz} \end{array}$$

Since 22 oz equals 1 lb 6 oz,

$$19 \text{ lb } 22 \text{ oz } = 19 \text{ lb } + \overbrace{1 \text{ lb } 6 \text{ oz}}$$
$$= 20 \text{ lb } 6 \text{ oz}$$

Bryan weighed 20 lb 6 oz on his first birthday.

■ Work Practice 6

Answers
4. 2 tons 900 lb **5.** 1 lb 6 oz
6. 6 lb 4 oz

Objective C Defining and Converting Metric System Units of Mass

In scientific and technical areas, a careful distinction is made between **weight** and **mass**. **Weight** is really a measure of the pull of gravity. The farther from Earth an object gets, the less it weighs. However, **mass** is a measure of the amount of substance in the object and does not change. Astronauts orbiting Earth weigh much less than they weigh on Earth, but they have the same mass in orbit as they do on Earth. Here on Earth, weight and mass are the same, so either term may be used.

The basic unit of mass in the metric system is the **gram.** It is defined as the mass of water contained in a cube 1 centimeter (cm) on each side.

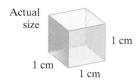

1 cubic centimeter

The following examples may help you get a feeling for metric masses:

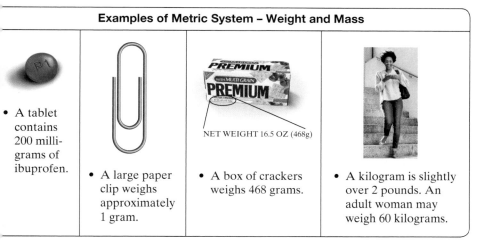

Examples of Metric System – Weight and Mass

- A tablet contains 200 milligrams of ibuprofen.
- A large paper clip weighs approximately 1 gram.
- A box of crackers weighs 468 grams.
- A kilogram is slightly over 2 pounds. An adult woman may weigh 60 kilograms.

The prefixes for units of mass in the metric system are the same as for units of length, as shown in the following table:

Metric Units of Mass
1 **kilo**gram (kg) = 1000 grams (g)
1 **hecto**gram (hg) = 100 g
1 **deka**gram (dag) = 10 g
1 **gram** (g) = 1 g
1 **deci**gram (dg) = 1/10 g or 0.1 g
1 **centi**gram (cg) = 1/100 g or 0.01 g
1 **milli**gram (mg) = 1/1000 g or 0.001 g

Concept Check True or false? A decigram is larger than a dekagram. Explain.

The **milligram,** the **gram,** and the **kilogram** are the three most commonly used units of mass in the metric system.

As with lengths, all units of mass are powers of 10 of the gram, so converting from one unit of mass to another only involves moving the decimal point. To convert from one unit of mass to another in the metric system, list the units of mass in order from largest to smallest.

✓ **Concept Check Answer**
false

Let's convert 4300 milligrams to grams. To convert from milligrams to grams, we move along the list 3 units to the left.

This means that we move the decimal point 3 places to the left to convert from milligrams to grams.

4300 mg = 4.3 g

Don't forget, the same conversion can be done with unit fractions.

$$4300 \text{ mg} = \frac{4300 \text{ mg}}{1} \cdot 1 = \frac{4300 \text{ mg}}{1} \cdot \frac{0.001 \text{ g}}{1 \text{ mg}}$$

$$= 4300 \cdot 0.001 \text{ g}$$

$$= 4.3 \text{ g} \quad \text{To multiply by 0.001, move the decimal point 3 places to the left.}$$

To see that this is reasonable, study the diagram:

Thus, 4300 mg = 4.3 g

Practice 7

Convert 3.41 g to milligrams.

Example 7 Convert 3.2 kg to grams.

Solution: First we convert by using a unit fraction.

$$3.2 \text{ kg} = 3.2 \text{ kg} \cdot 1 = 3.2 \text{ kg} \cdot \frac{\overset{\text{Unit fraction}}{1000 \text{ g}}}{1 \text{ kg}} = 3200 \text{ g}$$

Now let's list the units of mass in order from left to right and move from kilogram to grams.

kg hg dag g dg cg mg
     3 units to the right

3.200 kg = 3200. g
     3 places to the right

        1 kg   1 kg   1 kg   0.2 kg
                                        3.2 kg
                                        = 3200 g

       1000 g 1000 g 1000 g 200 g

■ Work Practice 7

Answer
7. 3410 mg

Example 8
Convert 2.35 cg to grams.

Solution: We list the units of mass in a chart and move from centigrams to grams.

kg hg dag g dg cg mg

2 units to the left

02.35 cg = 0.0235 g

2 places to the left

▶ Work Practice 8

Practice 8
Convert 56.2 cg to grams.

Objective D Performing Operations on Metric System Units of Mass

Arithmetic operations can be performed with metric units of mass just as we performed operations with metric units of length. We convert each number to the same unit of mass and add, subtract, multiply, or divide as with decimals.

Example 9
Subtract 5.4 dg from 1.6 g.

Solution: We convert both numbers to decigrams or to grams before subtracting.

5.4 dg = 0.54 g   or   1.6 g = 16 dg

$$\begin{array}{r} 1.60\text{ g} \\ -0.54\text{ g} \\ \hline 1.06\text{ g} \end{array}$$

$$\begin{array}{r} 16.0\text{ dg} \\ -5.4\text{ dg} \\ \hline 10.6\text{ g} \end{array}$$

The difference is 1.06 g or 10.6 dg.

▶ Work Practice 9

Practice 9
Subtract 3.1 dg from 2.5 g.

Example 10 Calculating Allowable Weight in an Elevator

An elevator has a weight limit of 1400 kg. A sign posted in the elevator indicates that the maximum capacity of the elevator is 17 persons. What is the average allowable weight for each passenger, rounded to the nearest kilogram?

Solution: To solve, notice that the total weight of 1400 kilograms ÷ 17 = average weight.

$$\begin{array}{r} 82.3\text{ kg} \approx 82\text{ kg} \\ 17\overline{)1400.0\text{ kg}} \\ -136 \\ \hline 40 \\ -34 \\ \hline 60 \\ -51 \\ \hline 9 \end{array}$$

Each passenger can weigh an average of 82 kg. (Recall that a kilogram is slightly over 2 pounds, so 82 kilograms is over 164 pounds.)

▶ Work Practice 10

Practice 10
Twenty-four bags of cement weigh a total of 550 kg. Find the average weight of 1 bag, rounded to the nearest kilogram.

Answers
8. 0.562 g   **9.** 2.19 g or 21.9 dg
10. 23 kg

Chapter 8 | Geometry and Measurement

Vocabulary, Readiness & Video Check

Use the choices below to fill in each blank.

mass   weight   gram

1. _____ is a measure of the amount of substance in an object. This measure does not change.
2. _____ is the measure of the pull of gravity.
3. The basic unit of mass in the metric system is the _____.

Fill in these blanks with the correct number. Choices for these blanks are not shown in the list of terms above.

4. One pound equals _____ ounces.
5. One ton equals _____ pounds.

See Video 8.5

Martin-Gay Interactive Videos   Watch the section lecture video and answer the following questions.

Objective A   6. In Example 2, what units are used in the numerator of the unit fraction and why was this decided?

Objective B   7. In Example 4, explain the first step taken to solve the problem.

Objective C   8. In Example 5, how many places is the decimal moved and in what direction? What is the final conversion?

Objective D   9. What is the answer to Example 7 in decigrams?

8.5 Exercise Set MyLab Math

Objective A *Convert as indicated. See Examples 1 through 3.*

1. 2 pounds to ounces
2. 5 pounds to ounces
3. 5 tons to pounds

4. 7 tons to pounds
5. 18,000 pounds to tons
6. 28,000 pounds to tons

7. 60 ounces to pounds
8. 90 ounces to pounds
9. 3500 pounds to tons

10. 11,000 pounds to tons
11. 12.75 pounds to ounces
12. 9.5 pounds to ounces

13. 4.9 tons to pounds
14. 8.3 tons to pounds
15. $4\frac{3}{4}$ pounds to ounces

16. $9\frac{1}{8}$ pounds to ounces
17. 2950 pounds to the nearest tenth of a ton
18. 51 ounces to the nearest tenth of a pound

19. $\frac{4}{5}$ oz to pounds
20. $\frac{1}{4}$ oz to pounds
21. $5\frac{3}{4}$ lb to ounces

Section 8.5 | Weight and Mass

22. $2\frac{1}{4}$ lb to ounces
23. 10 lb 1 oz to ounces
24. 7 lb 6 oz to ounces

25. 89 oz = ____ lb ____ oz
26. 100 oz = ____ lb ____ oz

Objective B *Perform each indicated operation. See Examples 4 through 6.*

27. 34 lb 12 oz + 18 lb 14 oz
28. 6 lb 10 oz + 10 lb 8 oz
29. 3 tons 1820 lb + 4 tons 930 lb

30. 1 ton 1140 lb + 5 tons 1200 lb
31. 5 tons 1050 lb − 2 tons 875 lb
32. 4 tons 850 lb − 1 ton 260 lb

33. 12 lb 4 oz − 3 lb 9 oz
34. 45 lb 6 oz − 26 lb 10 oz
35. 5 lb 3 oz × 6

36. 2 lb 5 oz × 5
37. 6 tons 1500 lb ÷ 5
38. 5 tons 400 lb ÷ 4

Objective C *Convert as indicated. See Examples 7 and 8.*

39. 500 g to kilograms
40. 820 g to kilograms
41. 4 g to milligrams
42. 9 g to milligrams

43. 25 kg to grams
44. 18 kg to grams
45. 48 mg to grams
46. 112 mg to grams

47. 6.3 g to kilograms
48. 4.9 g to kilograms
49. 15.14 g to milligrams
50. 16.23 g to milligrams

51. 6.25 kg to grams
52. 3.16 kg to grams
53. 35 hg to centigrams
54. 4.26 cg to dekagrams

Objective D *Perform each indicated operation. Remember to insert units when writing your answers. See Examples 9 and 10.*

55. 3.8 mg + 9.7 mg
56. 41.6 g + 9.8 g
57. 205 mg + 5.61 g
58. 2.1 g + 153 mg

59. 9 g − 7150 mg
60. 6.13 g − 418 mg
61. 1.61 kg − 250 g
62. 4 kg − 2410 g

63. 5.2 kg × 2.6
64. 4.8 kg × 9.3
65. 17 kg ÷ 8
66. 8.25 g ÷ 6

Objectives A B C D Mixed Practice *Solve. Remember to insert units when writing your answers. For Exercises 67 through 74, complete the chart. See Examples 1 through 10.*

	Object	Tons	Pounds	Ounces
67.	Statue of Liberty—weight of copper sheeting	100		
68.	Statue of Liberty—weight of steel	125		
69.	A 12-inch cube of osmium (heaviest metal)		1345	
70.	A 12-inch cube of lithium (lightest metal)		32	

Object	Grams	Kilograms	Milligrams	Centigrams
71. Capsule of amoxicillin (antibiotic)			500	
72. Tablet of Topamax (epilepsy and migraine uses)			25	
73. A six-year-old boy		21		
74. A golf ball	45			

75. A can of 7-Up weighs 336 grams. Find the weight in kilograms of 24 cans.

76. Guy Green normally weighs 73 kg, but he lost 2800 grams after being sick with the flu. Find Guy's new weight.

77. Sudafed is a decongestant that comes in two strengths. Regular strength contains 60 mg of medication. Extra strength contains 0.09 g of medication. How much extra medication is in the extra-strength tablet?

78. A small can of Planters sunflower seeds weighs 177 If each can contains 6 servings, find the weight of or serving.

79. Doris Johnson has two open containers of rice. If she combines 1 lb 10 oz from one container with 3 lb 14 oz from the other container, how much total rice does she have?

80. Dru Mizel maintains the records of the amount of coal delivered to his department in the steel mill. In January, 3 tons 1500 lb were delivered. In February, 2 tons 1200 lb were delivered. Find the total amoun delivered in these two months.

81. Carla Hamtini was amazed when she grew a 28-lb 10-oz zucchini in her garden, but later she learned that the heaviest zucchini ever grown weighed 64 lb 8 oz in Llanharry, Wales, by B. Lavery in 1990. How far below the record weight was Carla's zucchini? (*Source: Guinness World Records*)

82. The heaviest baby born in good health weighed an incredible 22 lb 8 oz. He was born in Italy in September 1955. How much heavier is this than a 7-lb 12-oz baby? (*Source: Guinness World Records*)

83. The smallest surviving premature baby weighed only 8 ounces, less than a can of soda. She was born in Witten, Germany, in 2016. How much lighter was she than an average European baby, who weighs about 7 lb 7.5 oz?

84. A large bottle of Hire's Root Beer weighs 1900 gran If a carton contains 6 large bottles of root beer, find the weight in kilograms of 5 cartons.

85. Three milligrams of preservatives are added to a 0.5-kg box of dried fruit. How many milligrams of preservatives are in 3 cartons of dried fruit if each carton contains 16 boxes?

86. One box of Swiss Miss Cocoa Mix weighs 0.385 kg, but 39 grams of this weight is the packaging. Find the actual weight of the cocoa in 8 boxes.

87. A carton of 12 boxes of Quaker Oats Oatmeal weighs 6.432 kg. Each box includes 26 grams of packaging material. What is the actual weight of the oatmeal in the carton?

88. The supermarket prepares hamburger in 85-gram market packages. When Leo Gonzalas gets home, he divides the package in half before refrigerating the meat. How much will each package weigh?

89. The Shop 'n Bag supermarket chain ships hamburger meat by placing 10 packages of hamburger in a box, with each package weighing 3 lb 4 oz. How much will 4 boxes of hamburger weigh?

90. The Quaker Oats Company ships its 1-lb 2-oz boxes of oatmeal in cartons containing 12 boxes of oatmeal. How much will 3 such cartons weigh?

91. A carton of Del Monte Pineapple weighs 55 lb 4 oz, but 2 lb 8 oz of this weight is due to packaging. Find the actual weight of the pineapple in 4 cartons.

92. The Hormel Corporation ships cartons of canned ham weighing 43 lb 2 oz each. Of this weight, 3 lb 4 oz is due to packaging. Find the actual weight of the ham found in 3 cartons.

Review

Write each fraction as a decimal. See Section 5.5.

93. $\dfrac{4}{25}$ **94.** $\dfrac{3}{5}$ **95.** $\dfrac{7}{8}$ **96.** $\dfrac{3}{16}$

Concept Extensions

Determine whether the measurement in each statement is reasonable.

97. The doctor prescribed a pill containing 2 kg of medication.

98. A full-grown cat weighs approximately 15 g.

99. A bag of flour weighs 4.5 kg.

100. A staple weighs 15 mg.

101. A professor weighs less than 150 g.

102. A car weighs 2000 mg.

103. Use a unit other than centigram and write a mass that is equivalent to 25 centigrams. (*Hint:* There are many possibilities.)

104. Use a unit other than pound and write a weight that is equivalent to 4000 pounds. (*Hint:* There are many possibilities.)

True or false? See the second Concept Check in this section.

105. A kilogram is larger than a gram.

106. A decigram is larger than a milligram.

107. Why is the decimal point moved to the right when grams are converted to milligrams?

108. To change 8 pounds to ounces, multiply by 16. Why is this the correct procedure?

8.6 Capacity

Objectives

A Define U.S. Units of Capacity and Convert from One Unit to Another.

B Perform Arithmetic Operations on U.S. Units of Capacity.

C Define Metric Units of Capacity and Convert from One Unit to Another.

D Perform Arithmetic Operations on Metric Units of Capacity.

Objective A Defining and Converting U.S. System Units of Capacity

Units of **capacity** are generally used to measure liquids. The number of gallons of gasoline needed to fill a gas tank in a car, the number of cups of water needed in a bread recipe, and the number of quarts of milk sold each day at a supermarket are all examples of using units of capacity. The following summary shows equivalencies between units of capacity:

U.S. Units of Capacity

8 fluid ounces (fl oz) = 1 cup (c)
2 cups = 1 pint (pt)
2 pints = 1 quart (qt)
4 quarts = 1 gallon (gal)

Just as with units of length and weight, we can form unit fractions to convert between different units of capacity. For instance,

$$\frac{2\text{ c}}{1\text{ pt}} = \frac{1\text{ pt}}{2\text{ c}} = 1 \quad \text{and} \quad \frac{2\text{ pt}}{1\text{ qt}} = \frac{1\text{ qt}}{2\text{ pt}} = 1$$

Practice 1

Convert 43 pints to quarts.

Example 1 Convert 9 quarts to gallons.

Solution: We multiply by the unit fraction $\frac{1\text{ gal}}{4\text{ qt}}$.

$$9\text{ qt} = \frac{9\text{ qt}}{1} \cdot 1$$
$$= \frac{9\text{ qt}}{1} \cdot \frac{1\text{ gal}}{4\text{ qt}}$$
$$= \frac{9\text{ gal}}{4}$$
$$= 2\frac{1}{4}\text{ gal}$$

Thus, 9 quarts is the same as $2\frac{1}{4}$ gallons, as shown in the diagram:

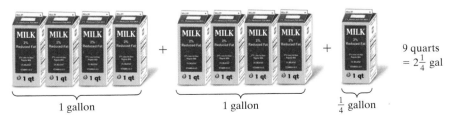

Work Practice 1

Answer

1. $21\frac{1}{2}$ qt

Example 2 Convert 14 cups to quarts.

Solution: Our equivalency table contains no direct conversion from cups to quarts. However, from this table we know that

$$1 \text{ qt} = 2 \text{ pt} = \frac{2 \text{ pt}}{1} \cdot 1 = \frac{2 \text{ pt}}{1} \cdot \frac{2 \text{ c}}{1 \text{ pt}} = 4 \text{ c}$$

so 1 qt = 4 c. Now we have the unit fraction $\frac{1 \text{ qt}}{4 \text{ c}}$. Thus,

$$14 \text{ c} = \frac{14 \text{ c}}{1} \cdot 1 = \frac{14 \text{ c}}{1} \cdot \frac{1 \text{ qt}}{4 \text{ c}} = \frac{14 \text{ qt}}{4} = \frac{7}{2} \text{ qt} \quad \text{or} \quad 3\frac{1}{2} \text{ qt}$$

Practice 2
Convert 26 quarts to cups.

Work Practice 2

✓ **Concept Check** If 50 cups is converted to quarts, will the equivalent number of quarts be less than or greater than 50? Explain.

Objective B Performing Operations on U.S. System Units of Capacity ▶

As is true of units of length and weight, units of capacity can be added, subtracted, multiplied, and divided.

Example 3 Subtract 3 qt from 4 gal 2 qt.

Solution: To subtract, we line up similar units.

```
    4 gal 2 qt
  −       3 qt
```

We cannot subtract 3 qt from 2 qt. We need to borrow 1 gallon from the 4 gallons, convert it to 4 quarts, and then combine it with the 2 quarts.

3 gal + (1 gal) 4 qt

```
    4 gal 2 qt   =     3 gal 6 qt
  −       3 qt   =   −       3 qt
                         3 gal 3 qt
```

To check, see that the sum of 3 gal 3 qt and 3 qt is 4 gal 2 qt.

Practice 3
Subtract 2 qt from 1 gal 1 qt.

Work Practice 3

Example 4 Finding the Amount of Water in an Aquarium

An aquarium contains 6 gal 3 qt of water. If 2 gal 2 qt of water is added, what is the total amount of water in the aquarium?

Solution:
```
  beginning water  →    6 gal 3 qt
+     water added  →  + 2 gal 2 qt
        total water →    8 gal 5 qt
```

(Continued on next page)

Practice 4
A large oil drum contains 15 gal 3 qt of oil. How much will be in the drum if an additional 4 gal 3 qt of oil is poured into it?

Answers
2. 104 c   **3.** 3 qt   **4.** 20 gal 2 qt

✓ **Concept Check Answer**
less than 50; answers may vary

Since 5 qt = 1 gal 1 qt, we have

$$= \overbrace{8 \text{ gal}}^{8 \text{ gal}} + \overbrace{1 \text{ gal } 1 \text{ qt}}^{5 \text{ qt}}$$
$$= 9 \text{ gal } 1 \text{ qt}$$

The total amount of water is 9 gal 1 qt.

■ Work Practice 4

Objective C Defining and Converting Metric System Units of Capacity

Thus far, we know that the basic unit of length in the metric system is the meter and that the basic unit of mass in the metric system is the gram. What is the basic unit of capacity? The **liter.** By definition, a **liter** is the capacity or volume of a cube measuring 10 centimeters on each side.

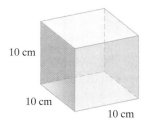

The following examples may help you get a feeling for metric capacities:

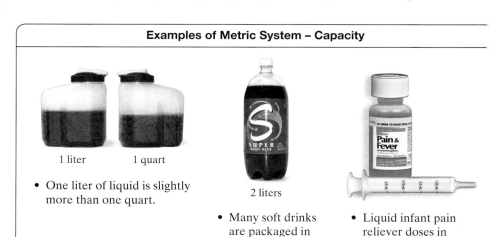

Examples of Metric System – Capacity

- 1 liter
- 1 quart
- One liter of liquid is slightly more than one quart.
- 2 liters
- Many soft drinks are packaged in 2-liter bottles.
- Liquid infant pain reliever doses in milliliters (ml or mL)

The metric system was designed to be a consistent system. Once again, the prefixes for metric units of capacity are the same as for metric units of length and mass, as summarized in the following table:

Metric Units of Capacity
1 **kilo**liter (kl) = 1000 liters (L)
1 **hecto**liter (hl) = 100 L
1 **deka**liter (dal) = 10 L
1 **liter** (L) = 1 L
1 **deci**liter (dl) = 1/10 L or 0.1 L
1 **centi**liter (cl) = 1/100 L or 0.01 L
1 **milli**liter (ml) = 1/1000 L or 0.001 L

The **milliliter** and the **liter** are the two most commonly used metric units of capacity.

Converting from one unit of capacity to another involves multiplying by powers of 10 or moving the decimal point to the left or to the right. Listing units of capacity in order from largest to smallest helps to keep track of how many places to move the decimal point when converting.

Let's convert 2.6 liters to milliliters. To convert from liters to milliliters, we move along the chart 3 units to the right.

kl   hl   dal   **L**   dl   cl   **ml**
                    $\underbrace{}_{\text{3 units to the right}}$

This means that we move the decimal point 3 places to the right to convert from liters to milliliters.

$2.600 \text{ L} = 2600. \text{ ml}$

This same conversion can be done with unit fractions.

$2.6 \text{ L} = \dfrac{2.6 \text{ L}}{1} \cdot 1$

$= \dfrac{2.6 \text{ L}}{1} \cdot \dfrac{1000 \text{ ml}}{1 \text{ L}}$

$= 2.6 \cdot 1000 \text{ ml}$

$= 2600 \text{ ml}$ To multiply by 1000, move the decimal point 3 places to the right.

To visualize the result, study the diagram below:

1000 ml   1000 ml   600 ml   2.6 L = 2600 ml

Thus, 2.6 L = 2600 ml.

Example 5   Convert 3210 ml to liters.

Solution: Let's use the unit fraction method first.

$3210 \text{ ml} = \dfrac{3210 \text{ ml}}{1} \cdot 1 = 3210 \text{ ml} \cdot \underbrace{\dfrac{1 \text{ L}}{1000 \text{ ml}}}_{\text{Unit fraction}} = 3.21 \text{ L}$

Now let's list the unit measures in order from left to right and move from milliliters to liters.

kl   hl   dal   L   dl   cl   ml
                    $\underbrace{}_{\text{3 units to the left}}$

$\underbrace{3210}_{\text{3 places to the left}} \text{ ml} = 3.210 \text{ L}$, the same results as before and shown below in the diagram.

1000 ml   1000 ml   1000 ml   210 ml

1 L   1 L   1 L   0.210 L   3210 ml = 3.210 L

Practice 5
Convert 2100 ml to liters.

Answer
5. 2.1 L

Work Practice 5

Practice 6
Convert 2.13 dal to liters.

Example 6 Convert 0.185 dl to milliliters.

Solution: We list the unit measures in order from left to right and move from deciliters to milliliters.

kl   hl   dal   L   dl   cl   ml

2 units to the right

0.185 dl = 18.5 ml

2 places to the right

Work Practice 6

Objective D Performing Operations on Metric System Units of Capacity

As was true for length and weight, arithmetic operations involving metric units of capacity can also be performed. Make sure that the metric units of capacity are the same before adding, subtracting, multiplying, or dividing.

Practice 7
Add 1250 ml to 2.9 L.

Example 7 Add 2400 ml to 8.9 L.

Solution: We must convert both to liters or both to milliliters before adding the capacities together.

2400 ml = 2.4 L   or   8.9 L = 8900 ml

  2.4 L   2400 ml
+ 8.9 L   + 8900 ml
 11.3 L   11,300 ml

The total is 11.3 L or 11,300 ml. They both represent the same capacity.

Work Practice 7

✓ **Concept Check** How could you estimate the following operation? Subtract 950 ml from 7.5 L.

Practice 8
If 28.6 L of water can be pumped every minute, how much water can be pumped in 85 minutes?

Example 8 Finding the Amount of Medication a Person Has Received

A patient hooked up to an IV unit in the hospital is to receive 12.5 ml of medication every hour. How much medication does the patient receive in 3.5 hours?

Solution: We multiply 12.5 ml by 3.5.

medication per hour   →   12.5 ml
×   hours   →   × 3.5
total medication   625
                           3750
                          43.75 ml

The patient receives 43.75 ml of medication.

Work Practice 8

Answers
6. 21.3 L   7. 4150 ml or 4.15 L
8. 2431 L

✓ **Concept Check Answer**
950 ml = 0.95 L; round 0.95 to 1;
7.5 − 1 = 6.5 L

Section 8.6 | Capacity

Vocabulary, Readiness & Video Check

Use the choices below to fill in each blank. Some choices may be used more than once.

cups   pints   liter
quarts   fluid ounces   capacity

1. Units of _____ are generally used to measure liquids.
2. The basic unit of capacity in the metric system is the _____.
3. One cup equals 8 _____.
4. One quart equals 2 _____.
5. One pint equals 2 _____.
6. One quart equals 4 _____.
7. One gallon equals 4 _____.

See Video 8.6

Martin-Gay Interactive Videos Watch the section lecture video and answer the following questions.

Objective A 8. Complete this statement based on Example 1: When using a unit fraction, we are not changing the _____, we are changing the _____.

Objective B 9. In Example 4, explain the first step taken to solve the exercise.

Objective C 10. In Example 5, how many places is the decimal moved and in what direction? What is the final conversion?

Objective D 11. What is the answer to Example 7 in dekaliters?

8.6 Exercise Set MyLab Math

Objective A Convert each measurement as indicated. See Examples 1 and 2.

1. 32 fluid ounces to cups
2. 16 quarts to gallons
3. 8 quarts to pints
4. 9 pints to quarts

5. 14 quarts to gallons
6. 11 cups to pints
7. 80 fluid ounces to pints
8. 18 pints to gallons

9. 2 quarts to cups
10. 3 pints to fluid ounces
11. 120 fluid ounces to quarts
12. 20 cups to gallons

13. 42 cups to quarts
14. 7 quarts to cups
15. $4\frac{1}{2}$ pints to cups
16. $6\frac{1}{2}$ gallons to quarts

17. 5 gal 3 qt to quarts
18. 4 gal 1 qt to quarts
19. $\frac{1}{2}$ cup to pints
20. $\frac{1}{2}$ pint to quarts

21. 58 qt = _____ gal _____ qt
22. 70 qt = _____ gal _____ qt

23. 39 pt = _____ gal _____ qt _____ pt

24. 29 pt = _____ gal _____ qt _____ pt

25. $2\frac{3}{4}$ gallons to pints

26. $3\frac{1}{4}$ quarts to cups

Objective **B** *Perform each indicated operation. See Examples 3 and 4.*

27. 5 gal 3 qt + 7 gal 3 qt

28. 2 gal 2 qt + 9 gal 3 qt

29. 1 c 5 fl oz + 2 c 7 fl oz

30. 2 c 3 fl oz + 2 c 6 fl oz

31. 3 gal − 1 gal 3 qt

32. 2 pt − 1 pt 1 c

33. 3 gal 1 qt − 1 qt 1 pt

34. 3 qt 1 c − 1 c 4 fl oz

35. 8 gal 2 qt × 2

36. 6 gal 1 pt × 2

37. 9 gal 2 qt ÷ 2

38. 5 gal 6 fl oz ÷ 2

Objective **C** *Convert as indicated. See Examples 5 and 6.*

39. 5 L to milliliters

40. 8 L to milliliters

41. 0.16 L to kiloliters

42. 0.127 L to kiloliters

43. 5600 ml to liters

44. 1500 ml to liters

45. 3.2 L to centiliters

46. 1.7 L to centiliters

47. 410 L to kiloliters

48. 250 L to kiloliters

49. 64 ml to liters

50. 39 ml to liters

51. 0.16 kl to liters

52. 0.48 kl to liters

53. 3.6 L to milliliters

54. 1.9 L to milliliters

Objective **D** *Perform each indicated operation. Remember to insert units when writing your answers. See Examples 7 and 8.*

55. 3.4 L + 15.9 L

56. 18.5 L + 4.6 L

57. 2700 ml + 1.8 L

58. 4.6 L + 1600 ml

59. 8.6 L − 190 ml

60. 4.8 L − 283 ml

61. 17,500 ml − 0.9 L

62. 6850 ml − 0.3 L

63. 480 ml × 8

64. 290 ml × 6

65. 81.2 L ÷ 0.5

66. 5.4 L ÷ 3.6

Section 8.6 | Capacity

Objectives A B C D Mixed Practice *Solve. Remember to insert units when writing your answers. For Exercises 67 through 70, complete the chart.*

	Capacity	Cups	Gallons	Quarts	Pints
67.	An average-size bath of water		21		
68.	A dairy cow's daily milk yield				38
69.	Your kidneys filter about this amount of blood every minute	4			
70.	The amount of water needed in a punch recipe	2			

71. Mike Schaferkotter drank 410 ml of Mountain Dew from a 2-liter bottle. How much Mountain Dew remains in the bottle?

72. The Werners' Volvo has a 54.5-L gas tank. Only 3.8 liters of gasoline still remain in the tank. How much is needed to fill it?

73. Margie Phitts added 354 ml of Prestone dry gas to the 18.6 L of gasoline in her car's tank. Find the total amount of gasoline in the tank.

74. Chris Peckaitis wishes to share a 2-L bottle of Coca-Cola equally with 7 of his friends. How much will each person get?

75. A garden tool engine requires a 30-to-1 gas-to-oil mixture. This means that $\frac{1}{30}$ of a gallon of oil should be mixed with 1 gallon of gas. Convert $\frac{1}{30}$ gallon to fluid ounces. Round to the nearest tenth.

76. Henning's Supermarket sells homemade soup in 1 qt 1 pt containers. How much soup is contained in three such containers?

77. Can 5 pt 1 c of fruit punch and 2 pt 1 c of ginger ale be poured into a 1-gal container without it overflowing?

78. Three cups of prepared Jell-O are poured into 6 dessert dishes. How many fluid ounces of Jell-O are in each dish?

79. Stanley Fisher paid $14 to fill his car with 44.3 liters of gasoline. Find the price per liter of gasoline to the nearest thousandth of a dollar.

80. A student carelessly misread the scale on a cylinder in the chemistry lab and added 40 cl of water to a mixture instead of 40 ml. Find the excess amount of water.

Review

Write each fraction in simplest form. See Section 4.2.

81. $\frac{20}{25}$ **82.** $\frac{75}{100}$ **83.** $\frac{27}{45}$ **84.** $\frac{56}{60}$ **85.** $\frac{72}{80}$ **86.** $\frac{18}{20}$

Concept Extensions

Determine whether the measurement in each statement is reasonable.

87. Clair took a dose of 2 L of cough medicine to cure her cough.

88. John drank 250 ml of milk for lunch.

89. Jeannie likes to relax in a tub filled with 3000 ml of hot water.

90. Sarah pumped 20 L of gasoline into her car yesterday.

Solve. See the Concept Checks in this section.

91. If 70 pints are converted to gallons, will the equivalent number of gallons be less than or greater than 70? Explain why.

92. If 30 gallons are converted to quarts, will the equivalent number of quarts be less than or greater than 30? Explain why.

93. Explain how to estimate the following operation: Add 986 ml to 6.9 L.

94. Explain how to borrow in order to subtract 1 gal 2 qt from 3 gal 1 qt.

95. Find the number of fluid ounces in 1 gallon.

96. Find the number of fluid ounces in 1.5 gallons.

A cubic centimeter (cc) is the amount of space that a volume of 1 ml occupies. Because of this, we will say that 1 cc = 1 ml.

A common syringe is one with a capacity of 3 cc. Use the diagram and give the measurement indicated by each arrow.

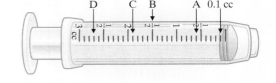

97. B **98.** A **99.** D **100.** C

In order to measure small dosages, such as for insulin, u-100 syringes are used. For these syringes, 1 cc has been divided into 100 equal units (u). Use the diagram and give the measurement indicated by each arrow in units (u) and then in cubic centimeters. Use 100 u = 1 cc.

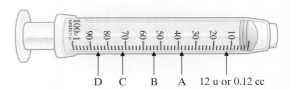

101. B **102.** A
103. D **104.** C

8.7 Temperature and Conversions Between the U.S. and Metric Systems

Objectives
A Convert Between the U.S. and Metric Systems.
B Convert Temperatures from Degrees Celsius to Degrees Fahrenheit.
C Convert Temperatures from Degrees Fahrenheit to Degrees Celsius.

Objective A Converting Between the U.S. and Metric Systems

The metric system probably had its beginnings in France in the 1600s, but it was the Metric Act of 1866 that made the use of this system legal (although not mandatory) in the United States. Other laws have followed that allow for a slow, but deliberate, transfer to the modernized metric system. In April 2001, for example, the U.S. Stock Exchanges completed their change to decimal trading instead of fractions. By the end of 2009, all products sold in Europe (with some exceptions) were required to have only metric units on their labels. (*Source*: U.S. Metric Association and National Institute of Standards and Technology)

You may be surprised at the number of everyday items we use that are already manufactured in metric units. We easily recognize 1 L and 2 L soda bottles, but what about the following?

- Pencil leads (0.5 mm or 0.7 mm)
- Camera film (35 mm)
- Sporting events (5-km or 10-km races)

Section 8.7 | Temperature and Conversions Between the U.S. and Metric Systems

- Medicines (500-mg capsules)

- Labels on retail goods (dual-labeled since 1994)

Since the United States has not completely converted to the metric system, we need to practice converting from one system to the other. Below is a table of mostly approximate conversions.

Length:		Capacity:		Weight (mass):	
Metric	U.S. System	Metric	U.S. System	Metric	U.S. System
1 m ≈ 1.09 yd		1 L ≈ 1.06 qt		1 kg ≈ 2.2 lb	
1 m ≈ 3.28 ft		1 L ≈ 0.26 gal		1 g ≈ 0.04 oz	
1 km ≈ 0.62 mi		3.79 L ≈ 1 gal		0.45 kg ≈ 1 lb	
2.54 cm = 1 in.		0.95 L ≈ 1 qt		28.35 g ≈ 1 oz	
0.30 m ≈ 1 ft		29.57 ml ≈ 1 fl oz			
1.61 km ≈ 1 mi					

There are many ways to perform these metric-to-U.S. conversions. We will do so by using unit fractions.

Example 1 Compact Discs

Standard-sized compact discs are 12 centimeters in diameter. Convert this length to inches. Round the result to two decimal places.

Solution: From our length conversion table, we know that 2.54 cm = 1 in. This fact gives us two unit fractions: $\frac{2.54 \text{ cm}}{1 \text{ in.}}$ and $\frac{1 \text{ in.}}{2.54 \text{ cm}}$. We use the unit fraction with cm in the denominator so that these units divide out.

$$12 \text{ cm} = \frac{12 \text{ cm}}{1} \cdot 1 = \frac{12 \text{ cm}}{1} \cdot \frac{1 \text{ in.}}{2.54 \text{ cm}}$$
← Units to convert to
← Original units

$$= \frac{12}{2.54} \text{ in.}$$

$$\approx 4.72 \text{ in.} \quad \text{Divide.}$$

Thus, the diameter of a standard compact disc is exactly 12 cm or approximately 4.72 inches. For a dimension this size, you can use a ruler to check. Another method is to approximate. Our result, 4.72 in., is close to 5 inches. Since 1 in. is about 2.5 cm, then 5 in. is about 5(2.5 cm) = 12.5 cm, which is close to 12 cm.

■ Work Practice 1

Practice 1

The center hole of a standard-sized compact disc is 1.5 centimeters in diameter. Convert this length to inches. Round the result to 2 decimal places.

Example 2 Liver

The liver is your largest internal organ. It weighs about 3.5 pounds in a grown man. Convert this weight to kilograms. Round to the nearest tenth. (*Source: Some Body!* by Dr. Pete Rowan)

Solution: $3.5 \text{ lb} \approx \frac{3.5 \text{ lb}}{1} \cdot \frac{0.45 \text{ kg}}{1 \text{ lb}} = 3.5(0.45 \text{ kg}) \approx 1.6 \text{ kg}$

Thus 3.5 pounds is approximately 1.6 kilograms. From the table of conversions, we know that 1 kg ≈ 2.2 lb. So that means 0.5 kg ≈ 1.1 lb and after adding, we have 1.5 kg ≈ 3.3 lb. Our result is reasonable.

■ Work Practice 2

Practice 2

A full-grown human heart weighs about 8 ounces. Convert this weight to grams. If necessary, round your result to the nearest tenth of a gram.

Answers
1. 0.59 in.   **2.** 226.8 g

Practice 3

Convert 237 ml to fluid ounces. Round to the nearest whole fluid ounce.

Example 3 Postage Stamp

Australia converted to the metric system in 1973. In that year, four postage stamps were issued to publicize this conversion. One such stamp is shown. Let's check the mathematics on the stamp by converting 7 fluid ounces to milliliters. Round to the nearest hundred.

Solution: $7 \text{ fl oz} \approx \dfrac{7 \text{ fl oz}}{1} \cdot \overbrace{\dfrac{29.57 \text{ ml}}{1 \text{ fl oz}}}^{\text{Unit fraction}} = 7(29.57 \text{ ml}) = 206.99 \text{ ml}$

Rounded to the nearest hundred, 7 fl oz ≈ 200 ml.

■ Work Practice 3

Now that we have practiced converting between two measurement systems, let's practice converting between two temperature scales.

Temperature When Gabriel Fahrenheit and Anders Celsius independently established units for temperature scales, each based his unit on the heat of water the moment it boils compared to the moment it freezes. One degree Celsius is $\dfrac{1}{100}$ of the difference in heat. One degree Fahrenheit is $\dfrac{1}{180}$ of the difference in heat. Celsius arbitrarily labeled the temperature at the freezing point at 0°C, making the boiling point 100°C; Fahrenheit labeled the freezing point 32°F, making the boiling point 212°F. Water boils at 212°F or 100°C.

By comparing the two scales in the figure, we see that a 20°C day is as warm as a 68°F day. Similarly, a sweltering 104°F day in the Mojave desert corresponds to a 40°C day.

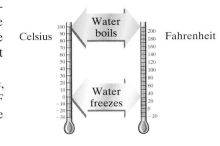

✓**Concept Check** Which of the following statements is correct? Explain.
a. 6°C is below the freezing point of water.
b. 6°F is below the freezing point of water.

Objective B Converting Degrees Celsius to Degrees Fahrenheit ▶

To convert from Celsius temperatures to Fahrenheit temperatures, see the box below. In this box, we use the symbol F to represent degrees Fahrenheit and the symbol C to represent degrees Celsius.

Converting Celsius to Fahrenheit

$$F = \dfrac{9}{5}C + 32 \quad \text{or} \quad F = 1.8C + 32$$

(To convert to Fahrenheit temperature, multiply the Celsius temperature by $\dfrac{9}{5}$ or 1.8, and then add 32.)

Answer
3. 8 fl oz

✓**Concept Check Answer**
b

Section 8.7 | Temperature and Conversions Between the U.S. and Metric Systems 635

Example 4 Convert 15°C to degrees Fahrenheit.

Solution: $F = \dfrac{9}{5}C + 32$

$= \dfrac{9}{5} \cdot 15 + 32$ Replace C with 15.

$= 27 + 32$ Simplify.

$= 59$ Add.

Thus, 15°C is equivalent to 59°F.

■ Work Practice 4

Practice 4

Convert 60°C to degrees Fahrenheit.

Example 5 Convert 29°C to degrees Fahrenheit.

Solution: $F = 1.8\,C + 32$

$= 1.8 \cdot 29 + 32$ Replace C with 29.

$= 52.2 + 32$ Multiply 1.8 by 29.

$= 84.2$ Add.

Therefore, 29°C is the same as 84.2°F.

■ Work Practice 5

Practice 5

Convert 32°C to degrees Fahrenheit.

Objective C Converting Degrees Fahrenheit to Degrees Celsius ▶

To convert from Fahrenheit temperatures to Celsius temperatures, see the box below. The symbol C represents degrees Celsius and the symbol F represents degrees Fahrenheit.

Converting Fahrenheit to Celsius

$C = \dfrac{5}{9}(F - 32)$

(To convert to Celsius temperature, subtract 32 from the Fahrenheit temperature, and then multiply by $\dfrac{5}{9}$.)

Example 6 Convert 59°F to degrees Celsius.

Solution: We evaluate the formula $C = \dfrac{5}{9}(F - 32)$ when F is 59.

$C = \dfrac{5}{9}(F - 32)$

$= \dfrac{5}{9} \cdot (59 - 32)$ Replace F with 59.

$= \dfrac{5}{9} \cdot (27)$ Subtract inside parentheses.

$= 15$ Multiply.

Therefore, 59°F is the same temperature as 15°C.

■ Work Practice 6

Practice 6

Convert 68°F to degrees Celsius.

Answers
4. 140°F **5.** 89.6°F **6.** 20°C

Practice 7
Convert 113°F to degrees Celsius. If necessary, round to the nearest tenth of a degree.

Example 7 Convert 114°F to degrees Celsius. If necessary, round to the nearest tenth of a degree.

Solution: $C = \frac{5}{9}(F - 32)$

$= \frac{5}{9}(114 - 32)$ Replace F with 114.

$= \frac{5}{9} \cdot (82)$ Subtract inside parentheses.

≈ 45.6 Multiply.

Therefore, 114°F is approximately 45.6°C.

■ Work Practice 7

Practice 8
During a bout with the flu, Albert's temperature reaches 102.8°F. What is his temperature measured in degrees Celsius? Round to the nearest tenth of a degree.

Example 8 Body Temperature

Normal body temperature is 98.6°F. What is this temperature in degrees Celsius?

Solution: We evaluate the formula $C = \frac{5}{9}(F - 32)$ when F is 98.6.

$C = \frac{5}{9}(F - 32)$

$= \frac{5}{9}(98.6 - 32)$ Replace F with 98.6.

$= \frac{5}{9} \cdot (66.6)$ Subtract inside parentheses.

$= 37$ Multiply.

Therefore, normal body temperature is 37°C.

■ Work Practice 8

✓**Concept Check** Clarissa must convert 40°F to degrees Celsius. What is wrong with her work shown below?

$F = 1.8 \cdot C + 32$
$F = 1.8 \cdot 40 + 32$
$F = 72 + 32$
$F = 104$

Answers
7. 45°C **8.** 39.3°C

✓**Concept Check Answer**
She used the conversion for Celsius to Fahrenheit instead of Fahrenheit to Celsius.

Vocabulary, Readiness & Video Check

See Video 8.7

Martin-Gay Interactive Videos Watch the section lecture video and answer the following questions.

Objective A 1. Write two conversions that may be used to solve Example 2.

2. Why isn't 0.1125 kg the final answer to Example 3?

Objective B 3. Which version of the formula is used to solve Example 4? What is the replacement value for C?

Objective C 4. In Example 5, what is the replacement value for F? What is the final conversion?

Section 8.7 | Temperature and Conversions Between the U.S. and Metric Systems

8.7 Exercise Set MyLab Math

Note: Because approximations are used, your answers may vary slightly from the answers given in the back of the book.

Objective A *Convert as indicated. If necessary, round answers to two decimal places. See Examples 1 through 3.*

1. 756 milliliters to fluid ounces
2. 18 liters to quarts
3. 86 inches to centimeters
4. 86 miles to kilometers
5. 1000 grams to ounces
6. 100 kilograms to pounds
7. 93 kilometers to miles
8. 9.8 meters to feet
9. 14.5 liters to gallons
10. 150 milliliters to fluid ounces
11. 30 pounds to kilograms
12. 15 ounces to grams

Fill in the chart. Give exact answers or round to one decimal place. See Examples 1 through 3.

	Meters	Yards	Centimeters	Feet	Inches
13. The height of a woman				5	
14. Statue of Liberty length of nose	1.37				
15. Leaning Tower of Pisa		60			
16. Blue whale		36			

Solve. If necessary, round answers to two decimal places. See Examples 1 through 3.

17. The balance beam for female gymnasts is 10 centimeters wide. Convert this width to inches.

18. In men's gymnastics, the rings are 250 centimeters from the floor. Convert this height to inches, then to feet.

19. In many states, the maximum speed limit for recreational vehicles is 50 miles per hour. Convert this to kilometers per hour.

20. In some states, the speed limit is 70 miles per hour. Convert this to kilometers per hour.

21. Ibuprofen comes in 200-milligram tablets. Convert this to ounces. (Round your answer to this exercise to 3 decimal places.)

22. Vitamin C tablets come in 500-milligram caplets. Convert this to ounces.

The 70-meter-diameter antenna is the largest and most sensitive Deep Space Network antenna. See the Chapter Opener and answer Exercises 23 through 26.

70-Meter Antenna

23. Convert 70 meters to feet.

24. The Deep Space Network sites also have a 26-meter antenna. Convert 26 meters to feet.

25. The 70-meter-diameter antenna can track a spacecraft traveling more than 16 billion kilometers from Earth. Convert this distance to miles.

26. The dish reflector and the mount atop the concrete pedestal of the 70-meter antenna weigh nearly 2.7 million kilograms. Convert this number to tons.

27. A stone is a unit in the British customary system. Use the conversion 14 pounds = 1 stone to check the equivalencies in this 1973 Australian stamp. Is 100 kilograms approximately 15 stone 10 pounds?

28. Convert 5 feet 11 inches to centimeters and check the conversion on this 1973 Australian stamp. Is it correct?

29. The Monarch butterfly migrates annually between the northern United States and central Mexico. The trip is about 4500 km long. Convert this to miles.

30. There is a species of African termite that builds nests up to 18 ft high. Convert this to meters.

31. A $3\frac{1}{2}$-inch diskette is not really $3\frac{1}{2}$ inches. To find its actual width, convert this measurement to centimeters, then to millimeters. Round the result to the nearest ten.

32. The average two-year-old is 84 centimeters tall. Convert this to feet and inches.

33. For an average adult, the weight of the right lung is greater than the weight of the left lung. If the right lung weighs 1.5 pounds and the left lung weighs 1.25 pounds, find the difference in grams. (*Source: Some Body!*)

34. The skin of an average adult weighs 9 pounds and is the heaviest organ. Find the weight in grams. (*Source: Some Body!*)

35. A fast sneeze has been clocked at about 167 kilometers per hour. Convert this to miles per hour. Round to the nearest whole.

36. A Boeing 747 has a cruising speed of about 980 kilometers per hour. Convert this to miles per hour. Round to the nearest whole.

37. The General Sherman giant sequoia tree has a diameter of about 8 meters at its base. Convert this to feet. (*Source: Fantastic Book of Comparisons*)

38. The largest crater on the near side of the moon is Billy Crater. It has a diameter of 303 kilometers. Convert this to miles. (*Source: Fantastic Book of Comparisons*)

Section 8.7 | Temperature and Conversions Between the U.S. and Metric Systems

39. The total length of the track on a CD is about 4.5 kilometers. Convert this to miles. Round to the nearest whole mile.

40. The distance between Mackinaw City, Michigan, and Cheyenne, Wyoming, is 2079 kilometers. Convert this to miles. Round to the nearest whole mile.

41. A doctor orders a dosage of 5 ml of medicine every 4 hours for 1 week. How many fluid ounces of medicine should be purchased? Round up to the next whole fluid ounce.

42. A doctor orders a dosage of 12 ml of medicine every 6 hours for 10 days. How many fluid ounces of medicine should be purchased? Round up to the next whole fluid ounce.

Without actually converting, choose the most reasonable answer.

43. This math book has a height of about _____.
   a. 28 mm
   b. 28 cm
   c. 28 m
   d. 28 km

44. A mile is _____ a kilometer.
   a. shorter than
   b. longer than
   c. the same length as

45. A liter has _____ capacity than a quart.
   a. less
   b. greater
   c. the same

46. A foot is _____ a meter.
   a. shorter than
   b. longer than
   c. the same length as

47. A kilogram weighs _____ a pound.
   a. the same as
   b. less than
   c. greater than

48. A football field is 100 yards, which is about _____.
   a. 9 m
   b. 90 m
   c. 900 m
   d. 9000 m

49. An $8\frac{1}{2}$-ounce glass of water has a capacity of about _____.
   a. 250 L
   b. 25 L
   c. 2.5 L
   d. 250 ml

50. A 5-gallon gasoline can has a capacity of about _____.
   a. 19 L
   b. 1.9 L
   c. 19 ml
   d. 1.9 ml

51. The weight of an average man is about _____.
   a. 700 kg
   b. 7 kg
   c. 0.7 kg
   d. 70 kg

52. The weight of a pill is about _____.
   a. 200 kg
   b. 20 kg
   c. 2 kg
   d. 200 mg

Objectives B C Mixed Practice *Convert as indicated. When necessary, round to the nearest tenth of a degree. See Examples 4 through 8.*

53. 77°F to degrees Celsius

54. 86°F to degrees Celsius

55. 104°F to degrees Celsius

56. 140°F to degrees Celsius

57. 50°C to degrees Fahrenheit

58. 80°C to degrees Fahrenheit

59. 115°C to degrees Fahrenheit

60. 225°C to degrees Fahrenheit

61. 20°F to degrees Celsius

62. 26°F to degrees Celsius

63. 142.1°F to degrees Celsius

64. 43.4°F to degrees Celsius

65. 92°C to degrees Fahrenheit

66. 75°C to degrees Fahrenheit

67. 12.4°C to degrees Fahrenheit

68. 48.6°C to degrees Fahrenheit

69. The hottest temperature ever recorded in the United States, in Death Valley, was 134°F. Convert this temperature to degrees Celsius. (*Source:* National Climatic Data Center)

70. The hottest temperature ever recorded in the United States in January was 95°F in Los Angeles. Convert this temperature to degrees Celsius. (*Source:* National Climatic Data Center)

71. A weather forecaster in Caracas predicts a high temperature of 27°C. Find this measurement in degrees Fahrenheit.

72. While driving to work, Alan Olda notices a temperature of 18°C flash on the local bank's temperature display. Find the corresponding temperature in degrees Fahrenheit.

73. At Mack Trucks' headquarters, the room temperature is to be set at 70°F, but the thermostat is calibrated in degrees Celsius. Find the temperature to be set.

74. The computer room at Merck, Sharp, and Dohm is normally cooled to 66°F. Find the corresponding temperature in degrees Celsius.

75. In a European cookbook, a recipe requires the ingredients for caramels to be heated to 118°C, but the cook has access only to a Fahrenheit thermometer. Find the temperature in degrees Fahrenheit that should be used to make the caramels.

76. The ingredients for divinity should be heated to 127°C, but the candy thermometer that Myung Kim has is calibrated to degrees Fahrenheit. Find how hot he should heat the ingredients.

77. The temperature of Earth's core is estimated to be 4000°C. Find the corresponding temperature in degrees Fahrenheit.

78. In 2012, the average temperature of Earth's surface was 58.3°F. Convert this temperature to degrees Celsius. (*Source:* NASA)

Review

Perform the indicated operations. See Section 1.7.

79. $6 \cdot 4 + 5 \div 1$

80. $10 \div 2 + 9(8)$

81. $3[(1 + 5) \cdot (8 - 6)]$

82. $5[(18 - 8) - 9]$

Concept Extensions

Determine whether the measurement in each statement is reasonable.

83. A 72°F room feels comfortable.

84. Water heated to 110°F will boil.

85. Josiah has a fever if a thermometer shows his temperature to be 40°F.

86. An air temperature of 20°F on a Vermont ski slope can be expected in the winter.

87. When the temperature is 30°C outside, an overcoat is needed.

88. An air-conditioned room at 60°C feels quite chilly.

89. Barbara has a fever when a thermometer records her temperature at 40°C.

90. Water cooled to 32°C will freeze.

Body surface area (BSA) is often used to calculate dosages for some drugs. BSA is calculated in square meters using a person's weight and height.

$$\text{BSA} = \sqrt{\frac{(\text{weight in kg}) \times (\text{height in cm})}{3600}}$$

For Exercises 91 through 96, calculate the BSA for each person. Round to the nearest hundredth. You will need to use the square root key on your calculator.

91. An adult whose height is 182 cm and weight is 90 kg

92. An adult whose height is 157 cm and weight is 63 kg

93. A child whose height is 40 in. and weight is 50 kg (*Hint:* Don't forget to first convert inches to centimeters.)

94. A child whose height is 26 in. and weight is 13 kg (*Hint:* Don't forget to first convert inches to centimeters.)

95. An adult whose height is 60 in. and weight is 150 lb

96. An adult whose height is 69 in. and weight is 172 lb

97. The most durable storage medium in the world is a nanostructured glass created at Southampton's Optoelectronics Research Center, United Kingdom. This remains stable for 13.8 billion years at a temperature of 371.9°F. Convert this temperature to degrees Celsius. Round your answer to the nearest tenth of a degree. (*Source: Guinness Book of World Records*)

98. The hottest-burning substance known is carbon subnitride. Its flame at one atmospheric pressure reaches 9010°F. Convert this temperature to degrees Celsius. (*Source: Guinness World Records*)

99. In your own words, describe how to convert from degrees Celsius to degrees Fahrenheit.

100. In your own words, describe how to convert from degrees Fahrenheit to degrees Celsius.

Chapter 8 Group Activity

Map Reading

Sections 8.1, 8.4, and 8.7

Materials:
- ruler
- string
- calculator

This activity may be completed by working in groups or individually.

Investigate the route you would take from Santa Rosa, New Mexico, to San Antonio, New Mexico. Use the map in the figure to answer the following questions. You may find that using string to match the roads on the map is useful when measuring distances.

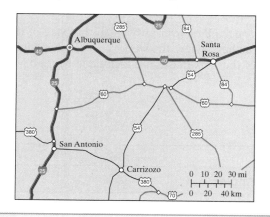

1. How many miles is it from Santa Rosa to San Antonio via Interstate 40 and Interstate 25? Convert this distance to kilometers.

2. How many miles is it from Santa Rosa to San Antonio via U.S. 54 and U.S. 380? Convert this distance to kilometers.

3. Assume that the speed limit on Interstates 40 and 25 is 65 miles per hour. How long would the trip take if you took this route and traveled 65 miles per hour the entire trip?

4. At what average speed would you have to travel on the U.S. routes to make the trip from Santa Rosa to San Antonio in the same amount of time that it would take on the interstate routes? Do you think this speed is reasonable on this route? Explain your reasoning.

5. Discuss in general the factors that might affect your decision between the different routes.

6. Explain which route you would choose in this case and why.

Chapter 8 Vocabulary Check

Fill in each blank with one of the words or phrases listed below.

transversal	line segment	obtuse	straight	adjacent	right	volume	area
acute	perimeter	vertical	supplementary	ray	angle	line	complementary
vertex	mass	unit fractions	gram	weight	meter	liter	surface area

1. _____ is a measure of the pull of gravity.
2. _____ is a measure of the amount of substance in an object. This measure does not change.
3. The basic unit of length in the metric system is the _____ .
4. To convert from one unit of length to another, _____ may be used.
5. The _____ is the basic unit of mass in the metric system.
6. The _____ is the basic unit of capacity in the metric system.
7. A(n) _____ is a piece of a line with two endpoints.
8. Two angles that have a sum of 90° are called _____ angles.
9. A(n) _____ is a set of points extending indefinitely in two directions.
10. The _____ of a polygon is the distance around the polygon.
11. A(n) _____ is made up of two rays that share the same endpoint. The common endpoint is called the _____ .
12. _____ measures the amount of surface of a region.
13. A(n) _____ is a part of a line with one endpoint. A ray extends indefinitely in one direction.
14. A line that intersects two or more lines at different points is called a(n) _____ .
15. An angle that measures 180° is called a(n) _____ angle.
16. The measure of the space of a solid is called its _____ .
17. When two lines intersect, four angles are formed. Two of these angles that are opposite each other are called _____ angles.
18. Two of the angles from Exercise 17 that share a common side are called _____ angles.
19. An angle whose measure is between 90° and 180° is called a(n) _____ angle.
20. An angle that measures 90° is called a(n) _____ angle.
21. An angle whose measure is between 0° and 90° is called a(n) _____ angle.
22. Two angles that have a sum of 180° are called _____ angles.
23. The _____ of a polyhedron is the sum of the areas of the faces of the polyhedron.

> **Helpful Hint**
> ▸ Are you preparing for your test? To help, don't forget to take these:
> - Chapter 8 Getting Ready for the Test on page 652
> - Chapter 8 Test on page 653
>
> The check all of your answers at the back of this text. For further review, the step-by-step video solutions to any of these exercises are located in MyLab Math.

8 Chapter Highlights

Definitions and Concepts	Examples
Section 8.1 Lines and Angles	
A **line** is a set of points extending indefinitely in two directions. A line has no width or height, but it does have length. We name a line by any two of its points.	Line AB or $\overleftrightarrow{AB}$
A **line segment** is a piece of a line with two endpoints.	Line segment AB or $\overline{AB}$

Definitions and Concepts	Examples

Section 8.1 Lines and Angles (continued)

A **ray** is a part of a line with one endpoint. A ray extends indefinitely in one direction.	Ray AB or $\overrightarrow{AB}$
An **angle** is made up of two rays that share the same endpoint. The common endpoint is called the **vertex**.	$\angle B$ or $\angle x$ or $\angle ABC$ or $\angle CBA$ 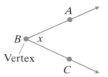

Section 8.2 Perimeter

Perimeter Formulas **Rectangle:** $P = 2l + 2w$ **Square:** $P = 4s$ **Triangle:** $P = a + b + c$ **Circumference of a Circle:** $C = 2\pi r$ or $C = \pi d$ where $\pi \approx 3.14$ or $\pi \approx \frac{22}{7}$	Find the perimeter of the rectangle.  $P = 2l + 2w$ $= 2 \cdot 28 \text{ meters} + 2 \cdot 15 \text{ meters}$ $= 56 \text{ meters} + 30 \text{ meters}$ $= 86 \text{ meters}$ The perimeter is 86 meters.

Section 8.3 Area, Volume, and Surface Area

Area Formulas **Rectangle:** $A = lw$ **Square:** $A = s^2$ **Triangle:** $A = \frac{1}{2}bh$ **Parallelogram:** $A = bh$ **Trapezoid:** $A = \frac{1}{2}(b + B)h$ **Circle:** $A = \pi r^2$ **Volume Formulas** **Rectangular Solid:** $V = lwh$ **Cube:** $V = s^3$ **Sphere:** $V = \frac{4}{3}\pi r^3$ **Right Circular Cylinder:** $V = \pi r^2 h$ **Cone:** $V = \frac{1}{3}\pi r^2 h$ **Square-Based Pyramid:** $V = \frac{1}{3}s^2 h$ **Surface Area Formulas:** See page 589.	Find the area of the square. $A = s^2$ $= (8 \text{ centimeters})^2$ $= 64 \text{ square centimeters}$ The area of the square is 64 square centimeters. Find the volume of the sphere. Use $\frac{22}{7}$ for π. $V = \frac{4}{3}\pi r^3$ $\approx \frac{4}{3} \cdot \frac{22}{7} \cdot (4 \text{ inches})^3$ $= \frac{4 \cdot 22 \cdot 64}{3 \cdot 7}$ cubic inches $= \frac{5632}{21}$ cubic inches or $268\frac{4}{21}$ cubic inches

Definitions and Concepts	Examples

Section 8.4 Linear Measurement

To convert from one unit of length to another, multiply by a **unit fraction** in the form

$$\frac{\text{units to convert to}}{\text{original units}}$$

Length: U.S. System of Measurement

12 inches (in.) = 1 foot (ft)
3 feet = 1 yard (yd)
5280 feet = 1 mile (mi)

Length: Metric System of Measurement

The basic unit of length in the metric system is the **meter**. A meter is slightly longer than a yard.

Metric Units of Length
1 **kilo**meter (km) = 1000 meters (m)
1 **hecto**meter (hm) = 100 m
1 **deka**meter (dam) = 10 m
1 meter (m) = 1 m
1 **deci**meter (dm) = 1/10 m or 0.1 m
1 **centi**meter (cm) = 1/100 m or 0.01 m
1 **milli**meter (mm) = 1/1000 m or 0.001 m

$$\frac{12 \text{ inches}}{1 \text{ foot}}, \frac{1 \text{ foot}}{12 \text{ inches}}, \frac{3 \text{ feet}}{1 \text{ yard}}$$

Convert 6 feet to inches.

$$6 \text{ ft} = \frac{6 \text{ ft}}{1} \cdot 1$$

$$= \frac{6 \text{ ft}}{1} \cdot \frac{12 \text{ in.}}{1 \text{ ft}} \quad \leftarrow \text{ units to convert to} \\ \leftarrow \text{ original units}$$

$$= 6 \cdot 12 \text{ in.}$$

$$= 72 \text{ in.}$$

Convert 3650 centimeters to meters.

$$3650 \text{ cm} = 3650 \text{ cm} \cdot 1$$

$$= \frac{3650 \text{ cm}}{1} \cdot \frac{0.01 \text{ m}}{1 \text{ cm}} = 36.5 \text{ m}$$

or

km hm dam m dm cm mm
                    ⎵
            2 units to the left

3650 cm = 36.5 m
        ⎵
  2 places to the left

Section 8.5 Weight and Mass

Weight is really a measure of the pull of gravity. **Mass** is a measure of the amount of substance in an object and does not change.

Weight: U.S. System of Measurement

16 ounces (oz) = 1 pound (lb)
2000 pounds = 1 ton

Mass: Metric System of Measurement

The **gram** is the basic unit of mass in the metric system. It is the mass of water contained in a cube 1 centimeter on each side. A paper clip weighs about 1 gram.

Metric Units of Mass
1 **kilo**gram (kg) = 1000 grams (g)
1 **hecto**gram (hg) = 100 g
1 **deka**gram (dag) = 10 g
1 gram (g) = 1 g
1 **deci**gram (dg) = 1/10 g or 0.1 g
1 **centi**gram (cg) = 1/100 g or 0.01 g
1 **milli**gram (mg) = 1/1000 g or 0.001 g

Convert 5 pounds to ounces.

$$5 \text{ lb} = 5 \text{ lb} \cdot 1 = \frac{5 \text{ lb}}{1} \cdot \frac{16 \text{ oz}}{1 \text{ lb}} = 80 \text{ oz}$$

Convert 260 grams to kilograms.

$$260 \text{ g} = \frac{260 \text{ g}}{1} \cdot 1 = \frac{260 \text{ g}}{1} \cdot \frac{1 \text{ kg}}{1000 \text{ g}} = 0.26 \text{ kg}$$

or

kg hg dag g dg cg mg
         ⎵
  3 units to the left

260 g = 0.26 kg
     ⎵
 3 places to the left

Definitions and Concepts	Examples

Section 8.6 Capacity

Capacity: U.S. System of Measurement

8 fluid ounces (fl oz) = 1 cup (c)
2 cups = 1 pint (pt)
2 pints = 1 quart (qt)
4 quarts = 1 gallon (gal)

Convert 5 pints to gallons.

1 gal = 4 qt = 8 pt

$$5 \text{ pt} = 5 \text{ pt} \cdot 1 = \frac{5 \text{ pt}}{1} \cdot \frac{1 \text{ gal}}{8 \text{ pt}} = \frac{5}{8} \text{ gal}$$

Capacity: Metric System of Measurement

The **liter** is the basic unit of capacity in the metric system. It is the capacity or volume of a cube measuring 10 centimeters on each side. A liter of liquid is slightly more than a quart.

Metric Units of Capacity

1 **kilo**liter (kl) = 1000 liters (L)
1 **hecto**liter (hl) = 100 L
1 **deka**liter (dal) = 10 L
1 liter (L) = 1 L
1 **deci**liter (dl) = 1/10 L or 0.1 L
1 **centi**liter (cl) = 1/100 L or 0.01 L
1 **milli**liter (ml) = 1/1000 L or 0.001 L

Convert 1.5 liters to milliliters.

$$1.5 \text{ L} = \frac{1.5 \text{ L}}{1} \cdot 1 = \frac{1.5 \text{ L}}{1} \cdot \frac{1000 \text{ ml}}{1 \text{ L}} = 1500 \text{ ml}$$

or

kl hl dal L dl cl ml

3 units to the left

1.500 L = 1500 ml

3 places to the right

Section 8.7 Temperature and Conversions Between the U.S. and Metric Systems

To convert between systems, use approximate unit fractions. See page 633.

Convert 7 feet to meters.

$$7 \text{ ft} \approx \frac{7 \text{ ft}}{1} \cdot \frac{0.30 \text{ m}}{1 \text{ ft}} = 2.1 \text{ m}$$

Convert 8 liters to quarts.

$$8 \text{ L} \approx \frac{8 \text{ L}}{1} \cdot \frac{1.06 \text{ qt}}{1 \text{ L}} = 8.48 \text{ qt}$$

Convert 363 grams to ounces.

$$363 \text{ g} \approx \frac{363 \text{ g}}{1} \cdot \frac{0.04 \text{ oz}}{1 \text{ g}} = 14.52 \text{ oz}$$

Celsius to Fahrenheit

$$F = \frac{9}{5}C + 32 \quad \text{or} \quad F = 1.8C + 32$$

Convert 35°C to degrees Fahrenheit.

$$F = \frac{9}{5} \cdot 35 + 32 = 63 + 32 = 95$$

35°C = 95°F

Fahrenheit to Celsius

$$C = \frac{5}{9}(F - 32)$$

Convert 50°F to degrees Celsius.

$$C = \frac{5}{9} \cdot (50 - 32) = \frac{5}{9} \cdot (18) = 10$$

50°F = 10°C

Chapter 8 Review

(8.1) *Classify each angle as acute, right, obtuse, or straight.*

1.
2.
3.
4.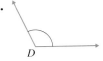

5. Find the complement of a 25° angle.

6. Find the supplement of a 105° angle.

Find the measure of angle x in each figure.

7.
8.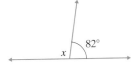
9.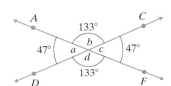
10.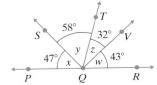

11. Identify the pairs of supplementary angles.

12. Identify the pairs of complementary angles.

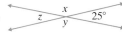

Find the measures of angles x, y, and z in each figure.

13.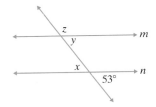
14.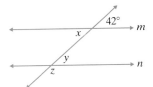

15. Given that $m \parallel n$.

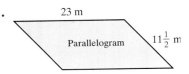

16. Given that $m \parallel n$.

(8.2) *Find the perimeter of each figure.*

17. Parallelogram, 23 m, $11\frac{1}{2}$ m

18. 11 cm, 7.6 cm, 12 cm

Chapter 8 Review   647

19.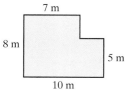

20.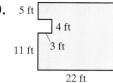

Solve.

21. Find the perimeter of a rectangular sign that measures 6 feet by 10 feet.

22. Find the perimeter of a town square that measures 110 feet on a side.

Find the circumference of each circle. Use $\pi \approx 3.14$.

23.

24.

(8.3) Find the area of each figure. For the circles, find the exact area and then use $\pi \approx 3.14$ to approximate the area.

25.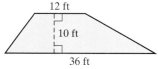

26.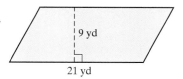

27.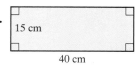

28.

29.

30.

31.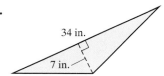

32.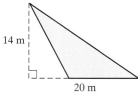

33.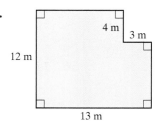

34.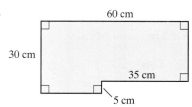

35. The amount of sealer necessary to seal a driveway depends on the area. Find the area of a rectangular driveway 36 feet by 12 feet.

36. Find how much carpet is necessary to cover the floor of the room shown.

Find the volume and surface area of the solids in Exercises 37 and 38. For Exercises 39 and 40, give the exact volume and an approximation.

37. 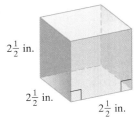 $2\frac{1}{2}$ in., $2\frac{1}{2}$ in., $2\frac{1}{2}$ in.

38. 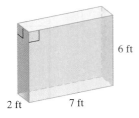 6 ft, 2 ft, 7 ft

39. Use $\pi \approx 3.14$. 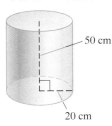 50 cm, 20 cm

40. Use $\pi \approx \frac{22}{7}$. 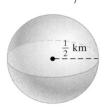 $\frac{1}{2}$ km

41. Find the volume of a pyramid with a square base 2 feet on a side and a height of 2 feet.

42. Approximate the volume of a tin can 8 inches high and 3.5 inches in radius. Use 3.14 for π.

43. A chest has 3 drawers. If each drawer has inside measurements of $2\frac{1}{2}$ feet by $1\frac{1}{2}$ feet by $\frac{2}{3}$ foot, find the total volume of the 3 drawers.

44. A cylindrical canister for a shop vacuum is 2 feet tall and 1 foot in *diameter*. Find its exact volume.

(8.4) *Convert.*

45. 108 in. to feet

46. 72 ft to yards

47. 1.5 mi to feet

48. $\frac{1}{2}$ yd to inches

49. 52 ft = _____ yd _____ ft

50. 46 in. = _____ ft _____ in.

51. 42 m to centimeters

52. 82 cm to millimeters

53. 12.18 mm to meters

54. 2.31 m to kilometers

Perform each indicated operation.

55. 4 yd 2 ft + 16 yd 2 ft

56. 7 ft 4 in. ÷ 2

57. 8 cm + 15 mm

58. 4 m − 126 cm

Solve.

59. A bolt of cloth contains 333 yd 1 ft of cotton ticking. Find the amount of material that remains after 163 yd 2 ft is removed from the bolt.

60. The student activities club is sponsoring a walk for hunger, and all students who participate will receive a sash with the name of the school to wear on the walk. If each sash requires 5 ft 2 in. of material and there are 50 students participating in the walk, how much material will the student activities club need?

Chapter 8 Review **649**

1. The trip from El Paso, TX, to Ontario, CA, is about 1235 km each way. Four friends agree to share the driving equally. How far must each drive on this round-trip vacation?

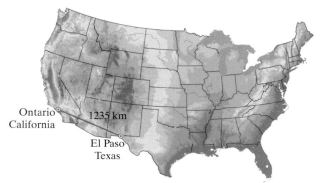

△ **62.** The college has ordered that NO SMOKING signs be placed above the doorway of each classroom. Each sign is 0.8 m long and 30 cm wide. Find the area of each sign. (*Hint:* Recall that the area of a rectangle = width · length.)

8.5) *Convert.*

63. 66 oz to pounds

64. 2.3 tons to pounds

65. 52 oz = _____ lb _____ oz

66. 10,300 lb = _____ tons _____ lb

67. 27 mg to grams

68. 40 kg to grams

69. 2.1 hg to dekagrams

70. 0.03 mg to decigrams

Perform each indicated operation.

71. 6 lb 5 oz − 2 lb 12 oz

72. 8 lb 6 oz × 4

73. 4.3 mg × 5

74. 4.8 kg − 4200 g

Solve.

75. Donshay Berry ordered 1 lb 12 oz of soft-center candies and 2 lb 8 oz of chewy-center candies for his party. Find the total weight of the candy ordered.

76. Four local townships jointly purchase 38 tons 300 lb of cinders to spread on their roads during an ice storm. Determine the weight of the cinders each township receives if they share the purchase equally.

8.6) *Convert.*

77. 28 pints to quarts

78. 40 fluid ounces to cups

79. 3 qt 1 pt to pints

80. 18 quarts to cups

81. 9 pt = _____ qt _____ pt

82. 15 qt = _____ gal _____ qt

83. 3.8 L to milliliters

84. 14 hl to kiloliters

85. 30.6 L to centiliters

86. 2.45 ml to liters

Perform each indicated operation.

87. 1 qt 1 pt + 3 qt 1 pt

88. 3 gal 2 qt × 2

89. 0.946 L − 210 ml

90. 6.1 L + 9400 ml

Solve.

91. Carlos Perez prepared 4 gal 2 qt of iced tea for a block party. During the first 30 minutes of the party, 1 gal 3 qt of the tea is consumed. How much iced tea remains?

92. A recipe for soup stock calls for 1 c 4 fl oz of beef broth. How much should be used if the recipe is cut in half?

93. Each bottle of Kiwi liquid shoe polish holds 85 ml of the polish. Find the number of liters of shoe polish contained in 8 boxes if each box contains 16 bottles.

94. Ivan Miller wants to pour three separate containers of saline solution into a single vat with a capacity of 10 liters. Will 6 liters of solution in the first container combined with 1300 milliliters in the second container and 2.6 liters in the third container fit into the larger vat?

(8.7) *Note: Because approximations are used in this section, your answers may vary slightly from the answers given in the back of the book.*

Convert as indicated. If necessary, round to two decimal places.

95. 7 meters to feet

96. 11.5 yards to meters

97. 17.5 liters to gallons

98. 7.8 liters to quarts

99. 15 ounces to grams

100. 23 pounds to kilograms

101. A compact disc is 1.2 mm thick. Find the height (in inches) of 50 discs.

102. If a person weighs 82 kilograms, how many pounds is this?

Convert. Round to the nearest tenth of a degree, if necessary.

103. 42°C to degrees Fahrenheit

104. 160°C to degrees Fahrenheit

105. 41.3°F to degrees Celsius

106. 80°F to degrees Celsius

Solve. Round to the nearest tenth of a degree, if necessary.

107. A sharp dip in the jet stream caused the temperature in New Orleans to drop to 35°F. Find the corresponding temperature in degrees Celsius.

108. A recipe for meat loaf calls for a 165°C oven. Find the setting used if the oven has a Fahrenheit thermometer.

Mixed Review

Find the following.

109. Find the supplement of a 72° angle.

110. Find the complement of a 1° angle.

Chapter 8   Review   **651**

Find the measure of angle x in each figure.

111.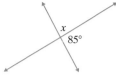

112.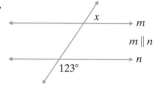

Find the perimeter of each figure.

113.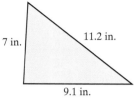

114.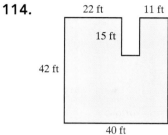

Find the area of each figure. For the circle, find the exact area and then use $\pi \approx 3.14$ to approximate the area.

115.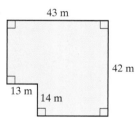

116.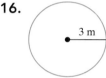

Find the volume of each solid.

117. Give an approximation using $\frac{22}{7}$ for π.

118. Find the surface area also.

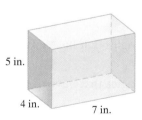

Convert the following.

119. 6.25 ft to inches

120. 8200 lb = _____ tons _____ lb

121. 5 m to centimeters

122. 286 mm to kilometers

123. 1400 mg to grams

124. 6.75 gallons to quarts

125. 86°C to degrees Fahrenheit

126. 51.8°F to degrees Celsius

Perform the indicated operations and simplify.

127. 9.3 km − 183 m

128. 35 L + 700 ml

129. 3 gal 3 qt + 4 gal 2 qt

130. 3.2 kg × 4

Chapter 8 — Getting Ready for the Test

MATCHING Match each word in the first column with its illustration in the second or third column.

1. line
2. line segment
3. ray
4. right angle
5. acute angle
6. obtuse angle

MULTIPLE CHOICE Exercises 7 through 16 are **Multiple Choice**. Choose the correct letter.

Use the given figure for Exercises 7 and 8.

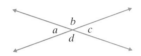

7. Choose two angles that have a sum of 180°.
   A. ∠a and ∠c **B.** ∠a and ∠d **C.** ∠b and ∠d

8. Choose two angles that have the same measure.
   A. ∠a and ∠b **B.** ∠c and ∠d **C.** ∠b and ∠d

Use the given figure for Exercises 9 and 10. For this figure, m ∥ n.

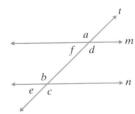

9. Choose two angles that have a sum of 180°.
   A. ∠a and ∠e **B.** ∠a and ∠d **C.** ∠a and ∠b

10. Choose two angles that have the same measure.
    A. ∠a and ∠e **B.** ∠a and ∠c **C.** ∠a and ∠f

For Exercises 11 through 16, the choices are below. Exercises 13 and 15 have two correct choices.

  A. perimeter **B.** area **C.** volume **D.** circumference

11. Which calculation is measured in square units?
12. Which calculation is measured in cubic units?
13. Which calculation is measured in units?

For Exercises 14 through 16, name the calculation (choices **A.**, **B.**, **C.**, **D.** above) to be used to solve each exercise.

14. The amount of material needed for a rectangular tablecloth.
15. The amount of trim needed to go around the edge of a tablecloth.
16. The amount of soil needed to fill a hole in the ground.

For additional practice go to your study plan in MyLab Math.

MyLab Math or **YouTube** Test

Chapter 8

1. Find the complement of a 78° angle.

2. Find the supplement of a 124° angle.

Answers

3. Find the measure of ∠x.

Find the measures of x, y, and z in each figure.

4.

5. Given that $m \parallel n$.

Find the unknown diameter or radius as indicated.

6.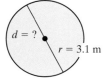

7.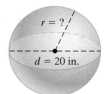

Find the perimeter (or circumference) and area of each figure. For the circle, give the exact value and an approximation using $\pi \approx 3.14$.

8.

9.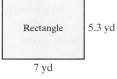

10.

1. _____

2. _____

3. _____

4. _____

5. _____

6. _____

7. _____

8. _____

9. _____

10. _____

653

Find the volume of each solid. For the cylinder, use $\pi \approx \dfrac{22}{7}$.

11.

12.

Solve.

13. Find the perimeter of a square photo with a side length of 4 inches.

14. How much soil is needed to fill a rectangular hole 3 feet by 3 feet by 2 feet?

15. Find how much baseboard is needed to go around a rectangular room that measures 18 feet by 13 feet. If baseboard costs $1.87 per foot, also calculate the total cost needed for materials.

Convert.

16. 280 in. = _____ ft _____ in.

17. $2\dfrac{1}{2}$ gal to quarts

18. 30 oz to pounds

19. 2.8 tons to pounds

20. 38 pt to gallons

21. 40 mg to grams

22. 2.4 kg to grams

23. 3.6 cm to millimeters

24. 4.3 dg to grams

25. 0.83 L to milliliters

Perform each indicated operation.

26. 3 qt 1 pt + 2 qt 1 pt

27. 8 lb 6 oz − 4 lb 9 oz

28. 2 ft 9 in. × 3

29. 5 gal 2 qt ÷ 2

30. 8 cm − 14 mm

31. 1.8 km + 456 m

Convert. Round to the nearest tenth of a degree, if necessary.

32. 84°F to degrees Celsius

33. 12.6°C to degrees Fahrenheit

34. The sugar maples in front of Bette MacMillan's house are 8.4 meters tall. Because they interfere with the phone lines, the telephone company plans to remove the top third of the trees. How tall will the maples be after they are shortened?

35. A total of 15 gal 1 qt of oil has been removed from a 20-gallon drum. How much oil still remains in the container

36. The engineer in charge of bridge construction said that the span of a certain bridge would be 88 m. But the actual construction required it to be 340 cm longer. Find the span of the bridge, in meters.

37. If 2 ft 9 in. of material is used to manufacture one scarf, how much material needed for 6 scarves?

Cumulative Review — Chapters 1–8

1. Solve: $3a - 6 = a + 4$

2. Solve: $2x + 1 = 3x - 5$

3. Evaluate:
  a. $\left(\dfrac{2}{5}\right)^4$
  b. $\left(-\dfrac{1}{4}\right)^2$

4. Evaluate:
  a. $\left(-\dfrac{1}{3}\right)^3$
  b. $\left(\dfrac{3}{7}\right)^2$

5. Add: $2\dfrac{4}{5} + 5 + 1\dfrac{1}{2}$

6. Add: $2\dfrac{1}{3} + 4\dfrac{2}{5} + 3$

7. Simplify by combining like terms:
$11.1x - 6.3 + 8.9x - 4.6$

8. Simplify by combining like terms:
$2.5y + 3.7 - 1.3y - 1.9$

9. Simplify: $\dfrac{5.68 + (0.9)^2 \div 100}{0.2}$

10. Simplify: $\dfrac{0.12 + 0.96}{0.5}$

11. Insert $<$, $>$, or $=$ to form a true statement.
$0.\overline{7} \quad \dfrac{7}{9}$

12. Insert $<$, $>$, or $=$ to form a true statement.
$0.43 \quad \dfrac{2}{5}$

13. Solve: $0.5y + 2.3 = 1.65$

14. Solve: $0.4x - 9.3 = 2.7$

15. An inner-city park is in the shape of a square that measures 300 feet on a side. Find the length of the diagonal of the park, rounded to the nearest whole foot.

16. A rectangular field is 200 feet by 125 feet. Find the length of the diagonal of the field, rounded to the nearest whole foot.

Answers

1. _____
2. _____
3. a. _____
   b. _____
4. a. _____
   b. _____
5. _____
6. _____
7. _____
8. _____
9. _____
10. _____
11. _____
12. _____
13. _____
14. _____
15. _____
16. _____

△ **17.** Given the rectangle shown:

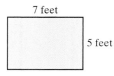

   a. Find the ratio of its width to its length.
   b. Find the ratio of its length to its perimeter.

△ **18.** A square is 9 inches by 9 inches.

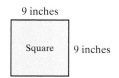

   a. Find the ratio of a side to its perimeter.
   b. Find the ratio of its perimeter to its area.

19. Determine whether -1 is a solution of the equation $3y + 1 = 3$.

20. Determine whether -5 is a solution of the equation $4(x + 7) = 8$.

21. Solve: $\dfrac{x}{3} = -2$

22. Solve: $9x = 8x + 1.02$

23. The standard dose of an antibiotic is 4 cc (cubic centimeters) for every 25 pounds (lb) of body weight. At this rate, find the standard dose for a 140-lb woman.

24. A recipe that makes 2 pie crusts calls for 3 cups of flour. How much flour is needed to make 5 pie crusts?

25. In a survey of 100 people, 17 people drive blue cars. What percent of the people drive blue cars?

26. Of 100 shoppers surveyed at a mall, 38 paid for their purchases using only cash. What percent of shoppers used only cash to pay for their purchases?

27. 13 is $6\dfrac{1}{2}$% of what number?

28. 54 is $4\dfrac{1}{2}$% of what number?

29. What number is 30% of 9?

30. What number is 42% of 30?

31. The number of applications for a mathematics scholarship at one university increased from 34 to 45 in one year. What is the percent of increase? Round to the nearest whole percent.

32. The price of a gallon of paint rose from $15 to $19. Find the percent of increase rounded to the nearest whole percent.

Cumulative Review

33. Find the sales tax and the total price on the purchase of an $85.50 atlas in a city where the sales tax rate is 7.5%.

34. A sofa has a purchase price of $375. If the sales tax rate is 8%, find the amount of sales tax and the total cost of the sofa.

35. Find the median of the list of numbers: 25, 54, 56, 57, 60, 71, 98

36. Find the median of the list of scores: 60, 95, 89, 72, 83

37. Find the probability of choosing a red marble from a box containing 1 red, 1 yellow, and 2 blue marbles.

38. Find the probability of choosing a nickel at random in a coin purse that contains 2 pennies, 2 nickels, and 3 quarters.

39. Find the complement of a 48° angle.

40. Find the supplement of a 137° angle.

41. Convert 8 feet to inches.

42. Convert 7 yards to feet.

43. Subtract 3 tons 1350 lb from 8 tons 1000 lb.

44. Add 8 lb 15 oz to 9 lb 3 oz.

45. Convert 59°F to degrees Celsius.

46. Convert 86°F to degrees Celsius.

33. _____
34. _____
35. _____
36. _____
37. _____
38. _____
39. _____
40. _____
41. _____
42. _____
43. _____
44. _____
45. _____
46. _____

9 Equations, Inequalities, and Problem Solving

In this chapter, we review solving equations and introduce solving inequalities. Solving equations and inequalities is necessary in order to solve word problems. Of course, problem solving is an integral topic in algebra and its discussion is continued throughout this text.

Sections

- 9.1 Symbols and Sets of Numbers
- 9.2 Properties of Real Numbers
- 9.3 Further Solving Linear Equations

  Integrated Review— Real Numbers and Solving Linear Equations

- 9.4 Further Problem Solving
- 9.5 Formulas and Problem Solving
- 9.6 Linear Inequalities and Problem Solving

Check Your Progress

Vocabulary Check
Chapter Highlights
Chapter Review
Getting Ready for the Test
Chapter Test
Cumulative Review

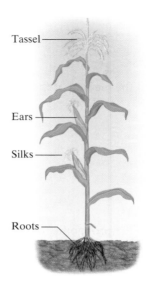

Maize Plant

Mature Maize Ear on a Stalk

Is It Corn or Maize?

In North America, Australia, and New Zealand, the words "corn" and "maize" mean the same thing, but that is not true in all countries. In many other countries, the word "corn" may refer to any cereal or grain crop. Because of this possible confusion, the term "maize" is preferred in scientific and international uses because it refers specifically to this one grain.

Maize is a staple food in many parts of the world, with more yearly production than that of wheat or rice. The bar graph below shows the yearly maize (corn) production in the United States, and the circle graph below shows the various uses of maize in the United States.

In Section 9.1, Exercises 69 through 72, we explore information about the corn crop from the top corn-producing states.

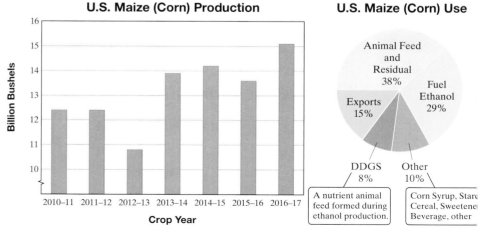

Source: USDA, NASS, Crop Production 2016 Summary, ERS Feed Outlook, Pro Exporter Network, Crop Year Ending August 31, 2017

9.1 Symbols and Sets of Numbers

Throughout the previous chapters, we have studied different sets of numbers. In this section, we review these sets of numbers. We also introduce a few new sets of numbers in order to show the relationships among these common sets of real numbers. We begin with a review of the set of natural numbers and the set of whole numbers and how we use symbols to compare these numbers. A **set** is a collection of objects, each of which is called a **member** or **element** of the set. A pair of brace symbols { } encloses the list of elements and is translated as "the set of" or "the set containing."

Objectives

A Define the Meaning of the Symbols $=$, $\neq$, $<$, $>$, $\leq$, and $\geq$.

B Translate Sentences into Mathematical Statements.

C Identify Integers, Rational Numbers, Irrational Numbers, and Real Numbers.

Natural numbers
$\{1, 2, 3, 4, 5, 6, \ldots\}$

Whole numbers
$\{0, 1, 2, 3, 4, 5, 6, \ldots\}$

> **Helpful Hint**
>
> The three dots (an ellipsis) at the end of the list of elements of a set means that the list continues in the same manner indefinitely.

Objective A Equality and Inequality Symbols

Picturing natural numbers and whole numbers on a number line helps us to see the order of the numbers. Symbols can be used to describe in writing the order of two quantities. We will use equality symbols and inequality symbols to compare quantities.

Below is a review of these symbols. The letters a and b are used to represent quantities. Recall from Section 1.8 that letters such as a and b that are used to represent numbers or quantities are called **variables.**

Equality and Inequality Symbols

		Meaning
Equality symbol:	$a = b$	a is equal to b.
Inequality symbols:	$a \neq b$	a is not equal to b.
	$a < b$	a is less than b.
	$a > b$	a is greater than b.
	$a \leq b$	a is less than or equal to b.
	$a \geq b$	a is greater than or equal to b.

These symbols may be used to form **mathematical statements** such as

$2 = 2$ and $2 \neq 6$

Recall that on a number line, we see that a number **to the right of** another number is **larger.** Similarly, a number **to the left of** another number is **smaller.** For example, 3 is to the left of 5 on the number line, which means that 3 is less than 5, or $3 < 5$. Similarly, 2 is to the right of 0 on the number line, which means that 2 is greater than 0, or $2 > 0$. Since 0 is to the left of 2, we can also say that 0 is less than 2, or $0 < 2$.

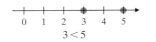
$3 < 5$

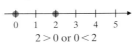
$2 > 0$ or $0 < 2$

Helpful Hint

Notice that $2 > 0$ has exactly the same meaning as $0 < 2$. Switching the order of the numbers and reversing the "direction" of the inequality symbol does not change the meaning of the statement.

$6 > 4$ has the same meaning as $4 < 6$.

Also notice that when the statement is true, the inequality arrow points to the smaller number.

Our discussion above can be generalized in the order property below.

Order Property for Real Numbers

For any two real numbers a and b, a is less than b if a is to the left of b on a number line.

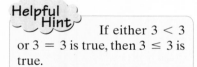

$a < b$ or also $b > a$

Practice 1–6

Determine whether each statement is true or false.
1. $8 < 6$
2. $100 > 10$
3. $21 \leq 21$
4. $21 \geq 21$
5. $0 \geq 5$
6. $25 \geq 22$

Examples
Determine whether each statement is true or false.

1. $2 < 3$ **True.**   Since 2 is to the left of 3 on a number line
2. $72 < 27$ **False.**   72 is to the right of 27 on a number line, so $72 > 27$.
3. $8 \geq 8$ **True.**   Since $8 = 8$ is true
4. $8 \leq 8$ **True.**   Since $8 = 8$ is true
5. $23 \leq 0$ **False.**   Since neither $23 < 0$ nor $23 = 0$ is true
6. $0 \leq 23$ **True.**   Since $0 < 23$ is true

■ Work Practice 1–6

Helpful Hint
If either $3 < 3$ or $3 = 3$ is true, then $3 \leq 3$ is true.

Objective B   Translating Sentences into Mathematical Statements

Now, let's use the symbols discussed on the previous page to translate sentences into mathematical statements.

Practice 7

Translate each sentence into a mathematical statement.
a. Fourteen is greater than or equal to fourteen.
b. Zero is less than five.
c. Nine is not equal to ten.

Example 7
Translate each sentence into a mathematical statement.

a. Nine is less than or equal to eleven.
b. Eight is greater than one.
c. Three is not equal to four.

Solution:

a. nine   is less than or equal to   eleven
    ↓   ↓   ↓
    9   ≤   11

b. eight   is greater than   one
    ↓   ↓   ↓
    8   >   1

c. three   is not equal to   four
    ↓   ↓   ↓
    3   ≠   4

■ Work Practice 7

Answers
1. false   2. true   3. true
4. true   5. false   6. true
7. a. $14 \geq 14$ b. $0 < 5$ c. $9 \neq 10$

Objective C Identifying Common Sets of Numbers

Whole numbers are not sufficient to describe many situations in the real world. For example, quantities smaller than zero must sometimes be represented, such as temperatures less than 0 degrees.

Recall that we can place numbers less than zero on a number line as follows: Numbers less than 0 are to the left of 0 and are labeled $-1, -2, -3,$ and so on. The numbers we have labeled on the number line below are called the set of **integers.**

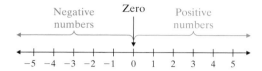

Integers to the left of 0 are called **negative integers;** integers to the right of 0 are called **positive integers.** The integer 0 is neither positive nor negative.

Integers

$$\{\ldots, -3, -2, -1, 0, 1, 2, 3, \ldots\}$$

Helpful Hint

A $-$ sign, such as the one in -2, tells us that the number is to the left of 0 on a number line.

-2 is read "negative two."

A $+$ sign or no sign tells us that the number lies to the right of 0 on a number line. For example, 3 and $+3$ both mean positive three.

Example 8 Use an integer to express the number in the following: "The lowest temperature ever recorded at South Pole Station, Antarctica, occurred during the month of June. The record-low temperature was 117 degrees Fahrenheit below zero." (*Source:* The National Oceanic and Atmospheric Administration)

Solution: The integer -117 represents 117 degrees Fahrenheit below zero.

Work Practice 8

Practice 8

Use an integer to express the number in the following: "The elevation of Laguna Salada in Mexico is 10 meters below sea level." (*Source: The World Almanac*)

Answer
8. -10

A problem with integers in real-life settings arises when quantities are smaller than some integer but greater than the next smallest integer. On a number line, these quantities may be visualized by points between integers. Some of these quantities between integers can be represented as quotients of integers. For example,

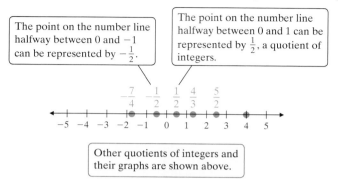

The point on the number line halfway between 0 and −1 can be represented by $-\frac{1}{2}$.

The point on the number line halfway between 0 and 1 can be represented by $\frac{1}{2}$, a quotient of integers.

Other quotients of integers and their graphs are shown above.

These numbers, each of which can be represented as a quotient of integers, are examples of **rational numbers.** It's not possible to list the set of rational numbers using the notation that we have been using. For this reason, we will use a different notation.

Rational Numbers

$$\left\{\frac{a}{b} \,\middle|\, a \text{ and } b \text{ are integers and } b \neq 0\right\}$$

We read this set as "the set of numbers $\frac{a}{b}$ such that a and b are integers and b **is not equal to 0.**"

Helpful Hint

We commonly refer to rational numbers as fractions.

Notice that every integer is also a rational number since each integer can be written as a quotient of integers. For example, the integer 5 is also a rational number since $5 = \frac{5}{1}$. For the rational number $\frac{5}{1}$, recall that the top number, 5, is called the numerator and the bottom number, 1, is called the denominator.

Let's practice **graphing** numbers on a number line.

Practice 9

Graph the numbers on the number line.

$-2\frac{1}{2}, \; -\frac{2}{3}, \; \frac{1}{5}, \; \frac{5}{4}, \; 2.25$

Example 9 Graph the numbers on a number line.

$-\frac{4}{3}, \; \frac{1}{4}, \; \frac{3}{2}, \; -2\frac{1}{8}, \; 3.5$

Solution: To help graph the improper fractions in the list, we first write them as mixed numbers.

■ Work Practice 9

Answer

9.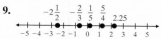

Every rational number has a point on the number line that corresponds to it. But not every point on the number line corresponds to a rational number. Those points that do not correspond to rational numbers correspond instead to **irrational numbers.**

Irrational Numbers

{Nonrational numbers that correspond to points on a number line}

An irrational number that you have probably seen is π. Also, $\sqrt{2}$, the length of the diagonal of the square shown below, is an irrational number.

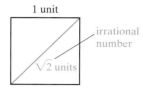

Both rational and irrational numbers can be written as decimal numbers. The decimal equivalent of a rational number will either terminate or repeat in a pattern. For example, upon dividing we find that

$\dfrac{3}{4} = 0.75$ (Decimal number terminates or ends.)

$\dfrac{2}{3} = 0.66666\ldots$ (Decimal number repeats in a pattern.)

The decimal representation of an irrational number will neither terminate nor repeat. (For further review of decimals, see Chapter 5.)

The set of numbers, each of which corresponds to a point on a number line, is called the set of **real numbers.** One and only one point on a number line corresponds to each real number.

Real Numbers

{All numbers that correspond to points on a number line}

Several different sets of numbers have been discussed in this section. The following diagram shows the relationships among these sets of real numbers. Notice that, together, the rational numbers and the irrational numbers make up the real numbers.

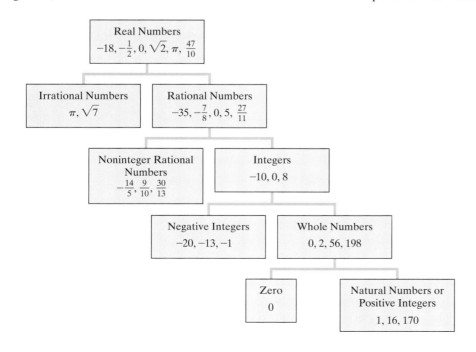

Now that other sets of numbers have been reviewed, let's continue our practice of comparing numbers.

Practice 10
Insert $<, >,$ or $=$ between the pairs of numbers to form true statements.

a. -11 $\quad$ -9
b. 4.511 $\quad$ 4.151
c. $\dfrac{7}{8}$ $\quad$ $\dfrac{2}{3}$

Example 10 Insert $<, >,$ or $=$ between the pairs of numbers to form true statements.

a. -5 $\quad$ -6
b. 3.195 $\quad$ 3.2
c. $\dfrac{1}{4}$ $\quad$ $\dfrac{1}{3}$

Solution:

a. $-5 > -6$ since -5 lies to the right of -6 on a number line.
b. By comparing digits in the same place values, we find that $3.195 < 3.2$, since $0.1 < 0.2$.
c. By dividing, we find that $\dfrac{1}{4} = 0.25$ and $\dfrac{1}{3} = 0.33\ldots$. Since $0.25 < 0.33\ldots$, $\dfrac{1}{4} < \dfrac{1}{3}$.

■ Work Practice 10

Practice 11
Given the set $\left\{-100, -\dfrac{2}{5}, 0, \pi, 6, 913\right\}$, list the numbers in this set that belong to the set of:

a. Natural numbers
b. Whole numbers
c. Integers
d. Rational numbers
e. Irrational numbers
f. Real numbers

Example 11 Given the set $\left\{-3, -2, 0, \dfrac{1}{4}, \sqrt{2}, 11, 112\right\}$, list the numbers in this set that belong to the set of:

a. Natural numbers $\quad$ b. Whole numbers $\quad$ c. Integers
d. Rational numbers $\quad$ e. Irrational numbers $\quad$ f. Real numbers

Solution:

a. The natural numbers are 11 and 112.
b. The whole numbers are 0, 11, and 112.
c. The integers are $-3, -2, 0, 11,$ and 112.
d. Recall that integers are rational numbers also. The rational numbers are $-3, -2, 0, \dfrac{1}{4}, 11,$ and 112.
e. The only irrational number is $\sqrt{2}$.
f. All numbers in the given set are real numbers.

■ Work Practice 11

Answers
10. a. $<$ $\quad$ b. $>$ $\quad$ c. $>$
11. a. $6, 913$ $\quad$ b. $0, 6, 913$
c. $-100, 0, 6, 913$
d. $-100, -\dfrac{2}{5}, 0, 6, 913$ $\quad$ e. π
f. all numbers in the given set

Vocabulary, Readiness & Video Check

Use the choices below to fill in each blank. Not all choices will be used.

| real | natural | whole | $\dfrac{1}{2}$ | $\dfrac{1}{4}$ |
| rational | inequality | integers | 0 | 1 |

1. The _____ numbers are $\{0, 1, 2, 3, 4, \ldots\}$.
2. The _____ numbers are $\{1, 2, 3, 4, 5, \ldots\}$.
3. The symbols $\neq, \leq,$ and $>$ are called _____ symbols.
4. The _____ are $\{\ldots, -3, -2, -1, 0, 1, 2, 3, \ldots\}$.
5. The _____ numbers are $\{$all numbers that correspond to points on a number line$\}$.

Section 9.1 | Symbols and Sets of Numbers

6. The _____ numbers are $\left\{\dfrac{a}{b} \mid a \text{ and } b \text{ are integers}, b \neq 0\right\}$.

7. The integer _____ is neither positive nor negative.

8. The point on a number line halfway between 0 and $\dfrac{1}{2}$ can be represented by _____.

Martin-Gay Interactive Videos Watch the section lecture video and answer the following questions.

Objective A 9. In Example 2, why is the symbol < inserted between the two numbers?

Objective B 10. Write the sentence given in Example 4 and translate it to a mathematical statement using symbols.

Objective C 11. Which sets of numbers does the number in Example 6 belong to? Why is this number not an irrational number?

See Video 9.1

9.1 Exercise Set MyLab Math

Objectives A C Mixed Practice *Insert* <, >, *or* = *in the space between the paired numbers to make each statement true. See Examples 1 through 6 and 10.*

1. 4   10
2. 8   5
3. 7   3
4. 9   15

5. 6.26   6.26
6. 1.13   1.13
7. 0   7
8. 20   0

9. The freezing point of water is 32° Fahrenheit. The boiling point of water is 212° Fahrenheit. Write an inequality statement using < or > comparing the numbers 32 and 212.

10. The freezing point of water is 0° Celsius. The boiling point of water is 100° Celsius. Write an inequality statement using < or > comparing the numbers 0 and 100.

11. An angle measuring 30° and an angle measuring 45° are shown. Write an inequality statement using ≤ or ≥ comparing the numbers 30 and 45.

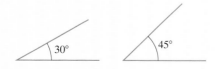

12. The sum of the measures of the angles of a parallelogram is 360°. The sum of the measures of the angles of a triangle is 180°. Write an inequality statement using ≤ or ≥ comparing the numbers 360 and 180.

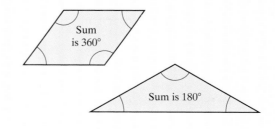

Determine whether each statement is true or false. See Examples 1 through 6 and 10.

13. $11 \leq 11$ **14.** $8 \geq 9$ **15.** $-11 > -10$ **16.** $-16 > -17$

17. $5.092 < 5.902$ **18.** $1.02 > 1.021$ **19.** $\dfrac{9}{10} \leq \dfrac{8}{9}$ **20.** $\dfrac{4}{5} \leq \dfrac{9}{11}$

Rewrite each inequality so that the inequality symbol points in the opposite direction and the resulting statement has the same meaning as the given one. See Examples 1 through 6 and 10.

21. $25 \geq 20$ **22.** $-13 \leq 13$ **23.** $0 < 6$

24. $5 > 3$ **25.** $-10 > -12$ **26.** $-4 < -2$

Objectives **B C** Mixed Practice—Translating *Write each sentence as a mathematical statement. See Example 7.*

27. Seven is less than eleven.

28. Twenty is greater than two.

29. Five is greater than or equal to four.

30. Negative ten is less than or equal to thirty-seven.

31. Fifteen is not equal to negative two.

32. Negative seven is not equal to seven.

Use integers to represent the value(s) in each statement. See Example 8.

33. The highest elevation in California is Mt. Whitney, with an altitude of 14,494 feet. The lowest elevation in California is Death Valley, with an altitude of 282 feet below sea level. (*Source:* U.S. Geological Survey)

34. Driskill Mountain, in Louisiana, has an altitude of 535 feet. New Orleans, Louisiana, lies 8 feet below sea level. (*Source:* U.S. Geological Survey)

35. The number of graduate students at the University of Texas at Austin was 28,288 fewer than the number of undergraduate students. (*Source:* University of Texas at Austin, 2016)

36. The number of students admitted to the class of 2020 at UCLA was 79,647 fewer students than the number that had applied. (*Source:* UCLA, 2016)

37. A community college student deposited $475 in her savings account. She later withdrew $195.

38. A deep-sea diver ascended 17 feet and later descended 15 feet.

Graph each set of numbers on the number line. See Example 9.

39. $-4, 0, 2, -2$

40. $-3, 0, 1, -5$

41. $-2, 4, \dfrac{1}{3}, -\dfrac{1}{4}$

42. $-5, 3, -\dfrac{1}{3}, \dfrac{7}{8}$

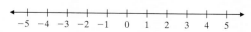

43. $-4.5, \dfrac{7}{4}, 3.25, -\dfrac{3}{2}$

44. $4.5, -\dfrac{9}{4}, 1.75, -\dfrac{7}{2}$

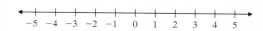

Tell which set or sets each number belongs to: natural numbers, whole numbers, integers, rational numbers, irrational numbers, or real numbers. See Example 11.

45. 0   **46.** $\frac{1}{4}$ **47.** -7 **48.** $-\frac{1}{7}$

49. 265   **50.** 7941   **51.** $\frac{2}{3}$ **52.** $\sqrt{3}$

Determine whether each statement is true or false.

53. Every rational number is also an integer.

54. Every natural number is positive.

55. 0 is a real number.

56. $\frac{1}{2}$ is an integer.

57. Every negative number is also a rational number.

58. Every rational number is also a real number.

59. Every real number is also a rational number.

60. Every whole number is an integer.

Review

Insert $<$, $>$, or $=$ in the appropriate space to make each statement true. See Section 2.1.

61. $|-5|$ -4 **62.** $|-12|$ $|0|$ **63.** $\left|-\frac{5}{8}\right|$ $\left|\frac{5}{8}\right|$ **64.** $\left|\frac{2}{5}\right|$ $\left|-\frac{2}{5}\right|$

65. $|-2|$ $|-2.7|$ **66.** $|-5.01|$ $|-5|$ **67.** $|0|$ $|-8|$ **68.** $|-12|$ $\frac{-24}{2}$

Concept Extensions

The bar graph shows corn production from the top six corn-producing states. (Source: National Agricultural Statistics Service)

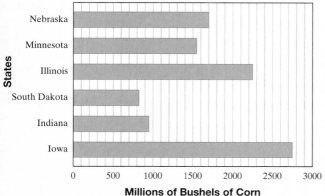

69. Write an inequality comparing the corn production in Illinois with the corn production in Iowa.

70. Write an inequality comparing the corn production in Minnesota with the corn production in South Dakota.

71. Determine the difference between the corn production in Nebraska and the corn production in Illinois.

72. Determine the difference between the corn production in Indiana and the corn production in Minnesota.

The apparent magnitude of a star is the measure of its brightness as seen by someone on Earth. The smaller the apparent magnitude, the brighter the star. Below, the apparent magnitudes of some stars are listed. Use this table to answer Exercises 73 through 78.

Star	Apparent Magnitude	Star	Apparent Magnitude
Arcturus	−0.04	Spica	0.98
Sirius	−1.46	Rigel	0.12
Vega	0.03	Regulus	1.35
Antares	0.96	Canopus	−0.72
Sun (Sol)	−26.7	Hadar	0.61

(*Source: Norton's 2000.0: Star Atlas and Reference Handbook*, 18th ed., Longman Group, UK, 1989)

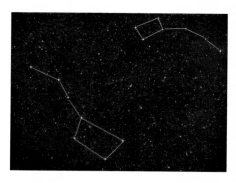

73. The apparent magnitude of the sun is −26.7. The apparent magnitude of the star Arcturus is −0.04. Write an inequality statement comparing the numbers −0.04 and −26.7.

74. The apparent magnitude of Antares is 0.96. The apparent magnitude of Spica is 0.98. Write an inequality statement comparing the numbers 0.96 and 0.98.

75. Which is brighter, the sun or Arcturus?

76. Which is dimmer, Antares or Spica?

77. Which star listed is the brightest?

78. Which star listed is the dimmest?

79. In your own words, explain how to find the absolute value of a number.

80. Give an example of a real-life situation that can be described with integers but not with whole numbers.

9.2 Properties of Real Numbers

Objectives

A Use the Commutative and Associative Properties.

B Use the Distributive Property.

C Use the Identity and Inverse Properties.

Objective A Using the Commutative and Associative Properties

In this section we review names of properties of real numbers with which we are already familiar. Throughout this section, the variables a, b, and c represent real numbers.

We know that order does not matter when adding numbers. For example, we know that $7 + 5$ is the same as $5 + 7$. This property is given a special name—the **commutative property of addition.** We also know that order does not matter when multiplying numbers. For example, we know that $-5(6) = 6(-5)$. This property means that multiplication is commutative also and is called the **commutative property of multiplication.**

Commutative Properties

Addition: $\quad a + b = b + a$

Multiplication: $\quad a \cdot b = b \cdot a$

These properties state that the *order* in which any two real numbers are added or multiplied does not change their sum or product. For example, if we let $a = 3$ and $b = 5$, then the commutative properties guarantee that

$$3 + 5 = 5 + 3 \quad \text{and} \quad 3 \cdot 5 = 5 \cdot 3$$

Helpful Hint

Is subtraction also commutative? Try an example. Is $3 - 2 = 2 - 3$? **No!** The left side of this statement equals 1; the right side equals -1. There is no commutative property of subtraction. Similarly, there is no commutative property of division. For example, $10 \div 2$ does not equal $2 \div 10$.

Example 1 Use a commutative property to complete each statement.

a. $x + 5 =$ _____
b. $3 \cdot x =$ _____

Solution:

a. $x + 5 = 5 + x$ By the commutative property of addition
b. $3 \cdot x = x \cdot 3$ By the commutative property of multiplication

Work Practice 1

Practice 1

Use a commutative property to complete each statement.

a. $7 \cdot y =$ _____
b. $4 + x =$ _____

✓**Concept Check** Which of the following pairs of actions are commutative?
a. "raking the leaves" and "bagging the leaves"
b. "putting on your left glove" and "putting on your right glove"
c. "putting on your coat" and "putting on your shirt"
d. "reading a novel" and "reading a newspaper"

Let's now discuss grouping numbers. When we add three numbers, the way in which they are grouped or associated does not change their sum. For example, we know that $2 + (3 + 4) = 2 + 7 = 9$. This result is the same if we group the numbers differently. In other words, $(2 + 3) + 4 = 5 + 4 = 9$, also. Thus, $2 + (3 + 4) = (2 + 3) + 4$. This property is called the **associative property of addition.**

In the same way, changing the grouping of numbers when multiplying does not change their product. For example, $2 \cdot (3 \cdot 4) = (2 \cdot 3) \cdot 4$ (check it). This is the **associative property of multiplication.**

Associative Properties

Addition: $(a + b) + c = a + (b + c)$
Multiplication: $(a \cdot b) \cdot c = a \cdot (b \cdot c)$

These properties state that the way in which three numbers are *grouped* does not change their sum or their product.

Answers
1. a. $y \cdot 7$ b. $x + 4$

✓**Concept Check Answer**
b, d

Practice 2

Use an associative property to complete each statement.

a. $5 \cdot (-3 \cdot 6) = $ _____
b. $(-2 + 7) + 3 = $ _____
c. $(q + r) + 17 = $ _____
d. $(ab) \cdot 21 = $ _____

Example 2 Use an associative property to complete each statement.

a. $5 + (4 + 6) = $ _____
b. $(-1 \cdot 2) \cdot 5 = $ _____
c. $(m + n) + 9 = $ _____
d. $(xy) \cdot 12 = $ _____

Solution:

a. $5 + (4 + 6) = (5 + 4) + 6$ By the associative property of addition
b. $(-1 \cdot 2) \cdot 5 = -1 \cdot (2 \cdot 5)$ By the associative property of multiplication
c. $(m + n) + 9 = m + (n + 9)$ By the associative property of addition
d. $(xy) \cdot 12 = x \cdot (y \cdot 12)$ Recall that xy means $x \cdot y$.

■ Work Practice 2

> **Helpful Hint**
> Remember the difference between the commutative properties and the associative properties. The commutative properties have to do with the *order* of numbers and the associative properties have to do with the *grouping* of numbers.

Practice 3–4

Determine whether each statement is true by an associative property or a commutative property.

3. $5 \cdot (4 \cdot 7) = 5 \cdot (7 \cdot 4)$
4. $-2 + (4 + 9) = (-2 + 4) + 9$

Examples Determine whether each statement is true by an associative property or a commutative property.

3. $(7 + 10) + 4 = (10 + 7) + 4$ Since the order of two numbers was changed and their grouping was not, this is true by the commutative property of addition.

4. $2 \cdot (3 \cdot 1) = (2 \cdot 3) \cdot 1$ Since the grouping of the numbers was changed and their order was not, this is true by the associative property of multiplication.

■ Work Practice 3–4

Let's now illustrate how these properties can help us simplify expressions.

Practice 5–6

Simplify each expression.

5. $(-3 + x) + 17$ 6. $4(5x)$

Examples Simplify each expression.

5. $10 + (x + 12) = 10 + (12 + x)$ By the commutative property of addition
$= (10 + 12) + x$ By the associative property of addition
$= 22 + x$ Add.

6. $-3(7x) = (-3 \cdot 7)x$ By the associative property of multiplication
$= -21x$ Multiply.

■ Work Practice 5–6

Objective B Using the Distributive Property

The **distributive property of multiplication over addition** is used repeatedly throughout algebra. It is useful because it allows us to write a product as a sum or a sum as a product.

We know that $7(2 + 4) = 7(6) = 42$. Compare that with

$$7(2) + 7(4) = 14 + 28 = 42$$

Since both original expressions equal 42, they must equal each other, or

$$7(2 + 4) = 7(2) + 7(4)$$

Answers

2. a. $(5 \cdot -3) \cdot 6$ b. $-2 + (7 + 3)$
c. $q + (r + 17)$ d. $a \cdot (b \cdot 21)$
3. commutative   4. associative
5. $14 + x$ 6. $20x$

This is an example of the distributive property. The product on the left side of the equal sign is equal to the sum on the right side. We can think of the 7 as being distributed to each number inside the parentheses.

Distributive Property of Multiplication Over Addition

$$a(b + c) = ab + ac$$

Since multiplication is commutative, this property can also be written as

$$(b + c)a = ba + ca$$

The distributive property can also be extended to more than two numbers inside the parentheses. For example,

$$3(x + y + z) = 3(x) + 3(y) + 3(z)$$
$$= 3x + 3y + 3z$$

Since we define subtraction in terms of addition, the distributive property is also true for subtraction. For example,

$$2(x - y) = 2(x) - 2(y)$$
$$= 2x - 2y$$

Examples Use the distributive property to write each expression without parentheses. Then simplify the result.

7. $2(x + y) = 2(x) + 2(y)$
$= 2x + 2y$

8. $-5(-3 + 2z) = -5(-3) + (-5)(2z)$
$= 15 - 10z$

9. $5(x + 3y - z) = 5(x) + 5(3y) - 5(z)$
$= 5x + 15y - 5z$

10. $-1(2 - y) = (-1)(2) - (-1)(y)$
$= -2 + y$

11. $-(3 + x - w) = -1(3 + x - w)$
$= (-1)(3) + (-1)(x) - (-1)(w)$
$= -3 - x + w$

12. $\frac{1}{2}(6x + 14) + 10 = \frac{1}{2}(6x) + \frac{1}{2}(14) + 10$ Apply the distributive property.
$= 3x + 7 + 10$ Multiply.
$= 3x + 17$ Add.

Helpful Hint
Notice in Example 11 that $-(3 + x - w)$ can be rewritten as $-1(3 + x - w)$.

Practice 7–12

Use the distributive property to write each expression without parentheses. Then simplify the result.

7. $5(x + y)$
8. $-3(2 + 7x)$
9. $4(x + 6y - 2z)$
10. $-1(3 - a)$
11. $-(8 + a - b)$
12. $\frac{1}{2}(2x + 4) + 9$

■ Work Practice 7–12

The distributive property can also be used to write a sum as a product.

Answers
7. $5x + 5y$ **8.** $-6 - 21x$
9. $4x + 24y - 8z$ **10.** $-3 + a$
11. $-8 - a + b$ **12.** $x + 11$

Practice 13–14

Use the distributive property to write each sum as a product.

13. $9 \cdot 3 + 9 \cdot y$

14. $6a + 6b$

Examples Use the distributive property to write each sum as a product.

13. $8 \cdot 2 + 8 \cdot x = 8(2 + x)$

14. $7s + 7t = 7(s + t)$

■ Work Practice 13–14

Objective C Using the Identity and Inverse Properties

Next, we look at the **identity properties.**

The number 0 is called the identity for addition because when 0 is added to any real number, the result is the same real number. In other words, the *identity* of the real number is not changed.

The number 1 is called the identity for multiplication because when a real number is multiplied by 1, the result is the same real number. In other words, the *identity* of the real number is not changed.

> **Identities for Addition and Multiplication**
>
> 0 is the identity element for addition.
>
> $$a + 0 = a \quad \text{and} \quad 0 + a = a$$
>
> 1 is the identity element for multiplication.
>
> $$a \cdot 1 = a \quad \text{and} \quad 1 \cdot a = a$$

Notice that 0 is the *only* number that can be added to any real number with the result that the sum is the same real number. Also, 1 is the *only* number that can be multiplied by any real number with the result that the product is the same real number.

Additive inverses or **opposites** were introduced in Section 2.1. Two numbers are called additive inverses or opposites if their sum is 0. The additive inverse or opposite of 6 is -6 because $6 + (-6) = 0$. The additive inverse or opposite of -5 is 5 because $-5 + 5 = 0$.

Reciprocals or **multiplicative inverses** were introduced in Section 4.3. Two nonzero numbers are called reciprocals or multiplicative inverses if their product is 1. The reciprocal or multiplicative inverse of $\frac{2}{3}$ is $\frac{3}{2}$ because $\frac{2}{3} \cdot \frac{3}{2} = 1$. Likewise, the reciprocal of -5 is $-\frac{1}{5}$ because $-5\left(-\frac{1}{5}\right) = 1$.

> **Additive or Multiplicative Inverses**
>
> The numbers a and $-a$ are additive inverses or opposites of each other because their sum is 0; that is,
>
> $$a + (-a) = 0$$
>
> The numbers b and $\frac{1}{b}$ (for $b \neq 0$) are reciprocals or multiplicative inverses of each other because their product is 1; that is,
>
> $$b \cdot \frac{1}{b} = 1$$

Answers

13. $9(3 + y)$ **14.** $6(a + b)$

Section 9.2 Properties of Real Numbers

✓**Concept Check** Which of the following is the
a. opposite of $-\dfrac{3}{10}$, and which is the
b. reciprocal of $-\dfrac{3}{10}$?

$$1, \quad -\dfrac{10}{3}, \quad \dfrac{3}{10}, \quad 0, \quad \dfrac{10}{3}, \quad -\dfrac{3}{10}$$

Examples Name the property illustrated by each true statement.

15. $3(x + y) = 3 \cdot x + 3 \cdot y$ Distributive property
16. $(x + 7) + 9 = x + (7 + 9)$ Associative property of addition (grouping changed)
17. $(b + 0) + 3 = b + 3$ Identity element for addition
18. $2 \cdot (z \cdot 5) = 2 \cdot (5 \cdot z)$ Commutative property of multiplication (order changed)
19. $-2 \cdot \left(-\dfrac{1}{2}\right) = 1$ Multiplicative inverse property
20. $-2 + 2 = 0$ Additive inverse property
21. $-6 \cdot (y \cdot 2) = (-6 \cdot 2) \cdot y$ Commutative and associative properties of multiplication (order and grouping changed)

■ Work Practice 15–21

Practice 15–21
Name the property illustrated by each true statement.
15. $7(a + b) = 7 \cdot a + 7 \cdot b$
16. $12 + y = y + 12$
17. $-4 \cdot (6 \cdot x) = (-4 \cdot 6) \cdot x$
18. $6 + (z + 2) = 6 + (2 + z)$
19. $3\left(\dfrac{1}{3}\right) = 1$
20. $(x + 0) + 23 = x + 23$
21. $(7 \cdot y) \cdot 10 = y \cdot (7 \cdot 10)$

Answers
15. distributive property
16. commutative property of addition
17. associative property of multiplication
18. commutative property of addition
19. multiplicative inverse property
20. identity element for addition
21. commutative and associative properties of multiplication

✓**Concept Check Answers**
a. $\dfrac{3}{10}$ b. $-\dfrac{10}{3}$

Vocabulary, Readiness & Video Check

Use the choices below to fill in each blank.

distributive property   associative property of multiplication   commutative property of addition
opposites or additive inverses   associative property of addition
reciprocals or multiplicative inverses   commutative property of multiplication

1. $x + 5 = 5 + x$ is a true statement by the _____.
2. $x \cdot 5 = 5 \cdot x$ is a true statement by the _____.
3. $3(y + 6) = 3 \cdot y + 3 \cdot 6$ is true by the _____.
4. $2 \cdot (x \cdot y) = (2 \cdot x) \cdot y$ is a true statement by the _____.
5. $x + (7 + y) = (x + 7) + y$ is a true statement by the _____.
6. The numbers $-\dfrac{2}{3}$ and $-\dfrac{3}{2}$ are called _____.
7. The numbers $-\dfrac{2}{3}$ and $\dfrac{2}{3}$ are called _____.

Martin-Gay Interactive Videos

See Video 9.2

Watch the section lecture video and answer the following questions.

Objective A 8. The commutative properties are discussed in Examples 1 and 2 and the associative properties are discussed in Examples 3–7. What's the one word used again and again to describe the commutative property? The associative property?

Objective B 9. In Example 10, what point is made about the term 2?

Objective C 10. Complete these statements based on the lecture given before Example 12:
- The identity element for addition is _____ because if we add _____ to any real number, the result is that real number.
- The identity element for multiplication is _____ because any real number times _____ gives a result of that original real number.

9.2 Exercise Set MyLab Math

Objective A *Use a commutative property to complete each statement. See Examples 1 and 3.*

1. $x + 16 =$ _____
2. $8 + y =$ _____
3. $-4 \cdot y =$ _____
4. $-2 \cdot x =$ _____
5. $xy =$ _____
6. $ab =$ _____
7. $2x + 13 =$ _____
8. $19 + 3y =$ _____

Use an associative property to complete each statement. See Examples 2 and 4.

9. $(xy) \cdot z =$ _____
10. $3 \cdot (x \cdot y) =$ _____
11. $2 + (a + b) =$ _____
12. $(y + 4) + z =$ _____
13. $4 \cdot (ab) =$ _____
14. $(-3y) \cdot z =$ _____
15. $(a + b) + c =$ _____
16. $6 + (r + s) =$ _____

Use the commutative and associative properties to simplify each expression. See Examples 5 and 6.

17. $8 + (9 + b)$
18. $(r + 3) + 11$
19. $4(6y)$
20. $2(42x)$
21. $\frac{1}{5}(5y)$
22. $\frac{1}{8}(8z)$
23. $(13 + a) + 13$
24. $7 + (x + 4)$
25. $-9(8x)$
26. $-3(12y)$
27. $\frac{3}{4}\left(\frac{4}{3}s\right)$
28. $\frac{2}{7}\left(\frac{7}{2}r\right)$
29. $-\frac{1}{2}(5x)$
30. $-\frac{1}{3}(7x)$

Section 9.2 Properties of Real Numbers   675

Objective B *Use the distributive property to write each expression without parentheses. Then simplify the result, if possible. See Examples 7 through 12.*

31. $4(x + y)$ **32.** $7(a + b)$ **33.** $9(x - 6)$ **34.** $11(y - 4)$

35. $2(3x + 5)$ **36.** $5(7 + 8y)$ **37.** $7(4x - 3)$ **38.** $3(8x - 1)$

39. $3(6 + x)$ **40.** $2(x + 5)$ **41.** $-2(y - z)$ **42.** $-3(z - y)$

43. $-\frac{1}{3}(3y + 5)$ **44.** $-\frac{1}{2}(2r + 11)$ **45.** $5(x + 4m + 2)$ **46.** $8(3y + z - 6)$

47. $-4(1 - 2m + n) + 4$ **48.** $-4(4 + 2p + 5) + 16$ **49.** $-(5x + 2)$ **50.** $-(9r + 5)$

51. $-(r - 3 - 7p)$ **52.** $-(q - 2 + 6r)$ **53.** $\frac{1}{2}(6x + 7) + \frac{1}{2}$ **54.** $\frac{1}{4}(4x - 2) - \frac{7}{2}$

55. $-\frac{1}{3}(3x - 9y)$ **56.** $-\frac{1}{5}(10a - 25b)$ **57.** $3(2r + 5) - 7$ **58.** $10(4s + 6) - 40$

59. $-9(4x + 8) + 2$ **60.** $-11(5x + 3) + 10$ **61.** $-0.4(4x + 5) - 0.5$ **62.** $-0.6(2x + 1) - 0.1$

Use the distributive property to write each sum as a product. See Examples 13 and 14.

63. $4 \cdot 1 + 4 \cdot y$ **64.** $14 \cdot z + 14 \cdot 5$ **65.** $11x + 11y$ **66.** $9a + 9b$

67. $(-1) \cdot 5 + (-1) \cdot x$ **68.** $(-3)a + (-3)y$ **69.** $30a + 30b$ **70.** $25x + 25y$

Objectives A C Mixed Practice *Name the property illustrated by each true statement. See Examples 15 through 21.*

71. $3 \cdot 5 = 5 \cdot 3$ **72.** $4(3 + 8) = 4 \cdot 3 + 4 \cdot 8$

73. $2 + (x + 5) = (2 + x) + 5$ **74.** $9 \cdot (x \cdot 7) = (9 \cdot x) \cdot 7$

75. $(x + 9) + 3 = (9 + x) + 3$ **76.** $1 \cdot 9 = 9$

77. $(4 \cdot y) \cdot 9 = 4 \cdot (y \cdot 9)$ **78.** $-4 \cdot (8 \cdot 3) = (8 \cdot 3) \cdot (-4)$

79. $0 + 6 = 6$

80. $(a + 9) + 6 = a + (9 + 6)$

81. $-4(y + 7) = -4 \cdot y + (-4) \cdot 7$

82. $(11 + r) + 8 = (r + 11) + 8$

83. $6 \cdot \dfrac{1}{6} = 1$

84. $r + 0 = r$

85. $-6 \cdot 1 = -6$

86. $-\dfrac{3}{4}\left(-\dfrac{4}{3}\right) = 1$

Review

Evaluate each expression for the given values. See Section 2.5.

87. If $x = -1$ and $y = 3$, find $y - x^2$.

88. If $g = 0$ and $h = -4$, find $gh - h^2$.

89. If $a = 2$ and $b = -5$, find $a - b^2$.

90. If $x = -3$, find $x^3 - x^2 + 4$.

91. If $y = -5$ and $z = 0$, find $yz - y^2$.

92. If $x = -2$, find $x^3 - x^2 - x$.

Concept Extensions

Fill in the table with the opposite (additive inverse), the reciprocal (multiplicative inverse), or the expression. Assume that the value of each expression is not 0.

	93.	94.	95.	96.	97.	98.
Expression	8	$-\dfrac{2}{3}$	x	$4y$		
Opposite						$7x$
Reciprocal					$\dfrac{1}{2x}$	

Decide whether each statement is true or false. See the second Concept Check in this section.

99. The opposite of $-\dfrac{a}{2}$ is $-\dfrac{2}{a}$.

100. The reciprocal of $-\dfrac{a}{2}$ is $\dfrac{a}{2}$.

Determine which pairs of actions are commutative. See the first Concept Check in this section.

101. "taking a test" and "studying for the test"

102. "putting on your shoes" and "putting on your socks"

103. "putting on your left shoe" and "putting on your right shoe"

104. "reading the sports section" and "reading the comics section"

105. "mowing the lawn" and "trimming the hedges"

106. "baking a cake" and "eating the cake"

107. "feeding the dog" and "feeding the cat"

108. "dialing a number" and "turning on the cell phone"

Name the property illustrated by each step.

109.
a. $\triangle + (\square + \bigcirc) = (\square + \bigcirc) + \triangle$
b. $ = (\bigcirc + \square) + \triangle$
c. $ = \bigcirc + (\square + \triangle)$

110.
a. $(x + y) + z = x + (y + z)$
b. $ = (y + z) + x$
c. $ = (z + y) + x$

111. Explain why 0 is called the identity element for addition.

112. Explain why 1 is called the identity element for multiplication.

113. Write an example that shows that division is not commutative.

114. Write an example that shows that subtraction is not commutative.

9.3 Further Solving Linear Equations

Objective A Solving Linear Equations

Since Chapter 2, we have been solving equations. In this section, we give a definition of a linear equation in one variable and review steps we may use to solve these equations.

Let's begin with a formal definition of a linear equation in one variable.

A **linear equation in one variable** can be written in the form

$$Ax + B = C$$

where A, B, and C are real numbers and $A \neq 0$.

We now combine our knowledge from the previous chapters and review solving linear equations.

Objectives

A Apply the General Strategy for Solving a Linear Equation.

B Solve Equations Containing Fractions or Decimals.

C Recognize Identities and Equations with No Solution.

To Solve Linear Equations in One Variable

Step 1: If an equation contains fractions or decimals, multiply both sides by the LCD to clear the equation of fractions or decimals.
Step 2: Use the distributive property to remove parentheses if they are present.
Step 3: Simplify each side of the equation by combining like terms.
Step 4: Get all variable terms on one side and all numbers on the other side by using the addition property of equality.
Step 5: Get the variable alone by using the multiplication property of equality.
Step 6: Check the solution by substituting it into the original equation.

Example 1 Solve: $4(2x - 3) + 7 = 3x + 5$

Solution: There are no fractions, so we begin with Step 2.

$$4(2x - 3) + 7 = 3x + 5$$

Step 2: $8x - 12 + 7 = 3x + 5$ Use the distributive property.
Step 3: $8x - 5 = 3x + 5$ Combine like terms.

(Continued on next page)

Practice 1
Solve:
$5(3x - 1) + 2 = 12x + 6$

Answer
1. $x = 3$

Step 4: Get all variable terms on one side of the equation and all numbers on the other side. One way to do this is by subtracting $3x$ from both sides and then adding 5 to both sides.

$8x - 5 - 3x = 3x + 5 - 3x$ Subtract $3x$ from both sides.
$5x - 5 = 5$ Simplify.
$5x - 5 + 5 = 5 + 5$ Add 5 to both sides.
$5x = 10$ Simplify.

Step 5: Use the multiplication property of equality to get x alone.

$\dfrac{5x}{5} = \dfrac{10}{5}$ Divide both sides by 5.
$x = 2$ Simplify.

Step 6: Check.

$4(2x - 3) + 7 = 3x + 5$ Original equation
$4[2(2) - 3] + 7 \stackrel{?}{=} 3(2) + 5$ Replace x with 2.
$4(4 - 3) + 7 \stackrel{?}{=} 6 + 5$
$4(1) + 7 \stackrel{?}{=} 11$
$4 + 7 \stackrel{?}{=} 11$
$11 = 11$ True

The solution is 2.

Helpful Hint: When checking solutions, use the original equation.

■ Work Practice 1

Practice 2

Solve: $9(5 - x) = -3x$

Example 2 Solve: $8(2 - t) = -5t$

Solution: First, we apply the distributive property.

$8(2 - t) = -5t$

Step 2: $16 - 8t = -5t$ Use the distributive property.
Step 4: $16 - 8t + 8t = -5t + 8t$ Add $8t$ to both sides.
$16 = 3t$ Combine like terms.
Step 5: $\dfrac{16}{3} = \dfrac{3t}{3}$ Divide both sides by 3.
$\dfrac{16}{3} = t$ Simplify.

Step 6: Check.

$8(2 - t) = -5t$ Original equation
$8\left(2 - \dfrac{16}{3}\right) \stackrel{?}{=} -5\left(\dfrac{16}{3}\right)$ Replace t with $\dfrac{16}{3}$.
$8\left(\dfrac{6}{3} - \dfrac{16}{3}\right) \stackrel{?}{=} -\dfrac{80}{3}$ The LCD is 3.
$8\left(-\dfrac{10}{3}\right) \stackrel{?}{=} -\dfrac{80}{3}$ Subtract fractions.
$-\dfrac{80}{3} = -\dfrac{80}{3}$ True

The solution is $\dfrac{16}{3}$.

■ Work Practice 2

Answer

2. $x = \dfrac{15}{2}$

Section 9.3 | Further Solving Linear Equations

Objective B Solving Equations Containing Fractions or Decimals

If an equation contains fractions, we can clear the equation of fractions by multiplying both sides by the LCD of all denominators. By doing this, we avoid working with time-consuming fractions.

Example 3 Solve: $\dfrac{x}{2} - 1 = \dfrac{2}{3}x - 3$

Solution: We begin by clearing fractions. To do this, we multiply both sides of the equation by the LCD, which is 6.

$$\dfrac{x}{2} - 1 = \dfrac{2}{3}x - 3$$

Step 1: $6\left(\dfrac{x}{2} - 1\right) = 6\left(\dfrac{2}{3}x - 3\right)$ Multiply both sides by the LCD, 6.

Step 2: $6\left(\dfrac{x}{2}\right) - 6(1) = 6\left(\dfrac{2}{3}x\right) - 6(3)$ Use the distributive property.

$$3x - 6 = 4x - 18$$ Simplify.

Helpful Hint: Don't forget to multiply *each* term by the LCD.

There are no longer grouping symbols and no like terms on either side of the equation, so we continue with Step 4.

$$3x - 6 = 4x - 18$$

Step 4: $3x - 6 - 3x = 4x - 18 - 3x$ Subtract $3x$ from both sides.

$$-6 = x - 18$$ Simplify.

$$-6 + 18 = x - 18 + 18$$ Add 18 to both sides.

$$12 = x$$ Simplify.

Step 5: The variable is now alone, so there is no need to apply the multiplication property of equality.

Step 6: Check.

$$\dfrac{x}{2} - 1 = \dfrac{2}{3}x - 3$$ Original equation

$$\dfrac{12}{2} - 1 \stackrel{?}{=} \dfrac{2}{3} \cdot 12 - 3$$ Replace x with 12.

$$6 - 1 \stackrel{?}{=} 8 - 3$$ Simplify.

$$5 = 5$$ True

The solution is 12.

Work Practice 3

Practice 3

Solve: $\dfrac{5}{2}x - 1 = \dfrac{3}{2}x - 4$

Example 4 Solve: $\dfrac{2(a + 3)}{3} = 6a + 2$

Solution: We clear the equation of fractions first.

$$\dfrac{2(a + 3)}{3} = 6a + 2$$

Step 1: $3 \cdot \dfrac{2(a + 3)}{3} = 3(6a + 2)$ Clear the fraction by multiplying both sides by the LCD, 3.

$$2(a + 3) = 3(6a + 2)$$ Simplify.

(Continued on next page)

Practice 4

Solve: $\dfrac{3(x - 2)}{5} = 3x + 6$

Answers

3. $x = -3$ 4. $x = -3$

Step 2: Next, we use the distributive property to remove parentheses.

$$2a + 6 = 18a + 6 \quad \text{Use the distributive property.}$$

Step 4: $2a + 6 - 18a = 18a + 6 - 18a \quad$ Subtract 18a from both sides.

$$-16a + 6 = 6 \quad \text{Simplify.}$$
$$-16a + 6 - 6 = 6 - 6 \quad \text{Subtract 6 from both sides.}$$
$$-16a = 0$$

Step 5: $\dfrac{-16a}{-16} = \dfrac{0}{-16} \quad$ Divide both sides by -16.

$$a = 0 \quad \text{Simplify.}$$

Step 6: To check, replace a with 0 in the original equation. The solution is 0.

■ Work Practice 4

Helpful Hint

Remember: When solving an equation, it makes no difference on which side of the equation variable terms lie. Just make sure that constant terms lie on the other side.

When solving a problem about money, you may need to solve an equation containing decimals. If you choose, you may multiply to clear the equation of decimals.

Practice 5

Solve:
$0.06x - 0.10(x - 2) = -0.16$

Helpful Hint

If you have trouble with this step, try removing parentheses first.

$0.25x + 0.10(x - 3) = 1.1$
$0.25x + 0.10x - 0.3 = 1.1$
$0.25x + 0.10x - 0.30 = 1.10$
$25x + 10x - 30 = 110$

Then continue.

Example 5
Solve: $0.25x + 0.10(x - 3) = 1.1$

Solution: First we clear this equation of decimals by multiplying both sides of the equation by 100. Recall that multiplying a decimal number by 100 has the effect of moving the decimal point 2 places to the right.

$$0.25x + 0.10(x - 3) = 1.1$$

Step 1: $0.25x + 0.10(x - 3) = 1.10 \quad$ Multiply both sides by 100.
$$25x + 10(x - 3) = 110$$

Step 2: $25x + 10x - 30 = 110 \quad$ Apply the distributive property.

Step 3: $35x - 30 = 110 \quad$ Combine like terms.

Step 4: $35x - 30 + 30 = 110 + 30 \quad$ Add 30 to both sides.
$$35x = 140 \quad \text{Combine like terms.}$$

Step 5: $\dfrac{35x}{35} = \dfrac{140}{35} \quad$ Divide both sides by 35.

$$x = 4$$

Step 6: To check, replace x with 4 in the original equation. The solution is 4.

■ Work Practice 5

Objective C Recognizing Identities and Equations with No Solution

So far, each equation that we have solved has had a single solution. However, not every equation in one variable has a single solution. Some equations have no solution, while others have an infinite number of solutions. For example,

$$x + 5 = x + 7$$

Answer
5. $x = 9$

has **no solution** since no matter which real number we replace x with, the equation is false.

real number + 5 = same real number + 7   FALSE

On the other hand,

$$x + 6 = x + 6$$

has infinitely many solutions since x can be replaced by any real number and the equation will always be true.

real number + 6 = same real number + 6   TRUE

The equation $x + 6 = x + 6$ is called an **identity.** The next two examples illustrate special equations like these.

Example 6 Solve: $-2(x - 5) + 10 = -3(x + 2) + x$

Solution:

$$-2(x - 5) + 10 = -3(x + 2) + x$$
$$-2x + 10 + 10 = -3x - 6 + x \quad \text{Apply the distributive property on both sides.}$$
$$-2x + 20 = -2x - 6 \quad \text{Combine like terms.}$$
$$-2x + 20 + 2x = -2x - 6 + 2x \quad \text{Add } 2x \text{ to both sides.}$$
$$20 = -6 \quad \text{Combine like terms.}$$

The final equation contains no variable terms, and the result is the false statement $20 = -6$. There is no value for x that makes $20 = -6$ a true equation. Thus, we conclude that there is **no solution** to this equation.

■ Work Practice 6

Practice 6

Solve:

$$5(2 - x) + 8x = 3(x - 6)$$

Example 7 Solve: $3(x - 4) = 3x - 12$

Solution: $3(x - 4) = 3x - 12$

$$3x - 12 = 3x - 12 \quad \text{Apply the distributive property.}$$

The left side of the equation is now identical to the right side. Every real number may be substituted for x and a true statement will result. We arrive at the same conclusion if we continue.

$$3x - 12 = 3x - 12$$
$$3x - 12 - 3x = 3x - 12 - 3x \quad \text{Subtract } 3x \text{ from both sides.}$$
$$-12 = -12 \quad \text{Combine like terms.}$$

Again, the final equation contains no variables, but this time the result is the true statement $-12 = -12$. This means that one side of the equation is identical to the other side. Thus, $3(x - 4) = 3x - 12$ is an **identity** and **all real numbers** are solutions.

■ Work Practice 7

Practice 7

Solve:

$$-6(2x + 1) - 14 = -10(x + 2) - 2x$$

✓ **Concept Check** Suppose you have simplified several equations and obtained the following results. What can you conclude about the solutions to the original equation?

a. $7 = 7$ **b.** $x = 0$ **c.** $7 = -4$

Answers

6. no solution
7. All real numbers are solutions.

✓ **Concept Check Answer**

a. All real numbers are solutions.
b. The solution is 0.
c. There is no solution.

Calculator Explorations — Checking Equations

We can use a calculator to check possible solutions of equations. To do this, replace the variable by the possible solution and evaluate each side of the equation separately.

Equation: $3x - 4 = 2(x + 6)$ Solution: $x = 16$

$$3x - 4 = 2(x + 6)$$
$$3(16) - 4 \stackrel{?}{=} 2(16 + 6)$$

Now evaluate each side with your calculator.

Evaluate left side: $\boxed{3}\ \boxed{\times}\ \boxed{16}\ \boxed{-}\ \boxed{4}\ \boxed{=}$

or

$\boxed{\text{ENTER}}$

Display: $\boxed{44}$

Evaluate right side: $\boxed{2}\ \boxed{(}\ \boxed{16}\ \boxed{+}\ \boxed{6}\ \boxed{)}\ \boxed{=}$

or

$\boxed{\text{ENTER}}$

Display: $\boxed{44}$

Since the left side equals the right side, the equation checks.

Use a calculator to check the possible solutions to each equation.

1. $2x = 48 + 6x$; $x = -12$
2. $-3x - 7 = 3x - 1$; $x = -1$
3. $5x - 2.6 = 2(x + 0.8)$; $x = 4.4$
4. $-1.6x - 3.9 = -6.9x - 25.6$; $x = 5$
5. $\dfrac{564x}{4} = 200x - 11(649)$; $x = 121$
6. $20(x - 39) = 5x - 432$; $x = 23.2$

Vocabulary, Readiness & Video Check

Throughout algebra, it is important to be able to distinguish between equations and expressions.

Remember,
- an equation contains an equal sign and
- an expression does not.

Among other things,
- we solve equations and
- we simplify or perform operations on expressions.

Identify each as an equation or an expression.

1. $x = -7$ _____
2. $x - 7$ _____
3. $4y - 6 + 9y + 1$ _____
4. $4y - 6 = 9y + 1$ _____
5. $\dfrac{1}{x} - \dfrac{x-1}{8}$ _____
6. $\dfrac{1}{x} - \dfrac{x-1}{8} = 6$ _____
7. $0.1x + 9 = 0.2x$ _____
8. $0.1x^2 + 9y - 0.2x^2$ _____

Section 9.3 | Further Solving Linear Equations   683

Martin-Gay Interactive Videos   Watch the section lecture video and answer the following questions.

See Video 9.3

Objective A 9. The general strategy for solving linear equations in one variable is discussed after Example 1. How many properties (not steps) are mentioned in this strategy and what are they?

Objective B 10. In the first step for solving Example 2, both sides of the equation are multiplied by the LCD. Why is the distributive property mentioned?

11. In Example 3, why is the number of decimal places in each term of the equation important?

Objective C 12. Complete each statement based on Examples 4 and 5. When solving an equation and all variable terms subtract out:
  a. If we have a true statement, then the equation has _____ solution(s).
  b. If we have a false statement, then the equation has _____ solution(s).

9.3 Exercise Set MyLab Math

Objective A *Solve each equation. See Examples 1 and 2.*

1. $-4y + 10 = -2(3y + 1)$
2. $-3x + 1 = -2(4x + 2)$
3. $15x - 8 = 10 + 9x$
4. $15x - 5 = 7 + 12x$
5. $-2(3x - 4) = 2x$
6. $-(5x - 10) = 5x$
7. $5(2x - 1) - 2(3x) = 1$
8. $3(2 - 5x) + 4(6x) = 12$
9. $-6(x - 3) - 26 = -8$
10. $-4(n - 4) - 23 = -7$
11. $8 - 2(a + 1) = 9 + a$
12. $5 - 6(2 + b) = b - 14$
13. $4x + 3 = -3 + 2x + 14$
14. $6y - 8 = -6 + 3y + 13$
15. $-2y - 10 = 5y + 18$
16. $-7n + 5 = 8n - 10$

Objective B *Solve each equation. See Examples 3 through 5.*

17. $\frac{2}{3}x + \frac{4}{3} = -\frac{2}{3}$
18. $\frac{4}{5}x - \frac{8}{5} = -\frac{16}{5}$
19. $\frac{3}{4}x - \frac{1}{2} = 1$
20. $\frac{2}{9}x - \frac{1}{3} = 1$
21. $0.50x + 0.15(70) = 35.5$
22. $0.40x + 0.06(30) = 9.8$
23. $\frac{2(x + 1)}{4} = 3x - 2$
24. $\frac{3(y + 3)}{5} = 2y + 6$
25. $x + \frac{7}{6} = 2x - \frac{7}{6}$
26. $\frac{5}{2}x - 1 = x + \frac{1}{4}$
27. $0.12(y - 6) + 0.06y = 0.08y - 0.7$
28. $0.60(z - 300) + 0.05z = 0.70z - 205$

Objective C *Solve each equation. See Examples 6 and 7.*

29. $4(3x + 2) = 12x + 8$

30. $14x + 7 = 7(2x + 1)$

31. $\dfrac{x}{4} + 1 = \dfrac{x}{4}$

32. $\dfrac{x}{3} - 2 = \dfrac{x}{3}$

33. $3x - 7 = 3(x + 1)$

34. $2(x - 5) = 2x + 10$

35. $-2(6x - 5) + 4 = -12x + 14$

36. $-5(4y - 3) + 2 = -20y + 17$

Objectives A B C Mixed Practice *Solve. See Examples 1 through 7.*

37. $\dfrac{6(3 - z)}{5} = -z$

38. $\dfrac{4(5 - w)}{3} = -w$

39. $-3(2t - 5) + 2t = 5t - 4$

40. $-(4a - 7) - 5a = 10 + a$

41. $5y + 2(y - 6) = 4(y + 1) - 2$

42. $9x + 3(x - 4) = 10(x - 5) + 7$

43. $\dfrac{3(x - 5)}{2} = \dfrac{2(x + 5)}{3}$

44. $\dfrac{5(x - 1)}{4} = \dfrac{3(x + 1)}{2}$

45. $0.7x - 2.3 = 0.5$

46. $0.9x - 4.1 = 0.4$

47. $5x - 5 = 2(x + 1) + 3x - 7$

48. $3(2x - 1) + 5 = 6x + 2$

49. $4(2n + 1) = 3(6n + 3) + 1$

50. $4(4y + 2) = 2(1 + 6y) + 8$

51. $x + \dfrac{5}{4} = \dfrac{3}{4}x$

52. $\dfrac{7}{8}x + \dfrac{1}{4} = \dfrac{3}{4}x$

53. $\dfrac{x}{2} - 1 = \dfrac{x}{5} + 2$

54. $\dfrac{x}{5} - 7 = \dfrac{x}{3} - 5$

55. $2(x + 3) - 5 = 5x - 3(1 + x)$

56. $4(2 + x) + 1 = 7x - 3(x - 2)$

57. $0.06 - 0.01(x + 1) = -0.02(2 - x)$

58. $-0.01(5x + 4) = 0.04 - 0.01(x + 4)$

59. $\dfrac{9}{2} + \dfrac{5}{2}y = 2y - 4$

60. $3 - \dfrac{1}{2}x = 5x - 8$

Review

Translating *Write each algebraic expression described. See Section 3.1.*

△ **61.** A plot of land is in the shape of a triangle. If one side is x meters, a second side is $(2x - 3)$ meters, and a third side is $(3x - 5)$ meters, express the perimeter of the lot as a simplified expression in x.

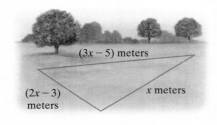

62. A portion of a board has length x feet. The other part has length $(7x - 9)$ feet. Express the total length of the board as a simplified expression in x.

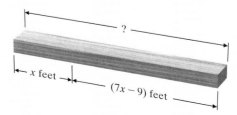

Section 9.3 Further Solving Linear Equations

Translating *Write each phrase as an algebraic expression. Use x for the unknown number. See Section 3.2.*

63. A number subtracted from -8
64. Three times a number
65. The sum of -3 and twice a number
66. The difference of 8 and twice a number
67. The product of 9 and the sum of a number and 20
68. The quotient of -12 and the difference of a number and 3

Concept Extensions

See the Concept Check in this section.

69. **a.** Solve: $x + 3 = x + 3$
    b. If you simplify an equation (such as the one in part **a**) and get a true statement such as $3 = 3$ or $0 = 0$, what can you conclude about the solution(s) of the original equation?
    c. On your own, construct an equation for which every real number is a solution.

70. **a.** Solve: $x + 3 = x + 5$
    b. If you simplify an equation (such as the one in part **a**) and get a false statement such as $3 = 5$ or $10 = 17$, what can you conclude about the solution(s) of the original equation?
    c. On your own, construct an equation that has no solution.

Match each equation in the first column with its solution in the second column. Items in the second column may be used more than once.

71. $5x + 1 = 5x + 1$
72. $3x + 1 = 3x + 2$
73. $2x - 6x - 10 = -4x + 3 - 10$
74. $x - 11x - 3 = -10x - 1 - 2$
75. $9x - 20 = 8x - 20$
76. $-x + 15 = x + 15$

a. all real numbers
b. no solution
c. 0

77. Explain the difference between simplifying an expression and solving an equation.

78. On your own, write an expression and then an equation. Label each.

*For Exercises 79 and 80, **a.** Write an equation for perimeter. **b.** Solve the equation in part (a). **c.** Find the length of each side.*

79. The perimeter of a geometric figure is the sum of the lengths of its sides. If the perimeter of the following pentagon (five-sided figure) is 28 centimeters, find the length of each side.

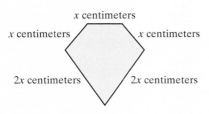

80. The perimeter of the following triangle is 35 meters. Find the length of each side.

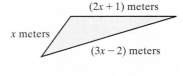

Fill in the blanks with numbers of your choice so that each equation has the given solution. Note: Each blank may be replaced by a different number.

81. $x + \underline{} = 2x - \underline{}$; solution: 9
82. $-5x - \underline{} = \underline{}$; solution: 2

Solve.

83. $1000(7x - 10) = 50(412 + 100x)$
84. $1000(x + 40) = 100(16 + 7x)$
85. $0.035x + 5.112 = 0.010x + 5.107$
86. $0.127x - 2.685 = 0.027x - 2.38$

Integrated Review — Sections 9.1–9.3

Real Numbers and Solving Linear Equations

Tell which set or sets each number belongs to: natural numbers, whole numbers, integers, rational numbers, irrational numbers, or real numbers.

1. 0
2. 143
3. $\dfrac{3}{8}$
4. 1
5. -13
6. $\dfrac{9}{10}$
7. $-\dfrac{1}{9}$
8. $\sqrt{5}$

Remove parentheses and simplify each expression.

9. $7(d-3)+10$
10. $9(z+7)-15$
11. $-4(3y-4)+12y$
12. $-3(2x+5)-6x$

Solve. Feel free to use the steps given in Section 9.3.

13. $2x - 7 = 6x - 27$
14. $3 + 8y = 3y - 2$
15. $-3a + 6 + 5a = 7a - 8a$
16. $4b - 8 - b = 10b - 3b$
17. $-\dfrac{2}{3}x = \dfrac{5}{9}$
18. $-\dfrac{3}{8}y = -\dfrac{1}{16}$
19. $10 = -6n + 16$
20. $-5 = -2m + 7$

Chapter 9 Integrated Review

1. $3(5c - 1) - 2 = 13c + 3$

22. $4(3t + 4) - 20 = 3 + 5t$

3. $\dfrac{2(z + 3)}{3} = 5 - z$

24. $\dfrac{3(w + 2)}{4} = 2w + 3$

5. $-2(2x - 5) = -3x + 7 - x + 3$

26. $-4(5x - 2) = -12x + 4 - 8x + 4$

7. $0.02(6t - 3) = 0.04(t - 2) + 0.02$

28. $0.03(m + 7) = 0.02(5 - m) + 0.11$

9. $-3y = \dfrac{4(y - 1)}{5}$

30. $-4x = \dfrac{5(1 - x)}{6}$

1. $\dfrac{1}{10}(3x - 7) = \dfrac{3}{10}x + 5$

32. $\dfrac{1}{7}(2x - 5) = \dfrac{2}{7}x + 1$

21. _____

22. _____

23. _____

24. _____

25. _____

26. _____

27. _____

28. _____

29. _____

30. _____

31. _____

32. _____

9.4 Further Problem Solving

Objectives

A Solve Problems Involving Direct Translations.

B Solve Problems Involving Relationships Among Unknown Quantities.

C Solve Problems Involving Consecutive Integers.

In this section, we review our problem-solving steps first introduced in Section 3 and continue to solve problems that are modeled by linear equations.

General Strategy for Problem Solving

1. UNDERSTAND the problem. During this step, become comfortable with the problem. Some ways of doing this are:
   Read and reread the problem.
   Choose a variable to represent the unknown.
   Construct a drawing.
   Propose a solution and check. Pay careful attention to how you check your proposed solution. This will help when writing an equation to model the problem.
2. TRANSLATE the problem into an equation.
3. SOLVE the equation.
4. INTERPRET the results: *Check* the proposed solution in the stated problem and *state* your conclusion.

Objective A Solving Direct Translation Problems

Much of problem solving involves a direct translation from a sentence to an equation.

Practice 1

Three times a number, minus 6, is the same as two times the number, plus 3. Find the number.

Example 1 Finding an Unknown Number

Twice a number, added to seven, is the same as three subtracted from the number. Find the number.

Solution: Translate the sentence into an equation and solve.

In words:	twice a number	added to	seven	is the same as	three subtracted from the number
Translate:	$2x$	$+$	7	$=$	$x - 3$

To solve, begin by subtracting x from both sides to isolate the variable term.

$$2x + 7 = x - 3$$
$$2x + 7 - x = x - 3 - x \quad \text{Subtract } x \text{ from both sides.}$$
$$x + 7 = -3 \quad \text{Combine like terms.}$$
$$x + 7 - 7 = -3 - 7 \quad \text{Subtract 7 from both sides.}$$
$$x = -10 \quad \text{Combine like terms.}$$

Check the solution in the problem as it was originally stated. To do so, replace "number" in the sentence with -10. Twice "-10" added to 7 is the same as 3 subtracted from "-10."

$$2(-10) + 7 \stackrel{?}{=} -10 - 3$$
$$-13 = -13$$

The unknown number is -10.

■ Work Practice 1

Answer
1. 9

Section 9.4 | Further Problem Solving

Helpful Hint

When checking solutions, go back to the original stated problem, rather than to your equation in case errors have been made in translating to an equation.

Example 2 Finding an Unknown Number

Twice the sum of a number and 4 is the same as four times the number decreased by 12. Find the number.

Solution:

1. UNDERSTAND. Read and reread the problem. If we let $x =$ the unknown number, then

   "the sum of a number and 4" translates to "$x + 4$" and
   "four times the number" translates to "$4x$"

2. TRANSLATE.

twice	the sum of a number and 4	is the same as	four times the number	decreased by	12
↓	↓	↓	↓	↓	↓
2	$(x + 4)$	=	$4x$	−	12

3. SOLVE

   $2(x + 4) = 4x - 12$
   $2x + 8 = 4x - 12$ Apply the distributive property.
   $2x + 8 - 4x = 4x - 12 - 4x$ Subtract $4x$ from both sides.
   $-2x + 8 = -12$
   $-2x + 8 - 8 = -12 - 8$ Subtract 8 from both sides.
   $-2x = -20$
   $\dfrac{-2x}{-2} = \dfrac{-20}{-2}$ Divide both sides by -2.
   $x = 10$

4. INTERPRET.

   Check: Check this solution in the problem as it was originally stated. To do so, replace "number" with 10. Twice the sum of "10" and 4 is 28, which is the same as 4 times "10" decreased by 12.

   State: The number is 10.

■ Work Practice 2

Practice 2

Three times the difference of a number and 5 is the same as twice the number decreased by 3. Find the number.

Objective B Solving Problems Involving Relationships Among Unknown Quantities

Example 3 Finding the Length of a Board

A 10-foot board is to be cut into two pieces so that the length of the longer piece is 4 times the length of the shorter. Find the length of each piece.

Solution:

1. UNDERSTAND the problem. To do so, read and reread the problem. You may also want to propose a solution. For example, if 3 feet represents the length of the shorter piece, then $4(3) = 12$ feet is the length of the longer piece, since it is 4 times the length of the shorter piece. This

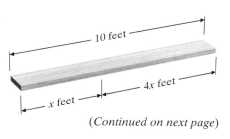

(*Continued on next page*)

Practice 3

An 18-foot wire is to be cut so that the length of the longer piece is 5 times the length of the shorter piece. Find the length of each piece.

Answers

2. 12
3. shorter piece: 3 feet; longer piece: 15 feet

guess gives a total board length of 3 feet + 12 feet = 15 feet, which is too long. However, the purpose of proposing a solution is not to guess correctly, but to help better understand the problem and how to model it.

In general, if we let

x = length of shorter piece, then

$4x$ = length of longer piece

2. **TRANSLATE** the problem. First, we write the equation in words.

length of shorter piece	added to	length of longer piece	equals	total length of board
↓	↓	↓	↓	↓
x	$+$	$4x$	$=$	10

3. **SOLVE.**

$x + 4x = 10$

$5x = 10$ Combine like terms.

$\dfrac{5x}{5} = \dfrac{10}{5}$ Divide both sides by 5.

$x = 2$

4. **INTERPRET.**

Check: Check the solution in the stated problem. If the length of the shorter piece of board is 2 feet, the length of the longer piece is $4 \cdot (2 \text{ feet}) = 8$ feet and the sum of the lengths of the two pieces is 2 feet + 8 feet = 10 feet.

State: The shorter piece of board is 2 feet and the longer piece of board is 8 feet.

■ Work Practice 3

Helpful Hint

Make sure that units are included in your answer, if appropriate.

Practice 4

Through the year 2020, the state of California will have 17 more electoral votes for president than the state of Texas. If the total electoral votes for these two states is 93, find the number of electoral votes for each state.

Example 4   Finding the Number of Republican and Democratic Representatives

As of January 2017, the total number of Democrats and Republicans in the U.S. House of Representatives was 435. There were 47 more Republican representatives than Democratic. Find the number of representatives from each party. (*Source:* Congressional Research Service)

Solution:

1. **UNDERSTAND** the problem. Read and re-read the problem. Let's suppose that there are 200 Democratic representatives. Since there are 47 more Republicans than Democrats, there must be 200 + 47 = 247 Republicans. The total number of Democrats and Republicans is then 200 + 247 = 447. This is incorrect since the total should be 435, but we now have a better understanding of the problem.

   In general, if we let

   x = number of Democrats, then

   $x + 47$ = number of Republicans

Answer

4. Texas: 38 electoral votes; California: 55 electoral votes

2. TRANSLATE the problem. First, we write the equation in words.

number of Democrats	added to	number of Republicans	equals	435
↓	↓	↓	↓	↓
x	$+$	$(x + 47)$	$=$	435

3. SOLVE.

$$x + (x + 47) = 435$$
$$2x + 47 = 435 \quad \text{Combine like terms.}$$
$$2x + 47 - 47 = 435 - 47 \quad \text{Subtract 47 from both sides.}$$
$$2x = 388$$
$$\frac{2x}{2} = \frac{388}{2} \quad \text{Divide both sides by 2.}$$
$$x = 194$$

4. INTERPRET.

Check: If there were 194 Democratic representatives, then there were $194 + 47 = 241$ Republican representatives. The total number of representatives is then $194 + 241 = 435$. The results check.

State: There were 194 Democratic and 241 Republican representatives in Congress in January 2017.

■ Work Practice 4

Example 5 Calculating Hours on the Job

A computer science major at a local university has a part-time job working on computers for his clients. He charges $20 to go to your home or office and then $25 per hour. During one month he visited 10 homes or offices and his total income was $575. How many hours did he spend working on computers?

Solution:

1. UNDERSTAND. Read and reread the problem. Let's propose that the student spent 20 hours working on computers. Pay careful attention as to how his income is calculated. For 20 hours and 10 visits, his income is $20(\$25) + 10(\$20) = \$700$, which is more than $575. We now have a better understanding of the problem and know that the time spent working on computers was less than 20 hours.

Let's let

$x =$ hours working on computers. Then

$25x =$ amount of money made while working on computers

2. TRANSLATE.

money made while working on computers	plus	money made for visits	is equal to	575
↓	↓	↓	↓	↓
$25x$	$+$	$10(20)$	$=$	575

3. SOLVE.

$$25x + 200 = 575$$
$$25x + 200 - 200 = 575 - 200 \quad \text{Subtract 200 from both sides.}$$
$$25x = 375 \quad \text{Simplify.}$$
$$\frac{25x}{25} = \frac{375}{25} \quad \text{Divide both sides by 25.}$$
$$x = 15 \quad \text{Simplify.}$$

(Continued on next page)

Practice 5

A car rental agency charges $28 a day and $0.15 a mile. If you rent a car for a day and your bill (before taxes) is $52, how many miles did you drive?

Answer
5. 160 miles

4. **INTERPRET.**

   Check: If the student works 15 hours and makes 10 visits, his income is $15(\$25) + 10(\$20) = \$575$.

   State: The student spent 15 hours working on computers.

◼ **Work Practice 5**

Practice 6

The measure of the second angle of a triangle is twice the measure of the smallest angle. The measure of the third angle of the triangle is three times the measure of the smallest angle. Find the measures of the angles.

Example 6 Finding Angle Measures

If the two walls of the Vietnam Veterans Memorial in Washington, D.C., were connected, an isosceles triangle would be formed. The measure of the third angle is 97.5° more than the measure of either of the two equal angles. Find the measure of the third angle. (*Source:* National Park Service)

Solution:

1. **UNDERSTAND.** Read and reread the problem. We then draw a diagram (recall that an isosceles triangle has two angles with the same measure) and let

   x = degree measure of one angle
   x = degree measure of the second, equal angle
   $x + 97.5$ = degree measure of the third angle

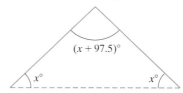

2. **TRANSLATE.** Recall that the sum of the measures of the angles of a triangle equals 180.

measure of first angle	+	measure of second angle	+	measure of third angle	equals	180
↓		↓		↓	↓	↓
x	+	x	+	$(x + 97.5)$	=	180

3. **SOLVE.**

   $x + x + (x + 97.5) = 180$
   $3x + 97.5 = 180$ Combine like terms.
   $3x + 97.5 - 97.5 = 180 - 97.5$ Subtract 97.5 from both sides.
   $3x = 82.5$
   $\dfrac{3x}{3} = \dfrac{82.5}{3}$ Divide both sides by 3.
   $x = 27.5$

4. **INTERPRET.**

   Check: If $x = 27.5$, then the measure of the third angle is $x + 97.5 = 125$. The sum of the angles is then $27.5 + 27.5 + 125 = 180$, the correct sum.

   State: The third angle measures 125°.*

◼ **Work Practice 6**

Answer

6. smallest: 30°; second: 60°; third: 90°

*The two walls actually meet at an angle of 125 degrees 12 minutes. The measurement of 97.5° given in the problem is an approximation.

Objective C Solving Consecutive Integer Problems

The next example has to do with consecutive integers.

	Example	General Representation
Consecutive Integers	11, 12, 13 (+1, +1)	Let x be an integer. $\quad x,\ x+1,\ x+2$ (+1, +1)
Consecutive Even Integers	38, 40, 42 (+2, +2)	Let x be an even integer. $\quad x,\ x+2,\ x+4$ (+2, +2)
Consecutive Odd Integers	57, 59, 61 (+2, +2)	Let x be an odd integer. $\quad x,\ x+2,\ x+4$ (+2, +2)

Example 7 Finding Area Codes

Some states have a single area code for the entire state. Two such states have area codes that are consecutive odd integers. If the sum of these integers is 1208, find the two area codes. (*Source: World Almanac*)

Solution:

1. UNDERSTAND. Read and reread the problem. If we let

   x = the first odd integer, then

   $x + 2$ = the next odd integer

2. TRANSLATE.

first odd integer	the sum of	next odd integer	is	1208
↓	↓	↓	↓	↓
x	$+$	$(x + 2)$	$=$	1208

3. SOLVE.

   $x + x + 2 = 1208$

   $2x + 2 = 1208$

   $2x + 2 - 2 = 1208 - 2$

   $2x = 1206$

   $\dfrac{2x}{2} = \dfrac{1206}{2}$

   $x = 603$

4. INTERPRET.

   Check: If $x = 603$, then the next odd integer $x + 2 = 603 + 2 = 605$. Notice their sum, $603 + 605 = 1208$, as needed.

   State: The area codes are 603 and 605.

   Note: New Hampshire's area code is 603 and South Dakota's area code is 605.

Work Practice 7

Practice 7

The sum of three consecutive even integers is 144. Find the integers.

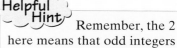

Remember, the 2 here means that odd integers are 2 units apart—for example, the odd integers 13 and $13 + 2 = 15$.

Answer
7. 46, 48, 50

Vocabulary, Readiness & Video Check

Fill in the table.

	A number:		Expression:		Expression:
1.	x	→	Double the number:	→	Double the number, decreased by 31:
2.	x	→	Three times the number:	→	Three times the number, increased by 17:
3.	x	→	The sum of the number and 5:	→	Twice the sum of the number and 5:
4.	x	→	The difference of the number and 11:	→	Seven times the difference of the number and 11:
5.	y	→	The difference of 20 and the number:	→	The difference of 20 and the number, divided by 3:
6.	y	→	The sum of -10 and the number:	→	The sum of -10 and the number, divided by 9:

Martin-Gay Interactive Videos Watch the section lecture video and answer the following questions.

See Video 9.4

Objective A 7. At the end of Example 1, where are we told is the best place to check the solution of an application problem?

Objective B 8. The solution of the equation for Example 3 is $x = 43$. Why is this not the solution to the application?

Objective C 9. What are two things that should be checked to make sure the solution of Example 4 is correct?

9.4 Exercise Set MyLab Math

Objective A *Solve. See Examples 1 and 2. For Exercises 1 through 4, write each of the following as equations. Then solve.*

1. The sum of twice a number and 7 is equal to the sum of the number and 6. Find the number.

2. The difference of three times a number and 1 is the same as twice the number. Find the number.

3. Three times a number, minus 6, is equal to two times the number, plus 8. Find the number.

4. The sum of 4 times a number and -2 is equal to the sum of 5 times the number and -2. Find the number.

 5. Twice the difference of a number and 8 is equal to three times the sum of the number and 3. Find the number.

6. Five times the sum of a number and -1 is the same as 6 times the number. Find the number.

7. The product of twice a number and 3 is the same as the difference of five times the number and $\frac{3}{4}$. Find the number.

8. If the difference of a number and 4 is doubled, the result is $\frac{1}{4}$ less than the number. Find the number.

Objective B *Solve. For Exercises 9 and 10, the solutions have been started for you. See Examples 3 and 4.*

9. A 25-inch piece of steel is cut into three pieces so that the second piece is twice as long as the first piece, and the third piece is one inch more than five times the length of the first piece. Find the lengths of the pieces.

Start the solution:

1. UNDERSTAND the problem. Reread it as many times as needed.
2. TRANSLATE into an equation. (Fill in the blanks below.)

total length of steel	equals	length of first piece	plus	length of second piece	plus	length of third piece
↓	↓	↓	↓	↓	↓	↓
25	=	___	+	___	+	___

Finish with:
3. SOLVE and 4. INTERPRET

10. A 46-foot piece of rope is cut into three pieces so that the second piece is three times as long as the first piece, and the third piece is two feet more than seven times the length of the first piece. Find the lengths of the pieces.

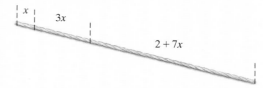

Start the solution:

1. UNDERSTAND the problem. Reread it as many times as needed.
2. TRANSLATE into an equation. (Fill in the blanks below.)

total length of rope	equals	length of first piece	plus	length of second piece	plus	length of third piece
↓	↓	↓	↓	↓	↓	↓
46	=	___	+	___	+	___

Finish with:
3. SOLVE and 4. INTERPRET

11. A 40-inch board is to be cut into three pieces so that the second piece is twice as long as the first piece and the third piece is 5 times as long as the first piece. If x represents the length of the first piece, find the lengths of all three pieces.

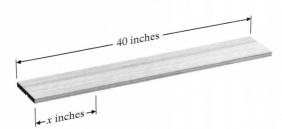

12. A 21-foot beam is to be divided so that the longer piece is 1 foot more than 3 times the length of the shorter piece. If x represents the length of the shorter piece, find the lengths of both pieces.

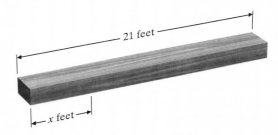

13. In 2016, New York produced 720 million pounds more apples than Pennsylvania. Together, the two states produced 1690 million pounds of apples. Find the amount of apples grown in New York and Pennsylvania in 2016. (*Source:* National Agriculture Statistics Service)

14. In the 2016 Summer Olympics, the U.S. team won 20 more gold medals than the Chinese team. If the total number of gold medals won by both teams was 72, find the number of gold medals won by each team. (*Source:* NBC Sports)

Solve. See Example 5.

15. A car rental agency advertised renting a Buick Century for $24.95 per day and $0.29 per mile. If you rent this car for 2 days, how many whole miles can you drive on a $100 budget?

16. A plumber gave an estimate for the renovation of a kitchen. Her hourly pay is $27 per hour and the plumbing parts will cost $80. If her total estimate is $404, how many hours does she expect this job to take?

17. In one U.S. city, the taxi cost is $3 plus $0.80 per mile. If you are traveling from the airport, there is an additional charge of $4.50 for tolls. How far can you travel from the airport by taxi for $27.50?

18. A professional carpet cleaning service charges $30 plus $25.50 per hour to come to your home. If your total bill from this company is $119.25 before taxes, for how many hours were you charged?

Solve. See Example 6.

△ 19. The flag of Equatorial Guinea contains an isosceles triangle. (Recall that an isosceles triangle contains two angles with the same measure.) If the measure of the third angle of the triangle is 30° more than twice the measure of either of the other two angles, find the measure of each angle of the triangle. (*Hint:* Recall that the sum of the measures of the angles of a triangle is 180°.)

△ 20. The flag of Brazil contains a parallelogram. One angle of the parallelogram is 15° less than twice the measure of the angle next to it. Find the measure of each angle of the parallelogram. (*Hint:* Recall that opposite angles of a parallelogram have the same measure and that the sum of the measures of the angles is 360°.)

21. The sum of the measures of the angles of a parallelogram is 360°. In the parallelogram below, angles A and D have the same measure as angles C and B. If the measure of angle C is twice the measure of angle A, find the measure of each angle.

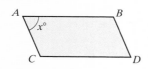

22. Recall that the sum of the measures of the angles of a triangle is 180°. In the triangle below, angle C has the same measure as angle B, and angle A measures 42° less than angle B. Find the measure of each angle.

Objective C *Solve. See Example 7. Fill in the table. Most of the first row has been completed for you.*

	First Integer	Next Integers		Indicated Sum
23. Three consecutive integers:	Integer: x	$x+1$	$x+2$	Sum of the three consecutive integers, simplified:
24. Three consecutive integers:	Integer: x			Sum of the second and third consecutive integers, simplified:
25. Three consecutive even integers:	Even integer: x			Sum of the first and third even consecutive integers, simplified:
26. Three consecutive odd integers:	Odd integer: x			Sum of the three consecutive odd integers, simplified:
27. Four consecutive integers:	Integer: x			Sum of the four consecutive integers, simplified:
28. Four consecutive integers:	Integer: x			Sum of the first and fourth consecutive integers, simplified:
29. Three consecutive odd integers:	Odd integer: x			Sum of the second and third consecutive odd integers, simplified:
30. Three consecutive even integers:	Even integer: x			Sum of the three consecutive even integers, simplified:

Solve. See Example 7

31. The left and right page numbers of an open book are two consecutive integers whose sum is 469. Find these page numbers.

32. The room numbers of two adjacent classrooms are two consecutive even numbers. If their sum is 654, find the classroom numbers.

33. To make an international telephone call, you need the code for the country you are calling. The codes for Belgium, France, and Spain are three consecutive integers whose sum is 99. Find the code for each country. (*Source:* The World Almanac and Book of Facts)

34. The code to unlock a student's combination lock happens to be three consecutive odd integers whose sum is 51. Find the integers.

Objectives A B C Mixed Practice *Solve. See Examples 1 through 7.*

35. A 17-foot piece of string is cut into two pieces so that the longer piece is 2 feet longer than twice the length of the shorter piece. Find the lengths of both pieces.

36. A 25-foot wire is to be cut so that the longer piece is one foot longer than 5 times the length of the shorter piece. Find the length of each piece.

37. Currently, the two fastest trains in the world are China's CRH380A and Germany's Transrapid TR-09. The sum of their fastest speeds is 581 miles per hour. If the maximum speed of the CRH380A is 23 miles per hour faster than the maximum speed of the Transrapid TR-09, find the speeds of each. (*Note:* The transrapid TR-09 is technically a monorail. *Source:* tiptoptens.com and telegraph.co.uk)

38. The Pentagon is the world's largest office building in terms of floor space. It has three times the amount of floor space as the Empire State Building. If the total floor space for these two buildings is approximately 8700 thousand square feet, find the floor space of each building.

39. Two angles are supplementary if their sum is 180°. The larger angle below measures eight degrees more than three times the measure of the smaller angle. If x represents the measure of the smaller angle and these two angles are supplementary, find the measure of each angle.

△ **40.** Two angles are complementary if their sum is 90°. Given the measures of the complementary angles shown, find the measure of each angle.

41. The measures of the angles of a triangle are 3 consecutive even integers. Find the measure of each angle.

42. A quadrilateral is a polygon with 4 sides. The sum of the measures of the 4 angles in a quadrilateral is 360°. If the measures of the angles of a quadrilateral are consecutive odd integers, find the measures.

43. The sum of $\frac{1}{5}$ and twice a number is equal to $\frac{4}{5}$ subtracted from three times the number. Find the number.

44. The sum of $\frac{2}{3}$ and four times a number is equal to $\frac{5}{6}$ subtracted from five times the number. Find the number.

45. Hertz Car Rental charges a daily rate of $39 plus $0.20 per mile for a certain car. Suppose that you rent that car for a day and your bill (before taxes) is $95. How many miles did you drive?

46. A woman's $15,000 estate is to be divided so that her husband receives twice as much as her son. Find the amount of money that her husband receives and the amount of money that her son receives.

47. One of the biggest rivalries in college football is the University of Michigan Wolverines and the Ohio State University Buckeyes. During their match-up in 2016, Ohio beat Michigan by 3 points. If their combined scores totaled 57, find the individual team scores.

48. In January 2016 there was one independent governor in the United States, and there were 17 more Republican governors than Democratic governors. How many Democrats and how many Republicans held governor's offices at that time? (*Source: Multistate.com*)

49. The number of counties in California and the number of counties in Montana are consecutive even integers whose sum is 114. If California has more counties than Montana, how many counties does each state have? (*Source: The World Almanac and Book of Facts*)

50. A student is building a bookcase with stepped shelves for her dorm room. She buys a 48-inch board and wants to cut the board into three pieces with lengths equal to three consecutive even integers. Find the three board lengths.

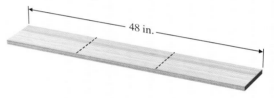

51. Scientists are continually updating information about the planets in our solar system, including the number of satellites orbiting each. Uranus is now believed to have 13 more satellites than Neptune. Also, Saturn is now believed to have 8 more than twice the number of satellites of Uranus. If the total number of satellites for these planets is 103, find the number of satellites for each planet. (*Source: National Space Science Data Center*)

52. The Apple iPhone 7 plus was introduced in 2016. The height of each iPhone 7 plus is 0.09 inch more than twice its width. If the sum of the height and the width is 9.30 inches, find each dimension. (*Source: Apple, Inc.*)

53. If the sum of a number and five is tripled, the result is one less than twice the number. Find the number.

54. Twice the sum of a number and six equals three times the sum of the number and four. Find the number.

55. The area of the Sahara Desert is 7 times the area of the Gobi Desert. If the sum of their areas is 4,000,000 square miles, find the area of each desert.

56. The largest meteorite in the world is the Hoba West, located in Namibia. Its weight is 3 times the weight of the Armanty meteorite, located in Outer Mongolia. If the sum of their weights is 88 tons, find the weight of each.

57. In the 2016 Summer Olympic Games, Brazil won more medals than New Zealand, which won more medals than Spain. If the numbers of medals won by these three countries are three consecutive integers whose sum is 54, find the number of medals won by each. (*Source:* NBS Sports)

58. To make an international telephone call, you need the code for the country you are calling. The codes for Mali Republic, Côte d'Ivoire, and Niger are three consecutive odd integers whose sum is 675. Find the code for each country.

59. In the 2015–2016 school year, there were 363 more male undergraduate students enrolled at MIT than female undergraduate students. If the total undergraduate enrollment was 4527 that year, find the numbers of female undergraduate students and male undergraduate students who were enrolled. (*Source:* MIT)

60. In 2015, approximately 2.4 million more trucks were sold in the United States than cars. If the total number of trucks and cars sold was 17.4 million, find the number of vehicles sold in each category. (*Source:* Alliance of Automobile Manufacturers)

61. A geodesic dome, based on the design by Buckminster Fuller, is composed of two different types of triangular panels. One of these is an isosceles triangle. In one geodesic dome, the measure of the third angle is 76.5° more than the measure of either of the two equal angles. Find the measure of the three angles. (*Source:* Buckminster Fuller Institute)

62. The measures of the angles of a particular triangle are such that the second and third angles are each four times the measure of the smallest angle. Find the measures of the angles of this triangle.

The graph below shows the states with the highest provisional tourism budgets in 2014–2015. Use the graph for Exercises 63 through 68.

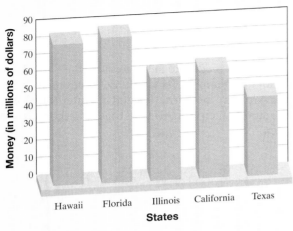

Source: U.S. Travel Association

63. Which state spent the most money on tourism?

64. Which state spent between $40 and $50 million on tourism?

65. The states of Florida and Illinois spent a total of $144.4 million on tourism. The state of Florida spent $24.8 million more than the state of Illinois. Find the amount that each state spent on tourism.

66. The states of California and Texas spent a total of $108.3 million on tourism. The state of Texas spent $16.5 million less than the state of California. Find the amount that each state spent on tourism.

Compare the heights of the bars in the graph with your results of the exercises below. Are your answers reasonable?

67. Exercise 65

68. Exercise 66

Review

Evaluate each expression for the given values. See Section 1.8.

69. $2W + 2L$; $W = 7$ and $L = 10$

70. $\frac{1}{2} Bh$; $B = 14$ and $h = 22$

71. πr^2; $r = 15$

72. $r \cdot t$; $r = 15$ and $t = 2$

Concept Extensions

73. A golden rectangle is a rectangle whose length is approximately 1.6 times its width. The early Greeks thought that a rectangle with these dimensions was the most pleasing to the eye, and examples of the golden rectangle are found in many early works of art. For example, the Parthenon in Athens contains many examples of golden rectangles.

Mike Hallahan would like to plant a rectangular garden in the shape of a golden rectangle. If he has 78 feet of fencing available, find the dimensions of the garden.

74. Dr. Dorothy Smith gave the students in her geometry class at the University of New Orleans the following question: Is it possible to construct a triangle such that the second angle of the triangle has a measure that is twice the measure of the first angle and the measure of the third angle is 5 times the measure of the first? If so, find the measure of each angle. (*Hint:* Recall that the sum of the measures of the angles of a triangle is 180°.)

75. Only male crickets chirp. They chirp at different rates depending on their species and the temperature of their environment. Suppose a certain species is currently chirping at a rate of 90 chirps per minute. At this rate, how many chirps occur in one hour? In one 24-hour day? In one year?

76. The human eye blinks once every 5 seconds on average. How many times does the average eye blink in one hour? In one 16-hour day while awake? In one year while awake?

77. In your own words, explain why a solution of a word problem should be checked using the original wording of the problem and not the equation written from the wording.

78. Give an example of how you recently solved a problem using mathematics.

*Recall from Exercise **73** that a golden rectangle is a rectangle whose length is approximately 1.6 times its width.*

79. It is thought that for about 75% of adults, a rectangle in the shape of the golden rectangle is the most pleasing to the eye. Draw three rectangles, one in the shape of the golden rectangle, and poll your class. Do the results agree with the percentage given above?

80. Examples of golden rectangles can be found today in architecture and manufacturing packaging. Find an example of a golden rectangle in your home. A few suggestions: the front face of a book, the floor of a room, the front of a box of food.

For Exercises 81 and 82, measure the dimensions of each rectangle and decide which one best approximates the shape of a golden rectangle.

81.

82.

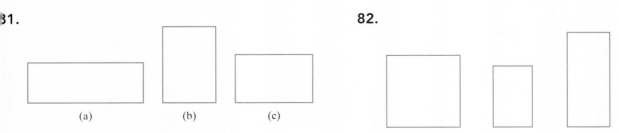

9.5 Formulas and Problem Solving

Objectives

A Use Formulas to Solve Problems.

B Solve a Formula or Equation for One of Its Variables.

Objective A Using Formulas to Solve Problems

A **formula** describes a known relationship among quantities. Many formulas are given as equations. For example, the formula

$$d = r \cdot t$$

stands for the relationship

$$\text{distance} = \text{rate} \cdot \text{time}$$

Let's look at one way that we can use this formula.

If we know we traveled a distance of 100 miles at a rate of 40 miles per hour, we can replace the variables d and r in the formula $d = rt$ and find our travel time, t.

$d = rt$ Formula
$100 = 40t$ Replace d with 100 and r with 40.

To solve for t, we divide both sides of the equation by 40.

$\dfrac{100}{40} = \dfrac{40t}{40}$ Divide both sides by 40.

$\dfrac{5}{2} = t$ Simplify.

The travel time was $\dfrac{5}{2}$ hours, or $2\dfrac{1}{2}$ hours, or 2.5 hours.

In this section, we solve problems that can be modeled by known formulas. We use the same problem-solving strategy that was used in the previous section.

Practice 1

A family is planning their vacation to visit relatives. They will drive from Cincinnati, Ohio, to Rapid City, South Dakota, a distance of 1180 miles. They plan to average a rate of 50 miles per hour. How much time will they spend driving?

Example 1 Finding Time Given Rate and Distance

A glacier is a giant mass of rocks and ice that flows downhill like a river. Portage Glacier in Alaska is about 6 miles, or 31,680 *feet*, long and moves 400 *feet* per year. Icebergs are created when the front end of the glacier flows into Portage Lake. How long does it take for ice at the head (beginning) of the glacier to reach the lake?

Solution:

1. UNDERSTAND. Read and reread the problem. The appropriate formula needed to solve this problem is the distance formula, $d = rt$. To become familiar with this formula, let's find the distance that ice traveling at a rate of 400 feet per year travels in 100 years. To do so, we let time t be 100 years and rate r be the given 400 feet per year, and substitute these values into the formula $d = rt$. We then have that distance $d = 400(100) = 40,000$ feet. Since we are interested in finding how long it takes ice to travel 31,680 feet, we now know that it is less than 100 years.

Answer
1. 23.6 hours

Since we are using the formula $d = rt$, we let

  t = the time in years for ice to reach the lake
  r = rate or speed of ice
  d = distance from beginning of glacier to lake

2. **TRANSLATE.** To translate to an equation, we use the formula $d = rt$ and let distance $d = 31{,}680$ feet and rate $r = 400$ feet per year.

$$d = r \cdot t$$
$$31{,}680 = 400 \cdot t \quad \text{Let } d = 31{,}680 \text{ and } r = 400.$$

3. **SOLVE.** Solve the equation for t. To solve for t, we divide both sides by 400.

$$\frac{31{,}680}{400} = \frac{400 \cdot t}{400} \quad \text{Divide both sides by 400.}$$
$$79.2 = t \quad \text{Simplify.}$$

4. **INTERPRET.**

   Check: To check, substitute 79.2 for t and 400 for r in the distance formula and check to see that the distance is 31,680 feet.

   State: It takes 79.2 years for the ice at the head of Portage Glacier to reach the lake.

Helpful Hint
Don't forget to include units, if appropriate.

Work Practice 1

Example 2 Calculating the Length of a Garden

Charles Pecot can afford enough fencing to enclose a rectangular garden with a perimeter of 140 feet. If the width of his garden is to be 30 feet, find the length.

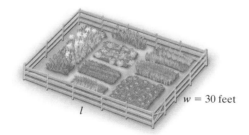

Practice 2
A wood deck is being built behind a house. The width of the deck must be 18 feet because of the shape of the house. If there is 450 square feet of decking material, find the length of the deck.

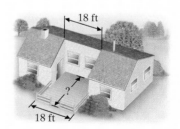

Solution:

1. **UNDERSTAND.** Read and reread the problem. The formula needed to solve this problem is the formula for the perimeter of a rectangle, $P = 2l + 2w$. Before continuing, let's become familar with this formula.

   l = the length of the rectangular garden
   w = the width of the rectangular garden
   P = perimeter of the garden

2. **TRANSLATE.** To translate to an equation, we use the formula $P = 2l + 2w$ and let perimeter $P = 140$ feet and width $w = 30$ feet.

$$P = 2l + 2w \quad \text{Let } P = 140 \text{ and } w = 30.$$
$$140 = 2l + 2(30)$$

(Continued on next page)

Answer
2. 25 feet

Practice 3

Convert the temperature 5°C to Fahrenheit.

3. SOLVE.

$$140 = 2l + 2(30)$$
$$140 = 2l + 60 \quad \text{Multiply 2(30).}$$
$$140 - 60 = 2l + 60 - 60 \quad \text{Subtract 60 from both sides.}$$
$$80 = 2l \quad \text{Combine like terms.}$$
$$40 = l \quad \text{Divide both sides by 2.}$$

4. INTERPRET.

Check: Substitute 40 for l and 30 for w in the perimeter formula and check to see that the perimeter is 140 feet.

State: The length of the rectangular garden is 40 feet.

■ Work Practice 2

Example 3 Finding an Equivalent Temperature

The average maximum temperature for January in Algiers, Algeria, is 59° Fahrenheit. Find the equivalent temperature in degrees Celsius.

Solution:

1. UNDERSTAND. Read and reread the problem. A formula that can be used to solve this problem is the formula for converting degrees Celsius to degrees Fahrenheit, $F = \frac{9}{5}C + 32$. Before continuing, become familiar with this formula. Using this formula, we let

   C = temperature in degrees Celsius, and
   F = temperature in degrees Fahrenheit.

2. TRANSLATE. To translate to an equation, we use the formula $F = \frac{9}{5}C + 32$ and let degrees Fahrenheit F = 59.

   Formula: $\quad F = \frac{9}{5}C + 32$

   Substitute: $\quad 59 = \frac{9}{5}C + 32 \quad$ Let F = 59.

3. SOLVE.

$$59 = \frac{9}{5}C + 32$$
$$59 - 32 = \frac{9}{5}C + 32 - 32 \quad \text{Subtract 32 from both sides.}$$
$$27 = \frac{9}{5}C \quad \text{Combine like terms.}$$
$$\frac{5}{9} \cdot 27 = \frac{5}{9} \cdot \frac{9}{5}C \quad \text{Multiply both sides by } \frac{5}{9}.$$
$$15 = C \quad \text{Simplify.}$$

4. INTERPRET.

Check: To check, replace C with 15 and F with 59 in the formula and see that a true statement results.

State: Thus, 59° Fahrenheit is equivalent to 15° Celsius.

■ Work Practice 3

Answer

3. 41°F

In the next example, we use the formula for perimeter of a rectangle as in Example 2. In Example 2, we knew the width of the rectangle. In this example, both the length and width are unknown.

Example 4 Finding Road Sign Dimensions

The length of a rectangular road sign is 2 feet less than three times its width. Find the dimensions if the perimeter is 28 feet.

Practice 4

The length of a rectangle is 1 meter more than 4 times its width. Find the dimensions if the perimeter is 52 meters.

Solution:

1. **UNDERSTAND.** Read and reread the problem. Recall that the formula for the perimeter of a rectangle is $P = 2l + 2w$. Draw a rectangle and guess the solution. If the width of the rectangular sign is 5 feet, its length is 2 feet less than 3 times the width, or $3(5 \text{ feet}) - 2 \text{ feet} = 13 \text{ feet}$. The perimeter P of the rectangle is then $2(13 \text{ feet}) + 2(5 \text{ feet}) = 36 \text{ feet}$, too large. We now know that the width is less than 5 feet.

   Proposed rectangle:

   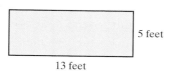

   Let
   w = the width of the rectangular sign; then
   $3w - 2$ = the length of the sign.

   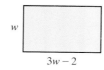

   Draw a rectangle and label it with the assigned variables.

2. **TRANSLATE.**
   Formula: $P = 2l + 2w$
   Substitute: $28 = 2(3w - 2) + 2w$

3. **SOLVE.**
   $28 = 2(3w - 2) + 2w$
   $28 = 6w - 4 + 2w$ Apply the distributive property.
   $28 = 8w - 4$
   $28 + 4 = 8w - 4 + 4$ Add 4 to both sides.
   $32 = 8w$
   $\dfrac{32}{8} = \dfrac{8w}{8}$ Divide both sides by 8.
   $4 = w$

4. **INTERPRET.**

   Check: If the width of the sign is 4 feet, the length of the sign is $3(4 \text{ feet}) - 2 \text{ feet} = 10 \text{ feet}$. This gives the rectangular sign a perimeter of $P = 2(4 \text{ feet}) + 2(10 \text{ feet}) = 28 \text{ feet}$, the correct perimeter.

   State: The width of the sign is 4 feet and the length of the sign is 10 feet.

Work Practice 4

Answer
4. length: 21 m; width: 5 m

Objective B Solving a Formula for a Variable

We say that the formula

$$d = rt$$

is solved for d because d is alone on one side of the equation and the other side contains no d's. Suppose that we have a large number of problems to solve where we are given distance d and rate r and asked to find time t. In this case, it may be easier to first solve the formula $d = rt$ for t. To solve for t, we divide both sides of the equation by r.

$$d = rt$$
$$\frac{d}{r} = \frac{rt}{r} \quad \text{Divide both sides by } r.$$
$$\frac{d}{r} = t \quad \text{Simplify.}$$

To solve a formula or an equation for a specified variable, we use the same steps as for solving a linear equation except that we treat the specified variable as the only variable in the equation. These steps are listed next.

Solving Equations for a Specified Variable

Step 1: Multiply on both sides to clear the equation of fractions if they appear.
Step 2: Use the distributive property to remove parentheses if they appear.
Step 3: Simplify each side of the equation by combining like terms.
Step 4: Get all terms containing the specified variable on one side and all other terms on the other side by using the addition property of equality.
Step 5: Get the specified variable alone by using the multiplication property of equality.

Practice 5

Solve $C = 2\pi r$ for r. (This formula is used to find the circumference, C, of a circle given its radius, r.)

Example 5 Solve $V = lwh$ for l.

Solution: This formula is used to find the volume of a box. To solve for l, we divide both sides by wh.

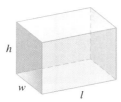

$$V = lwh$$
$$\frac{V}{wh} = \frac{lwh}{wh} \quad \text{Divide both sides by } wh.$$
$$\frac{V}{wh} = l \quad \text{Simplify.}$$

Since we have l alone on one side of the equation, we have solved for l in terms of V, w, and h. Remember that it does not matter on which side of the equation we get the variable alone.

■ Work Practice 5

Answer

5. $r = \dfrac{C}{2\pi}$

Example 6
Solve $y = mx + b$ for x.

Solution: First we get mx alone by subtracting b from both sides.

$$y = mx + b$$
$$y - b = mx + b - b \quad \text{Subtract } b \text{ from both sides.}$$
$$y - b = mx \quad \text{Combine like terms.}$$

Next we solve for x by dividing both sides by m.

$$\frac{y - b}{m} = \frac{mx}{m}$$
$$\frac{y - b}{m} = x \quad \text{Simplify.}$$

■ Work Practice 6

Practice 6
Solve $P = 2l + 2w$ for l.

✓ Concept Check Solve:
a. ● = ■ − ■ for ■
b. ● = ■ · ▲ − ■ for ■

Example 7
Solve $P = 2l + 2w$ for w.

Solution: This formula relates the perimeter of a rectangle to its length and width. Find the term containing the variable w. To get this term, $2w$, alone, subtract $2l$ from both sides.

$$P = 2l + 2w$$
$$P - 2l = 2l + 2w - 2l \quad \text{Subtract } 2l \text{ from both sides.}$$
$$P - 2l = 2w \quad \text{Combine like terms.}$$
$$\frac{P - 2l}{2} = \frac{2w}{2} \quad \text{Divide both sides by 2.}$$
$$\frac{P - 2l}{2} = w \quad \text{Simplify.}$$

■ Work Practice 7

Practice 7
Solve $P = 2a + b - c$ for a.

Helpful Hint The 2s may *not* be divided out here. Although 2 is a factor of the denominator, 2 is *not* a factor of the numerator since it is not a factor of both terms in the numerator.

The next example has an equation containing a fraction. We will first clear the equation of fractions and then solve for the specified variable.

Example 8
Solve $F = \frac{9}{5}C + 32$ for C.

Solution:
$$F = \frac{9}{5}C + 32$$
$$5(F) = 5\left(\frac{9}{5}C + 32\right) \quad \text{Clear the fraction by multiplying both sides by the LCD.}$$
$$5F = 9C + 160 \quad \text{Distribute the 5.}$$
$$5F - 160 = 9C + 160 - 160 \quad \text{To get the term containing the variable } C \text{ alone, subtract 160 from both sides.}$$
$$5F - 160 = 9C \quad \text{Combine like terms.}$$
$$\frac{5F - 160}{9} = \frac{9C}{9} \quad \text{Divide both sides by 9.}$$
$$\frac{5F - 160}{9} = C \quad \text{Simplify.}$$

■ Work Practice 8

Practice 8
Solve $A = \frac{a + b}{2}$ for b.

Answers
6. $l = \dfrac{P - 2w}{2}$ 7. $a = \dfrac{P - b + c}{2}$
8. $b = 2A - a$

✓ Concept Check Answer
a. ● + ■   b. $\dfrac{● + ■}{▲}$

Vocabulary, Readiness & Video Check

Martin-Gay Interactive Videos Watch the section lecture video and answer the following questions.

See Video 9.5

Objective A
1. Complete this statement based on the lecture given before Example 1: A formula is an equation that describes known _____ among quantities.
2. In Example 2, how are the units for the solution determined?

Objective B
3. In Example 4, what is the equation $5x = 30$ used to show?

9.5 Exercise Set MyLab Math

Objective A Substitute the given values into each given formula and solve for the unknown variable. See Examples 1 through 4.

1. $A = bh$; $A = 45, b = 15$ (Area of a parallelogram)

2. $d = rt$; $d = 195, t = 3$ (Distance formula)

3. $S = 4lw + 2wh$; $S = 102, l = 7, w = 3$ (Surface area of a special rectangular box)

4. $V = lwh$; $l = 14, w = 8, h = 3$ (Volume of a rectangular box)

5. $A = \frac{1}{2}h(B + b)$; $A = 180, B = 11, b = 7$ (Area of a trapezoid)

6. $A = \frac{1}{2}h(B + b)$; $A = 60, B = 7, b = 3$ (Area of a trapezoid)

7. $P = a + b + c$; $P = 30, a = 8, b = 10$ (Perimeter of a triangle)

8. $V = \frac{1}{3}Ah$; $V = 45, h = 5$ (Volume of a pyramid)

9. $C = 2\pi r$; $C = 15.7$ (Circumference of a circle) (Use the approximation 3.14 for π.)

10. $A = \pi r^2$; $r = 4$ (Area of a circle) (Use the approximation 3.14 for π.)

Objective B Solve each formula for the specified variable. See Examples 5 through 8.

11. $f = 5gh$ for h

12. $x = 4\pi y$ for y

13. $V = lwh$ for w

14. $T = mnr$ for n

15. $3x + y = 7$ for y

16. $-x + y = 13$ for y

17. $A = P + PRT$ for R

18. $A = P + PRT$ for T

19. $V = \dfrac{1}{3}Ah$ for A

20. $D = \dfrac{1}{4}fk$ for k

21. $P = a + b + c$ for a

22. $PR = x + y + z + w$ for z

23. $S = 2\pi rh + 2\pi r^2$ for h

△ 24. $S = 4lw + 2wh$ for h

Objective A *Solve. For Exercises 25 and 26, the solutions have been started for you. See Examples 1 through 4.*

25. The iconic NASDAQ sign in New York's Times Square has a width of 84 feet and an area of 10,080 square feet. Find the height (or length) of the sign. (*Source:* livedesignonline.com)

Start the solution:

1. UNDERSTAND the problem. Reread it as many times as needed.
2. TRANSLATE into an equation. (Fill in the blanks below.)

   Area = length times width
   ↓   ↓   ↓   ↓   ↓

   ___ = x · ___

Finish with:
3. SOLVE and 4. INTERPRET

△ 26. The world's largest sign for Coca-Cola is located in Arica, Chile. The rectangular sign has a length of 400 feet and an area of 52,400 square feet. Find the width of the sign. (*Source:* Fabulous Facts about Coca-Cola, Atlanta, GA)

Start the solution:

1. UNDERSTAND the problem. Reread it as many times as needed.
2. TRANSLATE into an equation. (Fill in the blanks below.)

   Area = length times width
   ↓   ↓   ↓   ↓   ↓

   ___ = ___ · w

Finish with:
3. SOLVE and 4. INTERPRET

27. A frame shop charges according to both the amount of framing needed to surround the picture and the amount of glass needed to cover the picture.
    a. Find the area and perimeter of the picture below.
    b. Identify whether the frame has to do with perimeter or area and the same with the glass.

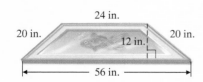

28. A decorator is painting and placing a border completely around a parallelogram-shaped wall.
    a. Find the area and perimeter of the wall below.
    b. Identify whether the border has to do with perimeter or area and the same with paint.

29. For the purpose of purchasing new baseboard and carpet,
    a. Find the area and perimeter of the room below (neglecting doors).
    b. Identify whether baseboard has to do with area or perimeter and the same with carpet.

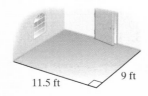

30. For the purpose of purchasing lumber for a new fence and seed to plant grass,
    a. Find the area and perimeter of the yard below.
    b. Identify whether a fence has to do with area or perimeter and the same with grass seed.
    $\left(A = \dfrac{1}{2}bh\right)$

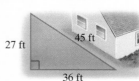

31. Convert Nome, Alaska's 14°F high temperature to Celsius.

32. Convert Paris, France's low temperature of −5°C to Fahrenheit.

33. The X-30 is a "space plane" that skims the edge of space at 4000 miles per hour. Neglecting altitude, if the circumference of the Earth is approximately 25,000 miles, how long does it take for the X-30 to travel around the Earth?

34. In the United States, a notable hang glider flight was a 303-mile, $8\frac{1}{2}$-hour flight from New Mexico to Kansas. What was the average rate during this flight?

35. An architect designs a rectangular flower garden such that the width is exactly two-thirds of the length. If 260 feet of antique picket fencing is to be used to enclose the garden, find the dimensions of the garden.

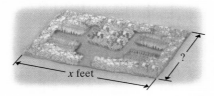

36. If the length of a rectangular parking lot is 10 meters less than twice its width, and the perimeter is 400 meters, find the length of the parking lot.

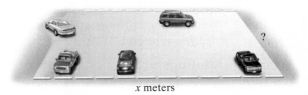

37. A flower bed is in the shape of a triangle with one side twice the length of the shortest side and the third side 30 feet more than the length of the shortest side. Find the dimensions if the perimeter is 102 feet.

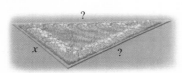

38. The perimeter of a yield sign in the shape of an isosceles triangle is 22 feet. If the shortest side is 2 feet less than the other two sides, find the length of the shortest side. (*Hint:* An isosceles triangle has two sides the same length.)

39. The Eurostar is a high-speed train that shuttles passengers between London, England, and Paris, France, via the Channel Tunnel. The Eurostar can make the trip in about $2\frac{1}{4}$ hours at an average speed of 136 mph. About how long is the Eurostar's route connecting London and Paris? (*Source: Eurostar*)

40. A family is planning their vacation to Disney World. They will drive from a small town outside New Orleans, Louisiana, to Orlando, Florida, a distance of 700 miles. They plan to average a rate of 56 mph. How long will this trip take?

Dolbear's Law states the relationship between the rate at which Snowy Tree Crickets chirp and the air temperature of the environment. The formula is

$$T = 50 + \frac{N - 40}{4}, \text{ where}$$

T = temperature in degrees Fahrenheit and
N = number of chirps per minute

41. If $N = 86$, find the temperature in degrees Fahrenheit, T.

42. If $N = 94$, find the temperature in degrees Fahrenheit, T.

43. If $T = 55°F$, find the number of chirps per minute.

44. If $T = 65°F$, find the number of chirps per minute.

Section 9.5 | Formulas and Problem Solving

Use the results of Exercises 41–44 to complete each sentence with "increases" or "decreases."

45. As the number of cricket chirps per minute increases, the air temperature of their environment _____.

46. As the air temperature of their environment decreases, the number of cricket chirps per minute _____.

Solve. See Examples 1 through 4.

47. Piranha fish require 1.5 cubic feet of water per fish to maintain a healthy environment. Find the maximum number of piranhas you could put in a tank measuring 8 feet by 3 feet by 6 feet.

△ 48. Find the maximum number of goldfish you can put in a cylindrical tank whose diameter is 8 meters and whose height is 3 meters, if each goldfish needs 2 cubic meters of water. ($V = \pi r^2 h$)

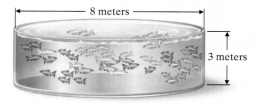

49. A lawn is in the shape of a trapezoid with a height of 60 feet and bases of 70 feet and 130 feet. How many bags of fertilizer must be purchased to cover the lawn if each bag covers 4000 square feet?

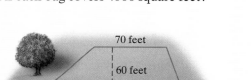

△ 50. If the area of a right-triangularly shaped sail is 20 square feet and its base is 5 feet, find the height of the sail. $\left(A = \frac{1}{2}bh \right)$

51. Maria's Pizza sells one 16-inch cheese pizza or two 10-inch cheese pizzas for $9.99. Determine which size gives more pizza.

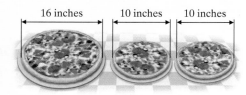

△ 52. Find how much rope is needed to wrap around the Earth at the equator, if the radius of the Earth is 4000 miles. (*Hint:* Use 3.14 for π and the formula for circumference.)

53. The perimeter of a geometric figure is the sum of the lengths of its sides. If the perimeter of the following pentagon (five-sided figure) is 48 meters, find the length of each side.

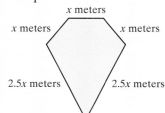

△ 54. The perimeter of the following triangle is 82 feet. Find the length of each side.

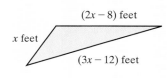

55. The Hawaiian volcano Kilauea is one of the world's most active volcanoes and has had continuous eruptive activity since 1983. Erupting lava flows through a tube system about 11 kilometers to the sea. Assume a lava flow speed of 0.5 kilometer per hour and calculate how long it takes to reach the sea.

56. The world's largest pink ribbon, the sign of the fight against breast cancer, was erected out of pink Post-it® notes on a billboard in New York City in October 2004. If the area of the rectangular billboard covered by the ribbon was approximately 3990 square feet, and the width of the billboard was approximately 57 feet, what was the height of this billboard?

△ 57. The perimeter of an equilateral triangle is 7 inches more than the perimeter of a square, and the sides of the triangle are 5 inches longer than the sides of the square. Find the side of the triangle. (*Hint:* An equilateral triangle has three sides the same length.)

△ 58. A square animal pen and a pen shaped like an equilateral triangle have equal perimeters. Find the length of the sides of each pen if the sides of the triangular pen are 15 less than twice a side of the square pen. (*Hint:* An equilateral triangle has three sides the same length.)

59. Find how long it takes Tran Nguyen to drive 135 miles on I-10 if he merges onto I-10 at 10 a.m. and drives nonstop with his cruise control set on 60 mph.

60. Beaumont, Texas, is about 150 miles from Toledo Bend. If Leo Miller leaves Beaumont at 4 a.m. and averages 45 mph, when should he arrive at Toledo Bend?

△ 61. The longest runway at Los Angeles International Airport has the shape of a rectangle and an area of 1,813,500 square feet. This runway is 150 feet wide. How long is the runway? (*Source:* Los Angeles World Airports)

62. Normal room temperature is about 78°F. Convert this temperature to Celsius.

63. The highest temperature ever recorded in Europe was 122°F in Seville, Spain, in August of 1881. Convert this record high temperature to Celsius. (*Source:* National Climatic Data Center)

64. The lowest temperature ever recorded in Oceania was −10°C at the Haleakala Summit in Maui, Hawaii, in January 1961. Convert this record low temperature to Fahrenheit. (*Source:* National Climatic Data Center)

△ 65. The IZOD IndyCar series is an open-wheeled race car competition based in the United States. An IndyCar car has a maximum length of 204 inches, a maximum width of 76.5 inches, and a maximum height of 44 inches. When the IZOD IndyCar series travels to another country for a grand prix, teams must ship their cars. Find the volume of the smallest shipping crate needed to ship an IndyCar car of maximum dimensions. (*Source:* Championship Auto Racing Teams, Inc.)

66. During a recent IndyCar road course race, the winner's average speed was 118 mph. Based on this speed, how long would it take an IndyCar driver to travel from Los Angeles to New York City, a distance of about 2810 miles by road, without stopping. Round to the nearest tenth of an hour.

IndyCar Racing Car

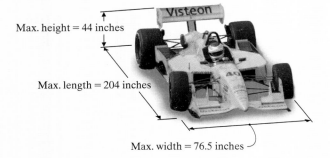

Max. height = 44 inches
Max. length = 204 inches
Max. width = 76.5 inches

67. The Hoberman Sphere is a toy ball that expands and contracts. When it is completely closed, it has a diameter of 9.5 inches. Find the volume of the Hoberman Sphere when it is completely closed. Use 3.14 for π. Round to the nearest whole cubic inch. (*Hint:* Volume of a sphere $= \frac{4}{3}\pi r^3$.) (*Source:* Hoberman Designs, Inc.)

68. When the Hoberman Sphere (see Exercise **67**) is completely expanded, its diameter is 30 inches. Find the volume of the Hoberman Sphere when it is completely expanded. Use 3.14 for π. (*Source:* Hoberman Designs, Inc.)

69. The average temperature on the planet Mercury is 167°C. Convert this temperature to degrees Fahrenheit. Round to the nearest degree. (*Source:* National Space Science Data Center)

70. The average temperature on the planet Jupiter is −227°F. Convert this temperature to degrees Celsius. Round to the nearest degree. (*Source:* National Space Science Data Center)

Review

Write each percent as a decimal. See Section 6.2.

71. 32%   **72.** 8%   **73.** 200%   **74.** 0.5%

Write each decimal as a percent. See Section 6.2.

75. 0.17   **76.** 0.03   **77.** 7.2   **78.** 5

Concept Extensions

Solve.

79. $N = R + \dfrac{V}{G}$ for V (Urban forestry: tree plantings per year)

80. $B = \dfrac{F}{P - V}$ for V (Business: break-even point)

81. The formula $V = lwh$ is used to find the volume of a box. If the length of a box is doubled, the width is doubled, and the height is doubled, how does this affect the volume? Explain your answer.

82. The formula $A = bh$ is used to find the area of a parallelogram. If the base of a parallelogram is doubled and its height is doubled, how does this affect the area? Explain your answer.

83. Use the Dolbear's Law formula from Exercises **41–46** and calculate when the number of cricket chirps per minute is the same as the temperature in degrees Fahrenheit. (*Hint:* Replace T with N and solve for N, or replace N with T and solve for T.)

84. Find the temperature at which the Celsius measurement and the Fahrenheit measurement are the same number.

Solve. See the Concept Check in this section.

85. ▲ − ● · ■ = ⬟ for ●

86. ⬢ · ■ + ▲ = ⬣ for ■

87. Flying fish do not *actually* fly, but glide. They have been known to travel a distance of 1300 feet at a rate of 20 miles per hour. How many seconds would it take to travel this distance? (*Hint:* First convert miles per hour to feet per second. Recall that 1 mile = 5280 feet.) Round to the nearest tenth of a second.

88. A glacier is a giant mass of rocks and ice that flows downhill like a river. Exit Glacier, near Seward, Alaska, moves at a rate of 20 inches a day. Find the distance in feet the glacier moves in a year. (Assume 365 days a year.) Round to two decimal places.

Substitute the given values into each given formula and solve for the unknown variable. If necessary, round to one decimal place.

89. $I = PRT$; $I = 1{,}056{,}000, R = 0.055, T = 6$ (Simple interest formula)

90. $I = PRT$; $I = 3750, P = 25{,}000, R = 0.05$ (Simple interest formula)

91. $V = \frac{4}{3}\pi r^3$; $r = 3$ (Volume of a sphere) (Use a calculator approximation for π.)

92. $V = \frac{1}{3}\pi r^2 h$; $V = 565.2, r = 6$ (Volume of a cone) (Use a calculator approximation for π.)

9.6 Linear Inequalities and Problem Solving

Objectives

A Graph Inequalities on a Number Line.

B Use the Addition Property of Inequality to Solve Inequalities.

C Use the Multiplication Property of Inequality to Solve Inequalities.

D Use Both Properties to Solve Inequalities.

E Solve Problems Modeled by Inequalities.

In Section 9.1, we reviewed these inequality symbols and their meanings:

$<$ means "is less than" $\leq$ means "is less than or equal to"
$>$ means "is greater than" $\geq$ means "is greater than or equal to"

An **inequality** is a statement that contains one of the symbols above.

Equations	Inequalities
$x = 3$	$x \leq 3$
$5n - 6 = 14$	$5n - 6 > 14$
$12 = 7 - 3y$	$12 \leq 7 - 3y$
$\frac{x}{4} - 6 = 1$	$\frac{x}{4} - 6 > 1$

Objective A Graphing Inequalities on a Number Line

Recall that the single solution to the equation $x = 3$ is 3. The solutions of the inequality $x \leq 3$ include 3 and *all real numbers less than 3* (for example, $-10, \frac{1}{2}$, and 2.9). Because we can't list all numbers less than 3, we instead show a picture of the solutions by graphing them on a number line.

To graph the solutions of $x \leq 3$, we shade the numbers to the left of 3 since they are less than 3. Then we place a closed circle on the point representing 3. The closed circle indicates that 3 *is* a solution: 3 *is* less than or equal to 3.

To graph the solutions of $x < 3$, we shade the numbers to the left of 3. Then we place an open circle on the point representing 3. The open circle indicates that 3 *is not* a solution: 3 *is not* less than 3.

Example 1 Graph: $x \geq -1$

Solution: To graph the solutions of $x \geq -1$, we place a closed circle at -1 since the inequality symbol is $\geq$ and -1 is greater than or equal to -1. Then we shade to the right of -1.

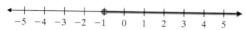

■ Work Practice 1

Practice 1
Graph: $x \geq -2$

Example 2 Graph: $-1 > x$

Solution: Recall from Section 9.1 that $-1 > x$ means the same as $x < -1$. The graph of the solutions of $x < -1$ is shown below.

■ Work Practice 2

Practice 2
Graph: $5 > x$

Example 3 Graph: $-4 < x \leq 2$

Solution: We read $-4 < x \leq 2$ as "-4 is less than x and x is less than or equal to 2," or as "x is greater than -4 and x is less than or equal to 2." To graph the solutions of this inequality, we place an open circle at -4 (-4 is not part of the graph), a closed circle at 2 (2 is part of the graph), and shade all numbers between -4 and 2. Why? All numbers between -4 and 2 are greater than -4 *and* also less than 2.

■ Work Practice 3

Practice 3
Graph: $-3 \leq x < 1$

Objective B Using the Addition Property

When solutions of a linear inequality are not immediately obvious, they are found through a process similar to the one used to solve a linear equation. Our goal is to get the variable alone on one side of the inequality. We use properties of inequality similar to properties of equality.

Addition Property of Inequality
If a, b, and c are real numbers, then
$$a < b \quad \text{and} \quad a + c < b + c$$
are equivalent inequalities.

This property also holds true for subtracting values, since subtraction is defined in terms of addition. In other words, adding or subtracting the same quantity from both sides of an inequality does not change the solutions of the inequality.

Answers

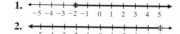

Practice 4

Solve $x - 6 \geq -11$. Graph the solutions.

Example 4
Solve $x + 4 \leq -6$. Graph the solutions.

Solution: To solve for x, subtract 4 from both sides of the inequality.

$x + 4 \leq -6$ Original inequality
$x + 4 - 4 \leq -6 - 4$ Subtract 4 from both sides.
$x \leq -10$ Simplify.

The graph of the solutions is shown below.

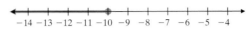

■ Work Practice 4

Helpful Hint

Notice that any number less than or equal to -10 is a solution to $x \leq -10$. For example, solutions include

$$-10, \quad -200, \quad -11\frac{1}{2}, \quad -\sqrt{130}, \quad \text{and} \quad -50.3$$

Objective C Using the Multiplication Property

An important difference between solving linear equations and solving linear inequalities is shown when we multiply or divide both sides of an inequality by a nonzero real number. For example, start with the true statement $6 < 8$ and multiply both sides by 2. As we see below, the resulting inequality is also true.

$6 < 8$ True
$2(6) < 2(8)$ Multiply both sides by 2.
$12 < 16$ True

But if we start with the same true statement $6 < 8$ and multiply both sides by -2, the resulting inequality is not a true statement.

$6 < 8$ True
$-2(6) < -2(8)$ Multiply both sides by -2.
$-12 < -16$ False

Notice, however, that if we reverse the direction of the inequality symbol, the resulting inequality is true.

$-12 < -16$ False
$-12 > -16$ True

This demonstrates the multiplication property of inequality.

Multiplication Property of Inequality

1. If $a, b,$ and c are real numbers, and c is **positive,** then
$$a < b \quad \text{and} \quad ac < bc$$
are equivalent inequalities.
2. If $a, b,$ and c are real numbers, and c is **negative,** then
$$a < b \quad \text{and} \quad ac > bc$$
are equivalent inequalities.

Answer
4. $x \geq -5$

Because division is defined in terms of multiplication, this property also holds true when dividing both sides of an inequality by a nonzero number: If we multiply or divide both sides of an inequality by a negative number, **the direction of the inequality sign must be reversed for the inequalities to remain equivalent.**

Concept Check Fill in each box with $<, >, \leq,$ or $\geq$.

a. Since $-8 < -4$, then $3(-8) \,\square\, 3(-4)$.

b. Since $5 \geq -2$, then $\dfrac{5}{-7} \,\square\, \dfrac{-2}{-7}$.

c. If $a < b$, then $2a \,\square\, 2b$.

d. If $a \geq b$, then $\dfrac{a}{-3} \,\square\, \dfrac{b}{-3}$.

Example 5 Solve $-2x \leq -4$. Graph the solutions.

Solution: Remember to reverse the direction of the inequality symbol when dividing by a negative number.

$$-2x \leq -4$$
$$\dfrac{-2x}{-2} \geq \dfrac{-4}{-2} \quad \text{Divide both sides by } -2 \text{ and reverse the inequality sign.}$$
$$x \geq 2 \quad \text{Simplify.}$$

The graph of the solutions is shown.

Work Practice 5

Practice 5

Solve $-3x \leq 12$. Graph the solutions.

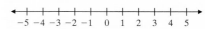

Example 6 Solve $2x < -4$. Graph the solutions.

Solution:

$$2x < -4$$
$$\dfrac{2x}{2} < \dfrac{-4}{2} \quad \text{Divide both sides by 2. Do not reverse the inequality sign.}$$
$$x < -2 \quad \text{Simplify.}$$

The graph of the solutions is shown.

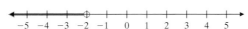

Work Practice 6

Since we cannot list all solutions to an inequality such as $x < -2$, we will use the set notation $\{x \mid x < -2\}$. Recall from Section 9.1 that this is read "the set of all x such that x is less than -2." We will use this notation when solving inequalities.

Objective D Using Both Properties of Inequality

The following steps may be helpful when solving inequalities in one variable. Notice that these steps are similar to the ones given in Section 9.3 for solving equations.

Practice 6

Solve $5x > -20$. Graph the solutions.

Answers

5. $x \geq -4$

6. $x > -4$

✓ **Concept Check Answers**

a. $<$ b. $\leq$ c. $<$ d. $\leq$

To Solve Linear Inequalities in One Variable

Step 1: If an inequality contains fractions or decimals, multiply both sides by the LCD to clear the inequality of fractions or decimals.

Step 2: Use the distributive property to remove parentheses if they appear.

Step 3: Simplify each side of the inequality by combining like terms.

Step 4: Get all variable terms on one side and all numbers on the other side by using the addition property of inequality.

Step 5: Get the variable alone by using the multiplication property of inequality.

Helpful Hint

Don't forget that if both sides of an inequality are multiplied or divided by a negative number, the direction of the inequality sign must be reversed.

Practice 7

Solve $-3x + 11 \leq -13$. Graph the solution set.

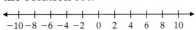

Example 7 Solve $-4x + 7 \geq -9$. Graph the solution set.

Solution:

$$-4x + 7 \geq -9$$
$$-4x + 7 - 7 \geq -9 - 7 \quad \text{Subtract 7 from both sides.}$$
$$-4x \geq -16 \quad \text{Simplify.}$$
$$\frac{-4x}{-4} \leq \frac{-16}{-4} \quad \text{Divide both sides by } -4 \text{ and reverse the direction of the inequality sign.}$$
$$x \leq 4 \quad \text{Simplify.}$$

The graph of the solution set $\{x \mid x \leq 4\}$ is shown.

■ Work Practice 7

Practice 8

Solve $2x - 3 > 4(x - 1)$. Graph the solution set.

Example 8 Solve $-5x + 7 < 2(x - 3)$. Graph the solution set.

Solution: $-5x + 7 < 2(x - 3)$

$$-5x + 7 < 2x - 6 \quad \text{Apply the distributive property.}$$
$$-5x + 7 - 2x < 2x - 6 - 2x \quad \text{Subtract } 2x \text{ from both sides.}$$
$$-7x + 7 < -6 \quad \text{Combine like terms.}$$
$$-7x + 7 - 7 < -6 - 7 \quad \text{Subtract 7 from both sides.}$$
$$-7x < -13 \quad \text{Combine like terms.}$$
$$\frac{-7x}{-7} > \frac{-13}{-7} \quad \text{Divide both sides by } -7 \text{ and reverse the direction of the inequality sign.}$$
$$x > \frac{13}{7} \quad \text{Simplify.}$$

The graph of the solution set $\left\{x \mid x > \frac{13}{7}\right\}$ is shown.

■ Work Practice 8

Answers

7. $\{x \mid x \geq 8\}$

8. $\left\{x \mid x < \frac{1}{2}\right\}$

Section 9.6 | Linear Inequalities and Problem Solving

Example 9 Solve: $2(x - 3) - 5 \leq 3(x + 2) - 18$

Solution:
$$2(x - 3) - 5 \leq 3(x + 2) - 18$$
$$2x - 6 - 5 \leq 3x + 6 - 18 \quad \text{Apply the distributive property.}$$
$$2x - 11 \leq 3x - 12 \quad \text{Combine like terms.}$$
$$-x - 11 \leq -12 \quad \text{Subtract } 3x \text{ from both sides.}$$
$$-x \leq -1 \quad \text{Add 11 to both sides.}$$
$$\frac{-x}{-1} \geq \frac{-1}{-1} \quad \text{Divide both sides by } -1 \text{ and reverse the direction of the inequality sign.}$$
$$x \geq 1 \quad \text{Simplify.}$$

The solution set is $\{x \mid x \geq 1\}$.

Work Practice 9

Practice 9

Solve:
$3(x + 5) - 1 \geq 5(x - 1) + 7$

Objective E Solving Problems Modeled by Inequalities

Problems containing words such as "at least," "at most," "between," "no more than," and "no less than" usually indicate that an inequality should be solved instead of an equation. In solving applications involving linear inequalities, we use the same procedure we used to solve applications involving linear equations.

Some Inequality Translations			
$\geq$	$\leq$	$<$	$>$
at least	at most	is less than	is greater than
no less than	no more than		

Example 10 12 subtracted from 3 times a number is less than 21. Find all numbers that make this statement true.

Solution:

1. UNDERSTAND. Read and reread the problem. This is a direct translation problem, and let's let

   x = the unknown number

2. TRANSLATE.

12	subtracted from	three times a number	is less than	21
		$3x$ − 12	$<$	21

3. SOLVE. $3x - 12 < 21$

   $3x < 33 \quad \text{Add 12 to both sides.}$

   $\dfrac{3x}{3} < \dfrac{33}{3} \quad \text{Divide both sides by 3 and do not reverse the direction of the inequality sign.}$

   $x < 11 \quad \text{Simplify.}$

4. INTERPRET.

   Check: Check the translation; then let's choose a number less than 11 to see if it checks. For example, let's check 10. 12 subtracted from 3 times 10 is 12 subtracted from 30, or 18. Since 18 is less than 21, the number 10 checks.

   State: All numbers less than 11 make the original statement true.

Work Practice 10

Practice 10

Twice a number, subtracted from 35 is greater than 15. Find all numbers that make this true.

Answers
9. $\{x \mid x \leq 6\}$
10. all numbers less than 10

Practice 11

Alex earns $600 per month plus 4% of all his sales. Find the minimum sales that will allow Alex to earn at least $3000 per month.

Example 11 Budgeting for a Wedding

A couple is having their wedding reception at the Gallery reception hall. They may spend at most $2000 for the reception. If the reception hall charges a $100 cleanup fee plus $36 per person, find the greatest number of people that they can invite and still stay within their budget.

Solution:

1. UNDERSTAND. Read and reread the problem. Suppose that 40 people attend the reception. The cost is then $100 + $36(40) = $100 + $1440 = $1540. Let x = the number of people who attend the reception.

2. TRANSLATE.

cleanup fee	+	cost per person	times	number of people	must be less than or equal to	$2000
↓	↓	↓	↓	↓	↓	↓
100	+	36	·	x	≤	2000

3. SOLVE.

$$100 + 36x \leq 2000$$
$$36x \leq 1900 \quad \text{Subtract 100 from both sides.}$$
$$x \leq 52\frac{7}{9} \quad \text{Divide both sides by 36.}$$

4. INTERPRET.

Check: Since x represents the number of people, we round down to the nearest whole, or 52. Notice that if 52 people attend, the cost is $100 + $36(52) = $1972. If 53 people attend, the cost is $100 + $36(53) = $2008, which is more than the given $2000.

State: The couple can invite at most 52 people to the reception.

■ Work Practice 11

Answer
11. $60,000

Vocabulary, Readiness & Video Check

Identify each as an equation, expression, or inequality.

1. $6x - 7(x + 9)$ _____
2. $6x = 7(x + 9)$ _____
3. $6x < 7(x + 9)$ _____
4. $5y - 2 \geq -38$ _____
5. $\dfrac{9}{7} = \dfrac{x + 2}{14}$ _____
6. $\dfrac{9}{7} - \dfrac{x + 2}{14}$ _____

Decide which number listed is not a solution to each given inequality.

7. $x \geq -3$; $-3, 0, -5, \pi$ _____
8. $x < 6$; $-6, |-6|, 0, -3.2$ _____
9. $x < 4.01$; $4, -4.01, 4.1, -4.1$ _____
10. $x \geq -3$; $-4, -3, -2, -(-2)$ _____

Section 9.6 | Linear Inequalities and Problem Solving

Martin-Gay Interactive Videos Watch the section lecture video and answer the following questions.

See Video 9.6

Objective A 11. From Example 1, when graphing an inequality, what inequality symbol(s) does an open circle indicate? What inequality symbol(s) does a closed circle indicate?

Objective B 12. From the lecture before Example 2, which property is the addition property of inequality very similar to?

Objective C 13. What is the answer to Example 3, written in solution set notation?

Objective D 14. When solving Example 4, why is special attention given to the coefficient of x in the last step?

Objective E 15. What is the phrase in Example 5 that tells us to translate to an inequality? What does this phrase translate to?

9.6 Exercise Set MyLab Math

Objective A *Graph each inequality on the number line. See Examples 1 and 2.*

1. $x \leq -1$

2. $y < 0$

3. $x > \dfrac{1}{2}$

4. $z \geq -\dfrac{2}{3}$

5. $y < 4$

6. $x > 3$

7. $-2 \leq m$

8. $-5 \geq x$

Graph each inequality on the number line. See Example 3.

9. $-1 < x < 3$

10. $-2 \leq x \leq 3$

11. $0 \leq y < 2$

12. $-4 < x \leq 0$

Objective B *Solve each inequality. Graph the solution set. Write each answer using set notation. See Example 4.*

13. $x - 2 \geq -7$

14. $x + 4 \leq 1$

15. $-9 + y < 0$

16. $-3 + m > 5$

17. $3x - 5 > 2x - 8$

18. $3 - 7x \geq 10 - 8x$

19. $4x - 1 \leq 5x - 2x$

20. $7x + 3 < 9x - 3x$

Objective C *Solve each inequality. Graph the solution set. Write each answer using set notation. See Examples 5 and 6.*

21. $2x < -6$

22. $3x > -9$

23. $-8x \leq 16$

24. $-5x < 20$

25. $-x > 0$

26. $-y \geq 0$

27. $\dfrac{3}{4}y \geq -2$

28. $\dfrac{5}{6}x \leq -8$

29. $-0.6y < -1.8$

30. $-0.3x > -2.4$

Objectives B C D Mixed Practice *Solve each inequality. Write each answer using set notation. See Examples 4 through 9.*

31. $-8 < x + 7$

32. $-11 > x + 4$

33. $7(x + 1) - 6x \geq -4$

34. $10(x + 2) - 9x \leq -$

35. $4x > 1$

36. $6x < 5$

37. $-\dfrac{2}{3}y \leq 8$

38. $-\dfrac{3}{4}y \geq 9$

39. $4(2z + 1) < 4$

40. $6(2 - z) \geq 12$

41. $3x - 7 < 6x + 2$

42. $2x - 1 \geq 4x - 5$

43. $5x - 7x \leq x + 2$

44. $4 - x < 8x + 2x$

45. $-6x + 2 \geq 2(5 - x)$

46. $-7x + 4 > 3(4 - x)$

47. $3(x - 5) < 2(2x - 1)$

48. $5(x - 2) \leq 3(2x - 1)$

49. $4(3x - 1) \leq 5(2x - 4)$

50. $3(5x - 4) \leq 4(3x - 2)$

51. $3(x + 2) - 6 > -2(x - 3) + 14$

52. $7(x - 2) + x \leq -4(5 - x) - 12$

53. $-5(1 - x) + x \leq -(6 - 2x) + 6$

54. $-2(x - 4) - 3x < -(4x + 1) + 2x$

55. $\frac{1}{4}(x + 4) < \frac{1}{5}(2x + 3)$

56. $\frac{1}{2}(x - 5) < \frac{1}{3}(2x - 1)$

57. $-5x + 4 \leq -4(x - 1)$

58. $-6x + 2 < -3(x + 4)$

Objective E *Solve the following. For Exercises 61 and 62, the solutions have been started for you. See Examples 10 and 11.*

59. Six more than twice a number is greater than negative fourteen. Find all numbers that make this statement true.

60. One more than five times a number is less than or equal to ten. Find all such numbers.

61. The perimeter of a rectangle is to be no greater than 100 centimeters and the width must be 15 centimeters. Find the maximum length of the rectangle.

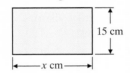

62. One side of a triangle is four times as long as another side, and the third side is 12 inches long. If the perimeter can be no longer than 87 inches, find the maximum lengths of the other two sides.

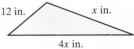

Start the solution:

1. UNDERSTAND the problem. Reread it as many times as needed.
2. TRANSLATE into an equation. (Fill in the blank below.)

the perimeter of the rectangle	is no greater than	100
↓	↓	↓
$x + 15 + x + 15$	____	100

Finish with:

3. SOLVE and 4. INTERPRET

Start the solution:

1. UNDERSTAND the problem. Reread it as many times as needed.
2. TRANSLATE into an equation. (Fill in the blank below.)

the perimeter of the triangle	is no greater than	87
↓	↓	↓
$12 + 4x + x$	____	87

Finish with:

3. SOLVE and 4. INTERPRET

63. Ben Holladay bowled 146 and 201 in his first two games. What must he bowl in his third game to have an average of at least 180? (*Hint:* The average of a list of numbers is their sum divided by the number of numbers in the list.)

64. On an NBA team the two forwards measure 6'8" and 6'6" tall and the two guards measure 6'0" and 5'9" tall. How tall should the center be if they wish to have a starting team average height of at least 6'5"?

65. Dennis and Nancy Wood are celebrating their 30th wedding anniversary by having a reception at Tiffany Oaks reception hall. They have budgeted $3000 for their reception. If the reception hall charges a $50.00 cleanup fee plus $34 per person, find the greatest number of people that they may invite and still stay within their budget.

66. A surprise retirement party is being planned for Pratap Puri. A total of $860 has been collected for the event, which is to be held at a local reception hall. This reception hall charges a cleanup fee of $40 and $15 per person for drinks and light snacks. Find the greatest number of people that may be invited and still stay within the $860 budget.

67. A 150-pound person uses 5.8 calories per minute when walking at a speed of 4 mph. How long must a person walk at this speed to use at least 200 calories? Round up to the nearest minute. (*Source: Home & Garden Bulletin* No. 72)

68. A 170-pound person uses 5.3 calories per minute when bicycling at a speed of 5.5 mph. How long must a person ride a bike at this speed in order to use at least 200 calories? Round up to the nearest minute. (*Source:* Same as Exercise **67**)

Review

Evaluate each expression. See Sections 1.7 and 4.3.

69. 3^4 **70.** 4^3 **71.** 1^8 **72.** 0^7 **73.** $\left(\dfrac{7}{8}\right)^2$ **74.** $\left(\dfrac{2}{3}\right)^3$

The graph shows the number of recreation visitors to Pecos National Historical Park in New Mexico from 2008 to 2016. Use this graph to answer Exercises 75 through 80. See Section 7.1.

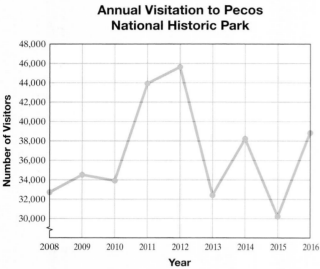

Source: National Park Service

75. How many people visited Pecos National Historical Park in 2008?

76. How many people visited Pecos National Historical Park in 2013?

77. Between which two years did the greatest increase in the number of visitors to Pecos National Historical Park occur?

78. In what year were there more than 45,000 visitors to Pecos National Historical Park?

79. In which year was the number of visitors to Pecos National Historical Park the lowest?

80. Between which two years did the greatest decrease in the number of visitors to Pecos National Historical Park occur?

Concept Extensions

Fill in the box with $<, >, \leq,$ or $\geq$. See the Concept Check in this section.

81. Since $3 < 5$, then $3(-4) \;\square\; 5(-4)$.

82. If $m \leq n$, then $2m \;\square\; 2n$.

83. If $m \leq n$, then $-2m \;\square\; -2n$.

84. If $-x < y$, then $x \;\square\; -y$.

85. When solving an inequality, when must you reverse the direction of the inequality symbol?

86. If both sides of the inequality $-3x < -30$ are divided by 3, do you reverse the direction of the inequality symbol? Why or why not?

Solve.

87. A history major has scores of 75, 83, and 85 on his history tests. Use an inequality to find the scores he can make on his final exam to receive a B in the class. The final exam counts as **two** tests, and a B is received if the final course average is greater than or equal to 80.

88. A mathematics major has scores of 85, 95, and 92 on her college geometry tests. Use an inequality to find the scores she can make on her final exam to receive an A in the course. The final exam counts as **three** tests, and an A is received if the final course average is greater than or equal to 90. Round to one decimal place.

89. Explain how solving a linear inequality is similar to solving a linear equation.

90. Explain how solving a linear inequality is different from solving a linear equation.

Chapter 9 Group Activity

Investigating Averages

Sections 9.1–9.6

Materials:
- small rubber ball or crumpled paper ball
- bucket or waste can

This activity may be completed by working in groups or individually.

1. Try shooting the ball into the bucket or waste can 5 times. Record your results below.

   Shots Made   **Shots Missed**

2. Find your shooting percent for the 5 shots (that is, the percent of the shots you actually made out of the number you tried).

3. Suppose you are going to try an additional 5 shots. How many of the next 5 shots will you have to make to have a 50% shooting percent for all 10 shots? An 80% shooting percent?

4. Did you solve an equation in Question 3? If so, explain what you did. If not, explain how you could use an equation to find the answers.

5. Suppose instead of trying 5 additional shots, you try 22 additional shots. How many of the next 22 shots will you have to make to have at least a 50% shooting percent for all 27 shots? At least a 70% shooting percent?

6. Choose one of the sports played at your college that is currently in season. How many regular-season games are scheduled? What is the team's current percent of games won?

7. Suppose the team has a goal of finishing the season with a winning percent better than 110% of their current wins. At least how many of the remaining games must they win to achieve their goal?

Chapter 9 Vocabulary Check

Fill in each blank with one of the words or phrases listed below.

no solution	all real numbers	linear equation in one variable
reciprocals	formula	reversed
equivalent inequalities	opposites	
linear inequality in one variable	the same	

1. A(n) _____ can be written in the form $ax + b = c$.
2. Inequalities that have the same solution are called _____.
3. An equation that describes a known relationship among quantities is called a(n) _____.
4. A(n) _____ can be written in the form $ax + b < c$ (or $>, \leq, \geq$).
5. The solution(s) to the equation $x + 5 = x + 5$ is/are _____.
6. The solution(s) to the equation $x + 5 = x + 4$ is/are _____.
7. If both sides of an inequality are multiplied or divided by the same positive number, the direction of the inequality symbol is _____.

8. If both sides of an inequality are multiplied by the same negative number, the direction of the inequality symbol is _____.
9. Two numbers whose sum is 0 are called _____.
10. Two numbers whose product is 1 are called _____.

Helpful Hint
Are you preparing for your test? To help, don't forget to take these:
- Chapter 9 Getting Ready for the Test on page 733
- Chapter 9 Test on page 734

Then check all of your answers at the back of this text. For further review, the step-by-step video solutions to any of these exercises are located in MyLab Math.

9 Chapter Highlights

Definitions and Concepts	Examples
Section 9.1 Symbols and Sets of Numbers	
A **set** is a collection of objects, called **elements**, enclosed in braces.	$\{a, c, e\}$
A line used to picture numbers is called a **number line**.	$\begin{array}{c}\leftarrow\!+\!+\!+\!+\!+\!+\!+\!+\!+\!+\!+\!\rightarrow\\ -5\ -4\ -3\ -2\ -1\ \ 0\ \ 1\ \ 2\ \ 3\ \ 4\ \ 5\end{array}$
Natural numbers: $\{1, 2, 3, 4, \ldots\}$ **Whole numbers:** $\{0, 1, 2, 3, 4, \ldots\}$ **Integers:** $\{\ldots, -3, -2, -1, 0, 1, 2, 3, \ldots\}$ **Rational numbers:** {real numbers that can be expressed as quotients of integers} **Irrational numbers:** {real numbers that cannot be expressed as quotients of integers} **Real numbers:** {all numbers that correspond to points on a number line}	Given the set $\left\{-3.4, \sqrt{3}, 0, \dfrac{2}{3}, 5, -4\right\}$, list the numbers that belong to the set of Natural numbers 5 Whole numbers 0, 5 Integers $-4, 0, 5$ Rational numbers $-4, -3.4, 0, \dfrac{2}{3}, 5$ Irrational numbers $\sqrt{3}$ Real numbers all numbers in the given set

Symbols: $=$ is equal to
               $\neq$ is not equal to
               $>$ is greater than
               $<$ is less than
               $\leq$ is less than or equal to
               $\geq$ is greater than or equal to
	$-7 = -7$
	 $3 \neq -3$
	 $4 > 1$
	 $1 < 4$
	 $6 \leq 6$
	 $18 \geq -\dfrac{1}{3}$

Section 9.2 Properties of Real Numbers	
Commutative Properties Addition: $a + b = b + a$ Multiplication: $a \cdot b = b \cdot a$	$3 + (-7) = -7 + 3$ $-8 \cdot 5 = 5 \cdot (-8)$
Associative Properties Addition: $(a + b) + c = a + (b + c)$ Multiplication: $(a \cdot b) \cdot c = a \cdot (b \cdot c)$	$(5 + 10) + 20 = 5 + (10 + 20)$ $(-3 \cdot 2) \cdot 11 = -3 \cdot (2 \cdot 11)$

Chapter 9 Highlights

Definitions and Concepts	Examples
Section 9.2 Properties of Real Numbers *(continued)*	

Two numbers whose product is 1 are called **multiplicative inverses** or **reciprocals**. The reciprocal of a nonzero number a is $\frac{1}{a}$ because $a \cdot \frac{1}{a} = 1$.

The reciprocal of 3 is $\frac{1}{3}$.

The reciprocal of $-\frac{2}{5}$ is $-\frac{5}{2}$.

Distributive Property

$a(b + c) = a \cdot b + a \cdot c$

$5(6 + 10) = 5 \cdot 6 + 5 \cdot 10$
$-2(3 + x) = -2 \cdot 3 + (-2)(x)$

Identities

$a + 0 = a \quad 0 + a = a$
$a \cdot 1 = a \quad 1 \cdot a = a$

$5 + 0 = 5 \quad\quad 0 + (-2) = -2$
$-14 \cdot 1 = -14 \quad\quad 1 \cdot 27 = 27$

Inverses

Additive or opposite: $a + (-a) = 0$

$7 + (-7) = 0$

Multiplicative or reciprocal: $b \cdot \frac{1}{b} = 1, \quad b \neq 0$

$3 \cdot \frac{1}{3} = 1$

Section 9.3 Further Solving Linear Equations

To Solve Linear Equations

Solve: $\dfrac{5(-2x + 9)}{6} + 3 = \dfrac{1}{2}$

1. Clear the equation of fractions.

   1. $6 \cdot \dfrac{5(-2x + 9)}{6} + 6 \cdot 3 = 6 \cdot \dfrac{1}{2}$

2. Remove any grouping symbols such as parentheses.

   2. $5(-2x + 9) + 18 = 3$ Apply the distributive property.
   $-10x + 45 + 18 = 3$

3. Simplify each side by combining like terms.

   3. $-10x + 63 = 3$ Combine like terms.

4. Get all variable terms on one side and all numbers on the other side by using the addition property of equality.

   4. $-10x + 63 - 63 = 3 - 63$ Subtract 63.
   $-10x = -60$

5. Get the variable alone by using the multiplication property of equality.

   5. $\dfrac{-10x}{-10} = \dfrac{-60}{-10}$ Divide by -10.
   $x = 6$

6. Check the solution by substituting it into the original equation.

Section 9.4 Further Problem Solving

Problem-Solving Steps

The height of the Hudson volcano in Chile is twice the height of the Kiska volcano in the Aleutian Islands. If the sum of their heights is 12,870 feet, find the height of each.

1. UNDERSTAND the problem.

   1. Read and reread the problem. Guess a solution and check your guess.

   Let x be the height of the Kiska volcano. Then $2x$ is the height of the Hudson volcano.

    x $2x$

(continued)

Definitions and Concepts	Examples
Section 9.4 Further Problem Solving (continued)	
2. TRANSLATE the problem.	2. height of Kiska added to height of Hudson is 12,870 $\quad\quad\quad x \quad\quad + \quad\quad 2x \quad\quad = \quad 12{,}870$
3. SOLVE the equation.	3. $x + 2x = 12{,}870$ $\quad\quad 3x = 12{,}870$ $\quad\quad\; x = 4290$
4. INTERPRET the results.	4. *Check:* If x is 4290, then $2x$ is $2(4290)$ or 8580. Their sum is $4290 + 8580$ or 12,870, the required amount. *State:* The Kiska volcano is 4290 feet tall, and the Hudson volcano is 8580 feet tall.
Section 9.5 Formulas and Problem Solving	
An equation that describes a known relationship among quantities is called a **formula.** **To solve a formula for a specified variable,** use the same steps as for solving a linear equation. Treat the specified variable as the only variable of the equation.	$A = lw$ (area of a rectangle) $I = PRT$ (simple interest) Solve: $P = 2l + 2w$ for l $\quad P = 2l + 2w$ $P - 2w = 2l + 2w - 2w \quad$ Subtract $2w$. $P - 2w = 2l$ $\dfrac{P - 2w}{2} = \dfrac{2l}{2} \quad$ Divide by 2. $\dfrac{P - 2w}{2} = l$
Section 9.6 Linear Inequalities and Problem Solving	
Properties of inequalities are similar to properties of equations. However, if you multiply or divide both sides of an inequality by the same *negative* number, you must reverse the direction of the inequality symbol.	$-2x \le 4$ $\dfrac{-2x}{-2} \ge \dfrac{4}{-2} \quad$ Divide by -2; reverse the inequality symbol. $x \ge -2$ 
To Solve Linear Inequalities 1. Clear the inequality of fractions. 2. Remove grouping symbols. 3. Simplify each side by combining like terms. 4. Write all variable terms on one side and all numbers on the other side using the addition property of inequality. 5. Get the variable alone by using the multiplication property of inequality.	Solve: $3(x + 2) \le -2 + 8$ 1. $3(x + 2) \le -2 + 8 \quad$ No fractions to clear. 2. $\quad 3x + 6 \le -2 + 8 \quad$ Apply the distributive property. 3. $\quad 3x + 6 \le 6 \quad$ Combine like terms. 4. $3x + 6 - 6 \le 6 - 6 \quad$ Subtract 6. $\quad 3x \le 0$ 5. $\quad \dfrac{3x}{3} \le \dfrac{0}{3} \quad$ Divide by 3. $\quad x \le 0$ The solution set is $\{x \mid x \le 0\}$.

Review   Chapter 9

(9.1) *Insert <, >, or = in the appropriate space to make each statement true. (See Section 2.1 for a review of absolute value.)*

1. 8   10
2. 7   2
3. −4   −5
4. $\dfrac{12}{2}$ −8
5. $|-7|$ $|-8|$
6. $|-9|$ −9
7. $-|-1|$ −1
8. $|-14|$ $-(-14)$
9. 1.2   1.02
10. $-\dfrac{3}{2}$ $-\dfrac{3}{4}$

TRANSLATING *Translate each statement into symbols.*

11. Four is greater than or equal to negative three.

12. Six is not equal to five.

13. 0.03 is less than 0.3.

14. For the 2014–2015 year, Missouri had an $18.9 million tourism budget and Virginia had a $20.8 million tourism budget. Write an inequality statement comparing the numbers 18.9 million and 20.8 million. (*Source:* U.S. Travel Association)

Given the sets of numbers below, list the numbers in each set that also belong to the set of:

a. Natural numbers   **b.** Whole numbers
c. Integers   **d.** Rational numbers
e. Irrational numbers   **f.** Real numbers

15. $\left\{-6, 0, 1, 1\dfrac{1}{2}, 3, \pi, 9.62\right\}$

16. $\left\{-3, -1.6, 2, 5, \dfrac{11}{2}, 15.1, \sqrt{5}, 2\pi\right\}$

The following chart shows the gains and losses in dollars of Density Oil and Gas stock for a particular week. Use this chart to answer Exercises 17 and 18.

Day	Gain or Loss (in dollars)
Monday	+1
Tuesday	−2
Wednesday	+5
Thursday	+1
Friday	−4

17. Which day showed the greatest loss?

18. Which day showed the greatest gain?

(9.2) *Name the property illustrated in each equation.*

19. $-6 + 5 = 5 + (-6)$

20. $6 \cdot 1 = 6$

21. $3(8 - 5) = 3 \cdot 8 + 3 \cdot (-5)$

729

22. $4 + (-4) = 0$

23. $2 + (3 + 9) = (2 + 3) + 9$

24. $2 \cdot 8 = 8 \cdot 2$

25. $6(8 + 5) = 6 \cdot 8 + 6 \cdot 5$

26. $(3 \cdot 8) \cdot 4 = 3 \cdot (8 \cdot 4)$

27. $4 \cdot \dfrac{1}{4} = 1$

28. $8 + 0 = 8$

(9.3) *Solve each equation.*

29. $\dfrac{5}{3}x + 4 = \dfrac{2}{3}x$

30. $\dfrac{7}{8}x + 1 = \dfrac{5}{8}x$

31. $-(5x + 1) = -7x + 3$

32. $-4(2x + 1) = -5x + 5$

33. $-6(2x - 5) = -3(9 + 4x)$

34. $3(8y - 1) = 6(5 + 4y)$

35. $\dfrac{3(2 - z)}{5} = z$

36. $\dfrac{4(n + 2)}{5} = -n$

37. $0.5(2n - 3) - 0.1 = 0.4(6 + 2n)$

38. $1.72y - 0.04y = 0.42$

39. $\dfrac{5(c + 1)}{6} = 2c - 3$

40. $-9 - 5a = 3(6a - 1)$

(9.4) *Solve each of the following.*

41. The height of the Washington Monument is 50.5 inches more than 10 times the length of a side of its square base. If the sum of these two dimensions is 7327 inches, find the height of the Washington Monument. (*Source:* National Park Service)

42. A 12-foot board is to be divided into two pieces so that one piece is twice as long as the other. If x represents the length of the shorter piece, find the length of each piece.

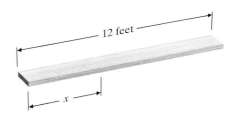

43. The National Park System in the United States includes a variety of park unit types. In 2016, there were a total of 41 national battlefields and national memorials. The number of national memorials is three less than three times the number of national battlefields. How many of each park unit were there? (*Source:* National Park System)

44. Find three consecutive integers whose sum is -114.

45. The quotient of a number and 3 is the same as the difference of the number and 2. Find the number.

46. Double the sum of a number and 6 is the opposite of the number. Find the number.

(9.5) *Substitute the given values into the given formulas and solve for the unknown variable.*

47. $P = 2l + 2w$; $P = 46, l = 14$

48. $V = lwh$; $V = 192, l = 8, w = 6$

Solve each equation for the indicated variable.

49. $y = mx + b$ for m

50. $r = vst - 5$ for s

51. $2y - 5x = 7$ for x

52. $3x - 6y = -2$ for y

△ **53.** $C = \pi d$ for d

△ **54.** $C = 2\pi r$ for r

55. A swimming pool holds 900 cubic meters of water. If its length is 20 meters and its height is 3 meters, find its width.

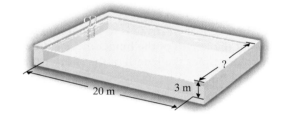

56. The perimeter of a rectangular billboard is 60 feet and has a length 6 feet longer than its width. Find the dimensions of the billboard.

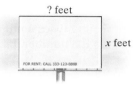

57. A charity 10K race is given annually to benefit a local hospice organization. How long would it take to run/walk a 10K race (10 kilometers or 10,000 meters) if your average pace is 125 **meters** per minute? Give your time in hours and minutes.

58. On September 14, 2013, the highest temperature recorded in the United States was 113°F, which occurred in Death Valley, California. Convert this temperature to degrees Celsius. (*Source:* National Weather Service)

(9.6) *Graph on the given number line.*

59. $x \leq -2$

60. $0 < x \leq 5$

Solve each inequality. Write each answer using set notation.

61. $x - 5 \leq -4$

62. $x + 7 > 2$

63. $-2x \geq -20$

64. $-3x > 12$

65. $5x - 7 > 8x + 5$

66. $x + 4 \geq 6x - 16$

67. $\dfrac{2}{3}y > 6$

68. $-0.5y \leq 7.5$ **69.** $-2(x - 5) > 2(3x - 2)$ **70.** $4(2x - 5) \leq 5x - 1$

71. A part-time salesperson earns $175 per week plus a 5% commission on all her sales. Find the minimum amount of sales she must make to ensure that she earns at least $300 per week.

72. An amateur golfer shot rounds of 76, 82, and 79 golfing. What must he shoot on his next round so that his average will be below 80?

Mixed Review

Solve each equation.

73. $6x + 2x - 1 = 5x + 11$ **74.** $2(3y - 4) = 6 + 7y$ **75.** $4(3 - a) - (6a + 9) = -12a$

76. $\dfrac{x}{3} - 2 = 5$ **77.** $2(y + 5) = 2y + 10$ **78.** $7x - 3x + 2 = 2(2x - 1)$

Solve.

79. The sum of six and twice a number is equal to seven less than the number. Find the number.

80. A 23-inch piece of string is to be cut into two pieces so that the length of the longer piece is three more than four times the length of the shorter piece. Find the lengths of both pieces.

Solve for the specified variable.

81. $V = \dfrac{1}{3}Ah$ for h

Solve each inequality. Graph the solution set. Write each answer using set notation.

82. $4x - 7 > 3x + 2$

83. $-5x < 20$

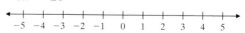

84. $-3(1 + 2x) + x \geq -(3 - x)$

Chapter 9   Getting Ready for the Test

MULTIPLE CHOICE Exercises 1 through 21 are **Multiple Choice**. Choose the correct letter.

For Exercises 1 and 2, choose two correct letters for each exercise.

1. Select the two numbers that are halfway between 2 and 3 on the number line.
   A. $3\frac{1}{2}$ **B.** 3.5   **C.** $2\frac{1}{2}$ **D.** 2.5

2. Select the two numbers that are halfway between -2 and -3 on the number line.
   A. $-3\frac{1}{2}$ **B.** -3.5 **C.** $-2\frac{1}{2}$ **D.** -2.5

For Exercises 3 through 5, choose the property that allows us to write the left side of each equal sign as the equivalent right side of the equal sign. Choices are below.

   A. distributive property   **B.** associative property   **C.** commutative property

3. $5(2x - 1) = 10x - 5$ 4. $9 + x = x + 9$
5. $5(4x) = (5 \cdot 4)x$

For Exercises 6 through 9, the exercise statement and the correct answer are given. Select the correct directions.

   A. Find the opposite.   **B.** Find the reciprocal.   **C.** Evaluate or simplify.

6. 5   Answer: $\frac{1}{5}$ 7. $3 + 2(-8)$ Answer: -13
8. 2^3 Answer: 8   9. -7 Answer: 7

For Exercises 10 through 13, identify each as an
   A. equation *or an*   **B.** expression

10. $6x + 2 + 4x - 10$ 11. $6x + 2 = 4x - 10$
12. $-2(x - 1) = 12$ 13. $-7\left(x + \frac{1}{2}\right) - 22$

14. Subtracting $100z$ from $8m$ translates to:
    A. $100z - 8m$ **B.** $8m - 100z$ **C.** $-800zm$ **D.** $92zm$

15. Subtracting $7x - 1$ from $9y$ translates to:
    A. $7x - 1 - 9y$ **B.** $9y - 7x - 1$ **C.** $9y - (7x - 1)$ **D.** $7x - 1 - (9y)$

For Exercises 16 through 19, an equation is given. Choose the correct solution.
   A. all real numbers   **B.** no solution   **C.** the solution is 0

16. $7x + 6 = 7x + 9$ 17. $2y - 5 = 2y - 5$
18. $11x - 13 = 10x - 13$ 19. $x + 15 = -x + 15$

20. To solve $5(3x - 2) = -(x + 20)$, we first use the distributive property and remove parentheses by multiplying. Once this is done, the equation is
    A. $15x - 2 = -x + 20$ **B.** $15x - 10 = -x - 20$ **C.** $15x - 10 = -x + 20$ **D.** $15x - 7 = -x - 20$

21. To solve $\frac{8x}{3} + 1 = \frac{x - 2}{10}$ we multiply through by the LCD, 30. Once this is done, the simplified equation is
    A. $80x + 1 = 3x - 6$ **B.** $80x + 6 = 3x - 6$ **C.** $8x + 1 = x - 2$ **D.** $80x + 30 = 3x - 6$

Chapter 9 Test

Translate each statement into symbols.

1. The absolute value of negative seven is greater than five.

2. The sum of nine and five is greater than or equal to four.

3. Given $\{-5, -1, \frac{1}{4}, 0, 1, 7, 11.6, \sqrt{7}, 3\pi\}$, list the numbers in this set that also belong to the set o

    a. Natural numbers   **b.** Whole numbers
    c. Integers   **d.** Rational numbers
    e. Irrational numbers   **f.** Real numbers

Identify the property illustrated by each expression.

4. $8 + (9 + 3) = (8 + 9) + 3$

5. $6 \cdot 8 = 8 \cdot 6$

6. $-6(2 + 4) = -6 \cdot 2 + (-6) \cdot 4$

7. $\frac{1}{6}(6) = 1$

8. Find the opposite of -9.

9. Find the reciprocal of $-\frac{1}{3}$.

Use the distributive property to write each expression without parentheses. Then simplify if possible.

10. $7 + 2(5y - 3)$

11. $4(x - 2) - 3(2x - 6)$

Solve each equation.

12. $4(n - 5) = -(4 - 2n)$

13. $-2(x - 3) = x + 5 - 3x$

14. $4z + 1 - z = 1 + z$

15. $\frac{2(x + 6)}{3} = x - 5$

16. $\frac{1}{2} - x + \frac{3}{2} = x - 4$

17. $-0.3(x - 4) + x = 0.5(3 - x)$

18. $-4(a + 1) - 3a = -7(2a - 3)$

734

Chapter 9 Test

Solve each application.

19. A number increased by two-thirds of the number is 35. Find the number.

20. A gallon of water seal covers 200 square feet. How many gallons are needed to paint two coats of water seal on a deck that measures 20 feet by 35 feet?

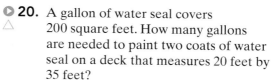

21. Find the value of x if $y = -14$, $m = -2$, and $b = -2$ in the formula $y = mx + b$.

Solve the equation for the indicated variable.

22. $V = \pi r^2 h$ for h

23. $3x - 4y = 10$ for y

Solve the inequality. Graph the solution set. Write the answer using set notation.

24. $3x - 5 > 7x + 3$

Solve each inequality. Write each answer using set notation.

25. $-0.3x \geq 2.4$

26. $-5(x - 1) + 6 \leq -3(x + 4) + 1$

27. $\dfrac{2(5x + 1)}{3} > 2$

28. California has more public libraries than any other state. It has 387 more public libraries than Ohio. If the total number of public libraries for these states is 1827, find the number of public libraries in California and the number in Ohio. (*Source:* Institute of Museum and Library Services)

19. _____

20. _____

21. _____

22. _____

23. _____

24. _____

25. _____

26. _____

27. _____

28. _____

Chapters 1–9 Cumulative Review

Answers

1. a. ____
   b. ____
   c. ____
2. a. ____
   b. ____
   c. ____
   d. ____
   e. ____
3. ____
4. ____
5. ____
6. ____
7. ____
8. ____
9. ____
10. ____
11. ____
12. ____
13. ____
14. ____
15. ____
16. ____
17. ____
18. ____
19. ____
20. ____
21. ____
22. ____

Simplify.

1. a. $-(-4)$ b. $-|-5|$ c. $-|6|$

2. Insert $<$, $>$, or $=$ in the appropriate space to make each statement true.
   a. $|0|\ \ 2$ b. $|-5|\ \ 5$
   c. $|-3|\ \ |-2|$ d. $|5|\ \ |6|$
   e. $|-7|\ \ |6|$

3. Simplify: $-2[-3 + 2(-1 + 6)] - 5$

4. Simplify: $\dfrac{3 + |4 - 3| + 2^2}{6 - 4}$

Add.

5. $-18 + 10$ **6.** $-8 + (-11)$ **7.** $12 + (-8)$

8. $-2 + 10$ **9.** $(-3) + 4 + (-11)$ **10.** $0.2 + (-0.5)$

Perform the indicated operation.

11. $\dfrac{2}{3} - \dfrac{10}{11}$ **12.** $\dfrac{2}{5} - \dfrac{39}{40}$ **13.** $2 - \dfrac{x}{3}$ **14.** $5 + \dfrac{x}{2}$

Evaluate. Let $x = -\dfrac{1}{2}$ and $y = \dfrac{1}{3}$.

15. $2x + y^2$ **16.** $2y + x^2$

Perform the indicated operation.

17. $-4\dfrac{2}{5} \cdot 1\dfrac{3}{11}$ **18.** $-2\dfrac{1}{2} \cdot \left(-2\dfrac{1}{2}\right)$ **19.** $-2\dfrac{1}{3} \div \left(-2\dfrac{1}{2}\right)$

20. $3\dfrac{3}{5} \div \left(-3\dfrac{1}{3}\right)$ **21.** $(-1.3)^2 + 2.4$ **22.** $1.2(7.3 - 9.3)$

23. $\sqrt{49}$ **24.** $\sqrt{64}$ **25.** $\sqrt{\dfrac{4}{25}}$ **26.** $\sqrt{\dfrac{9}{100}}$

Solve.

27. $1.2x + 5.8 = 8.2$

28. $1.3x - 2.6 = -9.1$

Determine whether each statement is true or false.

29. $8 \geq 8$ **30.** $-4 \leq -4$ **31.** $8 \leq 8$ **32.** $-4 \geq -4$

33. $23 \leq 0$ **34.** $-8 \leq 0$ **35.** $0 \leq 23$ **36.** $0 \leq -8$

Use the distributive property to write each expression without parentheses. Then simplify the result.

37. $-5(-3 + 2z)$ **38.** $-4(2x - 1)$ **39.** $\dfrac{1}{2}(6x + 14) + 10$ **40.** $9 + 2(5x + 6)$

Solve.

41. $\dfrac{2(a + 3)}{3} = 6a + 2$

42. $\dfrac{x}{2} + \dfrac{x}{5} = 3$

43. As of January 2017, the total number of Democrats and Republicans in the U.S. House of Representatives was 435. There were 47 more Republican representatives than Democratic. Find the number of representatives from each party. (*Source:* Congressional Research Service)

44. Three times the sum of a number and 2 is the same as 9 times the number. Find the number.

45. A glacier is a giant mass of rocks and ice that flows downhill like a river. Portage Glacier in Alaska is about 6 miles, or 31,680 feet, long and moves 400 feet per year. Icebergs are created when the front end of the glacier flows into Portage Lake. How long does it take for ice at the head (beginning) of the glacier to reach the lake?

46. Solve $V = lwh$ for w.

47. Graph $-1 > x$.

$\leftarrow \;|\;|\;|\;|\;|\;|\;|\;|\;|\;|\;|\; \rightarrow$
$\;\;-5\;-4\;-3\;-2\;-1\;\;0\;\;1\;\;2\;\;3\;\;4\;\;5$

48. Solve: $-3x < -30$

49. Solve: $2(x - 3) - 5 \leq 3(x + 2) - 18$

50. Solve: $10 + x < 6x - 10$

23. _____
24. _____
25. _____
26. _____
27. _____
28. _____
29. _____
30. _____
31. _____
32. _____
33. _____
34. _____
35. _____
36. _____
37. _____
38. _____
39. _____
40. _____
41. _____
42. _____
43. _____
44. _____
45. _____
46. _____
47. _____
48. _____
49. _____
50. _____

10 Exponents and Polynomials

Recall from Chapter 1 that an exponent is a shorthand notation for repeated factors. This chapter explores additional concepts about exponents and exponential expressions. An especially useful type of exponential expression is a polynomial. Polynomials model many real-world phenomena. Our goal in this chapter is to become proficient with operations on polynomials.

Sections

10.1 Exponents

10.2 Negative Exponents and Scientific Notation

10.3 Introduction to Polynomials

10.4 Adding and Subtracting Polynomials

10.5 Multiplying Polynomials

10.6 Special Products

Integrated Review— Exponents and Operations on Polynomials

10.7 Dividing Polynomials

Check Your Progress

Vocabulary Check
Chapter Highlights
Chapter Review
Getting Ready for the Test
Chapter Test
Cumulative Review

How Do You Listen to Music? Downloading? Streaming? A Physical CD or LP/Vinyl?

No matter how you listen to music, this industry is booming. In the United States, the music industry has grown to an estimated $15 billion. The number of paid subscribers to subscription streaming is increasing, and the number of digital music downloads is expected to decline. Interestingly enough, LP/vinyl albums are making a comeback. The bar graph below shows the increase in LP/vinyl album sales in the United States, but also study the circle graph to its right. Notice that although LP/vinyl album sales are increasing, they still represent a small part of the "total album sales pie". These sales are mostly through digital and CDs.

In Section 10.3, Exercises 25 and 26, we use the data below to predict future sales of LP/vinyl albums. (*Source:* IFPI.org)

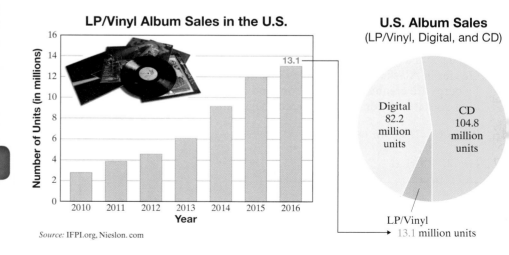

10.1 Exponents

Objective A Evaluating Exponential Expressions

In this section, we continue our work with integer exponents. Recall from Section 1.7 that repeated multiplication of the same factor can be written using exponents. For example,

$$2 \cdot 2 \cdot 2 \cdot 2 \cdot 2 = 2^5$$

The exponent 5 tells us how many times 2 is a factor. The expression 2^5 is called an **exponential expression**. It is also called the fifth **power** of 2, or we can say that 2 is **raised** to the fifth power.

$$5^6 = \underbrace{5 \cdot 5 \cdot 5 \cdot 5 \cdot 5 \cdot 5}_{\text{6 factors; each factor is 5}} \quad \text{and} \quad (-3)^4 = \underbrace{(-3) \cdot (-3) \cdot (-3) \cdot (-3)}_{\text{4 factors; each factor is } -3}$$

The base of an exponential expression is the repeated factor. The exponent is the number of times that the base is used as a factor.

$$\underset{\text{base}}{\uparrow} a^{n \leftarrow \text{exponent or power}} = \underbrace{a \cdot a \cdot a \cdots a}_{n \text{ factors; each factor is } a}$$

Objectives

A Evaluate Exponential Expressions.

B Use the Product Rule for Exponents.

C Use the Power Rule for Exponents.

D Use the Power Rules for Products and Quotients.

E Use the Quotient Rule for Exponents, and Define a Number Raised to the 0 Power.

F Decide Which Rule(s) to Use to Simplify an Expression.

Examples Evaluate each expression.

1. $2^3 = 2 \cdot 2 \cdot 2 = 8$
2. $3^1 = 3$. To raise 3 to the first power means to use 3 as a factor only once. When no exponent is shown, the exponent is assumed to be 1.
3. $(-4)^2 = (-4)(-4) = 16$
4. $-4^2 = -(4 \cdot 4) = -16$
5. $\left(\dfrac{1}{2}\right)^4 = \dfrac{1}{2} \cdot \dfrac{1}{2} \cdot \dfrac{1}{2} \cdot \dfrac{1}{2} = \dfrac{1}{16}$
6. $4 \cdot 3^2 = 4 \cdot 9 = 36$

■ Work Practice 1–6

Notice how similar -4^2 is to $(-4)^2$ in the examples above. The difference between the two is the parentheses. In $(-4)^2$, the parentheses tell us that the base, or the repeated factor, is -4. In -4^2, only 4 is the base.

Practice 1–6

Evaluate each expression.

1. 3^4 2. 7^1 3. $(-2)^3$
4. -2^3 5. $\left(\dfrac{2}{3}\right)^2$ 6. $5 \cdot 6^2$

Helpful Hint

Be careful when identifying the base of an exponential expression. Pay close attention to the use of parentheses.

$(-3)^2$ -3^2 $2 \cdot 3^2$

The base is -3.   The base is 3.   The base is 3.

$(-3)^2 = (-3)(-3) = 9$ $-3^2 = -(3 \cdot 3) = -9$ $2 \cdot 3^2 = 2 \cdot 3 \cdot 3 = 18$

An exponent has the same meaning whether the base is a number or a variable. If x is a real number and n is a positive integer, then x^n is the product of n factors, each of which is x.

$$x^n = \underbrace{x \cdot x \cdot x \cdot x \cdot x \cdots x}_{n \text{ factors; each factor is } x}$$

Answers

1. 81 2. 7 3. -8 4. -8 5. $\dfrac{4}{9}$
6. 180

Practice 7

Evaluate each expression for the given value of x.

a. $3x^2$ when x is 4

b. $\dfrac{x^4}{-8}$ when x is -2

Example 7 Evaluate each expression for the given value of x.

a. $2x^3$ when x is 5

b. $\dfrac{9}{x^2}$ when x is -3

Solution:

a. When x is 5, $2x^3 = 2 \cdot 5^3$
$= 2 \cdot (5 \cdot 5 \cdot 5)$
$= 2 \cdot 125$
$= 250$

b. When x is -3, $\dfrac{9}{x^2} = \dfrac{9}{(-3)^2}$
$= \dfrac{9}{(-3)(-3)}$
$= \dfrac{9}{9} = 1$

■ Work Practice 7

Objective B Using the Product Rule

Exponential expressions can be multiplied, divided, added, subtracted, and themselves raised to powers. Let's see if we can discover a shortcut method for multiplying exponential expressions with the same base. By our definition of an exponent,

$$5^4 \cdot 5^3 = \underbrace{(5 \cdot 5 \cdot 5 \cdot 5)}_{4 \text{ factors of } 5} \cdot \underbrace{(5 \cdot 5 \cdot 5)}_{3 \text{ factors of } 5}$$

$$= \underbrace{5 \cdot 5 \cdot 5 \cdot 5 \cdot 5 \cdot 5 \cdot 5}_{7 \text{ factors of } 5}$$

$$= 5^7$$

Also,

$$x^2 \cdot x^3 = (x \cdot x) \cdot (x \cdot x \cdot x)$$
$$= x \cdot x \cdot x \cdot x \cdot x$$
$$= x^5$$

In both cases, notice that the result is exactly the same if the exponents are added.

$$5^4 \cdot 5^3 = 5^{4+3} = 5^7 \quad \text{and} \quad x^2 \cdot x^3 = x^{2+3} = x^5$$

This suggests the following rule.

Product Rule for Exponents

If m and n are positive integers and a is a real number, then

$a^m \cdot a^n = a^{m+n}$ ← Add exponents.
         ↑
         └── Keep common base.

For example,

$3^5 \cdot 3^7 = 3^{5+7} = 3^{12}$ ← Add exponents.
         ↑
         └── Keep common base.

Answers

7. a. 48 b. -2

Helpful Hint

Don't forget that

$3^5 \cdot 3^7 \neq 9^{12}$ ← Add exponents.

Common base *not* kept.

$3^5 \cdot 3^7 = \underbrace{3 \cdot 3 \cdot 3 \cdot 3 \cdot 3}_{5 \text{ factors of } 3} \cdot \underbrace{3 \cdot 3 \cdot 3 \cdot 3 \cdot 3 \cdot 3 \cdot 3}_{7 \text{ factors of } 3}$

$= 3^{12}$ 12 factors of 3, *not* 9

In other words, to multiply two exponential expressions with the **same base,** we keep the base and add the exponents. We call this **simplifying** the exponential expression.

Examples Use the product rule to simplify each expression.

8. $4^2 \cdot 4^5 = 4^{2+5} = 4^7$ ← Add exponents.

   Keep common base.

9. $x^2 \cdot x^5 = x^{2+5} = x^7$

10. $y^3 \cdot y = y^3 \cdot y^1$
    $= y^{3+1}$
    $= y^4$

Helpful Hint
Don't forget that if no exponent is written, it is assumed to be 1.

11. $y^3 \cdot y^2 \cdot y^7 = y^{3+2+7} = y^{12}$

12. $(-5)^7 \cdot (-5)^8 = (-5)^{7+8} = (-5)^{15}$

Work Practice 8–12

Practice 8–12

Use the product rule to simplify each expression.
8. $7^3 \cdot 7^2$ 9. $x^4 \cdot x^9$
10. $r^5 \cdot r$ 11. $s^6 \cdot s^2 \cdot s^3$
12. $(-3)^9 \cdot (-3)$

✓**Concept Check** Where possible, use the product rule to simplify the expression.

a. $z^2 \cdot z^{14}$ b. $x^2 \cdot z^{14}$ c. $9^8 \cdot 9^3$ d. $9^8 \cdot 2^7$

Example 13 Use the product rule to simplify $(2x^2)(-3x^5)$.

Solution: Recall that $2x^2$ means $2 \cdot x^2$ and $-3x^5$ means $-3 \cdot x^5$.

$(2x^2)(-3x^5) = (2 \cdot x^2) \cdot (-3 \cdot x^5)$

$= (2 \cdot -3) \cdot (x^2 \cdot x^5)$ Group factors with common bases (using commutative and associative properties).

$= -6x^7$ Simplify.

Work Practice 13

Practice 13

Use the product rule to simplify $(6x^3)(-2x^9)$.

Examples Simplify.

14. $(x^2y)(x^3y^2) = (x^2 \cdot x^3) \cdot (y^1 \cdot y^2)$ Group like bases and write y as y^1.
    $= x^5 \cdot y^3$ or x^5y^3 Multiply.

15. $(-a^7b^4)(3ab^9) = (-1 \cdot 3) \cdot (a^7 \cdot a^1) \cdot (b^4 \cdot b^9)$
    $= -3a^8 b^{13}$

Work Practice 14–15

Practice 14–15

Simplify.
14. $(m^5n^{10})(mn^8)$
15. $(-x^9y)(4x^2y^{11})$

Answers
8. 7^5 9. x^{13} 10. r^6 11. s^{11}
12. $(-3)^{10}$ 13. $-12x^{12}$ 14. $m^6 n^{18}$
15. $-4x^{11}y^{12}$

✓**Concept Check Answers**

a. z^{16} b. cannot be simplified
c. 9^{11} d. cannot be simplified

Helpful Hint

These examples will remind you of the difference between adding and multiplying terms.

Addition

$$5x^3 + 3x^3 = (5 + 3)x^3 = 8x^3 \quad \text{By the distributive property}$$
$$7x + 4x^2 = 7x + 4x^2 \quad \text{Cannot be combined}$$

Multiplication

$$(5x^3)(3x^3) = 5 \cdot 3 \cdot x^3 \cdot x^3 = 15x^{3+3} = 15x^6 \quad \text{By the product rule}$$
$$(7x)(4x^2) = 7 \cdot 4 \cdot x \cdot x^2 = 28x^{1+2} = 28x^3 \quad \text{By the product rule}$$

Objective C   Using the Power Rule

Exponential expressions can themselves be raised to powers. Let's try to discover a rule that simplifies an expression like $(x^2)^3$. By the definition of a^n,

$$(x^2)^3 = (x^2)(x^2)(x^2) \quad (x^2)^3 \text{ means 3 factors of } (x^2).$$

which can be simplified by the product rule for exponents.

$$(x^2)^3 = (x^2)(x^2)(x^2) = x^{2+2+2} = x^6$$

Notice that the result is exactly the same if we multiply the exponents.

$$(x^2)^3 = x^{2 \cdot 3} = x^6$$

The following rule states this result.

Power Rule for Exponents

If m and n are positive integers and a is a real number, then

$$(a^m)^n = a^{mn} \leftarrow \text{Multiply exponents.}$$
$$\uparrow \text{ Keep the base.}$$

For example,

$$(7^2)^5 = 7^{2 \cdot 5} = 7^{10} \leftarrow \text{Multiply exponents.}$$
$$\uparrow \text{ Keep the base.}$$

$$[(-5)^3]^7 = (-5)^{3 \cdot 7} = (-5)^{21} \leftarrow \text{Multiply exponents.}$$
$$\uparrow \text{ Keep the base.}$$

In other words, to raise an exponential expression to a power, we keep the base and multiply the exponents.

Practice 16–17

Use the power rule to simplify each expression.

16. $(9^4)^{10}$ **17.** $(z^6)^3$

Examples   Use the power rule to simplify each expression.

16. $(5^3)^6 = 5^{3 \cdot 6} = 5^{18}$ **17.** $(y^8)^2 = y^{8 \cdot 2} = y^{16}$

■ Work Practice 16–17

Helpful Hint

Take a moment to make sure that you understand when to apply the product rule and when to apply the power rule.

Product Rule → Add Exponents	Power Rule → Multiply Exponents
$x^5 \cdot x^7 = x^{5+7} = x^{12}$	$(x^5)^7 = x^{5 \cdot 7} = x^{35}$
$y^6 \cdot y^2 = y^{6+2} = y^8$	$(y^6)^2 = y^{6 \cdot 2} = y^{12}$

Answers

16. 9^{40} **17.** z^{18}

Section 10.1 | Exponents

Objective D Using the Power Rules for Products and Quotients

When the base of an exponential expression is a product, the definition of a^n still applies. For example, simplify $(xy)^3$ as follows.

$(xy)^3 = (xy)(xy)(xy)$ $(xy)^3$ means 3 factors of (xy).
$ = x \cdot x \cdot x \cdot y \cdot y \cdot y$ Group factors with common bases.
$ = x^3 y^3$ Simplify.

Notice that to simplify the expression $(xy)^3$, we raise each factor within the parentheses to a power of 3.

$(xy)^3 = x^3 y^3$

In general, we have the following rule.

Power of a Product Rule

If n is a positive integer and a and b are real numbers, then

$$(ab)^n = a^n b^n$$

For example,

$$(3x)^5 = 3^5 x^5$$

In other words, to raise a product to a power, we raise each factor to the power.

Examples Simplify each expression.

8. $(st)^4 = s^4 \cdot t^4 = s^4 t^4$ Use the power of a product rule.
9. $(2a)^3 = 2^3 \cdot a^3 = 8a^3$ Use the power of a product rule.
10. $(-5x^2 y^3 z)^2 = (-5)^2 \cdot (x^2)^2 \cdot (y^3)^2 \cdot (z^1)^2$ Use the power of a product rule.
$ = 25 x^4 y^6 z^2$
11. $(-xy^3)^5 = (-1xy^3)^5 = (-1)^5 \cdot x^5 \cdot (y^3)^5$ Use the power of a product rule.
$ = -1 x^5 y^{15}$ or $-x^5 y^{15}$

Work Practice 18–21

Practice 18–21

Simplify each expression.
18. $(xy)^7$ **19.** $(3y)^4$
20. $(-2p^4 q^2 r)^3$ **21.** $(-a^4 b)^7$

Let's see what happens when we raise a quotient to a power. For example, we simplify $\left(\dfrac{x}{y}\right)^3$ as follows.

$\left(\dfrac{x}{y}\right)^3 = \left(\dfrac{x}{y}\right)\left(\dfrac{x}{y}\right)\left(\dfrac{x}{y}\right)$ $\left(\dfrac{x}{y}\right)^3$ means 3 factors of $\left(\dfrac{x}{y}\right)$.
$\phantom{\left(\dfrac{x}{y}\right)^3} = \dfrac{x \cdot x \cdot x}{y \cdot y \cdot y}$ Multiply fractions.
$\phantom{\left(\dfrac{x}{y}\right)^3} = \dfrac{x^3}{y^3}$ Simplify.

Notice that to simplify the expression $\left(\dfrac{x}{y}\right)^3$, we raise both the numerator and the denominator to a power of 3.

$\left(\dfrac{x}{y}\right)^3 = \dfrac{x^3}{y^3}$

Answers
18. $x^7 y^7$ **19.** $81 y^4$
20. $-8 p^{12} q^6 r^3$ **21.** $-a^{28} b^7$

Chapter 10 | Exponents and Polynomials

In general, we have the following rule.

> **Power of a Quotient Rule**
>
> If n is a positive integer and a and c are real numbers, then
> $$\left(\frac{a}{c}\right)^n = \frac{a^n}{c^n}, \quad c \neq 0$$
> For example,
> $$\left(\frac{y}{7}\right)^3 = \frac{y^3}{7^3}$$

In other words, to raise a quotient to a power, we raise both the numerator and the denominator to the power.

Practice 22–23

Simplify each expression.

22. $\left(\dfrac{r}{s}\right)^6$ 23. $\left(\dfrac{5x^6}{9y^3}\right)^2$

> **Examples** Simplify each expression.
>
> 22. $\left(\dfrac{m}{n}\right)^7 = \dfrac{m^7}{n^7}, \quad n \neq 0$ Use the power of a quotient rule.
>
> 23. $\left(\dfrac{2x^4}{3y^5}\right)^4 = \dfrac{2^4 \cdot (x^4)^4}{3^4 \cdot (y^5)^4}$ Use the power of a quotient rule.
>
> $\qquad = \dfrac{16x^{16}}{81y^{20}}, \quad y \neq 0$ Use the power rule for exponents.

■ Work Practice 22–23

Objective E Using the Quotient Rule and Defining the Zero Exponent

Another pattern for simplifying exponential expressions involves quotients.

$$\frac{x^5}{x^3} = \frac{x \cdot x \cdot x \cdot x \cdot x}{x \cdot x \cdot x}$$

$$= \frac{x \cdot x \cdot x \cdot x \cdot x}{x \cdot x \cdot x}$$

$$= 1 \cdot 1 \cdot 1 \cdot x \cdot x$$

$$= x \cdot x$$

$$= x^2$$

Notice that the result is exactly the same if we subtract exponents of the common base.

$$\frac{x^5}{x^3} = x^{5-3} = x^2$$

The following rule states this result in a general way.

> **Quotient Rule for Exponents**
>
> If m and n are positive integers and a is a real number, then
> $$\frac{a^m}{a^n} = a^{m-n}, \quad a \neq 0$$
> For example,
> $$\frac{x^6}{x^2} = x^{6-2} = x^4, \quad x \neq 0$$

Answers

22. $\dfrac{r^6}{s^6}, \; s \neq 0$ 23. $\dfrac{25x^{12}}{81y^6}, \; y \neq 0$

other words, to divide one exponential expression by another with a common base, we keep the base and subtract the exponents.

Examples Simplify each quotient.

24. $\dfrac{x^5}{x^2} = x^{5-2} = x^3$ Use the quotient rule.

25. $\dfrac{4^7}{4^3} = 4^{7-3} = 4^4 = 256$ Use the quotient rule.

26. $\dfrac{(-3)^5}{(-3)^2} = (-3)^3 = -27$ Use the quotient rule.

27. $\dfrac{2x^5y^2}{xy} = 2 \cdot \dfrac{x^5}{x^1} \cdot \dfrac{y^2}{y^1}$
$= 2 \cdot (x^{5-1}) \cdot (y^{2-1})$ Use the quotient rule.
$= 2x^4y^1$ or $2x^4y$

Work Practice 24–27

Practice 24–27

Simplify each quotient.

24. $\dfrac{y^7}{y^3}$ 25. $\dfrac{5^9}{5^6}$

26. $\dfrac{(-2)^{14}}{(-2)^{10}}$ 27. $\dfrac{7a^4b^{11}}{ab}$

Let's now give meaning to an expression such as x^0. To do so, we will simplify $\dfrac{x^3}{x^3}$ two ways and compare the results.

$\dfrac{x^3}{x^3} = x^{3-3} = x^0$ Apply the quotient rule.

$\dfrac{x^3}{x^3} = \dfrac{x \cdot x \cdot x}{x \cdot x \cdot x} = 1$ Apply the fundamental principle for fractions.

Since $\dfrac{x^3}{x^3} = x^0$ and $\dfrac{x^3}{x^3} = 1$, we define that $x^0 = 1$ as long as x is not 0.

Zero Exponent

$a^0 = 1$, as long as a is not 0.

For example, $5^0 = 1$.

other words, a base raised to the 0 power is 1, as long as the base is not 0.

Examples Simplify each expression.

28. $3^0 = 1$
29. $(5x^3y^2)^0 = 1$
30. $(-4)^0 = 1$
31. $-4^0 = -1 \cdot 4^0 = -1 \cdot 1 = -1$
32. $5x^0 = 5 \cdot x^0 = 5 \cdot 1 = 5$

Work Practice 28–32

Practice 28–32

Simplify each expression.
28. 8^0 29. $(2r^2s)^0$
30. $(-7)^0$ 31. -7^0
32. $7y^0$

Answers
24. y^4 25. 125   26. 16   27. $7a^3b^{10}$
28. 1   29. 1   30. 1   31. -1 32. 7

Chapter 10 | Exponents and Polynomials

Concept Check Suppose you are simplifying each expression. Tell whether you would *add* the exponents, *subtract* the exponents, *multiply* the exponents, *divide* the exponents, or *none of these*.

a. $(x^{63})^{21}$ b. $\dfrac{y^{15}}{y^3}$ c. $z^{16} + z^8$ d. $w^{45} \cdot w^9$

Objective F Deciding Which Rule to Use

Let's practice deciding which rule to use to simplify an expression. We will continue this discussion with more examples in the next section.

Example 33 Simplify each expression.

a. $x^7 \cdot x^4$ b. $\left(\dfrac{t}{2}\right)^4$ c. $(9y^5)^2$

Solution:

a. Here, we have a product, so we use the product rule to simplify.
$x^7 \cdot x^4 = x^{7+4} = x^{11}$

b. This is a quotient raised to a power, so we use the power of a quotient rule.
$\left(\dfrac{t}{2}\right)^4 = \dfrac{t^4}{2^4} = \dfrac{t^4}{16}$

c. This is a product raised to a power, so we use the power of a product rule.
$(9y^5)^2 = 9^2(y^5)^2 = 81y^{10}$

■ Work Practice 33

Practice 33
Simplify each expression.

a. $\dfrac{x^7}{x^4}$ b. $(3y^4)^4$ c. $\left(\dfrac{x}{4}\right)^3$

Answers
33. a. x^3 b. $81y^{16}$ c. $\dfrac{x^3}{64}$

✓ **Concept Check Answers**
a. multiply   b. subtract
c. none of these   d. add

Vocabulary, Readiness & Video Check

Use the choices below to fill in each blank. Some choices may be used more than once.

| 0 | base | add |
| 1 | exponent | multiply |

1. Repeated multiplication of the same factor can be written using a(n) _____.
2. In 5^2, the 2 is called the _____ and the 5 is called the _____.
3. To simplify $x^2 \cdot x^7$, keep the base and _____ the exponents.
4. To simplify $(x^3)^6$, keep the base and _____ the exponents.
5. The understood exponent on the term y is _____.
6. If $x^\square = 1$, the exponent is _____.

Each expression contains an exponent of 2. For each exercise, name the base for this exponent of 2.

7. 3^2 _____
8. $(-3)^2$ _____
9. -4^2 _____
10. $5 \cdot 3^2$ _____
11. $5x^2$ _____
12. $(5x)^2$ _____

Section 10.1 | Exponents 747

Martin-Gay Interactive Videos

See Video 10.1

Watch the section lecture video and answer the following questions.

Objective A 13. Examples 3 and 4 illustrate how to find the base of an exponential expression both with and without parentheses. Explain how identifying the base of Example 7 is similar to identifying the base of Example 4.

Objective B 14. Why were the commutative and associative properties applied in Example 12?

Objective C 15. What point is made at the end of Example 15?

Objective D 16. Although it's not especially emphasized in Example 20, what is helpful to remind ourself about the -2 in the problem?

Objective E 17. In Example 24, which exponent rule is used to show that any nonzero base raised to the power of zero is 1?

Objective F 18. When simplifying an exponential expression that's a fraction, will we always use the quotient rule? Refer to Example 30 to support your answer.

10.1 Exercise Set MyLab Math

Objective A *Evaluate each expression. See Examples 1 through 6.*

1. 7^2
2. -3^2
3. $(-5)^1$
4. $(-3)^2$
5. -2^4
6. -4^3
7. $(-2)^4$
8. $(-4)^3$
9. $\left(\dfrac{1}{3}\right)^3$
10. $\left(-\dfrac{1}{9}\right)^2$
11. $7 \cdot 2^4$
12. $9 \cdot 2^2$

Evaluate each expression for the given replacement values. See Example 7.

13. x^2 when $x = -2$
14. x^3 when $x = -2$
15. $5x^3$ when $x = 3$
16. $4x^2$ when $x = 5$
17. $2xy^2$ when $x = 3$ and $y = -5$
18. $-4x^2y^3$ when $x = 2$ and $y = -1$
19. $\dfrac{2z^4}{5}$ when $z = -2$
20. $\dfrac{10}{3y^3}$ when $y = -3$

Objective B *Use the product rule to simplify each expression. Write the results using exponents. See Examples 8 through 15.*

21. $x^2 \cdot x^5$
22. $y^2 \cdot y$
23. $(-3)^3 \cdot (-3)^9$
24. $(-5)^7 \cdot (-5)^6$
25. $(5y^4)(3y)$
26. $(-2z^3)(-2z^2)$
27. $(x^9y)(x^{10}y^5)$
28. $(a^2b)(a^{13}b^{17})$
29. $(-8mn^6)(9m^2n^2)$
30. $(-7a^3b^3)(7a^{19}b)$
31. $(4z^{10})(-6z^7)(z^3)$
32. $(12x^5)(-x^6)(x^4)$

33. The rectangle below has width $4x^2$ feet and length $5x^3$ feet. Find its area as an expression in x.

△ 34. The parallelogram below has base length $9y^7$ meters and height $2y^{10}$ meters. Find its area as an expression in y.

Objectives **C D** Mixed Practice *Use the power rule and the power of a product or quotient rule to simplify each expression. See Examples 16 through 23.*

35. $(x^9)^4$ **36.** $(y^7)^5$ **37.** $(pq)^8$ **38.** $(ab)^6$

39. $(2a^5)^3$ **40.** $(4x^6)^2$ **41.** $(x^2y^3)^5$ **42.** $(a^4b)^7$

43. $(-7a^2b^5c)^2$ **44.** $(-3x^7yz^2)^3$ **45.** $\left(\dfrac{r}{s}\right)^9$ **46.** $\left(\dfrac{q}{t}\right)^{11}$

47. $\left(\dfrac{mp}{n}\right)^9$ **48.** $\left(\dfrac{xy}{7}\right)^2$ **49.** $\left(\dfrac{-2xz}{y^5}\right)^2$ **50.** $\left(\dfrac{xy^4}{-3z^3}\right)^3$

△ 51. The square shown has sides of length $8z^5$ decimeters. Find its area.

△ 52. Given the circle below with radius $5y$ centimeters, find its area. Do not approximate π.

53. The vault below is in the shape of a cube. If each side is $3y^4$ feet, find its volume.

54. The silo shown is in the shape of a cylinder. If its radius is $4x$ meters and its height is $5x^3$ meters, find its volume. Do not approximate π.

Objective **E** *Use the quotient rule and simplify each expression. See Examples 24 through 27.*

55. $\dfrac{x^3}{x}$ **56.** $\dfrac{y^{10}}{y^9}$ **57.** $\dfrac{(-4)^6}{(-4)^3}$ **58.** $\dfrac{(-6)^{13}}{(-6)^{11}}$

59. $\dfrac{p^7q^{20}}{pq^{15}}$ **60.** $\dfrac{x^8y^6}{xy^5}$ **61.** $\dfrac{7x^2y^6}{14x^2y^3}$ **62.** $\dfrac{9a^4b^7}{27ab^2}$

Simplify each expression. See Examples 28 through 32.

63. 7^0 **64.** 23^0 **65.** $(2x)^0$ **66.** $(4y)^0$

67. $-7x^0$ **68.** $-2x^0$ **69.** $5^0 + y^0$ **70.** $-3^0 + 4^0$

Objectives A B C D E F Mixed Practice *Simplify each expression. See Examples 1 through 6 and through 33.*

71. -9^2
72. $(-9)^2$
73. $\left(\dfrac{1}{4}\right)^3$
74. $\left(\dfrac{2}{3}\right)^3$

75. $b^4 b^2$
76. $y^4 y$
77. $a^2 a^3 a^4$
78. $x^2 x^{15} x^9$

79. $(2x^3)(-8x^4)$
80. $(3y^4)(-5y)$
81. $(a^7 b^{12})(a^4 b^8)$
82. $(y^2 z^2)(y^{15} z^{13})$

83. $(-2mn^6)(-13m^8 n)$
84. $(-3s^5 t)(-7st^{10})$
85. $(z^4)^{10}$
86. $(t^5)^{11}$

87. $(4ab)^3$
88. $(2ab)^4$
89. $(-6xyz^3)^2$
90. $(-3xy^2 a^3)^3$

91. $\dfrac{z^{12}}{z^4}$
92. $\dfrac{b^6}{b^3}$
93. $\dfrac{3x^5}{x}$
94. $\dfrac{5x^9}{x}$

95. $(6b)^0$
96. $(5ab)^0$
97. $(9xy)^2$
98. $(2ab)^5$

99. $2^3 + 2^5$
100. $7^2 - 7^0$
101. $\left(\dfrac{3y^5}{6x^4}\right)^3$
102. $\left(\dfrac{2ab}{6yz}\right)^4$

103. $\dfrac{2x^3 y^2 z}{xyz}$
104. $\dfrac{5x^{12} y^{13}}{x^5 y^7}$

Review

Subtract. See Section 2.3.

105. $5 - 7$
106. $9 - 12$
107. $3 - (-2)$
108. $5 - (-10)$
109. $-11 - (-4)$
110. $-15 - (-21)$

Concept Extensions

Solve. See the Concept Checks in this section. For Exercises 111 through 114, match the expression with the operation needed to simplify each. A letter may be used more than once and a letter may not be used at all.

111. $(x^{14})^{23}$
112. $x^{14} \cdot x^{23}$
113. $x^{14} + x^{23}$
114. $\dfrac{x^{35}}{x^{17}}$

a. Add the exponents.
b. Subtract the exponents.
c. Multiply the exponents.
d. Divide the exponents.
e. None of these

Fill in the boxes so that each statement is true. (More than one answer is possible for each exercise.)

115. $x^{\square} \cdot x^{\square} = x^{12}$

116. $\left(x^{\square}\right)^{\square} = x^{20}$

117. $\dfrac{y^{\square}}{y^{\square}} = y^7$

118. $\left(y^{\square}\right)^{\square} \cdot \left(y^{\square}\right)^{\square} = y^{30}$

△ **119.** The formula $V = x^3$ can be used to find the volume V of a cube with side length x. Find the volume of a cube with side length 7 meters. (Volume is the capacity of a solid such as a cube and is measured in cubic units.)

△ **120.** The formula $S = 6x^2$ can be used to find the surface area S of a cube with side length x. Find the surface area of a cube with side length 5 meters. (Surface area is the area of the surface of the cube and is measured in square units.)

△ **121.** To find the amount of water that a swimming pool in the shape of a cube can hold, do we use the formula for volume of the cube or surface area of the cube? (See Exercises **119** and **120**.)

△ **122.** To find the amount of material needed to cover an ottoman in the shape of a cube, do we use the formula for volume of the cube or surface area of the cube? (See Exercises **119** and **120**.)

123. Explain why $(-5)^4 = 625$, while $-5^4 = -625$.

124. Explain why $5 \cdot 4^2 = 80$, while $(5 \cdot 4)^2 = 400$.

125. In your own words, explain why $5^0 = 1$.

126. In your own words, explain when $(-3)^n$ is positive and when it is negative.

Simplify each expression. Assume that variables represent positive integers.

127. $x^{5a}x^{4a}$ **128.** $b^{9a}b^{4a}$ **129.** $\left(a^b\right)^5$ **130.** $\left(2a^{4b}\right)^4$ **131.** $\dfrac{x^{9a}}{x^{4a}}$ **132.** $\dfrac{y^{15b}}{y^{6b}}$

10.2 Negative Exponents and Scientific Notation

Objective A Simplifying Expressions Containing Negative Exponents

Our work with exponential expressions so far has been limited to exponents that are positive integers or 0. Here we will also give meaning to an expression like x^{-3}.

Suppose that we wish to simplify the expression $\dfrac{x^2}{x^5}$. If we use the quotient rule for exponents, we subtract exponents:

$$\dfrac{x^2}{x^5} = x^{2-5} = x^{-3}, \quad x \neq 0$$

But what does x^{-3} mean? Let's simplify $\dfrac{x^2}{x^5}$ using the definition of a^n.

$$\dfrac{x^2}{x^5} = \dfrac{x \cdot x}{x \cdot x \cdot x \cdot x \cdot x}$$

$$= \dfrac{x \cdot x}{x \cdot x \cdot x \cdot x \cdot x} \qquad \text{Divide numerator and denominator by common factors by applying the fundamental principle for fractions.}$$

$$= \dfrac{1}{x^3}$$

If the quotient rule is to hold true for negative exponents, then x^{-3} must equal $\dfrac{1}{x^3}$.

From this example, we state the definition for negative exponents.

Objectives

A Simplify Expressions Containing Negative Exponents.

B Use the Rules and Definitions for Exponents to Simplify Exponential Expressions.

C Write Numbers in Scientific Notation.

D Convert Numbers in Scientific Notation to Standard Form.

E Perform Operations on Numbers Written in Scientific Notation.

Negative Exponents

If a is a real number other than 0 and n is an integer, then

$$a^{-n} = \dfrac{1}{a^n}$$

For example,

$$x^{-3} = \dfrac{1}{x^3}$$

In other words, another way to write a^{-n} is to take its reciprocal and change the sign of its exponent.

Examples Simplify by writing each expression with positive exponents only.

1. $3^{-2} = \dfrac{1}{3^2} = \dfrac{1}{9}$ Use the definition of negative exponents.

2. $2x^{-3} = 2^1 \cdot \dfrac{1}{x^3} = \dfrac{2^1}{x^3}$ or $\dfrac{2}{x^3}$ Use the definition of negative exponents.

3. $2^{-1} + 4^{-1} = \dfrac{1}{2} + \dfrac{1}{4} = \dfrac{2}{4} + \dfrac{1}{4} = \dfrac{3}{4}$

4. $(-2)^{-4} = \dfrac{1}{(-2)^4} = \dfrac{1}{(-2)(-2)(-2)(-2)} = \dfrac{1}{16}$

> **Helpful Hint** Don't forget that since there are no parentheses, only x is the base for the exponent -3.

Work Practice 1–4

Practice 1–4

Simplify by writing each expression with positive exponents only.

1. 5^{-3} **2.** $7x^{-4}$
3. $5^{-1} + 3^{-1}$ **4.** $(-3)^{-4}$

Answers

1. $\dfrac{1}{125}$ **2.** $\dfrac{7}{x^4}$ **3.** $\dfrac{8}{15}$ **4.** $\dfrac{1}{81}$

> **Helpful Hint**
>
> A negative exponent *does not affect* the sign of its base.
> Remember: Another way to write a^{-n} is to take its reciprocal and change the sign of its exponent: $a^{-n} = \frac{1}{a^n}$. For example,
>
> $$x^{-2} = \frac{1}{x^2}, \qquad 2^{-3} = \frac{1}{2^3} \text{ or } \frac{1}{8}$$
>
> $$\frac{1}{y^{-4}} = \frac{1}{\frac{1}{y^4}} = y^4, \qquad \frac{1}{5^{-2}} = 5^2 \text{ or } 25$$

From the preceding Helpful Hint, we know that $x^{-2} = \frac{1}{x^2}$ and $\frac{1}{y^{-4}} = y^4$. We ca use this to include another statement in our definition of negative exponents.

Negative Exponents

If a is a real number other than 0 and n is an integer, then

$$a^{-n} = \frac{1}{a^n} \quad \text{and} \quad \frac{1}{a^{-n}} = a^n$$

Practice 5–8

Simplify each expression. Write each result using positive exponents only.

5. $\left(\dfrac{6}{7}\right)^{-2}$ 6. $\dfrac{x}{x^{-4}}$

7. $\dfrac{y^{-9}}{z^{-5}}$ 8. $\dfrac{y^{-4}}{y^6}$

Examples Simplify each expression. Write each result using positive exponents only.

5. $\left(\dfrac{2}{x}\right)^{-3} = \dfrac{2^{-3}}{x^{-3}} = \dfrac{2^{-3}}{1} \cdot \dfrac{1}{x^{-3}} = \dfrac{1}{2^3} \cdot \dfrac{x^3}{1} = \dfrac{x^3}{2^3} = \dfrac{x^3}{8}$ Use the negative exponents rule.

6. $\dfrac{y}{y^{-2}} = \dfrac{y^1}{y^{-2}} = y^{1-(-2)} = y^3$ Use the quotient rule.

7. $\dfrac{p^{-4}}{q^{-9}} = p^{-4} \cdot \dfrac{1}{q^{-9}} = \dfrac{1}{p^4} \cdot q^9 = \dfrac{q^9}{p^4}$ Use the negative exponents rule.

8. $\dfrac{x^{-5}}{x^7} = x^{-5-7} = x^{-12} = \dfrac{1}{x^{12}}$

■ Work Practice 5–8

Objective B Simplifying Exponential Expressions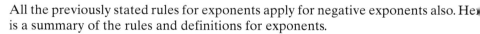

All the previously stated rules for exponents apply for negative exponents also. He is a summary of the rules and definitions for exponents.

Summary of Exponent Rules

If m and n are integers and a, b, and c are real numbers, then

Product rule for exponents: $a^m \cdot a^n = a^{m+n}$

Power rule for exponents: $(a^m)^n = a^{m \cdot n}$

Power of a product: $(ab)^n = a^n b^n$

Power of a quotient: $\left(\dfrac{a}{c}\right)^n = \dfrac{a^n}{c^n}, \quad c \neq 0$

Quotient rule for exponents: $\dfrac{a^m}{a^n} = a^{m-n}, \quad a \neq 0$

Zero exponent: $a^0 = 1, \quad a \neq 0$

Negative exponent: $a^{-n} = \dfrac{1}{a^n}, \quad a \neq 0$

Answers

5. $\dfrac{49}{36}$ 6. x^5 7. $\dfrac{z^5}{y^9}$ 8. $\dfrac{1}{y^{10}}$

Section 10.2 | Negative Exponents and Scientific Notation

Examples Simplify each expression. Write each result using positive exponents only.

9. $\dfrac{(x^3)^4 x}{x^7} = \dfrac{x^{12} \cdot x}{x^7} = \dfrac{x^{12+1}}{x^7} = \dfrac{x^{13}}{x^7} = x^{13-7} = x^6$ Use the power rule.

10. $\left(\dfrac{3a^2}{b}\right)^{-3} = \dfrac{3^{-3}(a^2)^{-3}}{b^{-3}}$ Raise each factor in the numerator and the denominator to the -3 power.

$= \dfrac{3^{-3}a^{-6}}{b^{-3}}$ Use the power rule.

$= \dfrac{b^3}{3^3 a^6}$ Use the negative exponent rule.

$= \dfrac{b^3}{27a^6}$ Write 3^3 as 27.

11. $(y^{-3}z^6)^{-6} = (y^{-3})^{-6}(z^6)^{-6}$ Raise each factor to the -6 power.

$= y^{18}z^{-36} = \dfrac{y^{18}}{z^{36}}$

12. $\dfrac{(2x)^5}{x^3} = \dfrac{2^5 \cdot x^5}{x^3} = 2^5 \cdot x^{5-3} = 32x^2$ Raise each factor in the numerator to the fifth power.

13. $\dfrac{x^{-7}}{(x^4)^3} = \dfrac{x^{-7}}{x^{12}} = x^{-7-12} = x^{-19} = \dfrac{1}{x^{19}}$

14. $(5y^3)^{-2} = 5^{-2}(y^3)^{-2} = 5^{-2}y^{-6} = \dfrac{1}{5^2 y^6} = \dfrac{1}{25y^6}$

15. $-\dfrac{22a^7 b^{-5}}{11a^{-2}b^3} = -\dfrac{22}{11} \cdot a^{7-(-2)}b^{-5-3} = -2a^9 b^{-8} = -\dfrac{2a^9}{b^8}$

16. $\dfrac{(2xy)^{-3}}{(x^2 y^3)^2} = \dfrac{2^{-3}x^{-3}y^{-3}}{(x^2)^2(y^3)^2} = \dfrac{2^{-3}x^{-3}y^{-3}}{x^4 y^6} = 2^{-3}x^{-3-4}y^{-3-6}$

$= 2^{-3}x^{-7}y^{-9} = \dfrac{1}{2^3 x^7 y^9}$ or $\dfrac{1}{8x^7 y^9}$

Work Practice 9–16

Practice 9–16
Simplify each expression. Write each result using positive exponents only.

9. $\dfrac{(x^5)^3 x}{x^4}$

10. $\left(\dfrac{9x^3}{y}\right)^{-2}$

11. $(a^{-4}b^7)^{-5}$

12. $\dfrac{(2x)^4}{x^8}$

13. $\dfrac{y^{-10}}{(y^5)^4}$

14. $(4a^2)^{-3}$

15. $-\dfrac{32x^{-3}y^{-6}}{8x^{-5}y^{-2}}$

16. $\dfrac{(3x^{-2}y)^{-2}}{(2x^7 y)^3}$

Objective C   Writing Numbers in Scientific Notation

Both very large and very small numbers frequently occur in many fields of science. For example, the distance between the sun and the dwarf planet Pluto is approximately 5,906,000,000 kilometers, and the mass of a proton is approximately 0.00000000000000000000000165 gram. It can be tedious to write these numbers in this standard decimal notation, so **scientific notation** is used as a convenient shorthand for expressing very large and very small numbers.

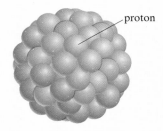

Mass of proton is approximately 0.00000000000000000000000165 gram

Answers
9. x^{12} 10. $\dfrac{y^2}{81x^6}$ 11. $\dfrac{a^{20}}{b^{35}}$ 12. $\dfrac{16}{x^4}$

13. $\dfrac{1}{y^{30}}$ 14. $\dfrac{1}{64a^6}$ 15. $-\dfrac{4x^2}{y^4}$

16. $\dfrac{1}{72x^{17}y^5}$

Scientific Notation

A positive number is written in scientific notation if it is written as the product of a number a, where $1 \leq a < 10$, and an integer power r of 10: $a \times 10^r$.

The following numbers are written in scientific notation. The $\times$ sign for multiplication is used as part of the notation.

$$2.03 \times 10^2 \quad 7.362 \times 10^7 \quad \underbrace{5.906 \times 10^9}_{} \quad \text{(Distance between the sun and Pluto)}$$

$$1 \times 10^{-3} \quad 8.1 \times 10^{-5} \quad \underbrace{1.65 \times 10^{-24}}_{} \quad \text{(Mass of a proton)}$$

The following steps are useful when writing positive numbers in scientific notation.

To Write a Number in Scientific Notation

Step 1: Move the decimal point in the original number so that the new number has a value between 1 and 10.

Step 2: Count the number of decimal places the decimal point is moved in Step 1. If the original number is 10 or greater, the count is positive. If the original number is less than 1, the count is negative.

Step 3: Multiply the new number in Step 1 by 10 raised to an exponent equal to the count found in Step 2.

Practice 17

Write each number in scientific notation.
a. 420,000   **b.** 0.00017
c. 9,060,000,000   **d.** 0.000007

Example 17

Write each number in scientific notation.

a. 367,000,000   **b.** 0.000003
c. 20,520,000,000   **d.** 0.00085

Solution:

a. Step 1: Move the decimal point until the number is between 1 and 10.
367,000,000.
  8 places

Step 2: The decimal point is moved 8 places and the original number is 10 or greater, so the count is positive 8.

Step 3: $367,000,000 = 3.67 \times 10^8$

b. Step 1: Move the decimal point until the number is between 1 and 10.
0.000003
  6 places

Step 2: The decimal point is moved 6 places and the original number is less than 1, so the count is -6.

Step 3: $0.000003 = 3.0 \times 10^{-6}$

c. $20,520,000,000 = 2.052 \times 10^{10}$

d. $0.00085 = 8.5 \times 10^{-4}$

■ Work Practice 17

Objective D Converting Numbers to Standard Form

A number written in scientific notation can be rewritten in standard form. For example, to write 8.63×10^3 in standard form, recall that $10^3 = 1000$.

$$8.63 \times 10^3 = 8.63(1000) = 8630$$

Answers
17. a. 4.2×10^5 **b.** 1.7×10^{-4}
c. 9.06×10^9 **d.** 7×10^{-6}

Section 10.2 | Negative Exponents and Scientific Notation 755

otice that the exponent on the 10 is positive 3, and we moved the decimal point 3 aces to the right.

To write 7.29×10^{-3} in standard form, recall that $10^{-3} = \frac{1}{10^3} = \frac{1}{1000}$.

$$7.29 \times 10^{-3} = 7.29\left(\frac{1}{1000}\right) = \frac{7.29}{1000} = 0.00729$$

he exponent on the 10 is negative 3, and we moved the decimal to the left 3 places.

In general, **to write a scientific notation number in standard form,** move the ecimal point the same number of places as the exponent on 10. If the exponent is ositive, move the decimal point to the right; if the exponent is negative, move the ecimal point to the left.

Example 18 Write each number in standard form, without exponents.

a. 1.02×10^5
b. 7.358×10^{-3}
c. 8.4×10^7
d. 3.007×10^{-5}

Solution:

a. Move the decimal point 5 places to the right.
$1.02 \times 10^5 = 102,000.$

b. Move the decimal point 3 places to the left.
$7.358 \times 10^{-3} = 0.007358$

c. $8.4 \times 10^7 = 84,000,000.$ 7 places to the right

d. $3.007 \times 10^{-5} = 0.00003007$ 5 places to the left

Work Practice 18

Practice 18
Write the numbers in standard form, without exponents.

a. 3.062×10^{-4}
b. 5.21×10^4
c. 9.6×10^{-5}
d. 6.002×10^6

✓**Concept Check** Which number in each pair is larger?
a. 7.8×10^3 or 2.1×10^5
b. 9.2×10^{-2} or 2.7×10^4
c. 5.6×10^{-4} or 6.3×10^{-5}

Objective E Performing Operations with Scientific Notation ▶

erforming operations on numbers written in scientific notation makes use of the iles and definitions for exponents.

Example 19 Perform each indicated operation. Write each result in standard decimal form.

a. $(8 \times 10^{-6})(7 \times 10^3)$
b. $\frac{12 \times 10^2}{6 \times 10^{-3}}$

Solution:

a. $(8 \times 10^{-6})(7 \times 10^3) = 8 \cdot 7 \cdot 10^{-6} \cdot 10^3$
$= 56 \times 10^{-3}$
$= 0.056$

b. $\frac{12 \times 10^2}{6 \times 10^{-3}} = \frac{12}{6} \times 10^{2-(-3)} = 2 \times 10^5 = 200,000$

Work Practice 19

Practice 19
Perform each indicated operation. Write each result in standard decimal form.

a. $(9 \times 10^7)(4 \times 10^{-9})$
b. $\frac{8 \times 10^4}{2 \times 10^{-3}}$

Answers
18. a. 0.0003062 b. 52,100
c. 0.000096 d. 6,002,000
19. a. 0.36 b. 40,000,000

✓**Concept Check Answers**
a. 2.1×10^5 b. 2.7×10^4
c. 5.6×10^{-4}

Calculator Explorations **Scientific Notation**

To enter a number written in scientific notation on a scientific calculator, locate the scientific notation key, which may be marked [EE] or [EXP]. To enter 3.1×10^7, press [3.1][EE][7]. The display should read [3.1 07].

Enter each number written in scientific notation on your calculator.

1. 5.31×10^3
2. -4.8×10^{14}
3. 6.6×10^{-9}
4. -9.9811×10^{-2}

Multiply each of the following on your calculator. Notice the form of the result.

5. $3,000,000 \times 5,000,000$
6. $230,000 \times 1000$

Multiply each of the following on your calculator. Write the product in scientific notation.

7. $(3.26 \times 10^6)(2.5 \times 10^{13})$
8. $(8.76 \times 10^{-4})(1.237 \times 10^9)$

Vocabulary, Readiness & Video Check

Fill in each blank with the correct choice.

1. The expression x^{-3} equals _____.
   a. $-x^3$ b. $\dfrac{1}{x^3}$ c. $\dfrac{-1}{x^3}$ d. $\dfrac{1}{x^{-3}}$

2. The expression 5^{-4} equals _____.
   a. -20 b. -625 c. $\dfrac{1}{20}$ d. $\dfrac{1}{625}$

3. The number 3.021×10^{-3} is written in _____.
   a. standard form b. expanded form
   c. scientific notation

4. The number 0.0261 is written in _____.
   a. standard form b. expanded form
   c. scientific notation

Write each expression using positive exponents only.

5. $5x^{-2}$
6. $3x^{-3}$
7. $\dfrac{1}{y^{-6}}$
8. $\dfrac{1}{x^{-3}}$
9. $\dfrac{4}{y^{-3}}$
10. $\dfrac{16}{y^{-7}}$

Martin-Gay Interactive Videos Watch the section lecture video and answer the following questions.

See Video 10.2

Objective A 11. What important reminder is given at the end of Example 1?

Objective B 12. Name all the rules and definitions used to simplify Example 8.

Objective C 13. From Examples 9 and 10, explain how the movement of the decimal point in Step 1 suggests the sign of the exponent on the number 10.

Objective D 14. From Example 11, what part of a number written in scientific notation is key in telling us how to write the number in standard form?

Objective E 15. In Example 13, what exponent rules were needed to evaluate the expression?

Section 10.2 | Negative Exponents and Scientific Notation

10.2 Exercise Set MyLab Math

Objective A *Simplify each expression. Write each result using positive exponents only. See Examples 1 through 8.*

1. 4^{-3}
2. 6^{-2}
3. $7x^{-3}$
4. $(7x)^{-3}$
5. $\left(-\dfrac{1}{4}\right)^{-3}$
6. $\left(-\dfrac{1}{8}\right)^{-2}$
7. $3^{-1} + 2^{-1}$
8. $4^{-1} + 4^{-2}$
9. $\dfrac{1}{p^{-3}}$
10. $\dfrac{1}{q^{-5}}$
11. $\dfrac{p^{-5}}{q^{-4}}$
12. $\dfrac{r^{-5}}{s^{-2}}$
13. $\dfrac{x^{-2}}{x}$
14. $\dfrac{y}{y^{-3}}$
15. $\dfrac{z^{-4}}{z^{-7}}$
16. $\dfrac{x^{-4}}{x^{-1}}$
17. $3^{-2} + 3^{-1}$
18. $4^{-2} - 4^{-3}$
19. $(-3)^{-2}$
20. $(-2)^{-6}$
21. $\dfrac{-1}{p^{-4}}$
22. $\dfrac{-1}{y^{-6}}$
23. $-2^0 - 3^0$
24. $5^0 + (-5)^0$

Objective B *Simplify each expression. Write each result using positive exponents only. See Examples 9 through 16.*

25. $\dfrac{x^2 x^5}{x^3}$
26. $\dfrac{y^4 y^5}{y^6}$
27. $\dfrac{p^2 p}{p^{-1}}$
28. $\dfrac{y^3 y}{y^{-2}}$
29. $\dfrac{(m^5)^4 m}{m^{10}}$
30. $\dfrac{(x^2)^8 x}{x^9}$
31. $\dfrac{r}{r^{-3} r^{-2}}$
32. $\dfrac{p}{p^{-3} q^{-5}}$
33. $(x^5 y^3)^{-3}$
34. $(z^5 x^5)^{-3}$
35. $\dfrac{(x^2)^3}{x^{10}}$
36. $\dfrac{(y^4)^2}{y^{12}}$
37. $\dfrac{(a^5)^2}{(a^3)^4}$
38. $\dfrac{(x^2)^5}{(x^4)^3}$
39. $\dfrac{8k^4}{2k}$
40. $\dfrac{27r^6}{3r^4}$
41. $\dfrac{-6m^4}{-2m^3}$
42. $\dfrac{15a^4}{-15a^5}$
43. $\dfrac{-24a^6 b}{6ab^2}$
44. $\dfrac{-5x^4 y^5}{15x^4 y^2}$
45. $\dfrac{6x^2 y^3}{-7x^2 y^5}$
46. $\dfrac{-8xa^2 b}{-5xa^5 b}$
47. $(3a^2 b^{-4})^3$
48. $(5x^3 y^{-2})^2$
49. $(a^{-5} b^2)^{-6}$
50. $(4^{-1} x^5)^{-2}$
51. $\left(\dfrac{x^{-2} y^4}{x^3 y^7}\right)^{-2}$
52. $\left(\dfrac{a^5 b}{a^7 b^{-2}}\right)^{-3}$
53. $\dfrac{4^2 z^{-3}}{4^3 z^{-5}}$
54. $\dfrac{5^{-1} z^7}{5^{-2} z^9}$
55. $\dfrac{3^{-1} x^4}{3^3 x^{-7}}$
56. $\dfrac{2^{-3} x^{-4}}{2^2 x}$
57. $\dfrac{7ab^{-4}}{7^{-1} a^{-3} b^2}$
58. $\dfrac{6^{-5} x^{-1} y^2}{6^{-2} x^{-4} y^4}$
59. $\dfrac{-12 m^5 n^{-7}}{4 m^{-2} n^{-3}}$
60. $\dfrac{-15 r^{-6} s}{5 r^{-4} s^{-3}}$
61. $\left(\dfrac{a^{-5} b}{ab^3}\right)^{-4}$
62. $\left(\dfrac{r^{-2} s^{-3}}{r^{-4} s^{-3}}\right)^{-3}$
63. $(5^2)(8)(2^0)$
64. $(3^4)(7^0)(2)$
65. $\dfrac{(xy^3)^5}{(xy)^{-4}}$
66. $\dfrac{(rs)^{-3}}{(r^2 s^3)^2}$
67. $\dfrac{(-2xy^{-3})^{-3}}{(xy^{-1})^{-1}}$
68. $\dfrac{(-3x^2 y^2)^{-2}}{(xyz)^{-2}}$
69. $\dfrac{(a^4 b^{-7})^{-5}}{(5a^2 b^{-1})^{-2}}$
70. $\dfrac{(a^6 b^{-2})^4}{(4a^{-3} b^{-3})^3}$

71. Find the volume of the cube.

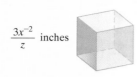

$\frac{3x^{-2}}{z}$ inches

72. Find the area of the triangle.

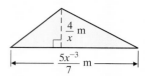

$\frac{4}{x}$ m

$\frac{5x^{-3}}{7}$ m

Objective C *Write each number in scientific notation. See Example 17.*

73. 78,000   **74.** 9,300,000,000   **75.** 0.00000167   **76.** 0.00000017

77. 0.00635   **78.** 0.00194   **79.** 1,160,000   **80.** 700,000

81. When it is completed in 2022, the Thirty Meter Telescope is expected to be the world's largest optical telescope. Located in an observatory complex at the summit of Mauna Kea in Hawaii, the elevation of the Thirty Meter Telescope will be roughly 4200 meters above sea level. Write 4200 in scientific notation.

82. The Thirty Meter Telescope (see Exercise **81**) will have the ability to view objects 13,000,000,000 light-years away. Write 13,000,000,000 in scientific notation.

Objective D *Write each number in standard form. See Example 18.*

83. 8.673×10^{-10} **84.** 9.056×10^{-4} **85.** 3.3×10^{-2}

86. 4.8×10^{-6} **87.** 2.032×10^{4} **88.** 9.07×10^{10}

89. Each second, the Sun converts 7.0×10^8 tons of hydrogen into helium and energy in the form of gamma rays. Write this number in standard form. (*Source:* Students for the Exploration and Development of Space)

90. In chemistry, Avogadro's number is the number of atoms in one mole of an element. Avogadro's number is $6.02214199 \times 10^{23}$. Write this number in standard form. (*Source:* National Institute of Standards and Technology)

Objectives C D Mixed Practice *See Examples 17 and 18. If a number is written in standard form, write it in scientific notation. If a number is written in scientific notation, write it in standard form. The bar graph below shows estimates of the top six national debts as of December, 2016.* (*Source:* The Economist)

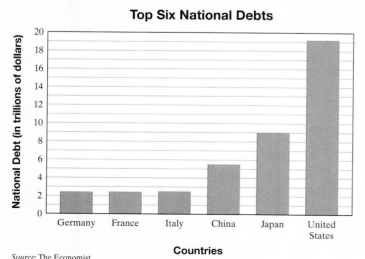

Top Six National Debts

Source: The Economist

91. Germany's national debt as of the end of 2016 was $2,415,000,000,000.

92. Italy's national debt as of the end of 2016 was $2,500,000,000,000.

93. China's national debt as of the end of 2016 was 5.5×10^{12}.

94. France's national debt as of the end of 2016 was 2.435×10^{12}.

95. Use the bar graph to estimate the national debt of Japan and then express it in both standard and scientific notation.

96. Use the bar graph to estimate the national debt of the United States and then express it in both standard and scientific notation.

Section 10.2 | Negative Exponents and Scientific Notation

Objective E *Evaluate each expression using exponential rules. Write each result in standard form. See Example 19.*

97. $(1.2 \times 10^{-3})(3 \times 10^{-2})$

98. $(2.5 \times 10^6)(2 \times 10^{-6})$

99. $(4 \times 10^{-10})(7 \times 10^{-9})$

100. $(5 \times 10^6)(4 \times 10^{-8})$

101. $\dfrac{8 \times 10^{-1}}{16 \times 10^5}$

102. $\dfrac{25 \times 10^{-4}}{5 \times 10^{-9}}$

103. $\dfrac{1.4 \times 10^{-2}}{7 \times 10^{-8}}$

104. $\dfrac{0.4 \times 10^5}{0.2 \times 10^{11}}$

105. Although the actual amount varies by season and time of day, the average volume of water that flows over Niagara Falls (the American and Canadian falls combined) each second is 7.5×10^5 gallons. How much water flows over Niagara Falls in an hour? Write the result in scientific notation. (*Hint:* 1 hour equals 3600 seconds.) (*Source:* http://niagarafallslive.com)

106. A beam of light travels 9.460×10^{12} kilometers per year. How far does light travel in 10,000 years? Write the result in scientific notation.

Review

Simplify each expression by combining any like terms. See Section 3.1.

107. $3x - 5x + 7$

108. $7w + w - 2w$

109. $y - 10 + y$

110. $-6z + 20 - 3z$

111. $7x + 2 - 8x - 6$

112. $10y - 14 - y - 14$

Concept Extensions

For Exercises 113 through 120, write each number in standard form. Then write the number in scientific notation.

113. The wireless subscriber connections in the United States at year's end 2015 were 377.9 million. (*Source:* CTIA—The Wireless Association)

114. Google hosted approximately 2 trillion searches for the first half of 2016. (*Source:* Google.com)

115. The surface of the Arctic Ocean encompasses 14.056 million square kilometers of water. (*Source:* CIA World Factbook)

116. The surface of the Pacific Ocean encompasses 155.557 million square kilometers of water. (*Source:* CIA World Factbook)

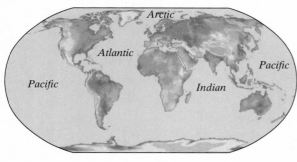

Solve.

117. A nanometer is one-billionth, or 10^{-9}, of a meter. A strand of DNA is about 2.5 nanometers in diameter. Use scientific notation to write the diameter of a DNA strand in terms of meters. (*Source:* United States National Nanotechnology Initiative)

118. A micrometer (sometimes referred to as a micron) is one-millionth, or 10^{-6}, of a meter. A single red blood cell is about 7 micrometers in diameter. Use scientific notation to write the diameter of a red blood cell in terms of meters. (*Source:* National Institute of Standards and Technology)

119. The Thirty Meter Telescope, described in Exercises **81–82**, will be capable of observing ultraviolet wavelengths measuring 310 nanometers. Express this wavelength in terms of meters using both standard form and scientific notation. (See Exercise **117** for a definition of nanometer.) (*Source:* TMT Observatory Corporation)

120. The Thirty Meter Telescope, described in Exercises **81–82**, will be capable of observing infrared wavelengths measuring 28 micrometers. Express this wavelength in terms of meters using both standard form and scientific notation. (See Exercise **118** for a definition of micrometer.) (*Source:* TMT Observatory Corporation)

Simplify.

121. $(2a^3)^3 a^4 + a^5 a^8$

122. $(2a^3)^3 a^{-3} + a^{11} a^{-5}$

Fill in the boxes so that each statement is true. (More than one answer is possible for these exercises.)

123. $x^\square = \dfrac{1}{x^5}$

124. $7^\square = \dfrac{1}{49}$

125. $z^\square \cdot z^\square = z^{-10}$

126. $(x^\square)^\square = x^{-15}$

127. Which is larger? See the Concept Check in this section.
  a. 9.7×10^{-2} or 1.3×10^1
  b. 8.6×10^5 or 4.4×10^7
  c. 6.1×10^{-2} or 5.6×10^{-4}

128. Determine whether each statement is true or false.
  a. $5^{-1} < 5^{-2}$
  b. $\left(\dfrac{1}{5}\right)^{-1} < \left(\dfrac{1}{5}\right)^{-2}$
  c. $a^{-1} < a^{-2}$ for all nonzero numbers.

129. It was stated earlier that for an integer n,
$$x^{-n} = \dfrac{1}{x^n}, \quad x \neq 0.$$
Explain why x may not equal 0.

130. The quotient rule states that
$$\dfrac{a^m}{a^n} = a^{m-n}, \ a \neq 0.$$
Explain why a may not equal 0.

Simplify each expression. Assume that variables represent positive integers.

131. $(x^{-3s})^3$

132. $a^{-4m} \cdot a^{5m}$

133. $a^{4m+1} \cdot a^4$

134. $(3y^{2z})^3$

10.3 Introduction to Polynomials

Objective A Defining Term and Coefficient

In this section, we introduce a special algebraic expression called a polynomial. Let's first review some definitions presented in Section 3.1.

Recall that a term is a number or the product of a number and variables raised to powers. The terms of an expression are separated by plus signs. The terms of the expression $4x^2 + 3x$ are $4x^2$ and $3x$. The terms of the expression $9x^4 - 7x - 1$, or $9x^4 + (-7x) + (-1)$, are $9x^4$, $-7x$, and -1.

Objectives

A Define Term and Coefficient of a Term.

B Define Polynomial, Monomial, Binomial, Trinomial, and Degree.

C Evaluate a Polynomial for Given Replacement Values.

D Simplify a Polynomial by Combining Like Terms.

E Simplify a Polynomial in Several Variables.

F Write a Polynomial in Descending Powers of the Variable and with No Missing Powers of the Variable.

Expression	Terms
$4x^2 + 3x$	$4x^2, 3x$
$9x^4 - 7x - 1$	$9x^4, -7x, -1$
$7y^3$	$7y^3$

The **numerical coefficient** of a term, or simply the **coefficient**, is the numerical factor of each term. If no numerical factor appears in the term, then the coefficient is understood to be 1. If the term is a number only, it is called a **constant term** or simply a **constant**.

Term	Coefficient
x^5	1
$3x^2$	3
$-4x$	-4
$-x^2 y$	-1
3 (constant)	3

Example 1 Complete the table for the expression $7x^5 - 8x^4 + x^2 - 3x + 5$.

Term	Coefficient
$7x^5$	
	-8
x^2	
	-3
5	

Practice 1

Complete the table for the expression $-6x^6 + 4x^5 + 7x^3 - 9x^2 - 1$.

Term	Coefficient
$-6x^6$	
	4
$7x^3$	
	-9
-1	

Solution: The completed table is shown below.

Term	Coefficient
$7x^5$	7
$-8x^4$	-8
x^2	1
$-3x$	-3
5	5

Work Practice 1

Answer
1. term: $4x^5$; $-9x^2$, coefficient: $-6, 7, -1$

Objective B Defining Polynomial, Monomial, Binomial, Trinomial, and Degree

Now we are ready to define what we mean by a polynomial.

Polynomial

A **polynomial in x** is a finite sum of terms of the form ax^n, where a is a real number and n is a whole number.

For example,

$$x^5 - 3x^3 + 2x^2 - 5x + 1$$

is a polynomial in x. Notice that this polynomial is written in **descending powers** of x, because the powers of x decrease from left to right. (Recall that the term 1 can be thought of as $1x^0$.)

On the other hand,

$$x^{-5} + 2x - 3$$

is **not** a polynomial because one of its terms contains a variable with an exponent -5, that is not a whole number.

Types of Polynomials

A **monomial** is a polynomial with exactly one term.
A **binomial** is a polynomial with exactly two terms.
A **trinomial** is a polynomial with exactly three terms.

The following are examples of monomials, binomials, and trinomials. Each of these examples is also a polynomial.

Polynomials			
Monomials	Binomials	Trinomials	More than Three Terms
ax^2	$x + y$	$x^2 + 4xy + y^2$	$5x^3 - 6x^2 + 3x - 6$
$-3z$	$3p + 2$	$x^5 + 7x^2 - x$	$-y^5 + y^4 - 3y^3 - y^2 + y$
4	$4x^2 - 7$	$-q^4 + q^3 - 2q$	$x^6 + x^4 - x^3 + 1$

Each term of a polynomial has a degree. The **degree of a term in one variable** is the exponent on the variable.

Practice 2

Identify the degree of each term of the trinomial
$-15x^3 + 2x^2 - 5$.

Example 2
Identify the degree of each term of the trinomial $12x^4 - 7x + 3$.

Solution: The term $12x^4$ has degree 4.
The term $-7x$ has degree 1 since $-7x$ is $-7x^1$.
The term 3 has degree 0 since 3 is $3x^0$.

■ Work Practice 2

Each polynomial also has a degree.

Degree of a Polynomial

The **degree of a polynomial** is the greatest degree of any term of the polynomial.

Answer
2. 3; 2; 0

Section 10.3 | Introduction to Polynomials

Example 3 Find the degree of each polynomial and tell whether the polynomial is a monomial, binomial, trinomial, or none of these.

a. $-2t^2 + 3t + 6$ b. $15x - 10$ c. $7x + 3x^3 + 2x^2 - 1$

Solution:

a. The degree of the trinomial $-2t^2 + 3t + 6$ is 2, the greatest degree of any of its terms.
b. The degree of the binomial $15x - 10$ or $15x^1 - 10$ is 1.
c. The degree of the polynomial $7x + 3x^3 + 2x^2 - 1$ is 3. The polynomial is neither a monomial, binomial, nor trinomial.

■ Work Practice 3

Practice 3

Find the degree of each polynomial and tell whether the polynomial is a monomial, binomial, trinomial, or none of these.

a. $-6x + 14$
b. $9x - 3x^6 + 5x^4 + 2$
c. $10x^2 - 6x - 6$

Objective C Evaluating Polynomials

Polynomials have different values depending on the replacement values for the variables. When we find the value of a polynomial for a given replacement value, we are evaluating the polynomial for that value.

Example 4 Evaluate each polynomial when $x = -2$.

a. $-5x + 6$ b. $3x^2 - 2x + 1$

Solution:

a. $-5x + 6 = -5(-2) + 6$ Replace x with -2.
$= 10 + 6$
$= 16$

b. $3x^2 - 2x + 1 = 3(-2)^2 - 2(-2) + 1$ Replace x with -2.
$= 3(4) - 2(-2) + 1$
$= 12 + 4 + 1$
$= 17$

■ Work Practice 4

Practice 4

Evaluate each polynomial when $x = -1$.

a. $-2x + 10$
b. $6x^2 + 11x - 20$

Many physical phenomena can be modeled by polynomials.

Example 5 Finding Free-Fall Time

The Swiss Re Building, completed in London in 2003, is a unique building. Londoners often refer to it as the "pickle building." The building is 592.1 feet tall. An object is dropped from the highest point of this building. Neglecting air resistance, the height in feet of the object above ground at time t seconds is given by the polynomial $-16t^2 + 592.1$. Find the height of the object when $t = 1$ second and when $t = 6$ seconds. (See next page for illustration.)

Solution: To find each height, we evaluate the polynomial when $t = 1$ and when $t = 6$.

$-16t^2 + 592.1 = -16(1)^2 + 592.1$ Replace t with 1.
$= -16(1) + 592.1$
$= -16 + 592.1$
$= 576.1$

The height of the object at 1 second is 576.1 feet.

$-16t^2 + 592.1 = -16(6)^2 + 592.1$ Replace t with 6.
$= -16(36) + 592.1$
$= -576 + 592.1 = 16.1$

(Continued on next page)

Practice 5

Find the height of the object in Example 5 when $t = 2$ seconds and $t = 4$ seconds.

Answers

3. a. binomial, 1 b. none of these, 6 c. trinomial, 2 4. a. 12 b. −25
5. 528.1 feet; 336.1 feet

The height of the object at 6 seconds is 16.1 feet.

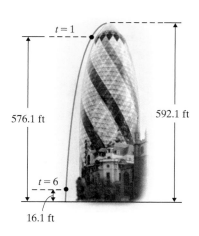

■ Work Practice 5

Objective D Simplifying Polynomials by Combining Like Terms

We can simplify polynomials with like terms by combining the like terms. Recall from Section 3.1 that like terms are terms that contain exactly the same variables raised to exactly the same powers.

Like Terms	Unlike Terms
$5x^2, -7x^2$	$3x, 3y$
$y, 2y$	$-2x^2, -5x$
$\frac{1}{2}a^2b, -a^2b$	$6st^2, 4s^2t$

Only like terms can be combined. We combine like terms by applying the distributive property.

Practice 6–10

Simplify each polynomial by combining any like terms.
6. $-6y + 8y$
7. $14y^2 + 3 - 10y^2 - 9$
8. $7x^3 + x^3$
9. $23x^2 - 6x - x - 15$
10. $\frac{2}{7}x^3 - \frac{1}{4}x + 2 - \frac{1}{2}x^3 + \frac{3}{8}x$

Examples Simplify each polynomial by combining any like terms.

6. $-3x + 7x = (-3 + 7)x = 4x$
7. $11x^2 + 5 + 2x^2 - 7 = 11x^2 + 2x^2 + 5 - 7$
   $= 13x^2 - 2$
8. $9x^3 + x^3 = 9x^3 + 1x^3$ Write x^3 as $1x^3$.
   $= 10x^3$
9. $5x^2 + 6x - 9x - 3 = 5x^2 - 3x - 3$ Combine like terms $6x$ and $-9x$.
10. $\frac{2}{5}x^4 + \frac{2}{3}x^3 - x^2 + \frac{1}{10}x^4 - \frac{1}{6}x^3$

$= \left(\frac{2}{5} + \frac{1}{10}\right)x^4 + \left(\frac{2}{3} - \frac{1}{6}\right)x^3 - x^2$

$= \left(\frac{4}{10} + \frac{1}{10}\right)x^4 + \left(\frac{4}{6} - \frac{1}{6}\right)x^3 - x^2$

$= \frac{5}{10}x^4 + \frac{3}{6}x^3 - x^2$

$= \frac{1}{2}x^4 + \frac{1}{2}x^3 - x^2$

■ Work Practice 6–10

Answers
6. $2y$ 7. $4y^2 - 6$ 8. $8x^3$
9. $23x^2 - 7x - 15$
10. $-\frac{3}{14}x^3 + \frac{1}{8}x + 2$

Section 10.3 | Introduction to Polynomials

Example 11
Write a polynomial that describes the total area of the squares and rectangles shown below. Then simplify the polynomial.

Solution: Recall that the area of a rectangle is length times width.

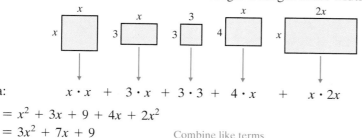

Area: $\quad x \cdot x \; + \; 3 \cdot x \; + \; 3 \cdot 3 \; + \; 4 \cdot x \; + \; x \cdot 2x$
$= x^2 + 3x + 9 + 4x + 2x^2$
$= 3x^2 + 7x + 9 \qquad$ Combine like terms.

■ Work Practice 11

Practice 11
Write a polynomial that describes the total area of the squares and rectangles shown below. Then simplify the polynomial.

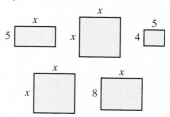

Objective E Simplifying Polynomials Containing Several Variables

A polynomial may contain more than one variable. One example is

$$5x + 3xy^2 - 6x^2y^2 + x^2y - 2y + 1$$

We call this expression a polynomial in several variables.

The **degree of a term** with more than one variable is the sum of the exponents on the variables. The **degree of a polynomial** in several variables is still the greatest degree of the terms of the polynomial.

Example 12
Identify the degrees of the terms and the degree of the polynomial $5x + 3xy^2 - 6x^2y^2 + x^2y - 2y + 1$.

Solution: To organize our work, we use a table.

Terms of Polynomial	Degree of Term	Degree of Polynomial
$5x$	1	
$3xy^2$	1 + 2, or 3	
$-6x^2y^2$	2 + 2, or 4	4 (greatest degree)
x^2y	2 + 1, or 3	
$-2y$	1	
1	0	

■ Work Practice 12

Practice 12
Identify the degrees of the terms and the degree of the polynomial $-2x^3y^2 + 4 - 8xy + 3x^3y + 5xy^2$.

To simplify a polynomial containing several variables, we combine any like terms.

Examples
Simplify each polynomial by combining any like terms.

13. $3xy - 5y^2 + 7yx - 9x^2 = (3 + 7)xy - 5y^2 - 9x^2$
$\qquad\qquad\qquad\qquad\qquad\quad = 10xy - 5y^2 - 9x^2$

14. $9a^2b - 6a^2 + 5b^2 + a^2b - 11a^2 + 2b^2$
$\quad = 10a^2b - 17a^2 + 7b^2$

Helpful Hint: This term can be written as $7yx$ or $7xy$.

■ Work Practice 13–14

Practice 13–14
Simplify each polynomial by combining any like terms.

13. $11ab - 6a^2 - ba + 8b^2$

14. $7x^2y^2 + 2y^2 - 4y^2x^2 + x^2 - y^2 + 5x^2$

Answers
11. $5x + x^2 + 20 + x^2 + 8x$;
$2x^2 + 13x + 20$
12. 5, 0, 2, 4, 3; 5
13. $10ab - 6a^2 + 8b^2$
14. $3x^2y^2 + y^2 + 6x^2$

Objective F Inserting "Missing" Terms

To prepare for dividing polynomials in Section 10.7, let's practice writing a polynomial in descending powers of the variable and with no "missing" powers.

Recall from Objective B that a polynomial such as

$$x^5 - 3x^3 + 2x^2 - 5x + 1$$

is written in descending powers of x because the powers of x decrease from left to right. Study the decreasing powers of x and notice that there is a "missing" power of x. This missing power is x^4. Writing a polynomial in decreasing powers of the variable helps you immediately determine important features of the polynomial, such as its degree. It is also sometimes helpful to write a polynomial so that there are no "missing" powers of x. For our polynomial above, if we simply insert a term of $0x^4$, which equals 0, we have an equivalent polynomial with no missing powers of x.

$$x^5 - 3x^3 + 2x^2 - 5x + 1 = x^5 + 0x^4 - 3x^3 + 2x^2 - 5x + 1$$

Practice 15

Write each polynomial in descending powers of the variable with no missing powers.

a. $x^2 + 9$
b. $9m^3 + m^2 - 5$
c. $-3a^3 + a^4$

Example 15 Write each polynomial in descending powers of the variable with no missing powers.

a. $x^2 - 4$ b. $3m^3 - m + 1$ c. $2x + x^4$

Solution:

a. $x^2 - 4 = x^2 + 0x^1 - 4$ or $x^2 + 0x - 4$ Insert a missing term of $0x^1$ or $0x$.
b. $3m^3 - m + 1 = 3m^3 + 0m^2 - m + 1$ Insert a missing term of $0m^2$.
c. $2x + x^4 = x^4 + 2x$ Write in descending powers of variable.
   $= x^4 + 0x^3 + 0x^2 + 2x + 0x^0$ Insert missing terms of $0x^3$, $0x^2$, and $0x^0$ (or 0).

■ Work Practice 15

Helpful Hint

Since there is no constant as a last term, we insert a $0x^0$. This $0x^0$ (or 0) is the final power of x in our polynomial.

Answers

15. a. $x^2 + 0x + 9$
b. $9m^3 + m^2 + 0m - 5$
c. $a^4 - 3a^3 + 0a^2 + 0a + 0a^0$

Vocabulary, Readiness & Video Check

Use the choices below to fill in each blank. Not all choices will be used.

| least | monomial | trinomial | coefficient |
| greatest | binomial | constant | |

1. A _____ is a polynomial with exactly two terms.
2. A _____ is a polynomial with exactly one term.
3. A _____ is a polynomial with exactly three terms.
4. The numerical factor of a term is called the _____.
5. A number term is also called a _____.
6. The degree of a polynomial is the _____ degree of any term of the polynomial.

Martin-Gay Interactive Videos

See Video 10.3

Watch the section lecture video and answer the following questions.

Objective A 7. How many terms does the polynomial in Example 1 have? What are they?

Objective B 8. For Example 2, why is the degree of each **term** found when the example asks for the degree of the **polynomial** only?

Objective C 9. From Example 3, what does the value of a polynomial depend on?

Objective D 10. When combining any like terms in a polynomial, as in Example 5, what are we doing to the polynomial?

Objective E 11. In Example 6, after combining like terms what is the degree of the binomial? Which term determines this?

Objective F 12. In Example 7, what power is "missing" from the original polynomial? What term is inserted to replace this missing power?

10.3 Exercise Set MyLab Math

Objective A *Complete each table for each polynomial. See Example 1.*

1. $x^2 - 3x + 5$

Term	Coefficient
x^2	
	-3
5	

2. $2x^3 - x + 4$

Term	Coefficient
	2
$-x$	
4	

3. $-5x^4 + 3.2x^2 + x - 5$

Term	Coefficient
$-5x^4$	
$3.2x^2$	
x	
-5	

4. $9.7x^7 - 3x^5 + x^3 - \frac{1}{4}x^2$

Term	Coefficient
$9.7x^7$	
$-3x^5$	
x^3	
$-\frac{1}{4}x^2$	

Objective B *Find the degree of each polynomial and determine whether it is a monomial, binomial, trinomial, or none of these. See Examples 2 and 3.*

5. $x + 2$
6. $-6y + 4$
7. $9m^3 - 5m^2 + 4m - 8$
8. $a + 5a^2 + 3a^3 - 4a^4$
9. $12x^4 - x^6 - 12x^2$
10. $7r^2 + 2r - 3r^5$
11. $3z - 5z^4$
12. $5y^6 + 2$

Objective C *Evaluate each polynomial when* **a.** $x = 0$ *and* **b.** $x = -1$. *See Examples 4 and 5.*

13. $5x - 6$
14. $2x - 10$
15. $x^2 - 5x - 2$
16. $x^2 + 3x - 4$
17. $-x^3 + 4x^2 - 15$
18. $-2x^3 + 3x^2 - 6$

A rocket is fired upward from the ground with an initial velocity of 200 feet per second. Neglecting air resistance, the height of the rocket in feet at any time t can be described by the polynomial $-16t^2 + 200t$. Find the height of the rocket at the times given in Exercises 19 through 22. See Example 5.

	Time, t (in seconds)	Height $-16t^2 + 200t$
19.	1	
20.	5	
21.	7.6	
22.	10.3	

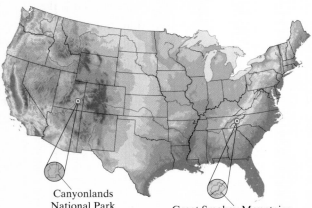

Canyonlands National Park

Great Smokey Mountains National Park

23. The polynomial $12x^2 - 26x + 454$ models the yearly number of visitors (in thousands) x years after 2010 at Canyonlands National Park. Use this polynomial to estimate the number of visitors to the park in 2020 ($x = 10$).

24. The polynomial $100x^2 - 238x + 9398$ models the yearly number of visitors (in thousands) x years after 2010 at Great Smoky Mountains National Park. Use this polynomial to estimate the number of visitors to the park in 2018 ($x = 8$).

25. The number of vinyl album sales (in millions) in the United States x years after 2010 is given by the polynomial $0.16x^2 + 0.9x + 2.7$ for 2010 through 2016. Use this model to predict the number of vinyl album sales in the United States in the year 2020 ($x = 10$). (See the Chapter Opener.)

26. Use the model in Exercise **25** to predict the number of vinyl album sales in the United States in the year 2022. (See the Chapter Opener.)

Objective D Simplify each expression by combining like terms. See Examples 6 through 10.

27. $9x - 20x$

28. $14y - 30y$

29. $14x^3 + 9x^3$

30. $18x^3 + 4x^3$

31. $7x^2 + 3 + 9x^2 - 10$

32. $8x^2 + 4 + 11x^2 - 20$

33. $15x^2 - 3x^2 - 13$

34. $12k^3 - 9k^3 + 11$

35. $8s - 5s + 4s$

36. $5y + 7y - 6y$

37. $0.1y^2 - 1.2y^2 + 6.7 - 1.9$

38. $7.6y + 3.2y^2 - 8y - 2.5y^2$

39. $\frac{2}{3}x^4 + 12x^3 + \frac{1}{6}x^4 - 19x^3 - 19$

40. $\frac{2}{5}x^4 - 23x^2 + \frac{1}{15}x^4 + 5x^2 - 5$

41. $\frac{3}{20}x^3 + \frac{1}{10} - \frac{3}{10}x - \frac{1}{5} - \frac{7}{20}x + 6x^2$

42. $\frac{5}{16}x^3 - \frac{1}{8} + \frac{3}{8}x + \frac{1}{4} - \frac{9}{16}x - 14x^2$

Write a polynomial that describes the total area of each set of rectangles and squares shown in Exercises 43 and 44. Then simplify the polynomial. See Example 11.

43.

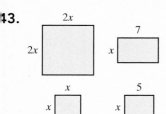

△ **44.**

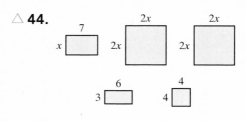

Recall that the perimeter of a figure such as the ones shown in Exercises 45 and 46 is the sum of the lengths of its sides. Write each perimeter as a polynomial. Then simplify the polynomial.

45.

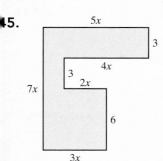

△ **46.**

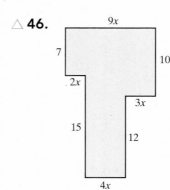

Objective E *Identify the degrees of the terms and the degree of the polynomial. See Example 12.*

47. $9ab - 6a + 5b - 3$

48. $y^4 - 6y^3x + 2x^2y^2 - 5y^2 + 3$

49. $x^3y - 6 + 2x^2y^2 + 5y^3$

50. $2a^2b + 10a^4b - 9ab + 6$

Simplify each polynomial by combining any like terms. See Examples 13 and 14.

51. $3ab - 4a + 6ab - 7a$

52. $-9xy + 7y - xy - 6y$

53. $4x^2 - 6xy + 3y^2 - xy$

54. $3a^2 - 9ab + 4b^2 - 7ab$

55. $5x^2y + 6xy^2 - 5yx^2 + 4 - 9y^2x$

56. $17a^2b - 16ab^2 + 3a^3 + 4ba^3 - b^2a$

57. $14y^3 - 9 + 3a^2b^2 - 10 - 19b^2a^2$

58. $18x^4 + 2x^3y^3 - 1 - 2y^3x^3 - 17x^4$

Objective F *Write each polynomial in descending powers of the variable and with no missing powers. See Example 15.*

59. $7x^2 + 3$

60. $5x^2 - 2$

61. $x^3 - 64$

62. $x^3 - 8$

63. $5y^3 + 2y - 10$

64. $6m^3 - 3m + 4$

65. $8y + 2y^4$

66. $11z + 4z^4$

67. $6x^5 + x^3 - 3x + 15$

68. $9y^5 - y^2 + 2y - 11$

Review

Simplify each expression. See Section 3.1.

69. $4 + 5(2x + 3)$

70. $9 - 6(5x + 1)$

71. $2(x - 5) + 3(5 - x)$

72. $-3(w + 7) + 5(w + 1)$

Concept Extensions

73. Describe how to find the degree of a term.

74. Describe how to find the degree of a polynomial.

75. Explain why xyz is a monomial while $x + y + z$ is a trinomial.

76. Explain why the degree of the term $5y^3$ is 3 and the degree of the polynomial $2y + y + 2y$ is 1.

Simplify, if possible.

77. $x^4 \cdot x^9$

78. $x^4 + x^9$

79. $a \cdot b^3 \cdot a^2 \cdot b^7$

80. $a + b^3 + a^2 + b^7$

81. $(y^5)^4 + (y^2)^{10}$

82. $x^5y^2 + y^2x^5$

Fill in the boxes so that the terms in each expression can be combined. Then simplify. Each exercise has more than one solution.

83. $7x^\square + 2x^\square$

84. $(3y^2)^\square + (4y^3)^\square$

85. Explain why the height of the rocket in Exercises **19** through **22** increases and then decreases as time passes.

86. Approximate (to the nearest tenth of a second) how long before the rocket in Exercises **19** through **22** hits the ground.

Simplify each polynomial by combining like terms.

87. $1.85x^2 - 3.76x + 9.25x^2 + 10.76 - 4.21x$

88. $7.75x + 9.16x^2 - 1.27 - 14.58x^2 - 18.34$

10.4 Adding and Subtracting Polynomials

Objective A Adding Polynomials

To add polynomials, we use commutative and associative properties and then combine like terms. To see if you are ready to add polynomials, try the Concept Check.

Objectives

A Add Polynomials.
B Subtract Polynomials.
C Add or Subtract Polynomials in One Variable.
D Add or Subtract Polynomials in Several Variables.

✓ **Concept Check** When combining like terms in the expression $5x - 8x^2 - 8x$, which of the following is the proper result?

a. $-11x^2$ **b.** $-3x - 8x^2$ **c.** $-11x$ **d.** $-11x^4$

To Add Polynomials

To add polynomials, combine all like terms.

Examples Add.

1. $(4x^3 - 6x^2 + 2x + 7) + (5x^2 - 2x)$
   $= 4x^3 - 6x^2 + 2x + 7 + 5x^2 - 2x$ Remove parentheses.
   $= 4x^3 + (-6x^2 + 5x^2) + (2x - 2x) + 7$ Group like terms.
   $= 4x^3 - x^2 + 7$ Simplify.

2. $(-2x^2 + 5x - 1) + (-2x^2 + x + 3)$
   $= -2x^2 + 5x - 1 - 2x^2 + x + 3$ Remove parentheses.
   $= (-2x^2 - 2x^2) + (5x + 1x) + (-1 + 3)$ Group like terms.
   $= -4x^2 + 6x + 2$ Simplify.

Practice 1–2

Add.
1. $(3x^5 - 7x^3 + 2x - 1) + (3x^3 - 2x)$
2. $(5x^2 - 2x + 1) + (-6x^2 + x - 1)$

Work Practice 1–2

Just as we can add numbers vertically, polynomials can be added vertically if we line up like terms underneath one another.

Example 3 Add $(7y^3 - 2y^2 + 7)$ and $(6y^2 + 1)$ using a vertical format.

Solution: Vertically line up like terms and add.

$7y^3 - 2y^2 + 7$
$ 6y^2 + 1$
$\overline{7y^3 + 4y^2 + 8}$

Practice 3

Add $(9y^2 - 6y + 5)$ and $(4y + 3)$ using a vertical format.

Work Practice 3

Objective B Subtracting Polynomials

To subtract one polynomial from another, recall the definition of subtraction. To subtract a number, we add its opposite: $a - b = a + (-b)$. To subtract a polynomial, we also add its opposite. Just as $-b$ is the opposite of b, $-(x^2 + 5)$ is the opposite of $(x^2 + 5)$.

Answers
1. $3x^5 - 4x^3 - 1$ 2. $-x^2 - x$
3. $9y^2 - 2y + 8$

To Subtract Polynomials

To subtract two polynomials, change the signs of the terms of the polynomial being subtracted and then add.

✓ **Concept Check Answer**
b

Chapter 10 | Exponents and Polynomials

Practice 4

Subtract:
$(9x + 5) - (4x - 3)$

Example 4 Subtract: $(5x - 3) - (2x - 11)$

Solution: From the definition of subtraction, we have

$(5x - 3) - (2x - 11) = (5x - 3) + [-(2x - 11)]$ Add the opposite.
$ = (5x - 3) + (-2x + 11)$ Apply the distributive property
$ = 5x - 3 - 2x + 11$ Remove parentheses.
$ = 3x + 8$ Combine like terms.

■ Work Practice 4

Practice 5

Subtract:
$(4x^3 - 10x^2 + 1)$
$-(-4x^3 + x^2 - 11)$

Example 5 Subtract: $(2x^3 + 8x^2 - 6x) - (2x^3 - x^2 + 1)$

Solution: First, we change the sign of each term of the second polynomial; then we add.

$(2x^3 + 8x^2 - 6x) - (2x^3 - x^2 + 1)$
$= (2x^3 + 8x^2 - 6x) + (-2x^3 + x^2 - 1)$
$= 2x^3 + 8x^2 - 6x - 2x^3 + x^2 - 1$
$= 2x^3 - 2x^3 + 8x^2 + x^2 - 6x - 1$
$= 9x^2 - 6x - 1$ Combine like terms.

■ Work Practice 5

Just as polynomials can be added vertically, so can they be subtracted vertically.

Practice 6

Subtract $(6y^2 - 3y + 2)$ from $(2y^2 - 2y + 7)$ using a vertical format.

Example 6 Subtract $(5y^2 + 2y - 6)$ from $(-3y^2 - 2y + 11)$ using vertical format.

Solution: Arrange the polynomials in a vertical format, lining up like terms.

$$\begin{array}{r} -3y^2 - 2y + 11 \\ -(5y^2 + 2y - 6) \end{array} \quad \begin{array}{r} -3y^2 - 2y + 11 \\ -5y^2 - 2y + 6 \\ \hline -8y^2 - 4y + 17 \end{array}$$

■ Work Practice 6

Helpful Hint

Don't forget to change the sign of each term in the polynomial being subtracted.

Objective C Adding and Subtracting Polynomials in One Variable

Let's practice adding and subtracting polynomials in one variable.

Practice 7

Subtract $(3x + 1)$ from the sum of $(4x - 3)$ and $(12x - 5)$.

Example 7 Subtract $(5z - 7)$ from the sum of $(8z + 11)$ and $(9z - 2)$.

Solution: Notice that $(5z - 7)$ is to be subtracted **from** a sum. The translation is

$[(8z + 11) + (9z - 2)] - (5z - 7)$
$= 8z + 11 + 9z - 2 - 5z + 7$ Remove grouping symbols.
$= 8z + 9z - 5z + 11 - 2 + 7$ Group like terms.
$= 12z + 16$ Combine like terms.

■ Work Practice 7

Answers

4. $5x + 8$ **5.** $8x^3 - 11x^2 + 12$
6. $-4y^2 + y + 5$ **7.** $13x - 9$

Section 10.4 | Adding and Subtracting Polynomials

Objective D Adding and Subtracting Polynomials in Several Variables

Now that we know how to add or subtract polynomials in one variable, we can also add and subtract polynomials in several variables.

Examples Add or subtract as indicated.

8. $(3x^2 - 6xy + 5y^2) + (-2x^2 + 8xy - y^2)$
$= 3x^2 - 6xy + 5y^2 - 2x^2 + 8xy - y^2$
$= x^2 + 2xy + 4y^2$ Combine like terms.

9. $(9a^2b^2 + 6ab - 3ab^2) - (5b^2a + 2ab - 3 - 9b^2)$
$= 9a^2b^2 + 6ab - 3ab^2 - 5b^2a - 2ab + 3 + 9b^2$
$= 9a^2b^2 + 4ab - 8ab^2 + 9b^2 + 3$ Combine like terms.

Work Practice 8–9

Practice 8–9
Add or subtract as indicated.
8. $(2a^2 - ab + 6b^2) + (-3a^2 + ab - 7b^2)$
9. $(5x^2y^2 + 3 - 9x^2y + y^2) - (-x^2y^2 + 7 - 8xy^2 + 2y^2)$

✓ Concept Check
If possible, simplify each expression by performing the indicated operation.
a. $2y + y$
b. $2y \cdot y$
c. $-2y - y$
d. $(-2y)(-y)$
e. $2x + y$

Answers
8. $-a^2 - b^2$
9. $6x^2y^2 - 4 - 9x^2y + 8xy^2 - y^2$

✓ Concept Check Answers
a. $3y$ **b.** $2y^2$ **c.** $-3y$ **d.** $2y^2$
e. cannot be simplified

Vocabulary, Readiness & Video Check

Simplify by combining like terms if possible.

1. $-9y - 5y$
2. $6m^5 + 7m^5$
3. $x + 6x$
4. $7z - z$
5. $5m^2 + 2m$
6. $8p^3 + 3p^2$

See Video 10.4

Martin-Gay Interactive Videos Watch the section lecture video and answer the following questions.

Objective A **7.** In Example 1, like terms are combined when adding the polynomials. What are the two sets of like terms?

Objective B **8.** In Example 2, why can't parentheses just be removed as they were in Example 1?

Objective C **9.** In Example 3, why are we told to be careful when translating to an expression?

Objective D **10.** In Example 5, why aren't any signs changed when parentheses are removed?

10.4 Exercise Set MyLab Math

Objective A *Add. See Examples 1 and 2.*

1. $(3x + 7) + (9x + 5)$

2. $(-y - 2) + (3y + 5)$

3. $(-7x + 5) + (-3x^2 + 7x + 5)$

4. $(3x - 8) + (4x^2 - 3x + 3)$

5. $(-5x^2 + 3) + (2x^2 + 1)$

6. $(3x^2 + 7) + (3x^2 + 9)$

7. $(-3y^2 - 4y) + (2y^2 + y - 1)$

8. $(7x^2 + 2x - 9) + (-3x^2 + 5)$

9. $(1.2x^3 - 3.4x + 7.9) + (6.7x^3 + 4.4x^2 - 10.9)$

10. $(9.6y^3 + 2.7y^2 - 8.6) + (1.1y^3 - 8.8y + 11.6)$

11. $\left(\dfrac{3}{4}m^2 - \dfrac{2}{5}m + \dfrac{1}{8}\right) + \left(-\dfrac{1}{4}m^2 - \dfrac{3}{10}m + \dfrac{11}{16}\right)$

12. $\left(-\dfrac{4}{7}n^2 + \dfrac{5}{6}n - \dfrac{1}{20}\right) + \left(\dfrac{3}{7}n^2 - \dfrac{5}{12}n - \dfrac{3}{10}\right)$

Add using a vertical format. See Example 3.

13. $\begin{array}{r} 3t^2 + 4 \\ \underline{5t^2 - 8} \end{array}$

14. $\begin{array}{r} 7x^3 + 3 \\ \underline{2x^3 - 7} \end{array}$

15. $\begin{array}{r} 10a^3 - 8a^2 + 4a + 9 \\ \underline{5a^3 + 9a^2 - 7a + 7} \end{array}$

16. $\begin{array}{r} 2x^3 - 3x^2 + x - 4 \\ \underline{5x^3 + 2x^2 - 3x + 2} \end{array}$

Objective B *Subtract. See Examples 4 and 5.*

17. $(2x + 5) - (3x - 9)$

18. $(4 + 5a) - (-a - 5)$

19. $(5x^2 + 4) - (-2x^2 + 4)$

20. $(-7y^2 + 5) - (-8y^2 + 12)$

21. $3x - (5x - 9)$

22. $4 - (-y - 4)$

23. $(2x^2 + 3x - 9) - (-4x + 7)$

24. $(-7x^2 + 4x + 7) - (-8x + 2)$

25. $(5x + 8) - (-2x^2 - 6x + 8)$

26. $(-6y^2 + 3y - 4) - (9y^2 - 3y)$

27. $(0.7x^2 + 0.2x - 0.8) - (0.9x^2 + 1.4)$

28. $(-0.3y^2 + 0.6y - 0.3) - (0.5y^2 + 0.3)$

29. $\left(\dfrac{1}{4}z^2 - \dfrac{1}{5}z\right) - \left(-\dfrac{3}{20}z^2 + \dfrac{1}{10}z - \dfrac{7}{20}\right)$

30. $\left(\dfrac{1}{3}x^2 - \dfrac{2}{7}x\right) - \left(\dfrac{4}{21}x^2 + \dfrac{1}{21}x - \dfrac{2}{3}\right)$

Section 10.4 | Adding and Subtracting Polynomials

...btract using a vertical format. See Example 6.

31. $4z^2 - 8z + 3$
$-(6z^2 + 8z - 3)$

32. $7a^2 - 9a + 6$
$-(11a^2 - 4a + 2)$

33. $5u^5 - 4u^2 + 3u - 7$
$-(3u^5 + 6u^2 - 8u + 2)$

34. $5x^3 - 4x^2 + 6x - 2$
$-(3x^3 - 2x^2 - x - 4)$

Objectives A B C Mixed Practice *Add or subtract as indicated. See Examples 1 through 7.*

35. $(3x + 5) + (2x - 14)$

36. $(2y + 20) + (5y - 30)$

37. $(9x - 1) - (5x + 2)$

38. $(7y + 7) - (y - 6)$

39. $(14y + 12) + (-3y - 5)$

40. $(26y + 17) + (-20y - 10)$

41. $(x^2 + 2x + 1) - (3x^2 - 6x + 2)$

42. $(5y^2 - 3y - 1) - (2y^2 + y + 1)$

43. $(3x^2 + 5x - 8) + (5x^2 + 9x + 12) - (8x^2 - 14)$

44. $(2x^2 + 7x - 9) + (x^2 - x + 10) - (3x^2 - 30)$

45. $(-a^2 + 1) - (a^2 - 3) + (5a^2 - 6a + 7)$

46. $(-m^2 + 3) - (m^2 - 13) + (6m^2 - m + 1)$

Translating *Perform each indicated operation. See Examples 3, 6, and 7.*

47. Subtract $4x$ from $(7x - 3)$.

48. Subtract y from $(y^2 - 4y + 1)$.

49. Add $(4x^2 - 6x + 1)$ and $(3x^2 + 2x + 1)$.

50. Add $(-3x^2 - 5x + 2)$ and $(x^2 - 6x + 9)$.

51. Subtract $(5x + 7)$ from $(7x^2 + 3x + 9)$.

52. Subtract $(5y^2 + 8y + 2)$ from $(7y^2 + 9y - 8)$.

53. Subtract $(4y^2 - 6y - 3)$ from the sum of $(8y^2 + 7)$ and $(6y + 9)$.

54. Subtract $(4x^2 - 2x + 2)$ from the sum of $(x^2 + 7x + 1)$ and $(7x + 5)$.

55. Subtract $(3x^2 - 4)$ from the sum of $(x^2 - 9x + 2)$ and $(2x^2 - 6x + 1)$.

56. Subtract $(y^2 - 9)$ from the sum of $(3y^2 + y + 4)$ and $(2y^2 - 6y - 10)$.

Objective D *Add or subtract as indicated. See Examples 8 and 9.*

57. $(9a + 6b - 5) + (-11a - 7b + 6)$

58. $(3x - 2 + 6y) + (7x - 2 - y)$

59. $(4x^2 + y^2 + 3) - (x^2 + y^2 - 2)$

60. $(7a^2 - 3b^2 + 10) - (-2a^2 + b^2 - 12)$

61. $(x^2 + 2xy - y^2) + (5x^2 - 4xy + 20y^2)$

62. $(a^2 - ab + 4b^2) + (6a^2 + 8ab - b^2)$

63. $(11r^2s + 16rs - 3 - 2r^2s^2) - (3sr^2 + 5 - 9r^2s^2)$

64. $(3x^2y - 6xy + x^2y^2 - 5) - (11x^2y^2 - 1 + 5yx^2)$

Objectives **A B C** Mixed Practice *For Exercises 65 through 68, find the perimeter of each figure.*

65.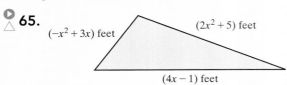

66.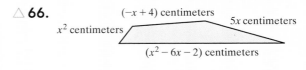

67.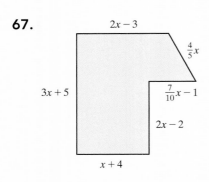

68.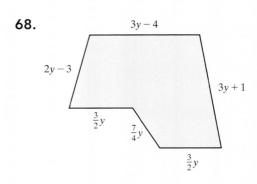

69. A wooden beam is $(4y^2 + 4y + 1)$ meters long. If a piece $(y^2 - 10)$ meters is cut off, express the length of the remaining piece of beam as a polynomial in y.

70. A piece of quarter-round molding is $(13x - 7)$ inches long. If a piece $(2x + 2)$ inches long is removed, express the length of the remaining piece of molding as a polynomial in x.

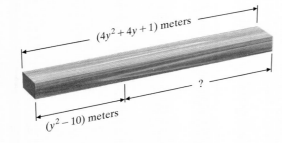

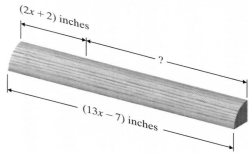

Perform each indicated operation.

71. $[(1.2x^2 - 3x + 9.1) - (7.8x^2 - 3.1 + 8)] + (1.2x - 6)$

72. $[(7.9y^4 - 6.8y^3 + 3.3y) + (6.1y^3 - 5)] - (4.2y^4 + 1.1y - 1)$

Review

Multiply. See Section 10.1.

73. $3x(2x)$ **74.** $-7x(x)$ **75.** $(12x^3)(-x^5)$ **76.** $6r^3(7r^{10})$ **77.** $10x^2(20xy^2)$ **78.** $-z^2y(11zy)$

Concept Extensions

Fill in the squares so that each is a true statement.

79. $3x^\square + 4x^2 = 7x^\square$

80. $9y^7 + 3y^\square = 12y^7$

81. $2x^\square + 3x^\square - 5x^\square + 4x^\square = 6x^4 - 2x^3$

82. $3y^\square + 7y^\square - 2y^\square - y^\square = 10y^5 - 3y^2$

Match each expression on the left with its simplification on the right. Not all letters on the right must be used and a letter may be used more than once.

83. $10y - 6y^2 - y$

84. $5x + 5x$

85. $(5x - 3) + (5x - 3)$

86. $(15x - 3) - (5x - 3)$

a. $3y$
b. $9y - 6y^2$
c. $10x$
d. $25x^2$
e. $10x - 6$
f. none of these

Simplify each expression by performing the indicated operation. Explain how you arrived at each answer. See the second Concept Check in this section.

87.
a. $z + 3z$
b. $z \cdot 3z$
c. $-z - 3z$
d. $(-z)(-3z)$

88.
a. $2y + y$
b. $2y \cdot y$
c. $-2y - y$
d. $(-2y)(-y)$

89.
a. $m \cdot m \cdot m$
b. $m + m + m$
c. $(-m)(-m)(-m)$
d. $-m - m - m$

90.
a. $x + x$
b. $x \cdot x$
c. $-x - x$
d. $(-x)(-x)$

91. The polynomial $437x^2 + 4868x + 3292$ represents the electricity generated (in thousand megawatthours) by photovoltaic solar sources in the United States during 2012–2015. The polynomial $-4489x^2 + 29{,}816x + 141{,}244$ represents the electricity generated (in thousand megawatthours) by wind power in the United States during 2012–2015. In both polynomials, x represents the number of years after 2012. Find a polynomial for the total electricity generated by both solar and wind power during 2012–2015. (*Source:* Based on information from the Energy Information Administration)

92. The polynomial $-43x^2 + 264x + 15{,}565$ represents the electricity generated (in thousand megawatt hours) by geothermal sources in the United States during 2012–2015. The polynomial $-603x^2 - 7199x + 276{,}272$ represents the electricity generated (in thousand megawatt hours) by conventional hydropower in the United States during 2012–2015. In both polynomials, x represents the number of years after 2012. Find a polynomial for the total electricity generated by both geothermal and hydropower during 2012–2015. (*Source:* based on data from the Energy Information Administration)

10.5 Multiplying Polynomials

Objectives
- **A** Multiply Monomials.
- **B** Multiply a Monomial by a Polynomial.
- **C** Multiply Two Polynomials.
- **D** Multiply Polynomials Vertically.

Objective A Multiplying Monomials

Recall from Section 10.1 that to multiply two monomials such as $(-5x^3)$ and $(-2x^4)$ we use the associative and commutative properties and regroup. Remember also that to multiply exponential expressions with a common base, we use the product rule for exponents and add exponents.

$(-5x^3)(-2x^4) = (-5)(-2)(x^3 \cdot x^4)$ Use the commutative and associative properties
$= 10x^7$ Multiply.

Examples Multiply.

1. $6x \cdot 4x = (6 \cdot 4)(x \cdot x)$ Use the commutative and associative properties
   $= 24x^2$ Multiply.
2. $-7x^2 \cdot 2x^5 = (-7 \cdot 2)(x^2 \cdot x^5)$
   $= -14x^7$
3. $(-12x^5)(-x) = (-12x^5)(-1x)$
   $= (-12)(-1)(x^5 \cdot x)$
   $= 12x^6$

■ Work Practice 1–3

Practice 1–3
Multiply.
1. $10x \cdot 9x$
2. $8x^3(-11x^7)$
3. $(-5x^4)(-x)$

✓ **Concept Check** Simplify.
a. $3x \cdot 2x$ b. $3x + 2x$

Objective B Multiplying Monomials by Polynomials

To multiply a monomial such as $7x$ by a trinomial such as $x^2 + 2x + 5$, we use the distributive property.

Examples Multiply.

4. $7x(x^2 + 2x + 5) = 7x(x^2) + 7x(2x) + 7x(5)$ Apply the distributive property
   $= 7x^3 + 14x^2 + 35x$ Multiply.
5. $5x(2x^3 + 6) = 5x(2x^3) + 5x(6)$ Apply the distributive property
   $= 10x^4 + 30x$ Multiply.
6. $-3x^2(5x^2 + 6x - 1)$
   $= (-3x^2)(5x^2) + (-3x^2)(6x) + (-3x^2)(-1)$ Apply the distributive property
   $= -15x^4 - 18x^3 + 3x^2$ Multiply.

■ Work Practice 4–6

Practice 4–6
Multiply.
4. $4x(x^2 + 4x + 3)$
5. $8x(7x^4 + 1)$
6. $-2x^3(3x^2 - x + 2)$

Answers
1. $90x^2$ 2. $-88x^{10}$ 3. $5x^5$
4. $4x^3 + 16x^2 + 12x$ 5. $56x^5 + 8x$
6. $-6x^5 + 2x^4 - 4x^3$

✓ **Concept Check Answers**
a. $6x^2$ b. $5x$

Objective C Multiplying Two Polynomials

We also use the distributive property to multiply two binomials.

Example 7 Multiply.

a. $(m + 4)(m + 6)$ **b.** $(3x + 2)(2x - 5)$

Solution:

a. $(m + 4)(m + 6) = m(m + 6) + 4(m + 6)$ Use the distributive property.
$= m \cdot m + m \cdot 6 + 4 \cdot m + 4 \cdot 6$ Use the distributive property.
$= m^2 + 6m + 4m + 24$ Multiply.
$= m^2 + 10m + 24$ Combine like terms.

b. $(3x + 2)(2x - 5) = 3x(2x - 5) + 2(2x - 5)$ Use the distributive property.
$= 3x(2x) + 3x(-5) + 2(2x) + 2(-5)$
$= 6x^2 - 15x + 4x - 10$ Multiply.
$= 6x^2 - 11x - 10$ Combine like terms.

Work Practice 7

Practice 7
Multiply:
a. $(x + 5)(x + 10)$
b. $(4x + 5)(3x - 4)$

This idea can be expanded so that we can multiply any two polynomials.

To Multiply Two Polynomials

Multiply each term of the first polynomial by each term of the second polynomial, and then combine like terms.

Examples Multiply.

8. $(2x - y)^2$
$= (2x - y)(2x - y)$ Using the meaning of an exponent, we have 2 factors of $(2x - y)$.
$= 2x(2x) + 2x(-y) + (-y)(2x) + (-y)(-y)$
$= 4x^2 - 2xy - 2xy + y^2$ Multiply.
$= 4x^2 - 4xy + y^2$ Combine like terms.

9. $(t + 2)(3t^2 - 4t + 2)$
$= t(3t^2) + t(-4t) + t(2) + 2(3t^2) + 2(-4t) + 2(2)$
$= 3t^3 - 4t^2 + 2t + 6t^2 - 8t + 4$
$= 3t^3 + 2t^2 - 6t + 4$ Combine like terms.

Work Practice 8–9

Practice 8–9
Multiply.
8. $(3x - 2y)^2$
9. $(x + 3)(2x^2 - 5x + 4)$

Concept Check Square where indicated. Simplify if possible.
a. $(4a)^2 + (3b)^2$ **b.** $(4a + 3b)^2$

Objective D Multiplying Polynomials Vertically

Another convenient method for multiplying polynomials is to multiply vertically, similar to the way we multiply real numbers. This method is shown in the next examples.

Answers
7. a. $x^2 + 15x + 50$
b. $12x^2 - x - 20$
8. $9x^2 - 12xy + 4y^2$
9. $2x^3 + x^2 - 11x + 12$

✓ **Concept Check Answers**
a. $16a^2 + 9b^2$ **b.** $16a^2 + 24ab + 9b^2$

Chapter 10 | Exponents and Polynomials

Practice 10

Multiply vertically:
$(3y^2 + 1)(y^2 - 4y + 5)$

Example 10 Multiply vertically: $(2y^2 + 5)(y^2 - 3y + 4)$

Solution:

$$\begin{array}{r} y^2 - 3y + 4 \\ 2y^2 + 5 \\ \hline 5y^2 - 15y + 20 \\ 2y^4 - 6y^3 + 8y^2 \\ \hline 2y^4 - 6y^3 + 13y^2 - 15y + 20 \end{array}$$

Multiply $y^2 - 3y + 4$ by 5.
Multiply $y^2 - 3y + 4$ by $2y^2$.
Combine like terms.

■ Work Practice 10

Practice 11

Find the product of $(4x^2 - x - 1)$ and $(3x^2 + 6x - 2)$ using a vertical format.

Example 11 Find the product of $(2x^2 - 3x + 4)$ and $(x^2 + 5x - 2)$ using a vertical format.

Solution: First, we arrange the polynomials in a vertical format. Then we multiply each term of the first polynomial by each term of the second polynomial.

$$\begin{array}{r} 2x^2 - 3x + 4 \\ x^2 + 5x - 2 \\ \hline -4x^2 + 6x - 8 \\ 10x^3 - 15x^2 + 20x \\ 2x^4 - 3x^3 + 4x^2 \\ \hline 2x^4 + 7x^3 - 15x^2 + 26x - 8 \end{array}$$

Multiply $2x^2 - 3x + 4$ by -2.
Multiply $2x^2 - 3x + 4$ by $5x$.
Multiply $2x^2 - 3x + 4$ by x^2.
Combine like terms.

Answers

10. $3y^4 - 12y^3 + 16y^2 - 4y + 5$
11. $12x^4 + 21x^3 - 17x^2 - 4x + 2$

■ Work Practice 11

Vocabulary, Readiness & Video Check

Fill in each blank with the correct choice.

1. The expression $5x(3x + 2)$ equals $5x \cdot 3x + 5x \cdot 2$ by the _____ property.
   a. commutative   **b.** associative   **c.** distributive

2. The expression $(x + 4)(7x - 1)$ equals $x(7x - 1) + 4(7x - 1)$ by the _____ property.
   a. commutative   **b.** associative   **c.** distributive

3. The expression $(5y - 1)^2$ equals _____.
   a. $2(5y - 1)$ **b.** $(5y - 1)(5y + 1)$ **c.** $(5y - 1)(5y - 1)$

4. The expression $9x \cdot 3x$ equals _____.
   a. $27x$ **b.** $27x^2$ **c.** $12x$ **d.** $12x^2$

Perform the indicated operation, if possible.

5. $x^3 \cdot x^5$ 6. $x^2 \cdot x^6$ 7. $x^3 + x^5$ 8. $x^2 + x^6$

9. $x^7 \cdot x^7$ 10. $x^{11} \cdot x^{11}$ 11. $x^7 + x^7$ 12. $x^{11} + x^{11}$

Martin-Gay Interactive Videos

See Video 10.5

Watch the section lecture video and answer the following questions.

Objective A 13. For Example 1, we use the product property to multiply the monomials. Is it possible to add the same two monomials? Why or why not?

Objective B 14. What property and what exponent rule are used in Examples 3 and 4?

Objective C 15. In Example 5, how many times is the distributive property actually applied? Explain.

Objective D 16. Would you say the vertical format used in Example 8 also applies the distributive property? Explain.

10.5 Exercise Set MyLab Math

Objective A Multiply. See Examples 1 through 3.

1. $8x^2 \cdot 3x$
2. $6x \cdot 3x^2$
3. $(-x^3)(-x)$
4. $(-x^6)(-x)$

5. $-4n^3 \cdot 7n^7$
6. $9t^6(-3t^5)$
7. $(-3.1x^3)(4x^9)$
8. $(-5.2x^4)(3x^4)$

9. $\left(-\frac{1}{3}y^2\right)\left(\frac{2}{5}y\right)$
10. $\left(-\frac{3}{4}y^7\right)\left(\frac{1}{7}y^4\right)$
11. $(2x)(-3x^2)(4x^5)$
12. $(x)(5x^4)(-6x^7)$

Objective B Multiply. See Examples 4 through 6.

13. $3x(2x + 5)$
14. $2x(6x + 3)$
15. $7x(x^2 + 2x - 1)$
16. $5y(y^2 + y - 10)$

17. $-2a(a + 4)$
18. $-3a(2a + 7)$
19. $3x(2x^2 - 3x + 4)$
20. $4x(5x^2 - 6x - 10)$

21. $3a^2(4a^3 + 15)$
22. $9x^3(5x^2 + 12)$
23. $-2a^2(3a^2 - 2a + 3)$
24. $-4b^2(3b^3 - 12b^2 - 6)$

25. $3x^2y(2x^3 - x^2y^2 + 8y^3)$
26. $4xy^2(7x^3 + 3x^2y^2 - 9y^3)$

27. $-y(4x^3 - 7x^2y + xy^2 + 3y^3)$
28. $-x(6y^3 - 5xy^2 + x^2y - 5x^3)$

29. $\frac{1}{2}x^2(8x^2 - 6x + 1)$
30. $\frac{1}{3}y^2(9y^2 - 6y + 1)$

Objective C *Multiply. See Examples 7 through 9.*

31. $(x + 4)(x + 3)$　　**32.** $(x + 2)(x + 9)$　　**33.** $(a + 7)(a - 2)$　　**34.** $(y - 10)(y + 11)$

35. $\left(x + \dfrac{2}{3}\right)\left(x - \dfrac{1}{3}\right)$　　**36.** $\left(x + \dfrac{3}{5}\right)\left(x - \dfrac{2}{5}\right)$　　**37.** $(3x^2 + 1)(4x^2 + 7)$　　**38.** $(5x^2 + 2)(6x^2 + 2)$

39. $(4x - 3)(3x - 5)$　　**40.** $(8x - 3)(2x - 4)$　　**41.** $(1 - 3a)(1 - 4a)$　　**42.** $(3 - 2a)(2 - a)$

43. $(2y - 4)^2$　　**44.** $(6x - 7)^2$　　**45.** $(x - 2)(x^2 - 3x + 7)$　　**46.** $(x + 3)(x^2 + 5x - 8)$

47. $(x + 5)(x^3 - 3x + 4)$　　**48.** $(a + 2)(a^3 - 3a^2 + 7)$　　**49.** $(2a - 3)(5a^2 - 6a + 4)$

50. $(3 + b)(2 - 5b - 3b^2)$　　**51.** $(7xy - y)^2$　　**52.** $(x^2 - 4)^2$

Objective D *Multiply vertically. See Examples 10 and 11.*

53. $(2x - 11)(6x + 1)$　　**54.** $(4x - 7)(5x + 1)$　　**55.** $(x + 3)(2x^2 + 4x - 1)$

56. $(4x - 5)(8x^2 + 2x - 4)$　　**57.** $(x^2 + 5x - 7)(2x^2 - 7x - 9)$　　**58.** $(3x^2 - x + 2)(x^2 + 2x + 1)$

Objectives A B C D Mixed Practice *Multiply. See Examples 1 through 11.*

59. $-1.2y(-7y^6)$　　**60.** $-4.2x(-2x^5)$　　**61.** $-3x(x^2 + 2x - 8)$　　**62.** $-5x(x^2 - 3x + 10)$

63. $(x + 19)(2x + 1)$　　**64.** $(3y + 4)(y + 11)$　　**65.** $\left(x + \dfrac{1}{7}\right)\left(x - \dfrac{3}{7}\right)$　　**66.** $\left(m + \dfrac{2}{9}\right)\left(m - \dfrac{1}{9}\right)$

67. $(3y + 5)^2$　　**68.** $(7y + 2)^2$　　**69.** $(a + 4)(a^2 - 6a + 6)$　　**70.** $(t + 3)(t^2 - 5t + 5$

Section 10.5 | Multiplying Polynomials   783

Express as the product of polynomials. Then multiply.

71. Find the area of the rectangle.

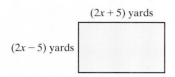

72. Find the area of the square field.

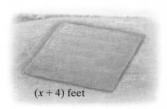

73. Find the area of the triangle.

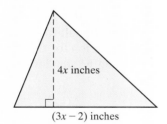

△ **74.** Find the volume of the cube-shaped glass block.

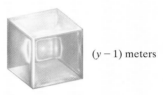

Review

In this section, we review operations on monomials. Study the box below, then proceed. See Sections 3.1, 10.1, and 10.2.

Operations on Monomials	
Multiply	Review the product rule for exponents.
Divide	Review the quotient rule for exponents.
Add or Subtract	Remember, we may only combine like terms.

Perform the operations on the monomials, if possible. The first two rows have been completed for you.

Monomials	Add	Subtract	Multiply	Divide
$6x, 3x$	$6x + 3x = 9x$	$6x - 3x = 3x$	$6x \cdot 3x = 18x^2$	$\frac{6x}{3x} = 2$
$-12x^2, 2x$	$-12x^2 + 2x$; can't be simplified	$-12x^2 - 2x$; can't be simplified	$-12x^2 \cdot 2x = -24x^3$	$\frac{-12x^2}{2x} = -6x$
75. $5a, 15a$				
76. $4y^3, 4y^7$				
77. $-3y^5, 9y^4$				
78. $-14x^2, 2x^2$				

Concept Extensions

79. Perform each indicated operation. Explain the difference between the two expressions.
  a. $(3x + 5) + (3x + 7)$
  b. $(3x + 5)(3x + 7)$

80. Perform each indicated operation. Explain the difference between the two expressions.
  a. $(8x - 3) - (5x - 2)$
  b. $(8x - 3)(5x - 2)$

Mixed Practice *Perform the indicated operations. See Sections 10.4 and 10.5.*

81. $(3x - 1) + (10x - 6)$

82. $(2x - 1) + (10x - 7)$

83. $(3x - 1)(10x - 6)$

84. $(2x - 1)(10x - 7)$

85. $(3x - 1) - (10x - 6)$

86. $(2x - 1) - (10x - 7)$

87. The area of the largest rectangle below is $x(x + 3)$. Find another expression for this area by finding the sum of the areas of the smaller rectangles.

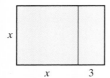

88. The area of the figure below is $(x + 2)(x + 3)$. Find another expression for this area by finding the sum of the areas of the smaller rectangles.

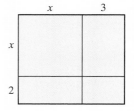

89. Write an expression for the area of the largest rectangle below in two different ways.

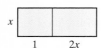

90. Write an expression for the area of the figure below in two different ways.

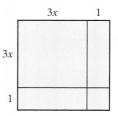

Simplify. See the Concept Checks in this section.

91. $5a + 6a$

92. $5a \cdot 6a$

Square where indicated. Simplify if possible.

93. $(5x)^2 + (2y)^2$

94. $(5x + 2y)^2$

95. Multiply each of the following polynomials.
  a. $(a + b)(a - b)$
  b. $(2x + 3y)(2x - 3y)$
  c. $(4x + 7)(4x - 7)$
  d. Can you make a general statement about all products of the form $(x + y)(x - y)$?

96. Evaluate each of the following.
  a. $(2 + 3)^2; 2^2 + 3^2$
  b. $(8 + 10)^2; 8^2 + 10^2$
  c. Does $(a + b)^2 = a^2 + b^2$ no matter what the values of a and b are? Why or why not?

10.6 Special Products

Objective A Using the FOIL Method

In this section, we multiply binomials using special products. First, we introduce a special order for multiplying binomials called the FOIL order or method. This order, or pattern, is a result of the distributive property. We demonstrate by multiplying $(3x + 1)$ by $(2x + 5)$.

Objectives

A Multiply Two Binomials Using the FOIL Method.

B Square a Binomial.

C Multiply the Sum and Difference of Two Terms.

D Use Special Products to Multiply Binomials.

The FOIL Method

F stands for the product of the **First** terms.

$(3x + 1)(2x + 5)$

$(3x)(2x) = 6x^2$ F

O stands for the product of the **Outer** terms.

$(3x + 1)(2x + 5)$

$(3x)(5) = 15x$ O

I stands for the product of the **Inner** terms.

$(3x + 1)(2x + 5)$

$(1)(2x) = 2x$ I

L stands for the product of the **Last** terms.

$(3x + 1)(2x + 5)$

$(1)(5) = 5$ L

$$(3x + 1)(2x + 5) = \overset{F}{6x^2} + \overset{O}{15x} + \overset{I}{2x} + \overset{L}{5}$$
$$= 6x^2 + 17x + 5 \quad \text{Combine like terms.}$$

Let's practice multiplying binomials using the FOIL method.

Example 1 Multiply: $(x - 3)(x + 4)$

Solution:

$$(x - 3)(x + 4) = \overset{F}{(x)(x)} + \overset{O}{(x)(4)} + \overset{I}{(-3)(x)} + \overset{L}{(-3)(4)}$$
$$= x^2 + 4x - 3x - 12$$
$$= x^2 + x - 12 \quad \text{Combine like terms.}$$

Work Practice 1

Practice 1

Multiply: $(x + 7)(x - 5)$

Helpful Hint Remember that the FOIL order for multiplying can be used only for the product of 2 binomials.

Example 2 Multiply: $(5x - 7)(x - 2)$

Solution:

$$(5x - 7)(x - 2) = \overset{F}{5x(x)} + \overset{O}{5x(-2)} + \overset{I}{(-7)(x)} + \overset{L}{(-7)(-2)}$$
$$= 5x^2 - 10x - 7x + 14$$
$$= 5x^2 - 17x + 14 \quad \text{Combine like terms.}$$

Work Practice 2

Practice 2

Multiply: $(6x - 1)(x - 4)$

Answers

1. $x^2 + 2x - 35$ **2.** $6x^2 - 25x + 4$

785

Practice 3

Multiply: $(2y^2 + 3)(y - 4)$

Example 3 Multiply: $(y^2 + 6)(2y - 1)$

Solution:
$$\overset{\text{F} \quad \text{O} \quad \text{I} \quad \text{L}}{(y^2 + 6)(2y - 1) = 2y^3 - 1y^2 + 12y - 6}$$

Notice in this example that there are no like terms that can be combined, so the product is $2y^3 - y^2 + 12y - 6$.

■ Work Practice 3

Objective B Squaring Binomials

An expression such as $(3y + 1)^2$ is called the square of a binomial. Since $(3y + 1)^2 = (3y + 1)(3y + 1)$, we can use the FOIL method to find this product.

Practice 4

Multiply: $(2x + 9)^2$

Example 4 Multiply: $(3y + 1)^2$

Solution:
$$(3y + 1)^2 = (3y + 1)(3y + 1)$$
$$\overset{\text{F} \quad \text{O} \quad \text{I} \quad \text{L}}{= (3y)(3y) + (3y)(1) + 1(3y) + 1(1)}$$
$$= 9y^2 + 3y + 3y + 1$$
$$= 9y^2 + 6y + 1$$

■ Work Practice 4

Notice the pattern that appears in Example 4.

$(3y + 1)^2 = 9y^2 + 6y + 1$

- $9y^2$ is the first term of the binomial squared: $(3y)^2 = 9y^2$.
- $6y$ is 2 times the product of both terms of the binomial: $(2)(3y)(1) = 6y$.
- 1 is the second term of the binomial squared: $(1)^2 = 1$.

This pattern leads to the formulas below, which can be used when squaring a binomial. We call these **special products**.

Squaring a Binomial

A binomial squared is equal to the square of the first term plus or minus twice the product of both terms plus the square of the second term.

$(a + b)^2 = a^2 + 2ab + b^2$
$(a - b)^2 = a^2 - 2ab + b^2$

This product can be visualized geometrically.

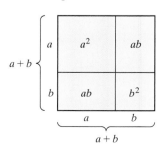

The area of the large square is side · side.
Area $= (a + b)(a + b) = (a + b)^2$
The area of the large square is also the sum of the areas of the smaller rectangles.
Area $= a^2 + ab + ab + b^2 = a^2 + 2ab + b^2$
Thus, $(a + b)^2 = a^2 + 2ab + b^2$.

Answers
3. $2y^3 - 8y^2 + 3y - 12$
4. $4x^2 + 36x + 81$

Examples
Use a special product to square each binomial.

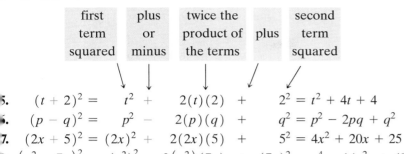

5. $(t + 2)^2 = t^2 + 2(t)(2) + 2^2 = t^2 + 4t + 4$
6. $(p - q)^2 = p^2 - 2(p)(q) + q^2 = p^2 - 2pq + q^2$
7. $(2x + 5)^2 = (2x)^2 + 2(2x)(5) + 5^2 = 4x^2 + 20x + 25$
8. $(x^2 - 7y)^2 = (x^2)^2 - 2(x^2)(7y) + (7y)^2 = x^4 - 14x^2y + 49y^2$

Work Practice 5–8

Practice 5–8

Use a special product to square each binomial.

5. $(y + 3)^2$
6. $(r - s)^2$
7. $(6x + 5)^2$
8. $(x^2 - 3y)^2$

Helpful Hint
Notice that
$(a + b)^2 \neq a^2 + b^2$ The middle term, $2ab$, is missing.
$(a + b)^2 = (a + b)(a + b) = a^2 + 2ab + b^2$

Likewise,
$(a - b)^2 \neq a^2 - b^2$
$(a - b)^2 = (a - b)(a - b) = a^2 - 2ab + b^2$

Objective C Multiplying the Sum and Difference of Two Terms

Another special product is the product of the sum and difference of the same two terms, such as $(x + y)(x - y)$. Finding this product by the FOIL method, we see a pattern emerge.

$(x + y)(x - y) = x^2 - xy + xy - y^2$
$ = x^2 - y^2$

Notice that the two middle terms subtract out. This is because the **O**uter product is the opposite of the **I**nner product. Only the **difference of squares** remains.

Multiplying the Sum and Difference of Two Terms
The product of the sum and difference of two terms is the square of the first term minus the square of the second term.
$(a + b)(a - b) = a^2 - b^2$

Answers
5. $y^2 + 6y + 9$
6. $r^2 - 2rs + s^2$
7. $36x^2 + 60x + 25$
8. $x^4 - 6x^2y + 9y^2$

Practice 9–13

Use a special product to multiply.

9. $(x + 9)(x - 9)$
10. $(5 + 4y)(5 - 4y)$
11. $\left(x - \dfrac{1}{3}\right)\left(x + \dfrac{1}{3}\right)$
12. $(3a - b)(3a + b)$
13. $(2x^2 - 6y)(2x^2 + 6y)$

Examples Use a special product to multiply.

$$\underset{\downarrow}{\text{first term squared}} \quad \underset{\downarrow}{\text{minus}} \quad \underset{\downarrow}{\text{second term squared}}$$

9. $(x + 4)(x - 4) = x^2 \;-\; 4^2 = x^2 - 16$
10. $(6t + 7)(6t - 7) = (6t)^2 \;-\; 7^2 = 36t^2 - 49$
11. $\left(x - \dfrac{1}{4}\right)\left(x + \dfrac{1}{4}\right) = x^2 \;-\; \left(\dfrac{1}{4}\right)^2 = x^2 - \dfrac{1}{16}$
12. $(2p - q)(2p + q) = (2p)^2 - q^2 = 4p^2 - q^2$
13. $(3x^2 - 5y)(3x^2 + 5y) = (3x^2)^2 - (5y)^2 = 9x^4 - 25y^2$

■ Work Practice 9–13

✓ **Concept Check** Match each expression on the left to the equivalent expression or expressions in the list on the right.

$(a + b)^2$ **a.** $(a + b)(a + b)$
$(a + b)(a - b)$ **b.** $a^2 - b^2$
    c. $a^2 + b^2$
    d. $a^2 - 2ab + b^2$
    e. $a^2 + 2ab + b^2$

Objective D Using Special Products

Let's now practice using our special products on a variety of multiplication problems. This practice will help us recognize when to apply what special product formula.

Practice 14–17

Use a special product to multiply, if possible.

14. $(7x - 1)^2$
15. $(5y + 3)(2y - 5)$
16. $(2a - 1)(2a + 1)$
17. $\left(5y - \dfrac{1}{9}\right)^2$

Examples Use a special product to multiply, if possible.

14. $(4x - 9)(4x + 9)$
    $= (4x)^2 - 9^2 = 16x^2 - 81$ This is the sum and difference of the same two terms.
15. $(3y + 2)^2$ This is a binomial squared.
    $= (3y)^2 + 2(3y)(2) + 2^2$
    $= 9y^2 + 12y + 4$
16. $(6a + 1)(a - 7)$ No special product applies.
         F   O   I   L   Use the FOIL method.
    $= 6a \cdot a + 6a(-7) + 1 \cdot a + 1(-7)$
    $= 6a^2 - 42a + a - 7$
    $= 6a^2 - 41a - 7$
17. $\left(4x - \dfrac{1}{11}\right)^2$ This is a binomial squared.
    $= (4x)^2 - 2(4x)\left(\dfrac{1}{11}\right) + \left(\dfrac{1}{11}\right)^2$
    $= 16x^2 - \dfrac{8}{11}x + \dfrac{1}{121}$

■ Work Practice 14–17

Answers

9. $x^2 - 81$ 10. $25 - 16y^2$
11. $x^2 - \dfrac{1}{9}$ 12. $9a^2 - b^2$
13. $4x^4 - 36y^2$
14. $49x^2 - 14x + 1$
15. $10y^2 - 19y - 15$
16. $4a^2 - 1$
17. $25y^2 - \dfrac{10}{9}y + \dfrac{1}{81}$

✓ **Concept Check Answer**
a and **e**, **b**

Section 10.6 | Special Products   789

> **Helpful Hint**
> - When multiplying two binomials, you may always use the FOIL order or method.
> - When multiplying any two polynomials, you may always use the distributive property to find the product.

Vocabulary, Readiness & Video Check

Answer each exercise true or false.

1. $(x + 4)^2 = x^2 + 16$ _____
2. For $(x + 6)(2x - 1)$, the product of the first terms is $2x^2$. _____
3. $(x + 4)(x - 4) = x^2 + 16$ _____
4. The product $(x - 1)(x^3 + 3x - 1)$ is a polynomial of degree 5. _____

Martin-Gay Interactive Videos Watch the section lecture video and answer the following questions.

See Video 10.6

Objective A 5. From Examples 1–3, for what type of multiplication problem is the FOIL order of multiplication used?

Objective B 6. Name at least one other method we can use to multiply Example 4.

Objective C 7. From Example 5, why does multiplying the sum and difference of the same two terms always give us a binomial answer?

Objective D 8. At the end of Example 8, what three special products for multiplying binomials are summarized?

10.6 Exercise Set MyLab Math

Objective A *Multiply using the FOIL method. See Examples 1 through 3.*

1. $(x + 3)(x + 4)$
2. $(x + 5)(x + 1)$
3. $(x - 5)(x + 10)$

4. $(y - 12)(y + 4)$
5. $(5x - 6)(x + 2)$
6. $(3y - 5)(2y + 7)$

7. $(y - 6)(4y - 1)$
8. $(2x - 9)(x - 11)$
9. $(2x + 5)(3x - 1)$

10. $(6x + 2)(x - 2)$
11. $(y^2 + 7)(6y + 4)$
12. $(y^2 + 3)(5y + 6)$

13. $\left(x - \dfrac{1}{3}\right)\left(x + \dfrac{2}{3}\right)$
14. $\left(x - \dfrac{2}{5}\right)\left(x + \dfrac{1}{5}\right)$
15. $(0.4 - 3a)(0.2 - 5a)$

16. $(0.3 - 2a)(0.6 - 5a)$
17. $(x + 5y)(2x - y)$
18. $(x + 4y)(3x - y)$

Objective B *Multiply. See Examples 4 through 8.*

19. $(x + 2)^2$
20. $(x + 7)^2$
21. $(2a - 3)^2$
22. $(7x - 3)^2$

23. $(3a - 5)^2$
24. $(5a - 2)^2$
25. $(x^2 + 0.5)^2$
26. $(x^2 + 0.3)^2$

27. $\left(y - \dfrac{2}{7}\right)^2$
28. $\left(y - \dfrac{3}{4}\right)^2$
29. $(2x - 1)^2$
30. $(5b - 4)^2$

31. $(5x + 9)^2$
32. $(6s + 2)^2$
33. $(3x - 7y)^2$
34. $(4s - 2y)^2$

35. $(4m + 5n)^2$
36. $(3n + 5m)^2$
37. $(5x^4 - 3)^2$
38. $(7x^3 - 6)^2$

Objective C *Multiply. See Examples 9 through 13.*

39. $(a - 7)(a + 7)$
40. $(b + 3)(b - 3)$
41. $(x + 6)(x - 6)$
42. $(x - 8)(x + 8)$

43. $(3x - 1)(3x + 1)$
44. $(7x - 5)(7x + 5)$
45. $(x^2 + 5)(x^2 - 5)$
46. $(a^2 + 6)(a^2 - 6)$

47. $(2y^2 - 1)(2y^2 + 1)$
48. $(3x^2 + 1)(3x^2 - 1)$
49. $(4 - 7x)(4 + 7x)$
50. $(8 - 7x)(8 + 7x)$

51. $\left(3x - \dfrac{1}{2}\right)\left(3x + \dfrac{1}{2}\right)$
52. $\left(10x + \dfrac{2}{7}\right)\left(10x - \dfrac{2}{7}\right)$
53. $(9x + y)(9x - y)$
54. $(2x - y)(2x + y)$

55. $(2m + 5n)(2m - 5n)$
56. $(5m + 4n)(5m - 4n)$

Objective D **Mixed Practice** *Multiply. See Examples 14 through 17.*

57. $(a + 5)(a + 4)$
58. $(a + 5)(a + 7)$
59. $(a - 7)^2$
60. $(b - 2)^2$

61. $(4a + 1)(3a - 1)$
62. $(6a + 7)(6a + 5)$
63. $(x + 2)(x - 2)$
64. $(x - 10)(x + 10)$

65. $(3a + 1)^2$
66. $(4a + 2)^2$
67. $(x + y)(4x - y)$
68. $(3x + 2)(4x - 2)$

69. $\left(\frac{1}{3}a^2 - 7\right)\left(\frac{1}{3}a^2 + 7\right)$ **70.** $\left(\frac{a}{2} + 4y\right)\left(\frac{a}{2} - 4y\right)$ ▶ **71.** $(3b + 7)(2b - 5)$ **72.** $(3y - 13)(y - 3)$

73. $(x^2 + 10)(x^2 - 10)$ **74.** $(x^2 + 8)(x^2 - 8)$ ▶ **75.** $(4x + 5)(4x - 5)$ **76.** $(3x + 5)(3x - 5)$

77. $(5x - 6y)^2$ **78.** $(4x - 9y)^2$ **79.** $(2r - 3s)(2r + 3s)$ **80.** $(6r - 2x)(6r + 2x)$

Express each as a product of polynomials in x. Then multiply and simplify.

81. Find the area of the square rug if its side is $(2x + 1)$ feet.

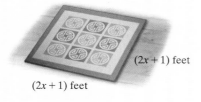
$(2x + 1)$ feet
$(2x + 1)$ feet

82. Find the area of the rectangular canvas if its length is $(3x - 2)$ inches and its width is $(x - 4)$ inches.

$(x - 4)$ inches
$(3x - 2)$ inches

Review

Simplify each expression. See Sections 10.1 and 10.2.

83. $\dfrac{50b^{10}}{70b^5}$ **84.** $\dfrac{60y^6}{80y^2}$ **85.** $\dfrac{8a^{17}b^5}{-4a^7b^{10}}$ **86.** $\dfrac{-6a^8y}{3a^4y}$ **87.** $\dfrac{2x^4y^{12}}{3x^4y^4}$ **88.** $\dfrac{-48ab^6}{32ab^3}$

Concept Extensions

Match each expression on the left to the equivalent expression on the right. See the Concept Check in this section. (Not all choices will be used.)

89. $(a - b)^2$

90. $(a - b)(a + b)$

91. $(a + b)^2$

92. $(a + b)^2(a - b)^2$

a. $a^2 - b^2$
b. $a^2 + b^2$
c. $a^2 - 2ab + b^2$
d. $a^2 + 2ab + b^2$
e. none of these

Fill in the squares so that a true statement forms.

93. $(x^\square + 7)(x^\square + 3) = x^4 + 10x^2 + 21$

94. $(5x^\square - 2)^2 = 25x^6 - 20x^3 + 4$

Find the area of the shaded figure. To do so, subtract the area of the smaller square(s) from the area of the larger geometric figure.

△ 95.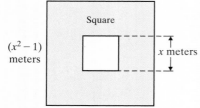

△ 96.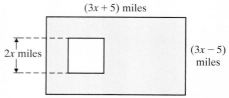

△ 97.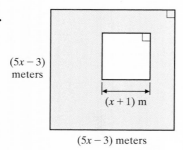

△ 98.

(3x − 4) centimeters

(3x + 4) centimeters

(with four small squares of side x)

99. In your own words, describe the different methods that can be used to find the product: $(2x - 5)(3x + 1)$.

100. In your own words, describe the different methods that can be used to find the product: $(5x + 1)^2$.

101. Suppose that a classmate asked you why $(2x + 1)^2$ is **not** $(4x^2 + 1)$. Write down your response to this classmate.

102. Suppose that a classmate asked you why $(2x + 1)^2$ **is** $(4x^2 + 4x + 1)$. Write down your response to this classmate.

Sections 10.1–10.6 Integrated Review

Exponents and Operations on Polynomials

Perform operations and simplify.

1. $(5x^2)(7x^3)$

2. $(4y^2)(-8y^7)$

3. -4^2

4. $(-4)^2$

5. $(x-5)(2x+1)$

6. $(3x-2)(x+5)$

7. $(x-5)+(2x+1)$

8. $(3x-2)+(x+5)$

9. $\dfrac{7x^9 y^{12}}{x^3 y^{10}}$

10. $\dfrac{20a^2 b^8}{14a^2 b^2}$

11. $(12m^7 n^6)^2$

12. $(4y^9 z^{10})^3$

13. $(4y-3)(4y+3)$

14. $(7x-1)(7x+1)$

15. $(x^{-7} y^5)^9$

16. 8^{-2}

17. $(3^{-1} x^9)^3$

18. $\dfrac{(r^7 s^{-5})^6}{(2r^{-4} s^{-4})^4}$

19. $(7x^2 - 2x + 3) - (5x^2 + 9)$

20. $(10x^2 + 7x - 9) - (4x^2 - 6x + 2)$

Answers

1. _____
2. _____
3. _____
4. _____
5. _____
6. _____
7. _____
8. _____
9. _____
10. _____
11. _____
12. _____
13. _____
14. _____
15. _____
16. _____
17. _____
18. _____
19. _____
20. _____

21. ___

22. ___

23. ___

24. ___

25. ___

26. ___

27. ___

28. ___

29. ___

30. ___

31. ___

32. ___

33. ___

34. ___

35. ___

36. ___

37. ___

38. ___

21. $0.7y^2 - 1.2 + 1.8y^2 - 6y + 1$

22. $7.8x^2 - 6.8x - 3.3 + 0.6x^2 - 0.9$

23. Subtract $(y^2 + 2)$ from $(3y^2 - 6y + 1)$.

24. $(z^2 + 5) - (3z^2 - 1) + \left(8z^2 + 2z - \dfrac{1}{2}\right)$

25. $(x + 4)^2$

26. $(y - 9)^2$

27. $(x + 4) + (x + 4)$

28. $(y - 9) + (y - 9)$

29. $7x^2 - 6xy + 4(y^2 - xy)$

30. $5a^2 - 3ab + 6(b^2 - a^2)$

31. $(x - 3)(x^2 + 5x - 1)$

32. $(x + 1)(x^2 - 3x - 2)$

33. $(2x - 7)(3x + 10)$

34. $(5x - 1)(4x + 5)$

35. $(2x - 7)(x^2 - 6x + 1)$

36. $(5x - 1)(x^2 + 2x - 3)$

37. $\left(2x + \dfrac{5}{9}\right)\left(2x - \dfrac{5}{9}\right)$

38. $\left(12y + \dfrac{3}{7}\right)\left(12y - \dfrac{3}{7}\right)$

10.7 Dividing Polynomials

Objective A Dividing by a Monomial

To divide a polynomial by a monomial, recall addition of fractions. Fractions that have a common denominator are added by adding the numerators:

$$\frac{a}{c} + \frac{b}{c} = \frac{a+b}{c}$$

If we read this equation from right to left and let a, b, and c be monomials, $c \neq 0$, we have the following.

Objectives
A Divide a Polynomial by a Monomial.
B Use Long Division to Divide a Polynomial by a Polynomial Other Than a Monomial.

> **To Divide a Polynomial by a Monomial**
>
> Divide each term of the polynomial by the monomial.
>
> $$\frac{a+b}{c} = \frac{a}{c} + \frac{b}{c}, \quad c \neq 0$$

Throughout this section, we assume that denominators are not 0.

Example 1 Divide: $(6m^2 + 2m) \div 2m$

Solution: We begin by writing the quotient in fraction form. Then we divide each term of the polynomial $6m^2 + 2m$ by the monomial $2m$ and use the quotient rule for exponents to simplify.

$$\frac{6m^2 + 2m}{2m} = \frac{6m^2}{2m} + \frac{2m}{2m}$$
$$= 3m + 1 \quad \text{Simplify.}$$

Check: To check, we multiply.

$$2m(3m + 1) = 2m(3m) + 2m(1) = 6m^2 + 2m$$

The quotient $3m + 1$ checks.

■ Work Practice 1

Practice 1
Divide: $(25x^3 + 5x^2) \div 5x^2$

✓**Concept Check** In which of the following is $\dfrac{x+5}{5}$ simplified correctly?

a. $\dfrac{x}{5} + 1$ **b.** x **c.** $x + 1$

Example 2 Divide: $\dfrac{9x^5 - 12x^2 + 3x}{3x^2}$

Solution: $\dfrac{9x^5 - 12x^2 + 3x}{3x^2} = \dfrac{9x^5}{3x^2} - \dfrac{12x^2}{3x^2} + \dfrac{3x}{3x^2}$ Divide each term by $3x^2$.

$$= 3x^3 - 4 + \dfrac{1}{x} \quad \text{Simplify.}$$

Notice that the quotient is not a polynomial because of the term $\dfrac{1}{x}$. This expression is called a rational expression—we will study rational expressions in Chapter 12. Although the quotient of two polynomials is not always a polynomial, we may still check by multiplying.

Practice 2
Divide: $\dfrac{24x^7 + 12x^2 - 4x}{4x^2}$

Answers
1. $5x + 1$ **2.** $6x^5 + 3 - \dfrac{1}{x}$

✓**Concept Check Answer**
a

(Continued on next page)

Check: $3x^2\left(3x^3 - 4 + \dfrac{1}{x}\right) = 3x^2(3x^3) - 3x^2(4) + 3x^2\left(\dfrac{1}{x}\right)$

$= 9x^5 - 12x^2 + 3x$

■ Work Practice 2

Practice 3

Divide: $\dfrac{12x^3y^3 - 18xy + 6y}{3xy}$

Example 3 Divide: $\dfrac{8x^2y^2 - 16xy + 2x}{4xy}$

Solution: $\dfrac{8x^2y^2 - 16xy + 2x}{4xy} = \dfrac{8x^2y^2}{4xy} - \dfrac{16xy}{4xy} + \dfrac{2x}{4xy}$ Divide each term by $4xy$.

$= 2xy - 4 + \dfrac{1}{2y}$ Simplify.

Check: $4xy\left(2xy - 4 + \dfrac{1}{2y}\right) = 4xy(2xy) - 4xy(4) + 4xy\left(\dfrac{1}{2y}\right)$

$= 8x^2y^2 - 16xy + 2x$

■ Work Practice 3

Objective B Dividing by a Polynomial Other Than a Monomial ▶

To divide a polynomial by a polynomial other than a monomial, we use a process known as long division. Polynomial long division is similar to number long division, so we review long division by dividing 13 into 3660.

$$\begin{array}{r} 281 \\ 13\overline{)3660} \\ -26 \\ \hline 106 \\ -104 \\ \hline 20 \\ -13 \\ \hline 7 \end{array}$$

$2 \cdot 13 = 26$

Subtract and bring down the next digit in the dividend.

$8 \cdot 13 = 104$

Subtract and bring down the next digit in the dividend.

$1 \cdot 13 = 13$

Subtract. There are no more digits to bring down, so the remainder is 7.

Helpful Hint Recall that 3660 is called the dividend.

The quotient is 281 R 7, which can be written as $281\dfrac{7}{13}$.   ← remainder / ← divisor

Recall that division can be checked by multiplication. To check this division problem, we see that

$13 \cdot 281 + 7 = 3660$, the dividend.

Now we demonstrate long division of polynomials.

Practice 4

Divide $x^2 + 12x + 35$ by $x + 5$ using long division.

Example 4 Divide $x^2 + 7x + 12$ by $x + 3$ using long division.

Solution:

To subtract, change the signs of these terms and add.

$$\begin{array}{r} x \\ x + 3\overline{)x^2 + 7x + 12} \\ \underline{x^2 + 3x}\downarrow \\ 4x + 12 \end{array}$$

How many times does x divide x^2? $\dfrac{x^2}{x} = x$.

Multiply: $x(x + 3)$

Subtract and bring down the next term.

Answers

3. $4x^2y^2 - 6 + \dfrac{2}{x}$ **4.** $x + 7$

Now we repeat this process.

$$\begin{array}{r} x + 4 \\ x + 3 \overline{)x^2 + 7x + 12} \\ \underline{x^2 + 3x} \\ 4x + 12 \\ \underline{4x + 12} \\ 0 \end{array}$$

How many times does x divide $4x$? $\frac{4x}{x} = 4$.

Multiply: $4(x + 3)$
Subtract. The remainder is 0.

To subtract, change the signs of these terms and add.

The quotient is $x + 4$.

Check: We check by multiplying.

or

divisor · quotient + remainder = dividend

$(x + 3) \cdot (x + 4) + 0 = x^2 + 7x + 12$

The quotient checks.

■ Work Practice 4

Example 5 Divide $6x^2 + 10x - 5$ by $3x - 1$ using long division.

Solution:

$$\begin{array}{r} 2x + 4 \\ 3x - 1 \overline{)6x^2 + 10x - 5} \\ \underline{6x^2 - 2x} \\ 12x - 5 \\ \underline{12x - 4} \\ -1 \end{array}$$

$\frac{6x^2}{3x} = 2x$, so $2x$ is a term of the quotient.

Multiply: $2x(3x - 1)$
Subtract and bring down the next term.
$\frac{12x}{3x} = 4$. Multiply: $4(3x - 1)$
Subtract. The remainder is -1.

Thus $(6x^2 + 10x - 5)$ divided by $(3x - 1)$ is $(2x + 4)$ with a remainder of -1. This can be written as follows.

$$\frac{6x^2 + 10x - 5}{3x - 1} = 2x + 4 + \frac{-1}{3x - 1} \quad \begin{array}{l} \leftarrow \text{remainder} \\ \leftarrow \text{divisor} \end{array}$$

or $2x + 4 - \frac{1}{3x - 1}$

Check: To check, we multiply $(3x - 1)(2x + 4)$. Then we add the remainder, -1, to this product.

$(3x - 1)(2x + 4) + (-1) = (6x^2 + 12x - 2x - 4) - 1$
$= 6x^2 + 10x - 5$

The quotient checks.

■ Work Practice 5

Notice that the division process is continued until the degree of the remainder polynomial is less than the degree of the divisor polynomial.

Recall that in Section 10.3 we practiced writing polynomials in descending order of powers and with no missing terms. For example, $2 - 4x^2$ written in this form is $-4x^2 + 0x + 2$. Writing the dividend and divisor in this form is helpful when dividing polynomials.

Practice 5
Divide: $8x^2 + 2x - 7$ by $2x - 1$

Answer
5. $4x + 3 + \frac{-4}{2x - 1}$ or $4x + 3 - \frac{4}{2x - 1}$

Practice 6
Divide: $(15 - 2x^2) \div (x - 3)$

Example 6 Divide: $(2 - 4x^2) \div (x + 1)$

Solution: We use the rewritten form of $2 - 4x^2$ from the previous page.

$$\begin{array}{r} -4x + 4 \\ x+1\overline{)-4x^2 + 0x + 2} \\ \underline{-4x^2 - 4x} \\ 4x + 2 \\ \underline{4x + 4} \\ -2 \end{array}$$

$\frac{-4x^2}{x} = -4x$, so $-4x$ is a term of the quotient.
Multiply: $-4x(x + 1)$
Subtract and bring down the next term.
$\frac{4x}{x} = 4$. Multiply: $4(x + 1)$
Remainder

Thus, $\dfrac{-4x^2 + 0x + 2}{x + 1}$ or $\dfrac{2 - 4x^2}{x + 1} = -4x + 4 + \dfrac{-2}{x + 1}$ or $-4x + 4 - \dfrac{2}{x + 1}$.

Check: To check, see that $(x + 1)(-4x + 4) + (-2) = 2 - 4x^2$.

■ Work Practice 6

Practice 7
Divide: $\dfrac{5 - x + 9x^3}{3x + 2}$

Example 7 Divide: $\dfrac{4x^2 + 7 + 8x^3}{2x + 3}$

Solution: Before we begin the division process, we rewrite $4x^2 + 7 + 8x^3$ as $8x^3 + 4x^2 + 0x + 7$. Notice that we have written the polynomial in descending order and have represented the missing x-term by $0x$.

$$\begin{array}{r} 4x^2 - 4x + 6 \\ 2x+3\overline{)8x^3 + 4x^2 + 0x + 7} \\ \underline{8x^3 + 12x^2} \\ -8x^2 + 0x \\ \underline{-8x^2 - 12x} \\ 12x + 7 \\ \underline{12x + 18} \\ -11 \quad \text{Remainder} \end{array}$$

Thus, $\dfrac{4x^2 + 7 + 8x^3}{2x + 3} = 4x^2 - 4x + 6 + \dfrac{-11}{2x + 3}$ or $4x^2 - 4x + 6 - \dfrac{11}{2x + 3}$.

■ Work Practice 7

Practice 8
Divide: $x^3 - 1$ by $x - 1$

Example 8 Divide $x^3 - 8$ by $x - 2$.

Solution: Notice that the polynomial $x^3 - 8$ is missing an x^2-term and an x-term. We'll represent these terms by inserting $0x^2$ and $0x$.

$$\begin{array}{r} x^2 + 2x + 4 \\ x-2\overline{)x^3 + 0x^2 + 0x - 8} \\ \underline{x^3 - 2x^2} \\ 2x^2 + 0x \\ \underline{2x^2 - 4x} \\ 4x - 8 \\ \underline{4x - 8} \\ 0 \end{array}$$

Thus, $\dfrac{x^3 - 8}{x - 2} = x^2 + 2x + 4$.

Check: To check, see that $(x^2 + 2x + 4)(x - 2) = x^3 - 8$.

■ Work Practice 8

Answers

6. $-2x - 6 + \dfrac{-3}{x - 3}$
or $-2x - 6 - \dfrac{3}{x - 3}$

7. $3x^2 - 2x + 1 + \dfrac{3}{3x + 2}$

8. $x^2 + x + 1$

Section 10.7 | Dividing Polynomials

Vocabulary, Readiness & Video Check

Use the choices below to fill in each blank. Choices may be used more than once.

dividend   divisor   quotient

1. In $6\overline{)18}$ (with quotient 3), the 18 is the _____, the 3 is the _____, and the 6 is the _____.

2. In $x + 1\overline{)x^2 + 3x + 2}$ (with quotient $x + 2$), the $x + 1$ is the _____, the $x^2 + 3x + 2$ is the _____, and the $x + 2$ is the _____.

Simplify each expression mentally.

3. $\dfrac{a^6}{a^4}$ 4. $\dfrac{p^8}{p^3}$ 5. $\dfrac{y^2}{y}$ 6. $\dfrac{a^3}{a}$

Martin-Gay Interactive Videos   Watch the section lecture video and answer the following questions.

See Video 10.7

Objective A 7. The lecture before Example 1 begins with adding two fractions with the same denominator. From there, the lecture continues to a method for dividing a polynomial by a monomial. What role does the monomial play in the fraction example?

Objective B 8. In Example 5, we're told that although we don't have to fill in missing powers in the divisor and the dividend, it really is a good idea to do so. Why?

10.7 Exercise Set MyLab Math

Objective A Perform each division. See Examples 1 through 3.

1. $\dfrac{12x^4 + 3x^2}{x}$

2. $\dfrac{15x^2 - 9x^5}{x}$

3. $\dfrac{20x^3 - 30x^2 + 5x + 5}{5}$

4. $\dfrac{8x^3 - 4x^2 + 6x + 2}{2}$

5. $\dfrac{15p^3 + 18p^2}{3p}$

6. $\dfrac{6x^5 + 3x^4}{3x^4}$

7. $\dfrac{-9x^4 + 18x^5}{6x^5}$

8. $\dfrac{14m^2 - 27m^3}{7m}$

9. $\dfrac{-9x^5 + 3x^4 - 12}{3x^3}$

10. $\dfrac{6a^2 - 4a + 12}{-2a^2}$

11. $\dfrac{4x^4 - 6x^3 + 7}{-4x^4}$

12. $\dfrac{-12a^3 + 36a - 15}{3a}$

Objective **B** *Find each quotient using long division. See Examples 4 and 5.*

13. $\dfrac{x^2 + 4x + 3}{x + 3}$

14. $\dfrac{x^2 + 7x + 10}{x + 5}$

15. $\dfrac{2x^2 + 13x + 15}{x + 5}$

16. $\dfrac{3x^2 + 8x + 4}{x + 2}$

17. $\dfrac{2x^2 - 7x + 3}{x - 4}$

18. $\dfrac{3x^2 - x - 4}{x - 1}$

19. $\dfrac{9a^3 - 3a^2 - 3a + 4}{3a + 2}$

20. $\dfrac{4x^3 + 12x^2 + x - 14}{2x + 3}$

21. $\dfrac{8x^2 + 10x + 1}{2x + 1}$

22. $\dfrac{3x^2 + 17x + 7}{3x + 2}$

23. $\dfrac{2x^3 + 2x^2 - 17x + 8}{x - 2}$

24. $\dfrac{4x^3 + 11x^2 - 8x - 10}{x + 3}$

Find each quotient using long division. Don't forget to write the polynomials in descending order and fill in any missing terms. See Examples 6 through 8.

25. $\dfrac{x^2 - 36}{x - 6}$

26. $\dfrac{a^2 - 49}{a - 7}$

27. $\dfrac{x^3 - 27}{x - 3}$

28. $\dfrac{x^3 + 64}{x + 4}$

29. $\dfrac{1 - 3x^2}{x + 2}$

30. $\dfrac{7 - 5x^2}{x + 3}$

31. $\dfrac{-4b + 4b^2 - 5}{2b - 1}$

32. $\dfrac{-3y + 2y^2 - 15}{2y + 5}$

Objectives **A B** **Mixed Practice** *Divide. If the divisor contains 2 or more terms, use long division. See Examples 1 through 8.*

33. $\dfrac{a^2b^2 - ab^3}{ab}$

34. $\dfrac{m^3n^2 - mn^4}{mn}$

35. $\dfrac{8x^2 + 6x - 27}{2x - 3}$

36. $\dfrac{18w^2 + 18w - 8}{3w + 4}$

37. $\dfrac{2x^2y + 8x^2y^2 - xy^2}{2xy}$

38. $\dfrac{11x^3y^3 - 33xy + x^2y^2}{11xy}$

39. $\dfrac{2b^3 + 9b^2 + 6b - 4}{b + 4}$

40. $\dfrac{2x^3 + 3x^2 - 3x + 4}{x + 2}$

41. $\dfrac{y^3 + 3y^2 + 4}{y - 2}$

42. $\dfrac{3x^3 + 11x + 12}{x + 4}$

43. $\dfrac{5 - 6x^2}{x - 2}$

44. $\dfrac{3 - 7x^2}{x - 3}$

45. $\dfrac{x^5 + x^2}{x^2 + x}$

46. $\dfrac{x^6 - x^3}{x^3 - x^2}$

Review

Fill in each blank. See Section 10.1.

47. $12 = 4 \cdot$ ___   **48.** $12 = 2 \cdot$ ___   **49.** $20 = -5 \cdot$ ___   **50.** $20 = -4 \cdot$ ___

51. $9x^2 = 3x \cdot$ ___   **52.** $9x^2 = 9x \cdot$ ___   **53.** $36x^2 = 4x \cdot$ ___   **54.** $36x^2 = 2x \cdot$ ___

Concept Extensions

Solve.

55. The perimeter of a square is $(12x^3 + 4x - 16)$ feet. Find the length of its side.

Perimeter is
$(12x^3 + 4x - 16)$ feet

56. The volume of the swimming pool shown is $(36x^5 - 12x^3 + 6x^2)$ cubic feet. If its height is $2x$ feet and its width is $3x$ feet, find its length.

3x feet

2x feet

57. The area of the parallelogram shown is $(10x^2 + 31x + 15)$ square meters. If its base is $(5x + 3)$ meters, find its height.

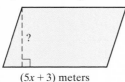

$(5x + 3)$ meters

58. The area of the top of the Ping-Pong table shown is $(49x^2 + 70x - 200)$ square inches. If its length is $(7x + 20)$ inches, find its width.

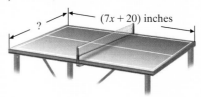

$(7x + 20)$ inches

59. Explain how to check a polynomial long division result when the remainder is 0.

60. Explain how to check a polynomial long division result when the remainder is not 0.

61. In which of the following is $\dfrac{a + 7}{7}$ simplified correctly? See the Concept Check in this section.

  a. $a + 1$
  b. a
  c. $\dfrac{a}{7} + 1$

62. In which of the following is $\dfrac{5x + 15}{5}$ simplified correctly? See the Concept Check in this section.

  a. $x + 15$
  b. $x + 3$
  c. $x + 1$

Chapter 10 Group Activity

Modeling with Polynomials

Materials:
- calculator

This activity may be completed by working in groups or individually.

Washington state is the leading producer of apples in the United States. The polynomial model $-312x^2 + 903x + 6227$ gives Washington's annual apple production (in million pounds) for the period 2013–2016. The polynomial model $-287x^2 + 730x + 10{,}531$ gives the total U.S. annual apple production (in million pounds) for the same period. In both models, x is the number of years after 2013. (Source: Based on data from the National Agricultural Statistics Service)

1. Use the given polynomials to complete the following table showing the annual apple production (both for Washington and all of the United States) over the period 2013–2016 by evaluating each polynomial at the given values of x. Then subtract each value in the fourth column from the corresponding value in the third column. Record the result in the last column, labeled "Difference." What do you think these values represent?

Year	x	Total U.S. Annual Apple Production (million pounds)	Washington's Annual Apple Production (million pounds)	Difference
2013	0			
2014	1			
2015	2			
2016	3			

2. Use the polynomial models to find a new polynomial model representing the annual apple production of *all other* U.S. states, excluding Washington. Then evaluate your new polynomial model to complete the accompanying table.

Year	x	Other Annual Apple Production (million pounds)
2013	0	
2014	1	
2015	2	
2016	3	

3. Compare the values in the last column of the table in Question **1** to the values in the last column of the table in Question **2**. What do you notice? What can you conclude?

4. Make a bar graph of the data in the table in Question **2**. Describe what you see.

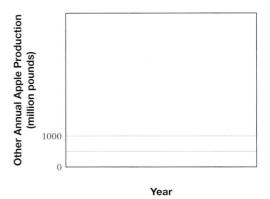

Chapter 10 Vocabulary Check

Fill in each blank with one of the words or phrases listed below.

| term | coefficient | monomial | binomial | trinomial |
| polynomials | degree of a term | degree of a polynomial | distributive | FOIL |

1. A _____ is a number or the product of a number and variables raised to powers.
2. The _____ method may be used when multiplying two binomials.
3. A polynomial with exactly 3 terms is called a _____.
4. The _____ is the greatest degree of any term of the polynomial.
5. A polynomial with exactly 2 terms is called a _____.
6. The _____ of a term is its numerical factor.

7. The _____ is the sum of the exponents on the variables in the term.
8. A polynomial with exactly 1 term is called a _____.
9. Monomials, binomials, and trinomials are all examples of _____.
10. The _____ property is used to multiply $2x(x - 4)$.

> **Helpful Hint**
> ▶ Are you preparing for your test? To help, don't forget to take these:
> - Chapter 10 Getting Ready for the Test on page 811
> - Chapter 10 Test on page 812
>
> Then check all of your answers at the back of this text. For further review, the step-by-step video solutions to any of these exercises are located in MyLab Math.

10 Chapter Highlights

Definitions and Concepts	Examples
Section 10.1 Exponents	
a^n means the product of n factors, each of which is a.	$3^2 = 3 \cdot 3 = 9$ $(-5)^3 = (-5)(-5)(-5) = -125$ $\left(\dfrac{1}{2}\right)^4 = \dfrac{1}{2} \cdot \dfrac{1}{2} \cdot \dfrac{1}{2} \cdot \dfrac{1}{2} = \dfrac{1}{16}$
Let m and n be integers and no denominators be 0. **Product Rule:** $a^m \cdot a^n = a^{m+n}$ **Power Rule:** $(a^m)^n = a^{mn}$ **Power of a Product Rule:** $(ab)^n = a^n b^n$ **Power of a Quotient Rule:** $\left(\dfrac{a}{b}\right)^n = \dfrac{a^n}{b^n}$ **Quotient Rule:** $\dfrac{a^m}{a^n} = a^{m-n}, a \neq 0$ **Zero Exponent:** $a^0 = 1, a \neq 0$	$x^2 \cdot x^7 = x^{2+7} = x^9$ $(5^3)^8 = 5^{3 \cdot 8} = 5^{24}$ $(7y)^4 = 7^4 y^4$ $\left(\dfrac{x}{8}\right)^3 = \dfrac{x^3}{8^3}$ $\dfrac{x^9}{x^4} = x^{9-4} = x^5, x \neq 0$ $5^0 = 1; x^0 = 1, x \neq 0$
Section 10.2 Negative Exponents and Scientific Notation	
If $a \neq 0$ and n is an integer, $a^{-n} = \dfrac{1}{a^n}$	$3^{-2} = \dfrac{1}{3^2} = \dfrac{1}{9}; 5x^{-2} = \dfrac{5}{x^2}$ Simplify: $\left(\dfrac{x^{-2}y}{x^5}\right)^{-2} = \dfrac{x^4 y^{-2}}{x^{-10}}$ $= x^{4-(-10)} y^{-2}$ $= \dfrac{x^{14}}{y^2}$
A positive number is written in scientific notation if it is written as the product of a number a, where $1 \leq a < 10$, and an integer power r of 10. $a \times 10^r$	$1200 = 1.2 \times 10^3$ $0.000000568 = 5.68 \times 10^{-7}$

Definitions and Concepts	Examples
Section 10.3 Introduction to Polynomials	
A **term** is a number or the product of a number and variables raised to powers.	$-5x, 7a^2b, \dfrac{1}{4}y^4, 0.2$
The **numerical coefficient,** or **coefficient,** of a term is its numerical factor.	**Term** **Coefficient** $7x^2$ 7 y 1 $-a^2b$ -1
A **polynomial** is a finite sum of terms of the form ax^n where a is a real number and n is a whole number.	$5x^3 - 6x^2 + 3x - 6$ (Polynomial)
A **monomial** is a polynomial with exactly 1 term.	$\dfrac{5}{6}y^3$ (Monomial)
A **binomial** is a polynomial with exactly 2 terms.	$-0.2a^2b - 5b^2$ (Binomial)
A **trinomial** is a polynomial with exactly 3 terms.	$3x^2 - 2x + 1$ (Trinomial)
The **degree of a polynomial** is the greatest degree of any term of the polynomial.	**Polynomial** **Degree** $5x^2 - 3x + 2$ 2 $7y + 8y^2z^3 - 12$ $2 + 3 = 5$
Section 10.4 Adding and Subtracting Polynomials	
To add polynomials, combine like terms.	Add. $(7x^2 - 3x + 2) + (-5x - 6)$ $= 7x^2 - 3x + 2 - 5x - 6$ $= 7x^2 - 8x - 4$
To subtract two polynomials, change the signs of the terms of the second polynomial, and then add.	Subtract. $(17y^2 - 2y + 1) - (-3y^3 + 5y - 6)$ $= (17y^2 - 2y + 1) + (3y^3 - 5y + 6)$ $= 17y^2 - 2y + 1 + 3y^3 - 5y + 6$ $= 3y^3 + 17y^2 - 7y + 7$
Section 10.5 Multiplying Polynomials	
To multiply two polynomials, multiply each term of one polynomial by each term of the other polynomial, and then combine like terms.	Multiply. $(2x + 1)(5x^2 - 6x + 2)$ $= 2x(5x^2 - 6x + 2) + 1(5x^2 - 6x + 2)$ $= 10x^3 - 12x^2 + 4x + 5x^2 - 6x + 2$ $= 10x^3 - 7x^2 - 2x + 2$

Chapter 10 Highlights

Definitions and Concepts	Examples
Section 10.6	**Special Products**

The **FOIL method** may be used when multiplying two binomials.	Multiply: $(5x - 3)(2x + 3)$ $(5x - 3)(2x + 3)$ with First, Outer, Inner, Last $\;\;\;\;\;\;\;\;\;\;\;\;\;\;F\;\;\;\;\;\;\;\;\;\;\;\;\;O\;\;\;\;\;\;\;\;\;\;\;\;\;\;I\;\;\;\;\;\;\;\;\;\;\;\;\;L$ $= (5x)(2x) + (5x)(3) + (-3)(2x) + (-3)(3)$ $= 10x^2 + 15x - 6x - 9$ $= 10x^2 + 9x - 9$
Squaring a Binomial $(a + b)^2 = a^2 + 2ab + b^2$ $(a - b)^2 = a^2 - 2ab + b^2$ **Multiplying the Sum and Difference of Two Terms** $(a + b)(a - b) = a^2 - b^2$	Square each binomial. $(x + 5)^2 = x^2 + 2(x)(5) + 5^2$ $ = x^2 + 10x + 25$ $(3x - 2y)^2 = (3x)^2 - 2(3x)(2y) + (2y)^2$ $ = 9x^2 - 12xy + 4y^2$ Multiply. $(6y + 5)(6y - 5) = (6y)^2 - 5^2$ $ = 36y^2 - 25$

Section 10.7	**Dividing Polynomials**

To divide a polynomial by a monomial, $\dfrac{a + b}{c} = \dfrac{a}{c} + \dfrac{b}{c}, c \neq 0$	Divide. $\dfrac{15x^5 - 10x^3 + 5x^2 - 2x}{5x^2}$ $= \dfrac{15x^5}{5x^2} - \dfrac{10x^3}{5x^2} + \dfrac{5x^2}{5x^2} - \dfrac{2x}{5x^2}$ $= 3x^3 - 2x + 1 - \dfrac{2}{5x}$
To divide a polynomial by a polynomial other than a monomial, use long division.	$\;\;\;\;\;\;\;\;5x - 1 + \dfrac{-4}{2x + 3}$ $2x + 3 \overline{)10x^2 + 13x - 7}$ $\underline{10x^2 + 15x}$ or $5x - 1 - \dfrac{4}{2x + 3}$ $-2x - 7$ $\underline{-2x - 3}$ -4

Chapter 10 Review

(10.1) *Each expression contains an exponent of 4. For each exercise, name the base for this exponent of 4.*

1. 3^4
2. $(-5)^4$
3. -5^4
4. x^4

Evaluate each expression.

5. 8^3
6. $(-6)^2$
7. -6^2
8. $-4^3 - 4^0$
9. $(3b)^0$
10. $\dfrac{8b}{8b}$

Simplify each expression.

11. $y^2 \cdot y^7$
12. $x^9 \cdot x^5$
13. $(2x^5)(-3x^6)$
14. $(-5y^3)(4y^4)$

15. $(x^4)^2$
16. $(y^3)^5$
17. $(3y^6)^4$
18. $(2x^3)^3$

19. $\dfrac{x^9}{x^4}$
20. $\dfrac{z^{12}}{z^5}$
21. $\dfrac{a^5 b^4}{ab}$
22. $\dfrac{x^4 y^6}{xy}$

23. $\dfrac{3x^4 y^{10}}{12xy^6}$
24. $\dfrac{2x^7 y^8}{8xy^2}$
25. $5a^7(2a^4)^3$
26. $(2x)^2(9x)$

27. $(-5a)^0 + 7^0 + 8^0$
28. $8x^0 + 9^0$

Simplify the given expression and choose the correct result.

29. $\left(\dfrac{3x^4}{4y}\right)^3$

   a. $\dfrac{27x^{64}}{64y^3}$
   b. $\dfrac{27x^{12}}{64y^3}$
   c. $\dfrac{9x^{12}}{12y^3}$
   d. $\dfrac{3x^{12}}{4y^3}$

30. $\left(\dfrac{5a^6}{b^3}\right)^2$

   a. $\dfrac{10a^{12}}{b^6}$
   b. $\dfrac{25a^{36}}{b^9}$
   c. $\dfrac{25a^{12}}{b^6}$
   d. $25a^{12}b^6$

(10.2) *Simplify each expression.*

31. 7^{-2}
32. -7^{-2}
33. $2x^{-4}$
34. $(2x)^{-4}$

35. $\left(\dfrac{1}{5}\right)^{-3}$
36. $\left(\dfrac{-2}{3}\right)^{-2}$
37. $2^0 + 2^{-4}$
38. $6^{-1} - 7^{-1}$

806

Simplify each expression. Write each answer using positive exponents only.

39. $\dfrac{x^5}{x^{-3}}$ **40.** $\dfrac{z^4}{z^{-4}}$ **41.** $\dfrac{r^{-3}}{r^{-4}}$ **42.** $\dfrac{y^{-2}}{y^{-5}}$

43. $\left(\dfrac{bc^{-2}}{bc^{-3}}\right)^4$ **44.** $\left(\dfrac{x^{-3}y^{-4}}{x^{-2}y^{-5}}\right)^{-3}$ **45.** $\dfrac{x^{-4}y^{-6}}{x^2 y^7}$ **46.** $\dfrac{a^5 b^{-5}}{a^{-5} b^5}$

Write each number in scientific notation.

47. 0.00027   **48.** 0.8868   **49.** 80,800,000   **50.** 868,000

51. In November 2016, approximately 137,000,000 people cast ballots in the U.S. presidential election. Write this number in scientific notation. (*Source:* U.S. election atlas)

52. The approximate diameter of the Milky Way galaxy is 150,000 light-years. Write this number in scientific notation. (*Source:* NASA IMAGE/POETRY Education and Public Outreach Program)

Write each number in standard form.

53. 8.67×10^5 **54.** 3.86×10^{-3} **55.** 8.6×10^{-4} **56.** 8.936×10^5

57. The volume of the planet Jupiter is 1.43128×10^{15} cubic kilometers. Write this number in standard form. (*Source:* National Space Science Data Center)

58. An angstrom is a unit of measure, equal to 1×10^{-10} meter, used for measuring wavelengths or the diameters of atoms. Write this number in standard form. (*Source:* National Institute of Standards and Technology)

Simplify. Express each result in standard form.

59. $(8 \times 10^4)(2 \times 10^{-7})$

60. $\dfrac{8 \times 10^4}{2 \times 10^{-7}}$

(10.3) *Find the degree of each polynomial.*

61. $y^5 + 7x - 8x^4$

62. $9y^2 + 30y + 25$

63. $-14x^2y - 28x^2y^3 - 42x^2y^2$

64. $6x^2y^2z^2 + 5x^2y^3 - 12xyz$

65. The Glass Bridge Skywalk is suspended 4000 feet over the Colorado River at the very edge of the Grand Canyon. Neglecting air resistance, the height of an object dropped from the Skywalk at time t seconds is given by the polynomial $-16t^2 + 4000$. Find the height of the object at the given times below.

t	0 seconds	1 second	3 seconds	5 seconds
$-16t^2 + 4000$				

△ **66.** The surface area of a box with a square base and a height of 5 units is given by the polynomial $2x^2 + 20x$. Fill in the table below by evaluating $2x^2 + 20x$ for the given values of x.

x	1	3	5.1	10
$2x^2 + 20x$				

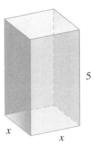

Combine like terms in each expression.

67. $7a^2 - 4a^2 - a^2$

68. $9y + y - 14y$

69. $6a^2 + 4a + 9a^2$

70. $21x^2 + 3x + x^2 + 6$

71. $4a^2b - 3b^2 - 8q^2 - 10a^2b + 7q^2$

72. $2s^{14} + 3s^{13} + 12s^{12} - s^{10}$

(10.4) *Add or subtract as indicated.*

73. $(3x^2 + 2x + 6) + (5x^2 + x)$

74. $(2x^5 + 3x^4 + 4x^3 + 5x^2) + (4x^2 + 7x + 6)$

75. $(-5y^2 + 3) - (2y^2 + 4)$

76. $(2m^7 + 3x^4 + 7m^6) - (8m^7 + 4m^2 + 6x^4)$

77. $(3x^2 - 7xy + 7y^2) - (4x^2 - xy + 9y^2)$

78. $(8x^6 - 5xy - 10y^2) - (7x^6 - 9xy - 12y^2)$

Translating *Perform the indicated operations.*

79. Add $(-9x^2 + 6x + 2)$ and $(4x^2 - x - 1)$.

80. Subtract $(4x^2 + 8x - 7)$ from the sum of $(x^2 + 7x + 9)$ and $(x^2 + 4)$.

(10.5) *Multiply each expression.*

81. $6(x + 5)$
82. $9(x - 7)$
83. $4(2a + 7)$
84. $9(6a - 3)$

85. $-7x(x^2 + 5)$
86. $-8y(4y^2 - 6)$
87. $-2(x^3 - 9x^2 + x)$
88. $-3a(a^2b + ab + b^2)$

89. $(3a^3 - 4a + 1)(-2a)$
90. $(6b^3 - 4b + 2)(7b)$
91. $(2x + 2)(x - 7)$

92. $(2x - 5)(3x + 2)$
93. $(4a - 1)(a + 7)$
94. $(6a - 1)(7a + 3)$

95. $(x + 7)(x^3 + 4x - 5)$
96. $(x + 2)(x^5 + x + 1)$
97. $(x^2 + 2x + 4)(x^2 + 2x - 4)$

98. $(x^3 + 4x + 4)(x^3 + 4x - 4)$
99. $(x + 7)^3$
100. $(2x - 5)^3$

(10.6) *Use special products to multiply each of the following.*

101. $(x + 7)^2$
102. $(x - 5)^2$
103. $(3x - 7)^2$
104. $(4x + 2)^2$

105. $(5x - 9)^2$
106. $(5x + 1)(5x - 1)$
107. $(7x + 4)(7x - 4)$
108. $(a + 2b)(a - 2b)$

109. $(2x - 6)(2x + 6)$
110. $(4a^2 - 2b)(4a^2 + 2b)$

Express each as a product of polynomials in x. Then multiply and simplify.

111. Find the area of the square if its side is $(3x - 1)$ meters.

$(3x - 1)$ meters

112. Find the area of the rectangle.

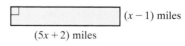

$(x - 1)$ miles
$(5x + 2)$ miles

(10.7) *Divide.*

113. $\dfrac{x^2 + 21x + 49}{7x^2}$

114. $\dfrac{5a^3b - 15ab^2 + 20ab}{-5ab}$

115. $(a^2 - a + 4) \div (a - 2)$

116. $(4x^2 + 20x + 7) \div (x + 5)$

117. $\dfrac{a^3 + a^2 + 2a + 6}{a - 2}$

118. $\dfrac{9b^3 - 18b^2 + 8b - 1}{3b - 2}$

119. $\dfrac{4x^4 - 4x^3 + x^2 + 4x - 3}{2x - 1}$

120. $\dfrac{-10x^2 - x^3 - 21x + 18}{x - 6}$

△ **121.** The area of the rectangle below is $(15x^3 - 3x^2 + 60)$ square feet. If its length is $3x^2$ feet, find its width.

Area is $(15x^3 - 3x^2 + 60)$ sq feet

122. The perimeter of the equilateral triangle below is $(21a^3b^6 + 3a - 3)$ units. Find the length of a side.

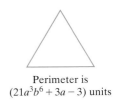

Perimeter is $(21a^3b^6 + 3a - 3)$ units

Mixed Review

Evaluate.

123. 3^3

124. $\left(-\dfrac{1}{2}\right)^3$

Simplify each expression. Write each answer using positive exponents only.

125. $(4xy^2)(x^3y^5)$

126. $\dfrac{18x^9}{27x^3}$

127. $\left(\dfrac{3a^4}{b^2}\right)^3$

128. $(2x^{-4}y^3)^{-4}$

129. $\dfrac{a^{-3}b^6}{9^{-1}a^{-5}b^{-2}}$

Perform the indicated operations and simplify.

130. $(-y^2 - 4) + (3y^2 - 6)$

131. $(6x + 2) + (5x - 7)$

132. $(5x^2 + 2x - 6) - (-x - 4)$

133. $(8y^2 - 3y + 1) - (3y^2 + 2)$

134. $(2x + 5)(3x - 2)$

135. $4x(7x^2 + 3)$

136. $(7x - 2)(4x - 9)$

137. $(x - 3)(x^2 + 4x - 6)$

Use special products to multiply.

138. $(5x + 4)^2$

139. $(6x + 3)(6x - 3)$

Divide.

140. $\dfrac{8a^4 - 2a^3 + 4a - 5}{2a^3}$

141. $\dfrac{x^2 + 2x + 10}{x + 5}$

142. $\dfrac{4x^3 + 8x^2 - 11x + 4}{2x - 3}$

Chapter 10 Getting Ready for the Test

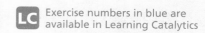

MATCHING *For Exercises 1 through 4, match the expression in the left column with the exponent operation needed to simplify in the right column. Letters may be used more than once or not at all.*

1. $x^2 \cdot x^5$
2. $(x^2)^5$
3. $x^2 + x^5$
4. $\dfrac{x^5}{x^2}$

A. multiply the exponents
B. divide the exponents
C. add the exponents
D. subtract the exponents
E. this expression will not simplify

MATCHING *For Exercises 5 through 8, match the operation in the left column with the result when the operation is performed on the given terms in the right columns. Letters may be used more than once or not at all.*

Given Terms: 20y and 4y

5. Add the terms
6. Subtract the terms
7. Multiply the terms
8. Divide the terms.

A. $80y$
B. $24y^2$
C. $16y$
D. 16
E. $80y^2$
F. $24y$
G. $16y^2$
H. $5y$
I. 5

9. **MULTIPLE CHOICE** *The expression 5^{-1} is equivalent to*

A. -5 B. 4 C. $\dfrac{1}{5}$ D. $-\dfrac{1}{5}$

10. **MULTIPLE CHOICE** *The expression 2^{-3} is equivalent to*

A. -6 B. -1 C. $-\dfrac{1}{6}$ D. $\dfrac{1}{8}$

MATCHING *For Exercises 11 through 14, match each expression in the left column with its simplified form in the right columns. Letters may be used more than once or not at all.*

11. $y + y + y$
12. $y \cdot y \cdot y$
13. $(-y)(-y)(-y)$
14. $-y - y - y$

A. $3y^3$
B. y^3
C. $3y$
D. $-3y$
E. $-3y^3$
F. $-y^3$

811

Chapter 10 Test

Evaluate each expression.

1. 2^5

2. $(-3)^4$

3. -3^4

4. 4^{-3}

Simplify each expression. Write the result using only positive exponents.

5. $(3x^2)(-5x^9)$

6. $\dfrac{y^7}{y^2}$

7. $\dfrac{r^{-8}}{r^{-3}}$

8. $\left(\dfrac{4x^2y^3}{x^3y^{-4}}\right)^2$

9. $\dfrac{6^2 x^{-4} y^{-1}}{6^3 x^{-3} y^7}$

Express each number in scientific notation.

10. 563,000

11. 0.0000863

Write each number in standard form.

12. 1.5×10^{-3}

13. 6.23×10^4

14. Simplify. Write the answer in standard form.
$(1.2 \times 10^5)(3 \times 10^{-7})$

15. a. Complete the table for the polynomial $4xy^2 + 7xyz + x^3y - 2$.

Term	Numerical Coefficient	Degree of Term
$4xy^2$		
$7xyz$		
x^3y		
-2		

b. What is the degree of the polynom

16. Simplify by combining like terms.
$5x^2 + 4x - 7x^2 + 11 + 8x$

Perform each indicated operation.

17. $(8x^3 + 7x^2 + 4x - 7) + (8x^3 - 7x - 6)$

18. $\begin{array}{r} 5x^3 + x^2 + 5x - 2 \\ -(8x^3 - 4x^2 + x - 7) \end{array}$

19. Subtract $(4x + 2)$ from the sum of $(8x^2 + 7x + 5)$ and $(x^3 - 8)$.

Chapter 10 Test 813

Multiply in Exercises 20 through 26.

20. $(3x + 7)(x^2 + 5x + 2)$

21. $3x^2(2x^2 - 3x + 7)$

22. $(x + 7)(3x - 5)$

23. $\left(3x - \dfrac{1}{5}\right)\left(3x + \dfrac{1}{5}\right)$

24. $(4x - 2)^2$

25. $(8x + 3)^2$

26. $(x^2 - 9b)(x^2 + 9b)$

27. The height of the Bank of China in Hong Kong is 1001 feet. Neglecting air resistance, the height of an object dropped from this building at time t seconds is given by the polynomial $-16t^2 + 1001$. Find the height of the object at the given times below.

t	0 seconds	1 second	3 seconds	5 seconds
$-16t^2 + 1001$				

28. Find the area of the top of the table. Express the area as a product, then multiply and simplify.

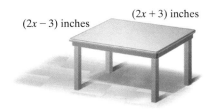

$(2x - 3)$ inches
$(2x + 3)$ inches

Divide.

29. $\dfrac{4x^2 + 2xy - 7x}{8xy}$

30. $(x^2 + 7x + 10) \div (x + 5)$

31. $\dfrac{27x^3 - 8}{3x + 2}$

20. _____
21. _____
22. _____
23. _____
24. _____
25. _____
26. _____
27. _____
28. _____
29. _____
30. _____
31. _____

Chapter 1–10 Cumulative Review

Answers

1. ____
2. ____
3. ____
4. ____
5. a. ____ b. ____
   c. ____
   d. ____
   e. ____
6. a. ____ b. ____
   c. ____ d. ____
   e. ____
7. ____
8. ____
9. ____
10. ____
11. ____
12. ____
13. ____
14. ____
15. ____
16. ____
17. ____
18. ____

Evaluate.

1. 9^2 **2.** 5^3 **3.** 3^4 **4.** 3^3

Write as an algebraic expression. Use x to represent "a number."

5. a. 7 increased by a number
   b. 15 decreased by a number
   c. the product of 2 and a number
   d. the quotient of a number and 5
   e. 2 subtracted from a number

6. a. the sum of a number and 3
   b. the product of 3 and a number
   c. twice a number
   d. 10 decreased by a number
   e. 5 times a number increased by 7

Use the distributive property to remove parentheses. Then simplify if possible.

7. $2(3 + 7x) - 15$

8. $5(x + 2)$

9. $-2(x - 5) + 4(2x + 2)$

10. $-2(y + 3z - 1)$

Solve.

11. $y - 5 = -2 - 6$

12. $\dfrac{y}{7} = 20$

13. $7(x - 2) = 9x - 6$

14. $6(2a - 1) - (11a + 6) = 7$

15. $\dfrac{3}{5}a = 9$

16. $\dfrac{2}{3}y = 16$

17. $3y = -\dfrac{2}{11}$

18. $5y = -\dfrac{1}{5}$

814

Cumulative Review

Write each decimal as a fraction or a mixed number.

19. 0.125

20. 0.250

21. 43.5

22. 10.75

23. −105.083

24. −31.07

Evaluate $x - y$ for the given replacement values.

25. $x = 2.8, y = 0.92$

26. $x = -1.2, y = 7.6$

Evaluate xy for the given replacement values.

27. $x = 2.3, y = 0.44$

28. $x = -6.1, y = 0.5$

29. Given the set $\left\{-3, -2, 0, \dfrac{1}{4}, \sqrt{2}, 11, 112\right\}$, list the numbers in this set that belong to the set of:

a. Natural numbers
b. Whole numbers
c. Integers
d. Rational numbers
e. Irrational numbers
f. Real numbers

30. Tell which set(s) the number $-2\dfrac{1}{2}$ belongs to. Use the sets of numbers from Exercise **29**.

Solve.

31. $0.25x + 0.10(x - 3) = 1.1$

32. $0.6x - 10 = 1.4x - 14$

33. Twice the sum of a number and 4 is the same as four times the number decreased by 12. Find the number.

34. Three times the difference of a number and 2 is the same as five times the number decreased by 10. Find the number.

19. _____
20. _____
21. _____
22. _____
23. _____
24. _____
25. _____
26. _____
27. _____
28. _____
29. a. _____
    b. _____
    c. _____
    d. _____
    e. _____
    f. _____
30. _____
31. _____
32. _____
33. _____
34. _____

35. Charles Pecot can afford enough fencing to enclose a rectangular garden with a perimeter of 140 feet. If the width of his garden is to be 30 feet, find the length.

36. A house is in the shape of a rectangle and has 2016 square feet of living area. If the length is 63 feet, find its width.

37. Solve: $-4x + 7 \geq -9$. Graph the solution set.

$$\begin{array}{c}\leftarrow\!\!+\!\!+\!\!+\!\!+\!\!+\!\!+\!\!+\!\!+\!\!+\!\!+\!\!\rightarrow\\-5\;-4\;-3\;-2\;-1\;\;0\;\;1\;\;2\;\;3\;\;4\;\;5\end{array}$$

38. Solve: $3x + 4 \geq 2x - 6$

39. Simplify each expression.

a. $x^7 \cdot x^4$ b. $\left(\dfrac{t}{2}\right)^4$ c. $(9y^5)^2$

40. Simplify each expression.

a. $y \cdot y^5$ b. $\left(\dfrac{2}{3}\right)^3$ c. $(8x^3)^2$

Simplify each expression. Write each result using positive exponents only.

41. $\left(\dfrac{3a^2}{b}\right)^{-3}$ **42.** $\left(\dfrac{2x^3}{y}\right)^{-2}$ **43.** $(5y^3)^{-2}$ **44.** $(3x^7)^{-3}$

Simplify each polynomial by combining any like terms.

45. $9x^3 + x^3$

46. $9x^3 - x^3$

47. $5x^2 + 6x - 9x - 3$

48. $2x - x^2 + 5x - 4x^2$

Multiply.

49. $7x(x^2 + 2x + 5)$

50. $-2x(x^2 - x + 1)$

Divide.

51. $\dfrac{9x^5 - 12x^2 + 3x}{3x^2}$

52. $\dfrac{4x^7 - 12x^2 + 2x}{2x}$

Factoring Polynomials

11

In Chapter 10, we multiplied polynomials. Now we will learn the reverse operation of multiplying—factoring. Factoring allows us to write a sum as a product. As we will see in this chapter, factoring can be used to solve equations other than linear equations. In Chapter 12, we will also use factoring to perform operations on rational expressions.

Why Are You in College?

There are probably as many answers as there are students. It may help you to know that college graduates have higher earnings and lower rates of unemployment. The double line graph below shows the increasing number of associate and bachelor degrees awarded over the years. It is also enlightening to know that an increasing number of high school graduates are interested in higher education.

In Exercise 99 of Section 11.1, we will explore how many students graduate from U.S. high schools each year and how many of those may expect to go to college.

Sections

- **11.1** The Greatest Common Factor and Factoring by Grouping
- **11.2** Factoring Trinomials of the Form $x^2 + bx + c$
- **11.3** Factoring Trinomials of the Form $ax^2 + bx + c$
- **11.4** Factoring Trinomials of the Form $ax^2 + bx + c$ by Grouping
- **11.5** Factoring Perfect Square Trinomials and the Difference of Two Squares

  Integrated Review— Choosing a Factoring Strategy

- **11.6** Solving Quadratic Equations by Factoring
- **11.7** Quadratic Equations and Problem Solving

Check Your Progress

Vocabulary Check
Chapter Highlights
Chapter Review
Getting Ready for the Test
Chapter Test
Cumulative Review

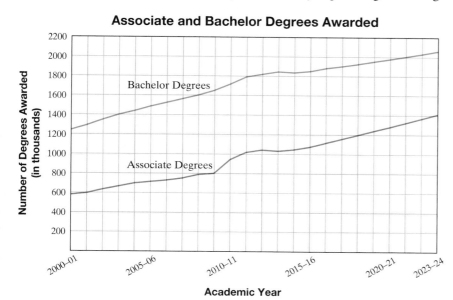

Source: National Center for Education Statistics (http://nces.ed.gov); U.S. Department of Education
Note: Some years are projected.

11.1 The Greatest Common Factor and Factoring by Grouping

Objectives

A Find the Greatest Common Factor of a List of Numbers.

B Find the Greatest Common Factor of a List of Terms.

C Factor Out the Greatest Common Factor from the Terms of a Polynomial.

D Factor a Polynomial by Grouping.

In the product $2 \cdot 3 = 6$, the numbers 2 and 3 are called **factors** of 6 and $2 \cdot 3$ a **factored form** of 6. This is true of polynomials also. Since $(x + 2)(x + 3)$ $x^2 + 5x + 6$, then $(x + 2)$ and $(x + 3)$ are factors of $x^2 + 5x + 6$, an $(x + 2)(x + 3)$ is a factored form of the polynomial.

a factored form of 6
$$2 \cdot 3 = 6$$
factor factor product

a factored form of x^5
$$x^2 \cdot x^3 = x^5$$
factor factor product

a factored form of $x^2 + 5x + 6$
$$(x + 2)(x + 3) = x^2 + 5x + 6$$
factor factor product

The process of writing a polynomial as a product is called **factoring** the polynomial.

Study the examples below and look for a pattern.

Multiplying: $5(x^2 + 3) = 5x^2 + 15$ $2x(x - 7) = 2x^2 - 14x$

Factoring: $5x^2 + 15 = 5(x^2 + 3)$ $2x^2 - 14x = 2x(x - 7)$

Do you see that factoring is the reverse process of multiplying?

$$x^2 + 5x + 6 \underset{\text{multiplying}}{\overset{\text{factoring}}{=}} (x + 2)(x + 3)$$

✓**Concept Check** Multiply: $2(x - 4)$
What do you think the result of factoring $2x - 8$ would be? Why?

Objective A Finding the Greatest Common Factor of a List of Numbers

The first step in factoring a polynomial is to see whether the terms of the polynomi have a common factor. If there is one, we can write the polynomial as a product b **factoring out** the common factor. We will usually factor out the *greatest* commo factor (GCF).

The GCF of a list of integers is the largest integer that is a factor of all the int gers in the list. For example, the GCF of 12 and 20 is 4 because 4 is the largest intege that is a factor of both 12 and 20. With large integers, the GCF may not be easi found by inspection. When this happens, use the following steps.

✓**Concept Check Answer**
$2x - 8$; the result would be $2(x - 4)$ because factoring is the reverse process of multiplying.

Section 11.1 | The Greatest Common Factor and Factoring by Grouping

Finding the GCF of a List of Integers

Step 1: Write each number as a product of prime numbers.
Step 2: Identify the common prime factors.
Step 3: The product of all common prime factors found in Step 2 is the greatest common factor. If there are no common prime factors, the greatest common factor is 1.

Recall from Section 4.2 that a prime number is a whole number other than 1 whose only factors are 1 and itself.

Example 1 Find the GCF of each list of numbers.

a. 28 and 40   b. 55 and 21   c. 15, 18, and 66

Solution:

a. Write each number as a product of primes.

$28 = 2 \cdot 2 \cdot 7 = 2^2 \cdot 7$
$40 = 2 \cdot 2 \cdot 2 \cdot 5 = 2^3 \cdot 5$

There are two common factors, each of which is 2, so the GCF is

$GCF = 2 \cdot 2 = 4$

b. $55 = 5 \cdot 11$
$21 = 3 \cdot 7$

There are no common prime factors; thus, the GCF is 1.

c. $15 = 3 \cdot 5$
$18 = 2 \cdot 3 \cdot 3 = 2 \cdot 3^2$
$66 = 2 \cdot 3 \cdot 11$

The only prime factor common to all three numbers is 3, so the GCF is

$GCF = 3$

Work Practice 1

Practice 1

Find the GCF of each list of numbers.

a. 45 and 75
b. 32 and 33
c. 14, 24, and 60

Objective B Finding the Greatest Common Factor of a List of Terms

The greatest common factor of a list of variables raised to powers is found in a similar way. For example, the GCF of x^2, x^3, and x^5 is x^2 because each term contains a factor of x^2 and no higher power of x is a factor of each term.

$x^2 = x \cdot x$
$x^3 = x \cdot x \cdot x$
$x^5 = x \cdot x \cdot x \cdot x \cdot x$

There are two common factors, each of which is x, so the $GCF = x \cdot x$ or x^2. From this example, we see that **the GCF of a list of common variables raised to powers is the variable raised to the smallest exponent in the list.**

Example 2 Find the GCF of each list of terms.

a. x^3, x^7, and x^5
b. y, y^4, and y^7

Practice 2

Find the GCF of each list of terms.

a. y^4, y^5, and y^8
b. x and x^{10}

Answers
1. a. 15 b. 1 c. 2
2. a. y^4 b. x

(*Continued on next page*)

Solution:

a. The GCF is x^3, since 3 is the smallest exponent to which x is raised.
b. The GCF is y^1 or y, since 1 is the smallest exponent on y.

■ Work Practice 2

The **greatest common factor (GCF) of a list of terms** is the product of the GCF of the numerical coefficients and the GCF of the variable factors.

$$20x^2y^2 = 2 \cdot 2 \cdot 5 \cdot x \cdot x \cdot y \cdot y$$
$$6xy^3 = 2 \cdot 3 \cdot x \cdot y \cdot y \cdot y$$
$$\text{GCF} = 2 \cdot x \cdot y \cdot y = 2xy^2$$

Helpful Hint

Remember that the GCF of a list of terms contains the smallest exponent on each common variable.

The GCF of x^5y^6, x^2y^7, and x^3y^4 is x^2y^4. — Smallest exponent on x
— Smallest exponent on y

Practice 3

Find the greatest common factor of each list of terms.
a. $6x^2$, $9x^4$, and $-12x^5$
b. $-16y$, $-20y^6$, and $40y^4$
c. a^5b^4, ab^3, and a^3b^2

Example 3 Find the greatest common factor of each list of terms.

a. $6x^2$, $10x^3$, and $-8x$
b. $-18y^2$, $-63y^3$, and $27y^4$
c. a^3b^2, a^5b, and a^6b^2

Solution:

a. $6x^2 = 2 \cdot 3 \cdot x^2$
$10x^3 = 2 \cdot 5 \cdot x^3$
$-8x = -1 \cdot 2 \cdot 2 \cdot 2 \cdot x^1$
$\text{GCF} = 2 \cdot x^1 \quad \text{or} \quad 2x$

→ The GCF of x^2, x^3, and x^1 is x^1 or x.

b. $-18y^2 = -1 \cdot 2 \cdot 3 \cdot 3 \cdot y^2$
$-63y^3 = -1 \cdot 3 \cdot 3 \cdot 7 \cdot y^3$
$27y^4 = 3 \cdot 3 \cdot 3 \cdot y^4$
$\text{GCF} = 3 \cdot 3 \cdot y^2 \quad \text{or} \quad 9y^2$

→ The GCF of y^2, y^3, and y^4 is y^2.

c. The GCF of a^3, a^5, and a^6 is a^3.
The GCF of b^2, b, and b^2 is b. Thus, the GCF of a^3b^2, a^5b, and a^6b^2 is a^3b.

■ Work Practice 3

Objective C Factoring Out the Greatest Common Factor

To factor a polynomial such as $8x + 14$, we first see whether the terms have a greatest common factor other than 1. In this case, they do: The GCF of $8x$ and 14 is 2.

We factor out 2 from each term by writing each term as the product of 2 and the term's remaining factors.

$$8x + 14 = 2 \cdot 4x + 2 \cdot 7$$

Using the distributive property, we can write

$$8x + 14 = 2 \cdot 4x + 2 \cdot 7$$
$$= 2(4x + 7)$$

Answers

3. a. $3x^2$ b. $4y$ c. ab^2

Section 11.1 | The Greatest Common Factor and Factoring by Grouping

Thus, a factored form of $8x + 14$ is $2(4x + 7)$. We can check by multiplying:

$$2(4x + 7) = 2 \cdot 4x + 2 \cdot 7 = 8x + 14$$

Helpful Hint

A factored form of $8x + 14$ is *not*

$$2 \cdot 4x + 2 \cdot 7$$

Although the *terms* have been factored (written as products), the *polynomial* $8x + 14$ has not been factored. A factored form of $8x + 14$ is the *product* $2(4x + 7)$.

 Concept Check Which of the following is/are factored form(s) of $6t + 18$?
a. 6 **b.** $6 \cdot t + 6 \cdot 3$ **c.** $6(t + 3)$ **d.** $3(t + 6)$

Example 4 Factor each polynomial by factoring out the greatest common factor (GCF).

a. $5ab + 10a$ **b.** $y^5 - y^{12}$

Solution:

a. The GCF of terms $5ab$ and $10a$ is $5a$. Thus,

$$5ab + 10a = 5a \cdot b + 5a \cdot 2$$
$$= 5a(b + 2) \quad \text{Apply the distributive property.}$$

We can check our work by multiplying $5a$ and $(b + 2)$.
$5a(b + 2) = 5a \cdot b + 5a \cdot 2 = 5ab + 10a$, the original polynomial.

b. The GCF of y^5 and y^{12} is y^5. Thus,

$$y^5 - y^{12} = y^5(1) - y^5(y^7)$$
$$= y^5(1 - y^7)$$

Helpful Hint
Don't forget the 1.

Work Practice 4

Practice 4

Factor each polynomial by factoring out the greatest common factor (GCF).

a. $10y + 25$
b. $x^4 - x^9$

Example 5 Factor: $-9a^5 + 18a^2 - 3a$

Solution:

$$-9a^5 + 18a^2 - 3a = 3a(-3a^4) + 3a(6a) + 3a(-1)$$
$$= 3a(-3a^4 + 6a - 1)$$

Work Practice 5

Helpful Hint
Don't forget the -1.

Practice 5

Factor: $-10x^3 + 8x^2 - 2x$

In Example 5, we could have chosen to factor out $-3a$ instead of $3a$. If we factor out $-3a$, we have

$$-9a^5 + 18a^2 - 3a = (-3a)(3a^4) + (-3a)(-6a) + (-3a)(1)$$
$$= -3a(3a^4 - 6a + 1)$$

Helpful Hint

Notice the changes in signs when factoring out $-3a$.

Answers
4. a. $5(2y + 5)$ **b.** $x^4(1 - x^5)$
5. $2x(-5x^2 + 4x - 1)$

✓ **Concept Check Answer**
c

Practice 6–8
Factor.
6. $4x^3 + 12x$
7. $\frac{2}{5}a^5 - \frac{4}{5}a^3 + \frac{1}{5}a^2$
8. $6a^3b + 3a^3b^2 + 9a^2b^4$

Examples Factor.

6. $6a^4 - 12a = 6a(a^3 - 2)$
7. $\frac{3}{7}x^4 + \frac{1}{7}x^3 - \frac{5}{7}x^2 = \frac{1}{7}x^2(3x^2 + x - 5)$
8. $15p^2q^4 + 20p^3q^5 + 5p^3q^3 = 5p^2q^3(3q + 4pq^2 + p)$

■ Work Practice 6–8

Practice 9
Factor: $7(p + 2) + q(p + 2)$

Example 9 Factor: $5(x + 3) + y(x + 3)$

Solution: The binomial $(x + 3)$ is present in both terms and is the greatest common factor. We use the distributive property to factor out $(x + 3)$.

$$5(x + 3) + y(x + 3) = (x + 3)(5 + y)$$

■ Work Practice 9

Practice 10
Factor $7xy^3(p + q) - (p + q)$

Example 10 Factor: $3m^2n(a + b) - (a + b)$

Solution: The greatest common factor is $(a + b)$.

$$3m^2n(a + b) - 1(a + b) = (a + b)(3m^2n - 1)$$

■ Work Practice 10

Objective D Factoring by Grouping

Once the GCF is factored out, we can often continue to factor the polynomial using a variety of techniques. We discuss here a technique called **factoring by grouping**. This technique can be used to factor some polynomials with four terms.

Practice 11
Factor $ab + 7a + 2b + 14$ by grouping.

 Helpful Hint Notice that this form, $x(y + 2) + 3(y + 2)$, is *not* a factored form of the original polynomial. It is a sum, not a product.

Example 11 Factor $xy + 2x + 3y + 6$ by grouping.

Solution: Notice that the first two terms of this polynomial have a common factor of x and that the second two terms have a common factor of 3. Because of this, group the first two terms, then the last two terms, and then factor out these common factors.

$$xy + 2x + 3y + 6 = (xy + 2x) + (3y + 6) \quad \text{Group terms.}$$
$$= x(y + 2) + 3(y + 2) \quad \text{Factor out GCF from each grouping.}$$

Next we factor out the common binomial factor, $(y + 2)$.

$$x(y + 2) + 3(y + 2) = (y + 2)(x + 3)$$

Now the result is a factored form because it is a product. We were able to write the polynomial as a product because of the common binomial factor, $(y + 2)$, that appeared. If this does not happen, try rearranging the terms of the original polynomial.

Check: Multiply $(y + 2)$ by $(x + 3)$.

$$(y + 2)(x + 3) = xy + 2x + 3y + 6,$$

the original polynomial.

Thus, a factored form of $xy + 2x + 3y + 6$ is the product $(y + 2)(x + 3)$.

■ Work Practice 11

You may want to try these steps when factoring by grouping.

Answers
6. $4x(x^2 + 3)$
7. $\frac{1}{5}a^2(2a^3 - 4a + 1)$
8. $3a^2b(2a + ab + 3b^3)$
9. $(p + 2)(7 + q)$
10. $(p + q)(7xy^3 - 1)$
11. $(b + 7)(a + 2)$

Section 11.1 | The Greatest Common Factor and Factoring by Grouping

To Factor a Four-Term Polynomial by Grouping
Step 1: Group the terms in two groups of two terms so that each group has a common factor.
Step 2: Factor out the GCF from each group.
Step 3: If there is a common binomial factor, factor it out.
Step 4: If not, rearrange the terms and try these steps again.

Examples Factor by grouping.

12. $15x^3 - 10x^2 + 6x - 4$
$= (15x^3 - 10x^2) + (6x - 4)$ Group the terms.
$= 5x^2(3x - 2) + 2(3x - 2)$ Factor each group.
$= (3x - 2)(5x^2 + 2)$ Factor out the common factor, $(3x - 2)$.

13. $3x^2 + 4xy - 3x - 4y$
$= (3x^2 + 4xy) + (-3x - 4y)$
$= x(3x + 4y) - 1(3x + 4y)$ Factor each group. A -1 is factored from the second pair of terms so that there is a common factor, $(3x + 4y)$.
$= (3x + 4y)(x - 1)$ Factor out the common factor, $(3x + 4y)$.

14. $2a^2 + 5ab + 2a + 5b$
$= (2a^2 + 5ab) + (2a + 5b)$ Factor each group. An understood 1 is written before $(2a + 5b)$ to help remember that
$= a(2a + 5b) + 1(2a + 5b)$ $(2a + 5b)$ is $1(2a + 5b)$.
$= (2a + 5b)(a + 1)$ Factor out the common factor, $(2a + 5b)$.

Practice 12–14
Factor by grouping.
12. $28x^3 - 7x^2 + 12x - 3$
13. $2xy + 5y^2 - 4x - 10y$
14. $3x^2 + 4xy + 3x + 4y$

Helpful Hint Notice that the factor of 1 is written when $(2a + 5b)$ is factored out.

Work Practice 12–14

Examples Factor by grouping.

15. $3x^3 - 2x - 9x^2 + 6$
$= x(3x^2 - 2) - 3(3x^2 - 2)$ Factor each group. A -3 is factored from the second pair of terms so that there is a common factor, $(3x^2 - 2)$.
$= (3x^2 - 2)(x - 3)$ Factor out the common factor, $(3x^2 - 2)$.

16. $3xy + 2 - 3x - 2y$

Notice that the first two terms have no common factor other than 1. However, if we rearrange these terms, a grouping emerges that does lead to a common factor.
$3xy + 2 - 3x - 2y$
$= (3xy - 3x) + (-2y + 2)$
$= 3x(y - 1) - 2(y - 1)$ Factor -2 from the second group.
$= (y - 1)(3x - 2)$ Factor out the common factor, $(y - 1)$.

17. $5x - 10 + x^3 - x^2 = 5(x - 2) + x^2(x - 1)$

There is no common binomial factor that can now be factored out. No matter how we rearrange the terms, no grouping will lead to a common factor. Thus, this polynomial is not factorable by grouping.

Practice 15–17
Factor by grouping.
15. $4x^3 + x - 20x^2 - 5$
16. $3xy - 4 + x - 12y$
17. $2x - 2 + x^3 - 3x^2$

Helpful Hint Throughout this chapter, we will be factoring polynomials. Even when the instructions do not so state, it is always a good idea to check your answers by multiplying.

Work Practice 15–17

Helpful Hint
One more reminder: When **factoring** a polynomial, make sure the polynomial is written as a **product**. For example, it is true that
$$3x^2 + 4xy - 3x - 4y = \underbrace{x(3x + 4y) - 1(3x + 4y)}_{\text{but this is not a factored form}},$$
since it is a **sum (difference)**, not a **product**.
A factored form of $3x^2 + 4xy - 3x - 4y$ is the **product** $(3x + 4y)(x - 1)$.

Answers
12. $(4x - 1)(7x^2 + 3)$
13. $(2x + 5y)(y - 2)$
14. $(3x + 4y)(x + 1)$
15. $(4x^2 + 1)(x - 5)$
16. $(3y + 1)(x - 4)$
17. cannot be factored by grouping

Vocabulary, Readiness & Video Check

Use the choices below to fill in each blank. Some choices may be used more than once and some may not be used at all.

greatest common factor   factors   factoring   true   false   least   greatest

1. Since $5 \cdot 4 = 20$, the numbers 5 and 4 are called _____ of 20.
2. The _____ of a list of integers is the largest integer that is a factor of all the integers in the list.
3. The greatest common factor of a list of common variables raised to powers is the variable raised to the _____ exponent in the list.
4. The process of writing a polynomial as a product is called _____.
5. True or false? A factored form of $7x + 21 + xy + 3y$ is $7(x + 3) + y(x + 3)$. _____
6. True or false? A factored form of $3x^3 + 6x + x^2 + 2$ is $3x(x^2 + 2)$. _____

Write the prime factorization of the following integers.

7. 14   8. 15

Write the GCF of the following pairs of integers.

9. 18, 3   10. 7, 35   11. 20, 15   12. 6, 15

See Video 11.1

Martin-Gay Interactive Videos   Watch the section lecture video and answer the following questions.

Objective A 13. Based on Example 1, give a general definition for the greatest common factor (GCF) of a list of numbers.
Objective B 14. When finding the GCF of the terms in Example 3, why are the numerical parts of the terms factored out, but not the variable parts?
Objective C 15. From Example 5, once we factor out the GCF, how can the number of terms in the other factor help us determine if our factorization is correct?
Objective D 16. In Examples 7 and 8, what are we reminded to always do first when factoring a polynomial? Also, a polynomial with how many terms suggests it might be factored by grouping?

11.1 Exercise Set MyLab Math

Objectives A B Mixed Practice *Find the GCF for each list. See Examples 1 through 3.*

1. 32, 36
2. 36, 90
3. 18, 42, 84
4. 30, 75, 135
5. 24, 14, 21
6. 15, 25, 27
7. y^2, y^4, y^7
8. x^3, x^2, x^5
9. z^7, z^9, z^{11}
10. y^8, y^{10}, y^{12}
11. $x^{10}y^2, xy^2, x^3y^3$
12. p^7q, p^8q^2, p^9q^3
13. $14x, 21$
14. $20y, 15$
15. $12y^4, 20y^3$
16. $32x^5, 18x^2$

Section 11.1 | The Greatest Common Factor and Factoring by Grouping

17. $-10x^2, 15x^3$ **18.** $-21x^3, 14x$ **19.** $12x^3, -6x^4, 3x^5$ **20.** $15y^2, 5y^7, -20y^3$

21. $-18x^2y, 9x^3y^3, 36x^3y$ **22.** $7x^3y^3, -21x^2y^2, 14xy^4$ **23.** $20a^6b^2c^8, 50a^7b$ **24.** $40x^7y^2z, 64x^9y$

Objective **C** Factor out the GCF from each polynomial. See Examples 4 through 10.

25. $3a + 6$ **26.** $18a + 12$ **27.** $30x - 15$ **28.** $42x - 7$ **29.** $x^3 + 5x^2$

30. $y^5 + 6y^4$ **31.** $6y^4 + 2y^3$ **32.** $5x^2 + 10x^6$ **33.** $32xy - 18x^2$ **34.** $10xy - 15x^2$

35. $4x - 8y + 4$ **36.** $7x + 21y - 7$ **37.** $6x^3 - 9x^2 + 12x$ **38.** $12x^3 + 16x^2 - 8x$

39. $a^7b^6 - a^3b^2 + a^2b^5 - a^2b^2$ **40.** $x^9y^6 + x^3y^5 - x^4y^3 + x^3y^3$ **41.** $5x^3y - 15x^2y + 10xy$

42. $14x^3y + 7x^2y - 7xy$ **43.** $8x^5 + 16x^4 - 20x^3 + 12$ **44.** $9y^6 - 27y^4 + 18y^2 + 6$

45. $\frac{1}{3}x^4 + \frac{2}{3}x^3 - \frac{4}{3}x^5 + \frac{1}{3}x$ **46.** $\frac{2}{5}y^7 - \frac{4}{5}y^5 + \frac{3}{5}y^2 - \frac{2}{5}y$ **47.** $y(x^2 + 2) + 3(x^2 + 2)$

48. $x(y^2 + 1) - 3(y^2 + 1)$ **49.** $z(y + 4) + 3(y + 4)$ **50.** $8(x + 2) - y(x + 2)$

51. $r(z^2 - 6) + (z^2 - 6)$ **52.** $q(b^3 - 5) + (b^3 - 5)$

Factor a negative number or a GCF with a negative coefficient from each polynomial. See Example 5.

53. $-2x - 14$ **54.** $-7y - 21$ **55.** $-2x^5 + x^7$

56. $-5y^3 + y^6$ **57.** $-6a^4 + 9a^3 - 3a^2$ **58.** $-5m^6 + 10m^5 - 5m^3$

Objective **D** Factor each four-term polynomial by grouping. If this is not possible, write "not factorable by grouping." See Examples 11 through 17.

59. $x^3 + 2x^2 + 5x + 10$ **60.** $x^3 + 4x^2 + 3x + 12$ **61.** $5x + 15 + xy + 3y$

62. $xy + y + 2x + 2$ **63.** $6x^3 - 4x^2 + 15x - 10$ **64.** $16x^3 - 28x^2 + 12x - 21$

65. $5m^3 + 6mn + 5m^2 + 6n$ **66.** $8w^2 + 7wv + 8w + 7v$ **67.** $2y - 8 + xy - 4x$

68. $6x - 42 + xy - 7y$ **69.** $2x^3 + x^2 + 8x + 4$ **70.** $2x^3 - x^2 - 10x + 5$

71. $3x - 3 + x^3 - 4x^2$ **72.** $7x - 21 + x^3 - 2x^2$ **73.** $4x^2 - 8xy - 3x + 6y$

74. $5xy - 15x - 6y + 18$ **75.** $5q^2 - 4pq - 5q + 4p$ **76.** $6m^2 - 5mn - 6m + 5n$

Objectives **C D** Mixed Practice *Factor out the GCF from each polynomial. Then factor by grouping.*

77. $12x^2y - 42x^2 - 4y + 14$

78. $90 + 15y^2 - 18x - 3xy^2$

79. $6a^2 + 9ab^2 + 6ab + 9b^3$

80. $16x^2 + 4xy^2 + 8xy + 2y^3$

Review

Multiply. See Section 10.5.

81. $(x + 2)(x + 5)$ **82.** $(y + 3)(y + 6)$ **83.** $(b + 1)(b - 4)$ **84.** $(x - 5)(x + 10)$

Fill in the chart by finding two numbers that have the given product and sum. The first column is filled in for you.

		85.	86.	87.	88.	89.	90.	91.	92.
Two Numbers	4, 7								
Their Product	28	12	20	8	16	−10	−9	−24	−36
Their Sum	11	8	9	−9	−10	3	0	−5	−5

Concept Extensions

See the Concept Checks in this section.

93. Which of the following is/are factored form(s) of $-2x + 14$?
  a. $-2(x + 7)$ **b.** $-2 \cdot x + 14$
  c. $-2(x - 14)$ **d.** $-2(x - 7)$

94. Which of the following is/are factored form(s) of $8a - 24$?
  a. $8 \cdot a - 24$ **b.** $8(a - 3)$
  c. $4(2a - 12)$ **d.** $8 \cdot a - 2 \cdot 12$

Which of the following expressions are factored?

95. $(a + 6)(a + 2)$

96. $(x + 5)(x + y)$

97. $5(2y + z) - b(2y + z)$

98. $3x(a + 2b) + 2(a + 2b)$

99. The number (in thousands) of students who graduated from U.S. high schools, both public and private, each year during 2000 through 2013 can be modeled by $-3x^2 + 78x + 2904$, where x is the number of years since 2000. (*Source:* National Center for Educational Statistics)
  a. Find the number of students who graduated from U.S. high schools in 2010. To do so, let $x = 10$ and evaluate $-3x^2 + 78x + 2904$.
  b. Use this expression to predict the number of students who will graduate from U.S. high schools in 2018.
  c. Factor the polynomial $-3x^2 + 78x + 2904$ by factoring -3 from each term.
  d. For the year 2010, the National Center for Higher Education determined that 62.5% of U.S. high school graduates went on to higher education. Using your answer from part **a**, determine how many of those graduating in 2010 pursued higher education.

100. The amount of bottled water consumed in the United States, in gallons per person, for the period 2010–2015 can be approximated by the polynomial $-\dfrac{9}{100}x^2 + \dfrac{246}{100}x + \dfrac{2594}{100}$, where x is the number of years after 2010. (*Source:* Bottledwater.org and *Fortune*)
  a. Find the approximate U.S. annual per capita consumption of bottled water in 2014. To do so, let $x = 4$ and evaluate $-\dfrac{9}{100}x^2 + \dfrac{246}{100}x + \dfrac{2594}{100}$.
  b. Find the approximate U.S. annual per capita consumption of bottled water in 2015.
  c. Suppose the annual per capita consumption of bottled water continues to be approximated by the polynomial $-\dfrac{9}{100}x^2 + \dfrac{246}{100}x + \dfrac{2594}{100}$. Use this polynomial to predict the per capita consumption of bottled water in 2020.
  d. Factor out the GCF $\dfrac{1}{100}$ from the polynomial $-\dfrac{9}{100}x^2 + \dfrac{246}{100}x + \dfrac{2594}{100}$.

101. The annual orange production (in thousand tons) in the United States for the period 2011–2016 can be approximated by the polynomial $87x^2 - 1131x + 9048$, where x is the number of years after 2011. (*Source:* Based on data from the National Agricultural Statistics Service and bloomberg.com)

  a. Find the approximate U.S. orange production in 2015. To do so, let $x = 4$ and evaluate $87x^2 - 1131x + 9048$.
  b. Find the approximate U.S. orange production in 2013.
  c. Factor out the GCF from the polynomial $87x^2 - 1131x + 9048$.

102. The polynomial $15x^2 + 12x + 354$ represents the approximate number of visitors (in thousands) per year to Redwoods National Park in California, during 2012–2015. In this polynomial, x represents the years since 2012. (*Source:* Based on data from the National Park Service)

  a. Find the approximate number of visitors to Redwoods National Park in 2013. To do so, let $x = 1$ and evaluate $15x^2 + 12x + 354$.
  b. Find the approximate number of visitors to Redwoods National Park in 2015.
  c. Factor out the GCF from the polynomial $15x^2 + 12x + 354$.

Write an expression for the area of each shaded region. Then write the expression as a factored polynomial.

103.

△ **104.**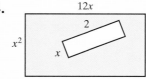

Write an expression for the length of each rectangle. (Hint: Factor the area binomial and recall that Area = width · length.)

105.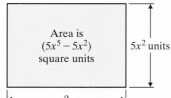

△ **106.**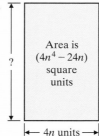

107. Construct a binomial whose greatest common factor is $5a^3$. (*Hint:* Multiply $5a^3$ by a binomial whose terms contain no common factor other than 1: $5a^3(\square + \square)$.)

108. Construct a trinomial whose greatest common factor is $2x^2$. See the hint for Exercise **107**.

109. Explain how you can tell whether a polynomial is written in factored form.

110. Construct a four-term polynomial that can be factored by grouping. Explain how you constructed the polynomial.

11.2 Factoring Trinomials of the Form $x^2 + bx + c$

Objectives

A Factor Trinomials of the Form $x^2 + bx + c$.

B Factor Out the Greatest Common Factor and Then Factor a Trinomial of the Form $x^2 + bx + c$.

Objective A Factoring Trinomials of the Form $x^2 + bx + c$

In this section, we factor trinomials of the form $x^2 + bx + c$, such as

$$x^2 + 7x + 12, \quad x^2 - 12x + 35, \quad x^2 + 4x - 12, \quad \text{and} \quad r^2 - r - 42$$

Notice that for these trinomials, the coefficient of the squared variable is 1.

Recall that factoring means to write as a product and that factoring and multiplying are reverse processes. Using the FOIL method of multiplying binomials, we have the following.

$$\begin{aligned} & \qquad\qquad\qquad \text{F} \quad \text{O} \quad \text{I} \quad \text{L} \\ (x+3)(x+1) &= x^2 + 1x + 3x + 3 \\ &= x^2 + 4x + 3 \end{aligned}$$

Thus, a factored form of $x^2 + 4x + 3$ is $(x + 3)(x + 1)$.

Notice that the product of the first terms of the binomials is $x \cdot x = x^2$, the first term of the trinomial. Also, the product of the last two terms of the binomials is $3 \cdot 1 = 3$, the third term of the trinomial. The sum of these same terms is $3 + 1 = 4$, the coefficient of the middle, x, term of the trinomial.

The product of these numbers is 3.

$$x^2 + 4x + 3 = (x + 3)(x + 1)$$

The sum of these numbers is 4.

Many trinomials, such as the one above, factor into two binomials. To factor $x^2 + 7x + 10$, let's assume that it factors into two binomials and begin by writing two pairs of parentheses. The first term of the trinomial is x^2, so we use x and x as the first terms of the binomial factors.

$$x^2 + 7x + 10 = (x + \square)(x + \square)$$

To determine the last term of each binomial factor, we look for two integers whose product is 10 and whose sum is 7. The integers are 2 and 5. Thus,

$$x^2 + 7x + 10 = (x + 2)(x + 5)$$

Check: To see if we have factored correctly, we multiply.

$$\begin{aligned} (x + 2)(x + 5) &= x^2 + 5x + 2x + 10 \\ &= x^2 + 7x + 10 \qquad \text{Combine like terms.} \end{aligned}$$

Helpful Hint

Since multiplication is commutative, the factored form of $x^2 + 7x + 10$ can be written as either $(x + 2)(x + 5)$ or $(x + 5)(x + 2)$.

To Factor a Trinomial of the Form $x^2 + bx + c$

The product of these numbers is c.

$$x^2 + bx + c = (x + \square)(x + \square)$$

The sum of these numbers is b.

Section 11.2 | Factoring Trinomials of the Form $x^2 + bx + c$

Example 1 Factor: $x^2 + 7x + 12$

Solution: We begin by writing the first terms of the binomial factors.

$(x + \square)(x + \square)$

Next we look for two numbers whose product is 12 and whose sum is 7. Since our numbers must have a positive product and a positive sum, we look at pairs of positive factors of 12 only.

Factors of 12	Sum of Factors
1, 12	13
2, 6	8
3, 4	7

Correct sum, so the numbers are 3 and 4.

Thus, $x^2 + 7x + 12 = (x + 3)(x + 4)$

Check: $(x + 3)(x + 4) = x^2 + 4x + 3x + 12 = x^2 + 7x + 12$

■ Work Practice 1

Practice 1

Factor: $x^2 + 12x + 20$

Example 2 Factor: $x^2 - 12x + 35$

Solution: Again, we begin by writing the first terms of the binomials.

$(x + \square)(x + \square)$

Now we look for two numbers whose product is 35 and whose sum is -12. Since our numbers must have a positive product and a negative sum, we look at pairs of negative factors of 35 only.

Factors of 35	Sum of Factors
$-1, -35$	-36
$-5, -7$	-12

Correct sum, so the numbers are -5 and -7.

$x^2 - 12x + 35 = (x - 5)(x - 7)$

Check: To check, multiply $(x - 5)(x - 7)$.

■ Work Practice 2

Practice 2

Factor each trinomial.
a. $x^2 - 23x + 22$
b. $x^2 - 27x + 50$

Example 3 Factor: $x^2 + 4x - 12$

Solution: $x^2 + 4x - 12 = (x + \square)(x + \square)$

We look for two numbers whose product is -12 and whose sum is 4. Since our numbers must have a negative product, we look at pairs of factors with opposite signs.

Factors of -12	Sum of Factors
$-1, 12$	11
$1, -12$	-11
$-2, 6$	4
$2, -6$	-4
$-3, 4$	1
$3, -4$	-1

Correct sum, so the numbers are -2 and 6.

$x^2 + 4x - 12 = (x - 2)(x + 6)$

■ Work Practice 3

Practice 3

Factor: $x^2 + 5x - 36$

Answers
1. $(x + 10)(x + 2)$
2. **a.** $(x - 1)(x - 22)$
   b. $(x - 2)(x - 25)$
3. $(x + 9)(x - 4)$

Practice 4
Factor each trinomial.
a. $q^2 - 3q - 40$
b. $y^2 + 2y - 48$

Example 4 Factor: $r^2 - r - 42$

Solution: Because the variable in this trinomial is r, the first term of each binomial factor is r.
$$r^2 - r - 42 = (r + \square)(r + \square)$$
Now we look for two numbers whose product is -42 and whose sum is -1, the numerical coefficient of r. The numbers are 6 and -7. Therefore,
$$r^2 - r - 42 = (r + 6)(r - 7)$$

■ Work Practice 4

Practice 5
Factor: $x^2 + 6x + 15$

Example 5 Factor: $a^2 + 2a + 10$

Solution: Look for two numbers whose product is 10 and whose sum is 2. Neither 1 and 10 nor 2 and 5 give the required sum, 2. We conclude that $a^2 + 2a + 10$ is not factorable with integers. A polynomial such as $a^2 + 2a + 10$ is called a **prime polynomial.**

■ Work Practice 5

Practice 6
Factor each trinomial.
a. $x^2 + 9xy + 14y^2$
b. $a^2 - 13ab + 30b^2$

Example 6 Factor: $x^2 + 5xy + 6y^2$

Solution: $x^2 + 5xy + 6y^2 = (x + \square)(x + \square)$
Recall that the middle term, $5xy$, is the same as $5yx$. Thus, we can see that $5y$ is the "coefficient" of x. We then look for two terms whose product is $6y^2$ and whose sum is $5y$. The terms are $2y$ and $3y$ because $2y \cdot 3y = 6y^2$ and $2y + 3y = 5y$. Therefore,
$$x^2 + 5xy + 6y^2 = (x + 2y)(x + 3y)$$

■ Work Practice 6

Practice 7
Factor: $x^4 + 8x^2 + 12$

Example 7 Factor: $x^4 + 5x^2 + 6$

Solution: As usual, we begin by writing the first terms of the binomials. Since the greatest power of x in this polynomial is x^4, we write
$$(x^2 + \square)(x^2 + \square) \quad \text{Since } x^2 \cdot x^2 = x^4$$
Now we look for two factors of 6 whose sum is 5. The numbers are 2 and 3. Thus,
$$x^4 + 5x^2 + 6 = (x^2 + 2)(x^2 + 3)$$

■ Work Practice 7

If the terms of a polynomial are not written in descending powers of the variable, you may want to rearrange the terms before factoring.

Practice 8
Factor: $48 - 14x + x^2$

Example 8 Factor: $40 - 13t + t^2$

Solution: First, we rearrange terms so that the trinomial is written in descending powers of t.
$$40 - 13t + t^2 = t^2 - 13t + 40$$
Next, try to factor.
$$t^2 - 13t + 40 = (t + \square)(t + \square)$$
Now we look for two factors of 40 whose sum is -13. The numbers are -8 and -5. Thus,
$$t^2 - 13t + 40 = (t - 8)(t - 5)$$

■ Work Practice 8

Answers
4. a. $(q - 8)(q + 5)$
b. $(y + 8)(y - 6)$
5. prime polynomial
6. a. $(x + 2y)(x + 7y)$
b. $(a - 3b)(a - 10b)$
7. $(x^2 + 6)(x^2 + 2)$
8. $(x - 6)(x - 8)$

Section 11.2 | Factoring Trinomials of the Form $x^2 + bx + c$

The following sign patterns may be useful when factoring trinomials.

Helpful Hint

A positive constant in a trinomial tells us to look for two numbers with the same sign. The sign of the coefficient of the middle term tells us whether the signs are both positive or both negative.

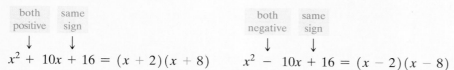

$$x^2 + 10x + 16 = (x + 2)(x + 8) \qquad x^2 - 10x + 16 = (x - 2)(x - 8)$$

(both positive → same sign) (both negative → same sign)

A negative constant in a trinomial tells us to look for two numbers with opposite signs.

$$x^2 + 6x - 16 = (x + 8)(x - 2) \qquad x^2 - 6x - 16 = (x - 8)(x + 2)$$

(opposite signs) (opposite signs)

Objective B Factoring Out the Greatest Common Factor

Remember that the first step in factoring any polynomial is to factor out the greatest common factor (if there is one other than 1 or -1).

Example 9 Factor: $3m^2 - 24m - 60$

Solution: First we factor out the greatest common factor, 3, from each term.

$$3m^2 - 24m - 60 = 3(m^2 - 8m - 20)$$

Now we factor $m^2 - 8m - 20$ by looking for two factors of -20 whose sum is -8. The factors are -10 and 2. Therefore, the complete factored form is

$$3m^2 - 24m - 60 = 3(m + 2)(m - 10)$$

■ Work Practice 9

Practice 9
Factor each trinomial.
a. $4x^2 - 24x + 36$
b. $x^3 + 3x^2 - 4x$

Helpful Hint

Remember to write the common factor, 3, as part of the factored form.

Example 10 Factor: $2x^4 - 26x^3 + 84x^2$

Solution:

$$2x^4 - 26x^3 + 84x^2 = 2x^2(x^2 - 13x + 42) \quad \text{Factor out common factor, } 2x^2.$$
$$= 2x^2(x - 6)(x - 7) \quad \text{Factor } x^2 - 13x + 42.$$

■ Work Practice 10

Practice 10
Factor: $5x^5 - 25x^4 - 30x^3$

Answers
9. a. $4(x - 3)(x - 3)$
b. $x(x + 4)(x - 1)$
10. $5x^3(x + 1)(x - 6)$

Vocabulary, Readiness & Video Check

Fill in each blank with "true" or "false."

1. To factor $x^2 + 7x + 6$, we look for two numbers whose product is 6 and whose sum is 7. _____
2. We can write the factorization $(y + 2)(y + 4)$ also as $(y + 4)(y + 2)$. _____
3. The factorization $(4x - 12)(x - 5)$ is completely factored. _____
4. The factorization $(x + 2y)(x + y)$ may also be written as $(x + 2y)^2$. _____

Complete each factored form.

5. $x^2 + 9x + 20 = (x + 4)(x \quad)$
6. $x^2 + 12x + 35 = (x + 5)(x \quad)$
7. $x^2 - 7x + 12 = (x - 4)(x \quad)$
8. $x^2 - 13x + 22 = (x - 2)(x \quad)$
9. $x^2 + 4x + 4 = (x + 2)(x \quad)$
10. $x^2 + 10x + 24 = (x + 6)(x \quad)$

Martin-Gay Interactive Videos Watch the section lecture video and answer the following questions.

Objective A 11. In Example 2, why are only negative factors of 15 considered?

Objective B 12. In Example 5, we know we need a positive and a negative factor of -10. How do we determine which factor is negative?

See Video 11.2

11.2 Exercise Set MyLab Math

Objective A *Factor each trinomial completely. If a polynomial can't be factored, write "prime." See Examples 1 through 8.*

1. $x^2 + 7x + 6$
2. $x^2 + 6x + 8$
3. $y^2 - 10y + 9$
4. $y^2 - 12y + 11$

5. $x^2 - 6x + 9$
6. $x^2 - 10x + 25$
7. $x^2 - 3x - 18$
8. $x^2 - x - 30$

9. $x^2 + 3x - 70$
10. $x^2 + 4x - 32$
11. $x^2 + 5x + 2$
12. $x^2 - 7x + 5$

13. $x^2 + 8xy + 15y^2$
14. $x^2 + 6xy + 8y^2$
15. $a^4 - 2a^2 - 15$
16. $y^4 - 3y^2 - 70$

17. $13 + 14m + m^2$
18. $17 + 18n + n^2$
19. $10t - 24 + t^2$
20. $6q - 27 + q^2$

21. $a^2 - 10ab + 16b^2$
22. $a^2 - 9ab + 18b^2$

Section 11.2 | Factoring Trinomials of the Form $x^2 + bx + c$

Objectives A B Mixed Practice *Factor each trinomial completely. Some of these trinomials contain a greatest common factor (other than 1). Don't forget to factor out the GCF first. See Examples 1 through 10.*

23. $2z^2 + 20z + 32$
24. $3x^2 + 30x + 63$
25. $2x^3 - 18x^2 + 40x$
26. $3x^3 - 12x^2 - 36x$

27. $x^2 - 3xy - 4y^2$
28. $x^2 - 4xy - 77y^2$
29. $x^2 + 15x + 36$
30. $x^2 + 19x + 60$

31. $x^4 - x^2 - 2$
32. $x^4 - 5x^2 - 14$
33. $r^2 - 16r + 48$
34. $r^2 - 10r + 21$

35. $x^2 + xy - 2y^2$
36. $x^2 - xy - 6y^2$
▶ 37. $3x^2 + 9x - 30$
38. $4x^2 - 4x - 48$

39. $3x^4 - 60x^2 + 108$
40. $2x^4 - 24x^2 + 70$
41. $x^2 - 18x - 144$
42. $x^2 + x - 42$

43. $r^2 - 3r + 6$
44. $x^2 + 4x - 10$
▶ 45. $x^2 - 8x + 15$
46. $x^2 - 9x + 14$

47. $6x^3 + 54x^2 + 120x$
48. $3x^3 + 3x^2 - 126x$
49. $4x^2y + 4xy - 12y$
50. $3x^2y - 9xy + 45y$

51. $x^2 - 4x - 21$
52. $x^2 - 4x - 32$
53. $x^2 + 7xy + 10y^2$
54. $x^2 - 3xy - 4y^2$

55. $64 + 24t + 2t^2$
56. $50 + 20t + 2t^2$
57. $x^3 - 2x^2 - 24x$
58. $x^3 - 3x^2 - 28x$

59. $2t^5 - 14t^4 + 24t^3$
60. $3x^6 + 30x^5 + 72x^4$
▶ 61. $5x^3y - 25x^2y^2 - 120xy^3$
62. $7a^3b - 35a^2b^2 + 42ab^3$

63. $162 - 45m + 3m^2$
64. $48 - 20n + 2n^2$
65. $-x^2 + 12x - 11$
(Factor out -1 first.)
66. $-x^2 + 8x - 7$
(Factor out -1 first.)

67. $\frac{1}{2}y^2 - \frac{9}{2}y - 11$
(Factor out $\frac{1}{2}$ first.)
68. $\frac{1}{3}y^2 - \frac{5}{3}y - 8$
(Factor out $\frac{1}{3}$ first.)
69. $x^3y^2 + x^2y - 20x$
70. $a^2b^3 + ab^2 - 30b$

Review

Multiply. See Section 10.5.

71. $(2x + 1)(x + 5)$
72. $(3x + 2)(x + 4)$
73. $(5y - 4)(3y - 1)$

74. $(4z - 7)(7z - 1)$
75. $(a + 3b)(9a - 4b)$
76. $(y - 5x)(6y + 5x)$

Concept Extensions

77. Write a polynomial that factors as $(x-3)(x+8)$.

78. To factor $x^2 + 13x + 42$, think of two numbers whose _____ is 42 and whose _____ is 13.

Complete each sentence in your own words.

79. If $x^2 + bx + c$ is factorable and c is negative, then the signs of the last-term factors of the binomials are opposite because ...

80. If $x^2 + bx + c$ is factorable and c is positive, then the signs of the last-term factors of the binomials are the same because ...

Remember that perimeter means distance around. Write the perimeter of each rectangle as a simplified polynomial. Then factor the polynomial completely.

△ **81.**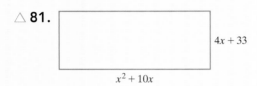

△ **82.**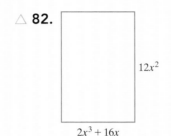

83. An object is thrown upward from the top of an 80-foot building with an initial velocity of 64 feet per second. Neglecting air resistance, the height of the object after t seconds is given by $-16t^2 + 64t + 80$. Factor this polynomial.

84. An object is thrown upward from the top of a 112-foot building with an initial velocity of 96 feet per second. Neglecting air resistance, the height of the object after t seconds is given by $-16t^2 + 96t + 112$. Factor this polynomial.

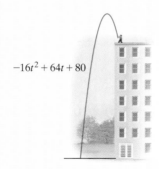

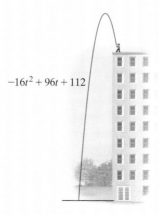

Factor each trinomial completely.

85. $x^2 + \dfrac{1}{2}x + \dfrac{1}{16}$

86. $x^2 + x + \dfrac{1}{4}$

87. $z^2(x+1) - 3z(x+1) - 70(x+1)$

88. $y^2(x+1) - 2y(x+1) - 15(x+1)$

Find a positive value of c so that each trinomial is factorable.

89. $n^2 - 16n + c$

90. $y^2 - 4y + c$

Find a positive value of b so that each trinomial is factorable.

91. $y^2 + by + 20$

92. $x^2 + bx + 15$

Factor each trinomial. (Hint: Notice that $x^{2n} + 4x^n + 3$ factors as $(x^n + 1)(x^n + 3)$. Remember: $x^n \cdot x^n = x^{n+n}$ or x^{2n}.)

93. $x^{2n} + 8x^n - 20$

94. $x^{2n} + 5x^n + 6$

11.3 Factoring Trinomials of the Form $ax^2 + bx + c$

Objective A Factoring Trinomials of the Form $ax^2 + bx + c$

Objectives

A Factor Trinomials of the Form $ax^2 + bx + c$, Where $a \neq 1$.

B Factor Out the GCF Before Factoring a Trinomial of the Form $ax^2 + bx + c$.

In this section, we factor trinomials of the form $ax^2 + bx + c$, such as

$$3x^2 + 11x + 6, \quad 8x^2 - 22x + 5, \quad \text{and} \quad 2x^2 + 13x - 7$$

Notice that the coefficient of the squared variable in these trinomials is a number other than 1. We will factor these trinomials using a trial-and-check method based on our work in the last section.

To begin, let's review the relationship between the numerical coefficients of the trinomial and the numerical coefficients of its factored form. For example, since

$$(2x + 1)(x + 6) = 2x^2 + 13x + 6,$$

a factored form of $2x^2 + 13x + 6$ is $(2x + 1)(x + 6)$.

Notice that $2x$ and x are factors of $2x^2$, the first term of the trinomial. Also, 6 and 1 are factors of 6, the last term of the trinomial, as shown:

$$2x^2 + 13x + 6 = (2x + 1)(x + 6)$$
$$\overbrace{}^{2x \cdot x}$$
$$\underbrace{}_{1 \cdot 6}$$

Also notice that $13x$, the middle term, is the sum of the following products:

$$2x^2 + 13x + 6 = (2x + 1)(x + 6)$$

$$\begin{array}{r} 1x \\ + 12x \\ \hline 13x \quad \text{Middle term} \end{array}$$

Let's use this pattern to factor $5x^2 + 7x + 2$. First, we find factors of $5x^2$. Since all numerical coefficients in this trinomial are positive, we will use factors with positive numerical coefficients only. Thus, the factors of $5x^2$ are $5x$ and x. Let's try these factors as first terms of the binomials. Thus far, we have

$$5x^2 + 7x + 2 = (5x + \square)(x + \square)$$

Next, we need to find positive factors of 2. Positive factors of 2 are 1 and 2. Now we try possible combinations of these factors as second terms of the binomials until we obtain a middle term of $7x$.

$$(5x + 1)(x + 2) = 5x^2 + 11x + 2$$

$1x$
$+10x$
$11x$ ⟶ **Incorrect** middle term

Let's try switching factors 2 and 1.

$$(5x + 2)(x + 1) = 5x^2 + 7x + 2$$

$2x$
$+5x$
$7x$ ⟶ **Correct** middle term

Thus a factored form of $5x^2 + 7x + 2$ is $(5x + 2)(x + 1)$. To check, we multiply $(5x + 2)$ and $(x + 1)$. The product is $5x^2 + 7x + 2$.

Practice 1

Factor each trinomial.
a. $5x^2 + 27x + 10$
b. $4x^2 + 12x + 5$

Example 1 Factor: $3x^2 + 11x + 6$

Solution: Since all numerical coefficients are positive, we use factors with positive numerical coefficients. We first find factors of $3x^2$.

Factors of $3x^2$: $3x^2 = 3x \cdot x$

If factorable, the trinomial will be of the form

$$3x^2 + 11x + 6 = (3x + \square)(x + \square)$$

Next we factor 6.

Factors of 6: $6 = 1 \cdot 6, \quad 6 = 2 \cdot 3$

Now we try combinations of factors of 6 until a middle term of $11x$ is obtained. Let's try 1 and 6 first.

$$(3x + 1)(x + 6) = 3x^2 + 19x + 6$$

$1x$
$+18x$
$19x$ ⟶ **Incorrect** middle term

Now let's next try 6 and 1.

$$(3x + 6)(x + 1)$$

Before multiplying, notice that the terms of the factor $3x + 6$ have a common factor of 3. The terms of the original trinomial $3x^2 + 11x + 6$ have no common factor other than 1, so the terms of its factors will also contain no common factor other than 1. This means that $(3x + 6)(x + 1)$ is not a factored form.

Next let's try 2 and 3 as last terms.

$$(3x + 2)(x + 3) = 3x^2 + 11x + 6$$

$2x$
$+9x$
$11x$ ⟶ **Correct** middle term

Helpful Hint

This is true in general: If the terms of a trinomial have no common factor (other than 1), then the terms of each of its binomial factors will contain no common factor (other than 1).

Answers
1. a. $(5x + 2)(x + 5)$
b. $(2x + 5)(2x + 1)$

Thus a factored form of $3x^2 + 11x + 6$ is $(3x + 2)(x + 3)$.

■ Work Practice 1

Section 11.3 | Factoring Trinomials of the Form $ax^2 + bx + c$

✓**Concept Check** Do the terms of $3x^2 + 29x + 18$ have a common factor? Without multiplying, decide which of the following factored forms could not be a factored form of $3x^2 + 29x + 18$.

a. $(3x + 18)(x + 1)$
b. $(3x + 2)(x + 9)$
c. $(3x + 6)(x + 3)$
d. $(3x + 9)(x + 2)$

Example 2 Factor: $8x^2 - 22x + 5$

Solution: Factors of $8x^2$: $\quad 8x^2 = 8x \cdot x, \quad 8x^2 = 4x \cdot 2x$

We'll try $8x$ and x.

$$8x^2 - 22x + 5 = (8x + \square)(x + \square)$$

Since the middle term, $-22x$, has a negative numerical coefficient, we factor 5 into negative factors.

Factors of 5: $\quad 5 = -1 \cdot -5$

Let's try -1 and -5.

$(8x - 1)(x - 5) = 8x^2 - 41x + 5$

$-1x$
$+(-40x)$
$-41x \longrightarrow$ **Incorrect** middle term

Now let's try -5 and -1.

$(8x - 5)(x - 1) = 8x^2 - 13x + 5$

$-5x$
$+(-8x)$
$-13x \longrightarrow$ **Incorrect** middle term

Don't give up yet! We can still try other factors of $8x^2$. Let's try $4x$ and $2x$ with -1 and -5.

$(4x - 1)(2x - 5) = 8x^2 - 22x + 5$

$-2x$
$+(-20x)$
$-22x \longrightarrow$ **Correct** middle term

A factored form of $8x^2 - 22x + 5$ is $(4x - 1)(2x - 5)$.

Work Practice 2

Practice 2

Factor each trinomial.
a. $2x^2 - 11x + 12$
b. $6x^2 - 5x + 1$

Example 3 Factor: $2x^2 + 13x - 7$

Solution: Factors of $2x^2$: $\quad 2x^2 = 2x \cdot x$

Factors of -7: $\quad -7 = 1 \cdot -7, \quad -7 = -1 \cdot 7$

We try possible combinations of these factors:

$(2x + 1)(x - 7) = 2x^2 - 13x - 7 \quad$ **Incorrect** middle term
$(2x - 1)(x + 7) = 2x^2 + 13x - 7 \quad$ **Correct** middle term

A factored form of $2x^2 + 13x - 7$ is $(2x - 1)(x + 7)$.

Work Practice 3

Practice 3

Factor each trinomial.
a. $3x^2 + 14x - 5$
b. $35x^2 + 4x - 4$

Answers
2. a. $(2x - 3)(x - 4)$
b. $(3x - 1)(2x - 1)$
3. a. $(3x - 1)(x + 5)$
b. $(5x + 2)(7x - 2)$

✓**Concept Check Answer**
no; **a, c, d**

Practice 4
Factor each trinomial.
a. $14x^2 - 3xy - 2y^2$
b. $12a^2 - 16ab - 3b^2$

Example 4 Factor: $10x^2 - 13xy - 3y^2$

Solution: Factors of $10x^2$: $10x^2 = 10x \cdot x$, $10x^2 = 2x \cdot 5x$
Factors of $-3y^2$: $-3y^2 = -3y \cdot y$, $-3y^2 = 3y \cdot -y$

We try some combinations of these factors:

$$
\begin{aligned}
(10x - 3y)(x + y) &= 10x^2 + 7xy - 3y^2 \\
(x + 3y)(10x - y) &= 10x^2 + 29xy - 3y^2 \\
(5x + 3y)(2x - y) &= 10x^2 + xy - 3y^2 \\
(2x - 3y)(5x + y) &= 10x^2 - 13xy - 3y^2 \quad \text{Correct middle term}
\end{aligned}
$$

A factored form of $10x^2 - 13xy - 3y^2$ is $(2x - 3y)(5x + y)$.

■ Work Practice 4

Practice 5
Factor: $2x^4 - 5x^2 - 7$

Example 5 Factor: $3x^4 - 5x^2 - 8$

Solution: Factors of $3x^4$: $3x^4 = 3x^2 \cdot x^2$
Factors of -8: $-8 = -2 \cdot 4, 2 \cdot -4, -1 \cdot 8, 1 \cdot -8$

Try combinations of these factors:

$$
\begin{aligned}
(3x^2 - 2)(x^2 + 4) &= 3x^4 + 10x^2 - 8 \\
(3x^2 + 4)(x^2 - 2) &= 3x^4 - 2x^2 - 8 \\
(3x^2 + 8)(x^2 - 1) &= 3x^4 + 5x^2 - 8 \quad \text{Incorrect sign on middle term, so switch signs in binomial factors.} \\
(3x^2 - 8)(x^2 + 1) &= 3x^4 - 5x^2 - 8 \quad \text{Correct middle term}
\end{aligned}
$$

■ Work Practice 5

Helpful Hint
Study the last two lines of Example 5. If a factoring attempt gives you a middle term whose numerical coefficient is the opposite of the desired numerical coefficient, try switching the signs of the last terms in the binomials.

Switched signs
$(3x^2 + 8)(x^2 - 1) = 3x^4 + 5x^2 - 8$ Middle term: $+5x$
$(3x^2 - 8)(x^2 + 1) = 3x^4 - 5x^2 - 8$ Middle term: $-5x$

Objective B Factoring Out the Greatest Common Factor

Don't forget that the first step in factoring any polynomial is to look for a common factor to factor out.

Practice 6
Factor each trinomial.
a. $3x^3 + 17x^2 + 10x$
b. $6xy^2 + 33xy - 18x$

Example 6 Factor: $24x^4 + 40x^3 + 6x^2$

Solution: Notice that all three terms have a common factor of $2x^2$. Thus factor out $2x^2$ first.

$$24x^4 + 40x^3 + 6x^2 = 2x^2(12x^2 + 20x + 3)$$

Answers
4. a. $(7x + 2y)(2x - y)$
b. $(6a + b)(2a - 3b)$
5. $(2x^2 - 7)(x^2 + 1)$
6. a. $x(3x + 2)(x + 5)$
b. $3x(2y - 1)(y + 6)$

Next we factor $12x^2 + 20x + 3$.

Factors of $12x^2$: $12x^2 = 4x \cdot 3x$, $\quad 12x^2 = 12x \cdot x$, $\quad 12x^2 = 6x \cdot 2x$

Since all terms in the trinomial have positive numerical coefficients, we factor 3 using positive factors only.

Factors of 3: $3 = 1 \cdot 3$

We try some combinations of the factors.

$2x^2(4x + 3)(3x + 1) = 2x^2(12x^2 + 13x + 3)$
$2x^2(12x + 1)(x + 3) = 2x^2(12x^2 + 37x + 3)$
$2x^2(2x + 3)(6x + 1) = 2x^2(12x^2 + 20x + 3)$ **Correct** middle term

A factored form of $24x^4 + 40x^3 + 6x^2$ is $2x^2(2x + 3)(6x + 1)$.

Helpful Hint Don't forget to include the common factor in the factored form.

Work Practice 6

When the term containing the squared variable has a negative coefficient, you may want to first factor out a common factor of -1.

Example 7 Factor: $-6x^2 - 13x + 5$

Solution: We begin by factoring out a common factor of -1.

$-6x^2 - 13x + 5 = -1(6x^2 + 13x - 5)$ Factor out -1.
$\qquad\qquad\qquad\;\; = -1(3x - 1)(2x + 5)$ Factor $6x^2 + 13x - 5$.

Work Practice 7

Practice 7
Factor: $-5x^2 - 19x + 4$

Answer
7. $-1(x + 4)(5x - 1)$

Vocabulary, Readiness & Video Check

Complete each factorization.

1. $2x^2 + 5x + 3$ factors as $(2x + 3)(\underline{\quad})$.
   a. $(x + 3)$ **b.** $(2x + 1)$ **c.** $(3x + 4)$ **d.** $(x + 1)$

2. $7x^2 + 9x + 2$ factors as $(7x + 2)(\underline{\quad})$.
   a. $(3x + 1)$ **b.** $(x + 1)$ **c.** $(x + 2)$ **d.** $(7x + 1)$

3. $3x^2 + 31x + 10$ factors as $\underline{\quad}$.
   a. $(3x + 2)(x + 5)$ **b.** $(3x + 5)(x + 2)$ **c.** $(3x + 1)(x + 10)$

4. $5x^2 + 61x + 12$ factors as $\underline{\quad}$.
   a. $(5x + 1)(x + 12)$ **b.** $(5x + 3)(x + 4)$ **c.** $(5x + 2)(x + 6)$

Martin-Gay Interactive Videos   Watch the section lecture video and answer the following questions.

Objective A   5. From Example 1, explain in general terms how we would go about factoring a trinomial with a first-term coefficient $\neq 1$.

Objective B   6. From Examples 3 and 5, how can factoring the GCF from a trinomial help us save time when trying to factor the remaining trinomial?

See Video 11.3

11.3 Exercise Set MyLab Math

Objective A *Complete each factored form. See Examples 1 through 5.*

1. $5x^2 + 22x + 8 = (5x + 2)$
2. $2y^2 + 15y + 25 = (2y + 5)$
3. $50x^2 + 15x - 2 = (5x + 2)$
4. $6y^2 + 11y - 10 = (2y + 5)$
5. $20x^2 - 7x - 6 = (5x + 2)$
6. $8y^2 - 2y - 55 = (2y + 5)$

Factor each trinomial completely. If a polynomial can't be factored, write "prime." See Examples 1 through 5.

7. $2x^2 + 13x + 15$
8. $3x^2 + 8x + 4$
9. $8y^2 - 17y + 9$
10. $21x^2 - 41x + 10$

11. $2x^2 - 9x - 5$
12. $36r^2 - 5r - 24$
13. $20r^2 + 27r - 8$
14. $3x^2 + 20x - 63$

15. $10x^2 + 31x + 3$
16. $12x^2 + 17x + 5$
17. $x + 3x^2 - 2$
18. $y + 8y^2 - 9$

19. $6x^2 - 13xy + 5y^2$
20. $8x^2 - 14xy + 3y^2$
21. $15m^2 - 16m - 15$
22. $25n^2 - 5n - 6$

23. $-9x + 20 + x^2$
24. $-7x + 12 + x^2$
25. $2x^2 - 7x - 99$
26. $2x^2 + 7x - 72$

27. $-27t + 7t^2 - 4$
28. $-3t + 4t^2 - 7$
29. $3a^2 + 10ab + 3b^2$
30. $2a^2 + 11ab + 5b^2$

31. $49p^2 - 7p - 2$
32. $3r^2 + 10r - 8$
33. $18x^2 - 9x - 14$
34. $42a^2 - 43a + 6$

35. $2m^2 + 17m + 10$
36. $3n^2 + 20n + 5$
37. $24x^2 + 41x + 12$
38. $24x^2 - 49x + 15$

Objectives A B Mixed Practice *Factor each trinomial completely. If a polynomial can't be factored, write "prime." See Examples 1 through 7.*

39. $12x^3 + 11x^2 + 2x$
40. $8a^3 + 14a^2 + 3a$
41. $21b^2 - 48b - 45$
42. $12x^2 - 14x - 10$

43. $7z + 12z^2 - 12$
44. $16t + 15t^2 - 15$
45. $6x^2y^2 - 2xy^2 - 60y^2$
46. $8x^2y + 34xy - 84y$

47. $4x^2 - 8x - 21$
48. $6x^2 - 11x - 10$
49. $3x^2 - 42x + 63$
50. $5x^2 - 75x + 60$

51. $8x^2 + 6xy - 27y^2$
52. $54a^2 + 39ab - 8b^2$
53. $-x^2 + 2x + 24$
54. $-x^2 + 4x + 21$

55. $4x^3 - 9x^2 - 9x$
56. $6x^3 - 31x^2 + 5x$
57. $24x^2 - 58x + 9$
58. $36x^2 + 55x - 14$

59. $40a^2b + 9ab - 9b$
60. $24y^2x + 7yx - 5x$
61. $30x^3 + 38x^2 + 12x$
62. $6x^3 - 28x^2 + 16x$

63. $6y^3 - 8y^2 - 30y$
64. $12x^3 - 34x^2 + 24x$
65. $10x^4 + 25x^3y - 15x^2y^2$
66. $42x^4 - 99x^3y - 15x^2y^2$

67. $-14x^2 + 39x - 10$
68. $-15x^2 + 26x - 8$
69. $16p^4 - 40p^3 + 25p^2$
70. $9q^4 - 42q^3 + 49q^2$

71. $-2x^2 + 9x + 5$
72. $-3x^2 + 8x + 16$
73. $-4 + 52x - 48x^2$
74. $-5 + 55x - 50x^2$

75. $2t^4 + 3t^2 - 27$
76. $4r^4 - 17r^2 - 15$
77. $5x^2y^2 + 20xy + 1$
78. $3a^2b^2 + 12ab + 1$

79. $6a^5 + 37a^3b^2 + 6ab^4$
80. $5m^5 + 26m^3h^2 + 5mh^4$

Review

Multiply. See Section 10.6.

81. $(x - 4)(x + 4)$
82. $(2x - 9)(2x + 9)$
83. $(x + 2)^2$

84. $(x + 3)^2$
85. $(2x - 1)^2$
86. $(3x - 5)^2$

The following graph shows the average number of texts sent per day by text message users in each age group. See Section (Source: Experion Marketing)

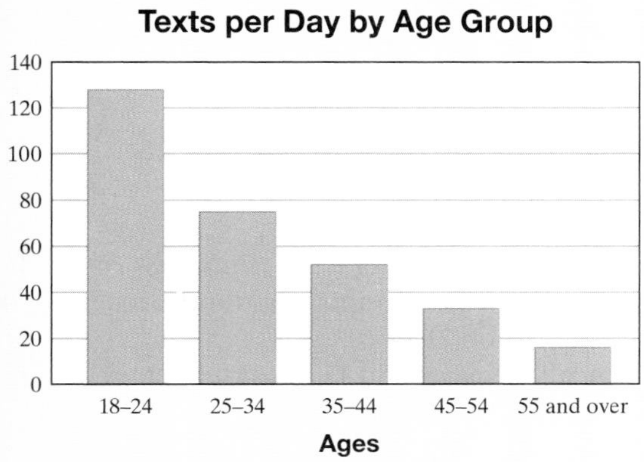

Texts per Day by Age Group

87. What range of ages sends and receives the greatest number of texts per day?

88. What range of ages sends and receives the fewest number of texts per day?

89. Describe any trend you see.

90. What do you think this graph would look like in 10 years? Explain your reasoning.

Concept Extensions

See the Concept Check in this section.

91. Do the terms of $4x^2 + 19x + 12$ have a common factor (other than 1)?

92. Without multiplying, decide which of the following factored forms is not a factored form of $4x^2 + 19x + 12$.
   a. $(2x + 4)(2x + 3)$
   b. $(4x + 4)(x + 3)$
   c. $(4x + 3)(x + 4)$
   d. $(2x + 2)(2x + 6)$

Write the perimeter of each figure as a simplified polynomial. Then factor the polynomial completely.

93.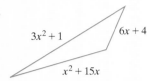

94.

Factor each trinomial completely.

95. $4x^2 + 2x + \dfrac{1}{4}$

96. $27x^2 + 2x - \dfrac{1}{9}$

97. $4x^2(y-1)^2 + 25x(y-1)^2 + 25(y-1)^2$

98. $3x^2(a+3)^3 - 28x(a+3)^3 + 25(a+3)^3$

Find a positive value of b so that each trinomial is factorable.

99. $3x^2 + bx - 5$

100. $2z^2 + bz - 7$

Find a positive value of c so that each trinomial is factorable.

101. $5x^2 + 7x + c$

102. $3x^2 - 8x + c$

103. In your own words, describe the steps you use to factor a trinomial.

104. A student in your class factored $6x^2 + 7x + 1$ a $(3x+1)(2x+1)$. Write down how you would explain the student's error.

11.4 Factoring Trinomials of the Form $ax^2 + bx + c$ by Grouping

Objective

A Use the Grouping Method to Factor Trinomials of the Form $ax^2 + bx + c$.

Objective A Using the Grouping Method

There is an alternative method that can be used to factor trinomials of the fo $ax^2 + bx + c, a \neq 1$. This method is called the **grouping method** because it u factoring by grouping as we learned in Section 11.1.

To see how this method works, recall from Section 11.2 that to factor a trinom such as $x^2 + 11x + 30$, we find two numbers such that

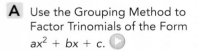

To factor a trinomial such as $2x^2 + 11x + 12$ by grouping, we use an extens of the method in Section 11.1. Here we look for two numbers such that

Section 11.4 | Factoring Trinomials of the Form $ax^2 + bx + c$ by Grouping

this time, we use the two numbers to write

$2x^2 + 11x + 12$ as

$= 2x^2 + \Box x + \Box x + 12$

Then we factor by grouping. Since we want a positive product, 24, and a positive sum, 11, we consider pairs of positive factors of 24 only.

Factors of 24	Sum of Factors
1, 24	25
2, 12	14
3, 8	11 Correct sum

The factors are 3 and 8. Now we use these factors to write the middle term, $11x$, as $3x + 8x$ (or $8x + 3x$). We replace $11x$ with $3x + 8x$ in the original trinomial and then we can factor by grouping.

$$2x^2 + 11x + 12 = 2x^2 + 3x + 8x + 12$$
$$= (2x^2 + 3x) + (8x + 12) \quad \text{Group the terms.}$$
$$= x(2x + 3) + 4(2x + 3) \quad \text{Factor each group.}$$
$$= (2x + 3)(x + 4) \quad \text{Factor out } (2x + 3).$$

In general, we have the following procedure.

To Factor Trinomials by Grouping

Step 1: Factor out a greatest common factor, if there is one other than 1.
Step 2: For the resulting trinomial $ax^2 + bx + c$, find two numbers whose product is $a \cdot c$ and whose sum is b.
Step 3: Write the middle term, bx, using the factors found in Step 2.
Step 4: Factor by grouping.

Example 1 Factor $8x^2 - 14x + 5$ by grouping.

Solution:

Step 1: The terms of this trinomial contain no greatest common factor other than 1.
Step 2: This trinomial is of the form $ax^2 + bx + c$, with $a = 8$, $b = -14$, and $c = 5$. Find two numbers whose product is $a \cdot c$ or $8 \cdot 5 = 40$ and whose sum is b or -14.

The numbers are -4 and -10.

Factors of 40	Sum of Factors
$-40, -1$	-41
$-20, -2$	-22
$-10, -4$	-14 Correct sum

Step 3: Write $-14x$ as $-4x - 10x$ so that
$8x^2 - 14x + 5 = 8x^2 - 4x - 10x + 5$

Step 4: Factor by grouping.
$8x^2 - 4x - 10x + 5 = 4x(2x - 1) - 5(2x - 1)$
$= (2x - 1)(4x - 5)$

Work Practice 1

Example 2 Factor $6x^2 - 2x - 20$ by grouping.

Solution:

Step 1: First factor out the greatest common factor, 2.
$6x^2 - 2x - 20 = 2(3x^2 - x - 10)$

(*Continued on next page*)

Practice 1

Factor each trinomial by grouping.

a. $3x^2 + 14x + 8$
b. $12x^2 + 19x + 5$

Practice 2

Factor each trinomial by grouping.

a. $30x^2 - 26x + 4$
b. $6x^2y - 7xy - 5y$

Answers
1. a. $(x + 4)(3x + 2)$
b. $(4x + 5)(3x + 1)$
2. a. $2(5x - 1)(3x - 2)$
b. $y(2x + 1)(3x - 5)$

Step 2: Next notice that $a = 3, b = -1$, and $c = -10$ in the resulting trinomial. Find two numbers whose product is $a \cdot c$ or $3(-10) = -30$ and whose sum is $b, -1$. The numbers are -6 and 5.

Step 3: $3x^2 - x - 10 = 3x^2 - 6x + 5x - 10$

Step 4: $3x^2 - 6x + 5x - 10 = 3x(x - 2) + 5(x - 2)$
$= (x - 2)(3x + 5)$

A factored form of $6x^2 - 2x - 20 = 2(x - 2)(3x + 5)$.

Don't forget to include the common factor of

■ Work Practice 2

Practice 3

Factor $12y^5 + 10y^4 - 42y^3$ by grouping.

Example 3 Factor $18y^4 + 21y^3 - 60y^2$ by grouping.

Solution:

Step 1: First factor out the greatest common factor, $3y^2$.

$18y^4 + 21y^3 - 60y^2 = 3y^2(6y^2 + 7y - 20)$

Step 2: Notice that $a = 6, b = 7$, and $c = -20$ in the resulting trinomial. Find two numbers whose product is $a \cdot c$ or $6(-20) = -120$ and whose sum is 7. It may help to factor -120 as a product of primes and -1.

$-120 = 2 \cdot 2 \cdot 2 \cdot 3 \cdot 5 \cdot (-1)$

Then choose pairings of factors until you have two pairings whose sum is

$2 \cdot 2 \cdot 2 \cdot 3 \cdot 5 \cdot (-1)$ The numbers are -8 and 15.

Step 3: $6y^2 + 7y - 20 = 6y^2 - 8y + 15y - 20$

Step 4: $6y^2 - 8y + 15y - 20 = 2y(3y - 4) + 5(3y - 4)$
$= (3y - 4)(2y + 5)$

A factored form of $18y^4 + 21y^3 - 60y^2$ is $3y^2(3y - 4)(2y + 5)$.

Don't forget to include the common factor of $3y^2$.

Answer

3. $2y^3(3y + 7)(2y - 3)$

■ Work Practice 3

Vocabulary, Readiness & Video Check

For each trinomial $ax^2 + bx + c$, choose two numbers whose product is $a \cdot c$ and whose sum is b.

1. $x^2 + 6x + 8$
   a. 4, 2   **b.** 7, 1   **c.** 6, 2   **d.** 6, 8

2. $x^2 + 11x + 24$
   a. 6, 4   **b.** 12, 2   **c.** 8, 3   **d.** 5, 6

3. $2x^2 + 13x + 6$
   a. 2, 6   **b.** 12, 1   **c.** 13, 1   **d.** 3, 4

4. $4x^2 + 8x + 3$
   a. 4, 3   **b.** 4, 4   **c.** 12, 1   **d.** 2, 6

Section 11.4 | Factoring Trinomials of the Form $ax^2 + bx + c$ by Grouping

Martin-Gay Interactive Videos Watch the section lecture video and answer the following questions.

See Video 11.4

Objective A 5. In the lecture following Example 1, why does writing a term as the sum or difference of two terms suggest we'd then try to factor by grouping?

11.4 Exercise Set MyLab Math

Objective A *Factor each polynomial by grouping. Notice that Step 3 has already been done in these exercises. See Examples 1 through 3.*

1. $x^2 + 3x + 2x + 6$
2. $x^2 + 5x + 3x + 15$
3. $y^2 + 8y - 2y - 16$
4. $z^2 + 10z - 7z - 70$
5. $8x^2 - 5x - 24x + 15$
6. $4x^2 - 9x - 32x + 72$
7. $5x^4 - 3x^2 + 25x^2 - 15$
8. $2y^4 - 10y^2 + 7y^2 - 35$

Factor each trinomial by grouping. Exercises 9 through 12 are broken into parts to help you get started. See Examples 1 through 3.

9. $6x^2 + 11x + 3$
   a. Find two numbers whose product is $6 \cdot 3 = 18$ and whose sum is 11.
   b. Write $11x$ using the factors from part **a**.
   c. Factor by grouping.

10. $8x^2 + 14x + 3$
    a. Find two numbers whose product is $8 \cdot 3 = 24$ and whose sum is 14.
    b. Write $14x$ using the factors from part **a**.
    c. Factor by grouping.

11. $15x^2 - 23x + 4$
    a. Find two numbers whose product is $15 \cdot 4 = 60$ and whose sum is -23.
    b. Write $-23x$ using the factors from part **a**.
    c. Factor by grouping.

12. $6x^2 - 13x + 5$
    a. Find two numbers whose product is $6 \cdot 5 = 30$ and whose sum is -13.
    b. Write $-13x$ using the factors from part **a**.
    c. Factor by grouping.

13. $21y^2 + 17y + 2$
14. $15x^2 + 11x + 2$
15. $7x^2 - 4x - 11$
16. $8x^2 - x - 9$
17. $10x^2 - 9x + 2$
18. $30x^2 - 23x + 3$
19. $2x^2 - 7x + 5$
20. $2x^2 - 7x + 3$
21. $12x + 4x^2 + 9$
22. $20x + 25x^2 + 4$
23. $4x^2 - 8x - 21$
24. $6x^2 - 11x - 10$
25. $10x^2 - 23x + 12$
26. $21x^2 - 13x + 2$
27. $2x^3 + 13x^2 + 15x$
28. $3x^3 + 8x^2 + 4x$
29. $16y^2 - 34y + 18$
30. $4y^2 - 2y - 12$
31. $-13x + 6 + 6x^2$
32. $-25x + 12 + 12x^2$

33. $54a^2 - 9a - 30$ **34.** $30a^2 + 38a - 20$ **35.** $20a^3 + 37a^2 + 8a$ **36.** $10a^3 + 17a^2 + 3a$

37. $12x^3 - 27x^2 - 27x$ **38.** $30x^3 - 155x^2 + 25x$ **39.** $3x^2y + 4xy^2 + y^3$ **40.** $6r^2t + 7rt^2 + t^3$

41. $20z^2 + 7z + 1$ **42.** $36z^2 + 6z + 1$

43. $24a^2 - 6ab - 30b^2$ **44.** $30a^2 + 5ab - 25b^2$

45. $15p^4 + 31p^3q + 2p^2q^2$ **46.** $20s^4 + 61s^3t + 3s^2t^2$

47. $35 + 12x + x^2$ **48.** $33 + 14x + x^2$

49. $6 - 11x + 5x^2$ **50.** $5 - 12x + 7x^2$

Review

Multiply. See Section 10.6.

51. $(x-2)(x+2)$ **52.** $(y-5)(y+5)$ **53.** $(y+4)(y+4)$ **54.** $(x+7)(x+7)$

55. $(9z+5)(9z-5)$ **56.** $(8y+9)(8y-9)$ **57.** $(4x-3)^2$ **58.** $(2z-1)^2$

Concept Extensions

Write the perimeter of each figure as a simplified polynomial. Then factor the polynomial.

59.
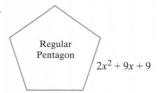
Regular Pentagon, $2x^2 + 9x + 9$

60.
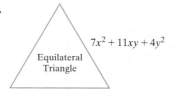
Equilateral Triangle, $7x^2 + 11xy + 4y^2$

Factor each polynomial by grouping.

61. $x^{2n} + 2x^n + 3x^n + 6$
(*Hint:* Don't forget that $x^{2n} = x^n \cdot x^n$.)

62. $x^{2n} + 6x^n + 10x^n + 60$

63. $3x^{2n} + 16x^n - 35$

64. $12x^{2n} - 40x^n + 25$

65. In your own words, explain how to factor a trinomial by grouping.

11.5 Factoring Perfect Square Trinomials and the Difference of Two Squares

Objective A Recognizing Perfect Square Trinomials

A trinomial that is the square of a binomial is called a **perfect square trinomial.** For example,

$$(x + 3)^2 = (x + 3)(x + 3)$$
$$= x^2 + 6x + 9$$

Thus $x^2 + 6x + 9$ is a perfect square trinomial.

In Chapter 10, we discovered special product formulas for squaring binomials.

$$(a + b)^2 = a^2 + 2ab + b^2 \quad \text{and} \quad (a - b)^2 = a^2 - 2ab + b^2$$

Because multiplication and factoring are reverse processes, we can now use these special products to help us factor perfect square trinomials. If we reverse these equations, we have the following.

Factoring Perfect Square Trinomials

$$a^2 + 2ab + b^2 = (a + b)^2$$
$$a^2 - 2ab + b^2 = (a - b)^2$$

Helpful Hint

Notice that for both given forms of a perfect square trinomial, the last term is positive. This is because the last term is a square.

To use these equations to help us factor, we must first be able to recognize a perfect square trinomial. A trinomial is a perfect square when

1. two terms, a^2 and b^2, are squares and
2. another term is $2 \cdot a \cdot b$ or $-2 \cdot a \cdot b$. That is, this term is twice the product of a and b, or its opposite.

Example 1 Decide whether $x^2 + 8x + 16$ is a perfect square trinomial.

Solution:
1. Two terms, x^2 and 16, are squares ($16 = 4^2$).
2. Twice the product of x and 4 is the other term of the trinomial.

$$2 \cdot x \cdot 4 = 8x$$

Thus, $x^2 + 8x + 16$ is a perfect square trinomial.

Work Practice 1

Practice 1

Decide whether each trinomial is a perfect square trinomial.
a. $x^2 + 12x + 36$
b. $x^2 + 20x + 100$

Example 2 Decide whether $4x^2 + 10x + 9$ is a perfect square trinomial.

Solution:
1. Two terms, $4x^2$ and 9, are squares.

$$4x^2 = (2x)^2 \quad \text{and} \quad 9 = 3^2$$

2. Twice the product of $2x$ and 3 is *not* the other term of the trinomial.

$$2 \cdot 2x \cdot 3 = 12x, \text{ not } 10x$$

The trinomial is *not* a perfect square trinomial.

Work Practice 2

Practice 2

Decide whether each trinomial is a perfect square trinomial.
a. $9x^2 + 20x + 25$
b. $4x^2 + 8x + 11$

Answers
1. a. yes b. yes
2. a. no b. no

Chapter 11 | Factoring Polynomials

Practice 3

Decide whether each trinomial is a perfect square trinomial.
a. $25x^2 - 10x + 1$
b. $9x^2 - 42x + 49$

Example 3 Decide whether $9x^2 - 12xy + 4y^2$ is a perfect square trinomial.

Solution:
1. Two terms, $9x^2$ and $4y^2$, are squares.

$$9x^2 = (3x)^2 \quad \text{and} \quad 4y^2 = (2y)^2$$

2. Twice the product of $3x$ and $2y$ is the opposite of the other term of the trinomial.

$$2 \cdot 3x \cdot 2y = 12xy, \text{ the opposite of } -12xy$$

Thus, $9x^2 - 12xy + 4y^2$ is a perfect square trinomial.

■ Work Practice 3

Objective B Factoring Perfect Square Trinomials

Now that we can recognize perfect square trinomials, we are ready to factor them.

Practice 4

Factor: $x^2 + 16x + 64$

Example 4 Factor: $x^2 + 12x + 36$

Solution:

$$x^2 + 12x + 36 = x^2 + 2 \cdot x \cdot 6 + 6^2 \quad 36 = 6^2 \text{ and } 12x = 2 \cdot x \cdot 6$$
$$a^2 + 2 \cdot a \cdot b + b^2$$
$$= (x + 6)^2$$
$$(a + b)^2$$

■ Work Practice 4

Practice 5

Factor: $9r^2 + 24rs + 16s^2$

Example 5 Factor: $25x^2 + 20xy + 4y^2$

Solution:

$$25x^2 + 20xy + 4y^2 = (5x)^2 + 2 \cdot 5x \cdot 2y + (2y)^2$$
$$= (5x + 2y)^2$$

■ Work Practice 5

Practice 6

Factor: $9n^4 - 6n^2 + 1$

Example 6 Factor: $4m^4 - 4m^2 + 1$

Solution:

$$4m^4 - 4m^2 + 1 = (2m^2)^2 - 2 \cdot 2m^2 \cdot 1 + 1^2$$
$$a^2 \quad - 2 \cdot a \cdot b + b^2$$
$$= (2m^2 - 1)^2$$
$$(a - b)^2$$

■ Work Practice 6

Answers
3. a. yes b. yes   4. $(x + 8)^2$
5. $(3r + 4s)^2$ 6. $(3n^2 - 1)^2$

Section 11.5 | Factoring Perfect Square Trinomials and the Difference of Two Squares

Example 7 Factor: $25x^2 + 50x + 9$

Solution: Notice that this trinomial is not a perfect square trinomial.

$25x^2 = (5x)^2, 9 = 3^2$

but

$2 \cdot 5x \cdot 3 = 30x$

and $30x$ is not the middle term, $50x$.

Although $25x^2 + 50x + 9$ is not a perfect square trinomial, it is factorable. Using techniques we learned in Section 11.3 or 11.4, we find that

$25x^2 + 50x + 9 = (5x + 9)(5x + 1)$

■ Work Practice 7

Example 8 Factor: $162x^3 - 144x^2 + 32x$

Solution: Don't forget to first look for a common factor. There is a greatest common factor of $2x$ in this trinomial.

$162x^3 - 144x^2 + 32x = 2x(81x^2 - 72x + 16)$
$= 2x[(9x)^2 - 2 \cdot 9x \cdot 4 + 4^2]$
$= 2x(9x - 4)^2$

■ Work Practice 8

Objective C Factoring the Difference of Two Squares

In Chapter 10, we discovered another special product, the product of the sum and difference of two terms a and b:

$(a + b)(a - b) = a^2 - b^2$

Reversing this equation gives us another factoring pattern, which we use to factor the difference of two squares.

Factoring the Difference of Two Squares
$a^2 - b^2 = (a + b)(a - b)$

To use this equation to help us factor, we must first be able to recognize the difference of two squares. A binomial is a difference of two squares if

1. both terms are squares and
2. the signs of the terms are different.

Let's practice using this pattern.

Examples Factor each binomial.

9. $z^2 - 4 = z^2 - 2^2 = (z + 2)(z - 2)$
$\ \uparrow\uparrow\uparrow\uparrow\uparrow\uparrow$
$\ a^2 - b^2 = (a + b)(a - b)$

10. $y^2 - 25 = y^2 - 5^2 = (y + 5)(y - 5)$

11. $y^2 - \dfrac{4}{9} = y^2 - \left(\dfrac{2}{3}\right)^2 = \left(y + \dfrac{2}{3}\right)\left(y - \dfrac{2}{3}\right)$

12. $x^2 + 4$

Practice 7
Factor: $9x^2 + 15x + 4$

Helpful Hint
A perfect square trinomial can also be factored by the methods found in Sections 11.2 through 11.4.

Practice 8
Factor:
a. $8n^2 + 40n + 50$
b. $12x^3 - 84x^2 + 147x$

Practice 9–12
Factor each binomial.
9. $x^2 - 9$ **10.** $a^2 - 16$
11. $c^2 - \dfrac{9}{25}$ **12.** $s^2 + 9$

Answers
7. $(3x + 1)(3x + 4)$
8. a. $2(2n + 5)^2$ **b.** $3x(2x - 7)^2$
9. $(x - 3)(x + 3)$
10. $(a - 4)(a + 4)$
11. $\left(c - \dfrac{3}{5}\right)\left(c + \dfrac{3}{5}\right)$
12. prime polynomial

Helpful Hint After the greatest common factor has been removed, the *sum* of two squares cannot be factored further using real numbers.

Note that the binomial $x^2 + 4$ is the *sum* of two squares since we can write $x^2 + 4$ as $x^2 + 2^2$. We might try to factor using $(x + 2)(x + 2)$ or $(x - 2)(x - 2)$. But when we multiply to check, we find that neither factoring is correct.

$$(x + 2)(x + 2) = x^2 + 4x + 4$$
$$(x - 2)(x - 2) = x^2 - 4x + 4$$

In both cases, the product is a trinomial, not the required binomial. In fac $x^2 + 4$ is a prime polynomial.

■ Work Practice 9–12

Practice 13–15

Factor each difference of two squares.
13. $9s^2 - 1$
14. $16x^2 - 49y^2$
15. $p^4 - 81$

Examples Factor each difference of two squares.

13. $4x^2 - 1 = (2x)^2 - 1^2 = (2x + 1)(2x - 1)$
14. $25a^2 - 9b^2 = (5a)^2 - (3b)^2 = (5a + 3b)(5a - 3b)$
15. $y^4 - 16 = (y^2)^2 - 4^2$
    $= (y^2 + 4)(y^2 - 4)$ Factor the difference of two squares.
    $= (y^2 + 4)(y + 2)(y - 2)$ Factor the difference of two squares.

■ Work Practice 13–15

Helpful Hint
1. Don't forget to first see whether there's a greatest common factor (other than 1) that can be factored out.
2. Factor completely. In other words, check to see whether any factors can be factored further (as in Example 15).

Practice 16–18

Factor each binomial.
16. $9x^3 - 25x$
17. $48x^4 - 3$
18. $-9x^2 + 100$

Examples Factor each binomial.

16. $4x^3 - 49x = x(4x^2 - 49)$ Factor out the common factor, x.
    $= x[(2x)^2 - 7^2]$
    $= x(2x + 7)(2x - 7)$ Factor the difference of two squares.
17. $162x^4 - 2 = 2(81x^4 - 1)$ Factor out the common factor, 2.
    $= 2(9x^2 + 1)(9x^2 - 1)$ Factor the difference of two squares.
    $= 2(9x^2 + 1)(3x + 1)(3x - 1)$ Factor the difference of two squares.
18. $-49x^2 + 16 = -1(49x^2 - 16)$ Factor out -1.
    $= -1(7x + 4)(7x - 4)$ Factor the difference of two squares.

■ Work Practice 16–18

Practice 19

Factor: $121 - m^2$

Example 19 Factor: $36 - x^2$

Solution: This is the difference of two squares. Factor as is or, if you like, firs write the binomial with the variable term first.

Factor as is: $36 - x^2 = 6^2 - x^2 = (6 + x)(6 - x)$
Rewrite binomial: $36 - x^2 = -x^2 + 36 = -1(x^2 - 36)$
$= -1(x + 6)(x - 6)$

Both factorizations are correct and are equal. To see this, factor -1 from $(6 - x$ in the first factorization.

■ Work Practice 19

Helpful Hint
When rearranging terms, keep in mind that the sign of a term is in front of the term

Answers
13. $(3s - 1)(3s + 1)$
14. $(4x - 7y)(4x + 7y)$
15. $(p^2 + 9)(p + 3)(p - 3)$
16. $x(3x - 5)(3x + 5)$
17. $3(4x^2 + 1)(2x + 1)(2x - 1)$
18. $-1(3x - 10)(3x + 10)$
19. $(11 + m)(11 - m)$ or $-1(m + 11)(m - 11)$

Section 11.5 | Factoring Perfect Square Trinomials and the Difference of Two Squares

Calculator Explorations Graphing

A graphing calculator is a convenient tool for evaluating an expression at a given replacement value. For example, let's evaluate $x^2 - 6x$ when $x = 2$. To do so, store the value 2 in the variable x and then enter and evaluate the algebraic expression.

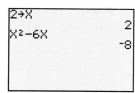

The value of $x^2 - 6x$ when $x = 2$ is -8. You may want to use this method for evaluating expressions as you explore the following.

We can use a graphing calculator to explore factoring patterns numerically. Use your calculator to evaluate $x^2 - 2x + 1$, $x^2 - 2x - 1$, and $(x - 1)^2$ for each value of x given in the table. What do you observe?

	$x^2 - 2x + 1$	$x^2 - 2x - 1$	$(x - 1)^2$
$x = 5$			
$x = -3$			
$x = 2.7$			
$x = -12.1$			
$x = 0$			

Notice in each case that $x^2 - 2x - 1 \neq (x - 1)^2$. Because for each x in the table the value of $x^2 - 2x + 1$ and the value of $(x - 1)^2$ are the same, we might guess that $x^2 - 2x + 1 = (x - 1)^2$. We can verify our guess algebraically with multiplication:
$$(x - 1)(x - 1) = x^2 - x - x + 1 = x^2 - 2x + 1$$

Vocabulary, Readiness & Video Check

Use the choices below to fill in each blank. Some choices may be used more than once and some choices may not be used at all.

| perfect square trinomial | true | $(5y)^2$ | $(x + 5y)^2$ |
| difference of two squares | false | $(x - 5y)^2$ | $5y^2$ |

1. A _____ is a trinomial that is the square of a binomial.
2. The term $25y^2$ written as a square is _____.
3. The expression $x^2 + 10xy + 25y^2$ is called a _____.
4. The expression $x^2 - 49$ is called a _____.
5. The factorization $(x + 5y)(x + 5y)$ may also be written as _____.
6. True or false? The factorization $(x - 5y)(x + 5y)$ may also be written as $(x - 5y)^2$. _____
7. The trinomial $x^2 - 6x - 9$ is a perfect square trinomial. _____
8. The binomial $y^2 + 9$ factors as $(y + 3)^2$. _____

Write each number or term as a square. For example, 16 written as a square is 4^2.

9. 64
10. 9
11. $121a^2$
12. $81b^2$
13. $36p^4$
14. $4q^4$

852 Chapter 11 | Factoring Polynomials

Martin-Gay Interactive Videos Watch the section lecture video and answer the following questions.

See Video 11.5

Objective A 15. The polynomial in Example 2 is shown to *not* be a perfect square trinomial. Does this necessarily mean it can't be factored?

Objective B 16. Describe in words the special patterns that the trinomials in Examples 3 and 4 have that identify them as perfect square trinomials.

Objective C 17. In Examples 5 and 6, what are two reasons the original binomial is rewritten so that each term is a square?

18. For Example 7, what is a prime polynomial?

11.5 Exercise Set MyLab Math

Objective A *Determine whether each trinomial is a perfect square trinomial. See Examples 1 through 3.*

1. $x^2 + 16x + 64$
2. $x^2 + 22x + 121$
3. $y^2 + 5y + 25$
4. $y^2 + 4y + 16$

5. $m^2 - 2m + 1$
6. $p^2 - 4p + 4$
7. $a^2 - 16a + 49$
8. $n^2 - 20n + 144$

9. $4x^2 + 12xy + 8y^2$
10. $25x^2 + 20xy + 2y^2$
11. $25a^2 - 40ab + 16b^2$
12. $36a^2 - 12ab + b^2$

Objective B *Factor each trinomial completely. See Examples 4 through 8.*

13. $x^2 + 22x + 121$
14. $x^2 + 18x + 81$
15. $x^2 - 16x + 64$
16. $x^2 - 12x + 36$

17. $16a^2 - 24a + 9$
18. $25x^2 - 20x + 4$
19. $x^4 + 4x^2 + 4$
20. $m^4 + 10m^2 + 25$

21. $2n^2 - 28n + 98$
22. $3y^2 - 6y + 3$
23. $16y^2 + 40y + 25$
24. $9y^2 + 48y + 64$

25. $x^2y^2 - 10xy + 25$
26. $4x^2y^2 - 28xy + 49$
27. $m^3 + 18m^2 + 81m$
28. $y^3 + 12y^2 + 36y$

29. $1 + 6x^2 + x^4$
30. $1 + 16x^2 + x^4$
31. $9x^2 - 24xy + 16y^2$
32. $25x^2 - 60xy + 36y^2$

Section 11.5 | Factoring Perfect Square Trinomials and the Difference of Two Squares

Objective C *Factor each binomial completely. See Examples 9 through 19.*

33. $x^2 - 4$ **34.** $x^2 - 36$ **35.** $81 - p^2$ **36.** $100 - t^2$

37. $-4r^2 + 1$ **38.** $-9t^2 + 1$ **39.** $9x^2 - 16$ **40.** $36y^2 - 25$

41. $16r^2 + 1$ **42.** $49y^2 + 1$ **43.** $-36 + x^2$ **44.** $-1 + y^2$

45. $m^4 - 1$ **46.** $n^4 - 16$ **47.** $x^2 - 169y^2$ **48.** $x^2 - 225y^2$

49. $18r^2 - 8$ **50.** $32t^2 - 50$ **51.** $9xy^2 - 4x$ **52.** $36x^2y - 25y$

53. $16x^4 - 64x^2$ **54.** $25y^4 - 100y^2$ ▶ **55.** $xy^3 - 9xyz^2$ **56.** $x^3y - 4xy^3$

57. $36x^2 - 64y^2$ **58.** $225a^2 - 81b^2$ **59.** $144 - 81x^2$ **60.** $12x^2 - 27$

61. $25y^2 - 9$ **62.** $49a^2 - 16$ ▶ **63.** $121m^2 - 100n^2$ **64.** $169a^2 - 49b^2$

65. $x^2y^2 - 1$ **66.** $a^2b^2 - 16$ **67.** $x^2 - \dfrac{1}{4}$

68. $y^2 - \dfrac{1}{16}$ ▶ **69.** $49 - \dfrac{9}{25}m^2$ **70.** $100 - \dfrac{4}{81}n^2$

Objectives B C Mixed Practice *Factor each binomial or trinomial completely. See Examples 4 through 19.*

71. $81a^2 - 25b^2$ **72.** $49y^2 - 100z^2$ **73.** $x^2 + 14xy + 49y^2$ **74.** $x^2 + 10xy + 25y^2$

75. $32n^4 - 112n^2 + 98$ **76.** $162a^4 - 72a^2 + 8$ **77.** $x^6 - 81x^2$

78. $n^9 - n^5$ **79.** $64p^3q - 81pq^3$ **80.** $100x^3y - 49xy^3$

Review

Solve each equation. See Section 9.3.

81. $x - 6 = 0$

82. $y + 5 = 0$

83. $2m + 4 = 0$

84. $3x - 9 = 0$

85. $5z - 1 = 0$

86. $4a + 2 = 0$

Concept Extensions

Factor each expression completely.

87. $x^2 - \frac{2}{3}x + \frac{1}{9}$

88. $x^2 - \frac{1}{25}$

89. $(x + 2)^2 - y^2$

90. $(y - 6)^2 - z^2$

91. $a^2(b - 4) - 16(b - 4)$

92. $m^2(n + 8) - 9(n + 8)$

93. $(x^2 + 6x + 9) - 4y^2$ (*Hint:* Factor the trinomial in parentheses first.)

94. $(x^2 + 2x + 1) - 36y^2$ (See the hint for Exercise **93**.)

95. $x^{2n} - 100$

96. $x^{2n} - 81$

97. Fill in the blank so that $x^2 + \underline{}x + 16$ is a perfect square trinomial.

98. Fill in the blank so that $9x^2 + \underline{}x + 25$ is a perfect square trinomial.

99. Describe a perfect square trinomial.

100. Write a perfect square trinomial that factors as $(x + 3y)^2$.

101. What binomial multiplied by $(x - 6)$ gives the difference of two squares?

102. What binomial multiplied by $(5 + y)$ gives the difference of two squares?

The area of the largest square in the figure is $(a + b)^2$. Use this figure to answer Exercises 103 and 104.

103. Write the area of the largest square as the sum of the areas of the smaller squares and rectangles.

104. What factoring formula from this section is visually represented by this square?

105. The Toroweap Overlook, on the North Rim of the Grand Canyon, lies 3000 vertical feet above the Colorado River. The view is spectacular, and the sheer drop is dramatic. A film crew creating a documentary about the Grand Canyon has suspended a camera platform 296 feet below the Overlook. A camera filter comes loose and falls to the river below. The height of the filter above the river, after t seconds, is given by the expression $2704 - 16t^2$.
  a. Find the height of the filter above the river after 3 seconds.
  b. Find the height of the filter above the river after 7 seconds.
  c. To the nearest whole second, estimate when the filter lands in the river.
  d. Factor $2704 - 16t^2$.

106. An object is dropped from the top of Pittsburgh's U.S. Steel Tower, which is 841 feet tall. (*Source: World Almanac* research) The height of the object after t seconds is given by the expression $841 - 16t^2$.
  a. Find the height of the object after 2 seconds.
  b. Find the height of the object after 5 seconds.
  c. To the nearest whole second, estimate when the object hits the ground.
  d. Factor $841 - 16t^2$.

841 feet

107. The world's tallest building is the Burj Khalifa in Dubai, United Arab Emirates, at a height of 2723 feet. (*Source:* Council on Tall Buildings and Urban Habitat) Suppose a worker is suspended 419 feet below the tip of the building, at a height of 2304 feet above the ground. If the worker accidentally drops a bolt, the height of the bolt after t seconds is given by the expression $2304 - 16t^2$.
  a. Find the height of the bolt after 3 seconds.
  b. Find the height of the bolt after 7 seconds.
  c. To the nearest whole second, estimate when the bolt hits the ground.
  d. Factor $2304 - 16t^2$.

108. A performer with the Moscow Circus is planning a stunt involving a free fall from the top of the MV Lomonosov State University building, which is 784 feet tall. (*Source:* Council on Tall Buildings and Urban Habitat) Neglecting air resistance, the performer's height above gigantic cushions positioned at ground level after t seconds is given by the expression $784 - 16t^2$.
  a. Find the performer's height after 2 seconds.
  b. Find the performer's height after 5 seconds.
  c. To the nearest whole second, estimate when the performer reaches the cushions positioned at ground level.
  d. Factor $784 - 16t^2$.

Integrated Review Sections 11.1–11.5

Choosing a Factoring Strategy

The following steps may be helpful when factoring polynomials.

To Factor a Polynomial

Step 1: Are there any common factors? If so, factor out the GCF.

Step 2: How many terms are in the polynomial?

  a. Two terms: Is it the difference of two squares? $a^2 - b^2 = (a-b)(a+b)$
  b. Three terms: Try one of the following.
      i. Perfect square trinomial: $a^2 + 2ab + b^2 = (a+b)^2$
         $a^2 - 2ab + b^2 = (a-b)^2$
      ii. If not a perfect square trinomial, factor using the methods presented in Sections 11.2 through 11.4.
  c. Four terms: Try factoring by grouping.

Step 3: See if any factors in the factored polynomial can be factored further.

Step 4: Check by multiplying.

Factor each polynomial completely.

1. $x^2 + x - 12$
2. $x^2 - 10x + 16$
3. $x^2 + 2x + 1$
4. $x^2 - 6x + 9$
5. $x^2 - x - 6$
6. $x^2 + x - 2$
7. $x^2 + x - 6$
8. $x^2 + 7x + 12$
9. $x^2 - 7x + 10$
10. $x^2 - x - 30$
11. $2x^2 - 98$
12. $3x^2 - 75$
13. $x^2 + 3x + 5x + 15$
14. $3y - 21 + xy - 7x$
15. $x^2 + 6x - 16$
16. $x^2 - 3x - 28$
17. $4x^3 + 20x^2 - 56x$
18. $6x^3 - 6x^2 - 120x$
19. $12x^2 + 34x + 24$
20. $24a^2 + 18ab - 15b^2$
21. $4a^2 - b^2$
22. $x^2 - 25y^2$
23. $28 - 13x - 6x^2$
24. $20 - 3x - 2x^2$
25. $4 - 2x + x^2$
26. $a + a^2 - 3$
27. $6y^2 + y - 15$
28. $4x^2 - x - 5$
29. $18x^3 - 63x^2 + 9x$
30. $12a^3 - 24a^2 + 4a$
31. $16a^2 - 56a + 49$
32. $25p^2 - 70p + 49$
33. $14 + 5x - x^2$
34. $3 - 2x - x^2$
35. $3x^4y + 6x^3y - 72x^2y$
36. $2x^3y + 8x^2y^2 - 10xy^3$

37. $12x^3y + 243xy$

38. $6x^3y^2 + 8xy^2$

39. $2xy - 72x^3y$

40. $2x^3 - 18x$

41. $x^3 + 6x^2 - 4x - 24$

42. $x^3 - 2x^2 - 36x + 72$

43. $6a^3 + 10a^2$

44. $4n^2 - 6n$

45. $3x^3 - x^2 + 12x - 4$

46. $x^3 - 2x^2 + 3x - 6$

47. $6x^2 + 18xy + 12y^2$

48. $12x^2 + 46xy - 8y^2$

49. $5(x + y) + x(x + y)$

50. $7(x - y) + y(x - y)$

51. $14t^2 - 9t + 1$

52. $3t^2 - 5t + 1$

53. $-3x^2 - 2x + 5$

54. $-7x^2 - 19x + 6$

55. $1 - 8a - 20a^2$

56. $1 - 7a - 60a^2$

57. $x^4 - 10x^2 + 9$

58. $x^4 - 13x^2 + 36$

59. $x^2 - 23x + 120$

60. $y^2 + 22y + 96$

61. $25p^2 - 70pq + 49q^2$

62. $16a^2 - 56ab + 49b^2$

63. $x^2 - 14x - 48$

64. $7x^2 + 24xy + 9y^2$

65. $-x^2 - x + 30$

66. $-x^2 + 6x - 8$

67. $3rs - s + 12r - 4$

68. $x^3 - 2x^2 + x - 2$

69. $4x^2 - 8xy - 3x + 6y$

70. $4x^2 - 2xy - 7yz + 14xz$

71. $x^2 + 9xy - 36y^2$

72. $3x^2 + 10xy - 8y^2$

73. $x^4 - 14x^2 - 32$

74. $x^4 - 22x^2 - 75$

75. Explain why it makes good sense to factor out the GCF first, before using other methods of factoring.

76. The sum of two squares usually does not factor. Is the sum of two squares $9x^2 + 81y^2$ factorable?

11.6 Solving Quadratic Equations by Factoring

Objectives

A Solve Quadratic Equations by Factoring.

B Solve Equations with Degree Greater Than Two by Factoring.

In this section, we introduce a new type of equation—the **quadratic equation.**

Quadratic Equation

A quadratic equation is one that can be written in the form

$$ax^2 + bx + c = 0$$

where a, b, and c are real numbers and $a \neq 0$.

Some examples of quadratic equations are shown below.

$$x^2 - 9x - 22 = 0 \qquad 4x^2 - 28 = -49 \qquad x(2x - 7) = 4$$

The form $ax^2 + bx + c = 0$ is called the **standard form** of a quadratic equation. The quadratic equation $x^2 - 9x - 22 = 0$ is the only equation above that is in standard form.

Quadratic equations model many real-life situations. For example, let's suppose we want to know how long before a person diving from a 144-foot cliff reaches the ocean. The answer to this question is found by solving the quadratic equation $-16t^2 + 144 = 0$. (See Example 1 in Section 11.7.)

144 feet

Objective A Solving Quadratic Equations by Factoring

Some quadratic equations can be solved by making use of factoring and the **zero-factor property.**

Zero-Factor Property

If a and b are real numbers and if $ab = 0$, then $a = 0$ or $b = 0$.

In other words, if the product of two numbers is 0, then at least one of the numbers must be 0.

Practice 1

Solve: $(x - 7)(x + 2) = 0$

Example 1 Solve: $(x - 3)(x + 1) = 0$

Solution: If this equation is to be a true statement, then either the factor $x - 3$ must be 0 or the factor $x + 1$ must be 0. In other words, either

$$x - 3 = 0 \qquad \text{or} \qquad x + 1 = 0$$

If we solve these two linear equations, we have

$$x = 3 \qquad \text{or} \qquad x = -1$$

Answer
1. 7 and -2

Thus, 3 and −1 are both solutions of the equation $(x - 3)(x + 1) = 0$. To check, we replace x with 3 in the original equation. Then we replace x with −1 in the original equation.

Check:

$(x - 3)(x + 1) = 0$ $\qquad$ $(x - 3)(x + 1) = 0$
$(3 - 3)(3 + 1) \stackrel{?}{=} 0$ Replace x with 3. $\qquad$ $(-1 - 3)(-1 + 1) \stackrel{?}{=} 0$ Replace x with −1.
$\quad\quad\quad 0(4) = 0$ True $\qquad\qquad\qquad\quad (-4)(0) = 0$ True

The solutions are 3 and −1.

■ Work Practice 1

Helpful Hint

The zero-factor property says that *if a product is 0, then a factor is 0.*

If $a \cdot b = 0$, then $a = 0$ or $b = 0$.
If $x(x + 5) = 0$, then $x = 0$ or $x + 5 = 0$.
If $(x + 7)(2x - 3) = 0$, then $x + 7 = 0$ or $2x - 3 = 0$.

Use this property only when the product is 0. For example, if $a \cdot b = 8$, we do not know the value of a or b. The values may be $a = 2$, $b = 4$ or $a = 8$, $b = 1$, or any other two numbers whose product is 8.

Example 2 Solve: $(x - 5)(2x + 7) = 0$

Solution: The product is 0. By the zero-factor property, this is true only when a factor is 0. To solve, we set each factor equal to 0 and solve the resulting linear equations.

$(x - 5)(2x + 7) = 0$
$x - 5 = 0 \quad \text{or} \quad 2x + 7 = 0$
$\quad x = 5 \qquad\qquad\quad 2x = -7$
$\qquad\qquad\qquad\qquad\quad x = -\dfrac{7}{2}$

Check: Let $x = 5$.

$(x - 5)(2x + 7) = 0$
$(5 - 5)(2 \cdot 5 + 7) \stackrel{?}{=} 0$ Replace x with 5.
$\quad\quad\quad 0 \cdot 17 \stackrel{?}{=} 0$
$\quad\quad\quad\quad 0 = 0$ True

Let $x = -\dfrac{7}{2}$.

$(x - 5)(2x + 7) = 0$
$\left(-\dfrac{7}{2} - 5\right)\left(2\left(-\dfrac{7}{2}\right) + 7\right) \stackrel{?}{=} 0$ Replace x with $-\dfrac{7}{2}$.
$\quad\quad\left(-\dfrac{17}{2}\right)(-7 + 7) \stackrel{?}{=} 0$
$\quad\quad\quad\quad\left(-\dfrac{17}{2}\right) \cdot 0 \stackrel{?}{=} 0$
$\quad\quad\quad\quad\quad\quad 0 = 0$ True

The solutions are 5 and $-\dfrac{7}{2}$.

■ Work Practice 2

Practice 2

Solve: $(x - 10)(3x + 1) = 0$

Answer

2. 10 and $-\dfrac{1}{3}$

Practice 3

Solve each equation.
a. $y(y + 3) = 0$
b. $x(4x - 3) = 0$

Example 3 Solve: $x(5x - 2) = 0$

Solution:

$$x(5x - 2) = 0$$
$$x = 0 \quad \text{or} \quad 5x - 2 = 0 \quad \text{Use the zero-factor property.}$$
$$5x = 2$$
$$x = \frac{2}{5}$$

Check these solutions in the original equation. The solutions are 0 and $\frac{2}{5}$.

■ Work Practice 3

Practice 4

Solve: $x^2 - 3x - 18 = 0$

Example 4 Solve: $x^2 - 9x - 22 = 0$

Solution: One side of the equation is 0. However, to use the zero-factor property, one side of the equation must be 0 *and* the other side must be written as product (must be factored). Thus, we must first factor this polynomial.

$$x^2 - 9x - 22 = 0$$
$$(x - 11)(x + 2) = 0 \quad \text{Factor.}$$

Now we can apply the zero-factor property.

$$x - 11 = 0 \quad \text{or} \quad x + 2 = 0$$
$$x = 11 \qquad\qquad x = -2$$

Check: Let $x = 11$. | Let $x = -2$.

$$x^2 - 9x - 22 = 0 \qquad\qquad x^2 - 9x - 22 = 0$$
$$11^2 - 9 \cdot 11 - 22 \stackrel{?}{=} 0 \qquad (-2)^2 - 9(-2) - 22 \stackrel{?}{=} 0$$
$$121 - 99 - 22 \stackrel{?}{=} 0 \qquad\qquad 4 + 18 - 22 \stackrel{?}{=} 0$$
$$22 - 22 \stackrel{?}{=} 0 \qquad\qquad\qquad 22 - 22 \stackrel{?}{=} 0$$
$$0 = 0 \quad \text{True} \qquad\qquad\qquad 0 = 0 \quad \text{True}$$

The solutions are 11 and -2.

■ Work Practice 4

Practice 5

Solve: $9x^2 - 24x = -16$

Example 5 Solve: $4x^2 - 28x = -49$

Solution: First we rewrite the equation in standard form so that one side is 0. Then we factor the polynomial.

$$4x^2 - 28x = -49$$
$$4x^2 - 28x + 49 = 0 \quad \text{Write in standard form by adding 49 to both sides.}$$
$$(2x - 7)(2x - 7) = 0 \quad \text{Factor.}$$

Next we use the zero-factor property and set each factor equal to 0. Since the factors are the same, the related equations will give the same solution.

$$2x - 7 = 0 \quad \text{or} \quad 2x - 7 = 0 \quad \text{Set each factor equal to 0.}$$
$$2x = 7 \qquad\qquad 2x = 7 \quad \text{Solve.}$$
$$x = \frac{7}{2} \qquad\qquad x = \frac{7}{2}$$

Check this solution in the original equation. The solution is $\frac{7}{2}$.

■ Work Practice 5

Answers
3. a. 0 and -3 **b.** 0 and $\frac{3}{4}$
4. 6 and -3 **5.** $\frac{4}{3}$

The following steps may be used to solve a quadratic equation by factoring.

To Solve Quadratic Equations by Factoring

Step 1: Write the equation in standard form so that one side of the equation is 0.
Step 2: Factor the quadratic equation completely.
Step 3: Set each factor containing a variable equal to 0.
Step 4: Solve the resulting equations.
Step 5: Check each solution in the original equation.

Since it is not always possible to factor a quadratic polynomial, not all quadratic equations can be solved by factoring. Other methods of solving quadratic equations are presented in Chapter 16.

Example 6 Solve: $x(2x - 7) = 4$

Solution: First we write the equation in standard form; then we factor.

$$x(2x - 7) = 4$$
$$2x^2 - 7x = 4 \quad \text{Multiply.}$$
$$2x^2 - 7x - 4 = 0 \quad \text{Write in standard form.}$$
$$(2x + 1)(x - 4) = 0 \quad \text{Factor.}$$
$$2x + 1 = 0 \quad \text{or} \quad x - 4 = 0 \quad \text{Set each factor equal to zero.}$$
$$2x = -1 \quad\quad\quad x = 4 \quad \text{Solve.}$$
$$x = -\frac{1}{2}$$

Check the solutions in the original equation. The solutions are $-\frac{1}{2}$ and 4.

■ Work Practice 6

Practice 6

Solve each equation.
a. $x(x - 4) = 5$
b. $x(3x + 7) = 6$

Helpful Hint

To solve the equation $x(2x - 7) = 4$, do **not** set each factor equal to 4. Remember that to apply the zero-factor property, one side of the equation must be 0 and the other side of the equation must be in factored form.

✓**Concept Check** Explain the error and solve the equation correctly.

$$(x - 3)(x + 1) = 5$$
$$x - 3 = 0 \quad \text{or} \quad x + 1 = 0$$
$$x = 3 \quad \text{or} \quad x = -1$$

Answers
6. a. 5 and -1 **b.** $\frac{2}{3}$ and -3

✓**Concept Check Answer**
To use the zero-factor property, one side of the equation must be 0, not 5. Correctly, $(x - 3)(x + 1) = 5$, $x^2 - 2x - 3 = 5$, $x^2 - 2x - 8 = 0$, $(x - 4)(x + 2) = 0$, $x - 4 = 0$ or $x + 2 = 0$, $x = 4$ or $x = -2$.

Objective B Solving Equations with Degree Greater Than Two by Factoring

Some equations with degree greater than 2 can be solved by factoring and then using the zero-factor property.

Chapter 11 | Factoring Polynomials

Practice 7

Solve: $2x^3 - 18x = 0$

Example 7 Solve: $3x^3 - 12x = 0$

Solution: To factor the left side of the equation, we begin by factoring out the greatest common factor, $3x$.

$$3x^3 - 12x = 0$$
$$3x(x^2 - 4) = 0 \quad \text{Factor out the GCF, } 3x.$$
$$3x(x + 2)(x - 2) = 0 \quad \text{Factor } x^2 - 4, \text{ a difference of two squares}$$
$$3x = 0 \quad \text{or} \quad x + 2 = 0 \quad \text{or} \quad x - 2 = 0 \quad \text{Set each factor equal to 0.}$$
$$x = 0 \qquad\qquad x = -2 \qquad\qquad x = 2 \quad \text{Solve.}$$

Thus, the equation $3x^3 - 12x = 0$ has three solutions: $0, -2,$ and 2.

Check: Replace x with each solution in the original equation.
Let $x = 0$.

$$3(0)^3 - 12(0) \stackrel{?}{=} 0$$
$$0 = 0 \quad \text{True}$$

Let $x = -2$.

$$3(-2)^3 - 12(-2) \stackrel{?}{=} 0$$
$$3(-8) + 24 \stackrel{?}{=} 0$$
$$0 = 0 \quad \text{True}$$

Let $x = 2$.

$$3(2)^3 - 12(2) \stackrel{?}{=} 0$$
$$3(8) - 24 \stackrel{?}{=} 0$$
$$0 = 0 \quad \text{True}$$

The solutions are $0, -2,$ and 2.

■ Work Practice 7

Practice 8

Solve:
$(x + 3)(3x^2 - 20x - 7) = 0$

Example 8 Solve: $(5x - 1)(2x^2 + 15x + 18) = 0$

Solution:

$$(5x - 1)(2x^2 + 15x + 18) = 0$$
$$(5x - 1)(2x + 3)(x + 6) = 0 \quad \text{Factor the trinomial.}$$
$$5x - 1 = 0 \quad \text{or} \quad 2x + 3 = 0 \quad \text{or} \quad x + 6 = 0 \quad \text{Set each factor equal to 0.}$$
$$5x = 1 \qquad\qquad 2x = -3 \qquad\qquad x = -6 \quad \text{Solve.}$$
$$x = \frac{1}{5} \qquad\qquad 3x = -\frac{3}{2}$$

Check each solution in the original equation. The solutions are $\frac{1}{5}, -\frac{3}{2},$ and -6.

■ Work Practice 8

Answers

7. $0, 3,$ and -3 **8.** $-3, -\frac{1}{3},$ and 7

Vocabulary, Readiness & Video Check

Use the choices below to fill in each blank. Not all choices will be used.

| $-3, 5$ | $a = 0$ or $b = 0$ | 0 | linear |
| 3, -5 | quadratic | 1 | |

1. An equation that can be written in the form $ax^2 + bx + c = 0$ (with $a \neq 0$) is called a _____ equation.

2. If the product of two numbers is 0, then at least one of the numbers must be _____.

3. The solutions to $(x - 3)(x + 5) = 0$ are _____.
4. If $a \cdot b = 0$, then _____.

Martin-Gay Interactive Videos Watch the section lecture video and answer the following questions.

Objective A 5. As shown in Examples 1–3, what two things have to be true in order to use the zero-factor theorem?

Objective B 6. Example 4 implies that the zero-factor theorem can be used with any number of factors on one side of the equation as long as the other side of the equation is zero. Why do you think this is true?

See Video 11.6

11.6 Exercise Set MyLab Math

Objective A Solve each equation. See Examples 1 through 3.

1. $(x - 2)(x + 1) = 0$
2. $(x + 3)(x + 2) = 0$
3. $(x - 6)(x - 7) = 0$
4. $(x + 4)(x - 10) = 0$
5. $(x + 9)(x + 17) = 0$
6. $(x - 11)(x - 1) = 0$
7. $x(x + 6) = 0$
8. $x(x - 7) = 0$
9. $3x(x - 8) = 0$
10. $2x(x + 12) = 0$
11. $(2x + 3)(4x - 5) = 0$
12. $(3x - 2)(5x + 1) = 0$
13. $(2x - 7)(7x + 2) = 0$
14. $(9x + 1)(4x - 3) = 0$
15. $\left(x - \dfrac{1}{2}\right)\left(x + \dfrac{1}{3}\right) = 0$
16. $\left(x + \dfrac{2}{9}\right)\left(x - \dfrac{1}{4}\right) = 0$
17. $(x + 0.2)(x + 1.5) = 0$
18. $(x + 1.7)(x + 2.3) = 0$

Solve. See Examples 4 through 6.

19. $x^2 - 13x + 36 = 0$
20. $x^2 + 2x - 63 = 0$
21. $x^2 + 2x - 8 = 0$
22. $x^2 - 5x + 6 = 0$
23. $x^2 - 7x = 0$
24. $x^2 - 3x = 0$
25. $x^2 + 20x = 0$
26. $x^2 + 15x = 0$
27. $x^2 = 16$
28. $x^2 = 9$
29. $x^2 - 4x = 32$
30. $x^2 - 5x = 24$
31. $(x + 4)(x - 9) = 4x$
32. $(x + 3)(x + 8) = x$
33. $x(3x - 1) = 14$
34. $x(4x - 11) = 3$

35. $3x^2 + 19x - 72 = 0$

36. $36x^2 + x - 21 = 0$

Objectives A B and Section 9.3 Mixed Practice *Solve each equation. See Examples 1 through 8. (A few exercises are linear equations.)*

37. $4x^3 - x = 0$

38. $4y^3 - 36y = 0$

39. $4(x - 7) = 6$

40. $5(3 - 4x) = 9$

41. $(4x - 3)(16x^2 - 24x + 9) = 0$

42. $(2x + 5)(4x^2 + 20x + 25) = 0$

43. $4y^2 - 1 = 0$

44. $4y^2 - 81 = 0$

45. $(2x + 3)(2x^2 - 5x - 3) = 0$

46. $(2x - 9)(x^2 + 5x - 36) = 0$

47. $x^2 - 15 = -2x$

48. $x^2 - 26 = -11x$

49. $30x^2 - 11x = 30$

50. $9x^2 + 7x = 2$

51. $5x^2 - 6x - 8 = 0$

52. $12x^2 + 7x - 12 = 0$

53. $6y^2 - 22y - 40 = 0$

54. $3x^2 - 6x - 9 = 0$

55. $(y - 2)(y + 3) = 6$

56. $(y - 5)(y - 2) = 28$

57. $x^3 - 12x^2 + 32x = 0$

58. $x^3 - 14x^2 + 49x = 0$

59. $x^2 + 14x + 49 = 0$

60. $x^2 + 22x + 121 = 0$

61. $12y = 8y^2$

62. $9y = 6y^2$

63. $7x^3 - 7x = 0$

64. $3x^3 - 27x = 0$

65. $3x^2 + 8x - 11 = 13 - 6x$

66. $2x^2 + 12x - 1 = 4 + 3x$

67. $3x^2 - 20x = -4x^2 - 7x - 6$

68. $4x^2 - 20x = -5x^2 - 6x - 5$

Review

Perform each indicated operation. Write all results in lowest terms. See Sections 4.3 and 4.5.

69. $\dfrac{3}{5} + \dfrac{4}{9}$

70. $\dfrac{2}{3} + \dfrac{3}{7}$

71. $\dfrac{7}{10} - \dfrac{5}{12}$

72. $\dfrac{5}{9} - \dfrac{5}{12}$

73. $\dfrac{4}{5} \cdot \dfrac{7}{8}$

74. $\dfrac{3}{7} \cdot \dfrac{12}{17}$

Concept Extensions

For Exercises 75 and 76, see the Concept Check in this section.

75. Explain the error and solve correctly:
$x(x - 2) = 8$
$x = 8$ or $x - 2 = 8$
     $x = 10$

76. Explain the error and solve correctly:
$(x - 4)(x + 2) = 0$
$x = -4$ or $x = 2$

77. Write a quadratic equation that has two solutions, 6 and −1. Leave the polynomial in the equation in factored form.

78. Write a quadratic equation that has two solutions, 0 and −2. Leave the polynomial in the equation in factored form.

79. Write a quadratic equation in standard form that has two solutions, 5 and 7.

80. Write an equation that has three solutions, 0, 1, and 2.

81. A compass is accidentally thrown upward and out of a hot-air balloon at a height of 300 feet. The height, y, of the compass at time x is given by the equation $y = -16x^2 + 20x + 300$.

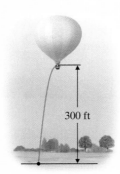

82. A rocket is fired upward from the ground with an initial velocity of 100 feet per second. The height, y, of the rocket at any time x is given by the equation $y = -16x^2 + 100x$.

a. Find the height of the compass at the given times by filling in the table below.

Time, x (in seconds)	0	1	2	3	4	5	6
Height, y (in feet)							

b. Use the table to determine when the compass strikes the ground.

c. Use the table to approximate the maximum height of the compass.

a. Find the height of the rocket at the given times by filling in the table below.

Time, x (in seconds)	0	1	2	3	4	5	6	7
Height, y (in feet)								

b. Use the table to determine between what two whole numbered seconds the rocket strikes the ground.

c. Use the table to approximate the maximum height of the rocket.

Solve each equation.

83. $(x - 3)(3x + 4) = (x + 2)(x - 6)$

84. $(2x - 3)(x + 6) = (x - 9)(x + 2)$

85. $(2x - 3)(x + 8) = (x - 6)(x + 4)$

86. $(x + 6)(x - 6) = (2x - 9)(x + 4)$

11.7 Quadratic Equations and Problem Solving

Objective

A Solve Problems That Can Be Modeled by Quadratic Equations.

Objective A Solving Problems Modeled by Quadratic Equations

Some problems may be modeled by quadratic equations. To solve these problems we use the same problem-solving steps that were introduced in Section 3.4. When solving these problems, keep in mind that a solution of an equation that models a problem may not be a solution to the problem. For example, a person's age or the length of a rectangle is always a positive number. Thus we discard solutions that do not make sense as solutions of the problem.

Practice 1

Cliff divers also frequent the falls at Waimea Falls Park in Oahu, Hawaii. Here, a diver can jump from a ledge 64 feet up the waterfall into a rocky pool below. Neglecting air resistance, the height of a diver above the pool after t seconds is $h = -16t^2 + 64$. Find how long it takes the diver to reach the pool.

Example 1 Finding Free-Fall Time

Since the 1940s, one of the top tourist attractions in Acapulco, Mexico, is watching the cliff divers off La Quebrada. The divers' platform is about 144 feet above the sea. These divers must time their descent just right, since they land in the crashing Pacific, in an inlet that is at most $9\frac{1}{2}$ feet deep. Neglecting air resistance, the height h in feet of a cliff diver above the ocean after t seconds is given by the quadratic equation $h = -16t^2 + 144$.

Find out how long it takes the diver to reach the ocean.

Solution:

1. UNDERSTAND. Read and reread the problem. Then draw a picture of the problem.

   The equation $h = -16t^2 + 144$ models the height of the falling diver at time t. Familiarize yourself with this equation by finding the height of the diver at time $t = 1$ second and $t = 2$ seconds.

   When $t = 1$ second, the height of the diver is $h = -16(1)^2 + 144 = 128$ feet.
   When $t = 2$ seconds, the height of the diver is $h = -16(2)^2 + 144 = 80$ feet.

2. TRANSLATE. To find out how long it takes the diver to reach the ocean, we want to know the value of t for which $h = 0$.

3. SOLVE. Solve the equation.

   $0 = -16t^2 + 144$
   $0 = -16(t^2 - 9)$ Factor out -16.
   $0 = -16(t - 3)(t + 3)$ Factor completely.
   $t - 3 = 0$ or $t + 3 = 0$ Set each factor containing a variable equal to 0.
   $t = 3$ or $t = -3$ Solve.

4. INTERPRET. Since the time t cannot be negative, the proposed solution is 3 seconds.

   Check: Verify that the height of the diver when t is 3 seconds is 0.

   When $t = 3$ seconds, $h = -16(3)^2 + 144 = -144 + 144 = 0$.

■ Work Practice 1

Answer
1. 2 sec

Example 2 Finding a Number

The square of a number plus three times the number is 70. Find the number.

Solution:

1. **UNDERSTAND.** Read and reread the problem. Suppose that the number is 5. The square of 5 is 5^2 or 25. Three times 5 is 15. Then $25 + 15 = 40$, not 70, so the number must be greater than 5. Remember, the purpose of proposing a number, such as 5, is to better understand the problem. Now that we do, we will let $x =$ the number.

2. **TRANSLATE.**

the square of a number	plus	three times the number	is	70
↓	↓	↓	↓	↓
x^2	$+$	$3x$	$=$	70

3. **SOLVE.**

   $x^2 + 3x = 70$
   $x^2 + 3x - 70 = 0$ Subtract 70 from both sides.
   $(x + 10)(x - 7) = 0$ Factor.
   $x + 10 = 0$ or $x - 7 = 0$ Set each factor equal to 0.
   $x = -10$ $x = 7$ Solve.

4. **INTERPRET.**

   Check: The square of -10 is $(-10)^2$, or 100. Three times -10 is $3(-10)$ or -30. Then $100 + (-30) = 70$, the correct sum, so -10 checks.
   The square of 7 is 7^2 or 49. Three times 7 is $3(7)$, or 21. Then $49 + 21 = 70$, the correct sum, so 7 checks.

   State: There are two numbers. They are -10 and 7.

Work Practice 2

Practice 2
The square of a number minus twice the number is 63. Find the number.

Example 3 Finding the Dimensions of a Sail

The height of a triangular sail is 2 meters less than twice the length of the base. If the sail has an area of 30 square meters, find the length of its base and the height.

Solution:

1. **UNDERSTAND.** Read and reread the problem. Since we are finding the length of the base and the height, we let

   $x =$ the length of the base

   Since the height is 2 meters less than twice the length of the base,

   $2x - 2 =$ the height

   An illustration is shown in the margin.

2. **TRANSLATE.** We are given that the area of the triangle is 30 square meters, so we use the formula for area of a triangle.

area of triangle	=	$\frac{1}{2}$	·	base	·	height
↓		↓		↓		↓
30	=	$\frac{1}{2}$	·	x	·	$(2x - 2)$

   (Continued on next page)

Practice 3
The length of a rectangular garden is 5 feet more than its width. The area of the garden is 176 square feet. Find the length and the width of the garden.

Height $= 2x - 2$
Base $= x$

Answers
2. 9 and -7
3. length: 16 ft; width: 11 ft

3. SOLVE. Now we solve the quadratic equation.

$$30 = \frac{1}{2}x(2x - 2)$$
$$30 = x^2 - x \quad \text{Multiply.}$$
$$0 = x^2 - x - 30 \quad \text{Write in standard form.}$$
$$0 = (x - 6)(x + 5) \quad \text{Factor.}$$
$$x - 6 = 0 \quad \text{or} \quad x + 5 = 0 \quad \text{Set each factor equal to 0.}$$
$$x = 6 \qquad\qquad x = -5$$

4. INTERPRET. Since x represents the length of the base, we discard the solution -5. The base of a triangle cannot be negative. The base is then 6 meters and the height is $2(6) - 2 = 10$ meters.

Check: To check this problem, we recall that

$$\text{area} = \frac{1}{2} \text{ base} \cdot \text{height or}$$

$$30 \stackrel{?}{=} \frac{1}{2}(6)(10)$$

$$30 = 30 \quad \text{True}$$

State: The base of the triangular sail is 6 meters and the height is 10 meters.

■ Work Practice 3

The next example has to do with consecutive integers. Study the following diagram for a review of consecutive integers.

Examples

If x is the first integer, then consecutive integers are
$x, x + 1, x + 2, \ldots$.

If x is the first even integer, then consecutive even integers are
$x, x + 2, x + 4, \ldots$.

If x is the first odd integer, then consecutive odd integers are
$x, x + 2, x + 4, \ldots$.

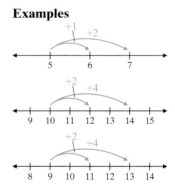

Practice 4

Find two consecutive odd integers whose product is 23 more than their sum.

Example 4 Finding Consecutive Even Integers

Find two consecutive even integers whose product is 34 more than their sum.

Solution:

1. UNDERSTAND. Read and reread the problem. Let's just choose two consecutive even integers to help us better understand the problem. Let's choose 10 and 12. Their product is $10(12) = 120$ and their sum is $10 + 12 = 22$. The product is $120 - 22$, or 98 greater than the sum. Thus our guess is incorrect but we have a better understanding of this example.
   Let's let x and $x + 2$ be the consecutive even integers.

2. TRANSLATE.

Product of integers	is	34	more than	sum of integers
↓	↓			
$x(x + 2)$	=	$x + (x + 2) + 34$		

Answer
4. 5 and 7 or -5 and -3

Section 11.7 | Quadratic Equations and Problem Solving

3. SOLVE. Now we solve the equation.

$$x(x + 2) = x + (x + 2) + 34$$
$$x^2 + 2x = x + x + 2 + 34 \quad \text{Multiply.}$$
$$x^2 + 2x = 2x + 36 \quad \text{Combine like terms.}$$
$$x^2 - 36 = 0 \quad \text{Write in standard form.}$$
$$(x + 6)(x - 6) = 0 \quad \text{Factor.}$$
$$x + 6 = 0 \quad \text{or} \quad x - 6 = 0 \quad \text{Set each factor equal to 0.}$$
$$x = -6 \qquad\qquad x = 6 \quad \text{Solve.}$$

4. INTERPRET. If $x = -6$, then $x + 2 = -6 + 2$, or -4.
If $x = 6$, then $x + 2 = 6 + 2$, or 8.

Check: $-6, -4$ $\qquad\qquad\qquad$ 6, 8

$-6(-4) \stackrel{?}{=} -6 + (-4) + 34 \qquad 6(8) \stackrel{?}{=} 6 + 8 + 34$
$\qquad 24 \stackrel{?}{=} -10 + 34 \qquad\qquad\qquad 48 \stackrel{?}{=} 14 + 34$
$\qquad 24 = 24 \qquad \text{True} \qquad\qquad 48 = 48 \qquad \text{True}$

State: The two consecutive even integers are -6 and -4 or 6 and 8.

Work Practice 4

The next example makes use of the **Pythagorean theorem.** Before we review this theorem, recall that a **right triangle** is a triangle that contains a 90° or right angle. The **hypotenuse** of a right triangle is the side opposite the right angle and the longest side of the triangle. The **legs** of a right triangle are the other sides of the triangle.

Pythagorean Theorem

In a right triangle, the sum of the squares of the lengths of the two legs is equal to the square of the length of the hypotenuse.

$$(\text{leg})^2 + (\text{leg})^2 = (\text{hypotenuse})^2 \quad \text{or} \quad a^2 + b^2 = c^2$$

Leg b, Hypotenuse c, Leg a

Helpful Hint: If you use this formula, don't forget that c represents the length of the hypotenuse.

Example 5 Finding the Dimensions of a Triangle

Find the lengths of the sides of a right triangle if the lengths can be expressed as three consecutive even integers.

Solution:

1. UNDERSTAND. Read and reread the problem. Let's suppose that the length of one leg of the right triangle is 4 units. Then the other leg is the next even integer, or 6 units, and the hypotenuse of the triangle is the next even integer, or 8 units. Remember that the hypotenuse is the longest side. Let's see if a triangle with sides of these lengths forms a right triangle. To do this, we check to see whether the Pythagorean theorem holds true.

$$4^2 + 6^2 \stackrel{?}{=} 8^2$$
$$16 + 36 \stackrel{?}{=} 64$$
$$52 = 64 \quad \text{False}$$

4 units, 8 units, 6 units

Our proposed numbers do not check, but we now have a better understanding of the problem.

(Continued on next page)

Practice 5

The length of one leg of a right triangle is 7 meters less than the length of the other leg. The length of the hypotenuse is 13 meters. Find the lengths of the legs.

Answer
5. 5 meters, 12 meters

We let x, $x + 2$, and $x + 4$ be three consecutive even integers. Since the integers represent lengths of the sides of a right triangle, we have the following

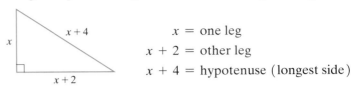

x = one leg
$x + 2$ = other leg
$x + 4$ = hypotenuse (longest side)

2. **TRANSLATE.** By the Pythagorean theorem, we have that

$$(\text{leg})^2 + (\text{leg})^2 = (\text{hypotenuse})^2$$
$$(x)^2 + (x + 2)^2 = (x + 4)^2$$

3. **SOLVE.** Now we solve the equation.

$x^2 + (x + 2)^2 = (x + 4)^2$	
$x^2 + x^2 + 4x + 4 = x^2 + 8x + 16$	Multiply.
$2x^2 + 4x + 4 = x^2 + 8x + 16$	Combine like terms.
$x^2 - 4x - 12 = 0$	Write in standard form.
$(x - 6)(x + 2) = 0$	Factor.
$x - 6 = 0 \quad \text{or} \quad x + 2 = 0$	Set each factor equal to 0.
$x = 6 \qquad\qquad x = -2$	

4. **INTERPRET.** We discard $x = -2$ since length cannot be negative. If $x =$ then $x + 2 = 8$ and $x + 4 = 10$.

Check: Verify that

$$(\text{leg})^2 + (\text{leg})^2 = (\text{hypotenuse})^2$$
$$6^2 + 8^2 \stackrel{?}{=} 10^2$$
$$36 + 64 \stackrel{?}{=} 100$$
$$100 = 100 \qquad \text{True}$$

State: The sides of the right triangle have lengths 6 units, 8 units, and 10 un

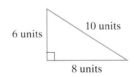

■ Work Practice 5

Vocabulary, Readiness & Video Check

See Video 11.7

Martin-Gay Interactive Videos Watch the section lecture video and answer the following question.

Objective A 1. In each of Examples 1–3, why aren't both solutions of the translated equation accepted as solutions to the application?

11.7 Exercise Set MyLab Math

Objective A See Examples 1 through 5 for all exercises.

Translating For Exercises 1 through 6, represent each given condition using a single variable, x.

1. The length and width of a rectangle whose length is 4 centimeters more than its width

2. The length and width of a rectangle whose length is twice its width

3. Two consecutive odd integers

4. Two consecutive even integers

5. The base and height of a triangle whose height is one more than four times its base

6. The base and height of a trapezoid whose base is three less than five times its height

Use the information given to find the dimensions of each figure.

7.

The *area* of the square is 121 square units. Find the length of its sides.

8.

The *area* of the rectangle is 84 square inches. Find its length and width.

9.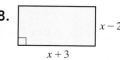

The *perimeter* of the quadrilateral is 120 centimeters. Find the lengths of its sides.

10.

The *perimeter* of the triangle is 29 feet. Find the lengths of its sides.

11.

The *area* of the parallelogram is 96 square miles. Find its base and height.

12.

The *area* of the circle is 25π square kilometers. Find its radius.

Solve.

13. An object is thrown upward from the top of an 80-foot building with an initial velocity of 64 feet per second. The height h of the object after t seconds is given by the quadratic equation $h = -16t^2 + 64t + 80$. When will the object hit the ground?

14. A hang glider accidentally drops her compass from the top of a 400-foot cliff. The height h of the compass after t seconds is given by the quadratic equation $h = -16t^2 + 400$. When will the compass hit the ground?

15. The width of a rectangle is 7 centimeters less than twice its length. Its area is 30 square centimeters. Find the dimensions of the rectangle.

16. The length of a rectangle is 9 inches more than its width. Its area is 112 square inches. Find the dimensions of the rectangle.

△ The equation $D = \frac{1}{2}n(n - 3)$ gives the number of diagonals D for a polygon with n sides. For example, a polygon with 6 sides has $D = \frac{1}{2} \cdot 6(6 - 3)$ or $D = 9$ diagonals. (See if you can count all 9 diagonals. Some are shown in the figure.) Use this equation, $D = \frac{1}{2}n(n - 3)$, for Exercises 17 through 20.

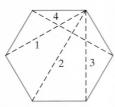

17. Find the number of diagonals for a polygon that has 12 sides.

18. Find the number of diagonals for a polygon that has 15 sides.

19. Find the number of sides n for a polygon that has 35 diagonals.

20. Find the number of sides n for a polygon that has 14 diagonals.

21. The sum of a number and its square is 132. Find the number.

22. The sum of a number and its square is 182. Find the number.

23. The product of two consecutive room numbers is 210. Find the room numbers.

24. The product of two consecutive page numbers is 420. Find the page numbers.

25. A ladder is leaning against a building so that the distance from the ground to the top of the ladder is one foot less than the length of the ladder. Find the length of the ladder if the distance from the bottom of the ladder to the building is 5 feet.

26. Use the given figure to find the length of the guy wire.

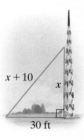

27. If the sides of a square are increased by 3 inches, the area becomes 64 square inches. Find the length of the sides of the original square.

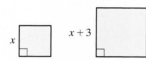

28. If the sides of a square are increased by 5 meters, the area becomes 100 square meters. Find the length of the sides of the original square.

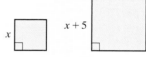

29. One leg of a right triangle is 4 millimeters longer than the shorter leg and the hypotenuse is 8 millimeters longer than the shorter leg. Find the lengths of the sides of the triangle.

30. One leg of a right triangle is 9 centimeters longer than the other leg and the hypotenuse is 45 centimeters. Find the lengths of the legs of the triangle.

31. The length of the base of a triangle is twice its height. If the area of the triangle is 100 square kilometers, find the height.

32. The height of a triangle is 2 millimeters less than the base. If the area is 60 square millimeters, find the base.

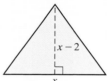

33. Find the length of the shorter leg of a right triangle if the longer leg is 12 feet more than the shorter leg and the hypotenuse is 12 feet less than twice the shorter leg.

34. Find the length of the shorter leg of a right triangle if the longer leg is 10 miles more than the shorter leg and the hypotenuse is 10 miles less than twice the shorter leg.

35. An object is dropped from 39 feet below the tip of the pinnacle atop one of the 1483-foot-tall Petronas Twin Towers in Kuala Lumpur, Malaysia. (*Source:* Council on Tall Buildings and Urban Habitat) The height h of the object after t seconds is given by the equation $h = -16t^2 + 1444$. Find how many seconds pass before the object reaches the ground.

36. An object is dropped from the top of 311 South Wacker Drive, a 961-foot-tall office building in Chicago. (*Source:* Council on Tall Buildings and Urban Habitat) The height h of the object after t seconds is given by the equation $h = -16t^2 + 961$. Find how many seconds pass before the object reaches the ground.

37. At the end of 2 years, P dollars invested at an interest rate r compounded annually increases to an amount, A dollars, given by

$$A = P(1 + r)^2$$

Find the interest rate if $100 increased to $144 in 2 years. Write your answer as a percent.

38. At the end of 2 years, P dollars invested at an interest rate r compounded annually increases to an amount, A dollars, given by

$$A = P(1 + r)^2$$

Find the interest rate if $2000 increased to $2420 in 2 years. Write your answer as a percent.

39. Find the dimensions of a rectangle whose width is 7 miles less than its length and whose area is 120 square miles.

40. Find the dimensions of a rectangle whose width is 2 inches less than half its length and whose area is 160 square inches.

41. If the cost, C, for manufacturing x units of a certain product is given by $C = x^2 - 15x + 50$, find the number of units manufactured at a cost of $9500.

42. If a switchboard handles n telephones, the number C of telephone connections it can make simultaneously is given by the equation $C = \dfrac{n(n-1)}{2}$. Find how many telephones are handled by a switchboard making 120 telephone connections simultaneously.

Review

The following double line graph shows a comparison of the number of annual visitors (in millions) to Acadia National Park and Grand Teton National Park for the years shown. Use this graph to answer Exercises 43 through 50. See Section 7.1.

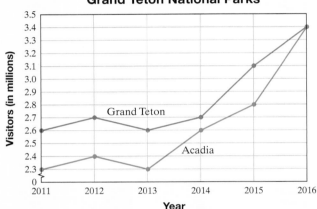

Source: National Park Service

43. Approximate the number of visitors to Acadia National Park in 2012.

44. Approximate the number of visitors to Grand Teton National Park in 2012.

45. Approximate the number of visitors to Acadia National Park in 2015.

46. Approximate the number of visitors to Grand Teton National Park in 2015.

47. Determine the year that the colored lines in this graph intersect.

48. For what year(s) on the graph is the number of visitors to Grand Teton National Park greater than 3 million?

49. In your own words, explain the meaning of the point of intersection in the graph.

50. Describe the trends shown in this graph and speculate as to why these trends have occurred.

Write each fraction in simplest form. See Section 4.2.

51. $\dfrac{20}{35}$ **52.** $\dfrac{24}{32}$ **53.** $\dfrac{27}{18}$ **54.** $\dfrac{15}{27}$ **55.** $\dfrac{14}{42}$ **56.** $\dfrac{45}{50}$

Concept Extensions

△ 57. The side of a square equals the width of a rectangle. The length of the rectangle is 6 meters longer than its width. The sum of the areas of the square and the rectangle is 176 square meters. Find the side of the square.

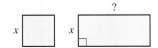

△ 58. Two boats travel at right angles to each other after leaving the same dock at the same time. One hour later the boats are 17 miles apart. If one boat travels 7 miles per hour faster than the other boat, find the rate of each boat.

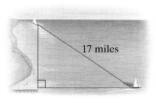

59. The sum of two numbers is 25, and the sum of their squares is 325. Find the numbers.

60. The sum of two numbers is 20, and the sum of their squares is 218. Find the numbers.

61. A rectangular pool is surrounded by a walk 4 meters wide. The pool is 6 meters longer than its width. If the total area of the pool and walk is 576 square meters more than the area of the pool, find the dimensions of the pool.

△ 62. A rectangular garden is surrounded by a walk of uniform width. The area of the garden is 180 square yards. If the dimensions of the garden plus the walk are 16 yards by 24 yards, find the width of the walk.

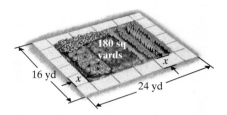

63. Write down two numbers whose sum is 10. Square each number and find the sum of the squares. Use this work to write a word problem like Exercise **59**. Then give the word problem to a classmate to solve.

64. Write down two numbers whose sum is 12. Square each number and find the sum of the squares. Use this work to write a word problem like Exercise **60**. Then give the word problem to a classmate to solve.

Chapter 11 Group Activity

Factoring polynomials can be visualized using areas of rectangles. To see this, let's first find the areas of the following squares and rectangles. (Recall that Area = Length · Width.)

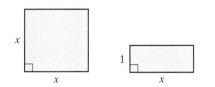

To use these areas to visualize factoring the polynomial $x^2 + 3x + 2$, for example, use the shapes below to form a rectangle. The factored form is found by reading the length and the width of the rectangle, as shown below.

Thus, $x^2 + 3x + 2 = (x + 2)(x + 1)$.

Try using this method to visualize the factored form of each polynomial below.

Work in a group and use tiles to find a factored form for the polynomials below. (Tiles can be handmade from index cards.)

1. $x^2 + 6x + 5$
2. $x^2 + 5x + 6$
3. $x^2 + 5x + 4$
4. $x^2 + 4x + 3$
5. $x^2 + 6x + 9$
6. $x^2 + 4x + 4$

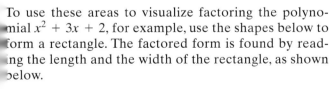
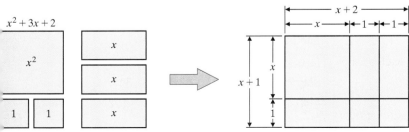

Chapter 11 Vocabulary Check

Fill in each blank with one of the words or phrases listed below. Some words or phrases may be used more than once.

| factoring | hypotenuse | quadratic equation |
| greatest common factor | leg | perfect square trinomial |

1. An equation that can be written in the form $ax^2 + bx + c = 0$ (with a not 0) is called a _____.
2. _____ is the process of writing an expression as a product.
3. The _____ of a list of terms is the product of all common factors.
4. A trinomial that is the square of some binomial is called a _____.
5. In a right triangle, the side opposite the right angle is called the _____.
6. In a right triangle, each side adjacent to the right angle is called a _____.
7. The Pythagorean theorem states that $(\text{leg})^2 + (\text{leg})^2 = (\text{_____})^2$.

> **Helpful Hint**
> ● Are you preparing for your test? To help, don't forget to take these:
> - Chapter 11 Getting Ready for the Test on page 883
> - Chapter 11 Test on page 884
>
> Then check all of your answers at the back of this text. For further review, the step-by-step video solutions to any of these exercises are located in MyLab Math.

11 Chapter Highlights

Definitions and Concepts	Examples
Section 11.1 The Greatest Common Factor and Factoring by Grouping	
Factoring is the process of writing an expression as a product.	Factor: $6 = 2 \cdot 3$ Factor: $x^2 + 5x + 6 = (x+2)(x+3)$
The GCF of a list of variable terms contains the smallest exponent on each common variable.	The GCF of z^5, z^3, and z^{10} is z^3.
The GCF of a list of terms is the product of all common factors.	Find the GCF of $8x^2y$, $10x^3y^2$, and $50x^2y^3$. $8x^2y = 2 \cdot 2 \cdot 2 \cdot x^2 \cdot y$ $10x^3y^2 = 2 \cdot 5 \cdot x^3 \cdot y^2$ $50x^2y^3 = 2 \cdot 5 \cdot 5 \cdot x^2 \cdot y^3$ $\text{GCF} = 2 \cdot x^2 \cdot y \text{ or } 2x^2y$
To Factor by Grouping **Step 1:** Group the terms in two groups so that each group has a common factor. **Step 2:** Factor out the GCF from each group. **Step 3:** If there is a common binomial factor, factor it out. **Step 4:** If not, rearrange the terms and try these steps again.	Factor: $10ax + 15a - 6xy - 9y$ **Step 1:** $(10ax + 15a) + (-6xy - 9y)$ **Step 2:** $5a(2x + 3) - 3y(2x + 3)$ **Step 3:** $(2x + 3)(5a - 3y)$

Definitions and Concepts	Examples

Section 11.2 Factoring Trinomials of the Form $x^2 + bx + c$

$x^2 + bx + c = (x + \Box)(x + \Box)$ The product of these numbers is c. The sum of these numbers is b.	Factor: $x^2 + 7x + 12$ $3 + 4 = 7 \quad 3 \cdot 4 = 12$ $x^2 + 7x + 12 = (x + 3)(x + 4)$

Section 11.3 Factoring Trinomials of the Form $ax^2 + bx + c$

To factor $ax^2 + bx + c$, try various combinations of factors of ax^2 and c until a middle term of bx is obtained when checking.	Factor: $3x^2 + 14x - 5$ Factors of $3x^2$: $3x, x$ Factors of -5: $-1, 5$ and $1, -5$ $(3x - 1)(x + 5)$ $-1x$ $+15x$ $14x$ Correct middle term

Section 11.4 Factoring Trinomials of the Form $ax^2 + bx + c$ by Grouping

To Factor $ax^2 + bx + c$ by Grouping **Step 1:** Find two numbers whose product is $a \cdot c$ and whose sum is b. **Step 2:** Rewrite bx, using the factors found in Step 1. **Step 3:** Factor by grouping.	Factor: $3x^2 + 14x - 5$ **Step 1:** Find two numbers whose product is $3 \cdot (-5)$ or -15 and whose sum is 14. They are 15 and -1. **Step 2:** $3x^2 + 14x - 5$ $= 3x^2 + 15x - 1x - 5$ **Step 3:** $= 3x(x + 5) - 1(x + 5)$ $= (x + 5)(3x - 1)$

Section 11.5 Factoring Perfect Square Trinomials and the Difference of Two Squares

A **perfect square trinomial** is a trinomial that is the square of some binomial.	**Perfect Square Trinomial = Square of Binomial** $x^2 + 4x + 4 = (x + 2)^2$ $25x^2 - 10x + 1 = (5x - 1)^2$
Factoring Perfect Square Trinomials $a^2 + 2ab + b^2 = (a + b)^2$ $a^2 - 2ab + b^2 = (a - b)^2$	Factor. $x^2 + 6x + 9 = x^2 + 2 \cdot x \cdot 3 + 3^2 = (x + 3)^2$ $4x^2 - 12x + 9 = (2x)^2 - 2 \cdot 2x \cdot 3 + 3^2$ $= (2x - 3)^2$
Difference of Two Squares $a^2 - b^2 = (a + b)(a - b)$	Factor. $x^2 - 9 = x^2 - 3^2 = (x + 3)(x - 3)$

Definitions and Concepts	Examples

Section 11.6 Solving Quadratic Equations by Factoring

A **quadratic equation** is an equation that can be written in the form $ax^2 + bx + c = 0$ with a not 0. The form $ax^2 + bx + c = 0$ is called the **standard form** of a quadratic equation.	**Quadratic Equation** **Standard Form** $x^2 = 16$ $x^2 - 16 = 0$ $y = -2y^2 + 5$ $2y^2 + y - 5 = 0$
Zero-Factor Property If a and b are real numbers and if $ab = 0$, then $a = 0$ or $b = 0$.	If $(x + 3)(x - 1) = 0$, then $x + 3 = 0$ or $x - 1 = 0$.
To Solve Quadratic Equations by Factoring **Step 1:** Write the equation in standard form so that one side of the equation is 0. **Step 2:** Factor completely. **Step 3:** Set each factor containing a variable equal to 0. **Step 4:** Solve the resulting equations. **Step 5:** Check solutions in the original equation.	Solve: $3x^2 = 13x - 4$ **Step 1:** $3x^2 - 13x + 4 = 0$ **Step 2:** $(3x - 1)(x - 4) = 0$ **Step 3:** $3x - 1 = 0$ or $x - 4 = 0$ **Step 4:** $\quad 3x = 1 \qquad\qquad x = 4$ $\qquad\quad x = \dfrac{1}{3}$ **Step 5:** Check both $\dfrac{1}{3}$ and 4 in the original equation.

Section 11.7 Quadratic Equations and Problem Solving

Problem-Solving Steps	A garden is in the shape of a rectangle whose length is two feet more than its width. If the area of the garden is 35 square feet, find its dimensions.
1. UNDERSTAND the problem.	1. Read and reread the problem. Guess a solution and check your guess. Draw a diagram. Let x be the width of the rectangular garden. Then $x + 2$ is the length. 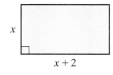
2. TRANSLATE.	2. length $\cdot$ width = area $\;\;\downarrow \qquad\qquad \downarrow \qquad\;\; \downarrow$ $(x + 2) \;\cdot\quad x \;\;=\;\; 35$
3. SOLVE.	3. $\qquad\qquad (x + 2)x = 35$ $\qquad\qquad x^2 + 2x - 35 = 0$ $\qquad\qquad (x - 5)(x + 7) = 0$ $x - 5 = 0 \qquad$ or $\qquad x + 7 = 0$ $\;\;\;\; x = 5 \qquad\qquad\qquad\;\; x = -7$
4. INTERPRET.	4. Discard the solution $x = -7$ since x represents width. *Check:* If x is 5 feet, then $x + 2 = 5 + 2 = 7$ feet. The area of a rectangle whose width is 5 feet and whose length is 7 feet is (5 feet)(7 feet) or 35 square feet. *State:* The garden is 5 feet by 7 feet.

Chapter 11 Review

(11.1) *Factor out the GCF from each polynomial.*

1. $5m + 30$

2. $6x^2 - 15x$

3. $4x^5 + 2x - 10x^4$

4. $20x^3 + 12x^2 + 24x$

5. $3x(2x + 3) - 5(2x + 3)$

6. $5x(x + 1) - (x + 1)$

Factor each polynomial by grouping.

7. $3x^2 - 3x + 2x - 2$

8. $3a^2 + 9ab + 3b^2 + ab$

9. $10a^2 + 5ab + 7b^2 + 14ab$

10. $6x^2 + 10x - 3x - 5$

(11.2) *Factor each trinomial.*

11. $x^2 + 6x + 8$

12. $x^2 - 11x + 24$

13. $x^2 + x + 2$

14. $x^2 - 5x - 6$

15. $x^2 + 2x - 8$

16. $x^2 + 4xy - 12y^2$

17. $x^2 + 8xy + 15y^2$

18. $72 - 18x - 2x^2$

19. $32 + 12x - 4x^2$

20. $5y^3 - 50y^2 + 120y$

21. To factor $x^2 + 2x - 48$, think of two numbers whose product is _____ and whose sum is _____.

22. What is the first step in factoring $3x^2 + 15x + 30$?

(11.3) or (11.4) *Factor each trinomial.*

23. $2x^2 + 13x + 6$

24. $4x^2 + 4x - 3$

25. $6x^2 + 5xy - 4y^2$

26. $x^2 - x + 2$

27. $2x^2 - 23x - 39$

28. $18x^2 - 9xy - 20y^2$

29. $10y^3 + 25y^2 - 60y$

30. $60y^3 - 39y^2 + 6y$

Write the perimeter of each figure as a simplified polynomial. Then factor each polynomial completely.

31.

△ 32.

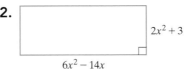

(11.5) *Determine whether each polynomial is a perfect square trinomial.*

33. $x^2 + 6x + 9$ **34.** $x^2 + 8x + 64$ **35.** $9m^2 - 12m + 16$ **36.** $4y^2 - 28y + 49$

Determine whether each binomial is a difference of two squares.

37. $x^2 - 9$ **38.** $x^2 + 16$ **39.** $4x^2 - 25y^2$ **40.** $9a^3 - 1$

Factor each polynomial completely.

41. $x^2 - 81$ **42.** $x^2 + 12x + 36$ **43.** $4x^2 - 9$ **44.** $9t^2 - 25s^2$

45. $16x^2 + y^2$ **46.** $n^2 - 18n + 81$ **47.** $3r^2 + 36r + 108$ **48.** $9y^2 - 42y + 49$

49. $5m^8 - 5m^6$ **50.** $4x^2 - 28xy + 49y^2$ **51.** $3x^2y + 6xy^2 + 3y^3$ **52.** $16x^4 - 1$

(11.6) *Solve each equation.*

53. $(x + 6)(x - 2) = 0$ **54.** $(x - 7)(x + 11) = 0$ **55.** $3x(x + 1)(7x - 2) = 0$

56. $4(5x + 1)(x + 3) = 0$ **57.** $x^2 + 8x + 7 = 0$ **58.** $x^2 - 2x - 24 = 0$ **59.** $x^2 + 10x = -25$

60. $x(x - 10) = -16$ **61.** $(3x - 1)(9x^2 + 3x + 1) = 0$ **62.** $56x^2 - 5x - 6 = 0$

63. $m^2 = 6m$ **64.** $r^2 = 25$ **65.** Write a quadratic equation in standard form that has the two solutions 4 and 5. **66.** Write a quadratic equation in standard form that has two solutions, both -1.

(11.7) *Use the given information to choose the correct dimensions.*

△ **67.** The perimeter of a rectangle is 24 inches. The length is twice the width. Find the dimensions of the rectangle.
   a. 5 inches by 7 inches **b.** 5 inches by 10 inches
   c. 4 inches by 8 inches **d.** 2 inches by 10 inches

△ **68.** The area of a rectangle is 80 meters. The length is one more than three times the width. Find the dimensions of the rectangle.
   a. 8 meters by 10 meters **b.** 4 meters by 13 mete
   c. 4 meters by 20 meters **d.** 5 meters by 16 mete

Use the given information to find the dimensions of each figure.

69. The *area* of the square is 81 square units. Find the length of a side.

△ 70. The *perimeter* of the quadrilateral is 47 units. Find the lengths of the sides.

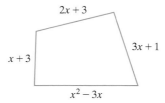

Solve.

71. A flag for a local organization is in the shape of a rectangle whose length is 15 inches less than twice its width. If the area of the flag is 500 square inches, find its dimensions.

△ 72. The base of a triangular sail is four times its height. If the area of the triangle is 162 square yards, find the base.

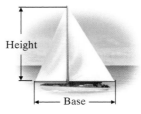

73. Find two consecutive positive integers whose product is 380.

74. Find two consecutive even positive integers whose product is 440.

75. A rocket is fired from the ground with an initial velocity of 440 feet per second. Its height h after t seconds is given by the equation $h = -16t^2 + 440t$.

  a. Find how many seconds pass before the rocket reaches a height of 2800 feet. Explain why two answers are obtained.
  b. Find how many seconds pass before the rocket reaches the ground again.

76. An architect's squaring instrument is in the shape of a right triangle. Find the length of the longer leg of the right triangle if the hypotenuse is 8 centimeters longer than the longer leg and the shorter leg is 8 centimeters shorter than the longer leg.

Mixed Review

Factor completely.

77. $6x + 24$

78. $7x - 63$

79. $11x(4x - 3) - 6(4x - 3)$

80. $2x(x - 5) - (x - 5)$

81. $3x^3 - 4x^2 + 6x - 8$

82. $xy + 2x - y - 2$

83. $2x^2 + 2x - 24$

84. $3x^3 - 30x^2 + 27x$

85. $4x^2 - 81$

86. $2x^2 - 18$

87. $16x^2 - 24x + 9$

88. $5x^2 + 20x + 20$

Solve.

89. $2x^2 - x - 28 = 0$

90. $x^2 - 2x = 15$

91. $2x(x + 7)(x + 4) = 0$

92. $x(x - 5) = -6$

93. $x^2 = 16x$

94. The perimeter of the following triangle is 48 inches. Find the lengths of its sides.

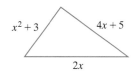

95. The width of a rectangle is 4 inches less than its length. Its area is 12 square inches. Find the dimensions of the rectangle.

Chapter 11 Getting Ready for the Test

MULTIPLE CHOICE All the exercises below are **Multiple Choice.** Choose the correct letter. Also, letters may be used more than once.

1. The greatest common factor of the terms of $10x^4 - 70x^3 + 2x^2 - 14x$ is
   A. $2x^2$ **B.** $2x$ **C.** $7x^2$ **D.** $7x$

2. Choose the expression that is NOT a factored form of $9y^3 - 18y^2$.
   A. $9(y^3 - 2y^2)$ **B.** $9y(y^2 - 2y)$ **C.** $9y^2(y - 2)$ **D.** $9 \cdot y^3 - 18 \cdot y^2$

For Exercises 3 through 6, identify each expression as:
   A. A factored expression or **B.** Not a factored expression

3. $(x - 1)(x + 5)$

4. $z(z + 12)(z - 12)$

5. $y(x - 6) + 1(x - 6)$

6. $m \cdot m - 5 \cdot 5$

For Exercises 7 through 9, choose the correct letter.

7. Choose the correct factored form for $4x^2 + 16$ or select "can't be factored."
   A. can't be factored **B.** $4(x^2 + 4)$ **C.** $4(x + 2)^2$ **D.** $4(x + 2)(x - 2)$

8. Which of the binomials can't be factored using real numbers?
   A. $x^2 + 64$ **B.** $x^2 - 64$ **C.** $x^3 + 64$ **D.** $x^3 - 64$

9. To solve $x(x + 2) = 15$, which is an incorrect next step?
   A. $x^2 + 2x = 15$ **B.** $x(x + 2) - 15 = 0$ **C.** $x = 15$ and $x + 2 = 15$

883

Chapter 11 Test

Factor each polynomial completely. If a polynomial cannot be factored, write "prime."

1. $9x^2 - 3x$
2. $x^2 + 11x + 28$
3. $49 - m^2$
4. $y^2 + 22y + 121$
5. $x^4 - 16$
6. $4(a + 3) - y(a + 3)$
7. $x^2 + 4$
8. $y^2 - 8y - 48$
9. $3a^2 + 3ab - 7a - 7b$
10. $3x^2 - 5x + 2$
11. $180 - 5x^2$
12. $3x^3 - 21x^2 + 30x$
13. $6t^2 - t - 5$
14. $xy^2 - 7y^2 - 4x + 28$
15. $x - x^5$
16. $x^2 + 14xy + 24y^2$

Solve each equation.

17. $(x - 3)(x + 9) = 0$
18. $x^2 + 5x = 14$
19. $x(x + 6) = 7$
20. $3x(2x - 3)(3x + 4) = 0$
21. $5t^3 - 45t = 0$
22. $t^2 - 2t - 15 = 0$
23. $6x^2 = 15x$

Solve.

24. A deck for a home is in the shape of a triangle. The length of the base of the triangle is 9 feet longer than its height. If the area of the triangle is 68 square feet, find the length of the base.

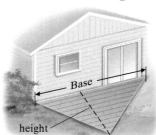

25. The *area* of the rectangle is 54 square units. Find the dimensions of the rectangle.

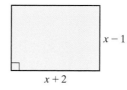

26. An object is dropped from the top of the Woolworth Building on Broadway in New York City. The height h of the object after t seconds is given by the equation

$$h = -16t^2 + 784$$

Find how many seconds pass before the object reaches the ground.

27. Find the lengths of the sides of a right triangle if the hypotenuse is 10 centimeters longer than the shorter leg and 5 centimeters longer than the longer leg.

28. A window washer is suspended 38 feet below the roof of the 1127-foot-tall John Hancock Center in Chicago. (*Source:* Council on Tall Buildings and Urban Habitat) If the window washer drops an object from this height, the object's height h after t seconds is given by the equation $h = -16t^2 + 1089$. Find how many seconds pass before the object reaches the ground.

24. _____

25. _____

26. _____

27. _____

28. _____

Chapters 1–11 Cumulative Review

Answers

1. Evaluate $x + 2y - z$ for $x = 3$, $y = -5$, and $z = -4$.

2. Evaluate $5x - y$ for $x = -2$ and $y = 4$.

3. Evaluate $7 - x^2$ for $x = -4$.

4. Evaluate $x^3 - y^3$ for $x = -2$ and $y = 4$.

Solve.

5. $-8 = n + 1$

6. $-8x = 72$

Simplify.

7. $2y - 6 + 4y + 8$

8. $-5a - 3 + a + 2$

9. $4x + 2 - 5x + 3$

10. $2x + 5x - 6$

Solve.

11. $3(3x - 5) = 10x$

12. $2(7x - 1) = 15x$

13. $3(2x - 6) + 6 = 0$

14. $4(x + 3) - 12 = 0$

15. $\dfrac{x - 5}{3} = \dfrac{x + 2}{5}$

16. $\dfrac{x + 1}{2} = \dfrac{x - 7}{3}$

17. Translate each sentence into a mathematical statement.
    a. Nine is less than or equal to eleven.
    b. Eight is greater than one.
    c. Three is not equal to four.

18. Translate each sentence into a mathematical statement.
    a. Five is greater than or equal to one.
    b. Two is not equal to negative four.

Solve.

19. $3(x - 4) = 3x - 12$

20. $2(x + 5) = 2x + 9$

21. Solve for l: $V = lwh$

22. Solve for x: $3x - 2y = 5$

Simplify each expression.

23. $(5^3)^6$
24. $(7^9)^2$
25. $(y^8)^2$
26. $(x^{11})^3$

Simplify the following expressions. Write each result using positive exponents only.

27. $\dfrac{(x^3)^4 x}{x^7}$
28. $\dfrac{(y^3)^9 \cdot y}{y^6}$
29. $(y^{-3} z^6)^{-6}$

30. $(xy^{-4})^{-2}$
31. $\dfrac{x^{-7}}{(x^4)^3}$
32. $\dfrac{(y^2)^5}{y^{-3}}$

Simplify each polynomial by combining any like terms.

33. $-3x + 7x$
34. $y + y$
35. $11x^2 + 5 + 2x^2 - 7$
36. $8y - y^2 + 4y - y^2$

Multiply.

37. $(2x - y)^2$
38. $(3x + 1)^2$

Use a special product to square each binomial.

39. $(t + 2)^2$
40. $(x - 4)^2$
41. $(x^2 - 7y)^2$
42. $(x^2 + 7y)^2$

Divide.

43. $\dfrac{8x^2 y^2 - 16xy + 2x}{4xy}$
44. $\dfrac{20a^2 b^3 - 5ab + 10b}{5ab}$

Factor completely.

45. $5(x + 3) + y(x + 3)$
46. $9(y - 2) + x(y - 2)$
47. $x^4 + 5x^2 + 6$
48. $x^4 - 4x^2 - 5$
49. $6x^2 - 2x - 20$
50. $10x^2 + 25x + 10$

51. Since the 1940s, one of the top tourist attractions in Acapulco, Mexico, is watching the cliff divers off La Quebrada. The divers' platform is about 144 feet above the sea. These divers must time their descent just right, since they land in the crashing Pacific, in an inlet that is at most $9\dfrac{1}{2}$ feet deep. Neglecting air resistance, the height h in feet of a cliff diver above the ocean after t seconds is given by the quadratic equation $h = -16t^2 + 144$.

Find out how long it takes the diver to reach the ocean.

52. The square of a number plus twice the number is 120. Find the number.

12 Rational Expressions

In this chapter, we expand our knowledge of algebraic expressions to include algebraic fractions, called *rational expressions*. We explore the operations of addition, subtraction, multiplication, and division using principles similar to the principles for numerical fractions.

Sections

- 12.1 Simplifying Rational Expressions
- 12.2 Multiplying and Dividing Rational Expressions
- 12.3 Adding and Subtracting Rational Expressions with the Same Denominator and Least Common Denominator
- 12.4 Adding and Subtracting Rational Expressions with Different Denominators
- 12.5 Solving Equations Containing Rational Expressions

   Integrated Review— Summary on Rational Expressions

- 12.6 Rational Equations and Problem Solving
- 12.7 Simplifying Complex Fractions

Check Your Progress

Vocabulary Check
Chapter Highlights
Chapter Review
Getting Ready for the Test
Chapter Test
Cumulative Review

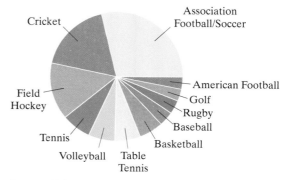

Source: worldatlas

What Do the Sports Above Have in Common?

All sports, from little league to professional, have numerous statistics kept on them. Although the final score of a match (or game) is a relatively easy statistic, determining the best player in a particular sport usually requires one or more agreed-upon formulas.

Let's focus on baseball and specifically what is called a player's slugging percentage. This is a popular measure of the hitting power of a player. Slugging percentage is different from batting average, for example, as it only deals with hits and not walks or being hit by pitches.

The bar graph below shows the all-time leaders in baseball slugging percentage. In Section 12.1, Exercises 75 through 82, we use the slugging percentage formula to calculate this particular statistic.

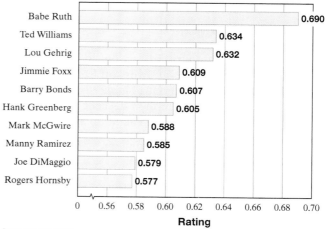

Source: Baseball Almanac

12.1 Simplifying Rational Expressions

Objective A Evaluating Rational Expressions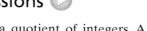

A rational number is a number that can be written as a quotient of integers. A *rational expression* is also a quotient; it is a quotient of polynomials. Examples are

$$\frac{2}{3}, \quad \frac{3y^3}{8}, \quad \frac{-4p}{p^3 + 2p + 1}, \quad \text{and} \quad \frac{5x^2 - 3x + 2}{3x + 7}$$

Objectives

A Find the Value of a Rational Expression Given a Replacement Number.

B Identify Values for Which a Rational Expression Is Undefined.

C Simplify, or Write Rational Expressions in Lowest Terms.

D Write Equivalent Rational Expressions of the Forms $-\frac{a}{b} = \frac{-a}{b} = \frac{a}{-b}$.

Rational Expression

A **rational expression** is an expression that can be written in the form

$$\frac{P}{Q}$$

where P and Q are polynomials and $Q \neq 0$.

Rational expressions have different numerical values depending on what values replace the variables.

Example 1 Find the numerical value of $\dfrac{x + 4}{2x - 3}$ for each replacement value.

a. $x = 5$ **b.** $x = -2$

Solution:

a. We replace each x in the expression with 5 and then simplify.

$$\frac{x + 4}{2x - 3} = \frac{5 + 4}{2(5) - 3} = \frac{9}{10 - 3} = \frac{9}{7}$$

b. We replace each x in the expression with -2 and then simplify.

$$\frac{x + 4}{2x - 3} = \frac{-2 + 4}{2(-2) - 3} = \frac{2}{-7} \quad \text{or} \quad -\frac{2}{7}$$

Practice 1
Find the value of $\dfrac{x - 3}{5x + 1}$ for each replacement value.

a. $x = 4$

b. $x = -3$

Work Practice 1

In the example above, we wrote $\dfrac{2}{-7}$ as $-\dfrac{2}{7}$. For a negative fraction such as $\dfrac{2}{-7}$, recall from Section 4.1 that

$$\frac{2}{-7} = \frac{-2}{7} = -\frac{2}{7}$$

In general, for any fraction,

$$\frac{-a}{b} = \frac{a}{-b} = -\frac{a}{b}, \quad b \neq 0$$

This is also true for rational expressions. For example,

$$\frac{-(x + 2)}{x} = \frac{x + 2}{-x} = -\frac{x + 2}{x}$$
↑
Notice the parentheses.

Answers

1. a. $\dfrac{1}{21}$ **b.** $\dfrac{3}{7}$

889

Helpful Hint

Do you recall why division by 0 is not defined? Remember, for example, that

$$\frac{8}{4} = 2 \text{ because } 2 \cdot 4 = 8.$$

Thus, if $\frac{8}{0} = a \text{ number},$

then *the number* $\cdot \, 0 = 8$. There is no number that when multiplied by 0 equals 8; thus $\frac{8}{0}$ is undefined. This is true in general for fractions and rational expressions.

Objective B Identifying When a Rational Expression Is Undefined

In the definition of rational expression (first "box" in this section), notice that we wrote $Q \neq 0$ for the denominator Q. The denominator of a rational expression must not equal 0 since division by 0 is not defined. (See the Helpful Hint to the left.) This means we must be careful when replacing the variable in a rational expression by a number. For example, suppose we replace x with 5 in the rational expression $\frac{3+x}{x-5}$. The expression becomes

$$\frac{3+x}{x-5} = \frac{3+5}{5-5} = \frac{8}{0}$$

But division by 0 is undefined. Therefore, in this expression we can allow x to be any real number *except* 5. **A rational expression is undefined for values that make the denominator 0.** Thus,

To find values for which a rational expression is undefined, find values for which the denominator is 0.

Practice 2

Are there any values for x for which each rational expression is undefined?

a. $\dfrac{x}{x+8}$

b. $\dfrac{x-3}{x^2+5x+4}$

c. $\dfrac{x^2-3x+2}{5}$

Example 2   Are there any values for x for which each expression is undefined?

a. $\dfrac{x}{x-3}$

b. $\dfrac{x^2+2}{x^2-3x+2}$

c. $\dfrac{x^3-6x^2-10x}{3}$

Solution: To find values for which a rational expression is undefined, we find values that make the denominator 0.

a. The denominator of $\dfrac{x}{x-3}$ is 0 when $x-3=0$ or when $x=3$. Thus, when $x=3$, the expression $\dfrac{x}{x-3}$ is undefined.

b. We set the denominator equal to 0.

$$x^2 - 3x + 2 = 0$$
$$(x-2)(x-1) = 0 \quad \text{Factor.}$$
$$x - 2 = 0 \quad \text{or} \quad x - 1 = 0 \quad \text{Set each factor equal to 0.}$$
$$x = 2 \qquad\qquad x = 1 \quad \text{Solve.}$$

Thus, when $x=2$ or $x=1$, the denominator $x^2 - 3x + 2$ is 0. So the rational expression $\dfrac{x^2+2}{x^2-3x+2}$ is undefined when $x=2$ or when $x=1$.

c. The denominator of $\dfrac{x^3-6x^2-10x}{3}$ is never 0, so there are no values of x for which this expression is undefined.

■ Work Practice 2

Note: Unless otherwise stated, we will now assume that variables in rational expressions are replaced only by values for which the expressions are defined.

Answers

2. a. $x = -8$ **b.** $x = -4, x = -1$
c. no

Objective C Simplifying Rational Expressions

A fraction is said to be written in lowest terms or simplest form when the numerator and denominator have no common factors other than 1 (or -1). For example, the fraction $\frac{7}{10}$ is written in lowest terms since the numerator and denominator have no common factors other than 1 (or -1).

The process of writing a rational expression in lowest terms or simplest form is called **simplifying** the rational expression.

Simplifying a rational expression is similar to simplifying a fraction. Recall from Section 4.2 that to simplify a fraction, we essentially "remove factors of 1." Our ability to do this comes from these facts:

- Any nonzero number over itself simplifies to 1 $\left(\frac{5}{5} = 1, \frac{-7.26}{-7.26} = 1, \text{ and } \frac{c}{c} = 1 \text{ as long as } c \text{ is not } 0\right)$, and
- The product of any number and 1 is that number $\left(19 \cdot 1 = 19, -8.9 \cdot 1 = -8.9, \frac{a}{b} \cdot 1 = \frac{a}{b}\right)$.

In other words, we have the following:

$$\frac{a \cdot c}{b \cdot c} = \frac{a}{b} \cdot \frac{c}{c} = \frac{a}{b}$$

Since $\frac{a}{b} \cdot 1 = \frac{a}{b}$

> **Helpful Hint**
> We use the Fundamental Principle of Fractions to simplify rational expressions. This process is also sometimes called
> - Dividing out common factors
>
> or
> - Removing a factor of 1
>
> (See Section 4.2 for a review.)

Simplify: $\frac{15}{20}$

$\frac{15}{20} = \frac{3 \cdot 5}{2 \cdot 2 \cdot 5}$ Factor the numerator and the denominator.

$= \frac{3 \cdot 5}{2 \cdot 2 \cdot 5}$ Look for common factors.

$= \frac{3}{2 \cdot 2} \cdot \frac{5}{5}$ Common factors in the numerator and denominator form factors of 1.

$= \frac{3}{2 \cdot 2} \cdot 1$ Write $\frac{5}{5}$ as 1.

$= \frac{3}{2 \cdot 2} = \frac{3}{4}$ Multiply to remove a factor of 1.

Before we use the same technique to simplify a rational expression, remember that as long as the denominator is not 0, $\frac{a^3b}{a^3b} = 1$, $\frac{x+3}{x+3} = 1$, and $\frac{7x^2 + 5x - 100}{7x^2 + 5x - 100} = 1$.

Simplify: $\frac{x^2 - 9}{x^2 + x - 6}$

$\frac{x^2 - 9}{x^2 + x - 6} = \frac{(x-3)(x+3)}{(x-2)(x+3)}$ Factor the numerator and the denominator.

$= \frac{(x-3)(x+3)}{(x-2)(x+3)}$ Look for common factors.

$= \frac{x-3}{x-2} \cdot \frac{x+3}{x+3}$

$= \frac{x-3}{x-2} \cdot 1$ Write $\frac{x+3}{x+3}$ as 1.

$= \frac{x-3}{x-2}$ Multiply to remove a factor of 1.

Just as for numerical fractions, we can use a shortcut notation. Remember that as long as exact factors in both the numerator and denominator are divided out, we are "removing a factor of 1." We will use the following notation to show this:

$$\frac{x^2 - 9}{x^2 + x - 6} = \frac{(x-3)(x+3)}{(x-2)(x+3)} \quad \text{A factor of 1 is identified by the shading.}$$

$$= \frac{x-3}{x-2} \quad \text{Remove a factor of 1.}$$

Thus, the rational expression $\frac{x^2-9}{x^2+x-6}$ has the same value as the rational expression $\frac{x-3}{x-2}$ for all values of x except 2 and -3. (Remember that when x is 2, the denominator of both rational expressions is 0 and that when x is -3, the original rational expression has a denominator of 0.)

As we simplify rational expressions, we will assume that the simplified rational expression is equal to the original rational expression for all real numbers except those for which either denominator is 0. The following steps may be used to simplify rational expressions.

To Simplify a Rational Expression

Step 1: Completely factor the numerator and denominator.

Step 2: Divide out factors common to the numerator and denominator. (This is the same as "removing a factor of 1.")

Practice 3

Simplify: $\dfrac{x^4 + x^3}{5x + 5}$

Example 3 Simplify: $\dfrac{5x - 5}{x^3 - x^2}$

Solution: To begin, we factor the numerator and denominator if possible. Then we look for common factors.

$$\frac{5x-5}{x^3-x^2} = \frac{5(x-1)}{x^2(x-1)} = \frac{5}{x^2}$$

■ Work Practice 3

Practice 4

Simplify: $\dfrac{x^2 + 11x + 18}{x^2 + x - 2}$

Example 4 Simplify: $\dfrac{x^2 + 8x + 7}{x^2 - 4x - 5}$

Solution: We factor the numerator and denominator and then look for common factors.

$$\frac{x^2+8x+7}{x^2-4x-5} = \frac{(x+7)(x+1)}{(x-5)(x+1)} = \frac{x+7}{x-5}$$

■ Work Practice 4

Practice 5

Simplify: $\dfrac{x^2 + 10x + 25}{x^2 + 5x}$

Example 5 Simplify: $\dfrac{x^2 + 4x + 4}{x^2 + 2x}$

Solution: We factor the numerator and denominator and then look for common factors.

$$\frac{x^2+4x+4}{x^2+2x} = \frac{(x+2)(x+2)}{x(x+2)} = \frac{x+2}{x}$$

■ Work Practice 5

Answers

3. $\dfrac{x^3}{5}$ **4.** $\dfrac{x+9}{x-1}$ **5.** $\dfrac{x+5}{x}$

Helpful Hint

When simplifying a rational expression, we look for **common *factors*, not common *terms*.**

$$\frac{x \cdot (x+2)}{x \cdot x} = \frac{x+2}{x}$$
Common factors. These can be divided out.

$$\frac{x+2}{x}$$
Common terms. There is no factor of 1 that can be generated.

✓ **Concept Check** Recall that we can remove only *factors* of 1. Which of the following are *not* true? Explain why.

a. $\frac{3-1}{3+5}$ simplifies to $-\frac{1}{5}$.

b. $\frac{2x+10}{2}$ simplifies to $x+5$.

c. $\frac{37}{72}$ simplifies to $\frac{3}{2}$.

d. $\frac{2x+3}{2}$ simplifies to $x+3$.

Example 6 Simplify: $\frac{x+9}{x^2-81}$

Solution: We factor and then apply the fundamental principle. Remember that this principle allows us to divide the numerator and denominator by all common factors.

$$\frac{x+9}{x^2-81} = \frac{x+9}{(x+9)(x-9)} = \frac{1}{x-9}$$

■ Work Practice 6

Practice 6

Simplify: $\frac{x+5}{x^2-25}$

Example 7 Simplify each rational expression.

a. $\frac{x+y}{y+x}$ b. $\frac{x-y}{y-x}$

Solution:

a. The expression $\frac{x+y}{y+x}$ can be simplified by using the commutative property of addition to rewrite the denominator $y+x$ as $x+y$.

$$\frac{x+y}{y+x} = \frac{x+y}{x+y} = 1$$

b. The expression $\frac{x-y}{y-x}$ can be simplified by recognizing that $y-x$ and $x-y$ are opposites. In other words, $y-x = -1(x-y)$. We proceed as follows:

$$\frac{x-y}{y-x} = \frac{1 \cdot (x-y)}{(-1)(x-y)} = \frac{1}{-1} = -1$$

■ Work Practice 7

Practice 7

Simplify each rational expression.

a. $\frac{x+4}{4+x}$ b. $\frac{x-4}{4-x}$

Answers

6. $\frac{1}{x-5}$ 7. a. 1 b. -1

✓ **Concept Check Answer**
a, c, d

Practice 8

Simplify: $\dfrac{2x^2 - 5x - 12}{16 - x^2}$

Example 8

Simplify: $\dfrac{4 - x^2}{3x^2 - 5x - 2}$

Solution:

$$\dfrac{4 - x^2}{3x^2 - 5x - 2} = \dfrac{(2 - x)(2 + x)}{(x - 2)(3x + 1)} \quad \text{Factor.}$$

$$= \dfrac{(-1)(x - 2)(2 + x)}{(x - 2)(3x + 1)} \quad \text{Write } 2 - x \text{ as } -1(x - 2).$$

$$= \dfrac{(-1)(2 + x)}{3x + 1} \text{ or } \dfrac{-2 - x}{3x + 1} \quad \text{Simplify.}$$

■ Work Practice 8

Objective D Writing Equivalent Forms of Rational Expressions

From Example 7(a), we have $y + x = x + y$. $\quad y + x$ and $x + y$ are equivalent.

From Example 7(b), we have $y - x = -1(x - y)$. $\quad y - x$ and $x - y$ are opposites.

Thus, $\dfrac{x + y}{y + x} = \dfrac{x + y}{x + y} = 1$ and $\dfrac{x - y}{y - x} = \dfrac{x - y}{-1(x - y)} = \dfrac{1}{-1} = -1$.

When performing operations on rational expressions, equivalent forms of answers often result. For this reason, it is very important to be able to recognize equivalent answers.

Practice 9

List 4 equivalent forms of $-\dfrac{3x + 7}{x - 6}$.

Helpful Hint Remember, a negative sign in front of a fraction or rational expression may be moved to the numerator or the denominator, but *not* both.

Example 9

List some equivalent forms of $-\dfrac{5x - 1}{x + 9}$.

Solution: To do so, recall that $-\dfrac{a}{b} = \dfrac{-a}{b} = \dfrac{a}{-b}$. Thus

$$-\dfrac{5x - 1}{x + 9} = \dfrac{-(5x - 1)}{x + 9} = \dfrac{-5x + 1}{x + 9} \text{ or } \dfrac{1 - 5x}{x + 9}$$

Also,

$$-\dfrac{5x - 1}{x + 9} = \dfrac{5x - 1}{-(x + 9)} = \dfrac{5x - 1}{-x - 9} \text{ or } \dfrac{5x - 1}{-9 - x}$$

Thus $-\dfrac{5x - 1}{x + 9} = \dfrac{-(5x - 1)}{x + 9} = \dfrac{-5x + 1}{x + 9} = \dfrac{5x - 1}{-(x + 9)} = \dfrac{5x - 1}{-x - 9}$

■ Work Practice 9

Keep in mind that many rational expressions may look different but in fact are equivalent.

Answers

8. $-\dfrac{2x + 3}{x + 4}$ or $\dfrac{-2x - 3}{x + 4}$

9. $\dfrac{-(3x + 7)}{x - 6}$; $\dfrac{-3x - 7}{x - 6}$; $\dfrac{3x + 7}{-(x - 6)}$; $\dfrac{3x + 7}{-x + 6}$

Section 12.1 | Simplifying Rational Expressions

Vocabulary, Readiness & Video Check

Use the choices below to fill in each blank. Not all choices will be used.

−1	0	simplifying	$\dfrac{-a}{-b}$	$\dfrac{-a}{b}$	$\dfrac{a}{-b}$
1	2	rational expression			

1. A _____ is an expression that can be written in the form $\dfrac{P}{Q}$, where P and Q are polynomials and $Q \neq 0$.

2. The expression $\dfrac{x+3}{3+x}$ simplifies to _____.

3. The expression $\dfrac{x-3}{3-x}$ simplifies to _____.

4. A rational expression is undefined for values that make the denominator _____.

5. The expression $\dfrac{7x}{x-2}$ is undefined for $x = $ _____.

6. The process of writing a rational expression in lowest terms is called _____.

7. For a rational expression, $-\dfrac{a}{b} = $ _____ = _____.

Decide which rational expression(s) can be simplified. (Do not actually simplify.)

8. $\dfrac{x}{x+7}$

9. $\dfrac{3+x}{x+3}$

10. $\dfrac{5-x}{x-5}$

11. $\dfrac{x+2}{x+8}$

See Video 12.1

Martin-Gay Interactive Videos Watch the section lecture video and answer the following questions.

Objective A 12. From the lecture before ▶ Example 1, what do the different values of a rational expression depend on? How are these different values found? ▶

Objective B 13. Why can't the denominators of rational expressions be zero? How can we find the numbers for which a rational expression is undefined? ▶

Objective C 14. In ▶ Example 7, why isn't a factor of x divided out of the expression at the end? ▶

Objective D 15. From ▶ Example 9, if we move a negative sign from in front of a rational expression to either the numerator or the denominator, when would we need to use parentheses and why? ▶

12.1 Exercise Set MyLab Math

Objective A Find the value of the following expressions when $x = 2$, $y = -2$, and $z = -5$. See Example 1.

1. $\dfrac{x+5}{x+2}$

2. $\dfrac{x+8}{x+1}$

3. $\dfrac{4z-1}{z-2}$

4. $\dfrac{7y-1}{y-1}$

5. $\dfrac{y^3}{y^2-1}$

6. $\dfrac{z}{z^2-5}$

7. $\dfrac{x^2+8x+2}{x^2-x-6}$

8. $\dfrac{x+5}{x^2+4x-8}$

Objective B *Find any numbers for which each rational expression is undefined. See Example 2.*

9. $\dfrac{7}{2x}$
10. $\dfrac{3}{5x}$
11. $\dfrac{x+3}{x+2}$
12. $\dfrac{5x+1}{x-9}$
13. $\dfrac{x-4}{2x-5}$

14. $\dfrac{x+1}{5x-2}$
15. $\dfrac{9x^3+4}{15x^2+30x}$
16. $\dfrac{19x^3+2}{x^2-x}$
17. $\dfrac{x^2-5x-2}{4}$
18. $\dfrac{9y^5+y^3}{9}$

19. $\dfrac{3x^2+9}{x^2-5x-6}$
20. $\dfrac{11x^2+1}{x^2-5x-14}$

21. $\dfrac{x}{3x^2+13x+14}$
22. $\dfrac{x}{2x^2+15x+27}$

Objective C *Simplify each expression. See Examples 3 through 8.*

23. $\dfrac{x+7}{7+x}$
24. $\dfrac{y+9}{9+y}$
25. $\dfrac{x-7}{7-x}$

26. $\dfrac{y-9}{9-y}$
27. $\dfrac{2}{8x+16}$
28. $\dfrac{3}{9x+6}$

29. $\dfrac{x-2}{x^2-4}$
30. $\dfrac{x+5}{x^2-25}$
31. $\dfrac{2x-10}{3x-30}$

32. $\dfrac{3x-9}{4x-16}$
33. $\dfrac{-5a-5b}{a+b}$
34. $\dfrac{-4x-4y}{x+y}$

35. $\dfrac{7x+35}{x^2+5x}$
36. $\dfrac{9x+99}{x^2+11x}$
37. $\dfrac{x+5}{x^2-4x-45}$

38. $\dfrac{x-3}{x^2-6x+9}$
39. $\dfrac{5x^2+11x+2}{x+2}$
40. $\dfrac{12x^2+4x-1}{2x+1}$

41. $\dfrac{x^3+7x^2}{x^2+5x-14}$
42. $\dfrac{x^4-10x^3}{x^2-17x+70}$
43. $\dfrac{14x^2-21x}{2x-3}$

44. $\dfrac{4x^2+24x}{x+6}$
45. $\dfrac{x^2+7x+10}{x^2-3x-10}$
46. $\dfrac{2x^2+7x-4}{x^2+3x-4}$

47. $\dfrac{3x^2+7x+2}{3x^2+13x+4}$
48. $\dfrac{4x^2-4x+1}{2x^2+9x-5}$
49. $\dfrac{2x^2-8}{4x-8}$

50. $\dfrac{5x^2 - 500}{35x + 350}$

51. $\dfrac{4 - x^2}{x - 2}$

52. $\dfrac{49 - y^2}{y - 7}$

53. $\dfrac{x^2 - 1}{x^2 - 2x + 1}$

54. $\dfrac{x^2 - 16}{x^2 - 8x + 16}$

Simplify each expression. Each exercise contains a four-term polynomial that should be factored by grouping. See Examples 3 through 8.

55. $\dfrac{x^2 + xy + 2x + 2y}{x + 2}$

56. $\dfrac{ab + ac + b^2 + bc}{b + c}$

57. $\dfrac{5x + 15 - xy - 3y}{2x + 6}$

58. $\dfrac{xy - 6x + 2y - 12}{y^2 - 6y}$

59. $\dfrac{2xy + 5x - 2y - 5}{3xy + 4x - 3y - 4}$

60. $\dfrac{2xy + 2x - 3y - 3}{2xy + 4x - 3y - 6}$

Objective D *Study Example 9. Then list four equivalent forms for each rational expression.*

61. $-\dfrac{x - 10}{x + 8}$

62. $-\dfrac{x + 11}{x - 4}$

63. $-\dfrac{5y - 3}{y - 12}$

64. $-\dfrac{8y - 1}{y - 15}$

Objectives C D Mixed Practice *Simplify each expression. Then determine whether the given answer is correct. See Examples 3 through 9.*

65. $\dfrac{9 - x^2}{x - 3}$; Answer: $-3 - x$?

66. $\dfrac{100 - x^2}{x - 10}$; Answer: $-10 - x$?

67. $\dfrac{7 - 34x - 5x^2}{25x^2 - 1}$; Answer: $\dfrac{x + 7}{-5x - 1}$?

68. $\dfrac{2 - 15x - 8x^2}{64x^2 - 1}$; Answer: $\dfrac{x + 2}{-8x - 1}$?

Review

Perform each indicated operation. See Section 4.3.

69. $\dfrac{1}{3} \cdot \dfrac{9}{11}$

70. $\dfrac{5}{27} \cdot \dfrac{2}{5}$

71. $\dfrac{1}{3} \div \dfrac{1}{4}$

72. $\dfrac{7}{8} \div \dfrac{1}{2}$

73. $\dfrac{13}{20} \div \dfrac{2}{9}$

74. $\dfrac{8}{15} \div \dfrac{5}{8}$

Concept Extensions

A baseball player's slugging percentage S can be calculated with the following formula: $S = \dfrac{H + D + 2T + 3R}{B}$, where H = number of hits, D = number of doubles, T = number of triples, R = number of home runs, and B = number of times at bat. *Use this formula to complete the table below and then rank players by their slugging percentage in 2016. Round answers to 3 decimal places.*

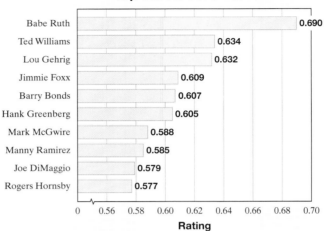

Source: Baseball Almanac

	Player Name (2016 Data)	B	H	D	T	R	$S = \dfrac{H + D + 2T + 3R}{B}$
75.	Miguel Cabrera (Detroit Tigers)	595	188	31	1	38	
76.	Nolan Arenado (Colorado Rockies)	618	182	35	6	41	
77.	David Ortiz (Boston Red Sox)	537	169	48	1	38	
78.	Nelson Cruz (Seattle Mariners)	589	169	27	1	43	
79.	Daniel Murphy (Washington Nationals)	531	184	47	5	25	
80.	Freddie Freeman (Atlanta Braves)	589	178	43	6	34	

81. Use your calculations above to name the player with the greatest slugging percentage.

82. Use your calculations above to name the player with the second-greatest slugging percentage.

Which of the following are incorrect and why? See the Concept Check in this section.

83. $\dfrac{5a - 15}{5}$ simplifies to $a - 3$?

84. $\dfrac{7m - 9}{7}$ simplifies to $m - 9$?

85. $\dfrac{1 + 2}{1 + 3}$ simplifies to $\dfrac{2}{3}$?

86. $\dfrac{46}{54}$ simplifies to $\dfrac{6}{5}$?

87. Explain how to write a fraction in lowest terms.

88. Explain how to write a rational expression in lowest terms.

89. Explain why the denominator of a fraction or a rational expression must not equal 0.

90. Does $\dfrac{(x-3)(x+3)}{x-3}$ have the same value as $x+3$ for all real numbers? Explain why or why not.

91. The average cost per DVD, in dollars, for a company to produce x DVDs on exercising is given by the formula $A = \dfrac{3x + 400}{x}$, where A is the average cost per DVD and x is the number of DVDs produced.
    a. Find the cost for producing 1 DVD.
    b. Find the average cost for producing 100 DVDs.
    c. Does the cost per DVD decrease or increase when more DVDs are produced? Explain your answer.

92. For a certain model of fax machine, the manufacturing cost C per machine is given by the equation
$$C = \dfrac{250x + 10{,}000}{x}$$
where x is the number of fax machines manufactured and cost C is in dollars per machine.
    a. Find the cost per fax machine when manufacturing 100 fax machines.
    b. Find the cost per fax machine when manufacturing 1000 fax machines.
    c. Does the cost per machine decrease or increase when more machines are manufactured? Explain why this is so.

93. The dose of medicine prescribed for a child depends on the child's age A in years and the adult dose D for the medication. Young's Rule is a formula used by pediatricians that gives a child's dose C as
$$C = \dfrac{DA}{A + 12}$$
Suppose that an 8-year-old child needs medication, and the normal adult dose is 1000 mg. What size dose should the child receive?

94. Calculating body-mass index is a way to gauge whether a person should lose weight. Doctors recommend that body-mass index values fall between 19 and 25. The formula for body-mass index B is
$$B = \dfrac{705w}{h^2}$$
where w is weight in pounds and h is height in inches. Should a 148-pound person who is 5 feet 6 inches tall lose weight?

95. Anthropologists and forensic scientists use a measure called the cephalic index to help classify skulls. The cephalic index of a skull with width W and length L from front to back is given by the formula
$$C = \dfrac{100W}{L}$$
A long skull has an index value less than 75, a medium skull has an index value between 75 and 85, and a broad skull has an index value over 85. Find the cephalic index of a skull that is 5 inches wide and 6.4 inches long. Classify the skull.

96. A company's gross profit margin P can be computed with the formula $P = \dfrac{R - C}{R}$, where $R =$ the company's revenue and $C =$ cost of goods sold. For the fiscal year 2012, Ford Motor Company had revenues of $134.25 billion and cost of goods sold $115.1 billion. (*Source:* Ford Motor Company) What was Ford's gross profit margin in 2012? Express the answer as a percent, rounded to the nearest tenth of a percent.

12.2 Multiplying and Dividing Rational Expressions

Objectives

A Multiply Rational Expressions.

B Divide Rational Expressions.

C Multiply and Divide Rational Expressions.

D Convert Between Units of Measure.

Objective A Multiplying Rational Expressions

Just as simplifying rational expressions is similar to simplifying number fractions, multiplying and dividing rational expressions is similar to multiplying and dividing number fractions.

Fractions	Rational Expressions
Multiply: $\dfrac{3}{5} \cdot \dfrac{10}{11}$	Multiply: $\dfrac{x-3}{x+5} \cdot \dfrac{2x+10}{x^2-9}$
Multiply numerators and then multiply denominators.	
$\dfrac{3}{5} \cdot \dfrac{10}{11} = \dfrac{3 \cdot 10}{5 \cdot 11}$	$\dfrac{x-3}{x+5} \cdot \dfrac{2x+10}{x^2-9} = \dfrac{(x-3) \cdot (2x+10)}{(x+5) \cdot (x^2-9)}$
Simplify by factoring numerators and denominators.	
$= \dfrac{3 \cdot 2 \cdot 5}{5 \cdot 11}$	$= \dfrac{(x-3) \cdot 2(x+5)}{(x+5)(x+3)(x-3)}$
Apply the fundamental principle.	
$= \dfrac{3 \cdot 2}{11}$ or $\dfrac{6}{11}$	$= \dfrac{2}{x+3}$

Multiplying Rational Expressions

If $\dfrac{P}{Q}$ and $\dfrac{R}{S}$ are rational expressions, then

$$\dfrac{P}{Q} \cdot \dfrac{R}{S} = \dfrac{PR}{QS}$$

To multiply rational expressions, multiply the numerators and then multiply the denominators.

Note: Recall that for Sections 12.1 through 12.4, we assume variables in rational expressions have only those replacement values for which the expressions are defined.

Practice 1

Multiply.

a. $\dfrac{16y}{3} \cdot \dfrac{1}{x^2}$

b. $\dfrac{-5a^3}{3b^3} \cdot \dfrac{2b^2}{15a}$

Example 1 Multiply.

a. $\dfrac{25x}{2} \cdot \dfrac{1}{y^3}$

b. $\dfrac{-7x^2}{5y} \cdot \dfrac{3y^5}{14x^2}$

Solution: To multiply rational expressions, we first multiply the numerators and then multiply the denominators of both expressions. Then we write the product in lowest terms.

a. $\dfrac{25x}{2} \cdot \dfrac{1}{y^3} = \dfrac{25x \cdot 1}{2 \cdot y^3} = \dfrac{25x}{2y^3}$

The expression $\dfrac{25x}{2y^3}$ is in lowest terms.

b. $\dfrac{-7x^2}{5y} \cdot \dfrac{3y^5}{14x^2} = \dfrac{-7x^2 \cdot 3y^5}{5y \cdot 14x^2}$ Multiply.

Answers

1. a. $\dfrac{16y}{3x^2}$ b. $-\dfrac{2a^2}{9b}$

Section 12.2 | Multiplying and Dividing Rational Expressions

The expression $\dfrac{-7x^2 \cdot 3y^5}{5y \cdot 14x^2}$ is not in lowest terms, so we factor the numerator and the denominator and apply the fundamental principle to "remove factors of 1."

$$= \dfrac{-1 \cdot 7 \cdot 3 \cdot x^2 \cdot y \cdot y^4}{5 \cdot 2 \cdot 7 \cdot x^2 \cdot y}$$ Common factors in the numerator and denominator form factors of 1.

$$= -\dfrac{3y^4}{10}$$ Divide out common factors. (This is the same as "removing a factor of 1.")

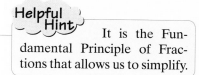

Helpful Hint It is the Fundamental Principle of Fractions that allows us to simplify.

■ Work Practice 1

When multiplying rational expressions, it is usually best to factor each numerator and denominator first. This will help us when we apply the fundamental principle to write the product in lowest terms.

Example 2 Multiply: $\dfrac{x^2 + x}{3x} \cdot \dfrac{6}{5x + 5}$

Solution:

$$\dfrac{x^2 + x}{3x} \cdot \dfrac{6}{5x + 5} = \dfrac{x(x+1)}{3x} \cdot \dfrac{2 \cdot 3}{5(x+1)}$$ Factor numerators and denominators.

$$= \dfrac{x(x+1) \cdot 2 \cdot 3}{3x \cdot 5 (x+1)}$$ Multiply.

$$= \dfrac{2}{5}$$ Divide out common factors.

Practice 2

Multiply: $\dfrac{3x + 6}{14} \cdot \dfrac{7x^2}{x^3 + 2x^2}$

■ Work Practice 2

The following steps may be used to multiply rational expressions.

To Multiply Rational Expressions

Step 1: Completely factor numerators and denominators.
Step 2: Multiply numerators and multiply denominators.
Step 3: Simplify or write the product in lowest terms by dividing out common factors.

Concept Check Which of the following is a true statement?

a. $\dfrac{1}{3} \cdot \dfrac{1}{2} \overset{?}{=} \dfrac{1}{5}$ b. $\dfrac{2}{x} \cdot \dfrac{5}{x} \overset{?}{=} \dfrac{10}{x}$ c. $\dfrac{3}{x} \cdot \dfrac{1}{2} \overset{?}{=} \dfrac{3}{2x}$ d. $\dfrac{x}{7} \cdot \dfrac{x+5}{4} \overset{?}{=} \dfrac{2x+5}{28}$

Example 3 Multiply: $\dfrac{3x + 3}{5x^2 - 5x} \cdot \dfrac{2x^2 + x - 3}{4x^2 - 9}$

Solution:

$$\dfrac{3x+3}{5x^2-5x} \cdot \dfrac{2x^2+x-3}{4x^2-9} = \dfrac{3(x+1)}{5x(x-1)} \cdot \dfrac{(2x+3)(x-1)}{(2x-3)(2x+3)}$$ Factor.

$$= \dfrac{3(x+1)\,(2x+3)(x-1)}{5x\,(x-1)\,(2x-3)\,(2x+3)}$$ Multiply.

$$= \dfrac{3(x+1)}{5x(2x-3)}$$ Simplify.

Practice 3

Multiply:

$\dfrac{4x + 8}{7x^2 - 14x} \cdot \dfrac{3x^2 - 5x - 2}{9x^2 - 1}$

Answers

2. $\dfrac{3}{2}$ 3. $\dfrac{4(x+2)}{7x(3x-1)}$

✓ **Concept Check Answer**

c

■ Work Practice 3

Objective B Dividing Rational Expressions

We can divide by a rational expression in the same way we divide by a number fraction. Recall that to divide by a fraction, we multiply by its reciprocal. For example, to divide $\frac{3}{2}$ by $\frac{7}{8}$, we multiply $\frac{3}{2}$ by $\frac{8}{7}$.

$$\frac{3}{2} \div \frac{7}{8} = \frac{3}{2} \cdot \frac{8}{7} = \frac{3 \cdot 4 \cdot 2}{2 \cdot 7} = \frac{12}{7}$$

Helpful Hint

Don't forget how to find reciprocals. The reciprocal of $\frac{a}{b}$ is $\frac{b}{a}$, $a \neq 0, b \neq 0$.

Dividing Rational Expressions

If $\frac{P}{Q}$ and $\frac{R}{S}$ are rational expressions and $\frac{R}{S}$ is not 0, then

$$\frac{P}{Q} \div \frac{R}{S} = \frac{P}{Q} \cdot \frac{S}{R} = \frac{PS}{QR}$$

To divide two rational expressions, multiply the first rational expression by the reciprocal of the second rational expression.

Example 4 Divide: $\frac{3x^3}{40} \div \frac{4x^3}{y^2}$

Solution:

$$\frac{3x^3}{40} \div \frac{4x^3}{y^2} = \frac{3x^3}{40} \cdot \frac{y^2}{4x^3} \quad \text{Multiply by the reciprocal of } \frac{4x^3}{y^2}.$$

$$= \frac{3\, x^3 \cdot y^2}{160\, x^3}$$

$$= \frac{3y^2}{160} \quad \text{Simplify.}$$

■ Work Practice 4

Practice 4

Divide: $\frac{7x^2}{6} \div \frac{x}{2y}$

Helpful Hint Remember, **to Divide by a Rational Expression,** multiply by its reciprocal.

Practice 5

Divide: $\frac{(x-4)^2}{6} \div \frac{3x-12}{2}$

Example 5 Divide: $\frac{(x+2)^2}{10} \div \frac{2x+4}{5}$

Solution:

$$\frac{(x+2)^2}{10} \div \frac{2x+4}{5} = \frac{(x+2)^2}{10} \cdot \frac{5}{2x+4} \quad \text{Multiply by the reciprocal of } \frac{2x+4}{5}$$

$$= \frac{(x+2)(x+2) \cdot 5}{5 \cdot 2 \cdot 2 \cdot (x+2)} \quad \text{Factor and multiply.}$$

$$= \frac{x+2}{4} \quad \text{Simplify.}$$

■ Work Practice 5

Answers

4. $\dfrac{7xy}{3}$ **5.** $\dfrac{x-4}{9}$

Section 12.2 | Multiplying and Dividing Rational Expressions

Example 6 Divide: $\dfrac{6x+2}{x^2-1} \div \dfrac{3x^2+x}{x-1}$

Solution:

$\dfrac{6x+2}{x^2-1} \div \dfrac{3x^2+x}{x-1} = \dfrac{6x+2}{x^2-1} \cdot \dfrac{x-1}{3x^2+x}$ Multiply by the reciprocal.

$= \dfrac{2(3x+1)(x-1)}{(x+1)(x-1) \cdot x(3x+1)}$ Factor and multiply.

$= \dfrac{2}{x(x+1)}$ Simplify.

■ Work Practice 6

Practice 6

Divide: $\dfrac{10x+4}{x^2-4} \div \dfrac{5x^3+2x^2}{x+2}$

Example 7 Divide: $\dfrac{2x^2-11x+5}{5x-25} \div \dfrac{4x-2}{10}$

Solution:

$\dfrac{2x^2-11x+5}{5x-25} \div \dfrac{4x-2}{10} = \dfrac{2x^2-11x+5}{5x-25} \cdot \dfrac{10}{4x-2}$ Multiply by the reciprocal.

$= \dfrac{(2x-1)(x-5) \cdot 2 \cdot 5}{5(x-5) \cdot 2(2x-1)}$ Factor and multiply.

$= \dfrac{1}{1} \text{ or } 1$ Simplify.

■ Work Practice 7

Practice 7

Divide:

$\dfrac{3x^2-10x+8}{7x-14} \div \dfrac{9x-12}{21}$

Objective C Multiplying and Dividing Rational Expressions ▶

Let's make sure that we understand the difference between multiplying and dividing rational expressions.

Rational Expressions	
Multiplication	Multiply the numerators and multiply the denominators.
Division	Multiply by the reciprocal of the divisor.

Example 8 Multiply or divide as indicated.

a. $\dfrac{x-4}{5} \cdot \dfrac{x}{x-4}$

b. $\dfrac{x-4}{5} \div \dfrac{x}{x-4}$

c. $\dfrac{x^2-4}{2x+6} \cdot \dfrac{x^2+4x+3}{2-x}$

Solution:

a. $\dfrac{x-4}{5} \cdot \dfrac{x}{x-4} = \dfrac{(x-4) \cdot x}{5 \cdot (x-4)} = \dfrac{x}{5}$

b. $\dfrac{x-4}{5} \div \dfrac{x}{x-4} = \dfrac{x-4}{5} \cdot \dfrac{x-4}{x} = \dfrac{(x-4)^2}{5x}$

c. $\dfrac{x^2-4}{2x+6} \cdot \dfrac{x^2+4x+3}{2-x} = \dfrac{(x-2)(x+2) \cdot (x+1)(x+3)}{2(x+3) \cdot (2-x)}$ Factor and multiply.

(Continued on next page)

Practice 8

Multiply or divide as indicated.

a. $\dfrac{x+3}{x} \cdot \dfrac{7}{x+3}$

b. $\dfrac{x+3}{x} \div \dfrac{7}{x+3}$

c. $\dfrac{3-x}{x^2+6x+5} \cdot \dfrac{2x+10}{x^2-7x+12}$

Answers

6. $\dfrac{2}{x^2(x-2)}$ 7. 1

8. a. $\dfrac{7}{x}$ b. $\dfrac{(x+3)^2}{7x}$

c. $-\dfrac{2}{(x+1)(x-4)}$

Recall from Section 12.1 that $x - 2$ and $2 - x$ are opposites. This means th $\frac{x-2}{2-x} = -1$. Thus,

$$\frac{(x-2)(x+2) \cdot (x+1)(x+3)}{2(x+3) \cdot (2-x)} = \frac{-1(x+2)(x+1)}{2}$$

$$= -\frac{(x+2)(x+1)}{2}$$

■ Work Practice 8

Objective D Converting Between Units of Measure

How many square inches are in 1 square foot?
How many cubic feet are in a cubic yard?

If you have trouble answering these questions, this section will be helpful to you.
Now that we know how to multiply fractions and rational expressions, we c use this knowledge to help us convert between units of measure. To do so, we w use **unit fractions.** A unit fraction is a fraction that equals 1. For example, sir 12 in. = 1 ft, we have the unit fractions

$$\frac{12 \text{ in.}}{1 \text{ ft}} = 1 \quad \text{and} \quad \frac{1 \text{ ft}}{12 \text{ in.}} = 1$$

Practice 9

288 square inches = _____ square feet

Example 9 18 square feet = _____ square yards

Solution: Let's multiply 18 square feet by a unit fraction that has square feet in t denominator and square yards in the numerator. From the diagram, you can see th

1 square yard = 9 square feet

Thus,

$$18 \text{ sq ft} = \frac{18 \text{ sq ft}}{1} \cdot 1 = \frac{\overset{2}{\cancel{18} \text{ sq ft}}}{1} \cdot \frac{1 \text{ sq yd}}{\underset{1}{\cancel{9} \text{ sq ft}}}$$

$$= \frac{2 \cdot 1}{1 \cdot 1} \text{ sq yd} = 2 \text{ sq yd}$$

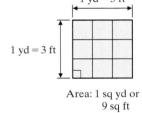

1 yd = 3 ft
1 yd = 3 ft
Area: 1 sq yd or 9 sq ft

Thus, 18 sq ft = 2 sq yd.
Draw a diagram of 18 sq ft to help you see that this is reasonable.

■ Work Practice 9

Practice 10

3.5 square feet = _____ square inches

Example 10 5.2 square yards = _____ square feet

Solution:

$$5.2 \text{ sq yd} = \frac{5.2 \text{ sq yd}}{1} \cdot 1 = \frac{5.2 \cancel{\text{ sq yd}}}{1} \cdot \frac{9 \text{ sq ft}}{1 \cancel{\text{ sq yd}}} \quad \begin{array}{l} \leftarrow \text{Units converting to} \\ \leftarrow \text{Units given} \end{array}$$

$$= \frac{5.2 \cdot 9}{1 \cdot 1} \text{ sq ft}$$

$$= 46.8 \text{ sq ft}$$

Thus, 5.2 sq yd = 46.8 sq ft.
Draw a diagram to see that this is reasonable.

■ Work Practice 10

Answers
9. 2 sq ft **10.** 504 sq in.

Example 11 Converting from Cubic Feet to Cubic Yards

The largest building in the world by volume is The Boeing Company's Everett, Washington, factory complex, where Boeing's wide-body jetliners, the 747, 767, and 777, are built. The volume of this factory complex is 472,370,319 cubic feet. Find the volume of this Boeing facility in cubic yards. (*Source:* The Boeing Company)

Solution: There are 27 cubic feet in 1 cubic yard. (See the diagram.)

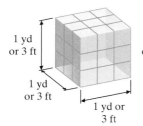

or (1 yd)(1 yd)(1 yd) = 1 cubic yard
(3 ft)(3 ft)(3 ft) = 27 cubic feet

$$472{,}370{,}319 \text{ cu ft} = 472{,}370{,}319 \text{ cu ft} \cdot \frac{1 \text{ cu yd}}{27 \text{ cu ft}}$$

$$= \frac{472{,}370{,}319}{27} \text{ cu yd}$$

$$= 17{,}495{,}197 \text{ cu yd}$$

Work Practice 11

Practice 11

The largest casino in the world is the Venetian, in Macau, on the southern tip of China. The gaming area for this casino is approximately 61,000 *square yards*. Find the size of the gaming area in *square feet*. (*Source: USA Today*)

Helpful Hint

When converting between units of measurement, if possible, write the unit fraction so that **the numerator contains the units you are converting to** and **the denominator contains the original units.**

$$48 \text{ in.} = \frac{48 \text{ in.}}{1} \cdot \frac{1 \text{ ft}}{12 \text{ in.}} \quad \begin{array}{l} \leftarrow \text{Units converting to} \\ \leftarrow \text{Original units} \end{array}$$

$$= \frac{48}{12} \text{ ft} = 4 \text{ ft}$$

Example 12

At the 2016 Summer Olympics, Jamaican athlete Usain Bolt won the gold medal in the men's 100-meter track event. He ran the distance at an average speed of 33.4 feet per second. Convert this speed to miles per hour. (*Source:* International Olympic Committee)

Solution: Recall that 1 mile = 5280 feet and 1 hour = 3600 seconds (60 · 60).

$$33.4 \text{ feet/second} = \frac{33.4 \text{ feet}}{1 \text{ second}} \cdot \frac{3600 \text{ seconds}}{1 \text{ hour}} \cdot \frac{1 \text{ mile}}{5280 \text{ feet}}$$

$$= \frac{33.4 \cdot 3600}{5280} \text{ miles/hour}$$

$$= 22.77 \text{ miles/hour}$$

Work Practice 12

Practice 12

The cheetah is the fastest land animal, being clocked at about 102.7 feet per second. Convert this to miles per hour. Round to the nearest tenth. (*Source: World Almanac and Book of Facts*)

Answers
11. 549,000 sq ft
12. 70.0 miles per hour

Vocabulary, Readiness & Video Check

Use the choices below to fill in each blank. Not all choices will be used.

opposites $\dfrac{a \cdot d}{b \cdot c}$ $\dfrac{a \cdot c}{b \cdot d}$ $\dfrac{x}{42}$ $\dfrac{x^2}{42}$ $\dfrac{2x}{42}$ $\dfrac{6}{7}$ $\dfrac{7}{6}$

reciprocals

1. The expressions $\dfrac{x}{2y}$ and $\dfrac{2y}{x}$ are called _____.

2. $\dfrac{a}{b} \cdot \dfrac{c}{d} =$ _____

3. $\dfrac{a}{b} \div \dfrac{c}{d} =$ _____

4. $\dfrac{x}{7} \cdot \dfrac{x}{6} =$ _____

5. $\dfrac{x}{7} \div \dfrac{x}{6} =$ _____

Martin-Gay Interactive Videos — Watch the section lecture video and answer the following questions.

See Video 12.2

Objective A 6. Would you say a person needs to be quite comfortable with factoring polynomials in order to be successful with multiplying rational expressions? Explain, referencing Example 2 in your answer.

Objective B 7. Based on the lecture before Example 3, complete the following statements: Dividing rational expressions is exactly like dividing _____. Therefore, to divide by a rational expression, multiply by its _____.

Objective C 8. In Examples 4 and 5, determining the operation is the first step in deciding how to perform the operation. Why is this so?

Objective D 9. In Example 6, why is the unit fraction $\dfrac{27 \text{ cu ft}}{1 \text{ cu yd}}$ used?

12.2 Exercise Set MyLab Math

Objective A Find each product and simplify if possible. See Examples 1 through 3.

1. $\dfrac{3x}{y^2} \cdot \dfrac{7y}{4x}$

2. $\dfrac{9x^2}{y} \cdot \dfrac{4y}{3x^3}$

3. $\dfrac{8x}{2} \cdot \dfrac{x^5}{4x^2}$

4. $\dfrac{6x^2}{10x^3} \cdot \dfrac{5x}{12}$

5. $-\dfrac{5a^2b}{30a^2b^2} \cdot b^3$

6. $-\dfrac{9x^3y^2}{18xy^5} \cdot y^3$

7. $\dfrac{x}{2x-14} \cdot \dfrac{x^2-7x}{5}$

8. $\dfrac{4x-24}{20x} \cdot \dfrac{5}{x-6}$

9. $\dfrac{6x+6}{5} \cdot \dfrac{10}{36x+36}$

10. $\dfrac{x^2+x}{8} \cdot \dfrac{16}{x+1}$

11. $\dfrac{(m+n)^2}{m-n} \cdot \dfrac{m}{m^2+mn}$

12. $\dfrac{(m-n)^2}{m+n} \cdot \dfrac{m}{m^2-mn}$

13. $\dfrac{x^2 - 25}{x^2 - 3x - 10} \cdot \dfrac{x + 2}{x}$

14. $\dfrac{a^2 - 4a + 4}{a^2 - 4} \cdot \dfrac{a + 3}{a - 2}$

15. $\dfrac{x^2 + 6x + 8}{x^2 + x - 20} \cdot \dfrac{x^2 + 2x - 15}{x^2 + 8x + 16}$

16. $\dfrac{x^2 + 9x + 20}{x^2 - 15x + 44} \cdot \dfrac{x^2 - 11x + 28}{x^2 + 12x + 35}$

Objective B *Find each quotient and simplify. See Examples 4 through 7.*

17. $\dfrac{5x^7}{2x^5} \div \dfrac{15x}{4x^3}$

18. $\dfrac{9y^4}{6y} \div \dfrac{y^2}{3}$

19. $\dfrac{8x^2}{y^3} \div \dfrac{4x^2 y^3}{6}$

20. $\dfrac{7a^2 b}{3ab^2} \div \dfrac{21a^2 b^2}{14ab}$

21. $\dfrac{(x - 6)(x + 4)}{4x} \div \dfrac{2x - 12}{8x^2}$

22. $\dfrac{(x + 3)^2}{5} \div \dfrac{5x + 15}{25}$

23. $\dfrac{3x^2}{x^2 - 1} \div \dfrac{x^5}{(x + 1)^2}$

24. $\dfrac{9x^5}{a^2 - b^2} \div \dfrac{27x^2}{3b - 3a}$

25. $\dfrac{m^2 - n^2}{m + n} \div \dfrac{m}{m^2 + nm}$

26. $\dfrac{(m - n)^2}{m + n} \div \dfrac{m^2 - mn}{m}$

27. $\dfrac{x + 2}{7 - x} \div \dfrac{x^2 - 5x + 6}{x^2 - 9x + 14}$

28. $\dfrac{x - 3}{2 - x} \div \dfrac{x^2 + 3x - 18}{x^2 + 2x - 8}$

29. $\dfrac{x^2 + 7x + 10}{x - 1} \div \dfrac{x^2 + 2x - 15}{x - 1}$

30. $\dfrac{x + 1}{2x^2 + 5x + 3} \div \dfrac{20x + 100}{2x + 3}$

Objective C **Mixed Practice** *Multiply or divide as indicated. See Example 8.*

31. $\dfrac{5x - 10}{12} \div \dfrac{4x - 8}{8}$

32. $\dfrac{6x + 6}{5} \div \dfrac{9x + 9}{10}$

33. $\dfrac{x^2 + 5x}{8} \cdot \dfrac{9}{3x + 15}$

34. $\dfrac{3x^2 + 12x}{6} \cdot \dfrac{9}{2x + 8}$

35. $\dfrac{7}{6p^2 + q} \div \dfrac{14}{18p^2 + 3q}$

36. $\dfrac{3x + 6}{20} \div \dfrac{4x + 8}{8}$

37. $\dfrac{3x + 4y}{x^2 + 4xy + 4y^2} \cdot \dfrac{x + 2y}{2}$

38. $\dfrac{x^2 - y^2}{3x^2 + 3xy} \cdot \dfrac{3x^2 + 6x}{3x^2 - 2xy - y^2}$

39. $\dfrac{(x + 2)^2}{x - 2} \div \dfrac{x^2 - 4}{2x - 4}$

40. $\dfrac{x + 3}{x^2 - 9} \div \dfrac{5x + 15}{(x - 3)^2}$

41. $\dfrac{x^2 - 4}{24x} \div \dfrac{2 - x}{6xy}$

42. $\dfrac{3y}{3 - x} \div \dfrac{12xy}{x^2 - 9}$

43. $\dfrac{a^2 + 7a + 12}{a^2 + 5a + 6} \cdot \dfrac{a^2 + 8a + 15}{a^2 + 5a + 4}$

44. $\dfrac{b^2 + 2b - 3}{b^2 + b - 2} \cdot \dfrac{b^2 - 4}{b^2 + 6b + 8}$

▶ 45. $\dfrac{5x - 20}{3x^2 + x} \cdot \dfrac{3x^2 + 13x + 4}{x^2 - 16}$

46. $\dfrac{9x + 18}{4x^2 - 3x} \cdot \dfrac{4x^2 - 11x + 6}{x^2 - 4}$

47. $\dfrac{8n^2 - 18}{2n^2 - 5n + 3} \div \dfrac{6n^2 + 7n - 3}{n^2 - 9n + 8}$

48. $\dfrac{36n^2 - 64}{3n^2 - 10n + 8} \div \dfrac{3n^2 - 5n - 12}{n^2 - 9n + 14}$

Objective D *Convert as indicated. See Examples 9 through 12.*

49. 10 square feet = _____ square inches.

50. 1008 square inches = _____ square feet.

51. 45 square feet = _____ square yards.

52. 2 square yards = _____ square inches.

▶ 53. 3 cubic yards = _____ cubic feet.

54. 2 cubic yards = _____ cubic inches.

55. 50 miles per hour = _____ feet per second (round to the nearest whole).

56. 10 feet per second = _____ miles per hour (round to the nearest tenth).

57. 6.3 square yards = _____ square feet.

58. 3.6 square yards = _____ square feet.

59. In January 2010, the Burj Khalifa Tower officially became the tallest building in the world. This tower has a curtain wall (the exterior skin of the building) that is approximately 133,500 square yards. Convert this to square feet. (*Source:* Burj Khalifa)

60. The Pentagon, headquarters for the Department of Defense, contains 3,705,793 square feet of office and storage space. Convert this to square yards. Round the nearest square yard. (*Source:* U.S. Department of Defense)

61. On January 7, 2011, Australian driver Barton Mawer set a new solar-powered-car land speed record of 80.9 feet per second in the Sunswift IV, built by a student team at the University of New South Wales. Convert this speed to miles per hour. Round to the nearest tenth. (*Source:* University of New South Wales)

62. Peregrine falcons are among the fastest birds in the world. When engaged in a high-speed dive for prey, a peregrine falcon can reach speeds over 200 miles per hour. Find this speed in feet per second. Round to the nearest tenth. (*Source:* Ohio Department of Natural Resources)

Review

Perform each indicated operation. See Section 4.4.

63. $\dfrac{1}{5} + \dfrac{4}{5}$

64. $\dfrac{3}{15} + \dfrac{6}{15}$

65. $\dfrac{9}{9} - \dfrac{19}{9}$

66. $\dfrac{4}{3} - \dfrac{8}{3}$

67. $\dfrac{6}{5} + \left(\dfrac{1}{5} - \dfrac{8}{5}\right)$

68. $-\dfrac{3}{2} + \left(\dfrac{1}{2} - \dfrac{3}{2}\right)$

Concept Extensions

Identify each statement as true or false. If false, correct the multiplication. See the Concept Check in this section.

69. $\dfrac{4}{a} \cdot \dfrac{1}{b} = \dfrac{4}{ab}$

70. $\dfrac{2}{3} \cdot \dfrac{2}{4} = \dfrac{2}{7}$

71. $\dfrac{x}{5} \cdot \dfrac{x+3}{4} = \dfrac{2x+3}{20}$

72. $\dfrac{7}{a} \cdot \dfrac{3}{a} = \dfrac{21}{a}$

73. Find the area of the rectangle.

$\dfrac{2x}{x^2-25}$ feet

$\dfrac{x+5}{9x}$ feet

△ 74. Find the area of the square.

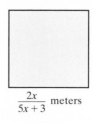

$\dfrac{2x}{5x+3}$ meters

Multiply or divide as indicated.

75. $\left(\dfrac{x^2-y^2}{x^2+y^2} \div \dfrac{x^2-y^2}{3x}\right) \cdot \dfrac{x^2+y^2}{6}$

76. $\left(\dfrac{x^2-9}{x^2-1} \cdot \dfrac{x^2+2x+1}{2x^2+9x+9}\right) \div \dfrac{2x+3}{1-x}$

77. $\left(\dfrac{2a+b}{b^2} \cdot \dfrac{3a^2-2ab}{ab+2b^2}\right) \div \dfrac{a^2-3ab+2b^2}{5ab-10b^2}$

78. $\left(\dfrac{x^2y^2-xy}{4x-4y} \div \dfrac{3y-3x}{8x-8y}\right) \cdot \dfrac{y-x}{8}$

79. In your own words, explain how you multiply rational expressions.

80. Explain how dividing rational expressions is similar to dividing rational numbers.

81. On December 22, 2016, 1 euro was equivalent to 1.0456 American dollars. If you had wanted to exchange $2000 U.S. for euros on that day for a European vacation, how many would you have received? Round to the nearest hundredth. (*Source:* Barclay's Bank)

82. An environmental technician finds that warm water from an industrial process is being discharged into a nearby pond at a rate of 30 gallons per minute. Plant regulations state that the flow rate should be no more than 0.1 cubic foot per second. Is the flow rate of 30 gallons per minute in violation of the plant regulations? (*Hint:* 1 cubic foot is equivalent to 7.48 gallons.)

12.3 Adding and Subtracting Rational Expressions with the Same Denominator and Least Common Denominator

Objectives

A Add and Subtract Rational Expressions with Common Denominators.

B Find the Least Common Denominator of a List of Rational Expressions.

C Write a Rational Expression as an Equivalent Expression Whose Denominator Is Given.

Objective A Adding and Subtracting Rational Expressions with the Same Denominator

Like multiplication and division, addition and subtraction of rational expressions are similar to addition and subtraction of rational numbers. In this section, we add and subtract rational expressions with a common denominator.

Add: $\dfrac{6}{5} + \dfrac{2}{5}$ | Add: $\dfrac{9}{x+2} + \dfrac{3}{x+2}$

Add the numerators and place the sum over the common denominator.

$\dfrac{6}{5} + \dfrac{2}{5} = \dfrac{6+2}{5}$ | $\dfrac{9}{x+2} + \dfrac{3}{x+2} = \dfrac{9+3}{x+2}$

$= \dfrac{8}{5}$ Simplify. | $= \dfrac{12}{x+2}$ Simplify.

Adding and Subtracting Rational Expressions with Common Denominators

If $\dfrac{P}{R}$ and $\dfrac{Q}{R}$ are rational expressions, then

$$\dfrac{P}{R} + \dfrac{Q}{R} = \dfrac{P+Q}{R} \quad \text{and} \quad \dfrac{P}{R} - \dfrac{Q}{R} = \dfrac{P-Q}{R}$$

To add or subtract rational expressions, add or subtract numerators and place the sum or difference over the common denominator.

Practice 1

Add: $\dfrac{8x}{3y} + \dfrac{x}{3y}$

Example 1 Add: $\dfrac{5m}{2n} + \dfrac{m}{2n}$

Solution:

$\dfrac{5m}{2n} + \dfrac{m}{2n} = \dfrac{5m + m}{2n}$ Add the numerators.

$= \dfrac{6m}{2n}$ Simplify the numerator by combining like terms.

$= \dfrac{3m}{n}$ Simplify by applying the fundamental principle.

■ Work Practice 1

Practice 2

Subtract: $\dfrac{3x}{3x-7} - \dfrac{7}{3x-7}$

Example 2 Subtract: $\dfrac{2y}{2y-7} - \dfrac{7}{2y-7}$

Solution:

$\dfrac{2y}{2y-7} - \dfrac{7}{2y-7} = \dfrac{2y-7}{2y-7}$ Subtract the numerators.

$= \dfrac{1}{1}$ or 1 Simplify.

■ Work Practice 2

Answers

1. $\dfrac{3x}{y}$ 2. 1

Section 12.3 | Adding and Subtracting Rational Expressions

Example 3 Subtract: $\dfrac{3x^2 + 2x}{x - 1} - \dfrac{10x - 5}{x - 1}$

Solution:

$$\dfrac{3x^2 + 2x}{x-1} - \dfrac{10x - 5}{x - 1} = \dfrac{3x^2 + 2x - (10x - 5)}{x - 1} \quad \text{Subtract the numerators. Notice the parentheses.}$$

$$= \dfrac{3x^2 + 2x - 10x + 5}{x - 1} \quad \text{Use the distributive property.}$$

$$= \dfrac{3x^2 - 8x + 5}{x - 1} \quad \text{Combine like terms.}$$

$$= \dfrac{(x - 1)(3x - 5)}{x - 1} \quad \text{Factor.}$$

$$= 3x - 5 \quad \text{Simplify.}$$

Work Practice 3

Practice 3

Subtract: $\dfrac{2x^2 + 5x}{x + 2} - \dfrac{4x + 6}{x + 2}$

Helpful Hint Parentheses are inserted so that the entire numerator, $10x - 5$, is subtracted.

Helpful Hint

Notice how the numerator $10x - 5$ was subtracted in Example 3.

This $-$ sign applies to the entire numerator $10x - 5$.
So parentheses are inserted here to indicate this.

$$\dfrac{3x^2 + 2x}{x - 1} - \dfrac{10x - 5}{x - 1} = \dfrac{3x^2 + 2x - (10x - 5)}{x - 1}$$

Objective B Finding the Least Common Denominator

Recall from Chapter 4 that to add and subtract fractions with different denominators, we first find the least common denominator (LCD). Then we write all fractions as equivalent fractions with the LCD.

For example, suppose we want to add $\dfrac{3}{8}$ and $\dfrac{1}{6}$. To find the LCD of the denominators, factor 8 and 6. Remember, the LCD is the same as the least common multiple, LCM. It is the smallest number that is a multiple of 6 and also 8.

$8 = 2 \cdot 2 \cdot 2$
$6 = 2 \cdot 3$

The LCM is a multiple of 6.
$\text{LCM} = 2 \cdot 2 \cdot 2 \cdot 3 = 24$
The LCM is a multiple of 8.

In the next section, we will find the sum $\dfrac{3}{8} + \dfrac{1}{6}$, but for now, let's concentrate on the LCD.

To add or subtract rational expressions with different denominators, we also first find the LCD and then write all rational expressions as equivalent expressions with the LCD. The **least common denominator (LCD) of a list of rational expressions** is a polynomial of least degree whose factors include all the factors of the denominators in the list.

To Find the Least Common Denominator (LCD)

Step 1: Factor each denominator completely.

Step 2: The least common denominator (LCD) is the product of all unique factors found in Step 1, each raised to a power equal to the greatest number of times that the factor appears in any one factored denominator.

Answer
3. $2x - 3$

Practice 4

Find the LCD for each pair.

a. $\dfrac{2}{9}, \dfrac{7}{15}$

b. $\dfrac{5}{6x^3}, \dfrac{11}{18x^5}$

Example 4 Find the LCD for each pair.

a. $\dfrac{1}{8}, \dfrac{3}{22}$

b. $\dfrac{7}{5x}, \dfrac{6}{15x^2}$

Solution:

a. We start by finding the prime factorization of each denominator.

$$8 = 2^3 \quad \text{and}$$
$$22 = 2 \cdot 11$$

Next we write the product of all the unique factors, each raised to a power equal to the greatest number of times that the factor appears.

The greatest number of times that the factor 2 appears is 3.

The greatest number of times that the factor 11 appears is 1.

$$\text{LCD} = 2^3 \cdot 11^1 = 8 \cdot 11 = 88$$

b. We factor each denominator.

$$5x = 5 \cdot x \quad \text{and}$$
$$15x^2 = 3 \cdot 5 \cdot x^2$$

The greatest number of times that the factor 5 appears is 1.
The greatest number of times that the factor 3 appears is 1.
The greatest number of times that the factor x appears is 2.

$$\text{LCD} = 3^1 \cdot 5^1 \cdot x^2 = 15x^2$$

■ Work Practice 4

Practice 5

Find the LCD of $\dfrac{3a}{a+5}$ and $\dfrac{7a}{a-5}$.

Example 5 Find the LCD of $\dfrac{7x}{x+2}$ and $\dfrac{5x^2}{x-2}$.

Solution: The denominators $x + 2$ and $x - 2$ are completely factored already. The factor $x + 2$ appears once and the factor $x - 2$ appears once.

$$\text{LCD} = (x+2)(x-2)$$

■ Work Practice 5

Practice 6

Find the LCD of $\dfrac{7x^2}{(x-4)^2}$ and $\dfrac{5x}{3x-12}$.

Example 6 Find the LCD of $\dfrac{6m^2}{3m+15}$ and $\dfrac{2}{(m+5)^2}$.

Solution: We factor each denominator.

$$3m + 15 = 3(m+5)$$
$$(m+5)^2 = (m+5)^2 \quad \text{This denominator is already factored.}$$

The greatest number of times that the factor 3 appears is 1.
The greatest number of times that the factor $m + 5$ appears *in any one denominator* is 2.

$$\text{LCD} = 3(m+5)^2$$

■ Work Practice 6

Answers

4. a. 45   b. $18x^5$
5. $(a+5)(a-5)$ 6. $3(x-4)^2$

✓ **Concept Check Answer**
b

✓ **Concept Check** Choose the correct LCD of $\dfrac{x}{(x+1)^2}$ and $\dfrac{5}{x+1}$.

a. $x + 1$ b. $(x+1)^2$ c. $(x+1)^3$ d. $5x(x+1)^2$

Section 12.3 | Adding and Subtracting Rational Expressions

Example 7 Find the LCD of $\dfrac{t-10}{2t^2+t-6}$ and $\dfrac{t+5}{t^2+3t+2}$.

Solution:
$$2t^2 + t - 6 = (2t-3)(t+2)$$
$$t^2 + 3t + 2 = (t+1)(t+2)$$
$$\text{LCD} = (2t-3)(t+2)(t+1)$$

■ Work Practice 7

Practice 7

Find the LCD of $\dfrac{y+5}{y^2+2y-3}$ and $\dfrac{y+4}{y^2-3y+2}$.

Example 8 Find the LCD of $\dfrac{2}{x-2}$ and $\dfrac{10}{2-x}$.

Solution: The denominators $x-2$ and $2-x$ are opposites. That is, $2-x = -1(x-2)$. We can use either $x-2$ or $2-x$ as the LCD.

$$\text{LCD} = x - 2 \quad \text{or} \quad \text{LCD} = 2 - x$$

■ Work Practice 8

Practice 8

Find the LCD of $\dfrac{6}{x-4}$ and $\dfrac{9}{4-x}$.

Objective C Writing Equivalent Rational Expressions

Next we practice writing a rational expression as an equivalent rational expression with a given denominator. To do this, we multiply by a form of 1. Recall that multiplying an expression by 1 produces an equivalent expression. In other words,

$$\frac{P}{Q} = \frac{P}{Q} \cdot 1 = \frac{P}{Q} \cdot \frac{R}{R} = \frac{PR}{QR}$$

Example 9 Write each rational expression as an equivalent rational expression with the given denominator.

a. $\dfrac{4b}{9a} = \dfrac{}{27a^2b}$ **b.** $\dfrac{7x}{2x+5} = \dfrac{}{6x+15}$

Solution:

a. We can ask ourselves: "What do we multiply $9a$ by to get $27a^2b$?" The answer is $3ab$, since $9a(3ab) = 27a^2b$. So we multiply by 1 in the form of $\dfrac{3ab}{3ab}$.

$$\frac{4b}{9a} = \frac{4b}{9a} \cdot 1 = \frac{4b}{9a} \cdot \frac{3ab}{3ab}$$
$$= \frac{4b(3ab)}{9a(3ab)} = \frac{12ab^2}{27a^2b}$$

b. First, factor the denominator on the right.

$$\frac{7x}{2x+5} = \frac{}{3(2x+5)}$$

To obtain the denominator on the right from the denominator on the left, we multiply by 1 in the form of $\dfrac{3}{3}$.

$$\frac{7x}{2x+5} = \frac{7x}{2x+5} \cdot \frac{3}{3} = \frac{7x \cdot 3}{(2x+5) \cdot 3} = \frac{21x}{3(2x+5)}$$

■ Work Practice 9

Practice 9

Write the rational expression as an equivalent rational expression with the given denominator.

$$\frac{2x}{5y} = \frac{}{20x^2y^2}$$

Answers
7. $(y+3)(y-1)(y-2)$
8. $x-4$ or $4-x$
9. $\dfrac{8x^3y}{20x^2y^2}$

914 Chapter 12 | Rational Expressions

Practice 10
Write the rational expression as an equivalent rational expression with the given denominator.
$$\frac{3}{x^2 - 25} = \frac{}{(x+5)(x-5)(x-3)}$$

Example 10 Write the rational expression as an equivalent rational expression with the given denominator.
$$\frac{5}{x^2 - 4} = \frac{}{(x-2)(x+2)(x-4)}$$

Solution: First we factor the denominator $x^2 - 4$ as $(x-2)(x+2)$. If we multiply the original denominator $(x-2)(x+2)$ by $x-4$, the result is the new denominator $(x-2)(x+2)(x-4)$. Thus, we multiply by 1 in the form of $\frac{x-4}{x-4}$.

$$\frac{5}{x^2 - 4} = \frac{5}{(x-2)(x+2)} = \frac{5}{(x-2)(x+2)} \cdot \frac{x-4}{x-4}$$

$$= \frac{5(x-4)}{(x-2)(x+2)(x-4)}$$

$$= \frac{5x - 20}{(x-2)(x+2)(x-4)}$$

■ Work Practice 10

Answer
10. $\dfrac{3x - 9}{(x+5)(x-5)(x-3)}$

Vocabulary, Readiness & Video Check

Use the choices below to fill in each blank. Not all choices will be used.

$\dfrac{9}{22}$ $\dfrac{5}{22}$ $\dfrac{9}{11}$ $\dfrac{5}{11}$ $\dfrac{ac}{b}$ $\dfrac{a-c}{b}$ $\dfrac{a+c}{b}$ $\dfrac{5-6+x}{x}$ $\dfrac{5-(6+x)}{x}$

1. $\dfrac{7}{11} + \dfrac{2}{11} =$ _____

2. $\dfrac{7}{11} - \dfrac{2}{11} =$ _____

3. $\dfrac{a}{b} + \dfrac{c}{b} =$ _____

4. $\dfrac{a}{b} - \dfrac{c}{b} =$ _____

5. $\dfrac{5}{x} - \dfrac{6+x}{x} =$ _____

Martin-Gay Interactive Videos

See Video 12.3

Watch the section lecture video and answer the following questions.

Objective A 6. In Example 3, why is it important to place parentheses around the second numerator when writing as one expression?

Objective B 7. In Examples 4 and 5, we factor the denominators completely. How does this help determine the LCD?

Objective C 8. Based on Example 6, complete the following statements: To write an equivalent rational expression, we multiply the _____ of a rational expression by the same expression as the denominator. This means we're multiplying the original rational expression by a factor of _____ and therefore not changing the _____ of the original expression.

12.3 Exercise Set MyLab Math

Objective A *Add or subtract as indicated. Simplify the result if possible. See Examples 1 through 3.*

1. $\dfrac{a}{13} + \dfrac{9}{13}$

2. $\dfrac{x+1}{7} + \dfrac{6}{7}$

3. $\dfrac{4m}{3n} + \dfrac{5m}{3n}$

4. $\dfrac{3p}{2q} + \dfrac{11p}{2q}$

5. $\dfrac{4m}{m-6} - \dfrac{24}{m-6}$

6. $\dfrac{8y}{y-2} - \dfrac{16}{y-2}$

7. $\dfrac{9}{3+y} + \dfrac{y+1}{3+y}$

8. $\dfrac{9}{y+9} + \dfrac{y-5}{y+9}$

9. $\dfrac{5x^2+4x}{x-1} - \dfrac{6x+3}{x-1}$

10. $\dfrac{x^2+9x}{x+7} - \dfrac{4x+14}{x+7}$

11. $\dfrac{4a}{a^2+2a-15} - \dfrac{12}{a^2+2a-15}$

12. $\dfrac{3y}{y^2+3y-10} - \dfrac{6}{y^2+3y-10}$

13. $\dfrac{2x+3}{x^2-x-30} - \dfrac{x-2}{x^2-x-30}$

14. $\dfrac{3x-1}{x^2+5x-6} - \dfrac{2x-7}{x^2+5x-6}$

15. $\dfrac{2x+1}{x-3} + \dfrac{3x+6}{x-3}$

16. $\dfrac{4p-3}{2p+7} + \dfrac{3p+8}{2p+7}$

17. $\dfrac{2x^2}{x-5} - \dfrac{25+x^2}{x-5}$

18. $\dfrac{6x^2}{2x-5} - \dfrac{25+2x^2}{2x-5}$

19. $\dfrac{5x+4}{x-1} - \dfrac{2x+7}{x-1}$

20. $\dfrac{7x+1}{x-4} - \dfrac{2x+21}{x-4}$

Objective B *Find the LCD for each list of rational expressions. See Examples 4 through 8.*

21. $\dfrac{19}{2x}, \dfrac{5}{4x^3}$

22. $\dfrac{17x}{4y^5}, \dfrac{2}{8y}$

23. $\dfrac{9}{8x}, \dfrac{3}{2x+4}$

24. $\dfrac{1}{6y}, \dfrac{3x}{4y+12}$

25. $\dfrac{2}{x+3}, \dfrac{5}{x-2}$

26. $\dfrac{-6}{x-1}, \dfrac{4}{x+5}$

27. $\dfrac{x}{x+6}, \dfrac{10}{3x+18}$

28. $\dfrac{12}{x+5}, \dfrac{x}{4x+20}$

29. $\dfrac{8x^2}{(x-6)^2}, \dfrac{13x}{5x-30}$

30. $\dfrac{9x^2}{7x-14}, \dfrac{6x}{(x-2)^2}$

31. $\dfrac{1}{3x+3}, \dfrac{8}{2x^2+4x+2}$

32. $\dfrac{19x+5}{4x-12}, \dfrac{3}{2x^2-12x+18}$

33. $\dfrac{5}{x-8}, \dfrac{3}{8-x}$

34. $\dfrac{2x+5}{3x-7}, \dfrac{5}{7-3x}$

35. $\dfrac{5x+1}{x^2+3x-4}, \dfrac{3x}{x^2+2x-3}$

36. $\dfrac{4}{x^2+4x+3}, \dfrac{4x-2}{x^2+10x+21}$

37. $\dfrac{2x}{3x^2+4x+1}, \dfrac{7}{2x^2-x-1}$

38. $\dfrac{3x}{4x^2+5x+1}, \dfrac{5}{3x^2-2x-1}$

39. $\dfrac{1}{x^2-16}, \dfrac{x+6}{2x^3-8x^2}$

40. $\dfrac{5}{x^2-25}, \dfrac{x+9}{3x^3-15x^2}$

Objective C *Rewrite each rational expression as an equivalent rational expression with the given denominator. See Examples 9 and 10.*

41. $\dfrac{3}{2x} = \dfrac{}{4x^2}$
42. $\dfrac{3}{9y^5} = \dfrac{}{72y^9}$
43. $\dfrac{6}{3a} = \dfrac{}{12ab^2}$
44. $\dfrac{5}{4y^2x} = \dfrac{}{32y^3x^2}$

45. $\dfrac{9}{2x+6} = \dfrac{}{2y(x+3)}$
46. $\dfrac{4x+1}{3x+6} = \dfrac{}{3y(x+2)}$
47. $\dfrac{9a+2}{5a+10} = \dfrac{}{5b(a+2)}$
48. $\dfrac{5+y}{2x^2+10} = \dfrac{}{4(x^2+5)}$

49. $\dfrac{x}{x^3+6x^2+8x} = \dfrac{}{x(x+4)(x+2)(x+1)}$
50. $\dfrac{5x}{x^3+2x^2-3x} = \dfrac{}{x(x-1)(x-5)(x+3)}$

51. $\dfrac{9y-1}{15x^2-30} = \dfrac{}{30x^2-60}$
52. $\dfrac{6m-5}{3x^2-9} = \dfrac{}{12x^2-36}$

Mixed Practice (Sections 12.2 and 12.3) *Perform the indicated operations.*

53. $\dfrac{5x}{7} + \dfrac{9x}{7}$
54. $\dfrac{5x}{7} \cdot \dfrac{9x}{7}$
55. $\dfrac{x+3}{4} \div \dfrac{2x-1}{4}$
56. $\dfrac{x+3}{4} - \dfrac{2x-1}{4}$

57. $\dfrac{x^2}{x-6} - \dfrac{5x+6}{x-6}$
58. $\dfrac{-2x}{x^3-8x} + \dfrac{3x}{x^3-8x}$
59. $\dfrac{x^2+5x}{x^2-25} \cdot \dfrac{3x-15}{x^2}$
60. $\dfrac{-2x}{x^3-8x} \div \dfrac{3x}{x^3-8x}$

61. $\dfrac{x^3+7x^2}{3x^3-x^2} \div \dfrac{5x^2+36x+7}{9x^2-1}$
62. $\dfrac{12x-6}{x^2+3x} \cdot \dfrac{4x^2+13x+3}{4x^2-1}$

Review

Perform each indicated operation. See Section 4.5.

63. $\dfrac{2}{3} + \dfrac{5}{7}$
64. $\dfrac{9}{10} - \dfrac{3}{5}$
65. $\dfrac{2}{6} - \dfrac{3}{4}$
66. $\dfrac{11}{15} + \dfrac{5}{9}$
67. $\dfrac{1}{12} + \dfrac{3}{20}$
68. $\dfrac{7}{30} + \dfrac{3}{18}$

Concept Extensions

For Exercises 69 and 70, see the Concept Check in this section.

69. Choose the correct LCD of $\dfrac{11a^3}{4a-20}$ and $\dfrac{15a^3}{(a-5)^2}$.
    a. $4a(a-5)(a+5)$ b. $a-5$
    c. $(a-5)^2$ d. $4(a-5)^2$
    e. $(4a-20)(a-5)^2$

70. Choose the correct LCD of $\dfrac{5}{14x^2}$ and $\dfrac{y}{6x^3}$.
    a. $84x^5$ b. $84x^3$
    c. $42x^3$ d. $42x^5$

For Exercises 71 and 72, an algebra student approaches you with each incorrect solution. Find the error and correct the work shown below.

71. $\dfrac{2x-6}{x-5} - \dfrac{x+4}{x-5}$
    $= \dfrac{2x-6-x+4}{x-5}$
    $= \dfrac{x-2}{x-5}$

72. $\dfrac{x}{x+3} + \dfrac{2}{x+3}$
    $= \dfrac{x+2}{x+3}$
    $= \dfrac{2}{3}$

Multiple choice. Select the correct result.

73. $\dfrac{3}{x} + \dfrac{y}{x} =$

a. $\dfrac{3+y}{x^2}$ b. $\dfrac{3+y}{2x}$ c. $\dfrac{3+y}{x}$

74. $\dfrac{3}{x} - \dfrac{y}{x} =$

a. $\dfrac{3-y}{x^2}$ b. $\dfrac{3-y}{2x}$ c. $\dfrac{3-y}{x}$

75. $\dfrac{3}{x} \cdot \dfrac{y}{x} =$

a. $\dfrac{3y}{x}$ b. $\dfrac{3y}{x^2}$ c. $3y$

76. $\dfrac{3}{x} \div \dfrac{y}{x} =$

a. $\dfrac{3}{y}$ b. $\dfrac{y}{3}$ c. $\dfrac{3}{x^2 y}$

Write each rational expression as an equivalent expression with a denominator of $x - 2$.

77. $\dfrac{5}{2-x}$

78. $\dfrac{8y}{2-x}$

79. $-\dfrac{7+x}{2-x}$

80. $\dfrac{x-3}{-(x-2)}$

81. A square has a side of length $\dfrac{5}{x-2}$ meters. Express its perimeter as a rational expression.

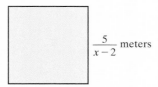

△ **82.** A trapezoid has sides of the indicated lengths. Find its perimeter.

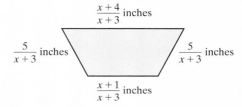

83. Write two rational expressions with the same denominator whose sum is $\dfrac{5}{3x-1}$.

84. Write two rational expressions with the same denominator whose difference is $\dfrac{x-7}{x^2+1}$.

85. The planet Mercury revolves around the Sun in 88 Earth days. It takes Jupiter 4332 Earth days to make one revolution around the Sun. (*Source:* National Space Science Data Center) If the two planets are aligned as shown in the figure, how long will it take for them to align again?

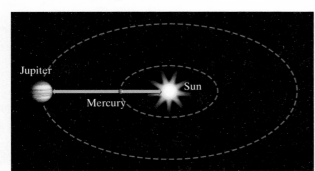

86. You are throwing a barbecue and you want to make sure that you purchase the same number of hot dogs as hot dog buns. Hot dogs come 8 to a package and hot dog buns come 12 to a package. What is the least number of each type of package you should buy?

87. Write some instructions to help a friend who is having difficulty finding the LCD of two rational expressions.

88. In your own words, describe how to add or subtract two rational expressions with the same denominator.

89. Explain why the LCD of the rational expressions $\dfrac{7}{x+1}$ and $\dfrac{9x}{(x+1)^2}$ is $(x+1)^2$ and not $(x+1)^3$.

90. Explain the similarities between subtracting $\dfrac{3}{8}$ from $\dfrac{7}{8}$ and subtracting $\dfrac{6}{x+3}$ from $\dfrac{9}{x+3}$.

12.4 Adding and Subtracting Rational Expressions with Different Denominators

Objective

A Add and Subtract Rational Expressions with Different Denominators.

Objective A Adding and Subtracting Rational Expressions with Different Denominators

Let's add $\frac{3}{8}$ and $\frac{1}{6}$. In the previous section, we found the LCD of 8 and 6 to be 24. Now let's write equivalent fractions with denominator 24 by multiplying by different forms of 1.

$$\frac{3}{8} = \frac{3}{8} \cdot 1 = \frac{3}{8} \cdot \frac{3}{3} = \frac{3 \cdot 3}{8 \cdot 3} = \frac{9}{24}$$

$$\frac{1}{6} = \frac{1}{6} \cdot 1 = \frac{1}{6} \cdot \frac{4}{4} = \frac{1 \cdot 4}{6 \cdot 4} = \frac{4}{24}$$

Now that the denominators are the same, we may add.

$$\frac{3}{8} + \frac{1}{6} = \frac{9}{24} + \frac{4}{24} = \frac{9 + 4}{24} = \frac{13}{24}$$

We add or subtract rational expressions the same way. You may want to use the steps below.

To Add or Subtract Rational Expressions with Different Denominators

Step 1: Find the LCD of the rational expressions.

Step 2: Rewrite each rational expression as an equivalent expression whose denominator is the LCD found in Step 1.

Step 3: Add or subtract numerators and write the sum or difference over the common denominator.

Step 4: Simplify or write the rational expression in lowest terms.

Practice 1

Perform each indicated operation.

a. $\frac{y}{5} - \frac{3y}{15}$ **b.** $\frac{5}{8x} + \frac{11}{10x^2}$

Example 1 Perform each indicated operation.

a. $\frac{a}{4} - \frac{2a}{8}$ **b.** $\frac{3}{10x^2} + \frac{7}{25x}$

Solution:

a. First, we must find the LCD. Since $4 = 2^2$ and $8 = 2^3$, the LCD $= 2^3 = 8$. Next we write each fraction as an equivalent fraction with the denominator 8, and then we subtract.

$$\frac{a}{4} = \frac{a}{4} \cdot 1 = \frac{a}{4} \cdot \frac{2}{2} = \frac{a \cdot 2}{4 \cdot 2} = \frac{2a}{8}$$

$$\frac{a}{4} - \frac{2a}{8} = \frac{2a}{8} - \frac{2a}{8} = \frac{2a - 2a}{8} = \frac{0}{8} = 0$$

Notice that we wrote $\frac{a}{4}$ as the equivalent expression $\frac{2a}{8}$. Multiplying by a form of 1 means we multiply the numerator and the denominator by the same number. Since this is so, we will start using the shorthand notation on the next page.

Answers

1. a. 0   **b.** $\frac{25x + 44}{40x^2}$

Section 12.4 | Adding and Subtracting Rational Expressions

$$\frac{a}{4} = \frac{a(2)}{4(2)} = \frac{2a}{8}$$

Multiplying the numerator and denominator by 2 is the same as multiplying by $\frac{2}{2}$ or 1.

b. Since $10x^2 = 2 \cdot 5 \cdot x \cdot x$ and $25x = 5 \cdot 5 \cdot x$, the LCD $= 2 \cdot 5^2 \cdot x^2 = 50x^2$. We write each fraction as an equivalent fraction with a denominator of $50x^2$.

$$\frac{3}{10x^2} + \frac{7}{25x} = \frac{3(5)}{10x^2(5)} + \frac{7(2x)}{25x(2x)}$$

$$= \frac{15}{50x^2} + \frac{14x}{50x^2}$$

$$= \frac{15 + 14x}{50x^2} \qquad \text{Add numerators. Write the sum over the common denominator.}$$

Work Practice 1

Example 2 Subtract: $\dfrac{6x}{x^2 - 4} - \dfrac{3}{x + 2}$

Solution: Since $x^2 - 4 = (x + 2)(x - 2)$, the LCD $= (x + 2)(x - 2)$. We write equivalent expressions with the LCD as denominators.

$$\frac{6x}{x^2 - 4} - \frac{3}{x + 2} = \frac{6x}{(x + 2)(x - 2)} - \frac{3(x - 2)}{(x + 2)(x - 2)}$$

$$= \frac{6x - 3(x - 2)}{(x + 2)(x - 2)} \qquad \text{Subtract numerators. Write the difference over the common denominator.}$$

$$= \frac{6x - 3x + 6}{(x + 2)(x - 2)} \qquad \text{Apply the distributive property in the numerator.}$$

$$= \frac{3x + 6}{(x + 2)(x - 2)} \qquad \text{Combine like terms in the numerator.}$$

Next we factor the numerator to see if this rational expression can be simplified.

$$\frac{3x + 6}{(x + 2)(x - 2)} = \frac{3\,(x + 2)}{(x + 2)\,(x - 2)} \qquad \text{Factor.}$$

$$= \frac{3}{x - 2} \qquad \text{Apply the fundamental principle to simplify.}$$

Work Practice 2

Practice 2

Subtract: $\dfrac{10x}{x^2 - 9} - \dfrac{5}{x + 3}$

Example 3 Add: $\dfrac{2}{3t} + \dfrac{5}{t + 1}$

Solution: The LCD is $3t(t + 1)$. We write each rational expression as an equivalent rational expression with a denominator of $3t(t + 1)$.

$$\frac{2}{3t} + \frac{5}{t + 1} = \frac{2(t + 1)}{3t(t + 1)} + \frac{5(3t)}{(t + 1)(3t)}$$

$$= \frac{2(t + 1) + 5(3t)}{3t(t + 1)} \qquad \text{Add numerators. Write the sum over the common denominator.}$$

$$= \frac{2t + 2 + 15t}{3t(t + 1)} \qquad \text{Apply the distributive property in the numerator.}$$

$$= \frac{17t + 2}{3t(t + 1)} \qquad \text{Combine like terms in the numerator.}$$

Work Practice 3

Practice 3

Add: $\dfrac{5}{7x} + \dfrac{2}{x + 1}$

Answers

2. $\dfrac{5}{x - 3}$ **3.** $\dfrac{19x + 5}{7x(x + 1)}$

Practice 4

Subtract: $\dfrac{10}{x-6} - \dfrac{15}{6-x}$

Example 4 Subtract: $\dfrac{7}{x-3} - \dfrac{9}{3-x}$

Solution: To find a common denominator, we notice that $x-3$ and $3-x$ are opposites. That is, $3-x = -(x-3)$. We write the denominator $3-x$ as $-(x-3)$ and simplify.

$$\dfrac{7}{x-3} - \dfrac{9}{3-x} = \dfrac{7}{x-3} - \dfrac{9}{-(x-3)}$$

$$= \dfrac{7}{x-3} - \dfrac{-9}{x-3} \qquad \text{Apply } \dfrac{a}{-b} = \dfrac{-a}{b}.$$

$$= \dfrac{7-(-9)}{x-3} \qquad \text{Subtract numerators. Write the difference over the common denominator.}$$

$$= \dfrac{16}{x-3}$$

Work Practice 4

Practice 5

Add: $2 + \dfrac{x}{x+5}$

Example 5 Add: $1 + \dfrac{m}{m+1}$

Solution: Recall that 1 is the same as $\dfrac{1}{1}$. The LCD of $\dfrac{1}{1}$ and $\dfrac{m}{m+1}$ is $m+1$.

$$1 + \dfrac{m}{m+1} = \dfrac{1}{1} + \dfrac{m}{m+1} \qquad \text{Write 1 as } \dfrac{1}{1}.$$

$$= \dfrac{1(m+1)}{1(m+1)} + \dfrac{m}{m+1} \qquad \text{Multiply both the numerator and the denominator of } \dfrac{1}{1} \text{ by } m+1.$$

$$= \dfrac{m+1+m}{m+1} \qquad \text{Add numerators. Write the sum over the common denominator.}$$

$$= \dfrac{2m+1}{m+1} \qquad \text{Combine like terms in the numerator.}$$

Work Practice 5

Practice 6

Subtract: $\dfrac{4}{3x^2+2x} - \dfrac{3x}{12x+8}$

Example 6 Subtract: $\dfrac{3}{2x^2+x} - \dfrac{2x}{6x+3}$

Solution: First, we factor the denominators.

$$\dfrac{3}{2x^2+x} - \dfrac{2x}{6x+3} = \dfrac{3}{x(2x+1)} - \dfrac{2x}{3(2x+1)}$$

The LCD is $3x(2x+1)$. We write equivalent expressions with denominator $3x(2x+1)$.

$$\dfrac{3}{x(2x+1)} - \dfrac{2x}{3(2x+1)} = \dfrac{3(3)}{x(2x+1)(3)} - \dfrac{2x(x)}{3(2x+1)(x)}$$

$$= \dfrac{9-2x^2}{3x(2x+1)} \qquad \text{Subtract numerators. Write the difference over the common denominator.}$$

Work Practice 6

Answers

4. $\dfrac{25}{x-6}$ **5.** $\dfrac{3x+10}{x+5}$ **6.** $\dfrac{16-3x^2}{4x(3x+2)}$

Section 12.4 | Adding and Subtracting Rational Expressions

Example 7 Add: $\dfrac{2x}{x^2 + 2x + 1} + \dfrac{x}{x^2 - 1}$

Solution: First we factor the denominators.

$\dfrac{2x}{x^2 + 2x + 1} + \dfrac{x}{x^2 - 1}$

$= \dfrac{2x}{(x + 1)(x + 1)} + \dfrac{x}{(x + 1)(x - 1)}$ Rewrite each expression with LCD $(x + 1)(x + 1)(x - 1)$.

$= \dfrac{2x(x - 1)}{(x + 1)(x + 1)(x - 1)} + \dfrac{x(x + 1)}{(x + 1)(x - 1)(x + 1)}$

$= \dfrac{2x(x - 1) + x(x + 1)}{(x + 1)^2(x - 1)}$ Add numerators. Write the sum over the common denominator.

$= \dfrac{2x^2 - 2x + x^2 + x}{(x + 1)^2(x - 1)}$ Apply the distributive property in the numerator.

$= \dfrac{3x^2 - x}{(x + 1)^2(x - 1)}$ or $\dfrac{x(3x - 1)}{(x + 1)^2(x - 1)}$

The numerator was factored as a last step to see if the rational expression could be simplified further. Since there are no factors common to the numerator and the denominator, we can't simplify further.

Work Practice 7

Practice 7

Add: $\dfrac{6x}{x^2 + 4x + 4} + \dfrac{x}{x^2 - 4}$

Answer

7. $\dfrac{x(7x - 10)}{(x + 2)^2(x - 2)}$

Vocabulary, Readiness & Video Check

Multiple choice. Choose the correct response.

1. $\dfrac{3}{7x} + \dfrac{5}{7} =$

   a. $\dfrac{3}{7x} + \dfrac{5}{7x} = \dfrac{8}{7x}$ **b.** $\dfrac{3}{7x} + \dfrac{5}{7} \cdot \dfrac{x}{x} = \dfrac{3 + 5x}{7x}$ **c.** $\dfrac{3}{7x} + \dfrac{5}{7} \cdot \dfrac{x}{x} = \dfrac{8x}{7x}$ or $\dfrac{8}{7}$

2. $\dfrac{1}{x} + \dfrac{2}{x^2} =$

   a. $\dfrac{1}{x} \cdot \dfrac{x}{x} + \dfrac{2}{x^2} = \dfrac{x + 2}{x^2}$ **b.** $\dfrac{3}{x^3}$ **c.** $\dfrac{1}{x} \cdot \dfrac{x}{x} + \dfrac{2}{x^2} = \dfrac{3x}{x^2}$ or $\dfrac{3}{x}$

Martin-Gay Interactive Videos   Watch the section lecture video and answer the following question.

See Video 12.4

Objective A   **3.** What special case is shown in Example 2, and what's the purpose of presenting it?

12.4 Exercise Set MyLab Math

Objective A *Perform each indicated operation. Simplify if possible. See Examples 1 through 7.*

1. $\dfrac{4}{2x} + \dfrac{9}{3x}$
2. $\dfrac{15}{7a} + \dfrac{8}{6a}$
3. $\dfrac{15a}{b} + \dfrac{6b}{5}$
4. $\dfrac{4c}{d} - \dfrac{8d}{5}$
5. $\dfrac{3}{x} + \dfrac{5}{2x^2}$
6. $\dfrac{14}{3x^2} + \dfrac{6}{x}$
7. $\dfrac{6}{x+1} + \dfrac{10}{2x+2}$
8. $\dfrac{8}{x+4} - \dfrac{3}{3x+12}$
9. $\dfrac{3}{x+2} - \dfrac{2x}{x^2-4}$
10. $\dfrac{5}{x-4} + \dfrac{4x}{x^2-16}$
11. $\dfrac{3}{4x} + \dfrac{8}{x-2}$
12. $\dfrac{5}{y^2} - \dfrac{y}{2y+1}$
13. $\dfrac{6}{x-3} + \dfrac{8}{3-x}$
14. $\dfrac{15}{y-4} + \dfrac{20}{4-y}$
15. $\dfrac{9}{x-3} + \dfrac{9}{3-x}$
16. $\dfrac{5}{a-7} + \dfrac{5}{7-a}$
17. $\dfrac{-8}{x^2-1} - \dfrac{7}{1-x^2}$
18. $\dfrac{-9}{25x^2-1} + \dfrac{7}{1-25x^2}$
19. $\dfrac{5}{x} + 2$
20. $\dfrac{7}{x^2} - 5x$
21. $\dfrac{5}{x-2} + 6$
22. $\dfrac{6y}{y+5} + 1$
23. $\dfrac{y+2}{y+3} - 2$
24. $\dfrac{7}{2x-3} - 3$
25. $\dfrac{-x+2}{x} - \dfrac{x-6}{4x}$
26. $\dfrac{-y+1}{y} - \dfrac{2y-5}{3y}$
27. $\dfrac{5x}{x+2} - \dfrac{3x-4}{x+2}$
28. $\dfrac{7x}{x-3} - \dfrac{4x+9}{x-3}$
29. $\dfrac{3x^4}{7} - \dfrac{4x^2}{21}$
30. $\dfrac{5x}{6} + \dfrac{11x^2}{2}$
31. $\dfrac{1}{x+3} - \dfrac{1}{(x+3)^2}$
32. $\dfrac{5x}{(x-2)^2} - \dfrac{3}{x-2}$
33. $\dfrac{4}{5b} + \dfrac{1}{b-1}$
34. $\dfrac{1}{y+5} + \dfrac{2}{3y}$
35. $\dfrac{2}{m} + 1$
36. $\dfrac{6}{x} - 1$
37. $\dfrac{2x}{x-7} - \dfrac{x}{x-2}$
38. $\dfrac{9x}{x-10} - \dfrac{x}{x-3}$
39. $\dfrac{6}{1-2x} - \dfrac{4}{2x-1}$
40. $\dfrac{10}{3n-4} - \dfrac{5}{4-3n}$

41. $\dfrac{7}{(x+1)(x-1)} + \dfrac{8}{(x+1)^2}$

42. $\dfrac{5}{(x+1)(x+5)} - \dfrac{2}{(x+5)^2}$

43. $\dfrac{x}{x^2-1} - \dfrac{2}{x^2-2x+1}$

44. $\dfrac{x}{x^2-4} - \dfrac{5}{x^2-4x+4}$

45. $\dfrac{3a}{2a+6} - \dfrac{a-1}{a+3}$

46. $\dfrac{1}{2x+2y} - \dfrac{y}{x+y}$

47. $\dfrac{y-1}{2y+3} + \dfrac{3}{(2y+3)^2}$

48. $\dfrac{x-6}{5x+1} + \dfrac{6}{(5x+1)^2}$

49. $\dfrac{5}{2-x} + \dfrac{x}{2x-4}$

50. $\dfrac{-1}{a-2} + \dfrac{4}{4-2a}$

51. $\dfrac{15}{x^2+6x+9} + \dfrac{2}{x+3}$

52. $\dfrac{2}{x^2+4x+4} + \dfrac{1}{x+2}$

53. $\dfrac{13}{x^2-5x+6} - \dfrac{5}{x-3}$

54. $\dfrac{-7}{y^2-3y+2} - \dfrac{2}{y-1}$

55. $\dfrac{70}{m^2-100} + \dfrac{7}{2(m+10)}$

56. $\dfrac{27}{y^2-81} + \dfrac{3}{2(y+9)}$

57. $\dfrac{x+8}{x^2-5x-6} + \dfrac{x+1}{x^2-4x-5}$

58. $\dfrac{x+4}{x^2+12x+20} + \dfrac{x+1}{x^2+8x-20}$

59. $\dfrac{5}{4n^2-12n+8} - \dfrac{3}{3n^2-6n}$

60. $\dfrac{6}{5y^2-25y+30} - \dfrac{2}{4y^2-8y}$

Mixed Practice (Sections 12.2, 12.3 and 12.4) *Perform the indicated operations. Addition, subtraction, multiplication, and division of rational expressions are included here.*

61. $\dfrac{15x}{x+8} \cdot \dfrac{2x+16}{3x}$

62. $\dfrac{9z+5}{15} \cdot \dfrac{5z}{81z^2-25}$

63. $\dfrac{8x+7}{3x+5} - \dfrac{2x-3}{3x+5}$

64. $\dfrac{2z^2}{4z-1} - \dfrac{z-2z^2}{4z-1}$

65. $\dfrac{5a+10}{18} \div \dfrac{a^2-4}{10a}$

66. $\dfrac{9}{x^2-1} \div \dfrac{12}{3x+3}$

67. $\dfrac{5}{x^2-3x+2} + \dfrac{1}{x-2}$

68. $\dfrac{4}{2x^2+5x-3} + \dfrac{2}{x+3}$

Review

Solve each linear or quadratic equation. See Sections 9.3 and 11.6.

69. $3x + 5 = 7$

70. $5x - 1 = 8$

71. $2x^2 - x - 1 = 0$

72. $4x^2 - 9 = 0$

73. $4(x + 6) + 3 = -3$

74. $2(3x + 1) + 15 = -7$

Concept Extensions

Perform each indicated operation.

75. $\dfrac{3}{x} - \dfrac{2x}{x^2 - 1} + \dfrac{5}{x + 1}$

76. $\dfrac{5}{x - 2} + \dfrac{7x}{x^2 - 4} - \dfrac{11}{x}$

77. $\dfrac{5}{x^2 - 4} + \dfrac{2}{x^2 - 4x + 4} - \dfrac{3}{x^2 - x - 6}$

78. $\dfrac{8}{x^2 + 6x + 5} - \dfrac{3x}{x^2 + 4x - 5} + \dfrac{2}{x^2 - 1}$

79. $\dfrac{9}{x^2 + 9x + 14} - \dfrac{3x}{x^2 + 10x + 21} + \dfrac{x + 4}{x^2 + 5x + 6}$

80. $\dfrac{x + 10}{x^2 - 3x - 4} - \dfrac{8}{x^2 + 6x + 5} - \dfrac{9}{x^2 + x - 20}$

81. A board of length $\dfrac{3}{x + 4}$ inches was cut into two pieces. If one piece is $\dfrac{1}{x - 4}$ inches, express the length of the other piece as a rational expression.

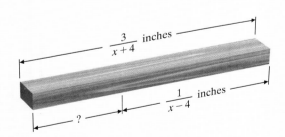

△ **82.** The length of a rectangle is $\dfrac{3}{y - 5}$ feet, while its width is $\dfrac{2}{y}$ feet. Find its perimeter and then find its area.

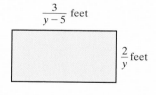

83. In ice hockey, penalty killing percentage is a statistic calculated as $1 - \dfrac{G}{P}$, where G = opponent's power play goals and P = opponent's power play opportunities. Simplify this expression.

84. The dose of medicine prescribed for a child depends on the child's age A in years and the adult dose D for the medication. Two expressions that give a child's dose are Young's Rule, $\dfrac{DA}{A + 12}$, and Cowling's Rule, $\dfrac{D(A + 1)}{24}$. Find an expression for the difference in the doses given by these expressions.

85. Explain when the LCD of the rational expressions in a sum is the product of the denominators.

86. Explain when the LCD is the same as one of the denominators of a rational expression to be added or subtracted.

87. Two angles are said to be complementary if the sum of their measures is 90°. If one angle measures $\dfrac{40}{x}$ degrees, find the measure of its complement.

△ 88. Two angles are said to be supplementary if the sum of their measures is 180°. If one angle measures $\dfrac{x+2}{x}$ degrees, find the measure of its supplement.

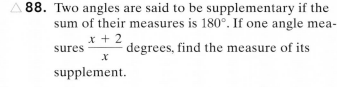

89. In your own words, explain how to add two rational expressions with different denominators.

90. In your own words, explain how to subtract two rational expressions with different denominators.

12.5 Solving Equations Containing Rational Expressions

Objective A Solving Equations Containing Rational Expressions

Objectives

A Solve Equations Containing Rational Expressions.

B Solve Equations Containing Rational Expressions for a Specified Variable.

In Chapters 4 and 9, we solved equations containing fractions. In this section, we continue the work we began in those chapters by solving equations containing rational expressions. For example,

$$\dfrac{x}{2} + \dfrac{8}{3} = \dfrac{1}{6} \quad \text{and} \quad \dfrac{4x}{x^2 + x - 30} + \dfrac{2}{x - 5} = \dfrac{1}{x + 6}$$

are equations containing rational expressions. To solve equations such as these, we use the multiplication property of equality to clear the equation of fractions by multiplying both sides of the equation by the LCD.

Example 1 Solve: $\dfrac{x}{2} + \dfrac{8}{3} = \dfrac{1}{6}$

Solution: The LCD of denominators 2, 3, and 6 is 6, so we multiply both sides of the equation by 6.

$$6\left(\dfrac{x}{2} + \dfrac{8}{3}\right) = 6\left(\dfrac{1}{6}\right)$$

$$6\left(\dfrac{x}{2}\right) + 6\left(\dfrac{8}{3}\right) = 6\left(\dfrac{1}{6}\right) \quad \text{Apply the distributive property.}$$

$$3 \cdot x + 16 = 1 \quad \text{Multiply and simplify.}$$

$$3x = -15 \quad \text{Subtract 16 from both sides.}$$

$$x = -5 \quad \text{Divide both sides by 3.}$$

(Continued on next page)

Practice 1

Solve: $\dfrac{x}{4} + \dfrac{4}{5} = \dfrac{1}{20}$

Helpful Hint Make sure that *each* term is multiplied by the LCD.

Answer
1. $x = -3$

Check: To check, we replace x with -5 in the original equation.

$$\frac{-5}{2} + \frac{8}{3} \overset{?}{=} \frac{1}{6} \quad \text{Replace } x \text{ with } -5.$$

$$\frac{1}{6} = \frac{1}{6} \quad \text{True}$$

This number checks, so the solution is -5.

■ Work Practice 1

Practice 2

Solve:

Helpful Hint
Multiply *each* term by 18.

Example 2 Solve: $\dfrac{t-4}{2} - \dfrac{t-3}{9} = \dfrac{5}{18}$

Solution: The LCD of denominators 2, 9, and 18 is 18, so we multiply both sides of the equation by 18.

$$18\left(\frac{t-4}{2} - \frac{t-3}{9}\right) = 18\left(\frac{5}{18}\right)$$

$$18\left(\frac{t-4}{2}\right) - 18\left(\frac{t-3}{9}\right) = 18\left(\frac{5}{18}\right) \quad \text{Apply the distributive property.}$$

$$9(t-4) - 2(t-3) = 5 \quad \text{Simplify.}$$

$$9t - 36 - 2t + 6 = 5 \quad \text{Use the distributive property.}$$

$$7t - 30 = 5 \quad \text{Combine like terms.}$$

$$7t = 35$$

$$t = 5 \quad \text{Solve for } t.$$

Check: $\dfrac{t-4}{2} - \dfrac{t-3}{9} = \dfrac{5}{18}$

$$\frac{5-4}{2} - \frac{5-3}{9} \overset{?}{=} \frac{5}{18} \quad \text{Replace } t \text{ with 5.}$$

$$\frac{1}{2} - \frac{2}{9} \overset{?}{=} \frac{5}{18} \quad \text{Simplify.}$$

$$\frac{5}{18} = \frac{5}{18} \quad \text{True}$$

The solution is 5.

■ Work Practice 2

Recall from Section 12.1 that a rational expression is defined for all real numbers except those that make the denominator of the expression 0. This means that if an equation contains *rational expressions with variables in the denominator*, we must be certain that the proposed solution does not make the denominator 0. If replacing the variable with the proposed solution makes the denominator 0, the rational expression is undefined and this proposed solution must be rejected.

Answer
2. $x = -6$

Section 12.5 | Solving Equations Containing Rational Expressions

Example 3 Solve: $3 - \dfrac{6}{x} = x + 8$

Helpful Hint: Notice that Example 3 contains our first equation with a variable in the denominator.

Solution: In this equation, 0 cannot be a solution because if x is 0, the rational expression $\dfrac{6}{x}$ is undefined. The LCD is x, so we multiply both sides of the equation by x.

$$x\left(3 - \dfrac{6}{x}\right) = x(x + 8)$$

$$x(3) - x\left(\dfrac{6}{x}\right) = x \cdot x + x \cdot 8 \quad \text{Apply the distributive property.}$$

$$3x - 6 = x^2 + 8x \quad \text{Simplify.}$$

Helpful Hint: Multiply *each* term by x.

Now we write the quadratic equation in standard form and solve for x.

$$0 = x^2 + 5x + 6$$
$$0 = (x + 3)(x + 2) \quad \text{Factor.}$$
$$x + 3 = 0 \quad \text{or} \quad x + 2 = 0 \quad \text{Set each factor equal to 0 and solve.}$$
$$x = -3 \qquad\qquad x = -2$$

Notice that neither -3 nor -2 makes the denominator in the original equation equal to 0.

Check: To check these solutions, we replace x in the original equation by -3, and then by -2.

If $x = -3$:
$$3 - \dfrac{6}{x} = x + 8$$
$$3 - \dfrac{6}{-3} \stackrel{?}{=} -3 + 8$$
$$3 - (-2) \stackrel{?}{=} 5$$
$$5 = 5 \quad \text{True}$$

If $x = -2$:
$$3 - \dfrac{6}{x} = x + 8$$
$$3 - \dfrac{6}{-2} \stackrel{?}{=} -2 + 8$$
$$3 - (-3) \stackrel{?}{=} 6$$
$$6 = 6 \quad \text{True}$$

Both -3 and -2 are solutions.

Work Practice 3

Practice 3

Solve: $2 + \dfrac{6}{x} = x + 7$

The following steps may be used to solve an equation containing rational expressions.

To Solve an Equation Containing Rational Expressions

Step 1: Multiply both sides of the equation by the LCD of all rational expressions in the equation.
Step 2: Remove any grouping symbols and solve the resulting equation.
Step 3: Check the solution in the original equation.

Answer
3. $x = -6, x = 1$

Practice 4

Solve:

$$\frac{2}{x+3} + \frac{3}{x-3} = \frac{-2}{x^2-9}$$

Example 4 Solve: $\dfrac{4x}{x^2+x-30} + \dfrac{2}{x-5} = \dfrac{1}{x+6}$

Solution: The denominator $x^2 + x - 30$ factors as $(x+6)(x-5)$. The LCD is then $(x+6)(x-5)$, so we multiply both sides of the equation by this LCD.

$$(x+6)(x-5)\left(\frac{4x}{x^2+x-30} + \frac{2}{x-5}\right) = (x+6)(x-5)\left(\frac{1}{x+6}\right) \quad \text{Multiply by the LCD.}$$

$$(x+6)(x-5) \cdot \frac{4x}{x^2+x-30} + (x+6)(x-5) \cdot \frac{2}{x-5} \quad \text{Apply the distributive property.}$$

$$= (x+6)(x-5) \cdot \frac{1}{x+6}$$

$$\begin{aligned} 4x + 2(x+6) &= x - 5 &&\text{Simplify.} \\ 4x + 2x + 12 &= x - 5 &&\text{Apply the distributive property.} \\ 6x + 12 &= x - 5 &&\text{Combine like terms.} \\ 5x &= -17 \\ x &= -\frac{17}{5} &&\text{Divide both sides by 5.} \end{aligned}$$

Check: Check by replacing x with $-\dfrac{17}{5}$ in the original equation. The solution is $-\dfrac{17}{5}$.

■ Work Practice 4

Practice 5

Solve: $\dfrac{5x}{x-1} = \dfrac{5}{x-1} + 3$

Example 5 Solve: $\dfrac{2x}{x-4} = \dfrac{8}{x-4} + 1$

Solution: Multiply both sides by the LCD, $x - 4$.

$$(x-4)\left(\frac{2x}{x-4}\right) = (x-4)\left(\frac{8}{x-4} + 1\right) \quad \text{Multiply by the LCD.}$$

$$(x-4) \cdot \frac{2x}{x-4} = (x-4) \cdot \frac{8}{x-4} + (x-4) \cdot 1 \quad \text{Use the distributive property.}$$

$$\begin{aligned} 2x &= 8 + (x-4) &&\text{Simplify.} \\ 2x &= 4 + x \\ x &= 4 \end{aligned}$$

Helpful Hint

As we can see from Example 5, it is important to check the proposed solution(s) in the original equation.

Notice that 4 makes a denominator 0 in the original equation. Therefore, 4 is *not* a solution and this equation has *no solution*.

■ Work Practice 5

✓ **Concept Check** When can we clear fractions by multiplying through by the LCD?

a. When adding or subtracting rational expressions
b. When solving an equation containing rational expressions
c. Both of these
d. Neither of these

Answers

4. $x = -1$ **5.** no solution

✓ Concept Check Answer

b

Example 6

Solve: $x + \dfrac{14}{x-2} = \dfrac{7x}{x-2} + 1$

Solution: Notice the denominators in this equation. We can see that 2 can't be a solution. The LCD is $x - 2$, so we multiply both sides of the equation by $x - 2$.

$$(x-2)\left(x + \dfrac{14}{x-2}\right) = (x-2)\left(\dfrac{7x}{x-2} + 1\right)$$

$$(x-2)(x) + (x-2)\left(\dfrac{14}{x-2}\right) = (x-2)\left(\dfrac{7x}{x-2}\right) + (x-2)(1)$$

$x^2 - 2x + 14 = 7x + x - 2$	Simplify.
$x^2 - 2x + 14 = 8x - 2$	Combine like terms.
$x^2 - 10x + 16 = 0$	Write the quadratic equation in standard form.
$(x-8)(x-2) = 0$	Factor.
$x - 8 = 0$ or $x - 2 = 0$	Set each factor equal to 0.
$x = 8 \qquad x = 2$	Solve.

As we have already noted, 2 can't be a solution of the original equation. So we need replace x only with 8 in the original equation. We find that 8 is a solution; the only solution is 8.

■ Work Practice 6

Practice 6

Solve:
$$x - \dfrac{6}{x+3} = \dfrac{2x}{x+3} + 2$$

Objective B Solving Equations for a Specified Variable ▶

The last example in this section is an equation containing several variables, and we are directed to solve for one of the variables. The steps used in the preceding examples can be applied to solve equations for a specified variable as well.

Example 7

Solve $\dfrac{1}{a} + \dfrac{1}{b} = \dfrac{1}{x}$ for x.

Solution: (This type of equation often models a work problem, as we shall see in the next section.) The LCD is abx, so we multiply both sides by abx.

$$abx\left(\dfrac{1}{a} + \dfrac{1}{b}\right) = abx\left(\dfrac{1}{x}\right)$$

$$abx\left(\dfrac{1}{a}\right) + abx\left(\dfrac{1}{b}\right) = abx \cdot \dfrac{1}{x}$$

$bx + ax = ab$	Simplify.
$x(b + a) = ab$	Factor out x from each term on the left side.
$\dfrac{x(b+a)}{b+a} = \dfrac{ab}{b+a}$	Divide both sides by $b + a$.
$x = \dfrac{ab}{b+a}$	Simplify.

This equation is now solved for x.

■ Work Practice 7

Practice 7

Solve $\dfrac{1}{a} + \dfrac{1}{b} = \dfrac{1}{x}$ for a.

Answers

6. $x = 4$ **7.** $a = \dfrac{bx}{b-x}$

Vocabulary, Readiness & Video Check

Multiple choice. Choose the correct response.

1. Multiply both sides of the equation $\frac{3x}{2} + 5 = \frac{1}{4}$ by 4. The result is:
   a. $3x + 5 = 1$ **b.** $6x + 5 = 1$ **c.** $6x + 20 = 1$ **d.** $6x + 9 = 1$

2. Multiply both sides of the equation $\frac{1}{x} - \frac{3}{5x} = 2$ by $5x$. The result is:
   a. $1 - 3 = 10x$ **b.** $5 - 3 = 10x$ **c.** $5x - 3 = 10x$ **d.** $5 - 3 = 7x$

Choose the correct LCD for the fractions in each equation.

3. Equation: $\frac{9}{x} + \frac{3}{4} = \frac{1}{12}$; LCD: _____
   a. $4x$ **b.** $12x$ **c.** $48x$ **d.** x

4. Equation: $\frac{8}{3x} - \frac{1}{x} = \frac{7}{9}$; LCD: _____
   a. x **b.** $3x$ **c.** $27x$ **d.** $9x$

5. Equation: $\frac{9}{x-1} = \frac{7}{(x-1)^2}$; LCD: _____
   a. $(x-1)^2$ **b.** $(x-1)$ **c.** $(x-1)^3$ **d.** 63

6. Equation: $\frac{1}{x-2} - \frac{3}{x^2-4} = 8$; LCD: _____
   a. $(x-2)$ **b.** $(x+2)$ **c.** (x^2-4) **d.** $(x-2)(x^2-4)$

Martin-Gay Interactive Videos

See Video 12.5

Watch the section lecture video and answer the following questions.

Objective A

7. After multiplying through by the LCD and then simplifying, why is it important to take a moment and determine whether we have a linear or a quadratic equation before we finish solving the problem?

8. From Examples 2–5, what extra step is needed when checking solutions to an equation containing rational expressions?

Objective B

9. The steps for solving Example 6 for a specified variable are the same as what other steps? How do we treat this specified variable?

Section 12.5 | Solving Equations Containing Rational Expressions

12.5 Exercise Set MyLab Math

Objective A *Solve each equation and check each solution. See Examples 1 through 3.*

1. $\dfrac{x}{5} + 3 = 9$
2. $\dfrac{x}{5} - 2 = 9$
3. $\dfrac{x}{2} + \dfrac{5x}{4} = \dfrac{x}{12}$
4. $\dfrac{x}{6} + \dfrac{4x}{3} = \dfrac{x}{18}$

5. $2 - \dfrac{8}{x} = 6$
6. $5 + \dfrac{4}{x} = 1$
7. $2 + \dfrac{10}{x} = x + 5$
8. $6 + \dfrac{5}{y} = y - \dfrac{2}{y}$

9. $\dfrac{a}{5} = \dfrac{a-3}{2}$
10. $\dfrac{b}{5} = \dfrac{b+2}{6}$
11. $\dfrac{x-3}{5} + \dfrac{x-2}{2} = \dfrac{1}{2}$
12. $\dfrac{a+5}{4} + \dfrac{a+5}{2} = \dfrac{a}{8}$

Solve each equation and check each proposed solution. See Examples 4 through 6.

13. $\dfrac{3}{2a-5} = -1$
14. $\dfrac{6}{4-3x} = -3$
15. $\dfrac{4y}{y-4} + 5 = \dfrac{5y}{y-4}$

16. $\dfrac{2a}{a+2} - 5 = \dfrac{7a}{a+2}$
17. $2 + \dfrac{3}{a-3} = \dfrac{a}{a-3}$
18. $\dfrac{2y}{y-2} - \dfrac{4}{y-2} = 4$

19. $\dfrac{1}{x+3} + \dfrac{6}{x^2-9} = 1$
20. $\dfrac{1}{x+2} + \dfrac{4}{x^2-4} = 1$
21. $\dfrac{2y}{y+4} + \dfrac{4}{y+4} = 3$

22. $\dfrac{5y}{y+1} - \dfrac{3}{y+1} = 4$
23. $\dfrac{2x}{x+2} - 2 = \dfrac{x-8}{x-2}$
24. $\dfrac{4y}{y-3} - 3 = \dfrac{3y-1}{y+3}$

Solve each equation. See Examples 1 through 6.

25. $\dfrac{2}{y} + \dfrac{1}{2} = \dfrac{5}{2y}$
26. $\dfrac{6}{3y} + \dfrac{3}{y} = 1$
27. $\dfrac{a}{a-6} = \dfrac{-2}{a-1}$

28. $\dfrac{5}{x-6} = \dfrac{x}{x-2}$
29. $\dfrac{11}{2x} + \dfrac{2}{3} = \dfrac{7}{2x}$
30. $\dfrac{5}{3} - \dfrac{3}{2x} = \dfrac{3}{2}$

31. $\dfrac{2}{x-2} + 1 = \dfrac{x}{x+2}$
32. $1 + \dfrac{3}{x+1} = \dfrac{x}{x-1}$
33. $\dfrac{x+1}{3} - \dfrac{x-1}{6} = \dfrac{1}{6}$

34. $\dfrac{3x}{5} - \dfrac{x-6}{3} = -\dfrac{2}{5}$
35. $\dfrac{t}{t-4} = \dfrac{t+4}{6}$
36. $\dfrac{15}{x+4} = \dfrac{x-4}{x}$

37. $\dfrac{y}{2y+2} + \dfrac{2y-16}{4y+4} = \dfrac{2y-3}{y+1}$
38. $\dfrac{1}{x+2} = \dfrac{4}{x^2-4} - \dfrac{1}{x-2}$

39. $\dfrac{4r-4}{r^2+5r-14} + \dfrac{2}{r+7} = \dfrac{1}{r-2}$

40. $\dfrac{3}{x+3} = \dfrac{12x+19}{x^2+7x+12} - \dfrac{5}{x+4}$

41. $\dfrac{x+1}{x+3} = \dfrac{x^2-11x}{x^2+x-6} - \dfrac{x-3}{x-2}$

42. $\dfrac{2t+3}{t-1} - \dfrac{2}{t+3} = \dfrac{5-6t}{t^2+2t-3}$

Objective B *Solve each equation for the indicated variable. See Example 7.*

43. $R = \dfrac{E}{I}$ for I (Electronics: resistance of a circuit)

44. $T = \dfrac{V}{Q}$ for Q (Water purification: settling time)

45. $T = \dfrac{2U}{B+E}$ for B (Merchandising: stock turnover rate)

46. $i = \dfrac{A}{t+B}$ for t (Hydrology: rainfall intensity)

47. $B = \dfrac{705w}{h^2}$ for w (Health: body-mass index)

48. $\dfrac{A}{W} = L$ for W (Geometry: area of a rectangle)

49. $N = R + \dfrac{V}{G}$ for G (Urban forestry: tree plantings per year)

50. $C = \dfrac{D(A+1)}{24}$ for A (Medicine: Cowling's Rule for child's dose)

51. $\dfrac{C}{\pi r} = 2$ for r (Geometry: circumference of a circle)

52. $W = \dfrac{CE^2}{2}$ for C (Electronics: energy stored in a capacitor)

53. $\dfrac{1}{y} + \dfrac{1}{3} = \dfrac{1}{x}$ for x

54. $\dfrac{1}{5} + \dfrac{2}{y} = \dfrac{1}{x}$ for x

Review

Translating *Write each phrase as an expression. See Sections 3.2 and 9.1.*

55. The reciprocal of x

56. The reciprocal of $x + 1$

57. The reciprocal of x, added to the reciprocal of 2

58. The reciprocal of x, subtracted from the reciprocal of 5

Answer each question.

59. If a tank is filled in 3 hours, what part of the tank is filled in 1 hour?

60. If a strip of beach is cleaned in 4 hours, what part of the beach is cleaned in 1 hour?

Concept Extensions

61. Explain the difference between solving an equation such as $\frac{x}{2} + \frac{3}{4} = \frac{x}{4}$ for x and performing an operation such as adding $\frac{x}{2} + \frac{3}{4}$.

62. When solving an equation such as $\frac{y}{4} = \frac{y}{2} - \frac{1}{4}$, we may multiply all terms by 4. When subtracting two rational expressions such as $\frac{y}{2} - \frac{1}{4}$, we may not. Explain why.

Determine whether each of the following is an equation or an expression. If it is an equation, then solve it for its variable. If it is an expression, perform the indicated operation.

63. $\frac{1}{x} + \frac{5}{9}$

64. $\frac{1}{x} + \frac{5}{9} = \frac{2}{3}$

65. $\frac{5}{x-1} - \frac{2}{x} = \frac{5}{x(x-1)}$

66. $\frac{5}{x-1} - \frac{2}{x}$

Recall that two angles are supplementary if the sum of their measures is 180°. Find the measures of the supplementary angles.

67.

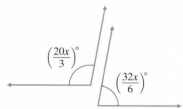

△ 68.

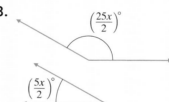

Recall that two angles are complementary if the sum of their measures is 90°. Find the measures of the complementary angles.

69.

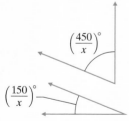

△ 70.

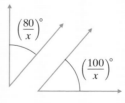

Solve each equation.

71. $\frac{4}{a^2 + 4a + 3} + \frac{2}{a^2 + a - 6} - \frac{3}{a^2 - a - 2} = 0$

72. $\frac{-4}{a^2 + 2a - 8} + \frac{1}{a^2 + 9a + 20} = \frac{-4}{a^2 + 3a - 10}$

Integrated Review — Sections 12.1–12.5

Summary on Rational Expressions

It is important to know the difference between performing operations with rational expressions and solving an equation containing rational expressions. Study the examples below.

Performing Operations with Rational Expressions

Adding: $\dfrac{1}{x} + \dfrac{1}{x+5} = \dfrac{1\cdot(x+5)}{x(x+5)} + \dfrac{1\cdot x}{x(x+5)} = \dfrac{x+5+x}{x(x+5)} = \dfrac{2x+5}{x(x+5)}$

Subtracting: $\dfrac{3}{x} - \dfrac{5}{x^2 y} = \dfrac{3\cdot xy}{x\cdot xy} - \dfrac{5}{x^2 y} = \dfrac{3xy - 5}{x^2 y}$

Multiplying: $\dfrac{2}{x}\cdot\dfrac{5}{x-1} = \dfrac{2\cdot 5}{x(x-1)} = \dfrac{10}{x(x-1)}$

Dividing: $\dfrac{4}{2x+1} \div \dfrac{x-3}{x} = \dfrac{4}{2x+1}\cdot\dfrac{x}{x-3} = \dfrac{4x}{(2x+1)(x-3)}$

Solving an Equation Containing Rational Expressions

To solve an equation containing rational expressions, we clear the equation of fractions by multiplying both sides by the LCD.

$$\dfrac{3}{x} - \dfrac{5}{x-1} = \dfrac{1}{x(x-1)} \quad \text{Note that } x \text{ can't be 0 or 1.}$$

$$x(x-1)\left(\dfrac{3}{x}\right) - x(x-1)\left(\dfrac{5}{x-1}\right) = x(x-1)\cdot\dfrac{1}{x(x-1)} \quad \text{Multiply both sides by the LCD.}$$

$$3(x-1) - 5x = 1 \quad \text{Simplify.}$$
$$3x - 3 - 5x = 1 \quad \text{Use the distributive property.}$$
$$-2x - 3 = 1 \quad \text{Combine like terms.}$$
$$-2x = 4 \quad \text{Add 3 to both sides.}$$
$$x = -2 \quad \text{Divide both sides by } -2.$$

Don't forget to check to make sure our proposed solution of -2 does not make any denominators 0. If it does, this proposed solution is *not* a solution of the equation. -2 checks and is the solution.

Determine whether each of the following is an equation or an expression. If it is an equation, solve it for its variable. If it is an expression, perform the indicated operation.

1. $\dfrac{1}{x} + \dfrac{2}{3}$

2. $\dfrac{3}{a} + \dfrac{5}{6}$

3. $\dfrac{1}{x} + \dfrac{2}{3} = \dfrac{3}{x}$

4. $\dfrac{3}{a} + \dfrac{5}{6} = 1$

5. $\dfrac{2}{x+1} - \dfrac{1}{x}$

6. $\dfrac{4}{x-3} - \dfrac{1}{x}$

7. $\dfrac{2}{x+1} - \dfrac{1}{x} = 1$

8. $\dfrac{4}{x-3} - \dfrac{1}{x} = \dfrac{6}{x(x-3)}$

9. $\dfrac{15x}{x+8}\cdot\dfrac{2x+16}{3x}$

10. $\dfrac{9z+5}{15}\cdot\dfrac{5z}{81z^2-25}$

1. $\dfrac{2x+1}{x-3} + \dfrac{3x+6}{x-3}$

12. $\dfrac{4p-3}{2p+7} + \dfrac{3p+8}{2p+7}$

3. $\dfrac{x+5}{7} = \dfrac{8}{2}$

14. $\dfrac{1}{2} = \dfrac{x+1}{8}$

5. $\dfrac{5a+10}{18} \div \dfrac{a^2-4}{10a}$

16. $\dfrac{9}{x^2-1} \div \dfrac{12}{3x+3}$

7. $\dfrac{x+2}{3x-1} + \dfrac{5}{(3x-1)^2}$

18. $\dfrac{4}{(2x-5)^2} + \dfrac{x+1}{2x-5}$

9. $\dfrac{x-7}{x} - \dfrac{x+2}{5x}$

20. $\dfrac{10x-9}{x} - \dfrac{x-4}{3x}$

21. $\dfrac{3}{x+3} = \dfrac{5}{x^2-9} - \dfrac{2}{x-3}$

22. $\dfrac{9}{x^2-4} + \dfrac{2}{x+2} = \dfrac{-1}{x-2}$

23. Explain the difference between solving an equation such as $\dfrac{x}{3} + \dfrac{1}{6} = \dfrac{x}{6}$ for x and performing an operation such as adding $\dfrac{x}{3} + \dfrac{1}{6}$.

24. When solving an equation such as $\dfrac{y}{6} = \dfrac{y}{3} - \dfrac{1}{6}$, we may multiply all terms by 6. When subtracting two rational expressions such as $\dfrac{y}{3} - \dfrac{1}{6}$, we may not. Explain why.

11. _____

12. _____

13. _____

14. _____

15. _____

16. _____

17. _____

18. _____

19. _____

20. _____

21. _____

22. _____

23. _____

24. _____

12.6 Rational Equations and Problem Solving

Objectives

A Solve Problems About Numbers.

B Solve Problems About Work.

C Solve Problems About Distance.

Objective A Solving Problems About Numbers

In this section, we solve problems that can be modeled by equations containing rational expressions. To solve these problems, we use the same problem-solving steps that were first introduced in Section 3.4. In our first example, our goal is to find an unknown number.

Example 1 Finding an Unknown Number

The quotient of a number and 6, minus $\frac{5}{3}$, is the quotient of the number and 2. Find the number.

Solution:

1. UNDERSTAND. Read and reread the problem. Suppose that the unknown number is 2; then we see if the quotient of 2 and 6, or $\frac{2}{6}$, minus $\frac{5}{3}$ is equal to the quotient of 2 and 2, or $\frac{2}{2}$.

$$\frac{2}{6} - \frac{5}{3} = \frac{1}{3} - \frac{5}{3} = -\frac{4}{3}, \text{ not } \frac{2}{2}$$

Don't forget that the purpose of a proposed solution is to better understand the problem.
Let $x =$ the unknown number.

2. TRANSLATE.

In words:	the quotient of x and 6	minus	$\frac{5}{3}$	is	the quotient of x and 2
	↓	↓	↓	↓	↓
Translate:	$\frac{x}{6}$	$-$	$\frac{5}{3}$	$=$	$\frac{x}{2}$

3. SOLVE. Here, we solve the equation $\frac{x}{6} - \frac{5}{3} = \frac{x}{2}$. We begin by multiplying both sides of the equation by the LCD, 6.

$$6\left(\frac{x}{6} - \frac{5}{3}\right) = 6\left(\frac{x}{2}\right)$$

$$6\left(\frac{x}{6}\right) - 6\left(\frac{5}{3}\right) = 6\left(\frac{x}{2}\right) \quad \text{Apply the distributive property.}$$

$$x - 10 = 3x \quad \text{Simplify.}$$

$$-10 = 2x \quad \text{Subtract } x \text{ from both sides.}$$

$$\frac{-10}{2} = \frac{2x}{2} \quad \text{Divide both sides by 2.}$$

$$-5 = x \quad \text{Simplify.}$$

4. INTERPRET.

Check: To check, we verify that "the quotient of -5 and 6 minus $\frac{5}{3}$ is the quotient of -5 and 2," or $-\frac{5}{6} - \frac{5}{3} = -\frac{5}{2}$.

State: The unknown number is -5.

■ Work Practice 1

Practice 1

The quotient of a number and 2, minus $\frac{1}{3}$, is the quotient of the number and 6. Find the number.

Answer
1. 1

Section 12.6 | Rational Equations and Problem Solving

Objective B Solving Problems About Work

The next example is often called a work problem. Work problems usually involve people or machines doing a certain task.

Example 2 Finding Work Rates

Sam Waterton and Frank Schaffer work in a plant that manufactures automobiles. Sam can complete a quality control tour of the plant in 3 hours while his assistant, Frank, needs 7 hours to complete the same job. The regional manager is coming to inspect the plant facilities, so both Sam and Frank are directed to complete a quality control tour together. How long will this take?

Practice 2

Guillaume Beauchesne and Greg Langacker volunteer at a local recycling plant. Guillaume can sort a batch of recyclables in 2 hours alone while his friend Greg needs 3 hours to complete the same job. If they work together, how long will it take them to sort one batch?

Solution:

1. UNDERSTAND. Read and reread the problem. The key idea here is the relationship between the **time** (hours) it takes to complete the job and the **part of the job** completed in 1 unit of time (hour). For example, if the **time** it takes Sam to complete the job is 3 hours, the **part of the job** he can complete in 1 hour is $\frac{1}{3}$. Similarly, Frank can complete $\frac{1}{7}$ of the job in 1 hour.

Let x = the **time** in hours it takes Sam and Frank to complete the job together. Then $\frac{1}{x}$ = the **part of the job** they complete in 1 hour.

	Hours to Complete Total Job	Part of Job Completed in 1 Hour
Sam	3	$\frac{1}{3}$
Frank	7	$\frac{1}{7}$
Together	x	$\frac{1}{x}$

2. TRANSLATE.

In words:	part of job Sam completes in 1 hour	added to	part of job Frank completes in 1 hour	is equal to	part of job they complete together in 1 hour
Translate:	$\frac{1}{3}$	$+$	$\frac{1}{7}$	$=$	$\frac{1}{x}$

3. SOLVE. Here, we solve the equation $\frac{1}{3} + \frac{1}{7} = \frac{1}{x}$. We begin by multiplying both sides of the equation by the LCD, $21x$.

$$21x\left(\frac{1}{3}\right) + 21x\left(\frac{1}{7}\right) = 21x\left(\frac{1}{x}\right)$$
$$7x + 3x = 21 \qquad \text{Simplify.}$$
$$10x = 21$$
$$x = \frac{21}{10} \quad \text{or} \quad 2\frac{1}{10} \text{ hours}$$

Answer

2. $1\frac{1}{5}$ hours

(Continued on next page)

938 Chapter 12 | Rational Expressions

4. **INTERPRET.**

   Check: Our proposed solution is $2\frac{1}{10}$ hours. This proposed solution is reasonable since $2\frac{1}{10}$ hours is more than half of Sam's time and less than half Frank's time. Check this solution in the originally *stated* problem.

   State: Sam and Frank can complete the quality control tour in $2\frac{1}{10}$ hours.

■ Work Practice 2

✓ **Concept Check** Solve $E = mc^2$

   a. for m **b.** for c^2

Objective C **Solving Problems About Distance**

Next we look at a problem solved by the distance formula,

$$d = r \cdot t$$

Example 3 Finding Speeds of Vehicles

A car travels 180 miles in the same time that a truck travels 120 miles. If the car's speed is 20 miles per hour faster than the truck's, find the car's speed and the truck's speed.

Solution:

1. **UNDERSTAND.** Read and reread the problem. Suppose that the truck's speed is 45 miles per hour. Then the car's speed is 20 miles per hour faster, or 65 miles per hour.

   We are given that the car travels 180 miles in the same time that the truck travels 120 miles. To find the time it takes the car to travel 180 miles, remember that since $d = rt$, we know that $\dfrac{d}{r} = t$.

      Car's Time   Truck's Time

   $t = \dfrac{d}{r} = \dfrac{180}{65} = 2\dfrac{50}{65} = 2\dfrac{10}{13}$ hours $t = \dfrac{d}{r} = \dfrac{120}{45} = 2\dfrac{30}{45} = 2\dfrac{2}{3}$ hours

   Since the times are not the same, our proposed solution is not correct. But we have a better understanding of the problem.

   Let $x = $ the speed of the truck.

   Since the car's speed is 20 miles per hour faster than the truck's, then

      $x + 20 = $ the speed of the car

   Use the formula $d = r \cdot t$ or **d**istance $= $ **r**ate $\cdot$ **t**ime. Prepare a chart to organize the information in the problem.

Practice 3

A car travels 600 miles in the same time that a motorcycle travels 450 miles. If the car's speed is 15 miles per hour faster than the motorcycle's, find the speed of the car and the speed of the motorcycle.

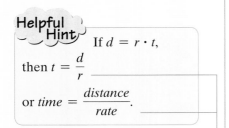

Helpful Hint

If $d = r \cdot t$,

then $t = \dfrac{d}{r}$

or time $= \dfrac{distance}{rate}$.

Answer

3. car: 60 mph; motorcycle: 45 mph

✓ **Concept Check Answers**

a. $m = \dfrac{E}{c^2}$ **b.** $c^2 = \dfrac{E}{m}$

	Distance	=	Rate	·	Time
Truck	120		x		$\dfrac{120}{x}$ ← distance / ← rate
Car	180		$x + 20$		$\dfrac{180}{x + 20}$ ← distance / ← rate

Section 12.6 | Rational Equations and Problem Solving

2. TRANSLATE. Since the car and the truck traveled the same amount of time, we have that

In words: car's time = truck's time

Translate: $\dfrac{180}{x+20} = \dfrac{120}{x}$

3. SOLVE. We begin by multiplying both sides of the equation by the LCD, $x(x+20)$, or cross multiplying.

$$\dfrac{180}{x+20} = \dfrac{120}{x}$$

$180x = 120(x+20)$

$180x = 120x + 2400$ Use the distributive property.

$60x = 2400$ Subtract $120x$ from both sides.

$x = 40$ Divide both sides by 60.

4. INTERPRET. The speed of the truck is 40 miles per hour. The speed of the car must then be $x + 20$ or 60 miles per hour.

Check: Find the time it takes the car to travel 180 miles and the time it takes the truck to travel 120 miles.

Car's Time
$t = \dfrac{d}{r} = \dfrac{180}{60} = 3$ hours

Truck's Time
$t = \dfrac{d}{r} = \dfrac{120}{40} = 3$ hours

Since both travel the same amount of time, the proposed solution is correct.

State: The car's speed is 60 miles per hour and the truck's speed is 40 miles per hour.

Work Practice 3

Vocabulary, Readiness & Video Check

Without solving algebraically, select the best choice for each exercise.

1. One person can complete a job in 7 hours. A second person can complete the same job in 5 hours. How long will it take them to complete the job if they work together?

  a. more than 7 hours
  b. between 5 and 7 hours
  c. less than 5 hours

2. One inlet pipe can fill a pond in 30 hours. A second inlet pipe can fill the same pond in 25 hours. How long before the pond is filled if both inlet pipes are on?

  a. less than 25 hours
  b. between 25 and 30 hours
  c. more than 30 hours

Fill in a Table Given the variable in the first column, use the phrase in the second column to translate to an expression, and then continue to the phrase in the third column to translate to another expression.

3.	A number: x	The reciprocal of the number:	The reciprocal of the number, decreased by 3:
4.	A number: y	The reciprocal of the number:	The reciprocal of the number, increased by 2:
5.	A number: z	The sum of the number and 5:	The reciprocal of the sum of the number and 5:
6.	A number: x	The difference of the number and 1:	The reciprocal of the difference of the number and 1:
7.	A number: y	Twice the number:	Eleven divided by twice the number:
8.	A number: z	Triple the number:	Negative ten divided by triple the number:

Martin-Gay Interactive Videos

See Video 12.6

Watch the section lecture video and answer the following questions.

Objective A 9. What words or phrases in Example 1 told you to translate to an equation containing rational expressions?

Objective B 10. From Example 2, how can you determine a somewhat reasonable answer to a work problem before you begin to solve it?

Objective C 11. The following problem is worded like Example 3 in the video, but uses different quantities.

A car travels 325 miles in the same time that a motorcycle travels 290 miles. If the car's speed is 7 miles per hour faster than the motorcycle's, find the speed of the car and the speed of the motorcycle. Fill in the table and set up an equation based on this problem (do not solve). Use Example 3 in the video as a model for your work.

	d	=	r	·	t
Car					
Motorcycle					

12.6 Exercise Set MyLab Math

Objective A *Solve the following. See Example 1.*

1. Three times the reciprocal of a number equals 9 times the reciprocal of 6. Find the number.

2. Twelve divided by the sum of x and 2 equals the quotient of 4 and the difference of x and 2. Find x.

3. If twice a number added to 3 is divided by the number plus 1, the result is three halves. Find the number.

4. A number added to the product of 6 and the reciprocal of the number equals -5. Find the number.

Objective B *See Example 2.*

5. Smith Engineering found that an experienced surveyor surveys a roadbed in 4 hours. An apprentice surveyor needs 5 hours to survey the same stretch of road. If the two work together, find how long it takes them to complete the job.

6. An experienced bricklayer constructs a small wall in 3 hours. The apprentice completes the job in 6 hours. Find how long it takes if they work together.

7. In 2 minutes, a conveyor belt moves 300 pounds of recyclable aluminum from the delivery truck to a storage area. A smaller belt moves the same amount of cans the same distance in 6 minutes. If both belts are used, find how long it takes to move the cans to the storage area.

8. Find how long it takes the conveyor belts described in Exercise 7 to move 1200 pounds of cans. (*Hint:* Think of 1200 pounds as four 300-pound jobs.)

Section 12.6 Rational Equations and Problem Solving

Objective C See Example 3.

9. A jogger begins her workout by jogging to the park, a distance of 12 miles. She rests, then jogs home at the same speed but along a different route. This return trip is 18 miles and her time is one hour longer. Find her jogging speed. Complete the accompanying chart and use it to find her jogging speed.

	Distance	=	Rate	·	Time
Trip to Park	12				
Return Trip	18				

10. A boat can travel 9 miles upstream in the same amount of time it takes to travel 11 miles downstream. If the current of the river is 3 miles per hour, complete the chart below and use it to find the speed of the boat in still water.

	Distance	=	Rate	·	Time
Upstream	9		$r - 3$		
Downstream	11		$r + 3$		

11. A cyclist rode the first 20-mile portion of his workout at a constant speed. For the 16-mile cooldown portion of his workout, he reduced his speed by 2 miles per hour. Each portion of the workout took the same time. Find the cyclist's speed during the first portion and find his speed during the cooldown portion.

12. A semi truck travels 300 miles through the flatland in the same amount of time that it travels 180 miles through mountains. The rate of the truck is 20 miles per hour slower in the mountains than in the flatland. Find both the flatland rate and the mountain rate.

Objectives A B C Mixed Practice Solve the following. See Examples 1 through 3. (Note: Some exercises can be modeled by equations without rational expressions.)

13. One-fourth equals the quotient of a number and 8. Find the number.

14. Four times a number added to 5 is divided by 6. The result is $\frac{7}{2}$. Find the number.

15. Marcus and Tony work for Lombardo's Pipe and Concrete. Mr. Lombardo is preparing an estimate for a customer. He knows that Marcus lays a slab of concrete in 6 hours. Tony lays the same-size slab in 4 hours. If both work on the job and the cost of labor is $45.00 per hour, decide what the labor estimate should be.

16. Mr. Dodson can paint his house by himself in 4 days. His son needs an additional day to complete the job if he works by himself. If they work together, find how long it takes to paint the house.

17. A pilot can travel 400 miles with the wind in the same amount of time as 336 miles against the wind. Find the speed of the wind if the pilot's speed in still air is 230 miles per hour.

18. A fisherman on Pearl River rows 9 miles downstream in the same amount of time he rows 3 miles upstream. If the current is 6 miles per hour, find how long it takes him to cover the 12 miles.

19. Two divided by the difference of a number and 3, minus 4 divided by the sum of the number and 3, equals 8 times the reciprocal of the difference of the number squared and 9. What is the number?

20. If 15 times the reciprocal of a number is added to the ratio of 9 times the number minus 7 and the number plus 2, the result is 9. What is the number?

21. A pilot flies 630 miles with a tail wind of 35 miles per hour. Against the wind, he flies only 455 miles in the same amount of time. Find the rate of the plane in still air.

22. A marketing manager travels 1080 miles in a corporate jet and then an additional 240 miles by car. If the car ride takes one hour longer than the jet ride takes and if the rate of the jet is 6 times the rate of the car, find the time the manager travels by jet and find the time the manager travels by car.

23. The quotient of a number and 3, minus 1, equals $\frac{5}{3}$. Find the number.

24. The quotient of a number and 5, minus 1, equals $\frac{7}{5}$. Find the number.

25. Two hikers are 11 miles apart and walking toward each other. They meet in 2 hours. Find the rate of each hiker if one hiker walks 1.1 mph faster than the other.

26. On a 255-mile trip, Gary Alessandrini traveled at an average speed of 70 mph, got a speeding ticket, and then traveled at 60 mph for the remainder of the trip. If the entire trip took 4.5 hours and the speeding ticket stop took 30 minutes, how long did Gary speed before getting stopped?

27. One custodian cleans a suite of offices in 3 hours. When a second worker is asked to join the regular custodian, the job takes only $1\frac{1}{2}$ hours. How long does it take the second worker to do the same job alone?

28. One person proofreads copy for a small newspaper in 4 hours. If a second proofreader is also employed, the job can be done in $2\frac{1}{2}$ hours. How long does it take the second proofreader to do the same job alone?

29. A boater travels 16 miles per hour on the water on a still day. During one particularly windy day, he finds that he travels 48 miles with the wind behind him in the same amount of time that he travels 16 miles into the wind. Find the rate of the wind.

    Let x be the rate of the wind.

	r	$\cdot$	t	$=$	d
With wind	$16 + x$				48
Into wind	$16 - x$				16

30. The current on a portion of the Mississippi River is 3 miles per hour. A barge can go 6 miles upstream in the same amount of time it takes to go 10 miles downstream. Find the speed of the boat in still water.

    Let x be the speed of the boat in still water.

	r	$\cdot$	t	$=$	d
Upstream	$x - 3$				6
Downstream	$x + 3$				10

31. Currently, the Toyota Corolla is the best-selling car in the world. A driver of this car took a day trip around the Maine coastline driving at two different speeds. He drove 70 miles at a slower speed and 300 miles at a speed 40 miles per hour faster. If the time spent driving at the faster speed was twice that spent driving at the slower speed, find the two speeds during the trip. (*Source: Forbes*)

32. The second best-selling car in the world is the Volkswagen Golf. Suppose that during a test drive of two Golfs, one car travels 224 miles in the same time that the second car travels 175 miles. If the speed of the first car is 14 miles per hour faster than the speed of the second car, find the speed of both cars. (*Source: Forbes*)

33. A pilot can fly an MD-11 2160 miles with the wind in the same time she can fly 1920 miles against the wind. If the speed of the wind is 30 mph, find the speed of the plane in still air. (*Source:* Air Transport Association of America)

34. A pilot can fly a DC-10 1365 miles against the wind in the same time he can fly 1575 miles with the wind. If the speed of the plane in still air is 490 miles per hour, find the speed of the wind. (*Source:* Air Transport Association of America)

35. A jet plane traveling at 500 mph overtakes a propeller plane traveling at 200 mph that had a 2-hour head start. How far from the starting point are the planes?

36. How long will it take a bus traveling at 60 miles per hour to overtake a car traveling at 40 mph if the car had a 1.5-hour head start?

Section 12.6 | Rational Equations and Problem Solving

7. One pipe fills a storage pool in 20 hours. A second pipe fills the same pool in 15 hours. When a third pipe is added and all three are used to fill the pool, it takes only 6 hours. Find how long it takes the third pipe to do the job.

38. One pump fills a tank 2 times as fast as another pump. If the pumps work together, they fill the tank in 18 minutes. How long does it take each pump to fill the tank?

9. A car travels 280 miles in the same time that a motorcycle travels 240 miles. If the car's speed is 10 miles per hour faster than the motorcycle's, find the speed of the car and the speed of the motorcycle.

40. A bus traveled on a level road for 3 hours at an average speed 20 miles per hour faster than it traveled on a winding road. The time spent on the winding road was 4 hours. Find the average speed on the level road if the entire trip was 305 miles.

1. In 6 hours, an experienced cook prepares enough pies to supply a local restaurant's daily order. Another cook prepares the same number of pies in 7 hours. Together with a third cook, they prepare the pies in 2 hours. Find how long it takes the third cook to prepare the pies alone.

42. Mrs. Smith balances the company books in 8 hours. It takes her assistant 12 hours to do the same job. If they work together, find how long it takes them to balance the books.

3. Suppose two trains leave Holbrook, Arizona, at the same time, traveling in opposite directions. One train travels 10 mph faster than the other. In 3.5 hours, the trains are 322 miles apart. Find the speed of each train.

44. Suppose two cars leave Brinkley, Arkansas, at the same time, traveling in opposite directions. One car travels 8 mph faster than the other car. In 2.5 hours, the cars are 280 miles apart. Find the speed of each car.

eview

mplify. Follow the circled steps in the order shown. See Sections 4.3 and 4.4.

5. $\dfrac{\left.\dfrac{3}{4} + \dfrac{1}{4}\right\} \leftarrow \text{① Add.}}{\left.\dfrac{3}{8} + \dfrac{13}{8}\right\} \leftarrow \text{② Add.}}$

46. $\dfrac{\left.\dfrac{9}{5} + \dfrac{6}{5}\right\} \leftarrow \text{① Add.}}{\left.\dfrac{17}{6} + \dfrac{7}{6}\right\} \leftarrow \text{② Add.}}$

7. $\dfrac{\left.\dfrac{2}{5} + \dfrac{1}{5}\right\} \leftarrow \text{① Add.}}{\left.\dfrac{7}{10} + \dfrac{7}{10}\right\} \leftarrow \text{② Add.}} \leftarrow \text{③ Divide.}$

48. $\dfrac{\left.\dfrac{1}{4} + \dfrac{5}{4}\right\} \leftarrow \text{① Add.}}{\left.\dfrac{3}{8} + \dfrac{7}{8}\right\} \leftarrow \text{② Add.}} \leftarrow \text{③ Divide.}$

oncept Extensions

9. One pump fills a tank 3 times as fast as another pump. If the pumps work together, they fill the tank in 21 minutes. How long does it take each pump to fill the tank?

50. It takes 9 hours for pump A to fill a tank alone. Pump B takes 15 hours to fill the same tank alone. If pumps A, B, and C are used, the tank fills in 5 hours. How long does it take pump C to fill the tank alone?

1. Person A can complete a job in 5 hours, and person B can complete the same job in 3 hours. Without solving algebraically, discuss reasonable and unreasonable answers for how long it would take them to complete the job together.

52. For which of the following equations can we immediately use cross products to solve for x?

a. $\dfrac{2-x}{5} = \dfrac{1+x}{3}$ **b.** $\dfrac{2}{5} - x = \dfrac{1+x}{3}$

Solve. See the Concept Check in this section.

Solve $D = RT$

53. for R

54. for T

55. A hyena spots a giraffe 0.5 mile away and begins running toward it. The giraffe starts running away from the hyena just as the hyena begins running toward it. A hyena can run at a speed of 40 mph and a giraffe can run at 32 mph. How long will it take the hyena to overtake the giraffe? (*Source: The World Almanac and Book of Facts*)

56. The two fastest cars in the world are the Hennessey Venom GT and the Bugatti Veyron 16.4 Super Sport. At an international auto demonstration, the Hennessey traveled 390 miles in the same time the Bugatti traveled 363 miles. If the speed of the Hennessey was 18 mph faster than the speed of the Bugatti, find the speed of both cars. (*Source: Top Ten of Everything*)

12.7 Simplifying Complex Fractions

Objectives

A Simplify Complex Fractions Using Method 1.

B Simplify Complex Fractions Using Method 2.

A rational expression whose numerator or denominator or both numerator and denominator contain fractions is called a **complex rational expression** or a **complex fraction**. Some examples are

$$\dfrac{4}{2 - \dfrac{1}{2}} \qquad \dfrac{\dfrac{3}{2}}{\dfrac{4}{7} - x} \qquad \dfrac{\dfrac{1}{x+2}}{x + 2 - \dfrac{1}{x}} \quad \begin{array}{l} \leftarrow \text{Numerator of complex fraction} \\ \leftarrow \text{Main fraction bar} \\ \leftarrow \text{Denominator of complex fraction} \end{array}$$

Our goal in this section is to write complex fractions in simplest form. A complex fraction is in simplest form when it is in the form $\dfrac{P}{Q}$, where P and Q are polynomials that have no common factors.

Objective A Simplifying Complex Fractions— Method 1

In this section, two methods of simplifying complex fractions are presented. The first method presented uses the fact that the main fraction bar indicates division.

Section 12.7 | Simplifying Complex Fractions

> **Method 1: To Simplify a Complex Fraction**
>
> **Step 1:** Add or subtract fractions in the numerator or denominator so that the numerator is a single fraction and the denominator is a single fraction.
>
> **Step 2:** Perform the indicated division by multiplying the numerator of the complex fraction by the reciprocal of the denominator of the complex fraction.
>
> **Step 3:** Write the rational expression in lowest terms.

Example 1 Simplify the complex fraction $\dfrac{\frac{5}{8}}{\frac{2}{3}}$.

Solution: Since the numerator and denominator of the complex fraction are already single fractions, we proceed to Step 2: Perform the indicated division by multiplying the numerator $\dfrac{5}{8}$ by the reciprocal of the denominator $\dfrac{2}{3}$.

$$\frac{\frac{5}{8}}{\frac{2}{3}} = \frac{5}{8} \div \frac{2}{3} = \frac{5}{8} \cdot \frac{3}{2} = \frac{15}{16}$$

The reciprocal of $\dfrac{2}{3}$ is $\dfrac{3}{2}$.

Practice 1

Simplify the complex fraction $\dfrac{\frac{3}{7}}{\frac{5}{9}}$.

■ Work Practice 1

Example 2 Simplify: $\dfrac{\frac{2}{3} + \frac{1}{5}}{\frac{2}{3} - \frac{2}{9}}$

Solution: We simplify the numerator and denominator of the complex fraction separately. First we add $\dfrac{2}{3}$ and $\dfrac{1}{5}$ to obtain a single fraction in the numerator. Then we subtract $\dfrac{2}{9}$ from $\dfrac{2}{3}$ to obtain a single fraction in the denominator.

$$\dfrac{\frac{2}{3} + \frac{1}{5}}{\frac{2}{3} - \frac{2}{9}} = \dfrac{\frac{2(5)}{3(5)} + \frac{1(3)}{5(3)}}{\frac{2(3)}{3(3)} - \frac{2}{9}}$$ The LCD of the numerator's fractions is 15.

The LCD of the denominator's fractions is 9.

$$= \dfrac{\frac{10}{15} + \frac{3}{15}}{\frac{6}{9} - \frac{2}{9}}$$ Simplify.

$$= \dfrac{\frac{13}{15}}{\frac{4}{9}}$$

Add the numerator's fractions.

Subtract the denominator's fractions.

Practice 2

Simplify: $\dfrac{\frac{3}{4} - \frac{2}{3}}{\frac{1}{2} + \frac{3}{8}}$

(Continued on next page)

Answers

1. $\dfrac{27}{35}$ **2.** $\dfrac{2}{21}$

Next we perform the indicated division by multiplying the numerator of the complex fraction by the reciprocal of the denominator of the complex fraction.

$$\dfrac{\dfrac{13}{15}}{\dfrac{4}{9}} = \dfrac{13}{15} \cdot \dfrac{9}{4} \quad \text{The reciprocal of } \dfrac{4}{9} \text{ is } \dfrac{9}{4}.$$

$$= \dfrac{13 \cdot 3 \cdot 3}{3 \cdot 5 \cdot 4} = \dfrac{39}{20}$$

■ Work Practice 2

Practice 3

Simplify: $\dfrac{\dfrac{2}{5} - \dfrac{1}{x}}{\dfrac{2x}{15} - \dfrac{1}{3}}$

Example 3 Simplify: $\dfrac{\dfrac{1}{z} - \dfrac{1}{2}}{\dfrac{1}{3} - \dfrac{z}{6}}$

Solution: Subtract to get a single fraction in the numerator and a single fraction in the denominator of the complex fraction.

$$\dfrac{\dfrac{1}{z} - \dfrac{1}{2}}{\dfrac{1}{3} - \dfrac{z}{6}} = \dfrac{\dfrac{2}{2z} - \dfrac{z}{2z}}{\dfrac{2}{6} - \dfrac{z}{6}} \quad \begin{array}{l}\text{The LCD of the numerator's fractions is } 2z.\\ \\ \text{The LCD of the denominator's fractions is } 6.\end{array}$$

$$= \dfrac{\dfrac{2-z}{2z}}{\dfrac{2-z}{6}}$$

$$= \dfrac{2-z}{2z} \cdot \dfrac{6}{2-z} \quad \text{Multiply by the reciprocal of } \dfrac{2-z}{6}.$$

$$= \dfrac{2 \cdot 3 \cdot (2-z)}{2 \cdot z \cdot (2-z)} \quad \text{Factor.}$$

$$= \dfrac{3}{z} \quad \text{Write in lowest terms.}$$

■ Work Practice 3

Objective B Simplifying Complex Fractions—Method 2

Next we study a second method for simplifying complex fractions. In this method, we multiply the numerator and the denominator of the complex fraction by the LCD of all fractions in the complex fraction.

Method 2: To Simplify a Complex Fraction

Step 1: Find the LCD of all the fractions in the complex fraction.

Step 2: Multiply both the numerator and the denominator of the complex fraction by the LCD from Step 1.

Step 3: Perform the indicated operations and write the result in lowest terms.

We use Method 2 to rework Example 2.

Answer

3. $\dfrac{3}{x}$

Section 12.7 | Simplifying Complex Fractions 947

Example 4 Simplify: $\dfrac{\frac{2}{3} + \frac{1}{5}}{\frac{2}{3} - \frac{2}{9}}$

Solution: The LCD of $\frac{2}{3}, \frac{1}{5}, \frac{2}{3}$, and $\frac{2}{9}$ is 45, so we multiply the numerator and the denominator of the complex fraction by 45. Then we perform the indicated operations, and write in lowest terms.

$$\dfrac{\frac{2}{3} + \frac{1}{5}}{\frac{2}{3} - \frac{2}{9}} = \dfrac{45\left(\frac{2}{3} + \frac{1}{5}\right)}{45\left(\frac{2}{3} - \frac{2}{9}\right)}$$

$$= \dfrac{45\left(\frac{2}{3}\right) + 45\left(\frac{1}{5}\right)}{45\left(\frac{2}{3}\right) - 45\left(\frac{2}{9}\right)} \quad \text{Apply the distributive property.}$$

$$= \dfrac{30 + 9}{30 - 10} = \dfrac{39}{20} \quad \text{Simplify.}$$

Work Practice 4

Practice 4

Use Method 2 to simplify the complex fraction in Practice 2:

$$\dfrac{\frac{3}{4} - \frac{2}{3}}{\frac{1}{2} + \frac{3}{8}}$$

Helpful Hint

The same complex fraction was simplified using two different methods in Examples 2 and 4. Notice that the simplified results are the same.

Example 5 Simplify: $\dfrac{\frac{x+1}{y}}{\frac{x}{y} + 2}$

Solution: The LCD of $\dfrac{x+1}{y}, \dfrac{x}{y}$, and $\dfrac{2}{1}$ is y, so we multiply the numerator and the denominator of the complex fraction by y.

$$\dfrac{\frac{x+1}{y}}{\frac{x}{y} + 2} = \dfrac{y\left(\frac{x+1}{y}\right)}{y\left(\frac{x}{y} + 2\right)}$$

$$= \dfrac{y\left(\frac{x+1}{y}\right)}{y\left(\frac{x}{y}\right) + y \cdot 2} \quad \text{Apply the distributive property in the denominator.}$$

$$= \dfrac{x+1}{x + 2y} \quad \text{Simplify.}$$

Work Practice 5

Practice 5

Simplify: $\dfrac{1 + \frac{x}{y}}{\frac{2x+1}{y}}$

Answers

4. $\dfrac{2}{21}$ **5.** $\dfrac{y+x}{2x+1}$

Chapter 12 | Rational Expressions

Practice 6

Simplify: $\dfrac{\dfrac{5}{6y} + \dfrac{y}{x}}{\dfrac{y}{3} - x}$

Example 6 Simplify: $\dfrac{\dfrac{x}{y} + \dfrac{3}{2x}}{\dfrac{x}{2} + y}$

Solution: The LCD of $\dfrac{x}{y}, \dfrac{3}{2x}, \dfrac{x}{2},$ and $\dfrac{y}{1}$ is $2xy$, so we multiply both the numerator and the denominator of the complex fraction by $2xy$.

$$\dfrac{\dfrac{x}{y} + \dfrac{3}{2x}}{\dfrac{x}{2} + y} = \dfrac{2xy\left(\dfrac{x}{y} + \dfrac{3}{2x}\right)}{2xy\left(\dfrac{x}{2} + y\right)}$$

$$= \dfrac{2xy\left(\dfrac{x}{y}\right) + 2xy\left(\dfrac{3}{2x}\right)}{2xy\left(\dfrac{x}{2}\right) + 2xy(y)} \quad \text{Apply the distributive property.}$$

$$= \dfrac{2x^2 + 3y}{x^2y + 2xy^2}$$

or $\dfrac{2x^2 + 3y}{xy(x + 2y)}$

■ Work Practice 6

Answer

6. $\dfrac{5x + 6y^2}{2xy^2 - 6x^2y}$ or $\dfrac{5x + 6y^2}{2xy(y - 3x)}$

Vocabulary, Readiness & Video Check

Complete the steps by writing the simplified complex fraction.

1. $\dfrac{\dfrac{y}{2}}{\dfrac{5x}{2}} = \dfrac{2\left(\dfrac{y}{2}\right)}{2\left(\dfrac{5x}{2}\right)} = \dfrac{?}{?}$

2. $\dfrac{\dfrac{10}{x}}{\dfrac{z}{x}} = \dfrac{x\left(\dfrac{10}{x}\right)}{x\left(\dfrac{z}{x}\right)} = \dfrac{?}{?}$

3. $\dfrac{\dfrac{3}{x}}{\dfrac{5}{x^2}} = \dfrac{x^2\left(\dfrac{3}{x}\right)}{x^2\left(\dfrac{5}{x^2}\right)} = \dfrac{?}{?}$

4. $\dfrac{\dfrac{a}{10}}{\dfrac{b}{20}} = \dfrac{20\left(\dfrac{a}{10}\right)}{20\left(\dfrac{b}{20}\right)} = \dfrac{?}{?}$

One method for simplifying a complex fraction is to multiply the fraction's numerator and denominator by the LCD of all fractions in the complex fraction. For each complex fraction, choose the LCD of its fractions.

5. $\dfrac{\dfrac{1}{4} + \dfrac{1}{2}}{\dfrac{1}{3} + \dfrac{1}{2}}$ The LCD of $\dfrac{1}{4}, \dfrac{1}{2}, \dfrac{1}{3},$ and $\dfrac{1}{2}$ is

a. 4 **b.** 2 **c.** 12 **d.** 6

6. $\dfrac{\dfrac{3}{5} + \dfrac{2}{3}}{\dfrac{1}{10} + \dfrac{1}{6}}$ The LCD of $\dfrac{3}{5}, \dfrac{2}{3}, \dfrac{1}{10},$ and $\dfrac{1}{6}$ is

a. 15 **b.** 30 **c.** 60 **d.** 180

7. $\dfrac{\dfrac{5}{2x^2} + \dfrac{3}{16x}}{\dfrac{x}{8} + \dfrac{3}{4x}}$ The LCD of $\dfrac{5}{2x^2}, \dfrac{3}{16x}, \dfrac{x}{8},$ and $\dfrac{3}{4x}$ is

a. $16x^2$ **b.** $32x^3$ **c.** $16x$ **d.** $16x^3$

8. $\dfrac{\dfrac{11}{6} + \dfrac{10}{x^2}}{\dfrac{7}{9} + \dfrac{5}{x}}$ The LCD of $\dfrac{11}{6}, \dfrac{10}{x^2}, \dfrac{7}{9},$ and $\dfrac{5}{x}$ is

a. 18 **b.** x^2 **c.** $18x^2$ **d.** $54x^3$

Section 12.7 | Simplifying Complex Fractions

Martin-Gay Interactive Videos Watch the section lecture video and answer the following questions.

See Video 12.7

Objective A 9. From Example 2, before we can rewrite the complex fraction as a division problem, what must we make sure we have?

Objective B 10. How does finding an LCD in Method 2, as in Examples 4 and 5, differ from finding an LCD in Method 1? Mention the purpose of the LCD in each method.

12.7 Exercise Set MyLab Math

Objectives A B Mixed Practice *Simplify each complex fraction. See Examples 1 through 6.*

1. $\dfrac{\frac{1}{2}}{\frac{3}{4}}$

2. $\dfrac{\frac{1}{8}}{-\frac{5}{12}}$

3. $\dfrac{-\frac{4x}{9}}{-\frac{2x}{3}}$

4. $\dfrac{-\frac{6y}{11}}{\frac{4y}{9}}$

5. $\dfrac{\frac{1+x}{6}}{\frac{1+x}{3}}$

6. $\dfrac{\frac{6x-3}{5x^2}}{\frac{2x-1}{10x}}$

7. $\dfrac{\frac{1}{2}+\frac{2}{3}}{\frac{5}{9}-\frac{5}{6}}$

8. $\dfrac{\frac{3}{4}-\frac{1}{2}}{\frac{3}{8}+\frac{1}{6}}$

9. $\dfrac{2+\frac{7}{10}}{1+\frac{3}{5}}$

10. $\dfrac{4-\frac{11}{12}}{5+\frac{1}{4}}$

11. $\dfrac{\frac{1}{3}}{\frac{1}{2}-\frac{1}{4}}$

12. $\dfrac{\frac{7}{10}-\frac{3}{5}}{\frac{1}{2}}$

13. $\dfrac{-\frac{2}{9}}{-\frac{14}{3}}$

14. $\dfrac{\frac{3}{8}}{\frac{4}{15}}$

15. $\dfrac{-\frac{5}{12x^2}}{\frac{25}{16x^3}}$

16. $\dfrac{-\frac{7}{8y}}{\frac{21}{4y}}$

17. $\dfrac{\frac{m}{n}-1}{\frac{m}{n}+1}$

18. $\dfrac{\frac{x}{2}+2}{\frac{x}{2}-2}$

19. $\dfrac{\frac{1}{5}-\frac{1}{x}}{\frac{7}{10}+\frac{1}{x^2}}$

20. $\dfrac{\frac{1}{y^2}+\frac{2}{3}}{\frac{1}{y}-\frac{5}{6}}$

21. $\dfrac{1+\frac{1}{y-2}}{y+\frac{1}{y-2}}$

22. $\dfrac{x-\frac{1}{2x+1}}{1-\frac{x}{2x+1}}$

23. $\dfrac{\frac{4y-8}{16}}{\frac{6y-12}{4}}$

24. $\dfrac{\frac{7y+21}{3}}{\frac{3y+9}{8}}$

25. $\dfrac{\dfrac{x}{y}+1}{\dfrac{x}{y}-1}$ 26. $\dfrac{\dfrac{3}{5y}+8}{\dfrac{3}{5y}-8}$ 27. $\dfrac{1}{2+\dfrac{1}{3}}$ 28. $\dfrac{3}{1-\dfrac{4}{3}}$

29. $\dfrac{\dfrac{ax+ab}{x^2-b^2}}{\dfrac{x+b}{x-b}}$ 30. $\dfrac{\dfrac{m+2}{m-2}}{\dfrac{2m+4}{m^2-4}}$ 31. $\dfrac{\dfrac{-3+y}{4}}{\dfrac{8+y}{28}}$ 32. $\dfrac{\dfrac{-x+2}{18}}{\dfrac{8}{9}}$

33. $\dfrac{3+\dfrac{12}{x}}{1-\dfrac{16}{x^2}}$ 34. $\dfrac{2+\dfrac{6}{x}}{1-\dfrac{9}{x^2}}$ 35. $\dfrac{\dfrac{8}{x+4}+2}{\dfrac{12}{x+4}-2}$ 36. $\dfrac{\dfrac{25}{x+5}+5}{\dfrac{3}{x+5}-5}$

37. $\dfrac{\dfrac{s}{r}+\dfrac{r}{s}}{\dfrac{s}{r}-\dfrac{r}{s}}$ 38. $\dfrac{\dfrac{2}{x}+\dfrac{x}{2}}{\dfrac{2}{x}-\dfrac{x}{2}}$

39. $\dfrac{\dfrac{6}{x-5}+\dfrac{x}{x-2}}{\dfrac{3}{x-6}-\dfrac{2}{x-5}}$ 40. $\dfrac{\dfrac{4}{x}+\dfrac{x}{x+1}}{\dfrac{1}{2x}+\dfrac{1}{x+6}}$

Review

Use the bar graph below to answer Exercises 41 through 44. See Section 7.1. Note: Some of these players are still competing; thus, their total prize money may increase.

Women's Tennis Career Prize Money Leaders

Players (top to bottom): Serena Williams, Maria Sharapova, Venus Williams, Victoria Azarenka, Agnieszka Radwanska

Career Prize Money (in millions): 0, 20, 40, 60, 80, 100

Source: Women's Tennis Association, December 2016

41. Which women's tennis player has earned the most prize money in her career?

42. Estimate how much more prize money Maria Sharapova has earned in her career than Victoria Azarenka.

43. What is the approximate spread in lifetime prize money between Agnieszka Radwanska and Venus Williams?

44. To date in her career, Serena Williams has won 94 doubles and singles tournament titles. Assuming her prize money is earned only for tournament titles, how much prize money has she earned, on average, per tournament title?

Concept Extensions

45. Explain how to simplify a complex fraction using Method 1.

46. Explain how to simplify a complex fraction using Method 2.

To find the average of two numbers, we find their sum and divide by 2. For example, the average of 65 and 81 is found by simplifying $\frac{65 + 81}{2}$. This simplifies to $\frac{146}{2} = 73$. Use this for Exercises 47 through 50.

47. Find the average of $\frac{1}{3}$ and $\frac{3}{4}$.

48. Write the average of $\frac{3}{n}$ and $\frac{5}{n^2}$ as a simplified rational expression.

49. A carpenter needs to drill a hole halfway between the two marked points. An intersecting board keeps him from measuring between the marked points, but he does have earlier measurements as shown. How far from the left side of the marked board should he drill?

50. Use the same diagram as for Exercise **49**. Suppose the measurements are 7.2 inches and 10.3 inches. How far from the left side of the marked board should he drill?

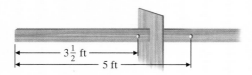

Solve.

51. In electronics, when two resistors R_1 (read R sub 1) and R_2 (read R sub 2) are connected in parallel, the total resistance is given by the complex fraction
$$\frac{1}{\frac{1}{R_1} + \frac{1}{R_2}}.$$
Simplify this expression.

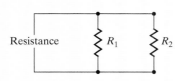

52. Astronomers occasionally need to know the day of the week a particular date fell on. The complex fraction
$$\frac{J + \frac{3}{2}}{7}$$
where J is the *Julian day number*, is used to make this calculation. Simplify this expression.

Simplify each of the following. First, write each expression with positive exponents. Then simplify the complex fraction. The first step has been completed for Exercise 53.

53. $\dfrac{x^{-1} + 2^{-1}}{x^{-2} - 4^{-1}} = \dfrac{\frac{1}{x} + \frac{1}{2}}{\frac{1}{x^2} - \frac{1}{4}}$

54. $\dfrac{3^{-1} - x^{-1}}{9^{-1} - x^{-2}}$

55. $\dfrac{y^{-2}}{1 - y^{-2}}$

56. $\dfrac{4 + x^{-1}}{3 + x^{-1}}$

57. If the distance formula $d = r \cdot t$ is solved for t, then $t = \dfrac{d}{r}$. Use this formula to find t if distance d is $\dfrac{20x}{3}$ miles and rate r is $\dfrac{5x}{9}$ miles per hour. Write t in simplified form.

△ 58. If the formula for the area of a rectangle, $A = l \cdot w$, is solved for w, then $w = \dfrac{A}{l}$. Use this formula to find w if area A is $\dfrac{4x - 2}{3}$ square meters and length l is $\dfrac{6x - 3}{5}$ meters. Write w in simplified form.

Chapter 12 Group Activity

Fast-Growing Careers

According to U.S. Bureau of Labor Statistics projections, the careers listed below will have the largest job growth in the years shown.

Occupation	Employment (number in thousands)		
	2014	2024	Change
1. Home health aides	913.5	1260.6	+347.1
2. Physical therapists	210.9	282.6	+71.7
3. Nurse practitioners	170.4	230.0	+59.6
4. Physical therapist assistants and aides	128.7	180.2	+51.5
5. Physician assistants	94.4	122.7	+28.3
6. Operations research analyst	91.3	118.7	+27.4
7. Occupational therapy assistants and aides	41.9	58.7	+16.8
8. Statisticians	30.0	40.2	+10.2
9. Ambulance drivers and attendants but not EMTs	20.0	26.5	+6.5
10. Wind turbine technician	4.4	9.2	+4.8

What do all of these in-demand occupations have in common? They all require a knowledge of math! For some careers, like nurse practitioners, statisticians, and operations research analysts, the ways math is used on the job may be obvious. For other occupations, the use of math may not be quite as obvious. However, tasks common to many jobs, such as filling in a time sheet or a medication log, writing up an expense report, planning a budget, figuring a bill, ordering supplies, and even making a work schedule, all require math.

Activity

Suppose that your college placement office is planning to publish an occupational handbook on math in popular occupations. Choose one of the occupations from the given list that interests you. Research the occupation. Then write a brief entry for the occupational handbook that describes how a person in that career would use math in his or her job. Include an example if possible.

Chapter 12 Vocabulary Check

Fill in each blank with one of the words or phrases listed below. Not all choices will be used.

least common denominator   simplifying   reciprocals   numerator $\dfrac{-a}{b}$

rational expression   unit   complex fraction   denominator $\dfrac{-a}{-b}$ $\dfrac{a}{-b}$

1. A _____, is an expression that can be written in the form $\dfrac{P}{Q}$, where P and Q are polynomials and Q is not 0.

2. In a _____ the numerator or denominator or both may contain fractions.

3. For a rational expression, $-\dfrac{a}{b} =$ _____ = _____.

4. A rational expression is undefined when the _____ is 0.

5. The process of writing a rational expression in lowest terms is called _____.

6. The expressions $\dfrac{2x}{7}$ and $\dfrac{7}{2x}$ are called _____.

7. The _____ of a list of rational expressions is a polynomial of least degree whose factors include all factors of the denominators in the list.

8. A _____ fraction is a fraction that equals 1.

> **Helpful Hint**   Are you preparing for your test? To help, don't forget to take these:
> - Chapter 12 Getting Ready for the Test on page 961
> - Chapter 12 Test on page 962
>
> Then check all of your answers at the back of this text. For further review, the step-by-step video solutions to any of these exercises are located in MyLab Math.

12 Chapter Highlights

Definitions and Concepts	Examples
Section 12.1 Simplifying Rational Expressions	
A **rational expression** is an expression that can be written in the form $\dfrac{P}{Q}$, where P and Q are polynomials and Q does not equal 0.	$\dfrac{7y^3}{4}, \dfrac{x^2 + 6x + 1}{x - 3}, \dfrac{-5}{s^3 + 8}$
To find values for which a rational expression is undefined, find values for which the denominator is 0.	Find any values for which the expression $\dfrac{5y}{y^2 - 4y + 3}$ is undefined. $y^2 - 4y + 3 = 0$ Set the denominator equal to 0. $(y - 3)(y - 1) = 0$ Factor. $y - 3 = 0$ or $y - 1 = 0$ Set each factor equal to 0. $y = 3$ $y = 1$ Solve. The expression is undefined when y is 3 and when y is 1.
To Simplify a Rational Expression **Step 1:** Factor the numerator and denominator. **Step 2:** Divide out factors common to the numerator and denominator. (This is the same as removing a factor of 1.)	Simplify: $\dfrac{4x + 20}{x^2 - 25}$ $\dfrac{4x + 20}{x^2 - 25} = \dfrac{4(x + 5)}{(x + 5)(x - 5)} = \dfrac{4}{x - 5}$
Section 12.2 Multiplying and Dividing Rational Expressions	
To Multiply Rational Expressions **Step 1:** Factor numerators and denominators. **Step 2:** Multiply numerators and multiply denominators. **Step 3:** Write the product in lowest terms. $\dfrac{P}{Q} \cdot \dfrac{R}{S} = \dfrac{PR}{QS}$	Multiply: $\dfrac{4x + 4}{2x - 3} \cdot \dfrac{2x^2 + x - 6}{x^2 - 1}$ $\dfrac{4x + 4}{2x - 3} \cdot \dfrac{2x^2 + x - 6}{x^2 - 1}$ $= \dfrac{4(x + 1)}{2x - 3} \cdot \dfrac{(2x - 3)(x + 2)}{(x + 1)(x - 1)}$ $= \dfrac{4(x + 1)(2x - 3)(x + 2)}{(2x - 3)(x + 1)(x - 1)}$ $= \dfrac{4(x + 2)}{x - 1}$

(Continued)

Definitions and Concepts	Examples
Section 12.2 Multiplying and Dividing Rational Expressions (*continued*)	

To divide by a rational expression, multiply by the reciprocal. $$\frac{P}{Q} \div \frac{R}{S} = \frac{P}{Q} \cdot \frac{S}{R} = \frac{PS}{QR}$$	Divide: $\dfrac{15x + 5}{3x^2 - 14x - 5} \div \dfrac{15}{3x - 12}$ $\dfrac{15x + 5}{3x^2 - 14x - 5} \div \dfrac{15}{3x - 12}$ $= \dfrac{5(3x + 1)}{(3x + 1)(x - 5)} \cdot \dfrac{3(x - 4)}{3 \cdot 5}$ $= \dfrac{x - 4}{x - 5}$

Section 12.3 Adding and Subtracting Rational Expressions with the Same Denominator and Least Common Denominator	

To add or subtract rational expressions with the same denominator, add or subtract numerators, and place the sum or difference over the common denominator. $$\frac{P}{R} + \frac{Q}{R} = \frac{P + Q}{R}$$ $$\frac{P}{R} - \frac{Q}{R} = \frac{P - Q}{R}$$	Perform each indicated operation. $\dfrac{5}{x + 1} + \dfrac{x}{x + 1} = \dfrac{5 + x}{x + 1}$ $\dfrac{2y + 7}{y^2 - 9} - \dfrac{y + 4}{y^2 - 9}$ $= \dfrac{(2y + 7) - (y + 4)}{y^2 - 9}$ $= \dfrac{2y + 7 - y - 4}{y^2 - 9}$ $= \dfrac{y + 3}{(y + 3)(y - 3)}$ $= \dfrac{1}{y - 3}$
To Find the Least Common Denominator (LCD) **Step 1:** Factor the denominators. **Step 2:** The LCD is the product of all unique factors, each raised to a power equal to the greatest number of times that it appears in any one factored denominator.	Find the LCD for $\dfrac{7x}{x^2 + 10x + 25}$ and $\dfrac{11}{3x^2 + 15x}$ $x^2 + 10x + 25 = (x + 5)(x + 5)$ $3x^2 + 15x = 3x(x + 5)$ $\text{LCD} = 3x(x + 5)(x + 5)$ or $3x(x + 5)^2$

Definitions and Concepts	Examples

Section 12.4 Adding and Subtracting Rational Expressions with Different Denominators

To Add or Subtract Rational Expressions with Different Denominators

Step 1: Find the LCD.

Step 2: Rewrite each rational expression as an equivalent expression whose denominator is the LCD.

Step 3: Add or subtract numerators and place the sum or difference over the common denominator.

Step 4: Write the result in lowest terms.

Perform the indicated operation.

$$\frac{9x+3}{x^2-9} - \frac{5}{x-3}$$

$$= \frac{9x+3}{(x+3)(x-3)} - \frac{5}{x-3}$$

LCD is $(x+3)(x-3)$.

$$= \frac{9x+3}{(x+3)(x-3)} - \frac{5(x+3)}{(x-3)(x+3)}$$

$$= \frac{9x+3-5(x+3)}{(x+3)(x-3)}$$

$$= \frac{9x+3-5x-15}{(x+3)(x-3)}$$

$$= \frac{4x-12}{(x+3)(x-3)}$$

$$= \frac{4(x-3)}{(x+3)(x-3)} = \frac{4}{x+3}$$

Section 12.5 Solving Equations Containing Rational Expressions

To Solve an Equation Containing Rational Expressions

Step 1: Multiply both sides of the equation by the LCD of all rational expressions in the equation.

Step 2: Remove any grouping symbols and solve the resulting equation.

Step 3: Check the solution in the original equation.

Solve: $\dfrac{5x}{x+2} + 3 = \dfrac{4x-6}{x+2}$ The LCD is $x+2$.

$$(x+2)\left(\frac{5x}{x+2} + 3\right) = (x+2)\left(\frac{4x-6}{x+2}\right)$$

$$(x+2)\left(\frac{5x}{x+2}\right) + (x+2)(3) = (x+2)\left(\frac{4x-6}{x+2}\right)$$

$$5x + 3x + 6 = 4x - 6$$

$$4x = -12$$

$$x = -3$$

The solution checks; the solution is -3.

Section 12.6 Rational Equations and Problem Solving

Problem-Solving Steps

1. UNDERSTAND. Read and reread the problem.

A small plane and a car leave Kansas City, Missouri, and head for Minneapolis, Minnesota, a distance of 450 miles. The speed of the plane is 3 times the speed of the car, and the plane arrives 6 hours ahead of the car. Find the speed of the car.

Let x = the speed of the car.
Then $3x$ = the speed of the plane.

	Distance =	Rate ·	Time
Car	450	x	$\dfrac{450}{x}$ $\left(\dfrac{\text{distance}}{\text{rate}}\right)$
Plane	450	$3x$	$\dfrac{450}{3x}$ $\left(\dfrac{\text{distance}}{\text{rate}}\right)$

(Continued)

Definitions and Concepts	Examples
Section 12.6 Rational Equations and Problem Solving (*continued*)	

2. TRANSLATE.	In words: $\dfrac{\text{plane's}}{\text{time}} + 6 \text{ hours} = \dfrac{\text{car's}}{\text{time}}$
3. SOLVE.	Translate: $\dfrac{450}{3x} + 6 = \dfrac{450}{x}$
	$\dfrac{450}{3x} + 6 = \dfrac{450}{x}$
	$3x\left(\dfrac{450}{3x}\right) + 3x(6) = 3x\left(\dfrac{450}{x}\right)$
	$450 + 18x = 1350$
	$18x = 900$
	$x = 50$
4. INTERPRET.	**Check** this solution in the originally stated problem. **State** the conclusion: The speed of the car is 50 miles per hour.

Definitions and Concepts	Examples
Section 12.7 Simplifying Complex Fractions	

Method 1: To Simplify a Complex Fraction

Step 1: Add or subtract fractions in the numerator and the denominator of the complex fraction.

Step 2: Perform the indicated division.

Step 3: Write the result in lowest terms.

Simplify:

$$\dfrac{\dfrac{1}{x} + 2}{\dfrac{1}{x} - \dfrac{1}{y}} = \dfrac{\dfrac{1}{x} + \dfrac{2x}{x}}{\dfrac{y}{xy} - \dfrac{x}{xy}}$$

$$= \dfrac{\dfrac{1 + 2x}{x}}{\dfrac{y - x}{xy}}$$

$$= \dfrac{1 + 2x}{x} \cdot \dfrac{xy}{y - x}$$

$$= \dfrac{y(1 + 2x)}{y - x}$$

Method 2: To Simplify a Complex Fraction

Step 1: Find the LCD of all fractions in the complex fraction.

Step 2: Multiply the numerator and the denominator of the complex fraction by the LCD.

Step 3: Perform the indicated operations and write the result in lowest terms.

$$\dfrac{\dfrac{1}{x} + 2}{\dfrac{1}{x} - \dfrac{1}{y}} = \dfrac{xy\left(\dfrac{1}{x} + 2\right)}{xy\left(\dfrac{1}{x} - \dfrac{1}{y}\right)}$$

$$= \dfrac{xy\left(\dfrac{1}{x}\right) + xy(2)}{xy\left(\dfrac{1}{x}\right) - xy\left(\dfrac{1}{y}\right)}$$

$$= \dfrac{y + 2xy}{y - x} \text{ or } \dfrac{y(1 + 2x)}{y - x}$$

Chapter 12 Review

12.1) *Find any real number(s) for which each rational expression is undefined.*

1. $\dfrac{x+5}{x^2-4}$

2. $\dfrac{5x+9}{4x^2-4x-15}$

Find the value of each rational expression when $x=5$, $y=7$, and $z=-2$.

3. $\dfrac{2-z}{z+5}$

4. $\dfrac{x^2+xy-y^2}{x+y}$

Simplify each rational expression.

5. $\dfrac{2x+6}{x^2+3x}$

6. $\dfrac{3x-12}{x^2-4x}$

7. $\dfrac{x+2}{x^2-3x-10}$

8. $\dfrac{x+4}{x^2+5x+4}$

9. $\dfrac{x^3-4x}{x^2+3x+2}$

10. $\dfrac{5x^2-125}{x^2+2x-15}$

11. $\dfrac{x^2-x-6}{x^2-3x-10}$

12. $\dfrac{x^2-2x}{x^2+2x-8}$

Simplify each expression. First, factor the four-term polynomials by grouping.

13. $\dfrac{x^2+xa+xb+ab}{x^2-xc+bx-bc}$

14. $\dfrac{x^2+5x-2x-10}{x^2-3x-2x+6}$

12.2) *Perform each indicated operation and simplify.*

15. $\dfrac{15x^3y^2}{z} \cdot \dfrac{z}{5xy^3}$

16. $\dfrac{-y^3}{8} \cdot \dfrac{9x^2}{y^3}$

17. $\dfrac{x^2-9}{x^2-4} \cdot \dfrac{x-2}{x+3}$

18. $\dfrac{2x+5}{x-6} \cdot \dfrac{2x}{-x+6}$

19. $\dfrac{x^2-5x-24}{x^2-x-12} \div \dfrac{x^2-10x+16}{x^2+x-6}$

20. $\dfrac{4x+4y}{xy^2} \div \dfrac{3x+3y}{x^2y}$

21. $\dfrac{x^2+x-42}{x-3} \cdot \dfrac{(x-3)^2}{x+7}$

22. $\dfrac{2a+2b}{3} \cdot \dfrac{a-b}{a^2-b^2}$

23. $\dfrac{2x^2-9x+9}{8x-12} \div \dfrac{x^2-3x}{2x}$

24. $\dfrac{x^2-y^2}{x^2+xy} \div \dfrac{3x^2-2xy-y^2}{3x^2+6x}$

957

(12.3) *Perform each indicated operation and simplify.*

25. $\dfrac{x}{x^2 + 9x + 14} + \dfrac{7}{x^2 + 9x + 14}$

26. $\dfrac{x}{x^2 + 2x - 15} + \dfrac{5}{x^2 + 2x - 15}$

27. $\dfrac{4x - 5}{3x^2} - \dfrac{2x + 5}{3x^2}$

28. $\dfrac{9x + 7}{6x^2} - \dfrac{3x + 4}{6x^2}$

Find the LCD of each pair of rational expressions.

29. $\dfrac{x + 4}{2x}, \dfrac{3}{7x}$

30. $\dfrac{x - 2}{x^2 - 5x - 24}, \dfrac{3}{x^2 + 11x + 24}$

Rewrite each rational expression as an equivalent expression whose denominator is the given polynomial.

31. $\dfrac{5}{7x} = \dfrac{}{14x^3y}$

32. $\dfrac{9}{4y} = \dfrac{}{16y^3x}$

33. $\dfrac{x + 2}{x^2 + 11x + 18} = \dfrac{}{(x + 2)(x - 5)(x + 9)}$

34. $\dfrac{3x - 5}{x^2 + 4x + 4} = \dfrac{}{(x + 2)^2(x + 3)}$

(12.4) *Perform each indicated operation and simplify.*

35. $\dfrac{4}{5x^2} + \dfrac{6}{y}$

36. $\dfrac{2}{x - 3} - \dfrac{4}{x - 1}$

37. $\dfrac{4}{x + 3} - 2$

38. $\dfrac{3}{x^2 + 2x - 8} + \dfrac{2}{x^2 - 3x + 2}$

39. $\dfrac{2x - 5}{6x + 9} - \dfrac{4}{2x^2 + 3x}$

40. $\dfrac{x - 1}{x^2 - 2x + 1} - \dfrac{x + 1}{x - 1}$

(12.5) *Solve each equation.*

41. $\dfrac{n}{10} = 9 - \dfrac{n}{5}$

42. $\dfrac{2}{x + 1} - \dfrac{1}{x - 2} = -\dfrac{1}{2}$

43. $\dfrac{y}{2y + 2} + \dfrac{2y - 16}{4y + 4} = \dfrac{y - 3}{y + 1}$

44. $\dfrac{2}{x - 3} - \dfrac{4}{x + 3} = \dfrac{8}{x^2 - 9}$

45. $\dfrac{x - 3}{x + 1} - \dfrac{x - 6}{x + 5} = 0$

46. $x + 5 = \dfrac{6}{x}$

Chapter 12 Review 959

12.6) *Solve.*

47. Five times the reciprocal of a number equals the sum of $\dfrac{3}{2}$ the reciprocal of the number and $\dfrac{7}{6}$. What is the number?

48. The reciprocal of a number equals the reciprocal of the difference of 4 and the number. Find the number.

49. A car travels 90 miles in the same time that a car traveling 10 miles per hour slower travels 60 miles. Find the speed of each car.

50. The current in a bayou near Lafayette, Louisiana, is 4 miles per hour. A paddle boat travels 48 miles upstream in the same amount of time it takes to travel 72 miles downstream. Find the speed of the boat in still water.

51. When Mark and Maria manicure Mr. Stergeon's lawn, it takes them 5 hours. If Mark works alone, it takes 7 hours. Find how long it takes Maria alone.

52. It takes pipe A 20 days to fill a fish pond. Pipe B takes 15 days. Find how long it takes both pipes together to fill the pond.

12.7) *Simplify each complex fraction.*

53. $\dfrac{\dfrac{5x}{27}}{\dfrac{10xy}{21}}$

54. $\dfrac{\dfrac{3}{5}+\dfrac{2}{7}}{\dfrac{1}{5}+\dfrac{5}{6}}$

55. $\dfrac{3-\dfrac{1}{y}}{2-\dfrac{1}{y}}$

56. $\dfrac{\dfrac{6}{x+2}+4}{\dfrac{8}{x+2}-4}$

Mixed Review

Simplify each rational expression.

57. $\dfrac{4x+12}{8x^2+24x}$

58. $\dfrac{x^3-6x^2+9x}{x^2+4x-21}$

Perform the indicated operations and simplify.

59. $\dfrac{x^2+9x+20}{x^2-25}\cdot\dfrac{x^2-9x+20}{x^2+8x+16}$

60. $\dfrac{x^2-x-72}{x^2-x-30}\div\dfrac{x^2+6x-27}{x^2-9x+18}$

61. $\dfrac{x}{x^2-36}+\dfrac{6}{x^2-36}$

62. $\dfrac{5x-1}{4x}-\dfrac{3x-2}{4x}$

63. $\dfrac{3x}{x^2+9x+14}-\dfrac{6x}{x^2+4x-21}$

64. $\dfrac{4}{3x^2+8x-3}+\dfrac{2}{3x^2-7x+2}$

Solve.

65. $\dfrac{4}{a-1} + 2 = \dfrac{3}{a-1}$

66. $\dfrac{x}{x+3} + 4 = \dfrac{x}{x+3}$

Solve.

67. The quotient of twice a number and three, minus one-sixth, is the quotient of the number and two. Find the number.

68. Mr. Crocker can paint his shed by himself in three days. His son will need an additional day to complete the job if he works alone. If they work together, find how long it takes to paint the shed.

Simplify each complex fraction.

69. $\dfrac{\frac{1}{4}}{\frac{1}{3} + \frac{1}{2}}$

70. $\dfrac{4 + \frac{2}{x}}{6 + \frac{3}{x}}$

Convert as indicated.

71. 1.8 square yards = _____ square feet

72. 135 cubic feet = _____ cubic yards

Chapter 12 — Getting Ready for the Test

MULTIPLE CHOICE Exercises 1 through 12 are **Multiple Choice**. Select the correct choice.

1. $\dfrac{x-8}{8-x}$ simplifies to

    A. 1   **B.** −1   **C.** −2   **D.** −8

2. $\dfrac{8}{x^2} \cdot \dfrac{4}{x^2} =$

    A. $\dfrac{32}{x^2}$ **B.** $\dfrac{2}{x^2}$ **C.** $\dfrac{32}{x^4}$ **D.** 2   **E.** $\dfrac{1}{2}$

3. $\dfrac{8}{x^2} \div \dfrac{4}{x^2} =$

    A. $\dfrac{32}{x^2}$ **B.** $\dfrac{2}{x^2}$ **C.** $\dfrac{32}{x^4}$ **D.** 2   **E.** $\dfrac{1}{2}$

4. $\dfrac{8}{x^2} + \dfrac{4}{x^2} =$

    A. $\dfrac{32}{x^2}$ **B.** $\dfrac{2}{x^2}$ **C.** $\dfrac{12}{x^4}$ **D.** $\dfrac{12}{x^2}$

5. $\dfrac{7x}{x-1} - \dfrac{5+2x}{x-1} =$

    A. 5   **B.** $\dfrac{9x-5}{x-1}$ **C.** $\dfrac{5}{x-1}$ **D.** $\dfrac{14}{x-1}$

6. The LCD of $\dfrac{9}{25x}$ and $\dfrac{z}{10x^3}$ is

    A. $250x^4$ **B.** $250x$ **C.** $50x^4$ **D.** $50x^3$

For Exercises 7 through 10, identify each as an

A. expression or   **B.** equation.

7. $\dfrac{5}{x} + \dfrac{1}{3}$ 8. $\dfrac{5}{x} + \dfrac{1}{3} = \dfrac{2}{x}$ 9. $\dfrac{a+5}{11} = 9$ 10. $\dfrac{a+5}{11} \cdot 9$

For Exercises 11 and 12, select the correct choice.

11. Multiply the given equation through by the LCD of its terms. Choose the correct equivalent equation once this is done and terms are simplified.   Given Equation: $\dfrac{x+3}{4} + \dfrac{5}{6} = 3$

    A. $(x+3) + 5 = 3$ **B.** $3(x+3) + 2 \cdot 5 = 3$ **C.** $3(x+3) + 2 \cdot 5 = 12 \cdot 3$ **D.** $6(x+3) + 4 \cdot 5 = 3$

12. Translate to an equation. Let x be the unknown number. "The quotient of a number and 5 equals the sum of that number and 12."

    A. $\dfrac{x}{5} = x + 12$ **B.** $\dfrac{5}{x} = x + 12$ **C.** $\dfrac{x}{5} = x \cdot 12$ **D.** $\dfrac{x}{5} \cdot (x + 12)$

Chapter 12 Test

MyLab Math or **YouTube** — For additional practice go to your study plan in MyLab Math.

Answers

1. Find any real numbers for which the following expression is undefined.

$$\frac{x+5}{x^2+4x+3}$$

2. For a certain computer desk, the average manufacturing cost C per desk (in dollars) is

$$C = \frac{100x+3000}{x}$$

where x is the number of desks manufactured.
   a. Find the average cost per desk when manufacturing 200 computer desks.
   b. Find the average cost per desk when manufacturing 1000 computer desks.

Simplify each rational expression.

3. $\dfrac{3x-6}{5x-10}$

4. $\dfrac{x+6}{x^2+12x+36}$

5. $\dfrac{7-x}{x-7}$

6. $\dfrac{y-x}{x^2-y^2}$

7. $\dfrac{2m^3-2m^2-12m}{m^2-5m+6}$

8. $\dfrac{ay+3a+2y+6}{ay+3a+5y+15}$

Perform each indicated operation and simplify if possible.

9. $\dfrac{x^2-13x+42}{x^2+10x+21} \div \dfrac{x^2-4}{x^2+x-6}$

10. $\dfrac{3}{x-1} \cdot (5x-5)$

11. $\dfrac{y^2-5y+6}{2y+4} \cdot \dfrac{y+2}{2y-6}$

12. $\dfrac{5}{2x+5} - \dfrac{6}{2x+5}$

13. $\dfrac{5a}{a^2-a-6} - \dfrac{2}{a-3}$

14. $\dfrac{6}{x^2-1} + \dfrac{3}{x+1}$

15. $\dfrac{x^2-9}{x^2-3x} \div \dfrac{x^2+4x+1}{2x+10}$

16. $\dfrac{x+2}{x^2+11x+18} + \dfrac{5}{x^2-3x-10}$

17. $\dfrac{4y}{y^2+6y+5} - \dfrac{3}{y^2+5y+4}$

Solve each equation.

18. $\dfrac{4}{y} - \dfrac{5}{3} = -\dfrac{1}{5}$

19. $\dfrac{5}{y+1} = \dfrac{4}{y+2}$

20. $\dfrac{a}{a-3} = \dfrac{3}{a-3} - \dfrac{3}{2}$

21. $\dfrac{10}{x^2 - 25} = \dfrac{3}{x+5} + \dfrac{1}{x-5}$

22. $x - \dfrac{14}{x-1} = 4 - \dfrac{2x}{x-1}$

Simplify each complex fraction.

23. $\dfrac{\dfrac{5x^2}{yz^2}}{\dfrac{10x}{z^3}}$

24. $\dfrac{\dfrac{b}{a} - \dfrac{a}{b}}{\dfrac{1}{b} + \dfrac{1}{a}}$

25. $\dfrac{5 - \dfrac{1}{y^2}}{\dfrac{1}{y} + \dfrac{2}{y^2}}$

26. One number plus five times its reciprocal is equal to six. Find the number.

27. A pleasure boat traveling down the Red River takes the same time to go 14 miles upstream as it takes to go 16 miles downstream. If the current of the river is 2 miles per hour, find the speed of the boat in still water.

28. An inlet pipe can fill a tank in 12 hours. A second pipe can fill the tank in 15 hours. If both pipes are used, find how long it takes to fill the tank.

18. _____
19. _____
20. _____
21. _____
22. _____
23. _____
24. _____
25. _____
26. _____
27. _____
28. _____

Chapters 1–12 Cumulative Review

Write each fraction or mixed number as a percent.

1. $\dfrac{7}{20}$
2. $\dfrac{4}{5}$
3. $\dfrac{2}{3}$
4. $\dfrac{1}{9}$
5. $2\dfrac{1}{4}$
6. $3\dfrac{3}{4}$

Name the property illustrated by each true statement.

7. $2 \cdot (z \cdot 5) = 2 \cdot (5 \cdot z)$
8. $3 + y = y + 3$
9. $(x + 7) + 9 = x + (7 + 9)$
10. $(x \cdot 7) \cdot 9 = x \cdot (7 \cdot 9)$

11. A 10-foot board is to be cut into two pieces so that the length of the longer piece is 4 times the length of the shorter. Find the length of each piece.

12. A 45-foot piece of rope is to be cut into three pieces. Two pieces are to have equal length and the third piece is to be 3 feet shorter than the longer piece. Find the length of each piece.

13. Solve $y = mx + b$ for x.
14. Solve $y = mx + b$ for b.

15. Solve $x + 4 \leq -6$. Graph the solutions.
16. Solve $x - 1 > -10$.

Simplify each quotient.

17. $\dfrac{x^5}{x^2}$
18. $\dfrac{x^9}{x}$
19. $\dfrac{4^7}{4^3}$
20. $\dfrac{7^{12}}{7^4}$

21. $\dfrac{(-3)^5}{(-3)^2}$
22. $\dfrac{(-4)^9}{(-4)^7}$
23. $\dfrac{2x^5 y^2}{xy}$
24. $\dfrac{13a^5 b}{a^4}$

Simplify by writing each expression with positive exponents only.

25. $2x^{-3}$ **26.** $9x^{-2}$ **27.** $(-2)^{-4}$ **28.** $(-3)^{-3}$

Multiply.

29. $5x(2x^3 + 6)$ **30.** $3y(4y^2 - 2)$

31. $-3x^2(5x^2 + 6x - 1)$ **32.** $-5y(7y^2 - 3y + 1)$

Divide.

33. $\dfrac{4x^2 + 7 + 8x^3}{2x + 3}$ **34.** $\dfrac{6x^2 - 7x + 4}{2x + 1}$

Factor.

35. $x^2 + 7x + 12$ **36.** $x^2 + 17x + 70$

37. $25x^2 + 20xy + 4y^2$ **38.** $36a^2 - 48ab + 16b^2$

39. Solve: $x^2 - 9x - 22 = 0$ **40.** Solve: $x^2 + 2x - 15 = 0$

41. Multiply: $\dfrac{x^2 + x}{3x} \cdot \dfrac{6}{5x + 5}$ **42.** Divide: $\dfrac{3x - 12}{2} \div \dfrac{5x - 20}{4x}$

43. Subtract: $\dfrac{3x^2 + 2x}{x - 1} - \dfrac{10x - 5}{x - 1}$ **44.** Add: $\dfrac{4x}{x + 2} + \dfrac{8}{x + 2}$

45. Subtract: $\dfrac{6x}{x^2-4} - \dfrac{3}{x+2}$

46. Add: $\dfrac{2}{x-3} + \dfrac{5x}{x^2-9}$

47. Solve: $\dfrac{t-4}{2} - \dfrac{t-3}{9} = \dfrac{5}{18}$

48. $\dfrac{y}{2} - \dfrac{y}{4} = \dfrac{1}{6}$

49. Sam Waterton and Frank Schaffer work in a plant that manufactures automobiles. Sam can complete a quality control tour of the plant in 3 hours while his assistant, Frank, needs 7 hours to complete the same job. The regional manager is coming to inspect the plant facilities, so both Sam and Frank are directed to complete a quality control tour together. How long will this take?

50. One machine can complete a task in 18 hours. A second machine can do the same task in 12 hours. How long would it take to complete the task if both machines are able to work on it?

Simplify.

51. $\dfrac{\dfrac{1}{z} - \dfrac{1}{2}}{\dfrac{1}{3} - \dfrac{z}{6}}$

52. $\dfrac{\dfrac{x}{9} - \dfrac{1}{x}}{1 + \dfrac{3}{x}}$

Graphing Equations and Inequalities

13

In Chapter 9 we learned to solve and graph the solutions of linear equations and inequalities in one variable on number lines. Now we define and present techniques for solving and graphing linear equations and inequalities in two variables on grids. Two-variable equations lead directly to the concept of *function*, perhaps the most important concept in all of mathematics. Functions are introduced in Section 13.6.

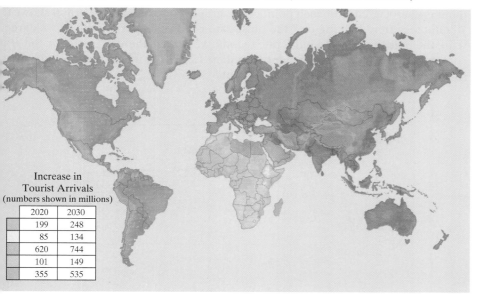

International Tourist Arrivals Forecast for 2020–2030 (numbers shown in millions)

Increase in Tourist Arrivals (numbers shown in millions)

2020	2030
199	248
85	134
620	744
101	149
355	535

Sections

- **13.1** The Rectangular Coordinate System
- **13.2** Graphing Linear Equations
- **13.3** Intercepts
- **13.4** Slope and Rate of Change
- **13.5** Equations of Lines
  - Integrated Review—Summary on Linear Equations
- **13.6** Introduction to Functions
- **13.7** Graphing Linear Inequalities in Two Variables
- **13.8** Direct and Inverse Variation

Check Your Progress

- Vocabulary Check
- Chapter Highlights
- Chapter Review
- Getting Ready for the Test
- Chapter Test
- Cumulative Review

What Is Tourism Toward 2030?

Tourism 2020 Vision was the World Tourism Organization's long-term forecast of world tourism through 2020. *Tourism Toward 2030* is its new program title for longer-term forecasts to 2030. The broken-line graph below shows the forecast for number of tourists, which is extremely important as these numbers greatly affect a country's economy. In Section 13.1, Exercises 45 through 50, we read a bar graph showing the top tourist destinations by country.

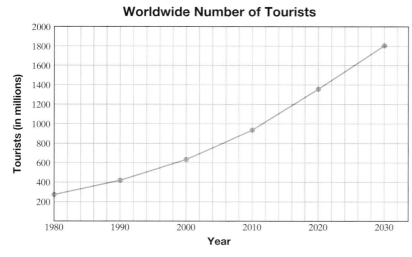

Data from World Tourism Organization (UNWTO)

13.1 The Rectangular Coordinate System

Objectives

A Plot Ordered Pairs of Numbers on the Rectangular Coordinate System.

B Graph Paired Data to Create a Scatter Diagram.

C Find the Missing Coordinate of an Ordered Pair Solution, Given One Coordinate of the Pair.

In Sections 7.1 and 7.2, we learned how to read graphs. The broken line graph below shows the relationship between the time before and after smoking a cigarette and pulse rate. The horizontal line or axis shows time in minutes and the vertical line or axis shows the pulse rate in heartbeats per minute. Notice that there are two numbers associated with each point of the graph. For example, the graph shows that 15 minutes after "lighting up," the pulse rate is 80 beats per minute. If we agree to write the time first and the pulse rate second, we can say there is a point on the graph corresponding to the **ordered pair** of numbers (15, 80). A few more ordered pairs are shown alongside their corresponding points.

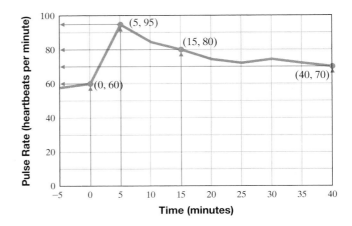

Objective A Plotting Ordered Pairs of Numbers

In general, we use the idea of ordered pairs to describe the location of a point in a plane (such as a piece of paper). We start with a horizontal and a vertical axis. Each axis is a number line, and for the sake of consistency we construct our axes to intersect at the 0 coordinate of both. This point of intersection is called the **origin**. Notice that these two number lines or axes divide the plane into four regions called **quadrants.** The quadrants are usually numbered with Roman numerals as shown. The axes are not considered to be in any quadrant.

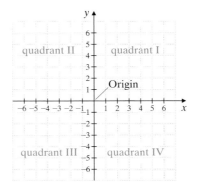

It is helpful to label axes, so we label the horizontal axis the **x-axis** and the vertical axis the **y-axis.** We call the system described above the **rectangular coordinate system,** or the **coordinate plane.** Just as with other graphs shown, we can then describe the locations of points by ordered pairs of numbers. We list the horizontal, **x-axis** measurement first and the vertical, **y-axis** measurement second.

To plot or graph the point corresponding to the ordered pair (a,b) we start at the origin. We then move a units left or right (right if a is positive, left if a is negative). From there, we move b units up or down (up if b is positive, down if b is negative). For example, to plot the point corresponding to the ordered pair $(3, 2)$, we start at the origin, move 3 units right, and from there move 2 units up. (See the figure below.) The x-value, 3, is also called the **x-coordinate** and the y-value, 2, is also called the **y-coordinate**. From now on, we will call the point with coordinates $(3, 2)$ simply the point $(3, 2)$. The point $(-2, 5)$ is also graphed below.

Practice 1

On a single coordinate system, plot each ordered pair. State in which quadrant, or on which axis, each point lies.

a. $(4, 2)$ b. $(-1, -3)$
c. $(2, -2)$ d. $(-5, 1)$
e. $(0, 3)$ f. $(3, 0)$
g. $(0, -4)$ h. $\left(-2\frac{1}{2}, 0\right)$
i. $\left(1, -3\frac{3}{4}\right)$

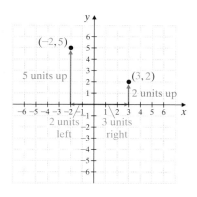

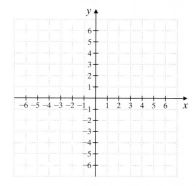

Helpful Hint

Don't forget that **each ordered pair corresponds to exactly one point in the plane and that each point in the plane corresponds to exactly one ordered pair.**

Concept Check Is the graph of the point $(-5, 1)$ in the same location as the graph of the point $(1, -5)$? Explain.

Example 1
On a single coordinate system, plot each ordered pair. State in which quadrant, or on which axis, each point lies.

a. $(5, 3)$ b. $(-2, -4)$ c. $(1, -2)$ d. $(-5, 3)$ e. $(0, 0)$
f. $(0, 2)$ g. $(-5, 0)$ h. $\left(0, -5\frac{1}{2}\right)$ i. $\left(4\frac{2}{3}, -3\right)$

Solution:

a. Point $(5, 3)$ lies in quadrant I.
b. Point $(-2, -4)$ lies in quadrant III.
c. Point $(1, -2)$ lies in quadrant IV.
d. Point $(-5, 3)$ lies in quadrant II.
e–h. Points $(0, 0)$, $(0, 2)$, and $\left(0, -5\frac{1}{2}\right)$ lie on the y-axis. Points $(0, 0)$ and $(-5, 0)$ lie on the x-axis.
i. Point $\left(4\frac{2}{3}, -3\right)$ lies in quadrant IV.

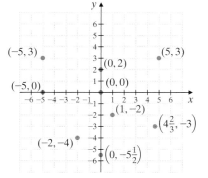

Answers
1.

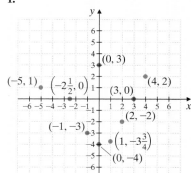

a. Point $(4, 2)$ lies in quadrant I.
b. Point $(-1, -3)$ lies in quadrant III.
c. Point $(2, -2)$ lies in quadrant IV.
d. Point $(-5, 1)$ lies in quadrant II.
e–h. Points $(3, 0)$ and $\left(-2\frac{1}{2}, 0\right)$ lie on the x-axis. Points $(0, 3)$ and $(0, -4)$ lie on the y-axis.
i. Point $\left(1, -3\frac{3}{4}\right)$ lies in quadrant IV.

✓ **Concept Check Answer**
The graph of point $(-5, 1)$ lies in quadrant II and the graph of point $(1, -5)$ lies in quadrant IV. They are *not* in the same location.

Work Practice 1

Helpful Hint

In Example 1, notice that the point $(0, 0)$ lies on both the x-axis and the y-axis. It is the only point in the entire rectangular coordinate system that has this feature. Why? It is the only point of intersection of the x-axis and the y-axis.

Practice 2

The table gives the number of tornadoes that occurred in the United States for the years shown. (*Source:* Storm Prediction Center, National Oceanic and Atmospheric Administration)

Year	Tornadoes
2010	1282
2011	1693
2012	939
2013	891
2014	881
2015	1183
2016	968

a. Write this paired data as a set of ordered pairs of the form (year, number of tornadoes).
b. Create a scatter diagram of the paired data.

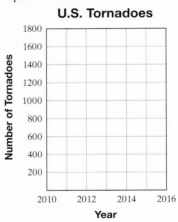

c. What trend in the paired data, if any, does the scatter diagram show?

Answers

2. a. (2010, 1282), (2011, 1693), (2012, 939), (2013, 891), (2014, 881), (2015, 1183), (2016, 968)

b.

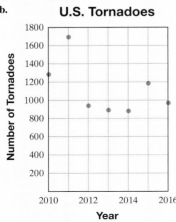

c. The number of tornadoes varies greatly from year to year.

✓ Concept Check Answers
a. $(-3, 5)$ **b.** $(0, -6)$

✓ **Concept Check** For each description of a point in the rectangular coordinate system, write an ordered pair that represents it.

a. Point A is located three units to the left of the y-axis and five units above the x-axis.

b. Point B is located six units below the origin.

From Example 1, notice that the y-coordinate of any point on the x-axis is 0. For example, the point $(-5, 0)$ lies on the x-axis. Also, the x-coordinate of any point on the y-axis is 0. For example, the point $(0, 2)$ lies on the y-axis.

Objective B Creating Scatter Diagrams

Data that can be represented as ordered pairs are called **paired data**. Many types of data collected from the real world are paired data. For instance, the annual measurements of a child's height can be written as ordered pairs of the form (year, height in inches) and are paired data. The graph of paired data as points in the rectangular coordinate system is called a **scatter diagram**. Scatter diagrams can be used to look for patterns and trends in paired data.

Example 2 The table gives the annual net sales for Dick's Sporting Goods for the years shown. (*Source:* Dick's Sporting Goods)

Year	Dick's Sporting Goods Net Sales (in billions of dollars)
2010	4.9
2011	5.2
2012	5.8
2013	6.2
2014	6.8
2015	7.3
2016	7.9

a. Write this paired data as a set of ordered pairs of the form (year, net sales in billions of dollars).
b. Create a scatter diagram of the paired data.
c. What trend in the paired data does the scatter diagram show?

Solution:

a. The ordered pairs are (2010, 4.9), (2011, 5.2), (2012, 5.8), (2013, 6.2), (2014, 6.8), (2015, 7.3), and (2016, 7.9).

b. We begin by plotting the ordered pairs. Because the x-coordinate in each ordered pair is a year, we label the x-axis "Year" and mark the horizontal axis with the years given. Then we label the y-axis or vertical axis "Net Sales (in billions of dollars)." In this case, it is convenient to mark the vertical axis in increments of 1, starting with 0.

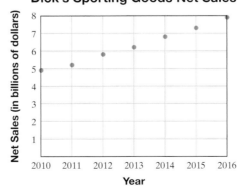

c. The scatter diagram shows that Dick's Sporting Goods net sales steadily increased over the years 2010–2016.

■ Work Practice 2

Objective C Completing Ordered Pair Solutions

Let's see how we can use ordered pairs to record solutions of equations containing two variables. An equation in one variable such as $x + 1 = 5$ has one solution, 4: The number 4 is the value of the variable x that makes the equation true.

An equation in two variables, such as $2x + y = 8$, has solutions consisting of two values, one for x and one for y. For example, $x = 3$ and $y = 2$ is a solution of $2x + y = 8$ because, if x is replaced with 3 and y with 2, we get a true statement.

$$2x + y = 8$$
$$2(3) + 2 \stackrel{?}{=} 8 \quad \text{Replace } x \text{ with 3 and } y \text{ with 2.}$$
$$8 = 8 \quad \text{True}$$

The solution $x = 3$ and $y = 2$ can be written as $(3, 2)$, an ordered pair of numbers.

In general, an ordered pair is a **solution** of an equation in two variables if replacing the variables by the values of the ordered pair results in a *true statement*.

For example, another ordered pair solution of $2x + y = 8$ is $(5, -2)$. Replacing x with 5 and y with -2 results in a true statement.

$$2x + y = 8$$
$$2(5) + (-2) \stackrel{?}{=} 8 \quad \text{Replace } x \text{ with 5 and } y \text{ with } -2.$$
$$10 - 2 \stackrel{?}{=} 8$$
$$8 = 8 \quad \text{True}$$

Example 3 Complete each ordered pair so that it is a solution to the equation $3x + y = 12$.

a. $(0,)$ **b.** $(, 6)$ **c.** $(-1,)$

Solution:

a. In the ordered pair $(0,)$, the x-value is 0. We let $x = 0$ in the equation and solve for y.

$$3x + y = 12$$
$$3(0) + y = 12 \quad \text{Replace } x \text{ with 0.}$$
$$0 + y = 12$$
$$y = 12$$

The completed ordered pair is $(0, 12)$.

b. In the ordered pair $(, 6)$, the y-value is 6. We let $y = 6$ in the equation and solve for x.

$$3x + y = 12$$
$$3x + 6 = 12 \quad \text{Replace } y \text{ with 6.}$$
$$3x = 6 \quad \text{Subtract 6 from both sides.}$$
$$x = 2 \quad \text{Divide both sides by 3.}$$

The ordered pair is $(2, 6)$.

c. In the ordered pair $(-1,)$, the x-value is -1. We let $x = -1$ in the equation and solve for y.

$$3x + y = 12$$
$$3(-1) + y = 12 \quad \text{Replace } x \text{ with } -1.$$
$$-3 + y = 12$$
$$y = 15 \quad \text{Add 3 to both sides.}$$

The ordered pair is $(-1, 15)$.

Work Practice 3

Practice 3

Complete each ordered pair so that it is a solution to the equation $x + 2y = 8$.

a. $(0,)$
b. $(, 3)$
c. $(-4,)$

Answers
3. a. $(0, 4)$ **b.** $(2, 3)$ **c.** $(-4, 6)$

Chapter 13 | Graphing Equations and Inequalities

Solutions of equations in two variables can also be recorded in a **table of paired values,** as shown in the next example.

Practice 4

Complete the table for the equation $y = -2x$.

	x	y
a.	-3	
b.		0
c.		10

Example 4 Complete the table for the equation $y = 3x$.

	x	y
a.	-1	
b.		0
c.		-9

Solution:

a. We replace x with -1 in the equation and solve for y.

$y = 3x$
$y = 3(-1)$ Let $x = -1$.
$y = -3$

The ordered pair is $(-1, -3)$.

b. We replace y with 0 in the equation and solve for x.

$y = 3x$
$0 = 3x$ Let $y = 0$.
$0 = x$ Divide both sides by 3.

The ordered pair is $(0, 0)$.

c. We replace y with -9 in the equation and solve for x.

$y = 3x$
$-9 = 3x$ Let $y = -9$.
$-3 = x$ Divide both sides by 3.

The ordered pair is $(-3, -9)$. The completed table is shown to the right.

x	y
-1	-3
0	0
-3	-9

■ Work Practice 4

Practice 5

Complete the table for the equation $y = \frac{1}{3}x - 1$.

	x	y
a.	-3	
b.	0	
c.		0

Example 5 Complete the table for the equation

$$y = \frac{1}{2}x - 5.$$

	x	y
a.	-2	
b.	0	
c.		0

Solution:

a. Let $x = -2$.

$y = \frac{1}{2}x - 5$
$y = \frac{1}{2}(-2) - 5$
$y = -1 - 5$
$y = -6$

b. Let $x = 0$.

$y = \frac{1}{2}x - 5$
$y = \frac{1}{2}(0) - 5$
$y = 0 - 5$
$y = -5$

c. Let $y = 0$.

$y = \frac{1}{2}x - 5$
$0 = \frac{1}{2}x - 5$ Now, solve for
$5 = \frac{1}{2}x$ Add 5.
$10 = x$ Multiply by 2

Ordered pairs: $(-2, -6)$ $(0, -5)$ $(10, 0)$

The completed table is

x	-2	0	10
y	-6	-5	0

■ Work Practice 5

Answers

4.

	x	y
a.	-3	6
b.	0	0
c.	-5	10

5.

	x	y
a.	-3	-2
b.	0	-1
c.	3	0

Notice in the previous example, Example 5, that a table showing ordered pair solutions may be written vertically or horizontally.

By now, you may also have noticed that equations in two variables often have more than one solution. We discuss this more in the next section.

Example 6 A small business purchased a computer for $2000. The business predicts that the computer will be used for 5 years and the value in dollars y of the computer in x years is $y = -300x + 2000$. Complete the table.

x	0	1	2	3	4	5
y						

Solution:

To find the value of y when x is 0, we replace x with 0 in the equation. We use this same procedure to find y when x is 1 and when x is 2.

When $x = 0$,
$y = -300x + 2000$
$y = -300 \cdot 0 + 2000$
$y = 0 + 2000$
$y = 2000$

When $x = 1$,
$y = -300x + 2000$
$y = -300 \cdot 1 + 2000$
$y = -300 + 2000$
$y = 1700$

When $x = 2$,
$y = -300x + 2000$
$y = -300 \cdot 2 + 2000$
$y = -600 + 2000$
$y = 1400$

We have the ordered pairs (0, 2000), (1, 1700), and (2, 1400). This means that in 0 years the value of the computer is $2000, in 1 year the value of the computer is $1700, and in 2 years the value is $1400. To complete the table of values, we continue the procedure for $x = 3$, $x = 4$, and $x = 5$.

When $x = 3$,
$y = -300x + 2000$
$y = -300 \cdot 3 + 2000$
$y = -900 + 2000$
$y = 1100$

When $x = 4$,
$y = -300x + 2000$
$y = -300 \cdot 4 + 2000$
$y = -1200 + 2000$
$y = 800$

When $x = 5$,
$y = -300x + 2000$
$y = -300 \cdot 5 + 2000$
$y = -1500 + 2000$
$y = 500$

The completed table is shown below.

x	0	1	2	3	4	5
y	2000	1700	1400	1100	800	500

Work Practice 6

Practice 6
A company purchased a portable scanner for $250. The business manager of the company predicts that the scanner will be used for 4 years and the value in dollars y of the machine in x years is $y = -50x + 250$. Complete the table.

x	1	2	3	4
y				

The ordered pair solutions recorded in the completed table for Example 6 are another set of paired data. They are graphed next. Notice that this scatter diagram gives a visual picture of the decrease in value of the computer.

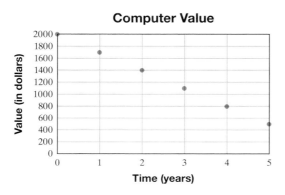

Answer

6.

x	1	2	3	4
y	200	150	100	50

Vocabulary, Readiness & Video Check

Use the choices below to fill in each blank. The exercises below all have to do with the rectangular coordinate system.

origin	x-coordinate	x-axis	scatter diagram	four
quadrants	y-coordinate	y-axis	solution	one

1. The horizontal axis is called the _____.
2. The vertical axis is called the _____.
3. The intersection of the horizontal axis and the vertical axis is a point called the _____.
4. The axes divide the plane into regions, called _____. There are _____ of these regions.
5. In the ordered pair of numbers $(-2, 5)$, the number -2 is called the _____ and the number 5 is called the _____.
6. Each ordered pair of numbers corresponds to _____ point in the plane.
7. An ordered pair is a(n) _____ of an equation in two variables if replacing the variables by the coordinates of the ordered pair results in a true statement.
8. The graph of paired data as points in a rectangular coordinate system is called a(n) _____.

Martin-Gay Interactive Videos Watch the section lecture video and answer the following questions.

See Video 13.1

Objective A 9. Several points are plotted in Examples 1–6. Where do we always start when plotting a point? How does the first coordinate tell us to move? How does the second coordinate tell us to move?

Objective B 10. From Example 7, what kind of data can be graphed in a scatter diagram?

Objective C 11. In Example 8, when finding the missing value in an ordered pair solution of a linear equation in two variables, how can we check our solution?

13.1 Exercise Set MyLab Math

Objective A *Plot each ordered pair. State in which quadrant or on which axis each point lies. See Example 1.*

1. a. $(1, 5)$ b. $(-5, -2)$ c. $(-3, 0)$ d. $(0, -1)$
   e. $(2, -4)$ f. $\left(-1, 4\frac{1}{2}\right)$ g. $(3.7, 2.2)$ h. $\left(\frac{1}{2}, -3\right)$

2. a. $(2, 4)$ b. $(0, 2)$ c. $(-2, 1)$ d. $(-3, -$
   e. $\left(3\frac{3}{4}, 0\right)$ f. $(5, -4)$ g. $(-3.4, 4.8)$ h. $\left(\frac{1}{3}, -5\right)$

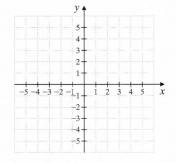
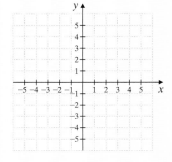

Find the x- and y-coordinates of each labeled point. See Example 1.

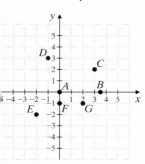

3. A

4. B

5. C

6. D

7. E

8. F

9. G

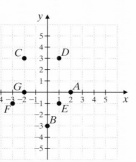

10. A

11. B

12. C

13. D

14. E

15. F

16. G

Objective B Solve. See Example 2.

17. The table shows the domestic box office (in billions of dollars) for the U.S. and Canadian movie industry during the years shown. (*Source:* Motion Picture Association of America)

Year	Box Office (in billions of dollars)
2010	10.6
2011	10.2
2012	10.8
2013	10.9
2014	10.4
2015	11.1
2016	11.4

a. Write this paired data as a set of ordered pairs of the form (year, box office).

b. In your own words, write the meaning of the ordered pair (2010, 10.6).

c. Create a scatter diagram of the paired data. Be sure to label the axes appropriately.

Domestic Box Office

d. What trend in the paired data does the scatter diagram show?

18. The table shows the amount of money (in billions of dollars) that Americans spent on their pets for the years shown. (*Source:* American Pet Products Association, Inc.)

Year	Pet-Related Expenditures (in billions of dollars)
2011	51
2012	53
2013	56
2014	58
2015	60
2016	63

a. Write this paired data as a set of ordered pairs of the form (year, pet-related expenditures).

b. In your own words, write the meaning of the ordered pair (2016, 63).

c. Create a scatter diagram of the paired data. Be sure to label the axes appropriately.

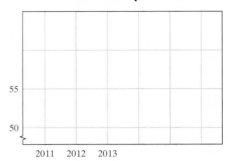

d. What trend in the paired data does the scatter diagram show?

19. Minh, a psychology student, kept a record of how much time she spent studying for each of her 20-point psychology quizzes and her score on each quiz.

Hours Spent Studying	0.50	0.75	1.00	1.25	1.50	1.50	1.75	2.00
Quiz Score	10	12	15	16	18	19	19	20

a. Write the data as ordered pairs of the form (hours spent studying, quiz score).

b. In your own words, write the meaning of the ordered pair (1.25, 16).

c. Create a scatter diagram of the paired data. Be sure to label the axes appropriately.

d. What might Minh conclude from the scatter diagram?

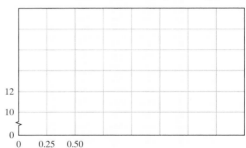

20. A local lumberyard uses quantity pricing. The table shows the price per board for different amounts of lumber purchased.

Price per Board (in dollars)	Number of Boards Purchased
8.00	1
7.50	10
6.50	25
5.00	50
2.00	100

a. Write the data as ordered pairs of the form (price per board, number of boards purchased).

b. In your own words, write the meaning of the ordered pair (2.00, 100).

c. Create a scatter diagram of the paired data. Be sure to label the axes appropriately.

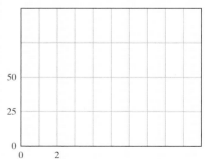

d. What trend in the paired data does the scatter diagram show?

Section 13.1 | The Rectangular Coordinate System

Objective C *Complete each ordered pair so that it is a solution to the given linear equation. See Example 3.*

21. $x - 4y = 4$; (, -2), (4,)   **22.** $x - 5y = -1$; (, -2), (4,)

23. $y = \frac{1}{4}x - 3$; (-8,), (, 1)   **24.** $y = \frac{1}{5}x - 2$; (-10,), (, 1)

Complete the table of ordered pairs for each linear equation. See Examples 4 and 5.

25. $y = -7x$

x	y
0	
-1	
	2

26. $y = -9x$

x	y
	0
-3	
	2

27. $x = -y + 2$

x	y
0	
	0
	-3

28. $x = -y + 4$

x	y
	0
0	
	-3

29. $y = \frac{1}{2}x$

x	y
0	
-6	
	1

30. $y = \frac{1}{3}x$

x	y
0	
-6	
	1

31. $x + 3y = 6$

x	y
0	
	0
	1

32. $2x + y = 4$

x	y
0	
	0
	2

33. $y = 2x - 12$

x	y
0	
	-2
3	

34. $y = 5x + 10$

x	y
	0
	5
0	

35. $2x + 7y = 5$

x	y
0	
	0
	1

36. $x - 6y = 3$

x	y
0	
1	
	-1

Objectives A B C Mixed Practice *Complete the table of ordered pairs for each equation. Then plot the ordered pair solutions. See Examples 1 through 5.*

37. $x = -5y$

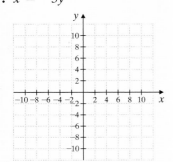

x	y
	0
	1
10	

38. $y = -3x$

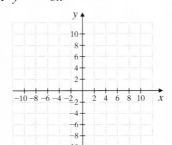

x	y
0	
-2	
	9

39. $y = \dfrac{1}{3}x + 2$

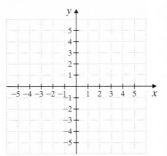

x	y
0	
−3	
	0

40. $y = \dfrac{1}{2}x + 3$

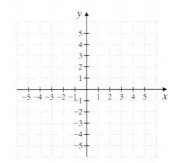

x	y
0	
−4	
	0

Solve. See Example 6.

41. The cost in dollars y of producing x computer desks is given by $y = 80x + 5000$.

   a. Complete the table.

x	100	200	300
y			

   b. Find the number of computer desks that can be produced for $8600. (*Hint:* Find x when $y = 8600$.)

42. The hourly wage y of an employee at a certain production company is given by $y = 0.25x + 9$, where x is the number of units produced by the employee in an hour.

   a. Complete the table.

x	0	1	5	10
y				

   b. Find the number of units that the employee must produce each hour to earn an hourly wage of $12.25. (*Hint:* Find x when $y = 12.25$.)

43. The average annual cinema admission price y (in dollars) from 2007 through 2016 is given by $y = 0.17x + 7.09$. In this equation, x represents the number of years after 2007. (*Source:* Motion Picture Association of America)

   a. Complete the table.

x	2	5	8
y			

   b. Find the year in which the average cinema admission price was approximately $8.00. (*Hint:* Find x when $y = 8.00$ and round to the nearest whole number.)

   c. Use the given equation to predict when the cinema admission price might be $10.00. (Use the hint for part **b.**)

44. The number of farms y in the United States from 20 through 2015 is given by $y = -14,300x + 2,112,000$ In the equation, x represents the number of years after 2012. (*Source:* Based on data from the Nationa Agricultural Statistics Service)

   a. Complete the table.

x	0	2	4
y			

   b. Find the year in which there were approximately 2,070,000 farms. (*Hint:* Find x when $y = 2,070,000$ and round to the nearest whole number.)

   c. Use the given equation to predict when the number of farms might be 2,000,000. (Use the hint for part **b.**)

Review

The following bar graph shows the top 10 tourist destinations and the number of tourists that visit each destination per year forecasted for 2020. Use this graph to answer Exercises 45 through 50. See Section 7.1.

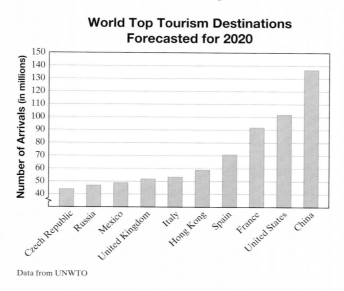

Data from UNWTO

45. Which location shown is predicted to be the most popular tourist destination?

46. Which location shown is predicted to be the least popular tourist destination?

47. Which locations shown are predicted to have more than 70 million tourists per year?

48. Which locations shown are predicted to have more than 100 million tourists per year?

49. Estimate the predicted number of tourists per year whose destination is Italy.

50. Estimate the predicted number of tourists per year whose destination is Mexico.

Solve each equation for y. See Section 9.5.

51. $x + y = 5$

52. $x - y = 3$

53. $2x + 4y = 5$

54. $5x + 2y = 7$

55. $10x = -5y$

56. $4y = -8x$

Concept Extensions

Answer each exercise with true or false.

57. Point $(-1, 5)$ lies in quadrant IV.

58. Point $(3, 0)$ lies on the y-axis.

59. For the point $\left(-\frac{1}{2}, 1.5\right)$, the first value, $-\frac{1}{2}$, is the x-coordinate and the second value, 1.5, is the y-coordinate.

60. The ordered pair $\left(2, \frac{2}{3}\right)$ is a solution of $2x - 3y = 6$.

For Exercises 61 through 65, fill in each blank with "0," "positive," or "negative." For Exercises 66 and 67, fill in each blank with "x" or "y."

	Point	Location
61.	(____, ____)	quadrant III
62.	(____, ____)	quadrant I
63.	(____, ____)	quadrant IV
64.	(____, ____)	quadrant II
65.	(____, ____)	origin
66.	(number, 0)	___-axis
67.	(0, number)	___-axis

68. Give an example of an ordered pair whose location is in (or on)
   a. quadrant I   **b.** quadrant II   **c.** quadrant III
   d. quadrant IV   **e.** x-axis   **f.** y-axis

Solve. See the Concept Checks in this section.

69. Is the graph of $(3, 0)$ in the same location as the graph of $(0, 3)$? Explain why or why not.

70. Give the coordinates of a point such that if the coordinates are reversed, their location is the same.

71. In general, what points can have coordinates reversed and still have the same location?

72. In your own words, describe how to plot or graph an ordered pair of numbers.

73. Discuss any similarities in the graphs of the ordered pair solutions for Exercises **37–40**.

74. Discuss any differences in the graphs of the ordered pair solutions for Exercises **37–40**.

Write an ordered pair for each point described.

75. Point C is four units to the right of the y-axis and seven units below the x-axis.

76. Point D is three units to the left of the origin.

77. Find the perimeter of the rectangle whose vertices are the points with coordinates $(-1, 5)$, $(3, 5)$, $(3, -4)$, and $(-1, -4)$.

78. Find the area of the rectangle whose vertices are the points with coordinates $(5, 2)$, $(5, -6)$, $(0, -6)$ and $(0, 2)$.

The scatter diagram below shows the annual number of people enrolled as Gold Star Members at Costco Wholesale. The horizontal axis represents the number of years after 2012.

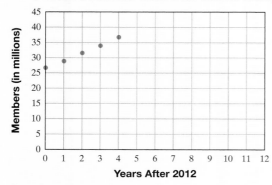

Costco's Annual Gold Star Membership

Source: Costco Wholesale Corporation

79. Estimate the annual Gold Star Membership for years 1, 2, 3, and 4.

80. Use a straightedge or ruler and this scatter diagram to predict Costco's Gold Star Membership in the year 2020.

13.2 Graphing Linear Equations

In the previous section, we found that equations in two variables may have more than one solution. For example, both $(2, 2)$ and $(0, 4)$ are solutions of the equation $x + y = 4$. In fact, this equation has an infinite number of solutions. Other solutions include $(-2, 6)$, $(4, 0)$, and $(6, -2)$. Notice the pattern that appears in the graph of these solutions.

Objective

A Graph a Linear Equation by Finding and Plotting Ordered Pair Solutions.

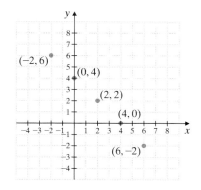

These solutions all appear to lie on the same line, as seen in the graph below. It can be shown that every ordered pair solution of the equation corresponds to a point on this line and that every point on this line corresponds to an ordered pair solution. Thus, we say that this line is the **graph of the equation** $x + y = 4$.

> **Helpful Hint**
> Notice that we can show only a part of a line on a graph. The arrowheads on each end of the line remind us that the line actually extends indefinitely in both directions.

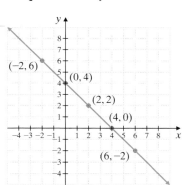

The equation $x + y = 4$ is called a *linear equation in two variables* and *the graph of every linear equation in two variables is a straight line.*

Linear Equation in Two Variables

A **linear equation in two variables** is an equation that can be written in the form

$$Ax + By = C$$

where A, B, and C are real numbers and A and B are not both 0. This form is called **standard form**. The graph of a linear equation in two variables is a straight line.

> **Helpful Hint**
> Notice from above that the form $Ax + By = C$
> - is called standard form, and
> - has an understood exponent of 1 on both x and y.

A linear equation in two variables may be written in many forms. Standard form $Ax + By = C$, is just one of these many forms.

Following are examples of linear equations in two variables.

$$2x + y = 8 \qquad -2x = 7y \qquad y = \frac{1}{3}x + 2 \qquad y = 7$$

(Standard form)

Objective A Graphing Linear Equations

From geometry, we know that a straight line is determined by just two points. Thus, to graph a linear equation in two variables, we need to find just two of its infinitely many solutions. Once we do so, we plot the solution points and draw the line connecting the points. Usually, we find a third solution as well, as a check.

Example 1 Graph the linear equation $2x + y = 5$.

Solution: To graph this equation, we find three ordered pair solutions of $2x + y = 5$. To do this, we choose a value for one variable, x or y, and solve for the other variable. For example, if we let $x = 1$, then $2x + y = 5$ becomes

$2x + y = 5$
$2(1) + y = 5$ Replace x with 1.
$2 + y = 5$ Multiply.
$y = 3$ Subtract 2 from both sides.

Since $y = 3$ when $x = 1$, the ordered pair $(1, 3)$ is a solution of $2x + y = 5$. Next, we let $x = 0$.

$2x + y = 5$
$2(0) + y = 5$ Replace x with 0.
$0 + y = 5$
$y = 5$

The ordered pair $(0, 5)$ is a second solution.

The two solutions found so far allow us to draw the straight line that is the graph of all solutions of $2x + y = 5$. However, we will find a third ordered pair as a check. Let $y = -1$.

$2x + y = 5$
$2x + (-1) = 5$ Replace y with -1.
$2x - 1 = 5$
$2x = 6$ Add 1 to both sides.
$x = 3$ Divide both sides by 2.

Helpful Hint All three points should fall on the same straight line. If not, check your ordered pair solutions for a mistake.

The third solution is $(3, -1)$. These three ordered pair solutions are listed in the table and plotted on the coordinate plane. The graph of $2x + y = 5$ is the line through the three points.

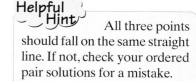

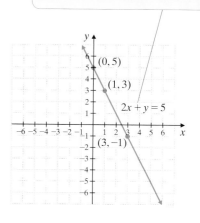

x	y
1	3
0	5
3	−1

■ Work Practice 1

Practice 1

Graph the linear equation $x + 3y = 6$.

Answer
1.
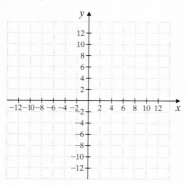

Example 2

Graph the linear equation $-5x + 3y = 15$.

Solution: We find three ordered pair solutions of $-5x + 3y = 15$.

Let $x = 0$.	Let $y = 0$.	Let $x = -2$.
$-5x + 3y = 15$	$-5x + 3y = 15$	$-5x + 3y = 15$
$-5 \cdot 0 + 3y = 15$	$-5x + 3 \cdot 0 = 15$	$-5 \cdot -2 + 3y = 15$
$0 + 3y = 15$	$-5x + 0 = 15$	$10 + 3y = 15$
$3y = 15$	$-5x = 15$	$3y = 5$
$y = 5$	$x = -3$	$y = \frac{5}{3}$ or $1\frac{2}{3}$

The ordered pairs are $(0, 5)$, $(-3, 0)$, and $\left(-2, 1\frac{2}{3}\right)$. The graph of $-5x + 3y = 15$ is the line through the three points.

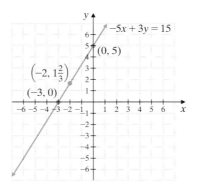

Work Practice 2

Example 3

Graph the linear equation $y = 3x$.

Solution: We find three ordered pair solutions. Since this equation is solved for y, we'll choose three x-values.

If $x = 2$, $y = 3 \cdot 2 = 6$.
If $x = 0$, $y = 3 \cdot 0 = 0$.
If $x = -1$, $y = 3 \cdot -1 = -3$.

Next, we plot the ordered pair solutions and draw a line through the plotted points. The line is the graph of $y = 3x$.

Think about the following for a moment: A line is made up of an infinite number of points. Every point on the line defined by $y = 3x$ represents an ordered pair solution of the equation, and every ordered pair solution is a point on this line.

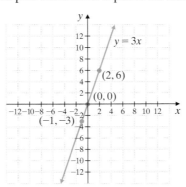

Work Practice 3

Practice 2

Graph the linear equation $-2x + 4y = 8$.

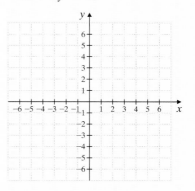

Practice 3

Graph the linear equation $y = 2x$.

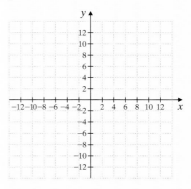

Answers

2.

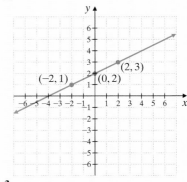

3.

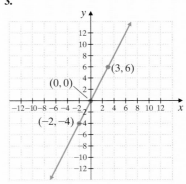

Practice 4

Graph the linear equation $y = -\dfrac{1}{2}x + 4$.

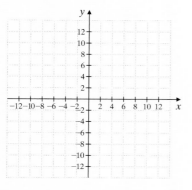

Practice 5

Graph the linear equation $x = 3$.

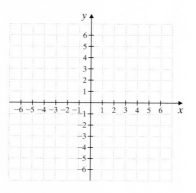

Answers

4.

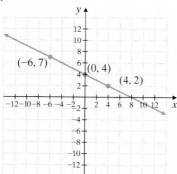

5.

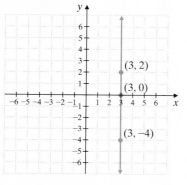

Helpful Hint

When graphing a linear equation in two variables, if it is
- solved for y, it may be easier to find ordered pair solutions by choosing x-values.
- solved for x, it may be easier to find ordered pair solutions by choosing y-values.

Example 4 Graph the linear equation $y = -\dfrac{1}{3}x + 2$.

Solution: We find three ordered pair solutions, plot the solutions, and draw a line through the plotted solutions. To avoid fractions, we'll choose x-values that are multiples of 3 to substitute into the equation.

If $x = 6$, then $y = -\dfrac{1}{3} \cdot 6 + 2 = -2 + 2 = 0$.

If $x = 0$, then $y = -\dfrac{1}{3} \cdot 0 + 2 = 0 + 2 = 2$.

If $x = -3$, then $y = -\dfrac{1}{3} \cdot -3 + 2 = 1 + 2 = 3$.

x	y
6	0
0	2
-3	3

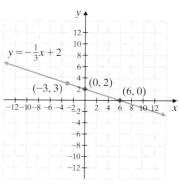

■ Work Practice 4

Let's take a moment and compare the graphs in Examples 3 and 4. The graph of $y = 3x$ tilts upward (as we follow the line from left to right) and the graph of $y = -\dfrac{1}{3}x + 2$ tilts downward (as we follow the line from left to right). We will learn more about the tilt, or slope, of a line in Section 13.4.

Example 5 Graph the linear equation $y = -2$.

Solution: The equation $y = -2$ can be written in standard form as $0x + y = -2$. No matter what value we replace x with, y is always -2.

x	y
0	-2
3	-2
-2	-2

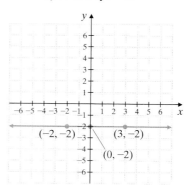

Notice that the graph of $y = -2$ is a horizontal line.

■ Work Practice 5

Linear equations are often used to model real data, as seen in the next example.

Example 6 Estimating the Number of Registered Nurses

One of the occupations expected to have the most growth in the next few years is registered nurse. The number of people y (in thousands) employed as registered nurses in the United States can be estimated by the linear equation $y = 43.9x + 2751$, where x is the number of years after the year 2014. (*Source: Based on data from the Bureau of Labor Statistics*)

a. Graph the equation.
b. Use the graph to predict the number of registered nurses in the year 2025.

Solution:

a. To graph $y = 43.9x + 2751$, choose x-values and substitute into the equation.

If $x = 0$, then $y = 43.9(0) + 2751 = 2751$.
If $x = 2$, then $y = 43.9(2) + 2751 = 2838.8$.
If $x = 5$, then $y = 43.9(5) + 2751 = 2970.5$.

x	y
0	2751
2	2838.8
5	2970.5

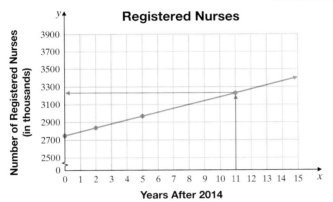

b. To use the graph to *predict* the number of registered nurses in the year 2025, we need to find the y-coordinate that corresponds to $x = 11$. (11 years after 2014 is the year 2025.) To do so, find 11 on the x-axis. Move vertically upward to the graphed line and then horizontally to the left. We approximate the number on the y-axis to be 3230. Thus, in the year 2025, we predict that there will be 3230 thousand registered nurses. (The value found by substituting 11 for x in the equation is 3223.9.)

■ Work Practice 6

Practice 6
Use the graph in Example 6 to predict the number of registered nurses in 2026.

Helpful Hint
From Example 5, we learned that equations such as $y = -2$ are linear equations since $y = -2$ can be written as $0x + y = -2$.

Helpful Hint
Make sure you understand that models are mathematical approximations of the data for the known years. (For example, see the model in Example 6.) Any number of unknown factors can affect future years, so be cautious when using models to make predictions.

Answer
6. 3275 thousand

Calculator Explorations Graphing

In this section, we begin an optional study of graphing calculators and graphing software packages for computers. These graphers use the same point plotting technique that was introduced in this section. The advantage of this graphing technology is, of course, that graphing calculators and computers can find and plot ordered pair solutions much faster than we can. Note, however, that the features described in these boxes may not be available on all graphing calculators.

The rectangular screen where a portion of the rectangular coordinate system is displayed is called a **window**. We call it a **standard window** for graphing when both the *x*- and *y*-axes show coordinates between -10 and 10. This information is often displayed in the window menu on a graphing calculator as follows.

Xmin = -10
Xmax = 10
  Xscl = 1 The scale on the *x*-axis is one unit per tick mark.
Ymin = -10
Ymax = 10
  Yscl = 1 The scale on the *y*-axis is one unit per tick mark.

To use a graphing calculator to graph the equation $y = 2x + 3$, press the $\boxed{Y=}$ key and enter the keystrokes $\boxed{2}\boxed{x}\boxed{+}\boxed{3}$. The top row should now read $Y_1 = 2x + 3$. Next press the $\boxed{GRAPH}$ key, and the display should look like this:

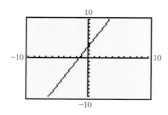

Graph the following linear equations. (Unless otherwise stated, use a standard window when graphing.)

1. $y = -3x + 7$

2. $y = -x + 5$

3. $y = 2.5x - 7.9$

4. $y = -1.3x + 5.2$

5. $y = -\dfrac{3}{10}x + \dfrac{32}{5}$

6. $y = \dfrac{2}{9}x - \dfrac{22}{3}$

Vocabulary, Readiness & Video Check

Martin-Gay Interactive Videos Watch the section lecture video and answer the following questions.

See Video 13.2

Objective A 1. In the lecture before Example 1, it's mentioned that we need only two points to determine a line. Why, then, are three ordered pair solutions found in Examples 1–3?

2. What does a graphed line represent, as discussed at the end of Examples 1 and 3?

13.2 Exercise Set MyLab Math

Objective A *For each equation, find three ordered pair solutions by completing the table. Then use the ordered pairs to graph the equation. See Examples 1 through 5.*

1. $x - y = 6$

x	y
	0
4	
	-1

2. $x - y = 4$

x	y
0	
	2
-1	

3. $y = -4x$

x	y
1	
0	
-1	

4. $y = -5x$

x	y
1	
0	
-1	

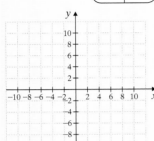
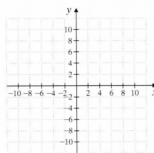
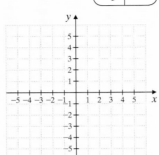
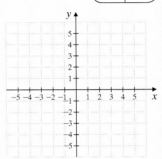

5. $y = \dfrac{1}{3}x$

x	y
0	
6	
-3	

6. $y = \dfrac{1}{2}x$

x	y
0	
-4	
2	

7. $y = -4x + 3$

x	y
0	
1	
2	

8. $y = -5x + 2$

x	y
0	
1	
2	

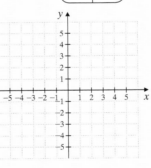
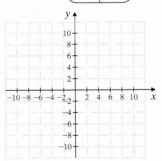

Graph each linear equation. See Examples 1 through 5.

9. $x + y = 1$

10. $x + y = 7$

11. $x - y = -2$

12. $-x + y = 6$

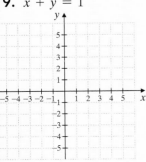
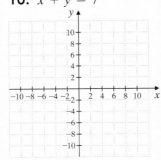
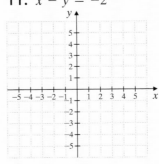
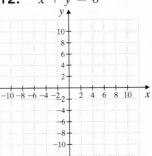

13. $x - 2y = 6$
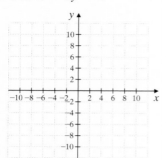

14. $-x + 5y = 5$
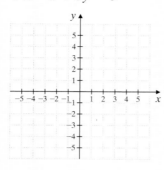

15. $y = 6x + 3$
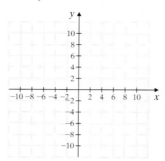

16. $y = -2x + 7$
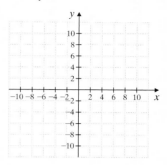

17. $x = -4$
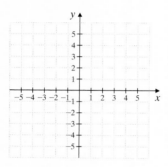

18. $y = 5$
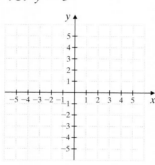

19. $y = 3$
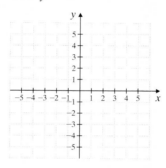

20. $x = -1$
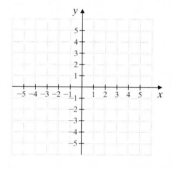

21. $y = x$
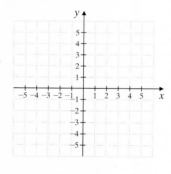

22. $y = -x$
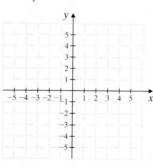

23. $x = -3y$
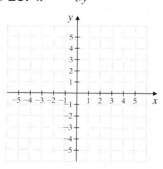

24. $x = 4y$
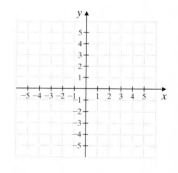

25. $x + 3y = 9$
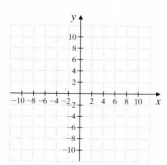

26. $2x + y = 2$
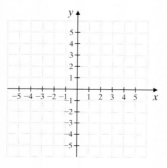

27. $y = \frac{1}{2}x + 2$
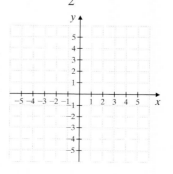

28. $y = \frac{1}{4}x + 3$
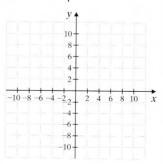

Section 13.2 | Graphing Linear Equations

29. $3x - 2y = 12$

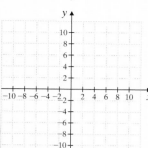

30. $2x - 7y = 14$

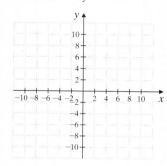

31. $y = -3.5x + 4$

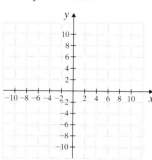

32. $y = -1.5x - 3$

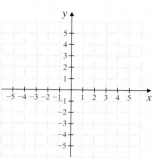

Solve. See Example 6.

33. The number of students y (in thousands) taking the SAT college entrance exam each year from 2011 through 2015 can be approximated by the linear equation $y = 11x + 1647$, where x is the number of years after 2011. (*Source:* Based on data from the College Board)

  a. Graph the linear equation.

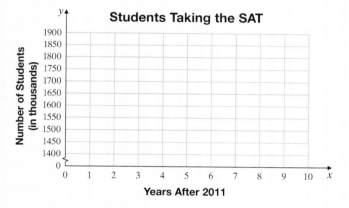

  b. Does the point $(7, 1724)$ lie on the line? If so, what does this ordered pair mean?

34. College is getting more expensive every year. The average cost for tuition and fees at a public two-year college y from 2001 through 2016 can be approximated by the linear equation $y = 90x + 2211$, where x is the number of years after 2001. (*Source:* The College Board: Trends in College Pricing 2016)

  a. Graph the linear equation.

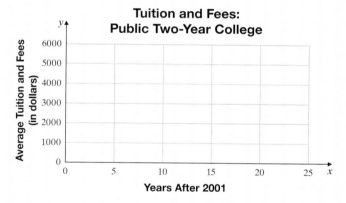

  b. Does the point $(20, 4011)$ lie on the line? If so, what does this ordered pair mean?

35. The total annual revenue y (in billions of euros) for IKEA from 2010 through 2016 can be approximated by the equation $y = 1.7x + 23.5$, where x is the number of years after 2010. (*Source:* Based on data from IKEA Group)

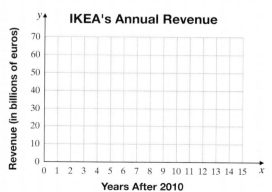

  a. Graph the linear equation.
  b. Complete the ordered pair $(6, \ )$.
  c. Write a sentence explaining the meaning of the ordered pair found in part **b**.

36. For the period 1970 through 2016, the annual food-and-drink sales for restaurants in the United States can be estimated by $y = 15.7x - 23.1$, where x is the number of years after 1970 and y is the food-and-drink sales in billions of dollars. (*Source:* Based on data from the National Restaurant Association)

a. Graph the linear equation.

b. Complete the ordered pair (43,).
c. Write a sentence explaining the meaning of the ordered pair found in part **b**.

Review

△ **37.** The coordinates of three vertices of a rectangle are $(-2, 5)$, $(4, 5)$, and $(-2, -1)$. Find the coordinates of the fourth vertex. See Section 13.1.

△ **38.** The coordinates of two vertices of a square are $(-3, -1)$ and $(2, -1)$. Find the coordinates of two pairs of possible points for the third and fourth vertices. See Section 13.1.

Complete each table. See Section 13.1.

39. $x - y = -3$

x	y
0	
	0

40. $y - x = 5$

x	y
0	
	0

41. $y = 2x$

x	y
0	
	0

42. $x = -3y$

x	y
0	
	0

Concept Extensions

Graph each pair of linear equations on the same set of axes. Discuss how the graphs are similar and how they are different.

43. $y = 5x$
$y = 5x + 4$

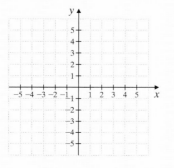

44. $y = 2x$
$y = 2x + 5$

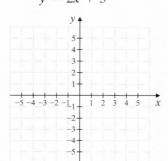

45. $y = -2x$
$y = -2x - 3$

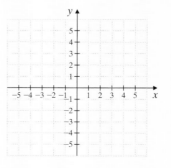

46. $y = x$
$y = x - 7$

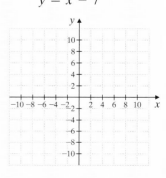

47. Graph the nonlinear equation $y = x^2$ by completing the table shown. Plot the ordered pairs and connect them with a smooth curve.

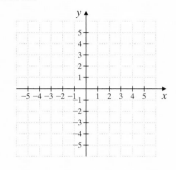

x	y
0	
1	
-1	
2	
-2	

48. Graph the nonlinear equation $y = |x|$ by completing the table shown. Plot the ordered pairs and connect them. This curve is "V" shaped.

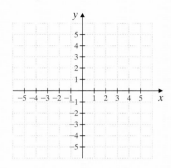

x	y
0	
1	
-1	
2	
-2	

49. The perimeter of the trapezoid below is 22 centimeters. Write a linear equation in two variables for the perimeter. Find y if x is 3 centimeters.

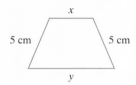

△ 50. The perimeter of the rectangle below is 50 miles. Write a linear equation in two variables for the perimeter. Use this equation to find x when y is 20 miles.

51. If (a, b) is an ordered pair solution of $x + y = 5$, is (b, a) also a solution? Explain why or why not.

52. If (a, b) is an ordered pair solution of $x - y = 5$, is (b, a) also a solution? Explain why or why not.

13.3 Intercepts

Objective A Identifying Intercepts

The graph of $y = 4x - 8$ is shown below. Notice that this graph crosses the y-axis at the point $(0, -8)$. This point is called the **y-intercept**. Likewise the graph crosses the x-axis at $(2, 0)$. This point is called the **x-intercept**.

Objectives

- **A** Identify Intercepts of a Graph.
- **B** Graph a Linear Equation by Finding and Plotting Intercept Points.
- **C** Identify and Graph Vertical and Horizontal Lines.

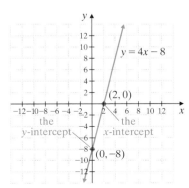

The intercepts are $(2, 0)$ and $(0, -8)$.

Practice 1–5

Identify the *x*- and *y*-intercepts.

1.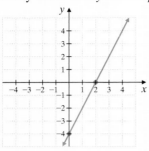

2.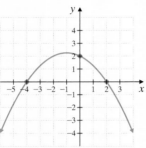

3.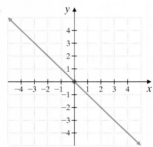

4.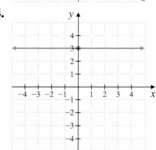

5.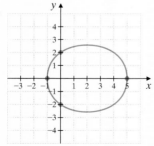

Answers
1. *x*-intercept: $(2, 0)$; *y*-intercept: $(0, -4)$
2. *x*-intercepts: $(-4, 0)$, $(2, 0)$; *y*-intercept: $(0, 2)$
3. *x*-intercept and *y*-intercept: $(0, 0)$
4. no *x*-intercept; *y*-intercept: $(0, 3)$
5. *x*-intercepts: $(-1, 0)$, $(5, 0)$; *y*-intercepts: $(0, 2)$, $(0, -2)$

Helpful Hint

If a graph crosses the *x*-axis at $(2, 0)$ and the *y*-axis at $(0, -8)$, then

$\underbrace{(2, 0)}_{x\text{-intercept}} \quad \underbrace{(0, -8)}_{y\text{-intercept}}$

Notice that for the *x*-intercept, the *y*-value is 0 and that for the *y*-intercept, the *x*-value is 0.

Note: Sometimes in mathematics, you may see just the number -8 stated as the *y*-intercept, and 2 stated as the *x*-intercept.

Examples Identify the *x*- and *y*-intercepts.

1.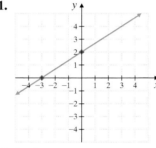

   Solution:
   x-intercept: $(-3, 0)$
   y-interceipt: $(0, 2)$

2.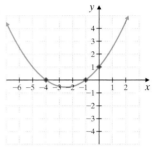

   Solution:
   x-intercepts: $(-4, 0)$, $(-1, 0)$
   y-intercept: $(0, 1)$

3.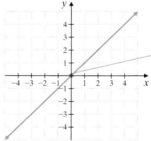

   Solution:
   x-intercept: $(0, 0)$
   y-intercept: $(0, 0)$

   Here, the *x*- and *y*-intercepts happen to be the same point.

   ### Helpful Hint
   Notice that any time $(0, 0)$ is a point of a graph, then it is an *x*-intercept and a *y*-intercept. Why? It is the *only* point that lies on both axes.

4.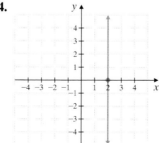

   Solution:
   x-intercept: $(2, 0)$
   y-intercept: none

5.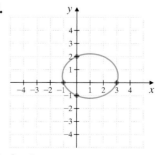

   Solution:
   x-intercepts: $(-1, 0)$, $(3, 0)$
   y-intercepts: $(0, 2)$, $(0, -1)$

■ Work Practice 1–5

Objective B Finding and Plotting Intercepts

Given an equation of a line, we can usually find intercepts easily since one coordinate is 0.

To find the *x*-intercept of a line from its equation, let $y = 0$, since a point on the *x*-axis has a *y*-coordinate of 0. To find the *y*-intercept of a line from its equation, let $x = 0$, since a point on the *y*-axis has an *x*-coordinate of 0.

Finding *x*- and *y*-Intercepts

To find the *x*-intercept, let $y = 0$ and solve for *x*.
To find the *y*-intercept, let $x = 0$ and solve for *y*.

Example 6 Graph $x - 3y = 6$ by finding and plotting its intercepts.

Solution: We let $y = 0$ to find the *x*-intercept and $x = 0$ to find the *y*-intercept.

Let $y = 0$.
$x - 3y = 6$
$x - 3(0) = 6$
$x - 0 = 6$
$x = 6$

Let $x = 0$.
$x - 3y = 6$
$0 - 3y = 6$
$-3y = 6$
$y = -2$

The *x*-intercept is $(6, 0)$ and the *y*-intercept is $(0, -2)$. We find a third ordered pair solution to check our work. If we let $y = -1$, then $x = 3$. We plot the points $(6, 0)$, $(0, -2)$, and $(3, -1)$. The graph of $x - 3y = 6$ is the line drawn through these points, as shown.

x	y
6	0
0	-2
3	-1

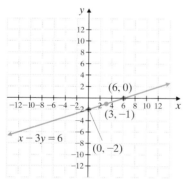

Work Practice 6

Example 7 Graph $x = -2y$ by finding and plotting its intercepts.

Solution: We let $y = 0$ to find the *x*-intercept and $x = 0$ to find the *y*-intercept.

Let $y = 0$.
$x = -2y$
$x = -2(0)$
$x = 0$

Let $x = 0$.
$x = -2y$
$0 = -2y$
$0 = y$

Both the *x*-intercept and *y*-intercept are $(0, 0)$. In other words, when $x = 0$, then $y = 0$, which gives the ordered pair $(0, 0)$. Also, when $y = 0$, then $x = 0$, which gives the same ordered pair, $(0, 0)$. This happens when the graph passes through the origin. Since two points are needed to determine a line, we must find at least one more ordered pair that satisfies $x = -2y$. Since the equation is solved for *x*, we

(Continued on next page)

Practice 6

Graph $2x - y = 4$ by finding and plotting its intercepts.

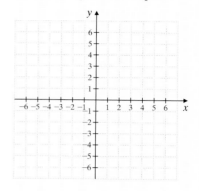

Practice 7

Graph $y = 3x$ by finding and plotting its intercepts.

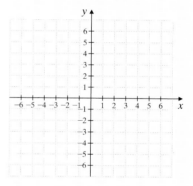

Answers

6.

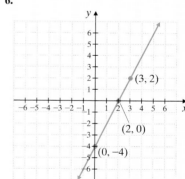

7.

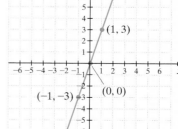

choose y-values so that there is no need to solve to find the corresponding x-values. We let $y = -1$ to find a second ordered pair solution and let $y = 1$ as a check point.

Let $y = -1$.
$$x = -2(-1)$$
$$x = 2 \quad \text{Multiply.}$$

Let $y = 1$.
$$x = -2(1)$$
$$x = -2 \quad \text{Multiply.}$$

The ordered pairs are $(0, 0)$, $(2, -1)$, and $(-2, 1)$. We plot these points to graph $x = -2y$.

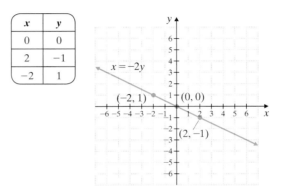

■ Work Practice 7

Practice 8

Graph: $3x = 2y + 4$

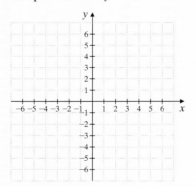

Example 8 Graph: $4x = 3y - 9$

Solution: Find the x- and y-intercepts, and then choose $x = 2$ to find a checkpoint.

Let $y = 0$.
$$4x = 3(0) - 9$$
$$4x = -9$$
Solve for x.
$$x = -\frac{9}{4} \text{ or } -2\frac{1}{4}$$

Let $x = 0$.
$$4 \cdot 0 = 3y - 9$$
$$9 = 3y$$
Solve for y.
$$3 = y$$

Let $x = 2$.
$$4(2) = 3y - 9$$
$$8 = 3y - 9$$
Solve for y.
$$17 = 3y$$
$$\frac{17}{3} = y \text{ or } y = 5\frac{2}{3}$$

The ordered pairs are $\left(-2\frac{1}{4}, 0\right)$, $(0, 3)$, and $\left(2, 5\frac{2}{3}\right)$. The equation $4x = 3y - 9$ is graphed as follows.

x	y
$-2\frac{1}{4}$	0
0	3
2	$5\frac{2}{3}$

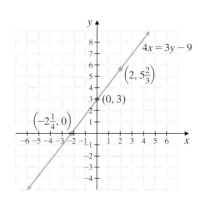

■ Work Practice 8

Answer

8.

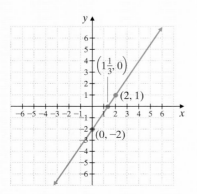

Objective C Graphing Vertical and Horizontal Lines

From Section 13.2, recall that the equation $x = 2$, for example, is a linear equation in two variables because it can be written in the form $x + 0y = 2$. The graph of this equation is a vertical line, as reviewed in the next example.

Example 9 Graph: $x = 2$

Solution: The equation $x = 2$ can be written as $x + 0y = 2$. For any y-value chosen, notice that x is 2. No other value for x satisfies $x + 0y = 2$. Any ordered pair whose x-coordinate is 2 is a solution of $x + 0y = 2$. We will use the ordered pair solutions $(2, 3), (2, 0)$, and $(2, -3)$ to graph $x = 2$.

x	y
2	3
2	0
2	-3

The graph is a vertical line with x-intercept $(2, 0)$. Note that this graph has no y-intercept because x is never 0.

■ Work Practice 9

In general, we have the following.

Vertical Lines

The graph of $x = c$, where c is a real number, is a **vertical line** with x-intercept $(c, 0)$.

Example 10 Graph: $y = -3$

Solution: The equation $y = -3$ can be written as $0x + y = -3$. For any x-value chosen, y is -3. If we choose 4, 1, and -2 as x-values, the ordered pair solutions are $(4, -3), (1, -3)$, and $(-2, -3)$. We use these ordered pairs to graph $y = -3$. The graph is a horizontal line with y-intercept $(0, -3)$ and no x-intercept.

x	y
4	-3
1	-3
-2	-3

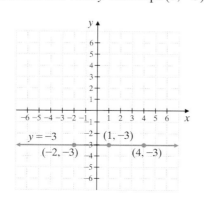

■ Work Practice 10

Practice 9
Graph: $x = -3$

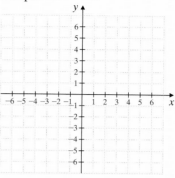

Practice 10
Graph: $y = 4$

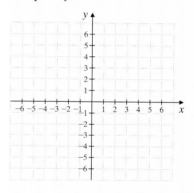

Answers

9.
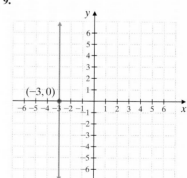

10.
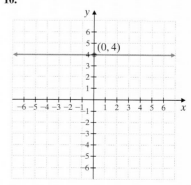

In general, we have the following.

Horizontal Lines

The graph of $y = c$, where c is a real number, is a **horizontal line** with y-intercept $(0, c)$.

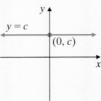

Calculator Explorations Graphing

You may have noticed that to use the $\boxed{Y=}$ key on a graphing calculator to graph an equation, the equation must be solved for y. For example, to graph $2x + 3y = 7$, we solve the equation for y.

$$2x + 3y = 7$$
$$3y = -2x + 7 \quad \text{Subtract } 2x \text{ from both sides.}$$
$$\frac{3y}{3} = -\frac{2x}{3} + \frac{7}{3} \quad \text{Divide both sides by 3.}$$
$$y = -\frac{2}{3}x + \frac{7}{3} \quad \text{Simplify.}$$

To graph $2x + 3y = 7$ or $y = -\frac{2}{3}x + \frac{7}{3}$, press the $\boxed{Y=}$ key and enter

$$Y_1 = -\frac{2}{3}x + \frac{7}{3}$$

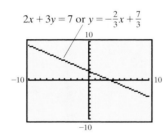

2. $-2.61y = x$

3. $3x + 7y = 21$

4. $-4x + 6y = 12$

5. $-2.2x + 6.8y = 15.5$

6. $5.9x - 0.8y = -10.4$

Graph each linear equation.
1. $x = 3.78y$

Vocabulary, Readiness & Video Check

Use the choices below to fill in each blank. Some choices may be used more than once. Exercises 1 and 2 come from Section 13.2.

| x | vertical | x-intercept | linear |
| y | horizontal | y-intercept | standard |

1. An equation that can be written in the form $Ax + By = C$ is called a(n) _____ equation in two variables.
2. The form $Ax + By = C$ is called _____ form.
3. The graph of the equation $y = -1$ is a(n) _____ line.
4. The graph of the equation $x = 5$ is a(n) _____ line.
5. A point where a graph crosses the y-axis is called a(n) _____.
6. A point where a graph crosses the x-axis is called a(n) _____.
7. Given an equation of a line, to find the x-intercept (if there is one), let _____ = 0 and solve for _____.
8. Given an equation of a line, to find the y-intercept (if there is one), let _____ = 0 and solve for _____.

Answer the following true or false.

9. All lines have an x-intercept *and* a y-intercept. _____
10. The graph of $y = 4x$ contains the point $(0, 0)$. _____
11. The graph of $x + y = 5$ has an x-intercept of $(5, 0)$ and a y-intercept of $(0, 5)$. _____
12. The graph of $y = 5x$ contains the point $(5, 1)$. _____

Martin-Gay Interactive Videos

See Video 13.3

Watch the section lecture video and answer the following questions.

Objective A 13. At the end of ▶ Example 2, patterns are discussed. What reason is given for why x-intercepts have y-values of 0? For why y-intercepts have x-values of 0?

Objective B 14. In ▶ Example 3, the goal is to use the x- and y-intercepts to graph a line. Yet once the two intercepts are found, a third point is also found before the line is graphed. Why do you think this practice of finding a third point is continued?

Objective C 15. From ▶ Examples 5 and 6, what is the coefficient of x when the equation of a horizontal line is written as $Ax + By = C$? What is the coefficient of y when the equation of a vertical line is written as $Ax + By = C$?

13.3 Exercise Set MyLab Math

Objective A *Identify the intercepts. See Examples 1 through 5.*

1.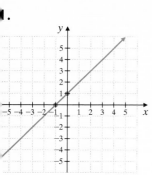

2.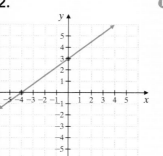

3.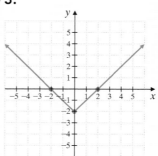

4.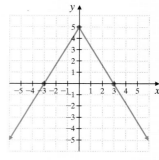

5.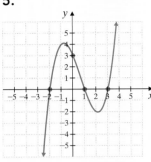

6.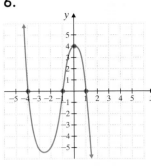

7.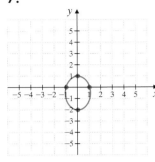

8.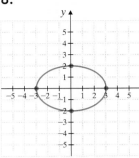

Objective **B** *Graph each linear equation by finding and plotting its intercepts. See Examples 6 through 8.*

9. $x - y = 3$
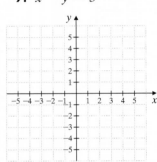

10. $x - y = -4$
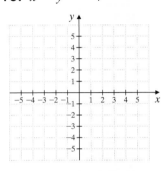

11. $x = 5y$
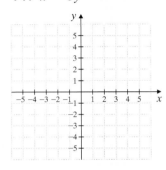

12. $x = 2y$
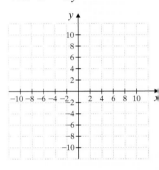

13. $-x + 2y = 6$
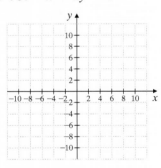

14. $x - 2y = -8$
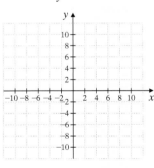

15. $2x - 4y = 8$
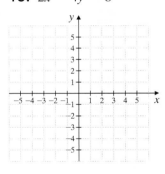

16. $2x + 3y = 6$
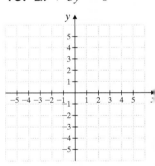

17. $y = 2x$
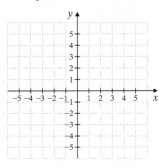

18. $y = -2x$
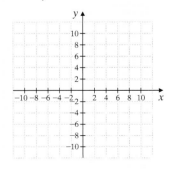

19. $y = 3x + 6$
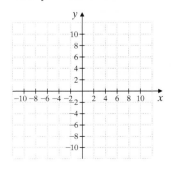

20. $y = 2x + 10$
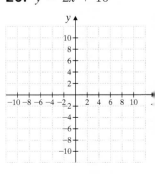

Section 13.3 | Intercepts   999

Objective C *Graph each linear equation. See Examples 9 and 10.*

21. $x = -1$
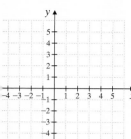

22. $y = 5$
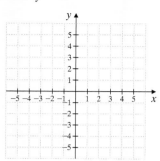

23. $y = 0$
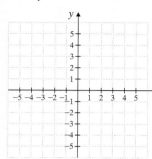

24. $x = 0$
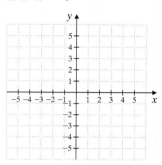

25. $y + 7 = 0$
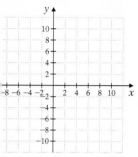

26. $x - 2 = 0$
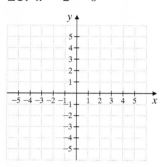

27. $x + 3 = 0$
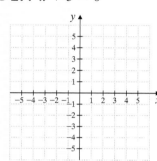

28. $y - 6 = 0$
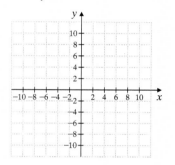

Objectives B C Mixed Practice *Graph each linear equation. See Examples 6 through 10.*

29. $x = y$
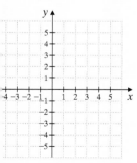

30. $x = -y$
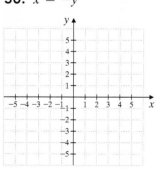

31. $x + 8y = 8$
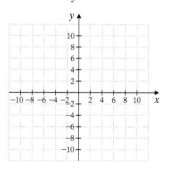

32. $x + 3y = 9$
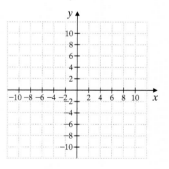

33. $5 = 6x - y$
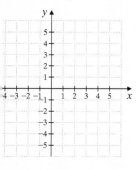

34. $4 = x - 3y$
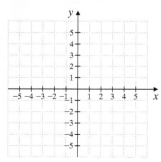

35. $-x + 10y = 11$
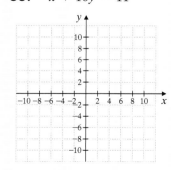

36. $-x + 9y = 10$
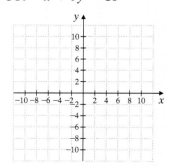

37. $x = -4\frac{1}{2}$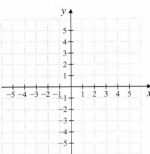

38. $x = -1\frac{3}{4}$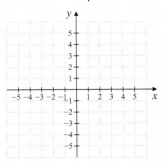

39. $y = 3\frac{1}{4}$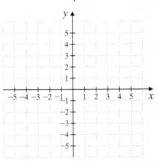

40. $y = 2\frac{1}{2}$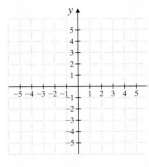

41. $y = -\frac{2}{3}x + 1$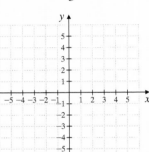

42. $y = -\frac{3}{5}x + 3$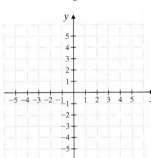

43. $4x - 6y + 2 = 0$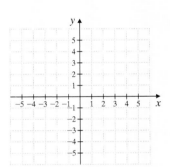

44. $9x - 6y + 3 = 0$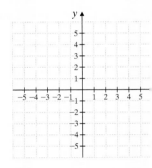

Review

Simplify. See Sections 2.2, 2.3, and 4.2.

45. $\dfrac{-6 - 3}{2 - 8}$

46. $\dfrac{4 - 5}{-1 - 0}$

47. $\dfrac{-8 - (-2)}{-3 - (-2)}$

48. $\dfrac{12 - 3}{10 - 9}$

49. $\dfrac{0 - 6}{5 - 0}$

50. $\dfrac{2 - 2}{3 - 5}$

Concept Extensions

Match each equation with its graph.

51. $y = 3$
a.

52. $y = 2x + 2$
b.

53. $x = 3$
c.

54. $y = 2x + 3$
d.

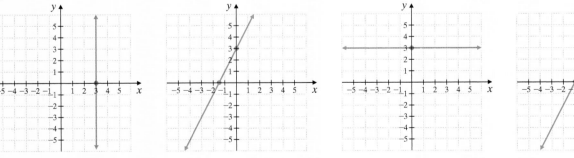

55. What is the greatest number of x- and y-intercepts that a line can have?

56. What is the smallest number of x- and y-intercepts that a line can have?

57. What is the smallest number of x- and y-intercepts that a circle can have?

58. What is the greatest number of x- and y-intercepts that a circle can have?

59. Discuss whether a vertical line ever has a y-intercept.

60. Discuss whether a horizontal line ever has an x-intercept.

The production supervisor at Alexandra's Office Products finds that it takes 3 hours to manufacture a particular office chair and 6 hours to manufacture an office desk. A total of 1200 hours is available to produce office chairs and desks of this style. The linear equation that models this situation is $3x + 6y = 1200$, where x represents the number of chairs produced and y represents the number of desks manufactured.

61. Complete the ordered pair solution $(0, \;)$ of this equation. Describe the manufacturing situation that corresponds to this solution.

62. Complete the ordered pair solution $(\;, 0)$ of this equation. Describe the manufacturing situation that corresponds to this solution.

63. If 50 desks are manufactured, find the greatest number of chairs that can be made.

64. If 50 chairs are manufactured, find the greatest number of desks that can be made.

*Two lines in the same plane that do not intersect are called **parallel lines.***

65. Use your own graph paper to draw a line parallel to the line $y = -1$ that intersects the y-axis at -4. What is the equation of this line?

66. Use your own graph paper to draw a line parallel to the line $x = 5$ that intersects the x-axis at 1. What is the equation of this line?

Solve.

67. As print newspaper sales decline, the number of employees in the print newspaper business is also declining. Employment in the print newspaper industry from 2009 to 2015 can be modeled by the equation $y = -1950x + 45{,}564$, where x represents the number of years after 2009. (*Source:* American Society of News Editors)
   a. Find the x-intercept of this equation (round to the nearest tenth).
   b. What does this x-intercept mean?

68. The number y of Barnes & Noble retail stores in operation for the years 2012 to 2016 can be modeled by the equation $y = -12.9x + 689$, where x represents the number of years after 2012. (*Source:* Based on data from Barnes & Noble, Inc.)
   a. Find the y-intercept of this equation.
   b. What does this y-intercept mean?

13.4 Slope and Rate of Change

Objective A Finding the Slope of a Line Given Two Points

Objectives

A Find the Slope of a Line Given Two Points of the Line.

B Find the Slope of a Line Given Its Equation.

C Find the Slopes of Horizontal and Vertical Lines.

D Compare the Slopes of Parallel and Perpendicular Lines.

E Interpret Slope as a Rate of Change.

Thus far, much of this chapter has been devoted to graphing lines. You have probably noticed by now that a key feature of a line is its slant or steepness. In mathematics, the slant or steepness of a line is formally known as its **slope.** We measure the slope of a line by the ratio of vertical change (rise) to the corresponding horizontal change (run) as we move along the line.

On the line at the top of the next page, for example, suppose that we begin at the point $(1, 2)$ and move to the point $(4, 6)$. The vertical change is the change in y-coordinates: $6 - 2$ or 4 units. The corresponding horizontal change is the change in x-coordinates: $4 - 1 = 3$ units. The ratio of these changes is

$$\text{slope} = \frac{\text{change in } y \text{ (vertical change or rise)}}{\text{change in } x \text{ (horizontal change or run)}} = \frac{4}{3}$$

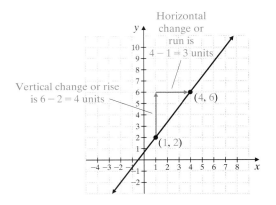

The slope of this line, then, is $\frac{4}{3}$. This means that for every 4 units of change in y-coordinates, there is a corresponding change of 3 units in x-coordinates.

Helpful Hint

It makes no difference what two points of a line are chosen to find its slope. The slope of a line is the same everywhere on the line.

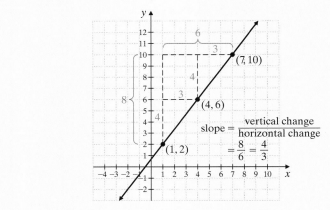

To find the slope of a line, then, choose two points of the line. Label the two x-coordinates of the two points x_1 and x_2 (read "x sub one" and "x sub two"), and label the corresponding y-coordinates y_1 and y_2.

The vertical change or **rise** between these points is the difference in the y-coordinates: $y_2 - y_1$. The horizontal change or **run** between the points is the difference of the x-coordinates: $x_2 - x_1$. The slope of the line is the ratio of $y_2 - y_1$ to $x_2 - x_1$, and we traditionally use the letter m to denote slope: $m = \dfrac{y_2 - y_1}{x_2 - x_1}$.

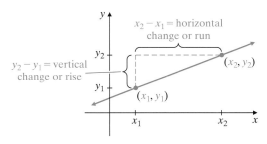

Section 13.4 | Slope and Rate of Change

Slope of a Line

The slope m of the line containing the points (x_1, y_1) and (x_2, y_2) is given by

$$m = \frac{\text{rise}}{\text{run}} = \frac{\text{change in } y}{\text{change in } x} = \frac{y_2 - y_1}{x_2 - x_1}, \quad \text{as long as } x_2 \neq x_1$$

Example 1 Find the slope of the line through $(-1, 5)$ and $(2, -3)$. Graph the line.

Solution: Let (x_1, y_1) be $(-1, 5)$ and (x_2, y_2) be $(2, -3)$. Then, by the definition of slope, we have the following.

$$m = \frac{y_2 - y_1}{x_2 - x_1}$$
$$= \frac{-3 - 5}{2 - (-1)}$$
$$= \frac{-8}{3} = -\frac{8}{3}$$

The slope of the line is $-\frac{8}{3}$.

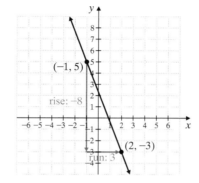

Work Practice 1

Helpful Hint

When finding slope, it makes no difference which point is identified as (x_1, y_1) and which is identified as (x_2, y_2). Just remember that whatever y-value is first in the numerator, its corresponding x-value is first in the denominator. Another way to calculate the slope in Example 1 is

$$m = \frac{y_2 - y_1}{x_2 - x_1} = \frac{5 - (-3)}{-1 - 2} = \frac{8}{-3} \text{ or } -\frac{8}{3} \leftarrow \text{Same slope as found in Example 1}$$

Concept Check The points $(-2, -5)$, $(0, -2)$, $(4, 4)$, and $(10, 13)$ all lie on the same line. Work with a partner and verify that the slope is the same no matter which points are used to find slope.

Example 2 Find the slope of the line through $(-1, -2)$ and $(2, 4)$. Graph the line.

Solution: Let (x_1, y_1) be $(2, 4)$ and (x_2, y_2) be $(-1, -2)$.

$$m = \frac{y_2 - y_1}{x_2 - x_1}$$
$$= \frac{-2 - 4}{-1 - 2} \quad \begin{array}{l}y\text{-value} \\ \text{corresponding } x\text{-value}\end{array}$$
$$= \frac{-6}{-3} = 2$$

The slope is 2.

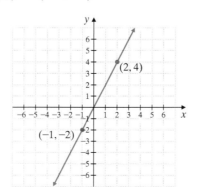

Work Practice 2

Practice 1
Find the slope of the line through $(-2, 3)$ and $(4, -1)$. Graph the line.

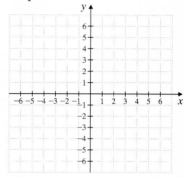

Practice 2
Find the slope of the line through $(-2, 1)$ and $(3, 5)$. Graph the line. (Answer on following page.)

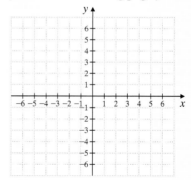

Answer

1. $m = -\frac{2}{3}$

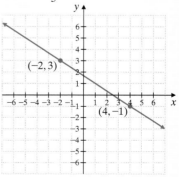

✓ **Concept Check Answers**

$m = \frac{3}{2}$

Chapter 13 | Graphing Equations and Inequalities

✓ **Concept Check** What is wrong with the following slope calculation for points $(3, 5)$ and $(-2, 6)$?

$$m = \frac{5 - 6}{-2 - 3} = \frac{-1}{-5} = \frac{1}{5}$$

Notice that the slope of the line in Example 1 is negative and that the slope of line in Example 2 is positive. Let your eye follow the line with negative slope from to right and notice that the line "goes down." If you follow the line with positive sl from left to right, you will notice that the line "goes up." This is true in general.

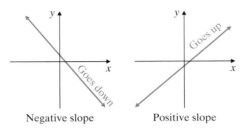

Negative slope   Positive slope

Helpful Hint To decide whether a li "goes up" or "goes down," alwa **follow the line from left to right.**

Objective B Finding the Slope of a Line Given Its Equation

As we have seen, the slope of a line is defined by two points on the line. Thus, if know the equation of a line, we can find its slope by finding two of its points. example, let's find the slope of the line

$$y = 3x - 2$$

To find two points, we can choose two values for x and substitute to find responding y-values. If $x = 0$, for example, $y = 3 \cdot 0 - 2$ or $y = -2$. If $x =$ $y = 3 \cdot 1 - 2$ or $y = 1$. This gives the ordered pairs $(0, -2)$ and $(1, 1)$. Using definition for slope, we have

$$m = \frac{1 - (-2)}{1 - 0} = \frac{3}{1} = 3 \quad \text{The slope is 3.}$$

Notice that the slope, 3, is the same as the coefficient of x in the equa $y = 3x - 2$. This is true in general.

If a linear equation is solved for y, the coefficient of x is the line's slope. In oth words, the slope of the line given by $y = mx + b$ is m, the coefficient of x.

$$y = m\underset{\text{slope}}{x} + b$$

Practice 3

Find the slope of the line $5x + 4y = 10$.

✓ **Concept Check Answers**
$m = \dfrac{5 - 6}{3 - (-2)} = \dfrac{-1}{5} = -\dfrac{1}{5}$

Answers

2. $m = \dfrac{4}{5}$

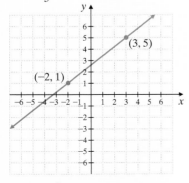

3. $m = -\dfrac{5}{4}$

Example 3 Find the slope of the line $-2x + 3y = 11$.

Solution: When we solve for y, the coefficient of x is the slope.

$$-2x + 3y = 11$$
$$3y = 2x + 11 \quad \text{Add } 2x \text{ to both sides.}$$
$$y = \frac{2}{3}x + \frac{11}{3} \quad \text{Divide both sides by 3.}$$

The slope is $\dfrac{2}{3}$.

■ Work Practice 3

Section 13.4 | Slope and Rate of Change

Example 4 Find the slope of the line $-y = 5x - 2$.

Solution: Remember, the equation must be solved for y (not $-y$) in order for the coefficient of x to be the slope.

To solve for y, let's divide both sides of the equation by -1.

$$-y = 5x - 2$$
$$\frac{-y}{-1} = \frac{5x}{-1} - \frac{2}{-1} \quad \text{Divide both sides by } -1.$$
$$y = -5x + 2 \quad \text{Simplify.}$$

The slope is -5.

Work Practice 4

Practice 4
Find the slope of the line $-y = -2x + 7$.

Objective C Finding Slopes of Horizontal and Vertical Lines

Example 5 Find the slope of the line $y = -1$.

Solution: Recall that $y = -1$ is a horizontal line with y-intercept -1. To find the slope, we find two ordered pair solutions of $y = -1$, knowing that solutions of $y = -1$ must have a y-value of -1. We will use $(2, -1)$ and $(-3, -1)$. We let (x_1, y_1) be $(2, -1)$ and (x_2, y_2) be $(-3, -1)$.

$$m = \frac{y_2 - y_1}{x_2 - x_1} = \frac{-1 - (-1)}{-3 - 2} = \frac{0}{-5} = 0$$

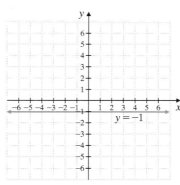

The slope of the line $y = -1$ is 0. Since the y-values will have a difference of 0 for every horizontal line, we can say that all **horizontal lines have a slope of 0.**

Work Practice 5

Practice 5
Find the slope of $y = 3$.

Example 6 Find the slope of the line $x = 5$.

Solution: Recall that the graph of $x = 5$ is a vertical line with x-intercept 5. To find the slope, we find two ordered pair solutions of $x = 5$. Ordered pair solutions of $x = 5$ must have an x-value of 5. We will use $(5, 0)$ and $(5, 4)$. We let $(x_1, y_1) = (5, 0)$ and $(x_2, y_2) = (5, 4)$.

$$m = \frac{y_2 - y_1}{x_2 - x_1} = \frac{4 - 0}{5 - 5} = \frac{4}{0}$$

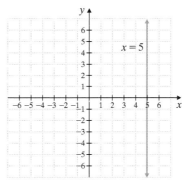

(*Continued on next page*)

Practice 6
Find the slope of the line $x = -2$.

Answers
4. $m = 2$ **5.** $m = 0$
6. undefined slope

Helpful Hint Slope of 0 and undefined slope are not the same. Vertical lines have undefined slope, while horizontal lines have a slope of 0.

Since $\frac{4}{0}$ is undefined, we say that the slope of the vertical line $x = 5$ is undefined.

Since the x-values will have a difference of 0 for every vertical line, we can say that all **vertical lines have undefined slope.**

■ Work Practice 6

Here is a general review of slope.

Summary of Slope

Slope m of the line through (x_1, y_1) and (x_2, y_2) is given by the equation

$$m = \frac{y_2 - y_1}{x_2 - x_1}.$$

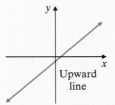

Upward line

Positive slope: $m > 0$

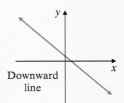

Downward line

Negative slope: $m < 0$

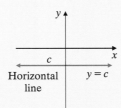

Horizontal line $y = c$

Zero slope: $m = 0$

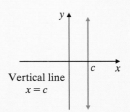

Vertical line $x = c$

No slope or undefined slope

Objective D Comparing Slopes of Parallel and Perpendicular Lines ▶

Two lines in the same plane are **parallel** if they do not intersect. Slopes of lines can help us determine whether lines are parallel. Since parallel lines have the same steepness, it follows that they have the same slope.

For example, the graphs of

$$y = -2x + 4$$

and

$$y = -2x - 3$$

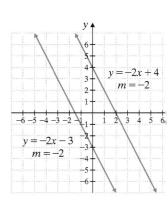

are shown. These lines have the same slope, -2. They also have different y-intercepts, so the lines are parallel. (If the y-intercepts were the same also, the lines would be the same.)

Parallel Lines

Nonvertical parallel lines have the same slope and different y-intercepts.

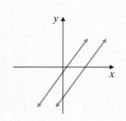

Two lines are **perpendicular** if they lie in the same plane and meet at a 90° (right) angle. How do the slopes of perpendicular lines compare? The product of the slopes of two perpendicular lines is -1.

For example, the graphs of

$$y = 4x + 1$$

and

$$y = -\frac{1}{4}x - 3$$

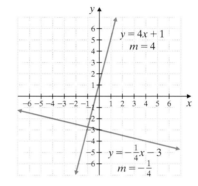

are shown. The slopes of the lines are 4 and $-\frac{1}{4}$. Their product is $4\left(-\frac{1}{4}\right) = -1$, so the lines are perpendicular.

Perpendicular Lines

If the product of the slopes of two lines is -1, then the lines are perpendicular.

(Two nonvertical lines are perpendicular if the slope of one is the negative reciprocal of the slope of the other.)

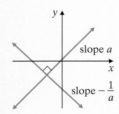

Helpful Hint

Here are examples of numbers that are negative (opposite) reciprocals.

Number	Negative Reciprocal	Their product is -1.
$\frac{2}{3}$	$-\frac{3}{2}$	$\frac{2}{3} \cdot -\frac{3}{2} = -\frac{6}{6} = -1$
-5 or $-\frac{5}{1}$	$\frac{1}{5}$	$-5 \cdot \frac{1}{5} = -\frac{5}{5} = -1$

Here are a few important points about vertical and horizontal lines.
- Two distinct vertical lines are parallel.
- Two distinct horizontal lines are parallel.
- A horizontal line and a vertical line are always perpendicular.

Practice 7

Determine whether each pair of lines is parallel, perpendicular, or neither.

a. $x + y = 5$
$2x + y = 5$

b. $5y = 2x - 3$
$5x + 2y = 1$

c. $y = 2x + 1$
$4x - 2y = 8$

Example 7

Determine whether each pair of lines is parallel, perpendicular, or neither.

a. $y = -\frac{1}{5}x + 1$
$2x + 10y = 3$

b. $x + y = 3$
$-x + y = 4$

c. $3x + y = 5$
$2x + 3y = 6$

Solution:

a. The slope of the line $y = -\frac{1}{5}x + 1$ is $-\frac{1}{5}$. We find the slope of the second line by solving its equation for y.

$$2x + 10y = 3$$
$$10y = -2x + 3 \quad \text{Subtract } 2x \text{ from both sides.}$$
$$y = \frac{-2}{10}x + \frac{3}{10} \quad \text{Divide both sides by 10.}$$
$$y = -\frac{1}{5}x + \frac{3}{10} \quad \text{Simplify.}$$

The slope of this line is $-\frac{1}{5}$ also. Since the lines have the same slope and different y-intercepts, they are parallel, as shown below on the left graph.

Helpful Hint

Note: To find the y-intercept of a line, let $x = 0$.

- For $y = -\frac{1}{5}x + 1$, the y-intercept is $(0, 1)$.
- For $y = -\frac{1}{5}x + \frac{3}{10}$, the y-intercept is $\left(0, \frac{3}{10}\right)$.

Thus, the y-intercepts are different.

b. To find each slope, we solve each equation for y.

$x + y = 3$ $\qquad$ $-x + y = 4$
$y = -x + 3$ $\qquad$ $y = x + 4$
↑ $\qquad\qquad\qquad$ ↑
The slope is -1. $\qquad$ The slope is 1.

The slopes are not the same, so the lines are not parallel. Next we check the product of the slopes: $(-1)(1) = -1$. Since the product is -1, the lines are perpendicular, as shown below on the right.

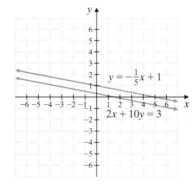
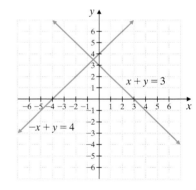

c. We solve each equation for y to find each slope. The slopes are -3 and $-\frac{2}{3}$. The slopes are not the same and their product is not -1. Thus, the lines are neither parallel nor perpendicular.

■ Work Practice 7

✓ **Concept Check** Consider the line $-6x + 2y = 1$.

a. Write the equations of two lines parallel to this line.
b. Write the equations of two lines perpendicular to this line.

Answers

7. a. neither b. perpendicular c. parallel

✓ **Concept Check Answers**

Answers may vary; for example,

a. $y = 3x - 3$, $y = 3x - 1$

b. $y = -\frac{1}{3}x$, $y = -\frac{1}{3}x + 1$

Section 13.4 | Slope and Rate of Change

Objective E Interpreting Slope as a Rate of Change

Slope can also be interpreted as a rate of change. In other words, slope tells us how fast y is changing with respect to x. To see this, let's look at a few of the many real-world applications of slope. For example, the pitch of a roof, used by builders and architects, is its slope. The pitch of the roof on the right is $\frac{7}{10}\left(\frac{\text{rise}}{\text{run}}\right)$. This means that the roof rises vertically 7 feet for every horizontal 10 feet. The rate of change for the roof is 7 vertical feet (y) per 10 horizontal feet (x).

The grade of a road is its slope written as a percent. A 7% grade, as shown below, means that the road rises (or falls) 7 feet for every horizontal 100 feet. (Recall that $7\% = \frac{7}{100}$.) Here, the slope of $\frac{7}{100}$ gives us the rate of change. The road rises (in our diagram) 7 vertical feet (y) for every 100 horizontal feet (x).

Example 8 Finding the Grade of a Road

At one part of the road to the summit of Pike's Peak, the road rises 15 feet for a horizontal distance of 250 feet. Find the grade of the road.

Solution: Recall that the grade of a road is its slope written as a percent.

$$\text{grade} = \frac{\text{rise}}{\text{run}} = \frac{15}{250} = 0.06 = 6\%$$

The grade is 6%.

Work Practice 8

Example 9 Finding the Slope of a Line

The following graph shows the number y of movie screens in the United States, where x is the number of years after 2011. Find the slope of the line and attach the proper units for the rate of change. Then write a sentence explaining the meaning of slope in this application. (*Source:* Motion Picture Association of America)

Solution: Use $(1, 39.764)$ and $(3, 40.276)$ to calculate slope.

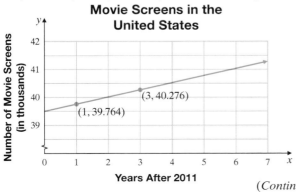

Practice 8
Find the grade of the road shown.

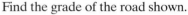

Practice 9
Find the slope of the line and write the slope as a rate of change. This graph represents annual restaurant-industry employment y (in billions of workers) for year x. Write a sentence explaining the meaning of slope in this application.

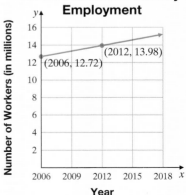

Source: National Restaurant Association

Answers
8. 15% **9.** $m = 0.21$; Each year the number of workers employed in the U.S. restaurant industry increases by 0.21 million, or 210,000, workers per year.

(*Continued on next page*)

$$m = \frac{40.276 - 39.764}{3 - 1} = \frac{0.512}{2} = \frac{0.256 \text{ thousand screens}}{1 \text{ year}}$$

This means that the rate of change of the number of movie screens is 0.256 thousand screens per year, or each year 256 movie screens are added in the United States.

Work Practice 9

Calculator Explorations Graphing

It is possible to use a graphing calculator to sketch the graph of more than one equation on the same set of axes. This feature can be used to see parallel lines with the same slope. For example, graph the equations $y = \frac{2}{5}x$, $y = \frac{2}{5}x + 7$, and $y = \frac{2}{5}x - 4$ on the same set of axes. To do so, press the $\boxed{Y =}$ key and enter the equations on the first three lines.

$Y_1 = \frac{2}{5}x$

$Y_2 = \frac{2}{5}x + 7$

$Y_3 = \frac{2}{5}x - 4$

The displayed equations should look like this:

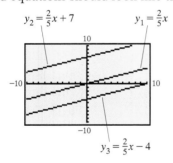

These lines are parallel, as expected, since they all have a slope of $\frac{2}{5}$. The graph of $y = \frac{2}{5}x + 7$ is the graph of $y = \frac{2}{5}x$ moved 7 units upward with a y-intercept of 7. Also, the graph of $y = \frac{2}{5}x - 4$ is the graph of $y = \frac{2}{5}x$ moved 4 units downward with a y-intercept of -4.

Graph the parallel lines on the same set of axes. Describe the similarities and differences in their graphs.

1. $y = 3.8x$, $y = 3.8x - 3$, $y = 3.8x + 9$

2. $y = -4.9x$, $y = -4.9x + 1$, $y = -4.9x + 8$

3. $y = \frac{1}{4}x$, $y = \frac{1}{4}x + 5$, $y = \frac{1}{4}x - 8$

4. $y = -\frac{3}{4}x$, $y = -\frac{3}{4}x - 5$, $y = -\frac{3}{4}x + 6$

Vocabulary, Readiness & Video Check

Use the choices below to fill in each blank. Not all choices will be used.

m	x	0	positive	undefined
b	y	slope	negative	

1. The measure of the steepness or tilt of a line is called _____.
2. If an equation is written in the form $y = mx + b$, the value of the letter _____ is the value of the slope of the graph.
3. The slope of a horizontal line is _____.
4. The slope of a vertical line is _____.
5. If the graph of a line moves upward from left to right, the line has _____ slope.
6. If the graph of a line moves downward from left to right, the line has _____ slope.
7. Given two points of a line, slope = $\dfrac{\text{change in } \underline{\qquad}}{\text{change in } \underline{\qquad}}$.

State whether the slope of the line is positive, negative, 0, or undefined.

8.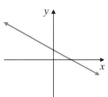
9.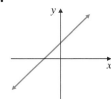
10.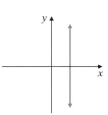
11.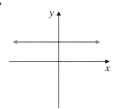

Decide whether a line with the given slope slants upward or downward or is horizontal or vertical.

12. $m = \dfrac{7}{6}$ _____
13. $m = -3$ _____
14. $m = 0$ _____
15. m is undefined. _____

Martin-Gay Interactive Videos

See Video 13.4

Watch the section lecture video and answer the following questions.

Objective A 16. What important point is made during Example 1 having to do with the order of the points in the slope formula?

Objective B 17. From Example 5, how do we write an equation in "slope-intercept form"? Once the equation is in slope-intercept form, how do we identify the slope?

Objective C 18. In the lecture after Example 8, different slopes are summarized. What is the difference between zero slope and undefined slope? What does "no slope" mean?

Objective D 19. From Example 10, what form of the equation is best to determine if two lines are parallel or perpendicular? Why?

Objective E 20. Writing the slope as a rate of change in Example 11 gave real-life meaning to the slope. What step in the general strategy for problem solving does this correspond to?

13.4 Exercise Set MyLab Math

Objective A *Find the slope of the line that passes through the given points. See Examples 1 and 2.*

1. $(-1, 5)$ and $(6, -2)$
2. $(-1, 16)$ and $(3, 4)$
3. $(1, 4)$ and $(5, 3)$
4. $(3, 1)$ and $(2, 6)$

5. $(5, 1)$ and $(-2, 1)$
6. $(-8, 3)$ and $(-2, 3)$
7. $(-4, 3)$ and $(-4, 5)$
8. $(-2, -3)$ and $(-2, 5)$

Use the points shown on each graph to find the slope of each line. See Examples 1 and 2.

9.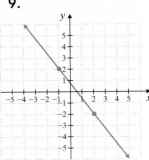
10.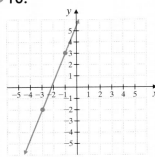
11.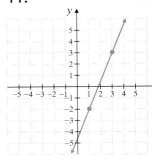
12.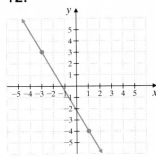

For each graph, determine which line has the greater slope.

13.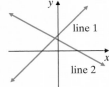
14.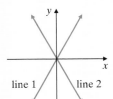
15.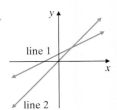
16.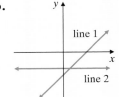

Section 13.4 | Slope and Rate of Change

Objectives B C Mixed Practice *Find the slope of each line. See Examples 3 through 6.*

17. $y = 5x - 2$
18. $y = -2x + 6$
19. $y = -0.3x + 2.5$
20. $y = -7.6x - 0.1$
21. $2x + y = 7$
22. $-5x + y = 10$

23.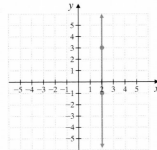
24.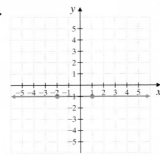

25. $2x - 3y = 10$
26. $3x - 5y = 1$
27. $x = 1$
28. $y = -2$

29. $x = 2y$
30. $x = -4y$
31. $y = -3$
32. $x = 5$

33. $-3x - 4y = 6$
34. $-4x - 7y = 9$
35. $20x - 5y = 1.2$
36. $24x - 3y = 5.7$

Objective D *Find the slope of a line that is (a) parallel and (b) perpendicular to the line through each pair of points. See Example 7.*

37. $(-3, -3)$ and $(0, 0)$
38. $(6, -2)$ and $(1, 4)$
39. $(-8, -4)$ and $(3, 5)$
40. $(6, -1)$ and $(-4, -10)$

Determine whether each pair of lines is parallel, perpendicular, or neither. See Example 7.

41. $y = \frac{2}{9}x + 3$
    $y = -\frac{2}{9}x$

42. $y = \frac{1}{5}x + 20$
    $y = -\frac{1}{5}x$

43. $x - 3y = -6$
    $y = 3x - 9$

44. $y = 4x - 2$
    $4x + y = 5$

45. $6x = 5y + 1$
    $-12x + 10y = 1$

46. $-x + 2y = -2$
    $2x = 4y + 3$

47. $6 + 4x = 3y$
    $3x + 4y = 8$

48. $10 + 3x = 5y$
    $5x + 3y = 1$

Objective E *The pitch of a roof is its slope. Find the pitch of each roof shown. See Example 8.*

49.

50.

The grade of a road is its slope written as a percent. Find the grade of each road shown. See Example 8.

51.

52.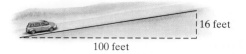

53. One of Japan's superconducting "bullet" trains is researched and tested at the Yamanashi Maglev Test Line near Otsuki City. The steepest section of the track has a rise of 2580 meters for a horizontal distance of 6450 meters. What is the grade (slope written as a percent) of this section of track? (*Source:* Japan Railways Central Co.)

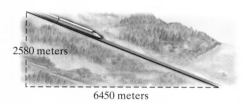

54. Professional plumbers suggest that a sewer pipe rise 0.25 inch for every horizontal foot. Find the recommended slope for a sewer pipe and write the slope as a grade, or percent. Round to the nearest percent.

55. There has been controversy over the past few years about the world's steepest street. *The Guinness Book of Records* listed Baldwin Street, in Dunedin, New Zealand, as the world's steepest street, but Canton Avenue in the Pittsburgh neighborhood of Beechview may actually be steeper. Calculate each grade to the nearest percent.

Canton Avenue	For every 30 meters of horizontal distance, the vertical change is 11 meters.	
Baldwin Street	For every 2.86 meters of horizontal distance, the vertical change is 1 meter.	

56. According to federal regulations, a wheelchair ramp should rise no more than 1 foot for a horizontal distance of 12 feet. Write the slope as a grade. Round to the nearest tenth of a percent.

Find the slope of each line and write a sentence using the slope as a rate of change. Don't forget to attach the proper units. See Example 9.

57. This graph approximates the number of U.S. households that have televisions y (in millions) for year x.

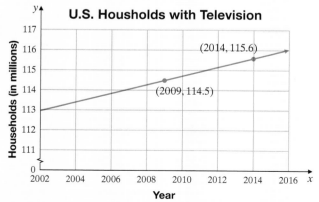

58. The graph approximates the amount of money y (in billions of dollars) spent worldwide on tourism for year x.

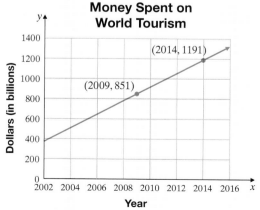

59. Americans are keeping their cars longer. The graph below shows the median age y (in years) of automobiles in the United States for the years shown.

60. The graph below shows the total cost y (in dollars) of owning and operating an SUV in the United States in 2016, where x is the annual number of miles driven.

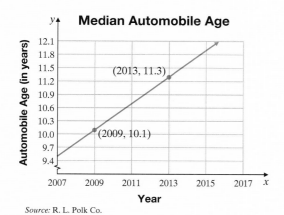

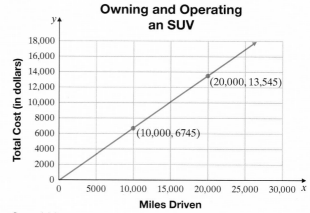

Review

Solve each equation for y. See Section 9.5.

61. $y - (-6) = 2(x - 4)$

62. $y - 7 = -9(x - 6)$

63. $y - 1 = -6(x - (-2))$

64. $y - (-3) = 4(x - (-5))$

Concept Extensions

Match each line with its slope.

A. $m = 0$

B. undefined slope

C. $m = 3$

D. $m = 1$

E. $m = -\dfrac{1}{2}$

F. $m = -\dfrac{3}{4}$

65.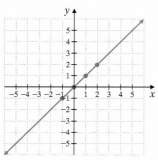

66.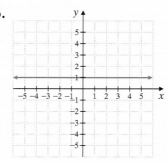

67.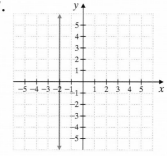

68. 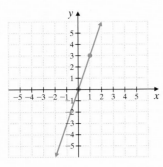 **69.** 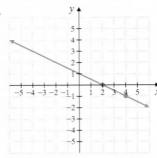 **70.**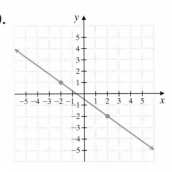

Solve. See a Concept Check in this section.

71. Verify that the points $(2, 1), (0, 0), (-2, -1)$, and $(-4, -2)$ are all on the same line by computing the slope between each pair of points. (See the first Concept Check.)

72. Given the points $(2, 3)$ and $(-5, 1)$, can the slope of the line through these points be calculated by $\dfrac{1 - 3}{2 - (-5)}$? Why or why not? (See the second Concept Check.)

73. Write the equations of three lines parallel to $10x - 5y = -7$. (See the third Concept Check.)

74. Write the equations of two lines perpendicular to $10x - 5y = -7$. (See the third Concept Check.)

The following line graph shows the average fuel economy (in miles per gallon) of passenger automobiles produced during each of the model years shown. Use this graph to answer Exercises 83 through 88.

75. What was the average fuel economy (in miles per gallon) for automobiles produced during 2008?

76. Find the decrease in average fuel economy for automobiles between the years 2010 and 2011.

77. During which of the model years shown was average fuel economy the lowest?
What was the average fuel economy for that year?

78. During which of the model years shown was average fuel economy the highest?
What was the average fuel economy for that year?

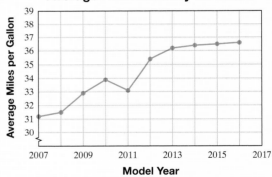

Average Fuel Economy for Autos

Source: Bureau of Transportation Statistics

79. Of the following line segments, which has the greatest slope: from 2007 to 2008, from 2009 to 2010, or from 2011 to 2012?

80. What line segment has a slope of 1?

81. Find x so that the pitch of the roof is $\dfrac{2}{5}$.

82. Find x so that the pitch of the roof is $\dfrac{1}{3}$.

83. There were approximately 2322 heart transplants performed in the United States in 2011. In 2016, the number of heart transplants in the United States rose to 2924. (*Source:* Organ Procurement and Transplantation Network)

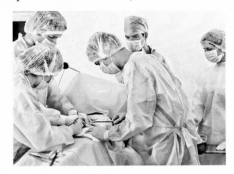

a. Write two ordered pairs of the form (year, number of heart transplants).
b. Find the slope of the line between the two points.
c. Write a sentence explaining the meaning of the slope as a rate of change.

84. The average price of an acre of U.S. cropland was $2980 in 2011. In 2015, the price of an acre rose to $4130. (*Source:* National Agricultural Statistics Service)

a. Write two ordered pairs of the form (year, price of acre).
b. Find the slope of the line through the two points.
c. Write a sentence explaining the meaning of the slope as a rate of change.

85. Show that the quadrilateral with vertices $(1, 3)$, $(2, 1)$, $(-4, 0)$, and $(-3, -2)$ is a parallelogram.

86. Show that a triangle with vertices at the points $(1, 1)$, $(-4, 4)$, and $(-3, 0)$ is a right triangle.

Find the slope of the line through the given points.

87. $(-3.8, 1.2)$ and $(-2.2, 4.5)$

88. $(2.1, 6.7)$ and $(-8.3, 9.3)$

89. $(14.3, -10.1)$ and $(9.8, -2.9)$

90. $(2.3, 0.2)$ and $(7.9, 5.1)$

91. The graph of $y = \frac{1}{2}x$ has a slope of $\frac{1}{2}$. The graph of $y = 3x$ has a slope of 3. The graph of $y = 5x$ has a slope of 5. Graph all three equations on a single coordinate system. As the slope becomes larger, how does the steepness of the line change?

92. The graph of $y = -\frac{1}{3}x + 2$ has a slope of $-\frac{1}{3}$. The graph of $y = -2x + 2$ has a slope of -2. The graph of $y = -4x + 2$ has a slope of -4. Graph all three equations on a single coordinate system. As the absolute value of the slope becomes larger, how does the steepness of the line change?

13.5 Equations of Lines

Objectives

A Use the Slope-Intercept Form to Write an Equation of a Line.

B Use the Slope-Intercept Form to Graph a Linear Equation.

C Use the Point-Slope Form to Find an Equation of a Line Given Its Slope and a Point of the Line.

D Use the Point-Slope Form to Find an Equation of a Line Given Two Points of the Line.

E Find Equations of Vertical and Horizontal Lines.

F Use the Point-Slope Form to Solve Problems.

We know that when a linear equation is solved for y, the coefficient of x is the slope of the line. For example, the slope of the line whose equation is $y = 3x + 1$ is 3. In the equation $y = 3x + 1$, what does 1 represent? To find out, let $x = 0$ and watch what happens.

$$y = 3x + 1$$
$$y = 3 \cdot 0 + 1 \quad \text{Let } x = 0.$$
$$y = 1$$

We now have the ordered pair $(0, 1)$, which means that 1 is the y-intercept. This is true in general. To see this, let $x = 0$ and solve for y in $y = mx + b$.

$$y = m \cdot 0 + b \quad \text{Let } x = 0.$$
$$y = b$$

We obtain the ordered pair $(0, b)$, which means that point is the y-intercept. The form $y = mx + b$ is appropriately called the *slope-intercept form* of a linear equation.

$$y = \underset{\uparrow}{m}x + \underset{\uparrow}{b}$$
$$\text{slope} \quad \text{y-intercept is } (0, b).$$

Slope-Intercept Form

When a linear equation in two variables is written in **slope-intercept form**,

$$y = \underset{\uparrow}{m}x + \underset{\uparrow}{b}$$
$$\text{slope} \quad (0, b), \text{y-intercept}$$

then m is the slope of the line and $(0, b)$ is the y-intercept of the line.

Objective A Using the Slope-Intercept Form to Write an Equation

The slope-intercept form can be used to write the equation of a line when we know its slope and y-intercept.

Practice 1

Find an equation of the line with y-intercept $(0, -2)$ and slope of $\dfrac{3}{5}$.

Example 1 Find an equation of the line with y-intercept $(0, -3)$ and slope of $\dfrac{1}{4}$.

Solution: We are given the slope and the y-intercept. We let $m = \dfrac{1}{4}$ and $b = -3$ and write the equation in slope-intercept form, $y = mx + b$.

$$y = mx + b$$
$$y = \dfrac{1}{4}x + (-3) \quad \text{Let } m = \dfrac{1}{4} \text{ and } b = -3.$$
$$y = \dfrac{1}{4}x - 3 \quad \text{Simplify.}$$

■ Work Practice 1

Objective B Using the Slope-Intercept Form to Graph an Equation

Answer

1. $y = \dfrac{3}{5}x - 2$

We also can use the slope-intercept form of the equation of a line to graph a linear equation.

Section 13.5 | Equations of Lines 1019

Example 2 Use the slope-intercept form to graph the equation $y = \frac{3}{5}x - 2$.

Solution: Since the equation $y = \frac{3}{5}x - 2$ is written in slope-intercept form $y = mx + b$, the slope of its graph is $\frac{3}{5}$ and the y-intercept is $(0, -2)$. To graph this equation, we begin by plotting the point $(0, -2)$. From this point, we can find another point of the graph by using the slope $\frac{3}{5}$ and recalling that slope is $\frac{\text{rise}}{\text{run}}$. We start at the y-intercept and move 3 units up since the numerator of the slope is 3; then we move 5 units to the right since the denominator of the slope is 5. We stop at the point $(5, 1)$. The line through $(0, -2)$ and $(5, 1)$ is the graph of $y = \frac{3}{5}x - 2$.

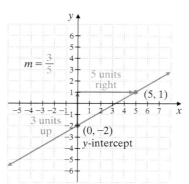

■ Work Practice 2

Example 3 Use the slope-intercept form to graph the equation $4x + y = 1$.

Solution: First we write the given equation in slope-intercept form.

$$4x + y = 1$$
$$y = -4x + 1$$

The graph of this equation will have slope -4 and y-intercept $(0, 1)$. To graph this line, we first plot the point $(0, 1)$. To find another point of the graph, we use the slope -4, which can be written as $\frac{-4}{1}\left(\frac{4}{-1}\text{ could also be used}\right)$. We start at the point $(0, 1)$ and move 4 units down (since the numerator of the slope is -4), and then 1 unit to the right (since the denominator of the slope is 1).

We arrive at the point $(1, -3)$. The line through $(0, 1)$ and $(1, -3)$ is the graph of $4x + y = 1$.

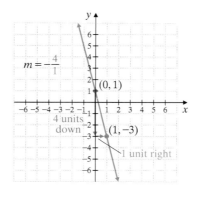

■ Work Practice 3

Helpful Hint

In Example 3, if we interpret the slope of -4 as $\frac{4}{-1}$, we arrive at $(-1, 5)$ for a second point. Notice that this point is also on the line.

Objective C Writing an Equation Given Its Slope and a Point ▶

Thus far, we have seen that we can write an equation of a line if we know its slope and y-intercept. We can also write an equation of a line if we know its slope and any

Practice 2

Use the slope-intercept form to graph the equation $y = \frac{2}{3}x - 4$.

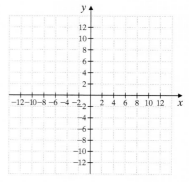

Practice 3

Use the slope-intercept form to graph $3x + y = 2$.

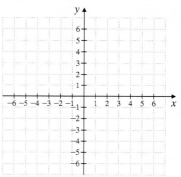

Answers

2.
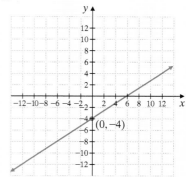

3.
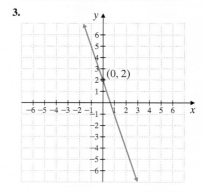

point on the line. To see how we do this, let m represent slope and (x_1, y_1) represent a point on the line. Then if (x, y) is any other point on the line, we have that

$$\frac{y - y_1}{x - x_1} = m$$

$$y - y_1 = m(x - x_1) \quad \text{Multiply both sides by } (x - x_1).$$

↑ slope

This is the *point-slope form* of the equation of a line.

Point-Slope Form of the Equation of a Line

The **point-slope form** of the equation of a line is

$$y - y_1 = m(x - x_1)$$

↑ slope   ↑ (x_1, y_1) point on the line

where m is the slope of the line and (x_1, y_1) is a point on the line.

Practice 4

Find an equation of the line with slope -3 that passes through $(2, -4)$. Write the equation in slope-intercept form, $y = mx + b$.

Example 4 Find an equation of the line with slope -2 that passes through $(-1, 5)$. Write the equation in slope-intercept form, $y = mx + b$, and in standard form, $Ax + By = C$.

Solution: Since the slope and a point on the line are given, we use point-slope form, $y - y_1 = m(x - x_1)$, to write the equation. Let $m = -2$ and $(-1, 5) = (x_1, y_1)$.

$$y - y_1 = m(x - x_1)$$
$$y - 5 = -2[x - (-1)] \quad \text{Let } m = -2 \text{ and } (x_1, y_1) = (-1, 5).$$
$$y - 5 = -2(x + 1) \quad \text{Simplify.}$$
$$y - 5 = -2x - 2 \quad \text{Use the distributive property.}$$

To write the equation in slope-intercept form, $y = mx + b$, we simply solve the equation for y. To do this, we add 5 to both sides.

$$y - 5 = -2x - 2$$
$$y = -2x + 3 \quad \text{Slope-intercept form}$$
$$2x + y = 3 \quad \text{Add } 2x \text{ to both sides and we have standard form.}$$

■ Work Practice 4

Objective D Writing an Equation Given Two Points

We can also find an equation of a line when we are given any two points of the line.

Practice 5

Find an equation of the line through $(1, 3)$ and $(5, -2)$. Write the equation in the form $Ax + By = C$.

Example 5 Find an equation of the line through $(2, 5)$ and $(-3, 4)$. Write the equation in the form $Ax + By = C$.

Solution: First, use the two given points to find the slope of the line.

$$m = \frac{4 - 5}{-3 - 2} = \frac{-1}{-5} = \frac{1}{5}$$

Next we use the slope $\frac{1}{5}$ and either one of the given points to write the equation in point-slope form. We use $(2, 5)$. Let $x_1 = 2$, $y_1 = 5$, and $m = \frac{1}{5}$.

$$y - y_1 = m(x - x_1) \quad \text{Use point-slope form.}$$
$$y - 5 = \frac{1}{5}(x - 2) \quad \text{Let } x_1 = 2, y_1 = 5, \text{ and } m = \frac{1}{5}.$$

Answers

4. $y = -3x + 2$ **5.** $5x + 4y = 17$

$5(y-5) = 5 \cdot \frac{1}{5}(x-2)$ Multiply both sides by 5 to clear fractions.
$5y - 25 = x - 2$ Use the distributive property and simplify.
$-x + 5y - 25 = -2$ Subtract x from both sides.
$-x + 5y = 23$ Add 25 to both sides.

■ Work Practice 5

Helpful Hint

When you multiply both sides of the equation from Example 5, $-x + 5y = 23$, by -1, it becomes $x - 5y = -23$.
Both $-x + 5y = 23$ and $x - 5y = -23$ are in the form $Ax + By = C$ and both are equations of the same line.

Objective E Finding Equations of Vertical and Horizontal Lines

Recall from Section 13.3 that:

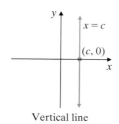
Vertical line

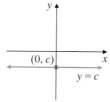
Horizontal line

Example 6 Find an equation of the vertical line through $(-1, 5)$.

Solution: The equation of a vertical line can be written in the form $x = c$, so an equation for the vertical line passing through $(-1, 5)$ is $x = -1$.

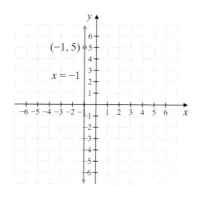

■ Work Practice 6

Practice 6

Find an equation of the vertical line through $(3, -2)$.

Example 7 Find an equation of the line parallel to the line $y = 5$ and passing through $(-2, -3)$.

Solution: Since the graph of $y = 5$ is a horizontal line, any line parallel to it is also horizontal. The equation of a horizontal line can be written in the form $y = c$. An equation for the horizontal line passing through $(-2, -3)$ is $y = -3$.

Practice 7

Find an equation of the line parallel to the line $y = -2$ and passing through $(4, 3)$.

Answers
6. $x = 3$ **7.** $y = 3$

(Continued on next page)

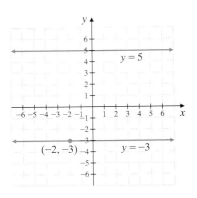

■ Work Practice 7

Objective F Using the Point-Slope Form to Solve Problems

Problems occurring in many fields can be modeled by linear equations in two variables. The next example is from the field of marketing and shows how consumer demand for a product depends on the price of the product.

Example 8 The Whammo Company has learned that by pricing a newly released Frisbee at $6, sales will reach 2000 Frisbees per day. Raising the price to $8 will cause the sales to fall to 1500 Frisbees per day.

a. Assume that the relationship between sales price and number of Frisbees sold is linear, and write an equation describing this relationship. Write the equation in slope-intercept form. Use ordered pairs of the form (sales price, number sold).

b. Predict the daily sales of Frisbees if the price is $7.50.

Solution:

a. We use the given information and write two ordered pairs. Our ordered pairs are $(6, 2000)$ and $(8, 1500)$. To use the point-slope form to write an equation, we find the slope of the line that contains these points.

$$m = \frac{2000 - 1500}{6 - 8} = \frac{500}{-2} = -250$$

Next we use the slope and either one of the points to write the equation in point-slope form. We use $(6, 2000)$.

$$y - y_1 = m(x - x_1) \quad \text{Use point-slope form.}$$
$$y - 2000 = -250(x - 6) \quad \text{Let } x_1 = 6, y_1 = 2000, \text{ and } m = -250.$$
$$y - 2000 = -250x + 1500 \quad \text{Use the distributive property.}$$
$$y = -250x + 3500 \quad \text{Write in slope-intercept form.}$$

Practice 8

The Pool Entertainment Company learned that by pricing a new pool toy at $10, local sales will reach 200 a week. Lowering the price to $9 will cause sales to rise to 250 a week.

a. Assume that the relationship between sales price and number of toys sold is linear, and write an equation describing this relationship. Write the equation in slope-intercept form. Use ordered pairs of the form (sales price, number sold).

b. Predict the weekly sales of the toy if the price is $7.50.

Answer
8. a. $y = -50x + 700$ **b.** 325

b. To predict the sales if the price is $7.50, we find y when $x = 7.50$.

$y = -250x + 3500$
$y = -250(7.50) + 3500$ Let $x = 7.50$.
$y = -1875 + 3500$
$y = 1625$

If the price is $7.50, sales will reach 1625 Frisbees per day.

Work Practice 8

We also could have solved Example 8 by using ordered pairs of the form (number sold, sales price).

Here is a summary of our discussion on linear equations thus far.

Forms of Linear Equations

$Ax + By = C$	**Standard form** of a linear equation. A and B are not both 0.
$y = mx + b$	**Slope-intercept form** of a linear equation. The slope is m and the y-intercept is $(0, b)$.
$y - y_1 = m(x - x_1)$	**Point-slope form** of a linear equation. The slope is m and (x_1, y_1) is a point on the line.
$y = c$	**Horizontal line** The slope is 0 and the y-intercept is $(0, c)$.
$x = c$	**Vertical line** The slope is undefined and the x-intercept is $(c, 0)$.

Parallel and Perpendicular Lines

Nonvertical parallel lines have the same slope.
The product of the slopes of two nonvertical perpendicular lines is -1.

Calculator Explorations Graphing

A graphing calculator is a very useful tool for discovering patterns. To discover the change in the graph of a linear equation caused by a change in slope, try the following: Use a standard window and graph a linear equation in the form $y = mx + b$. Recall that the graph of such an equation will have slope m and y-intercept $(0, b)$.

First graph $y = x + 3$. To do so, press the $\boxed{Y =}$ key and enter $Y_1 = x + 3$. Notice that this graph has slope 1 and that the y-intercept is 3. Next, on the same set of axes, graph $y = 2x + 3$ and $y = 3x + 3$ by pressing $\boxed{Y =}$ and entering $Y_2 = 2x + 3$ and $Y_3 = 3x + 3$.

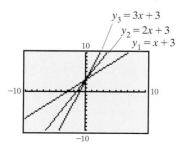

Notice the difference in the graph of each equation as the slope changes from 1 to 2 to 3. How would the graph of $y = 5x + 3$ appear? To see the change in the graph caused by a change to negative slope, try graphing $y = -x + 3$, $y = -2x + 3$, and $y = -3x + 3$ on the same set of axes.

Use a graphing calculator to graph the following equations. For each exercise, graph the first equation and use its graph to predict the appearance of the other equations. Then graph the other equations on the same set of axes and check your prediction.

1. $y = x$; $y = 6x$, $y = -6x$

2. $y = -x$; $y = -5x$, $y = -10x$

3. $y = \dfrac{1}{2}x + 2;\ y = \dfrac{3}{4}x + 2,\ y = x + 2$

4. $y = x + 1;\ y = \dfrac{5}{4}x + 1,\ y = \dfrac{5}{2}x + 1$

Vocabulary, Readiness & Video Check

Use the choices below to fill in each blank. Some choices may be used more than once and some not at all.

b	(y_1, x_1)	point-slope	vertical	standard
m	(x_1, y_1)	slope-intercept	horizontal	

1. The form $y = mx + b$ is called _____ form. When a linear equation in two variables is written in this form, _____ is the slope of its graph and $(0,$ _____$)$ is its y-intercept.

2. The form $y - y_1 = m(x - x_1)$ is called _____ form. When a linear equation in two variables is written in this form, _____ is the slope of its graph and _____ is a point on the graph.

For Exercises 3 through 6, identify the form that the linear equation in two variables is written in. For Exercises 7 and 8, identify the appearance of the graph of the equation.

3. $y - 7 = 4(x + 3);$ _____ form

4. $5x - 9y = 11;$ _____ form

5. $y = \dfrac{3}{4}x - \dfrac{1}{3};$ _____ form

6. $y + 2 = \dfrac{-1}{3}(x - 2)$ _____ form

7. $y = \dfrac{1}{2};$ _____ line

8. $x = -17;$ _____ line

Martin-Gay Interactive Videos

See Video 13.5

Watch the section lecture video and answer the following questions.

Objective A   **9.** In Example 1, what is the y-intercept?

Objective B   **10.** We can use the slope-intercept form to graph a line. Complete these statements based on Example 2: Start by graphing the _____. From this point, find another point by applying the slope—if necessary, rewrite the slope as a(n) _____.

Objective C   **11.** In Example 4, we use the point-slope form to find an equation of a line given the slope and a point. How do we then write this equation in standard form?

Objective D   **12.** The lecture before Example 5 discusses how to find the equation of a line given two points. Is there any circumstance when we might want to use the slope–intercept form to find the equation of a line given two points? If so, when?

Objective E 13. Solve Examples 6 and 7 again, this time using the point $(-2, -3)$ in each exercise.

Objective F 14. In Example 8, we are told to use ordered pairs of the form (time, speed). Explain why it is important to keep track of how we define our ordered pairs and/or our variables.

13.5 Exercise Set MyLab Math

Objective A *Write an equation of the line with each given slope, m, and y-intercept, (0, b). See Example 1.*

1. $m = 5, b = 3$
2. $m = -3, b = -3$
3. $m = -4, b = -\dfrac{1}{6}$
4. $m = 2, b = \dfrac{3}{4}$
5. $m = \dfrac{2}{3}, b = 0$
6. $m = -\dfrac{4}{5}, b = 0$
7. $m = 0, b = -8$
8. $m = 0, b = -2$
9. $m = -\dfrac{1}{5}, b = \dfrac{1}{9}$
10. $m = \dfrac{1}{2}, b = -\dfrac{1}{3}$

Objective B *Use the slope-intercept form to graph each equation. See Examples 2 and 3.*

11. $y = 2x + 1$
12. $y = -4x - 1$
13. $y = \dfrac{2}{3}x + 5$
14. $y = \dfrac{1}{4}x - 3$

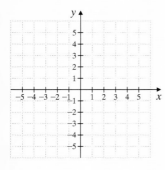
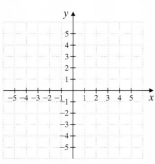
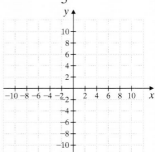
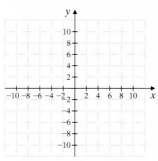

15. $y = -5x$
16. $y = -6x$
17. $4x + y = 6$
18. $-3x + y = 2$

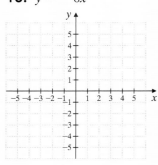
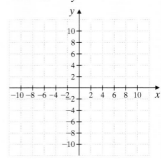
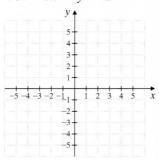

19. $4x - 7y = -14$ **20.** $3x - 4y = 4$ **21.** $x = \dfrac{5}{4}y$ **22.** $x = \dfrac{3}{2}y$

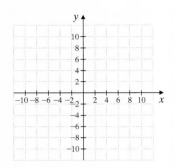 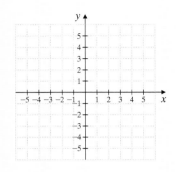 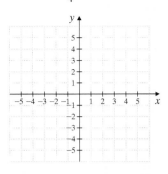 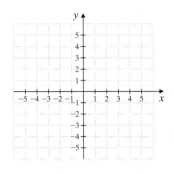

Objective C *Find an equation of each line with the given slope that passes through the given point. Write the equation in the form $Ax + By = C$. See Example 4.*

23. $m = 6;\ (2, 2)$

24. $m = 4;\ (1, 3)$

25. $m = -8;\ (-1, -5)$

26. $m = -2;\ (-11, -12)$

27. $m = \dfrac{3}{2};\ (5, -6)$

28. $m = \dfrac{2}{3};\ (-8, 9)$

29. $m = -\dfrac{1}{2};\ (-3, 0)$

30. $m = -\dfrac{1}{5};\ (4, 0)$

Objective D *Find an equation of the line passing through each pair of points. Write the equation in the form $Ax + By = C$. See Example 5.*

31. $(3, 2)$ and $(5, 6)$

32. $(6, 2)$ and $(8, 8)$

33. $(-1, 3)$ and $(-2, -5)$

34. $(-4, 0)$ and $(6, -1)$

35. $(2, 3)$ and $(-1, -1)$

36. $(7, 10)$ and $(-1, -1)$

37. $(0, 0)$ and $\left(-\dfrac{1}{8}, \dfrac{1}{13}\right)$

38. $(0, 0)$ and $\left(-\dfrac{1}{2}, \dfrac{1}{3}\right)$

Objective E *Find an equation of each line. See Example 6.*

39. Vertical line through $(0, 2)$

40. Horizontal line through $(1, 4)$

41. Horizontal line through $(-1, 3)$

42. Vertical line through $(-1, 3)$

43. Vertical line through $\left(-\dfrac{7}{3}, -\dfrac{2}{5}\right)$

44. Horizontal line through $\left(\dfrac{2}{7}, 0\right)$

Find an equation of each line. See Example 7.

45. Parallel to $y = 5$, through $(1, 2)$

46. Perpendicular to $y = 5$, through $(1, 2)$

47. Perpendicular to $x = -3$, through $(-2, 5)$

48. Parallel to $y = -4$, through $(0, -3)$

49. Parallel to $x = 0$, through $(6, -8)$

50. Perpendicular to $x = 7$, through $(-5, 0)$

Objectives **A C D E** Mixed Practice *See Examples 1 and 4 through 7. Find an equation of each line described. Write each equation in slope-intercept form when possible.*

51. With slope $-\dfrac{1}{2}$, through $\left(0, \dfrac{5}{3}\right)$

52. With slope $\dfrac{5}{7}$, through $(0, -3)$

53. Through $(10, 7)$ and $(7, 10)$

54. Through $(5, -6)$ and $(-6, 5)$

55. With undefined slope, through $\left(-\dfrac{3}{4}, 1\right)$

56. With slope 0, through $(6.7, 12.1)$

57. Slope 1, through $(-7, 9)$

58. Slope 5, through $(6, -8)$

59. Slope -5, y-intercept $(0, 7)$

60. Slope -2, y-intercept $(0, -4)$

61. Through $(-8, 11)$, parallel to $x = 5$

62. Through $(9, -12)$, parallel to the y-axis

63. Through $(2, 3)$ and $(0, 0)$

64. Through $(4, 7)$ and $(0, 0)$

65. Through $(-2, -3)$, perpendicular to the y-axis

66. Through $(0, 12)$, perpendicular to the x-axis

67. Slope $-\dfrac{4}{7}$, through $(-1, -2)$

68. Slope $-\dfrac{3}{5}$, through $(4, 4)$

Objective **F** *Solve. Assume each exercise describes a linear relationship. Write the equations in slope-intercept form. See Example 8.*

69. In 2007, a total of 6809 different magazines were in print in the United States. By 2015, this number was 7293. (*Source:* MPA—The Association of Magazine Media)
   a. Write two ordered pairs of the form (years after 2007, number of magazines in print) for this situation.
   b. Assume the relationship between years after 2007 and number of magazines in print is linear over this period. Use the ordered pairs from part **a** to write an equation for the line relating year after 2007 to number of magazines in print.
   c. Use the linear equation in part **b** to estimate the number of magazines in print in 2013.

70. The number of digital magazines published in the United States is rapidly expanding the magazine market. In 2011, there were approximately 3.4 million digital magazine subscriptions. By 2015, this had grown to about 16 million digital magazine subscriptions. (*Source:* MPA—The Association of Magazine Media)
   a. Write two ordered pairs of the form (years after 2011, millions of digital magazine subscriptions).
   b. Assume the relationship between years after 2011 and millions of digital magazine subscriptions is linear over this period. Use the ordered pairs from part **a** to write an equation for the line relating years after 2011 to millions of digital magazine subscriptions.
   c. Use the linear equation from part **b** to estimate the millions of digital magazine subscriptions in 2018 if this trend were to continue.

71. A rock is dropped from the top of a 400-foot cliff. After 1 second, the rock is traveling 32 feet per second. After 3 seconds, the rock is traveling 96 feet per second.

  a. Assume that the relationship between time and speed is linear and write an equation describing this relationship. Use ordered pairs of the form (time, speed).

  b. Use this equation to determine the speed of the rock 4 seconds after it is dropped.

72. A Hawaiian fruit company is studying the sales of a pineapple sauce to see if this product is to be continued. At the end of its first year, profits on this product amounted to $30,000. At the end of the fourth year, profits were $66,000.

  a. Assume that the relationship between years on the market and profit is linear and write an equation describing this relationship. Use ordered pairs of the form (years on the market, profit).

  b. Use this equation to predict the profit at the end of 7 years.

73. In 2012 there were approximately 434,000 hybrid vehicles sold in the United States. In 2015, there were approximately 491,000 such vehicles sold. (*Source:* Oak Ridge National Laboratory)

  a. Write an equation describing the relationship between time and the number of hybrid vehicles sold. Use ordered pairs of the form (years past 2012, number of vehicles sold).

  b. Use this equation to estimate the number of hybrid sales in 2014.

74. In 2012, there were approximately 980 thousand restaurants in the United States. In 2016, there were 1046 thousand restaurants. (*Source:* National Restaurant Association)

  a. Write an equation describing the relationship between time and the number of restaurants. Use ordered pairs of the form (years past 2012, number of restaurants in thousands).

  b. Use this equation to predict the number of eating establishments in 2018.

75. In 2012 there were 5320 indoor cinema sites in the United States. In 2016, there were approximately 5470 indoor cinema sites. (*Source:* National Association of Theater Owners)

  a. Write an equation describing this relationship. Use ordered pairs of the form (years past 2012, number of indoor cinema sites).

  b. Use this equation to predict the number of indoor cinema sites in 2018.

76. In 2015, the U.S. population per square mile of land area was approximately 90.7. In 2010, this person-per-square-mile population was 87.4. (*Source:* World Bank)

  a. Write an equation describing the relationship between year and persons per square mile. Use ordered pairs of the form (years past 2010, persons per square mile).

  b. Use this equation to predict the person-per-square-mile population in 2018.

77. The Pool Fun Company has learned that, by pricing a newly released Fun Noodle at $3, sales will reach 10,000 Fun Noodles per day during the summer. Raising the price to $5 will cause the sales to fall to 8000 Fun Noodles per day.

    a. Assume that the relationship between sales price and number of Fun Noodles sold is linear and write an equation describing this relationship. Use ordered pairs of the form (sales price, number sold).

    b. Predict the daily sales of Fun Noodles if the price is $3.50.

78. The value of a building bought in 1995 may be depreciated (or decreased) as time passes for income tax purposes. Seven years after the building was bought, this value was $225,000 and 12 years after it was bought, this value was $195,000.

    a. If the relationship between number of years past 1995 and the depreciated value of the building is linear, write an equation describing this relationship. Use ordered pairs of the form (years past 1995, value of building).

    b. Use this equation to estimate the depreciated value of the building in 2013.

Review

Find the value of $x^2 - 3x + 1$ for each given value of x. See Sections 1.7, 1.8, and Chapter 2.

79. 2   **80.** 5   **81.** -1 **82.** -3

Concept Extensions

Match each linear equation with its graph.

83. $y = 2x + 1$ **84.** $y = -x + 1$ **85.** $y = -3x - 2$ **86.** $y = \frac{5}{3}x - 2$

A.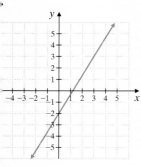
B.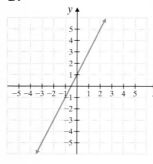
C.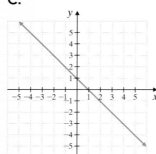
D.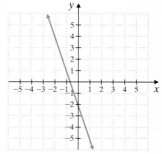

87. Write an equation in standard form of the line that contains the point $(-2, 4)$ and has the same slope as the line $y = 2x + 5$.

88. Write an equation in standard form of the line that contains the point $(3, 0)$ and has the same slope as the line $y = -3x - 1$.

89. Write an equation in standard form of the line that contains the point $(-1, 2)$ and is

    a. parallel to the line $y = 3x - 1$.

    b. perpendicular to the line $y = 3x - 1$.

△ **90.** Write an equation in standard form of the line that contains the point $(4, 0)$ and is

    a. parallel to the line $y = -2x + 3$.

    b. perpendicular to the line $y = -2x + 3$.

Integrated Review — Sections 13.1–13.5

Summary on Linear Equations

Answers

1. _____

2. _____

3. _____

4. _____

5. _____

6. _____

7. _____

8. _____

Find the slope of each line.

1.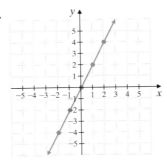

2.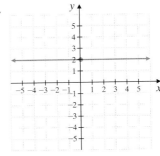

3.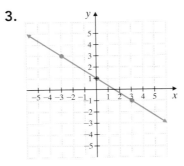

4.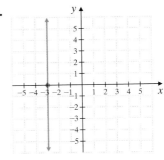

Graph each linear equation. For Exercises 11 and 12, label the intercepts.

5. $y = -2x$

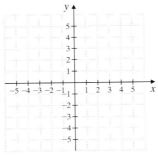

6. $x + y = 3$

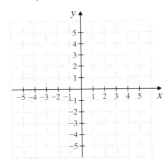

7. $x = -1$

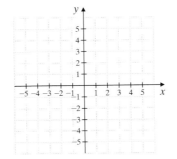

8. $y = 4$

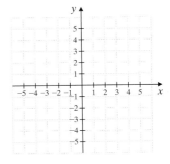

9. $x - 2y = 6$

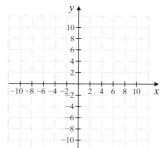

10. $y = 3x + 2$

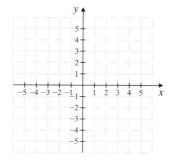

11. $y = -\dfrac{3}{4}x + 3$

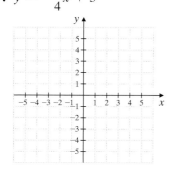

12. $5x - 2y = 8$

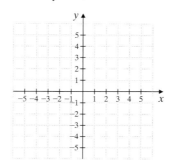

Find the slope of each line by writing the equation in slope-intercept form.

13. $y = 3x - 1$ **14.** $y = -6x + 2$ **15.** $7x + 2y = 11$ **16.** $2x - y = 0$

Find the slope of each line.

17. $x = 2$

18. $y = -4$

19. Write an equation of the line with slope $m = 2$ and y-intercept $\left(0, -\dfrac{1}{3}\right)$. Write the equation in the form $y = mx + b$.

20. Find an equation of the line with slope $m = -4$ that passes through the point $(-1, 3)$. Write the equation in the form $y = mx + b$.

21. Find an equation of the line that passes through the points $(2, 0)$ and $(-1, -3)$. Write the equation in the form $Ax + By = C$.

Determine whether each pair of lines is parallel, perpendicular, or neither.

22. $6x - y = 7$
$2x + 3y = 4$

23. $3x - 6y = 4$
$y = -2x$

24. Yogurt is an ever more popular food item. In 2012, U.S. production of yogurt stood at approximately 4416 million pounds. In 2015, this number rose to 4743 million pounds of yogurt. (*Source:* United States Department of Agriculture)

  a. Write two ordered pairs of the form (year, millions of pounds of yogurt produced).
  b. Find the slope of the line between these two points.
  c. Write a sentence explaining the meaning of the slope as a rate of change.

9. _____
10. _____
11. _____
12. _____
13. _____
14. _____
15. _____
16. _____
17. _____
18. _____
19. _____
20. _____
21. _____
22. _____
23. _____
24. a. _____
    b. _____
    c. _____

13.6 Introduction to Functions

Objectives

A. Identify Relations, Domains, and Ranges.

B. Identify Functions.

C. Use the Vertical Line Test.

D. Use Function Notation.

Objective A Identifying Relations, Domains, and Ranges

In previous sections, we have discussed the relationships between two quantities. For example, the relationship between the length of the side of a square x and its area is described by the equation $y = x^2$. Ordered pairs can be used to write down solutions of this equation. For example, (2, 4) is a solution of $y = x^2$, and this notation tells us that the x-value 2 is related to the y-value 4 for this equation. In other words, when the length of the side of a square is 2 units, its area is 4 square units.

Examples of Relationships Between Two Quantities

Area of Square: $y = x^2$	Equation of Line: $y = x + 2$	Online Advertising Spending

Some Ordered Pairs

x	y
2	4
5	25
7	49
12	144

Some Ordered Pairs

x	y
−3	−1
0	2
2	4
9	11

Ordered Pairs

Year	Billions of Dollars
2016	68.1
2017	75.3
2018	81.9
2019	88.1
2020	93.5

A set of ordered pairs is called a **relation.** The set of all x-coordinates is called the **domain** of a relation, and the set of all y-coordinates is called the **range** of a relation. Equations such as $y = x^2$ are also called relations since equations in two variables define a set of ordered pair solutions.

Practice 1

Find the domain and range of the relation $\{(-3, 5), (-3, 1), (4, 6), (7, 0)\}$.

Example 1 Find the domain and the range of the relation $\{(0, 2), (3, 3), (-1, 0), (3, -2)\}$.

Solution: The domain is the set of all x-coordinates, or $\{-1, 0, 3\}$, and the range is the set of all y-coordinates, or $\{-2, 0, 2, 3\}$.

■ Work Practice 1

Objective B Identifying Functions

Paired data occur often in real-life applications. Some special sets of paired data, or ordered pairs, are called *functions*.

Answer

1. domain: $\{-3, 4, 7\}$; range: $\{0, 1, 5, 6\}$

Section 13.6 | Introduction to Functions   1033

Function

A **function** is a set of ordered pairs in which each *x*-coordinate has exactly one *y*-coordinate.

In other words, a function cannot have two ordered pairs with the same *x*-coordinate but different *y*-coordinates.

Example 2 Determine whether each relation is also a function.

a. $\{(-1, 1), (2, 3), (7, 3), (8, 6)\}$
b. $\{(0, -2), (1, 5), (0, 3), (7, 7)\}$

Solution:

a. Although the ordered pairs (2, 3) and (7, 3) have the same *y*-value, each *x*-value is assigned to only one *y*-value, so this set of ordered pairs is a function.
b. The *x*-value 0 is paired with two *y*-values, −2 and 3, so this set of ordered pairs is not a function.

Work Practice 2

Practice 2

Are the following relations also functions?

a. $\{(2, 5), (-3, 7), (4, 5), (0, -1)\}$
b. $\{(1, 4), (6, 6), (1, -3), (7, 5)\}$

Relations and functions can be described by graphs of their ordered pairs.

Example 3 Which graph is the graph of a function?

a.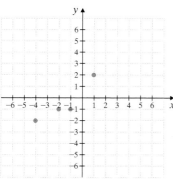
b.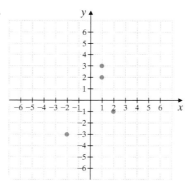

Practice 3

Is each graph the graph of a function?

a.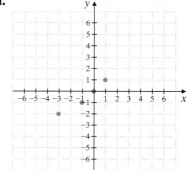

b.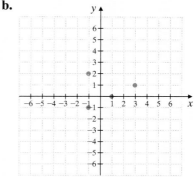

Solution:

a. This is the graph of the relation $\{(-4, -2), (-2, -1), (-1, -1), (1, 2)\}$. Each *x*-coordinate has exactly one *y*-coordinate, so this is the graph of a function.
b. This is the graph of the relation $\{(-2, -3), (1, 2), (1, 3), (2, -1)\}$. The *x*-coordinate 1 is paired with two *y*-coordinates, 2 and 3, so this is not the graph of a function.

Work Practice 3

Objective C Using the Vertical Line Test

The graph in Example 3(b) was not the graph of a function because the *x*-coordinate 1 was paired with two *y*-coordinates, 2 and 3. Notice that when an *x*-coordinate is paired with more than one *y*-coordinate, a vertical line can be drawn that

Answers

2. **a.** a function   **b.** not a function
3. **a.** a function   **b.** not a function

1034 Chapter 13 | Graphing Equations and Inequalities

will intersect the graph at more than one point. We can use this fact to determine whether a relation is also a function. We call this the vertical line test.

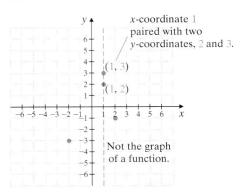

Vertical Line Test

If a vertical line can be drawn so that it intersects a graph more than once, the graph is not the graph of a function. (If no such vertical line can be drawn, the graph is that of a function.)

This vertical line test works for all types of graphs on the rectangular coordinate system.

Example 4 Use the vertical line test to determine whether each graph is the graph of a function.

a. b. c. d.

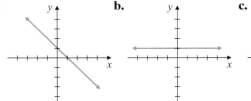 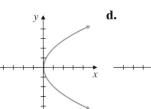

Solution:

a. This graph is the graph of a function since no vertical line will intersect the graph more than once.
b. This graph is also the graph of a function; no vertical line will intersect it more than once.
c. This graph is not the graph of a function. Vertical lines can be drawn that intersect the graph in two points. An example of one such line is shown.

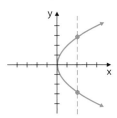

d. This graph is not the graph of a function. A vertical line can be drawn that intersects this line at every point.

Work Practice 4

Practice 4

Determine whether each graph is the graph of a function.

a.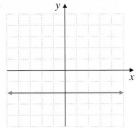

b.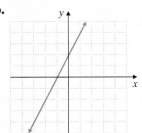

c.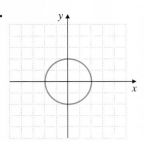

d.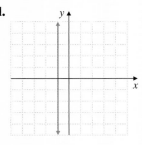

Answers
4. a. a function b. a function
c. not a function d. not a function

Recall that the graph of a linear equation is a line, and a line that is not vertical will pass the vertical line test. **Thus, all linear equations are functions except those of the form $x = c$, which are vertical lines.**

Example 5 Which of the following linear equations are functions?

a. $y = x$ **b.** $y = 2x + 1$ **c.** $y = 5$ **d.** $x = -1$

Solution: **a, b,** and **c** are functions because their graphs are nonvertical lines. **d** is not a function because its graph is a vertical line.

Work Practice 5

Practice 5

Which of the following linear equations are functions?

a. $y = 2x$ **b.** $y = -3x - 1$
c. $y = 8$ **d.** $x = 2$

Examples of functions can often be found in magazines, newspapers, books, and other printed material in the form of tables or graphs such as that in Example 6.

Example 6 The graph shows the sunrise time for Indianapolis, Indiana, for the first of each month for one year. Use this graph to answer the questions.

a. Approximate the time of sunrise on February 1.
b. Approximate the date(s) when the sun rises at 5 a.m.

Practice 6

Use the graph in Example 6 to answer the questions.

a. Approximate the time of sunrise on March 1.
b. Approximate the date(s) when the sun rises at 6 a.m.

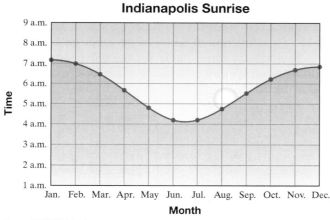

Indianapolis Sunrise

Source: Wolff World Atlas

c. Is this the graph of a function?

Solution:

a. As shown on the next page, to approximate the time of sunrise on February 1, we find the mark on the horizontal axis that corresponds to February 1. From this mark, we move vertically upward (shown in blue) until the graph is reached. From that point on the graph, we move horizontally to the left until the vertical axis is reached. The vertical axis there reads 7 a.m.

b. To approximate the date(s) when the sun rises at 5 a.m., we find 5 a.m. on the time axis and move horizontally to the right (shown in red). Notice that we will hit the graph at two points, corresponding to two dates for which the sun rises at 5 a.m. We follow both points on the graph vertically downward until the horizontal axis is reached. The sun rises at 5 a.m. at approximately the end of the month of April and early in the month of August.

(Continued on next page)

Answers
5. a, b, and **c** are functions.
6. a. 6:30 a.m. **b.** middle of March and middle of September

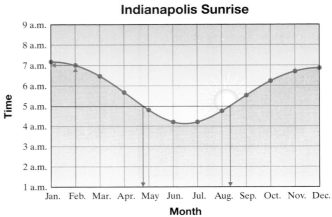

Source: Wolff World Atlas

c. The graph is the graph of a function since it passes the vertical line test. In other words, for every day of the year in Indianapolis, there is exactly one sunrise time.

■ Work Practice 6

Objective D Using Function Notation

The graph of the linear equation $y = 2x + 1$ passes the vertical line test, so we see that $y = 2x + 1$ is a function. In other words, $y = 2x + 1$ gives us a rule for writing ordered pairs where every x-coordinate is paired with at most one y-coordinate.

The variable y is a function of the variable x. For each value of x, there is only one value of y. Thus, we say the variable x is the **independent variable** because any value in the domain can be assigned to x. The variable y is the **dependent variable** because its value depends on x.

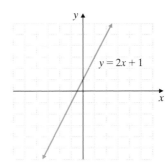

We often use letters such as f, g, and h to name functions. For example, the symbol $f(x)$ means *function of x* and is read "f of x." This notation is called **function notation.** The equation $y = 2x + 1$ can be written as $f(x) = 2x + 1$ using function notation, and these equations mean the same thing. In other words, $y = f(x)$.

The notation $f(1)$ means to replace x with 1 and find the resulting y or function value. Since

$$f(x) = 2x + 1$$

then

$$f(1) = 2(1) + 1 = 3$$

is means that, when $x = 1$, y or $f(x) = 3$, and we have the ordered pair $(1, 3)$.
w let's find $f(2), f(0)$, and $f(-1)$.

$f(x) = 2x + 1$	$f(x) = 2x + 1$	$f(x) = 2x + 1$
$f(2) = 2(2) + 1$	$f(0) = 2(0) + 1$	$f(-1) = 2(-1) + 1$
$= 4 + 1$	$= 0 + 1$	$= -2 + 1$
$= 5$	$= 1$	$= -1$

dered Pair: $(2, 5)$ $(0, 1)$ $(-1, -1)$

> **Helpful Hint**
> Note that, for example, if $f(2) = 5$, the corresponding ordered pair is $(2, 5)$.

> **Helpful Hint**
> Note that $f(x)$ is a special symbol in mathematics used to denote a function. The symbol $f(x)$ is read "f of x." It does **not** mean $f \cdot x$ (f times x).

Example 7 Given $g(x) = x^2 - 3$, find the following and list the corresponding ordered pairs generated.

a. $g(2)$ **b.** $g(-2)$ **c.** $g(0)$

Solution:

a.
$g(x) = x^2 - 3$
$g(2) = 2^2 - 3$
$= 4 - 3$
$= 1$

b.
$g(x) = x^2 - 3$
$g(-2) = (-2)^2 - 3$
$= 4 - 3$
$= 1$

c.
$g(x) = x^2 - 3$
$g(0) = 0^2 - 3$
$= 0 - 3$
$= -3$

Ordered Pairs:	$g(2) = 1$ gives $(2, 1)$	$g(-2) = 1$ gives $(-2, 1)$	$g(0) = -3$ gives $(0, -3)$

Practice 7
Given $f(x) = x^2 + 1$, find the following and list the corresponding ordered pairs.

a. $f(1)$ **b.** $f(-3)$ **c.** $f(0)$

Work Practice 7

We now practice finding the domain and the range of a function. The domain of r functions will be the set of all possible real numbers that x can be replaced by. e range is the set of corresponding y-values.

Example 8 Find the domain of each function.

a. $g(x) = \dfrac{1}{x}$ **b.** $f(x) = 2x + 1$

Solution:

a. Recall that we cannot divide by 0, so the domain of $g(x)$ is the set of all real numbers except 0.

b. In this function, x can be any real number. The domain of $f(x)$ is the set of all real numbers.

Work Practice 8

Practice 8
Find the domain of each function.

a. $h(x) = 6x + 3$

b. $f(x) = \dfrac{1}{x^2}$

Concept Check Suppose that the value of f is -7 when the function is evaluated at 2. Write this situation in function notation.

Answers
7. a. $2; (1, 2)$ **b.** $10; (-3, 10)$ **c.** $1; (0, 1)$
8. a. Domain: all real numbers
   b. Domain: all real numbers except 0

✓ **Concept Check Answer**
$f(2) = -7$

Practice 9

Find the domain and the range of each function graphed.

a.
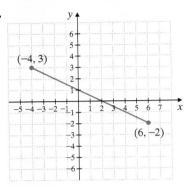

b.
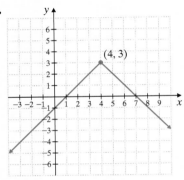

Example 9
Find the domain and the range of each function graphed.

a.
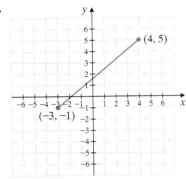

b.
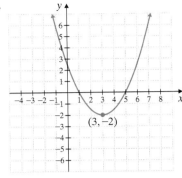

Solution:

a.
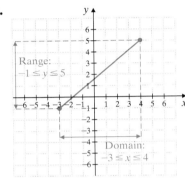

b.
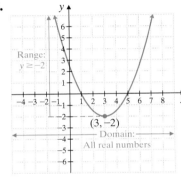

■ Work Practice 9

Answers

9. a. Domain: $-4 \leq x \leq 6$
   Range: $-2 \leq y \leq 3$
   b. Domain: all real numbers
   Range: $y \leq 3$

Vocabulary, Readiness & Video Check

Use the choices below to fill in each blank. Some choices may not be used.

| $x = c$ | horizontal | domain | relation | (7, 3) | x | $\{x \mid x \leq 5\}$ |
| $y = c$ | vertical | range | function | (3, 7) | y | all real numbers |

1. A set of ordered pairs is called a(n) _____.
2. A set of ordered pairs that assigns to each x-value exactly one y-value is called a(n) _____.
3. The set of all y-coordinates of a relation is called the _____.
4. The set of all x-coordinates of a relation is called the _____.
5. All linear equations are functions except those whose graphs are _____ lines.
6. All linear equations are functions except those whose equations are of the form _____.
7. If $f(3) = 7$, the corresponding ordered pair is _____.
8. The domain of $f(x) = x + 5$ is _____.
9. For the function $y = mx + b$, the dependent variable is _____ and the independent variable is _____.

Section 13.6 | Introduction to Functions

Martin-Gay Interactive Videos

See Video 13.6

Watch the section lecture video and answer the following questions.

Objective A 10. In the lecture before Example 1, relations are discussed. Why can an equation in two variables define a relation?

Objective B 11. Based on Examples 2 and 3, can a set of ordered pairs with no repeated *x*-values, but with repeated *y*-values, be a function? For example: $\{(0, 4), (-3, 4), (2, 4)\}$.

Objective C 12. After reviewing Example 8, explain why the vertical line test works.

Objective D 13. Using Example 10, write the three function values found and their corresponding ordered pairs. One example is: $f(0) = 2$ corresponds to $(0, 2)$.

13.6 Exercise Set MyLab Math

Objective A *Find the domain and the range of each relation. See Example 1.*

1. $\{(2, 4), (0, 0), (-7, 10), (10, -7)\}$

2. $\{(3, -6), (1, 4), (-2, -2)\}$

3. $\{(0, -2), (1, -2), (5, -2)\}$

4. $\{(5, 0), (5, -3), (5, 4), (5, 3)\}$

Objective B *Determine whether each relation is also a function. See Example 2.*

5. $\{(1, 1), (2, 2), (-3, -3), (0, 0)\}$

6. $\{(11, 6), (-1, -2), (0, 0), (3, -2)\}$

7. $\{(-1, 0), (-1, 6), (-1, 8)\}$

8. $\{(1, 2), (3, 2), (1, 4)\}$

Objectives B C Mixed Practice *Determine whether each graph is the graph of a function. For Exercises 9 through 12, either write down the ordered pairs or use the vertical line test. See Examples 3 and 4.*

9.
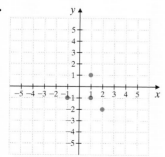

10.
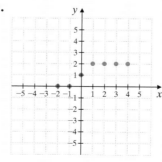

11.
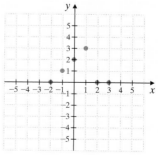

12. 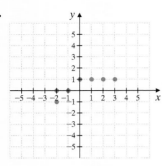 **13.** 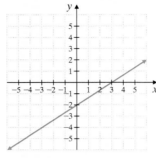 **14.**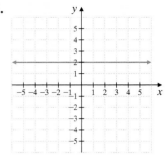

15. 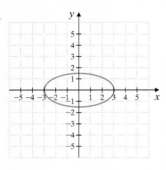 **16.**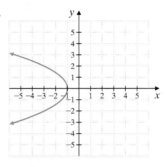

For each exercise, choose the value of x so that the relation is NOT also a function.

17. $\{(2, 3), (-1, 7), (x, 9)\}$
   a. -1 b. 1 c. 9 d. 7

18. $\{(-8, 0), (x, 1), (5, -3)\}$
   a. 0 b. -3 c. -5 d. 5

Decide whether the equation describes a function. See Example 5.

19. $y - x = 7$
20. $2x - 3y = 9$
21. $y = 6$
22. $x = 3$
23. $x = -2$
24. $y = -9$
25. $x = y^2$
26. $y = x^2 - 3$

(*Hint:* For Exercises **25** and **26**, check to see whether each x-value pairs with exactly one y-value.)

The graph shows the sunset times for Seward, Alaska for the first of each month for one year. Use this graph to answer Exercises 27 through 32. See Example 6.

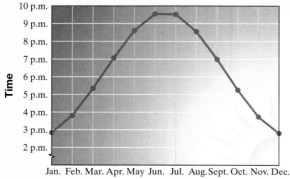

27. Approximate the time of sunset on June 1.

28. Approximate the time of sunset on November 1.

29. Approximate the date(s) when the sunset is at 3 p.m.

30. Approximate the date(s) when the sunset is at 9 p.m.

31. Is this graph the graph of a function? Why or why not?

32. Do you think a graph of sunset times for any location will always be a function? Why or why not?

Section 13.6 | Introduction to Functions

This graph shows the U.S. hourly minimum wage for each year shown. Use this graph to answer Exercises 33 through 38. See Example 6.

Source: U.S. Department of Labor

33. Approximate the minimum wage before October 1996.

34. Approximate the minimum wage in 2006.

35. Find the year when the minimum wage increased to over $7.00 per hour.

36. According to the graph, what hourly wage was in effect for the greatest number of years?

37. Is this graph the graph of a function? Why or why not?

38. Do you think that a similar graph of your hourly wage on January 1 of every year (whether you are working or not) would be the graph of a function? Why or why not?

This graph shows the cost of mailing a large envelope through the U.S. Postal Service by weight. Use this graph to answer Exercises 39 through 44. See Example 6.

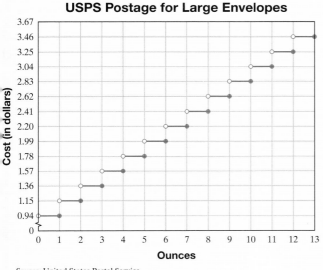

Source: United States Postal Service

39. Approximate the postage to mail a large envelope weighing more than 4 ounces but not more than 5 ounces.

40. Approximate the postage to mail a large envelope weighing more than 7 ounces but not more than 8 ounces.

41. Give the weight of a large envelope that costs $1.15 to mail.

42. If you have $3.00, what is the weight of the largest envelope you can mail for that amount of money?

43. Is this graph a function? Why or why not?

44. Do you think that a similar graph of postage to mail a first-class letter would be the graph of a function? Why or why not?

Objective D Find $f(-2)$, $f(0)$, and $f(3)$ for each function. See Example 7.

45. $f(x) = 2x - 5$ **46.** $f(x) = 3 - 7x$ **47.** $f(x) = x^2 + 2$ **48.** $f(x) = x^2 - 4$

49. $f(x) = 3x$ **50.** $f(x) = -3x$ **51.** $f(x) = |x|$ **52.** $f(x) = |2 - x|$

Find $h(-1), h(0),$ and $h(4)$ for each function. See Example 7.

53. $h(x) = -5x$ **54.** $h(x) = -3x$ **55.** $h(x) = 2x^2 + 3$ **56.** $h(x) = 3x^2$

For each given function value, write a corresponding ordered pair.

57. $f(3) = 6$ **58.** $f(7) = -2$ **59.** $g(0) = -\dfrac{1}{2}$

60. $g(0) = -\dfrac{7}{8}$ **61.** $h(-2) = 9$ **62.** $h(-10) = 1$

Objectives **A** **D** Mixed Practice Find the domain of each function. See Example 8.

63. $f(x) = 3x - 7$ **64.** $g(x) = 5 - 2x$ **65.** $h(x) = \dfrac{1}{x + 5}$ **66.** $f(x) = \dfrac{1}{x - 6}$

Objectives **A** **D** Mixed Practice Find the domain and the range of each relation graphed. See Example 9.

67. 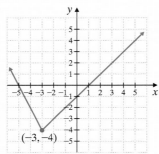   **68.** 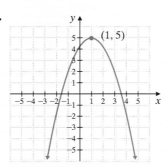   **69.**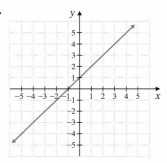

70. 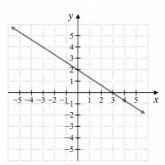   **71.** 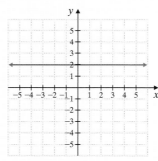   **72.**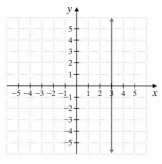

Use the graph of f below to answer Exercises 73 through 78.

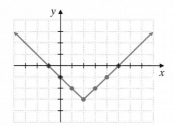

73. Complete the ordered pair solution for f. $(0, \ \ )$
74. Complete the ordered pair solution for f. $(3, \ \ )$
75. $f(0) = $ _____?
76. $f(3) = $ _____?
77. If $f(x) = 0$, find the value(s) of x.
78. If $f(x) = -1$, find the value(s) of x.

Review

Solve each inequality. See Section 9.6.

79. $2x + 5 < 7$ **80.** $3x - 1 \geq 11$ **81.** $-x + 6 \leq 9$ **82.** $-2x + 3 > 3$

Find the perimeter of each figure. See Sections 12.3 and 12.4.

83.
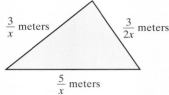

△ 84.
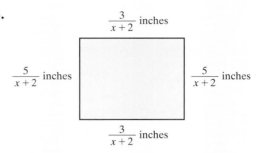

Concept Extensions

Solve. See the Concept Check in this section.

85. If a function f is evaluated at -5, the value of the function is 12. Write this situation using function notation.

86. Suppose $(9, 20)$ is an ordered pair solution for the function g. Write this situation using function notation.

The graph of the function f is below. Use this graph to answer Exercises 87 through 90.

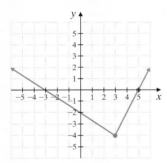

87. Write the coordinates of the lowest point of the graph.

88. Write the answer to Exercise **87** in function notation.

89. An x-intercept of this graph is $(5, 0)$. Write this using function notation.

90. Write the other x-intercept of this graph (see Exercise **89**) using function notation.

91. In your own words define (a) function; (b) domain; (c) range.

92. Explain the vertical line test and how it is used.

93. Since $y = x + 7$ is a function, rewrite the equation using function notation.

94. Since $y = 3$ is a function, rewrite the equation using function notation.

95. The dosage in milligrams of Ivermectin, a heartworm preventive for a dog who weighs x pounds, is given by the function

$$f(x) = \frac{136}{25} x$$

  a. Find the proper dosage for a dog who weighs 35 pounds.
  b. Find the proper dosage for a dog who weighs 70 pounds.

96. Forensic scientists use the function

$$f(x) = 2.59x + 47.24$$

to estimate the height of a woman, in centimeters, given the length x of her femur bone in centimeters.

  a. Estimate the height of a woman whose femur measures 46 centimeters.
  b. Estimate the height of a woman whose femur measures 39 centimeters.

13.7 Graphing Linear Inequalities in Two Variables

Objectives

A Determine Whether an Ordered Pair Is a Solution of a Linear Inequality in Two Variables.

B Graph a Linear Inequality in Two Variables.

Recall that a linear equation in two variables is an equation that can be written in the form $Ax + By = C$, where A, B, and C are real numbers and A and B are not both 0. A **linear inequality in two variables** is an inequality that can be written in one of the forms

$$Ax + By < C \qquad Ax + By \leq C$$
$$Ax + By > C \qquad Ax + By \geq C$$

where A, B, and C are real numbers and A and B are not both 0.

Objective A Determining Solutions of Linear Inequalities in Two Variables

Just as for linear equations in x and y, an ordered pair is a **solution** of an inequality in x and y if replacing the variables with the coordinates of the ordered pair results in a true statement.

Example 1 Determine whether each ordered pair is a solution of the inequality $2x - y < 6$.

a. $(5, -1)$ **b.** $(2, 7)$

Solution:

a. We replace x with 5 and y with -1 and see if a true statement results.

$$2x - y < 6$$
$$2(5) - (-1) < 6 \quad \text{Replace } x \text{ with 5 and } y \text{ with } -1.$$
$$10 + 1 < 6$$
$$11 < 6 \quad \text{False}$$

The ordered pair $(5, -1)$ is not a solution since $11 < 6$ is a false statement.

b. We replace x with 2 and y with 7 and see if a true statement results.

$$2x - y < 6$$
$$2(2) - (7) < 6 \quad \text{Replace } x \text{ with 2 and } y \text{ with 7.}$$
$$4 - 7 < 6$$
$$-3 < 6 \quad \text{True}$$

The ordered pair $(2, 7)$ is a solution since $-3 < 6$ is a true statement.

■ Work Practice 1

Practice 1

Determine whether each ordered pair is a solution of $x - 4y > 8$.

a. $(-3, 2)$ **b.** $(9, 0)$

Objective B Graphing Linear Inequalities in Two Variables

The linear equation $x - y = 1$ is graphed next. Recall that all points on the line correspond to ordered pairs that satisfy the equation $x - y = 1$.

Notice that the line defined by $x - y = 1$ divides the rectangular coordinate system plane into 2 sides. All points on one side of the line satisfy the inequality $x - y < 1$, and all points on the other side satisfy the inequality $x - y > 1$. The graph on the next page shows a few examples of this.

Answers

1. a. no   **b.** yes

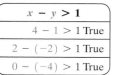
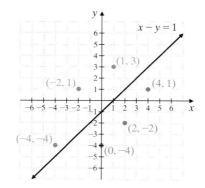

The graph of $x - y < 1$ is the region shaded blue and the graph of $x - y > 1$ is the region shaded red below.

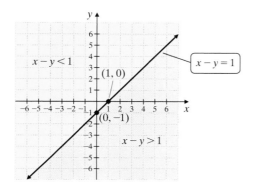

The region to the left of the line and the region to the right of the line are called **half-planes.** Every line divides the plane (similar to a sheet of paper extending indefinitely in all directions) into two half-planes; the line is called the **boundary.**

Recall that the inequality $x - y \leq 1$ means

$x - y = 1$ or $x - y < 1$

Thus, the graph of $x - y \leq 1$ is the blue half-plane $x - y < 1$ along with the boundary line $x - y = 1$.

To Graph a Linear Inequality in Two Variables

Step 1: Graph the boundary line found by replacing the inequality sign with an equal sign. If the inequality sign is $>$ or $<$, graph a dashed boundary line (indicating that the points on the line are not solutions of the inequality). If the inequality sign is $\geq$ or $\leq$, graph a solid boundary line (indicating that the points on the line are solutions of the inequality).

Step 2: Choose a point *not* on the boundary line as a test point. Substitute the coordinates of this test point into the *original* inequality.

Step 3: If a true statement is obtained in Step 2, shade the half-plane that contains the test point. If a false statement is obtained, shade the half-plane that does not contain the test point.

Practice 2

Graph: $x - y > 3$

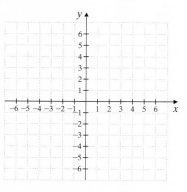

Practice 3

Graph: $x - 4y \le 4$

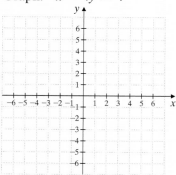

Answers

2.

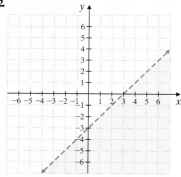

3.

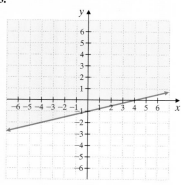

✓ **Concept Check Answers**
a. no   **b.** yes   **c.** yes

Example 2 Graph: $x + y < 7$

Solution:

Step 1: First we graph the boundary line by graphing the equation $x + y = 7$. We graph this boundary as a *dashed line* because the inequality sign is $<$, and thus the points on the line are not solutions of the inequality $x + y < 7$.

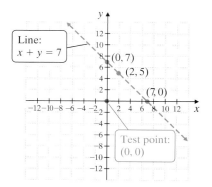

Step 2: Next we choose a test point, being careful *not* to choose a point on the boundary line. We choose $(0, 0)$, and substitute the coordinates of $(0, 0)$ into $x + y < 7$.

$x + y < 7$ Original inequality
$0 + 0 < 7$ Replace x with 0 and y with 0.
$0 < 7$ True

Step 3: Since the result is a true statement, $(0, 0)$ is a solution of $x + y < 7$, and every point in the same half-plane as $(0, 0)$ is also a solution. To indicate this, we shade the entire half-plane containing $(0, 0)$, as shown.

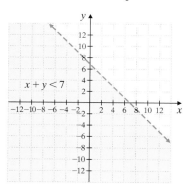

Graph of $x + y < 7$

▪ Work Practice 2

✓ **Concept Check** Determine whether $(0, 0)$ is included in the graph of
a. $y \ge 2x + 3$ **b.** $x < 7$ **c.** $2x - 3y < 6$

Example 3 Graph: $2x - y \ge 3$

Solution:

Step 1: We graph the boundary line by graphing $2x - y = 3$. We draw this line as a solid line because the inequality sign is $\ge$, and thus the points on the line are solutions of $2x - y \ge 3$.

Step 2: Once again, $(0, 0)$ is a convenient test point since it is not on the boundary line.

We substitute 0 for x and 0 for y into the original inequality.

$$2x - y \geq 3$$
$$2(0) - 0 \geq 3 \quad \text{Let } x = 0 \text{ and } y = 0.$$
$$0 \geq 3 \quad \text{False}$$

Step 3: Since the statement is false, no point in the half-plane containing $(0, 0)$ is a solution. Therefore, we shade the half-plane that does not contain $(0, 0)$. Every point in the shaded half-plane and every point on the boundary line is a solution of $2x - y \geq 3$.

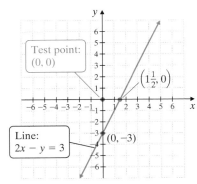

Step 1 and Step 2 above

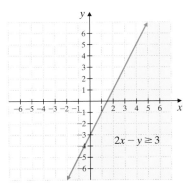

Graph of $2x - y \geq 3$

Work Practice 3

Helpful Hint

When graphing an inequality, make sure the test point is substituted into the **original inequality**. For Example 3, we substituted the test point $(0, 0)$ into the **original inequality** $2x - y \geq 3$, *not* $2x - y = 3$.

Example 4 Graph: $x > 2y$

Solution:

Step 1: We find the boundary line by graphing $x = 2y$. The boundary line is a dashed line since the inequality symbol is $>$.

Step 2: We cannot use $(0, 0)$ as a test point because it is a point on the boundary line. We choose instead $(0, 2)$.

$$x > 2y$$
$$0 > 2(2) \quad \text{Let } x = 0 \text{ and } y = 2.$$
$$0 > 4 \quad \text{False}$$

Step 3: Since the statement is false, we shade the half-plane that does not contain the test point $(0, 2)$, as shown.

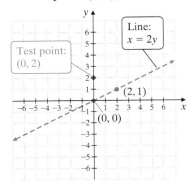

Step 1 and Step 2 above

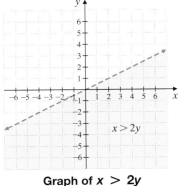

Graph of $x > 2y$

Practice 4

Graph: $y < 3x$

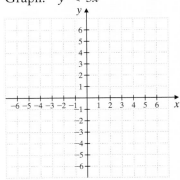

Answer

4.

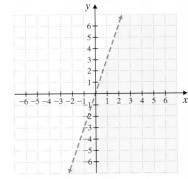

Work Practice 4

Practice 5
Graph: $3x + 2y \geq 12$

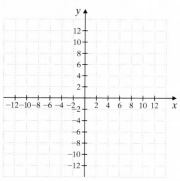

Practice 6
Graph: $x < 2$

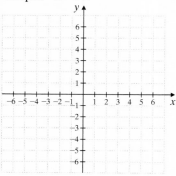

Example 5
Graph: $5x + 4y \leq 20$

Solution: We graph the solid boundary line $5x + 4y = 20$ and choose $(0, 0)$ as the test point.

$$5x + 4y \leq 20$$
$$5(0) + 4(0) \leq 20 \quad \text{Let } x = 0 \text{ and } y = 0.$$
$$0 \leq 20 \quad \text{True}$$

We shade the half-plane that contains $(0, 0)$, as shown.

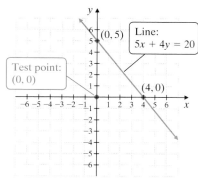

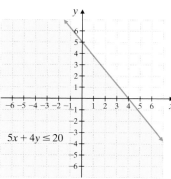

Steps 1 and 2 to graph $5x + 4y \leq 20$

Graph of $5x + 4y \leq 20$

■ Work Practice 5

Example 6
Graph: $y > 3$

Solution: We graph the dashed boundary line $y = 3$ and choose $(0, 0)$ as the test point. (Recall that the graph of $y = 3$ is a horizontal line with y-intercept 3.)

$$y > 3$$
$$0 > 3 \quad \text{Let } y = 0.$$
$$0 > 3 \quad \text{False}$$

We shade the half-plane that does not contain $(0, 0)$, as shown.

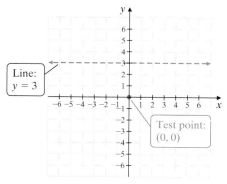

Steps 1 and 2 to graph $y > 3$

Graph of $y > 3$

■ Work Practice 6

Answers
5.

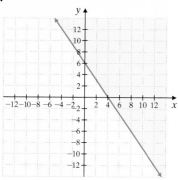

6.

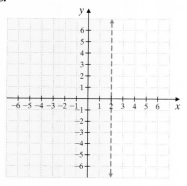

Example 7

Graph: $y \leq \dfrac{2}{3}x - 4$

Solution: Graph the solid boundary line $y = \dfrac{2}{3}x - 4$. This equation is in slope-intercept form, with slope $\dfrac{2}{3}$ and y-intercept -4.

We use this information to graph the line. Then we choose $(0, 0)$ as our test point.

Check the test point.

$y \leq \dfrac{2}{3}x - 4$

$0 \stackrel{?}{\leq} \dfrac{2}{3} \cdot 0 - 4$

$0 \leq -4$ False

Since false, we shade the half-plane that does not contain $(0, 0)$, as shown.

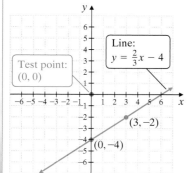

Steps 1 and 2 to graph $y \leq \dfrac{2}{3}x - 4$

Graph of $y \leq \dfrac{2}{3}x - 4$

■ Work Practice 7

Practice 7

Graph: $y \geq \dfrac{1}{4}x + 3$

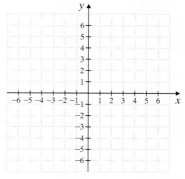

Answer

7.

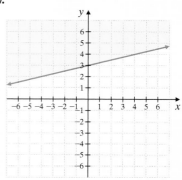

Vocabulary, Readiness & Video Check

Use the choices below to fill in each blank. Some choices may be used more than once, and some not at all.

true	$x < 2$	$y < 2$	half-planes
false	$x \leq 2$	$y \leq 2$	linear inequality in two variables

1. The statement $5x - 6y < 7$ is an example of a _____.
2. A boundary line divides a plane into two regions called _____.
3. True or false? The graph of $5x - 6y < 7$ includes its corresponding boundary line. _____
4. True or false? When graphing a linear inequality, to determine which side of the boundary line to shade, choose a point *not* on the boundary line. _____
5. True or false? The boundary line for the inequality $5x - 6y < 7$ is the graph of $5x - 6y = 7$. _____
6. The graph of _____ is

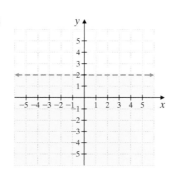

Martin-Gay Interactive Videos

See Video 13.7

Watch the section lecture video and answer the following questions.

Objective A 7. From Example 1, how do we determine whether an ordered pair in x and y is a solution of an inequality in x and y?

Objective B 8. From Example 3, how do we find the equation of the boundary line? How do we determine if the points on the boundary line are solutions of the inequality?

13.7 Exercise Set MyLab Math

Objective A *Determine whether the ordered pairs given are solutions of the linear inequality in two variables. See Example 1.*

1. $x - y > 3$; $(0, 3)$, $(2, -1)$

2. $y - x < -2$; $(2, 1)$, $(5, -1)$

3. $3x - 5y \leq -4$; $(2, 3)$, $(-1, -1)$

4. $2x + y \geq 10$; $(0, 11)$, $(5, 0)$

5. $x < -y$; $(0, 2)$, $(-5, 1)$

6. $y > 3x$; $(0, 0)$, $(1, 4)$

Objective B *Graph each inequality. See Examples 2 through 7.*

7. $x + y \leq 1$
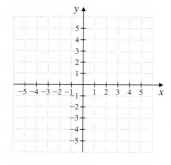

8. $x + y \geq -2$
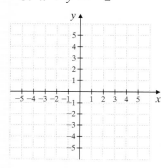

9. $2x - y > -4$
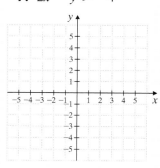

10. $x - 3y < 3$
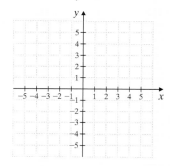

11. $y \geq 2x$
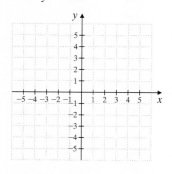

12. $y \leq 3x$
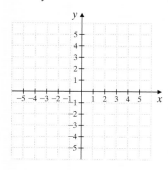

13. $x < -3y$
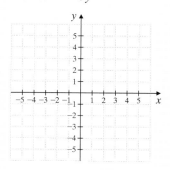

14. $x > -2y$
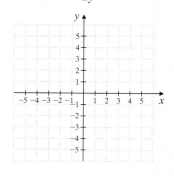

15. $y \geq x + 5$

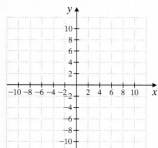

16. $y \leq x + 1$

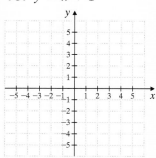

17. $y < 4$

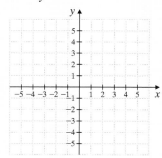

18. $y > 2$

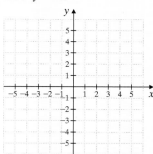

19. $x \geq -3$

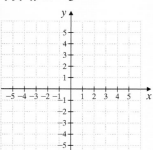

20. $x \leq -1$

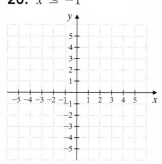

21. $5x + 2y \leq 10$

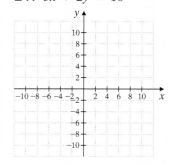

22. $4x + 3y \geq 12$

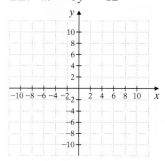

23. $x > y$

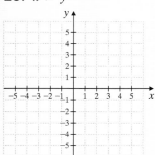

24. $x \leq -y$

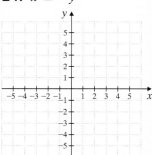

25. $x - y \leq 6$

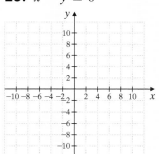

26. $x - y > 10$

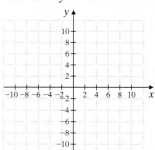

27. $x \geq 0$

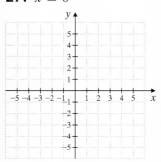

28. $y \leq 0$

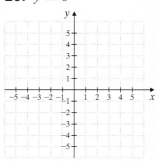

29. $2x + 7y > 5$

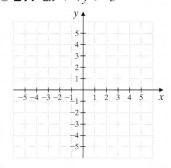

30. $3x + 5y \leq -2$

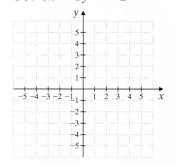

31. $y \geq \dfrac{1}{2}x - 4$ **32.** $y < \dfrac{2}{5}x - 3$ **33.** $y < -\dfrac{3}{4}x + 2$ **34.** $y > -\dfrac{1}{3}x + 4$

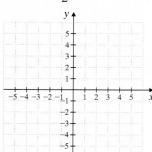 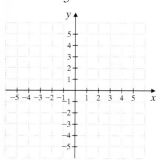 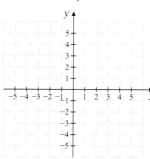 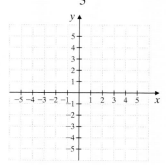

Review

Approximate the coordinates of each point of intersection. See Section 13.1.

35.   **36.**   **37.**   **38.**

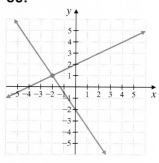 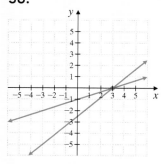 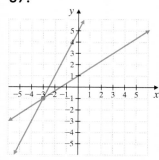 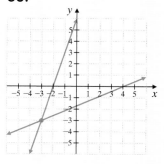

Concept Extensions

Match each inequality with its graph.

A. $x > 2$ **B.** $y < 2$ **C.** $y \leq 2x$ **D.** $y \leq -3x$

39.   **40.**   **41.**   **42.**

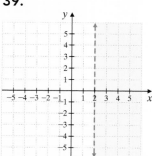 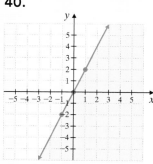 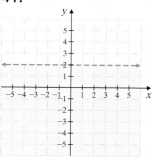 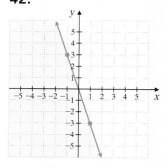

43. Explain why a point on the boundary line should not be chosen as the test point.

44. Write an inequality whose solutions are all points with coordinates whose sum is at least 13.

Determine whether (1, 1) is included in each graph. See the Concept Check in this section.

45. $3x + 4y < 8$ **46.** $y > 5x$ **47.** $y \geq -\dfrac{1}{2}x$ **48.** $x > 3$

49. It's the end of the budgeting period for Dennis Fernandes and he has $500 left in his budget for car rental expenses. He plans to spend this budget on a sales trip throughout southern Texas. He will rent a car that costs $30 per day and $0.15 per mile and he can spend no more than $500.

  a. Write an inequality describing this situation. Let x = number of days and let y = number of miles.

  b. Graph this inequality below.

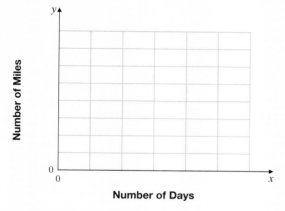

  c. Why is the grid showing quadrant I only?

50. Scott Sambracci and Sara Thygeson are planning their wedding. They have calculated that they want the cost of their wedding ceremony x plus the cost of their reception y to be no more than $5000.

  a. Write an inequality describing this relationship.

  b. Graph this inequality below.

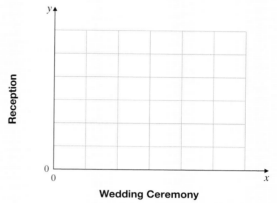

  c. Why is the grid showing quadrant I only?

13.8 Direct and Inverse Variation

Thus far, we have studied linear equations in two variables. Recall that such an equation can be written in the form $Ax + By = C$, where A and B are not both 0. Also recall that the graph of a linear equation in two variables is a line. In this section, we begin by looking at a particular family of linear equations—those that can be written in the form

$$y = kx$$

where k is a constant. This family of equations is called *direct variation*.

Objectives

A Solve Problems Involving Direct Variation.

B Solve Problems Involving Inverse Variation.

C Solve Problems Involving Other Types of Direct and Inverse Variation.

D Solve Applications of Variation.

Objective **A** Solving Direct Variation Problems

Let's suppose that you are earning minimum wage, $7.25 per hour, at a part-time job. The amount of money you earn depends on the number of hours you work. This is illustrated by the following table:

Hours Worked	0	1	2	3	4
Money Earned (before deductions)	0	7.25	14.50	21.75	29.00

and so on

In general, to calculate your earnings (before deductions), multiply the constant $7.25 by the number of hours you work. If we let y represent the amount of money earned and x represent the number of hours worked, we get the direct variation equation

$$y = 7.25 \cdot x$$

earnings = $7.25 · hours worked

Notice that in this direct variation equation, as the number of hours increases, the pay increases as well.

> ### Direct Variation
>
> **y varies directly as x,** or **y is directly proportional to x,** if there is a nonzero constant k such that
>
> $$y = kx$$
>
> The number k is called the **constant of variation** or the **constant of proportionality.**

In our direct variation example, $y = 7.25x$, the constant of variation is 7.25.

Let's use the previous table to graph $y = 7.25x$. We begin our graph at the ordered pair solution $(0, 0)$. Why? We assume that the least amount of hours worked is 0. If 0 hours are worked, then the pay is $0.

As illustrated in this graph, a direct variation equation $y = kx$ is linear. Also notice that $y = 7.25x$ is a function since its graph passes the vertical line test.

Practice 1

Write a direct variation equation that satisfies:

x	4	$\frac{1}{2}$	1.5	6
y	8	1	3	12

Example 1 Write a direct variation equation of the form $y = kx$ that satisfies the ordered pairs in the table below.

x	2	9	1.5	-1
y	6	27	4.5	-3

Solution: We are given that there is a direct variation relationship between x and y. This means that

$$y = kx$$

By studying the given values, you may be able to mentally calculate k. If not, to find k, we simply substitute one given ordered pair into this equation and solve for k. We'll use the given pair $(2, 6)$.

$$y = kx$$
$$6 = k \cdot 2$$
$$\frac{6}{2} = \frac{k \cdot 2}{2}$$
$$3 = k \qquad \text{Solve for } k.$$

Since $k = 3$, we have the equation $y = 3x$.
To check, see that each given y is 3 times the given x.

■ Work Practice 1

Answer
1. $y = 2x$

Let's try another type of direct variation example.

Example 2 Suppose that y varies directly as x. If y is 17 when x is 34, find the constant of variation and the direct variation equation. Then find y when x is 12.

Solution: Let's use the same method as in Example 1 to find k. Since we are told that y varies directly as x, we know the relationship is of the form

$$y = kx$$

Let $y = 17$ and $x = 34$ and solve for k.

$$17 = k \cdot 34$$

$$\frac{17}{34} = \frac{k \cdot 34}{34}$$

$$\frac{1}{2} = k \qquad \text{Solve for } k.$$

Thus, the constant of variation is $\frac{1}{2}$ and the equation is $y = \frac{1}{2}x$.

To find y when $x = 12$, use $y = \frac{1}{2}x$ and replace x with 12.

$$y = \frac{1}{2}x$$

$$y = \frac{1}{2} \cdot 12 \qquad \text{Replace } x \text{ with 12.}$$

$$y = 6$$

Thus, when x is 12, y is 6.

■ Work Practice 2

Practice 2
Suppose that y varies directly as x. If y is 15 when x is 45, find the constant of variation and the direct variation equation. Then find y when x is 3.

Let's review a few facts about linear equations of the form $y = kx$.

Direct Variation: $y = kx$

- There is a direct variation relationship between x and y.
- The graph is a line.
- The line will always go through the origin $(0, 0)$. Why?
    Let $x = 0$. Then $y = k \cdot 0$ or $y = 0$.
- The slope of the graph of $y = kx$ is k, the constant of variation. Why? Remember that the slope of an equation of the form $y = mx + b$ is m, the coefficient of x.
- The equation $y = kx$ describes a function. Each x has a unique y and its graph passes the vertical line test.

Practice 3
Find the constant of variation and the direct variation equation for the line below.

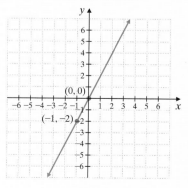

Example 3 The line is the graph of a direct variation equation. Find the constant of variation and the direct variation equation.

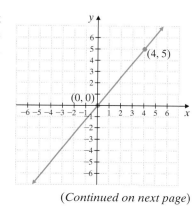

(*Continued on next page*)

Answers
2. $k = \frac{1}{3}; y = \frac{1}{3}x; y = 1$
3. $k = 2; y = 2x$

Solution: Recall that k, the constant of variation, is the same as the slope of the line. Thus, to find k, we use the slope formula and find slope.

Using the given points $(0, 0)$ and $(4, 5)$, we have

$$\text{slope} = \frac{5 - 0}{4 - 0} = \frac{5}{4}$$

Thus, $k = \frac{5}{4}$ and the variation equation is $y = \frac{5}{4}x$.

■ Work Practice 3

Objective B Solving Inverse Variation Problems

In this section, we introduce another type of variation called inverse variation.

Let's suppose you need to drive a distance of 40 miles. You know that the faster you drive the distance, the sooner you arrive at your destination. Recall that there a mathematical relationship (formula) between distance, rate, and time. It is $d = r \cdot$

In our example, distance is a constant 40 miles, so we have $40 = r \cdot t$ or $t = \frac{40}{r}$.

For example, if you drive 10 mph, the time to drive the 40 miles is

$$t = \frac{40}{r} = \frac{40}{10} = 4 \text{ hours}$$

If you drive 20 mph, the time is

$$t = \frac{40}{r} = \frac{40}{20} = 2 \text{ hours}$$

Again, notice that as speed increases, time decreases. Below are some ordered pa solutions of $t = \frac{40}{r}$ and its graph.

Rate (mph)	r	5	10	20	40	60	80
Time (hr)	t	8	4	2	1	$\frac{2}{3}$	$\frac{1}{2}$

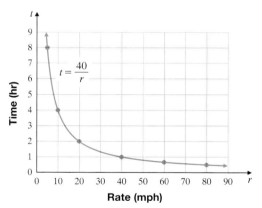

Notice that the graph of this variation is not a line, but it passes the vertical lin test so $t = \frac{40}{r}$ does describe a function. This is an example of inverse variation.

Inverse Variation

y varies inversely as x, or **y is inversely proportional to x,** if there is a nonzero constant k such that

$$y = \frac{k}{x}$$

The number k is called the **constant of variation** or the **constant of proportionality**

In our inverse variation example, $t = \dfrac{40}{r}$ or $y = \dfrac{40}{x}$, the constant of variation is 40.

We can immediately see differences and similarities in direct variation and inverse variation.

Direct Variation	$y = kx$	linear equation	both functions
Inverse Variation	$y = \dfrac{k}{x}$	rational equation	

Remember from Chapter 12 that $y = \dfrac{k}{x}$ is a rational equation and not a linear equation. Also notice that because x is in the denominator, x can be any value except 0. We can still derive an inverse variation equation from a table of values.

Example 4 Write an inverse variation equation of the form $y = \dfrac{k}{x}$ that satisfies the ordered pairs in the table below.

x	2	4	$\frac{1}{2}$
y	6	3	24

Solution: Since there is an inverse variation relationship between x and y, we know that $y = \dfrac{k}{x}$.

To find k, choose one given ordered pair and substitute the values into the equation. We'll use (2, 6).

$y = \dfrac{k}{x}$

$6 = \dfrac{k}{2}$

$2 \cdot 6 = 2 \cdot \dfrac{k}{2}$ Multiply both sides by 2.

$12 = k$ Solve.

Since $k = 12$, we have the equation $y = \dfrac{12}{x}$.

■ Work Practice 4

Practice 4

Write an inverse variation equation of the form $y = \dfrac{k}{x}$ that satisfies:

x	4	10	40	-2
y	5	2	$\frac{1}{2}$	-10

Helpful Hint

Multiply both sides of the inverse variation relationship equation $y = \dfrac{k}{x}$ by x (as long as x is not 0), and we have $xy = k$. This means that if y varies inversely as x, their product is always the constant of variation k. For an example of this, check the table from Example 4:

x	2	4	$\frac{1}{2}$
y	6	3	24

$2 \cdot 6 = 12$ $4 \cdot 3 = 12$ $\frac{1}{2} \cdot 24 = 12$

Answer

4. $y = \dfrac{20}{x}$

Practice 5

Suppose that y varies inversely as x. If y is 4 when x is 0.8, find the constant of variation and the inverse variation equation. Then find y when x is 20.

Example 5 Suppose that y varies inversely as x. If $y = 0.02$ when $x = 75$, find the constant of variation and the inverse variation equation. Then find y when x is 30.

Solution: Since y varies inversely as x, the constant of variation may be found by simply finding the product of the given x and y.

$$k = xy = 75(0.02) = 1.5$$

To check, we will use the inverse variation equation

$$y = \frac{k}{x}$$

Let $y = 0.02$ and $x = 75$ and solve for k.

$$0.02 = \frac{k}{75}$$

$$75(0.02) = 75 \cdot \frac{k}{75} \quad \text{Multiply both sides by 75.}$$

$$1.5 = k \quad \text{Solve for } k.$$

Thus, the constant of variation is 1.5 and the equation is $y = \frac{1.5}{x}$. To find y when $x = 30$, use $y = \frac{1.5}{x}$ and replace x with 30.

$$y = \frac{1.5}{x}$$

$$y = \frac{1.5}{30} \quad \text{Replace } x \text{ with 30.}$$

$$y = 0.05$$

Thus, when x is 30, y is 0.05.

■ Work Practice 5

Objective C Solving Other Types of Direct and Inverse Variation Problems

It is possible for y to vary directly or inversely as powers of x.

> **Direct and Inverse Variation as nth Powers of x**
>
> **y varies directly as a power of x** if there is a nonzero constant k and a natural number n such that
>
> $$y = kx^n$$
>
> **y varies inversely as a power of x** if there is a nonzero constant k and a natural number n such that
>
> $$y = \frac{k}{x^n}$$

Practice 6

The area of a circle varies directly as the square of its radius. A circle with radius 7 inches has an area of 49π square inches. Find the area of a circle whose radius is 4 feet.

Example 6 The surface area of a cube A varies directly as the square of length of its sides. If A is 54 when s is 3, find A when $s = 4.2$.

s

Answers

5. $k = 3.2; y = \frac{3.2}{x}; y = 0.16$

6. 16π sq ft

Solution: Since the surface area A varies directly as the square of side s, we have

$A = ks^2$

To find k, let $A = 54$ and $s = 3$.

$A = k \cdot s^2$
$54 = k \cdot 3^2$ Let $A = 54$ and $s = 3$.
$54 = 9k$ $3^2 = 9$.
$6 = k$ Divide by 9.

The formula for surface area of a cube is then
$A = 6s^2$, where s is the length of a side.

To find the surface area when $s = 4.2$, substitute.

$A = 6s^2$
$A = 6 \cdot (4.2)^2$
$A = 105.84$

The surface area of a cube whose side measures 4.2 units is 105.84 sq units.

■ Work Practice 6

Objective D Solving Applications of Variation

There are many real-life applications of direct and inverse variation.

Example 7 The weight of a body w varies inversely with the square of its distance from the center of Earth, d. If a person weighs 160 pounds on the surface of Earth, what is the person's weight 200 miles above the surface? (Assume that the radius of Earth is 4000 miles.)

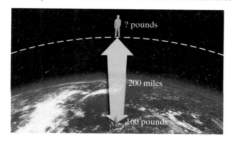

Practice 7

The distance d that an object falls is directly proportional to the square of the time of the fall, t. If an object falls 144 feet in 3 seconds, find how far the object falls in 5 seconds.

Solution:
1. UNDERSTAND. Make sure you read and reread the problem.
2. TRANSLATE. Since we are told that weight w varies inversely with the square of its distance from the center of Earth, d, we have

$w = \dfrac{k}{d^2}$

3. SOLVE. To solve the problem, we first find k. To do so, we use the fact that the person weighs 160 pounds on Earth's surface, which is a distance of 4000 miles from the Earth's center.

$w = \dfrac{k}{d^2}$

$160 = \dfrac{k}{(4000)^2}$

$2{,}560{,}000{,}000 = k$

Thus, we have $w = \dfrac{2{,}560{,}000{,}000}{d^2}$.

(Continued on next page)

Answer
7. 400 feet

Since we want to know the person's weight 200 miles above the Earth's surface, we let $d = 4200$ and find w.

$$w = \frac{2,560,000,000}{d^2}$$

$$w = \frac{2,560,000,000}{(4200)^2} \quad \text{A person 200 miles above the Earth's surface is 4200 miles from the Earth's center.}$$

$$w \approx 145 \quad \text{Simplify.}$$

4. **INTERPRET.** *Check:* Your answer is reasonable since the farther a person is from Earth, the less the person weighs. *State:* Thus, 200 miles above the surface of the Earth, a 160-pound person weighs approximately 145 pounds.

■ Work Practice 7

Vocabulary, Readiness & Video Check

State whether each equation represents direct or inverse variation.

1. $y = \dfrac{k}{x}$, where k is a constant. _____

2. $y = kx$, where k is a constant. _____

3. $y = 5x$ _____

4. $y = \dfrac{5}{x}$ _____

5. $y = \dfrac{7}{x^2}$ _____

6. $y = 6.5x^4$ _____

7. $y = \dfrac{11}{x}$ _____

8. $y = 18x$ _____

9. $y = 12x^2$ _____

10. $y = \dfrac{20}{x^3}$ _____

Martin-Gay Interactive Videos

See Video 13.8

Watch the section lecture video and answer the following questions.

Objective A 11. Based on the lecture before ▭ Example 1, what kind of equation is a direct variation equation? What does k represent in this equation? ▶

Objective B 12. In ▭ Example 4, why is it not necessary to place the given values of x and y into the inverse variation equation in order to find k? ▶

Objective C 13. From ▭ Examples 5–7, does a power on x change the basic direct and inverse variation formula relationships? ▶

Objective D 14. In ▭ Example 8, why is it reasonable to expect our answer to be a greater distance than the original distance given in the problem? ▶

Section 13.8 Exercise Set MyLab Math

Objective A *Write a direct variation equation, $y = kx$, that satisfies the ordered pairs in each table. See Example 1.*

1.
x	0	6	10
y	0	3	5

2.
x	0	2	−1	3
y	0	14	−7	21

3.
x	−2	2	4	5
y	−12	12	24	30

4.
x	3	9	−2	12
y	1	3	$-\frac{2}{3}$	4

Write a direct variation equation, $y = kx$, that describes each graph. See Example 3.

5. 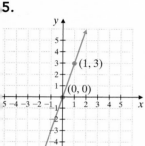   6.   7.   8.

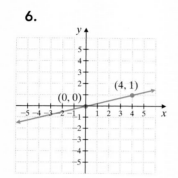
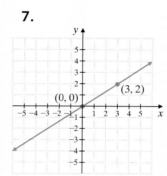
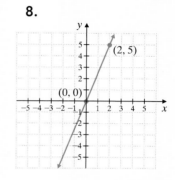

Objective B *Write an inverse variation equation, $y = \dfrac{k}{x}$, that satisfies the ordered pairs in each table. See Example 4.*

9.
x	1	−7	3.5	−2
y	7	−1	2	−3.5

10.
x	2	−11	4	−4
y	11	−2	5.5	−5.5

11.
x	10	$\frac{1}{2}$	$-\frac{1}{4}$
y	0.05	1	−2

12.
x	4	$\frac{1}{5}$	−8
y	0.1	2	−0.05

Objectives A B C Translating *Write an equation to describe each variation. Use k for the constant of proportionality. See Examples 1 through 6.*

13. y varies directly as x.   **14.** a varies directly as b.   **15.** h varies inversely as t.   **16.** s varies inversely as t.

17. z varies directly as x^2.   **18.** p varies inversely as x^2.   **19.** y varies inversely as z^3.   **20.** x varies directly as y^4.

21. x varies inversely as $\sqrt{y}$.   **22.** y varies directly as d^2.

Objectives A B C *Solve. See Examples 2, 5, and 6.*

23. y varies directly as x. If $y = 20$ when $x = 5$, find y when x is 10.

24. y varies directly as x. If $y = 27$ when $x = 3$, find y when x is 2.

25. y varies inversely as x. If $y = 5$ when $x = 60$, find y when x is 100.

26. y varies inversely as x. If $y = 200$ when $x = 5$, find y when x is 4.

27. z varies directly as x^2. If $z = 96$ when $x = 4$, find z when $x = 3$.

28. s varies directly as t^3. If $s = 270$ when $t = 3$, find s when $t = 1$.

29. a varies inversely as b^3. If $a = \frac{3}{2}$ when $b = 2$, find a when b is 3.

30. p varies inversely as q^2. If $p = \frac{5}{16}$ when $q = 8$, find p when $q = \frac{1}{2}$.

Objectives **A B C D** Solve. See Examples 1 through 7.

31. Your paycheck (before deductions) varies directly as the number of hours you work. If your paycheck is $166.50 for 18 hours, find your pay for 10 hours.

32. If your paycheck (before deductions) is $304.50 for 30 hours, find your pay for 34 hours. (See Exercise **31**.)

33. The cost of manufacturing a certain type of headphone varies inversely as the number of headphones increases. If 5000 headphones can be manufactured for $9.00 each, find the cost per headphone to manufacture 7500 headphones.

34. The cost of manufacturing a certain composition notebook varies inversely as the number of notebooks increases. If 10,000 notebooks can be manufactured for $0.50 each, find the cost per notebook to manufacture 18,000 notebooks. Round your answer to the nearest cent.

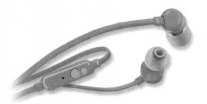

35. The distance a spring stretches varies directly with the weight attached to the spring. If a 60-pound weight stretches the spring 4 inches, find the distance that an 80-pound weight stretches the spring.

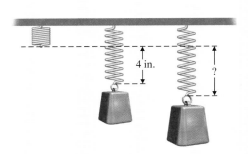

36. If a 30-pound weight stretches a spring 10 inches, find the distance a 20-pound weight stretches the spring. (See Exercise **35**.)

37. The weight of an object varies inversely as the square of its distance from the center of Earth. If a person weighs 180 pounds on Earth's surface, what is his weight 10 miles above the surface of Earth? (Assume that Earth's radius is 4000 miles and round your answer to one decimal place.)

38. For a constant distance, the rate of travel varies inversely as the time traveled. If a family travels 55 mph and arrives at a destination in 4 hours, how long will the return trip take traveling at 60 mph?

39. The distance d that an object falls is directly proportional to the square of the time of the fall, t. A person who is parachuting for the first time is told to wait 10 seconds before opening the parachute. If the person falls 64 feet in 2 seconds, find how far he falls in 10 seconds.

40. The distance needed for a car to stop, d, is directly proportional to the square of its rate of travel, r. Under certain driving conditions, a car traveling 60 mph needs 300 feet to stop. With these same driving conditions, how long does it take a car to stop if the car is traveling 30 mph when the brakes are applied?

Review

Add the equations by adding the left sides of the equations, bringing down an equal sign, and then adding the right sides of the equations. See Section 10.4.

41. $-3x + 4y = 7$
    $3x - 2y = 9$

42. $x - y = -9$
    $-x - y = -14$

43. $5x - 0.4y = 0.7$
    $-9x + 0.4y = -0.2$

44. $1.9x - 2y = 5.7$
    $-1.9x - 0.1y = 2.3$

Concept Extensions

45. Suppose that y varies directly as x. If x is tripled, what is the effect on y?

46. Suppose that y varies directly as x^2. If x is tripled, what is the effect on y?

47. The period of a pendulum p (the time of one complete back-and-forth swing) varies directly with the square root of its length, ℓ. If the length of the pendulum is quadrupled, what is the effect on the period, p?

48. For a constant distance, the rate of travel r varies inversely with the time traveled, t. If a car traveling 100 mph completes a test track in 6 minutes, find the rate needed to complete the same test track in 4 minutes. (*Hint:* Convert minutes to hours.)

Chapter 13 Group Activity

Finding a Linear Model

This activity may be completed by working in groups or individually.

The following table shows the actual number of international tourist arrivals to the United States for the years 2012 through 2015.

Year	International Tourist Arrivals to the United States (in millions)
2012	67
2013	70
2014	75
2015	79

Source: World Tourism Organization

1. Make a scatter diagram of the paired data in the table.

2. Use what you have learned in this chapter to write an equation of the line representing the paired data in the table. Explain how you found the equation, and what each variable represents.

3. What is the slope of your line? What does the slope mean in this context?

4. Use your linear equation to predict the number of international tourist arrivals to the United States in 2022.

5. Compare your linear equation to that found by other students or groups. Is it the same, similar, or different? How?

6. Compare your prediction from question **4** to that of other students or groups. Describe what you find.

7. Suppose that the number of international tourist arrivals to the United States for 2020 was estimated to be 107 million. If this data point is added to the chart, how does it affect your results?

Chapter 13 Vocabulary Check

Fill in each blank with one of the words listed below.

y-axis	x-axis	solution	linear	standard	point-slope
x-intercept	y-intercept	y	x	slope	relation
domain	range	direct	inverse	slope-intercept	function

1. An ordered pair is a(n) _____ of an equation in two variables if replacing the variables by the coordinates of the ordered pair results in a true statement.
2. The vertical number line in the rectangular coordinate system is called the _____.
3. A(n) _____ equation can be written in the form $Ax + By = C$.
4. A(n) _____ is a point of the graph where the graph crosses the x-axis.
5. The form $Ax + By = C$ is called _____ form.
6. A(n) _____ is a point of the graph where the graph crosses the y-axis.
7. A set of ordered pairs that assigns to each x-value exactly one y-value is called a(n) _____.
8. The equation $y = 7x - 5$ is written in _____ form.
9. The set of all x-coordinates of a relation is called the _____ of the relation.
10. The set of all y-coordinates of a relation is called the _____ of the relation.
11. A set of ordered pairs is called a(n) _____.
12. The equation $y + 1 = 7(x - 2)$ is written in _____ form.
13. To find an x-intercept of a graph, let _____ = 0.
14. The horizontal number line in the rectangular coordinate system is called the _____.
15. To find a y-intercept of a graph, let _____ = 0.
16. The _____ of a line measures the steepness or tilt of the line.
17. The equation $y = kx$ is an example of _____ variation.
18. The equation $y = \dfrac{k}{x}$ is an example of _____ variation.

> **Helpful Hint**
>
> ▶ Are you preparing for your test? To help, don't forget to take these:
> - Chapter 13 Getting Ready for the Test on page 1075
> - Chapter 13 Test on page 1076
>
> Then check all of your answers at the back of this text. For further review, the step-by-step video solutions to any of these exercises are located in MyLab Math.

13 Chapter Highlights

Definitions and Concepts	Examples

Section 13.1 The Rectangular Coordinate System

The **rectangular coordinate system** consists of a plane and a vertical and a horizontal number line intersecting at their 0 coordinates. The vertical number line is called the **y-axis** and the horizontal number line is called the **x-axis.** The point of intersection of the axes is called the **origin.**

To **plot** or **graph** an ordered pair means to find its corresponding point on a rectangular coordinate system.

To plot or graph an ordered pair such as $(3, -2)$, start at the origin. Move 3 units to the right and from there, 2 units down.

To plot or graph $(-3, 4)$, start at the origin. Move 3 units to the left and from there, 4 units up.

An ordered pair is a **solution** of an equation in two variables if replacing the variables with the coordinates of the ordered pair results in a true statement.

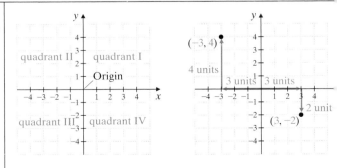

Chapter 13 Highlights

Definitions and Concepts	Examples

Section 13.1 The Rectangular Coordinate System (continued)

If one coordinate of an ordered pair solution of an equation is known, the other value can be determined by substitution.	Complete the ordered pair (0,) for the equation $x - 6y = 12$. $x - 6y = 12$ $0 - 6y = 12$ Let $x = 0$. $\dfrac{-6y}{-6} = \dfrac{12}{-6}$ Divide by -6. $y = -2$ The ordered pair solution is $(0, -2)$.

Section 13.2 Graphing Linear Equations

A **linear equation in two variables** is an equation that can be written in the form $Ax + By = C$, where A and B are not both 0. The form $Ax + By = C$ is called **standard form**. To graph a linear equation in two variables, find three ordered pair solutions. Plot the solution points and draw the line connecting the points.	$3x + 2y = -6 \qquad x = -5$ $\qquad\quad y = 3 \qquad y = -x + 10$ $x + y = 10$ is in standard form. Graph: $x - 2y = 5$ 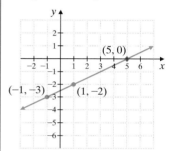 | x | y | |---|---| | 5 | 0 | | 1 | -2 | | -1 | -3 |

Section 13.3 Intercepts

An **intercept** of a graph is a point where the graph intersects an axis. If a graph intersects the x-axis at a, then $(a, 0)$ is an **x-intercept**. If a graph intersects the y-axis at b, then $(0, b)$ is a **y-intercept**.	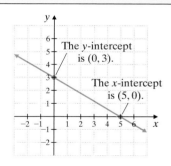 The y-intercept is $(0, 3)$. The x-intercept is $(5, 0)$.
To find the **x-intercept(s)**, let $y = 0$ and solve for x. To find the **y-intercept(s)**, let $x = 0$ and solve for y.	Find the intercepts for $2x - 5y = -10$. If $y = 0$, then $\qquad$ If $x = 0$, then $2x - 5 \cdot 0 = -10 \qquad 2 \cdot 0 - 5y = -10$ $2x = -10 \qquad\qquad -5y = -10$ $\dfrac{2x}{2} = \dfrac{-10}{2} \qquad\qquad \dfrac{-5y}{-5} = \dfrac{-10}{-5}$ $x = -5 \qquad\qquad\quad y = 2$

(continued)

Definitions and Concepts	Examples
Section 13.3 Intercepts (continued)	

The x-intercept is $(-5, 0)$. The y-intercept is $(0, 2)$.

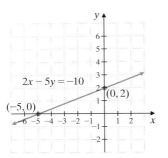

The graph of $x = c$ is a vertical line with x-intercept $(c, 0)$.

The graph of $y = c$ is a horizontal line with y-intercept $(0, c)$.

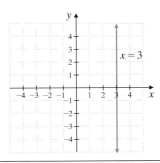 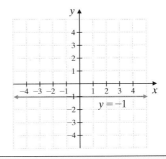

Section 13.4 Slope and Rate of Change

The **slope m** of the line through points (x_1, y_1) and (x_2, y_2) is given by

$$m = \frac{y_2 - y_1}{x_2 - x_1} \quad \text{as long as } x_2 \neq x_1$$

A horizontal line has slope 0.
The slope of a vertical line is undefined.
Nonvertical parallel lines have the same slope.
Two nonvertical lines are perpendicular if the slope of one is the negative reciprocal of the slope of the other.

The slope of the line through points $(-1, 6)$ and $(-5, 8)$ is

$$m = \frac{y_2 - y_1}{x_2 - x_1} = \frac{8 - 6}{-5 - (-1)} = \frac{2}{-4} = -\frac{1}{2}$$

The slope of the line $y = -5$ is 0.
The line $x = 3$ has undefined slope.

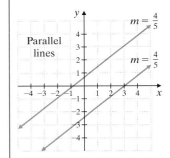 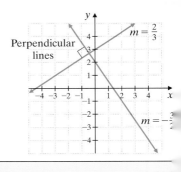

Section 13.5 Equations of Lines

Slope-Intercept Form

$$y = mx + b$$

m is the slope of the line.
$(0, b)$ is the y-intercept.

Find the slope and the y-intercept of the line $2x + 3y = 6$.
Solve for y:

$$2x + 3y = 6$$
$$3y = -2x + 6 \quad \text{Subtract } 2x.$$
$$y = -\frac{2}{3}x + 2 \quad \text{Divide by 3.}$$

The slope of the line is $-\frac{2}{3}$ and the y-intercept is $(0, 2)$.

Definitions and Concepts	Examples
Section 13.5 Equations of Lines *(continued)*	

Point-Slope Form

$$y - y_1 = m(x - x_1)$$

m is the slope.
(x_1, y_1) is a point of the line.

Find an equation of the line with slope $\frac{3}{4}$ that contains the point $(-1, 5)$.

$$y - 5 = \frac{3}{4}[x - (-1)]$$

$4(y - 5) = 3(x + 1)$ Multiply by 4.
$4y - 20 = 3x + 3$ Distribute.
$-3x + 4y = 23$ Subtract $3x$ and add 20.

Section 13.6 Introduction to Functions	

A set of ordered pairs is a **relation**. The set of all x-coordinates is called the **domain** of the relation and the set of all y-coordinates is called the **range** of the relation.

A **function** is a set of ordered pairs that assigns to each x-value exactly one y-value.

Vertical Line Test

If a vertical line can be drawn so that it intersects a graph more than once, the graph is not the graph of a function. (If no such line can be drawn, the graph is that of a function.)

The domain of the relation

$$\{(0, 5), (2, 5), (4, 5), (5, -2)\}$$

is $\{0, 2, 4, 5\}$. The range is $\{-2, 5\}$.

Which are graphs of functions?

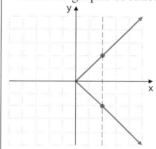 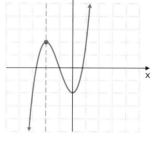

This graph is not the graph of a function.   This graph is the graph of a function.

The symbol $f(x)$ means **function of x**. This notation is called **function notation**.

If $f(x) = 3x - 7$, then
$$f(-1) = 3(-1) - 7$$
$$= -3 - 7$$
$$= -10$$

Section 13.7 Graphing Linear Inequalities in Two Variables	

A **linear inequality in two variables** is an inequality that can be written in one of these forms:

$Ax + By < C$ $Ax + By \leq C$
$Ax + By > C$ $Ax + By \geq C$

where A and B are not both 0.

$2x - 5y < 6$ $x \geq -5$
$y > -8x$ $y \leq 2$

(continued)

Definitions and Concepts	Examples
Section 13.7 Graphing Linear Inequalities in Two Variables *(continued)*	
To Graph a Linear Inequality 1. Graph the boundary line by graphing the related equation. Draw the line solid if the inequality symbol is $\leq$ or $\geq$. Draw the line dashed if the inequality symbol is $<$ or $>$. 2. Choose a test point not on the line. Substitute its coordinates into the original inequality. 3. If the resulting inequality is true, shade the half-plane that contains the test point. If the inequality is not true, shade the half-plane that does not contain the test point.	Graph: $2x - y \leq 4$ 1. Graph $2x - y = 4$. Draw a solid line because the inequality symbol is $\leq$. 2. Check the test point $(0, 0)$ in the original inequality, $2x - y \leq 4$. $2 \cdot 0 - 0 \leq 4$ Let $x = 0$ and $y = 0$. $0 \leq 4$ True 3. The inequality is true, so shade the half-plane containing $(0, 0)$, as shown. 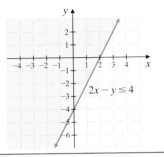
Section 13.8 Direct and Inverse Variation	
y **varies directly as** x, or y **is directly proportional to** x, if there is a nonzero constant k such that $y = kx$ y **varies inversely as** x, or y **is inversely proportional to** x, if there is a nonzero constant k such that $y = \dfrac{k}{x}$	The circumference of a circle C varies directly as its radius r. $C = \underbrace{2\pi}_{k} r$ Pressure P varies inversely with volume V. $P = \dfrac{k}{V}$

Chapter 13   Review

(13.1) *Plot each point on the same rectangular coordinate system.*

1. $(-7, 0)$

2. $\left(0, 4\dfrac{4}{5}\right)$

3. $(-2, -5)$

4. $(1, -3)$

5. $(0.7, 0.7)$

6. $(-6, 4)$

Complete each ordered pair so that it is a solution of the given equation.

7. $-2 + y = 6x; (7,)$

8. $y = 3x + 5; (, -8)$

Complete the table of values for each given equation.

9. $9 = -3x + 4y$

x	y
	0
	3
9	

10. $y = 5$

x	y
7	
-7	
0	

11. $x = 2y$

x	y
	0
	5
	-5

12. The cost in dollars of producing x compact disc holders is given by $y = 5x + 2000$.

  a. Complete the table.

x	1	100	1000
y			

  b. Find the number of compact disc holders that can be produced for $6430.

3.2) Graph each linear equation.

13. $x - y = 1$

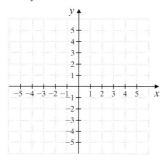

14. $x + y = 6$

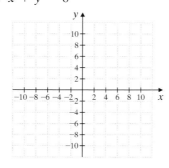

15. $x - 3y = 12$

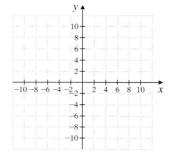

16. $5x - y = -8$

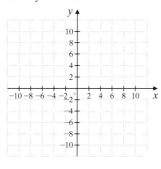

17. $x = 3y$

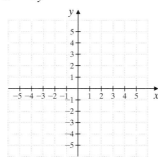

18. $y = -2x$

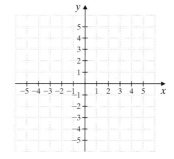

3.3) Identify the intercepts in each graph.

19.

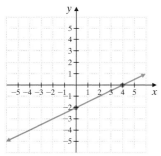

20.

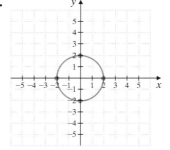

Graph each linear equation.

21. $y = -3$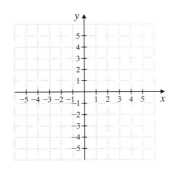

22. $x = 5$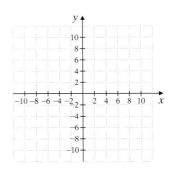

Find the intercepts of each equation.

23. $x - 3y = 12$

24. $-4x + y = 8$

(13.4) *Find the slope of each line.*

25.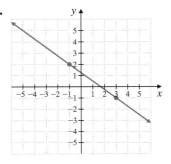

26.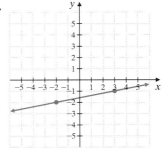

Match each line with its slope.

a.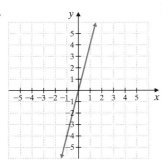
b.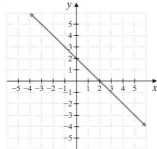
c.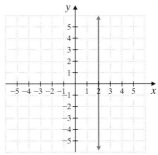
d.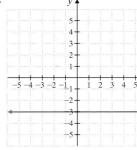

27. $m = 0$

28. $m = -1$

29. undefined slope

30. $m = 4$

Find the slope of the line that passes through each pair of points.

31. $(2, 5)$ and $(6, 8)$

32. $(4, 7)$ and $(1, 2)$

33. $(1, 3)$ and $(-2, -9)$

34. $(-4, 1)$ and $(3, -6)$

Find the slope of each line.

35. $y = 3x + 7$

36. $x - 2y = 4$

37. $y = -2$

38. $x = 0$

Determine whether each pair of lines is parallel, perpendicular, or neither.

39. $x - y = -6$
$x + y = 3$

40. $3x + y = 7$
$-3x - y = 10$

41. $y = 4x + \dfrac{1}{2}$
$4x + 2y = 1$

42. $y = 6x - \dfrac{1}{3}$
$x + 6y = 6$

Find the slope of each line and write the slope as a rate of change. Don't forget to attach the proper units.

43. The graph below approximates the total number of U.S. magazines in print for each year x.

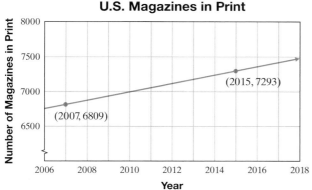

Source: MPA—The Association of Magazine Media

44. The graph below approximates the number of lung transplants y in the United States for year x. Round to the nearest whole.

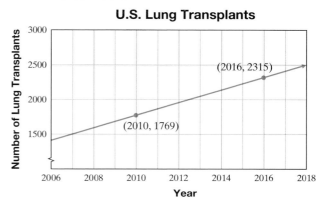

Source: Organ Procurement and Transplantation Network

(13.5) Determine the slope and the y-intercept of the graph of each equation.

45. $x - 6y = -1$

46. $3x + y = 7$

Write an equation of each line.

47. slope -5; y-intercept $\left(0, \dfrac{1}{2}\right)$

48. slope $\dfrac{2}{3}$; y-intercept $(0, 6)$

Match each equation with its graph.

49. $y = 2x + 1$

50. $y = -4x$

51. $y = 2x$

52. $y = 2x - 1$

a.

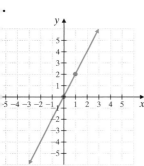

b.

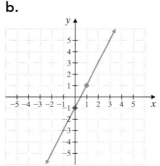

c.

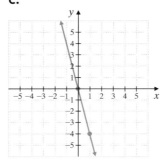

d.

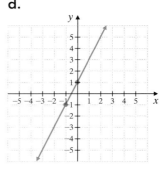

Write an equation of the line with the given slope that passes through the given point. Write the equation in the form $Ax + By = C$.

53. $m = 4; (2, 0)$
54. $m = -3; (0, -5)$
55. $m = \dfrac{3}{5}; (1, 4)$
56. $m = -\dfrac{1}{3}; (-3, 3)$

Write an equation of the line passing through each pair of points. Write the equation in the form $y = mx + b$.

57. $(1, 7)$ and $(2, -7)$
58. $(-2, 5)$ and $(-4, 6)$

(13.6) *Determine whether each relation or graph is a function.*

59. $\{(7, 1), (7, 5), (2, 6)\}$
60. $\{(0, -1), (5, -1), (2, 2)\}$

61.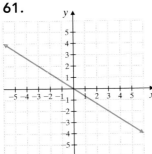
62.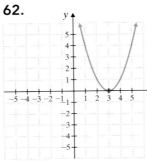
63.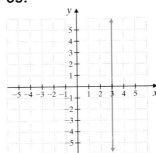
64.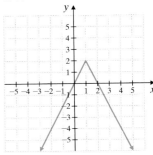

Find each indicated function value for the function $f(x) = -2x + 6$.

65. $f(0)$
66. $f(-2)$
67. $f\left(\dfrac{1}{2}\right)$
68. $f\left(-\dfrac{1}{2}\right)$

(13.7) *Graph each inequality.*

69. $x + 6y < 6$
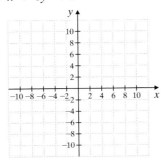

70. $x + y > -2$
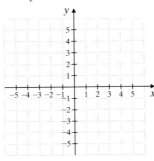

71. $y \geq -7$
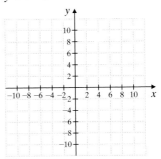

72. $y \leq -4$
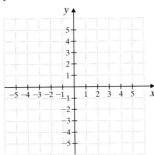

73. $-x \leq y$
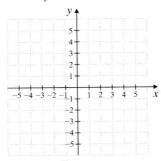

74. $x \geq -y$
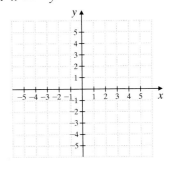

3.8) Solve.

75. y varies directly as x. If $y = 40$ when $x = 4$, find y when x is 11.

76. y varies inversely as x. If $y = 4$ when $x = 6$, find y when x is 48.

77. y varies inversely as x^3. If $y = 12.5$ when $x = 2$, find y when x is 3.

78. y varies directly as x^2. If $y = 175$ when $x = 5$, find y when $x = 10$.

79. The cost of manufacturing a certain medicine varies inversely as the amount of medicine manufactured increases. If 3000 milliliters can be manufactured for $6600, find the cost to manufacture 5000 milliliters.

80. The distance a spring stretches varies directly with the weight attached to the spring. If a 150-pound weight stretches the spring 8 inches, find the distance that a 90-pound weight stretches the spring.

Mixed Review

Complete the table of values for each given equation.

81. $2x - 5y = 9$

x	y
	1
2	
	-3

82. $x = -3y$

x	y
0	
	1
6	

Find the intercepts for each equation.

83. $2x - 3y = 6$

84. $-5x + y = 10$

Graph each linear equation.

85. $x - 5y = 10$

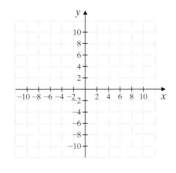

86. $x + y = 4$

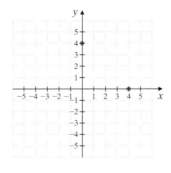

87. $y = -4x$

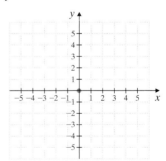

88. $2x + 3y = -6$

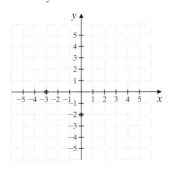

89. $x = 3$

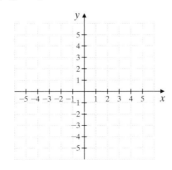

90. $y = -2$

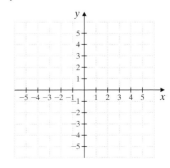

Find the slope of the line that passes through each pair of points.

91. $(3, -5)$ and $(-4, 2)$

92. $(1, 3)$ and $(-6, -8)$

Find the slope of each line.

93.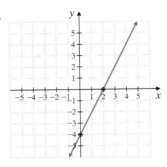

94.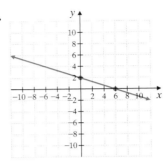

Determine the slope and y-intercept of the graph of each equation.

95. $-2x + 3y = -15$

96. $6x + y - 2 = 0$

Write an equation of the line with the given slope that passes through the given point. Write the equation in the form $Ax + By = C$.

97. $m = -5; (3, -7)$

98. $m = 3; (0, 6)$

Write an equation of the line passing through each pair of points. Write the equation in the form $Ax + By = C$.

99. $(-3, 9)$ and $(-2, 5)$

100. $(3, 1)$ and $(5, -9)$

Chapter 13 Getting Ready for the Test

MULTIPLE CHOICE Exercises 1 through 4 are **Multiple Choice**. Choose the correct letter.

For Exercises 1 and 2, choose the ordered pair that is NOT a solution of the linear equation.

1. $x - y = 5$
   A. $(7, 2)$
   B. $(0, -5)$
   C. $(-2, 3)$
   D. $(-2, -7)$

2. $y = 4$
   A. $(4, 0)$
   B. $(0, 4)$
   C. $(2, 4)$
   D. $(100, 4)$

3. What is the most and then the fewest number of intercepts a line may have?
   A. most: 2; fewest: 1
   B. most: infinite number; fewest: 1
   C. most: 2; fewest: 0
   D. most: infinite number; fewest: 0

4. Choose the linear equation:
   A. $\sqrt{x} - 3y = 7$
   B. $2x = 6^2$
   C. $4x^3 + 6y^3 = 5^3$
   D. $y = |x|$

MATCHING For Exercises 5 through 8, **Match** each numbered line in the rectangular system with its slope to the right. Each slope may be used only once.

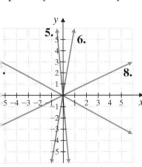

5.   A. $m = 5$

6.   B. $m = -10$

7.   C. $m = \dfrac{1}{2}$

8.   D. $m = -\dfrac{4}{7}$

MULTIPLE CHOICE Exercises 9 through 14 are **Multiple Choice**. Choose the correct letter.

9. An ordered pair solution for the function $f(x)$ is $(0, 5)$. This solution using function notation is:
   A. $f(5) = 0$
   B. $f(5) = f(0)$
   C. $f(0) = 5$
   D. $0 = 5$

10. Given: $(2, 3)$ and $(0, 9)$. Final Answer: $y = -3x + 9$. Select the correct directions:
    A. Find the slope of the line through the two points.
    B. Find an equation of the line through the two points. Write the equation in standard form.
    C. Find an equation of the line through the two points. Write the equation in slope-intercept form.

For Exercises 11 through 14, use the graph to fill in each blank using the choices below.
   A. -2 B. 2 C. 4 D. 0 E. 3

11. $f(0) =$ _____.

12. $f(4) =$ _____.

13. If $f(x) = 0$, then $x =$ _____ or $x =$ _____.

14. $f(1) =$ _____.

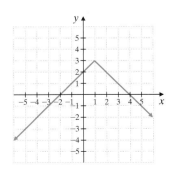

1075

Chapter 13 Test

Complete each ordered pair so that it is a solution of the given equation.

1. $12y - 7x = 5$; $(1, \ )$

2. $y = 17$; $(-4, \ )$

Find the slope of each line.

3.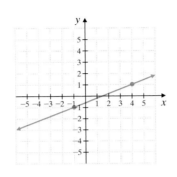

4.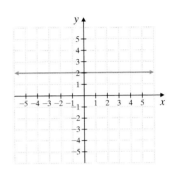

5. Passes through $(6, -5)$ and $(-1, 2)$

6. Passes through $(0, -8)$ and $(-1, -1)$

7. $-3x + y = 5$

8. $x = 6$

Graph.

9. $2x + y = 8$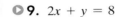

10. $-x + 4y = 5$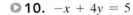

11. $x - y \geq -2$

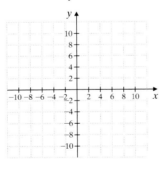
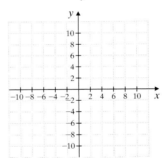
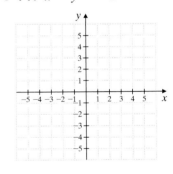

12. $y \geq -4x$

13. $5x - 7y = 10$

14. $2x - 3y > -6$

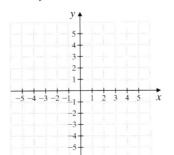
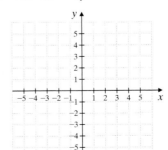
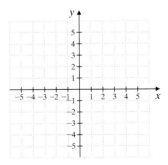

Chapter 13 Test

15. $6x + y > -1$
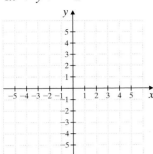

16. $y = -1$
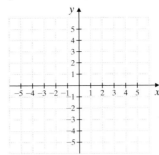

17. Determine whether the graphs of $y = 2x - 6$ and $-4x = 2y$ are parallel lines, perpendicular lines, or neither.

Find the equation of each line. Write the equation in the form $Ax + By = C$.

18. Slope $-\dfrac{1}{4}$, passes through $(2, 2)$

19. Passes through the origin and $(6, -7)$

20. Passes through $(2, -5)$ and $(1, 3)$

21. Slope $\dfrac{1}{8}$; y-intercept $(0, 12)$

Determine whether each relation is a function.

22. $\{(-1, 2), (-2, 4), (-3, 6), (-4, 8)\}$

23. $\{(-3, -3), (0, 5), (-3, 2), (0, 0)\}$

24. The graph shown in Exercise 3

25. The graph shown in Exercise 4

Find the indicated function values for each function.

26. $f(x) = 2x - 4$
  a. $f(-2)$
  b. $f(0.2)$
  c. $f(0)$

27. $f(x) = x^3 - x$
  a. $f(-1)$
  b. $f(0)$
  c. $f(4)$

28. The perimeter of the parallelogram below is 42 meters. Write a linear equation in two variables for the perimeter. Use this equation to find x when y is 8 meters.

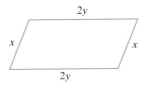

15. _____
16. _____
17. _____
18. _____
19. _____
20. _____
21. _____
22. _____
23. _____
24. _____
25. _____
26. a. _____
    b. _____
    c. _____
27. a. _____
    b. _____
    c. _____
28. _____

29. The table gives the percent of total U.S. music revenue derived from streaming music for the years shown. (*Source:* Recording Industry Association of America)

Year	Percent of Music Revenue from Streaming
2011	9
2012	15
2013	21
2014	27
2015	34

a. Write this data as a set of ordered pairs of the form (year, percent of music revenue from streaming).

b. Create a scatter diagram of the data. Be sure to label the axes properly.

Percent of Music Revenue from Streaming

30. This graph approximates the gross box office sales y (in billions) for Canada and the U.S. for the year x. Find the slope of the line and write the slope as a rate of change. Don't forget to attach the proper units.

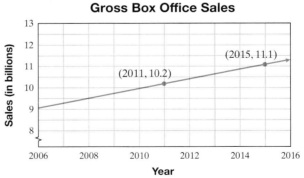

Gross Box Office Sales

Source: National Association of Theater Owners

31. y varies directly as x. If $y = 10$ when $x = 15$, find y when x is 42.

32. y varies inversely as x^2. If $y = 8$ when $x = 5$, find y when x is 15.

Cumulative Review — Chapters 1–13

1. The state of Colorado is in the shape of a rectangle whose length is 380 miles and whose width is 280 miles. Find its area.

2. In a pecan orchard, there are 21 trees in each row and 7 rows of trees. How many pecan trees are there?

3. Add: $1 + (-10) + (-8) + 9$

4. Add: $-2 + (-7) + 3 + (-4)$

5. Write $\dfrac{8}{3x}$ as an equivalent fraction whose denominator is $24x$.

6. Write $\dfrac{3}{2c}$ as an equivalent fraction with denominator $8c$.

7. Subtract: $14 - 8\dfrac{3}{7}$

8. Subtract: $15 - 4\dfrac{2}{5}$

9. Evaluate $-2x + 5$ for $x = 3.8$.

10. Evaluate $6x - 1$ for $x = -2.1$.

11. Write $\dfrac{22}{7}$ as a decimal. Round to the nearest hundredth.

12. Write $\dfrac{37}{19}$ as a decimal. Round to the nearest thousandth.

13. Solve $2x < -4$. Graph the solutions.

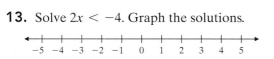

14. Solve $3x \leq -9$. Graph the solutions.

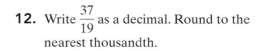

Find the degree of each polynomial and tell whether the polynomial is a monomial, binomial, trinomial, or none of these.

15.
 a. $-2t^2 + 3t + 6$
 b. $15x - 10$
 c. $7x + 3x^3 + 2x^2 - 1$

16.
 a. $-7y + 2$
 b. $8x - x^2 - 1$
 c. $9y^3 - 6y + 2 + y^2$

Perform the indicated operation.

17. Add: $(-2x^2 + 5x - 1)$ and $(-2x^2 + x + 3)$

18. Add: $(9x - 5)$ and $(x^2 - 6x + 5)$

19. Multiply: $(3y + 1)^2$

20. Multiply: $(2x - 5)^2$

21. Factor: $-9a^5 + 18a^2 - 3a$

22. Factor: $2x^5 - x^3$

23. Factor: $x^2 + 4x - 12$

24. Factor: $x^2 + 4x - 21$

25. Factor: $8x^2 - 22x + 5$

26. Factor: $15x^2 + x - 2$

27. Solve: $4x^2 - 28x = -49$

28. Solve: $x^2 - 9x = -14$

29. Divide: $\dfrac{2x^2 - 11x + 5}{5x - 25} \div \dfrac{4x - 2}{10}$

30. Multiply: $\dfrac{3x^2 + 17x - 6}{5x + 5} \cdot \dfrac{2x + 2}{4x + 24}$

Write the rational expression as an equivalent rational expression with the given denominator.

31. $\dfrac{4b}{9a} = \dfrac{}{27a^2 b}$

32. $\dfrac{7x}{11y} = \dfrac{}{99x^2 y^2}$

33. Add: $1 + \dfrac{m}{m + 1}$

34. Subtract: $1 - \dfrac{m}{m + 1}$

35. Solve: $3 - \dfrac{6}{x} = x + 8$

36. Solve: $2 + \dfrac{10}{x} = x + 5$

37. Simplify: $\dfrac{\dfrac{x + 1}{y}}{\dfrac{x}{y} + 2}$

38. Simplify: $\dfrac{\dfrac{x}{2} + 2}{\dfrac{x}{2} - 2}$

Complete each ordered pair solution so that it is a solution to the equation.

39. $3x + y = 12$
 a. $(0,\ )$
 b. $(\ , 6)$
 c. $(-1,\ )$

40. $-x + 4y = -20$
 a. $(0,\ )$
 b. $(\ , 0)$
 c. $(\ , -2)$

41. Graph: $2x + y = 5$

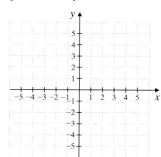

42. Graph: $y = -2x$

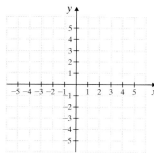

43. Find the slope of the line $-2x + 3y = 11$.

44. Find the slope of the line $7x + 4y = 10$.

45. Find an equation of the line with slope -2 that passes through $(-1, 5)$. Write the equation in standard form, $Ax + By = C$.

46. Find an equation of the line passing through $(2, -7)$ with slope -5. Write the equation in the form $Ax + By = C$.

47. Given $g(x) = x^2 - 3$, find the following and list the corresponding ordered pairs generated.
  a. $g(2)$
  b. $g(-2)$
  c. $g(0)$

48. Given $f(x) = 3x^2 + 2$, find the following and list the corresponding ordered pairs generated.
  a. $f(0)$
  b. $f(4)$
  c. $f(-1)$

14 Systems of Equations

In Chapter 13, we graphed equations containing two variables. As we have seen, equations like these are often needed to represent relationships between two different quantities. There are also many opportunities to compare and contrast two such equations, called a *system of equations*. This chapter presents *linear systems* and ways we solve these systems and apply them to real-life situations.

Sections

14.1 Solving Systems of Linear Equations by Graphing

14.2 Solving Systems of Linear Equations by Substitution

14.3 Solving Systems of Linear Equations by Addition

  Integrated Review—Summary on Solving Systems of Equations

14.4 Systems of Linear Equations and Problem Solving

Check Your Progress

Vocabulary Check
Chapter Highlights
Chapter Review
Getting Ready for the Test
Chapter Test
Cumulative Review

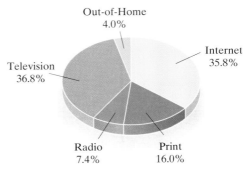

Source: Strategy Analytics Advertising Forecast

Where Are You Exposed to More Advertising: Internet or Television?

Advertising is big business. Notice the largest sectors above correspond to TV and Internet (or digital). This means that most advertising money spent in the United States is spent on TV advertising and Internet (or digital) advertising. In fact, the 2016 total for these two types of advertising is estimated to be $140 billion. As you may imagine, the fastest-growing U.S. consumer segment is Internet users. Since advertisers follow consumers, Internet advertising is the fastest-growing sector above.

Budgets for Internet advertising are increasing at a faster rate than budgets for TV advertising as shown on the double line graph below. In Section 14.3, Exercise 59, we study these two types of advertising further.

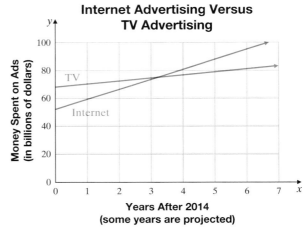

Source: PricewaterhouseCoopers Global

14.1 Solving Systems of Linear Equations by Graphing

A **system of linear equations** consists of two or more linear equations. In this section, we focus on solving systems of linear equations containing two equations in two variables. Examples of such linear systems are

$$\begin{cases} 3x - 3y = 0 \\ x = 2y \end{cases} \quad \begin{cases} x - y = 0 \\ 2x + y = 10 \end{cases} \quad \begin{cases} y = 7x - 1 \\ y = 4 \end{cases}$$

Objectives

A Decide Whether an Ordered Pair Is a Solution of a System of Linear Equations.

B Solve a System of Linear Equations by Graphing.

C Without Graphing, Determine the Number of Solutions of a System.

Objective A Deciding Whether an Ordered Pair Is a Solution

A **solution** of a system of two equations in two variables is an ordered pair of numbers that is a solution of both equations in the system.

Example 1 Determine whether $(12, 6)$ is a solution of the system

$$\begin{cases} 2x - 3y = 6 \\ x = 2y \end{cases}$$

Solution: To determine whether $(12, 6)$ is a solution of the system, we replace x with 12 and y with 6 in both equations.

$2x - 3y = 6$ First equation
$2(12) - 3(6) \stackrel{?}{=} 6$ Let $x = 12$ and $y = 6$.
$24 - 18 \stackrel{?}{=} 6$ Simplify.
$6 = 6$ True

$x = 2y$ Second equation
$12 \stackrel{?}{=} 2(6)$ Let $x = 12$ and $y = 6$.
$12 = 12$ True

Since $(12, 6)$ is a solution of both equations, it is a solution of the system.

■ Work Practice 1

Practice 1

Determine whether $(3, 9)$ is a solution of the system

$$\begin{cases} 5x - 2y = -3 \\ y = 3x \end{cases}$$

Example 2 Determine whether $(-1, 2)$ is a solution of the system

$$\begin{cases} x + 2y = 3 \\ 4x - y = 6 \end{cases}$$

Solution: We replace x with -1 and y with 2 in both equations.

$x + 2y = 3$ First equation
$-1 + 2(2) \stackrel{?}{=} 3$ Let $x = -1$ and $y = 2$.
$-1 + 4 \stackrel{?}{=} 3$ Simplify.
$3 = 3$ True

$4x - y = 6$ Second equation
$4(-1) - 2 \stackrel{?}{=} 6$ Let $x = -1$ and $y = 2$.
$-4 - 2 \stackrel{?}{=} 6$ Simplify.
$-6 = 6$ False

$(-1, 2)$ is not a solution of the second equation, $4x - y = 6$, so it is not a solution of the system.

■ Work Practice 2

Practice 2

Determine whether $(3, -2)$ is a solution of the system

$$\begin{cases} 2x - y = 8 \\ x + 3y = 4 \end{cases}$$

Objective B Solving Systems of Equations by Graphing

Since a solution of a system of two equations in two variables is a solution common to both equations, it is also a point common to the graphs of both equations. Let's practice finding solutions of both equations in a system—that is, solutions of the system—by graphing and identifying points of intersection.

Answers

1. $(3, 9)$ is a solution of the system.
2. $(3, -2)$ is not a solution of the system.

1083

Practice 3

Solve the system of equations by graphing.

$$\begin{cases} -3x + y = -10 \\ x - y = 6 \end{cases}$$

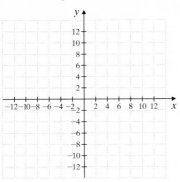

Practice 4

Solve the system of equations by graphing.

$$\begin{cases} x + 3y = -1 \\ y = 1 \end{cases}$$

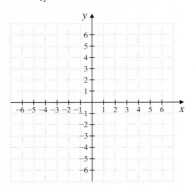

Answers

3. $(2, -4)$;

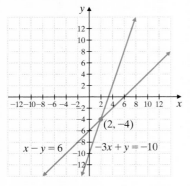

4. See page 1086.

Example 3 Solve the system of equations by graphing.

$$\begin{cases} -x + 3y = 10 \\ x + y = 2 \end{cases}$$

Solution: On a single set of axes, graph each linear equation.

$-x + 3y = 10$

x	y
0	$\frac{10}{3}$
-4	2
2	4

Helpful Hint The point of intersection gives the solution of the system.

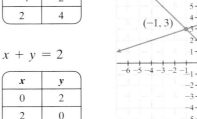

$x + y = 2$

x	y
0	2
2	0
1	1

The two lines appear to intersect at the point $(-1, 3)$. To check, we replace x with -1 and y with 3 in both equations.

$-x + 3y = 10$ First equation
$-(-1) + 3(3) \stackrel{?}{=} 10$ Let $x = -1$ and $y = 3$.
$1 + 9 \stackrel{?}{=} 10$ Simplify.
$10 = 10$ True

$x + y = 2$ Second equation
$-1 + 3 \stackrel{?}{=} 2$ Let $x = -1$ and $y = 3$.
$2 = 2$ True

$(-1, 3)$ checks, so it is the solution of the system.

■ Work Practice 3

Helpful Hint

Neatly drawn graphs can help when "guessing" the solution of a system of linear equations by graphing.

Example 4 Solve the system of equations by graphing.

$$\begin{cases} 2x + 3y = -2 \\ x = 2 \end{cases}$$

Solution: We graph each linear equation on a single set of axes.

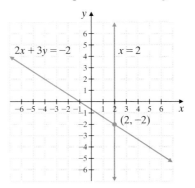

The two lines appear to intersect at the point $(2, -2)$. To determine whether $(2, -2)$ is the solution, we replace x with 2 and y with -2 in both equations.

$2x + 3y = -2$ First equation
$2(2) + 3(-2) \stackrel{?}{=} -2$ Let $x = 2$ and $y = -2$.
$4 + (-6) \stackrel{?}{=} -2$ Simplify.
$-2 = -2$ True

$x = 2$ Second equation
$2 \stackrel{?}{=} 2$ Let $x = 2$.
$2 = 2$ True

Since a true statement results in both equations, $(2, -2)$ is the solution of the system.

■ Work Practice 4

Not all systems of linear equations have a single solution. Some systems have no solution and some have an infinite number of solutions.

Example 5 Solve the system of equations by graphing.

$$\begin{cases} 2x + y = 7 \\ 2y = -4x \end{cases}$$

Solution: We graph the two equations in the system. The equations in slope-intercept form are $y = -2x + 7$ and $y = -2x$. Notice from the equations that the lines have the same slope, -2, and different y-intercepts. This means that the lines are parallel.

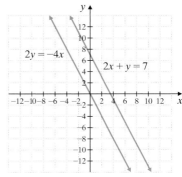

Since the lines are parallel, they do not intersect. This means that the system has *no solution*.

■ Work Practice 5

Practice 5
Solve the system of equations by graphing.
$$\begin{cases} 3x - y = 6 \\ 6x = 2y \end{cases}$$

Example 6 Solve the system of equations by graphing.

$$\begin{cases} x - y = 3 \\ -x + y = -3 \end{cases}$$

Solution: We graph each equation. The graphs of the equations are the same line. To see this, notice that if both sides of the first equation in the system are multiplied by -1, the result is the second equation.

$x - y = 3$ First equation
$-1(x - y) = -1(3)$ Multiply both sides by -1.
$-x + y = -3$ Simplify. This is the second equation.

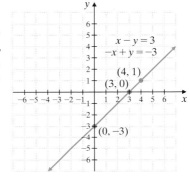

Any ordered pair that is a solution of one equation is a solution of the other equation and is then a solution of the system. This means that the system has an infinite number of solutions.

■ Work Practice 6

Practice 6
Solve the system of equations by graphing.
$$\begin{cases} x + y = -4 \\ -2x - 2y = 8 \end{cases}$$

Answers
5. See page 1086.   6. See page 1086.

Practice 7

Without graphing, determine the number of solutions of the system.

$$\begin{cases} 5x + 4y = 6 \\ x - y = 3 \end{cases}$$

Answers

4. $(-4, 1)$;

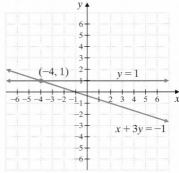

5. no solution;

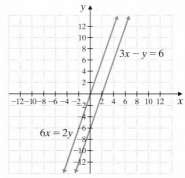

6. infinite number of solutions;

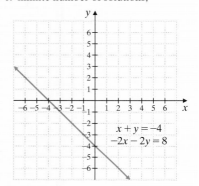

7. one solution

Examples 5 and 6 are special cases of systems of linear equations. A system that has no solution is said to be an **inconsistent system.** If the graphs of the two equations of a system are identical, we call the equations **dependent equations.** Thus, the system in Example 5 is an inconsistent system and the equations in the system in Example 6 are dependent equations.

As we have seen, three different situations can occur when graphing the two lines associated with the equations in a linear system. These situations are shown in the figures.

One point of intersection: one solution
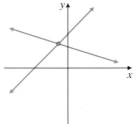
Consistent system
(at least one solution)
Independent equations
(graphs of equations differ)

Parallel lines: no solution
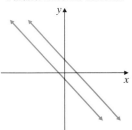
Inconsistent system
(no solution)
Independent equations
(graphs of equations differ)

Same line: infinite number of solutions
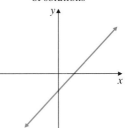
Consistent system
(at least one solution)
Dependent equations
(graphs of equations identical)

Objective C Finding the Number of Solutions of a System Without Graphing

You may have suspected by now that graphing alone is not an accurate way to solve a system of linear equations. For example, a solution of $\left(\dfrac{1}{2}, \dfrac{2}{9}\right)$ is unlikely to be read correctly from a graph. The next two sections present two accurate methods of solving these systems. In the meantime, we can decide how many solutions a system has by writing each equation in slope-intercept form.

Example 7 Without graphing, determine the number of solutions of the system

$$\begin{cases} \dfrac{1}{2}x - y = 2 \\ x = 2y + 5 \end{cases}$$

Solution: First write each equation in slope-intercept form.

$\dfrac{1}{2}x - y = 2$ First equation

$\dfrac{1}{2}x = y + 2$ Add y to both sides.

$\dfrac{1}{2}x - 2 = y$ Subtract 2 from both sides.

$x = 2y + 5$ Second equation

$x - 5 = 2y$ Subtract 5 from both sides.

$\dfrac{x}{2} - \dfrac{5}{2} = \dfrac{2y}{2}$ Divide both sides by 2.

$\dfrac{1}{2}x - \dfrac{5}{2} = y$ Simplify.

The slope of each line is $\dfrac{1}{2}$, but they have different y-intercepts. This tells us that the lines representing these equations are parallel. Since the lines are parallel, the system has no solution and is inconsistent.

Work Practice 7

Example 8

Without graphing, determine the number of solutions of the system.

$$\begin{cases} 3x - y = 4 \\ x + 2y = 8 \end{cases}$$

Solution: Once again, the slope-intercept form helps determine how many solutions this system has.

$3x - y = 4$ First equation	$x + 2y = 8$ Second equation
$3x = y + 4$ Add y to both sides.	$x = -2y + 8$ Subtract $2y$ from both sides.
$3x - 4 = y$ Subtract 4 from both sides.	$x - 8 = -2y$ Subtract 8 from both sides.
	$\dfrac{x}{-2} - \dfrac{8}{-2} = \dfrac{-2y}{-2}$ Divide both sides by -2.
	$-\dfrac{1}{2}x + 4 = y$ Simplify.

The slope of the second line is $-\dfrac{1}{2}$, whereas the slope of the first line is 3. Since the slopes are not equal, the two lines are neither parallel nor identical and must intersect. Therefore, this system has one solution and is consistent.

■ Work Practice 8

Practice 8

Without graphing, determine the number of solutions of the system.

$$\begin{cases} -\dfrac{2}{3}x + y = 6 \\ 3y = 2x + 5 \end{cases}$$

Answer
8. no solution

🖩 Calculator Explorations   Graphing

A graphing calculator may be used to approximate solutions of systems of equations. For example, to approximate the solution of the system

$$\begin{cases} y = -3.14x - 1.35 \\ y = 4.88x + 5.25, \end{cases}$$

first graph each equation on the same set of axes. Then use the Intersect feature of your calculator to approximate the point of intersection.

The approximate point of intersection is $(-0.82, 1.23)$.

Solve each system of equations. Approximate the solutions to two decimal places.

1. $\begin{cases} y = -2.68x + 1.21 \\ y = 5.22x - 1.68 \end{cases}$

2. $\begin{cases} y = 4.25x + 3.89 \\ y = -1.88x + 3.21 \end{cases}$

3. $\begin{cases} 4.3x - 2.9y = 5.6 \\ 8.1x + 7.6y = -14.1 \end{cases}$

4. $\begin{cases} -3.6x - 8.6y = 10 \\ -4.5x + 9.6y = -7.7 \end{cases}$

Vocabulary, Readiness & Video Check

Fill in each blank with one of the words or phrases listed below.

system of linear equations   solution   consistent
dependent   inconsistent   independent

1. In a system of linear equations in two variables, if the graphs of the equations are the same, the equations are _____ equations.
2. Two or more linear equations are called a(n) _____.
3. A system of equations that has at least one solution is called a(n) _____ system.
4. A(n) _____ of a system of two equations in two variables is an ordered pair of numbers that is a solution of both equations in the system.
5. A system of equations that has no solution is called a(n) _____ system.
6. In a system of linear equations in two variables, if the graphs of the equations are different, the equations are _____ equations.

Each rectangular coordinate system shows the graph of the equations in a system of equations. Use each graph to determine the number of solutions for each associated system. If the system has only one solution, give its coordinates.

7.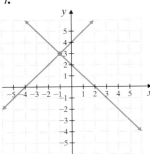

8.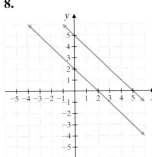

9.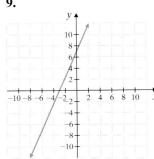

10.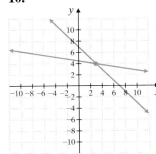

See Video 14.1

Martin-Gay Interactive Videos   Watch the section lecture video and answer the following questions.

Objective A   11. In Example 1, the first ordered pair is a solution of the first equation of the system. Why is this not enough to determine whether this ordered pair is a solution of the system?

Objective B   12. From Examples 2 and 3, why is finding the solution of a system of equations from a graph considered "guessing" and this proposed solution checked algebraically?

Objective C   13. From Examples 5–7, explain how the slope-intercept form tells us how many solutions a system of equations has.

14.1 Exercise Set MyLab Math

Objective A *Determine whether each ordered pair is a solution of the system of linear equations. See Examples 1 and 2.*

1. $\begin{cases} x + y = 8 \\ 3x + 2y = 21 \end{cases}$
   a. $(2, 4)$
   b. $(5, 3)$

2. $\begin{cases} 2x + y = 5 \\ x + 3y = 5 \end{cases}$
   a. $(5, 0)$
   b. $(2, 1)$

3. $\begin{cases} 3x - y = 5 \\ x + 2y = 11 \end{cases}$
   a. $(3, 4)$
   b. $(0, -5)$

4. $\begin{cases} 2x - 3y = 8 \\ x - 2y = 6 \end{cases}$
   a. $(-2, -4)$
   b. $(7, 2)$

5. $\begin{cases} 2y = 4x + 6 \\ 2x - y = -3 \end{cases}$
   a. $(-3, -3)$
   b. $(0, 3)$

6. $\begin{cases} x + 5y = -4 \\ -2x = 10y + 8 \end{cases}$
   a. $(-4, 0)$
   b. $(6, -2)$

7. $\begin{cases} -2 = x - 7y \\ 6x - y = 13 \end{cases}$
   a. $(-2, 0)$
   b. $\left(\dfrac{1}{2}, \dfrac{5}{14}\right)$

8. $\begin{cases} 4x = 1 - y \\ x - 3y = -8 \end{cases}$
   a. $(0, 1)$
   b. $\left(\dfrac{1}{6}, \dfrac{1}{3}\right)$

Objective B *Solve each system of linear equations by graphing. See Examples 3 through 6.*

9. $\begin{cases} x + y = 4 \\ x - y = 2 \end{cases}$

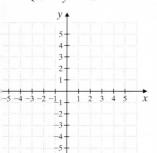

10. $\begin{cases} x + y = 3 \\ x - y = 5 \end{cases}$

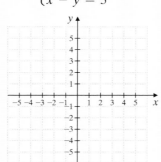

11. $\begin{cases} x + y = 6 \\ -x + y = -6 \end{cases}$

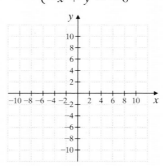

12. $\begin{cases} x + y = 1 \\ -x + y = -3 \end{cases}$

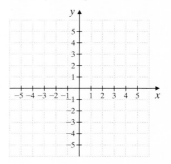

13. $\begin{cases} y = 2x \\ 3x - y = -2 \end{cases}$

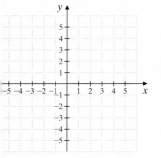

14. $\begin{cases} y = -3x \\ 2x - y = -5 \end{cases}$

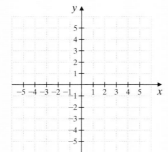

15. $\begin{cases} y = x + 1 \\ y = 2x - 1 \end{cases}$

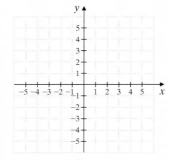

16. $\begin{cases} y = 3x - 4 \\ y = x + 2 \end{cases}$

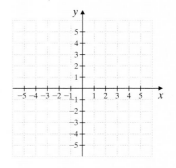

17. $\begin{cases} 2x + y = 0 \\ 3x + y = 1 \end{cases}$
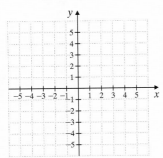

18. $\begin{cases} 2x + y = 1 \\ 3x + y = 0 \end{cases}$
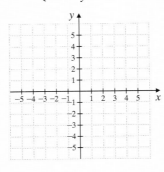

19. $\begin{cases} y = -x - 1 \\ y = 2x + 5 \end{cases}$
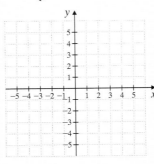

20. $\begin{cases} y = x - 1 \\ y = -3x - 5 \end{cases}$
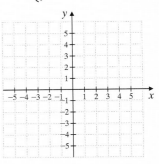

21. $\begin{cases} x + y = 5 \\ x + y = 6 \end{cases}$
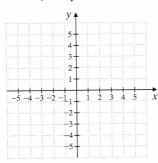

22. $\begin{cases} x - y = 4 \\ x - y = 1 \end{cases}$
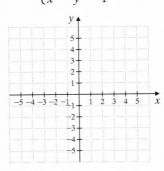

23. $\begin{cases} 2x - y = 6 \\ y = 2 \end{cases}$
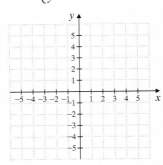

24. $\begin{cases} x + y = 5 \\ x = 4 \end{cases}$
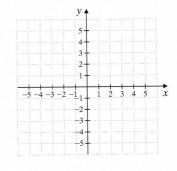

25. $\begin{cases} x - 2y = 2 \\ 3x + 2y = -2 \end{cases}$
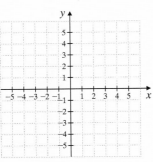

26. $\begin{cases} x + 3y = 7 \\ 2x - 3y = -4 \end{cases}$
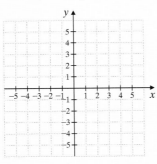

27. $\begin{cases} 2x + y = 4 \\ 6x = -3y + 6 \end{cases}$
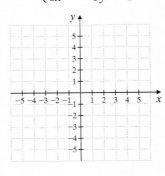

28. $\begin{cases} y + 2x = 3 \\ 4x = 2 - 2y \end{cases}$
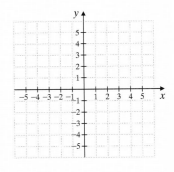

29. $\begin{cases} y - 3x = -2 \\ 6x - 2y = 4 \end{cases}$
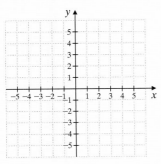

30. $\begin{cases} x - 2y = -6 \\ -2x + 4y = 12 \end{cases}$
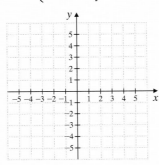

31. $\begin{cases} x = 3 \\ y = -1 \end{cases}$
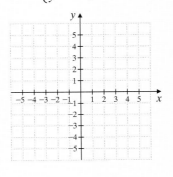

32. $\begin{cases} x = -5 \\ y = 3 \end{cases}$
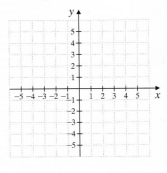

33. $\begin{cases} y = x - 2 \\ y = 2x + 3 \end{cases}$

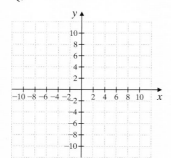

34. $\begin{cases} y = x + 5 \\ y = -2x - 4 \end{cases}$

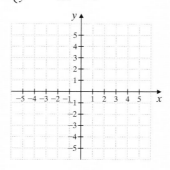

35. $\begin{cases} 2x - 3y = -2 \\ -3x + 5y = 5 \end{cases}$

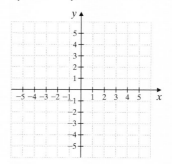

36. $\begin{cases} 4x - y = 7 \\ 2x - 3y = -9 \end{cases}$

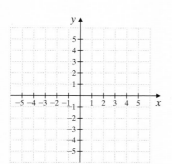

37. $\begin{cases} 6x - y = 4 \\ \dfrac{1}{2}y = -2 + 3x \end{cases}$

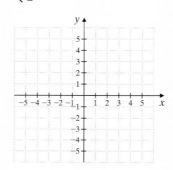

38. $\begin{cases} 3x - y = 6 \\ \dfrac{1}{3}y = -2 + x \end{cases}$

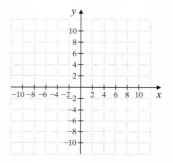

Objective C *Without graphing, decide:*

Are the graphs of the equations identical lines, parallel lines, or lines intersecting at a single point?
How many solutions does the system have? See Examples 7 and 8.

39. $\begin{cases} 4x + y = 24 \\ x + 2y = 2 \end{cases}$

40. $\begin{cases} 3x + y = 1 \\ 3x + 2y = 6 \end{cases}$

41. $\begin{cases} 2x + y = 0 \\ 2y = 6 - 4x \end{cases}$

42. $\begin{cases} 3x + y = 0 \\ 2y = -6x \end{cases}$

43. $\begin{cases} 6x - y = 4 \\ \dfrac{1}{2}y = -2 + 3x \end{cases}$

44. $\begin{cases} 3x - y = 2 \\ \dfrac{1}{3}y = -2 + 3x \end{cases}$

45. $\begin{cases} x = 5 \\ y = -2 \end{cases}$

46. $\begin{cases} y = 3 \\ x = -4 \end{cases}$

47. $\begin{cases} 3y - 2x = 3 \\ x + 2y = 9 \end{cases}$

48. $\begin{cases} 2y = x + 2 \\ y + 2x = 3 \end{cases}$

49. $\begin{cases} 6y + 4x = 6 \\ 3y - 3 = -2x \end{cases}$

50. $\begin{cases} 8y + 6x = 4 \\ 4y - 2 = 3x \end{cases}$

51. $\begin{cases} x + y = 4 \\ x + y = 3 \end{cases}$

52. $\begin{cases} 2x + y = 0 \\ y = -2x + 1 \end{cases}$

Review

Solve each equation. See Section 9.3.

53. $5(x - 3) + 3x = 1$

54. $-2x + 3(x + 6) = 17$

55. $4\left(\dfrac{y + 1}{2}\right) + 3y = 0$

56. $-y + 12\left(\dfrac{y - 1}{4}\right) = 3$

57. $8a - 2(3a - 1) = 6$

58. $3z - (4z - 2) = 9$

Concept Extensions

59. Draw a graph of two linear equations whose associated system has the solution $(-1, 4)$.

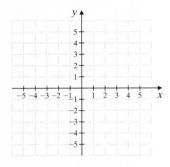

60. Draw a graph of two linear equations whose associated system has the solution $(3, -2)$.

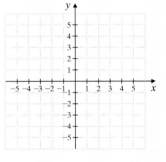

61. Draw a graph of two linear equations whose associated system has no solution.

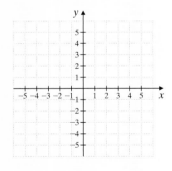

62. Draw a graph of two linear equations whose associated system has an infinite number of solutions.

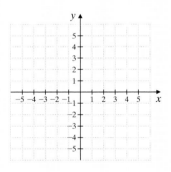

The double line graph below shows the number of digital 3-D and analog movie screens in U.S. cinemas for the years shown. Use this graph to answer Exercises 63 and 64. (Source: Motion Picture Association of America, Inc.)

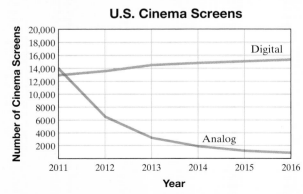

Source: Motion Picture Association of America, Inc.

63. Between what pairs of years did the number of digital 3-D cinema screens equal the number of analog cinema screens?

64. For what year was the number of digital 3-D cinema screens less than the number of analog cinema screens?

The double line graph below shows the number of pounds of fresh Pacific salmon imported to or exported from the United States during the given years. Use this graph to answer Exercises 65 and 66.

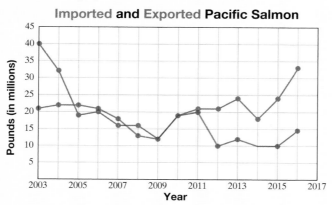

Imported and Exported Pacific Salmon

Source: USDA Economic Research Service

65. During which year(s) did the number of pounds of imported Pacific salmon equal the number of pounds of exported Pacific salmon?

66. For what year(s) was the number of pounds of exported Pacific salmon less than the number of pounds of imported Pacific salmon?

67. Construct a system of two linear equations that has $(2, 5)$ as a solution.

68. Construct a system of two linear equations that has $(0, 1)$ as a solution.

69. The ordered pair $(-2, 3)$ is a solution of the three linear equations below:

$$x + y = 1$$
$$2x - y = -7$$
$$x + 3y = 7$$

If each equation has a distinct graph, describe the graph of all three equations on the same axes.

70. Explain how to use a graph to determine the number of solutions of a system.

71. Below are tables of values for two linear equations.
  a. Find a solution of the corresponding system.
  b. Graph several ordered pairs from each table and sketch the two lines.

72. Below are tables of values for two linear equations.
  a. Find a solution of the corresponding system.
  b. Graph several ordered pairs from each table and sketch the two lines.

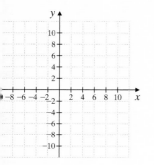

x	y
1	3
2	5
3	7
4	9
5	11

x	y
1	6
2	7
3	8
4	9
5	10

x	y
-3	5
-1	1
0	-1
1	-3
2	-5

x	y
-3	7
-1	1
0	-2
1	-5
2	-8

c. Does your graph confirm the solution from part **a**?

c. Does your graph confirm the solution from part **a**?

14.2 Solving Systems of Linear Equations by Substitution

Objective

A Use the Substitution Method to Solve a System of Linear Equations.

Objective A Using the Substitution Method

You may have suspected by now that graphing alone is not an accurate way to solve a system of linear equations. For example, a solution of $\left(\frac{1}{2}, \frac{2}{9}\right)$ is unlikely to be read correctly from a graph. In this section, we discuss a second, more accurate method for solving systems of equations. This method is called the **substitution method** and is introduced in the next example.

Practice 1

Use the substitution method to solve the system:
$$\begin{cases} 2x + 3y = 13 \\ x = y + 4 \end{cases}$$

Example 1 Solve the system:
$$\begin{cases} 2x + y = 10 & \text{First equation} \\ x = y + 2 & \text{Second equation} \end{cases}$$

Solution: The second equation in this system is $x = y + 2$. This tells us that x and $y + 2$ have the same value. This means that we may substitute $y + 2$ for x in the first equation.

$$2x + y = 10 \quad \text{First equation}$$

$$2(y + 2) + y = 10 \quad \text{Substitute } y + 2 \text{ for } x \text{ since } x = y + 2.$$

Notice that this equation now has one variable, y. Let's now solve this equation for y.

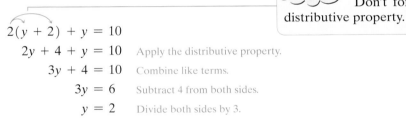

Helpful Hint Don't forget the distributive property.

$$2(y + 2) + y = 10$$
$$2y + 4 + y = 10 \quad \text{Apply the distributive property.}$$
$$3y + 4 = 10 \quad \text{Combine like terms.}$$
$$3y = 6 \quad \text{Subtract 4 from both sides.}$$
$$y = 2 \quad \text{Divide both sides by 3.}$$

Now we know that the y-value of the ordered pair solution of the system is 2. To find the corresponding x-value, we replace y with 2 in the second equation, $x = y + 2$, and solve for x.

$$x = y + 2 \quad \text{Second equation}$$
$$x = 2 + 2 \quad \text{Let } y = 2.$$
$$x = 4$$

The solution of the system is the ordered pair $(4, 2)$. Since an ordered pair solution must satisfy both linear equations in the system, we could have chosen the equation $2x + y = 10$ to find the corresponding x-value. The resulting x-value is the same.

Check: We check to see that $(4, 2)$ satisfies both equations of the original system.

First Equation	Second Equation
$2x + y = 10$	$x = y + 2$
$2(4) + 2 \stackrel{?}{=} 10$	$4 \stackrel{?}{=} 2 + 2$ Let $x = 4$ and $y = 2$.
$10 = 10$ True	$4 = 4$ True

Answer
1. $(5, 1)$

The solution of the system is $(4, 2)$.

A graph of the two equations shows the two lines intersecting at the point $(4, 2)$.

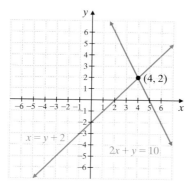

■ Work Practice 1

Example 2 Solve the system:
$$\begin{cases} 5x - y = -2 \\ y = 3x \end{cases}$$

Solution: The second equation is solved for y in terms of x. We substitute $3x$ for y in the first equation.

$5x - y = -2$ First equation

$5x - (3x) = -2$ Substitute $3x$ for y.

Now we solve for x.

$5x - 3x = -2$

$2x = -2$ Combine like terms.

$x = -1$ Divide both sides by 2.

The x-value of the ordered pair solution is -1. To find the corresponding y-value, we replace x with -1 in the second equation, $y = 3x$.

$y = 3x$ Second equation

$y = 3(-1)$ Let $x = -1$.

$y = -3$

Check to see that the solution of the system is $(-1, -3)$.

■ Work Practice 2

To solve a system of equations by substitution, we first need an equation solved for one of its variables, as in Examples 1 and 2. If neither equation in a system is solved for x or y, this will be our first step.

Example 3 Solve the system:
$$\begin{cases} x + 2y = 7 \\ 2x + 2y = 13 \end{cases}$$

Solution: Notice that neither equation is solved for x or y. Thus, we choose one of the equations and solve for x or y. We will solve the first equation for x so that we will not introduce tedious fractions when solving. To solve the first equation for x, we subtract $2y$ from both sides.

$x + 2y = 7$ First equation

$x = 7 - 2y$ Subtract $2y$ from both sides.

(Continued on next page)

Practice 2

Use the substitution method to solve the system:
$$\begin{cases} 4x - y = 2 \\ y = 5x \end{cases}$$

Practice 3

Solve the system:
$$\begin{cases} 3x + y = 5 \\ 3x - 2y = -7 \end{cases}$$

Answers

2. $(-2, -10)$ **3.** $\left(\dfrac{1}{3}, 4\right)$

Helpful Hint Don't forget to insert parentheses when substituting $7 - 2y$ for x.

Since $x = 7 - 2y$, we now substitute $7 - 2y$ for x in the second equation and solve for y.

$$2x + 2y = 13 \quad \text{Second equation}$$
$$2(7 - 2y) + 2y = 13 \quad \text{Let } x = 7 - 2y.$$
$$14 - 4y + 2y = 13 \quad \text{Apply the distributive property.}$$
$$14 - 2y = 13 \quad \text{Simplify.}$$
$$-2y = -1 \quad \text{Subtract 14 from both sides.}$$
$$y = \frac{1}{2} \quad \text{Divide both sides by } -2.$$

Helpful Hint To find x, any equation in two variables equivalent to the original equations of the system may be used. We used this equation since it is solved for x.

To find x, we let $y = \frac{1}{2}$ in the equation $x = 7 - 2y$.

$$x = 7 - 2y$$
$$x = 7 - 2\left(\frac{1}{2}\right) \quad \text{Let } y = \frac{1}{2}.$$
$$x = 7 - 1$$
$$x = 6$$

Check the solution in both equations of the original system. The solution is $\left(6, \frac{1}{2}\right)$.

■ Work Practice 3

The following steps summarize how to solve a system of equations by the substitution method.

To Solve a System of Two Linear Equations by the Substitution Method

Step 1: Solve one of the equations for one of its variables.
Step 2: Substitute the expression for the variable found in Step 1 into the other equation.
Step 3: Solve the equation from Step 2 to find the value of one variable.
Step 4: Substitute the value found in Step 3 into any equation containing both variables to find the value of the other variable.
Step 5: Check the proposed solution in the original system.

✓ **Concept Check** As you solve the system

$$\begin{cases} 2x + y = -5 \\ x - y = 5 \end{cases}$$

you find that $y = -5$. Is this the solution of the system?

Practice 4

Solve the system:
$$\begin{cases} 5x - 2y = 6 \\ -3x + y = -3 \end{cases}$$

Answer
4. $(0, -3)$

✓ **Concept Check Answer**
no, the solution will be an ordered pair

Example 4 Solve the system:
$$\begin{cases} 7x - 3y = -14 \\ -3x + y = 6 \end{cases}$$

Solution: Since the coefficient of y is 1 in the second equation, we will solve the second equation for y. This way, we avoid introducing tedious fractions.

$$-3x + y = 6 \quad \text{Second equation}$$
$$y = 3x + 6$$

Next, we substitute $3x + 6$ for y in the first equation.

$$7x - 3y = -14 \quad \text{First equation}$$
$$7x - 3(3x + 6) = -14 \quad \text{Let } y = 3x + 6.$$
$$7x - 9x - 18 = -14 \quad \text{Use the distributive property.}$$
$$-2x - 18 = -14 \quad \text{Simplify.}$$
$$-2x = 4 \quad \text{Add 18 to both sides.}$$
$$x = -2 \quad \text{Divide both sides by } -2.$$

To find the corresponding y-value, we substitute -2 for x in the equation $y = 3x + 6$. Then $y = 3(-2) + 6$ or $y = 0$. The solution of the system is $(-2, 0)$. Check this solution in both equations of the system.

■ Work Practice 4

✓ **Concept Check** To avoid fractions, which of the equations below would you use to solve for x?

a. $3x - 4y = 15$ **b.** $14 - 3y = 8x$ **c.** $7y + x = 12$

Helpful Hint

When solving a system of equations by the substitution method, begin by solving an equation for one of its variables. If possible, solve for a variable that has a coefficient of 1 or −1 to avoid working with time-consuming fractions.

Example 5 Solve the system: $\begin{cases} \dfrac{1}{2}x - y = 3 \\ x = 6 + 2y \end{cases}$

Solution: The second equation is already solved for x in terms of y. Thus we substitute $6 + 2y$ for x in the first equation and solve for y.

$$\dfrac{1}{2}x - y = 3 \quad \text{First equation}$$
$$\dfrac{1}{2}(6 + 2y) - y = 3 \quad \text{Let } x = 6 + 2y.$$
$$3 + y - y = 3 \quad \text{Apply the distributive property.}$$
$$3 = 3 \quad \text{Simplify.}$$

Arriving at a true statement such as $3 = 3$ indicates that the two linear equations in the original system are equivalent. This means that their graphs are identical, as shown in the figure. There is an infinite number of solutions to the system, and any solution of one equation is also a solution of the other.

Practice 5
Solve the system:
$$\begin{cases} -x + 3y = 6 \\ y = \dfrac{1}{3}x + 2 \end{cases}$$

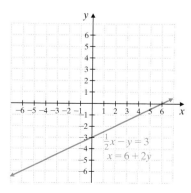

■ Work Practice 5

Answer
5. infinite number of solutions

✓ **Concept Check Answer**
c

Helpful Hint

Know that an infinite number of solutions does *not* mean that any ordered pair is a solution of both equations of the system.

An infinite number of solutions for Example 5 means that any of the infinite number of ordered pairs that is a solution of one equation in the system is also a solution of the other and is thus a solution of the system.

For Example 5,
$(2, 0)$ is *not* a solution of the system, but
$(6, 0)$ is a solution of the system.

Practice 6

Solve the system:
$$\begin{cases} 2x - 3y = 6 \\ -4x + 6y = 12 \end{cases}$$

Example 6 Solve the system:
$$\begin{cases} 6x + 12y = 5 \\ -4x - 8y = 0 \end{cases}$$

Solution: We choose the second equation and solve for y. (*Note:* Although you might not see this beforehand, if you solve the second equation for x, the result is $x = -2y$ and no fractions are introduced. Either way will lead to the correct solution.)

$$-4x - 8y = 0 \quad \text{Second equation}$$
$$-8y = 4x \quad \text{Add } 4x \text{ to both sides.}$$
$$\frac{-8y}{-8} = \frac{4x}{-8} \quad \text{Divide both sides by } -8.$$
$$y = -\frac{1}{2}x \quad \text{Simplify.}$$

Now we replace y with $-\frac{1}{2}x$ in the first equation.

$$6x + 12y = 5 \quad \text{First equation}$$
$$6x + 12\left(-\frac{1}{2}x\right) = 5 \quad \text{Let } y = -\frac{1}{2}x.$$
$$6x + (-6x) = 5 \quad \text{Simplify.}$$
$$0 = 5 \quad \text{Combine like terms.}$$

The false statement $0 = 5$ indicates that this system has no solution. The graph of the linear equations in the system is a pair of parallel lines, as shown in the figure.

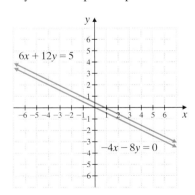

■ Work Practice 6

Answer

6. no solution

✓ Concept Check Answer

a. parallel lines b. intersect at one point c. identical graphs

✓ Concept Check Describe how the graphs of the equations in a system appear if the system has

a. no solution   b. one solution   c. an infinite number of solutions

Section 14.2 Solving Systems of Linear Equations by Substitution

Vocabulary, Readiness & Video Check

Give the solution of each system. If the system has no solution or an infinite number of solutions, say so. If the system has one solution, find it.

1. $\begin{cases} y = 4x \\ -3x + y = 1 \end{cases}$
When solving, you obtain $x = 1$.

2. $\begin{cases} 4x - y = 17 \\ -8x + 2y = 0 \end{cases}$
When solving, you obtain $0 = 34$.

3. $\begin{cases} 4x - y = 17 \\ -8x + 2y = -34 \end{cases}$
When solving, you obtain $0 = 0$.

4. $\begin{cases} 5x + 2y = 25 \\ x = y + 5 \end{cases}$
When solving, you obtain $y = 0$.

5. $\begin{cases} x + y = 0 \\ 7x - 7y = 0 \end{cases}$
When solving, you obtain $x = 0$.

6. $\begin{cases} y = -2x + 5 \\ 4x + 2y = 10 \end{cases}$
When solving, you obtain $0 = 0$.

Martin-Gay Interactive Videos Watch the section lecture video and answer the following question.

See Video 14.2

Objective A 7. The systems in Examples 2–4 all need one of their equations solved for a variable as a first step. What important part of the substitution method is emphasized in each example?

4.2 Exercise Set MyLab Math

Objective A *Solve each system of equations by the substitution method. See Examples 1 and 2.*

1. $\begin{cases} x + y = 3 \\ x = 2y \end{cases}$

2. $\begin{cases} x + y = 20 \\ x = 3y \end{cases}$

3. $\begin{cases} x + y = 6 \\ y = -3x \end{cases}$

4. $\begin{cases} x + y = 6 \\ y = -4x \end{cases}$

5. $\begin{cases} y = 3x + 1 \\ 4y - 8x = 12 \end{cases}$

6. $\begin{cases} y = 2x + 3 \\ 5y - 7x = 18 \end{cases}$

7. $\begin{cases} y = 2x + 9 \\ y = 7x + 10 \end{cases}$

8. $\begin{cases} y = 5x - 3 \\ y = 8x + 4 \end{cases}$

Solve each system of equations by the substitution method. See Examples 1 through 6.

9. $\begin{cases} 3x - 4y = 10 \\ y = x - 3 \end{cases}$

10. $\begin{cases} 4x - 3y = 10 \\ y = x - 5 \end{cases}$

11. $\begin{cases} x + 2y = 6 \\ 2x + 3y = 8 \end{cases}$

12. $\begin{cases} x + 3y = -5 \\ 2x + 2y = 6 \end{cases}$

13. $\begin{cases} 3x + 2y = 16 \\ x = 3y - 2 \end{cases}$

14. $\begin{cases} 2x + 3y = 18 \\ x = 2y - 5 \end{cases}$

15. $\begin{cases} 2x - 5y = 1 \\ 3x + y = -7 \end{cases}$

16. $\begin{cases} 3y - x = 6 \\ 4x + 12y = 0 \end{cases}$

17. $\begin{cases} 4x + 2y = 5 \\ -2x = y + 4 \end{cases}$

18. $\begin{cases} 2y = x + 2 \\ 6x - 12y = 0 \end{cases}$
19. $\begin{cases} 4x + y = 11 \\ 2x + 5y = 1 \end{cases}$
20. $\begin{cases} 3x + y = -14 \\ 4x + 3y = -22 \end{cases}$

21. $\begin{cases} x + 2y + 5 = -4 + 5y - x \\ 2x + x = y + 4 \end{cases}$
(*Hint:* First simplify each equation.)

22. $\begin{cases} 5x + 4y - 2 = -6 + 7y - 3x \\ 3x + 4x = y + 3 \end{cases}$
(*Hint:* See Exercise **21**.)

23. $\begin{cases} 6x - 3y = 5 \\ x + 2y = 0 \end{cases}$
24. $\begin{cases} 10x - 5y = -21 \\ x + 3y = 0 \end{cases}$
▶ 25. $\begin{cases} 3x - y = 1 \\ 2x - 3y = 10 \end{cases}$

26. $\begin{cases} 2x - y = -7 \\ 4x - 3y = -11 \end{cases}$
27. $\begin{cases} -x + 2y = 10 \\ -2x + 3y = 18 \end{cases}$
28. $\begin{cases} -x + 3y = 18 \\ -3x + 2y = 19 \end{cases}$

29. $\begin{cases} 5x + 10y = 20 \\ 2x + 6y = 10 \end{cases}$
30. $\begin{cases} 6x + 3y = 12 \\ 9x + 6y = 15 \end{cases}$
▶ 31. $\begin{cases} 3x + 6y = 9 \\ 4x + 8y = 16 \end{cases}$

32. $\begin{cases} 2x + 4y = 6 \\ 5x + 10y = 16 \end{cases}$
▶ 33. $\begin{cases} \frac{1}{3}x - y = 2 \\ x - 3y = 6 \end{cases}$
34. $\begin{cases} \frac{1}{4}x - 2y = 1 \\ x - 8y = 4 \end{cases}$

35. $\begin{cases} x = \frac{3}{4}y - 1 \\ 8x - 5y = -6 \end{cases}$
36. $\begin{cases} x = \frac{5}{6}y - 2 \\ 12x - 5y = -9 \end{cases}$

Review

Write equivalent equations by multiplying both sides of each given equation by the given nonzero number. See Section 3.

37. $3x + 2y = 6$ by -2
38. $-x + y = 10$ by 5
39. $-4x + y = 3$ by 3
40. $5a - 7b = -4$ by -4

Add the binomials. See Section 10.4.

41. $\begin{array}{r} 3n + 6m \\ 2n - 6m \end{array}$
42. $\begin{array}{r} -2x + 5y \\ 2x + 11y \end{array}$
43. $\begin{array}{r} -5a - 7b \\ 5a - 8b \end{array}$
44. $\begin{array}{r} 9q + p \\ -9q - p \end{array}$

Concept Extensions

Solve each system by the substitution method. First simplify each equation by combining like terms.

45. $\begin{cases} -5y + 6y = 3x + 2(x - 5) - 3x + 5 \\ 4(x + y) - x + y = -12 \end{cases}$

46. $\begin{cases} 5x + 2y - 4x - 2y = 2(2y + 6) - 7 \\ 3(2x - y) - 4x = 1 + 9 \end{cases}$

47. Explain how to identify a system with no solution when using the substitution method.

48. Occasionally, when using the substitution method, we obtain the equation $0 = 0$. Explain how this result indicates that the graphs of the equations in the system are identical.

Solve. See a Concept Check in this section.

49. As you solve the system $\begin{cases} 3x - y = -6 \\ -3x + 2y = 7 \end{cases}$, you find that $y = 1$. Is this the solution of the system?

50. As you solve the system $\begin{cases} x = 5y \\ y = 2x \end{cases}$, you find that $x = 0$ and $y = 0$. What is the solution of this system?

Section 14.2 | Solving Systems of Linear Equations by Substitution 1101

51. To avoid fractions, which of the equations below would you use if solving for *y*? Explain why.
   a. $\frac{1}{2}x - 4y = \frac{3}{4}$ b. $8x - 5y = 13$ c. $7x - y = 19$

52. Give the number of solutions for a system if the graphs of the equations in the system are
   a. lines intersecting in one point
   b. parallel lines
   c. same line

Use a graphing calculator to solve each system.

53. $\begin{cases} y = 5.1x + 14.56 \\ y = -2x - 3.9 \end{cases}$ 54. $\begin{cases} y = 3.1x - 16.35 \\ y = -9.7x + 28.45 \end{cases}$ 55. $\begin{cases} 3x + 2y = 14.04 \\ 5x + y = 18.5 \end{cases}$ 56. $\begin{cases} x + y = -15.2 \\ -2x + 5y = -19.3 \end{cases}$

57. U.S. consumer spending *y* (in billions of dollars) on DVD- or Blu-ray-format home entertainment from 2014 to 2016 is given by $y = -0.6x + 6.8$, where *x* is the number of years after 2014. U.S. consumer spending *y* (in billions of dollars) on streaming services home entertainment from 2014 to 2016 is given by $y = 0.9x + 4.6$, where *x* is the number of years after 2014. (*Source:* Based on data from Variety)

   a. Use the substitution method to solve this system of equations.
   $$\begin{cases} y = -0.6x + 6.8 \\ y = 0.9x + 4.6 \end{cases}$$
   Round *x* to the nearest tenth and *y* to the nearest whole.
   b. Explain the meaning of your answer to part **a**.
   c. Sketch a graph of the system of equations. Write a sentence describing the trends in the popularity of these types of home entertainment formats.

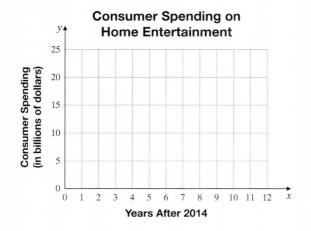

58. Consumption of diet soda in the United States continues to decline as the consumption of bottled water increases. For that reason, more beverage companies are adding bottled water to their stable of products. From 2012 through 2016, the amount *y* (in gallons per person) of diet soda consumed in the United States is given by the equation $y = -1.5x + 36$, where *x* is the number of years after 2012. For the same period, the consumption of bottled water (in gallons per person) in the United States can be represented by the equation $y = 2.1x + 30$, where *x* is the number of years after 2012. (*Source: Beverage Digest* and *Forbes*)

   a. Use the substitution method to solve this system of equations.
   $$\begin{cases} y = -1.5x + 36 \\ y = 2.1x + 30 \end{cases}$$
   Round *x* and *y* to the nearest tenth.
   b. Explain the meaning of your answer to part **a**.
   c. Sketch a graph of the system of equations. Write a sentence describing the trends in diet soda and bottled water consumption in the United States between 2012 and 2016.

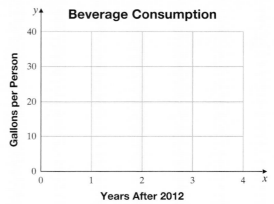

14.3 Solving Systems of Linear Equations by Addition

Objective

A Use the Addition Method to Solve a System of Linear Equations.

Objective A Using the Addition Method

We have seen that substitution is an accurate method for solving a system of linear equations. Another accurate method is the **addition** or **elimination method.** The addition method is based on the addition property of equality: Adding equal quantities to both sides of an equation does not change the solution of the equation. In symbols,

$$\text{if } A = B \text{ and } C = D, \text{ then } A + C = B + D$$

To see how we use this to solve a system of equations, study Example 1.

Example 1 Solve the system: $\begin{cases} x + y = 7 \\ x - y = 5 \end{cases}$

Solution: Since the left side of each equation is equal to its right side, we are adding equal quantities when we add the left sides of the equations together and add the right sides of the equations together. This adding eliminates the variable y and gives us an equation in one variable, x. We can then solve for x.

$$\begin{array}{ll} x + y = 7 & \text{First equation} \\ x - y = 5 & \text{Second equation} \\ \hline 2x = 12 & \text{Add the equations to eliminate } y. \\ x = 6 & \text{Divide both sides by 2.} \end{array}$$

The x-value of the solution is 6. To find the corresponding y-value, we let $x = 6$ in either equation of the system. We will use the first equation.

$$\begin{array}{ll} x + y = 7 & \text{First equation} \\ 6 + y = 7 & \text{Let } x = 6. \\ y = 1 & \text{Solve for } y. \end{array}$$

The solution is $(6, 1)$.

Helpful Hint Notice in Example 1 that our goal when solving a system of equations by the addition method is to eliminate a variable when adding the equations.

Check: Check the solution in both equations of the original system.

First Equation
$x + y = 7$
$6 + 1 \stackrel{?}{=} 7$ Let $x = 6$ and $y = 1$.
$7 = 7$ True

Second Equation
$x - y = 5$
$6 - 1 \stackrel{?}{=} 5$ Let $x = 6$ and $y = 1$
$5 = 5$ True

Thus, the solution of the system is $(6, 1)$.

If we graph the two equations in the system, we have two lines that intersect at the point $(6, 1)$, as shown.

Practice 1

Use the addition method to solve the system:
$\begin{cases} x + y = 13 \\ x - y = 5 \end{cases}$

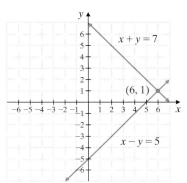

■ Work Practice 1

Answer
1. $(9, 4)$

Section 14.3 | Solving Systems of Linear Equations by Addition

Example 2 Solve the system: $\begin{cases} -2x + y = 2 \\ -x + 3y = -4 \end{cases}$

Solution: If we simply add these two equations, the result is still an equation in two variables. However, from Example 1, remember that our goal is to eliminate one of the variables so that we have an equation in the other variable. To do this, notice what happens if we multiply *both sides* of the first equation by -3. We are allowed to do this by the multiplication property of equality. Then the system

$\begin{cases} -3(-2x + y) = -3(2) \\ -x + 3y = -4 \end{cases}$ simplifies to $\begin{cases} 6x - 3y = -6 \\ -x + 3y = -4 \end{cases}$

When we add the resulting equations, the y-variable is eliminated.

$$\begin{aligned} 6x - 3y &= -6 \\ -x + 3y &= -4 \\ \hline 5x &= -10 \quad \text{Add.} \\ x &= -2 \quad \text{Divide both sides by 5.} \end{aligned}$$

To find the corresponding y-value, we let $x = -2$ in either of the original equations. We use the first equation of the original system.

$$\begin{aligned} -2x + y &= 2 \quad \text{First equation} \\ -2(-2) + y &= 2 \quad \text{Let } x = -2. \\ 4 + y &= 2 \\ y &= -2 \end{aligned}$$

Check the ordered pair $(-2, -2)$ in both equations of the *original* system. The solution is $(-2, -2)$.

Work Practice 2

Practice 2
Solve the system:
$\begin{cases} 2x - y = -6 \\ -x + 4y = 17 \end{cases}$

Helpful Hint
When finding the second value of an ordered pair solution, any equation equivalent to one of the original equations in the system may be used.

In Example 2, the decision to multiply the first equation by -3 was no accident. **To eliminate a variable** when adding two equations, **the coefficient of the variable in one equation must be the opposite of its coefficient in the other equation.**

Helpful Hint
Be sure to multiply *both sides* of an equation by the chosen number when solving by the addition method. A common mistake is to multiply only the side containing the variables.

Example 3 Solve the system: $\begin{cases} 2x - y = 7 \\ 8x - 4y = 1 \end{cases}$

Solution: When we multiply both sides of the first equation by -4, the resulting coefficient of x is -8. This is the opposite of 8, the coefficient of x in the second equation. Then the system

$\begin{cases} -4(2x - y) = -4(7) \\ 8x - 4y = 1 \end{cases}$ simplifies to

Helpful Hint: Don't forget to multiply both sides by -4.

$$\begin{aligned} \begin{cases} -8x + 4y = -28 \\ 8x - 4y = 1 \end{cases} \\ \hline 0 = -27 \quad \text{Add the equations.} \end{aligned}$$

(Continued on next page)

Practice 3
Solve the system:
$\begin{cases} x - 3y = -2 \\ -3x + 9y = 5 \end{cases}$

Answers
2. $(-1, 4)$ **3.** no solution

When we add the equations, both variables are eliminated and we have $0 = -2$, a false statement. This means that the system has no solution. The equations, graphed, would represent parallel lines.

■ Work Practice 3

Practice 4

Solve the system:
$$\begin{cases} 2x + 5y = 1 \\ -4x - 10y = -2 \end{cases}$$

Example 4 Solve the system: $\begin{cases} 3x - 2y = 2 \\ -9x + 6y = -6 \end{cases}$

Solution: First we multiply both sides of the first equation by 3 and then we add the resulting equations.

$$\begin{cases} 3(3x - 2y) = 3(2) \\ -9x + 6y = -6 \end{cases} \text{ simplifies to } \begin{cases} 9x - 6y = 6 \\ \underline{-9x + 6y = -6} \\ 0 = 0 \end{cases} \text{ Add the equations.}$$

Both variables are eliminated and we have $0 = 0$, a true statement. This means that the system has an infinite number of solutions. The equations, if graphed, would be the same line.

■ Work Practice 4

✓**Concept Check** Suppose you are solving the system

$$\begin{cases} 3x + 8y = -5 \\ 2x - 4y = 3 \end{cases}$$

You decide to use the addition method by multiplying both sides of the second equation by 2. In which of the following was the multiplication performed correctly? Explain.

a. $4x - 8y = 3$ **b.** $4x - 8y = 6$

In the next example, we multiply both equations by numbers so that coefficients of a variable are opposites.

Practice 5

Solve the system:
$$\begin{cases} 4x + 5y = 14 \\ 3x - 2y = -1 \end{cases}$$

Example 5 Solve the system: $\begin{cases} 3x + 4y = 13 \\ 5x - 9y = 6 \end{cases}$

Solution: We can eliminate the variable y by multiplying the first equation by 9 and the second equation by 4. Then we add the resulting equations.

$$\begin{cases} 9(3x + 4y) = 9(13) \\ 4(5x - 9y) = 4(6) \end{cases} \text{ simplifies to } \begin{cases} 27x + 36y = 117 \\ \underline{20x - 36y = 24} \\ 47x = 141 \\ x = 3 \end{cases} \begin{array}{l} \text{Add the equations.} \\ \text{Solve for } x. \end{array}$$

To find the corresponding y-value, we let $x = 3$ in one of the original equations of the system. Doing so in either of these equations will give $y = 1$. Check to see that $(3, 1)$ satisfies each equation in the original system. The solution is $(3, 1)$.

■ Work Practice 5

Answers

4. infinite number of solutions
5. $(1, 2)$

✓**Concept Check Answer**

b; answers may vary

If we had decided to eliminate x instead of y in Example 5, the first equation could have been multiplied by 5 and the second by -3. Try solving the original system this way to check that the solution is $(3, 1)$.

The following steps summarize how to solve a system of linear equations by the addition method.

To Solve a System of Two Linear Equations by the Addition Method

Step 1: Rewrite each equation in standard form, $Ax + By = C$.
Step 2: If necessary, multiply one or both equations by a nonzero number so that the coefficients of the chosen variable in the system are opposites.
Step 3: Add the equations.
Step 4: Find the value of one variable by solving the resulting equation from Step 3.
Step 5: Find the value of the second variable by substituting the value found in Step 4 into either of the original equations.
Step 6: Check the proposed solution in the original system.

Concept Check Suppose you are solving the system
$$\begin{cases} -4x + 7y = 6 \\ x + 2y = 5 \end{cases}$$
by the addition method.
What step(s) should you take if you wish to eliminate x when adding the equations?
What step(s) should you take if you wish to eliminate y when adding the equations?

Example 6 Solve the system: $\begin{cases} -x - \dfrac{y}{2} = \dfrac{5}{2} \\ \dfrac{x}{6} - \dfrac{y}{2} = 0 \end{cases}$

Solution: We begin by clearing each equation of fractions. To do so, we multiply both sides of the first equation by the LCD, 2, and both sides of the second equation by the LCD, 6. Then the system

$$\begin{cases} 2\left(-x - \dfrac{y}{2}\right) = 2\left(\dfrac{5}{2}\right) \\ 6\left(\dfrac{x}{6} - \dfrac{y}{2}\right) = 6(0) \end{cases} \text{ simplifies to } \begin{cases} -2x - y = 5 \\ x - 3y = 0 \end{cases}$$

We can now eliminate the variable x by multiplying the second equation by 2.

$$\begin{cases} -2x - y = 5 \\ 2(x - 3y) = 2(0) \end{cases} \text{ simplifies to } \begin{cases} -2x - y = 5 \\ \underline{2x - 6y = 0} \\ -7y = 5 \end{cases}$$ Add the equations.

$$y = -\dfrac{5}{7}$$ Solve for y.

To find x, we could replace y with $-\dfrac{5}{7}$ in one of the equations with two variables. Instead, let's go back to the simplified system and multiply by appropriate factors to eliminate the variable y and solve for x. To do this, we multiply the first equation by -3. Then the system

$$\begin{cases} -3(-2x - y) = -3(5) \\ x - 3y = 0 \end{cases} \text{ simplifies to } \begin{cases} 6x + 3y = -15 \\ \underline{x - 3y = 0} \\ 7x = -15 \end{cases}$$ Add the equations.

$$x = -\dfrac{15}{7}$$ Solve for x.

Check the ordered pair $\left(-\dfrac{15}{7}, -\dfrac{5}{7}\right)$ in both equations of the original system. The solution is $\left(-\dfrac{15}{7}, -\dfrac{5}{7}\right)$.

Work Practice 6

Practice 6
Solve the system:
$$\begin{cases} -\dfrac{x}{3} + y = \dfrac{4}{3} \\ \dfrac{x}{2} - \dfrac{5}{2}y = -\dfrac{1}{2} \end{cases}$$

Answer
6. $\left(-\dfrac{17}{2}, -\dfrac{3}{2}\right)$

✓ **Concept Check Answer**
a. multiply the second equation by 4
b. possible answer: multiply the first equation by -2 and the second equation by 7

Vocabulary, Readiness & Video Check

Given the system $\begin{cases} 3x - 2y = -9 \\ x + 5y = 14 \end{cases}$ read each row (Step 1, Step 2, and Result). Then answer whether the result is true or false.

	Step 1	Step 2	Result	True or False?
1.	Multiply 2nd equation through by -3.	Add the resulting equation to the 1st equation.	The y's are eliminated.	
2.	Multiply 2nd equation through by -3.	Add the resulting equation to the 1st equation.	The x's are eliminated.	
3.	Multiply 1st equation by 5 and 2nd equation by 2.	Add the two new equations.	The y's are eliminated.	
4.	Multiply 1st equation by 5 and 2nd equation by -2.	Add the two new equations.	The y's are eliminated.	

Martin-Gay Interactive Videos Watch the section lecture video and answer the following question.

See Video 14.3

Objective A 5. For the addition/elimination method, sometimes we need to multiply an equation through by a nonzero number so that the coefficients of a variable are opposites, as is shown in Example 2. What property allows us to do this? What important reminder is made at this step?

14.3 Exercise Set MyLab Math

Objective A Solve each system of equations by the addition method. See Example 1.

1. $\begin{cases} 3x + y = 5 \\ 6x - y = 4 \end{cases}$
2. $\begin{cases} 4x + y = 13 \\ 2x - y = 5 \end{cases}$
3. $\begin{cases} x - 2y = 8 \\ -x + 5y = -17 \end{cases}$
4. $\begin{cases} x - 2y = -11 \\ -x + 5y = 23 \end{cases}$

Solve each system of equations by the addition method. If a system contains fractions or decimals, you may want to first clear each equation of the fractions or decimals. See Examples 1 through 6.

5. $\begin{cases} 3x + y = -11 \\ 6x - 2y = -2 \end{cases}$
6. $\begin{cases} 4x + y = -13 \\ 6x - 3y = -15 \end{cases}$
7. $\begin{cases} 3x + 2y = 11 \\ 5x - 2y = 29 \end{cases}$

8. $\begin{cases} 4x + 2y = 2 \\ 3x - 2y = 12 \end{cases}$
9. $\begin{cases} x + 5y = 18 \\ 3x + 2y = -11 \end{cases}$
10. $\begin{cases} x + 4y = 14 \\ 5x + 3y = 2 \end{cases}$

11. $\begin{cases} x + y = 6 \\ x - y = 6 \end{cases}$
12. $\begin{cases} x - y = 1 \\ -x + 2y = 0 \end{cases}$
13. $\begin{cases} 2x + 3y = 0 \\ 4x + 6y = 3 \end{cases}$

14. $\begin{cases} 3x + y = 4 \\ 9x + 3y = 6 \end{cases}$
15. $\begin{cases} -x + 5y = -1 \\ 3x - 15y = 3 \end{cases}$
16. $\begin{cases} 2x + y = 6 \\ 4x + 2y = 12 \end{cases}$

17. $\begin{cases} 3x - 2y = 7 \\ 5x + 4y = 8 \end{cases}$
18. $\begin{cases} 6x - 5y = 25 \\ 4x + 15y = 13 \end{cases}$
19. $\begin{cases} 8x = -11y - 16 \\ 2x + 3y = -4 \end{cases}$

20. $\begin{cases} 10x + 3y = -12 \\ 5x = -4y - 16 \end{cases}$

21. $\begin{cases} 4x - 3y = 7 \\ 7x + 5y = 2 \end{cases}$

22. $\begin{cases} -2x + 3y = 10 \\ 3x + 4y = 2 \end{cases}$

23. $\begin{cases} 4x - 6y = 8 \\ 6x - 9y = 12 \end{cases}$

24. $\begin{cases} 9x - 3y = 12 \\ 12x - 4y = 18 \end{cases}$

25. $\begin{cases} 2x - 5y = 4 \\ 3x - 2y = 4 \end{cases}$

26. $\begin{cases} 6x - 5y = 7 \\ 4x - 6y = 7 \end{cases}$

27. $\begin{cases} \dfrac{x}{3} + \dfrac{y}{6} = 1 \\ \dfrac{x}{2} - \dfrac{y}{4} = 0 \end{cases}$

28. $\begin{cases} \dfrac{x}{2} + \dfrac{y}{8} = 3 \\ x - \dfrac{y}{4} = 0 \end{cases}$

29. $\begin{cases} \dfrac{10}{3}x + 4y = -4 \\ 5x + 6y = -6 \end{cases}$

30. $\begin{cases} \dfrac{3}{2}x + 4y = 1 \\ 9x + 24y = 5 \end{cases}$

31. $\begin{cases} x - \dfrac{y}{3} = -1 \\ -\dfrac{x}{2} + \dfrac{y}{8} = \dfrac{1}{4} \end{cases}$

32. $\begin{cases} 2x - \dfrac{3y}{4} = -3 \\ x + \dfrac{y}{9} = \dfrac{13}{3} \end{cases}$

33. $\begin{cases} -4(x + 2) = 3y \\ 2x - 2y = 3 \end{cases}$

34. $\begin{cases} -9(x + 3) = 8y \\ 3x - 3y = 8 \end{cases}$

35. $\begin{cases} \dfrac{x}{3} - y = 2 \\ -\dfrac{x}{2} + \dfrac{3y}{2} = -3 \end{cases}$

36. $\begin{cases} \dfrac{x}{2} + \dfrac{y}{4} = 1 \\ -\dfrac{x}{4} - \dfrac{y}{8} = 1 \end{cases}$

37. $\begin{cases} \dfrac{3}{5}x - y = -\dfrac{4}{5} \\ 3x + \dfrac{y}{2} = -\dfrac{9}{5} \end{cases}$

38. $\begin{cases} 3x + \dfrac{7}{2}y = \dfrac{3}{4} \\ -\dfrac{x}{2} + \dfrac{5}{3}y = -\dfrac{5}{4} \end{cases}$

39. $\begin{cases} 3.5x + 2.5y = 17 \\ -1.5x - 7.5y = -33 \end{cases}$

40. $\begin{cases} -2.5x - 6.5y = 47 \\ 0.5x - 4.5y = 37 \end{cases}$

41. $\begin{cases} 0.02x + 0.04y = 0.09 \\ -0.1x + 0.3y = 0.8 \end{cases}$

42. $\begin{cases} 0.04x - 0.05y = 0.105 \\ 0.2x - 0.6y = 1.05 \end{cases}$

Review

Translating *Rewrite each sentence using mathematical symbols. Do not solve the equations. See Sections 9.3 and 9.4.*

43. Twice a number, added to 6, is 3 less than the number.

44. The sum of three consecutive integers is 66.

45. Three times a number, subtracted from 20, is 2.

46. Twice the sum of 8 and a number is the difference of the number and 20.

47. The product of 4 and the sum of a number and 6 is twice the number.

48. If the quotient of twice a number and 7 is subtracted from the reciprocal of the number, the result is 2.

Concept Extensions

Solve. See a Concept Check in this section.

49. To solve this system by the addition method and eliminate the variable y,
$$\begin{cases} 4x + 2y = -7 \\ 3x - y = -12 \end{cases}$$
by what value would you multiply the second equation? What do you get when you complete the multiplication?

Given the system of linear equations $\begin{cases} 3x - y = -8 \\ 5x + 3y = 2 \end{cases}$:

50. Use the addition method and
  a. Solve the system by eliminating x.
  b. Solve the system by eliminating y.

51. Suppose you are solving the system
$$\begin{cases} 3x + 8y = -5 \\ 2x - 4y = 3 \end{cases}$$
You decide to use the addition method by multiplying both sides of the second equation by 2. In which of the following was the multiplication performed correctly? Explain.
  a. $4x - 8y = 3$
  b. $4x - 8y = 6$

52. Suppose you are solving the system
$$\begin{cases} -2x - y = 0 \\ -2x + 3y = 6 \end{cases}$$
You decide to use the addition method by multiplying both sides of the first equation by 3, then adding the resulting equation to the second equation. Which of the following is the correct sum? Explain.
  a. $-8x = 6$
  b. $-8x = 9$

53. When solving a system of equations by the addition method, how do we know when the system has no solution?

54. Explain why the addition method might be preferred over the substitution method for solving the system $\begin{cases} 2x - 3y = 5 \\ 5x + 2y = 6 \end{cases}$.

55. Use the system of linear equations below to answer the questions.
$$\begin{cases} x + y = 5 \\ 3x + 3y = b \end{cases}$$
  a. Find the value of b so that the system has an infinite number of solutions.
  b. Find a value of b so that there are no solutions to the system.

56. Use the system of linear equations below to answer the questions.
$$\begin{cases} x + y = 4 \\ 2x + by = 8 \end{cases}$$
  a. Find the value of b so that the system has an infinite number of solutions.
  b. Find a value of b so that the system has a single solution.

Solve each system by the addition method.

57. $\begin{cases} 2x + 3y = 14 \\ 3x - 4y = -69.1 \end{cases}$

58. $\begin{cases} 5x - 2y = -19.8 \\ -3x + 5y = -3.7 \end{cases}$

59. In recent years, budgets for advertising on the Internet have been increasing at a faster rate than budgets for advertising on television. The amount of money y (in billions) budgeted for Internet advertising from 2014 through 2017 can be approximated by $-43x + 5y = 251$, where x is the number of years after 2014. The amount of money y (in billions) budgeted for television advertising can be approximated by $21x - 10y = -685$, where x is the number of years after 2014.

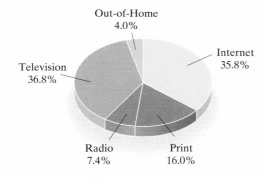

U.S. Ad Spending by Media Type

Out-of-Home 4.0%
Internet 35.8%
Television 36.8%
Radio 7.4%
Print 16.0%

a. Use the addition method to solve this system of equations.

$$\begin{cases} -43x + 5y = 251 \\ 21x - 10y = -685 \end{cases}$$

(Eliminate y first and solve for x. Round x to the nearest whole number. Because of rounding, the y-value of your ordered pair solution may vary.)

b. Interpret your solution from part **a**.

c. Use the year in your answer to part **b** to find how much money was spent on each type of advertising that year.

60. As the economy and job marketplace change, demand for certain types of workers changes. The number of jobs for audiologists that is predicted for 2014 through 2024 can be approximated by $-38x + 10y = 112$. The number of jobs for exercise physiologists that is predicted for the same period can be approximated by $15x - 10y = -155$. For both equations, x is the number of years after 2014, and y is the number of jobs in thousands. (*Source:* Based on data from the U.S. Bureau of Labor Statistics)

a. Use the addition method to solve this system of equations.

$$\begin{cases} -38x + 10y = 112 \\ 15x - 10y = -155 \end{cases}$$

(Eliminate y first and solve for x. Round x to the nearest whole number. Because of rounding, the y-value of your ordered pair solution may vary.)

b. Interpret your solution from part **a**.

c. Using the year in your answer to part **b**, estimate the number of audiologist jobs and exercise physiologist jobs in that year.

Integrated Review — Sections 14.1–14.3

Summary on Solving Systems of Equations

Solve each system by either the addition method or the substitution method.

1. $\begin{cases} 2x - 3y = -11 \\ y = 4x - 3 \end{cases}$

2. $\begin{cases} 4x - 5y = 6 \\ y = 3x - 10 \end{cases}$

3. $\begin{cases} x + y = 3 \\ x - y = 7 \end{cases}$

4. $\begin{cases} x - y = 20 \\ x + y = -8 \end{cases}$

5. $\begin{cases} x + 2y = 1 \\ 3x + 4y = -1 \end{cases}$

6. $\begin{cases} x + 3y = 5 \\ 5x + 6y = -2 \end{cases}$

7. $\begin{cases} y = x + 3 \\ 3x = 2y - 6 \end{cases}$

8. $\begin{cases} y = -2x \\ 2x - 3y = -16 \end{cases}$

9. $\begin{cases} y = 2x - 3 \\ y = 5x - 18 \end{cases}$

10. $\begin{cases} y = 6x - 5 \\ y = 4x - 11 \end{cases}$

11. $\begin{cases} x + \dfrac{1}{6}y = \dfrac{1}{2} \\ 3x + 2y = 3 \end{cases}$

12. $\begin{cases} x + \dfrac{1}{3}y = \dfrac{5}{12} \\ 8x + 3y = 4 \end{cases}$

13. $\begin{cases} x - 5y = 1 \\ -2x + 10y = 3 \end{cases}$

14. $\begin{cases} -x + 2y = 3 \\ 3x - 6y = -9 \end{cases}$

15. $\begin{cases} 0.2x - 0.3y = -0.95 \\ 0.4x + 0.1y = 0.55 \end{cases}$

16. $\begin{cases} 0.08x - 0.04y = -0.11 \\ 0.02x - 0.06y = -0.09 \end{cases}$

17. $\begin{cases} x = 3y - 7 \\ 2x - 6y = -14 \end{cases}$

18. $\begin{cases} y = \dfrac{x}{2} - 3 \\ 2x - 4y = 0 \end{cases}$

19. $\begin{cases} 2x + 5y = -1 \\ 3x - 4y = 33 \end{cases}$

20. $\begin{cases} 7x - 3y = 2 \\ 6x + 5y = -21 \end{cases}$

21. Which method, substitution or addition, would you prefer to use to solve the system below? Explain your reasoning.

$$\begin{cases} 3x + 2y = -2 \\ y = -2x \end{cases}$$

22. Which method, substitution or addition, would you prefer to use to solve the system below? Explain your reasoning.

$$\begin{cases} 3x - 2y = -3 \\ 6x + 2y = 12 \end{cases}$$

14.4 Systems of Linear Equations and Problem Solving

Objective A Using a System of Equations for Problem Solving

Many of the word problems solved earlier with one-variable equations can also be solved with two equations in two variables. We use the same problem-solving steps that we have used throughout this text. The only difference is that two variables are assigned to represent the two unknown quantities and that the problem is translated into two equations.

Objective

A Use a System of Equations to Solve Problems.

Problem-Solving Steps

1. **UNDERSTAND** the problem. During this step, become comfortable with the problem. Some ways of doing this are to

   Read and reread the problem.
   Choose two variables to represent the two unknowns.
   Construct a drawing.
   Propose a solution and check. Pay careful attention to how you check your proposed solution. This will help when writing equations to model the problem.

2. **TRANSLATE** the problem into two equations.
3. **SOLVE** the system of equations.
4. **INTERPRET** the results: *Check* the proposed solution in the stated problem and *state* your conclusion.

Example 1 Finding Unknown Numbers

Find two numbers whose sum is 37 and whose difference is 21.

Solution:

1. **UNDERSTAND.** Read and reread the problem. Suppose that one number is 20. If their sum is 37, the other number is 17 because $20 + 17 = 37$. Is their difference 21? No; $20 - 17 = 3$. Our proposed solution is incorrect, but we now have a better understanding of the problem.

   Since we are looking for two numbers, we let

   $x =$ first number and
   $y =$ second number

2. **TRANSLATE.** Since we have assigned two variables to this problem, we translate our problem into two equations.

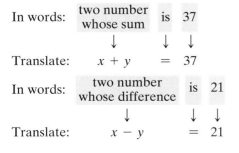

(*Continued on next page*)

Practice 1

Find two numbers whose sum is 50 and whose difference is 22.

Answer
1. 36 and 14

3. SOLVE. Now we solve the system.

$$\begin{cases} x + y = 37 \\ x - y = 21 \end{cases}$$

Notice that the coefficients of the variable y are opposites. Let's then solve by the addition method and begin by adding the equations.

$$\begin{aligned} x + y &= 37 \\ \underline{x - y} &= \underline{21} \quad \text{Add the equations.} \\ 2x &= 58 \\ x &= 29 \quad \text{Divide both sides by 2.} \end{aligned}$$

Now we let $x = 29$ in the first equation to find y.

$$\begin{aligned} x + y &= 37 \quad \text{First equation} \\ 29 + y &= 37 \\ y &= 8 \quad \text{Subtract 29 from both sides.} \end{aligned}$$

4. INTERPRET. The solution of the system is $(29, 8)$.

Check: Notice that the sum of 29 and 8 is $29 + 8 = 37$, the required sum. Their difference is $29 - 8 = 21$, the required difference.

State: The numbers are 29 and 8.

Work Practice 1

Practice 2

Admission prices at a local weekend fair were $5 for children and $7 for adults. The total money collected was $3379, and 587 people attended the fair. How many children and how many adults attended the fair?

Example 2 Solving a Problem About Prices

The Cirque du Soleil show Varekai is performing locally. Matinee admission for 4 adults and 2 children is $374, while admission for 2 adults and 3 children is $285.

a. What is the price of an adult's ticket?
b. What is the price of a child's ticket?
c. Suppose that a special rate of $1000 is offered for groups of 20 persons. Should a group of 4 adults and 16 children use the group rate? Why or why not?

Solution:

1. UNDERSTAND. Read and reread the problem and guess a solution. Let's suppose that the price of an adult's ticket is $50 and the price of a child's ticket is $40. To check our proposed solution, let's see if admission for 4 adults and 2 children is $374. Admission for 4 adults is 4($50) or $200 and admission for 2 children is 2($40) or $80. This gives a total admission of $200 + $80 = $280, not the required $374. Again, though, we have accomplished the purpose of this process: We have a better understanding of the problem. To continue, we let

$A = $ the price of an adult's ticket and
$C = $ the price of a child's ticket

Answer
2. 365 children and 222 adults

2. **TRANSLATE.** We translate the problem into two equations using both variables.

In words: admission for 4 adults and admission for 2 children is $374

Translate: $4A + 2C = 374$

In words: admission for 2 adults and admission for 3 children is $285

Translate: $2A + 3C = 285$

3. **SOLVE.** We solve the system.

$$\begin{cases} 4A + 2C = 374 \\ 2A + 3C = 285 \end{cases}$$

Since both equations are written in standard form, we solve by the addition method. First we multiply the second equation by -2 so that when we add the equations, we eliminate the variable A. Then the system

$$\begin{cases} 4A + 2C = 374 \\ -2(2A + 3C) = -2(285) \end{cases}$$ simplifies to $$\begin{cases} 4A + 2C = 374 \\ -4A - 6C = -570 \end{cases}$$

Add the equations.

$$-4C = -196$$
$$C = 49 \text{ or } \$49, \text{ the child's ticket price}$$

To find A, we replace C with 49 in the first equation.

$4A + 2C = 374$ First equation
$4A + 2(49) = 374$ Let $C = 49$.
$4A + 98 = 374$
$4A = 276$
$A = 69$ or $69, the adult's ticket price

4. **INTERPRET.**

Check: Notice that 4 adults and 2 children will pay $4(\$69) + 2(\$49) = \$276 + \$98 = \$374$, the required amount. Also, the price for 2 adults and 3 children is $2(\$69) + 3(\$49) = \$138 + \$147 = \$285$, the required amount.

State: Answer the three original questions.
a. Since $A = 69$, the price of an adult's ticket is $69.
b. Since $C = 49$, the price of a child's ticket is $49.
c. The regular admission price for 4 adults and 16 children is

$$4(\$69) + 16(\$49) = \$276 + \$784$$
$$= \$1060$$

This is $60 more than the special group rate of $1000, so they should request the group rate.

Work Practice 2

Practice 3

Two cars are 440 miles apart and traveling toward each other. They meet in 3 hours. If one car's speed is 10 miles per hour faster than the other car's speed, find the speed of each car.

	$r \cdot t = d$		
Faster Car			
Slower Car			

Example 3   Finding Rates

As part of an exercise program, two students, Louisa and Alfredo, start walking each morning. They live 15 miles away from each other. They decide to meet one day by walking toward one another. After 2 hours they meet. If Louisa walks one mile per hour faster than Alfredo, find both walking speeds.

Solution:

1. **UNDERSTAND.** Read and reread the problem. Let's propose a solution and use the formula $d = r \cdot t$ to check. Suppose that Louisa's rate is 4 miles per hour. Since Louisa's rate is 1 mile per hour faster, Alfredo's rate is 3 miles per hour. To check, see if they can walk a total of 15 miles in 2 hours. Louisa's distance is rate · time = $4(2) = 8$ miles and Alfredo's distance is rate · time = $3(2) = 6$ miles. Their total distance is 8 miles + 6 miles = 14 miles, not the required 15 miles. Now that we have a better understanding of the problem, let's model it with a system of equations.

   First, we let

   $x =$ Alfredo's rate in miles per hour and
   $y =$ Louisa's rate in miles per hour

   Now we use the facts stated in the problem and the formula $d = rt$ to fill in the following chart.

	$r \cdot t = d$		
Alfredo	x	2	$2x$
Louisa	y	2	$2y$

2. **TRANSLATE.** We translate the problem into two equations using both variables.

   In words: Alfredo's distance + Louisa's distance = 15 miles
   Translate: $2x + 2y = 15$

   In words: Louisa's rate is 1 mile per hour faster than Alfredo's
   Translate: $y = x + 1$

3. **SOLVE.** The system of equations we are solving is

   $$\begin{cases} 2x + 2y = 15 \\ y = x + 1 \end{cases}$$

Answer

3. One car's speed is $68\frac{1}{3}$ mph and the other car's speed is $78\frac{1}{3}$ mph.

Let's use substitution to solve the system since the second equation is solved for y.

$$2x + 2y = 15 \quad \text{First equation}$$
$$2x + 2(x + 1) = 15 \quad \text{Replace } y \text{ with } x + 1.$$
$$2x + 2x + 2 = 15$$
$$4x = 13$$
$$x = \frac{13}{4} = 3\frac{1}{4} \text{ or } 3.25$$
$$y = x + 1 = 3\frac{1}{4} + 1 = 4\frac{1}{4} \text{ or } 4.25$$

4. INTERPRET. Alfredo's proposed rate is $3\frac{1}{4}$ miles per hour and Louisa's proposed rate is $4\frac{1}{4}$ miles per hour.

Check: Use the formula $d = rt$ and find that in 2 hours, Alfredo's distance is $(3.25)(2)$ miles or 6.5 miles. In 2 hours, Louisa's distance is $(4.25)(2)$ miles or 8.5 miles. The total distance walked is 6.5 miles + 8.5 miles or 15 miles, the given distance.

State: Alfredo walks at a rate of 3.25 miles per hour and Louisa walks at a rate of 4.25 miles per hour.

■ Work Practice 3

Example 4 Finding Amounts of Solutions

A chemistry teaching assistant needs 10 liters of a 20% saline solution (salt water) for his 2 p.m. laboratory class. Unfortunately, the only mixtures on hand are a 5% saline solution and a 25% saline solution. How much of each solution should he mix to produce the 20% solution?

Solution:

1. UNDERSTAND. Read and reread the problem. Suppose that we need 4 liters of the 5% solution. Then we need $10 - 4 = 6$ liters of the 25% solution. To see if this gives us 10 liters of a 20% saline solution, let's find the amount of pure salt in each solution.

	concentration rate	×	amount of solution	=	amount of pure salt
	↓		↓		↓
5% solution:	0.05	×	4 liters	=	0.2 liter
25% solution:	0.25	×	6 liters	=	1.5 liters
20% solution:	0.20	×	10 liters	=	2 liters

Since 0.2 liter + 1.5 liters = 1.7 liters, not 2 liters, our proposed solution is incorrect. But we have gained some insight into how to model and check this problem.

We let

x = number of liters of 5% solution and
y = number of liters of 25% solution

5% saline solution   25% saline solution   20% saline solution

(Continued on next page)

Practice 4

A pharmacist needs 50 liters of a 60% alcohol solution. She currently has available a 20% solution and a 70% solution. How many liters of each must she use to make the needed 50 liters of 60% alcohol solution?

Answer
4. 10 liters of the 20% alcohol solution and 40 liters of the 70% alcohol solution

Now we use a table to organize the given data.

	Concentration Rate	Liters of Solution	Liters of Pure Salt
First Solution	5%	x	$0.05x$
Second Solution	25%	y	$0.25y$
Mixture Needed	20%	10	$(0.20)(10)$

2. **TRANSLATE.** We translate into two equations using both variables.

In words: liters of 5% solution + liters of 25% solution = 10 liters

Translate: $x + y = 10$

In words: salt in 5% solution + salt in 25% solution = salt in mixture

Translate: $0.05x + 0.25y = (0.20)(10)$

3. **SOLVE.** Here we solve the system

$$\begin{cases} x + y = 10 \\ 0.05x + 0.25y = 2 \end{cases}$$

To solve by the addition method, we first multiply the first equation by -25 and the second equation by 100. Then the system

$$\begin{cases} -25(x + y) = -25(10) \\ 100(0.05x + 0.25y) = 100(2) \end{cases} \text{ simplifies to } \begin{cases} -25x - 25y = -250 \\ 5x + 25y = 200 \end{cases}$$

$$-20x = -50 \quad \text{Add.}$$
$$x = 2.5$$

To find y, we let $x = 2.5$ in the first equation of the original system.

$$x + y = 10$$
$$2.5 + y = 10 \quad \text{Let } x = 2.5.$$
$$y = 7.5$$

4. **INTERPRET.** Thus, we propose that he needs to mix 2.5 liters of 5% saline solution with 7.5 liters of 25% saline solution.

Check: Notice that $2.5 + 7.5 = 10$, the required number of liters. Also, the sum of the liters of salt in the two solutions equals the liters of salt in the required mixture:

$$0.05(2.5) + 0.25(7.5) = 0.20(10)$$
$$0.125 + 1.875 = 2$$

State: He needs 2.5 liters of the 5% saline solution and 7.5 liters of the 25% saline solution.

■ Work Practice 4

✓ **Concept Check** Suppose you mix an amount of a 30% acid solution with an amount of a 50% acid solution. Which of the following acid strengths would be possible for the resulting acid mixture?
 a. 22%   **b.** 44%   **c.** 63%

✓ Concept Check Answer
b

Section 14.4 Systems of Linear Equations and Problem Solving

Vocabulary, Readiness & Video Check

Martin-Gay Interactive Videos Watch the section lecture video and answer the following question.

See Video 14.4

Objective A 1. In the lecture before Example 1, the problem-solving steps for solving applications involving systems are discussed. How do these steps differ from the general problem-solving strategy steps?

14.4 Exercise Set MyLab Math

Without actually solving each problem, choose the correct solution by deciding which choice satisfies the given conditions.

1. The length of a rectangle is 3 feet longer than the width. The perimeter is 30 feet. Find the dimensions of the rectangle.
   a. length = 8 feet; width = 5 feet
   b. length = 8 feet; width = 7 feet
   c. length = 9 feet; width = 6 feet

2. An isosceles triangle, a triangle with two sides of equal length, has a perimeter of 20 inches. Each of the equal sides is one inch longer than the third side. Find the lengths of the three sides.
   a. 6 inches, 6 inches, and 7 inches
   b. 7 inches, 7 inches, and 6 inches
   c. 6 inches, 7 inches, and 8 inches

3. Two computer disks and three notebooks cost $17. However, five computer disks and four notebooks cost $32. Find the price of each.
   a. notebook = $4; computer disk = $3
   b. notebook = $3; computer disk = $4
   c. notebook = $5; computer disk = $2

4. Two music CDs and four DVDs cost a total of $40. However, three music CDs and five DVDs cost $55. Find the price of each.
   a. CD = $12; DVD = $4
   b. CD = $15; DVD = $2
   c. CD = $10; DVD = $5

5. Kesha has a total of 100 coins, all of which are either dimes or quarters. The total value of the coins is $13.00. Find the number of each type of coin.
   a. 80 dimes; 20 quarters
   b. 20 dimes; 44 quarters
   c. 60 dimes; 40 quarters

6. Samuel has 28 gallons of saline solution available in two large containers at his pharmacy. One container holds three times as much as the other container. Find the capacity of each container.
   a. 15 gallons; 5 gallons
   b. 20 gallons; 8 gallons
   c. 21 gallons; 7 gallons

Objective A *Write a system of equations describing each situation. Do not solve the system. See Example 1.*

7. Two numbers add up to 15 and have a difference of 7.

8. The total of two numbers is 16. The first number plus 2 more than 3 times the second equals 18.

9. Keiko has a total of $6500, which she has invested in two accounts. The larger account is $800 greater than the smaller account.

10. Dominique has four times as much money in his savings account as in his checking account. The total amount is $2300.

Solve. See Examples 1 through 4.

11. Two numbers total 83 and have a difference of 17. Find the two numbers.

12. The sum of two numbers is 76 and their difference is 52. Find the two numbers.

13. A first number plus twice a second number is 8. Twice the first number plus the second totals 25. Find the numbers.

14. One number is 4 more than twice a second number. Their total is 25. Find the numbers.

15. Nolan Arenado of the Colorado Rockies led Major League Baseball in runs batted in for the 2016 regular season. David Ortiz of the Boston Red Sox, who came in second to Arenado, had 6 fewer runs batted in for the 2016 regular season. Together, these two players brought home 260 runs during the 2016 regular season. How many runs batted in did Arenado and Ortiz each account for? (*Source:* Major League Baseball)

16. The highest scorer during the WNBA 2016 regular season was Tina Charles of the New York Liberty. Over the season, Charles scored 32 more points than the second-highest scorer, Maya Moore of the Minnesota Lynx. Together, Charles and Moore scored 1344 points during the regular 2016 season. How many points did each player score over the course of the season? (*Source:* Women's National Basketball Association)

17. Ann Marie Jones has been pricing Amtrak train fares for a group trip to New York. Three adults and four children must pay $159. Two adults and three children must pay $112. Find the price of an adult's ticket, and find the price of a child's ticket.

18. Last month, Jerry Papa purchased five DVDs and two CDs at Wall-to-Wall Sound for $65. This month he bought three DVDs and four CDs for $81. Find the price of each DVD, and find the price of each CD.

19. Johnston and Betsy Waring have a jar containing 80 coins, all of which are either quarters or nickels. The total value of the coins is $14.60. How many of each type of coin do they have?

20. Sarah and Keith Robinson purchased 40 stamps, a mixture of 47¢ and 34¢ stamps. Find the number of each type of stamp if they spent $18.15.

21. Norman and Suzanne Scarpulla own 30 shares of McDonald's Corp. stock and 68 shares of Toys R Us stock. As the New York Stock Exchange opened on December 30, 2016, their stock portfolio consisting of these two stocks was worth $8811.20. The McDonald's stock was $47.40 more per share than the Toys R Us stock. What was the price of each stock on that day? (*Source:* YAHOO Finance)

22. Saralee Rose has investments in Google and Facebook stock. As the NASDAQ exchange opened on December 30, 2016, Google stock was at $782.75 per share, and Facebook stock was at $116.60 per share. Saralee's portfolio made up of these two stocks was worth $30,627.60 at that time. If Saralee owns 15 more shares of Google stock than she owns of Facebook stock, how many shares of each type of stock does she own? (*Source:* Scottrade, Inc.)

23. Twice last month, Judy Carter rented a car from Enterprise in Fresno, California, and traveled around the Southwest on business. Enterprise rents this car for a daily fee, plus an additional charge per mile driven. Judy recalls that her first trip lasted 4 days, she drove 450 miles, and the rental cost her $240.50. On her second business trip she drove the same level of car 200 miles in 3 days, and paid $146.00 for the rental. Find the daily fee and the mileage charge.

24. Joan Gundersen rented the same car model twice from Hertz, which rents this car model for a daily fee plus an additional charge per mile driven. Joan recalls that the car rented for 5 days and driven for 300 miles cost her $178, while the same model car rented for 4 days and driven for 500 miles cost $197. Find the daily fee and the mileage charge.

25. Pratap Puri rowed 18 miles down the Delaware River in 2 hours, but the return trip took him $4\frac{1}{2}$ hours. Find the rate Pratap can row in still water and find the rate of the current.

Let x = rate Pratap can row in still water and
y = rate of the current

	$d =$	r $\cdot$	t
Downstream		$x + y$	
Upstream		$x - y$	

26. The Jonathan Schultz family took a canoe 10 miles down the Allegheny River in $1\frac{1}{4}$ hours. After lunch it took them 4 hours to return. Find the rate of the current.

Let x = rate the family can row in still water and
y = rate of the current

	$d =$	r $\cdot$	t
Downstream		$x + y$	
Upstream		$x - y$	

27. Dave and Sandy Hartranft are frequent flyers with Delta Airlines. They often fly from Philadelphia to Chicago, a distance of 780 miles. On one particular trip they fly into the wind, and the flight takes 2 hours. The return trip, with the wind behind them, takes only $1\frac{1}{2}$ hours. Find the speed of the wind and find the speed of the plane in still air.

28. With a strong wind behind it, a United Airlines jet flies 2400 miles from Los Angeles to Orlando in $4\frac{3}{4}$ hours. The return trip takes 6 hours, as the plane flies into the wind. Find the speed of the plane in still air and find the wind speed to the nearest tenth of a mile per hour.

29. Kevin Briley began a 186-mile bicycle trip to build up stamina for a triathlon competition. Unfortunately, his bicycle chain broke, so he finished the trip walking. The whole trip took 6 hours. If Kevin walks at a rate of 4 miles per hour and rides at 40 miles per hour, find the amount of time he spent on the bicycle.

30. In Canada, eastbound and westbound trains travel along the same track, with sidings to pull onto to avoid accidents. Two trains are now 150 miles apart, with the westbound train traveling twice as fast as the eastbound train. A warning must be issued to pull one train onto a siding, or else the trains will crash in $1\frac{1}{4}$ hours. Find the speed of the eastbound train and the speed of the westbound train.

31. Dorren Schmidt is a chemist with Gemco Pharmaceutical. She needs to prepare 12 ounces of a 9% hydrochloric acid solution. Find the amount of a 4% solution and the amount of a 12% solution she should mix to get this solution.

Concentration Rate	Liters of Solution	Liters of Pure Acid
0.04	x	$0.04x$
0.12	y	?
0.09	12	?

32. Elise Everly is preparing 15 liters of a 25% saline solution. Elise has two other saline solutions with strengths of 40% and 10%. Find the amount of 40% solution and the amount of 10% solution she should mix to get 15 liters of a 25% solution.

Concentration Rate	Liters of Solution	Liters of Pure Salt
0.40	x	$0.40x$
0.10	y	?
0.25	15	?

33. Wayne Osby blends coffee for a local coffee café. He needs to prepare 200 pounds of blended coffee beans selling for $3.95 per pound. He intends to do this by blending together a high-quality bean costing $4.95 per pound and a cheaper bean costing $2.65 per pound. To the nearest pound, find how much of the high-quality coffee beans and how much of the cheaper coffee beans he should blend.

34. Macadamia nuts cost an astounding $16.50 per pound, but research by an independent firm says that mixed nuts sell better if macadamias are included. The standard mix costs $9.25 per pound. Find how many pounds of macadamias and how many pounds of the standard mix should be combined to produce 40 pounds that will cost $10 per pound. Find the amounts to the nearest tenth of a pound.

35. Recall that two angles are complementary if the sum of their measures is 90°. Find the measures of two complementary angles if one angle is twice the other.

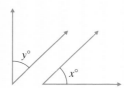

36. Recall that two angles are supplementary if the sum of their measures is 180°. Find the measures of two supplementary angles if one angle is 20° more than four times the other.

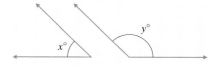

37. Find the measures of two complementary angles if one angle is 10° more than three times the other.

38. Find the measures of two supplementary angles if one angle is 18° more than twice the other.

39. Kathi and Robert Hawn had a pottery stand at the annual Skippack Craft Fair. They sold some of their pottery at the original price of $9.50 each, but later decreased the price of each by $2. If they sold all 90 pieces and took in $721, find how many they sold at the original price and how many they sold at the reduced price.

40. A charity fundraiser consisted of a spaghetti supper where a total of 387 people were fed. They charged $6.80 for adults and half price for children. If they took in $2444.60, find how many adults and how many children attended the supper.

41. The Santa Fe National Historic Trail is approximately 1200 miles between Old Franklin, Missouri, and Santa Fe, New Mexico. Suppose that a group of hikers start from each town and walk the trail toward each other. They meet after a total hiking time of 240 hours. If one group travels $\frac{1}{2}$ mile per hour slower than the other group, find the rate of each group. (*Source:* National Park Service)

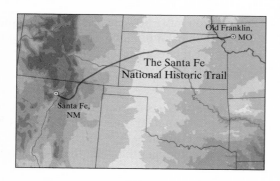

42. California 1 South is a historic highway that stretches 123 miles along the coast from Monterey to Morro Bay. Suppose that two cars start driving this highway, one from each town. They meet after 3 hours. Find the rate of each car if one car travels 1 mile per hour faster than the other car. (*Source: National Geographic*)

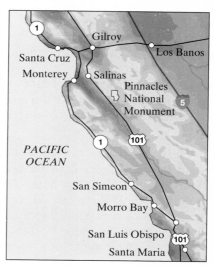

43. A 30% solution of fertilizer is to be mixed with a 60% solution of fertilizer in order to get 150 gallons of a 50% solution. How many gallons of the 30% solution and 60% solution should be mixed?

44. A 10% acid solution is to be mixed with a 50% acid solution in order to get 120 ounces of a 20% acid solution. How many ounces of the 10% solution and 50% solution should be mixed?

45. Traffic signs are regulated by the *Manual on Uniform Traffic Control Devices* (MUTCD). According to this manual, if the sign below is placed on a freeway, its perimeter must be 144 inches. Also, its length must be 12 inches longer than its width. Find the dimensions of this sign.

46. According to the MUTCD (see Exercise **45**), this sign must have a perimeter of 60 inches. Also, its length must be 6 inches longer than its width. Find the perimeter of this sign.

Review

Find the square of each expression. For example, the square of 7 is 7^2 or 49. The square of 5x is $(5x)^2$ or $25x^2$. See Section 10.

47. 4   **48.** 3   **49.** $6x$ **50.** $11y$ **51.** $10y^3$ **52.** $8x^5$

Concept Extensions

Solve. See the Concept Check in this section.

53. Suppose you mix an amount of candy costing $0.49 a pound with candy costing $0.65 a pound. Which of the following costs per pound could result?
  a. $0.58   **b.** $0.72   **c.** $0.29

54. Suppose you mix a 50% acid solution with pure acid (100%). Which of the following acid strengths are possible for the resulting acid mixture?
  a. 25%   **b.** 150%   **c.** 62%   **d.** 90%

△ **55.** Dale and Sharon Mahnke have decided to fence off a garden plot behind their house, using their house as the "fence" along one side of the garden. The length (which runs parallel to the house) is 3 feet less than twice the width. Find the dimensions if 33 feet of fencing is used along the three sides requiring it.

△ **56.** Judy McElroy plans to erect 152 feet of fencing to make a rectangular horse pasture. A river bank serves as one side length of the rectangle. If each width is 4 feet longer than half the length, find the dimensions.

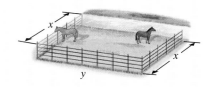

Chapter 14 Group Activity

Break-Even Point

Sections 14.1, 14.2, 14.3, 14.4

When a business sells a new product, it generally does not start making a profit right away. There are usually many expenses associated with creating a new product. These expenses might include an advertising blitz to introduce the product to the public. These start-up expenses might also include the cost of market research and product development or any brand-new equipment needed to manufacture the product. Start-up costs like these are generally called *fixed costs* because they don't depend on the number of items manufactured. Expenses that do depend on the number of items manufactured, such as the cost of materials and shipping, are called *variable costs*. The total cost of manufacturing the new product is given by the cost equation Total cost = Fixed costs + Variable costs.

For instance, suppose a greeting card company is launching a new line of greeting cards. The company spent $7000 doing product research and development for the new line and spent $15,000 advertising the new line. The company does not need to buy any new equipment to manufacture the cards, but the paper and ink needed to make each card will cost $0.20 per card. The total cost y in dollars for manufacturing x cards is $y = 22,000 + 0.20x$.

Once a business sets a price for a new product, the company can find the product's expected *revenue*. Revenue is the amount of money the company takes in from the sales of its product. The revenue from selling a product is given by the revenue equation Revenue = Price per item × Number of items sold.

For instance, suppose that the card company plans to sell its new cards for $1.50 each. The revenue y, in dollars, that the company can expect to receive from the sales of x cards is $y = 1.50x$.

If the total cost and revenue equations are graphed on the same coordinate system, the graphs should intersect. The point of intersection is where total cost equals revenue and is called the *break-even point*. The break-even point gives the number of items x that must be manufactured and sold for the company to recover its expenses. If fewer than this number of items are produced and sold, the company loses money. If more than this number of items are produced and sold, the company makes a profit. In the case of the greeting card company, approximately 16,923 cards must be manufactured and sold for the company to break even on this new card line. The total cost and revenue of producing and selling 16,923 cards is the same. It is approximately $25,385.

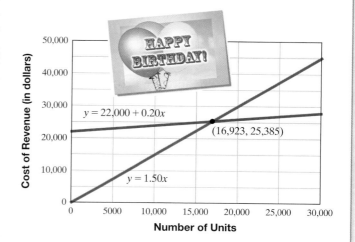

Group Activity

Suppose your group is starting a small business near your campus.

a. Choose a business and decide what campus-related product or service you will provide.

b. Research the fixed costs of starting up such a business.

c. Research the variable costs of producing such a product or providing such a service.

d. Decide how much you will charge per unit of your product or service.

e. Find a system of equations for the total cost and revenue of your product or service.

f. How many units of your product or service must be sold before your business will break even?

Chapter 14 Vocabulary Check

Fill in each blank with one of the words or phrases listed below.

system of linear equations   solution   consistent   independent
dependent   inconsistent   substitution   addition

1. In a system of linear equations in two variables, if the graphs of the equations are the same, the equations are _____ equations.
2. Two or more linear equations are called a(n) _____.
3. A system of equations that has at least one solution is called a(n) _____ system.
4. A(n) _____ of a system of two equations in two variables is an ordered pair of numbers that is a solution of both equations in the system.
5. Two algebraic methods for solving systems of equations are _____ and _____.
6. A system of equations that has no solution is called a(n) _____ system.
7. In a system of linear equations in two variables, if the graphs of the equations are different, the equations are _____ equations.

> **Helpful Hint**
> ▶ Are you preparing for your test?
> To help, don't forget to take these:
> - Chapter 14 Getting Ready for the Test on page 11
> - Chapter 14 Test on page 1131
>
> Then check all of your answers at the back of the te
> For further review, the step-by-step video solutions
> any of these exercises are located in MyLab Math.

14 Chapter Highlights

Definitions and Concepts	Examples
Section 14.1 Solving Systems of Linear Equations by Graphing	
A **system of linear equations** consists of two or more linear equations.	$\begin{cases} 2x + y = 6 \\ x = -3y \end{cases}$ $\begin{cases} -3x + 5y = 10 \\ x - 4y = -2 \end{cases}$
A **solution** of a system of two equations in two variables is an ordered pair of numbers that is a solution of both equations in the system.	Determine whether $(-1, 3)$ is a solution of the system. $$\begin{cases} 2x - y = -5 \\ x = 3y - 10 \end{cases}$$ Replace x with -1 and y with 3 in both equations. $$2x - y = -5$$ $$2(-1) - 3 \stackrel{?}{=} -5$$ $$-5 = -5 \quad \text{True}$$ $$x = 3y - 10$$ $$-1 \stackrel{?}{=} 3(3) - 10$$ $$-1 = -1 \quad \text{True}$$ $(-1, 3)$ is a solution of the system.
Graphically, a solution of a system is a point common to the graphs of both equations.	Solve by graphing: $\begin{cases} 3x - 2y = -3 \\ x + y = 4 \end{cases}$ 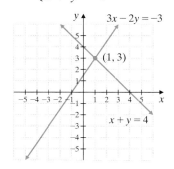

Chapter 14 Highlights

Definitions and Concepts	Examples
Section 14.1 Solving Systems of Linear Equations by Graphing *(continued)*	
Three different situations can occur when graphing the two lines associated with the equations in a linear system.	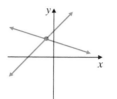 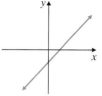 One point of intersection; one solution Same line; infinite number of solutions 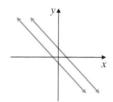 Parallel lines; no solution

Section 14.2 Solving Systems of Linear Equations by Substitution

To Solve a System of Linear Equations by the Substitution Method

Step 1: Solve one equation for a variable.

Step 2: Substitute the expression for the variable into the other equation.

Step 3: Solve the equation from Step 2 to find the value of one variable.

Step 4: Substitute the value from Step 3 in either original equation to find the value of the other variable.

Step 5: Check the solution in both original equations.

Solve by substitution.
$$\begin{cases} 3x + 2y = 1 \\ x = y - 3 \end{cases}$$

Substitute $y - 3$ for x in the first equation.

$$3x + 2y = 1$$
$$3(y - 3) + 2y = 1$$
$$3y - 9 + 2y = 1$$
$$5y = 10$$
$$y = 2 \quad \text{Divide by 5.}$$

To find x, substitute 2 for y in $x = y - 3$ so that $x = 2 - 3$ or -1. The solution $(-1, 2)$ checks.

Section 14.3 Solving Systems of Linear Equations by Addition

To Solve a System of Linear Equations by the Addition Method

Step 1: Rewrite each equation in standard form, $Ax + By = C$.

Step 2: Multiply one or both equations by a nonzero number so that the coefficients of a variable are opposites.

Step 3: Add the equations.

Step 4: Find the value of one variable by solving the resulting equation.

Step 5: Substitute the value from Step 4 into either original equation to find the value of the other variable.

Step 6: Check the solution in both original equations.

Solve by addition.
$$\begin{cases} x - 2y = 8 \\ 3x + y = -4 \end{cases}$$

Multiply both sides of the first equation by -3.
$$\begin{cases} -3x + 6y = -24 \\ 3x + y = -4 \end{cases}$$
$$7y = -28 \quad \text{Add.}$$
$$y = -4 \quad \text{Divide by 7.}$$

To find x, let $y = -4$ in an original equation.
$$x - 2(-4) = 8 \quad \text{First equation}$$
$$x + 8 = 8$$
$$x = 0$$

The solution $(0, -4)$ checks.

(continued)

Definitions and Concepts	Examples
Section 14.3 Solving Systems of Linear Equations by Addition (*continued*)	
If solving a system of linear equations by substitution or addition yields a true statement such as $-2 = -2$, then the graphs of the equations in the system are identical and the system has an infinite number of solutions.	Solve: $\begin{cases} 2x - 6y = -2 \\ x = 3y - 1 \end{cases}$ Substitute $3y - 1$ for x in the first equation. $2(3y - 1) - 6y = -2$ $6y - 2 - 6y = -2$ $-2 = -2$ True The system has an infinite number of solutions.
If solving a system of linear equations yields a false statement such as $0 = 3$, the graphs of the equations in the system are parallel lines and the system has no solution.	Solve: $\begin{cases} 5x - 2y = 6 \\ -5x + 2y = -3 \end{cases}$ $ 0 = 3$ False The system has no solution.
Section 14.4 Systems of Linear Equations and Problem Solving	

Problem-Solving Steps
 1. UNDERSTAND. Read and reread the problem.
	Two angles are supplementary if the sum of their measures is 180°. The larger of two supplementary angles is three times the smaller, decreased by twelve. Find the measure of each angle. Let
	     x = measure of smaller angle and
	     y = measure of larger angle

2. TRANSLATE.	In words: the sum of supplementary angles is 180° Translate: $ x + y = 180$ In words: larger angle is 3 times smaller decreased by 12 Translate: $ y = 3x - 12$
3. SOLVE.	Solve the system. $\begin{cases} x + y = 180 \\ y = 3x - 12 \end{cases}$ Use the substitution method and replace y with $3x - 12$ in the first equation. $x + y = 180$ $x + (3x - 12) = 180$ $4x = 192$ $x = 48$
4. INTERPRET.	Since $y = 3x - 12$, then $y = 3 \cdot 48 - 12$ or 132. The solution checks. The smaller angle measures 48° and the larger angle measures 132°.

Chapter 14 Review

4.1) *Determine whether each ordered pair is a solution of the system of linear equations.*

1. $\begin{cases} 2x - 3y = 12 \\ 3x + 4y = 1 \end{cases}$
   a. $(12, 4)$
   b. $(3, -2)$

2. $\begin{cases} 2x + 3y = 1 \\ 3y - x = 4 \end{cases}$
   a. $(2, 2)$
   b. $(-1, 1)$

3. $\begin{cases} 5x - 6y = 18 \\ 2y - x = -4 \end{cases}$
   a. $(-6, -8)$
   b. $\left(3, \dfrac{5}{2}\right)$

4. $\begin{cases} 4x + y = 0 \\ -8x - 5y = 9 \end{cases}$
   a. $\left(\dfrac{3}{4}, -3\right)$
   b. $(-2, 8)$

Solve each system of equations by graphing.

5. $\begin{cases} x + y = 5 \\ x - y = 1 \end{cases}$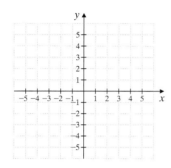

6. $\begin{cases} x + y = 3 \\ x - y = -1 \end{cases}$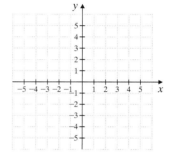

7. $\begin{cases} x = 5 \\ y = -1 \end{cases}$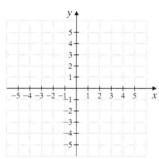

8. $\begin{cases} x = -3 \\ y = 2 \end{cases}$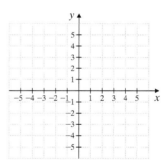

9. $\begin{cases} 2x + y = 5 \\ x = -3y \end{cases}$

10. $\begin{cases} 3x + y = -2 \\ y = -5x \end{cases}$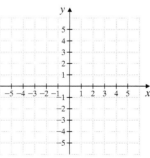

11. $\begin{cases} y = 3x \\ -6x + 2y = 6 \end{cases}$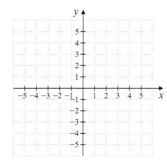

12. $\begin{cases} x - 2y = 2 \\ -2x + 4y = -4 \end{cases}$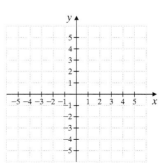

1127

(14.2) *Solve each system of equations by the substitution method.*

13. $\begin{cases} y = 2x + 6 \\ 3x - 2y = -11 \end{cases}$
14. $\begin{cases} y = 3x - 7 \\ 2x - 3y = 7 \end{cases}$

15. $\begin{cases} x + 3y = -3 \\ 2x + y = 4 \end{cases}$
16. $\begin{cases} 3x + y = 11 \\ x + 2y = 12 \end{cases}$

17. $\begin{cases} 4y = 2x + 6 \\ x - 2y = -3 \end{cases}$
18. $\begin{cases} 9x = 6y + 3 \\ 6x - 4y = 2 \end{cases}$

19. $\begin{cases} x + y = 6 \\ y = -x - 4 \end{cases}$
20. $\begin{cases} -3x + y = 6 \\ y = 3x + 2 \end{cases}$

(14.3) *Solve each system of equations by the addition method.*

21. $\begin{cases} 2x + 3y = -6 \\ x - 3y = -12 \end{cases}$
22. $\begin{cases} 4x + y = 15 \\ -4x + 3y = -19 \end{cases}$
23. $\begin{cases} 2x - 3y = -15 \\ x + 4y = 31 \end{cases}$

24. $\begin{cases} x - 5y = -22 \\ 4x + 3y = 4 \end{cases}$
25. $\begin{cases} 2x - 6y = -1 \\ -x + 3y = \dfrac{1}{2} \end{cases}$
26. $\begin{cases} 0.6x - 0.3y = -1.5 \\ 0.04x - 0.02y = -0.1 \end{cases}$

27. $\begin{cases} \dfrac{3}{4}x + \dfrac{2}{3}y = 2 \\ x + \dfrac{y}{3} = 6 \end{cases}$
28. $\begin{cases} 10x + 2y = 0 \\ 3x + 5y = 33 \end{cases}$

(14.4) *Solve each problem by writing and solving a system of linear equations.*

29. The sum of two numbers is 16. Three times the larger number decreased by the smaller number is 72. Find the two numbers.

30. The Forrest Theater can seat a total of 360 people. They take in $15,150 when every seat is sold. If orchestra section tickets cost $45 and balcony tickets cost $35, find the number of seats in the orchestra section and the number of seats in the balcony.

31. A riverboat can go 340 miles upriver in 19 hours, but the return trip takes only 14 hours. Find the current of the river and find the speed of the riverboat in still water to the nearest tenth of a mile.

	$d =$	r	$\cdot\; t$
Upriver		$x - y$	
Downriver		$x + y$	

32. Find the amount of a 6% acid solution and the amount of a 14% acid solution Pat Mayfield should combine to prepare 50 cc (cubic centimeters) of a 12% solution.

33. A deli charges $3.80 for a breakfast of three eggs and four strips of bacon. The charge is $2.75 for two eggs and three strips of bacon. Find the cost of each egg and the cost of each strip of bacon.

34. An exercise enthusiast alternates between jogging and walking. He traveled 15 miles during the past 3 hours. He jogs at a rate of 7.5 miles per hour and walks at a rate of 4 miles per hour. Find how much time, to the nearest hundredth of an hour, he actually spent jogging and how much time he spent walking.

Mixed Review

Solve each system of equations by graphing.

35. $\begin{cases} x - 2y = 1 \\ 2x + 3y = -12 \end{cases}$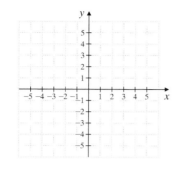

36. $\begin{cases} 3x - y = -4 \\ 6x - 2y = -8 \end{cases}$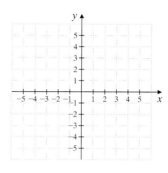

Solve each system of equations.

37. $\begin{cases} x + 4y = 11 \\ 5x - 9y = -3 \end{cases}$

38. $\begin{cases} x + 9y = 16 \\ 3x - 8y = 13 \end{cases}$

39. $\begin{cases} y = -2x \\ 4x + 7y = -15 \end{cases}$

40. $\begin{cases} 3y = 2x + 15 \\ -2x + 3y = 21 \end{cases}$

41. $\begin{cases} 3x - y = 4 \\ 4y = 12x - 16 \end{cases}$

42. $\begin{cases} x + y = 19 \\ x - y = -3 \end{cases}$

43. $\begin{cases} x - 3y = -11 \\ 4x + 5y = -10 \end{cases}$

44. $\begin{cases} -x - 15y = 44 \\ 2x + 3y = 20 \end{cases}$

45. $\begin{cases} 2x + y = 3 \\ 6x + 3y = 9 \end{cases}$

46. $\begin{cases} -3x + y = 5 \\ -3x + y = -2 \end{cases}$

Solve each problem by writing and solving a system of linear equations.

47. The sum of two numbers is 12. Three times the smaller number increased by the larger number is 20. Find the numbers.

48. The difference of two numbers is −18. Twice the smaller decreased by the larger is −23. Find the two numbers.

49. Emma Hodges has a jar containing 65 coins, all of which are either nickels or dimes. The total value of the coins is $5.30. How many of each type does she have?

50. Sarah and Owen Hebert purchased 26 stamps, a mixture of $1.15 and 47¢ stamps. Find the number of each type of stamp if they spent $19.02.

Chapter 14 — Getting Ready for the Test

1. MULTIPLE CHOICE The ordered pair $(-1, 2)$ is a solution to what system?

A. $\begin{cases} 5x - y = -7 \\ x - y = 3 \end{cases}$
B. $\begin{cases} 3x - y = -5 \\ x + y = 1 \end{cases}$
C. $\begin{cases} x = 2 \\ x + y = 1 \end{cases}$
D. $\begin{cases} y = -1 \\ x + y = -3 \end{cases}$

MATCHING For Exercises 2 through 5, **Match** each system with the graph of its equations.

2. $\begin{cases} x = 3 \\ y = -3 \end{cases}$

3. $\begin{cases} x = -3 \\ y = 3 \end{cases}$

4. $\begin{cases} x = 3 \\ y = 3 \end{cases}$

5. $\begin{cases} x = -3 \\ y = -3 \end{cases}$

A.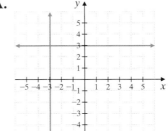

B.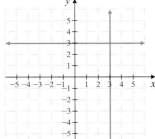

C.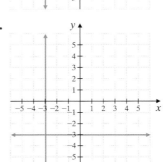

D.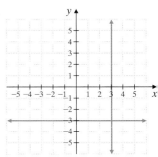

6. MULTIPLE CHOICE When solving a system of two linear equations in two variables, all variables subtract out and the resulting equation is $0 = 5$. What does this mean?

A. the solution is $(0, 5)$ B. the system has an infinite number of solutions   C. the system has no solution

MATCHING For Exercises 7 through 10, **Match** each system with its solution. Letter choices may be used more than once or not at all.

7. $\begin{cases} y = 5x + 2 \\ y = -5x + 2 \end{cases}$

8. $\begin{cases} y = \dfrac{1}{2}x - 3 \\ y = \dfrac{1}{2}x + 7 \end{cases}$

9. $\begin{cases} y = 4x + 2 \\ 8x - 2y = -4 \end{cases}$

10. $\begin{cases} y = 6x \\ y = -\dfrac{1}{6}x \end{cases}$

A. no solution
B. one solution
C. two solutions
D. an infinite number of solutions

MULTIPLE CHOICE Choose the correct choice for Exercises 11 and 12. The system for these exercises is:

$$\begin{cases} 5x - y = -8 \\ 2x + 3y = 1 \end{cases}$$

11. When solving, if we decide to multiply the first equation above by 3, the result of the multiplication is:

A. $15x - 3y = -8$ B. $6x + 9y = 1$ C. $6x + 9y = 3$ D. $15x - 3y = -24$

12. When solving, if we decide to multiply the second equation above by -5, the result of the multiplication is:

A. $-10x - 15y = 1$ B. $-25x + 5y = 40$ C. $-10x - 15y = -5$ D. $-25x + 5y = -8$

Chapter 14 Test

Answer each question true or false.

1. A system of two linear equations in two variables can have exactly two solutions.

2. Although $(1, 4)$ is not a solution of $x + 2y = 6$, it can still be a solution of the system $\begin{cases} x + 2y = 6 \\ x + y = 5 \end{cases}$.

3. If the two equations in a system of linear equations are added and the result is $3 = 0$, the system has no solution.

4. If the two equations in a system of linear equations are added and the result is $3x = 0$, the system has no solution.

Is the ordered pair a solution of the given linear system?

5. $\begin{cases} 2x - 3y = 5 \\ 6x + y = 1 \end{cases}$; $(1, -1)$

6. $\begin{cases} 4x - 3y = 24 \\ 4x + 5y = -8 \end{cases}$; $(3, -4)$

Solve each system by graphing.

7. $\begin{cases} x - y = 2 \\ 3x - y = -2 \end{cases}$

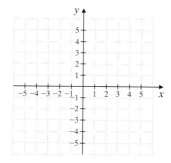

8. $\begin{cases} y = -3x \\ 3x + y = 6 \end{cases}$

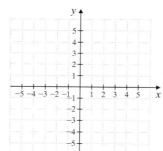

Solve each system by the substitution method.

9. $\begin{cases} 3x - 2y = -14 \\ y = x + 5 \end{cases}$

10. $\begin{cases} \dfrac{1}{2}x + 2y = -\dfrac{15}{4} \\ 4x = -y \end{cases}$

Solve each system by the addition method.

11. $\begin{cases} x + y = 28 \\ x - y = 12 \end{cases}$

12. $\begin{cases} 4x - 6y = 7 \\ -2x + 3y = 0 \end{cases}$

Solve each system using the substitution method or the addition method.

13. $\begin{cases} 3x + y = 7 \\ 4x + 3y = 1 \end{cases}$

14. $\begin{cases} 3(2x + y) = 4x + 20 \\ x - 2y = 3 \end{cases}$

Answers

1. _____
2. _____
3. _____
4. _____
5. _____
6. _____
7. _____
8. _____
9. _____
10. _____
11. _____
12. _____
13. _____
14. _____

Chapter 14 | Systems of Equations

15. $\begin{cases} \dfrac{x-3}{2} = \dfrac{2-y}{4} \\ \dfrac{7-2x}{3} = \dfrac{y}{2} \end{cases}$

16. $\begin{cases} 8x - 4y = 12 \\ y = 2x - 3 \end{cases}$

17. $\begin{cases} 0.01x - 0.06y = -0.23 \\ 0.2x + 0.4y = 0.2 \end{cases}$

18. $\begin{cases} x - \dfrac{2}{3}y = 3 \\ -2x + 3y = 10 \end{cases}$

Solve each problem by writing and using a system of linear equations.

19. Two numbers have a sum of 124 and a difference of 32. Find the numbers.

20. Find the amount of a 12% saline solution a lab assistant should add to 80 cc (cubic centimeters) of a 22% saline solution in order to have a 16% solution.

21. Texas and Missouri are the states with the most farms. Texas has 140 thousand more farms than Missouri and the total number of farms for these two states is 356 thousand. Find the number of farms for each state.

22. Two hikers start at opposite ends of the St. Tammany Trails and walk toward each other. The trail is 36 miles long and they meet in 4 hours. If one hiker is twice as fast as the other, find both hiking speeds.

The double line graph below shows the average attendance per game for the years shown for the Minnesota Twins and the Texas Rangers baseball teams. Use this for Exercises 23 and 24.

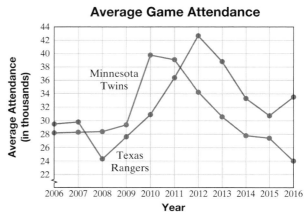

Source: Baseball Almanac

23. In what year(s) was the average attendance per game for the Texas Rangers greater than the average attendance per game for the Minnesota Twins?

24. In what year was the average attendance per game for the Texas Rangers closest to the average attendance per game for the Minnesota Twins, 2011 or 2016?

Cumulative Review — Chapters 1–14

Subtract.

1. $8 - 15$

2. $4 - 7$

3. $-4 - (-5)$

4. $3 - (-2)$

5. Solve: $x = -60 + 4 + 10$

6. Solve: $4x = -2 + 3x$

7. Translate to an equation: 1.2 is 30% of what number?

8. Translate to an equation: 9 is 45% of what number?

9. What percent of 50 is 8?

10. What percent of 16 is 4?

11. Mr. Percy, the principal at Slidell High School, counted 31 freshmen absent during a particular day. If this is 4% of the total number of freshmen, how many freshmen are there at Slidell High School?

12. 2% of the apples in a shipment are rotten. If there are 29 rotten apples, how many apples are in the shipment?

Solve.

13. $-2(x - 5) + 10 = -3(x + 2) + x$

14. $4(4y + 2) = 2(1 + 6y) + 8$

15. Solve $-5x + 7 < 2(x - 3)$. Graph the solution set.

$\leftarrow\!\!+\!\!+\!\!+\!\!+\!\!+\!\!+\!\!+\!\!+\!\!+\!\!+\!\!+\!\!\rightarrow$
$-5\ -4\ -3\ -2\ -1\ \ 0\ \ 1\ \ 2\ \ 3\ \ 4\ \ 5$

16. Solve $-7x + 4 \leq 3(4 - x)$. Graph the solution set.

$\leftarrow\!\!+\!\!+\!\!+\!\!+\!\!+\!\!+\!\!+\!\!+\!\!+\!\!+\!\!+\!\!\rightarrow$
$-5\ -4\ -3\ -2\ -1\ \ 0\ \ 1\ \ 2\ \ 3\ \ 4\ \ 5$

Simplify each expression.

17. $\left(\dfrac{m}{n}\right)^7$

18. $(-5x^3)(-7x^4)$

19. $\left(\dfrac{2x^4}{3y^5}\right)^4$

20. $\left(\dfrac{5x^2}{4y^3}\right)^2$

Answers

1. _____
2. _____
3. _____
4. _____
5. _____
6. _____
7. _____
8. _____
9. _____
10. _____
11. _____
12. _____
13. _____
14. _____
15. _____
16. _____
17. _____
18. _____
19. _____
20. _____

Chapter 14 | Systems of Equations

21. Subtract: $(2x^3 + 8x^2 - 6x) - (2x^3 - x^2 + 1)$

22. Subtract: $(7x + 1) - (-x - 3)$

23. Divide $6x^2 + 10x - 5$ by $3x - 1$ using long division.

24. Divide $3x^2 - x - 4$ by $x - 1$ using long division.

25. Solve: $x(2x - 7) = 4$

26. Solve: $x(x - 5) = 24$

△ **27.** Find the lengths of the sides of a right triangle if the lengths can be expressed as three consecutive even integers.

28. The sum of a number and its square is 132. Find the number(s).

29. Subtract: $\dfrac{2y}{2y - 7} - \dfrac{7}{2y - 7}$

30. Add: $\dfrac{x^2 + 3}{x + 9} + \dfrac{9x - 3}{x + 9}$

31. Find the slope of the line $y = -1$.

32. Find the slope of the line $x = 2$.

33. Find an equation of the line through $(2, 5)$ and $(-3, 4)$. Write the equation in the form $Ax + By = C$.

34. Find an equation of the line through $(5, -6)$ and $(-6, 5)$. Write the equation in the form $Ax + By = C$.

35. Find the domain and the range of the relation $\{(0, 2), (3, 3), (-1, 0), (3, -2)\}$.

36. Find the domain and the range of the relation $\{(2, 3), (2, 0), (2, -2), (2, 4)\}$.

37. Solve the system: $\begin{cases} x + 2y = 7 \\ 2x + 2y = 13 \end{cases}$

38. Solve the system: $\begin{cases} 3y = x + 6 \\ 4x + 12y = 0 \end{cases}$

39. Solve the system: $\begin{cases} -x - \dfrac{y}{2} = \dfrac{5}{2} \\ \dfrac{x}{6} - \dfrac{y}{2} = 0 \end{cases}$

40. Solve the system: $\begin{cases} x - \dfrac{3y}{8} = -\dfrac{3}{2} \\ x + \dfrac{y}{9} = \dfrac{13}{3} \end{cases}$

41. Find two numbers whose sum is 37 and whose difference is 21.

42. The sum of two numbers is 75 and their difference is 9. Find the two numbers.

37. _____

38. _____

39. _____

40. _____

41. _____

42. _____

15 Roots and Radicals

Having spent the last chapter studying equations, we return now to algebraic expressions. We expand on our skills of operating on expressions—adding, subtracting, multiplying, dividing, and raising to powers—to include finding roots. Just as subtraction is defined by addition and division by multiplication, finding roots is defined by raising to powers. As we master finding roots, we will work with equations that contain roots and solve problems that can be modeled by such equations.

Sections

- 15.1 Introduction to Radicals
- 15.2 Simplifying Radicals
- 15.3 Adding and Subtracting Radicals
- 15.4 Multiplying and Dividing Radicals

    Integrated Review—Simplifying Radicals

- 15.5 Solving Equations Containing Radicals
- 15.6 Radical Equations and Problem Solving

Check Your Progress

Vocabulary Check

Chapter Highlights

Chapter Review

Getting Ready for the Test

Chapter Test

Cumulative Review

Did you know that pendulums can be used to demonstrate that the earth rotates on its axis?

In 1851, French physicist Léon Foucault developed a special pendulum in an experiment to demonstrate that the earth rotates on its axis. He connected his tall pendulum, capable of running for many hours, to the roof of the Paris Observatory. The pendulum's bob was able to swing back and forth in one plane but not to twist in other directions. So, when the pendulum bob appeared to move in a circle over time, he demonstrated that it was not the pendulum but the building that moved. And since the building was firmly attached to the earth, it must be the earth rotating that created the apparent circular motion of the bob. In Section 15.1, Exercise 9, roots are used to explore the time it takes Foucault's pendulum to complete one swing of its bob.

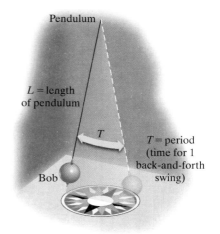

15.1 Introduction to Radicals

Objective A Finding Square Roots

In this section, we define finding the **root** of a number by its reverse operation, raising a number to a power. We begin with squares and square roots.

The *square* of 5 is $5^2 = 25$.
The *square* of -5 is $(-5)^2 = 25$.
The *square* of $\frac{1}{2}$ is $\left(\frac{1}{2}\right)^2 = \frac{1}{4}$.

The reverse operation of squaring a number is finding a **square root** of a number. For example,

A *square root* of 25 is 5, because $5^2 = 25$.
A *square root* of 25 is also -5, because $(-5)^2 = 25$.
A *square root* of $\frac{1}{4}$ is $\frac{1}{2}$, because $\left(\frac{1}{2}\right)^2 = \frac{1}{4}$.

Objectives

- **A** Find Square Roots.
- **B** Find Cube Roots.
- **C** Find nth Roots.
- **D** Approximate Square Roots.
- **E** Simplify Radicals Containing Variables.

In general, the number b is a square root of a number a if $b^2 = a$.

The symbol $\sqrt{}$ is used to denote the **positive** or **principal square root** of a number. For example,

$\sqrt{25} = 5$ only, since $5^2 = 25$ and 5 is positive.

The symbol $-\sqrt{}$ is used to denote the **negative square root.** For example,

$-\sqrt{25} = -5$

The symbol $\sqrt{}$ is called a **radical** or **radical sign.** The expression within or under a radical sign is called the **radicand.** An expression containing a radical is called a **radical expression.**

$\sqrt{a}$ — radical sign
— radicand

Square Root

If a is a positive number, then

$\sqrt{a}$ is the **positive square root** of a and
$-\sqrt{a}$ is the **negative square root** of a.

Also, $\sqrt{0} = 0$.

Examples Find each square root.

1. $\sqrt{36} = 6$, because $6^2 = 36$ and 6 is positive.
2. $-\sqrt{16} = -4$. The negative sign in front of the radical indicates the negative square root of 16.
3. $\sqrt{\frac{9}{100}} = \frac{3}{10}$ because $\left(\frac{3}{10}\right)^2 = \frac{9}{100}$ and $\frac{3}{10}$ is positive.
4. $\sqrt{0} = 0$ because $0^2 = 0$.
5. $\sqrt{0.64} = 0.8$ because $(0.8)^2 = 0.64$ and 0.8 is positive.

Work Practice 1–5

Practice 1–5

Find each square root.
1. $\sqrt{100}$
2. $-\sqrt{81}$
3. $\sqrt{\dfrac{25}{81}}$
4. $\sqrt{1}$
5. $\sqrt{0.81}$

Answers

1. 10   2. -9 3. $\dfrac{5}{9}$ 4. 1   5. 0.9

Is the square root of a negative number a real number? For example, is $\sqrt{-4}$ a real number? To answer this question, we ask ourselves, is there a real number whose square is -4? Since there is no real number whose square is -4, we say that $\sqrt{-4}$ is not a real number. In general,

A square root of a negative number is not a real number.

Study the following table to make sure you understand the differences discussed earlier.

Number	Square Roots of Number	$\sqrt{\text{number}}$	$-\sqrt{\text{number}}$
25	$-5, 5$	$\sqrt{25} = 5$ only	$-\sqrt{25} = -5$
$\dfrac{1}{4}$	$-\dfrac{1}{2}, \dfrac{1}{2}$	$\sqrt{\dfrac{1}{4}} = \dfrac{1}{2}$ only	$-\sqrt{\dfrac{1}{4}} = -\dfrac{1}{2}$
-9	No real square roots.	$\sqrt{-9}$ is not a real number.	

Objective B Finding Cube Roots

We can find roots other than square roots. For example, since $2^3 = 8$, we call 2 the **cube root** of 8. In symbols, we write

$\sqrt[3]{8} = 2$ The number 3 is called the **index.**

Also,

$\sqrt[3]{-64} = -4$ Since $(-4)^3 = -64$

Notice that unlike the square root of a negative number, the cube root of a negative number is a real number. This is so because while we cannot find a real number whose *square* is negative, we *can* find a real number whose *cube* is negative. In fact, the cube of a negative number is a negative number. Therefore, the cube root of a negative number is a negative number.

Practice 6–8

Find each cube root.

6. $\sqrt[3]{27}$
7. $\sqrt[3]{-8}$
8. $\sqrt[3]{\dfrac{1}{64}}$

Examples Find each cube root.

6. $\sqrt[3]{1} = 1$ because $1^3 = 1$.
7. $\sqrt[3]{-27} = -3$ because $(-3)^3 = -27$.
8. $\sqrt[3]{\dfrac{1}{125}} = \dfrac{1}{5}$ because $\left(\dfrac{1}{5}\right)^3 = \dfrac{1}{125}$.

■ Work Practice 6–8

Objective C Finding *n*th Roots

Just as we can raise a real number to powers other than 2 or 3, we can find roots other than square roots and cube roots. In fact, we can take the *n*th root of a number when *n* is any natural number. An ***n*th root** of a number *a* is a number whose *n*th power is *a*.

In symbols, the *n*th root of *a* is written as $\sqrt[n]{a}$. Recall that *n* is called the **index.** The index 2 is usually omitted for square roots.

Helpful Hint

If the index is even, as it is in $\sqrt{}, \sqrt[4]{}, \sqrt[6]{}$, and so on, the radicand must be nonnegative for the root to be a real number. For example,

$\sqrt[4]{81} = 3$ but $\sqrt[4]{-81}$ is not a real number.
$\sqrt[6]{64} = 2$ but $\sqrt[6]{-64}$ is not a real number.

Answers

6. 3   7. -2 8. $\dfrac{1}{4}$

✓**Concept Check** Which of the following is a real number?
a. $\sqrt{-64}$ b. $\sqrt[4]{-64}$ c. $\sqrt[5]{-64}$ d. $\sqrt[6]{-64}$

Examples Find each root.

9. $\sqrt[4]{16} = 2$ because $2^4 = 16$ and 2 is positive.
10. $\sqrt[5]{-32} = -2$ because $(-2)^5 = -32$.
11. $-\sqrt[6]{1} = -1$ because $\sqrt[6]{1} = 1$.
12. $\sqrt[4]{-81}$ is not a real number since the index, 4, is even and the radicand, -81, is negative. In other words, there is no real number that when raised to the 4th power gives -81.

Work Practice 9–12

Practice 9–12
Find each root.
9. $\sqrt[4]{-16}$
10. $\sqrt[5]{-1}$
11. $\sqrt[4]{256}$
12. $\sqrt[6]{-1}$

Objective D Approximating Square Roots

Recall that numbers such as 1, 4, 9, 25, and $\frac{4}{25}$ are called **perfect squares**, since $1^2 = 1, 2^2 = 4, 3^2 = 9, 5^2 = 25$, and $\left(\frac{2}{5}\right)^2 = \frac{4}{25}$. Square roots of perfect square radicands simplify to rational numbers.

What happens when we try to simplify a root such as $\sqrt{3}$? Since 3 is not a perfect square, $\sqrt{3}$ is not a rational number. It cannot be written as a quotient of integers. It is called an **irrational number** and we can find a decimal **approximation** of it. To find decimal approximations, use a calculator or Appendix A.4. (For calculator help, see the next example or the box at the end of this section.)

Example 13 Use a calculator or Appendix A.4 to approximate $\sqrt{3}$ to three decimal places.

Solution: We may use Appendix A.4 or a calculator to approximate $\sqrt{3}$. To use a calculator, find the square root key $\boxed{\sqrt{}}$.

$\sqrt{3} \approx 1.732050808$

To three decimal places, $\sqrt{3} \approx 1.732$.

Work Practice 13

From Example 13, we found that
$\sqrt{3} \approx 1.732$

To see if the approximation is reasonable, notice that since
$1 < 3 < 4$, then
$\sqrt{1} < \sqrt{3} < \sqrt{4}$, or
$1 < \sqrt{3} < 2$.

Since $\sqrt{3}$ is a number between 1 and 2, our result of $\sqrt{3} \approx 1.732$ is reasonable.

Practice 13
Use a calculator or Appendix A.4 to approximate $\sqrt{22}$ to three decimal places.

Objective E Simplifying Radicals Containing Variables

Radicals can also contain variables. To simplify radicals containing variables, special care must be taken. To see how we simplify $\sqrt{x^2}$, let's look at a few examples in this form.

If $x = 3$, we have $\sqrt{3^2} = \sqrt{9} = 3$, or x.
If x is 5, we have $\sqrt{5^2} = \sqrt{25} = 5$, or x.

From these two examples, you may think that $\sqrt{x^2}$ simplifies to x. Let's now look at an example where x is a negative number. If $x = -3$, we have $\sqrt{(-3)^2} = \sqrt{9} = 3$, not -3, our original x. To make sure that $\sqrt{x^2}$ simplifies to a nonnegative number, we have the following.

Answers
9. not a real number 10. -1 11. 4
12. not a real number 13. 4.690

Concept Check Answer
c

For any real number a,
$$\sqrt{a^2} = |a|$$

Thus,
$$\sqrt{x^2} = |x|,$$
$$\sqrt{(-8)^2} = |-8| = 8$$
$$\sqrt{(7y)^2} = |7y|, \quad \text{and so on.}$$

To avoid this confusion, for the rest of the chapter we assume that **if a variable appears in the radicand of a radical expression, it represents positive numbers only.** Then

$$\sqrt{x^2} = |x| = x \text{ since } x \text{ is a positive number.}$$
$$\sqrt{y^2} = y \qquad \text{Because } (y)^2 = y^2$$
$$\sqrt{x^8} = x^4 \qquad \text{Because } (x^4)^2 = x^8$$
$$\sqrt{9x^2} = 3x \qquad \text{Because } (3x)^2 = 9x^2$$

Practice 14–19

Simplify each expression. Assume that all variables represent positive numbers.

14. $\sqrt{z^8}$
15. $\sqrt{x^{20}}$
16. $\sqrt{4x^6}$
17. $\sqrt[3]{8y^{12}}$
18. $\sqrt{\dfrac{z^8}{81}}$
19. $\sqrt[3]{-64x^9 y^{24}}$

Answers
14. z^4 15. x^{10} 16. $2x^3$
17. $2y^4$ 18. $\dfrac{z^4}{9}$ 19. $-4x^3 y^8$

Examples Simplify each expression. Assume that all variables represent positive numbers.

14. $\sqrt{z^2} = z$ because $(z)^2 = z^2$.
15. $\sqrt{x^6} = x^3$ because $(x^3)^2 = x^6$.
16. $\sqrt[3]{27y^6} = 3y^2$ because $(3y^2)^3 = 27y^6$.
17. $\sqrt{16x^{16}} = 4x^8$ because $(4x^8)^2 = 16x^{16}$.
18. $\sqrt{\dfrac{x^4}{25}} = \dfrac{x^2}{5}$ because $\left(\dfrac{x^2}{5}\right)^2 = \dfrac{x^4}{25}$.
19. $\sqrt[3]{-125a^{12}b^{15}} = -5a^4 b^5$ because $(-5a^4 b^5)^3 = -125a^{12}b^{15}$.

■ Work Practice 14–19

Calculator Explorations Simplifying Square Roots

To simplify or approximate square roots using a calculator, locate the key marked $\boxed{\sqrt{}}$. To simplify $\sqrt{25}$ using a scientific calculator, press $\boxed{25}$ $\boxed{\sqrt{}}$. The display should read $\boxed{5}$. To simplify $\sqrt{25}$ using a graphing calculator, press $\boxed{\sqrt{}}$ $\boxed{25}$ $\boxed{\text{ENTER}}$.

To approximate $\sqrt{30}$, press $\boxed{30}$ $\boxed{\sqrt{}}$ (or $\boxed{\sqrt{}}$ $\boxed{30}$). The display should read $\boxed{5.477225575}$. This is an approximation for $\sqrt{30}$. A three-decimal-place approximation is

$$\sqrt{30} \approx 5.477$$

Is this answer reasonable? Since 30 is between perfect squares 25 and 36, $\sqrt{30}$ is between $\sqrt{25} = 5$ and $\sqrt{36} = 6$. The calculator result is then reasonable since 5.477225575 is between 5 and 6.

Use a calculator to approximate each expression to three decimal places. Decide whether each result is reasonable.

1. $\sqrt{6}$
2. $\sqrt{14}$
3. $\sqrt{11}$
4. $\sqrt{200}$
5. $\sqrt{82}$
6. $\sqrt{46}$

Many scientific calculators have a key, such as $\boxed{\sqrt[x]{y}}$, that can be used to approximate roots other than square roots. To approximate these roots using a graphing calculator, look under the $\boxed{\text{MATH}}$ menu or consult your manual. To use a $\boxed{\sqrt[x]{y}}$ key to find $\sqrt[3]{8}$, press $\boxed{3}$ $\boxed{\sqrt[x]{y}}$ $\boxed{8}$ (press $\boxed{\text{ENTER}}$ if needed). The display should read $\boxed{2}$.

Use a calculator to approximate each expression to three decimal places. Decide whether each result is reasonable.

7. $\sqrt[3]{40}$
8. $\sqrt[3]{71}$
9. $\sqrt[4]{20}$
10. $\sqrt[4]{15}$
11. $\sqrt[5]{18}$
12. $\sqrt[6]{2}$

Section 15.1 Introduction to Radicals

Vocabulary, Readiness & Video Check

Use the choices below to fill in each blank.

| positive | index | radical sign | power |
| negative | principal | square root | radicand |

1. The symbol $\sqrt{}$ is used to denote the positive, or _____, square root.
2. In the expression $\sqrt[4]{16}$, the number 4 is called the _____, the number 16 is called the _____, and $\sqrt{}$ is called the _____.
3. The reverse operation of squaring a number is finding a(n) _____ of a number.
4. For a positive number a,
   $-\sqrt{a}$ is the _____ square root of a and
   $\sqrt{a}$ is the _____ square root of a.
5. An nth root of a number a is a number whose nth _____ is a.

Answer each true or false.

6. $\sqrt{4} = -2$ _____
7. $\sqrt{-9} = -3$ _____
8. $\sqrt{1000} = 100$ _____
9. $\sqrt{1} = 1$ and $\sqrt{0} = 0$ _____
10. $\sqrt{64} = 8$ and $\sqrt[3]{64} = 4$ _____

Martin-Gay Interactive Videos — Watch the section lecture video and answer the following questions.

See Video 15.1

Objective A 11. Explain the differences between Examples 1 and 2, including how we know which one simplifies to a positive number and which one simplifies to a negative number.

Objective B 12. From Example 11, what is an important difference between the square root and the cube root of a negative number?

Objective C 13. From Examples 12–15, given a negative radicand, what kind of index must we have to be a real number?

Objective D 14. From Example 16, how do we determine if an approximate answer is reasonable?

Objective E 15. As explained in Example 19, when simplifying radicals containing variables, what is a shortcut you can use when dealing with exponents?

15.1 Exercise Set  MyLab Math

Objective A *Find each square root. See Examples 1 through 5.*

1. $\sqrt{16}$
2. $\sqrt{64}$
3. $\sqrt{\dfrac{1}{25}}$
4. $\sqrt{\dfrac{1}{64}}$
5. $-\sqrt{100}$
6. $-\sqrt{36}$
7. $\sqrt{-4}$
8. $\sqrt{-25}$
9. $-\sqrt{121}$
10. $-\sqrt{49}$

11. $\sqrt{\dfrac{9}{25}}$ 12. $\sqrt{\dfrac{4}{81}}$ 13. $\sqrt{900}$ 14. $\sqrt{400}$ 15. $\sqrt{144}$

16. $\sqrt{169}$ 17. $\sqrt{\dfrac{1}{100}}$ 18. $\sqrt{\dfrac{1}{121}}$ 19. $\sqrt{0.25}$ 20. $\sqrt{0.49}$

Objective **B** *Find each cube root. See Examples 6 through 8.*

21. $\sqrt[3]{125}$ 22. $\sqrt[3]{64}$ 23. $\sqrt[3]{-64}$ 24. $\sqrt[3]{-27}$ 25. $-\sqrt[3]{8}$

26. $-\sqrt[3]{27}$ 27. $\sqrt[3]{\dfrac{1}{8}}$ 28. $\sqrt[3]{\dfrac{1}{64}}$ 29. $\sqrt[3]{-125}$ 30. $\sqrt[3]{-1}$

Objectives **A B C** Mixed Practice *Find each root. See Examples 1 through 12.*

31. $\sqrt[5]{32}$ 32. $\sqrt[4]{81}$ 33. $\sqrt{81}$ 34. $\sqrt{49}$

35. $\sqrt[4]{-16}$ 36. $\sqrt{-9}$ 37. $\sqrt[3]{-\dfrac{27}{64}}$ 38. $\sqrt[3]{-\dfrac{8}{27}}$

39. $-\sqrt[4]{625}$ 40. $-\sqrt[5]{32}$ 41. $\sqrt[6]{1}$ 42. $\sqrt[5]{1}$

Objective **D** *Approximate each square root to three decimal places. See Example 13.*

43. $\sqrt{7}$ 44. $\sqrt{10}$ 45. $\sqrt{37}$ 46. $\sqrt{27}$ 47. $\sqrt{136}$ 48. $\sqrt{8}$

49. A standard baseball diamond is a square with 90-foot sides connecting the bases. The distance from home plate to second base is $90 \cdot \sqrt{2}$ feet. Approximate $\sqrt{2}$ to two decimal places and use your result to approximate the distance $90 \cdot \sqrt{2}$ feet.

50. The roof of the warehouse shown needs to be shingled. The total area of the roof is exactly $480 \cdot \sqrt{29}$ square feet. Approximate $\sqrt{29}$ to two decimal places and use your result to approximate the area $480 \cdot \sqrt{29}$ square feet. Approximate this area to the nearest whole number.

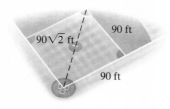

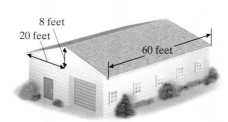

Objective **E** *Find each root. Assume that all variables represent positive numbers. See Examples 14 through 19.*

51. $\sqrt{m^2}$ 52. $\sqrt{y^{10}}$ 53. $\sqrt{x^4}$ 54. $\sqrt{x^6}$

55. $\sqrt{9x^8}$ 56. $\sqrt{36x^{12}}$ 57. $\sqrt{81x^2}$ 58. $\sqrt{100z^4}$

59. $\sqrt{a^2b^4}$ 60. $\sqrt{x^{12}y^{20}}$ 61. $\sqrt{16a^6b^4}$ 62. $\sqrt{4m^{14}n^2}$

63. $\sqrt[3]{a^6 b^{18}}$
64. $\sqrt[3]{x^{12} y^{18}}$
65. $\sqrt[3]{-8x^3 y^{27}}$
66. $\sqrt[3]{-27a^6 b^{30}}$
67. $\sqrt{\dfrac{x^6}{36}}$
68. $\sqrt{\dfrac{y^8}{49}}$
69. $\sqrt{\dfrac{25y^2}{9}}$
70. $\sqrt{\dfrac{4x^2}{81}}$

Review

Write each integer as a product of two integers such that one of the factors is a perfect square. For example, we can write $18 = 9 \cdot 2$, where 9 is a perfect square. See Section 4.2.

71. 50
72. 8
73. 32
74. 75
75. 28
76. 44
77. 27
78. 90

Concept Extensions

Solve. See the Concept Check in this section.

79. Which of the following is a real number?
    a. $\sqrt[7]{-1}$
    b. $\sqrt[3]{-125}$
    c. $\sqrt[6]{-128}$
    d. $\sqrt[8]{-1}$

80. Which of the following is a real number?
    a. $\sqrt{-1}$
    b. $\sqrt[3]{-1}$
    c. $\sqrt[4]{-1}$
    d. $\sqrt[5]{-1}$

The length of a side of a square is given by the expression $\sqrt{A}$, where A is the square's area. Use this expression for Exercises 81 through 84. Be sure to attach the appropriate units.

81. The area of a square is 49 square miles. Find the length of a side of the square.

82. The area of a square is $\dfrac{1}{81}$ square meters. Find the length of a side of the square.

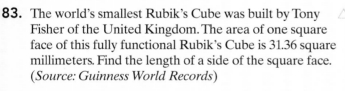

83. The world's smallest Rubik's Cube was built by Tony Fisher of the United Kingdom. The area of one square face of this fully functional Rubik's Cube is 31.36 square millimeters. Find the length of a side of the square face. (Source: Guinness World Records)

84. A parking lot is in the shape of a square with area 2500 square yards. Find the length of a side.

85. Simplify $\sqrt{\sqrt{81}}$.
86. Simplify $\sqrt[3]{\sqrt[3]{1}}$.
87. Simplify $\sqrt{\sqrt{10{,}000}}$.
88. Simplify $\sqrt{\sqrt{1{,}600{,}000{,}000}}$.

For each square root below, give two whole numbers that the square root lies between. For example, since 11 is between 9 and 16, then $\sqrt{11}$ is between 3 and 4.

89. $\sqrt{18}$
90. $\sqrt{28}$
91. $\sqrt{80}$
92. $\sqrt{98}$

93. The formula for calculating the period (one back-and-forth swing) of a pendulum is $T = 2\pi\sqrt{\dfrac{L}{g}}$, where T is time of the period of the swing, L is the length of the pendulum, and g is the acceleration of gravity. At the California Academy of Sciences, one can see a Foucault's pendulum with length = 30 ft and $g = 32$ ft/sec^2. Using $\pi \approx 3.14$, find the period of this pendulum. (Round to the nearest tenth of a second.)

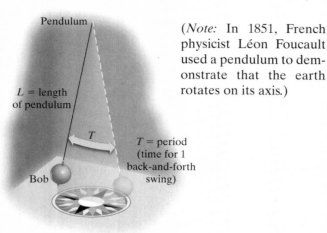

(*Note:* In 1851, French physicist Léon Foucault used a pendulum to demonstrate that the earth rotates on its axis.)

94. If the amount of gold discovered by humankind could be assembled in one place, it would be a cube with a volume of 195,112 cubic feet. Each side of the cube would be $\sqrt[3]{195{,}112}$ feet long. How long would one side of the cube be? (*Source: Reader's Digest*)

95. Explain why the square root of a negative number is not a real number.

96. Explain why the cube root of a negative number is a real number.

97. Graph $y = \sqrt{x}$. (Complete the table below, plot the ordered pair solutions, and draw a smooth curve through the points. Remember that since the radicand cannot be negative, this particular graph begins at the point with coordinates $(0, 0)$.)

x	y
0	0
1	
3	(approximate)
4	
9	

98. Graph $y = \sqrt[3]{x}$. (Complete the table below, plot the ordered pair solutions, and draw a smooth curve through the points.)

x	y
-8	
-2	(approximate)
-1	
0	
1	
2	(approximate)
8	

Recall from this section that $\sqrt{a^2} = |a|$ for any real number a. Simplify the following given that x represents any real number.

99. $\sqrt{x^2}$

100. $\sqrt{4x^2}$

101. $\sqrt{(x+2)^2}$

102. $\sqrt{x^2 + 6x + 9}$
(*Hint:* First factor $x^2 + 6x + 9$.)

Use a graphing calculator and graph each function. Observe the graph from left to right and give the ordered pair that corresponds to the "beginning" of the graph. Then tell why the graph starts at that point.

103. $y = \sqrt{x - 2}$

104. $y = \sqrt{x + 3}$

105. $y = \sqrt{x + 4}$

106. $y = \sqrt{x - 5}$

15.2 Simplifying Radicals

Objective A Simplifying Radicals Using the Product Rule

A square root is simplified when the radicand contains no perfect square factors (other than 1). For example, $\sqrt{20}$ is not simplified because $\sqrt{20} = \sqrt{4 \cdot 5}$ and 4 is a perfect square.

To begin simplifying square roots, we notice the following pattern.

$$\sqrt{9 \cdot 16} = \sqrt{144} = 12$$
$$\sqrt{9} \cdot \sqrt{16} = 3 \cdot 4 = 12$$

Since both expressions simplify to 12, we can write

$$\sqrt{9 \cdot 16} = \sqrt{9} \cdot \sqrt{16}$$

This suggests the following product rule for square roots.

Objectives

A Use the Product Rule to Simplify Radicals.

B Use the Quotient Rule to Simplify Radicals.

C Use Both Rules to Simplify Radicals Containing Variables.

D Simplify Cube Roots.

Product Rule for Square Roots

If $\sqrt{a}$ and $\sqrt{b}$ are real numbers, then

$$\sqrt{a \cdot b} = \sqrt{a} \cdot \sqrt{b}$$

In other words, the square root of a product is equal to the product of the square roots.

To simplify $\sqrt{45}$, for example, we factor 45 so that one of its factors is a perfect square factor.

$$\sqrt{45} = \sqrt{9 \cdot 5} \quad \text{Factor 45.}$$
$$= \sqrt{9} \cdot \sqrt{5} \quad \text{Use the product rule.}$$
$$= 3\sqrt{5} \quad \text{Write } \sqrt{9} \text{ as 3.}$$

The notation $3\sqrt{5}$ means $3 \cdot \sqrt{5}$. Since the radicand 5 has no perfect square factor other than 1, the expression $3\sqrt{5}$ is in simplest form.

> **Helpful Hint**
> Remember that the notation
> $3\sqrt{5}$ means $3 \cdot \sqrt{5}$.

> **Helpful Hint**
> A radical expression in simplest form *does not mean* a decimal approximation. The simplest form of a radical expression is an exact form and may still contain a radical.
> $\underbrace{\sqrt{45} = 3\sqrt{5}}_{\text{exact}} \quad \underbrace{\sqrt{45} \approx 6.71}_{\text{decimal approximation}}$

Examples Simplify.

1. $\sqrt{54} = \sqrt{9 \cdot 6}$ Factor 54 so that one factor is a perfect square. 9 is a perfect square.
$\phantom{\sqrt{54}} = \sqrt{9} \cdot \sqrt{6}$ Use the product rule.
$\phantom{\sqrt{54}} = 3\sqrt{6}$ Write $\sqrt{9}$ as 3.

2. $\sqrt{12} = \sqrt{4 \cdot 3}$ Factor 12 so that one factor is a perfect square. 4 is a perfect square.
$\phantom{\sqrt{12}} = \sqrt{4} \cdot \sqrt{3}$ Use the product rule.
$\phantom{\sqrt{12}} = 2\sqrt{3}$ Write $\sqrt{4}$ as 2.

Practice 1–4

Simplify.

1. $\sqrt{40}$ **2.** $\sqrt{18}$
3. $\sqrt{500}$ **4.** $\sqrt{15}$

Answers
1. $2\sqrt{10}$ **2.** $3\sqrt{2}$ **3.** $10\sqrt{5}$
4. $\sqrt{15}$

1145

3. $\sqrt{200} = \sqrt{100 \cdot 2}$ Factor 200 so that one factor is a perfect square. 100 is a perfect square.
$= \sqrt{100} \cdot \sqrt{2}$ Use the product rule.
$= 10\sqrt{2}$ Write $\sqrt{100}$ as 10.

4. $\sqrt{35}$ The radicand 35 contains no perfect square factors other than 1. Thus $\sqrt{35}$ is in simplest form.

■ Work Practice 1–4

In Example 3, 100 is the largest perfect square factor of 200. What happens if we don't use the largest perfect square factor? Although using the largest perfect square factor saves time, the result is the same no matter what perfect square factor is used. For example, it is also true that $200 = 4 \cdot 50$. Then

$$\sqrt{200} = \sqrt{4} \cdot \sqrt{50}$$
$$= 2 \cdot \sqrt{50}$$

Since $\sqrt{50}$ is not in simplest form, we continue.

$$\sqrt{200} = 2 \cdot \sqrt{50}$$
$$= 2 \cdot \sqrt{25 \cdot 2}$$
$$= 2 \cdot \sqrt{25} \cdot \sqrt{2}$$
$$= 2 \cdot 5 \cdot \sqrt{2}$$
$$= 10\sqrt{2}$$

Practice 5
Simplify $7\sqrt{75}$.

Example 5   Simplify $3\sqrt{8}$.

Solution: Remember that $3\sqrt{8}$ means $3 \cdot \sqrt{8}$.

$3 \cdot \sqrt{8} = 3 \cdot \sqrt{4 \cdot 2}$ Factor 8 so that one factor is a perfect square.
$= 3 \cdot \sqrt{4} \cdot \sqrt{2}$ Use the product rule.
$= 3 \cdot 2 \cdot \sqrt{2}$ Write $\sqrt{4}$ as 2.
$= 6 \cdot \sqrt{2}$ or $6\sqrt{2}$ Write $3 \cdot 2$ as 6.

■ Work Practice 5

Objective B   Simplifying Radicals Using the Quotient Rule ▶

Next, let's examine the square root of a quotient.

$$\sqrt{\frac{16}{4}} = \sqrt{4} = 2$$

Also,

$$\frac{\sqrt{16}}{\sqrt{4}} = \frac{4}{2} = 2$$

Since both expressions equal 2, we can write

$$\sqrt{\frac{16}{4}} = \frac{\sqrt{16}}{\sqrt{4}}$$

This suggests the following quotient rule.

Answer
5. $35\sqrt{3}$

Section 15.2 | Simplifying Radicals 1147

Quotient Rule for Square Roots

If $\sqrt{a}$ and $\sqrt{b}$ are real numbers and $b \neq 0$, then

$$\sqrt{\frac{a}{b}} = \frac{\sqrt{a}}{\sqrt{b}}$$

In other words, the square root of a quotient is equal to the quotient of the square roots.

Examples Use the quotient rule to simplify.

6. $\sqrt{\dfrac{25}{36}} = \dfrac{\sqrt{25}}{\sqrt{36}} = \dfrac{5}{6}$

7. $\sqrt{\dfrac{3}{64}} = \dfrac{\sqrt{3}}{\sqrt{64}} = \dfrac{\sqrt{3}}{8}$

8. $\sqrt{\dfrac{40}{81}} = \dfrac{\sqrt{40}}{\sqrt{81}}$ Use the quotient rule.

$\phantom{8.\ \sqrt{\dfrac{40}{81}}} = \dfrac{\sqrt{4} \cdot \sqrt{10}}{9}$ Use the product rule and write $\sqrt{81}$ as 9.

$\phantom{8.\ \sqrt{\dfrac{40}{81}}} = \dfrac{2\sqrt{10}}{9}$ Write $\sqrt{4}$ as 2.

■ Work Practice 6–8

Practice 6–8

Use the quotient rule to simplify.

6. $\sqrt{\dfrac{16}{81}}$

7. $\sqrt{\dfrac{2}{25}}$

8. $\sqrt{\dfrac{45}{49}}$

Objective C Simplifying Radicals Containing Variables

Recall that $\sqrt{x^6} = x^3$ because $(x^3)^2 = x^6$. If a variable radicand in a square root has an odd exponent, we write the exponential expression so that one factor is the greatest even power contained in the expression. Then we use the product rule to simplify.

Examples Simplify each radical. Assume that all variables represent positive numbers.

9. $\sqrt{x^5} = \sqrt{x^4 \cdot x} = \sqrt{x^4} \cdot \sqrt{x} = x^2\sqrt{x}$

10. $\sqrt{8y^2} = \sqrt{4 \cdot 2 \cdot y^2} = \sqrt{4y^2 \cdot 2} = \sqrt{4y^2} \cdot \sqrt{2} = 2y\sqrt{2}$; 4 and y^2 are both perfect square factors so we grouped them under one radical.

11. $\sqrt{\dfrac{45}{x^6}} = \dfrac{\sqrt{45}}{\sqrt{x^6}} = \dfrac{\sqrt{9 \cdot 5}}{x^3} = \dfrac{\sqrt{9} \cdot \sqrt{5}}{x^3} = \dfrac{3\sqrt{5}}{x^3}$

12. $\sqrt{\dfrac{5p^3}{9}} = \dfrac{\sqrt{5p^3}}{\sqrt{9}} = \dfrac{\sqrt{p^2 \cdot 5p}}{3} = \dfrac{\sqrt{p^2} \cdot \sqrt{5p}}{3} = \dfrac{p\sqrt{5p}}{3}$

■ Work Practice 9–12

Practice 9–12

Simplify each radical. Assume that all variables represent positive numbers.

9. $\sqrt{x^{11}}$ 10. $\sqrt{18x^4}$

11. $\sqrt{\dfrac{27}{x^8}}$ 12. $\sqrt{\dfrac{7y^7}{25}}$

Answers

6. $\dfrac{4}{9}$ 7. $\dfrac{\sqrt{2}}{5}$ 8. $\dfrac{3\sqrt{5}}{7}$ 9. $x^5\sqrt{x}$

10. $3x^2\sqrt{2}$ 11. $\dfrac{3\sqrt{3}}{x^4}$ 12. $\dfrac{y^3\sqrt{7y}}{5}$

Objective D Simplifying Cube Roots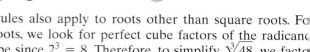

The product and quotient rules also apply to roots other than square roots. For example, to simplify cube roots, we look for perfect cube factors of the radicand. Recall that 8 is a perfect cube since $2^3 = 8$. Therefore, to simplify $\sqrt[3]{48}$, we factor 48 as $8 \cdot 6$.

$$\sqrt[3]{48} = \sqrt[3]{8 \cdot 6} \qquad \text{Factor 48.}$$
$$= \sqrt[3]{8} \cdot \sqrt[3]{6} \qquad \text{Use the product rule.}$$
$$= 2\sqrt[3]{6} \qquad \text{Write } \sqrt[3]{8} \text{ as 2.}$$

$2\sqrt[3]{6}$ is in simplest form since the radicand, 6, contains no perfect cube factors other than 1.

Practice 13–16

Simplify each radical.
13. $\sqrt[3]{88}$
14. $\sqrt[3]{50}$
15. $\sqrt[3]{\dfrac{10}{27}}$
16. $\sqrt[3]{\dfrac{81}{8}}$

Answers
13. $2\sqrt[3]{11}$ 14. $\sqrt[3]{50}$ 15. $\dfrac{\sqrt[3]{10}}{3}$
16. $\dfrac{3\sqrt[3]{3}}{2}$

Examples Simplify each radical.

13. $\sqrt[3]{54} = \sqrt[3]{27 \cdot 2} = \sqrt[3]{27} \cdot \sqrt[3]{2} = 3\sqrt[3]{2}$
14. $\sqrt[3]{18}$ The number 18 contains no perfect cube factors, so $\sqrt[3]{18}$ cannot be simplified further.
15. $\sqrt[3]{\dfrac{7}{8}} = \dfrac{\sqrt[3]{7}}{\sqrt[3]{8}} = \dfrac{\sqrt[3]{7}}{2}$
16. $\sqrt[3]{\dfrac{40}{27}} = \dfrac{\sqrt[3]{40}}{\sqrt[3]{27}} = \dfrac{\sqrt[3]{8 \cdot 5}}{3} = \dfrac{\sqrt[3]{8} \cdot \sqrt[3]{5}}{3} = \dfrac{2\sqrt[3]{5}}{3}$

■ Work Practice 13–16

Vocabulary, Readiness & Video Check

Use the choices below to fill in each blank. Not all choices will be used.

$a \cdot b$ $\qquad \dfrac{a}{b} \qquad \dfrac{\sqrt{a}}{\sqrt{b}} \qquad \sqrt{a} \cdot \sqrt{b}$

1. If $\sqrt{a}$ and $\sqrt{b}$ are real numbers, then $\sqrt{a \cdot b} =$ _____.
2. If $\sqrt{a}$ and $\sqrt{b}$ are real numbers, then $\sqrt{\dfrac{a}{b}} =$ _____.

For Exercises 3 and 4, fill in the blanks using the example: $\sqrt{4 \cdot 9} = \sqrt{4} \cdot \sqrt{9} = 2 \cdot 3 = 6$.

3. $\sqrt{16 \cdot 25} = \sqrt{\_\_} \cdot \sqrt{\_\_} = \_\_ \cdot \_\_ = \_\_$.
4. $\sqrt{36 \cdot 3} = \sqrt{\_\_} \cdot \sqrt{\_\_} = \_\_ \cdot \sqrt{\_\_} = \_\_$.

True or False? Decide whether each radical is completely simplified.

5. $\sqrt{48} = 2\sqrt{12}$ Completely simplified: _____
6. $\sqrt[3]{48} = 2\sqrt[3]{6}$ Completely simplified: _____

Section 15.2 | Simplifying Radicals

Martin-Gay Interactive Videos

See Video 15.2

Watch the section lecture video and answer the following questions.

Objective A 7. From Example 3, if we have trouble finding a perfect square factor in the radicand, what is recommended?

Objective B 8. Based on the lecture before Example 5, complete the following statement: In words, the quotient rule for square roots says that the square root of a quotient is equal to the square root of the _____ over the square root of the _____.

Objective C 9. From Examples 6–8, we know that even powers of a variable are perfect square factors of the variable. Therefore, what must be true about the power of any variable left in the radicand of a simplified square root? Explain.

Objective D 10. From Example 9, how does factoring the radicand as a product of primes help simplify higher roots also?

15.2 Exercise Set MyLab Math

Objective A *Use the product rule to simplify each radical. See Examples 1 through 4.*

1. $\sqrt{20}$
2. $\sqrt{44}$
3. $\sqrt{50}$
4. $\sqrt{28}$
5. $\sqrt{33}$

6. $\sqrt{21}$
7. $\sqrt{98}$
8. $\sqrt{125}$
9. $\sqrt{60}$
10. $\sqrt{90}$

11. $\sqrt{180}$
12. $\sqrt{150}$
13. $\sqrt{52}$
14. $\sqrt{75}$

Use the product rule to simplify each radical. See Example 5.

15. $3\sqrt{25}$
16. $9\sqrt{36}$
17. $7\sqrt{63}$

18. $11\sqrt{99}$
19. $-5\sqrt{27}$
20. $-6\sqrt{75}$

Objective B *Use the quotient rule and the product rule to simplify each radical. See Examples 6 through 8.*

21. $\sqrt{\dfrac{8}{25}}$
22. $\sqrt{\dfrac{63}{16}}$
23. $\sqrt{\dfrac{27}{121}}$
24. $\sqrt{\dfrac{24}{169}}$

25. $\sqrt{\dfrac{9}{4}}$
26. $\sqrt{\dfrac{100}{49}}$
27. $\sqrt{\dfrac{125}{9}}$
28. $\sqrt{\dfrac{27}{100}}$

29. $\sqrt{\dfrac{11}{36}}$
30. $\sqrt{\dfrac{30}{49}}$
31. $-\sqrt{\dfrac{27}{144}}$
32. $-\sqrt{\dfrac{84}{121}}$

Objective **C** Simplify each radical. Assume that all variables represent positive numbers. See Examples 9 through 12.

33. $\sqrt{x^7}$
34. $\sqrt{y^3}$
35. $\sqrt{x^{13}}$
36. $\sqrt{y^{17}}$

37. $\sqrt{36a^3}$
38. $\sqrt{81b^5}$
39. $\sqrt{96x^4}$
40. $\sqrt{40y^{10}}$

41. $\sqrt{\dfrac{12}{m^2}}$
42. $\sqrt{\dfrac{63}{p^2}}$
43. $\sqrt{\dfrac{9x}{y^{10}}}$
44. $\sqrt{\dfrac{6y^2}{z^{16}}}$

45. $\sqrt{\dfrac{88}{x^{12}}}$
46. $\sqrt{\dfrac{500}{y^{22}}}$

Objectives **A B C** Mixed Practice Simplify each radical. See Examples 1 through 12.

47. $8\sqrt{4}$
48. $6\sqrt{49}$
49. $\sqrt{\dfrac{36}{121}}$
50. $\sqrt{\dfrac{25}{144}}$

51. $\sqrt{175}$
52. $\sqrt{700}$
53. $\sqrt{\dfrac{20}{9}}$
54. $\sqrt{\dfrac{45}{64}}$

55. $\sqrt{24m^7}$
56. $\sqrt{50n^{13}}$
57. $\sqrt{\dfrac{23y^3}{4x^6}}$
58. $\sqrt{\dfrac{41x^5}{9y^8}}$

Objective **D** Simplify each radical. See Examples 13 through 16.

59. $\sqrt[3]{24}$
60. $\sqrt[3]{81}$
61. $\sqrt[3]{250}$
62. $\sqrt[3]{56}$

63. $\sqrt[3]{\dfrac{5}{64}}$
64. $\sqrt[3]{\dfrac{32}{125}}$
65. $\sqrt[3]{\dfrac{23}{8}}$
66. $\sqrt[3]{\dfrac{37}{27}}$

67. $\sqrt[3]{\dfrac{15}{64}}$
68. $\sqrt[3]{\dfrac{4}{27}}$
69. $\sqrt[3]{270}$
70. $\sqrt[3]{108}$

Review

Perform each indicated operation. See Sections 10.4 and 10.5.

71. $6x + 8x$
72. $(6x)(8x)$
73. $(2x + 3)(x - 5)$

74. $(2x + 3) + (x - 5)$
75. $9y^2 - 9y^2$
76. $(9y^2)(-8y^2)$

Concept Extensions

Simplify each radical. Assume that all variables represent positive numbers.

77. $\sqrt{x^6 y^3}$
78. $\sqrt{a^{13} b^{14}}$
79. $\sqrt{98x^5 y^4}$

80. $\sqrt{27x^8 y^{11}}$
81. $\sqrt[3]{-8x^6}$
82. $\sqrt[3]{27x^{12}}$

3. If a cube is to have a volume of 80 cubic inches, then each side must be $\sqrt[3]{80}$ inches long. Simplify the radical representing the side length.

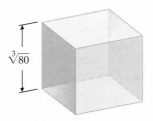

84. Jeannie Boswell is swimming across a 40-foot-wide river, trying to head straight across to the opposite shore. However, the current is strong enough to move her downstream 100 feet by the time she reaches land. (See the figure.) Because of the current, the actual distance she swims is $\sqrt{11{,}600}$ feet. Simplify this radical.

5. By using replacement values for a and b, show that $\sqrt{a^2 + b^2}$ does not equal $a + b$.

86. By using replacement values for a and b, show that $\sqrt{a + b}$ does not equal $\sqrt{a} + \sqrt{b}$.

7. The "Water Cube" was the swimming and diving venue for the 2008 Beijing Summer Olympics. It is not actually a cube, because it is only 31 meters tall, which is not the same as its width and length. However, the roof of it is a square. If the area of the roof of the Water Cube is 31,329 square meters, find the dimensions of the roof of the Water Cube.

8. The competition diving pool in the Water Cube at the Beijing Summer Olympics is not a cube either. It has a square footprint, but is only 5 meters deep. If the volume of the diving pool is 3125 cubic meters, find the length and width of the competition diving pool.

The length of a side of a cube is given by the expression $\dfrac{\sqrt{6A}}{6}$, where A is the cube's surface area. Use this expression for Exercises 89 through 92. Be sure to attach the appropriate units.

9. The surface area of a cube is 120 square inches. Find the exact length of a side of the cube.

0. The surface area of a cube is 594 square feet. Find the exact length of a side of the cube.

Solve.

91. Rubik's Cube, named after its inventor, Erno Rubik, was first imagined by him in 1974, and by 1980 was a worldwide phenomenon. A standard Rubik's Cube has a surface area of 30.375 square inches. Find the length of one side of a Rubik's Cube. (A few world records are listed below. *Source: Guinness World Records*)
- Fastest time to solve 1 Rubik's Cube: 4.73 sec by Feliks Zemdegs (Australia) in 2016.
- Most Rubik's Cubes solved in 1 hour: 65 by Aashik Madhav G.S. (India) in 2017.

92. In 2011, Apple renovated its flagship Apple Store on Fifth Avenue in New York City, taking advantage of advances in glass manufacturing to simplify the giant glass cube that serves as the store's entrance. A cube of this size has a surface area of 6144 square feet. Find the length of one side of the Apple Store glass cube. (*Source:* Based on data from AppleInsider.com)

The cost C in dollars per day to operate a small delivery service is given by $C = 100\sqrt[3]{n} + 700$, where n is the number of deliveries per day.

93. Find the cost if the number of deliveries is 1000.

94. Approximate the cost if the number of deliveries is 500.

The Mosteller formula for calculating body surface area is $B = \sqrt{\dfrac{hw}{3600}}$, where B is an individual's body surface area in square meters, h is the individual's height in centimeters, and w is the individual's weight in kilograms. Use this formula in Exercises 95 and 96. Round answers to the nearest tenth.

95. Find the body surface area of a person who is 169 cm tall and weighs 64 kilograms.

96. Approximate the body surface area of a person who is 183 cm tall and weighs 85 kilograms.

15.3 Adding and Subtracting Radicals

Objective A Adding and Subtracting Radicals

Recall that to combine like terms, we use the distributive property.

$5x + 3x = (5 + 3)x = 8x$

The distributive property can also be applied to expressions containing the same radicals. For example,

$5\sqrt{2} + 3\sqrt{2} = (5 + 3)\sqrt{2} = 8\sqrt{2}$

Also,

$9\sqrt{5} - 6\sqrt{5} = (9 - 6)\sqrt{5} = 3\sqrt{5}$

Radical terms such as $5\sqrt{2}$ and $3\sqrt{2}$ are **like radicals,** as are $9\sqrt{5}$ and $6\sqrt{5}$. Like radicals have the same index and the same radicand.

Objectives

A Add or Subtract Like Radicals.

B Simplify Square Root Radical Expressions, and Then Add or Subtract Any Like Radicals.

C Simplify Cube Root Radical Expressions, and Then Add or Subtract Any Like Radicals.

Examples Add or subtract as indicated.

1. $4\sqrt{5} + 3\sqrt{5} = (4 + 3)\sqrt{5} = 7\sqrt{5}$
2. $\sqrt{10} - 6\sqrt{10} = 1\sqrt{10} - 6\sqrt{10} = (1 - 6)\sqrt{10} = -5\sqrt{10}$
3. $2\sqrt{6} + 2\sqrt{5}$ cannot be simplified further since the radicands are not the same.
4. $\sqrt{15} + \sqrt{15} - \sqrt{2} = 1\sqrt{15} + 1\sqrt{15} - \sqrt{2}$
   $= (1 + 1)\sqrt{15} - \sqrt{2}$
   $= 2\sqrt{15} - \sqrt{2}$

This expression cannot be simplified further since the radicands are not the same.

■ Work Practice 1–4

Practice 1–4

Add or subtract as indicated.
1. $6\sqrt{11} + 9\sqrt{11}$
2. $\sqrt{7} - 3\sqrt{7}$
3. $\sqrt{2} + \sqrt{2} - \sqrt{15}$
4. $3\sqrt{3} - 3\sqrt{2}$

✓**Concept Check** Which is true?
a. $2 + 3\sqrt{5} = 5\sqrt{5}$
b. $2\sqrt{3} + 2\sqrt{7} = 2\sqrt{10}$
c. $\sqrt{3} + \sqrt{5} = \sqrt{8}$
d. $\sqrt{3} + \sqrt{3} = 3$
e. None of the above is true. In each case, the left-hand side cannot be simplified further.

Objective B Simplifying Square Root Radicals Before Adding or Subtracting

At first glance, it appears that the expression $\sqrt{50} + \sqrt{8}$ cannot be simplified further because the radicands are different. However, the product rule can be used to simplify each radical, and then further simplification might be possible.

Examples Simplify each radical expression.

5. $\sqrt{50} + \sqrt{8} = \sqrt{25 \cdot 2} + \sqrt{4 \cdot 2}$ Factor radicands.
   $= \sqrt{25} \cdot \sqrt{2} + \sqrt{4} \cdot \sqrt{2}$ Use the product rule.
   $= 5\sqrt{2} + 2\sqrt{2}$ Simplify $\sqrt{25}$ and $\sqrt{4}$.
   $= 7\sqrt{2}$ Add like radicals.

Practice 5–8

Simplify each radical expression.
5. $\sqrt{27} + \sqrt{75}$
6. $3\sqrt{20} - 7\sqrt{45}$
7. $\sqrt{36} - \sqrt{48} - 4\sqrt{3} - \sqrt{9}$
8. $\sqrt{9x^4} - \sqrt{36x^3} + \sqrt{x^3}$

Answers
1. $15\sqrt{11}$ 2. $-2\sqrt{7}$ 3. $2\sqrt{2} - \sqrt{15}$
4. $3\sqrt{3} - 3\sqrt{2}$ 5. $8\sqrt{3}$ 6. $-15\sqrt{5}$
7. $3 - 8\sqrt{3}$ 8. $3x^2 - 5x\sqrt{x}$

✓**Concept Check Answer**
e

1153

6. $7\sqrt{12} - 2\sqrt{75} = 7\sqrt{4 \cdot 3} - 2\sqrt{25 \cdot 3}$ Factor radicands.
$\phantom{7\sqrt{12} - 2\sqrt{75}} = 7\sqrt{4} \cdot \sqrt{3} - 2\sqrt{25} \cdot \sqrt{3}$ Use the product rule.
$\phantom{7\sqrt{12} - 2\sqrt{75}} = 7 \cdot 2\sqrt{3} - 2 \cdot 5\sqrt{3}$ Simplify $\sqrt{4}$ and $\sqrt{25}$.
$\phantom{7\sqrt{12} - 2\sqrt{75}} = 14\sqrt{3} - 10\sqrt{3}$ Multiply.
$\phantom{7\sqrt{12} - 2\sqrt{75}} = 4\sqrt{3}$ Subtract like radicals.

7. $\sqrt{25} - \sqrt{27} - 2\sqrt{18} - \sqrt{16}$
$= 5 - \sqrt{9 \cdot 3} - 2\sqrt{9 \cdot 2} - 4$ Factor radicands and simplify $\sqrt{25}$ and $\sqrt{16}$.
$= 5 - \sqrt{9} \cdot \sqrt{3} - 2\sqrt{9} \cdot \sqrt{2} - 4$ Use the product rule.
$= 5 - 3\sqrt{3} - 2 \cdot 3\sqrt{2} - 4$ Simplify $\sqrt{9}$.
$= 1 - 3\sqrt{3} - 6\sqrt{2}$ Write $5 - 4$ as 1 and $2 \cdot 3$ as 6.

8. $2\sqrt{x^2} - \sqrt{25x^5} + \sqrt{x^5}$
$= 2x - \sqrt{25x^4 \cdot x} + \sqrt{x^4 \cdot x}$ Factor radicands so that one factor is a perfect square. Simplify $\sqrt{x^2}$.
$= 2x - \sqrt{25x^4} \cdot \sqrt{x} + \sqrt{x^4} \cdot \sqrt{x}$ Use the product rule.
$= 2x - 5x^2\sqrt{x} + x^2\sqrt{x}$ Write $\sqrt{25x^4}$ as $5x^2$ and $\sqrt{x^4}$ as x^2.
$= 2x - 4x^2\sqrt{x}$ Add like radicals.

■ Work Practice 5–8

Objective C Simplifying Cube Root Radicals Before Adding or Subtracting

Practice 9

Simplify the radical expression.
$10\sqrt[3]{81p^6} - \sqrt[3]{24p^6}$

Example 9 Simplify the radical expression.

$5\sqrt[3]{16x^3} - \sqrt[3]{54x^3}$
$= 5\sqrt[3]{8x^3 \cdot 2} - \sqrt[3]{27x^3 \cdot 2}$ Factor radicands so that one factor is a perfect cube.
$= 5 \cdot \sqrt[3]{8x^3} \cdot \sqrt[3]{2} - \sqrt[3]{27x^3} \cdot \sqrt[3]{2}$ Use the product rule.
$= 5 \cdot 2x\sqrt[3]{2} - 3x\sqrt[3]{2}$ Write $\sqrt[3]{8x^3}$ as $2x$ and $\sqrt[3]{27x^3}$ as $3x$.
$= 10x\sqrt[3]{2} - 3x\sqrt[3]{2}$ Write $5 \cdot 2x$ as $10x$.
$= 7x\sqrt[3]{2}$ Subtract like radicals.

■ Work Practice 9

Answer
9. $28p^2\sqrt[3]{3}$

Vocabulary, Readiness & Video Check

Fill in each blank.

1. Radicals that have the same index and same radicand are called _____.

2. The expressions $7\sqrt[3]{2x}$ and $-\sqrt[3]{2x}$ are called _____.

3. $11\sqrt{2} + 6\sqrt{2} =$ _____.
  a. $66\sqrt{2}$ **b.** $17\sqrt{2}$ **c.** $17\sqrt{4}$

4. $\sqrt{5}$ is the same as _____.
  a. $0\sqrt{5}$ **b.** $1\sqrt{5}$ **c.** $5\sqrt{5}$

5. $\sqrt{5} + \sqrt{5} =$ _____
  a. $\sqrt{10}$ **b.** 5 **c.** $2\sqrt{5}$

6. $9\sqrt{7} - \sqrt{7} =$ _____
  a. $8\sqrt{7}$ **b.** 9 **c.** 0

Section 15.3 | Adding and Subtracting Radicals   1155

Martin-Gay Interactive Videos   Watch the section lecture video and answer the following questions.

See Video 15.3

Objective A 7. From Examples 1–4, how is combining like radicals similar to combining like terms?

Objective B 8. From Example 5, why should we always check to see if all radical terms in our expression are simplified before attempting to add or subtract the radicals?

Objective C 9. In Example 8, what property is used during the simplification of the expression?

15.3 Exercise Set MyLab Math

Objective A *Add or subtract as indicated. See Examples 1 through 4.*

1. $4\sqrt{3} - 8\sqrt{3}$
2. $\sqrt{5} - 9\sqrt{5}$
3. $3\sqrt{6} + 8\sqrt{6} - 2\sqrt{6} - 5$
4. $12\sqrt{2} - 3\sqrt{2} + 8\sqrt{2} + 10$
5. $6\sqrt{5} - 5\sqrt{5} + \sqrt{2}$
6. $4\sqrt{3} + \sqrt{5} - 3\sqrt{3}$
7. $2\sqrt{3} + 5\sqrt{3} - \sqrt{2}$
8. $8\sqrt{14} + 2\sqrt{14} + \sqrt{5}$
9. $2\sqrt{2} - 7\sqrt{2} - 6$
10. $5\sqrt{7} + 2 - 11\sqrt{7}$

Objective B *Add or subtract by first simplifying each radical and then combining any like radicals. Assume that all variables represent positive numbers. See Examples 5 through 8.*

11. $\sqrt{12} + \sqrt{27}$
12. $\sqrt{50} + \sqrt{18}$
13. $\sqrt{45} + 3\sqrt{20}$
14. $5\sqrt{32} - \sqrt{72}$
15. $2\sqrt{54} - \sqrt{20} + \sqrt{45} - \sqrt{24}$
16. $2\sqrt{8} - \sqrt{128} + \sqrt{48} + \sqrt{18}$
17. $4x - 3\sqrt{x^2} + \sqrt{x}$
18. $x - 6\sqrt{x^2} + 2\sqrt{x}$
19. $\sqrt{25x} + \sqrt{36x} - 11\sqrt{x}$
20. $\sqrt{9x} - \sqrt{16x} + 2\sqrt{x}$
21. $\sqrt{\dfrac{5}{9}} + \sqrt{\dfrac{5}{81}}$
22. $\sqrt{\dfrac{3}{64}} + \sqrt{\dfrac{3}{16}}$
23. $\sqrt{\dfrac{3}{4}} - \sqrt{\dfrac{3}{64}}$
24. $\sqrt{\dfrac{2}{25}} + \sqrt{\dfrac{2}{9}}$

Objectives A B Mixed Practice *See Examples 1 through 8.*

25. $12\sqrt{5} - \sqrt{5} - 4\sqrt{5}$
26. $\sqrt{6} + 3\sqrt{6} + \sqrt{6}$
27. $\sqrt{75} + \sqrt{48}$
28. $2\sqrt{80} - \sqrt{45}$
29. $\sqrt{5} + \sqrt{15}$
30. $\sqrt{5} + \sqrt{5}$
31. $3\sqrt{x^3} - x\sqrt{4x}$
32. $x\sqrt{16x} - \sqrt{x^3}$
33. $\sqrt{8} + \sqrt{9} + \sqrt{18} + \sqrt{81}$
34. $\sqrt{6} + \sqrt{16} + \sqrt{24} + \sqrt{25}$
35. $4 + 8\sqrt{2} - 9$
36. $11 - 5\sqrt{7} - 8$

37. $2\sqrt{45} - 2\sqrt{20}$ **38.** $5\sqrt{18} + 2\sqrt{32}$ **39.** $\sqrt{35} - \sqrt{140}$ **40.** $\sqrt{15} - \sqrt{135}$

41. $6 - 2\sqrt{3} - \sqrt{3}$ **42.** $8 - \sqrt{2} - 5\sqrt{2}$ **43.** $3\sqrt{9x} + 2\sqrt{x}$ **● 44.** $5\sqrt{2x} + \sqrt{98x}$

45. $\sqrt{9x^2} + \sqrt{81x^2} - 11\sqrt{x}$ **46.** $\sqrt{100x^2} + 3\sqrt{x} - \sqrt{36x^2}$ **47.** $\sqrt{3x^3} + 3x\sqrt{x}$

48. $x\sqrt{4x} + \sqrt{9x^3}$ **49.** $\sqrt{32x^2} + \sqrt{32x^2} + \sqrt{4x^2}$ **50.** $\sqrt{18x^2} + \sqrt{24x^3} + \sqrt{2x^2}$

51. $\sqrt{40x} + \sqrt{40x^4} - 2\sqrt{10x} - \sqrt{5x^4}$ **52.** $\sqrt{72x^2} + \sqrt{54x} - x\sqrt{50} - 3\sqrt{2x}$

Objective C *Simplify each radical expression. See Example 9.*

53. $2\sqrt[3]{9} + 5\sqrt[3]{9} - \sqrt[3]{25}$ **54.** $8\sqrt[3]{4} + 2\sqrt[3]{4} - \sqrt[3]{49}$ **● 55.** $2\sqrt[3]{2} - 7\sqrt[3]{2} - 6$ **56.** $5\sqrt[3]{9} + 2 - 11\sqrt[3]{9}$

57. $\sqrt[3]{81} + \sqrt[3]{24}$ **58.** $\sqrt[3]{32} + \sqrt[3]{4}$ **● 59.** $\sqrt[3]{8} + \sqrt[3]{54} - 5$ **60.** $\sqrt[3]{64} + \sqrt[3]{14} - 9$

61. $2\sqrt[3]{8x^3} + 2\sqrt[3]{16x^3}$ **62.** $3\sqrt[3]{27z^3} + 3\sqrt[3]{81z^3}$ **63.** $12\sqrt[3]{y^7} - y^2\sqrt[3]{8y}$ **64.** $19\sqrt[3]{z^{11}} - z^3\sqrt[3]{125z^2}$

65. $\sqrt{40x} + x\sqrt[3]{40} - 2\sqrt{10x} - x\sqrt[3]{5}$ **66.** $\sqrt{72x^2} + \sqrt[3]{54} - x\sqrt{50} - 3\sqrt[3]{2}$

Review

Square each binomial. See Section 10.6.

67. $(x + 6)^2$ **68.** $(3x + 2)^2$ **69.** $(2x - 1)^2$ **70.** $(x - 5)^2$

Concept Extensions

71. In your own words, describe like radicals.

72. In the expression $\sqrt{5} + 2 - 3\sqrt{5}$, explain why 2 and -3 cannot be combined.

△ **73.** Find the perimeter of the rectangular picture frame.

△ **74.** Find the perimeter of the plot of land.

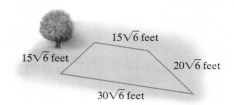

75. A water trough is to be made of wood. Each of the two triangular end pieces has an area of $\frac{3\sqrt{27}}{4}$ square feet. The two side panels are both rectangular. In simplest radical form, find the total area of the wood needed.

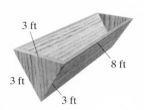

76. Eight wooden braces are to be attached along the diagonals of the vertical sides of a storage bin. Each of four of these diagonals has a length of $\sqrt{52}$ feet, while each of the other four has a length of $\sqrt{80}$ feet. In simplest radical form, find the total length of the wood needed for these braces.

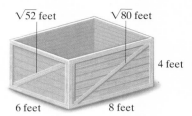

Determine whether each expression can be simplified. If yes, then simplify. See the Concept Check in this section.

77. $4\sqrt{2} + 3\sqrt{2}$

78. $3\sqrt{7} + 3\sqrt{6}$

79. $6 + 7\sqrt{6}$

80. $5x\sqrt{2} + 8x\sqrt{2}$

81. $\sqrt{7} + \sqrt{7} + \sqrt{7}$

82. $6\sqrt{5} - \sqrt{5}$

Simplify.

83. $\sqrt{\dfrac{x^3}{16}} - x\sqrt{\dfrac{9x}{25}} + \dfrac{\sqrt{81x^3}}{2}$

84. $7\sqrt{x^{11}y^7} - x^2y\sqrt{25x^7y^5} + \sqrt{8x^8y^2}$

15.4 Multiplying and Dividing Radicals

Objective A Multiplying Radicals

In Section 15.2, we used the product and quotient rules for radicals to help us simplify radicals. In this section, we use these rules to simplify products and quotients of radicals.

Objectives

A Multiply Radicals.
B Divide Radicals.
C Rationalize Denominators.
D Rationalize Denominators Using Conjugates.

Product Rule for Radicals

If $\sqrt{a}$ and $\sqrt{b}$ are real numbers, then
$$\sqrt{a} \cdot \sqrt{b} = \sqrt{a \cdot b}$$

In other words, the product of the square roots of two numbers is the square root of the product of the two numbers. For example,

$$\sqrt{3} \cdot \sqrt{2} = \sqrt{3 \cdot 2} = \sqrt{6}$$

Practice 1–4

Multiply. Then simplify each product if possible.

1. $\sqrt{5} \cdot \sqrt{2}$
2. $\sqrt{7} \cdot \sqrt{7}$
3. $\sqrt{6} \cdot \sqrt{3}$
4. $\sqrt{10x} \cdot \sqrt{2x}$

Examples Multiply. Then simplify each product if possible.

1. $\sqrt{5} \cdot \sqrt{3} = \sqrt{5 \cdot 3}$
   $= \sqrt{15}$
2. $\sqrt{3} \cdot \sqrt{3} = \sqrt{3 \cdot 3} = \sqrt{9} = 3$
3. $\sqrt{3} \cdot \sqrt{15} = \sqrt{45}$ Use the product rule.
   $= \sqrt{9 \cdot 5}$ Factor the radicand.
   $= \sqrt{9} \cdot \sqrt{5}$ Use the product rule.
   $= 3\sqrt{5}$ Simplify $\sqrt{9}$.
4. $\sqrt{2x^3} \cdot \sqrt{6x} = \sqrt{2x^3 \cdot 6x}$ Use the product rule.
   $= \sqrt{12x^4}$ Multiply.
   $= \sqrt{4x^4 \cdot 3}$ Write $12x^4$ so that one factor is a perfect square.
   $= \sqrt{4x^4} \cdot \sqrt{3}$ Use the product rule.
   $= 2x^2\sqrt{3}$ Simplify.

■ Work Practice 1–4

From Example 2, we found that

$\sqrt{3} \cdot \sqrt{3} = 3$ or $(\sqrt{3})^2 = 3$

This is true in general.

If a is a positive number,

$\sqrt{a} \cdot \sqrt{a} = a$ or $(\sqrt{a})^2 = a$

✓**Concept Check** Identify the true statement(s).

   a. $\sqrt{7} \cdot \sqrt{7} = 7$ b. $\sqrt{2} \cdot \sqrt{3} = 6$
   c. $(\sqrt{131})^2 = 131$ d. $\sqrt{5x} \cdot \sqrt{5x} = 5x$ (Here x is a positive number.)

When multiplying radical expressions containing more than one term, we u the same techniques we use to multiply other algebraic expressions with more th one term.

Practice 5

Multiply.

a. $\sqrt{7}(\sqrt{7} - \sqrt{3})$
b. $\sqrt{5x}(\sqrt{x} - 3\sqrt{5})$
c. $(\sqrt{x} + \sqrt{5})(\sqrt{x} - \sqrt{3})$

Example 5 Multiply.

a. $\sqrt{5}(\sqrt{5} - \sqrt{2})$ b. $\sqrt{3x}(\sqrt{x} - 5\sqrt{3})$
c. $(\sqrt{x} + \sqrt{2})(\sqrt{x} - \sqrt{7})$

Solution:

a. Using the distributive property, we have

$\sqrt{5}(\sqrt{5} - \sqrt{2}) = \sqrt{5} \cdot \sqrt{5} - \sqrt{5} \cdot \sqrt{2}$
$= 5 - \sqrt{10}$ Since $\sqrt{5} \cdot \sqrt{5} = 5$ and $\sqrt{5} \cdot \sqrt{2} = \sqrt{10}$

b. $\sqrt{3x}(\sqrt{x} - 5\sqrt{3}) = \sqrt{3x} \cdot \sqrt{x} - \sqrt{3x} \cdot 5\sqrt{3}$ Use the distributive property.
$= \sqrt{3x \cdot x} - 5\sqrt{3x \cdot 3}$ Use the product rule.
$= \sqrt{3 \cdot x^2} - 5\sqrt{9 \cdot x}$ Factor each radicand so that one factor is a perfect square.
$= \sqrt{3} \cdot \sqrt{x^2} - 5 \cdot \sqrt{9} \cdot \sqrt{x}$ Use the product rule.
$= x\sqrt{3} - 5 \cdot 3 \cdot \sqrt{x}$ Simplify.
$= x\sqrt{3} - 15\sqrt{x}$ Simplify.

Answers

1. $\sqrt{10}$ 2. 7   3. $3\sqrt{2}$ 4. $2x\sqrt{5}$
5. a. $7 - \sqrt{21}$ b. $x\sqrt{5} - 15\sqrt{x}$
c. $x - \sqrt{3x} + \sqrt{5x} - \sqrt{15}$

✓**Concept Check Answer**
a, c, d

c. Using the FOIL method of multiplication, we have

$(\sqrt{x} + \sqrt{2})(\sqrt{x} - \sqrt{7})$

FO$$I$$L

$= \sqrt{x} \cdot \sqrt{x} - \sqrt{x} \cdot \sqrt{7} + \sqrt{2} \cdot \sqrt{x} - \sqrt{2} \cdot \sqrt{7}$

$= x - \sqrt{7x} + \sqrt{2x} - \sqrt{14}$ Use the product rule.

Work Practice 5

The special product formulas also can be used to multiply expressions containing radicals.

Example 6 Multiply.

a. $(\sqrt{5} - 7)(\sqrt{5} + 7)$ **b.** $(\sqrt{7x} + 2)^2$

Solution:

a. $(\sqrt{5} - 7)(\sqrt{5} + 7) = (\sqrt{5})^2 - 7^2$ Recall that $(a - b)(a + b) = a^2 - b^2$.

$\phantom{(\sqrt{5} - 7)(\sqrt{5} + 7)} = 5 - 49$

$\phantom{(\sqrt{5} - 7)(\sqrt{5} + 7)} = -44$

b. $(\sqrt{7x} + 2)^2$

$= (\sqrt{7x})^2 + 2(\sqrt{7x})(2) + (2)^2$ Recall that $(a + b)^2 = a^2 + 2ab + b^2$.

$= 7x + 4\sqrt{7x} + 4$

Work Practice 6

Practice 6

Multiply.

a. $(\sqrt{3} + 8)(\sqrt{3} - 8)$

b. $(\sqrt{5x} + 4)^2$

Objective B Dividing Radicals

To simplify quotients of rational expressions, we use the quotient rule.

Quotient Rule for Radicals

If $\sqrt{a}$ and $\sqrt{b}$ are real numbers and $b \neq 0$, then

$$\frac{\sqrt{a}}{\sqrt{b}} = \sqrt{\frac{a}{b}}$$

Examples Divide. Then simplify the quotient if possible.

7. $\dfrac{\sqrt{14}}{\sqrt{2}} = \sqrt{\dfrac{14}{2}} = \sqrt{7}$

8. $\dfrac{\sqrt{100}}{\sqrt{5}} = \sqrt{\dfrac{100}{5}} = \sqrt{20} = \sqrt{4 \cdot 5} = \sqrt{4} \cdot \sqrt{5} = 2\sqrt{5}$

9. $\dfrac{\sqrt{12x^3}}{\sqrt{3x}} = \sqrt{\dfrac{12x^3}{3x}} = \sqrt{4x^2} = 2x$

Work Practice 7–9

Practice 7–9

Divide. Then simplify the quotient if possible.

7. $\dfrac{\sqrt{21}}{\sqrt{3}}$

8. $\dfrac{\sqrt{90}}{\sqrt{2}}$

9. $\dfrac{\sqrt{125x^3}}{\sqrt{5x}}$

Objective C Rationalizing Denominators

It is sometimes easier to work with radical expressions if the denominator does not contain a radical. To rewrite an expression so that the denominator does not contain a radical expression, we use the fact that we can multiply the numerator and the denominator of a fraction by the same nonzero number without changing the value

Answers

6. a. -61 **b.** $5x + 8\sqrt{5x} + 16$

7. $\sqrt{7}$ **8.** $3\sqrt{5}$ **9.** $5x$

of the expression. This is the same as multiplying the fraction by 1. For example, to get rid of the radical in the denominator of $\dfrac{\sqrt{5}}{\sqrt{2}}$, we multiply by 1 in the form of $\dfrac{\sqrt{2}}{\sqrt{2}}$. Th

$$\dfrac{\sqrt{5}}{\sqrt{2}} = \dfrac{\sqrt{5}}{\sqrt{2}} \cdot 1 = \dfrac{\sqrt{5}}{\sqrt{2}} \cdot \dfrac{\sqrt{2}}{\sqrt{2}} = \dfrac{\sqrt{5} \cdot \sqrt{2}}{\sqrt{2} \cdot \sqrt{2}} = \dfrac{\sqrt{10}}{2}$$

This process is called **rationalizing** the denominator.

Practice 10

Rationalize the denominator of $\dfrac{5}{\sqrt{3}}$.

Example 10 Rationalize the denominator of $\dfrac{2}{\sqrt{7}}$.

Solution: To rewrite $\dfrac{2}{\sqrt{7}}$ so that there is no radical in the denominator, multiply by 1 in the form of $\dfrac{\sqrt{7}}{\sqrt{7}}$.

$$\dfrac{2}{\sqrt{7}} = \dfrac{2}{\sqrt{7}} \cdot \dfrac{\sqrt{7}}{\sqrt{7}} = \dfrac{2 \cdot \sqrt{7}}{\sqrt{7} \cdot \sqrt{7}} = \dfrac{2\sqrt{7}}{7}$$

■ Work Practice 10

Practice 11

Rationalize the denominator of $\dfrac{\sqrt{7}}{\sqrt{20}}$.

Example 11 Rationalize the denominator of $\dfrac{\sqrt{5}}{\sqrt{12}}$.

Solution: We can multiply by $\dfrac{\sqrt{12}}{\sqrt{12}}$, but see what happens if we simplify first

$$\dfrac{\sqrt{5}}{\sqrt{12}} = \dfrac{\sqrt{5}}{\sqrt{4 \cdot 3}} = \dfrac{\sqrt{5}}{2\sqrt{3}}$$

To rationalize the denominator now, we multiply by $\dfrac{\sqrt{3}}{\sqrt{3}}$.

$$\dfrac{\sqrt{5}}{2\sqrt{3}} = \dfrac{\sqrt{5}}{2\sqrt{3}} \cdot \dfrac{\sqrt{3}}{\sqrt{3}} = \dfrac{\sqrt{5} \cdot \sqrt{3}}{2\sqrt{3} \cdot \sqrt{3}} = \dfrac{\sqrt{15}}{2 \cdot 3} = \dfrac{\sqrt{15}}{6}$$

■ Work Practice 11

Practice 12

Rationalize the denominator of $\sqrt{\dfrac{2}{45x}}$.

Example 12 Rationalize the denominator of $\sqrt{\dfrac{1}{18x}}$.

Solution: First we simplify.

$$\sqrt{\dfrac{1}{18x}} = \dfrac{\sqrt{1}}{\sqrt{18x}} = \dfrac{1}{\sqrt{9} \cdot \sqrt{2x}} = \dfrac{1}{3\sqrt{2x}}$$

Now to rationalize the denominator, we multiply by $\dfrac{\sqrt{2x}}{\sqrt{2x}}$.

$$\dfrac{1}{3\sqrt{2x}} = \dfrac{1}{3\sqrt{2x}} \cdot \dfrac{\sqrt{2x}}{\sqrt{2x}} = \dfrac{1 \cdot \sqrt{2x}}{3\sqrt{2x} \cdot \sqrt{2x}} = \dfrac{\sqrt{2x}}{3 \cdot 2x} = \dfrac{\sqrt{2x}}{6x}$$

■ Work Practice 12

Objective D Rationalizing Denominators Using Conjugates ▶

To rationalize a denominator that is a sum or a difference, such as the denominator

$$\dfrac{2}{4 + \sqrt{3}}$$

we multiply the numerator and the denominator by $4 - \sqrt{3}$. The expressi $4 + \sqrt{3}$ and $4 - \sqrt{3}$ are called conjugates of each other. When a radical express

Answers

10. $\dfrac{5\sqrt{3}}{3}$ **11.** $\dfrac{\sqrt{35}}{10}$ **12.** $\dfrac{\sqrt{10x}}{15x}$

Section 15.4 | Multiplying and Dividing Radicals

such as $4 + \sqrt{3}$ is multiplied by its conjugate, $4 - \sqrt{3}$, the product simplifies to an expression that contains no radicals.

In general, the expressions $a + b$ and $a - b$ are **conjugates** of each other.

$$(a + b)(a - b) = a^2 - b^2$$
$$(4 + \sqrt{3})(4 - \sqrt{3}) = 4^2 - (\sqrt{3})^2 = 16 - 3 = 13$$

Then

$$\frac{2}{4 + \sqrt{3}} = \frac{2(4 - \sqrt{3})}{(4 + \sqrt{3})(4 - \sqrt{3})} = \frac{2(4 - \sqrt{3})}{13}$$

Example 13 Rationalize the denominator of $\frac{2}{1 + \sqrt{3}}$.

Solution: We multiply the numerator and the denominator of this fraction by the conjugate of $1 + \sqrt{3}$, that is, by $1 - \sqrt{3}$.

$$\frac{2}{1 + \sqrt{3}} = \frac{2(1 - \sqrt{3})}{(1 + \sqrt{3})(1 - \sqrt{3})}$$
$$= \frac{2(1 - \sqrt{3})}{1^2 - (\sqrt{3})^2}$$
$$= \frac{2(1 - \sqrt{3})}{1 - 3}$$
$$= \frac{2(1 - \sqrt{3})}{-2}$$
$$= -\frac{2(1 - \sqrt{3})}{2} \qquad \frac{a}{-b} = -\frac{a}{b}$$
$$= -1(1 - \sqrt{3}) \qquad \text{Simplify.}$$
$$= -1 + \sqrt{3} \qquad \text{Multiply.}$$

Helpful Hint
Don't forget that $(\sqrt{3})^2 = 3$.

Practice 13
Rationalize the denominator of $\frac{3}{2 + \sqrt{7}}$.

■ Work Practice 13

Example 14 Rationalize the denominator of $\frac{\sqrt{5} + 4}{\sqrt{5} - 1}$.

Solution:

$$\frac{\sqrt{5} + 4}{\sqrt{5} - 1} = \frac{(\sqrt{5} + 4)(\sqrt{5} + 1)}{(\sqrt{5} - 1)(\sqrt{5} + 1)} \qquad \text{Multiply the numerator and denominator by } \sqrt{5} + 1, \text{ the conjugate of } \sqrt{5} - 1.$$
$$= \frac{5 + \sqrt{5} + 4\sqrt{5} + 4}{5 - 1} \qquad \text{Multiply.}$$
$$= \frac{9 + 5\sqrt{5}}{4} \qquad \text{Simplify.}$$

Practice 14
Rationalize the denominator of $\frac{\sqrt{2} + 5}{\sqrt{2} - 1}$.

■ Work Practice 14

Example 15 Rationalize the denominator of $\frac{3}{1 + \sqrt{x}}$.

Solution:

$$\frac{3}{1 + \sqrt{x}} = \frac{3(1 - \sqrt{x})}{(1 + \sqrt{x})(1 - \sqrt{x})} \qquad \text{Multiply the numerator and denominator by } 1 - \sqrt{x}, \text{ the conjugate of } 1 + \sqrt{x}.$$
$$= \frac{3(1 - \sqrt{x})}{1 - x}$$

Practice 15
Rationalize the denominator of $\frac{7}{2 - \sqrt{x}}$.

Answers
13. $-2 + \sqrt{7}$ **14.** $7 + 6\sqrt{2}$
15. $\frac{7(2 + \sqrt{x})}{4 - x}$

■ Work Practice 15

Vocabulary, Readiness & Video Check

Fill in each blank.

1. $\sqrt{7} \cdot \sqrt{3} = $ _____
2. $\sqrt{10} \cdot \sqrt{10} = $ _____
3. $\dfrac{\sqrt{15}}{\sqrt{3}} = $ _____

4. The process of eliminating the radical in the denominator of a radical expression is called _____.
5. The conjugate of $2 + \sqrt{3}$ is _____.

Martin-Gay Interactive Videos Watch the section lecture video and answer the following questions.

See Video 15.4

Objective A
6. In ▯ Examples 1 and 3, the product rule for radicals is applied twice, but in different ways. Explain.
7. Starting with ▯ Example 2, what important reminder is made repeatedly about the square root of a positive number that is squared?

Objective B
8. From ▯ Examples 5 and 6, when we're looking at a quotient of two radicals, what would make us think to apply the quotient rule in order to simplify?

Objective C
9. From the lecture before ▯ Example 7, what is the goal of rationalizing a denominator?

Objective D
10. From ▯ Example 9, why will multiplying a denominator by its conjugate rationalize the denominator?

15.4 Exercise Set MyLab Math

Objective A *Multiply and simplify. Assume that all variables represent positive real numbers. See Examples 1 through*

1. $\sqrt{8} \cdot \sqrt{2}$
2. $\sqrt{3} \cdot \sqrt{12}$
3. $\sqrt{10} \cdot \sqrt{5}$
4. $\sqrt{2} \cdot \sqrt{14}$

5. $(\sqrt{6})^2$
6. $(\sqrt{10})^2$
7. $\sqrt{2x} \cdot \sqrt{2x}$
8. $\sqrt{5y} \cdot \sqrt{5y}$

9. $(2\sqrt{5})^2$
10. $(3\sqrt{10})^2$
11. $(6\sqrt{x})^2$
12. $(8\sqrt{y})^2$

13. $\sqrt{3x^5} \cdot \sqrt{6x}$
14. $\sqrt{21y^7} \cdot \sqrt{3y}$
15. $\sqrt{2xy^2} \cdot \sqrt{8xy}$
16. $\sqrt{18x^2y^2} \cdot \sqrt{2x^2y}$

17. $\sqrt{6}(\sqrt{5} + \sqrt{7})$
18. $\sqrt{10}(\sqrt{3} - \sqrt{7})$
19. $\sqrt{10}(\sqrt{2} + \sqrt{5})$

20. $\sqrt{6}(\sqrt{3} + \sqrt{2})$
21. $\sqrt{7y}(\sqrt{y} - 2\sqrt{7})$
22. $\sqrt{5b}(2\sqrt{b} + \sqrt{5})$

23. $(\sqrt{3} + 6)(\sqrt{3} - 6)$
24. $(\sqrt{5} + 2)(\sqrt{5} - 2)$
25. $(\sqrt{3} + \sqrt{5})(\sqrt{2} - \sqrt{5})$

26. $(\sqrt{7} + \sqrt{5})(\sqrt{2} - \sqrt{5})$
27. $(2\sqrt{11} + 1)(\sqrt{11} - 6)$
28. $(5\sqrt{3} + 2)(\sqrt{3} - 1)$

29. $(\sqrt{x}+6)(\sqrt{x}-6)$ 30. $(\sqrt{y}+5)(\sqrt{y}-5)$ 31. $(\sqrt{x}-7)^2$

32. $(\sqrt{x}+4)^2$ 33. $(\sqrt{6y}+1)^2$ 34. $(\sqrt{3y}-2)^2$

Objective B Divide and simplify. Assume that all variables represent positive real numbers. See Examples 7 through 9.

35. $\dfrac{\sqrt{32}}{\sqrt{2}}$ 36. $\dfrac{\sqrt{40}}{\sqrt{10}}$ 37. $\dfrac{\sqrt{21}}{\sqrt{3}}$ 38. $\dfrac{\sqrt{55}}{\sqrt{5}}$ 39. $\dfrac{\sqrt{90}}{\sqrt{5}}$

40. $\dfrac{\sqrt{96}}{\sqrt{8}}$ 41. $\dfrac{\sqrt{75y^5}}{\sqrt{3y}}$ 42. $\dfrac{\sqrt{24x^7}}{\sqrt{6x}}$ 43. $\dfrac{\sqrt{150}}{\sqrt{2}}$ 44. $\dfrac{\sqrt{120}}{\sqrt{3}}$

45. $\dfrac{\sqrt{72y^5}}{\sqrt{3y^3}}$ 46. $\dfrac{\sqrt{54x^3}}{\sqrt{2x}}$ 47. $\dfrac{\sqrt{24x^3y^4}}{\sqrt{2xy}}$ 48. $\dfrac{\sqrt{96x^5y^3}}{\sqrt{3x^2y}}$

Objective C Rationalize each denominator and simplify. Assume that all variables represent positive real numbers. See Examples 10 through 12.

49. $\dfrac{\sqrt{3}}{\sqrt{5}}$ 50. $\dfrac{\sqrt{2}}{\sqrt{3}}$ 51. $\dfrac{7}{\sqrt{2}}$ 52. $\dfrac{8}{\sqrt{11}}$

53. $\dfrac{1}{\sqrt{6y}}$ 54. $\dfrac{1}{\sqrt{10z}}$ 55. $\sqrt{\dfrac{5}{18}}$ 56. $\sqrt{\dfrac{7}{12}}$

57. $\sqrt{\dfrac{3}{x}}$ 58. $\sqrt{\dfrac{5}{x}}$ 59. $\sqrt{\dfrac{1}{8}}$ 60. $\sqrt{\dfrac{1}{27}}$

61. $\sqrt{\dfrac{2}{15}}$ 62. $\sqrt{\dfrac{11}{14}}$ 63. $\sqrt{\dfrac{3}{20}}$ 64. $\sqrt{\dfrac{3}{50}}$

65. $\dfrac{3x}{\sqrt{2x}}$ 66. $\dfrac{5y}{\sqrt{3y}}$ 67. $\dfrac{8y}{\sqrt{5}}$ 68. $\dfrac{7x}{\sqrt{2}}$

69. $\sqrt{\dfrac{x}{36y}}$ 70. $\sqrt{\dfrac{z}{49y}}$ 71. $\sqrt{\dfrac{y}{12x}}$ 72. $\sqrt{\dfrac{x}{20y}}$

Objective D Rationalize each denominator and simplify. Assume that all variables represent positive real numbers. See Examples 13 through 15.

73. $\dfrac{3}{\sqrt{2}+1}$ 74. $\dfrac{6}{\sqrt{5}+2}$

75. $\dfrac{4}{2-\sqrt{5}}$ 76. $\dfrac{2}{\sqrt{10}-3}$

77. $\dfrac{\sqrt{5}+1}{\sqrt{6}-\sqrt{5}}$ 78. $\dfrac{\sqrt{3}+1}{\sqrt{3}-\sqrt{2}}$

79. $\dfrac{\sqrt{3}+1}{\sqrt{2}-1}$ 80. $\dfrac{\sqrt{2}-2}{2-\sqrt{3}}$

81. $\dfrac{5}{2 + \sqrt{x}}$

82. $\dfrac{9}{3 + \sqrt{x}}$

83. $\dfrac{3}{\sqrt{x} - 4}$

84. $\dfrac{4}{\sqrt{x} - 1}$

Review

Solve each equation. See Section 9.3.

85. $x + 5 = 7^2$

86. $2y - 1 = 3^2$

87. $4z^2 + 6z - 12 = (2z)^2$

88. $16x^2 + x + 9 = (4x)^2$

89. $9x^2 + 5x + 4 = (3x + 1)^2$

90. $x^2 + 3x + 4 = (x + 2)^2$

Concept Extensions

△ 91. Find the area of a rectangular room whose length is $13\sqrt{2}$ meters and width is $5\sqrt{6}$ meters.

△ 92. Find the volume of a microwave oven whose length is $\sqrt{3}$ feet, width is $\sqrt{2}$ feet, and height is $\sqrt{2}$ feet.

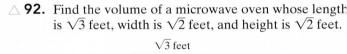

△ 93. If a circle has area A, then the formula for the radius r of the circle is

$$r = \sqrt{\dfrac{A}{\pi}}$$

Rationalize the denominator of this expression.

△ 94. If the surface area of a round ball is S, then the formula for the radius r of the ball is

$$r = \sqrt{\dfrac{S}{4\pi}}$$

Simplify this expression by rationalizing the denominator.

Identify each statement as true or false. See the Concept Check in this section.

95. $\sqrt{5} \cdot \sqrt{5} = 5$

96. $\sqrt{5} \cdot \sqrt{3} = 15$

97. $\sqrt{3x} \cdot \sqrt{3x} = 2\sqrt{3x}$

98. $\sqrt{3x} + \sqrt{3x} = 2\sqrt{3x}$

99. $\sqrt{11} + \sqrt{2} = \sqrt{13}$

100. $\sqrt{11} \cdot \sqrt{2} = \sqrt{22}$

101. When rationalizing the denominator of $\frac{\sqrt{2}}{\sqrt{3}}$, explain why both the numerator and the denominator must be multiplied by $\sqrt{3}$.

102. In your own words, explain why $\sqrt{6} + \sqrt{2}$ cannot be simplified further, but $\sqrt{6} \cdot \sqrt{2}$ can be.

103. To rationalize the denominator of $\frac{\sqrt[3]{2}}{\sqrt[3]{3}}$, multiply the numerator and the denominator by $\sqrt[3]{9}$. Then simplify. Explain why this works.

104. When rationalizing the denominator of $\frac{5}{1+\sqrt{2}}$, explain why multiplying by $\frac{\sqrt{2}}{\sqrt{2}}$ will not accomplish this, but multiplying by $\frac{1-\sqrt{2}}{1-\sqrt{2}}$ will.

It is often more convenient to work with a radical expression whose numerator is rationalized. Rationalize the numerator of each expression by multiplying the numerator and denominator by the conjugate of the numerator.

105. $\frac{\sqrt{3}+1}{\sqrt{2}-1}$

106. $\frac{\sqrt{2}-2}{2-\sqrt{3}}$

Integrated Review
Sections 15.1–15.4

Simplifying Radicals

Simplify. Assume that all variables represent positive numbers.

1. $\sqrt{36}$
2. $\sqrt{48}$
3. $\sqrt{x^4}$

4. $\sqrt{y^7}$
5. $\sqrt{16x^2}$
6. $\sqrt{18x^{11}}$

7. $\sqrt[3]{8}$
8. $\sqrt[4]{81}$
9. $\sqrt[3]{-27}$

10. $\sqrt{-4}$
11. $\sqrt{\dfrac{11}{9}}$
12. $\sqrt[3]{\dfrac{7}{64}}$

13. $-\sqrt{16}$
14. $-\sqrt{25}$
15. $\sqrt{\dfrac{9}{49}}$

16. $\sqrt{\dfrac{1}{64}}$
17. $\sqrt{a^8 b^2}$
18. $\sqrt{x^{10} y^{20}}$

19. $\sqrt{25m^6}$
20. $\sqrt{9n^{16}}$

Add or subtract as indicated.

21. $5\sqrt{7} + \sqrt{7}$
22. $\sqrt{50} - \sqrt{8}$

Chapter 15 Integrated Review   **1167**

23. $5\sqrt{2} - 5\sqrt{3}$

24. $2\sqrt{x} + \sqrt{25x} - \sqrt{36x} + 3x$

Multiply and simplify if possible.

25. $\sqrt{2} \cdot \sqrt{15}$

26. $\sqrt{3} \cdot \sqrt{3}$

27. $(2\sqrt{7})^2$

28. $(3\sqrt{5})^2$

29. $\sqrt{3}(\sqrt{11} + 1)$

30. $\sqrt{6}(\sqrt{3} - 2)$

31. $\sqrt{8y} \cdot \sqrt{2y}$

32. $\sqrt{15x^2} \cdot \sqrt{3x^2}$

33. $(\sqrt{x} - 5)(\sqrt{x} + 2)$

34. $(3 + \sqrt{2})^2$

Divide and simplify if possible.

35. $\dfrac{\sqrt{8}}{\sqrt{2}}$

36. $\dfrac{\sqrt{45}}{\sqrt{15}}$

37. $\dfrac{\sqrt{24x^5}}{\sqrt{2x}}$

38. $\dfrac{\sqrt{75a^4b^5}}{\sqrt{5ab}}$

Rationalize each denominator.

39. $\sqrt{\dfrac{1}{6}}$

40. $\dfrac{x}{\sqrt{20}}$

41. $\dfrac{4}{\sqrt{6} + 1}$

42. $\dfrac{\sqrt{2} + 1}{\sqrt{x} - 5}$

23. _____

24. _____

25. _____

26. _____

27. _____

28. _____

29. _____

30. _____

31. _____

32. _____

33. _____

34. _____

35. _____

36. _____

37. _____

38. _____

39. _____

40. _____

41. _____

42. _____

15.5 Solving Equations Containing Radicals

Objectives

A Solve Radical Equations by Using the Squaring Property of Equality Once.

B Solve Radical Equations by Using the Squaring Property of Equality Twice.

Objective A Using the Squaring Property of Equality Once

In this section, we solve **radical equations** such as
$$\sqrt{x+3} = 5 \quad \text{and} \quad \sqrt{2x+1} = \sqrt{3x}$$

Radical equations contain variables in the radicand. To solve these equations, we re on the following squaring property.

> **The Squaring Property of Equality**
> If $a = b$, then $a^2 = b^2$.

Unfortunately, this squaring property does not guarantee that all solutions the new equation are solutions of the original equation. For example, if we squa both sides of the equation
$$x = 2$$
we have
$$x^2 = 4$$

This new equation has two solutions, 2 and -2, while the original equation, $x =$ has only one solution. For this reason, we must **always check proposed solutions radical equations in the original equation.**

Practice 1
Solve: $\sqrt{x-2} = 7$

Example 1 Solve: $\sqrt{x+3} = 5$

Solution: To solve this radical equation, we use the squaring property of equ ity and square both sides of the equation.

$$\sqrt{x+3} = 5$$
$$(\sqrt{x+3})^2 = 5^2 \quad \text{Square both sides.}$$
$$x + 3 = 25 \quad \text{Simplify.}$$
$$x = 22 \quad \text{Subtract 3 from both sides.}$$

Check: We replace x with 22 in the original equation.

$$\sqrt{x+3} = 5 \quad \text{Original equation}$$
$$\sqrt{22+3} \stackrel{?}{=} 5 \quad \text{Let } x = 22.$$
$$\sqrt{25} \stackrel{?}{=} 5$$
$$5 = 5 \quad \text{True}$$

Since a true statement results, 22 is the solution.

> **Helpful Hint**
> Don't forget to check the proposed solutions of radical equations in the original equation.

■ Work Practice 1

When solving radical equations, if possible, move radicals so that at least one radic is by itself on one side of the equation.

Example 2 Solve: $\sqrt{x} = \sqrt{5x-2}$

Practice 2
Solve: $\sqrt{6x-1} = \sqrt{x}$

Solution: Each radical is by itself on one side of the equation. Let's begin sol ing by squaring both sides.

$$\sqrt{x} = \sqrt{5x-2} \quad \text{Original equation}$$
$$(\sqrt{x})^2 = (\sqrt{5x-2})^2 \quad \text{Square both sides.}$$
$$x = 5x - 2 \quad \text{Simplify.}$$
$$-4x = -2 \quad \text{Subtract } 5x \text{ from both sides.}$$
$$x = \frac{-2}{-4} = \frac{1}{2} \quad \text{Divide both sides by } -4 \text{ and simplify.}$$

Answers
1. $x = 51$ **2.** $x = \dfrac{1}{5}$

Check: We replace x with $\frac{1}{2}$ in the original equation.

$\sqrt{x} = \sqrt{5x - 2}$ Original equation

$\sqrt{\frac{1}{2}} \stackrel{?}{=} \sqrt{5 \cdot \frac{1}{2} - 2}$ Let $x = \frac{1}{2}$.

$\sqrt{\frac{1}{2}} \stackrel{?}{=} \sqrt{\frac{5}{2} - 2}$ Multiply.

$\sqrt{\frac{1}{2}} \stackrel{?}{=} \sqrt{\frac{5}{2} - \frac{4}{2}}$ Write 2 as $\frac{4}{2}$.

$\sqrt{\frac{1}{2}} = \sqrt{\frac{1}{2}}$ True

This statement is true, so the solution is $\frac{1}{2}$.

■ Work Practice 2

Example 3 Solve: $\sqrt{x} + 6 = 4$

Practice 3
Solve: $\sqrt{x} + 9 = 2$

Solution: First we write the equation so that the radical is by itself on one side of the equation.

$\sqrt{x} + 6 = 4$
$\sqrt{x} = -2$ Subtract 6 from both sides to get the radical by itself.

Normally we would now square both sides. Recall, however, that $\sqrt{x}$ is the principal or nonnegative square root of x, so $\sqrt{x}$ cannot equal -2 and thus this equation has no solution. We arrive at the same conclusion if we continue by applying the squaring property.

$\sqrt{x} = -2$
$(\sqrt{x})^2 = (-2)^2$ Square both sides.
$x = 4$ Simplify.

Check: We replace x with 4 in the original equation.

$\sqrt{x} + 6 = 4$ Original equation
$\sqrt{4} + 6 \stackrel{?}{=} 4$ Let $x = 4$.
$2 + 6 = 4$ False

Since 4 *does not* satisfy the original equation, this equation has no solution.

■ Work Practice 3

Example 3 makes it very clear that we *must* check proposed solutions in the original equation to determine if they are truly solutions. If a proposed solution does not work, we say that the value is an **extraneous solution.**

The following steps can be used to solve radical equations containing square roots.

To Solve a Radical Equation Containing Square Roots

Step 1: Arrange terms so that one radical is by itself on one side of the equation. That is, isolate a radical.
Step 2: Square both sides of the equation.
Step 3: Simplify both sides of the equation.
Step 4: If the equation still contains a radical term, repeat Steps 1 through 3.
Step 5: Solve the equation.
Step 6: Check all solutions in the original equation for extraneous solutions.

Answer
3. no solution

Practice 4
Solve: $\sqrt{9y^2 + 2y - 10} = 3y$

Example 4 Solve: $\sqrt{4y^2 + 5y - 15} = 2y$

Solution: The radical is already isolated, so we start by squaring both sides.

$$\sqrt{4y^2 + 5y - 15} = 2y$$
$$(\sqrt{4y^2 + 5y - 15})^2 = (2y)^2 \quad \text{Square both sides.}$$
$$4y^2 + 5y - 15 = 4y^2 \quad \text{Simplify.}$$
$$5y - 15 = 0 \quad \text{Subtract } 4y^2 \text{ from both sides.}$$
$$5y = 15 \quad \text{Add 15 to both sides.}$$
$$y = 3 \quad \text{Divide both sides by 5.}$$

Check: We replace y with 3 in the original equation.

$$\sqrt{4y^2 + 5y - 15} = 2y \quad \text{Original equation}$$
$$\sqrt{4 \cdot 3^2 + 5 \cdot 3 - 15} \stackrel{?}{=} 2 \cdot 3 \quad \text{Let } y = 3.$$
$$\sqrt{4 \cdot 9 + 15 - 15} \stackrel{?}{=} 6 \quad \text{Simplify.}$$
$$\sqrt{36} \stackrel{?}{=} 6$$
$$6 = 6 \quad \text{True}$$

This statement is true, so the solution is 3.

■ Work Practice 4

Practice 5
Solve: $\sqrt{x + 1} - x = -5$

Example 5 Solve: $\sqrt{x + 3} - x = -3$

Solution: First we isolate the radical by adding x to both sides. Then we square both sides.

$$\sqrt{x + 3} - x = -3$$
$$\sqrt{x + 3} = x - 3 \quad \text{Add } x \text{ to both sides.}$$
$$(\sqrt{x + 3})^2 = (x - 3)^2 \quad \text{Square both sides.}$$
$$x + 3 = x^2 - 6x + 9 \quad \text{Simplify.}$$

Helpful Hint Don't forget that $(x - 3)^2 = (x - 3)(x - 3) = x^2 - 6x + 9$.

To solve the resulting quadratic equation, we write the equation in standard form by subtracting x and 3 from both sides.

$$x + 3 = x^2 - 6x + 9$$
$$3 = x^2 - 7x + 9 \quad \text{Subtract } x \text{ from both sides.}$$
$$0 = x^2 - 7x + 6 \quad \text{Subtract 3 from both sides.}$$
$$0 = (x - 6)(x - 1) \quad \text{Factor.}$$
$$0 = x - 6 \quad \text{or} \quad 0 = x - 1 \quad \text{Set each factor equal to zero.}$$
$$6 = x \quad\quad\quad 1 = x \quad \text{Solve for } x.$$

Check: We replace x with 6 and then x with 1 in the original equation.

Let $x = 6$.
$$\sqrt{x + 3} - x = -3$$
$$\sqrt{6 + 3} - 6 \stackrel{?}{=} -3$$
$$\sqrt{9} - 6 \stackrel{?}{=} -3$$
$$3 - 6 \stackrel{?}{=} -3$$
$$-3 = -3 \quad \text{True}$$

Let $x = 1$.
$$\sqrt{x + 3} - x = -3$$
$$\sqrt{1 + 3} - 1 \stackrel{?}{=} -3$$
$$\sqrt{4} - 1 \stackrel{?}{=} -3$$
$$2 - 1 \stackrel{?}{=} -3$$
$$1 = -3 \quad \text{False}$$

Since replacing x with 1 resulted in a false statement, 1 is an extraneous solution. The only solution is 6.

■ Work Practice 5

Answers
4. $y = 5$ **5.** $x = 8$

Objective B Using the Squaring Property of Equality Twice

If a radical equation contains two radicals, we may need to use the squaring property twice.

Example 6 Solve: $\sqrt{x-4} = \sqrt{x} - 2$

Solution:

$$\sqrt{x-4} = \sqrt{x} - 2$$
$$(\sqrt{x-4})^2 = (\sqrt{x} - 2)^2 \quad \text{Square both sides.}$$
$$x - 4 = x - 4\sqrt{x} + 4$$

$-8 = -4\sqrt{x}$ To get the radical term alone, subtract x and 4 from both sides.

$2 = \sqrt{x}$ Divide both sides by -4.

$4 = x$ Square both sides again.

Check the proposed solution in the original equation. The solution is 4.

■ Work Practice 6

Practice 6

Solve: $\sqrt{x} + 3 = \sqrt{x + 15}$

Helpful Hint

Don't forget:

$$(\sqrt{x} - 2)^2 = (\sqrt{x} - 2)(\sqrt{x} - 2)$$
$$= \sqrt{x} \cdot \sqrt{x} - 2\sqrt{x} - 2\sqrt{x} + 4$$
$$= x - 4\sqrt{x} + 4$$

Answer
6. $x = 1$

Vocabulary, Readiness & Video Check

Martin-Gay Interactive Videos

See Video 15.5

Watch the section lecture video and answer the following questions.

Objective A 1. From Examples 1 and 2, why must we be sure to check our proposed solution(s) in the original equation?

Objective B 2. Solving Example 5 requires using the squaring property twice. Is anything else done differently to solve these equations as compared to equations where the property is used only once?

15.5 Exercise Set MyLab Math

Objective A *Solve each equation. See Examples 1 through 3.*

1. $\sqrt{x} = 9$
2. $\sqrt{x} = 4$
3. $\sqrt{x+5} = 2$
4. $\sqrt{x+12} = 3$
5. $\sqrt{x} - 2 = 5$
6. $4\sqrt{x} - 7 = 5$
7. $3\sqrt{x} + 5 = 2$
8. $3\sqrt{x} + 8 = 5$
9. $\sqrt{x} = \sqrt{3x-8}$
10. $\sqrt{x} = \sqrt{4x-3}$
11. $\sqrt{4x-3} = \sqrt{x+3}$
12. $\sqrt{5x-4} = \sqrt{x+8}$

Solve each equation. See Examples 4 and 5.

13. $\sqrt{9x^2 + 2x - 4} = 3x$
14. $\sqrt{4x^2 + 3x - 9} = 2x$
15. $\sqrt{x} = x - 6$
16. $\sqrt{x} = x - 2$
17. $\sqrt{x+7} = x + 5$
18. $\sqrt{x+5} = x - 1$
19. $\sqrt{3x+7} - x = 3$
20. $x = \sqrt{4x-7} + 1$
21. $\sqrt{16x^2 + 2x + 2} = 4x$
22. $\sqrt{4x^2 + 3x + 2} = 2x$
23. $\sqrt{2x^2 + 6x + 9} = 3$
24. $\sqrt{3x^2 + 6x + 4} = 2$

Objective B *Solve each equation. See Example 6.*

25. $\sqrt{x-7} = \sqrt{x} - 1$
26. $\sqrt{x-8} = \sqrt{x} - 2$
27. $\sqrt{x} + 2 = \sqrt{x+24}$
28. $\sqrt{x} + 5 = \sqrt{x+55}$
29. $\sqrt{x+8} = \sqrt{x} + 2$
30. $\sqrt{x} + 1 = \sqrt{x+15}$

Objectives A B Mixed Practice *Solve each equation. See Examples 1 through 6.*

31. $\sqrt{2x+6} = 4$
32. $\sqrt{3x+7} = 5$
33. $\sqrt{x+6} + 1 = 3$
34. $\sqrt{x+5} + 2 = 5$
35. $\sqrt{x+6} + 5 = 3$
36. $\sqrt{2x-1} + 7 = 1$
37. $\sqrt{16x^2 - 3x + 6} = 4x$
38. $\sqrt{9x^2 - 2x + 8} = 3x$
39. $-\sqrt{x} = -6$
40. $-\sqrt{y} = -8$
41. $\sqrt{x+9} = \sqrt{x} - 3$
42. $\sqrt{x} - 6 = \sqrt{x+36}$
43. $\sqrt{2x+1} + 3 = 5$
44. $\sqrt{3x-1} + 1 = 4$
45. $\sqrt{x} + 3 = 7$
46. $\sqrt{x} + 5 = 10$
47. $\sqrt{4x} = \sqrt{2x+6}$
48. $\sqrt{5x+6} = \sqrt{8x}$
49. $\sqrt{2x+1} = x - 7$
50. $\sqrt{2x+5} = x - 5$
51. $x = \sqrt{2x-2} + 1$
52. $\sqrt{2x-4} + 2 = x$
53. $\sqrt{1-8x} - x = 4$
54. $\sqrt{2x+5} - 1 = x$

Review

Translating *Translate each sentence into an equation and then solve. See Section 9.4.*

55. If 8 is subtracted from the product of 3 and x, the result is 19. Find x.
56. If 3 more than x is subtracted from twice x, the result is 11. Find x.

57. The length of a rectangle is twice the width. The perimeter is 24 inches. Find the length.

58. The length of a rectangle is 2 inches longer than the width. The perimeter is 24 inches. Find the length.

Concept Extensions

Solve each equation.

59. $\sqrt{x-3} + 3 = \sqrt{3x+4}$

60. $\sqrt{2x+3} = \sqrt{x-2} + 2$

61. Explain why proposed solutions of radical equations must be checked in the original equation.

62. Is 8 a solution of the equation $\sqrt{x-4} - 5 = \sqrt{x+1}$? Explain why or why not.

63. The formula $b = \sqrt{\dfrac{V}{2}}$ can be used to determine the length b of a side of the base of a square-based pyramid with height 6 units and volume V cubic units.

  a. Find the length of the side of the base that produces a pyramid with each volume. (Round to the nearest tenth of a unit.)

V	20	200	2000
b			

  b. Notice in the table that volume V has been increased by a factor of 10 each time. Does the corresponding length b of a side increase by a factor of 10 each time also?

64. The formula $r = \sqrt{\dfrac{V}{2\pi}}$ can be used to determine the radius r of a cylinder with height 2 units and volume V cubic units.

  a. Find the radius needed to manufacture a cylinder with each volume. (Round to the nearest tenth of a unit.)

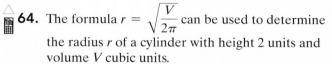

V	10	100	1000
r			

  b. Notice in the table that volume V has been increased by a factor of 10 each time. Does the corresponding radius increase by a factor of 10 each time also?

The formula for the radius of a sphere is $r = \sqrt[3]{\dfrac{3V}{4\pi}}$, where V is the volume in cubic millimeters and r is the radius in millimeters. Use 3.14 as an approximation for π, and round Exercises 65 and 66 to the nearest millimeter.

65. Find the radius of a table tennis ball whose volume is 33,494 cubic millimeters.

66. Find the radius of a baseball whose volume is 212,067 cubic millimeters.

Graphing calculators can be used to solve equations. To solve $\sqrt{x-2} = x - 5$, for example, graph $y_1 = \sqrt{x-2}$ and $y_2 = x - 5$ on the same set of axes. Use the Trace and Zoom features or an Intersect feature to find the point of intersection of the graphs. The x-value of the point is the solution of the equation. Use a graphing calculator to solve the equations below. Approximate solutions to the nearest hundredth.

67. $\sqrt{x-2} = x - 5$

68. $\sqrt{x+1} = 2x - 3$

69. $-\sqrt{x+4} = 5x - 6$

70. $-\sqrt{x+5} = -7x + 1$

15.6 Radical Equations and Problem Solving

Objectives

A Use the Pythagorean Theorem to Solve Problems.

B Solve Problems Using Formulas Containing Radicals.

Objective A Using the Pythagorean Theorem

Applications of radicals can be found in geometry, finance, science, and other areas of technology. Our first application involves the Pythagorean theorem, which gives a formula that relates the lengths of the three sides of a right triangle. We studied the Pythagorean theorem in Chapters 7 and 11, and we review it here.

The Pythagorean Theorem

If a and b are lengths of the legs of a right triangle and c is the length of the hypotenuse, then $a^2 + b^2 = c^2$.

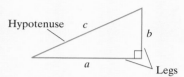

Practice 1

Find the length of the hypotenuse of the right triangle shown.

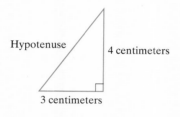

Example 1 Find the length of the hypotenuse of a right triangle whose legs are 6 inches and 8 inches long.

Solution: Because this is a right triangle, we use the Pythagorean theorem. We let $a = 6$ inches and $b = 8$ inches. Length c must be the length of the hypotenuse.

$$a^2 + b^2 = c^2 \quad \text{Use the Pythagorean theorem.}$$
$$6^2 + 8^2 = c^2 \quad \text{Substitute the lengths of the legs.}$$
$$36 + 64 = c^2 \quad \text{Simplify.}$$
$$100 = c^2$$

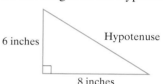

Since c represents a length, we know that c is positive and is the principal square root of 100.

$$100 = c^2$$
$$\sqrt{100} = c \quad \text{Use the definition of principal square root.}$$
$$10 = c \quad \text{Simplify.}$$

The hypotenuse has a length of 10 inches.

■ Work Practice 1

Practice 2

Find the length of the leg of the right triangle shown. Give the exact length and a two-decimal-place approximation.

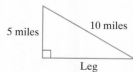

Example 2 Find the length of the leg of the right triangle shown. Give the exact length and a two-decimal-place approximation.

Solution: We let $a = 2$ meters and b be the unknown length of the other leg. The hypotenuse is $c = 5$ meters.

$$a^2 + b^2 = c^2 \quad \text{Use the Pythagorean theorem.}$$
$$2^2 + b^2 = 5^2 \quad \text{Let } a = 2 \text{ and } c = 5.$$
$$4 + b^2 = 25$$
$$b^2 = 21$$
$$b = \sqrt{21} \approx 4.58 \text{ meters}$$

The length of the leg is exactly $\sqrt{21}$ meters or approximately 4.58 meters.

■ Work Practice 2

Answers

1. 5 cm **2.** $5\sqrt{3}$ mi; 8.66 mi

Example 3 Finding a Distance

A surveyor must determine the distance across a lake at points P and Q, as shown in the figure. To do this, she finds a third point, R, such that line QR is perpendicular to line PQ. If the length of $\overline{PR}$ is 320 feet and the length of $\overline{QR}$ is 240 feet, what is the distance across the lake? Approximate this distance to the nearest whole foot.

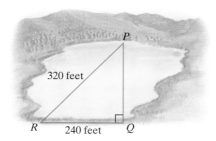

Practice 3
Evan Saacks wants to determine the distance at certain points across a pond on his property. He is able to measure the distances shown on the following diagram. Find how wide the pond is to the nearest tenth of a foot.

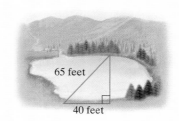

Solution:
1. **UNDERSTAND.** Read and reread the problem. We will set up the problem using the Pythagorean theorem. By creating a line perpendicular to line PQ, the surveyor deliberately constructed a right triangle. The hypotenuse, $\overline{PR}$, has a length of 320 feet, so we let $c = 320$ in the Pythagorean theorem. The side $\overline{QR}$ is one of the legs, so we let $a = 240$ and $b =$ the unknown length.

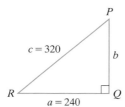

2. **TRANSLATE.**

   $a^2 + b^2 = c^2$ Use the Pythagorean theorem.

   $240^2 + b^2 = 320^2$ Let $a = 240$ and $c = 320$.

3. **SOLVE.**

   $57{,}600 + b^2 = 102{,}400$

   $\quad\quad\quad\; b^2 = 44{,}800$ Subtract 57,600 from both sides.

   $\quad\quad\quad\;\; b = \sqrt{44{,}800}$ Use the definition of principal square root.

   $\quad\quad\quad\quad\;\; = 80\sqrt{7}$ Simplify.

4. **INTERPRET.**

   Check: See that $240^2 + (\sqrt{44{,}800})^2 = 320^2$.

   State: The distance across the lake is *exactly* $\sqrt{44{,}800}$ or $80\sqrt{7}$ feet. The surveyor can now use a calculator to find that $80\sqrt{7}$ feet is *approximately* 211.6601 feet, so the distance across the lake is roughly 212 feet.

   Work Practice 3

Objective B Using Formulas Containing Radicals

The Pythagorean theorem is an extremely important formula in mathematics and should be memorized. But there are other applications involving formulas containing radicals that are not quite as well known, such as the velocity formula used in the next example.

Answer
3. 51.2 feet

Practice 4

Use the formula from Example 4 to find the velocity of an object after it has fallen 20 feet. Round to the nearest tenth.

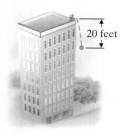

Answer
4. $16\sqrt{5}$ ft per sec ≈ 35.8 ft per sec

Example 4 Finding the Velocity of an Object

A formula used to determine the velocity v, in feet per second, of an object after it has fallen a certain height (neglecting air resistance) is $v = \sqrt{2gh}$, where g is the acceleration due to gravity and h is the height the object has fallen. On Earth, the acceleration g due to gravity is approximately 32 feet per second per second. Find the velocity of a person after falling 5 feet. Round to the nearest tenth.

Solution: We are told that $g = 32$ feet per second per second. To find the velocity v when $h = 5$ feet, we use the velocity formula.

$$v = \sqrt{2gh} \quad \text{Use the velocity formula.}$$
$$= \sqrt{2 \cdot 32 \cdot 5} \quad \text{Substitute known values.}$$
$$= \sqrt{320}$$
$$= 8\sqrt{5} \quad \text{Simplify the radicand.}$$

The velocity of the person after falling 5 feet is *exactly* $8\sqrt{5}$ feet per second, or *approximately* 17.9 feet per second.

■ Work Practice 4

Vocabulary, Readiness & Video Check

Martin-Gay Interactive Videos Watch the section lecture video and answer the following questions.

See Video 15.6

Objective A 1. From ▶ Examples 1 and 2, when solving exercises using the Pythagorean theorem, what two things must we keep in mind? ▶

2. What very important point is made as the final answer to ▶ Example 1 is being found? ▶

Objective B 3. In ▶ Example 4, how do we know to give an estimated answer instead of an exact answer? In what form would the exact answer be given? ▶

15.6 Exercise Set MyLab Math

Objective A *Use the Pythagorean theorem to find the length of the unknown side of each right triangle. Give the exact answer and a two-decimal-place approximation. See Examples 1 and 2.*

1.

2.

3.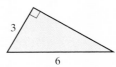

4.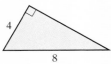

5.

6.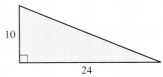

7.

8.

9.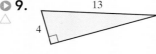

10.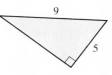

Find the length of the unknown side of each right triangle with sides a, b, and c, where c is the hypotenuse. See Examples 1 and 2. Give the exact answer and a two-decimal-place approximation.

11. $a = 4, b = 5$

12. $a = 2, b = 7$

13. $b = 2, c = 6$

14. $b = 1, c = 5$

15. $a = \sqrt{10}, c = 10$

16. $a = \sqrt{7}, c = \sqrt{35}$

Solve each problem. See Example 3.

17. A wire is used to anchor a 20-foot-tall pole. One end of the wire is attached to the top of the pole. The other end is fastened to a stake five feet away from the bottom of the pole. Find the length of the wire rounded to the nearest tenth of a foot.

18. Jim Spivey needs to connect two underground pipelines, which are offset by 3 feet, as pictured in the diagram. Neglecting the joints needed to join the pipes, find the length of the shortest possible connecting pipe rounded to the nearest hundredth of a foot.

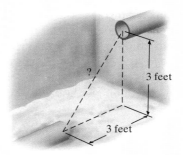

△ **19.** Robert Weisman needs to attach a diagonal brace to a rectangular frame in order to make it structurally sound. If the framework is 6 feet by 10 feet, find how long the brace needs to be to the nearest tenth of a foot.

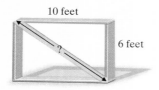

△ **20.** Elizabeth Kaster is flying a kite. She let out 80 feet of string and attached the string to a stake in the ground. The kite is now directly above her brother Mike, who is 32 feet away from the stake. Find the height of the kite to the nearest foot.

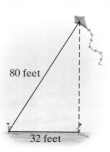

Objective B *Solve each problem. See Example 4.*

△ **21.** For a square-based pyramid, the formula $b = \sqrt{\dfrac{3V}{h}}$ describes the relationship among the length b of one side of the base, the volume V, and the height h. Find the volume if each side of the base is 6 feet long, and the pyramid is 2 feet high.

22. The formula $t = \dfrac{\sqrt{d}}{4}$ relates the distance d, in feet, that an object falls in t seconds, assuming that air resistance does not slow down the object. Find how long, to the nearest hundredth of a second, it takes an object to reach the ground from the top of the Willis Tower in Chicago, a distance of 1730 feet. (*Source:* Council on Tall Buildings and Urban Habitat)

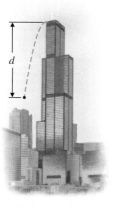

23. Police use the formula $s = \sqrt{30fd}$ to estimate the speed s of a car just before it skidded. In this formula, the speed s is measured in miles per hour, d represents the distance the car skidded in feet, and f represents the coefficient of friction. The value of f depends on the type of road surface, and for wet concrete f is 0.35. Find how fast a car was moving if it skidded 280 feet on wet concrete. Round your result to the nearest mile per hour.

24. The coefficient of friction of a certain dry road is 0.95. Use the formula in Exercise **23** to find how far a car will skid on this dry road if it is traveling at a rate of 60 mph. Round the length to the nearest foot.

25. The formula $v = \sqrt{2.5r}$ can be used to estimate the maximum safe velocity, v, in miles per hour, at which a car can travel if it is driven along a curved road with a **radius of curvature** r in feet. Find the maximum safe speed to the nearest whole number if a cloverleaf exit on an expressway has a radius of curvature of 300 feet.

26. Use the formula from Exercise **25** to find the radius of curvature if the safe velocity is 30 mph.

The maximum distance d in kilometers that you can see from a height of h meters is given by $d = 3.5\sqrt{h}$. Use this equation for Exercises 27 through 30.

27. Find how far you can see from the top of the Comcast Building in New York City, a height of 259.1 meters. Round to the nearest tenth of a kilometer. (*Source:* Council on Tall Buildings and Urban Habitat)

28. Find how far you can see from the top of Great American Tower at Queen City Square in Cincinnati, Ohio, a height of 202.7 meters. Round to the nearest tenth of a kilometer. (*Source:* Council on Tall Buildings and Urban Habitat)

29. The newly built One World Trade Center, in New York City, is the tallest building in the Western Hemisphere. Its height, including the spire at the top of the building, is 541.3 meters. Find how far you could see from the top of One World Trade Center's spire. Round to the nearest tenth of a kilometer. (*Source:* Council on Tall Buildings and Urban Habitat)

30. Guests can take in the views from One World Trade Center by visiting the building's One World Observatory, located at a height of 386.1 meters. Find how far a visitor to One World Observatory could see. Round to the nearest tenth of a kilometer. (*Source:* Council on Tall Buildings and Urban Habitat)

Review

Find two numbers whose square is the given number. See Section 15.1.

31. 9   **32.** 25   **33.** 100

34. 49   **35.** 64   **36.** 121

Concept Extensions

For each triangle, find the length of y, then x.

△ 37.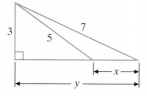

△ 38.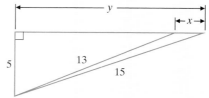

△ 39. Mike and Sandra Hallahan leave the seashore at the same time. Mike drives northward at a rate of 30 miles per hour, while Sandra drives west at 60 mph. Find how far apart they are after 3 hours to the nearest mile.

△ 40. Railroad tracks are invariably made up of relatively short sections of rail connected by expansion joints. To see why this construction is necessary, consider a single rail 100 feet long (or 1200 inches). On an extremely hot day, suppose it expands 1 inch in the hot sun to a new length of 1201 inches. Theoretically, the track would bow upward as pictured.

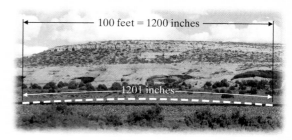

Let us approximate the bulge in the railroad this way.

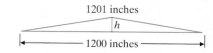

Calculate the height h of the bulge to the nearest tenth of an inch.

✎ 41. Based on the results of Exercise **40**, explain why railroads use short sections of rail connected by expansion joints.

Chapter 15 Group Activity

Graphing and the Distance Formula

One application of radicals is finding the distance between two points in the coordinate plane. This can be very useful in graphing.

The distance d between two points with coordinates (x_1, y_1) and (x_2, y_2) is given by the **distance formula** $d = \sqrt{(x_2 - x_1)^2 + (y_2 - y_1)^2}$.

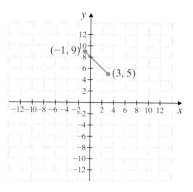

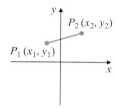

Suppose we want to find the distance between the two points $(-1, 9)$ and $(3, 5)$. We can use the distance formula with $(x_1, y_1) = (-1, 9)$ and $(x_2, y_2) = (3, 5)$. Then we have

$$d = \sqrt{(x_2 - x_1)^2 + (y_2 - y_1)^2}$$
$$= \sqrt{[3 - (-1)]^2 + (5 - 9)^2}$$
$$= \sqrt{(4)^2 + (-4)^2}$$
$$= \sqrt{16 + 16}$$
$$= \sqrt{32} = 4\sqrt{2}$$

The distance between the two points is exactly $4\sqrt{2}$ units or approximately 5.66 units.

Group Activity

Brainstorm to come up with several disciplines or activities in which the distance formula might be useful. Make up an example that shows how the distance formula would be used in one of the activities on your list. Then present your example to the rest of the class.

Chapter 15 Vocabulary Check

Fill in each blank with one of the words or phrases listed below. Not all choices will be used.

index	radicand	like radicals
rationalizing the denominator	conjugate	leg
principal square root	radical	hypotenuse

1. The expressions $5\sqrt{x}$ and $7\sqrt{x}$ are examples of _____.
2. In the expression $\sqrt[3]{45}$, the number 3 is the _____, the number 45 is the _____, and $\sqrt{}$ is called the _____ sign.
3. The _____ of $a + b$ is $a - b$.
4. The _____ of 25 is 5.
5. The process of eliminating the radical in the denominator of a radical expression is called _____.
6. The Pythagorean theorem states that for a right triangle, $(\text{leg})^2 + (\text{leg})^2 = (\underline{})^2$.

Helpful Hint

▶ Are you preparing for your test?
To help, don't forget to take these:
- Chapter 15 Getting Ready for the Test on page 1188
- Chapter 15 Test on page 1189

Then check all of your answers at the back of this text. For further review, the step-by-step video solutions to any of these exercises are located in MyLab Math.

15 Chapter Highlights

Definitions and Concepts	Examples
Section 15.1 Introduction to Radicals	
The **positive** or **principal square root** of a positive number a is written as $\sqrt{a}$. The **negative square root** of a is written as $-\sqrt{a}$. $\sqrt{a} = b$ only if $b^2 = a$ and $b > 0$.	$\sqrt{25} = 5 \qquad \sqrt{100} = 10$ $-\sqrt{9} = -3 \qquad \sqrt{\dfrac{4}{49}} = \dfrac{2}{7}$
A square root of a negative number is not a real number.	$\sqrt{-4}$ is not a real number.
The **cube root** of a real number a is written as $\sqrt[3]{a}$. $\sqrt[3]{a} = b$ only if $b^3 = a$. The **nth root** of a number a is written as $\sqrt[n]{a}$. $\sqrt[n]{a} = b$ only if $b^n = a$. In $\sqrt[n]{a}$, the natural number n is called the **index**, the symbol $\sqrt{}$ is called a **radical**, and the expression within the radical is called the **radicand**. (*Note:* If the index is even, the radicand must be nonnegative for the root to be a real number.)	$\sqrt[3]{64} = 4 \qquad \sqrt[3]{-8} = -2$ $\sqrt[4]{81} = 3$ $\sqrt[5]{-32} = -2$ index ↓ $\sqrt[n]{a}$ ↖ radicand
Section 15.2 Simplifying Radicals	
Product Rule for Radicals If $\sqrt{a}$ and $\sqrt{b}$ are real numbers, then $\sqrt{a \cdot b} = \sqrt{a} \cdot \sqrt{b}$	
A square root is in **simplified form** if the radicand contains no perfect square factors other than 1. To simplify a square root, factor the radicand so that one of its factors is a perfect square factor.	$\sqrt{45} = \sqrt{9 \cdot 5}$ $\phantom{\sqrt{45}} = \sqrt{9} \cdot \sqrt{5}$ $\phantom{\sqrt{45}} = 3\sqrt{5}$

Definitions and Concepts	Examples
Section 15.2 Simplifying Radicals *(continued)*	
Quotient Rule for Radicals If $\sqrt{a}$ and $\sqrt{b}$ are real numbers and $b \neq 0$, then $$\sqrt{\frac{a}{b}} = \frac{\sqrt{a}}{\sqrt{b}}$$	$$\sqrt{\frac{18}{x^6}} = \frac{\sqrt{9 \cdot 2}}{\sqrt{x^6}} = \frac{\sqrt{9} \cdot \sqrt{2}}{x^3} = \frac{3\sqrt{2}}{x^3}$$
Section 15.3 Adding and Subtracting Radicals	
Like radicals are radical expressions that have the same index and the same radicand. To **combine like radicals,** use the distributive property.	$5\sqrt{2}, -7\sqrt{2}, \sqrt{2}$ $2\sqrt{7} - 13\sqrt{7} = (2-13)\sqrt{7} = -11\sqrt{7}$ $\sqrt{8} + \sqrt{50} = 2\sqrt{2} + 5\sqrt{2} = 7\sqrt{2}$
Section 15.4 Multiplying and Dividing Radicals	
The product and quotient rules for radicals may be used to simplify products and quotients of radicals.	Perform each indicated operation and simplify. Multiply. $\sqrt{2} \cdot \sqrt{8} = \sqrt{16} = 4$ $(\sqrt{3x} + 1)(\sqrt{5} - \sqrt{3})$ $= \sqrt{15x} - \sqrt{9x} + \sqrt{5} - \sqrt{3}$ $= \sqrt{15x} - 3\sqrt{x} + \sqrt{5} - \sqrt{3}$ Divide. $$\frac{\sqrt{20}}{\sqrt{2}} = \sqrt{\frac{20}{2}} = \sqrt{10}$$
The process of eliminating the radical in the denominator of a radical expression is called **rationalizing the denominator.**	Rationalize the denominator. $$\frac{5}{\sqrt{11}} = \frac{5 \cdot \sqrt{11}}{\sqrt{11} \cdot \sqrt{11}} = \frac{5\sqrt{11}}{11}$$
The **conjugate** of $a + b$ is $a - b$. To rationalize a denominator that is a sum or difference of radicals, multiply the numerator and the denominator by the conjugate of the denominator.	The conjugate of $2 + \sqrt{3}$ is $2 - \sqrt{3}$. Rationalize the denominator. $$\frac{5}{6 - \sqrt{5}} = \frac{5(6 + \sqrt{5})}{(6 - \sqrt{5})(6 + \sqrt{5})}$$ $$= \frac{5(6 + \sqrt{5})}{36 - 5}$$ $$= \frac{5(6 + \sqrt{5})}{31}$$

Definitions and Concepts	Examples
Section 15.5 Solving Equations Containing Radicals	
To Solve a Radical Equation Containing Square Roots **Step 1:** Get one radical by itself on one side of the equation. **Step 2:** Square both sides of the equation. **Step 3:** Simplify both sides of the equation. **Step 4:** If the equation still contains a radical term, repeat Steps 1 through 3. **Step 5:** Solve the equation. **Step 6:** Check solutions in the original equation.	Solve: $$\sqrt{2x-1} - x = -2$$ $$\sqrt{2x-1} = x - 2$$ $$(\sqrt{2x-1})^2 = (x-2)^2 \quad \text{Square both sides}$$ $$2x - 1 = x^2 - 4x + 4$$ $$0 = x^2 - 6x + 5$$ $$0 = (x-1)(x-5) \quad \text{Factor.}$$ $$x - 1 = 0 \quad \text{or} \quad x - 5 = 0$$ $$x = 1 \qquad\qquad x = 5 \quad \text{Solve.}$$ Check both proposed solutions in the original equation. Here, 5 checks but 1 does not. The only solution is 5.
Section 15.6 Radical Equations and Problem Solving	
Problem-Solving Steps 1. UNDERSTAND. Read and reread the problem.	A rain gutter is to be mounted on the eaves of a house 15 feet above the ground. A garden is adjacent to the house, so the closest a ladder can be placed to the house is 6 feet. How long a ladder is needed for installing the gutter? Let $x =$ the length of the ladder. 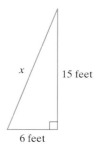 Here, we use the Pythagorean theorem. The unknown length x is the hypotenuse. In words: $$(\text{leg})^2 + (\text{leg})^2 = (\text{hypotenuse})^2$$
2. TRANSLATE. 3. SOLVE.	Translate: $$6^2 + 15^2 = x^2$$ $$36 + 225 = x^2$$ $$261 = x^2$$ $$\sqrt{261} = x \quad \text{or} \quad x = 3\sqrt{29}$$
4. INTERPRET.	Check and state. The ladder needs to be $3\sqrt{29}$ feet or approximately 16.2 feet long.

Chapter 15   Review

15.1) *Find each root.*

1. $\sqrt{81}$
2. $-\sqrt{49}$
3. $\sqrt[3]{27}$
4. $\sqrt[4]{81}$

5. $-\sqrt{\dfrac{9}{64}}$
6. $\sqrt{\dfrac{36}{81}}$
7. $\sqrt[4]{16}$
8. $\sqrt[3]{-8}$

9. Which radical(s) is not a real number?
   a. $\sqrt{4}$ b. $-\sqrt{4}$ c. $\sqrt{-4}$ d. $\sqrt[3]{-4}$

10. Which radical(s) is not a real number?
    a. $\sqrt{-5}$ b. $\sqrt[3]{-5}$ c. $\sqrt[4]{-5}$ d. $\sqrt[5]{-5}$

Find each root. Assume that all variables represent positive numbers.

11. $\sqrt{x^{12}}$
12. $\sqrt{x^8}$
13. $\sqrt{9y^2}$
14. $\sqrt{25x^4}$

15.2) *Simplify each expression using the product rule. Assume that all variables represent positive numbers.*

15. $\sqrt{40}$
16. $\sqrt{24}$
17. $\sqrt{54}$
18. $\sqrt{88}$
19. $\sqrt{x^5}$

20. $\sqrt{y^7}$
21. $\sqrt{20x^2}$
22. $\sqrt{50y^4}$
23. $\sqrt[3]{54}$
24. $\sqrt[3]{88}$

Simplify each expression using the quotient rule. Assume that all variables represent positive numbers.

25. $\sqrt{\dfrac{18}{25}}$
26. $\sqrt{\dfrac{75}{64}}$
27. $-\sqrt{\dfrac{50}{9}}$
28. $-\sqrt{\dfrac{12}{49}}$

29. $\sqrt{\dfrac{11}{x^2}}$
30. $\sqrt{\dfrac{7}{y^4}}$
31. $\sqrt{\dfrac{y^5}{100}}$
32. $\sqrt{\dfrac{x^3}{81}}$

15.3) *Add or subtract by combining like radicals.*

33. $5\sqrt{2} - 8\sqrt{2}$
34. $\sqrt{3} - 6\sqrt{3}$

35. $6\sqrt{5} + 3\sqrt{6} - 2\sqrt{5} + \sqrt{6}$
36. $-\sqrt{7} + 8\sqrt{2} - \sqrt{7} - 6\sqrt{2}$

Add or subtract by simplifying each radical and then combining like terms. Assume that all variables represent positive numbers.

37. $\sqrt{28} + \sqrt{63} + \sqrt{56}$
38. $\sqrt{75} + \sqrt{48} - \sqrt{16}$

39. $\sqrt{\dfrac{5}{9}} - \sqrt{\dfrac{5}{36}}$
40. $\sqrt{\dfrac{11}{25}} + \sqrt{\dfrac{11}{16}}$

41. $\sqrt{45x^2} + 3\sqrt{5x^2} - 7x\sqrt{5} + 10$
42. $\sqrt{50x} - 9\sqrt{2x} + \sqrt{72x} - \sqrt{3x}$

15.4) *Multiply and simplify if possible. Assume that all variables represent positive numbers.*

43. $\sqrt{3} \cdot \sqrt{6}$
44. $\sqrt{5} \cdot \sqrt{15}$

1185

45. $\sqrt{2}(\sqrt{5} - \sqrt{7})$

46. $\sqrt{5}(\sqrt{11} + \sqrt{3})$

47. $(\sqrt{3} + 2)(\sqrt{6} - 5)$

48. $(\sqrt{5} + 1)(\sqrt{5} - 3)$

49. $(\sqrt{x} - 2)^2$

50. $(\sqrt{y} + 4)^2$

Divide and simplify if possible. Assume that all variables represent positive numbers.

51. $\dfrac{\sqrt{27}}{\sqrt{3}}$

52. $\dfrac{\sqrt{20}}{\sqrt{5}}$

53. $\dfrac{\sqrt{160}}{\sqrt{8}}$

54. $\dfrac{\sqrt{96}}{\sqrt{3}}$

55. $\dfrac{\sqrt{30x^6}}{\sqrt{2x^3}}$

56. $\dfrac{\sqrt{54x^5y^2}}{\sqrt{3xy^2}}$

Rationalize each denominator and simplify.

57. $\dfrac{\sqrt{2}}{\sqrt{11}}$

58. $\dfrac{\sqrt{3}}{\sqrt{13}}$

59. $\sqrt{\dfrac{5}{6}}$

60. $\sqrt{\dfrac{7}{10}}$

61. $\dfrac{1}{\sqrt{5x}}$

62. $\dfrac{5}{\sqrt{3y}}$

63. $\sqrt{\dfrac{3}{x}}$

64. $\sqrt{\dfrac{6}{y}}$

65. $\dfrac{3}{\sqrt{5} - 2}$

66. $\dfrac{8}{\sqrt{10} - 3}$

67. $\dfrac{\sqrt{2} + 1}{\sqrt{3} - 1}$

68. $\dfrac{\sqrt{3} - 2}{\sqrt{5} + 2}$

69. $\dfrac{10}{\sqrt{x} + 5}$

70. $\dfrac{8}{\sqrt{x} - 1}$

(15.5) *Solve each radical equation.*

71. $\sqrt{2x} = 6$

72. $\sqrt{x + 3} = 4$

73. $\sqrt{x} + 3 = 8$

74. $\sqrt{x} + 8 = 3$

75. $\sqrt{2x + 1} = x - 7$

76. $\sqrt{3x + 1} = x - 1$

77. $\sqrt{x + 3} = \sqrt{x + 15}$

78. $\sqrt{x - 5} = \sqrt{x} - 1$

(15.6) *Use the Pythagorean theorem to find the length of each unknown side. Give the exact answer and a two-decimal-place approximation.*

 79.
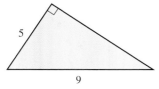

 80.
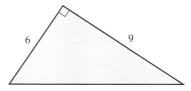

 81. Romeo is standing 20 feet away from the wall below Juliet's balcony during a school play. Juliet is on the balcony, 12 feet above the ground. Find how far apart Romeo and Juliet are.

 82. The diagonal of a rectangle is 10 inches long. If the width of the rectangle is 5 inches, find the length of the rectangle.

Use the formula $r = \sqrt{\dfrac{S}{4\pi}}$, where r = the radius of a sphere and S = the surface area of the sphere, for Exercises 83 and 84.

△ **83.** Find the radius of a sphere to the nearest tenth of an inch if the surface area is 72 square inches.

△ **84.** Find the exact surface area of a sphere if its radius is 6 inches. (Do not approximate π.)

Mixed Review

Find each root. Assume all variables represent positive numbers.

85. $\sqrt{144}$ **86.** $-\sqrt[3]{64}$ **87.** $\sqrt{16x^{16}}$ **88.** $\sqrt{4x^{24}}$

Simplify each expression. Assume all variables represent positive numbers.

89. $\sqrt{18x^7}$ **90.** $\sqrt{48y^6}$ **91.** $\sqrt{\dfrac{y^4}{81}}$ **92.** $\sqrt{\dfrac{x^9}{9}}$

Add or subtract by simplifying and then combining like terms. Assume all variables represent positive numbers.

93. $\sqrt{12} + \sqrt{75}$

94. $\sqrt{63} + \sqrt{28} - \sqrt{9}$

95. $\sqrt{\dfrac{3}{16}} - \sqrt{\dfrac{3}{4}}$

96. $\sqrt{45x^3} + x\sqrt{20x} - \sqrt{5x^3}$

Multiply and simplify if possible. Assume all variables represent positive numbers.

97. $\sqrt{7} \cdot \sqrt{14}$ **98.** $\sqrt{3}(\sqrt{9} - \sqrt{2})$ **99.** $(\sqrt{2} + 4)(\sqrt{5} - 1)$ **100.** $(\sqrt{x} + 3)^2$

Divide and simplify if possible. Assume all variables represent positive numbers.

101. $\dfrac{\sqrt{120}}{\sqrt{5}}$

102. $\dfrac{\sqrt{60x^9}}{\sqrt{15x^7}}$

Rationalize each denominator and simplify.

103. $\sqrt{\dfrac{2}{7}}$

104. $\dfrac{3}{\sqrt{2x}}$

105. $\dfrac{3}{\sqrt{x} - 6}$

106. $\dfrac{\sqrt{7} - 5}{\sqrt{5} + 3}$

Solve each radical equation.

107. $\sqrt{4x} = 2$ **108.** $\sqrt{x - 4} = 3$ **109.** $\sqrt{4x + 8} + 6 = x$ **110.** $\sqrt{x - 8} = \sqrt{x} - 2$

111. Use the Pythagorean theorem to find the length of the unknown side. Give the exact answer and a two-decimal-place approximation.

112. The diagonal of a rectangle is 6 inches long. If the width of the rectangle is 2 inches, find the length of the rectangle.

Chapter 15 — Getting Ready for the Test

MULTIPLE CHOICE Exercises 1 through 11 are **Multiple Choice**. Select the correct choice.

1. Choose the expression that simplifies to -4.
   A. $\sqrt{-16}$
   B. $-\sqrt{16}$
   C. $\sqrt[3]{8}$
   D. $\sqrt[3]{-8}$

2. $7\sqrt{3} - \sqrt{3} =$
   A. 7
   B. 6
   C. $6\sqrt{3}$
   D. cannot be simplified

3. $7\sqrt{3} \cdot \sqrt{2} =$
   A. 35
   B. $7\sqrt{6}$
   C. 42
   D. $8\sqrt{6}$
   E. cannot be simplified

4. $(4\sqrt{5})^2 =$
   A. 40
   B. 80
   C. $8\sqrt{5}$
   D. $16\sqrt{5}$
   E. cannot be simplified

5. Simplify: $\sqrt{\dfrac{28}{25}}$
   A. $\dfrac{14}{5}$
   B. $\dfrac{14}{\sqrt{25}}$
   C. $\dfrac{2\sqrt{7}}{5}$
   D. cannot be simplified

6. Simplify: $\sqrt{18x^{16}}$
   A. $9x^8$
   B. $9x^4$
   C. $3x^4\sqrt{2}$
   D. $3x^8\sqrt{2}$

7. Simplify: $\sqrt[3]{64}$
   A. 8
   B. 4
   C. 192
   D. cannot be simplified

8. Simplify: $\sqrt[3]{x^{27}}$
   A. x^3
   B. x^9
   C. $x^{13}\sqrt[3]{x}$
   D. cannot be simplified

9. To rationalize the denominator of $\dfrac{\sqrt{5}}{\sqrt{2}}$, we multiply by:
   A. $\dfrac{\sqrt{5}}{\sqrt{2}}$
   B. $\dfrac{\sqrt{10}}{\sqrt{10}}$
   C. $\dfrac{\sqrt{2}}{\sqrt{2}}$
   D. $\dfrac{\sqrt{5}}{\sqrt{5}}$

10. Square both sides of the equation $3\sqrt{x} = \sqrt{10x - 9}$. The result is:
    A. $3x = 10x - 9$
    B. $3x^2 = 10x - 9$
    C. $9x = 10x - 9$
    D. $3x^2 = 100x^2 - 38x + 81$

11. Square both sides of the equation $x + 1 = \sqrt{9x - 9}$. The result is:
    A. $x^2 + 2x + 1 = 9x - 9$
    B. $x^2 + 1 = 9x - 9$
    C. $x^2 + x + 1 = 9x - 9$
    D. $x + 1 = 9x - 9$

Chapter 15 Test

Simplify each radical. Indicate if the radical is not a real number. Assume that x represents a positive number.

1. $\sqrt{16}$
2. $\sqrt[3]{125}$
3. $\sqrt[4]{81}$

4. $\sqrt{\dfrac{9}{16}}$
5. $\sqrt[4]{-81}$
6. $\sqrt{x^{10}}$

Simplify each radical. Assume that all variables represent positive numbers.

7. $\sqrt{54}$
8. $\sqrt{92}$
9. $\sqrt{y^7}$
10. $\sqrt{24x^8}$

11. $\sqrt[3]{27}$
12. $\sqrt[3]{16}$
13. $\sqrt{\dfrac{5}{16}}$
14. $\sqrt{\dfrac{y^3}{25}}$

Perform each indicated operation. Assume that all variables represent positive numbers.

15. $\sqrt{13} + \sqrt{13} - 4\sqrt{13}$
16. $\sqrt{18} - \sqrt{75} + 7\sqrt{3} - \sqrt{8}$

17. $\sqrt{\dfrac{3}{4}} + \sqrt{\dfrac{3}{25}}$
18. $\sqrt{7} \cdot \sqrt{14}$
19. $\sqrt{2}(\sqrt{6} - \sqrt{5})$

Answers

1. _____
2. _____
3. _____
4. _____
5. _____
6. _____
7. _____
8. _____
9. _____
10. _____
11. _____
12. _____
13. _____
14. _____
15. _____
16. _____
17. _____
18. _____
19. _____

20. $(\sqrt{x}+2)(\sqrt{x}-3)$

21. $\dfrac{\sqrt{50}}{\sqrt{10}}$

22. $\dfrac{\sqrt{40x^4}}{\sqrt{2x}}$

Rationalize each denominator. Assume that all variables represent positive numbers.

23. $\sqrt{\dfrac{2}{3}}$

24. $\dfrac{8}{\sqrt{5y}}$

25. $\dfrac{8}{\sqrt{6}+2}$

26. $\dfrac{1}{3-\sqrt{x}}$

Solve each radical equation.

27. $\sqrt{x}+8=11$

28. $\sqrt{3x-6}=\sqrt{x+4}$

29. $\sqrt{2x-2}=x-5$

30. Find the length of the unknown leg of the right triangle shown. Give the exact answer.

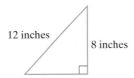

31. The formula $r=\sqrt{\dfrac{A}{\pi}}$ can be used to find the radius r of a circle given its area A. Use this formula to approximate the radius of the given circle. Round to two decimal places.

Area is 15 square meters.

Cumulative Review — Chapters 1–15

1. Round 736.2359 to the nearest tenth.

2. Round 328.174 to the nearest tenth.

3. Add: $23.85 + 1.604$

4. Add: $12.762 + 4.29$

5. Is -9 a solution of the equation $3.7y = -3.33$?

6. Is 6 a solution of the equation $2.8x = 16.8$?

7. Find: $\sqrt{\dfrac{1}{36}}$

8. Find: $\sqrt{\dfrac{4}{25}}$

9. Solve: $4(2x - 3) + 7 = 3x + 5$

10. Solve: $3(2 - 5x) + 24x = 12$

11. Write each number in standard form, without exponents.
    a. 1.02×10^5
    b. 7.358×10^{-3}
    c. 8.4×10^7
    d. 3.007×10^{-5}

12. Write each number in standard form, without exponents.
    a. 8.26×10^4
    b. 9.9×10^{-2}
    c. 1.002×10^5
    d. 8.039×10^{-3}

13. Multiply: $(3x + 2)(2x - 5)$

14. Multiply: $(5x - 1)(4x + 1)$

15. Factor $xy + 2x + 3y + 6$ by grouping.

16. Factor: $16x^3 - 28x^2 + 12x - 21$

17. Factor: $3x^2 + 11x + 6$

18. Factor: $9x^2 - 5x - 4$

Are there any values of x for which each expression is undefined?

19. a. $\dfrac{x}{x - 3}$
    b. $\dfrac{x^2 + 2}{x^2 - 3x + 2}$
    c. $\dfrac{x^3 - 6x^2 - 10x}{3}$

20. a. $\dfrac{x - 3}{x}$
    b. $\dfrac{x + 1}{5}$
    c. $\dfrac{x^2 - 3}{x^2 - 4}$

Answers

1.
2.
3.
4.
5.
6.
7.
8.
9.
10.
11. a.
    b.
    c.
    d.
12. a.
    b.
    c.
    d.
13.
14.
15.
16.
17.
18.
19. a.
    b.
    c.
20. a.
    b.
    c.

21. Simplify: $\dfrac{x^2 + 4x + 4}{x^2 + 2x}$

22. Simplify: $\dfrac{16x^2 - 4y^2}{4x - 2y}$

Perform each indicated operation.

23. a. $\dfrac{a}{4} - \dfrac{2a}{8}$

b. $\dfrac{3}{10x^2} + \dfrac{7}{25x}$

24. a. $\dfrac{x}{5} - \dfrac{3x}{10}$

b. $\dfrac{9}{12a^2} + \dfrac{5}{16a}$

25. Solve: $\dfrac{4x}{x^2 + x - 30} + \dfrac{2}{x - 5} = \dfrac{1}{x + 6}$

26. Solve: $\dfrac{3}{x + 3} = \dfrac{12x + 19}{x^2 + 7x + 12} - \dfrac{5}{x +}$

27. Graph $y = -3$.

28. Graph $x = 2$.

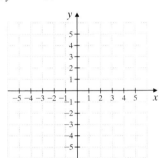
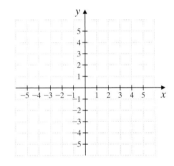

Find each cube root.

29. $\sqrt[3]{1}$

30. $\sqrt[3]{8}$

31. $\sqrt[3]{-27}$

32. $\sqrt[3]{-8}$

33. $\sqrt[3]{\dfrac{1}{125}}$

34. $\sqrt[3]{\dfrac{27}{64}}$

Simplify.

35. $\sqrt{54}$

36. $\sqrt{40}$

37. $\sqrt{200}$

38. $\sqrt{125}$

Add or subtract by first simplifying each radical.

39. $7\sqrt{12} - 2\sqrt{75}$

40. $\sqrt{75} + \sqrt{48}$

41. $2\sqrt{x^2} - \sqrt{25x^5} + \sqrt{x^5}$

42. $5\sqrt{x^2} + \sqrt{36x} + \sqrt{49x^2}$

Rationalize each denominator.

43. $\dfrac{2}{\sqrt{7}}$

44. $\dfrac{4}{\sqrt{5}}$

45. Solve: $\sqrt{x} = \sqrt{5x - 2}$

46. Solve: $\sqrt{x + 5} = x - 1$

Quadratic Equations

16

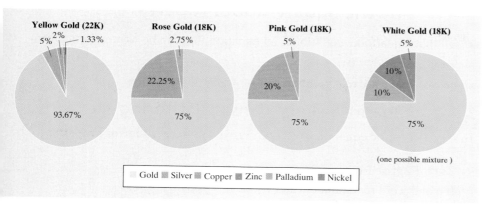

Does Gold Really Come in Colors?

How are rose gold, pink gold, and white gold formed? As we see above, they come from mixing gold with other metals. Most of the gold jewelry that we buy is a mixture (or an alloy) of metals. The purity of gold is given by its karat (K). Twenty-four karat gold is called fine gold and is greater than 99.7% pure gold.

Gold has many interesting qualities. For example, one ounce of gold can be stretched into a golden thread of about 5 miles in length, or it can be made into a thin sheet of about 300 square feet in area.

Below is a bar graph of the price of gold at the beginning of each year shown. In Section 16.2, Exercise 48, we will explore the current and possible future price of gold.

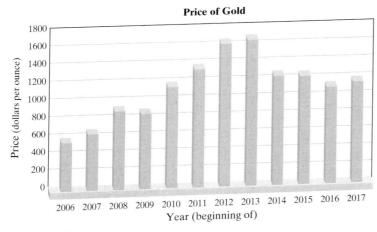

Source: goldprice.org

An important part of the study of algebra is learning to use methods for solving equations. Starting in Chapter 2, we presented techniques for solving linear equations in one variable. In Chapter 11, we solved quadratic equations in one variable by factoring the quadratic expressions. We now present other methods for solving quadratic equations in one variable.

Sections

16.1 Solving Quadratic Equations by the Square Root Property

16.2 Solving Quadratic Equations by Completing the Square

16.3 Solving Quadratic Equations by the Quadratic Formula

Integrated Review—Summary on Solving Quadratic Equations

16.4 Graphing Quadratic Equations in Two Variables

Check Your Progress

Vocabulary Check

Chapter Highlights

Chapter Review

Getting Ready for the Test

Chapter Test

Cumulative Review

16.1 Solving Quadratic Equations by the Square Root Property

Objectives
A Use the Square Root Property to Solve Quadratic Equations.

B Use the Square Root Property to Solve Applications.

Recall that a quadratic equation is an equation that can be written in the form
$$ax^2 + bx + c = 0$$
where a, b, and c are real numbers and $a \neq 0$.

Solving Quadratic Equations by Factoring

To solve quadratic equations by factoring, we use the **zero-factor property**:

> **Zero Factor Property**
> If a and b are real numbers and
> if $ab = 0$, then $a = 0$ or $b = 0$.

Examples 1 and 2 review the process of solving quadratic equations by factoring.

Example 1 Solve: $x^2 - 4 = 0$

Solution:
$$x^2 - 4 = 0$$
$(x + 2)(x - 2) = 0$ Factor.
$x + 2 = 0$ or $x - 2 = 0$ Use the zero-factor property.
$x = -2$ $x = 2$ Solve each equation.

The solutions are -2 and 2.

■ Work Practice 1

Practice 1
Solve: $x^2 - 25 = 0$

Example 2 Solve: $3y^2 + 13y = 10$

Solution: Recall that to use the zero-factor property, one side of the equation must be 0 and the other side must be factored.
$$3y^2 + 13y = 10$$
$3y^2 + 13y - 10 = 0$ Subtract 10 from both sides.
$(3y - 2)(y + 5) = 0$ Factor.
$3y - 2 = 0$ or $y + 5 = 0$ Use the zero-factor property.
$3y = 2$ $y = -5$ Solve each equation.
$y = \dfrac{2}{3}$

The solutions are $\dfrac{2}{3}$ and -5.

■ Work Practice 2

Practice 2
Solve: $2x^2 - 3x = 9$

Objective A Using the Square Root Property

Consider solving Example 1, $x^2 - 4 = 0$, another way. First, add 4 to both sides of the equation.
$$x^2 - 4 = 0$$
$x^2 = 4$ Add 4 to both sides.

Now we see that the value for x must be a number whose square is 4. Therefore $x = \sqrt{4} = 2$ or $x = -\sqrt{4} = -2$. This reasoning is an example of the square root property.

Answers
1. 5 and -5 **2.** $-\dfrac{3}{2}$ and 3

Section 16.1 | Solving Quadratic Equations by the Square Root Property

Square Root Property

If $x^2 = a$ for $a \geq 0$, then

$x = \sqrt{a}$ or $x = -\sqrt{a}$

Example 3 Use the square root property to solve $x^2 - 9 = 0$.

Solution: First we solve for x^2 by adding 9 to both sides.

$x^2 - 9 = 0$
$x^2 = 9$ Add 9 to both sides.

Next we use the square root property.

$x = \sqrt{9}$ or $x = -\sqrt{9}$
$x = 3$ $x = -3$

Check:

$x^2 - 9 = 0$ Original equation $x^2 - 9 = 0$ Original equation
$3^2 - 9 \stackrel{?}{=} 0$ Let $x = 3$. $(-3)^2 - 9 \stackrel{?}{=} 0$ Let $x = -3$.
$0 = 0$ True $0 = 0$ True

The solutions are 3 and -3.

Work Practice 3

Practice 3

Use the square root property to solve $x^2 - 16 = 0$.

Example 4 Use the square root property to solve $2x^2 = 7$.

Solution: First we solve for x^2 by dividing both sides by 2. Then we use the square root property.

$2x^2 = 7$

$x^2 = \dfrac{7}{2}$ Divide both sides by 2.

$x = \sqrt{\dfrac{7}{2}}$ or $x = -\sqrt{\dfrac{7}{2}}$ Use the square root property.

$x = \dfrac{\sqrt{7} \cdot \sqrt{2}}{\sqrt{2} \cdot \sqrt{2}}$ $x = -\dfrac{\sqrt{7} \cdot \sqrt{2}}{\sqrt{2} \cdot \sqrt{2}}$ Rationalize the denominator.

$x = \dfrac{\sqrt{14}}{2}$ $x = -\dfrac{\sqrt{14}}{2}$ Simplify.

Remember to check both solutions in the original equation. The solutions are $\dfrac{\sqrt{14}}{2}$ and $-\dfrac{\sqrt{14}}{2}$.

Work Practice 4

Practice 4

Use the square root property to solve $3x^2 = 11$.

Example 5 Use the square root property to solve $(x - 3)^2 = 16$.

Solution: Instead of x^2, here we have $(x - 3)^2$. But the square root property can still be used.

$(x - 3)^2 = 16$

$x - 3 = \sqrt{16}$ or $x - 3 = -\sqrt{16}$ Use the square root property.
$x - 3 = 4$ $x - 3 = -4$ Write $\sqrt{16}$ as 4 and $-\sqrt{16}$ as -4.
$x = 7$ $x = -1$ Solve.

(Continued on next page)

Practice 5

Use the square root property to solve $(x - 4)^2 = 49$.

Answers

3. 4 and -4 **4.** $\dfrac{\sqrt{33}}{3}$ and $-\dfrac{\sqrt{33}}{3}$

5. 11 and -3

Chapter 16 | Quadratic Equations

Check:

$(x-3)^2 = 16$ Original equation $(x-3)^2 = 16$ Original equation
$(7-3)^2 \stackrel{?}{=} 16$ Let $x = 7$. $(-1-3)^2 \stackrel{?}{=} 16$ Let $x = -1$.
$4^2 \stackrel{?}{=} 16$ Simplify. $(-4)^2 \stackrel{?}{=} 16$ Simplify.
$16 = 16$ True $16 = 16$ True

Both 7 and -1 are solutions.

■ Work Practice 5

Practice 6

Use the square root property to solve $(x-5)^2 = 18$.

Example 6 Use the square root property to solve $(x+1)^2 = 8$.

Solution: $(x+1)^2 = 8$

$x + 1 = \sqrt{8}$ or $x + 1 = -\sqrt{8}$ Use the square root property.
$x + 1 = 2\sqrt{2}$ $x + 1 = -2\sqrt{2}$ Simplify the radical.
$x = -1 + 2\sqrt{2}$ $x = -1 - 2\sqrt{2}$ Solve for x.

Check both solutions in the original equation. The solutions are $-1 + 2\sqrt{2}$ a $-1 - 2\sqrt{2}$. This can be written compactly as $-1 \pm 2\sqrt{2}$. The notation $\pm$ is re as "plus or minus."

Helpful Hint

read "plus or minus"
↓
The notation $-1 \pm \sqrt{5}$, for example, is just a shorthand notation for both $-1 + \sqrt{5}$ and $-1 - \sqrt{5}$.

■ Work Practice 6

Practice 7

Use the square root property to solve $(x+3)^2 = -5$.

Example 7 Use the square root property to solve $(x-1)^2 = -2$.

Solution: This equation has no real solution because the square root of -2 not a real number.

■ Work Practice 7

Practice 8

Use the square root property to solve $(4x+1)^2 = 15$.

Example 8 Use the square root property to solve $(5x-2)^2 = 10$.

Solution: $(5x-2)^2 = 10$

$5x - 2 = \sqrt{10}$ or $5x - 2 = -\sqrt{10}$ Use the square root property.
$5x = 2 + \sqrt{10}$ $5x = 2 - \sqrt{10}$ Add 2 to both sides.
$x = \dfrac{2 + \sqrt{10}}{5}$ $x = \dfrac{2 - \sqrt{10}}{5}$ Divide both sides by 5.

Check both solutions in the original equation. The solutions are $\dfrac{2 + \sqrt{10}}{5}$ a $\dfrac{2 - \sqrt{10}}{5}$, which can be written as $\dfrac{2 \pm \sqrt{10}}{5}$.

■ Work Practice 8

Helpful Hint

For some applications and graphing purposes, decimal approximations of exac solutions to quadratic equations may be desired.

Exact Solutions from Example 8	Decimal Approximations
$\dfrac{2 + \sqrt{10}}{5} \approx$	1.032
$\dfrac{2 - \sqrt{10}}{5} \approx$	-0.232

Answers
6. $5 \pm 3\sqrt{2}$ **7.** no real solution
8. $\dfrac{-1 \pm \sqrt{15}}{4}$

Objective B Using the Square Root Property to Solve Applications

Many real-world applications are modeled by quadratic equations. In the next example, we use the quadratic formula $h = 16t^2$. This formula gives the distance h traveled by a free-falling object in time t. One important note is that this formula does not take into account any air resistance.

Example 9 Finding the Length of Time of a Dive

The record for the highest dive into a lake was made by Harry Froboess of Switzerland. In 1936 he dove 394 feet from the airship Hindenburg into Lake Constance. To the nearest tenth of a second, how long did his dive take? (*Source: Guinness World Records*)

Solution:

1. UNDERSTAND. To approximate the time of the dive, we use the formula* $h = 16t^2$, where t is time in seconds and h is the distance in feet traveled by a free-falling body or object. For example, to find the distance traveled in 1 second, or 3 seconds, we let $t = 1$ and then $t = 3$.

   If $t = 1$, $h = 16(1)^2 = 16 \cdot 1 = 16$ feet.
   If $t = 3$, $h = 16(3)^2 = 16 \cdot 9 = 144$ feet.

   Since a body travels 144 feet in 3 seconds, we now know the dive of 394 feet lasted longer than 3 seconds.

2. TRANSLATE. Use the formula $h = 16t^2$, let the distance $h = 394$, and we have the equation $394 = 16t^2$.

3. SOLVE. To solve $394 = 16t^2$ for t, we will use the square root property.

   $394 = 16t^2$

   $\dfrac{394}{16} = t^2$ Divide both sides by 16.

   $24.625 = t^2$ Simplify.

   $\sqrt{24.625} = t$ or $-\sqrt{24.625} = t$ Use the square root property.

   $5.0 \approx t$ or $-5.0 \approx t$ Approximate.

4. INTERPRET.

   Check: We reject the solution -5.0 since the length of the dive is not a negative number.

   State: The dive lasted approximately 5 seconds.

Work Practice 9

Practice 9

Use the formula $h = 16t^2$ (see Example 9) to find how long, to the nearest tenth of a second, it takes a free-falling body to fall 650 feet.

Answer
9. 6.4 sec

*The formula $h = 16t^2$ does not take into account air resistance.

Vocabulary, Readiness & Video Check

Martin-Gay Interactive Videos

See Video 16.1

Watch the section lecture video and answer the following questions.

Objective A 1. As explained in ▯ Example 2, why is $a \geq 0$ in the statement of the square root property?

Objective B 2. In ▯ Example 6, how can we tell by looking at the translated equation that the square root property can be used to solve it? Why is the negative square root not considered?

16.1 Exercise Set MyLab Math

Solve each equation by factoring. See Examples 1 and 2.

1. $k^2 - 49 = 0$
2. $k^2 - 9 = 0$
3. $m^2 + 2m = 15$
4. $m^2 + 6m = 7$
5. $2x^2 - 32 = 0$

6. $2x^2 - 98 = 0$
7. $4a^2 - 36 = 0$
8. $7a^2 - 175 = 0$
9. $x^2 + 7x = -10$
10. $x^2 + 10x = -$

Objective A *Use the square root property to solve each quadratic equation. See Examples 3 and 4.*

11. $x^2 = 64$
12. $x^2 = 121$
13. $x^2 = 21$
14. $x^2 = 22$
15. $x^2 = \dfrac{1}{25}$

16. $x^2 = \dfrac{1}{16}$
17. $x^2 = -4$
18. $x^2 = -25$
19. $3x^2 = 13$

20. $5x^2 = 2$
21. $7x^2 = 4$
22. $2x^2 = 9$
23. $2x^2 - 10 = 0$
24. $3x^2 - 45 = 0$

Use the square root property to solve each quadratic equation. See Examples 5 through 8.

25. $(x - 5)^2 = 49$
26. $(x + 2)^2 = 25$
27. $(x + 2)^2 = 7$
28. $(x - 7)^2 = 2$

29. $\left(m - \dfrac{1}{2}\right)^2 = \dfrac{1}{4}$
30. $\left(m + \dfrac{1}{3}\right)^2 = \dfrac{1}{9}$
31. $(p + 2)^2 = 10$
32. $(p - 7)^2 = 13$

33. $(3y + 2)^2 = 100$
34. $(4y - 3)^2 = 81$
35. $(z - 4)^2 = -9$
36. $(z + 7)^2 = -20$

37. $(2x - 11)^2 = 50$
38. $(3x - 17)^2 = 28$
39. $(3x - 7)^2 = 32$
40. $(5x - 11)^2 = 54$

Use the square root property to solve. See Examples 3 through 8.

41. $x^2 - 29 = 0$

42. $x^2 - 35 = 0$

43. $(x + 6)^2 = 24$

44. $(x + 5)^2 = 20$

45. $\frac{1}{2}n^2 = 5$

46. $\frac{1}{5}y^2 = 2$

47. $(4x - 1)^2 = 5$

48. $(7x - 2)^2 = 11$

49. $3z^2 = 36$

50. $3z^2 = 24$

51. $(8 - 3x)^2 - 45 = 0$

52. $(10 - 9x)^2 - 75 = 0$

Objective B *The formula for the area of a square is $A = s^2$, where s is the length of a side. Use this formula for Exercises 53 through 56. For each exercise, give the exact answer and a two-decimal-place approximation.*

53. If the area of a square is 20 square inches, find the length of a side.

54. If the area of a square is 32 square meters, find the length of a side.

55. The "Water Cube" National Swimming Center was constructed in Beijing for the 2008 Summer Olympics. Its square base has an area of 31,329 square meters. Find the length of a side of this building. (*Source: ARUP East Asia*)

56. The Washington Monument has a square base whose area is approximately 3039 square feet. Find the length of a side. (*Source: The World Almanac*)

Note: The Beijing Water Cube was converted to an indoor Water Park and recently reopened.

Solve. For Exercises 57 through 60, use the formula $h = 16t^2$. See Example 9. Round each answer to the nearest tenth of a second.

57. The highest regularly performed dives are made by professional divers from La Quebrada. If this cliff in Acapulco has a height of 87.6 feet, determine the time of a dive. (*Source: Guinness World Records*)

58. In 1988, Eddie Turner saved Frank Fanan, who became unconscious after an injury while jumping out of an airplane. Fanan fell 11,136 feet before Turner pulled his ripcord. Determine the time of Fanan's unconscious free fall.

59. The Hualapai Indian Tribe allowed the Grand Canyon Skywalk to be built over the rim of the Grand Canyon on its tribal land. The skywalk extends 70 feet beyond the canyon's edge and is 4000 feet above the canyon floor. Determine the time, to the nearest tenth of a second, it would take an object, dropped off the skywalk, to land at the bottom of the Grand Canyon. (*Source: Boston Globe*)

60. If a sandblaster drops his goggles from a bridge 400 feet from the water below, find how long it takes for the goggles to hit the water.

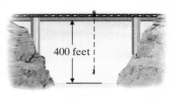

△ **61.** The area of a circle is found by the equation $A = \pi r^2$. If the area A of a certain circle is 36π square inches, find its radius r.

62. If the area of the circle below is 10π square units, find its exact radius. (See Exercise **61**.)

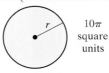

Review

Factor each perfect square trinomial. See Section 11.5.

63. $x^2 + 6x + 9$ **64.** $y^2 + 10y + 25$ **65.** $x^2 - 4x + 4$ **66.** $x^2 - 20x + 100$

Concept Extensions

67. Explain why the equation $x^2 = -9$ has no real solution.

68. Explain why the equation $x^2 = 9$ has two solutions.

Solve each quadratic equation by first factoring the perfect square trinomial on the left side. Then apply the square root property.

69. $x^2 + 4x + 4 = 16$ **70.** $y^2 - 10y + 25 = 11$

Solve each quadratic equation by using the square root property. Use a calculator and round each solution to the nearest hundredth.

71. $x^2 = 1.78$ **72.** $(x - 1.37)^2 = 5.71$

73. The number of U.S. highway bridges for the years 2006 to 2016 can be modeled by the equation $y = 110(x - 2)^2 + 596{,}680$, where $x = 0$ represents the year 2006. Assume that this trend continued and find the first year in which there were 600,000 highway bridges in the United States. (*Hint:* Replace y with 600,000 in the equation and solve for x. Round to the nearest year.) (*Source:* Based on data from the U.S. Department of Transportation, Federal Highway Administration)

74. The annual wireless data usage (in trillions of megabytes) in the United States for the years 2005 through 2015 can be estimated by $y = 0.5(x - 6)^2 + 1$, where $x = 0$ represents the year 2005. Assume that this trend continues, and determine the first year in which the annual wireless data usage will surpass 20 trillion megabytes. (*Hint:* Replace y with 20 in the equation and solve for x. Round to the nearest year.) (*Source:* Based on data from CTIA — The Wireless Association)

16.2 Solving Quadratic Equations by Completing the Square

Objective A Completing the Square to Solve $x^2 + bx + c = 0$

Objectives

A Solve Quadratic Equations of the Form $x^2 + bx + c = 0$ by Completing the Square.

B Solve Quadratic Equations of the Form $ax^2 + bx + c = 0$ by Completing the Square.

In the last section, we used the square root property to solve equations such as

$$(x + 1)^2 = 8 \quad \text{and} \quad (5x - 2)^2 = 3$$

Notice that one side of each equation is a quantity squared and that the other side is a constant. To solve

$$x^2 + 2x = 4$$

notice that if we add 1 to both sides of the equation, the left side is a perfect square trinomial that can be factored.

$$x^2 + 2x + 1 = 4 + 1 \quad \text{Add 1 to both sides.}$$
$$(x + 1)^2 = 5 \quad \text{Factor.}$$

Now we can solve this equation as we did in the previous section, by using the square root property.

$$x + 1 = \sqrt{5} \quad \text{or} \quad x + 1 = -\sqrt{5} \quad \text{Use the square root property.}$$
$$x = -1 + \sqrt{5} \quad \quad x = -1 - \sqrt{5} \quad \text{Solve.}$$

The solutions are $-1 \pm \sqrt{5}$.

Adding a number to $x^2 + 2x$ to form a perfect square trinomial is called **completing the square** on $x^2 + 2x$.

In general, we have the following:

Completing the Square

To complete the square on $x^2 + bx$, add $\left(\dfrac{b}{2}\right)^2$. To find $\left(\dfrac{b}{2}\right)^2$, **find half the coefficient of x, and then square the result.**

Example 1 Solve $x^2 + 6x + 3 = 0$ by completing the square.

Solution: First we get the variable terms alone by subtracting 3 from both sides of the equation.

$$x^2 + 6x + 3 = 0$$
$$x^2 + 6x = -3 \quad \text{Subtract 3 from both sides.}$$

Next we find half the coefficient of the x-term, and then square it. We add this result to *both sides* of the equation. This will make the left side a perfect square trinomial. The coefficient of x is 6, and half of 6 is 3. So we add 3^2 or 9 to both sides.

$$x^2 + 6x + 9 = -3 + 9 \quad \text{Complete the square.}$$
$$(x + 3)^2 = 6 \quad \text{Factor the trinomial } x^2 + 6x + 9.$$
$$x + 3 = \sqrt{6} \quad \text{or} \quad x + 3 = -\sqrt{6} \quad \text{Use the square root property.}$$
$$x = -3 + \sqrt{6} \quad \quad x = -3 - \sqrt{6} \quad \text{Subtract 3 from both sides.}$$

Check by substituting $-3 + \sqrt{6}$ and $-3 - \sqrt{6}$ in the original equation. The solutions are $-3 \pm \sqrt{6}$.

Work Practice 1

Practice 1

Solve $x^2 + 8x + 1 = 0$ by completing the square.

Answer
1. $-4 \pm \sqrt{15}$

Chapter 16 | Quadratic Equations

> **Helpful Hint**
> Remember, when completing the square, add the number that completes the square to **both sides of the equation.**

Practice 2
Solve $x^2 - 14x = -32$ by completing the square.

Example 2 Solve $x^2 - 10x = -14$ by completing the square.

Solution: The variable terms are already alone on one side of the equation. The coefficient of x is -10. Half of -10 is -5, and $(-5)^2 = 25$. So we add 25 to both sides.

$$x^2 - 10x = -14$$
$$x^2 - 10x + 25 = -14 + 25$$

> **Helpful Hint** Add 25 to *both* sides of the equation.

$$(x - 5)^2 = 11 \quad \text{Factor the trinomial and simplify } -14 + 25.$$
$$x - 5 = \sqrt{11} \quad \text{or} \quad x - 5 = -\sqrt{11} \quad \text{Use the square root property.}$$
$$x = 5 + \sqrt{11} \qquad x = 5 - \sqrt{11} \quad \text{Add 5 to both sides.}$$

The solutions are $5 \pm \sqrt{11}$.

■ Work Practice 2

Objective B Completing the Square to Solve $ax^2 + bx + c = 0$

The method of completing the square can be used to solve *any* quadratic equation whether the coefficient of the squared variable is 1 or not. When the coefficient of the squared variable is not 1, we first divide both sides of the equation by the coefficient of the squared variable so that the new coefficient is 1. Then we complete the square.

Practice 3
Solve $4x^2 - 16x - 9 = 0$ by completing the square.

Example 3 Solve $4x^2 - 8x - 5 = 0$ by completing the square.

Solution: Since the coefficient of x^2 is 4, not 1, we first divide both sides of the equation by 4 so that the coefficient of x^2 is 1.

$$4x^2 - 8x - 5 = 0$$
$$x^2 - 2x - \frac{5}{4} = 0 \quad \text{Divide both sides by 4.}$$
$$x^2 - 2x = \frac{5}{4} \quad \text{Get the variable terms alone on one side of the equation.}$$

The coefficient of x is -2. Half of -2 is -1, and $(-1)^2 = 1$. So we add 1 to both sides.

$$x^2 - 2x + 1 = \frac{5}{4} + 1$$
$$(x - 1)^2 = \frac{9}{4} \quad \text{Factor } x^2 - 2x + 1 \text{ and simplify } \frac{5}{4} + 1.$$
$$x - 1 = \sqrt{\frac{9}{4}} \quad \text{or} \quad x - 1 = -\sqrt{\frac{9}{4}} \quad \text{Use the square root property.}$$
$$x = 1 + \frac{3}{2} \qquad x = 1 - \frac{3}{2} \quad \text{Add 1 to both sides and simplify the radical.}$$
$$x = \frac{5}{2} \qquad x = -\frac{1}{2} \quad \text{Simplify.}$$

Both $\frac{5}{2}$ and $-\frac{1}{2}$ are solutions.

■ Work Practice 3

Answers
2. $7 \pm \sqrt{17}$ **3.** $\frac{9}{2}$ and $-\frac{1}{2}$

Section 16.2 | Solving Quadratic Equations by Completing the Square

The following steps may be used to solve a quadratic equation in x by completing the square.

To Solve a Quadratic Equation in x by Completing the Square

Step 1: If the coefficient of x^2 is 1, go to Step 2. If not, divide both sides of the equation by the coefficient of x^2.
Step 2: Get all terms with variables on one side of the equation and constants on the other side.
Step 3: Find half the coefficient of x and then square the result. Add this number to both sides of the equation.
Step 4: Factor the resulting perfect square trinomial.
Step 5: Use the square root property to solve the equation.

Example 4 Solve $2x^2 + 6x = -7$ by completing the square.

Solution: The coefficient of x^2 is not 1. We divide both sides by 2, the coefficient of x^2.

$$2x^2 + 6x = -7$$
$$x^2 + 3x = -\frac{7}{2} \quad \text{Divide both sides by 2.}$$
$$x^2 + 3x + \frac{9}{4} = -\frac{7}{2} + \frac{9}{4} \quad \text{Add } \left(\frac{3}{2}\right)^2 \text{ or } \frac{9}{4} \text{ to both sides.}$$
$$\left(x + \frac{3}{2}\right)^2 = -\frac{5}{4} \quad \text{Factor the left side and simplify the right.}$$

There is no real solution to this equation since the square root of a negative number is not a real number.

Work Practice 4

Practice 4
Solve $2x^2 + 10x = -13$ by completing the square.

Example 5 Solve $2x^2 = 10x + 1$ by completing the square.

Solution: First we divide both sides of the equation by 2, the coefficient of x^2.

$$2x^2 = 10x + 1$$
$$x^2 = 5x + \frac{1}{2} \quad \text{Divide both sides by 2.}$$

Next we get the variable terms alone by subtracting $5x$ from both sides.

$$x^2 - 5x = \frac{1}{2}$$
$$x^2 - 5x + \frac{25}{4} = \frac{1}{2} + \frac{25}{4} \quad \text{Add } \left(-\frac{5}{2}\right)^2 \text{ or } \frac{25}{4} \text{ to both sides.}$$
$$\left(x - \frac{5}{2}\right)^2 = \frac{27}{4} \quad \text{Factor the left side and simplify the right side.}$$
$$x - \frac{5}{2} = \sqrt{\frac{27}{4}} \quad \text{or} \quad x - \frac{5}{2} = -\sqrt{\frac{27}{4}} \quad \text{Use the square root property.}$$
$$x - \frac{5}{2} = \frac{3\sqrt{3}}{2} \qquad x - \frac{5}{2} = -\frac{3\sqrt{3}}{2} \quad \text{Simplify.}$$
$$x = \frac{5}{2} + \frac{3\sqrt{3}}{2} \qquad x = \frac{5}{2} - \frac{3\sqrt{3}}{2}$$

The solutions are $\dfrac{5 \pm 3\sqrt{3}}{2}$.

Work Practice 5

Practice 5
Solve $2x^2 = -6x + 5$ by completing the square.

Answers

4. no real solution **5.** $\dfrac{-3 \pm \sqrt{19}}{2}$

Vocabulary, Readiness & Video Check

Use the choices below to fill in each blank. Not all choices will be used, and these exercises come from Sections 16.1 and 16.2

| $\sqrt{a}$ | linear equation | zero | $\left(\dfrac{b}{2}\right)^2$ | $\dfrac{b}{2}$ | 6 |
| $\pm\sqrt{a}$ | quadratic equation | one | completing the square | 9 | 3 |

1. By the zero-factor property, if the product of two numbers is zero, then at least one of these two numbers must be _____.
2. If a is a positive number, and if $x^2 = a$, then $x = $ _____.
3. An equation that can be written in the form $ax^2 + bx + c = 0$ where $a, b,$ and c are real numbers and a is not zero called a(n) _____.
4. The process of solving a quadratic equation by writing it in the form $(x + a)^2 = c$ is called _____.
5. To complete the square on $x^2 + 6x$, add _____.
6. To complete the square on $x^2 + bx$, add _____.

Fill in the blank with the number needed to make each expression a perfect square trinomial. See Example 1.

7. $p^2 + 8p + $ _____
8. $p^2 + 6p + $ _____
9. $x^2 + 20x + $ _____
10. $x^2 + 18x + $ _____
11. $y^2 + 14y + $ _____
12. $y^2 + 2y + $ _____

Martin-Gay Interactive Videos Watch the section lecture video and answer the following questions.

Objective A 13. In Examples 3 and 4, explain why the constant that completes the square is added to both sides of the equation.

Objective B 14. In Example 5, why is the equation first divided through by 2?

See Video 16.2

16.2 Exercise Set MyLab Math

Objective A *Solve each quadratic equation by completing the square. See Examples 1 and 2.*

1. $x^2 + 8x = -12$
2. $x^2 - 10x = -24$
3. $x^2 + 2x - 7 = 0$
4. $z^2 + 6z - 9 = 0$
5. $x^2 - 6x = 0$
6. $y^2 + 4y = 0$
7. $y^2 + 5y + 4 = 0$
8. $y^2 - 5y + 6 = 0$
9. $x^2 - 2x - 1 = 0$
10. $x^2 - 4x + 2 = 0$
11. $z^2 + 5z = 7$
12. $x^2 - 7x = 5$

Objective B *Solve each quadratic equation by completing the square. See Examples 3 through 5.*

13. $3x^2 - 6x = 24$
14. $2x^2 + 18x = -40$
15. $5x^2 + 10x + 6 = 0$
16. $3x^2 - 12x + 14 = 0$

Section 16.2 | Solving Quadratic Equations by Completing the Square

17. $2x^2 = 6x + 5$
18. $4x^2 = -20x + 3$
19. $2y^2 + 8y + 5 = 0$
20. $4z^2 - 8z + 1 = 0$

Objectives A B Mixed Practice *Solve each quadratic equation by completing the square. See Examples 1 through 5.*

21. $x^2 + 6x - 25 = 0$
22. $x^2 - 6x + 7 = 0$
23. $x^2 - 3x - 3 = 0$

24. $x^2 - 9x + 3 = 0$
25. $2y^2 - 3y + 1 = 0$
26. $2y^2 - y - 1 = 0$

27. $x(x + 3) = 18$ (*Hint:* First use the distributive property and multiply.)
28. $x(x - 3) = 18$ (See hint for Exercise **27**.)
29. $3z^2 + 6z + 4 = 0$

30. $2y^2 + 8y + 9 = 0$
31. $4x^2 + 16x = 48$
32. $6x^2 - 30x = -36$

Review

Simplify each expression. See Section 15.2.

33. $\dfrac{3}{4} - \sqrt{\dfrac{25}{16}}$
34. $\dfrac{3}{5} + \sqrt{\dfrac{16}{25}}$
35. $\dfrac{1}{2} + \sqrt{\dfrac{9}{4}}$
36. $\dfrac{9}{10} - \sqrt{\dfrac{49}{100}}$

Simplify each expression. See Section 15.4.

37. $\dfrac{6 + 4\sqrt{5}}{2}$
38. $\dfrac{10 + 20\sqrt{3}}{2}$
39. $\dfrac{3 - 9\sqrt{2}}{6}$
40. $\dfrac{12 - 8\sqrt{7}}{16}$

Concept Extensions

41. In your own words, describe a perfect square trinomial.

42. Describe how to find the number to add to $x^2 - 7x$ to make a perfect square trinomial.

43. Write your own quadratic equation to be solved by completing the square. Write it in the form

perfect square trinomial = a number that is not a perfect square

$$x^2 + 6x + 9 = 11$$

Complete part **a** as an example.
  a. Solve $x^2 + 6x + 9 = 11$.
  b. Solve your quadratic equation by completing the square.

44. Follow the directions of Exercise **43**, except write your equation in the form

perfect square trinomial = negative number

Solve your quadratic equation by completing the square.

45. Find a value of k that will make $x^2 + kx + 16$ a perfect square trinomial.

46. Find a value of k that will make $x^2 + kx + 25$ a perfect square trinomial.

47. The revenues from product sales y (in millions of dollars) of Abiomed, Inc., maker of the AbioCor artificial heart, during fiscal years 2012 through 2016 can be modeled by the equation $y = 12x^2 + 3x + 132$, where $x = 0$ represents 2012. Assume that this trend continues and predict the year after 2012 in which Abiomed's revenues from product sales will be $540 million. (Round to the nearest whole number.) (*Source:* Based on data from Abiomed, Inc.)

48. The average price y of gold (in dollars per ounce) from 2013 through 2017 is given by the equation $y = 63x^2 - 364x + 1622$, where x is the number of years after 2013. Assume that this trend continued and find the year after 2013 in which the price of gold was $1706 per ounce. (*Source:* Based on data from goldprice.org)

Recall that a graphing calculator may be used to solve an equation. For example, to solve $x^2 + 8x = -12$ (Exercise 1), graph

$y_1 = x^2 + 8x$
$y_2 = -12$

The x-coordinate of the point of intersection of the graphs is the solution. Use a graphing calculator to solve each equation. Round solutions to the nearest hundredth.

49. Exercise 1 **50.** Exercise 2 **51.** Exercise 17 **52.** Exercise 12

16.3 Solving Quadratic Equations by the Quadratic Formula

Objectives

A Use the Quadratic Formula to Solve Quadratic Equations.

B Approximate Solutions to Quadratic Equations.

Objective A Using the Quadratic Formula

We can use the technique of completing the square to develop a formula to find solutions of any quadratic equation. We develop and use the **quadratic formula** in this section.

Recall that a quadratic equation in **standard form** is

$$ax^2 + bx + c = 0, \quad a \neq 0$$

To develop the quadratic formula, let's complete the square for this quadratic equation in standard form.

First we divide both sides of the equation by the coefficient of x^2 and then get the variable terms alone on one side of the equation.

$$x^2 + \frac{b}{a}x + \frac{c}{a} = 0 \quad \text{Divide by } a; \text{ recall that } a \text{ cannot be 0.}$$

$$x^2 + \frac{b}{a}x = -\frac{c}{a} \quad \text{Get the variable terms alone on one side of the equation.}$$

Section 16.3 | Solving Quadratic Equations by the Quadratic Formula

The coefficient of x is $\frac{b}{a}$. Half of $\frac{b}{a}$ is $\frac{b}{2a}$ and $\left(\frac{b}{2a}\right)^2 = \frac{b^2}{4a^2}$. So we add $\frac{b^2}{4a^2}$ to both sides of the equation.

$$x^2 + \frac{b}{a}x + \frac{b^2}{4a^2} = -\frac{c}{a} + \frac{b^2}{4a^2} \quad \text{Add } \frac{b^2}{4a^2} \text{ to both sides.}$$

$$\left(x + \frac{b}{2a}\right)^2 = -\frac{c}{a} + \frac{b^2}{4a^2} \quad \text{Factor the left side.}$$

$$\left(x + \frac{b}{2a}\right)^2 = -\frac{4ac}{4a^2} + \frac{b^2}{4a^2} \quad \text{Multiply } -\frac{c}{a} \text{ by } \frac{4a}{4a} \text{ so that the terms on the right side have a common denominator.}$$

$$\left(x + \frac{b}{2a}\right)^2 = \frac{b^2 - 4ac}{4a^2} \quad \text{Simplify the right side.}$$

Now we use the square root property.

$$x + \frac{b}{2a} = \sqrt{\frac{b^2 - 4ac}{4a^2}} \quad \text{or} \quad x + \frac{b}{2a} = -\sqrt{\frac{b^2 - 4ac}{4a^2}} \quad \text{Use the square root property.}$$

$$x + \frac{b}{2a} = \frac{\sqrt{b^2 - 4ac}}{2a} \qquad x + \frac{b}{2a} = -\frac{\sqrt{b^2 - 4ac}}{2a} \quad \text{Simplify the radical.}$$

$$x = -\frac{b}{2a} + \frac{\sqrt{b^2 - 4ac}}{2a} \qquad x = -\frac{b}{2a} - \frac{\sqrt{b^2 - 4ac}}{2a} \quad \text{Subtract } \frac{b}{2a} \text{ from both sides.}$$

$$x = \frac{-b + \sqrt{b^2 - 4ac}}{2a} \qquad x = \frac{-b - \sqrt{b^2 - 4ac}}{2a} \quad \text{Simplify.}$$

The solutions are $\frac{-b \pm \sqrt{b^2 - 4ac}}{2a}$. This final equation is called the **quadratic formula** and gives the solutions of any quadratic equation.

Quadratic Formula

If a, b, and c are real numbers and $a \neq 0$, a quadratic equation written in the standard form $ax^2 + bx + c = 0$ has solutions

$$x = \frac{-b \pm \sqrt{b^2 - 4ac}}{2a}$$

Helpful Hint

Don't forget that to correctly identify a, b, and c in the quadratic formula, you should write the equation in standard form.

Quadratic Equations in Standard Form

$5x^2 - 6x + 2 = 0$	$a = 5, b = -6, c = 2$
$4y^2 - 9 = 0$	$a = 4, b = 0, c = -9$
$x^2 + x = 0$	$a = 1, b = 1, c = 0$
$\sqrt{2}x^2 + \sqrt{5}x + \sqrt{3} = 0$	$a = \sqrt{2}, b = \sqrt{5}, c = \sqrt{3}$

Chapter 16 | Quadratic Equations

Practice 1
Solve $2x^2 - x - 5 = 0$ using the quadratic formula.

Example 1 Solve $3x^2 + x - 3 = 0$ using the quadratic formula.

Solution: This equation is in standard form with $a = 3$, $b = 1$, and $c = -3$. By the quadratic formula, we have

$$x = \frac{-b \pm \sqrt{b^2 - 4ac}}{2a}$$

$$x = \frac{-1 \pm \sqrt{1^2 - 4 \cdot 3 \cdot (-3)}}{2 \cdot 3} \quad \text{Let } a = 3, b = 1, \text{ and } c = -3.$$

$$= \frac{-1 \pm \sqrt{1 + 36}}{6} \quad \text{Simplify.}$$

$$= \frac{-1 \pm \sqrt{37}}{6}$$

Check both solutions in the original equation. The solutions are $\frac{-1 + \sqrt{37}}{6}$ and $\frac{-1 - \sqrt{37}}{6}$.

■ Work Practice 1

Practice 2
Solve $3x^2 + 8x = 3$ using the quadratic formula.

Example 2 Solve $2x^2 - 9x = 5$ using the quadratic formula.

Solution: First we write the equation in standard form by subtracting 5 from both sides.

$$2x^2 - 9x = 5$$
$$2x^2 - 9x - 5 = 0$$

Next we note that $a = 2$, $b = -9$, and $c = -5$. We substitute these values into the quadratic formula.

$$x = \frac{-b \pm \sqrt{b^2 - 4ac}}{2a}$$

$$x = \frac{-(-9) \pm \sqrt{(-9)^2 - 4 \cdot 2 \cdot (-5)}}{2 \cdot 2} \quad \text{Substitute into the formula.}$$

$$= \frac{9 \pm \sqrt{81 + 40}}{4} \quad \text{Simplify.}$$

$$= \frac{9 \pm \sqrt{121}}{4} = \frac{9 \pm 11}{4}$$

Helpful Hint: Notice that the fraction bar is under the entire numerator $-b \pm \sqrt{b^2 - 4ac}$.

Then,

$$x = \frac{9 - 11}{4} = -\frac{1}{2} \quad \text{or} \quad x = \frac{9 + 11}{4} = 5$$

Check $-\frac{1}{2}$ and 5 in the original equation. Both $-\frac{1}{2}$ and 5 are solutions.

■ Work Practice 2

Answers
1. $\frac{1 + \sqrt{41}}{4}$ and $\frac{1 - \sqrt{41}}{4}$
2. $\frac{1}{3}$ and -3

The following steps may be useful when solving a quadratic equation by the quadratic formula.

Section 16.3 | Solving Quadratic Equations by the Quadratic Formula

To Solve a Quadratic Equation by the Quadratic Formula
Step 1: Write the quadratic equation in standard form: $ax^2 + bx + c = 0$.
Step 2: If necessary, clear the equation of fractions to simplify calculations.
Step 3: Identify $a, b,$ and c.
Step 4: Replace $a, b,$ and c in the quadratic formula with the identified values, and simplify.

✓ **Concept Check** For the quadratic equation $2x^2 - 5 = 7x$, if $a = 2$ and $c = -5$ in the quadratic formula, the value of b is which of the following?

a. $\dfrac{7}{2}$ **b.** 7   **c.** -5 **d.** -7

Example 3 Solve $7x^2 = 1$ using the quadratic formula.

Solution: First we write the equation in standard form by subtracting 1 from both sides.

$$7x^2 = 1$$
$$7x^2 - 1 = 0$$

Helpful Hint
$7x^2 - 1 = 0$ can be written as $7x^2 + 0x - 1 = 0$. This form helps you see that $b = 0$.

Next we replace $a, b,$ and c with the identified values: $a = 7, b = 0, c = -1$.

$$x = \frac{0 \pm \sqrt{0^2 - 4 \cdot 7 \cdot (-1)}}{2 \cdot 7} \quad \text{Substitute into the formula.}$$

$$= \frac{\pm \sqrt{28}}{14} \quad \text{Simplify.}$$

$$= \frac{\pm 2\sqrt{7}}{14}$$

$$= \pm \frac{2\sqrt{7}}{2 \cdot 7}$$

$$= \pm \frac{\sqrt{7}}{7}$$

The solutions are $\dfrac{\sqrt{7}}{7}$ and $-\dfrac{\sqrt{7}}{7}$.

■ Work Practice 3

Practice 3
Solve $5x^2 = 2$ using the quadratic formula.

Notice that we could have solved the equation $7x^2 = 1$ in Example 3 by dividing both sides by 7 and then using the square root property. We solved the equation by the quadratic formula to show that this formula can be used to solve any quadratic equation.

Example 4 Solve $x^2 = -x - 1$ using the quadratic formula.

Solution: First we write the equation in standard form.

$$x^2 + x + 1 = 0$$

Next we replace $a, b,$ and c in the quadratic formula with $a = 1, b = 1,$ and $c = 1$.

(Continued on next page)

Practice 4
Solve $x^2 = -2x - 3$ using the quadratic formula.

Answers
3. $\dfrac{\sqrt{10}}{5}$ and $-\dfrac{\sqrt{10}}{5}$
4. no real solution

✓ **Concept Check Answer**
d

$$x = \frac{-1 \pm \sqrt{1^2 - 4 \cdot 1 \cdot 1}}{2 \cdot 1}$$ Substitute into the formula.

$$= \frac{-1 \pm \sqrt{-3}}{2}$$ Simplify.

There is no real number solution because $\sqrt{-3}$ is not a real number.

■ **Work Practice 4**

Practice 5

Solve $\frac{1}{3}x^2 - x = 1$ using the quadratic formula.

Example 5 Solve $\frac{1}{2}x^2 - x = 2$ using the quadratic formula.

Solution: We write the equation in standard form and then clear the equation of fractions by multiplying both sides by the LCD, 2.

$$\frac{1}{2}x^2 - x = 2$$

$$\frac{1}{2}x^2 - x - 2 = 0 \quad \text{Write in standard form.}$$

$$x^2 - 2x - 4 = 0 \quad \text{Multiply both sides by 2.}$$

Here, $a = 1$, $b = -2$, and $c = -4$, so we substitute these values into the quadratic formula.

$$x = \frac{-(-2) \pm \sqrt{(-2)^2 - 4 \cdot 1 \cdot (-4)}}{2 \cdot 1}$$

$$= \frac{2 \pm \sqrt{20}}{2} = \frac{2 \pm 2\sqrt{5}}{2} \quad \text{Simplify.}$$

$$= \frac{2(1 \pm \sqrt{5})}{2} = 1 \pm \sqrt{5} \quad \text{Factor and simplify.}$$

The solutions are $1 - \sqrt{5}$ and $1 + \sqrt{5}$.

■ **Work Practice 5**

Notice that in Example 5, although we cleared the equation of fractions, the coefficients $a = \frac{1}{2}$, $b = -1$, and $c = -2$ will give the same results.

Helpful Hint

When simplifying an expression such as

$$\frac{3 \pm 6\sqrt{2}}{6}$$

first factor out a common factor from the terms of the numerator and then simplify.

$$\frac{3 \pm 6\sqrt{2}}{6} = \frac{3(1 \pm 2\sqrt{2})}{2 \cdot 3} = \frac{1 \pm 2\sqrt{2}}{2}$$

Answer

5. $\frac{3 + \sqrt{21}}{2}$ and $\frac{3 - \sqrt{21}}{2}$

Objective B Approximating Solutions to Quadratic Equations

Sometimes approximate solutions for quadratic equations are appropriate.

Example 6 Approximate the exact solutions of the quadratic equation in Example 1. Round the approximations to the nearest tenth.

Solution: From Example 1, we have exact solutions $\dfrac{-1 \pm \sqrt{37}}{6}$. Thus,

$\dfrac{-1 + \sqrt{37}}{6} \approx 0.847127088 \approx 0.8$ to the nearest tenth.

$\dfrac{-1 - \sqrt{37}}{6} \approx -1.180460422 \approx -1.2$ to the nearest tenth.

Thus approximate solutions to the quadratic equation in Example 1 are 0.8 and -1.2.

Work Practice 6

Practice 6
Approximate the exact solutions of the quadratic equation in Practice 1. Round the approximations to the nearest tenth.

Answer
6. $\dfrac{1 + \sqrt{41}}{4} \approx 1.9, \dfrac{1 - \sqrt{41}}{4} \approx -1.4$

Vocabulary, Readiness & Video Check

Fill in each blank.

1. The quadratic formula is _____.

Identify the values of a, b, and c in each quadratic equation.

2. $5x^2 - 7x + 1 = 0$; $a = $ _____, $b = $ _____, $c = $ _____
3. $x^2 + 3x - 7 = 0$; $a = $ _____, $b = $ _____, $c = $ _____
4. $x^2 - 6 = 0$; $a = $ _____, $b = $ _____, $c = $ _____
5. $x^2 + x - 1 = 0$; $a = $ _____, $b = $ _____, $c = $ _____
6. $9x^2 - 4 = 0$; $a = $ _____, $b = $ _____, $c = $ _____

Simplify the following.

7. $\dfrac{-1 \pm \sqrt{1^2 - 4(1)(-2)}}{2(1)}$

8. $\dfrac{-(-5) \pm \sqrt{(-5)^2 - 4(2)(3)}}{2(2)}$

9. $\dfrac{-5 \pm \sqrt{5^2 - 4(1)(2)}}{2(1)}$

10. $\dfrac{-7 \pm \sqrt{7^2 - 4(2)(1)}}{2(2)}$

Martin-Gay Interactive Videos *Watch the section lecture video and answer the following questions.*

See Video 16.3

Objective A **11.** Based on the lectures and Examples 1–3, answer the following.
  a. Must a quadratic equation be written in standard form in order to use the quadratic formula? Why or why not?
  b. Must fractions be cleared from the equation before using the quadratic formula? Why or why not?

Objective B **12.** From Example 4, how are approximate solutions found?

16.3 Exercise Set MyLab Math

Objective A *Use the quadratic formula to solve each quadratic equation. See Examples 1 through 4.*

1. $x^2 - 3x + 2 = 0$
2. $x^2 - 5x - 6 = 0$
3. $3k^2 + 7k + 1 = 0$
4. $7k^2 + 3k - 1 = 0$

5. $4x^2 - 3 = 0$
6. $25x^2 - 15 = 0$
7. $5z^2 - 4z + 3 = 0$
8. $3x^2 + 2x + 1 = 0$

9. $y^2 = 7y + 30$
10. $y^2 = 5y + 36$
11. $2x^2 = 10$
12. $5x^2 = 15$

13. $m^2 - 12 = m$
14. $m^2 - 14 = 5m$
15. $3 - x^2 = 4x$
16. $10 - x^2 = 2x$

17. $6x^2 + 9x = 2$
18. $3x^2 - 9x = 8$
19. $7p^2 + 2 = 8p$
20. $11p^2 + 2 = 10p$

21. $x^2 - 6x + 2 = 0$
22. $x^2 - 10x + 19 = 0$
23. $2x^2 - 6x + 3 = 0$
24. $5x^2 - 8x + 2 = 0$

25. $3x^2 = 1 - 2x$
26. $5y^2 = 4 - y$
27. $4y^2 = 6y + 1$
28. $6z^2 = 2 - 3z$

29. $x^2 + x + 2 = 0$
30. $k^2 + 2k + 5 = 0$
31. $20y^2 = 3 - 11y$
32. $2z^2 = z + 3$

33. $x^2 - 5x - 2 = 0$
34. $x^2 - 2x - 5 = 0$
35. $3x^2 - x - 14 = 0$
36. $5x^2 - 13x - 6 = 0$

Use the quadratic formula to solve each quadratic equation. See Example 5.

37. $\dfrac{m^2}{2} = m + \dfrac{1}{2}$
38. $\dfrac{m^2}{2} = 3m - 1$
39. $3p^2 - \dfrac{2}{3}p + 1 = 0$
40. $\dfrac{5}{2}p^2 - p + \dfrac{1}{2} = 0$

41. $4p^2 + \dfrac{3}{2} = -5p$
42. $4p^2 + \dfrac{3}{2} = 5p$
43. $5x^2 = \dfrac{7}{2}x + 1$
44. $2x^2 = \dfrac{5}{2}x + \dfrac{7}{2}$

45. $x^2 - \dfrac{11}{2}x - \dfrac{1}{2} = 0$
46. $\dfrac{2}{3}x^2 - 2x - \dfrac{2}{3} = 0$
47. $5z^2 - 2z = \dfrac{1}{5}$
48. $9z^2 + 12z = -1$

Objectives A B **Mixed Practice** *Use the quadratic formula to solve each quadratic equation. Find the exact solutions; then approximate these solutions to the nearest tenth. See Examples 1 through 6.*

49. $3x^2 = 21$
50. $2x^2 = 26$
51. $x^2 + 6x + 1 = 0$
52. $x^2 + 4x + 2 = 0$

53. $x^2 = 9x + 4$
54. $x^2 = 7x + 5$
55. $3x^2 - 2x - 2 = 0$
56. $5x^2 - 3x - 1 = 0$

Review

Graph the following linear equations in two variables. See Sections 13.2 and 13.3.

57. $y = -3$

58. $x = 4$

59. $y = 3x - 2$

60. $y = 2x + 3$

Concept Extensions

Solve. See the Concept Check in this section. For the quadratic equation $5x^2 + 2 = x$, if $a = 5$,

61. What is the value of b?
a. $\dfrac{1}{5}$ b. 0   c. -1 d. 1

62. What is the value of c?
a. 5   b. x c. -2 d. 2

For the quadratic equation $7y^2 = 3y$, if $b = 3$,

63. What is the value of a?
a. 7   b. -7 c. 0   d. 1

64. What is the value of c?
a. 7   b. 3   c. 0   d. 1

65. In a recent year, Nestle created a chocolate bar that the company claimed weighed more than 2 tons. The rectangular bar had a base area of approximately 34.65 square feet, and its length was 0.6 foot shorter than three times its width. Find the length and width of the bar. (*Source:* Nestle)

△ 66. The area of a rectangular conference room table is 95 square feet. If its length is six feet longer than its width, find the dimensions of the table. Round each dimension to the nearest tenth.

Solve each equation using the quadratic formula.

67. $x^2 + 3\sqrt{2}x - 5 = 0$

68. $y^2 - 2\sqrt{5}y - 1 = 0$

69. Explain how to identify a, b, and c correctly when solving a quadratic equation by the quadratic formula.

70. Explain how the quadratic formula is developed and why it is useful.

Use the quadratic formula and a calculator to solve each equation. Round solutions to the nearest tenth.

71. $7.3z^2 + 5.4z - 1.1 = 0$

72. $1.2x^2 - 5.2x - 3.9 = 0$

A rocket is launched from the top of an 80-foot cliff with an initial velocity of 120 feet per second. The height, h, of the rocket after t seconds is given by the equation $h = -16t^2 + 120t + 80$.

73. How long after the rocket is launched will it be 30 feet from the ground? Round to the nearest tenth of a second.

74. How long after the rocket is launched will it strike the ground? Round to the nearest tenth of a second. (*Hint:* The rocket will strike the ground when its height $h = 0$.)

75. Restaurant industry food and drink sales y (in billions of dollars) in the United States from 2008 through 2017 can be approximated by the equation $y = 2x^2 + 9.6x + 565$, where x is the number of years since 2008. Assume this trend continues and predict the year in which the restaurant industry food and drink sales will be $1160 billion. (Round to the nearest whole number.) (*Source:* National Restaurant Association)

76. Retail sales y (in billions of dollars) for Target Corporation for the years 2012 through 2016 is approximated by the equation $y = -0.06x^2 + x + 70$, where $x = 0$ represents the year 2012. Assume that this trend continues and predict the year after 2012 in which Target's retail sales will be approximately $68 billion. (Round to the nearest whole number.) (*Source:* Based on data from Target Corporation)

Sections 16.1–16.3 Integrated Review

Summary on Solving Quadratic Equations

An important skill in mathematics is learning when to use one technique in favor of another. We now practice this by deciding which method to use when solving quadratic equations. Although both the quadratic formula and completing the square can be used to solve any quadratic equation, the quadratic formula is usually less tedious and thus preferred. The following steps may be used to solve a quadratic equation.

To Solve a Quadratic Equation

Step 1: If the equation is in the form $ax^2 = c$ or $(ax + b)^2 = c$, use the square root property and solve. If not, go to Step 2.
Step 2: Write the equation in standard form: $ax^2 + bx + c = 0$.
Step 3: Try to solve the equation by factoring. If not possible, go to Step 4.
Step 4: Solve the equation by the quadratic formula.

Study the examples below to help you review these steps.

Example 1 Solve $m^2 - 2m - 7 = 0$.

Solution: The equation is in standard form, but the quadratic expression $m^2 - 2m - 7$ is not factorable, so use the quadratic formula with $a = 1, b = -2$, and $c = -7$.

$$m^2 - 2m - 7 = 0$$

$$m = \frac{-(-2) \pm \sqrt{(-2)^2 - 4 \cdot 1 \cdot (-7)}}{2 \cdot 1} = \frac{2 \pm \sqrt{32}}{2}$$

$$m = \frac{2 \pm 4\sqrt{2}}{2} = \frac{2(1 \pm 2\sqrt{2})}{2} = 1 \pm 2\sqrt{2}$$

The solutions are $1 - 2\sqrt{2}$ and $1 + 2\sqrt{2}$.

■ Work Practice 1

Practice 1
Solve $y^2 - 4y - 6 = 0$.

Example 2 Solve $(3x + 1)^2 = 20$.

Solution: This equation is in a form that makes the square root property easy to apply.

$$(3x + 1)^2 = 20$$
$$3x + 1 = \pm\sqrt{20} \quad \text{Apply the square root property.}$$
$$3x + 1 = \pm 2\sqrt{5} \quad \text{Simplify } \sqrt{20}.$$
$$3x = -1 \pm 2\sqrt{5}$$
$$x = \frac{-1 \pm 2\sqrt{5}}{3}$$

The solutions are $\dfrac{-1 - 2\sqrt{5}}{3}$ and $\dfrac{-1 + 2\sqrt{5}}{3}$.

■ Work Practice 2

Practice 2
Solve $(2x + 5)^2 = 45$.

Answers
1. $2 \pm \sqrt{10}$
2. $\dfrac{-5 \pm 3\sqrt{5}}{2}$

Chapter 16 | Quadratic Equations

Practice 3

Solve $x^2 - \frac{5}{2}x = -\frac{3}{2}$.

Example 3 Solve $x^2 - \frac{11}{2}x = -\frac{5}{2}$.

Solution: The fractions make factoring more difficult and complicate the calculations for using the quadratic formula. Clear the equation of fractions by multiplying both sides of the equation by the LCD, 2.

$$x^2 - \frac{11}{2}x = -\frac{5}{2}$$

$$x^2 - \frac{11}{2}x + \frac{5}{2} = 0 \qquad \text{Write in standard form.}$$

$$2x^2 - 11x + 5 = 0 \qquad \text{Multiply both sides by 2.}$$

$$(2x - 1)(x - 5) = 0 \qquad \text{Factor.}$$

$$2x - 1 = 0 \quad \text{or} \quad x - 5 = 0 \qquad \text{Apply the zero factor theorem.}$$

$$2x = 1 \qquad\qquad x = 5$$

$$x = \frac{1}{2} \qquad\qquad x = 5$$

The solutions are $\frac{1}{2}$ and 5.

■ Work Practice 3

Answer

3. $\frac{3}{2}, 1$

Answers

1. _____
2. _____
3. _____
4. _____
5. _____
6. _____
7. _____
8. _____
9. _____
10. _____
11. _____
12. _____
13. _____
14. _____
15. _____
16. _____

Choose and use a method to solve each equation.

1. $5x^2 - 11x + 2 = 0$

2. $5x^2 + 13x - 6 = 0$

3. $x^2 - 1 = 2x$

4. $x^2 + 7 = 6x$

5. $a^2 = 20$

6. $a^2 = 72$

7. $x^2 - x + 4 = 0$

8. $x^2 - 2x + 7 = 0$

9. $3x^2 - 12x + 12 = 0$

10. $5x^2 - 30x + 45 = 0$

11. $9 - 6p + p^2 = 0$

12. $49 - 28p + 4p^2 = 0$

13. $4y^2 - 16 = 0$

14. $3y^2 - 27 = 0$

15. $x^2 - 3x + 2 = 0$

16. $x^2 + 7x + 12 = 0$

17. $(2z + 5)^2 = 25$

18. $(3z - 4)^2 = 16$

19. $30x = 25x^2 + 2$

20. $12x = 4x^2 + 4$

21. $\frac{2}{3}m^2 - \frac{1}{3}m - 1 = 0$

22. $\frac{5}{8}m^2 + m - \frac{1}{2} = 0$

23. $x^2 - \frac{1}{2}x - \frac{1}{5} = 0$

24. $x^2 + \frac{1}{2}x - \frac{1}{8} = 0$

25. $4x^2 - 27x + 35 = 0$

26. $9x^2 - 16x + 7 = 0$

27. $(7 - 5x)^2 = 18$

28. $(5 - 4x)^2 = 75$

29. $3z^2 - 7z = 12$

30. $6z^2 + 7z = 6$

31. $x = x^2 - 110$

32. $x = 56 - x^2$

33. $\frac{3}{4}x^2 - \frac{5}{2}x - 2 = 0$

34. $x^2 - \frac{6}{5}x - \frac{8}{5} = 0$

35. $x^2 - 0.6x + 0.05 = 0$

36. $x^2 - 0.1x - 0.06 = 0$

37. $10x^2 - 11x + 2 = 0$

38. $20x^2 - 11x + 1 = 0$

39. $\frac{1}{2}z^2 - 2z + \frac{3}{4} = 0$

40. $\frac{1}{5}z^2 - \frac{1}{2}z - 2 = 0$

41. Explain how you will decide what method to use when solving quadratic equations.

16.4 Graphing Quadratic Equations in Two Variables

Objectives

A Graph Quadratic Equations of the Form $y = ax^2$.

B Graph Quadratic Equations of the Form $y = ax^2 + bx + c$.

C Use the Vertex Formula to Determine the Vertex of a Parabola.

Recall from Section 13.2 that the graph of a linear equation in two variables $Ax + By = C$, is a straight line. In this section, we will find that the graph of a quadratic equation in the form $y = ax^2 + bx + c$ is a parabola.

Objective A Graphing $y = ax^2$

We begin our work by graphing $y = x^2$. To do so, we will find and plot ordered pairs solutions of this equation. Let's select a few values for x, find the corresponding y-values, and record them in a table of values to keep track. Then we can plot the points corresponding to these solutions on a coordinate plane.

If $x = -3$, then $y = (-3)^2$, or 9.
If $x = -2$, then $y = (-2)^2$, or 4.
If $x = -1$, then $y = (-1)^2$, or 1.
If $x = 0$, then $y = 0^2$, or 0.
If $x = 1$, then $y = 1^2$, or 1.
If $x = 2$, then $y = 2^2$, or 4.
If $x = 3$, then $y = 3^2$, or 9.

x	y
-3	9
-2	4
-1	1
0	0
1	1
2	4
3	9

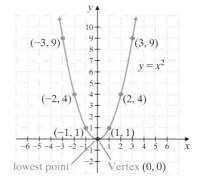

Practice 1
Graph: $y = -3x^2$

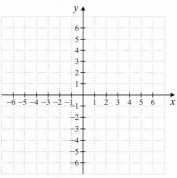

The graph of $y = x^2$ is a smooth curve through the plotted points. This curve is called a **parabola.** The lowest point on a parabola opening upward is called the **vertex.** The vertex is $(0, 0)$ for the parabola $y = x^2$. If we fold the graph along the y-axis, the two pieces of the parabola match perfectly. For this reason, we say the graph is **symmetric about the y-axis,** and we call the y-axis the **axis of symmetry.**

Notice that the parabola that corresponds to the equation $y = x^2$ opens upward. This happens when the coefficient of x^2 is positive. In the equation $y = x^2$, the coefficient of x^2 is 1. Example 1 shows the graph of a quadratic equation where the coefficient of x^2 is negative.

Answer
1.
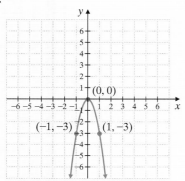

Example 1 Graph: $y = -2x^2$

Solution: We begin by selecting x-values and calculating the corresponding y-values. Then we plot the ordered pairs found and draw a smooth curve through those points. Notice that when the coefficient of x^2 is negative, the corresponding parabola opens downward. When a parabola opens downward, the vertex is the highest point of the parabola. The vertex of this parabola is (0, 0), and the axis of symmetry is again the y-axis.

$y = -2x^2$

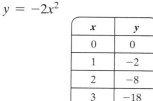

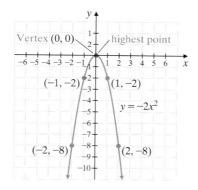

■ Work Practice 1

Objective B Graphing $y = ax^2 + bx + c$

Just as for linear equations, we can use x- and y-intercepts to help graph quadratic equations. Recall from Chapter 13 that an x-intercept is the point where the graph crosses the x-axis. A y-intercept is the point where the graph crosses the y-axis. We find intercepts just as we did in Chapter 13.

> **Helpful Hint**
> Recall that:
> To find x-intercepts, let $y = 0$ and solve for x.
> To find y-intercepts, let $x = 0$ and solve for y.

Example 2 Graph: $y = x^2 - 4$

Solution: If we write this equation as $y = x^2 + 0x + (-4)$, we can see that it is in the form $y = ax^2 + bx + c$. To graph it, we first find the intercepts. To find the y-intercept, we let $x = 0$. Then

$y = 0^2 - 4 = -4$

To find x-intercepts, we let $y = 0$.

$0 = x^2 - 4$
$0 = (x - 2)(x + 2)$
$x - 2 = 0$ or $x + 2 = 0$
$x = 2$ $x = -2$

Thus far, we have the y-intercept $(0, -4)$ and the x-intercepts $(2, 0)$ and $(-2, 0)$. Now we can select additional x-values, find the corresponding y-values, plot the points, and draw a smooth curve through the points.

(Continued on next page)

Practice 2
Graph: $y = x^2 - 9$

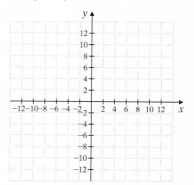

Answer
2.
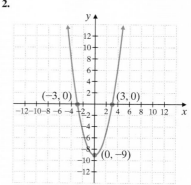

$y = x^2 - 4$

x	y
0	-4
1	-3
2	0
3	5
-1	-3
-2	0
-3	5

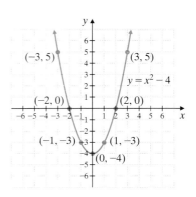

Notice that the vertex of this parabola is $(0, -4)$.

■ Work Practice 2

✓ **Concept Check** Tell whether the graph of each equation opens upward or downward.

a. $y = 2x^2$ **b.** $y = 3x^2 + 4x - 5$ **c.** $y = -5x^2 + 2$

Helpful Hint

For the graph of $y = ax^2 + bx + c$,

If a is positive, the parabola opens upward.
If a is negative, the parabola opens downward.

✓ **Concept Check** For which of the following graphs of $y = ax^2 + bx + c$ would the value of a be negative?

a. 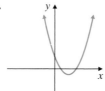 **b.**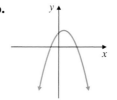

Objective C Using the Vertex Formula

Thus far, we have accidentally stumbled upon the vertex of each parabola that we have graphed. However, our choice of values for x may not yield an ordered pair for the vertex of the parabola. It would be helpful if we could first find the vertex of a parabola. Next we would determine whether the parabola opens upward or downward. Finally we would calculate additional points such as x- and y-intercepts as needed. In fact, there is a formula that may be used to find the vertex of a parabola.

Vertex Formula

The vertex of the parabola $y = ax^2 + bx + c$ has x-coordinate

$$\frac{-b}{2a}$$

The corresponding y-coordinate of the vertex is obtained by substituting the x-coordinate into the equation and finding y.

✓ First Concept Check Answer
a. upward b. upward c. downward

✓ Second Concept Check Answer
b

One way to develop this formula is to notice that the x-value of the vertex of the parabolas that we are considering lies halfway between its x-intercepts. Another way to develop this formula is to complete the square on the general form of a quadratic equation: $y = ax^2 + bx + c$. We will not show the development of this formula here.

Example 3 Graph: $y = x^2 - 6x + 8$

Solution: In the equation $y = x^2 - 6x + 8$, $a = 1$ and $b = -6$.

Vertex: The x-coordinate of the vertex is

$$\frac{-b}{2a} = \frac{-(-6)}{2 \cdot 1} = 3 \quad \text{Use the vertex formula, } \frac{-b}{2a}.$$

To find the corresponding y-coordinate, we let $x = 3$ in the original equation.

$$y = x^2 - 6x + 8 = 3^2 - 6 \cdot 3 + 8 = -1$$

The vertex is $(3, -1)$ and the parabola opens upward since a is positive. We now find and plot the intercepts.

Intercepts: To find the x-intercepts, we let $y = 0$.

$$0 = x^2 - 6x + 8$$

We factor the expression $x^2 - 6x + 8$ to find $(x - 4)(x - 2) = 0$. The x-intercepts are $(4, 0)$ and $(2, 0)$.

If we let $x = 0$ in the original equation, then $y = 8$ gives us the y-intercept $(0, 8)$. Now we plot the vertex $(3, -1)$ and the intercepts $(4, 0)$, $(2, 0)$, and $(0, 8)$. Then we can sketch the parabola.

These and two additional points are shown in the table.

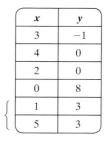

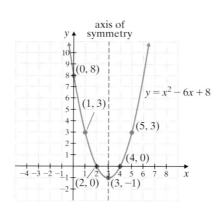

Work Practice 3

Study Example 3 and let's use it to write down a general procedure for graphing quadratic equations.

Graphing Parabolas Defined by $y = ax^2 + bx + c$

1. **Find the vertex by using the formula $x = \dfrac{-b}{2a}$.** Don't forget to find the y-value of the vertex.
2. **Find the intercepts.**
   - Let $x = 0$ and solve for y to find the y-intercept. There will be only one.
   - Let $y = 0$ and solve for x to find any x-intercepts. There may be 0, 1, or 2.
3. **Plot the vertex and the intercepts.**
4. **Find and plot additional points on the graph.** Then draw a smooth curve through the plotted points. Keep in mind that if $a > 0$, the parabola opens upward and that if $a < 0$, the parabola opens downward.

Practice 3
Graph: $y = x^2 - 2x - 3$

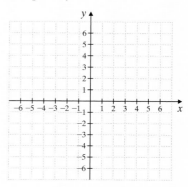

Answer

3.

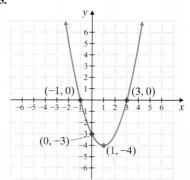

Practice 4

Graph: $y = x^2 - 4x + 1$

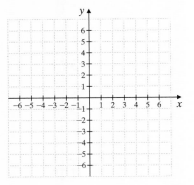

Example 4 Graph: $y = x^2 + 2x - 5$

Solution: In the equation $y = x^2 + 2x - 5$, $a = 1$ and $b = 2$. Using the vertex formula, we find that the x-coordinate of the vertex is

$$x = \frac{-b}{2a} = \frac{-2}{2 \cdot 1} = -1$$

The y-coordinate is

$$y = (-1)^2 + 2(-1) - 5 = -6$$

Thus the vertex is $(-1, -6)$.

To find the x-intercepts, we let $y = 0$.

$$0 = x^2 + 2x - 5$$

This cannot be solved by factoring, so we use the quadratic formula.

$$x = \frac{-2 \pm \sqrt{2^2 - 4(1)(-5)}}{2 \cdot 1} \quad \text{Let } a = 1, b = 2, \text{ and } c = -5.$$

$$x = \frac{-2 \pm \sqrt{24}}{2}$$

$$x = \frac{-2 \pm 2\sqrt{6}}{2} \quad \text{Simplify the radical.}$$

$$x = \frac{2(-1 \pm \sqrt{6})}{2} = -1 \pm \sqrt{6}$$

The x-intercepts are $(-1 + \sqrt{6}, 0)$ and $(-1 - \sqrt{6}, 0)$. We use a calculator approximate these so that we can easily graph these intercepts.

$$-1 + \sqrt{6} \approx 1.4 \quad \text{and} \quad -1 - \sqrt{6} \approx -3.4$$

To find the y-intercept, we let $x = 0$ in the original equation and find that $y = -$ Thus the y-intercept is $(0, -5)$. You will find, because of symmetry, that $(-2, -5)$ is also an ordered pair solution.

x	y
−1	−6
$-1 + \sqrt{6} \approx 1.4$	0
$-1 - \sqrt{6} \approx -3.4$	0
0	−5
−2	−5

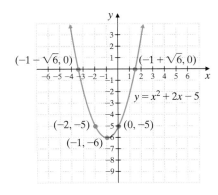

Work Practice 4

Answer
4.

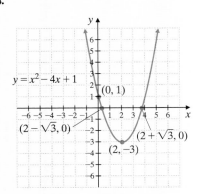

Helpful Hint

Notice that the number of x-intercepts of the graph of the parabola $y = ax^2 + bx + c$ is the same as the number of real solutions of $0 = ax^2 + bx + c$.

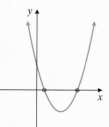

$y = ax^2 + bx + c$
$a > 0$
Two x-intercepts
Two real solutions of
$0 = ax^2 + bx + c$

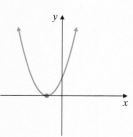

$y = ax^2 + bx + c$
$a > 0$
One x-intercept
One real solution of
$0 = ax^2 + bx + c$

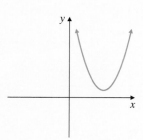

$y = ax^2 + bx + c$
$a > 0$
No x-intercepts
No real solutions of
$0 = ax^2 + bx + c$

Calculator Explorations — Graphing

Recall that a graphing calculator may be used to solve quadratic equations. The x-intercepts of the graph of $y = ax^2 + bx + c$ are solutions of $0 = ax^2 + bx + c$. To solve $x^2 - 7x - 3 = 0$, for example, graph $y = x^2 - 7x - 3$. The x-intercepts of the graph are the solutions of the equation.

Use a graphing calculator to solve each quadratic equation. Round solutions to two decimal places.

1. $x^2 - 7x - 3 = 0$
2. $2x^2 - 11x - 1 = 0$
3. $-1.7x^2 + 5.6x - 3.7 = 0$
4. $-5.8x^2 + 2.3x - 3.9 = 0$
5. $5.8x^2 - 2.6x - 1.9 = 0$
6. $7.5x^2 - 3.7x - 1.1 = 0$

Vocabulary, Readiness & Video Check

Martin-Gay Interactive Videos Watch the section lecture video and answer the following questions.

See Video 16.4

Objective A 1. In Example 1, how are the vertex and line of symmetry of a parabola explained?

Objective B 2. In Example 2, what important point was accidentally found? Why would it be useful to have an algebraic way to find this point?

Objective C 3. From Example 3, how can finding the vertex and noting whether the parabola opens up or down possibly help save us time and work? Explain using an example.

16.4 Exercise Set MyLab Math

Objective **A** *Graph each quadratic equation by finding and plotting ordered pair solutions. See Example 1.*

1. $y = 2x^2$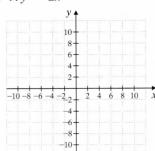

2. $y = 3x^2$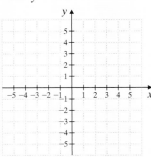

3. $y = -x^2$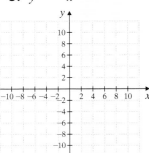

4. $y = -4x^2$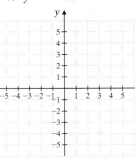

Objective **B** *Sketch the graph of each equation. Label the vertex and the intercepts. See Example 2.*

5. $y = x^2 - 1$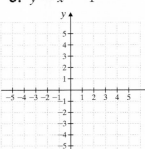

6. $y = x^2 - 16$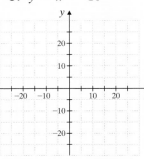

7. $y = x^2 + 4$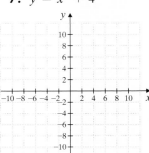

8. $y = x^2 + 9$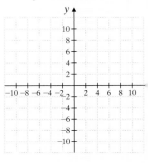

Objectives **A** **B** **C** *Sketch the graph of each equation. Label the vertex and the intercepts. See Examples 1 through*

9. $y = -x^2 + 4x - 4$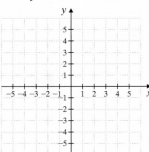

10. $y = -x^2 - 2x - 1$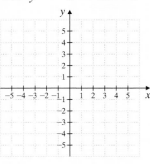

11. $y = x^2 + 5x + 4$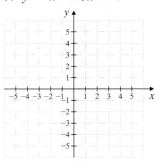

12. $y = x^2 + 7x + 10$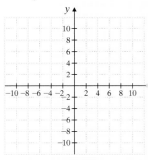

13. $y = x^2 - 4x + 5$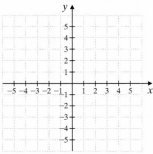

14. $y = x^2 - 6x + 10$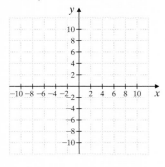

15. $y = 2 - x^2$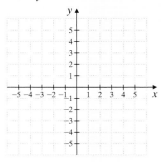

16. $y = 3 - x^2$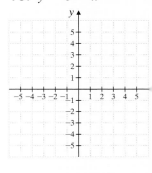

17. $y = \dfrac{1}{3}x^2$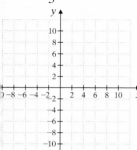

18. $y = \dfrac{1}{2}x^2$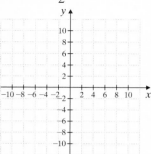

19. $y = x^2 + 6x$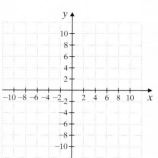

20. $y = x^2 - 4x$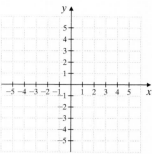

21. $y = x^2 + 2x - 8$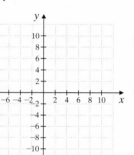

22. $y = x^2 - 2x - 3$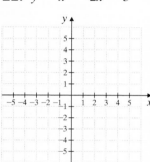

23. $y = -\dfrac{1}{2}x^2$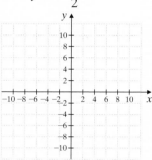

24. $y = -\dfrac{1}{3}x^2$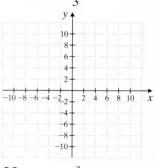

25. $y = 2x^2 - 11x + 5$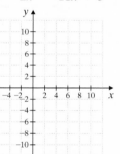

26. $y = 2x^2 + x - 3$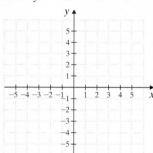

27. $y = -x^2 + 4x - 3$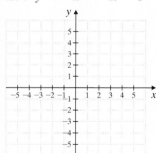

28. $y = -x^2 + 6x - 8$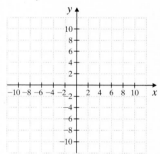

Review

Simplify each complex fraction. See Section 12.7.

29. $\dfrac{\frac{1}{7}}{\frac{2}{5}}$

30. $\dfrac{\frac{3}{8}}{\frac{1}{7}}$

31. $\dfrac{\frac{1}{x}}{\frac{2}{x^2}}$

32. $\dfrac{\frac{x}{5}}{\frac{2}{x}}$

33. $\dfrac{2x}{1 - \frac{1}{x}}$

34. $\dfrac{x}{x - \frac{1}{x}}$

35. $\dfrac{\frac{a-b}{2b}}{\frac{b-a}{8b^2}}$

36. $\dfrac{\frac{2a^2}{a-3}}{\frac{a}{3-a}}$

Concept Extensions

For Exercises 37 through 40, sketch the graph of each equation. Label the vertex and intercepts. Use the quadratic formula to locate the exact x-intercepts.

37. $y = x^2 + 2x - 2$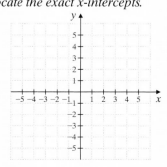

38. $y = x^2 - 4x - 3$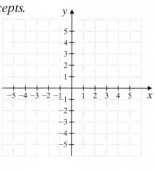

39. $y = x^2 - 3x + 1$

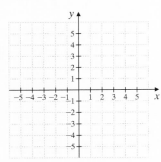

40. $y = x^2 - 2x - 5$

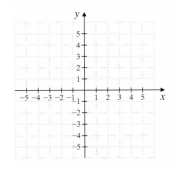

41. The height h of a fireball launched from a Roman candle with an initial velocity of 128 feet per second is given by the equation $h = -16t^2 + 128t$, where t is time in seconds after launch.
   Use the graph of this equation to answer each question.
   a. Estimate the maximum height of the fire ball.
   b. Estimate the time when the fireball is at its maximum height.
   c. Estimate the time when the fireball would return to the ground.

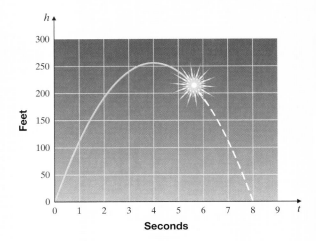

42. Determine the maximum number and the minimum number of x-intercepts for a parabola. Explain your answers.

Match the graph of each equation of the form $y = ax^2 + bx + c$ with the given description.

43. $a > 0$, two x-intercepts

44. $a < 0$, one x-intercept

45. $a < 0$, no x-intercept

46. $a > 0$, no x-intercept

47. $a > 0$, one x-intercept

48. $a < 0$, two x-intercepts

A
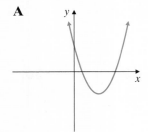

B
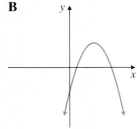

C
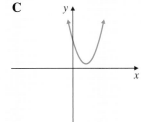

D
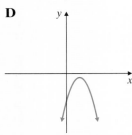

E
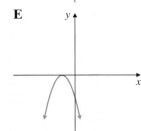

F
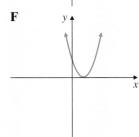

Chapter 16 Group Activity

Uses of Parabolas

In this chapter, we learned that the graph of a quadratic equation in two variables of the form $y = ax^2 + bx + c$ is a shape called a **parabola**. The figure to the right shows the general shape of a parabola.

The shape of a parabola shows up in many situations, both natural and human-made, in the world around us.

Natural Situations

- **Hurricanes** The paths of many hurricanes are roughly shaped like parabolas. In the Northern Hemisphere, hurricanes generally begin moving to the northwest. Then, as they move farther from the equator, they swing around to head in a northeasterly direction.

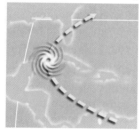

- **Projectiles** The force of the Earth's gravity acts on a projectile launched into the air. The resulting path of the projectile, anything from a bullet to a football, is generally shaped like a parabola.

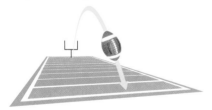

- **Orbits** There are several different possible shapes for orbits of satellites, planets, moons, and comets in outer space. One of the possible types of orbits is in the shape of a parabola. A parabolic orbit is most often seen with comets.

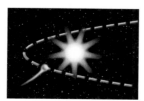

Human-Made Situations

- **Telescopes** Because a parabola has nice reflecting properties, its shape is used in many kinds of telescopes. The largest nonsteerable radio telescope is the Arecibo Observatory in Puerto Rico. This telescope consists of a huge parabolic dish built into a valley. The dish is about 1000 feet across.

- **Training Astronauts** Astronauts must be able to work in zero-gravity conditions on missions in space. However, it's nearly impossible to escape the force of gravity on Earth. To help astronauts train to work in weightlessness, a specially modified jet can be flown in a parabolic path. At the top of the parabola, weightlessness can be simulated for up to 30 seconds at a time.

- **Architecture** The reinforced concrete arches used in many modern buildings are based on the shape of a parabola.

- **Music** The design of the modern flute incorporates a parabolic head joint.

Group Activity

There are many other physical applications of parabolas. For example, satellite dishes often have parabolic shapes. Choose a physical example of a parabola given here or use one of your own and write a report (with diagrams).

Chapter 16 Vocabulary Check

Fill in each blank with one of the words or phrases listed below. Some choices may be used more than once and some may not be used at all.

square root	vertex	one	parabola
completing the square	quadratic	zero	

1. If $x^2 = a$, then $x = \sqrt{a}$ or $x = -\sqrt{a}$. This property is called the _____ property.
2. The graph of $y = x^2$ is called a _____.
3. The formula $\dfrac{-b}{2a}$, where $y = ax^2 + bx + c$, is called the _____ formula.
4. The process of solving a quadratic equation by writing it in the form $(x + a)^2 = c$ is called _____.
5. The formula $x = \dfrac{-b \pm \sqrt{b^2 - 4ac}}{2a}$ is called the _____ formula.
6. The lowest point on a parabola that opens upward is called the _____.
7. The zero-factor property states that if the product of two numbers is zero, then at least one of the two numbers is _____.

Helpful Hint
▶ Are you preparing for your test? To help, don't forget to take these:
- Chapter 16 Getting Ready for the Test on page 123
- Chapter 16 Test on page 1235

Then check all of your answers at the back of this text. For further review, the step-by-step video solutions to any of these exercises are located in MyLab Math.

16 Chapter Highlights

Definitions and Concepts	Examples
Section 16.1 Solving Quadratic Equations by the Square Root Property	
Square Root Property If $x^2 = a$ for $a \geq 0$, then $x = \sqrt{a}$ or $x = -\sqrt{a}$.	Solve the equation. $(x - 1)^2 = 15$ $x - 1 = \sqrt{15}$ or $x - 1 = -\sqrt{15}$ $x = 1 + \sqrt{15}$ $\quad x = 1 - \sqrt{15}$
Section 16.2 Solving Quadratic Equations by Completing the Square	
To Solve a Quadratic Equation by Completing the Square **Step 1:** If the coefficient of x^2 is not 1, divide both sides of the equation by the coefficient. **Step 2:** Get all terms with variables alone on one side. **Step 3:** Complete the square by adding the square of half of the coefficient of x to both sides. **Step 4:** Factor the perfect square trinomial. **Step 5:** Use the square root property to solve.	Solve $2x^2 + 12x - 10 = 0$ by completing the square. $\dfrac{2x^2}{2} + \dfrac{12x}{2} - \dfrac{10}{2} = \dfrac{0}{2}$ Divide by 2. $x^2 + 6x - 5 = 0$ Simplify. $x^2 + 6x = 5$ Add 5. The coefficient of x is 6. Half of 6 is 3 and $3^2 = 9$. Add 9 to both sides. $x^2 + 6x + 9 = 5 + 9$ $(x + 3)^2 = 14$ Factor. $x + 3 = \sqrt{14}$ or $x + 3 = -\sqrt{14}$ $x = -3 + \sqrt{14}$ $\quad x = -3 - \sqrt{14}$

Chapter 16 Highlights

Definitions and Concepts	Examples
Section 16.3 Solving Quadratic Equations by the Quadratic Formula	

Quadratic Formula

If a, b, and c are real numbers and $a \neq 0$, the quadratic equation $ax^2 + bx + c = 0$ has solutions

$$x = \frac{-b \pm \sqrt{b^2 - 4ac}}{2a}$$

To Solve a Quadratic Equation by the Quadratic Formula

Step 1: Write the equation in standard form: $ax^2 + bx + c = 0$.

Step 2: If necessary, clear the equation of fractions.

Step 3: Identify a, b, and c.

Step 4: Replace a, b, and c in the quadratic formula with the identified values, and simplify.

Identify a, b, and c in the quadratic equation

$$4x^2 - 6x = 5$$

First, subtract 5 from both sides.

$$4x^2 - 6x - 5 = 0$$

$a = 4$, $b = -6$, and $c = -5$

Solve $3x^2 - 2x - 2 = 0$.

In this equation, $a = 3$, $b = -2$, and $c = -2$.

$$x = \frac{-(-2) \pm \sqrt{(-2)^2 - 4(3)(-2)}}{2 \cdot 3}$$

$$= \frac{2 \pm \sqrt{4 - (-24)}}{6}$$

$$= \frac{2 \pm \sqrt{28}}{6} = \frac{2 \pm \sqrt{4 \cdot 7}}{6} = \frac{2 \pm 2\sqrt{7}}{6}$$

$$= \frac{2(1 \pm \sqrt{7})}{2 \cdot 3} = \frac{1 \pm \sqrt{7}}{3}$$

Section 16.4 Graphing Quadratic Equations in Two Variables

The graph of a quadratic equation $y = ax^2 + bx + c$, $a \neq 0$, is called a **parabola**. The lowest point on a parabola opening upward or the highest point on a parabola opening downward is called the **vertex**. The vertical line through the vertex is the **axis of symmetry**.

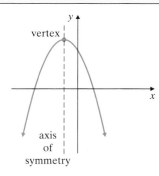

Vertex Formula

The vertex of the parabola $y = ax^2 + bx + c$ has x-coordinate $\dfrac{-b}{2a}$. To find the corresponding y-coordinate, substitute the x-coordinate into the original equation and solve for y.

Graph: $y = 2x^2 - 6x + 4$

The x-coordinate of the vertex is

$$x = \frac{-b}{2a} = \frac{-(-6)}{2(2)} = \frac{6}{4} = \frac{3}{2}$$

The y-coordinate is

$$y = 2\left(\frac{3}{2}\right)^2 - 6\left(\frac{3}{2}\right) + 4 = 2\left(\frac{9}{4}\right) - 9 + 4 = -\frac{1}{2}$$

The vertex is $\left(\dfrac{3}{2}, -\dfrac{1}{2}\right)$.

The y-intercept is

$$y = 2 \cdot 0^2 - 6 \cdot 0 + 4 = 4$$

The x-intercepts are

$$0 = 2x^2 - 6x + 4$$
$$0 = 2(x - 2)(x - 1)$$
$$x = 2 \quad \text{or} \quad x = 1$$

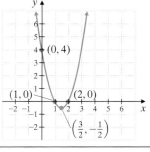

Chapter 16 Review

(16.1) *Solve each quadradic equation by factoring.*

1. $x^2 - 121 = 0$
2. $y^2 - 100 = 0$
3. $3m^2 - 5m = 2$
4. $7m^2 + 2m = 5$

Use the square root property to solve each quadratic equation.

5. $x^2 = 36$
6. $x^2 = 81$
7. $k^2 = 50$
8. $k^2 = 45$
9. $(x - 11)^2 = 49$
10. $(x + 3)^2 = 100$
11. $(4p + 5)^2 = 41$
12. $(3p + 7)^2 = 37$

Solve. For Exercises 13 and 14, use the formula $h = 16t^2$, where h is the height in feet at time t seconds.

13. If Kara Washington dives from a height of 100 feet, how long before she hits the water?
14. How long does a 5-mile free fall take? Round your result to the nearest tenth of a second. (*Hint:* 1 mi = 5280 ft)

(16.2) *Solve each quadratic equation by completing the square.*

15. $x^2 - 9x = -8$
16. $x^2 + 8x = 20$
17. $x^2 + 4x = 1$
18. $x^2 - 8x = 3$
19. $x^2 - 6x + 7 = 0$
20. $x^2 + 6x + 7 = 0$
21. $2y^2 + y - 1 = 0$
22. $4y^2 + 3y - 1 = 0$

(16.3) *Use the quadratic formula to solve each quadratic equation.*

23. $9x^2 + 30x + 25 = 0$
24. $16x^2 - 72x + 81 = 0$
25. $7x^2 = 35$
26. $11x^2 = 33$
27. $x^2 - 10x + 7 = 0$
28. $x^2 + 4x - 7 = 0$
29. $3x^2 + x - 1 = 0$
30. $x^2 + 3x - 1 = 0$
31. $2x^2 + x + 5 = 0$
32. $7x^2 - 3x + 1 = 0$

For the exercise numbers given, approximate the exact solutions to the nearest tenth.

33. Exercise 29
34. Exercise 30

35. The annual number of visitors y (in thousands) to Yosemite National Park in California is modeled by the equation $y = 78x^2 - 267x + 3975$. In this equation, x is the number of years since 2011. Assume that this trend continued and find the year after 2011 in which 3882 thousand people visited Yosemite National Park. (*Source:* Based on data from the National Park Service)

36. The amount y of electricity generated by solar power (in thousand megawatt hours per day) in the United States from 2012 through 2016 is modeled by the equation $y = 2x^2 + 13x + 12$, where x represents the number of years after 2012. Assume that this trend continues and find the year after 2012 in which the amount of electricity generated by solar power is 127 thousand megawatt hours per day. (*Source:* Based on information from the Energy Information Administration)

(16.4) *Graph each quadratic equation and find and plot any intercepts.*

37. $y = 5x^2$

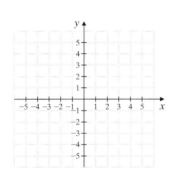

38. $y = -\dfrac{1}{2}x^2$

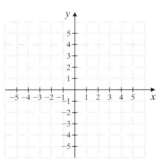

Graph each quadratic equation. Label the vertex and the intercepts with their coordinates.

39. $y = x^2 - 25$

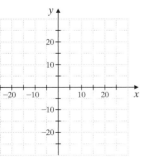

40. $y = x^2 - 36$

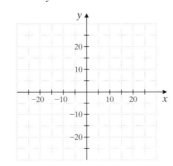

41. $y = x^2 + 3$

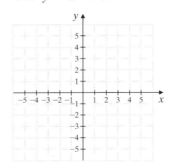

42. $y = x^2 + 8$

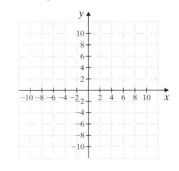

43. $y = -4x^2 + 8$

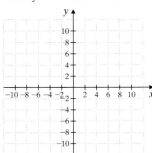

44. $y = -3x^2 + 9$

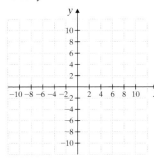

45. $y = x^2 + 3x - 10$

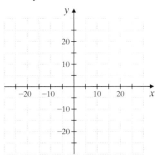

46. $y = x^2 + 3x - 4$

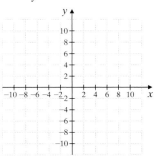

47. $y = -x^2 - 5x - 6$

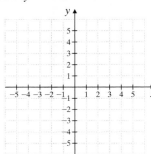

48. $y = 3x^2 - x - 2$

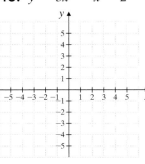

49. $y = 2x^2 - 11x - 6$

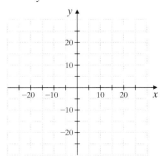

50. $y = -x^2 + 4x + 8$

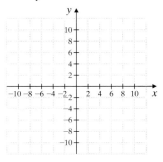

Match each quadratic equation with its graph.

51. $y = 2x^2$
A

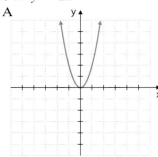

52. $y = -x^2$
B

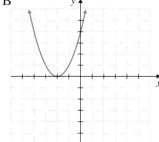

53. $y = x^2 + 4x + 4$
C
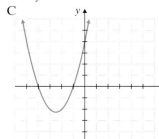

54. $y = x^2 + 5x + 4$
D
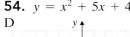
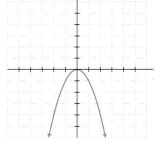

Quadratic equations in the form $y = ax^2 + bx + c$ are graphed below. Determine the number of real solutions for the related equation $0 = ax^2 + bx + c$ from each graph.

55.

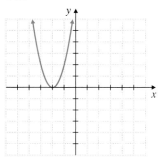

56.

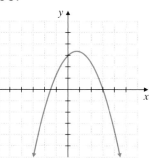

57.

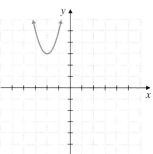

58.

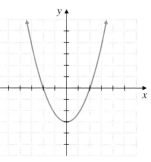

Mixed Review

Use the square root property to solve each quadratic equation.

59. $x^2 = 49$ **60.** $y^2 = 75$ **61.** $(x - 7)^2 = 64$

Solve each quadratic equation by completing the square.

62. $x^2 + 4x = 6$ **63.** $3x^2 + x = 2$ **64.** $4x^2 - x - 2 = 0$

Use the quadratic formula to solve each quadratic equation.

65. $4x^2 - 3x - 2 = 0$ **66.** $5x^2 + x - 2 = 0$

67. $4x^2 + 12x + 9 = 0$ **68.** $2x^2 + x + 4 = 0$

Graph each quadratic equation. Label the vertex and the intercepts with their coordinates.

69. $y = 4 - x^2$ **70.** $y = x^2 + 4$ **71.** $y = x^2 + 6x + 8$ **72.** $y = x^2 - 2x - 4$

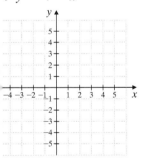 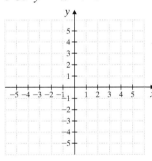 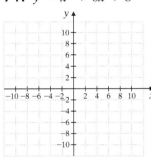 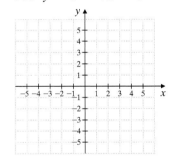

Chapter 16 — Getting Ready for the Test

MULTIPLE CHOICE All the exercises below are **Multiple Choice**. Choose the correct letter.

1. To solve $x^2 - 6x = -1$ by completing the square, choose the next correct step.
   A. $x^2 - 6x + 9 = -1$ **B.** $x^2 - 6x + 9 = -1 + 9$ **C.** $x^2 - 6x - 9 = -1 - 9$ **D.** $(x-6)^2 = (-1)^2$

2. To solve $x^2 + 5x = 4$ by completing the square, choose the next correct step.
   A. $x^2 + 5x + 25 = 4$ **B.** $x^2 + 5x + 25 = 4 + 25$ **C.** $x^2 + 5x + \frac{25}{4} = 4$ **D.** $x^2 + 5x + \frac{25}{4} = 4 + \frac{25}{4}$

3. The expression $\dfrac{12 \pm 3\sqrt{7}}{9}$ simplifies to:
   A. $\dfrac{\pm 15\sqrt{7}}{9}$ **B.** $\dfrac{4 \pm 3\sqrt{7}}{3}$ **C.** $\dfrac{4 \pm \sqrt{7}}{3}$ **D.** $4 \pm \sqrt{7}$

4. The expression $\dfrac{5 \pm 10\sqrt{2}}{5}$ simplifies to:
   A. $1 \pm 2\sqrt{2}$ **B.** $\pm 10\sqrt{2}$ **C.** $\pm 3\sqrt{2}$ **D.** $1 \pm 10\sqrt{2}$

For Exercises 5 and 6, the quadratic equation is $7x^2 = 3 - x$, with $a = 7$ in the quadratic formula.

5. Choose the value of c.
   A. 3 **B.** -3 **C.** 1 **D.** -1 **E.** x

6. Choose the value of b.
   A. 3 **B.** -3 **C.** 1 **D.** -1 **E.** x

For Exercises 7 through 10, choose the correct letter.

7. Select the vertex of the graph of $y = x^2 + 3$.
   A. $(0, 0)$ **B.** $(0, 3)$ **C.** $(3, 0)$ **D.** $\left(\dfrac{9}{2}, 3\right)$

8. Select the vertex of the graph of $y = x^2 - 2x$.
   A. $(1, -2)$ **B.** $(-1, 3)$ **C.** $(2, 0)$ **D.** $(1, -1)$

9. Select the vertex of the graph of $y = -x^2 + 2x - 4$.
   A. $(-1, -7)$ **B.** $(2, -4)$ **C.** $(1, -1)$ **D.** $(1, -3)$

10. Select the intercept(s) of the graph of $y = x^2 + x + 1$.
    A. $(0, 1), (-1, 0)$ **B.** $(0, 1)$ **C.** $(1, 0), (-1, 0)$ **D.** $(0, 1), (1, 0), (-1, 0)$

Chapter 16 Test

Solve by factoring.

1. $x^2 - 400 = 0$
2. $2x^2 - 11x = 21$

Solve using the square root property.

3. $5k^2 = 80$
4. $(3m - 5)^2 = 8$

Solve by completing the square.

5. $x^2 - 26x + 160 = 0$
6. $3x^2 + 12x - 4 = 0$

Solve using the quadratic formula.

7. $x^2 - 3x - 10 = 0$
8. $p^2 - \frac{5}{3}p - \frac{1}{3} = 0$

Solve by the most appropriate method.

9. $(3x - 5)(x + 2) = -6$
10. $(3x - 1)^2 = 16$
11. $3x^2 - 7x - 2 = 0$
12. $x^2 - 4x - 5 = 0$
13. $3x^2 - 7x + 2 = 0$
14. $2x^2 - 6x + 1 = 0$

15. The height of a triangle is 4 times the length of the base. The area of the triangle is 18 square feet. Find the height and base of the triangle.

Answers

1. _____
2. _____
3. _____
4. _____
5. _____
6. _____
7. _____
8. _____
9. _____
10. _____
11. _____
12. _____
13. _____
14. _____
15. _____

Graph each quadratic equation. Label the vertex and the intercepts with their coordinates.

16. $y = -5x^2$

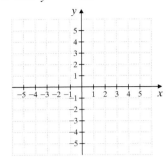

17. $y = x^2 - 4$

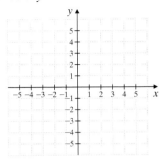

18. $y = x^2 - 7x + 10$

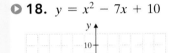

19. $y = 2x^2 + 4x - 1$

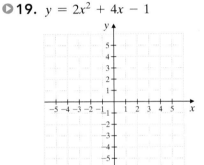

20. The number of diagonals d that a polygon with n sides has is given by the formula

$$d = \frac{n^2 - 3n}{2}$$

Find the number of sides of a polygon if it has 9 diagonals.

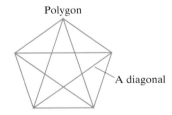

Solve.

21. The highest dive from a diving board was made by Laso Schaller of Switzerland in August 2015. He dove from a height of 193.83 feet at Maggia, Ticlno, Switzerland. To the nearest tenth of a second, how long did the dive take? Use the formula $h = 16t^2$. (*Source: Guinness Book of World Records*)

Cumulative Review — Chapters 1–16

Divide.

1. $\dfrac{786.1}{1000}$

2. $\dfrac{818}{1000}$

3. $\dfrac{0.12}{10}$

4. $-\dfrac{5.03}{100}$

5. Using the circle graph shown, determine the percent of visitors who came to the United States from Mexico or Canada.

6. Using the circle graph shown, determine the percent of visitors to the United States that came from Europe, Asia, or South America.

Visitors to U.S. by Region

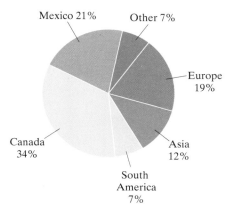

Source: Office of Travel and Tourism Industries, 2012

7. Find the perimeter of a rectangle with a length of 11 inches and a width of 3 inches.

△ 8. Find the perimeter of a triangular yard whose sides are 6 feet, 8 feet, and 11 feet.

9. Find the area of the parallelogram.

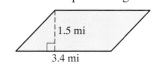

△ 10. Find the area of the triangle.

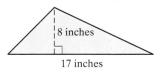

11. Convert 3210 ml to liters.

12. Convert 4321 cl to liters.

13. Solve: $8(2 - t) = -5t$

14. Solve: $\dfrac{5}{2}x - 1 = x + \dfrac{1}{4}$

Simplify the following expressions.

15. 3^0

16. -2^0

17. Factor: $r^2 - r - 42$

18. Factor: $y^2 + 3y - 70$

19. Factor: $10x^2 - 13xy - 3y^2$

20. Factor: $72x^2 - 35xy + 3y^2$

1237

21. Factor $8x^2 - 14x + 5$ by grouping.

22. Factor $15x^2 - 4x - 4$ by grouping.

23. Factor completely.
   a. $4x^3 - 49x$
   b. $162x^4 - 2$

24. Factor completely.
   a. $9x^3 - x$
   b. $5x^4 - 5$

25. Solve: $(5x - 1)(2x^2 + 15x + 18) = 0$

26. Solve: $(x + 4)(40x^2 - 34x + 3) =$

27. Simplify: $\dfrac{x^2 + 8x + 7}{x^2 - 4x - 5}$

28. Simplify: $\dfrac{x^2 - 6x + 5}{x^2 + 6x - 7}$

29. The quotient of a number and 6, minus $\dfrac{5}{3}$, is the quotient of the number and 2. Find the number.

30. Four times a number, added to 5, is divided by 6. The result is $\dfrac{7}{2}$. Find the number.

31. Complete the table for the equation $y = 3x$.

	x	y
a.	-1	
b.		0
c.		-9

32. Complete the table for the equation $y = -2x + 7$.

	x	y
a.	0	
b.		0
c.	5	

Determine whether each pair of lines is parallel, perpendicular, or neither.

33.
   a. $y = -\dfrac{1}{5}x + 1$
      $2x + 10y = 3$
   b. $x + y = 3$
      $-x + y = 4$
   c. $3x + y = 5$
      $2x + 3y = 6$

34.
   a. $y = -2x + 3$
      $y = -2x + 5$
   b. $-2x + y = 3$
      $x + 2y = 9$
   c. $3x - 2y = -8$
      $3x + 2y = 1$

35. Determine whether each relation is also a function.
   a. $\{(-1, 1), (2, 3), (7, 3), (8, 6)\}$
   b. $\{(0, -2), (1, 5), (0, 3), (7, 7)\}$

36. Determine whether each relation is also a function.
   a. $\{(-2, 3), (-5, 7), (9, 3), (0, 0)\}$
   b. $\left\{(9, -1), \left(0, \dfrac{1}{2}\right), (2, -1), (9, 0)\right\}$

37. Solve the system:
$\begin{cases} 2x + y = 10 \\ x = y + 2 \end{cases}$

38. Solve the system:
$\begin{cases} 8x - 3y = -4 \\ y = 7x - 3 \end{cases}$

Cumulative Review

Find each square root.

39. $\sqrt{36}$ **40.** $\sqrt{81}$ **41.** $\sqrt{\dfrac{9}{100}}$ **42.** $\sqrt{\dfrac{16}{25}}$

Rationalize the denominator.

43. $\dfrac{2}{1 + \sqrt{3}}$ **44.** $\dfrac{7}{\sqrt{5} - 2}$

Use the square root property to solve.

45. $(x - 3)^2 = 16$ **46.** $(x + 4)^2 = 9$

47. Solve $\dfrac{1}{2}x^2 - x = 2$ using the quadratic formula.

48. Solve $2x^2 = \dfrac{5}{2}x + \dfrac{7}{2}$ using the quadratic formula.

39. _____

40. _____

41. _____

42. _____

43. _____

44. _____

45. _____

46. _____

47. _____

48. _____

Appendix A: Tables

A.1 Table of Geometric Figures

Plane Figures Have Length and Width but No Thickness or Depth		
Name	Description	Figure
Polygon	Union of three or more coplanar line segments that intersect with each other only at each end point, with each end point shared by two segments.	
Triangle	Polygon with three sides (sum of measures of three angles is 180°).	
Scalene Triangle	Triangle with no sides of equal length.	
Isosceles Triangle	Triangle with two sides of equal length.	
Equilateral Triangle	Triangle with all sides of equal length.	
Right Triangle	Triangle that contains a right angle.	leg, hypotenuse, leg
Quadrilateral	Polygon with four sides (sum of measures of four angles is 360°).	
Trapezoid	Quadrilateral with exactly one pair of opposite sides parallel.	base, leg, parallel sides, leg, base
Isosceles Trapezoid	Trapezoid with legs of equal length.	
Parallelogram	Quadrilateral with both pairs of opposite sides parallel.	
Rhombus	Parallelogram with all sides of equal length.	
Rectangle	Parallelogram with four right angles.	

Plane Figures Have Length and Width but No Thickness or Depth (*continued*)		
Name	Description	Figure
Square	Rectangle with all sides of equal length.	
Circle	All points in a plane the same distance from a fixed point called the **center.**	

Solid Figures Have Length, Width, and Height or Depth		
Name	Description	Figure
Rectangular Solid	A solid with six sides, all of which are rectangles.	
Cube	A rectangular solid whose six sides are squares.	
Sphere	All points the same distance from a fixed point called the **center.**	
Right Circular Cylinder	A cylinder having two circular bases that are perpendicular to its altitude.	
Right Circular Cone	A cone with a circular base that is perpendicular to its altitude.	

A.2 Table of Percents, Decimals, and Fraction Equivalents

Percent	Decimal	Fraction
1%	0.01	$\frac{1}{100}$
5%	0.05	$\frac{1}{20}$
10%	0.1	$\frac{1}{10}$
12.5% or $12\frac{1}{2}$%	0.125	$\frac{1}{8}$
$16.\overline{6}$% or $16\frac{2}{3}$%	$0.1\overline{6}$	$\frac{1}{6}$
20%	0.2	$\frac{1}{5}$
25%	0.25	$\frac{1}{4}$
30%	0.3	$\frac{3}{10}$
$33.\overline{3}$% or $33\frac{1}{3}$%	$0.\overline{3}$	$\frac{1}{3}$
37.5% or $37\frac{1}{2}$%	0.375	$\frac{3}{8}$
40%	0.4	$\frac{2}{5}$
50%	0.5	$\frac{1}{2}$
60%	0.6	$\frac{3}{5}$
62.5% or $62\frac{1}{2}$%	0.625	$\frac{5}{8}$
$66.\overline{6}$% or $66\frac{2}{3}$%	$0.\overline{6}$	$\frac{2}{3}$
70%	0.7	$\frac{7}{10}$
75%	0.75	$\frac{3}{4}$
80%	0.8	$\frac{4}{5}$
$83.\overline{3}$% or $83\frac{1}{3}$%	$0.8\overline{3}$	$\frac{5}{6}$
87.5% or $87\frac{1}{2}$%	0.875	$\frac{7}{8}$
90%	0.9	$\frac{9}{10}$
100%	1.0	1
110%	1.1	$1\frac{1}{10}$
125%	1.25	$1\frac{1}{4}$
$133.\overline{3}$% or $133\frac{1}{3}$%	$1.\overline{3}$	$1\frac{1}{3}$
150%	1.5	$1\frac{1}{2}$
$166.\overline{6}$% or $166\frac{2}{3}$%	$1.\overline{6}$	$1\frac{2}{3}$
175%	1.75	$1\frac{3}{4}$
200%	2.0	2

A.3 Table on Finding Common Percents of a Number

Common Percent Equivalences*	Shortcut Method for Finding Percent	Examples
$1\% = 0.01$ (or $\frac{1}{100}$)	To find 1% of a number, multiply by 0.01. To do so, move the decimal point 2 places to the left.	1% of 210 is 2.10 or 2.1. 1% of 1500 is 15. 1% of 8.6 is 0.086.
$10\% = 0.1$ (or $\frac{1}{10}$)	To find 10% of a number, multiply by 0.1, or move the decimal point of the number 1 place to the left.	10% of 140 is 14. 10% of 30 is 3. 10% of 17.6 is 1.76.
$25\% = \frac{1}{4}$	To find 25% of a number, find $\frac{1}{4}$ of the number, or divide the number by 4.	25% of 20 is $\frac{20}{4}$ or 5. 25% of 8 is 2. 25% of 10 is $\frac{10}{4}$ or $2\frac{1}{2}$.
$50\% = \frac{1}{2}$	To find 50% of a number, find $\frac{1}{2}$ of the number, or divide the number by 2.	50% of 64 is $\frac{64}{2}$ or 32. 50% of 1000 is 500. 50% of 9 is $\frac{9}{2}$ or $4\frac{1}{2}$.
$100\% = 1$	To find 100% of a number, multiply the number by 1. In other words, 100% of a number is the number.	100% of 98 is 98. 100% of 1407 is 1407. 100% of 18.4 is 18.4.
$200\% = 2$	To find 200% of a number, multiply the number by 2.	200% of 31 is 31 · 2 or 62. 200% of 750 is 1500. 200% of 6.5 is 13.

*See Appendix A.2.

A.4 Table of Squares and Square Roots

n	n^2	$\sqrt{n}$	n	n^2	$\sqrt{n}$
1	1	1.000	51	2601	7.141
2	4	1.414	52	2704	7.211
3	9	1.732	53	2809	7.280
4	16	2.000	54	2916	7.348
5	25	2.236	55	3025	7.416
6	36	2.449	56	3136	7.483
7	49	2.646	57	3249	7.550
8	64	2.828	58	3364	7.616
9	81	3.000	59	3481	7.681
10	100	3.162	60	3600	7.746
11	121	3.317	61	3721	7.810
12	144	3.464	62	3844	7.874
13	169	3.606	63	3969	7.937
14	196	3.742	64	4096	8.000
15	225	3.873	65	4225	8.062
16	256	4.000	66	4356	8.124
17	289	4.123	67	4489	8.185
18	324	4.243	68	4624	8.246
19	361	4.359	69	4761	8.307
20	400	4.472	70	4900	8.367
21	441	4.583	71	5041	8.426
22	484	4.690	72	5184	8.485
23	529	4.796	73	5329	8.544
24	576	4.899	74	5476	8.602
25	625	5.000	75	5625	8.660
26	676	5.099	76	5776	8.718
27	729	5.196	77	5929	8.775
28	784	5.292	78	6084	8.832
29	841	5.385	79	6241	8.888
30	900	5.477	80	6400	8.944
31	961	5.568	81	6561	9.000
32	1024	5.657	82	6724	9.055
33	1089	5.745	83	6889	9.110
34	1156	5.831	84	7056	9.165
35	1225	5.916	85	7225	9.220
36	1296	6.000	86	7396	9.274
37	1369	6.083	87	7569	9.327
38	1444	6.164	88	7744	9.381
39	1521	6.245	89	7921	9.434
40	1600	6.325	90	8100	9.487
41	1681	6.403	91	8281	9.539
42	1764	6.481	92	8464	9.592
43	1849	6.557	93	8649	9.644
44	1936	6.633	94	8836	9.695
45	2025	6.708	95	9025	9.747
46	2116	6.782	96	9216	9.798
47	2209	6.856	97	9409	9.849
48	2304	6.928	98	9604	9.899
49	2401	7.000	99	9801	9.950
50	2500	7.071	100	10,000	10.000

Factoring Sums and Differences of Cubes

Although the sum of two squares usually does not factor, the sum or difference of two cubes can be factored and reveal factoring patterns. The pattern for the sum of cubes can be checked by multiplying the binomial $x + y$ and the trinomial $x^2 - xy + y^2$. The pattern for the difference of two cubes can be checked by multiplying the binomial $x - y$ by the trinomial $x^2 + xy + y^2$.

Sum or Difference of Two Cubes
$$a^3 + b^3 = (a + b)(a^2 - ab + b^2)$$
$$a^3 - b^3 = (a - b)(a^2 + ab + b^2)$$

Example 1 Factor $x^3 + 8$.

Solution:

First, write the binomial in the form $a^3 + b^3$.

$x^3 + 8 = x^3 + 2^3$ Write in the form $a^3 + b^3$.

If we replace a with x and b with 2 in the formula above, we have

$$x^3 + 2^3 = (x + 2)[x^2 - (x)(2) + 2^2]$$
$$= (x + 2)(x^2 - 2x + 4)$$

Helpful Hint

When factoring sums or differences of cubes, notice the sign patterns.

$$x^3 + y^3 = (x + y)(x^2 - xy + y^2)$$
same sign — opposite signs — always positive

$$x^3 - y^3 = (x - y)(x^2 + xy + y^2)$$
same sign — opposite signs — always positive

Example 2 Factor $y^3 - 27$.

Solution:

$$\begin{aligned} y^3 - 27 &= y^3 - 3^3 &\text{Write in the form } a^3 - b^3. \\ &= (y - 3)\left[y^2 + (y)(3) + 3^2\right] \\ &= (y - 3)(y^2 + 3y + 9) \end{aligned}$$

Example 3 Factor $64x^3 + 1$.

Solution:

$$\begin{aligned} 64x^3 + 1 &= (4x)^3 + 1^3 \\ &= (4x + 1)\left[(4x)^2 - (4x)(1) + 1^2\right] \\ &= (4x + 1)(16x^2 - 4x + 1) \end{aligned}$$

Example 4 Factor $54a^3 - 16b^3$.

Solution: Remember to factor out common factors first before using other factoring methods.

$$\begin{aligned} 54a^3 - 16b^3 &= 2(27a^3 - 8b^3) &\text{Factor out the GCF 2.} \\ &= 2\left[(3a)^3 - (2b)^3\right] &\text{Difference of two cubes.} \\ &= 2(3a - 2b)\left[(3a)^2 + (3a)(2b) + (2b)^2\right] \\ &= 2(3a - 2b)(9a^2 + 6ab + 4b^2) \end{aligned}$$

Appendix B | Factoring Sums and Differences of Cubes

B Exercise Set MyLab Math

Factor the binomials completely. See Examples 1 through 4.

1. $a^3 + 27$
2. $b^3 - 8$
3. $8a^3 + 1$
4. $64x^3 - 1$
5. $5k^3 + 40$
6. $6r^3 - 162$
7. $x^3y^3 - 64$
8. $8x^3 - y^3$
9. $x^3 + 125$
10. $a^3 - 216$
11. $24x^4 - 81xy^3$
12. $375y^6 - 24y^3$
13. $27 - t^3$
14. $125 + r^3$
15. $8r^3 - 64$
16. $54r^3 + 2$
17. $t^3 - 343$
18. $s^3 + 216$
19. $s^3 - 64t^3$
20. $8t^3 + s^3$

Appendix C

Mixture and Uniform Motion Problem Solving

Objectives

A Solve Mixture Problems.

B Solve Distance or Uniform Motion Problems.

This appendix is devoted to solving problems in the categories listed. The same problem-solving steps used in previous sections are also followed in this appendix. They are listed below for review.

General Strategy for Problem Solving

1. UNDERSTAND the problem. During this step, become comfortable with the problem. Some ways of doing this are as follows:

   Read and reread the problem.
   Choose a variable to represent the unknown.
   Construct a drawing whenever possible.
   Propose a solution and check. Pay careful attention to how you check your proposed solution. This will help when writing an equation to model the problem.

2. TRANSLATE the problem into an equation.
3. SOLVE the equation.
4. INTERPRET the results: *Check* the proposed solution in the stated problem and *state* your conclusion.

Objective A Solving Mixture Problems

Mixture problems involve two or more different quantities being combined to form a new mixture. These applications range from Dow Chemical's need to form a chemical mixture of a required strength to Planter's Peanut Company's need to find the correct mixture of peanuts and cashews given taste and price constraints.

Practice 1

How much 20% dye solution and 50% dye solution should be mixed to obtain 6 liters of a 40% solution?

Example 1 Calculating Percent for a Lab Experiment

A chemist working on his doctoral degree at Massachusetts Institute of Technology needs 12 liters of a 50% acid solution for a lab experiment. The stockroom has only 40% and 70% solutions. How much of each solution should be mixed together to form 12 liters of a 50% solution?

Solution:

1. UNDERSTAND. First, read and reread the problem a few times. Next, guess a solution. Suppose that we need 7 liters of the 40% solution. Then we need $12 - 7 = 5$ liters of the 70% solution. To see if this is indeed the answer, find the amount of pure acid in 7 liters of the 40% solution, in 5 liters of the 70% solution, and in 12 liters of a 50% solution, the required amount and strength.

number of liters	×	acid strength	=	amount of pure acid
7 liters	×	40%	=	7(0.40) or 2.8 liters
5 liters	×	70%	=	5(0.70) or 3.5 liters
12 liters	×	50%	=	12(0.50) or 6 liters

Answer

1. 2 liters of the 20% solution; 4 liters of the 50% solution

Since 2.8 liters + 3.5 liters = 6.3 liters and not 6, our guess is incorrect, but we have gained some valuable insight into how to model and check this problem.
Let

x = number of liters of 40% solution; then
$12 - x$ = number of liters of 70% solution.

2. **TRANSLATE.** To help us translate to an equation, the following table summarizes the information given. Recall that the amount of acid in each solution is found by multiplying the acid strength of each solution by the number of liters.

	No. of Liters	· Acid Strength	= Amount of Acid
40% Solution	x	40%	$0.40x$
70% Solution	$12 - x$	70%	$0.70(12 - x)$
50% Solution Needed	12	50%	$0.50(12)$

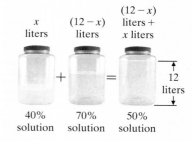

The amount of acid in the final solution is the sum of the amounts of acid in the two beginning solutions.

In words: acid in 40% solution + acid in 70% solution = acid in 50% mixture

Translate: $\quad 0.40x \quad + \quad 0.70(12 - x) \quad = \quad 0.50(12)$

3. **SOLVE.**

$$0.40x + 0.70(12 - x) = 0.50(12)$$
$$0.4x + 8.4 - 0.7x = 6 \quad \text{Apply the distributive property.}$$
$$-0.3x + 8.4 = 6 \quad \text{Combine like terms.}$$
$$-0.3x = -2.4 \quad \text{Subtract 8.4 from both sides.}$$
$$x = 8 \quad \text{Divide both sides by } -0.3.$$

4. **INTERPRET.**

Check: To check, recall how we checked our guess.

State: If 8 liters of the 40% solution are mixed with $12 - 8$ or 4 liters of the 70% solution, the result is 12 liters of a 50% solution.

Work Practice 1

The next example has to do with finding an unknown number of a certain denomination of coin or bill. These problems are extremely useful in that they help you understand the difference between the number of coins or bills and the total value of the money.

For example, suppose there are seven $5 bills. The *number* of $5 bills is 7 and the *total value* of the money is $5(7) = $35.

Study the table below for more examples.

Denomination of Coin or Bill	Number of Coins or Bills	Value of Coins or Bills
20-dollar bills	17	$20(17) = $340
nickels	31	$0.05(31) = $1.55
quarters	x	$0.25(x) = $0.25x

Practice 2

A stack of $5 and $20 bills was counted by the treasurer of an organization. The total value of the money was $1710 and there were 47 more $5 bills than $20 bills. Find the number of each type of bill.

Example 2 Finding Numbers of Denominations

Part of the proceeds from a local talent show was $2420 worth of $10 and $20 bills. If there were 37 more $20 bills than $10 bills, find the number of each denomination.

Solution:

1. UNDERSTAND the problem. To do so, read and reread the problem. If you'd like, let's guess a solution. Suppose that there are 25 $10 bills. Since there are 37 more $20 bills, we have $25 + 37 = 62$ $20 bills. The total amount of money is $10(25) + $20(62) = $1490, less than the given amount of $2420. Remember that our purpose for guessing is to help us better understand the problem.

   We are looking for the number of each denomination, so we let

   $x =$ number of $10 bills

   There are 37 more $20 bills, so

   $x + 37 =$ number of $20 bills

2. TRANSLATE. To help us translate to an equation, study the table below.

Denomination	Number of Bills	Value of Bills (in dollars)
$10 bills	x	$10x$
$20 bills	$x + 37$	$20(x + 37)$

   Since the total value of these bills is $2420, we have

   In words: value of $10 bills plus value of $20 bills is 2420

   Translate: $10x + 20(x + 37) = 2420$

3. SOLVE:
$$10x + 20x + 740 = 2420 \quad \text{Use the distributive property.}$$
$$30x + 740 = 2420 \quad \text{Add like terms.}$$
$$30x = 1680 \quad \text{Subtract 740 from both sides.}$$
$$\frac{30x}{30} = \frac{1680}{30} \quad \text{Divide both sides by 30.}$$
$$x = 56$$

4. INTERPRET the results.

   Check: Since x represents the number of $10 bills, we have 56 $10 bills and $56 + 37$, or 93, $20 bills. The total amount of these bills is $10(56) + $20(93) = $2420, the correct total.

   State: There are 56 $10 bills and 93 $20 bills.

■ Work Practice 2

Answer
2. 106 $5 bills; 59 $20 bills

Objective B Solving Distance or Uniform Motion Problems

This example involves the distance formula $d = r \cdot t$.

Example 3 Finding Time Given Rate and Distance

Marie Antonio, a bicycling enthusiast, rode her 21-speed at an average speed of 18 miles per hour on level roads and then slowed down to an average speed of 10 mph on the hilly roads of the trip. If she covered a distance of 98 miles, how long did the entire trip take if traveling the level roads took the same time as traveling the hilly roads?

Solution:

1. UNDERSTAND the problem. To do so, read and reread the problem. The formula $d = r \cdot t$ is needed. At this time, let's guess a solution. Suppose that she spent 2 hours traveling on the level roads. This means that she also spent 2 hours traveling on the hilly roads, since the times spent were the same. What is her total distance? Her distance on the level road is rate $\cdot$ time $= 18(2) = 36$ miles. Her distance on the hilly roads is rate $\cdot$ time $= 10(2) = 20$ miles. This gives a total distance of 36 miles + 20 miles = 56 miles, not the correct distance of 98 miles. Remember that the purpose of guessing a solution is not to guess correctly (although this may happen) but to help better understand the problem and how to model it with an equation. We are looking for the length of the entire trip, so we begin by letting

    x = the time spent on level roads

    Because the same amount of time was spent on hilly roads, then also

    x = the time spent on hilly roads

2. TRANSLATE. To help us translate to an equation, we now summarize the information from the problem in the following table. Fill in the rates given and the variables used to represent the times, and use the formula $d = r \cdot t$ to fill in the distance column.

	Rate	· Time =	Distance
Level	18	x	$18x$
Hilly	10	x	$10x$

    We'll draw a diagram to help visualize distance.

    Since the entire trip covered 98 miles, we have that

    In words: total distance = level distance + hilly distance
    Translate: 98 = 18x + 10x

3. SOLVE.

    $98 = 28x$ Add like terms.

    $\dfrac{98}{28} = \dfrac{28x}{28}$ Divide both sides by 28.

    $3.5 = x$

    (Continued on next page)

Practice 3

Sat Tranh took a short hike with his friends up Mt. Wachusett. They hiked uphill at a steady pace of 1.5 miles per hour, and downhill at a rate of 4 miles per hour. If the time to climb the mountain took an hour more than the time to hike down, how long did the entire hike take?

Answer
3. 2.2 hours

Practice 4

The Kansas City Southern Railway has a station in Mexico City, Mexico. Suppose two trains leave Mexico City at the same time. One travels east and the other west at a speed that is 10 mph slower. In 1.5 hours, the trains are 171 miles apart. Find the speed of each train.

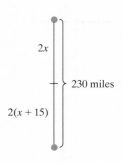

Answer
4. eastbound: 62 mph; westbound: 52 mph

4. INTERPRET the results.

Check: Recall that x represents the time spent on the level portion of the trip and also the time spent on the hilly portion. If Marie rode for 3.5 hours at 18 mph, her distance was $18(3.5) = 63$ miles. If Marie rode for 3.5 hours at 10 mph, her distance was $10(3.5) = 35$ miles. The total distance was 63 miles + 35 miles = 98 miles, the required distance.

State: The time of the entire trip was then 3.5 hours + 3.5 hours or 7 hours.

■ Work Practice 3

Example 4 Finding Train Speeds

The Kansas City Southern Railway operates in 10 states and Mexico. Suppose two trains leave Neosho, Missouri, at the same time. One travels north and the other travels south at a speed that is 15 miles per hour faster. In 2 hours, the trains are 230 miles apart. Find the speed of each train.

Solution:

1. UNDERSTAND the problem. Read and reread the problem. Guess a solution and check. Let's let

   x = speed of train traveling north

   Because the train traveling south is 15 mph faster, we have

   $x + 15$ = speed of train traveling south

2. TRANSLATE. Just as for Example 3, let's summarize our information in a table. Use the formula $d = r \cdot t$ to fill in the distance column.

	r	$\cdot$ t $=$	d
North Train	x	2	$2x$
South Train	$x + 15$	2	$2(x + 15)$

   Since the total distance between the trains is 230 miles, we have

   In words: north train distance + south train distance = total distance

   Translate: $2x + 2(x + 15) = 230$

3. SOLVE. $2x + 2x + 30 = 230$ Use the distributive property.
   $4x + 30 = 230$ Combine like terms.
   $4x = 200$ Subtract 30 from both sides.
   $\dfrac{4x}{4} = \dfrac{200}{4}$ Divide both sides by 4.
   $x = 50$ Simplify.

4. INTERPRET the results.

   Check: Recall that x is the speed of the train traveling north, or 50 mph. In 2 hours, this train travels a distance of $2(50) = 100$ miles. The speed of the train traveling south is $x + 15$ or $50 + 15 = 65$ mph. In 2 hours, the train travels $2(65) = 130$ miles. The total distance between the trains is 100 miles + 130 miles = 230 miles, the required distance.

   State: The northbound train's speed is 50 mph and the southbound train's speed is 65 mph.

■ Work Practice 4

Appendix C | Mixture and Uniform Motion Problem Solving

C Exercise Set MyLab Math

Objective A *Solve. For each exercise, a table is given for you to complete and use to write an equation that models the situation. See Example 1.*

1. How much pure acid should be mixed with 2 gallons of a 40% acid solution in order to get a 70% acid solution?

	Number of Gallons	·	Acid Strength	=	Amount of Acid
Pure Acid			100%		
40% Acid Solution					
70% Acid Solution Needed					

2. How many cubic centimeters (cc) of a 25% antibiotic solution should be added to 10 cubic centimeters of a 60% antibiotic solution in order to get a 30% antibiotic solution?

	Number of Cubic cm	·	Antibiotic Strength	=	Amount of Antibiotic
25% Antibiotic Solution					
60% Antibiotic Solution					
30% Antibiotic Solution Needed					

3. Community Coffee Company wants to create a new flavor of Cajun coffee. How many pounds of coffee worth $7 per pound should be added to 14 pounds of coffee worth $4 per pound to get a mixture worth $5 a pound?

	Number of Pounds	·	Cost per Pound	=	Value
$7 per lb Coffee					
$4 per lb Coffee					
$5 per lb Coffee Wanted					

4. Planter's Peanut Company wants to mix 20 pounds of peanuts worth $3 a pound with cashews worth $5 a pound in order to try out a mix worth $3.50 a pound. How many pounds of cashews should be added to the peanuts?

	Number of Pounds	·	Cost per Pound	=	Value
$3 per lb Peanuts					
$5 per lb Cashews					
$3.50 per lb Mixture Wanted					

Complete the table. The first and sixth rows have been completed for you. See Example 2.

		Number of Coins or Bills	Value of Coins or Bills (in dollars)
	pennies	x	$0.01x$
5.	dimes	y	
6.	quarters	z	
7.	nickels	$(x + 7)$	
8.	half-dollars	$(20 - z)$	
	$5 bills	$9x$	$5(9x)$
9.	$20 bills	$4y$	
10.	$100 bills	$97z$	
11.	$50 bills	$(35 - x)$	
12.	$10 bills	$(15 - y)$	

Solve. See Examples 1 and 2.

13. Part of the proceeds from a garage sale was $280 worth of $5 and $10 bills. If there were 20 more $5 bills than $10 bills, find the number of each denomination.

	Number of Bills	Value of Bills
$5 bills		
$10 bills		
Total		

14. A bank teller is counting $20 and $50 bills. If there are six times as many $20 bills as $50 bills and the total amount of money is $3910, find the number of each denomination.

	Number of Bills	Value of Bills
$20 bills		
$50 bills		
Total		

15. How much of an alloy that is 20% copper should be mixed with 200 ounces of an alloy that is 50% copper in order to get an alloy that is 30% copper?

16. How much water should be added to 30 gallons of a solution that is 70% antifreeze in order to get a mixture that is 60% antifreeze?

17. A new self-tanning lotion for everyday use is to be sold. First, a speculative lotion mixture is made by mixing 800 ounces of everyday moisturizing lotion worth $0.30 per ounce with self-tanning lotion worth $3 per ounce. If the speculative lotion is to cost $1.20 per ounce, how many ounces of the self-tanning lotion should be in the mixture?

18. The owner of a local chocolate shop wants to develop a new trail mix. How many pounds of chocolate-covered peanuts worth $5 a pound should be mixed with 10 pounds of granola bites worth $2 a pound to get a mixture worth $3 a pound?

Objective B *Solve. See Example 3.*

19. Two cars leave Richmond, Virginia, at the same time after visiting the nearby Richmond International Speedway. The cars travel in opposite directions, one traveling north at 56 mph and one traveling south at 47 mph. When will the two cars be 206 miles apart?

20. Two cars leave Las Vegas, Nevada, at the same time after visiting the Las Vegas Motor Speedway. The cars travel in opposite directions, one traveling northeast at 65 mph and one traveling southwest at 41 mph. When will the two cars be 530 miles apart?

Solve. See Example 4.

21. Suppose two trains leave Corpus Christi, Texas, at the same time, traveling in opposite directions. One train travels 10 mph faster than the other. In 2.5 hours, the trains are 205 miles apart. Find the speed of each train.

22. Suppose two trains leave Cleveland, Ohio, at the same time, traveling in opposite directions. One train travels 8 mph faster than the other. In 1.5 hours, the trains are 162 miles apart. Find the speed of each train.

Solve. See Examples 3 and 4.

23. A truck and a van leave the same location at the same time and travel in opposite directions. The truck's speed is 52 mph and the van's speed is 63 mph. When will the truck and the van be 460 miles apart?

24. Two cars leave the same location at the same time and travel in opposite directions. One car's speed is 65 mph and the other car's speed is 45 mph. When will the two cars be 330 miles apart?

25. Two cars leave Pecos, Texas, at the same time and both travel east on Interstate 20. The first car's speed is 70 mph and the second car's speed is 58 mph. When will the cars be 30 miles apart?

26. Two cars leave Savannah, Georgia, at the same time and both travel north on Interstate 95. The first car's speed is 40 mph and the second car's speed is 50 mph. When will the cars be 20 miles apart?

7. A jet plane traveling at 500 mph overtakes a propeller plane traveling at 200 mph that had a 2-hour head start. How far from the starting point are the planes?

9. A bus traveled on a level road for 3 hours at an average speed 20 miles per hour faster than it traveled on a winding road. The time spent on the winding road was 4 hours. Find the average speed on the level road if the entire trip was 305 miles.

1. Alan and Dave Schaferkötter leave from the same point bicycling in opposite directions, Alan bicycling at 5.5 miles per hour and Dave at 6.5 mph. How long will they be able to talk on their long-range walkie-talkies if the walkie-talkies have a 35-mile range?

3. Two hikers are 11 miles apart and walking toward each other. They meet in 2 hours. Find the rate of each hiker if one hiker walks 1.1 mph faster than the other.

5. Mark Martin can row upstream at 5 mph and downstream at 11 mph. If Mark starts rowing upstream until he gets tired and then rows downstream to his starting point, how far did Mark row if the entire trip took 4 hours?

7. A stack of $20, $50, and $100 bills was retrieved as part of an FBI investigation. There were 46 more $50 bills than $100 bills. Also, the number of $20 bills was 7 times the number of $100 bills. If the total value of the money was $9550, find the number of each type of bill.

28. How long will it take a bus traveling at 60 miles per hour to overtake a car traveling at 40 mph if the car had a 1.5-hour head start?

30. The Jones family drove to Disneyland at 50 miles per hour and returned on the same route at 40 mph. Find the distance to Disneyland if the total driving time was 7.2 hours.

32. Kathleen and Cade Williams leave simultaneously from the same point hiking in opposite directions, Kathleen walking at 4 miles per hour and Cade at 5 mph. How long can they talk on their walkie-talkies if the walkie-talkies have a 20-mile radius?

34. Nedra and Latonya Dominguez are 12 miles apart hiking toward each other. How long will it take them to meet if Nedra walks at 3 mph and Latonya walks 1 mph faster?

36. On a 255-mile trip, Gary Alessandrini traveled at an average speed of 70 mph, got a speeding ticket, and then traveled at 60 mph for the remainder of the trip. If the entire trip took 4.5 hours and the speeding ticket stop took 30 minutes, how long did Gary speed before getting stopped?

38. A man places his pocket change in a jar every day. The jar is full and his children have counted the change. The total value is $44.86. Let x represent the number of quarters, and use the information below to find the number of each type of coin.

There are:
136 more dimes than quarters
8 times as many nickels as quarters
32 more than 16 times as many pennies as quarters

Appendix D

Systems of Linear Inequalities

A **solution of a system of linear inequalities** is an ordered pair that satisfies each inequality in the system. The set of all such ordered pairs is the solution set of the system. Graphing this set gives us a picture of the solution set. We can graph a system of inequalities by graphing each inequality in the system and identifying the region of overlap.

Example 1 Graph the solutions of the system: $\begin{cases} 3x \geq y \\ x + 2y \leq 8 \end{cases}$

Solution: We begin by graphing each inequality on the *same* set of axes. The graph of the solutions of the system is the region contained in the graphs of both inequalities. In other words, it is their intersection.

First let's graph $3x \geq y$. The boundary line is the graph of $3x = y$. We sketch a solid boundary line since the inequality $3x \geq y$ means $3x > y$ or $3x = y$. The test point $(1, 0)$ satisfies the inequality, so we shade the half-plane that includes $(1, 0)$.

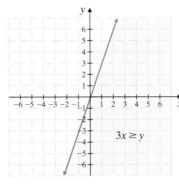
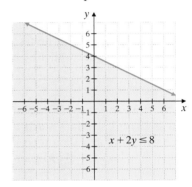

Next we sketch a solid boundary line $x + 2y = 8$ on the same set of axes. The test point $(0, 0)$ satisfies the inequality $x + 2y \leq 8$, so we shade the half-plane that includes $(0, 0)$. (For clarity, the graph of $x + 2y \leq 8$ is shown here on a separate set of axes.) An ordered pair solution of the system must satisfy both inequalities. These solutions are points that lie in both shaded regions. The solution of the system is the darkest shaded region. This solution includes parts of both boundary lines.

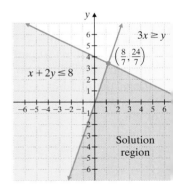

In linear programming, it is sometimes necessary to find the coordinates of the **corner point**: the point at which the two boundary lines intersect. To find the corner point for the system of Example 1, we solve the related linear system

$$\begin{cases} 3x = y \\ x + 2y = 8 \end{cases}$$

using either the substitution or the elimination method. The lines intersect at $\left(\dfrac{8}{7}, \dfrac{24}{7}\right)$, the corner point of the graph.

Graphing the Solutions of a System of Linear Inequalities

Step 1: Graph each inequality in the system on the same set of axes.

Step 2: The solutions of the system are the points common to the graphs of all the inequalities in the system.

Example 2 Graph the solutions of the system: $\begin{cases} x - y < 2 \\ x + 2y > -1 \end{cases}$

Solution: Graph both inequalities on the same set of axes. Both boundary lines are dashed lines since the inequality symbols are $<$ and $>$. The solutions of the system are the regions shown by the darkest shading. In this example, the boundary lines are not a part of the solution.

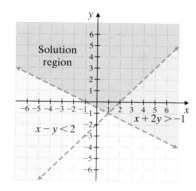

Example 3 Graph the solutions of the system: $\begin{cases} -3x + 4y < 12 \\ x \geq 2 \end{cases}$

Solution: Graph both inequalities on the same set of axes.

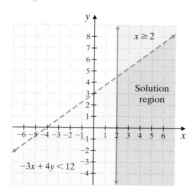

The solutions of the system are the darkest shaded regions, including a portion of the line $x = 2$.

D Exercise Set MyLab Math

Graph the solutions of each system of linear inequalities. See Examples 1 through 3.

1. $\begin{cases} y \geq x + 1 \\ y \geq 3 - x \end{cases}$

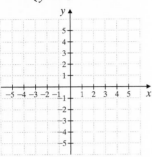

2. $\begin{cases} y \geq x - 3 \\ y \geq -1 - x \end{cases}$

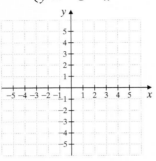

3. $\begin{cases} y < 3x - 4 \\ y \leq x + 2 \end{cases}$

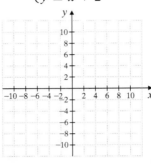

4. $\begin{cases} y \leq 2x + 1 \\ y > x + 2 \end{cases}$

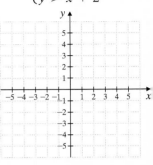

5. $\begin{cases} y < -2x - 2 \\ y > x + 4 \end{cases}$

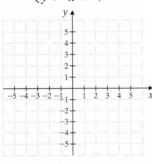

6. $\begin{cases} y \leq 2x + 4 \\ y \geq -x - 5 \end{cases}$

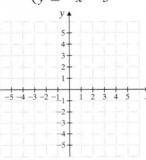

7. $\begin{cases} y \geq -x + 2 \\ y \leq 2x + 5 \end{cases}$

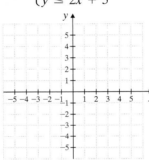

8. $\begin{cases} y \geq x - 5 \\ y \leq -3x + 3 \end{cases}$

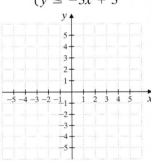

9. $\begin{cases} x \geq 3y \\ x + 3y \leq 6 \end{cases}$

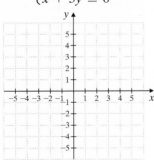

10. $\begin{cases} -2x < y \\ x + 2y < 3 \end{cases}$

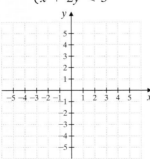

11. $\begin{cases} y + 2x \geq 0 \\ 5x - 3y \leq 12 \end{cases}$

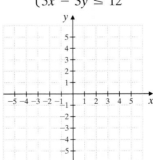

12. $\begin{cases} y + 2x \leq 0 \\ 5x + 3y \geq -2 \end{cases}$

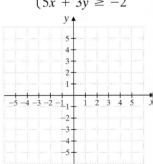

Appendix D | Systems of Linear Inequalities

13. $\begin{cases} 3x - 4y \geq -6 \\ 2x + y \leq 7 \end{cases}$

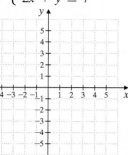

14. $\begin{cases} 4x - y \geq -2 \\ 2x + 3y \leq -8 \end{cases}$

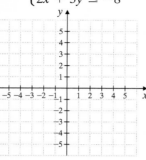

15. $\begin{cases} x \leq 2 \\ y \geq -3 \end{cases}$

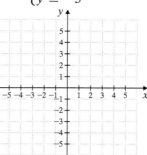

16. $\begin{cases} x \geq -3 \\ y \geq -2 \end{cases}$

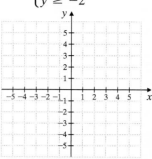

17. $\begin{cases} y \geq 1 \\ x < -3 \end{cases}$

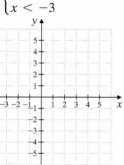

18. $\begin{cases} y > 2 \\ x \geq -1 \end{cases}$

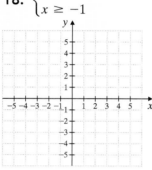

19. $\begin{cases} 2x + 3y < -8 \\ x \geq -4 \end{cases}$

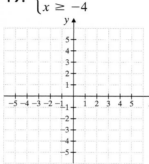

20. $\begin{cases} 3x + 2y \leq 6 \\ x < 2 \end{cases}$

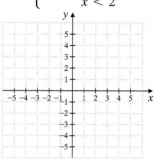

21. $\begin{cases} 2x - 5y \leq 9 \\ y \leq -3 \end{cases}$

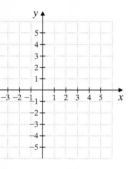

22. $\begin{cases} 2x + 5y \leq -10 \\ y \geq 1 \end{cases}$

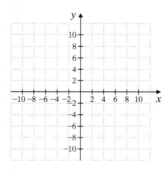

23. $\begin{cases} y \geq \dfrac{1}{2}x + 2 \\ y \leq \dfrac{1}{2}x - 3 \end{cases}$

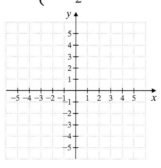

24. $\begin{cases} y \geq -\dfrac{3}{2}x + 3 \\ y < -\dfrac{3}{2}x + 6 \end{cases}$

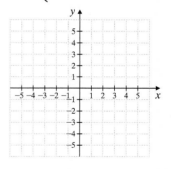

Match each system of inequalities to the corresponding graph.

A 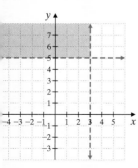 **B** 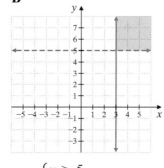 **C** 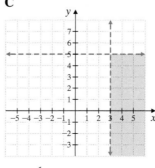 **D**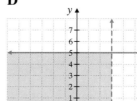

25. $\begin{cases} y < 5 \\ x > 3 \end{cases}$

26. $\begin{cases} y > 5 \\ x < 3 \end{cases}$

27. $\begin{cases} y \leq 5 \\ x < 3 \end{cases}$

28. $\begin{cases} y > 5 \\ x \geq 3 \end{cases}$

Appendix E: Geometric Formulas

Rectangle
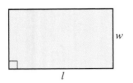
Perimeter: $P = 2l + 2w$
Area: $A = lw$

Square

Perimeter: $P = 4s$
Area: $A = s^2$

Triangle
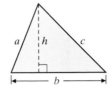
Perimeter: $P = a + b + c$
Area: $A = \frac{1}{2}bh$

Sum of Angles of Triangle
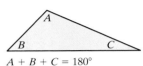
$A + B + C = 180°$
The sum of the measures of the three angles is 180°.

Pythagorean Theorem (for right triangles)
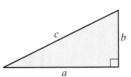
Perimeter: $P = a + b + c$
Area: $A = \frac{1}{2}ab$
One 90° (right) angle

Isosceles Triangle

Triangle has:
two equal sides and
two equal angles.

Equilateral Triangle

Triangle has:
three equal sides and
three equal angles.
Measure of each angle is 60°.

Trapezoid
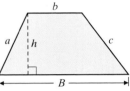
Perimeter: $P = a + b + c + B$
Area: $A = \frac{1}{2}h(B + b)$

Parallelogram
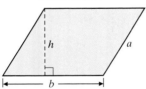
Perimeter: $P = 2a + 2b$
Area: $A = bh$

Circle

Circumference: $C = \pi d$
$C = 2\pi r$
Area: $A = \pi r^2$

Rectangular Solid

Volume: $V = LWH$
Surface Area:
$S = 2LW + 2HL + 2HW$

Cube

Volume: $V = s^3$
Surface Area: $S = 6s^2$

Cone

Volume: $V = \frac{1}{3}\pi r^2 h$
Surface Area:
$S = \pi r \sqrt{r^2 + h^2} + \pi r^2$

Right Circular Cylinder

Volume: $V = \pi r^2 h$
Surface Area: $S = 2\pi r^2 + 2\pi rh$

Sphere

Volume: $V = \frac{4}{3}\pi r^3$
Surface Area: $S = 4\pi r^2$

Square-Based Pyramid

Volume: $V = \frac{1}{3}s^2 h$

Contents of Student Resources

Study Skills Builders

Attitude and Study Tips:
1. Have You Decided to Complete This Course Successfully?
2. Tips for Studying for an Exam
3. What to Do the Day of an Exam
4. Are You Satisfied with Your Performance on a Particular Quiz or Exam?
5. How Are You Doing?
6. Are You Preparing for Your Final Exam?

Organizing Your Work:
7. Learning New Terms
8. Are You Organized?
9. Organizing a Notebook
10. How Are Your Homework Assignments Going?

MyLab Math and MathXL:
11. Tips for Turning in Your Homework on Time
12. Tips for Doing Your Homework Online
13. Organizing Your Work
14. Getting Help with Your Homework Assignments
15. Tips for Preparing for an Exam
16. How Well Do You Know the Resources Available to You in MyLab Math?

Additional Help Inside and Outside Your Textbook:
17. How Well Do You Know Your Textbook?
18. Are You Familiar with Your Textbook Supplements?
19. Are You Getting All the Mathematics Help That You Need?

Bigger Picture—Study Guide Outline

Practice Final Exam

Answers to Selected Exercises

Student Resources

Study Skills Builders

Attitude and Study Tips

Study Skills Builder 1

Have You Decided to Complete This Course Successfully?

Ask yourself if one of your current goals is to complete this course successfully.

If it is not a goal of yours, ask yourself why. One common reason is fear of failure. Amazingly enough, fear of failure alone can be strong enough to keep many of us from doing our best in any endeavor.

Another common reason is that you simply haven't taken the time to think about or write down your goals for this course. To help accomplish this, answer the questions below.

Exercises

1. Write down your goal(s) for this course.

2. Now list steps you will take to make sure your goal(s) in Exercise **1** are accomplished.

3. Rate your commitment to this course with a number between 1 and 5. Use the diagram below to help.

High Commitment		Average Commitment		Not Committed at All
5	4	3	2	1

4. If you have rated your personal commitment level (from the exercise above) as a 1, 2, or 3, list the reasons why this is so. Then determine whether it is possible to increase your commitment level to a 4 or 5.

   Good luck, and don't forget that a positive attitude will make a big difference.

Study Skills Builder 2

Tips for Studying for an Exam
To prepare for an exam, try the following study techniques:

- Start the study process days before your exam.
- Make sure that you are up to date on your assignments.
- If there is a topic that you are unsure of, use one of the many resources that are available to you. For example,

   See your instructor.
   View a lecture video on the topic.
   Visit a learning resource center on campus.
   Read the textbook material and examples on the topic.

- Reread your notes and carefully review the Chapter Highlights at the end of any chapter.
- Work the review exercises at the end of the chapter.
- Find a quiet place to take the Chapter Test found at the end of the chapter. Do not use any resources when taking this sample test. This way, you will have a clear indication of how prepared you are for your exam. Check your answers and use the Chapter Test Prep Videos to make sure that you correct any missed exercises.

Good luck, and keep a positive attitude.

Exercises

Let's see how you did on your last exam.

1. How many days before your last exam did you start studying for that exam?
2. Were you up to date on your assignments at that time or did you need to catch up on assignments?
3. List the most helpful text supplement (if you used one).
4. List the most helpful campus supplement (if you used one).
5. List your process for preparing for a mathematics test.
6. Was this process helpful? In other words, were you satisfied with your performance on your exam?
7. If not, what changes can you make in your process that will make it more helpful to you?

Study Skills Builder 3

What to Do the Day of an Exam
Your first exam may be soon. On the day of an exam, don't forget to try the following:

- Allow yourself plenty of time to arrive.
- Read the directions on the test carefully.
- Read each problem carefully as you take your test. Make sure that you answer the question asked.
- Watch your time and pace yourself so that you may attempt each problem on your test.
- Check your work and answers.
- **Do not turn your test in early.** If you have extra time, spend it double-checking your work.

Good luck!

Exercises

Answer the following questions based on your most recent mathematics exam, whenever that was.

1. How soon before class did you arrive?
2. Did you read the directions on the test carefully?
3. Did you make sure you answered the question asked for each problem on the exam?
4. Were you able to attempt each problem on your exam?
5. If your answer to Exercise **4** is no, list reasons why.
6. Did you have extra time on your exam?
7. If your answer to Exercise **6** is yes, describe how you spent that extra time.

Study Skills Builder 4

Are You Satisfied with Your Performance on a Particular Quiz or Exam?

If not, don't forget to analyze your quiz or exam and look for common errors. Were most of your errors a result of:

- *Carelessness?* Did you turn in your quiz or exam before the allotted time expired? If so, resolve to use any extra time to check your work.
- *Running out of time?* Answer the questions you are sure of first. Then attempt the questions you are unsure of, and delay checking your work until all questions have been answered.
- *Not understanding a concept?* If so, review that concept and correct your work so that you make sure you understand it before the next quiz or the final exam.
- *Test conditions?* When studying for a quiz or exam, make sure you place yourself in conditions similar to test conditions. For example, before your next quiz or exam, take a sample test without the aid of your notes or text.

(For a sample test, see your instructor or use the Chapter Test at the end of each chapter.)

Exercises

1. Have you corrected all your previous quizzes and exams?
2. List any errors you have found common to two or more of your graded papers.
3. Is one of your common errors not understanding a concept? If so, are you making sure you understand all the concepts for the next quiz or exam?
4. Is one of your common errors making careless mistakes? If so, are you now taking all the time allotted to check over your work so that you can minimize the number of careless mistakes?
5. Are you satisfied with your grades thus far on quizzes and tests?
6. If your answer to Exercise 5 is no, are there any more suggestions you can make to your instructor or yourself to help? If so, list them here and share them with your instructor.

Study Skills Builder 5

How Are You Doing?

If you haven't done so yet, take a few moments and think about how you are doing in this course. Are you working toward your goal of successfully completing this course? Is your performance on homework, quizzes, and tests satisfactory? If not, you might want to see your instructor to see if he/she has any suggestions on how you can improve your performance. Reread Section 1.1 for ideas on places to get help with your mathematics course.

Exercises

Answer the following.

1. List any textbook supplements you are using to help you through this course.
2. List any campus resources you are using to help you through this course.
3. Write a short paragraph describing how you are doing in your mathematics course.
4. If improvement is needed, list ways that you can work toward improving your situation as described in Exercise 3.

Study Skills Builder 6

Are You Preparing for Your Final Exam?

To prepare for your final exam, try the following study techniques:

- Review the material that you will be responsible for on your exam. This includes material from your textbook, your notebook, and any handouts from your instructor.
- Review any formulas that you may need to memorize.
- Check to see if your instructor or mathematics department will be conducting a final exam review.
- Check with your instructor to see whether final exams from previous semesters/quarters are available to students for review.
- Use your previously taken exams as a practice final exam. To do so, rewrite the test questions in mixed order on blank sheets of paper. This will help you prepare for exam conditions.
- If you are unsure of a few concepts, see your instructor or visit a learning lab for assistance. Also, view the video segment of any troublesome sections.
- If you need further exercises to work, try the Cumulative Reviews at the end of the chapters.

Once again, good luck! I hope you are enjoying this textbook and your mathematics course.

Organizing Your Work

Study Skills Builder 7

Learning New Terms

Many of the terms used in this text may be new to you. It will be helpful to make a list of new mathematical terms and symbols as you encounter them and to review them frequently. Placing these new terms (including page references) on 3 × 5 index cards might help you later when you're preparing for a quiz.

Exercises

1. Name one way you might place a word and its definition on a 3 × 5 card.

2. How do new terms stand out in this text so that they can be found?

Study Skills Builder 8

Are You Organized?

Have you ever had trouble finding a completed assignment? When it's time to study for a test, are your notes neat and organized? Have you ever had trouble reading your own mathematics handwriting? (Be honest—I have.)

When any of these things happen, it's time to get organized. Here are a few suggestions:

- Write your notes and complete your homework assignments in a notebook with pockets (spiral or ring binder).
- Take class notes in this notebook, and then follow the notes with your completed homework assignment.
- When you receive graded papers or handouts, place them in the notebook pocket so that you will not lose them.
- Mark (possibly with an exclamation point) any note(s) that seem extra important to you.
- Mark (possibly with a question mark) any notes or homework that you are having trouble with.
- See your instructor or a math tutor for help with the concepts or exercises that you are having trouble understanding.
- If you are having trouble reading your own handwriting, *slow down* and write your mathematics work clearly!

Exercises

1. Have you been completing your assignments on time?

2. Have you been correcting any exercises you may be having difficulty with?

3. If you are having trouble understanding a mathematical concept or correcting any homework exercises, have you visited your instructor, a tutor, or your campus math lab?

4. Are you taking lecture notes in your mathematics course? (By the way, these notes should include worked-out examples solved by your instructor.)

5. Is your mathematics course material (handouts, graded papers, lecture notes) organized?

6. If your answer to Exercise **5** is no, take a moment and review your course material. List at least two ways that you might better organize it.

Study Skills Builder 9

Organizing a Notebook

It's never too late to get organized. If you need ideas about organizing a notebook for your mathematics course, try some of these:

- Use a spiral or ring binder notebook with pockets and use it for mathematics only.
- Start each page by writing the book's section number you are working on at the top.
- When your instructor is lecturing, take notes. *Always* include any examples your instructor works for you.
- Place your worked-out homework exercises in your notebook immediately after the lecture notes from that section. This way, a section's worth of material is together.
- Homework exercises: Attempt and check all assigned homework.
- Place graded quizzes in the pockets of your notebook or a special section of your binder.

Exercises

Check your notebook organization by answering the following questions.

1. Do you have a spiral or ring binder notebook for your mathematics course only?
2. Have you ever had to flip through several sheets of notes and work in your mathematics notebook to determine what section's work you are in?
3. Are you now writing the textbook's section number at the top of each notebook page?
4. Have you ever lost or had trouble finding a graded quiz or test?
5. Are you now placing all your graded work in a dedicated place in your notebook?
6. Are you attempting all of your homework and placing all of your work in your notebook?
7. Are you checking and correcting your homework in your notebook? If not, why not?
8. Are you writing in your notebook the examples your instructor works for you in class?

Study Skills Builder 10

How Are Your Homework Assignments Going?

It is very important in mathematics to keep up with homework. Why? Many concepts build on each other. Often your understanding of a day's concepts depends on an understanding of the previous day's material.

Remember that completing your homework assignment involves a lot more than attempting a few of the problems assigned.

To complete a homework assignment, remember these four things:

- Attempt all of it.
- Check it.
- Correct it.
- If needed, ask questions about it.

Exercises

Take a moment and review your completed homework assignments. Answer the questions below based on this review.

1. Approximate the fraction of your homework you have attempted.
2. Approximate the fraction of your homework you have checked (if possible).
3. If you are able to check your homework, have you corrected it when errors have been found?
4. When working homework, if you do not understand a concept, what do you do?

MyLab Math and MathXL

Study Skills Builder 11

Tips for Turning in Your Homework on Time

It is very important to keep up with your mathematics homework assignments. Why? Many concepts in mathematics build upon each other.

Remember these four tips to help ensure your work is completed on time:

- Know the assignments and due dates set by your instructor.
- Do not wait until the last minute to submit your homework.
- Set a goal to submit your homework 6–8 hours before the scheduled due date in case you have unexpected technology trouble.
- Schedule enough time to complete each assignment.

Following the tips above will also help you avoid potentially losing points for late or missed assignments.

Exercises

Take a moment to consider your work on your homework assignments to date and answer the following questions:

1. What percentage of your assignments have you turned in on time?
2. Why might it be a good idea to submit your homework 6–8 hours before the scheduled deadline?
3. If you have missed submitting any homework by the due date, list some of the reasons why this occurred.
4. What steps do you plan to take in the future to ensure your homework is submitted on time?

Study Skills Builder 12

Tips for Doing Your Homework Online

Practice is one of the main keys to success in any mathematics course. Did you know that MyLab Math/MathXL provides you with **immediate feedback** for each exercise? If you are incorrect, you are given hints to work the exercise correctly. You have **unlimited practice opportunities** and can rework any exercises you have trouble with until you master them, and submit homework assignments unlimited times before the deadline.

Remember these success tips when doing your homework online:

- Attempt all assigned exercises.
- Write down (neatly) your step-by-step work for each exercise before entering your answer.
- Use the immediate feedback provided by the program to help you check and correct your work for each exercise.
- Rework any exercises you have trouble with until you master them.
- Work through your homework assignment as many times as necessary until you are satisfied.

Exercises

Take a moment to think about your homework assignments to date and answer the following:

1. Have you attempted all assigned exercises?
2. Of the exercises attempted, have you also written out your work before entering your answer—so that you can check it?
3. Are you familiar with how to enter answers using the MathXL player so that you avoid answer-entry-type errors?
4. List some ways the immediate feedback and practice supports have helped you with your homework. If you have not used these supports, how do you plan to use them with the success tips above on your next assignment?

Study Skills Builder 13

Organizing Your Work

Have you ever used any readily available paper (such as the back of a flyer, another course assignment, Post-its, etc.) to work out your homework exercises before entering the answer in MathXL? To save time, have you ever entered answers directly into MathXL without working the exercises on paper? When it's time to study, have you ever been unable to find your completed work or read and follow your own mathematics handwriting?

When any of these things happen, it's time to get organized. Here are some suggestions:

- Write your step-by-step work for each homework exercise (neatly) on lined, loose-leaf paper and keep this in a 3-ring binder.
- Refer to your step-by-step work when you receive feedback that your answer is incorrect in MathXL. Double-check using the steps and hints provided by the program and correct your work accordingly.
- Keep your written homework with your class notes for that section.
- Identify any exercises you are having trouble with and ask questions about them.
- Keep all graded quizzes and tests in this binder as well to study later.

If you follow the suggestions above, you and your instructor or tutor will be able to follow your steps and correct any mistakes. You will also have a written copy of your work to refer to later to ask questions and study for tests.

Exercises

1. Why is it important that you write out your step-by-step work for homework exercises and keep a hard copy of all work submitted online?
2. If you have gotten an incorrect answer, are you able to follow your steps and find your error?
3. If you were asked today to review your previous homework assignments and first test, could you find them? If not, list some ways you might better organize your work.

Study Skills Builder 14

Getting Help with Your Homework Assignments

There are many resources available to you through MathXL to help you work through any homework exercises you may have trouble with. It is important that you know what these resources are and know when and how to use them.

Let's review the features found in the homework exercises:

- **Help Me Solve This**—provides step-by-step help for the exercise you are working. You must work an additional exercise of the same type (without this help) before you can get credit for having worked it correctly.
- **View an Example**—allows you to view a correctly worked exercise similar to the one you are having trouble with. You can then go back to your original exercise and work it on your own.
- **E-Book**—allows you to read examples from your text and find similar exercises.
- **Video**—your text author, Elayn Martin-Gay, works an exercise similar to the one you need help with. **Not all exercises have an accompanying video clip.
- **Ask My Instructor**—allows you to e-mail your instructor for help with an exercise.

Exercises

1. How does the "Help Me Solve This" feature work?
2. If the "View an Example" feature is used, is it necessary to work an additional problem before continuing the assignment?
3. When might be a good time to use the "Video" feature? Do all exercises have an accompanying video clip?
4. Which of the features above have you used? List those you found the most helpful to you.
5. If you haven't used the features discussed, list those you plan to try on your next homework assignment.

Study Skills Builder 15

Tips for Preparing for an Exam

Did you know that you can rework your previous homework assignments in MyLab Math and MathXL? This is a great way to prepare for tests. To do this, open a previous homework assignment and click "similar exercise." This will generate new exercises similar to the homework you have submitted. You can then rework the exercises and assignments until you feel confident that you understand them.

To prepare for an exam, follow these tips:

- Review your written work for your previous homework assignments along with your class notes.
- Identify any exercises or topics that you have questions on or have difficulty understanding.
- Rework your previous assignments in MyLab Math and MathXL until you fully understand them and can do them without help.
- Get help for any topics you feel unsure of or for which you have questions.

Exercises

1. Are your current homework assignments up to date and is your written work for them organized in a binder or notebook? If the answer is no, it's time to get organized. For tips on this, see Study Skills Builder 13—Organizing Your Work.

2. How many days in advance of an exam do you usually start studying?

3. List some ways you think that working previous homework assignments can help you prepare for your test.

4. List two or three resources you can use to get help for any topics you are unsure of or have questions on.

Good luck!

Study Skills Builder 16

How Well Do You Know the Resources Available to You in MyLab Math?

There are many helpful resources available to you in MyLab Math. Let's take a moment to locate and explore a few of them now. Go into your MyLab Math course, and visit the Multimedia Library, Tools for Success, and E-Book.

Let's see what you found.

Exercises

1. List the resources available to you in the Multimedia Library.

2. List the resources available to you in the Tools for Success folder.

3. Where did you find the English/Spanish Audio Glossary?

4. Can you view videos from the E-Book?

5. Did you find any resources you did not know about? If so, which ones?

6. Which resources have you used most often or find most helpful?

Additional Help Inside and Outside Your Textbook

Study Skills Builder 17

How Well Do You Know Your Textbook?

The questions below will help determine whether you are familiar with your textbook. For additional information, see Section 1.1 in this text.

1. What does the ⊙ icon mean?
2. What does the ⟍ icon mean?
3. What does the △ icon mean?
4. Where can you find a review for each chapter? What answers to this review can be found in the back of your text?
5. Each chapter contains an overview of the chapter along with examples. What is this feature called?
6. Each chapter contains a review of vocabulary. What is this feature called?
7. There are practice exercises that are contained in this text. What are they and how can they be used?
8. This text contains a student section in the back entitled Student Resources. List the contents of this section and how they might be helpful.
9. What exercise answers are available in this text? Where are they located?

Study Skills Builder 18

Are You Familiar with Your Textbook Supplements?
Below is a review of some of the student supplements available for additional study. Check to see if you are using the ones most helpful to you.

- Chapter Test Prep Videos. These videos provide video clip solutions to the Chapter Test exercises in this text. You will find them extremely useful when studying for tests or exams.
- Interactive DVD Lecture Series. These are keyed to each section of the text. The material is presented by me, Elayn Martin-Gay, and I have placed a ▶ by the exercises in the text that I have worked on the video.
- The *Student Solutions Manual*. This contains worked-out solutions to odd-numbered exercises as well as every exercise in the Integrated Reviews, Chapter Reviews, Chapter Tests, and Cumulative Reviews.
- Pearson Tutor Center. Mathematics questions may be phoned, faxed, or e-mailed to this center.
- MyLab Math is a text-specific online course. MathXL is an online homework, tutorial, and assessment system.

Take a moment and determine whether these are available to you.

As usual, your instructor is your best source of information.

Exercises

Let's see how you are doing with textbook supplements.

1. Name one way the Lecture Videos can be helpful to you.
2. Name one way the Chapter Test Prep Videos can help you prepare for a chapter test.
3. List any textbook supplements that you have found useful.
4. Have you located and visited a learning resource lab located on your campus?
5. List the textbook supplements that are currently housed in your campus' learning resource lab.

Study Skills Builder 19

Are You Getting All the Mathematics Help That You Need?
Remember that, in addition to your instructor, there are many places to get help with your mathematics course. For example:

- This text has an accompanying video lesson for every section, and the CD in this text contains worked-out solutions to every Chapter Test exercise.
- The back of the book contains answers to odd-numbered exercises.
- A *Student Solutions Manual* is available that contains worked-out solutions to odd-numbered exercises as well as solutions to every exercise in the Integrated Reviews, Chapter Reviews, Chapter Tests, and Cumulative Reviews.
- Don't forget to check with your instructor for other local resources available to you, such as a tutor center.

Exercises

1. List items you find helpful in the text and all student supplements to this text.
2. List all the campus help that is available to you for this course.
3. List any help (besides the textbook) from Exercises **1** and **2** above that you are using.
4. List any help (besides the textbook) that you feel you should try.
5. Write a goal for yourself that includes trying everything you listed in Exercise **4** during the next week.

Bigger Picture— Study Guide Outline

OUTLINE: PART 1

Operations on Sets of Numbers and Solving Equations

I. Operations on Sets of Numbers

  A. Whole Numbers

    1. **Add or Subtract:** (Sec. 1.3)

$$\begin{array}{r} 14 \\ +39 \\ \hline 53 \end{array} \qquad \begin{array}{r} 300 \\ -27 \\ \hline 273 \end{array}$$

    2. **Multiply or Divide:** (Sec. 1.5, 1.6)

$$\begin{array}{r} 238 \\ \times\ 47 \\ \hline 1666 \\ 9520 \\ \hline 11{,}186 \end{array} \qquad \begin{array}{r} 127\ \text{R}\ 2 \\ 7\overline{)891} \\ \underline{-7} \\ 19 \\ \underline{-14} \\ 51 \\ \underline{-49} \\ 2 \end{array}$$

    3. **Exponent:** (Sec. 1.7)

$$3^4 = \overbrace{3 \cdot 3 \cdot 3 \cdot 3}^{4\ \text{factors of 3}} = 81$$

    4. **Order of Operations:** (Sec. 1.7)

$$\begin{aligned} 24 \div 3 \cdot 2 - (2 + 8) &= 24 \div 3 \cdot 2 - (10) &&\text{Simplify within parentheses.} \\ &= 8 \cdot 2 - 10 &&\text{Multiply or divide from left to right.} \\ &= 16 - 10 &&\text{Multiply or divide from left to right.} \\ &= 6 &&\text{Add or subtract from left to right.} \end{aligned}$$

    5. **Square Root:** (Sec. 7.3)

$$\sqrt{25} = 5 \text{ because } 5 \cdot 5 = 25 \text{ and 5 is a positive number.}$$

  B. Integers

    1. **Add:** (Sec. 2.2) $-5 + (-2) = -7$ Adding like signs: Add absolute values. Attach the common sign.

      $-5 + 2 = -3$ Adding unlike signs: Subtract absolute values. Attach the sign of the number with the larger absolute value.

    2. **Subtract:** Add the first number to the opposite of the second number. (Sec. 2.3)

$$7 - 10 = 7 + (-10) = -3$$

    3. **Multiply or Divide:** Multiply or divide as usual. If the signs of the two numbers are the same, the answer is positive. If the signs of the two numbers are different, the answer is negative. (Sec. 2.4)

$$-5 \cdot 5 = -25, \quad \frac{-32}{-8} = 4$$

C. **Fractions**
   1. **Simplify:** Factor the numerator and denominator. Then divide out factors of 1 by dividing out common factors in the numerator and denominator. (Sec. 4.2)

   Simplify: $\dfrac{20}{28} = \dfrac{4 \cdot 5}{4 \cdot 7} = \dfrac{5}{7}$

   2. **Multiply:** Numerator times numerator over denominator times denominator. (Sec. 4.3)

   $\dfrac{5}{9} \cdot \dfrac{2}{7} = \dfrac{10}{63}$

   3. **Divide:** First fraction times the reciprocal of the second fraction. (Sec. 4.3)

   $\dfrac{2}{11} \div \dfrac{3}{4} = \dfrac{2}{11} \cdot \dfrac{4}{3} = \dfrac{8}{33}$

   4. **Add or Subtract:** Must have same denominators. If not, find the LCD, and write each fraction as an equivalent fraction with the LCD as denominator. (Sec. 4.4, Sec. 4.5)

   $\dfrac{2}{5} + \dfrac{1}{15} = \dfrac{2}{5} \cdot \dfrac{3}{3} + \dfrac{1}{15} = \dfrac{6}{15} + \dfrac{1}{15} = \dfrac{7}{15}$

D. **Decimals**
   1. **Add or Subtract:** Line up decimal points. (Sec. 5.2)

   $\begin{array}{r} 1.27 \\ +\ 0.6 \\ \hline 1.87 \end{array}$

   2. **Multiply:** (Sec. 5.3)

   $\begin{array}{r} 2.56 \\ \times\ 3.2 \\ \hline 512 \\ 7680 \\ \hline 8.192 \end{array}$ 2 decimal places
   1 decimal place
   $2 + 1 = 3$
   3 decimal places

   3. **Divide:** (Sec. 5.4) $8\overline{)5.6}$ gives 0.7; $0.6\overline{)0.7\,86}$ gives 1.36

II. **Solving Equations**

   A. **Equations in General:** Simplify both sides of the equation by removing parentheses and adding any like terms. Then use the addition property to write variable terms on one side and constants (or numbers) on the other side. Then use the multiplication property to solve for the variable by dividing both sides of the equation by the coefficient of the variable. (Sec. 3.3)

   Solve: $2(x - 5) = 80$
   $2x - 10 = 80$ Use the distributive property.
   $2x - 10 + 10 = 80 + 10$ Add 10 to both sides.
   $2x = 90$ Simplify.
   $\dfrac{2x}{2} = \dfrac{90}{2}$ Divide both sides by 2.
   $x = 45$ Simplify.

   B. **Proportions:** Set cross products equal to each other. Then solve. (Sec. 6.1)

   $\dfrac{14}{3} = \dfrac{2}{n}$ or $14 \cdot n = 3 \cdot 2$ or $14 \cdot n = 6$ or $n = \dfrac{6}{14} = \dfrac{3}{7}$

C. **Percent Problems**
1. **Solved by Equations:** Remember that "of" means multiplication and "i[s]" means equals. (Sec. 6.3)
   "12% of some number is 6" translates to
   $$12\% \cdot n = 6 \text{ or } 0.12 \cdot n = 6 \text{ or } n = \frac{6}{0.12} \text{ or } n = 50$$
2. **Solved by Proportions:** Remember that percent, p, is identified by % o[r] "percent"; base, b, usually appears after "of"; and amount, a, is the pa[rt] compared to the whole. (Sec. 6.4)
   "12% of some number is 6" translates to
   $$\frac{6}{b} = \frac{12}{100} \text{ or } 6 \cdot 100 = b \cdot 12 \text{ or } \frac{600}{12} = b \text{ or } 50 = b$$

OUTLINE: PART 2

Simplifying Expressions and Solving Equations and Inequalities

I. **Simplifying Expressions**

A. **Exponents** (Sec. 10.1, 10.2)

$$x^7 \cdot x^5 = x^{12}; \quad (x^7)^5 = x^{35}; \quad \frac{x^7}{x^5} = x^2; \quad x^0 = 1; \quad 8^{-2} = \frac{1}{8^2} = \frac{1}{64}$$

B. **Polynomials**
1. **Add:** Combine like terms. (Sec. 10.4)
   $$(3y^2 + 6y + 7) + (9y^2 - 11y - 15) = 3y^2 + 6y + 7 + 9y^2 - 11y -$$
   $$= 12y^2 - 5y - 8$$
2. **Subtract:** Change the sign of the terms of the polynomial being subtracte[d] then add. (Sec. 10.4)
   $$(3y^2 + 6y + 7) - (9y^2 - 11y - 15) = 3y^2 + 6y + 7 - 9y^2 + 11y +$$
   $$= -6y^2 + 17y + 22$$
3. **Multiply:** Multiply each term of one polynomial by each term of the othe[r] polynomial. (Sec. 10.5)
   $$(x + 5)(2x^2 - 3x + 4) = x(2x^2 - 3x + 4) + 5(2x^2 - 3x + 4)$$
   $$= 2x^3 - 3x^2 + 4x + 10x^2 - 15x + 20$$
   $$= 2x^3 + 7x^2 - 11x + 20$$
4. **Divide:** (Sec. 10.7)
   a. To divide by a monomial, divide each term of the polynomial by the monomial.
   $$\frac{8x^2 + 2x - 6}{2x} = \frac{8x^2}{2x} + \frac{2x}{2x} - \frac{6}{2x} = 4x + 1 - \frac{3}{x}$$

b. To divide by a polynomial other than a monomial, use long division.

$$2x + 5 \overline{)2x^2 - 7x + 10} \quad \begin{array}{c} x - 6 + \dfrac{40}{2x + 5} \end{array}$$

$$\begin{array}{r} 2x^2 + 5x \\ \hline -12x + 10 \\ -12x - 30 \\ \hline 40 \end{array}$$

C. Factoring Polynomials

See the Chapter 11 Integrated Review for steps.

$$3x^4 - 78x^2 + 75 = 3(x^4 - 26x^2 + 25) \quad \text{Factor out GCF—always first step.}$$
$$= 3(x^2 - 25)(x^2 - 1) \quad \text{Factor trinomial.}$$
$$= 3(x + 5)(x - 5)(x + 1)(x - 1) \quad \text{Factor further—each a difference of squares.}$$

D. Rational Expressions

1. **Simplify:** Factor the numerator and denominator. Then divide out factors of 1 by dividing out common factors in the numerator and denominator. (Sec. 12.1)

$$\frac{x^2 - 9}{7x^2 - 21x} = \frac{(x + 3)(x - 3)}{7x(x - 3)} = \frac{x + 3}{7x}$$

2. **Multiply:** Multiply numerators, then multiply denominators. (Sec. 12.2)

$$\frac{5z}{2z^2 - 9z - 18} \cdot \frac{22z + 33}{10z} = \frac{5 \cdot z}{(2z + 3)(z - 6)} \cdot \frac{11(2z + 3)}{2 \cdot 5 \cdot z} = \frac{11}{2(z - 6)}$$

3. **Divide:** First fraction times the reciprocal of the second fraction. (Sec. 12.2)

$$\frac{14}{x + 5} \div \frac{x + 1}{2} = \frac{14}{x + 5} \cdot \frac{2}{x + 1} = \frac{28}{(x + 5)(x + 1)}$$

4. **Add or Subtract:** Must have same denominator. If not, find the LCD and write each fraction as an equivalent fraction with the LCD as denominator. (Sec. 12.3, 12.4)

$$\frac{9}{10} - \frac{x + 1}{x + 5} = \frac{9(x + 5)}{10(x + 5)} - \frac{10(x + 1)}{10(x + 5)}$$
$$= \frac{9x + 45 - 10x - 10}{10(x + 5)} = \frac{-x + 35}{10(x + 5)}$$

E. Radicals

1. **Simplify Square Roots:** If possible, factor the radicand so that one factor is a perfect square. Then use the product rule and simplify. (Sec. 15.2)

$$\sqrt{75} = \sqrt{25 \cdot 3} = \sqrt{25} \cdot \sqrt{3} = 5\sqrt{3}$$

2. **Add or Subtract:** Only like radicals (same index and radicand) can be added or subtracted. (Sec. 15.3)

$$8\sqrt{10} - \sqrt{40} + \sqrt{5} = 8\sqrt{10} - 2\sqrt{10} + \sqrt{5} = 6\sqrt{10} + \sqrt{5}$$

3. **Multiply or Divide:** (Sec. 15.4) $\sqrt{a} \cdot \sqrt{b} = \sqrt{ab}; \dfrac{\sqrt{a}}{\sqrt{b}} = \sqrt{\dfrac{a}{b}}.$

$$\sqrt{11} \cdot \sqrt{3} = \sqrt{33}; \quad \frac{\sqrt{140}}{\sqrt{7}} = \sqrt{\frac{140}{7}} = \sqrt{20} = \sqrt{4 \cdot 5} = 2\sqrt{5}$$

4. **Rationalizing the Denominator:** (Sec. 15.4)

   a. If denominator is one term,
   $$\frac{5}{\sqrt{11}} = \frac{5 \cdot \sqrt{11}}{\sqrt{11} \cdot \sqrt{11}} = \frac{5\sqrt{11}}{11}$$

   b. If denominator is two terms, multiply by 1 in the form of $\frac{\text{conjugate of denominator}}{\text{conjugate of denominator}}$.
   $$\frac{13}{3 + \sqrt{2}} = \frac{13}{3 + \sqrt{2}} \cdot \frac{3 - \sqrt{2}}{3 - \sqrt{2}} = \frac{13(3 - \sqrt{2})}{9 - 2} = \frac{13(3 - \sqrt{2})}{7}$$

II. Solving Equations and Inequalities

A. **Linear Equations:** Power on variable is 1 and there are no variables in denominator. (Sec. 9.3)

$7(x - 3) = 4x + 6$	Linear equation (If fractions, multiply by LCD.)
$7x - 21 = 4x + 6$	Use the distributive property.
$3x - 21 = 6$	Subtract $4x$ from both sides.
$3x = 27$	Add 21 to both sides.
$x = 9$	Divide both sides by 3.

B. **Linear Inequalities:** Same as linear equation except if you multiply or divide by a negative number, then reverse direction of inequality. (Sec. 9.6)

$-4x + 11 \leq -1$	Linear inequality
$-4x \leq -12$	Subtract 11 from both sides.
$\frac{-4x}{-4} \geq \frac{-12}{-4}$	Divide both sides by -4 and reverse the direction of the inequality symbol.
$x \geq 3$	Simplify.

C. **Quadratic and Higher Degree Equations:** Solve: First write the equation in standard form (one side is 0).

   1. If the polynomial on one side factors, solve by factoring. (Sec. 11.6)
   2. If the polynomial does not factor, solve by the quadratic formula. (Sec. 16.3)

By Factoring:	**By Quadratic Formula:**
$x^2 + x = 6$	$x^2 + x = 5$
$x^2 + x - 6 = 0$	$x^2 + x - 5 = 0$
$(x - 2)(x + 3) = 0$	$a = 1, b = 1, c = -5$
$x - 2 = 0$ or $x + 3 = 0$	$x = \frac{-1 \pm \sqrt{1^2 - 4(1)(-5)}}{2 \cdot 1}$
$x = 2$ or $x = -3$	$= \frac{-1 \pm \sqrt{21}}{2}$

D. **Equations with Rational Expressions:** Make sure the proposed solution does not make a denominator 0. (Sec. 12.5)

$$\frac{3}{x} - \frac{1}{x - 1} = \frac{4}{x - 1} \quad \text{Equation with rational expressions}$$

$$x(x - 1) \cdot \frac{3}{x} - x(x - 1) \cdot \frac{1}{x - 1} = x(x - 1) \cdot \frac{4}{x - 1} \quad \text{Multiply through by } x(x - 1).$$

$$3(x-1) - x \cdot 1 = x \cdot 4 \quad \text{Simplify.}$$
$$3x - 3 - x = 4x \quad \text{Use the distributive property.}$$
$$-3 = 2x \quad \text{Simplify and move variable terms to right side.}$$
$$-\frac{3}{2} = x \quad \text{Divide both sides by 2.}$$

E. **Equations with Radicals:** To solve, isolate a radical, then square both sides. You may have to repeat this. Check possible solution(s) in the original equation. (Sec. 15.5)

$$\sqrt{x + 49} + 7 = x$$
$$\sqrt{x + 49} = x - 7 \quad \text{Subtract 7 from both sides.}$$
$$x + 49 = x^2 - 14x + 49 \quad \text{Square both sides.}$$
$$0 = x^2 - 15x \quad \text{Set terms equal to 0.}$$
$$0 = x(x - 15) \quad \text{Factor.}$$
$$\cancel{x = 0} \text{ or } x = 15 \quad \text{Set each factor equal to 0 and solve.}$$

Practice Final Exam

Chapters 1–8

Simplify by performing the indicated operations.

1. $2^3 \cdot 5^2$

2. $16 + 9 \div 3 \cdot 4 - 7$

3. $18 - 24$

4. $5 \cdot (-20)$

5. $\sqrt{49}$

6. $(-5)^3 - 24 \div (-3)$

7. $0 \div 49$

8. $62 \div 0$

9. $-\dfrac{8}{15y} - \dfrac{2}{15y}$

10. $\dfrac{11}{12} - \dfrac{3}{8} + \dfrac{5}{24}$

11. $\dfrac{3a}{8} \cdot \dfrac{16}{6a^3}$

12. $-\dfrac{16}{3} \div -\dfrac{3}{12}$

13. $\begin{array}{r} 19 \\ -2\frac{3}{11} \\ \hline \end{array}$

14. $\dfrac{0.23 + 1.63}{-0.3}$

15. 10.2×4.01

16. Write 0.6% as a decimal.

17. Write 6.1 as a percent.

18. Write $\dfrac{3}{8}$ as a percent.

19. Write 0.345 as a fraction.

20. Write $-\dfrac{13}{26}$ as a decimal.

21. Round 34.8923 to the nearest tenth.

Evaluate each expression for the given replacement value(s).

22. $5(x^3 - 2)$ for $x = 2$

23. $10 - y^2$ for $y = -3$

24. $x \div y$ for $x = \dfrac{1}{2}$ and $y = 3\dfrac{7}{8}$

25. Simplify: $-(3z + 2) - 5z - 18$

△ 26. Write an expression that represents the perimeter of the equilateral triangle (a triangle with three sides of equal length). Then simplify the expression.

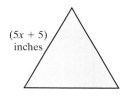

$(5x + 5)$ inches

Solve each equation.

27. $\dfrac{n}{-7} = 4$

28. $-4x + 7 = 15$

29. $-4(x - 11) - 34 = 10 - 12$

30. $\dfrac{x}{5} + x = -\dfrac{24}{5}$

31. $2(x + 5.7) = 6x - 3.4$

32. $\dfrac{5}{y + 1} = \dfrac{4}{y + 2}$

33. Find the perimeter and area.

20 yards
Rectangle 10 yards

34. Find the average of $-12, -13, 0,$ and 9.

Solve.

35. The difference of three times a number and five times the same number is 4. Find the number.

36. During a 258-mile trip, a car used $10\dfrac{3}{4}$ gallons of gas. How many miles would we expect the car to travel on 1 gallon of gas?

37. In a 10-kilometer race, there are 112 more men entered than women. Find the number of female runners if the total number of runners in the race is 600.

38. In a sample of 85 fluorescent bulbs, 3 were found to be defective. At this rate, how many defective bulbs should be found in 510 bulbs?

39. A $120 framed picture is on sale for 15% off. Find the discount and the sale price.

40. Approximate to the nearest hundredth of a centimeter the length of the missing side of a right triangle with legs of 4 centimeters each.

41. Find the complement of a 78° angle.

42. Given that $m \parallel n$, find the measures of $x, y,$ and z.

Practice Final Exam

43. Find the perimeter and area.

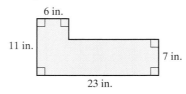

44. Find the circumference and area. Give the exact values and then approximations using $\pi \approx 3.14$.

Convert.

45. $2\frac{1}{2}$ gallons to quarts

46. 2.4 kilograms to grams

Chapters 9–16

Evaluate.

47. -3^4

48. 4^{-3}

Perform the indicated operations and simplify if possible.

49. $7 + 2(5y - 3)$

50. $\begin{array}{r} 5x^3 + x^2 + 5x - 2 \\ -(8x^3 - 4x^2 + x - 7) \end{array}$

51. $(4x - 2)^2$

52. $(3x + 7)(x^2 + 5x + 2)$

Factor.

53. $6t^2 - t - 5$

54. $180 - 5x^2$

55. $3a^2 + 3ab - 7a - 7b$

56. $3x^3 - 21x^2 + 30x$

Simplify. Write answers with positive exponents only.

57. $\left(\dfrac{4x^2y^3}{x^3y^{-4}}\right)^2$

58. $\dfrac{5 - \dfrac{1}{y^2}}{\dfrac{1}{y} + \dfrac{2}{y^2}}$

Perform the indicated operations and simplify if possible.

59. $\dfrac{x^2 - 9}{x^2 - 3x} \div \dfrac{x^2 + 4x + 1}{2x + 10}$

60. $\dfrac{5a}{a^2 - a - 6} - \dfrac{2}{a - 3}$

61. $(x^2 + 7x + 10) \div (x + 5)$
  (Long division)

Practice Final Exam   **1281**

Solve each equation or inequality.

62. $4(n - 5) = -(4 - 2n)$

63. $(3x - 5)(x + 2) = -6$

64. $-5(x - 1) + 6 \leq -3(x + 4) + 1$

65. $2x^2 - 6x + 1 = 0$

66. $\dfrac{4}{y} - \dfrac{5}{3} = -\dfrac{1}{5}$

67. $\dfrac{a}{a - 3} = \dfrac{3}{a - 3} - \dfrac{3}{2}$

68. $\sqrt{2x - 2} = x - 5$

Graph the following.

69. $5x - 7y = 10$

70. $y = -1$

71. $y \geq -4x$

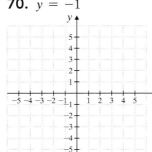
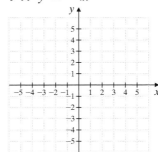

Find the slope of each line.

72. through $(6, -5)$ and $(-1, 2)$

73. $-3x + y = 5$

Write equations of the following lines. Write each equation in standard form.

74. through $(2, -5)$ and $(1, 3)$

75. slope $\dfrac{1}{8}$; y-intercept $(0, 12)$

Solve each system of equations.

76. $\begin{cases} 3x - 2y = -14 \\ y = x + 5 \end{cases}$

77. $\begin{cases} 4x - 6y = 7 \\ -2x + 3y = 0 \end{cases}$

Answer the questions about functions.

78. If $f(x) = x^3 - x$, find
  a. $f(-1)$
  b. $f(0)$
  c. $f(4)$

79. Decide whether the following is a function:
$\{(-3, -3), (0, 5), (-3, 2), (0, 0)\}$

Simplify.

80. $\sqrt{54}$

81. $\sqrt{24x^8}$

62. _____
63. _____
64. _____
65. _____
66. _____
67. _____
68. _____
69. _____
70. _____
71. _____
72. _____
73. _____
74. _____
75. _____
76. _____
77. _____
    a.
    b.
78. c.
79. _____
80. _____
81. _____

Practice Final Exam

Perform the indicated operations and simplify if possible.

82. $\sqrt{18} - \sqrt{75} + 7\sqrt{3} - \sqrt{8}$

83. $\dfrac{\sqrt{40x^4}}{\sqrt{2x}}$

84. $\sqrt{2}(\sqrt{6} - \sqrt{5})$

Rationalize each denominator.

85. $\dfrac{8}{\sqrt{5y}}$

86. $\dfrac{8}{\sqrt{6} + 2}$

Graph the quadratic equation. Label the vertex and the intercepts with their coordinates.

87. $y = x^2 - 7x + 10$

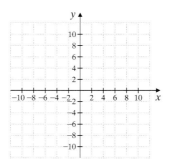

Solve each application.

88. One number plus five times its reciprocal is equal to six. Find the number.

89. A number increased by two-thirds of the number is 35. Find the number.

90. California has more public libraries than any other state. It has 387 more public libraries than Ohio. If the total number of public libraries for these states is 1827, find the number of public libraries in California and the number in Ohio. (*Source*: Institute of Museum and Library Services)

91. A pleasure boat traveling down the Red River takes the same time to go 14 miles upstream as it takes to go 16 miles downstream. If the current of the river is 2 miles per hour, find the speed of the boat in still water.

Answers to Selected Exercises

Chapter 1 The Whole Numbers

Section 1.2

Vocabulary, Readiness & Video Check **1.** whole **3.** words **5.** period **7.** hundreds **9.** 80,000

Exercise Set 1.2 **1.** tens **3.** thousands **5.** hundred-thousands **7.** millions **9.** three hundred fifty-four **11.** eight thousand, two hundred seventy-nine **13.** twenty-six thousand, nine hundred ninety **15.** two million, three hundred eighty-eight thousand **17.** twenty-four million, three hundred fifty thousand, one hundred eighty-five **19.** thirty-seven thousand, nine hundred seventeen **21.** two thousand, seven hundred twenty **23.** eighteen million, three hundred thousand **25.** fourteen thousand, four hundred thirty-three **27.** seventeen thousand, four hundred ten **29.** 6587 **31.** 59,800 **33.** 13,601,011 **35.** 7,000,017 **37.** 260,997 **39.** 250 **41.** 1024 **43.** $247,966,675 **45.** 565 **47.** 200 + 9 **49.** 3000 + 400 + 70 **51.** 80,000 + 700 + 70 + 4 **53.** 60,000 + 6000 + 40 + 9 **55.** 30,000,000 + 9,000,000 + 600,000 + 80,000 **57.** 1786 **59.** Mount Baker **61.** Glacier Peak **63.** German shepherd **65.** Labrador retriever; seventy-five **67.** 70 pounds **69.** 9861 **71.** no; one hundred five **73.** answers may vary **75.** 1,000,000,000,000

Section 1.3

Calculator Explorations **1.** 134 **3.** 340 **5.** 2834 **7.** 770 **9.** 109 **11.** 8978

Vocabulary, Readiness & Video Check **1.** number **3.** 0 **5.** minuend; subtrahend; difference **7.** order; commutative **9.** place; right; left **11.** triangle; 3

Exercise Set 1.3 **1.** 36 **3.** 292 **5.** 49 **7.** 5399 **9.** 209,078 **11.** 25 **13.** 212 **15.** 11,926 **17.** 16,717 **19.** 35,901 **21.** 632,389 **23.** 600 **25.** 25 **27.** 288 **29.** 168 **31.** 5723 **33.** 504 **35.** 79 **37.** 32,711 **39.** 5041 **41.** 31,213 **43.** 1034 **45.** 9 **47.** 8518 **49.** 22,876 **51.** 25 ft **53.** 24 in. **55.** 29 in. **57.** 44 m **59.** 2093 **61.** 266 **63.** 20 **65.** 544 **67.** 72 **69.** 88 **71.** 3170 thousand **73.** $619 **75.** 264,000 sq mi **77.** 283,000 sq mi **79.** 340 ft **81.** 264 pages **83.** 27,441,402 **85.** 100 dB **87.** 58 dB **89.** 3444 **91.** 124 ft **93.** California **95.** 2282 **97.** Florida and Georgia **99.** 5894 mi **101.** minuend: 48; subtrahend: 1 **103.** minuend: 70; subtrahend: 7 **105.** answers may vary **107.** correct **109.** incorrect; 530 **111.** incorrect; 685 **113.** correct
115. $52\underline{6}9$
$-23\underline{8}5$
$288\underline{4}$
117. answers may vary **119.** no; 1089 more pages

Section 1.4

Vocabulary, Readiness & Video Check **1.** graph **3.** 70; 60 **5.** 3 is the place we're rounding to (tens), and the digit to the right of this place is 5 or greater, so we need to add 1 to the 3. **7.** Each circled digit is to the right of the place value being rounded to and is used to determine whether or not we add 1 to the digit in the place value being rounded to.

Exercise Set 1.4 **1.** 420 **3.** 640 **5.** 2800 **7.** 500 **9.** 21,000 **11.** 34,000 **13.** 328,500 **15.** 36,000 **17.** 39,990 **19.** 30,000,000 **21.** 5280; 5300; 5000 **23.** 9440; 9400; 9000 **25.** 14,880; 14,900; 15,000 **27.** 311,000 miles **29.** 60,100 days **31.** $150,000,000,000 **33.** $4,200,000 **35.** USA: 220,000,000; India: 292,000,000 **37.** 130 **39.** 80 **41.** 5700 **43.** 300 **45.** 11,400 **47.** incorrect **49.** correct **51.** correct **53.** $3400 **55.** 900 mi **57.** 6000 ft **59.** Springfield was larger by approximately 40,000. **61.** The decrease was 182,000 **63.** 74,900,000; 75,000,000; 80,000,000 **65.** 64,100,000; 64,000,000; 60,000,000 **67.** 5723, for example **69. a.** 8550 **b.** 8649 **71.** answers may vary **73.** 140 m

Section 1.5

Calculator Explorations **1.** 3456 **3.** 15,322 **5.** 272,291

Vocabulary, Readiness & Video Check **1.** 0 **3.** product; factor **5.** grouping; associative **7.** length **9.** distributive **11.** Area is measured in square units, and here we have meters times meters, or square meters; the correct answer is 63 *square* meters, so the correct units are square meters.

Exercise Set 1.5 **1.** 24 **3.** 0 **5.** 0 **7.** 87 **9.** $6 \cdot 3 + 6 \cdot 8$ **11.** $4 \cdot 3 + 4 \cdot 9$ **13.** $20 \cdot 14 + 20 \cdot 6$ **15.** 512 **17.** 3678 **19.** 1662 **21.** 6444 **23.** 1157 **25.** 24,418 **27.** 24,786 **29.** 15,600 **31.** 0 **33.** 6400 **35.** 48,126 **37.** 142,506 **39.** 2,369,826 **41.** 64,790 **43.** 3,949,935 **45.** area: 63 sq m; perimeter: 32 m **47.** area: 680 sq ft; perimeter: 114 ft **49.** 240,000 **51.** 300,000 **53.** c **55.** c **57.** 880 **59.** 4200 **61.** 4480 **63.** 375 cal **65.** $3290 **67. a.** 20 **b.** 100 **c.** 2000 lb **69.** 8800 sq ft **71.** 56,000 sq ft **73.** 5828 pixels **75.** 2100 characters **77.** 1280 cal **79.** $10, $60; $10, $200; $12, $36; $12, $36: total cost $372 **81.** 1,440,000 tea bags **83.** 135 **85.** 2144 **87.** 23 **89.** 15 **91.** $5 \cdot 6$ or $6 \cdot 5$ **93. a.** $5 + 5 + 5$ or $3 + 3 + 3 + 3 + 3$ **b.** answers may vary **95.** 203
97. $4\underline{2}$
$\times 9\underline{3}$
99. answers may vary **101.** 506 windows
$\times14$
812
$\underline{2030}$
2842

A1

Section 1.6

Calculator Explorations **1.** 53 **3.** 62 **5.** 261 **7.** 0

Vocabulary, Readiness & Video Check **1.** quotient; dividend; divisor **3.** 1 **5.** undefined **7.** 0 **9.** $202 \cdot 102 + 15 = 20{,}619$ **11.** addition and division

Exercise Set 1.6 **1.** 6 **3.** 12 **5.** 0 **7.** 31 **9.** 1 **11.** 8 **13.** undefined **15.** 1 **17.** 0 **19.** 9 **21.** 29 **23.** 74 **25.** 338 **27.** undefined **29.** 9 **31.** 25 **33.** 68 R 3 **35.** 236 R 5 **37.** 38 R 1 **39.** 326 R 4 **41.** 13 **43.** 49 **45.** 97 R 8 **47.** 209 R 11 **49.** 506 **51.** 202 R 7 **53.** 54 **55.** 99 R 100 **57.** 202 R 15 **59.** 579 R 72 **61.** 17 **63.** 511 R 3 **65.** 2132 R 32 **67.** 6080 **69.** 23 R 2 **71.** 5 R 25 **73.** 20 R 2 **75.** 33 students **77.** 165 lb **79.** 310 yd **81.** 89 bridges **83.** 11 light poles **85.** 5 mi **87.** 1760 yd **89.** 20 **91.** 387 **93.** 79 **95.** 74° **97.** 9278 **99.** 15,288 **101.** 679 **103.** undefined **105.** 9 R 12 **107.** c **109.** b **111.** $3,090,000,000 **113.** increase; answers may vary **115.** no; answers may vary **117.** 12 ft **119.** answers may vary **121.** 5 R 1

Integrated Review **1.** 194 **2.** 6555 **3.** 4524 **4.** 562 **5.** 67 **6.** undefined **7.** 1 **8.** 5 **9.** 0 **10.** 0 **11.** 0 **12.** 3 **13.** 63 **14.** 9 **15.** 138 **16.** 276 **17.** 1169 **18.** 9826 **19.** 182 R 4 **20.** 79,317 **21.** 1099 R 2 **22.** 111 R 1 **23.** 663 R 24 **24.** 1076 R 60 **25.** 1037 **26.** 9899 **27.** 30,603 **28.** 47,500 **29.** 71 **30.** 558 **31.** 6 R 8 **32.** 53 **33.** 183 **34.** 231 **35.** 9740; 9700; 10,000 **36.** 1430; 1400; 1000 **37.** 20,800; 20,800; 21,000 **38.** 432,200; 432,200; 432,000 **39.** perimeter: 24 ft; area: 36 sq ft **40.** perimeter: 42 in.; area: 98 sq in. **41.** 28 mi **42.** 26 m **43.** 24 **44.** 124 **45.** Lake Pontchartrain Bridge; 2175 ft **46.** $5562

Section 1.7

Calculator Explorations **1.** 4096 **3.** 3125 **5.** 2048 **7.** 2526 **9.** 4295 **11.** 8

Vocabulary, Readiness & Video Check **1.** base; exponent **3.** addition **5.** division **7.** 1 **9.** The area of a rectangle is length · width. A square is a special rectangle where length = width. Thus, the area of a square is side · side or $(\text{side})^2$.

Exercise Set 1.7 **1.** 4^3 **3.** 7^6 **5.** 12^3 **7.** $6^2 \cdot 5^3$ **9.** $9 \cdot 8^2$ **11.** $3 \cdot 2^4$ **13.** $3 \cdot 2^4 \cdot 5^5$ **15.** 64 **17.** 125 **19.** 32 **21.** 1 **23.** 7 **25.** 128 **27.** 256 **29.** 256 **31.** 729 **33.** 144 **35.** 100 **37.** 20 **39.** 729 **41.** 192 **43.** 162 **45.** 21 **47.** 7 **49.** 5 **51.** 16 **53.** 46 **55.** 8 **57.** 64 **59.** 83 **61.** 2 **63.** 48 **65.** 4 **67.** undefined **69.** 59 **71.** 52 **73.** 44 **75.** 12 **77.** 21 **79.** 3 **81.** 43 **83.** 8 **85.** 16 **87.** area: 49 sq m; perimeter: 28 m **89.** area: 529 sq mi; perimeter: 92 mi **91.** true **93.** false **95.** $(2 + 3) \cdot 6 - 2$ **97.** $24 \div (3 \cdot 2) + 2 \cdot 5$ **99.** 1260 ft **101.** 6,384,814 **103.** answers may vary

Section 1.8

Vocabulary, Readiness & Video Check **1.** expression **3.** expression; variables **5.** equation **7.** multiplication **9.** decreased by

Exercise Set 1.8 **1.** 28; 14; 147; 3 **3.** 152; 152; 0; undefined **5.** 57; 55; 56; 56 **7.** 9 **9.** 8 **11.** 6 **13.** 5 **15.** 117 **17.** 94 **19.** 5 **21.** 34 **23.** 20 **25.** 4 **27.** 4 **29.** 0 **31.** 33 **33.** 125 **35.** 121 **37.** 100 **39.** 60 **41.** 4 **43.** 16; 64; 144; 256 **45.** yes **47.** no **49.** no **51.** yes **53.** no **55.** yes **57.** 12 **59.** 6 **61.** 4 **63.** none **65.** 11 **67.** $x + 8$ **69.** $x + 8$ **71.** $20 - x$ **73.** $512x$ **75.** $8 \div x$ or $\dfrac{8}{x}$ **77.** $5x + (17 + x)$ **79.** $5x$ **81.** $11 - x$ **83.** $x - 5$ **85.** $6 \div x$ or $\dfrac{6}{x}$ **87.** $50 - 8x$ **89.** correct answer: 21 **91.** correct answer: 46 **93.** 274,657 **95.** 777 **97.** $5x$; answers may vary **99.** As t gets larger $16t^2$ gets larger.

Chapter 1 Vocabulary Check **1.** whole numbers **2.** perimeter **3.** place value **4.** exponent **5.** area **6.** digits **7.** variable **8.** equation **9.** expression **10.** solution **11.** set **12.** sum **13.** divisor **14.** dividend **15.** quotient **16.** factor **17.** product **18.** minuend **19.** subtrahend **20.** difference **21.** addend

Chapter 1 Review **1.** tens **2.** ten-millions **3.** seven thousand, six hundred forty **4.** forty-six million, two hundred thousand, one hundred twenty **5.** $3000 + 100 + 50 + 8$ **6.** $400{,}000{,}000 + 3{,}000{,}000 + 200{,}000 + 20{,}000 + 5000$ **7.** 81,900 **8.** 6,304,000,000 **9.** 615,000,000 **10.** 28,000,000 **11.** Oceania/Australia **12.** Asia **13.** 67 **14.** 67 **15.** 65 **16.** 304 **17.** 449 **18.** 840 **19.** 391 **20.** 7908 **21.** 4211 **22.** 1967 **23.** 1334 **24.** 886 **25.** 17,897 **26.** 34,658 **27.** 7523 mi **28.** $197,699 **29.** 216 ft **30.** 66 km **31.** 82 million or 82,000,000 **32.** 113 million or 113,000,000 **33.** May **34.** August **35.** $110 **36.** $240 **37.** 40 **38.** 50 **39.** 880 **40.** 500 **41.** 3800 **42.** 58,000 **43.** 40,000,000 **44.** 800,000 **45.** 7300 **46.** 4100 **47.** 2700 mi **48.** Europe: 832,000,000; Latin America/Caribbean: 626,000,000; difference: 206,000,000 **49.** 2208 **50.** 1396 **51.** 2280 **52.** 2898 **53.** 560 **54.** 900 **55.** 0 **56.** 0 **57.** 16,994 **58.** 8954 **59.** 113,634 **60.** 44,763 **61.** 411,426 **62.** 636,314 **63.** 1500 **64.** 4920 **65.** $898 **66.** $403,980 **67.** 91 sq mi **68.** 500 sq cm **69.** 7 **70.** 4 **71.** 5 R 2 **72.** 4 R 2 **73.** undefined **74.** 0 **75.** 33 R 2 **76.** 19 R 7 **77.** 24 R 2 **78.** 35 R 15 **79.** 506 R 10 **80.** 907 R 40 **81.** 2793 R 140 **82.** 2012 R 60 **83.** 18 R 2 **84.** 21 R 2 **85.** 27 boxes **86.** 13 miles **87.** 51 **88.** 59 **89.** 64 **90.** 125 **91.** 405 **92.** 400 **93.** 16 **94.** 10 **95.** 15 **96.** 7 **97.** 12 **98.** 9 **99.** 42 **100.** 33 **101.** 9 **102.** 2 **103.** 6 **104.** 29 **105.** 40 **106.** 72 **107.** 5 **108.** 7 **109.** 49 sq m **110.** 9 sq in. **111.** 5 **112.** 17 **113.** undefined **114.** 0 **115.** 121 **116.** 2 **117.** 4 **118.** 20 **119.** $x - 5$ **120.** $x + 7$ **121.** $10 \div x$ or $\dfrac{10}{x}$ **122.** $5x$ **123.** yes **124.** no **125.** no **126.** yes **127.** 11 **128.** 175 **129.** 14 **130.** none **131.** 417 **132.** 682 **133.** 2196 **134.** 2516 **135.** 1101 **136.** 1411 **137.** 458 R 8 **138.** 237 R 1 **139.** 70,848 **140.** 95,832 **141.** 1644 **142.** 8481 **143.** 840 **144.** 300,000 **145.** 12 **146.** 6 **147.** no **148.** yes **149.** 53 full boxes with 18 left over **150.** $86

Chapter 1 Getting Ready for the Test 1. D 2. E 3. F 4. B 5. C 6. B 7. D 8. A 9. C 10. B 11. D 12. A 13. C

Chapter 1 Test 1. eighty-two thousand, four hundred twenty-six 2. 402,550 3. 141 4. 113 5. 14,880 6. 766 R 42 7. 200 8. 98 9. 0 10. undefined 11. 33 12. 21 13. 48 14. 36 15. 5,698,000 16. 82 17. 52,000 18. 13,700 19. 1600 20. 92 21. 122 22. 1605 23. 7 R 2 24. $17 25. $126 26. 360 cal 27. $7905 28. 20 cm; 25 sq cm 29. 60 yd; 200 sq yd 30. 30 31. 1 32. **a.** $x \div 17$ or $\frac{x}{17}$ **b.** $2x - 20$ 33. yes 34. 10

Chapter 2 Integers and Introduction to Solving Equations

Section 2.1

Vocabulary, Readiness & Video Check 1. integers 3. inequality symbols 5. is less than; is greater than 7. absolute value 9. number of feet a miner works underground 11. negative 13. opposite of

Exercise Set 2.1 1. -1235 3. $-35{,}814$ 5. $+120$ 7. $-11{,}810$ 9. -1270 million 11. $-160, -147$; Guillermo 13. $-60°$ or $-60°$F

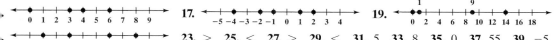

23. > 25. < 27. > 29. < 31. 5 33. 8 35. 0 37. 55 39. -5 41. 4 43. -23 45. 85 47. 7 49. -20 51. -3 53. 43 55. 15 57. 33 59. 6 61. -2 63. 32 65. -7 67. < 69. < 71. = 73. < 75. > 77. < 79. > 81. < 83. 31; -31 85. 28; 28 87. Caspian Sea 89. Lake Superior 91. iodine 93. oxygen 95. 13 97. 35 99. 360 101. $-|-8|, -|3|, 2^2, -(-5)$ 103. $-|-6|, -|1|, |-1|, -(-6)$ 105. $-10, -|-9|, -(-2), |-12|, 5^2$ 107. a, d 109. 8 111. false 113. true 115. false 117. answers may vary 119. no; answers may vary

Section 2.2

Calculator Explorations 1. -159 3. 44 5. $-894{,}855$

Vocabulary, Readiness & Video Check 1. 0 3. a 5. Negative; the numbers have different signs and the sign of the sum is the same as the sign of the number with the larger absolute value, -6. 7. The diver's current depth is 231 feet below the surface.

Exercise Set 2.2

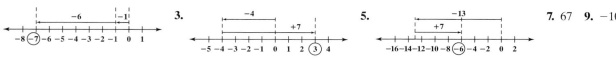

7. 67 9. -10 11. 0 13. 4 15. -6 17. -2 19. -9 21. -24 23. -840 25. 7 27. -3 29. -30 31. 40 33. -20 35. -125 37. -7 39. -246 41. 16 43. 13 45. -28 47. -11 49. 20 51. -34 53. -1 55. 0 57. -42 59. -70 61. 3 63. -21 65. $-6 + 25$; 19 67. $-31 + (-9) + 30$; -10 69. $0 + (-215) + (-16) = -231$; 231 ft below the surface 71. Walker: -3; Matsuyama: -2 73. $45{,}687{,}000{,}000 75. $-$7{,}707{,}000{,}000 77. 2°C 79. $13{,}609 81. $-31°$F 83. -7535 m 85. 44 87. 141 89. answers may vary 91. -3 93. -22 95. true 97. false 99. answers may vary

Section 2.3

Vocabulary, Readiness & Video Check 1. b 3. d 5. additive inverse 7. to follow the order of operations

Exercise Set 2.3 1. 0 3. 3 5. -5 7. 22 9. 3 11. -20 13. -12 15. -13 17. 508 19. -14 21. -4 23. -12 25. $-25 - 17$; -42 27. $-22 - (-3)$; -19 29. $2 - (-12)$; 14 31. -56 33. -5 35. -145 37. -37 39. 3 41. 1 43. -1 45. -31 47. 44 49. -32 51. -9 53. 14 55. -11 57. 31 59. 12 61. 20 63. 111°F 65. 180°F 67. 263°F 69. 26 strokes 71. $-10°$C 73. 154 ft 75. 69 ft 77. 652 ft 79. 144 ft 81. 1197°F 83. $-$45 billion 85. $-5 + x$ 87. $-20 - x$ 89. 5 91. 1058 93. answers may vary 95. 16 97. -20 99. -4 101. 0 103. -14 105. false 107. answers may vary

Section 2.4

Vocabulary, Readiness & Video Check 1. negative 3. positive 5. 0 7. undefined 9. multiplication 11. The phrase "lost four yards" in the example translates to the negative number -4.

Exercise Set 2.4 1. 12 3. -36 5. -81 7. 0 9. 48 11. -12 13. 80 15. 0 17. -15 19. -9 21. -27 23. -36 25. -64 27. -8 29. -5 31. 7 33. 0 35. undefined 37. -14 39. 0 41. -15 43. -63 45. 42 47. -24 49. 49 51. -5 53. -9 55. -6 57. 120 59. -1080 61. undefined 63. -6 65. -7 67. 3 69. -1 71. -32 73. 180 75. 1 77. -30 79. -1104 81. -2870 83. -56 85. -18 87. 35 89. -1 91. undefined 93. 6 95. 16; 4 97. 0; 0 99. $-54 \div 9$; -6 101. $-42(-6)$; 252 103. $-71 \cdot x$ or $-71x$ 105. $-16 - x$ 107. $-29 + x$ 109. $\frac{x}{-33}$ or $x \div (-33)$ 111. $3 \cdot (-4) = -12$; a loss of 12 yd

113. $5 \cdot (-20) = -100$; a depth of 100 feet   **115.** $-210°C$ **117.** $-189°C$ **119.** $-\$132$ million per month   **121. a.** $-23,705$ scree
b. -4741 screens per year   **123.** 109   **125.** 8   **127.** -19 **129.** -28 **131.** -8 **133.** negative   **135.** $(-5)^{17}, (-2)^{17}, (-2)^{12}, (-5$
137. answers may vary

Integrated Review   1. $-50; +122$ or 122 **2.** [number line with points at $-5, -2, 1, 3$]   **3.** $>$ **4.** $<$ **5.** $<$ **6.** $>$ **7.** 3   **8.** 9   **9.** -4
10. 5   **11.** -11 **12.** 3   **13.** -64 **14.** 0   **15.** 12   **16.** -20 **17.** -48 **18.** -9 **19.** 10   **20.** -2 **21.** 106   **22.** -3 **23.** 0
24. 4   **25.** 42   **26.** 6   **27.** 19   **28.** -900 **29.** -12 **30.** -19 **31.** undefined   **32.** 0   **33.** $-12 - (-8); -4$ **34.** $-17 + (-27);$ -
35. $-5(-25); 125$ **36.** $-100 \div (-5); 20$ **37.** $\dfrac{x}{-17}$ or $x \div (-17)$ **38.** $-3 + x$ **39.** $x - (-18)$ **40.** $-7 \cdot x$ or $-7x$ **41.** 9
42. -15 **43.** 27   **44.** 33   **45.** -15 **46.** -4

Section 2.5

Calculator Explorations   1. 48   **3.** -258

Vocabulary, Readiness & Video Check   1. division   **3.** average   **5.** subtraction   **7.** A fraction bar means divided by and it is a
grouping symbol.   **9.** Finding the average is a good application of both order of operations and adding and dividing integers.

Exercise Set 2.5   1. -125 **3.** -64 **5.** 32   **7.** -8 **9.** -11 **11.** -43 **13.** -8 **15.** 17   **17.** -1 **19.** 4   **21.** -3 **23.** 16
25. 13   **27.** -77 **29.** 80   **31.** 256   **33.** 53   **35.** 4   **37.** -64 **39.** 4   **41.** 16   **43.** -27 **45.** 34   **47.** 65   **49.** -7 **51.** 36   **53.** -1
55. 30   **57.** -3 **59.** -30 **61.** 1   **63.** -12 **65.** 0   **67.** -20 **69.** 9   **71.** -16 **73.** -128 **75.** 1   **77.** -50 **79.** -2 **81.** -19
83. 18   **85.** 0   **87.** no; answers may vary   **89.** 4050   **91.** 45   **93.** 32 in.   **95.** 30 ft   **97.** $2 \cdot (7 - 5) \cdot 3$ **99.** $-6 \cdot (10 - 4)$
101. answers may vary   **103.** answers may vary   **105.** 20,736   **107.** 8900   **109.** 9

Section 2.6

Vocabulary, Readiness & Video Check   1. expression   **3.** equation; expression   **5.** solution   **7.** addition
9. an equal sign   **11.** original; true

Exercise Set 2.6   1. yes   **3.** no   **5.** yes   **7.** yes   **9.** 18   **11.** -12 **13.** 9   **15.** -17 **17.** 4   **19.** -4 **21.** -14 **23.** -17 **25.** 0
27. 1   **29.** -7 **31.** -50 **33.** -25 **35.** 36   **37.** 21   **39.** 12   **41.** -80 **43.** -2 **45.** $x - (-2)$ **47.** $-6 \cdot x$ or $-6x$
49. $-15 + x$ **51.** $-8 \div x$ or $\dfrac{-8}{x}$ **53.** 41,574   **55.** -409 **57.** answers may vary   **59.** answers may vary

Chapter 2 Vocabulary Check   1. opposites   **2.** absolute value   **3.** integers   **4.** negative   **5.** positive   **6.** inequality
symbols   **7.** solution   **8.** average   **9.** expression   **10.** equation   **11.** is less than; is greater than   **12.** addition   **13.** multiplicati

Chapter 2 Review   1. -1572 **2.** $+11,239$ **3.** [number line with points at $-5, -3, 7$]   **4.** [number line with points at $-1, 5$]
5. 11   **6.** 0   **7.** -8 **8.** 9   **9.** -16 **10.** 2   **11.** $>$ **12.** $<$ **13.** $>$ **14.** $>$ **15.** 18   **16.** -42 **17.** false   **18.** true   **19.** true
20. true   **21.** 2   **22.** 3   **23.** -5 **24.** -10 **25.** Elevator D   **26.** Elevator B   **27.** 2   **28.** 14   **29.** 4   **30.** 17   **31.** -23 **32.** -2
33. -21 **34.** -70 **35.** 0   **36.** 0   **37.** -151 **38.** -606 **39.** $-20°C$ **40.** -150 ft   **41.** -19 **42.** 11   **43.** 8   **44.** -16 **45.** -2
46. -10 **47.** 20   **48.** 8   **49.** 0   **50.** -32 **51.** 0   **52.** 7   **53.** -10 **54.** -27 **55.** 692 ft   **56.** -25 **57.** -14 or 14 ft below grou
58. 82 ft   **59.** true   **60.** false   **61.** 21   **62.** -18 **63.** -64 **64.** 60   **65.** 25   **66.** -1 **67.** 0   **68.** 24   **69.** -5 **70.** 3   **71.** 0
72. undefined   **73.** -20 **74.** -9 **75.** 38   **76.** -5 **77.** $(-5)(2) = -10$ **78.** $(-50)(4) = -200$ **79.** $-1024 \div 4 = -256$
80. $-45 \div 9 = -5$ **81.** 49   **82.** -49 **83.** 0   **84.** -8 **85.** -16 **86.** 35   **87.** -32 **88.** -8 **89.** 7   **90.** -14 **91.** 39   **92.** -1
93. -2 **94.** -12 **95.** -3 **96.** -35 **97.** -5 **98.** 5   **99.** -1 **100.** -7 **101.** no   **102.** yes   **103.** -13 **104.** -20 **105.** -3
106. -9 **107.** -13 **108.** -31 **109.** 44   **110.** -26 **111.** -19 **112.** 38   **113.** 6   **114.** -5 **115.** -15 **116.** -19 **117.** 48
118. -21 **119.** 21   **120.** -5 **121.** $-27°C$ **122.** $6°C$ **123.** 13,118 ft   **124.** -9 **125.** 2   **126.** 3   **127.** -5 **128.** -25 **129.** $-$
130. 17   **131.** -21 **132.** -17 **133.** 12   **134.** -9 **135.** -200 **136.** 3

Chapter 2 Getting Ready for the Test   1. A   **2.** A   **3.** A   **4.** B   **5.** C   **6.** C   **7.** A   **8.** D   **9.** C   **10.** D   **11.** B   **12.** B
13. D   **14.** A

Chapter 2 Test   1. 3   **2.** -6 **3.** -100 **4.** 4   **5.** -30 **6.** 12   **7.** 65   **8.** 5   **9.** 12   **10.** -6 **11.** 50   **12.** -2 **13.** -11
14. -46 **15.** -117 **16.** 3456   **17.** 28   **18.** -213 **19.** -2 **20.** 2   **21.** -5 **22.** -32 **23.** -17 **24.** 1   **25.** -1
26. 88 ft below sea level   **27.** 45   **28.** 31,642   **29.** 3820 ft below sea level   **30.** -4 **31. a.** $17 \cdot x$ or $17x$ **b.** $20 - x$ **32.** 5
33. -28 **34.** -20 **35.** -4

Cumulative Review   1. hundred-thousands; Sec. 1.2, Ex. 1   **2.** hundreds; Sec. 1.2   **3.** thousands; Sec. 1.2, Ex. 2   **4.** thousands; Sec. 1
5. ten-millions; Sec. 1.2, Ex. 3   **6.** hundred-thousands; Sec. 1.2   **7. a.** $<$ **b.** $>$ **c.** $>$; Sec. 2.1, Ex. 3   **8. a.** $>$ **b.** $>$ **c.** $<$; Sec. 2.1
9. 39; Sec. 1.3, Ex. 3   **10.** 39; Sec. 1.3   **11.** 7321; Sec. 1.3, Ex. 6   **12.** 3013; Sec. 1.3   **13.** 36,184 mi; Sec. 1.3, Ex. 11   **14.** \$525; Sec. 1.
15. 570; Sec. 1.4, Ex. 1   **16.** 600; Sec. 1.4   **17.** 1800; Sec. 1.4, Ex. 5   **18.** 5000; Sec. 1.4   **19. a.** $5 \cdot 6 + 5 \cdot 5$ **b.** $20 \cdot 4 + 20 \cdot 7$
c. $2 \cdot 7 + 2 \cdot 9$; Sec. 1.5, Ex. 2   **20. a.** $5 \cdot 2 + 5 \cdot 12$ **b.** $9 \cdot 3 + 9 \cdot 6$ **c.** $4 \cdot 8 + 4 \cdot 1$; Sec. 1.5   **21.** 78,875; Sec. 1.5, Ex. 5   **22.** 31,096; Sec.

a. 6 **b.** 8 **c.** 7; Sec. 1.6, Ex. 1   **24. a.** 7 **b.** 8 **c.** 12; Sec. 1.6   **25.** 741; Sec. 1.6, Ex. 4   **26.** 456; Sec. 1.6   **27.** 12 cards each; 10 cards left
; Sec. 1.6, Ex. 11   **28.** $9; Sec. 1.6   **29.** 81; Sec. 1.7, Ex. 5   **30.** 125; Sec. 1.7   **31.** 6; Sec. 1.7, Ex. 6   **32.** 4; Sec. 1.7   **33.** 180; Sec. 1.7,
. 8   **34.** 56; Sec. 1.7   **35.** 2; Sec. 1.7, Ex. 13   **36.** 5; Sec. 1.7   **37.** 14; Sec. 1.8, Ex. 1   **38.** 14; Sec. 1.8   **39. a.** 9 **b.** 8 **c.** 0; Sec. 2.1,
. 4   **40. a.** 4 **b.** 7; Sec. 2.1   **41.** 23; Sec. 2.2, Ex. 7   **42.** 5; Sec. 2.2   **43.** 22; Sec. 2.3, Ex. 12   **44.** 5; Sec. 2.3   **45.** -21; Sec. 2.4, Ex. 1
, -10; Sec. 2.4   **47.** 0; Sec. 2.4, Ex. 3   **48.** -54; Sec. 2.4   **49.** -16; Sec. 2.5, Ex. 8   **50.** -27; Sec. 2.5

Chapter 3 Solving Equations and Problem Solving
Section 3.1

Vocabulary, Readiness & Video Check   **1.** expression; term   **3.** combine like terms   **5.** variable; constant   **7.** associative
 numerical coefficient   **11.** distributive property   **13.** addition; multiplication; $P =$ perimeter, $A =$ area

Exercise Set 3.1   **1.** $8x$ **3.** $-n$ **5.** $-2c$ **7.** $-6x$ **9.** $12a - 5$ **11.** $42x$ **13.** $-33y$ **15.** $72a$ **17.** $2y + 6$ **19.** $3a - 18$
$-12x - 28$ **23.** $2x + 1$ **25.** $15c + 3$ **27.** $-21n + 20$ **29.** $7w + 15$ **31.** $11x - 8$ **33.** $-2y + 16$ **35.** $-2y$ **37.** $-7z$
$8d - 3c$ **41.** $6y - 14$ **43.** $-q$ **45.** $2x + 22$ **47.** $-3x - 35$ **49.** $-3z - 15$ **51.** $-6x + 6$ **53.** $3x - 30$ **55.** $-r + 8$
$-7n + 3$ **59.** $9z - 14$ **61.** -6 **63.** $-4x + 10$ **65.** $2x + 20$ **67.** $7a + 12$ **69.** $3y + 5$ **71.** $(14y + 22)$ m
$(11a + 12)$ ft   **75.** $(125x + 55)$ in.   **77.** $36y$ sq in.   **79.** $(32x - 64)$ sq km   **81.** $(60y + 20)$ sq mi   **83.** 4700 sq ft
64 ft   **87.** $(3x + 6)$ ft   **89.** -3 **91.** 8   **93.** 0   **95.** incorrect; $15x - 10$ **97.** incorrect; $2xy$ **99.** correct
. incorrect; $4y - 12 + 11$ or $4y - 1$ **103.** distributive   **105.** associative   **107.** $(20x + 16)$ sq mi   **109.** $4824q + 12{,}274$
. answers may vary   **113.** answers may vary

Section 3.2

Vocabulary, Readiness & Video Check   **1.** equivalent   **3.** simplifying   **5.** addition   **7.** Simplify the left side of the equation
combining like terms.   **9.** addition property of equality.

Exercise Set 3.2   **1.** 6   **3.** 8   **5.** -4 **7.** 6   **9.** -1 **11.** 18   **13.** -8 **15.** -50 **17.** 3   **19.** -22 **21.** -6 **23.** 24   **25.** -30
12   **29.** 4   **31.** 3   **33.** -3 **35.** 1   **37.** 5   **39.** -11 **41.** -4 **43.** -3 **45.** -1 **47.** 0   **49.** 3   **51.** -6 **53.** -35 **55.** 10
-2 **59.** 0   **61.** 28   **63.** -5 **65.** -28 **67.** -28 **69.** 5   **71.** -4 **73.** $-7 + x$ **75.** $x - 11$ **77.** $-13x$ **79.** $\frac{x}{-12}$ or $-\frac{x}{12}$
$-11x + 5$ **83.** $-10 - 7x$ **85.** $4x + 7$ **87.** $2x - 17$ **89.** $-6(x + 15)$ **91.** $\frac{45}{-5x}$ or $-\frac{45}{5x}$ **93.** $\frac{17}{x} + (-15)$ or $\frac{17}{x} - 15$
Caracas   **97.** $1535   **99.** answers may vary   **101.** no; answers may vary   **103.** answers may vary   **105.** 67,896   **107.** -48 **109.** 42

Integrated Review   **1.** expression   **2.** equation   **3.** equation   **4.** expression   **5.** simplify   **6.** solve   **7.** $8x$ **8.** $-4y$ **9.** $-2a - 2$
$5a - 26$ **11.** $-8x - 14$ **12.** $-6x + 30$ **13.** $5y - 10$ **14.** $15x - 31$ **15.** $(12x - 6)$ sq m   **16.** $(2x + 9)$ ft   **17.** -4 **18.** -3
-10 **20.** 6   **21.** -15 **22.** -120 **23.** -5 **24.** -13 **25.** -24 **26.** -54 **27.** 12   **28.** -42 **29.** 2   **30.** 2   **31.** -3 **32.** 5
-5 **34.** 6   **35.** $x - 10$ **36.** $-20 + x$ **37.** $10x$ **38.** $\frac{10}{x}$ **39.** $-2x + 5$ **40.** $-4(x - 1)$

Section 3.3

Calculator Explorations   **1.** yes   **3.** no   **5.** yes

Vocabulary, Readiness & Video Check   **1.** $3x - 9 + x - 16$; $5(2x + 6) - 1 = 39$ **3.** addition   **5.** distributive
the addition property of equality; to make sure we get an equivalent equation   **9.** gives; amounts to

Exercise Set 3.3   **1.** -12 **3.** -3 **5.** 1   **7.** -45 **9.** -9 **11.** 6   **13.** 8   **15.** 5   **17.** 0   **19.** -5 **21.** -22 **23.** 6   **25.** -11
-7 **29.** -5 **31.** 270   **33.** 5   **35.** 3   **37.** 9   **39.** -6 **41.** 11   **43.** 3   **45.** 4   **47.** -4 **49.** 3   **51.** -1 **53.** -4 **55.** -5
0   **59.** 4   **61.** 1   **63.** -30 **65.** $-42 + 16 = -26$ **67.** $-5(-29) = 145$ **69.** $3(-14 - 2) = -48$ **71.** $\frac{100}{2(50)} = 1$
134,000,000 returns   **75.** 39,000,000 returns   **77.** 33   **79.** -37 **81.** b   **83.** a   **85.** $6x - 10 = 5x - 7$; $6x = 5x + 3$; $x = 3$
0   **89.** -4 **91.** no; answers may vary

Section 3.4

Vocabulary, Readiness & Video Check   **1.** The phrase is "a number subtracted from -20" so -20 goes first and we subtract the
number from that.   **3.** The original application asks for the fastest speeds of a pheasant and a falcon. The value of x is the speed in
h for a pheasant, so the falcon's speed still needs to be found.

A6 Answers to Selected Exercises

Exercise Set 3.4 **1.** $-5 + x = -7$ **3.** $3x = 27$ **5.** $-20 - x = 104$ **7.** $2x = 108$ **9.** $5(-3 + x) = -20$ **11.** $9 + 3x = 33$; 8 **13.** $3 + 4 + x = 16$; 9 **15.** $x - 3 = \dfrac{10}{5}$; 5 **17.** $30 - x = 3(x + 6)$; 3 **19.** $5x - 40 = x + 8$; 12 **21.** $3(x - 5) = \dfrac{108}{12}$; 8 **23.** $4x = 30 - 2x$; 5 **25.** California: 55 votes; Florida: 29 votes **27.** falcon: 185 mph; pheasant: 37 mph **29.** *USA Today*: 2283 thousand; *The Wall Street Journal*: 2057 thousand **31.** Apple: $154 billion; Google: $83 billion **33.** Xbox: $330; games $110 **35.** 5470 mi **37.** Beaver Stadium: 107,282; Michigan Stadium: 106,201 **39.** China: 140 million; Spain: 70 million **41.** United States: 12,200,000; Germany: 6,100,000 **43.** $225 **45.** Olympic: 120 yards; Under-10 field: 70 yards **47.** 77 points **49.** TV: $74.7 billion; Internet: $75.3 billion **51.** 590 **53.** 1000 **55.** 3000 **57.** yes; answers may vary **59.** $216,200 **61.** $549

Chapter 3 Vocabulary Check **1.** simplified; combined. **2.** like **3.** variable **4.** algebraic expression **5.** terms **6.** numerical coefficient **7.** evaluating the expression **8.** constant **9.** equation **10.** solution **11.** distributive **12.** multiplication **13.** addition

Chapter 3 Review **1.** $10y - 15$ **2.** $-6y - 10$ **3.** $-6a - 7$ **4.** $-8y + 2$ **5.** $2x + 10$ **6.** $-3y - 24$ **7.** $11x - 12$ **8.** $-4m -$ **9.** $-5a + 4$ **10.** $12y - 9$ **11.** $16y - 5$ **12.** $x - 2$ **13.** $(4x + 6)$ yd **14.** $20y$ m **15.** $(6x - 3)$ sq yd **16.** $(45x + 8)$ sq cm **17.** -2 **18.** 10 **19.** -7 **20.** 5 **21.** -12 **22.** 45 **23.** -6 **24.** -1 **25.** -25 **26.** -8 **27.** -2 **28.** -2 **29.** -8 **30.** -45 **31.** 5 **32.** -5 **33.** -63 **34.** -15 **35.** 5 **36.** 12 **37.** -6 **38.** 4 **39.** $-5x$ **40.** $x - 3$ **41.** $-5 + x$ **42.** $\dfrac{-2}{x}$ or $-\dfrac{2}{x}$ **43.** $-5x - 50$ **44.** $2x + 11$ **45.** $\dfrac{70}{x + 6}$ **46.** $2(x - 13)$ **47.** 21 **48.** -10 **49.** 2 **50.** 2 **51.** 11 **52.** -5 **53.** -15 **54.** 10 **55.** -2 **56.** -6 **57.** -1 **58.** 1 **59.** 0 **60.** 20 **61.** $20 - (-8) = 28$ **62.** $-2 - 19 = -21$ **63.** $\dfrac{-75}{5 + 20} = -3$ **64.** $5[2 + (-6)] = -20$ **65.** $2x - 8 = 40$ **66.** $6x = x + 20$ **67.** $\dfrac{x}{2} - 12 = 10$ **68.** $x - 3 = \dfrac{x}{4}$ **69.** 5 **70.** -16 **71.** 2386 vo **72.** 84 DVDs **73.** $-11x$ **74.** $-35x$ **75.** $22x - 19$ **76.** $-9x - 32$ **77.** -1 **78.** -25 **79.** 13 **80.** -6 **81.** -22 **82.** -6 **83.** -15 **84.** 18 **85.** -5 **86.** 11 **87.** 2 **88.** -1 **89.** 0 **90.** -6 **91.** 5 **92.** 1 **93.** Vermont: 108.1 trillion BTU; District of Columbia: 132.4 trillion BTU **94.** North Dakota 86,843 mi; South Dakota: 82,354 mi

Chapter 3 Getting Ready for the Test **1.** A **2.** B **3.** A **4.** C **5.** C **6.** B **7.** B **8.** A **9.** C **10.** A **11.** D **12.** E **13.** B **14.** C **15.** E **16.** F

Chapter 3 Test **1.** $-5x + 5$ **2.** $-6y - 14$ **3.** $-8z - 20$ **4.** $(15x + 15)$ in. **5.** $(12x - 4)$ sq m **6.** -6 **7.** -6 **8.** 24 **9.** -2 **10.** 6 **11.** 3 **12.** -2 **13.** 0 **14.** 4 **15.** $-23 + x$ **16.** $-2 - 3x$ **17.** $2 \cdot 5 + (-15) = -5$ **18.** $3x + 6 = -30$ **19.** -2 **20.** 8 free throws **21.** 244 women

Cumulative Review Chapters 1–3 **1.** three hundred eight million, sixty-three thousand, five hundred fifty-seven; Sec. 1.2, Ex. 7 **2.** two hundred seventy-six thousand, four; Sec. 1.2 **3.** 13 in.; Sec. 1.3, Ex. 9 **4.** 18 in.; Sec. 1.3 **5.** 726; Sec. 1.3, Ex. 8 **6.** 9585; Sec. 1.3 **7.** 249,000; Sec. 1.4, Ex. 3 **8.** 844,000; Sec. 1.4 **9.** 200; Sec. 1.5, Ex. 3a **10.** 29,230; Sec. 1.5 **11.** 208; Sec. 1.6, Ex. 5 **12.** 86; Sec. 1.6 **13.** 7; Sec. 1.7, Ex. 9 **14.** 35; Sec. 1.7 **15.** 26; Sec. 1.8, Ex. 4 **16.** 10; Sec. 1.8 **17.** 20 is a solution; 26 and 40 are not solutions.; Sec. 1.8, Ex. 7 **18. a.** $<$ **b.** $>$; Sec. 2.1 **19.** 3; Sec. 2.2, Ex. 1 **20.** -7; Sec. 2.2 **21.** -25; Sec. 2.2, Ex. 5 **22.** -4; Sec. 2.2 **23.** 23; Sec. 2.2, Ex. 7 **24.** 17; Sec. 2.2 **25.** -14; Sec. 2.3, Ex. 2 **26.** -5; Sec. 2.3 **27.** 11; Sec. 2.3, Ex. 3 **28.** 29; Sec. 2.3 **29.** -4; Sec. 2.3, Ex. 4 **30.** -3; Sec. 2.3 **31.** -2; Sec. 2.4, Ex. 10 **32.** 6; Sec. 2.4 **33.** 5; Sec. 2.4, Ex. 11 **34.** -13; Sec. 2.4 **35.** -16; Sec. 2.4, Ex. 12 **36.** -10; Sec. 2.4 **37.** 9; Sec. 2.5, Ex. 1 **38.** -32; Sec. 2.5 **39.** -9; Sec. 2.5, Ex. 2 **40.** 25; Sec. 2.5 **41.** $6y + 2$; Sec. 3.1, Ex. 2 **42.** $3x + 9$; Sec. 3.1 **43.** not a solution; Sec. 2.6, Ex. 1 **44.** solution; Sec. 2.6 **45.** 3; Sec. 2.6, Ex. 7 **46.** -5; Sec. 2.6 **47.** 12; Sec. 3.2, Ex. 7 **48.** -2; Sec. 3.2 **49.** software: $420; computer system: $1680; Sec. 3.4, Ex. 4 **50.** 11; Sec. 3.4

Chapter 4 Fractions and Mixed Numbers

Section 4.1

Vocabulary, Readiness & Video Check **1.** fraction; denominator; numerator **3.** improper; proper; mixed number **5.** equal; improper **7.** how many equal parts to divide each whole number into **9.** addition; +

Exercise Set 4.1 **1.** numerator: 1; denominator: 2; proper **3.** numerator: 10; denominator: 3; improper **5.** numerator: 15; denominator: 15; improper **7.** $\dfrac{1}{3}$ **9. a.** $\dfrac{11}{4}$ **b.** $2\dfrac{3}{4}$ **11. a.** $\dfrac{23}{6}$ **b.** $3\dfrac{5}{6}$ **13.** $\dfrac{7}{12}$ **15.** $\dfrac{3}{7}$ **17.** $\dfrac{4}{9}$ **19. a.** $\dfrac{4}{3}$ **b.** $1\dfrac{1}{3}$ **21. a.** $\dfrac{11}{2}$ **b.** $5\dfrac{1}{2}$ **23.** $\dfrac{1}{6}$ **25.** $\dfrac{5}{8}$ **27.** **29.** **31.** **33.** $\dfrac{42}{131}$ of the students **35. a.** 89 **b.** $\dfrac{89}{131}$ of the students **37.** $\dfrac{7}{44}$ of the presidents **39.** $\dfrac{10}{15}$ of the tropical storms **41.** $\dfrac{11}{31}$ of the month

$\frac{10}{31}$ of the class **45. a.** $\frac{33}{50}$ of the states **b.** 17 states **c.** $\frac{17}{50}$ of the states **47. a.** $\frac{21}{50}$ of the marbles **b.** 29 **c.** $\frac{29}{50}$ of the marbles **49.** $\frac{1}{4}$ [number line] **51.** $\frac{4}{7}$ [number line] **53.** $\frac{8}{5}$ [number line] **55.** $\frac{7}{3}$ [number line] **57.** 1 **59.** −5 **61.** 0 **63.** 1 **65.** undefined **67.** 3 **69.** $\frac{7}{3}$ **71.** $\frac{18}{5}$ **73.** $\frac{53}{8}$ **75.** $\frac{83}{7}$ **77.** $\frac{187}{20}$ **79.** $\frac{500}{3}$ **81.** $3\frac{2}{5}$ **83.** $4\frac{5}{8}$ **85.** $3\frac{2}{15}$ **87.** 15 **89.** $1\frac{7}{175}$ **91.** $6\frac{65}{112}$ **93.** 9 **95.** 125 **97.** $\frac{-11}{2};\frac{11}{-2}$ **99.** $\frac{13}{-15};-\frac{13}{15}$ **101.** answers may vary **103.** ●●●●○○○○○○ **105.** 7 **107.** $\frac{947}{2247}$ of the stores **109.** $\frac{1500}{1580}$ of the affiliates

Section 4.2

Calculator Explorations 1. $\frac{4}{7}$ **3.** $\frac{20}{27}$ **5.** $\frac{8}{15}$ **7.** $\frac{2}{9}$

Vocabulary, Readiness & Video Check 1. prime factorization **3.** prime **5.** equivalent **7.** Check that every factor is a prime number and check that the product of the factors is the original number **9.** You can simplify the two fractions and then compare them. $\frac{3}{9}$ and $\frac{6}{18}$ both simplify to $\frac{1}{3}$ so the original fractions are equivalent.

Exercise Set 4.2 1. $2^2 \cdot 5$ **3.** $2^4 \cdot 3$ **5.** 3^4 **7.** $2 \cdot 3^4$ **9.** $2 \cdot 5 \cdot 11$ **11.** $5 \cdot 17$ **13.** $2^4 \cdot 3 \cdot 5$ **15.** $2^2 \cdot 3^2 \cdot 23$ **17.** $\frac{1}{4}$ **19.** $\frac{2x}{21}$ **21.** $\frac{7}{8}$ **23.** $\frac{2}{3}$ **25.** $\frac{7}{10}$ **27.** $-\frac{7}{9}$ **29.** $\frac{5x}{6}$ **31.** $\frac{27}{64}$ **33.** $\frac{5x}{8}$ **35.** $-\frac{5}{8}$ **37.** $\frac{3x^2y}{2}$ **39.** $\frac{3}{4}$ **41.** $\frac{5}{8z}$ **43.** $\frac{3}{14}$ **45.** $-\frac{11}{17y}$ **47.** $\frac{7}{8}$ **49.** $14a^2$ **51.** equivalent **53.** not equivalent **55.** equivalent **57.** equivalent **59.** not equivalent **61.** not equivalent **63.** $\frac{1}{4}$ of a shift **65.** $\frac{1}{2}$ mi **67. a.** $\frac{8}{25}$ **b.** 34 states **c.** $\frac{17}{25}$ **69.** $\frac{5}{12}$ of the wall **71. a.** 12 states **b.** $\frac{1}{4}$ **73.** $\frac{15}{122}$ of U.S astronauts **75.** −3 **77.** −14 **79.** answers may vary **81.** $\frac{3}{5}$ **83.** $\frac{9}{25}$ **85.** $\frac{1}{25}$ **87.** $2^2 \cdot 3^5 \cdot 5 \cdot 7$ **89.** answers may vary **91.** no; answers may vary **93.** $\frac{3}{50}$ **95.** answers may vary **97.** $\frac{6}{25}$ **99.** answers may vary **101.** 786, 222, 900, 1470 **103.** 6; answers may vary

Section 4.3

Vocabulary, Readiness & Video Check 1. $\frac{a \cdot c}{b \cdot d}$ **3.** $\frac{2 \cdot 2 \cdot 2}{7}; \frac{2}{7} \cdot \frac{2}{7} \cdot \frac{2}{7}$ **5.** $\frac{a \cdot d}{b \cdot c}$ **7.** We have a negative fraction times a positive fraction and a negative number times a positive number is a negative number. **9.** numerator; denominator **11.** radius $= \frac{1}{2} \cdot$ diameter

Exercise Set 4.3 1. $\frac{18}{77}$ **3.** $-\frac{5}{28}$ **5.** $\frac{1}{15}$ **7.** $\frac{18x}{55}$ **9.** $\frac{3a^2}{4}$ **11.** $\frac{x^2}{y}$ **13.** 0 **15.** $-\frac{17}{25}$ **17.** $\frac{1}{56}$ **19.** $\frac{1}{125}$ **21.** $\frac{4}{9}$ **23.** $-\frac{4}{27}$ **25.** $\frac{4}{5}$ **27.** $-\frac{1}{6}$ **29.** $-\frac{16}{9x}$ **31.** $\frac{121y}{60}$ **33.** $-\frac{1}{6}$ **35.** $\frac{x}{25}$ **37.** $\frac{10}{27}$ **39.** $\frac{18x^2}{35}$ **41.** $\frac{10}{9}$ **43.** $-\frac{1}{4}$ **45.** $\frac{9}{16}$ **47.** xy^2 **49.** $\frac{77}{2}$ **51.** $-\frac{36}{x}$ **53.** $\frac{3}{49}$ **55.** $-\frac{19y}{7}$ **57.** $\frac{4}{11}$ **59.** $\frac{8}{3}$ **61.** $\frac{15x}{4}$ **63.** $\frac{8}{9}$ **65.** $\frac{1}{60}$ **67.** b **69.** $\frac{2}{5}$ **71.** $-\frac{5}{3}$ **73. a.** $\frac{1}{3}$ **b.** $\frac{12}{25}$ **75. a.** $-\frac{36}{55}$ **b.** $-\frac{44}{45}$ **77.** yes **79.** no **81.** 50 **83.** 20 **85.** 112 **87.** 84 million astronauts **89.** 868 mi **91.** $\frac{3}{16}$ in. **93.** $1838 **95.** 50 libraries **97.** $\frac{1}{14}$ sq ft **99.** 3840 mi **101.** 2400 mi **103.** 201 **105.** 196 **107.** answers may vary **109.** 5 **111.** 40,630,000 people **113.** 8505

Section 4.4

Vocabulary, Readiness & Video Check 1. like; unlike **3.** $-\frac{a}{b}$ **5.** least common denominator (LCD) **7.** numerators; denominator **9.** $P = \frac{5}{12} + \frac{7}{12} + \frac{5}{12} + \frac{7}{12}$; 2 meters **11.** Multiplying by 1 does not change the value of a fraction.

A8 Answers to Selected Exercises

Exercise Set 4.4   **1.** $\dfrac{7}{11}$ **3.** $\dfrac{2}{3}$ **5.** $-\dfrac{1}{4}$ **7.** $-\dfrac{1}{2}$ **9.** $\dfrac{2}{3x}$ **11.** $-\dfrac{x}{9}$ **13.** $\dfrac{6}{11}$ **15.** $\dfrac{3}{4}$ **17.** $-\dfrac{3}{y}$ **19.** $-\dfrac{19}{33}$ **21.** $-\dfrac{1}{3}$ **23.** $\dfrac{7a-3}{4}$ **25.** $\dfrac{9}{10}$ **27.** $-\dfrac{13x}{14}$ **29.** $\dfrac{9x+1}{15}$ **31.** $-\dfrac{x}{2}$ **33.** $-\dfrac{2}{3}$ **35.** $\dfrac{3x}{4}$ **37.** $\dfrac{5}{4}$ **39.** $\dfrac{2}{5}$ **41.** 1 in.   **43.** 2 m   **45.** $\dfrac{7}{10}$ mi   **47.** $\dfrac{8}{25}$ **49.** $\dfrac{7}{25}$ **51.** $\dfrac{1}{50}$ **53.** 45   **55.** 72   **57.** 150   **59.** 24x   **61.** 126   **63.** 168   **65.** 14   **67.** 20   **69.** 25   **71.** 56x   **73.** 8y   **75.** 20a   **77.** $\dfrac{54}{100}; \dfrac{50}{100}; \dfrac{46}{100}; \dfrac{50}{100}; \dfrac{15}{100}; \dfrac{65}{100}; \dfrac{45}{100}; \dfrac{52}{100}; \dfrac{60}{100}; \dfrac{61}{100}; \dfrac{48}{100}; \dfrac{50}{100}$ **79.** drugs, health and beauty aids   **81.** 9   **83.** 125   **85.** 49   **87.** 24   **89.** $\dfrac{2}{7}+\dfrac{9}{7}=\dfrac{11}{7}$ **91.** answers may vary   **93.** 1; answers may vary   **95.** 814x   **97.** answers may vary   **99.** a, b, and d

Section 4.5

Calculator Explorations   **1.** $\dfrac{37}{80}$ **3.** $\dfrac{95}{72}$ **5.** $\dfrac{394}{323}$

Vocabulary, Readiness & Video Check   **1.** equivalent; least common denominator   **3.** $\dfrac{4}{24}; \dfrac{15}{24}; \dfrac{19}{24}$ **5.** expression; equation   **7.** They are unlike terms and so cannot be combined.   **9.** $\dfrac{3}{2}+\dfrac{1}{3}; 6$

Exercise Set 4.5   **1.** $\dfrac{5}{6}$ **3.** $\dfrac{1}{6}$ **5.** $-\dfrac{4}{33}$ **7.** $-\dfrac{3}{14}$ **9.** $\dfrac{3x}{5}$ **11.** $\dfrac{24-y}{12}$ **13.** $\dfrac{11}{36}$ **15.** $-\dfrac{44}{7}$ **17.** $\dfrac{89a}{99}$ **19.** $\dfrac{4y-1}{6}$ **21.** $\dfrac{x+6}{2x}$ **23.** $-\dfrac{8}{33}$ **25.** $\dfrac{3}{14}$ **27.** $\dfrac{11y-10}{35}$ **29.** $-\dfrac{11}{36}$ **31.** $\dfrac{1}{20}$ **33.** $\dfrac{33}{56}$ **35.** $\dfrac{17}{16}$ **37.** $-\dfrac{2}{9}$ **39.** $-\dfrac{11}{30}$ **41.** $\dfrac{15+11y}{33}$ **43.** $-\dfrac{53}{42}$ **45.** $\dfrac{7x}{8}$ **47.** $\dfrac{11}{18}$ **49.** $\dfrac{44a}{39}$ **51.** $-\dfrac{11}{60}$ **53.** $\dfrac{5y+9}{9y}$ **55.** $\dfrac{19}{20}$ **57.** $\dfrac{56}{45}$ **59.** $\dfrac{40+9x}{72x}$ **61.** $-\dfrac{5}{24}$ **63.** $\dfrac{37x-20}{56}$ **65.** <   **67.** >   **69.** >   **71.** $\dfrac{13}{12}$ **73.** $\dfrac{1}{4}$ **75.** $\dfrac{11}{6}$ **77.** $\dfrac{34}{15}$ cm or $2\dfrac{4}{15}$ cm   **79.** $\dfrac{17}{10}$ m or $1\dfrac{7}{10}$ m   **81.** $x+\dfrac{1}{2}$ **83.** $-\dfrac{3}{8}-x$ **85.** $\dfrac{7}{100}$ mph   **87.** $\dfrac{5}{8}$ in.   **89.** $\dfrac{49}{100}$ of students   **91.** $\dfrac{47}{32}$ in.   **93.** $\dfrac{19}{25}$ **95.** $\dfrac{3}{40}$ **97.** $\dfrac{63}{100}$ **99.** -20 **101.** -6 **103.** $\dfrac{3}{5}+\dfrac{4}{5}=\dfrac{7}{5}$ **105.** $\dfrac{223}{540}$ **107.** $\dfrac{49}{44}$ **109.** answers may vary   **111.** standard mail

Integrated Review   **1.** $\dfrac{3}{7}$ **2.** $\dfrac{5}{4}$ or $1\dfrac{1}{4}$ **3.** $\dfrac{73}{85}$ **4.** [bar diagram]   **5.** -1 **6.** 17   **7.** 0   **8.** undefined   **9.** $5\cdot 13$ **10.** $2\cdot 5\cdot 7$ **11.** $3^2\cdot 5\cdot 7$ **12.** $3^2\cdot 7^2$ **13.** $\dfrac{1}{7}$ **14.** $\dfrac{6}{5}$ **15.** $-\dfrac{14}{15}$ **16.** $-\dfrac{9}{10}$ **17.** $\dfrac{2x}{5}$ **18.** $\dfrac{3}{8y}$ **19.** $\dfrac{11z^2}{14}$ **20.** $\dfrac{7}{11ab^2}$ **21.** not equivalent   **22.** equivalent   **23. a.** $\dfrac{1}{25}$ **b.** 48   **c.** $\dfrac{24}{25}$ **24. a.** $\dfrac{11}{47}$ **b.** 108,000   **c.** $\dfrac{36}{47}$ **25.** 30   **26.** 14   **27.** 90   **28.** $\dfrac{28}{36}$ **29.** $\dfrac{55}{75}$ **30.** $\dfrac{40}{48}$ **31.** $\dfrac{4}{5}$ **32.** $-\dfrac{2}{5}$ **33.** $\dfrac{3}{25}$ **34.** $\dfrac{1}{3}$ **35.** $\dfrac{4}{5}$ **36.** $\dfrac{5}{9}$ **37.** $-\dfrac{1}{6y}$ **38.** $\dfrac{3x}{2}$ **39.** $\dfrac{1}{18}$ **40.** $\dfrac{4}{21}$ **41.** $-\dfrac{7}{48z}$ **42.** $-\dfrac{9}{50}$ **43.** $\dfrac{37}{40}$ **44.** $\dfrac{11}{36}$ **45.** $\dfrac{11}{18}$ **46.** $\dfrac{25y+12}{50}$ **47.** $\dfrac{2}{3}\cdot x$ or $\dfrac{2}{3}x$ **48.** $x\div\left(-\dfrac{1}{5}\right)$ **49.** $-\dfrac{8}{9}-x$ **50.** $\dfrac{6}{11}+x$ **51.** 1020   **52.** 24 lots   **53.** $\dfrac{3}{4}$ ft

Section 4.6

Vocabulary, Readiness & Video Check   **1.** complex   **3.** division   **5.** addition   **7.** distributive property   **9.** Since x is squared and the replacement value is negative, we use parentheses to make sure the whole value of x is squared. Without parentheses, the exponent would not apply to the negative sign.

Exercise Set 4.6   **1.** $\dfrac{1}{6}$ **3.** $\dfrac{7}{3}$ **5.** $\dfrac{x}{6}$ **7.** $\dfrac{23}{22}$ **9.** $\dfrac{2x}{13}$ **11.** $\dfrac{17}{60}$ **13.** $\dfrac{5}{8}$ **15.** $\dfrac{35}{9}$ **17.** $-\dfrac{17}{45}$ **19.** $\dfrac{11}{8}$ **21.** $\dfrac{29}{10}$ **23.** $\dfrac{27}{32}$ **25.** $\dfrac{1}{100}$ **27.** $\dfrac{9}{64}$ **29.** $\dfrac{7}{6}$ **31.** $-\dfrac{2}{5}$ **33.** $-\dfrac{2}{9}$ **35.** $\dfrac{11}{9}$ **37.** $\dfrac{5a}{2}$ **39.** $\dfrac{7}{2}$ **41.** $\dfrac{9}{20}$ **43.** $-\dfrac{13}{2}$ **45.** $\dfrac{9}{25}$ **47.** $-\dfrac{5}{32}$ **49.** 1   **51.** $-\dfrac{2}{5}$ **53.** $-\dfrac{11}{40}$ **55.** $\dfrac{x+6}{16}$ **57.** $\dfrac{7}{2}$ or $3\dfrac{1}{2}$ **59.** $\dfrac{49}{6}$ or $8\dfrac{1}{6}$ **61.** no; answers may very   **63.** $\dfrac{5}{8}$ **65.** $\dfrac{11}{56}$ **67.** halfway between a and b **69.** false   **71.** true   **73.** true   **75.** addition: answers may vary   **77.** subtraction, multiplication, addition, division   **79.** division, multiplication, subtraction, addition   **81.** $-\dfrac{77}{16}$ **83.** $-\dfrac{55}{16}$

Section 4.7

Calculator Explorations 1. $\frac{280}{11}$ 3. $\frac{3776}{35}$ 5. $26\frac{1}{14}$ 7. $92\frac{3}{10}$

Vocabulary, Readiness & Video Check 1. mixed number 3. round 5. The denominator of the mixed number we're graphing, $-3\frac{4}{5}$, is 5. 7. The fractional part of a mixed number should always be a proper fraction. 9. We're adding two mixed numbers with unlike signs, so the answer has the sign of the mixed number with the larger absolute value, which in this case is negative.

Exercise Set 4.7 1. [number line with $-2\frac{2}{3}$, $-\frac{1}{3}$, $\frac{7}{8}$] 3. [number line with $-3\frac{4}{5}$, $\frac{1}{3}$, $1\frac{1}{3}$] 5. b 7. a 9. $\frac{8}{21}$ 11. $4\frac{3}{8}$ 13. $7\frac{7}{10}$; 8 15. $23\frac{31}{35}$; 24 17. $12\frac{1}{2}$ 19. $5\frac{1}{2}$ 21. $18\frac{2}{3}$ 23. a 25. b 27. $6\frac{2}{3}$; 7 29. $13\frac{11}{14}$; 14 31. $17\frac{7}{25}$ 33. $47\frac{53}{84}$ 35. $25\frac{5}{14}$ 37. $13\frac{13}{24}$ 39. $2\frac{3}{5}$; 3 41. $7\frac{5}{14}$; 7 43. $\frac{24}{25}$ 45. $3\frac{5}{9}$ 47. $15\frac{3}{4}$ 49. 4 51. $5\frac{11}{14}$ 53. $6\frac{2}{9}$ 55. $\frac{25}{33}$ 57. $35\frac{13}{18}$ 59. $2\frac{1}{2}$ 61. $72\frac{19}{30}$ 63. $\frac{11}{14}$ 65. $5\frac{4}{7}$ 67. $13\frac{16}{33}$ 69. $-5\frac{2}{7} - x$ 71. $1\frac{9}{10} \cdot x$ 73. $3\frac{3}{16}$ mi 75. $9\frac{2}{5}$ in. 77. $7\frac{13}{20}$ in. 79. $3\frac{1}{2}$ sq yd 81. $\frac{15}{16}$ sq in. 83. $21\frac{5}{24}$ m 85. no; she will be short $\frac{1}{2}$ ft 87. $4\frac{2}{3}$ m 89. $9\frac{3}{4}$ min 91. $1\frac{4}{5}$ min 93. $-10\frac{3}{25}$ 95. $-24\frac{7}{8}$ 97. $-13\frac{59}{60}$ 99. $4\frac{2}{7}$ 101. $-1\frac{23}{24}$ 103. $\frac{73}{1000}$ 105. $1x$ or x 107. $1a$ or a 109. a, b, c 111. Incorrect; to divide mixed numbers, first write each mixed number as an improper fraction. 113. answers may vary 115. answers may vary 117. answers may vary

Section 4.8

Vocabulary, Readiness & Video Check 1. 6 3. 15 5. addition property of equality 7. We multiply by 12 because it is the LCD of all fractions in the equation. The equation no longer contains fractions.

Exercise Set 4.8 1. $-\frac{2}{3}$ 3. $\frac{1}{13}$ 5. $\frac{4}{5}$ 7. $\frac{11}{12}$ 9. $-\frac{7}{10}$ 11. $\frac{11}{16}$ 13. $\frac{11}{18}$ 15. $\frac{2}{7}$ 17. 12 19. -27 21. $\frac{27}{8}$ 23. $\frac{1}{21}$ 25. $\frac{2}{11}$ 27. $-\frac{3}{10}$ 29. 1 31. 10 33. -1 35. -15 37. $\frac{3x-28}{21}$ 39. $\frac{y+10}{2}$ 41. $\frac{7x}{15}$ 43. $\frac{4}{3}$ 45. 2 47. $\frac{21}{10}$ 49. $-\frac{1}{14}$ 51. -3 53. $-\frac{1}{9}$ 55. 50 57. $-\frac{3}{7}$ 59. 4 61. $\frac{3}{5}$ 63. $-\frac{1}{24}$ 65. $-\frac{5}{14}$ 67. 4 69. -36 71. 57,200 73. 330 75. answers may vary 77. $\frac{112}{11}$ 79. area: $\frac{3}{16}$ sq in.; perimeter: 2 in.

Chapter 4 Vocabulary Check 1. reciprocals 2. composite number 3. equivalent 4. improper fraction 5. prime number 6. simplest form 7. proper fraction 8. mixed number 9. numerator; denominator 10. prime factorization 11. undefined 12. 0 13. like 14. least common denominator 15. complex fraction 16. cross products

Chapter 4 Review 1. $\frac{2}{6}$ 2. $\frac{4}{7}$ 3. $\frac{7}{3}$ or $2\frac{1}{3}$ 4. $\frac{13}{4}$ or $3\frac{1}{4}$ 5. $\frac{11}{12}$ 6. a. 108 b. $\frac{108}{131}$ 7. -1 8. 1 9. 0 10. undefined 11. [number line $\frac{7}{9}$] 12. [number line $\frac{4}{7}$] 13. [number line $\frac{5}{4}$] 14. [number line $\frac{7}{5}$] 15. $3\frac{3}{4}$ 16. 3 17. $\frac{11}{5}$ 18. $\frac{35}{9}$ 19. $\frac{3}{7}$ 20. $\frac{5}{9}$ 21. $-\frac{1}{3x}$ 22. $-\frac{y^2}{2}$ 23. $\frac{29}{32c}$ 24. $\frac{18z}{23}$ 25. $\frac{5x}{3y^2}$ 26. $\frac{7b}{5c^2}$ 27. $\frac{2}{3}$ of a foot 28. $\frac{3}{5}$ of the cars 29. no 30. yes 31. $\frac{3}{10}$ 32. $-\frac{5}{14}$ 33. $\frac{9}{x^2}$ 34. $\frac{y}{2}$ 35. $-\frac{1}{27}$ 36. $-\frac{25}{144}$ 37. -2 38. $\frac{15}{4}$ 39. $\frac{27}{2}$ 40. $-\frac{5}{6y}$ 41. $\frac{12}{7}$ 42. $-\frac{63}{10}$ 43. $\frac{77}{48}$ sq ft 44. $\frac{4}{9}$ sq m 45. $\frac{10}{11}$ 46. $\frac{2}{3}$ 47. $-\frac{1}{3}$ 48. $\frac{4x}{5}$ 49. $\frac{4y-3}{21}$ 50. $-\frac{1}{15}$ 51. $3x$ 52. 24 53. 20 54. 35 55. $49a$ 56. $45b$ 57. 40 58. 10 59. $\frac{3}{4}$ of his homework 60. $\frac{3}{2}$ mi 61. $\frac{11}{18}$ 62. $\frac{7}{26}$ 63. $-\frac{1}{12}$ 64. $-\frac{5}{12}$ 65. $\frac{25x+2}{55}$ 66. $\frac{4+3b}{15}$ 67. $\frac{7y}{36}$ 68. $\frac{11x}{18}$ 69. $\frac{4y+45}{9y}$ 70. $\frac{9y-33}{11y}$ 71. $\frac{91}{150}$ 72. $\frac{5}{18}$ 73. $\frac{19}{9}$ m 74. $\frac{3}{2}$ ft 75. $\frac{21}{50}$ of the donors 76. $\frac{1}{4}$ yd 77. $\frac{4x}{7}$ 78. $\frac{3y}{11}$ 79. -2 80. $-7y$ 81. $\frac{15}{4}$ 82. $-\frac{5}{24}$ 83. $\frac{8}{13}$ 84. $-\frac{1}{27}$ 85. $\frac{29}{110}$ 86. $-\frac{1}{7}$ 87. $20\frac{7}{24}$ 88. $2\frac{51}{55}$; 3 89. $5\frac{1}{5}$; 6 90. $5\frac{1}{4}$ 91. 22 mi 92. $36\frac{2}{3}$ g 93. each measurement is $4\frac{1}{4}$ in. 94. $\frac{7}{10}$ yd 95. $-27\frac{5}{14}$

96. $-\frac{33}{40}$ **97.** $1\frac{5}{27}$ **98.** $-3\frac{15}{16}$ **99.** $\frac{5}{6}$ **100.** $-\frac{9}{10}$ **101.** -10 **102.** -6 **103.** 15 **104.** 1 **105.** 5 **106.** -4 **107.** $\frac{1}{4}$ **108.** $\frac{y^2}{2x}$ **109.** $\frac{1}{5}$ **110.** $-\frac{7}{12x}$ **111.** $\frac{11x}{12}$ **112.** $-\frac{23}{55}$ **113.** $-6\frac{2}{5}$ **114.** $12\frac{3}{8}$; 13 **115.** $2\frac{19}{35}$; 2 **116.** $\frac{22}{7}$ or $3\frac{1}{7}$ **117.** $-\frac{1}{12}$ **118.** $-\frac{9}{14}$ **119.** $-\frac{4}{9}$ **120.** -19 **121.** $44\frac{1}{2}$ yd **122.** $40\frac{1}{2}$ sq ft

Chapter 4 Getting Ready for the Test
1. A **2.** B **3.** D **4.** C **5.** C **6.** B **7.** C **8.** D **9.** B **10.** A **11.** F **12.** H **13.** B **14.** D **15.** C **16.** A **17.** D **18.** C **19.** D **20.** A **21.** B

Chapter 4 Test
1. $\frac{7}{16}$ **2.** $\frac{23}{3}$ **3.** $18\frac{3}{4}$ **4.** $\frac{4}{35}$ **5.** $-\frac{3x}{5}$ **6.** not equivalent **7.** equivalent **8.** $2^2 \cdot 3 \cdot 7$ **9.** $3^2 \cdot 5 \cdot 11$ **10.** $\frac{4}{3}$ **11.** $-\frac{4}{3}$ **12.** $\frac{8x}{9}$ **13.** $\frac{x-21}{7x}$ **14.** y^2 **15.** $\frac{16}{45}$ **16.** $\frac{9a+4}{10}$ **17.** $-\frac{2}{3y}$ **18.** $\frac{1}{a^2}$ **19.** $\frac{3}{4}$ **20.** $14\frac{1}{40}$ **21.** $16\frac{8}{11}$ **22.** $\frac{64}{3}$ or $21\frac{1}{3}$ **23.** $22\frac{1}{2}$ **24.** $-\frac{5}{3}$ or $-1\frac{2}{3}$ **25.** $\frac{9}{16}$ **26.** $\frac{3}{8}$ **27.** $\frac{11}{12}$ **28.** $\frac{3}{4x}$ **29.** $\frac{76}{21}$ or $3\frac{13}{21}$ **30.** -2 **31.** -4 **32.** 1 **33.** $\frac{5}{2}$ or $2\frac{1}{2}$ **34.** $\frac{4}{31}$ **35.** $3\frac{3}{4}$ ft **36.** $\frac{23}{50}$ **37.** $\frac{13}{50}$ **38.** $\$2820$ **39.** perimeter: $3\frac{1}{3}$ ft; area: $\frac{2}{3}$ sq ft **40.** 24 mi

Cumulative Review Chapters 1–4
1. five hundred forty-six; Sec. 1.2 . Ex. 5 **2.** one hundred fifteen; Sec. 1.2 **3.** twenty-seven thousand, thirty-four; Sec 1.2, Ex. 6 **4.** six thousand, five hundred seventy-three; Sec. 1.2 **5.** 759; Sec. 1.3, Ex. 1 **6.** 631; Sec. 1.3 **7.** 514; Sec. 1.3, Ex. 7 **8.** 933; Sec. 1.3 **9.** 278,000; Sec. 1.4, Ex. 2 **10.** 1440; Sec. 1.4 **11.** 57,600 megabytes; Sec. 1.5, Ex. 7 **12.** 1305 mi; Sec. 1.5 **13.** 7089 R 5; Sec. 1.6, Ex. 7 **14.** 379 R 10; Sec. 1.6 **15.** 7^3; Sec. 1.7, Ex. 1 **16.** 7^2; Sec. 1.7 **17.** $3^4 \cdot 9^3$; Sec. 1.7, Ex. 4 **18.** $9^4 \cdot 5^2$; Sec 1.7 **19.** 6; Sec. 1.8, Ex. 2 **20.** 52; Sec. 1.8 **21.** -7188; Sec. 2.1, Ex. 1 **22.** -21; Sec. 2.1 **23.** -4; Sec. 2.2, Ex. 3 **24.** 5; Sec. 2.2 **25.** 3; Sec. 2.3, Ex. 9 **26.** 10; Sec. 2.3 **27.** 25; Sec. 2.4, Ex. 8 **28.** -16; Sec. 2.4 **29.** -16; Sec. 2.5, Ex. 8 **30.** 25; Sec. 2.5 **31.** $6y + 2$; Sec. 3.1, Ex. 2 **32.** $6x + 9$; Sec. 3.1 **33.** -14; Sec. 3.2, Ex. 4 **34.** -18; Sec. 3.2 **35.** -1; Sec. 3.3, Ex. 2 **36.** -11; Sec. 3.3 **37.** $\frac{2}{5}$; Sec. 4.1, Ex. 3 **38.** $2^2 \cdot 3 \cdot 13$; Sec. 4.2 **39. a.** $\frac{38}{9}$ **b.** $\frac{19}{11}$; Sec. 4.1, Ex. 20 **40.** $7\frac{4}{5}$; Sec. 4.1 **41.** $\frac{7x}{11}$; Sec. 4.2, Ex. 5 **42.** $\frac{2}{3y}$; Sec. 4.2 **43.** $2\frac{11}{12}$; Sec. 4.7, Ex. 2 **44.** $2\frac{2}{3}$; Sec. 4.7 **45.** $\frac{5}{12}$; Sec. 4.3, Ex. 11 **46.** $\frac{11}{56}$; Sec. 4.7

Chapter 5 Decimals
Section 5.1

Vocabulary, Readiness & Video Check **1.** words; standard form **3.** decimals **5.** tenths; tens **7.** as "and" **9.** Reading a decimal correctly gives you the correct place value, which tells you the denominator of your equivalent fraction. **11.** When rounding, look to the digit to the right of the place value we're rounding to. In this case we look to the hundredths-place digit, which is 7.

Exercise Set 5.1 **1.** five and sixty-two hundredths **3.** sixteen and twenty-three hundredths **5.** negative two hundred five thousandt **7.** one hundred sixty-seven and nine thousandths **9.** three thousand and four hundredths **11.** one hundred five and six tenths **13.** two and forty-three hundredths

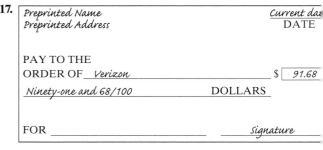

19. 2.8 **21.** 9.08 **23.** -705.625 **25.** 0.0046 **27.** $\frac{7}{10}$ **29.** $\frac{27}{100}$ **31.** $\frac{2}{5}$ **33.** $5\frac{2}{5}$ **35.** $-\frac{29}{500}$ **37.** $7\frac{1}{125}$ **39.** $15\frac{401}{500}$ **41.** $\frac{601}{2000}$ **43.** $0.8; \frac{8}{10}$ or $\frac{4}{5}$ **45.** seventy-seven thousandths; $\frac{77}{1000}$ **47.** < **49.** < **51.** < **53.** = **55.** < **57.** > **59.** < **61.** > **63.** 0.6 **65.** 98,210 **67.** -0.23 **69.** 0.594 **71.** 3.1 **73.** 3.142 **75.** $\$27$ **77.** $\$0.20$ **79.** 0.7 in. **81.** 1.92 min **83.** $\$68$ **85.** 225 days **87.** 57 **89.** 35 **91.** b **93.** a **95.** answers may vary **97.** 7.12 **99.** $\frac{26,849,577}{100,000,000,000}$ **101.** answers may vary **103.** answers may vary **105.** 0.26499, 0.25786 **107.** 0.10299, 0.1037, 0.1038, 0.9 **109.** $\$4200$ million

Section 5.2

Calculator Explorations 1. 328.742 3. 5.2414 5. 865.392

Vocabulary, Readiness & Video Check 1. last 3. like 5. false 7. Lining up the decimal points also lines up place values, so we only add or subtract digits in the same place values. 9. So the subtraction can be written vertically with decimal points lined up. 11. perimeter

Exercise Set 5.2 1. 7.7 3. 10.35 5. 27.0578 7. −8.57 9. 10.33 11. 465.56; 230 + 230 / 460 13. 115.123; 100 + 6 + 9 / 115 15. 50.409 17. 4.4 19. 15.3 21. 598.23 23. 1.83; 6 − 4 = 2 25. 876.6; 1000 − 100 / 900 27. 194.4 29. −6.32 31. −6.15 33. 3.1 35. 2.9988 37. 16.3 39. 3.1 41. −5.62 43. 363.36 45. −549.8 47. 861.6 49. 216.114 51. 0.088 53. −180.44 55. −1.1 57. 3.81 59. 3.39 61. 1.61 63. no 65. yes 67. no 69. $6.9x + 6.9$ 71. $3.47y − 10.97$ 73. $7.52 75. −$0.42 77. 28.56 m 79. 16.16 in. 81. 195.8 mph 83. 42.7 texts per day 85. $2356 million 87. 326.3 in. 89. 67.44 ft 91. 4.234 mph 93. Switzerland 95. 7.9 lb 97.

Country	Pounds of Chocolate per Person
Switzerland	19.8
Germany	17.4
UK	16.3
Ireland	16.2
Norway	14.6
Sweden	11.9

99. 138 101. $\frac{4}{9}$ 103. incorrect; 9.200 + 8.630 + 4.005 / 21.835 105. 6.08 in. 107. $1.20 109. 1 nickel, 1 dime, and 2 pennies; 3 nickels and 2 pennies; 1 dime and 7 pennies; 2 nickels and 7 pennies 111. answers may vary 113. answers may vary 115. $22.181x − 22.984$

Section 5.3

Vocabulary, Readiness & Video Check 1. sum 3. right; zeros 5. circumference 7. Knowing whether we placed the decimal point correctly in our product 9. $3(5.7) − (−0.2)$ 11. This is an application problem and needs units attached. The complete answer is 24.8 grams.

Exercise Set 5.3 1. 1.36 3. 0.6 5. −17.595 7. 55.008 9. 28.56; 7 × 4 = 28 11. 0.1041 13. 8.23854; 1 15. 11.2746 17. 65; × 8 / 8 19. 0.83 21. −7093 23. 70 25. 0.0983 27. 0.02523 29. 0.0492 31. 14,790 33. 1.29 35. −9.3762 37. 0.5623 39. 36.024 41. 1,500,000,000 43. 49,800,000 45. −0.6 47. 17.3 49. no 51. yes 53. 10π cm ≈ 31.4 cm 55. 18.2π yd ≈ 57.148 yd 57. $715.20 59. 24.8 g 61. 14.144 sq in. 63. 250π ft ≈ 785 ft 65. 135π m ≈ 423.9 m 67. 64.9605 in. 69. a. 62.8 m and 125.6 m b. yes 71. $404.25 73. 977.4 Canadian dollars 75. 16,406.16 Mexican pesos 77. 486 79. −9 81. 3.64 83. 3.56 85. −0.1105 87. 3,831,600 mi 89. answers may vary 91. answers may vary

Section 5.4

Calculator Explorations 1. not reasonable 3. reasonable

Vocabulary, Readiness & Video Check 1. quotient; divisor; dividend 3. left; zeros 5. a whole number 7. We just need to know how to move the decimal point. 1000 has three zeros, so we move the decimal point in the decimal number three places to the left. 9. We want the answer rounded to the nearest tenth, so we go to one extra place value, to the hundredths place, in order to round.

Exercise Set 5.4 1. 4.6 3. 0.094 5. 300 7. 7.3 9. 6.6; 6)36̄ with 6 above 11. 0.413 13. −600 15. 7 17. 4.8 19. 2100 21. 5.8 23. 5.5 25. 9.8; 7)70̄ with 10 above 27. 9.6 29. 45 31. 54.592 33. 0.0055 35. 23.87 37. 114.0 39. 0.83397 41. 2.687 43. −0.0129 45. 12.6 47. 1.31 49. 0.045625 51. 0.413 53. −8 55. −7.2 57. 1400 59. 30 61. −58,000 63. −0.69 65. 0.024 67. 65 69. −5.65 71. −7.0625 73. yes 75. no 77. 11 qt 79. 5.1 m 81. 11.4 boxes 83. 24 tsp 85. 8 days 87. 146.6 mi per week 89. 345.5 thousand books per hr 91. $\frac{21}{50}$ 93. $-\frac{1}{10}$ 95. 4.26 97. 1.578 99. −26.66 101. 904.29 103. c 105. b 107. 85.5 109. 8.6 ft 111. answers may vary 113. 65.2–82.6 knots 115. 27.3 m

Integrated Review 1. 2.57 2. 4.05 3. 8.9 4. 3.5 5. 0.16 6. 0.24 7. 0.27 8. 0.52 9. −4.8 10. 6.09 11. 75.56 12. 289.1
13. −24.974 14. −43.875 15. −8.6 16. 5.4 17. −280 18. 1600 19. 224.938 20. 145.079 21. 0.56 22. −0.63 23. 27.60
24. 145.6312 25. 5.4 26. −17.74 27. −414.44 28. −1295.03 29. −34 30. −28 31. 116.81 32. 18.79 33. 156.2 34. 1.56
35. 25.62 36. 5.62 37. Exact: 204.1 mi; Estimate: 200 mi 38. $1.89 39. $8.8 billion or $8,800,000,000

Section 5.5

Vocabulary, Readiness & Video Check 1. false 3. false 5. We place a bar over just the repeating digits and only 6 repeats in our decimal answer. 7. The fraction bar serves as a grouping symbol 9. $4(0.3) - (-2.4)$

Exercise Set 5.5 1. 0.2 3. 0.68 5. 0.75 7. −0.08 9. 2.25 11. $0.91\overline{6}$ 13. 0.425 15. 0.45 17. $-0.\overline{3}$ 19. 0.4375 21. $0.\overline{63}$
23. 5.85 25. 0.624 27. −0.33 29. 0.44 31. 0.6 33. 0.62 35. 0.86 37. 0.02 39. < 41. = 43. < 45. < 47. < 49.
51. < 53. < 55. 0.32, 0.34, 0.35 57. 0.49, 0.491, 0.498 59. $5.23, \frac{42}{8}, 5.34$ 61. $0.612, \frac{5}{8}, 0.649$ 63. 0.59 65. −3 67. 5.29
69. 9.24 71. 0.2025 73. −1.29 75. −15.4 77. −3.7 79. 25.65 sq in. 81. 0.248 sq yd 83. 5.76 85. 5.7 87. 3.6 89. $\frac{77}{50}$
91. $\frac{5}{2}$ 93. = 1 95. > 1 97. < 1 99. 0.057 101. 8200 stations 103. answers may vary

Section 5.6

Vocabulary, Readiness & Video Check 1. So that we are no longer working with decimals.

Exercise Set 5.6 1. 5.9 3. −0.43 5. 10.2 7. −4.5 9. 4 11. 0.45 13. 4.2 15. −4 17. 1.8 19. 10 21. 7.6 23. 60
25. −0.07 27. 20 29. 0.0148 31. −8.13 33. 1.5 35. −1 37. −7 39. 7 41. 53.2 43. $3x - 16$ 45. $\frac{3}{5x}$ 47. $\frac{13x}{21}$ 49. 3.
51. $6x - 0.61$ 53. $-2y + 6.8$ 55. 9.1 57. −3 59. $-4z + 16.67$ 61. 15.7 63. 5.85 65. $-2.1z - 10.1$ 67. answers may va
69. answers may vary 71. 7.683 73. 4.683

Section 5.7

Vocabulary, Readiness & Video Check 1. average 3. mean (or average) 5. grade point average 7. Place the data numbers in numerical order (or verify that they already are)

Exercise Set 5.7 1. mean: 21; median: 23; no mode 3. mean: 8.1; median: 8.2; mode: 8.2 5. mean: 0.5; median: 0.5; mode: 0.2 and
7. mean: 370.9; median: 313.5; no mode 9. 2109.2 ft 11. 1968.5 ft 13. answers may vary 15. 2.79 17. 3.64 19. 6.8 21. 6.9
23. 88.5 25. 73 27. 70 and 71 29. 9 rates 31. mean: 3773 mi; median: 3812 mi 33. $\frac{3}{5y}$ 35. $\frac{11}{15}$ 37. 35, 35, 37, 43
39. yes; answers may vary

Chapter 5 Vocabulary Check 1. decimal 2. numerator; denominator 3. vertically 4. and 5. sum 6. mode 7. circumferen
8. median; mean 9. mean 10. standard form

Chapter 5 Review 1. tenths 2. hundred-thousandths 3. negative twenty-three and forty-five hundredths 4. three hundred forty-five hundred-thousandths 5. one hundred nine and twenty-three hundredths 6. two hundred and thirty-two millionths
7. 8.06 8. −503.102 9. 16,025.0014 10. 14.011 11. $\frac{4}{25}$ 12. $-12\frac{23}{1000}$ 13. 0.00231 14. 25.25 15. > 16. = 17. < 18.
19. 0.6 20. 0.94 21. −42.90 22. 16.349 23. 890,500,000 24. 600,000 25. 18.1 26. 5.1 27. −7.28 28. −12.04 29. 320.3
30. 148.74236 31. 1.7 32. 2.49 33. −1324.5 34. −10.136 35. 65.02 36. 199.99802 37. 52.6 mi 38. −5.7 39. 22.2 in.
40. 38.9 ft 41. 72 42. 9345 43. −78.246 44. 73,246.446 45. 14π m ≈ 43.96 m 46. 20π in. ≈ 62.8 in. 47. 0.0877 48. 15.8
49. 70 50. −0.21 51. 8.059 52. 30.4 53. 0.02365 54. −9.3 55. 7.3 m 56. 45 months 57. 0.8 58. −0.923 59. $2.\overline{3}$ or 2.3
60. $0.21\overline{6}$ or 0.217 61. = 62. < 63. < 64. < 65. 0.832, 0.837, 0.839 66. $\frac{5}{8}, 0.626, 0.685$ 67. $0.42, \frac{3}{7}, 0.43$ 68. $\frac{19}{12}, 1.63, \frac{18}{11}$
69. −11.94 70. 3.89 71. 7.26 72. 0.81 73. 55 74. −129 75. 6.9 sq ft 76. 5.46 sq in. 77. 0.3 78. 92.81 79. 8.6 80. −
81. 1.98 82. −1.5 83. −20 84. 1 85. mean: 17.8; median: 14; no mode 86. mean: 58.1; median: 60; mode: 45 and 86 87. mea
24,500; median: 20,000; mode: 20,000 88. mean: 447.3; median: 420; mode: 400 89. 3.25 90. 2.57 91. two hundred and thirty-two ten-thousandths 92. −16.09 93. $\frac{847}{10,000}$ 94. $0.75, \frac{6}{7}, \frac{8}{9}$ 95. −0.07 96. 0.1125 97. 51.057 98. > 99. < 100. 86.91
101. 3.115 102. $123.00 103. $3646.00 104. −1.7 105. 5.26 106. −12.76 107. −14.907 108. 8.128 109. −7.245 110. 49
111. 23.904 112. 9600 sq ft 113. yes 114. 0.1024 115. 3.6 116. mean: 74.4; median: 73; mode: none 117. mean: 619.17; median: 647.5; mode: 327

Chapter 5 Getting Ready for the Test 1. C 2. A 3. B 4. E 5. C 6. A 7. B 8. D 9. B 10. B 11. C 12. B
13. C 14. A

Chapter 5 Test 1. forty-five and ninety-two thousandths 2. 3000.059 3. 17.595 4. −51.20 or −51.2 5. −20.42 6. 40.902
7. 0.037 8. 34.9 9. 0.862 10. < 11. < 12. $\frac{69}{200}$ 13. $-24\frac{73}{100}$ 14. −0.5 15. 0.941 16. 1.93 17. −6.2 18. $0.5x - 13.4$

19. −3 **20.** 3.7 **21.** mean: 38.4; median: 42; no mode **22.** mean: 12.625; median: 12.5; mode: 12 and 16 **23.** 3.07 **24.** 4,583,000,000
25. 2.31 sq mi **26.** 18π mi ≈ 56.52 mi **27. a.** 9904 sq ft **b.** 198.08 oz **28.** 54 mi

Chapters 1–5 Cumulative Review **1.** seventy-two; Sec. 1.2, Ex. 4 **2.** one hundred seven; Sec. 1.2 **3.** five hundred forty-six; Sec. 1.2, Ex. 5 **4.** five thousand, twenty-six; Sec. 1.2 **5.** 759; Sec. 1.3, Ex. 1 **6.** 19 in.; Sec. 1.3 **7.** 514; Sec. 1.3, Ex. 7 **8.** 121 R 1; Sec. 1.6 **9.** 278,000; Sec. 1.4, Ex. 2 **10.** $2 \cdot 3 \cdot 5$; Sec. 4.2 **11.** 20,296; Sec. 1.5, Ex. 4 **12.** 0; Sec. 1.5 **13. a.** 7 **b.** 12 **c.** 1 **d.** 1 **e.** 20 **f.** 1; Sec. 1.6, Ex. 2 **14.** 25; Sec. 1.6 **15.** 7; Sec. 1.7, Ex. 9 **16.** 49; Sec. 1.7 **17.** 81; Sec. 1.7, Ex. 5 **18.** 125; Sec. 1.7 **19.** 81; Sec. 1.7, Ex. 7 **20.** 1000; Sec. 1.7 **21.** 2; Sec. 1.8, Ex. 3 **22.** 6; Sec. 1.8 **23. a.** −13 **b.** 2 **c.** 0; Sec. 2.1, Ex. 5 **24. a.** 7 **b.** −4 **c.** 1; Sec. 2.1 **25.** −23; Sec. 2.2, Ex. 4 **26.** −22; Sec. 2.2 **27.** 180; Sec. 1.7, Ex. 8 **28.** 32; Sec. 1.7 **29.** −49; Sec. 2.4, Ex. 9 **30.** −32; Sec. 2.4 **31.** 25; Sec. 2.4, Ex. 8 **32.** −9; Sec. 2.4 **33.** $\frac{4}{3}; 1\frac{1}{3}$; Sec. 4.1, Ex. 10 **34.** $\frac{7}{4}; 1\frac{3}{4}$; Sec. 4.1 **35.** $\frac{11}{4}; 2\frac{3}{4}$; Sec. 4.1, Ex. 11 **36.** $\frac{14}{3}; 4\frac{2}{3}$; Sec. 4.1 **37.** $2^2 \cdot 3^2 \cdot 7$; Sec. 4.2, Ex. 3 **38.** 62; Sec. 1.3 **39.** $-\frac{36}{13}$; Sec. 4.2, Ex. 8 **40.** $\frac{79}{8}$; Sec. 4.1 **41.** equivalent; Sec. 4.2, Ex. 10 **42.** >; Sec. 4.5 **43.** $\frac{10}{33}$; Sec. 4.3, Ex. 1 **44.** $1\frac{1}{2}$; Sec. 4.7 **45.** $\frac{1}{8}$; Sec. 4.3, Ex. 2 **46.** 37; Sec. 4.7 **47.** −24; Sec. 3.2, Ex. 3 **48.** −8; Sec. 3.2 **49.** 829.6561; Sec. 5.2, Ex. 2 **50.** 230.8628; Sec. 5.2 **51.** 18.408; Sec. 5.3, Ex. 1 **52.** 28.251; Sec. 5.3

Chapter 6 Ratio, Proportion, and Percent

Section 6.1

Vocabulary, Readiness & Video Check **1.** proportion, ratio **3.** In order to simplify the ratio, quantities must have the same units of measurement to divide out. **5.** There are also many ways to set up an incorrect proportion, so just checking your solution in your proportion isn't enough. You need to determine if your solution is reasonable from the relationships given in the stated application.

Exercise Set 6.1 **1.** $\frac{2}{15}$ **3.** $\frac{5}{6}$ **5.** $\frac{1}{10}$ **7.** $\frac{19}{18}$ **9.** $\frac{77}{100}$ **11.** $\frac{2}{7}$ **13.** $\frac{47}{25}$ **15.** $\frac{17}{40}$ **17.** $\frac{15}{1}$ **19.** $\frac{2}{3}$ **21.** $\frac{10}{21}$ **23.** 4 **25.** $\frac{50}{9}$ **27.** $\frac{21}{4}$ **29.** 7 **31.** −3 **33.** $\frac{14}{9}$ **35.** 5 **37.** −20 **39.** 10,250 lb **41.** 165 cal **43.** 150 cal **45. a.** 18 tsp **b.** 6 tbsp **47.** 6 people **49.** 112 ft; 11-in. difference **51.** 102.9 mg **53.** 1257 ft **55.** 1350 sales **57.** 55 million **59.** 2.4 c **61. a.** 0.1 gal **b.** 13 fl oz **63. a.** 2062.5 mg **b.** no **65.** $2^3 \cdot 5^2$ **67.** 2^5 **69.** answers may vary **71.** 0.8 ml **73.** 1.25 ml

Section 6.2

Vocabulary, Readiness & Video Check **1.** Percent **3.** percent **5.** 0.01 **7.** Percent means "per 100." **9.** 1

Exercise Set 6.2 **1.** 96% **3.** football; 37% **5.** 50% **7.** 0.41 **9.** 0.06 **11.** 1.00 or 1 **13.** 0.736 **15.** 0.028 **17.** 0.006 **19.** 3.00 or 3 **21.** 0.3258 **23.** $\frac{2}{25}$ **25.** $\frac{1}{25}$ **27.** $\frac{9}{200}$ **29.** $\frac{7}{4}$ or $1\frac{3}{4}$ **31.** $\frac{1}{16}$ **33.** $\frac{31}{300}$ **35.** $\frac{179}{800}$ **37.** 22% **39.** 0.6% **41.** 530% **43.** 5.6% **45.** 22.28% **47.** 300% **49.** 70% **51.** 70% **53.** 80% **55.** 68% **57.** $37\frac{1}{2}$% **59.** $33\frac{1}{3}$% **61.** 450% **63.** 190% **65.** 81.82% **67.** 26.67% **69.** 0.6; $\frac{3}{5}$; $23\frac{1}{2}$%; $\frac{47}{200}$; 80%; 0.8; $0.333\overline{3}$; $\frac{1}{3}$; 87.5%; 0.875; 0.075; $\frac{3}{40}$ **71.** 2; 2; 280%; $2\frac{4}{5}$; 7.05; 7$\frac{1}{20}$; 454%; 4.54 **73.** 0.38; $\frac{19}{50}$ **75.** 0.99; $\frac{99}{100}$ **77.** 0.48; $\frac{12}{25}$ **79.** 0.005; $\frac{1}{200}$ **81.** 0.162; $\frac{81}{500}$ **83.** 78.1% **85.** 0.7% **87.** 0.47 **89.** $\frac{11}{36}$ **91.** $1\frac{5}{6}$ **93. a.** 52.9% **b.** 52.86% **95. b, d** **97.** 4% **99.** 75% **101.** greater **103.** 0.266; 26.6% **105.** occupational therapy assistant **107.** 0.30 or 0.3 **109.** answers may vary

Section 6.3

Vocabulary, Readiness & Video Check **1.** is **3.** amount; base; percent **5.** greater **7.** "of" translates to multiplication; "is" (or something equivalent) translates to an equal sign; "what" or "unknown" translates to our variable

Exercise Set 6.3 **1.** $18\% \cdot 81 = x$ **3.** $20\% \cdot x = 105$ **5.** $0.6 = 40\% \cdot x$ **7.** $x \cdot 80 = 3.8$ **9.** $x = 9\% \cdot 43$ **11.** $x \cdot 250 = 150$ **13.** 3.5 **15.** 28.7 **17.** 10 **19.** 600 **21.** 110% **23.** 34% **25.** 1 **27.** 645 **29.** 500 **31.** 5.16% **33.** 25.2 **35.** 35% **37.** 35 **39.** 0.624 **41.** 0.5% **43.** 145 **45.** 63% **47.** 4% **49.** 30 **51.** $3\frac{7}{11}$ **53.** $\frac{17}{12} = \frac{x}{20}$ **55.** $\frac{8}{9} = \frac{14}{x}$ **57. c** **59. b** **61.** answers may vary **63. b** **65. c** **67. c** **69. a** **71. a** **73.** answers may vary **75.** 686.625 **77.** 12,285

Section 6.4

Vocabulary, Readiness & Video Check 1. amount; base; percent 3. amount 5. 45 follows the word "of" so it is the base

Exercise Set 6.4 1. $\dfrac{a}{45} = \dfrac{98}{100}$ 3. $\dfrac{a}{150} = \dfrac{4}{100}$ 5. $\dfrac{14.3}{b} = \dfrac{26}{100}$ 7. $\dfrac{84}{b} = \dfrac{35}{100}$ 9. $\dfrac{70}{400} = \dfrac{p}{100}$ 11. $\dfrac{8.2}{82} = \dfrac{p}{100}$ 13. 26 15. 18.9 17. 600 19. 10 21. 120% 23. 28% 25. 37 27. 1.68 29. 1000 31. 210% 33. 55.18 33. 45% 37. 75 39. 0.864 41. 0.5% 43. 140 45. 9.6 47. 113% 49. $-\dfrac{7}{8}$ 51. $3\dfrac{2}{15}$ 53. 0.7 55. 2.19 57. answers may vary 59. no; $a = 16$ 61. yes 63. answers may vary 65. 12,011.2 67. 7270.6

Integrated Review 1. $\dfrac{9}{10}$ 2. $\dfrac{9}{25}$ 3. $\dfrac{43}{50}$ 4. $\dfrac{8}{23}$ 5. $\dfrac{2}{3}$ 6. $\dfrac{5}{3}$ 7. 25 8. 35 9. 86 weeks 10. 7 boxes 11. 94% 12. 17% 13. 37.5% 14. 350% 15. 470% 16. 800% 17. 45% 18. 106% 19. 675% 20. 325% 21. 2% 22. 6% 23. 0.71 24. 0.3 25. 0.03 26. 0.04 27. 2.24 28. 7 29. 0.029 30. 0.066 31. $0.07; \dfrac{7}{100}$ 32. $0.05; \dfrac{1}{20}$ 33. $0.068; \dfrac{17}{250}$ 34. $0.1125; \dfrac{9}{80}$ 35. $0.74; \dfrac{37}{50}$ 36. $0.45; \dfrac{9}{20}$ 37. $0.163; \dfrac{49}{300}$ 38. $0.127; \dfrac{19}{150}$ 39. 13.5 40. 100 41. 350 42. 120% 43. 28% 44. 76 45. 34 46. 130% 47. 46% 48. 37.8 49. 150 50. 62

Section 6.5

Vocabulary, Readiness & Video Check 1. The price of the home is $175,000.

Exercise Set 6.5 1. 1600 bolts 3. 8.8 lb 5. 14% 7. 16,388 screens 9. 38% 11. 496 chairs; 5704 chairs 13. 59,917 occupational therapy assistants 15. 1,843,920 17. 30% 19. 50% 21. 12.5% 23. 29.2% 25. $175,000 27. 31.2 hr 29. $867.87; $20,153.87 31. 35 ft 33. increase: $1043; tuition in 2016–2017: $10,037 35. increase: 1,494,100 enrolled in associate degree programs; 8,194,100 projected enrollment in 2024–2025 37. 30; 60% 39. 52; 80% 41. 2; 25% 43. 120; 75% 45. 44% 47. 137.5% 49. 374.6% 51. 9.8% 53. 14.8% 55. 1.9% 57. 141.4% 59. 51.4% 61. 4.56 63. 11.18 65. $\dfrac{1}{24}$ 67. $\dfrac{28}{39}$ 69. The increased number is double the original number. 71. 331,778 73. 93.0% 75. 324,228,700 77. answers may vary 79. percent of increase $= \dfrac{30}{150} = 20\%$ 81. False; the percents are different.

Section 6.6

Vocabulary, Readiness & Video Check 1. sales tax 3. commission 5. sale price 7. We write the commission rate as a percent

Exercise Set 6.6 1. $7.50 3. $858.93 5. 7% 7. a. $120 b. $130.20 9. $117; $1917 11. $485 13. 6% 15. $16.10; $246.10 17. $53,176.04 19. 14% 21. $4888.50 23. $185,500 25. $8.90; $80.10 27. $98.25; $98.25 29. $143.50; $266.50 31. $3255; $18,445 33. $45; $255 35. $27.45; $332.45 37. $3.08; $59.08 39. $7074 41. 8% 43. 1200 45. 132 47. 16 49. d 51. $4.00; $6.00; $8.00 53. $7.20; $10.80; $14.40 55. a discount of 60% is better; answers may vary 57. $26,838.45

Section 6.7

Calculator Explorations 1. $936.31 3. $9674.77 5. $634.49

Vocabulary, Readiness & Video Check 1. simple 3. Compound 5. Total amount 7. principal

Exercise Set 6.7 1. $32 3. $750 5. $700 7. $593.75 9. $77.50 11. $101,562.50; $264,062.50 13. $5562.50 15. $14,280 17. $46,815.37 19. $2327.14 21. $58,163.65 23. $2915.75 25. $2938.66 27. $2971.89 29. 32 yd 31. 35 m 33. $\dfrac{9x}{20}$ 35. $-\dfrac{131}{225}$ 37. answers may vary 39. answers may vary

Chapter 6 Vocabulary Check 1. of 2. is 3. Percent 4. Compound interest 5. amount; base 6. 100% 7. 0.01 8. $\dfrac{1}{100}$ 9. base; amount 10. percent of decrease 11. percent of increase 12. sales tax 13. total price 14. commission 15. amount of discount 16. sale price 17. proportion 18. ratio

Chapter 6 Review 1. $\dfrac{1}{5}$ 2. $\dfrac{2}{3}$ 3. 6 4. 500 5. 312.5 6. 50 7. 9 8. −17 9. 3 10. −19 11. 675 parts 12. $33.75 13. 37% 14. 77% 15. 0.26 16. 0.75 17. 0.035 18. 0.015 19. 2.75 20. 4.00 or 4 21. 0.4785 22. 0.8534 23. 160%

Answers to Selected Exercises **A15**

5.5% **25.** 7.6% **26.** 8.5% **27.** 71% **28.** 65% **29.** 600% **30.** 900% **31.** $\frac{7}{100}$ **32.** $\frac{3}{20}$ **33.** $\frac{1}{4}$ **34.** $\frac{17}{200}$ **35.** $\frac{51}{500}$
$\frac{1}{6}$ **37.** $\frac{1}{3}$ **38.** $1\frac{1}{10}$ **39.** 40% **40.** 70% **41.** $58\frac{1}{3}$% **42.** $166\frac{2}{3}$% **43.** 125% **44.** 60% **45.** 6.25% **46.** 62.5% **47.** 100,000
8000 **49.** 23% **50.** 114.5 **51.** 108.8 **52.** 150% **53.** 418 **54.** 300 **55.** 159.6 **56.** 180% **57.** 110% **58.** 165 **59.** 66%
16% **61.** 20.9% **62.** 106.25% **63.** $206,400 **64.** $13.23 **65.** $273.75 **66.** $2.17 **67.** $5000 **68.** $300.38 **69.** discount:
$00; sale price: $2100 **70.** discount: $9; sale price: $81 **71.** $160 **72.** $325 **73.** $30,104.61 **74.** $17,506.54 **75.** $180.61
$33,830.10 **77.** 0.038 **78.** 1.245 **79.** 54% **80.** 9520% **81.** $\frac{47}{100}$ **82.** $\frac{7}{125}$ **83.** $37\frac{1}{2}$% **84.** 120% **85.** 268.75 **86.** 110%
708.48 **88.** 134% **89.** 300% **90.** 38.4 **91.** 560 **92.** 325% **93.** 26% **94.** $6786.50 **95.** $617.70 **96.** $3.45 **97.** 12.5%
$1491 **99.** $11,687.50

apter 6 Getting Ready for the Test **1.** A **2.** B **3.** D **4.** A **5.** B **6.** C **7.** B **8.** D **9.** A **10.** C **11.** C **12.** A
D **14.** C

apter 6 Test **1.** 0.85 **2.** 5 **3.** 0.006 **4.** 5.6% **5.** 610% **6.** 35% **7.** $1\frac{1}{5}$ **8.** $\frac{77}{200}$ **9.** $\frac{1}{500}$ **10.** 55% **11.** 37.5% **12.** 175%
25% **14.** $\frac{9}{20}$ **15.** 33.6 **16.** 1250 **17.** 75% **18.** 38.4 lb **19.** $56,750 **20.** $383.21 **21.** 5% **22.** discount: $18; sale price:
$02 **23.** $395 **24.** 9% **25.** $647.50 **26.** $2005.63 **27.** $427 **28.** 2.3% **29.** $\frac{15}{2}$ **30.** −6 **31.** 18 bulbs

mulative Review Chapters 1–6 **1.** 20,296; Sec. 1.5, Ex. 4 **2.** 31,084; Sec. 1.5 **3.** −10; Sec. 2.3, Ex. 8 **4.** 10; Sec. 2.3
1; Sec. 2.6, Ex. 2 **6.** −1; Sec. 2.6 **7.** 2; Sec. 3.3, Ex. 4 **8.** 5; Sec. 3.3 **9.** $\frac{21}{7}$; Sec. 4.4, Ex. 20 **10.** $\frac{40}{5}$; Sec. 4.4
$-\frac{10}{27}$; Sec. 4.2, Ex. 6 **12.** $\frac{5y}{16}$; Sec. 4.2 **13.** $\frac{7}{10}$; Sec. 4.3, Ex. 13 **14.** $-\frac{4}{7}$; Sec. 4.3 **15.** $-\frac{1}{2}$; Sec. 4.4, Ex. 9 **16.** $\frac{1}{5}$; Sec. 4.4
$\frac{1}{28}$; Sec. 4.5, Ex. 5 **18.** $\frac{16}{45}$; Sec. 4.5 **19.** $\frac{3}{2}$; Sec. 4.6, Ex. 2 **20.** $\frac{50}{9}$; Sec. 4.6 **21.** 3; Sec. 4.8, Ex. 9 **22.** 4; Sec. 4.8
a. $\frac{38}{9}$ b. $\frac{19}{11}$; Sec. 4.1, Ex. 20 **24.** a. $\frac{17}{5}$ b. $\frac{44}{7}$; Sec. 4.1 **25.** $\frac{1}{8}$; Sec. 5.1, Ex. 9 **26.** $\frac{17}{20}$; Sec. 5.1 **27.** $-105\frac{83}{1000}$;
c. 5.1, Ex. 11 **28.** $17\frac{3}{200}$; Sec. 5.1 **29.** 67.69; Sec. 5.2, Ex. 6 **30.** 27.94; Sec. 5.2 **31.** 76.8; Sec. 5.3, Ex. 5 **32.** 1248.3; Sec. 5.3
−76,300; Sec. 5.3, Ex. 7 **34.** −8537.5; Sec. 5.3 **35.** 50; Sec. 5.4, Ex. 10 **36.** no; Sec. 5.4 **37.** 80.5; Sec. 5.7, Ex. 4
48; Sec. 5.7 **39.** $\frac{50}{63}$; Sec. 6.1, Ex. 2 **40.** $\frac{29}{38}$; Sec. 6.1 **41.** 63; Sec. 6.1, Ex. 4 **42.** −11; Sec. 6.1 **43.** 17.5 mi; Sec. 6.1, Ex. 6
35 problems; Sec. 6.1 **45.** $\frac{19}{1000}$; Sec. 6.2, Ex. 9 **46.** $\frac{23}{1000}$; Sec. 6.2 **47.** $\frac{1}{3}$; Sec. 6.2, Ex. 11 **48.** $1\frac{2}{25}$; Sec. 6.2
21; Sec. 6.3, Ex. 7 **50.** 35.7; Sec. 6.3

hapter 7 Graphs and Triangle Applications

ction 7.1

cabulary, Readiness & Video Check **1.** bar **3.** line **5.** range **7.** Count the number of symbols and multiply this number by
w much each symbol stands for (from the key). **9.** bar graph **11.** Answers may vary. For example, 6, 6, 6, 6.

ercise Set 7.1 **1.** Kansas **3.** 5.5 million or 5,500,000 acres **5.** South Dakota, Colorado, and Washington **7.** Montana
48,000 **11.** 2016 **13.** 18,000 **15.** 60,000 wildfires/year **17.** September **19.** 78 **21.** $\frac{1}{39}$ **23.** Tokyo, Japan; about 38 million
38,000,000 **25.** New York; 20.6 million or 20,600,000 **27.** approximately 3 million

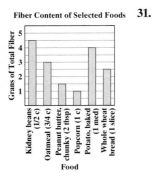

31.

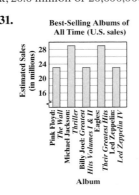

33. 15 adults **35.** 61 adults **37.** 24 adults **39.** 12 adults **41.** $\frac{9}{100}$ **43.** 55–64
45. 24 million householders **47.** 47 million householders **49.** 2 million
householders **51.** |; 1 **53.** ||||; 8 **55.** ||||; 6 **57.** ||||; 6 **59.** ||; 2
61. 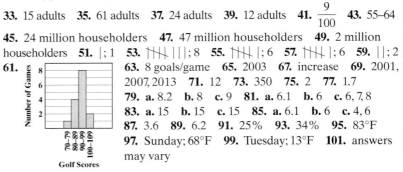 **63.** 8 goals/game **65.** 2003 **67.** increase **69.** 2001,
2007, 2013 **71.** 12 **73.** 350 **75.** 2 **77.** 1.7
79. a. 8.2 b. 8 c. 9 **81.** a. 6.1 b. 6 c. 6, 7, 8
83. a. 15 b. 15 c. 15 **85.** a. 6.1 b. 6 c. 4, 6
87. 3.6 **89.** 6.2 **91.** 25% **93.** 34% **95.** 83°F
97. Sunday; 68°F **99.** Tuesday; 13°F **101.** answers
may vary

Section 7.2

Vocabulary, Readiness & Video Check 1. circle 3. 360 5. 100%

Exercise Set 7.2 1. parent or guardian's home 3. $\frac{9}{35}$ 5. $\frac{9}{16}$ 7. Asia 9. 37% 11. 17,100,000 sq mi 13. 2,850,000 sq mi
15. 55% 17. nonfiction 19. 31,400 books 21. 27,632 books 23. 25,120 books
25. 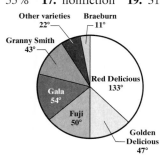 27. 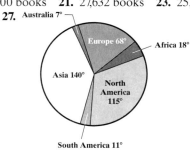 29. $2^2 \times 5$ 31. $2^3 \times 5$ 33. 5×17 35. answers may vary 37. 129,600,002 sq km 39. 55,542,858 sq km 41. 672 respondents 43. 2408 respondents 45. $\frac{12}{31}$ 47. no; answers may vary

Integrated Review 1. 320,000 2. 440,000 3. personal care aides 4. nursing assistants 5. Oroville Dam; 755 ft
6. New Bullards Bar Dam; 635 ft 7. 15 ft 8. 4 dams 9. Thursday and Saturday; 100°F 10. Monday; 82°F 11. Sunday, Monday, and Tuesday 12. Wednesday, Thursday, Friday, and Saturday 13. 70 qt containers 14. 52 qt containers
15. 2 qt containers 16. 6 qt containers 17. ||; 2 18. |; 1 19. |||; 3 20. ||||·|; 6 21. ||||; 5 22.

Section 7.3

Calculator Explorations 1. 32 3. 3.873 5. 9.849

Vocabulary, Readiness & Video Check 1. 10; −10 3. radical 5. perfect squares 7. c^2; b^2 9. The square roots of 49 are 7 and −7 since $7^2 = 49$ and $(-7)^2 = 49$. The radical sign means the positive square root only, so $\sqrt{49} = 7$. 11. The hypotenuse is the side across from the right angle.

Exercise Set 7.3 1. 2 3. 11 5. $\frac{1}{9}$ 7. $\frac{4}{8} = \frac{1}{2}$ 9. 1.732 11. 3.873 13. 5.568 15. 5.099 17. 6, 7 19. 10, 11 21. 16
23. 9.592 25. $\frac{7}{12}$ 27. 8.426 29. 13 in. 31. 6.633 cm 33. 52.802 m 35. 117 mm 37. 5 39. 12 41. 17.205 43. 44.822
45. 42.426 47. 1.732 49. 8.5 51. 141.42 yd 53. 25.0 ft 55. 340 ft 57. $\frac{5}{6}$ 59. $\frac{x}{30}$ 61. $\frac{21}{13y}$ 63. $\frac{9x}{64}$ 65. 6 67. 10
69. answers may vary 71. yes 73. $(\sqrt{80} - 6)$ in. ≈ 2.94 in.

Section 7.4

Vocabulary, Readiness & Video Check 1. false 3. true 5. false 7. The ratios of corresponding sides are the same.

Exercise Set 7.4 1. congruent; SSS 3. not congruent 5. congruent; ASA 7. congruent; SAS 9. $\frac{2}{1}$ 11. $\frac{3}{2}$ 13. 4.5 15. 6
17. 5 19. 13.5 21. 17.5 23. 10 25. 28.125 27. 10 29. 520 ft 31. 500 ft 33. 14.4 ft 35. 52 neon tetras 37. 381 ft 39. 4.
41. −1.23 43. $3\frac{8}{9}$ in.; no 45. 8.4 47. answers may vary 49. 200 ft, 300 ft, 425 ft

Section 7.5

Vocabulary, Readiness & Video Check 1. outcome 3. probability 5. 0 7. The number of outcomes equal the ending numb of branches drawn.

Answers to Selected Exercises

Exercise Set 7.5

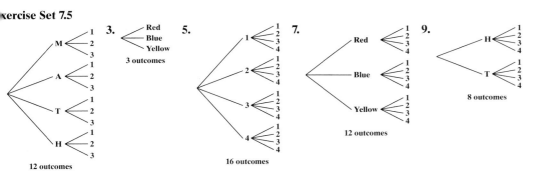

11. $\frac{1}{6}$ **13.** $\frac{1}{3}$ **15.** $\frac{1}{2}$ **17.** $\frac{2}{3}$ **19.** $\frac{1}{3}$ **21.** 1 **23.** $\frac{2}{3}$ **25.** $\frac{1}{7}$ **27.** $\frac{2}{7}$ **29.** $\frac{4}{7}$ **31.** $\frac{19}{100}$ **33.** $\frac{1}{20}$ **35.** $\frac{5}{6}$ **37.** $\frac{1}{6}$ **39.** $6\frac{2}{3}$ **41.** $\frac{1}{52}$ **43.** $\frac{1}{13}$ **45.** $\frac{1}{4}$ **47.** $\frac{1}{2}$ **49.** $\frac{5}{36}$ **51.** 0 **53.** answers may vary

Chapter 7 Vocabulary Check **1.** Congruent **2.** Similar **3.** leg **4.** leg **5.** hypotenuse **6.** right **7.** Pythagorean **8.** bar **9.** outcomes **10.** pictograph **11.** tree diagram **12.** experiment **13.** circle **14.** probability **15.** histogram; class interval; class frequency

Chapter 7 Review **1.** 2,500,000 **2.** 7,500,000 **3.** South **4.** Northeast **5.** South, West **6.** Northeast, Midwest **7.** 30% **8.** 2015 **9.** 1990, 2000, 2010, 2015 **10.** answers may vary **11.** 962 (exact number) **12.** 927 (exact number) **13.** 971 (exact number) **14.** 842 (exact number) **15.** 31 **16.** answers may vary **17.** 1 employee **18.** 4 employees **19.** 18 employees **20.** 9 employees **21.** ||||; 5 **22.** |||; 3 **23.** ||||; 4

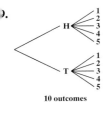

25. 3 **26.** 7 **27.** 35 **28.** 19 **29.** 5 **30.** 2.84 **31. a.** 62.1 **b.** 62 **c.** 63 **32. a.** 62.5 **b.** 63 **c.** 64 **33. a.** 12.9 **b.** 12 **c.** 11 and 12 **34. a.** 18.1 **b.** 18 **c.** 18 and 20 **35. a.** 59.8 **b.** 60 **c.** 60 **36. a.** 13.9 **b.** 14 **c.** 10 and 18 **37.** mortgage payment **38.** utilities **39.** $1225 **40.** $700 **41.** $\frac{39}{160}$ **42.** $\frac{7}{40}$ **43.** 78 **44.** 21 **45.** 2 **46.** 7 **47.** 8 **48.** 12 **49.** 3.464 **50.** 3.873 **51.** 0 **52.** 1 **53.** 7.071 **54.** 8.062 **55.** $\frac{2}{5}$ **56.** $\frac{1}{10}$ **57.** 13 **58.** 29 **59.** 10.7 **60.** 55.1 **61.** 28.28 cm **62.** 88.2 ft **63.** congruent; ASA **64.** not congruent **65.** $13\frac{1}{3}$ **66.** 17.4 **67.** approximately 33 ft **68.** $x = \frac{5}{6}$ in.; $y = 2\frac{1}{6}$ in.

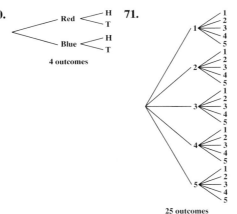

73. $\frac{1}{6}$ **74.** $\frac{1}{6}$ **75.** $\frac{1}{5}$ **76.** $\frac{1}{5}$ **77.** $\frac{3}{5}$ **78.** $\frac{2}{5}$ **79.** $\frac{1}{4}$ **80.** $\frac{3}{8}$ **81.** $\frac{1}{4}$ **82.** $\frac{1}{8}$ **83.** Audi A3 Spartback; 85 mpg **84.** Honda CR-Z; 31 mpg **85.** Chevrolet Malibu and Honda Accord; 49 mpg, or Ford C-Max and Toyota Camry; 42 mpg **86.** Toyota Prius; 54 mpg **87.** 6 **88.** $\frac{4}{9}$ **89.** 10.247 **90.** 5.657 **91.** 86.6 **92.** 20.8 **93.** 12 **94.** $6\frac{1}{2}$ **95. a.** 14.44 **b.** 14 **c.** 18 **d.** 8 **96. a.** 13 **b.** 15 **c.** 5 **d.** 20

Chapter 7 Getting Ready for the Test **1.** C **2.** C **3.** D **4.** C **5.** D **6.** D **7.** A **8.** B **9.** B **10.** A **11.** C **12.** C **13.** A **14.** B **15.** D **16.** D **17.** A and C **18.** A and B **19.** C and D **20.** A and C

Chapter 7 Test **1.** $225 **2.** 3rd week; $350 **3.** $1100 **4.** June, August, September **5.** February; 3 cm **6.** March and November **7.** 8.** 1.6% **9.** 2008, 2011 **10.** 2008–2009, 2011–2012, 2012–2013, 2014–2015 **11.** $\frac{17}{40}$ **12.** $\frac{31}{22}$ **13.** 74 million **14.** 90 million **15.** 9 students **16.** 11 students **17.** |; 1; |||; 3; ||||; 4; ||||; 5; |||| |||; 8; ||||; 4

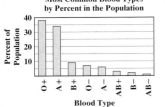

18. 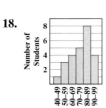 **19.** 8 **20.** 56 **21. a.** 92.8 **b.** 93 **c.** 93 and 94 **d.** 4 **22. a.** 26.3 **b.** 25 **c.** 35 **d.** 20 **23.** 7 **24.** 12.530 **25.** $\frac{8}{10} = \frac{4}{5}$ **26.** 5.66 cm **27.** 7.5 **28.** 69 ft **29.** 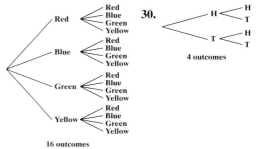 **30.** 4 outcomes

31. $\frac{1}{10}$ **32.** $\frac{1}{5}$

Cumulative Review Chapters 1–7 **1.** 47; Sec. 1.7, Ex. 12 **2.** 180; Sec. 1.7 **3.** −12; Sec. 2.3, Ex. 11 **4.** 9; Sec. 2.3 **5.** −3; Sec. 3.2, Ex. 2 **6.** −6; Sec. 3.2 **7.** 2; Sec. 4.8, Ex. 7 **8.** −10; Sec. 4.8 **9.** $7\frac{17}{24}$; Sec. 4.7, Ex. 9 **10.** $8\frac{3}{20}$; Sec. 4.7 **11.** $5\frac{9}{10}$; Sec. 5.1, Ex. 8 **12.** $2\frac{4}{5}$; Sec. 5.1 **13.** 3.432; Sec. 5.2, Ex. 5 **14.** 7.327; Sec. 5.2 **15.** 0.8496; Sec. 5.3, Ex. 2 **16.** 0.0294; Sec. 5.3 **17.** −0.052; Sec. 5.4, Ex. 3 **18.** 0.136; Sec. 5.4 **19.** 4.09; Sec. 5.5, Ex. 13 **20.** 7.29; Sec. 5.5 **21.** 0.25; Sec. 5.5, Ex. 1 **22.** 0.375; Sec. 5.5 **23.** 0.7; Sec. 5.6, Ex. 5 **24.** 1.68; Sec. 5.6 **25.** 8.944; Sec. 7.3, Ex. 7b **26.** 7.746; Sec. 7.3 **27.** $\frac{1}{6}$; Sec. 7.3, Ex. 5 **28.** $\frac{4}{7}$; Sec. 7.3 **29.** 4.90 in.; Sec. 7.3, Ex. 11 **30.** 16.12; Sec. 7.3 **31.** $\frac{31}{2}$; Sec. 6.1, Ex. 5 **32.** 16; Sec. 6.1 **33.** $\frac{12}{19}$; Sec. 7.4, Ex. 2 **34.** $\frac{4}{9}$; Sec. 7.4 **35.** 0.046; Sec. 6.2, Ex. 4 **36.** 0.32; Sec. 6.2 **37.** 0.0074; Sec. 6.2, Ex. 6 **38.** 0.027; Sec. 6.2 **39.** 21; Sec. 6.3, Ex. 7 **40.** 14.4; Sec. 6.3 **41.** 52; Sec. 6.4, Ex. 9 **42.** 38; Sec. 6.4 **43.** 8.5%; Sec. 6.6, Ex. 2 **44.** 6.5%; Sec. 6.6 **45.** $6772.12; Sec. 6.7, Ex. 5 **46.** $144.05; Sec. 6.7 **47.** 57; Sec. 5.7, Ex. 4 **48.** 48.5; Sec. 5.7 **49.** $\frac{1}{3}$; Sec. 7.5, Ex. 4 **50.** $\frac{1}{2}$; Sec. 7.5

Chapter 8 Geometry and Measurement
Section 8.1

Vocabulary, Readiness & Video Check **1.** plane **3.** Space **5.** ray **7.** straight **9.** acute **11.** Parallel; intersecting **13.** degrees **15.** vertical **17.** ∠WUV, ∠VUW, ∠U, ∠x **19.** 180° − 17° = 163°

Exercise Set 8.1 **1.** line; line CD or line l or $\overleftrightarrow{CD}$ **3.** line segment; line segment MN or $\overline{MN}$ **5.** angle; ∠GHI or ∠IHG or ∠H **7.** ray; ray UW or $\overrightarrow{UW}$ **9.** ∠CPR, ∠RPC **11.** ∠TPM, ∠MPT **13.** straight **15.** right **17.** obtuse **19.** acute **21.** 67° **23.** 163° **25.** 32° **27.** 30° **29.** ∠MNP and ∠RNO; ∠PNQ and ∠QNR **31.** ∠SPT and ∠TPQ; ∠SPR and ∠RPQ; ∠SPT and ∠SPR; ∠TPQ and ∠QPR **33.** 27° **35.** 132° **37.** m∠x = 30°; m∠y = 150°; m∠z = 30° **39.** m∠x = 77°; m∠y = 103°; m∠z = 77° **41.** m∠x = 100°; m∠y = 80°; m∠z = 100° **43.** m∠x = 134°; m∠y = 46°; m∠z = 134° **45.** ∠ABC or ∠CBA **47.** ∠DBE or ∠EBD **49.** 15° **51.** 50° **53.** 65° **55.** 95° **57.** $\frac{9}{8}$ or $1\frac{1}{8}$ **59.** $\frac{7}{32}$ **61.** $\frac{5}{6}$ **63.** $1\frac{1}{3}$ **65.** 360° **67.** 45° **69.** 54.8° **71.** false; answers may vary **73.** true **75.** m∠a = 60°; m∠b = 50°; m∠c = 110°; m∠d = 70°; m∠e = 120° **77.** no; answers may vary **79.** 45°; 45°

Section 8.2

Vocabulary, Readiness & Video Check **1.** perimeter **3.** π **5.** $\frac{22}{7}$ (or 3.14); 3.14 (or $\frac{22}{7}$) **7.** Opposite sides of a rectangle have the same measure, so we can just find the sum of the measures of all four sides.

Exercise Set 8.2 **1.** 64 ft **3.** 120 cm **5.** 21 in. **7.** 48 ft **9.** 42 in. **11.** 155 cm **13.** 21 ft **15.** 624 ft **17.** 346 yd **19.** 22 ft **21.** $55 **23. a.** 8 **b.** 72 in. **25.** 28 in. **27.** $36.12 **29.** 96 m **31.** 66 ft **33.** 74 cm **35.** 17π cm; 53.38 cm **37.** 16π mi; 50.24 mi **39.** 26π m; 81.64 m **41.** 15π ft; 47.1 ft **43.** 12,560 ft **45.** 30.7 mi **47.** 14π cm ≈ 43.96 cm **49.** 40 mm **51.** 84 ft **53.** 23 **55.** 1 **57.** 6 **59.** 10 **61. a.** width: 30 yd; length: 40 yd **b.** 140 yd **63.** b **65. a.** 62.8 m; 125.6 m **b.** yes **67.** answers may vary **69.** 27.4 m **71.** 75.4 m

Section 8.3

Vocabulary, Readiness & Video Check 1. surface area 3. Area 5. square 7. We don't have a formula for an L-shaped figure, so we divide it into two rectangles, use the formula to find the area of each, and then add these two areas.

Exercise Set 8.3 1. 7 sq m 3. $9\frac{3}{4}$ sq yd 5. 15 sq yd 7. 2.25π sq in. ≈ 7.065 sq in. 9. 36.75 sq ft 11. 28 sq m 13. 22 sq yd
15. $13\frac{3}{5}$ sq ft 17. $22\frac{1}{2}$ sq in. 19. 25 sq cm 21. 86 sq mi 23. 24 sq cm 25. 36π sq in. ≈ $113\frac{1}{7}$ sq in. 27. $V = 72$ cu in.;
$SA = 108$ sq in. 29. $V = 512$ cu cm; $SA = 384$ sq cm 31. $V = 4\pi$ cu yd ≈ $12\frac{4}{7}$ cu yd; $SA = (2\pi\sqrt{13} + 4\pi)$ sq yd ≈ 35.20 sq yd
33. $V = \frac{500}{3}\pi$ cu in. ≈ $523\frac{17}{21}$ cu in.; $SA = 100\pi$ sq in. ≈ $314\frac{2}{7}$ sq in. 35. $V = 9\pi$ cu in. ≈ $28\frac{2}{7}$ cu in. 37. $V = 75$ cu cm
39. $2\frac{10}{27}$ cu in. 41. $V = 8.4$ cu ft; $SA = 26$ sq ft 43. 113,625 sq ft 45. 168 sq ft 47. 960 cu cm 49. 9200 sq ft
51. $V = \frac{1372}{3}\pi$ cu in. or $457\frac{1}{3}\pi$ cu in.; $SA = 196\pi$ sq in. 53. a. 381 sq ft b. 4 squares 55. $V = 5.25\pi$ cu in. 57. 4π sq ft ≈ 12.56 sq ft
59. $V = 4.5\pi$ cu m; 14.13 cu m 61. 168 sq ft 63. $10\frac{5}{6}$ cu in. 65. 8.8 cu in. 67. 25 69. 9 71. 5 73. 20 75. perimeter
77. area 79. area 81. perimeter 83. 12-in. pizza 85. 2093.33 cu m 87. no; answers may vary 89. 7.74 sq in. 91. 298.5 sq m
93. no; answers may vary

Integrated Review 1. 153°; 63° 2. $m\angle x = 75°; m\angle y = 105°; m\angle z = 75°$ 3. $m\angle x = 128°; m\angle y = 52°; m\angle z = 128°$
4. $m\angle x = 52°$ 5. 4.6 in. 6. $4\frac{1}{4}$ in. 7. 20 m; 25 sq m 8. 12 ft; 6 sq ft 9. 10π cm ≈ 31.4 cm; 25π sq cm ≈ 78.5 sq cm
10. 32 mi; 44 sq mi 11. 54 cm; 143 sq cm 12. 62 ft; 238 sq ft 13. 9800 ft; 4,900,000 sq ft 14. $V = 64$ cu in.; $SA = 96$ sq in.
15. $V = 30.6$ cu ft; $SA = 63$ sq ft 16. $V = 400$ cu cm 17. $V = 4\frac{1}{2}\pi$ cu mi ≈ $14\frac{1}{7}$ cu mi

Section 8.4

Vocabulary, Readiness & Video Check 1. meter 3. yard 5. feet 7. feet 9. Both mean addition; $5\frac{2}{5} = 5 + \frac{2}{5}$ and
5 ft 2 in. = 5 ft + 2 in. 11. Since the metric system is based on base 10, we just need to move the decimal point to convert from one unit to another.

Exercise Set 8.4 1. 5 ft 3. 36 ft 5. 8 mi 7. 102 in. 9. $3\frac{1}{3}$ yd 11. 33,792 ft 13. 4.5 yd 15. 0.25 ft 17. 13 yd 1 ft 19. 7 ft 1 in.
21. 1 mi 4720 ft 23. 62 in. 25. 26 ft 27. 84 in. 29. 11 ft 2 in. 31. 22 yd 1 ft 33. 6 ft 5 in. 35. 7 ft 6 in. 37. 14 ft 4 in.
39. 83 yd 1 ft 41. 6000 cm 43. 4 cm 45. 0.5 km 47. 1.7 m 49. 15 m 51. 42,000 cm 53. 7000 m 55. 83 mm 57. 0.201 dm
59. 40 mm 61. 8.94 m 63. 2.94 m or 2940 mm 65. 1.29 cm or 12.9 mm 67. 12.64 km or 12,640 m 69. 54.9 m 71. 1.55 km
73. $348\frac{2}{3}$; 12,552 75. $11\frac{2}{3}$; 420 77. 5000; 0.005; 500 79. 0.065; 65; 0.000065 81. 342,000; 342,000,000; 34,200,000 83. 10 ft 6 in.
85. 5100 ft 87. 5.0 times 89. 26.7 mm 91. 15 ft 9 in. 93. 3.35 m 95. $72\frac{2}{3}$ yd 97. $\frac{21}{100}$ 99. 0.13 101. 0.25 103. no 105. yes
107. no 109. Estimate: 13 yd 111. answers may vary; for example, $1\frac{1}{3}$ yd or 48 in. 113. answers may vary 115. 334.89 sq m

Section 8.5

Vocabulary, Readiness & Video Check 1. Mass 3. gram 5. 2000 7. We can't subtract 9 oz from 4 oz, so we borrow
1 lb (= 16 oz) from 12 lb to add to the 4 oz; 12 lb 4 oz becomes 11 lb 20 oz. 9. 18.50 dg

Exercise Set 8.5 1. 32 oz 3. 10,000 lb 5. 9 tons 7. $3\frac{3}{4}$ lb 9. $1\frac{3}{4}$ tons 11. 204 oz 13. 9800 lb 15. 76 oz 17. 1.5 tons
19. $\frac{1}{20}$ lb 21. 92 oz 23. 161 oz 25. 5 lb 9 oz 27. 53 lb 10 oz 29. 8 tons 750 lb 31. 3 tons 175 lb 33. 8 lb 11 oz 35. 31 lb 2 oz
37. 1 ton 700 lb 39. 0.5 kg 41. 4000 mg 43. 25,000 g 45. 0.048 g 47. 0.0063 kg 49. 15,140 mg 51. 6250 g 53. 350,000 cg
55. 13.5 mg 57. 5.815 g or 5815 mg 59. 1850 mg or 1.85 g 61. 1360 g or 1.36 kg 63. 13.52 kg 65. 2.125 kg 67. 200,000; 3,200,000
69. $\frac{269}{400}$ or 0.6725; 21,520 71. 0.5; 0.0005; 50 73. 21,000; 21,000,000; 2,100,000 75. 8.064 kg 77. 30 mg 79. 5 lb 8 oz
81. 35 lb 14 oz 83. 6 lb 15.5 oz 85. 144 mg 87. 6.12 kg 89. 130 lb 91. 211 lb 93. 0.16 95. 0.875 97. no 99. yes
101. no 103. answers may vary; for example, 250 mg or 0.25 g 105. true 107. answers may vary

Section 8.6

Vocabulary, Readiness & Video Check   **1.** capacity   **3.** fluid ounces   **5.** cups   **7.** quarts   **9.** We can't subtract 3 qt from 0 qt, so we borrow 1 gal (= 4 qt) from 3 gal to get 2 gal 4 qt.   **11.** 0.45 dal

Exercise Set 8.6   **1.** 4 c   **3.** 16 pt   **5.** $3\frac{1}{2}$ gal   **7.** 5 pt   **9.** 8 c   **11.** $3\frac{3}{4}$ qt   **13.** $10\frac{1}{2}$ qt   **15.** 9 c   **17.** 23 qt   **19.** $\frac{1}{4}$ pt   **21.** 14 gal 2 c   **23.** 4 gal 3 qt 1 pt   **25.** 22 pt   **27.** 13 gal 2 qt   **29.** 4 c 4 fl oz   **31.** 1 gal 1 qt   **33.** 2 gal 3 qt 1 pt   **35.** 17 gal   **37.** 4 gal 3 qt   **39.** 5000 ml   **41.** 0.00016 kl   **43.** 5.6 L   **45.** 320 cl   **47.** 0.41 kl   **49.** 0.064 L   **51.** 160 L   **53.** 3600 ml   **55.** 19.3 L   **57.** 4.5 L or 4500 ml   **59.** 8410 ml or 8.41 L   **61.** 16,600 ml or 16.6 L   **63.** 3840 ml   **65.** 162.4 L   **67.** 336; 84; 168   **69.** $\frac{1}{4}$; 1; 2   **71.** 1.59 L   **73.** 18.954 L   **75.** 4.3 fl oz   **77.** yes   **79.** $0.316   **81.** $\frac{4}{5}$ **83.** $\frac{3}{5}$ **85.** $\frac{9}{10}$ **87.** no   **89.** no   **91.** less than; answers may vary   **93.** answers may vary   **95.** 128 fl oz   **97.** 1.5 cc   **99.** 2.7 cc   **101.** 54 u or 0.54 cc   **103.** 86 u or 0.86 cc

Section 8.7

Vocabulary, Readiness & Video Check   **1.** 1 L ≈ 0.26 gal or 3.79 L ≈ 1 gal   **3.** F = 1.8C + 32; 27

Exercise Set 8.7   **1.** 25.57 fl oz   **3.** 218.44 cm   **5.** 40 oz   **7.** 57.66 mi   **9.** 3.77 gal   **11.** 13.5 kg   **13.** 1.5; $1\frac{2}{3}$; 150; 60   **15.** 55; 5500; 180; 2160   **17.** 3.94 in.   **19.** 80.5 kph   **21.** 0.008 oz   **23.** 229.6 ft   **25.** 9.92 billion mi   **27.** yes   **29.** 2790 mi   **31.** 90 mm   **33.** 112.5   **35.** 104 mph   **37.** 26.24 ft   **39.** 3 mi   **41.** 8 fl oz   **43.** b   **45.** b   **47.** c   **49.** d   **51.** d   **53.** 25°C   **55.** 40°C   **57.** 122°F   **59.** 239°F   **61.** −6.7°C   **63.** 61.2°C   **65.** 197.6°F   **67.** 54.3°F   **69.** 56.7°C   **71.** 80.6°F   **73.** 21.1°C   **75.** 244.4°F   **77.** 7232°F   **79.** 29   **81.** 3   **83.** yes   **85.** no   **87.** no   **89.** yes   **91.** 2.13 sq m   **93.** 1.19 sq m   **95.** 1.69 sq m   **97.** 188.8°C   **99.** answers may vary

Chapter 8 Vocabulary Check   **1.** Weight   **2.** Mass   **3.** meter   **4.** unit fractions   **5.** gram   **6.** liter   **7.** line segment   **8.** complementary   **9.** line   **10.** perimeter   **11.** angle; vertex   **12.** Area   **13.** ray   **14.** transversal   **15.** straight   **16.** volume   **17.** vertical   **18.** adjacent   **19.** obtuse   **20.** right   **21.** acute   **22.** supplementary   **23.** surface area

Chapter 8 Review   **1.** right   **2.** straight   **3.** acute   **4.** obtuse   **5.** 65°   **6.** 75°   **7.** 58°   **8.** 98°   **9.** 90°   **10.** 25°   **11.** ∠a and ∠b; ∠b and ∠c; ∠c and ∠d; ∠d and ∠a   **12.** ∠x and ∠w; ∠y and ∠z   **13.** $m\angle x = 100°$; $m\angle y = 80°$; $m\angle z = 80°$ **14.** $m\angle x = 155°$; $m\angle y = 155°$; $m\angle z = 25°$ **15.** $m\angle x = 53°$; $m\angle y = 53°$; $m\angle z = 127°$ **16.** $m\angle x = 42°$; $m\angle y = 42°$; $m\angle z = 138°$ **17.** 69 m   **18.** 30.6 cm   **19.** 36 m   **20.** 90 ft   **21.** 32 ft   **22.** 440 ft   **23.** 5.338 in.   **24.** 31.4 yd   **25.** 240 sq ft   **26.** 189 sq yd   **27.** 600 sq cm   **28.** 82.81 sq m   **29.** 49π sq ft ≈ 153.86 sq ft   **30.** 9π sq in. ≈ 28.26 sq in.   **31.** 119 sq in.   **32.** 140 sq m   **33.** 144 sq m   **34.** 1625 sq cm   **35.** 432 sq ft   **36.** 130 sq ft   **37.** $V = 15\frac{5}{8}$ cu in.; $SA = 37\frac{1}{2}$ sq in.   **38.** $V = 84$ cu ft; $SA = 136$ sq ft   **39.** $V = 20{,}000\pi$ cu cm ≈ 62,800 cu cm   **40.** $V = \frac{1}{6}\pi$ cu km ≈ $\frac{11}{21}$ cu km   **41.** $2\frac{2}{3}$ cu ft   **42.** 307.72 cu in.   **43.** $7\frac{1}{2}$ cu ft   **44.** 0.5π cu ft or $\frac{1}{2}\pi$ cu ft   **45.** 9 ft   **46.** 24 yd   **47.** 7920 ft   **48.** 18 in.   **49.** 17 yd 1 ft   **50.** 3 ft 10 in.   **51.** 4200 cm   **52.** 820 mm   **53.** 0.01218 m   **54.** 0.00231 km   **55.** 21 yd 1 ft   **56.** 3 ft 8 in.   **57.** 9.5 cm or 95 mm   **58.** 2.74 m or 274 cm   **59.** 169 yd 2 ft   **60.** 258 ft 4 in.   **61.** 617.5 km   **62.** 0.24 sq m   **63.** $4\frac{1}{8}$ lb   **64.** 4600 lb   **65.** 3 lb 4 oz   **66.** 5 tons 300 lb   **67.** 0.027 g   **68.** 40,000 g   **69.** 21 dag   **70.** 0.0003 dg   **71.** 3 lb 9 oz   **72.** 33 lb 8 oz   **73.** 21.5 mg   **74.** 0.6 kg or 600 g   **75.** 4 lb 4 oz   **76.** 9 tons 1075 lb   **77.** 14 qt   **78.** 5 c   **79.** 7 pt   **80.** 72 c   **81.** 4 qt 1 pt   **82.** 3 gal 3 qt   **83.** 3800 ml   **84.** 1.4 kl   **85.** 3060 cl   **86.** 0.00245 L   **87.** 1 gal 1 qt   **88.** 7 gal   **89.** 736 ml or 0.736 L   **90.** 15.5 L or 15,500 ml   **91.** 2 gal 3 qt   **92.** 6 fl oz   **93.** 10.88 L   **94.** yes   **95.** 22.96 ft   **96.** 10.55 m   **97.** 4.55 gal   **98.** 8.27 qt   **99.** 425.25 g   **100.** 10.35 kg   **101.** 2.36 in.   **102.** 180.4 lb   **103.** 107.6°F   **104.** 320°F   **105.** 5.2°C   **106.** 26.7°C   **107.** 1.7°C   **108.** 329°F   **109.** 108°   **110.** 89°   **111.** 95°   **112.** 57°   **113.** 27.3 in.   **114.** 194 ft   **115.** 1624 sq m   **116.** 9π sq m ≈ 28.26 sq m   **117.** $346\frac{1}{2}$ cu in.   **118.** $V = 140$ cu in.; $SA = 166$ sq in.   **119.** 75 in.   **120.** 4 tons 200 lb   **121.** 500 cm   **122.** 0.000286 km   **123.** 1.4 g   **124.** 27 qt   **125.** 186.8°F   **126.** 11°C   **127.** 9117 m or 9.117 km   **128.** 35.7 L or 35,700 ml   **129.** 8 gal 1 qt   **130.** 12.8 kg

Chapter 8 Getting Ready for the Test   **1.** D   **2.** B   **3.** E   **4.** A   **5.** F   **6.** C   **7.** B   **8.** C   **9.** A   **10.** B   **11.** B   **12.** C   **13.** A and D   **14.** B   **15.** A or D   **16.** C

Chapter 8 Test   **1.** 12°   **2.** 56°   **3.** 50°   **4.** $m\angle x = 118°$; $m\angle y = 62°$; $m\angle z = 118°$ **5.** $m\angle x = 73°$; $m\angle y = 73°$; $m\angle z = 73°$ **6.** 6.2 m   **7.** 10 in.   **8.** circumference = 18π in. ≈ 56.52 in.; area = 81π sq in. ≈ 254.34 sq in.   **9.** perimeter = 24.6 yd; area = 37.1 sq yd   **10.** perimeter = 68 in.; area = 185 sq in.   **11.** $62\frac{6}{7}$ cu in.   **12.** 30 cu ft   **13.** 16 in.   **14.** 18 cu ft   **15.** 62 ft; $115.94   **16.** 23 ft 4 in.   **17.** 10 qt   **18.** $1\frac{7}{8}$ lb   **19.** 5600 lb   **20.** $4\frac{3}{4}$ gal   **21.** 0.04 g   **22.** 2400 g   **23.** 36 mm   **24.** 0.43 g   **25.** 830 ml   **26.** 1 gal 2 qt

27. 3 lb 13 oz **28.** 8 ft 3 in. **29.** 2 gal 3 qt **30.** 66 mm or 6.6 cm **31.** 2.256 km or 2256 m **32.** 28.9°C **33.** 54.7°F **34.** 5.6 m
35. 4 gal 3 qt **36.** 91.4 m **37.** 16 ft 6 in.

Cumulative Review Chapters 1–8 **1.** 5; Sec. 3.3, Ex. 1 **2.** 6; Sec. 3.3 **3. a.** $\frac{16}{625}$ **b.** $\frac{1}{16}$; Sec. 4.3, Ex. 10 **4. a.** $-\frac{1}{27}$ **b.** $\frac{9}{49}$;
Sec. 4.3 **5.** $9\frac{3}{10}$; Sec. 4.7, Ex. 11 **6.** $9\frac{11}{15}$; Sec. 4.7 **7.** $20x - 10.9$; Sec. 5.2, Ex. 13 **8.** $1.2y + 1.8$; Sec. 5.2 **9.** 28.4405; Sec. 5.5,
Ex. 14 **10.** 2.16; Sec. 5.5 **11.** =; Sec. 5.5, Ex. 9 **12.** >; Sec. 5.5 **13.** -1.3; Sec. 5.6, Ex. 6 **14.** 30; Sec. 5.6 **15.** 424 ft; Sec. 7.3,
Ex. 12 **16.** 236 ft; Sec. 7.3 **17. a.** $\frac{5}{7}$ **b.** $\frac{7}{24}$; Sec. 6.1, Ex. 3 **18. a.** $\frac{1}{4}$ **b.** $\frac{4}{9}$; Sec. 6.1 **19.** no; Sec. 2.6, Ex. 1 **20.** yes; Sec. 2.6
21. -6; Sec. 2.6, Ex. 8 **22.** 1.02; Sec. 2.6 **23.** 22.4 cc; Sec. 6.1, Ex. 7 **24.** 7.5 cups; Sec. 6.1 **25.** 17%; Sec. 6.2, Ex. 1 **26.** 38%; Sec.
6.2 **27.** 200; Sec. 6.3, Ex. 10 **28.** 1200; Sec. 6.3 **29.** 2.7; Sec. 6.4, Ex. 7 **30.** 12.6; Sec. 6.4 **31.** 32%; Sec. 6.5, Ex. 5 **32.** 27%; Sec.
6.5 **33.** sales tax: $6.41; total price: $91.91; Sec. 6.6, Ex. 1 **34.** sales tax: $30; total price: $405; Sec. 6.6 **35.** 57; Sec. 5.7, Ex. 4 **36.** 83;
Sec. 5.7 **37.** $\frac{1}{4}$; Sec. 7.5, Ex. 5 **38.** $\frac{2}{7}$; Sec. 7.5 **39.** 42°; Sec. 8.1, Ex. 4 **40.** 43°; Sec. 8.1 **41.** 96 in.; Sec. 8.4, Ex. 1 **42.** 21 feet; Sec. 8.4
43. 4 tons 1650 lb; Sec. 8.4, Ex. 4 **44.** 18 lb 2 oz; Sec. 8.5 **45.** 15°C; Sec. 8.7, Ex. 6 **46.** 30°C; Sec. 8.7

Chapter 9 Equations, Inequalities, and Problem Solving

Vocabulary, Readiness & Video Check **1.** whole **3.** inequality **5.** real **7.** 0 **9.** To form a true statement: $0 < 7$.
11. 0 belongs to the whole numbers, the integers, the rational numbers, and the real numbers; because 0 is a rational number, it cannot also be an irrational number.

Exercise Set 9.1 **1.** < **3.** > **5.** = **7.** < **9.** $32 < 212$ **11.** $30 \leq 45$ **13.** true **15.** false **17.** true **19.** false **21.** $20 \leq 25$
23. $6 > 0$ **25.** $-12 < -10$ **27.** $7 < 11$ **29.** $5 \geq 4$ **31.** $15 \neq -2$ **33.** 14,494; -282 **35.** $-28,288$ **37.** 475; -195
39. [number line from -4 to 5] **41.** [number line showing $-\frac{1}{4}, \frac{1}{3}$] **43.** [number line showing $-4.5, -\frac{3}{2}, \frac{7}{4}, 3.25$]
45. whole, integers, rational, real **47.** integers, rational, real **49.** natural, whole, integers, rational, real **51.** rational, real **53.** false
55. true **57.** false **59.** false **61.** > **63.** = **65.** < **67.** < **69.** 2250 million < 2750 million or 2,250,000,000 < 2,750,000,000
71. 550 million bushels less or $-550,000,000$ **73.** $-0.04 > -26.7$ **75.** sun **77.** sun **79.** answers may vary

Vocabulary, Readiness & Video Check **1.** commutative property of addition **3.** distributive property **5.** associative property
of addition **7.** opposites or additive inverses **9.** 2 is outside the parentheses, so the point is made that you should distribute
the -9 only to the terms within the parentheses and not also to the 2.

Exercise Set 9.2 **1.** $16 + x$ **3.** $y \cdot (-4)$ **5.** yx **7.** $13 + 2x$ **9.** $x \cdot (yz)$ **11.** $(2 + a) + b$ **13.** $(4a) \cdot b$ **15.** $a + (b + c)$
17. $17 + b$ **19.** $24y$ **21.** y **23.** $26 + a$ **25.** $-72x$ **27.** s **29.** $-\frac{5}{2}x$ **31.** $4x + 4y$ **33.** $9x - 54$ **35.** $6x + 10$ **37.** $28x - 21$
39. $18 + 3x$ **41.** $-2y + 2z$ **43.** $-y - \frac{5}{3}$ **45.** $5x + 20m + 10$ **47.** $8m - 4n$ **49.** $-5x - 2$ **51.** $-r + 3 + 7p$ **53.** $3x + 4$
55. $-x + 3y$ **57.** $6r + 8$ **59.** $-36x - 70$ **61.** $-1.6x - 2.5$ **63.** $4(1 + y)$ **65.** $11(x + y)$ **67.** $-1(5 + x)$ **69.** $30(a + b)$
71. commutative property of multiplication **73.** associative property of addition **75.** commutative property of addition
77. associative property of multiplication **79.** identity element for addition **81.** distributive property **83.** multiplicative
inverse property **85.** identity element for multiplication **87.** 2 **89.** -23 **91.** -25 **93.** $-8; \frac{1}{8}$ **95.** $-x; \frac{1}{x}$ **97.** $2x; -2x$
99. false **101.** no **103.** yes **105.** yes **107.** yes **109. a.** commutative property of addition **b.** commutative property of addition
c. associative property of addition **111.** answers may vary **113.** answers may vary

Calculator Explorations **1.** solution **3.** not a solution **5.** solution

Vocabulary, Readiness & Video Check **1.** equation **3.** expression **5.** expression **7.** equation **9.** 3; distributive property, addition property of equality, multiplication property of equality **11.** The number of decimal places in each number helps us determine the smallest power of 10 we can multiply through by so we are no longer dealing with decimals.

Exercise Set 9.3 **1.** -6 **3.** 3 **5.** 1 **7.** $\frac{3}{2}$ **9.** 0 **11.** -1 **13.** 4 **15.** -4 **17.** -3 **19.** 2 **21.** 50 **23.** 1 **25.** $\frac{7}{3}$ **27.** 0.2
29. all real numbers **31.** no solution **33.** no solution **35.** all real numbers **37.** 18 **39.** $\frac{19}{9}$ **41.** $\frac{14}{3}$ **43.** 13 **45.** 4

47. all real numbers **49.** $-\dfrac{3}{5}$ **51.** -5 **53.** 10 **55.** no solution **57.** 3 **59.** -17 **61.** $(6x - 8)$ m **63.** $-8 - x$ **65.** $-3 + 2x$ **67.** $9(x + 20)$ **69. a.** all real numbers **b.** answers may vary **c.** answers may vary **71.** a **73.** b **75.** c **77.** answers may vary **79. a.** $x + x + x + 2x + 2x = 28$ **b.** $x = 4$ **c.** x cm $= 4$ cm; $2x$ cm $= 8$ cm **81.** answers may vary **83.** 15.3 **85.** -0.2

Integrated Review **1.** whole, integers, rational, real **2.** natural, whole, integers, rational, real **3.** rational, real **4.** natural, whole, integers, rational, real **5.** integers, rational, real **6.** rational, real **7.** rational, real **8.** irrational, real **9.** $7d - 11$ **10.** $9z + 48$ **11.** 16 **12.** $-12x - 15$ **13.** 5 **14.** -1 **15.** -2 **16.** -2 **17.** $-\dfrac{5}{6}$ **18.** $\dfrac{1}{6}$ **19.** 1 **20.** 6 **21.** 4 **22.** 1 **23.** $\dfrac{9}{5}$ **24.** $-\dfrac{6}{5}$ **25.** all real numbers **26.** all real numbers **27.** 0 **28.** 0 **29.** $\dfrac{4}{19}$ **30.** $-\dfrac{5}{19}$ **31.** no solution **32.** no solution

Vocabulary, Readiness & Video Check **1.** $2x$; $2x - 31$ **3.** $x + 5$; $2(x + 5)$ **5.** $20 - y$; $\dfrac{20 - y}{3}$ or $(20 - y) \div 3$ **7.** in the statement of the application **9.** That the three angle measures are consecutive even integers and that they sum to 180°.

Exercise Set 9.4 **1.** $2x + 7 = x + 6$; -1 **3.** $3x - 6 = 2x + 8$; 14 **5.** -25 **7.** $-\dfrac{3}{4}$ **9.** 3 in.; 6 in.; 16 in. **11.** 1st piece: 5 in.; 2nd piece: 10 in.; 3rd piece: 25 in. **13.** Pennsylvania: 485 million pounds; New York: 1205 million pounds **15.** 172 mi **17.** 25 mi **19.** 1st angle: 37.5°; 2nd angle: 37.5°; 3rd angle: 105° **21.** A: 60°; B: 120°; C: 120°; D: 60° **23.** $3x + 3$ **25.** $x + 2$; $x + 4$; $2x + 4$ **27.** $x + 1$; $x + 2$; $x + 3$; $4x + 6$ **29.** $x + 2$; $x + 4$; $2x + 6$ **31.** 234, 235 **33.** Belgium: 32; France: 33; Spain: 34 **35.** 5 ft, 12 ft **37.** CRH380A: 302 mph; Transrapid TR-09: 279 mph **39.** 43°, 137° **41.** 58°, 60°, 62° **43.** 1 **45.** 280 mi **47.** Michigan: 27; Ohio: 30 **49.** Montana: 56 counties; California: 58 counties **51.** Neptune: 14 satellites; Uranus: 27 satellites; Saturn: 62 satellites **53.** -16 **55.** Sahara: 3,500,000 sq mi; Gobi: 500,000 sq mi **57.** Brazil: 19; New Zealand: 18; Spain: 17 **59.** females: 2082; males: 2445 **61.** 34.5°; 34.5°; 111° **63.** Florida **65.** Florida: $84.6 million; Illinois: $59.8 million **67.** answers may vary **69.** 34 **71.** 225π **73.** 15 ft by 24 ft **75.** 5400 chirps per hour; 129,600 chirps per day; 47,304,000 chirps per year **77.** answers may vary **79.** answers may vary **81.** c

Vocabulary, Readiness & Video Check **1.** relationships **3.** That the process of solving this equation for x—dividing both sides by 5, the coefficient of x—is the same process used to solve a formula for a specific variable. Treat whatever is multiplied by that specific variable as the coefficient—the coefficient is all the factors except that specific variable.

Exercise Set 9.5 **1.** $h = 3$ **3.** $h = 3$ **5.** $h = 20$ **7.** $c = 12$ **9.** $r = 2.5$ **11.** $h = \dfrac{f}{5g}$ **13.** $w = \dfrac{V}{lh}$ **15.** $y = 7 - 3x$ **17.** $R = \dfrac{A - P}{PT}$ **19.** $A = \dfrac{3V}{h}$ **21.** $a = P - b - c$ **23.** $h = \dfrac{S - 2\pi r^2}{2\pi r}$ **25.** 120 ft **27. a.** area: 480 sq in.; perimeter: 120 in. **b.** frame: perimeter; glass: area **29. a.** area: 103.5 sq ft; perimeter: 41 ft **b.** baseboard: perimeter; carpet: area **31.** $-10°C$ **33.** 6.25 hr **35.** length: 78 ft; width: 52 ft **37.** 18 ft, 36 ft, 48 ft **39.** 306 mi **41.** 61.5°F **43.** 60 chirps per minute **45.** increases **47.** 96 piranhas **49.** 2 bags **51.** one 16-in. pizza **53.** x m $= 6$ m; $2.5x$ m $= 15$ m **55.** 22 hr **57.** 13 in. **59.** 2.25 hr **61.** 12,090 ft **63.** 50°C **65.** 686,664 cu in. **67.** 449 cu in. **69.** 333°F **71.** 0.32 **73.** 2.00 or 2 **75.** 17% **77.** 720% **79.** $V = G(N - R)$ **81.** multiplies the volume by 8; answers may vary **83.** $53\dfrac{1}{3}$ **85.** $\bigcirc = \dfrac{\triangle - \square}{\square}$ **87.** 44.3 sec **89.** $P = 3,200,000$ **91.** $V = 113.1$

Vocabulary, Readiness & Video Check **1.** expression **3.** inequality **5.** equation **7.** -5 **9.** 4.1 **11.** An open circle indicates $>$ or $<$; a closed circle indicates $\geq$ or $\leq$. **13.** $\{x \mid x \geq -2\}$ **15.** is greater than; $>$

Exercise Set 9.6 **1.** **3.** **5.** **7.** **9.** **11.** **13.** $\{x \mid x \geq -5\}$ **15.** $\{y \mid y < 9\}$ **17.** $\{x \mid x > -3\}$ **19.** $\{x \mid x \leq 1\}$ **21.** $\{x \mid x < -3\}$ **23.** $\{x \mid x \geq -2\}$ **25.** $\{x \mid x < 0\}$ **27.** $\left\{y \mid y \geq -\dfrac{8}{3}\right\}$ **29.** $\{y \mid y > 3\}$ **31.** $\{x \mid x > -15\}$ **33.** $\{x \mid x \geq -11\}$ **35.** $\left\{x \mid x > \dfrac{1}{4}\right\}$ **37.** $\{y \mid y \geq -12\}$ **39.** $\{z \mid z < 0\}$ **41.** $\{x \mid x > -3\}$ **43.** $\left\{x \mid x \geq -\dfrac{2}{3}\right\}$ **45.** $\{x \mid x \leq -2\}$ **47.** $\{x \mid x > -13\}$ **49.** $\{x \mid x \leq -8\}$ **51.** $\{x \mid x > 4\}$ **53.** $\left\{x \mid x \leq \dfrac{5}{4}\right\}$ **55.** $\left\{x \mid x > \dfrac{8}{3}\right\}$ **57.** $\{x \mid x \geq 0\}$ **59.** all numbers greater than -10 **61.** 35 cm **63.** at least 193 **65.** 86 people **67.** at least 35 min **69.** 81 **71.** 1 **73.** $\dfrac{49}{64}$ **75.** about 32,700 (exact number) **77.** 2010 and 2011 **79.** 2015 **81.** $>$ **83.** $\geq$ **85.** when multiplying or dividing by a negative number **87.** final exam score ≥ 78.5 **89.** answers may vary

Answers to Selected Exercises **A23**

Chapter 9 Vocabulary Check **1.** linear equation in one variable **2.** equivalent inequalities **3.** formula **4.** linear inequality in one variable **5.** all real numbers **6.** no solution **7.** the same **8.** reversed **9.** opposites **10.** reciprocals

Chapter 9 Review **1.** < **2.** > **3.** > **4.** > **5.** < **6.** > **7.** = **8.** = **9.** > **10.** < **11.** $4 \geq -3$ **12.** $6 \neq 5$
13. $0.03 < 0.3$ **14.** 18.9 million < 20.8 million **15. a.** 1, 3 **b.** 0, 1, 3 **c.** $-6, 0, 1, 3$ **d.** $-6, 0, 1, 1\frac{1}{2}, 3, 9.62$ **e.** π **f.** all numbers in set
16. a. 2, 5 **b.** 2, 5 **c.** $-3, 2, 5$ **d.** $-3, -1.6, 2, 5, \frac{11}{2}, 15.1$ **e.** $\sqrt{5}, 2\pi$ **f.** all numbers in set **17.** Friday **18.** Wednesday
19. commutative property of addition **20.** identity element for multiplication **21.** distributive property **22.** additive inverse property
23. associative property of addition **24.** commutative property of multiplication **25.** distributive property **26.** associative property of multiplication **27.** multiplicative inverse property **28.** identity element for addition **29.** -4 **30.** -4 **31.** 2 **32.** -3
33. no solution **34.** no solution **35.** $\frac{3}{4}$ **36.** $-\frac{8}{9}$ **37.** 20 **38.** 0.25 **39.** $\frac{23}{7}$ **40.** $-\frac{6}{23}$ **41.** 6665.5 in. **42.** shorter piece: 4 ft; longer piece: 8 ft **43.** national battlefields: 11; national memorials: 30 **44.** $-39, -38, -37$ **45.** 3 **46.** -4 **47.** $w = 9$ **48.** $h = 4$
49. $m = \dfrac{y - b}{x}$ **50.** $s = \dfrac{r + 5}{vt}$ **51.** $x = \dfrac{2y - 7}{5}$ **52.** $y = \dfrac{2 + 3x}{6}$ **53.** $d = \dfrac{C}{\pi}$ **54.** $r = \dfrac{C}{2\pi}$ **55.** 15 m **56.** 18 ft by 12 ft
57. 1 hr and 20 min **58.** 45°C **59.** ←——○——→ at -2 **60.** ○——●→ from 0 to 5 **61.** $\{x \mid x \leq 1\}$ **62.** $\{x \mid x > -5\}$
63. $\{x \mid x \leq 10\}$ **64.** $\{x \mid x < -4\}$ **65.** $\{x \mid x < -4\}$ **66.** $\{x \mid x \leq 4\}$ **67.** $\{y \mid y > 9\}$ **68.** $\{y \mid y \geq -15\}$
69. $\left\{x \mid x < \dfrac{7}{4}\right\}$ **70.** $\left\{x \mid x \leq \dfrac{19}{3}\right\}$ **71.** at least \$2500 **72.** score must be less than 83 **73.** 4 **74.** -14 **75.** $-\dfrac{3}{2}$ **76.** 21
77. all real numbers **78.** no solution **79.** -13 **80.** shorter piece: 4 in.; longer piece: 19 in. **81.** $h = \dfrac{3V}{A}$ **82.** $\{x \mid x > 9\}$
○——→ at 9 **83.** $\{x \mid x > -4\}$ ←——● at -4 **84.** $\{x \mid x \leq 0\}$ ←——● at 0

Chapter 9 Getting Ready for the Test **1.** C and D **2.** C and D **3.** A **4.** C **5.** B **6.** B **7.** C **8.** C **9.** A
10. B **11.** A **12.** A **13.** B **14.** B **15.** C **16.** B **17.** A **18.** C **19.** C **20.** B **21.** D

Chapter 9 Test **1.** $|-7| > 5$ **2.** $9 + 5 \geq 4$ **3. a.** 1, 7 **b.** 0, 1, 7 **c.** $-5, -1, 0, 1, 7$ **d.** $-5, -1, 0, \dfrac{1}{4}, 1, 7, 11.6$ **e.** $\sqrt{7}, 3\pi$
f. all numbers in set **4.** associative property of addition **5.** commutative property of multiplication **6.** distributive property
7. multiplicative inverse property **8.** 9 **9.** -3 **10.** $10y + 1$ **11.** $-2x + 10$ **12.** 8 **13.** no solution **14.** 0 **15.** 27 **16.** 3
17. 0.25 **18.** $\dfrac{25}{7}$ **19.** 21 **20.** 7 gal **21.** $x = 6$ **22.** $h = \dfrac{V}{\pi r^2}$ **23.** $y = \dfrac{3x - 10}{4}$ **24.** $\{x \mid x < -2\}$ ←——○ at -2
25. $\{x \mid x \leq -8\}$ **26.** $\{x \mid x \geq 11\}$ **27.** $\left\{x \mid x > \dfrac{2}{5}\right\}$ **28.** California: 1107; Ohio: 720

Chapters 1–9 Cumulative Review **1. a.** 4 **b.** -5 **c.** -6; Sec. 2.1, Ex. 6 **2. a.** < **b.** = **c.** > **d.** < **e.** >; Sec. 2.1
3. -19; Sec. 2.5, Ex. 10 **4.** $\dfrac{4}{3}$; Sec. 2.5 **5.** -8; Sec. 2.2, Ex. 9 **6.** -19; Sec. 2.2 **7.** 4; Sec. 2.2, Ex. 10 **8.** 8; Sec. 2.2 **9.** -10; Sec. 2.2, Ex. 14 **10.** -0.3; Sec. 5.2 **11.** $-\dfrac{8}{33}$; Sec. 4.5, Ex. 4 **12.** $-\dfrac{23}{40}$; Sec. 4.5 **13.** $\dfrac{6 - x}{3}$; Sec. 4.5, Ex. 6 **14.** $\dfrac{10 + x}{2}$; Sec. 4.5 **15.** $-\dfrac{8}{9}$;
Sec. 4.6, Ex. 7 **16.** $\dfrac{11}{12}$; Sec. 4.6 **17.** $-5\dfrac{3}{5}$; Sec. 4.7, Ex. 21 **18.** $6\dfrac{1}{4}$; Sec. 4.7 **19.** $\dfrac{14}{15}$; Sec. 4.7, Ex. 22 **20.** $-1\dfrac{2}{25}$; Sec. 4.7 **21.** 4.09;
Sec. 5.5, Ex. 13 **22.** -2.4; Sec. 5.5 **23.** 7; Sec. 7.3, Ex. 1 **24.** 8; Sec. 7.3 **25.** $\dfrac{2}{5}$; Sec. 7.3, Ex. 6 **26.** $\dfrac{3}{10}$; Sec. 7.3 **27.** 2; Sec. 5.6, Ex. 3
28. -5; Sec. 5.6 **29.** true; Sec. 9.1, Ex. 3 **30.** true; Sec. 9.1 **31.** true; Sec. 9.1, Ex. 4 **32.** true; Sec. 9.1 **33.** false; Sec. 9.1, Ex. 5
34. true; Sec. 9.1 **35.** true; Sec. 9.1, Ex. 6 **36.** false; Sec. 9.1 **37.** $15 - 10z$; Sec. 9.2, Ex. 8 **38.** $-8x + 4$; Sec. 9.2 **39.** $3x + 17$;
Sec. 9.2, Ex. 12 **40.** $10x + 21$; Sec. 9.2 **41.** 0; Sec. 9.3, Ex. 4 **42.** $\dfrac{30}{7}$; Sec. 9.3 **43.** Democrats: 194; Republicans: 241; Sec. 9.4, Ex. 4
44. 1; Sec. 9.4 **45.** 79.2 yr; Sec. 9.5, Ex. 1 **46.** $w = \dfrac{V}{lh}$; Sec. 9.5 **47.** ←——○ at -1; Sec. 9.6, Ex. 2 **48.** $\{x \mid x > 10\}$;
Sec. 9.6 **49.** $\{x \mid x \geq 1\}$; Sec. 9.6, Ex. 9 **50.** $\{x \mid x > 4\}$; Sec. 9.6

Chapter 10 Exponents and Polynomials

Vocabulary, Readiness & Video Check 1. exponent 3. add 5. 1 7. base: 3; exponent: 2 9. base: 4; exponent: 2 11. base: x; exponent: 2 13. Example 4 can be written as $-4^2 = -1 \cdot 4^2$, which is similar to Example 7, $4 \cdot 3^2$, and shows why the negative sign should not be considered part of the base when there are no parentheses. 15. Be careful not to confuse the power rule with the product rule. The power rule involves a power raised to a power (exponents are multiplied), and the product rule involves a product (exponents are added). 17. the quotient rule

Exercise Set 10.1 1. 49 3. -5 5. -16 7. 16 9. $\dfrac{1}{27}$ 11. 112 13. 4 15. 135 17. 150 19. $\dfrac{32}{5}$ 21. x^7 23. $(-3)^{12}$ 25. $15y^5$ 27. $x^{19}y^6$ 29. $-72m^3n^8$ 31. $-24z^{20}$ 33. $20x^5$ sq ft 35. x^{36} 37. p^8q^8 39. $8a^{15}$ 41. $x^{10}y^{15}$ 43. $49a^4b^{10}c^2$ 45. $\dfrac{r^9}{s^9}$ 47. $\dfrac{m^9p^9}{n^9}$ 49. $\dfrac{4x^2z^2}{y^{10}}$ 51. $64z^{10}$ sq dm 53. $27y^{12}$ cu ft 55. x^2 57. -64 59. p^6q^5 61. $\dfrac{y^3}{2}$ 63. 1 65. 1 67. -7 69. 2 71. -81 73. $\dfrac{1}{64}$ 75. b^6 77. a^9 79. $-16x^7$ 81. $a^{11}b^{20}$ 83. $26m^9n^7$ 85. z^{40} 87. $64a^3b^3$ 89. $36x^2y^2z^6$ 91. z^8 93. $3x^4$ 95. 1 97. $81x^2y^2$ 99. 40 101. $\dfrac{y^{15}}{8x^{12}}$ 103. $2x^2y$ 105. -2 107. 5 109. -7 111. c 113. e 115. answers may vary 117. answers may vary 119. 343 cu m 121. volume 123. answers may vary 125. answers may vary 127. x^{9a} 129. a^{5b} 131. x^{5a}

Calculator Explorations 1. 5.31 EE 3 3. 6.6 EE -9 5. 1.5×10^{13} 7. 8.15×10^{19}

Vocabulary, Readiness & Video Check 1. $\dfrac{1}{x^3}$ 3. scientific notation 5. $\dfrac{5}{x^2}$ 7. y^6 9. $4y^3$ 11. A negative exponent has nothing to do with the sign of the simplified result. 13. When you move the decimal point to the left, the sign of the exponent will be positive; when you move the decimal point to the right, the sign of the exponent will be negative. 15. the quotient rule

Exercise Set 10.2 1. $\dfrac{1}{64}$ 3. $\dfrac{7}{x^3}$ 5. -64 7. $\dfrac{5}{6}$ 9. p^3 11. $\dfrac{q^4}{p^5}$ 13. $\dfrac{1}{x^3}$ 15. z^3 17. $\dfrac{4}{9}$ 19. $\dfrac{1}{9}$ 21. $-p^4$ 23. -2 25. x^4 27. p^4 29. m^{11} 31. r^6 33. $\dfrac{1}{x^{15}y^9}$ 35. $\dfrac{1}{x^4}$ 37. $\dfrac{1}{a^2}$ 39. $4k^3$ 41. $3m$ 43. $-\dfrac{4a^5}{b}$ 45. $-\dfrac{6}{7y^2}$ 47. $\dfrac{27a^6}{b^{12}}$ 49. $\dfrac{a^{30}}{b^{12}}$ 51. $x^{10}y^6$ 53. $\dfrac{z^2}{4}$ 55. $\dfrac{x^{11}}{81}$ 57. $\dfrac{49a^4}{b^6}$ 59. $-\dfrac{3m^7}{n^4}$ 61. $a^{24}b^8$ 63. 200 65. x^9y^{19} 67. $-\dfrac{y^8}{8x^2}$ 69. $\dfrac{25b^{33}}{a^{16}}$ 71. $\dfrac{27}{z^3x^6}$ cu in. 73. 7.8×10^4 75. 1.67×10^{-6} 77. 6.35×10^{-3} 79. 1.16×10^6 81. 4.2×10^3 83. 0.0000000008673 85. 0.033 87. 20,320 89. 700,000,000 91. 2.415×10^{12} 93. 5,500,000,000,000 95. 9,000,000,000,000; 9×10^{12} 97. 0.000036 99. 0.00000000000000028 101. 0.0000005 103. 200,000 105. 2.7×10^9 gal 107. $-2x + 7$ 109. $2y - 10$ 111. $-x - 4$ 113. 377,900,000; 3.779×10^8 115. 14,056,000; 1.4056×10^7 117. 2.5×10^{-9} m 119. 0.00000031 m; 3.1×10^{-7} m 121. $9a^{13}$ 123. -5 125. answers may vary 127. **a.** 1.3×10^1 **b.** 4.4×10^7 **c.** 6.1×10^{-2} 129. answers may vary 131. $\dfrac{1}{x^{9s}}$ 133. a^{4m+5}

Vocabulary, Readiness & Video Check 1. binomial 3. trinomial 5. constant 7. 3; x^2, $-3x$, 5 9. the replacement value for the variable 11. 2; $9ab$

Exercise Set 10.3 1. 1; $-3x$; 5 3. -5; 3.2; 1; -5 5. 1; binomial 7. 3; none of these 9. 6; trinomial 11. 4; binomial 13. **a.** -6 **b.** -11 15. **a.** -2 **b.** 4 17. **a.** -15 **b.** -10 19. 184 ft 21. 595.84 ft 23. 1349 thousand 25. 27.7 million 27. $-11x$ 29. $23x^3$ 31. $16x^2 - 7$ 33. $12x^2 - 13$ 35. $7s$ 37. $-1.1y^2 + 4.8$ 39. $\dfrac{5}{6}x^4 - 7x^3 - 19$ 41. $\dfrac{3}{20}x^3 + 6x^2 - \dfrac{13}{20}x - \dfrac{1}{10}$ 43. $4x^2 + 7x + x^2 + 5x$; $5x^2 + 12x$ 45. $5x + 3 + 4x + 3 + 2x + 6 + 3x + 7x$; $21x + 12$ 47. 2, 1, 1, 0; 2 49. 4, 0, 4, 3; 4 51. $9ab - 11a$ 53. $4x^2 - 7xy + 3y^2$ 55. $-3xy^2 + 4$ 57. $14y^3 - 19 - 16a^2b^2$ 59. $7x^2 + 0x + 3$ 61. $x^3 + 0x^2 + 0x - 64$ 63. $5y^3 + 0y^2 + 2y - 10$ 65. $2y^4 + 0y^3 + 0y^2 + 8y + 0y^0$ or $2y^4 + 0y^3 + 0y^2 + 8y + 0$ 67. $6x^5 + 0x^4 + x^3 + 0x^2 - 3x + 15$ 69. $10x + 19$ 71. $-x + 5$ 73. answers may vary 75. answers may vary 77. x^{13} 79. a^3b^{10} 81. $2y^{20}$ 83. answers may vary 85. answers may vary 87. $11.1x^2 - 7.97x + 10.76$

Vocabulary, Readiness & Video Check 1. $-14y$ 3. $7x$ 5. $5m^2 + 2m$ 7. $-3y^2$ and $2y^2$; $-4y$ and y 9. We're translating a subtraction problem. Order matters when subtracting, so we need to be careful that the order of the expressions is correct.

Exercise Set 10.4 1. $12x + 12$ 3. $-3x^2 + 10$ 5. $-3x^2 + 4$ 7. $-y^2 - 3y - 1$ 9. $7.9x^3 + 4.4x^2 - 3.4x - 3$ 11. $\dfrac{1}{2}m^2 - \dfrac{7}{10}m + \dfrac{13}{16}$ 13. $8t^2 - 4$ 15. $15a^3 + a^2 - 3a + 16$ 17. $-x + 14$ 19. $7x^2$ 21. $-2x + 9$ 23. $2x^2 + 7x - 16$ 25. $2x^2 + 11x$ 27. $-0.2x^2 + 0.2x - 2.2$ 29. $\dfrac{2}{5}z^2 - \dfrac{3}{10}z + \dfrac{7}{20}$ 31. $-2z^2 - 16z + 6$ 33. $2u^5 - 10u^2 + 11u - 9$ 35. $5x - 9$

37. $4x - 3$ **39.** $11y + 7$ **41.** $-2x^2 + 8x - 1$ **43.** $14x + 18$ **45.** $3a^2 - 6a + 11$ **47.** $3x - 3$ **49.** $7x^2 - 4x + 2$ **51.** $7x^2 - 2x + 2$ **53.** $4y^2 + 12y + 19$ **55.** $-15x + 7$ **57.** $-2a - b + 1$ **59.** $3x^2 + 5$ **61.** $6x^2 - 2xy + 19y^2$ **63.** $8r^2s + 16rs - 8 + 7r^2s^2$ **65.** $(x^2 + 7x + 4)$ ft **67.** $\left(\dfrac{19}{2}x + 3\right)$ units **69.** $(3y^2 + 4y + 11)$ m **71.** $-6.6x^2 - 1.8x - 1.8$ **73.** $6x^2$ **75.** $-12x^8$ **77.** $200x^3y^2$ **79.** $2;2$ **81.** $4;3;3;4$ **83. b** **85. e** **87. a.** $4z$ **b.** $3z^2$ **c.** $-4z$ **d.** $3z^2$; answers may vary **89. a.** m^3 **b.** $3m$ **c.** $-m^3$ **d.** $-3m$; answers may vary **91.** $-4052x^2 + 34{,}684x + 144{,}536$

Vocabulary, Readiness & Video Check **1.** distributive **3.** $(5y - 1)(5y - 1)$ **5.** x^8 **7.** cannot simplify **9.** x^{14} **11.** $2x^7$ **13.** No. The monomials are unlike terms. **15.** Three times: First $(a - 2)$ is distributed to a and 7, and then a is distributed to $(a - 2)$ and 7 is distributed to $(a - 2)$.

Exercise Set 10.5 **1.** $24x^3$ **3.** x^4 **5.** $-28n^{10}$ **7.** $-12.4x^{12}$ **9.** $-\dfrac{2}{15}y^3$ **11.** $-24x^8$ **13.** $6x^2 + 15x$ **15.** $7x^3 + 14x^2 - 7x$ **17.** $-2a^2 - 8a$ **19.** $6x^3 - 9x^2 + 12x$ **21.** $12a^5 + 45a^2$ **23.** $-6a^4 + 4a^3 - 6a^2$ **25.** $6x^5y - 3x^4y^3 + 24x^2y^4$ **27.** $-4x^3y + 7x^2y^2 - xy^3 - 3y^4$ **29.** $4x^4 - 3x^3 + \dfrac{1}{2}x^2$ **31.** $x^2 + 7x + 12$ **33.** $a^2 + 5a - 14$ **35.** $x^2 + \dfrac{1}{3}x - \dfrac{2}{9}$ **37.** $12x^4 + 25x^2 + 7$ **39.** $12x^2 - 29x + 15$ **41.** $1 - 7a + 12a^2$ **43.** $4y^2 - 16y + 16$ **45.** $x^3 - 5x^2 + 13x - 14$ **47.** $x^4 + 5x^3 - 3x^2 - 11x + 20$ **49.** $10a^3 - 27a^2 + 26a - 12$ **51.** $49x^2y^2 - 14xy^2 + y^2$ **53.** $12x^2 - 64x - 11$ **55.** $2x^3 + 10x^2 + 11x - 3$ **57.** $2x^4 + 3x^3 - 58x^2 + 4x + 63$ **59.** $8.4y^7$ **61.** $-3x^3 - 6x^2 + 24x$ **63.** $2x^2 + 39x + 19$ **65.** $x^2 - \dfrac{2}{7}x - \dfrac{3}{49}$ **67.** $9y^2 + 30y + 25$ **69.** $a^3 - 2a^2 - 18a + 24$ **71.** $(4x^2 - 25)$ sq yd **73.** $(6x^2 - 4x)$ sq in. **75.** $5a + 15a = 20a$; $5a - 15a = -10a$; $5a \cdot 15a = 75a^2$; $\dfrac{5a}{15a} = \dfrac{1}{3}$ **77.** $-3y^5 + 9y^4$, cannot be simplified; $-3y^5 - 9y^4$, cannot be simplified; $-3y^5 \cdot 9y^4 = -27y^9$; $\dfrac{-3y^5}{9y^4} = -\dfrac{y}{3}$ **79. a.** $6x + 12$ **b.** $9x^2 + 36x + 35$; answers may vary **81.** $13x - 7$ **83.** $30x^2 - 28x + 6$ **85.** $-7x + 5$ **87.** $x^2 + 3x$ **89.** $x + 2x^2$; $x(1 + 2x)$ **91.** $11a$ **93.** $25x^2 + 4y^2$ **95. a.** $a^2 - b^2$ **b.** $4x^2 - 9y^2$ **c.** $16x^2 - 49$ **d.** answers may vary

Vocabulary, Readiness & Video Check **1.** false **3.** false **5.** a binomial times a binomial **7.** Multiplying gives you four terms, and the two like terms will always subtract out.

Exercise Set 10.6 **1.** $x^2 + 7x + 12$ **3.** $x^2 + 5x - 50$ **5.** $5x^2 + 4x - 12$ **7.** $4y^2 - 25y + 6$ **9.** $6x^2 + 13x - 5$ **11.** $6y^3 + 4y^2 + 42y + 28$ **13.** $x^2 + \dfrac{1}{3}x - \dfrac{2}{9}$ **15.** $0.08 - 2.6a + 15a^2$ **17.** $2x^2 + 9xy - 5y^2$ **19.** $x^2 + 4x + 4$ **21.** $4a^2 - 12a + 9$ **23.** $9a^2 - 30a + 25$ **25.** $x^4 + x^2 + 0.25$ **27.** $y^2 - \dfrac{4}{7}y + \dfrac{4}{49}$ **29.** $4x^2 - 4x + 1$ **31.** $25x^2 + 90x + 81$ **33.** $9x^2 - 42xy + 49y^2$ **35.** $16m^2 + 40mn + 25n^2$ **37.** $25x^8 - 30x^4 + 9$ **39.** $a^2 - 49$ **41.** $x^2 - 36$ **43.** $9x^2 - 1$ **45.** $x^4 - 25$ **47.** $4y^4 - 1$ **49.** $16 - 49x^2$ **51.** $9x^2 - \dfrac{1}{4}$ **53.** $81x^2 - y^2$ **55.** $4m^2 - 25n^2$ **57.** $a^2 + 9a + 20$ **59.** $a^2 - 14a + 49$ **61.** $12a^2 - a - 1$ **63.** $x^2 - 4$ **65.** $9a^2 + 6a + 1$ **67.** $4x^2 + 3xy - y^2$ **69.** $\dfrac{1}{9}a^4 - 49$ **71.** $6b^2 - b - 35$ **73.** $x^4 - 100$ **75.** $16x^2 - 25$ **77.** $25x^2 - 60xy + 36y^2$ **79.** $4r^2 - 9s^2$ **81.** $(4x^2 + 4x + 1)$ sq ft **83.** $\dfrac{5b^5}{7}$ **85.** $-\dfrac{2a^{10}}{b^5}$ **87.** $\dfrac{2y^8}{3}$ **89. c** **91. d** **93.** 2 **95.** $(x^4 - 3x^2 + 1)$ sq m **97.** $(24x^2 - 32x + 8)$ sq m **99.** answers may vary **101.** answers may vary

Integrated Review **1.** $35x^5$ **2.** $-32y^9$ **3.** -16 **4.** 16 **5.** $2x^2 - 9x - 5$ **6.** $3x^2 + 13x - 10$ **7.** $3x - 4$ **8.** $4x + 3$ **9.** $7x^6y^2$ **10.** $\dfrac{10b^6}{7}$ **11.** $144m^{14}n^{12}$ **12.** $64y^{27}z^{30}$ **13.** $16y^2 - 9$ **14.** $49x^2 - 1$ **15.** $\dfrac{y^{45}}{x^{63}}$ **16.** $\dfrac{1}{64}$ **17.** $\dfrac{x^{27}}{27}$ **18.** $\dfrac{r^{58}}{16s^{14}}$ **19.** $2x^2 - 2x - 6$ **20.** $6x^2 + 13x - 11$ **21.** $2.5y^2 - 6y - 0.2$ **22.** $8.4x^2 - 6.8x - 4.2$ **23.** $2y^2 - 6y - 1$ **24.** $6z^2 + 2z + \dfrac{11}{2}$ **25.** $x^2 + 8x + 16$ **26.** $y^2 - 18y + 81$ **27.** $2x + 8$ **28.** $2y - 18$ **29.** $7x^2 - 10xy + 4y^2$ **30.** $-a^2 - 3ab + 6b^2$ **31.** $x^3 + 2x^2 - 16x + 3$ **32.** $x^3 - 2x^2 - 5x - 2$ **33.** $6x^2 - x - 70$ **34.** $20x^2 + 21x - 5$ **35.** $2x^3 - 19x^2 + 44x - 7$ **36.** $5x^3 + 9x^2 - 17x + 3$ **37.** $4x^2 - \dfrac{25}{81}$ **38.** $144y^2 - \dfrac{9}{49}$

Vocabulary, Readiness & Video Check **1.** dividend; quotient; divisor **3.** a^2 **5.** y **7.** the common denominator

Exercise Set 10.7 **1.** $12x^3 + 3x$ **3.** $4x^3 - 6x^2 + x + 1$ **5.** $5p^2 + 6p$ **7.** $-\dfrac{3}{2x} + 3$ **9.** $-3x^2 + x - \dfrac{4}{x^3}$ **11.** $-1 + \dfrac{3}{2x} - \dfrac{7}{4x^4}$ **13.** $x + 1$ **15.** $2x + 3$ **17.** $2x + 1 + \dfrac{7}{x - 4}$ **19.** $3a^2 - 3a + 1 + \dfrac{2}{3a + 2}$ **21.** $4x + 3 - \dfrac{2}{2x + 1}$ **23.** $2x^2 + 6x - 5 - \dfrac{2}{x - 2}$

25. $x + 6$ **27.** $x^2 + 3x + 9$ **29.** $-3x + 6 - \dfrac{11}{x+2}$ **31.** $2b - 1 - \dfrac{6}{2b-1}$ **33.** $ab - b^2$ **35.** $4x + 9$ **37.** $x + 4xy - \dfrac{y}{2}$ **39.** $2b^2 + b + 2 - \dfrac{12}{b+4}$ **41.** $y^2 + 5y + 10 + \dfrac{24}{y-2}$ **43.** $-6x - 12 - \dfrac{19}{x-2}$ **45.** $x^3 - x^2 + x$ **47.** 3 **49.** -4 **51.** $3x$ **53.** $9x$ **55.** $(3x^3 + x - 4)$ ft **57.** $(2x + 5)$ m **59.** answers may vary **61. c**

Chapter 10 Vocabulary Check **1.** term **2.** FOIL **3.** trinomial **4.** degree of a polynomial **5.** binomial **6.** coefficient **7.** degree of a term **8.** monomial **9.** polynomials **10.** distributive

Chapter 10 Review **1.** base: 3; exponent: 4 **2.** base: -5; exponent: 4 **3.** base: 5; exponent: 4 **4.** base: x; exponent: 4 **5.** 512 **6.** 36 **7.** -36 **8.** -65 **9.** 1 **10.** 1 **11.** y^9 **12.** x^{14} **13.** $-6x^{11}$ **14.** $-20y^7$ **15.** x^8 **16.** y^{15} **17.** $81y^{24}$ **18.** $8x^9$ **19.** x^5 **20.** z^7 **21.** a^4b^3 **22.** x^3y^5 **23.** $\dfrac{x^3y^4}{4}$ **24.** $\dfrac{x^6y^6}{4}$ **25.** $40a^{19}$ **26.** $36x^3$ **27.** 3 **28.** 9 **29.** b **30.** c **31.** $\dfrac{1}{49}$ **32.** $-\dfrac{1}{49}$ **33.** $\dfrac{2}{x^4}$ **34.** $\dfrac{1}{16x^4}$ **35.** 125 **36.** $\dfrac{9}{4}$ **37.** $\dfrac{17}{16}$ **38.** $\dfrac{1}{42}$ **39.** x^8 **40.** z^8 **41.** r **42.** y^3 **43.** c^4 **44.** $\dfrac{x^3}{y^3}$ **45.** $\dfrac{1}{x^6y^{13}}$ **46.** $\dfrac{a^{10}}{b^{10}}$ **47.** 2.7×10^{-4} **48.** 8.868×10^{-1} **49.** 8.08×10^7 **50.** 8.68×10^5 **51.** 1.37×10^8 **52.** 1.5×10^5 **53.** 867,000 **54.** 0.00386 **55.** 0.00086 **56.** 893,600 **57.** 1,431,280,000,000,000 **58.** 0.0000000001 **59.** 0.016 **60.** 400,000,000,000 **61.** 5 **62.** 2 **63.** 5 **64.** 6 **65.** 4000 ft; 3984 ft; 3856 ft; 3600 ft **66.** 22; 78; 154.02; 400 **67.** $2a^2$ **68.** $-4y$ **69.** $15a^2 + 4a$ **70.** $22x^2 + 3x + 6$ **71.** $-6a^2b - 3b^2 - q^2$ **72.** cannot be combined **73.** $8x^2 + 3x + 6$ **74.** $2x^5 + 3x^4 + 4x^3 + 9x^2 + 7x + 6$ **75.** $-7y^2 - 1$ **76.** $-6m^7 - 3x^4 + 7m^6 - 4m^2$ **77.** $-x^2 - 6xy - 2y^2$ **78.** $x^6 + 4xy + 2y^2$ **79.** $-5x^2 + 5x + 1$ **80.** $-2x^2 - x + 20$ **81.** $6x + 30$ **82.** $9x - 63$ **83.** $8a + 28$ **84.** $54a - 27$ **85.** $-7x^3 - 35x$ **86.** $-32y^3 + 48y$ **87.** $-2x^3 + 18x^2 - 2x$ **88.** $-3a^3b - 3a^2b - 3ab^2$ **89.** $-6a^4 + 8a^2 - 2a$ **90.** $42b^4 - 28b^2 + 14b$ **91.** $2x^2 - 12x - 14$ **92.** $6x^2 - 11x - 10$ **93.** $4a^2 + 27a - 7$ **94.** $42a^2 + 11a - 3$ **95.** $x^4 + 7x^3 + 4x^2 + 23x - 35$ **96.** $x^6 + 2x^5 + x^2 + 3x + 2$ **97.** $x^4 + 4x^3 + 4x^2 - 16$ **98.** $x^6 + 8x^4 + 16x^2 - 16$ **99.** $x^3 + 21x^2 + 147x + 343$ **100.** $8x^3 - 60x^2 + 150x - 125$ **101.** $x^2 + 14x + 49$ **102.** $x^2 - 10x + 25$ **103.** $9x^2 - 42x + 49$ **104.** $16x^2 + 16x + 4$ **105.** $25x^2 - 90x + 81$ **106.** $25x^2 - 1$ **107.** $49x^2 - 16$ **108.** $a^2 - 4b^2$ **109.** $4x^2 - 36$ **110.** $16a^4 - 4b^2$ **111.** $(9x^2 - 6x + 1)$ sq m **112.** $(5x^2 - 3x - 2)$ sq mi **113.** $\dfrac{1}{7} + \dfrac{3}{x} + \dfrac{7}{x^2}$ **114.** $-a^2 + 3b - 4$ **115.** $a + 1 + \dfrac{6}{a-2}$ **116.** $4x + \dfrac{7}{x+5}$ **117.** $a^2 + 3a + 8 + \dfrac{22}{a-2}$ **118.** $3b^2 - 4b - \dfrac{1}{3b-2}$ **119.** $2x^3 - x^2 + 2 - \dfrac{1}{2x-1}$ **120.** $-x^2 - 16x - 117 - \dfrac{684}{x-6}$ **121.** $\left(5x - 1 + \dfrac{20}{x^2}\right)$ ft **122.** $(7a^3b^6 + a - 1)$ units **123.** 27 **124.** $-\dfrac{1}{8}$ **125.** $4x^4y^7$ **126.** $\dfrac{2x^6}{3}$ **127.** $\dfrac{27a^{12}}{b^6}$ **128.** $\dfrac{x^{16}}{16y^{12}}$ **129.** $9a^2b^8$ **130.** $2y^2 - 10$ **131.** $11x - 5$ **132.** $5x^2 + 3x - 2$ **133.** $5y^2 - 3y - 1$ **134.** $6x^2 + 11x - 10$ **135.** $28x^3 + 12x$ **136.** $28x^2 - 71x + 18$ **137.** $x^3 + x^2 - 18x + 18$ **138.** $25x^2 + 40x + 16$ **139.** $36x^2 - 9$ **140.** $4a - 1 + \dfrac{2}{a^2} - \dfrac{5}{2a^3}$ **141.** $x - 3 + \dfrac{25}{x+5}$ **142.** $2x^2 + 7x + 5 + \dfrac{19}{2x-3}$

Chapter 10 Getting Ready for the Test **1.** C **2.** A **3.** E **4.** D **5.** F **6.** C **7.** E **8.** I **9.** C **10.** D **11.** C **12.** B **13.** F **14.** D

Chapter 10 Test **1.** 32 **2.** 81 **3.** -81 **4.** $\dfrac{1}{64}$ **5.** $-15x^{11}$ **6.** y^5 **7.** $\dfrac{1}{r^5}$ **8.** $\dfrac{16y^{14}}{x^2}$ **9.** $\dfrac{1}{6xy^8}$ **10.** 5.63×10^5 **11.** 8.63×10^{-5} **12.** 0.0015 **13.** 62,300 **14.** 0.036 **15. a.** 4, 3; 7, 3; 1, 4; -2, 0 **b.** 4 **16.** $-2x^2 + 12x + 11$ **17.** $16x^3 + 7x^2 - 3x - 13$ **18.** $-3x^3 + 5x^2 + 4x + 5$ **19.** $x^3 + 8x^2 + 3x - 5$ **20.** $3x^3 + 22x^2 + 41x + 14$ **21.** $6x^4 - 9x^3 + 21x^2$ **22.** $3x^2 + 16x - 35$ **23.** $9x^2 - \dfrac{1}{25}$ **24.** $16x^2 - 16x + 4$ **25.** $64x^2 + 48x + 9$ **26.** $x^4 - 81b^2$ **27.** 1001 ft; 985 ft; 857 ft; 601 ft **28.** $(4x^2 - 9)$ sq in. **29.** $\dfrac{x}{2y} + \dfrac{1}{4} - \dfrac{7}{8y}$ **30.** $x + 2$ **31.** $9x^2 - 6x + 4 - \dfrac{16}{3x+2}$

Cumulative Review Chapters 1–10 **1.** 81; Sec. 1.7, Ex. 5 **2.** 125; Sec. 1.7 **3.** 81; Sec. 1.7, Ex. 7 **4.** 27; Sec. 1.7 **5. a.** $7 + x$ **b.** $15 - x$ **c.** $2x$ **d.** $x \div 5$ or $\dfrac{x}{5}$ **e.** $x - 2$; Sec. 1.8, Ex. 8 **6. a.** $x + 3$ **b.** $3x$ **c.** $2x$ **d.** $10 - x$ **e.** $5x + 7$; Sec. 1.8 **7.** $14x - 9$; Sec. 3.1, Ex. 11 **8.** $5x + 10$; Sec. 3.1 **9.** $6x + 18$; Sec. 3.1, Ex. 12 **10.** $-2y - 6z + 2$; Sec. 3.1 **11.** -3; Sec. 3.2, Ex. 1 **12.** 140; Sec. 3.2 **13.** -4; Sec. 3.3, Ex. 3 **14.** 19; Sec. 3.3 **15.** 15; Sec. 4.8, Ex. 3 **16.** 24; Sec. 4.8 **17.** $-\dfrac{2}{33}$; Sec. 4.8, Ex. 5 **18.** $-\dfrac{1}{25}$; Sec. 4.8 **19.** $\dfrac{1}{8}$; Sec. 5.1, Ex. 9 **20.** $\dfrac{1}{4}$; Sec. 5.1 **21.** $43\dfrac{1}{2}$; Sec. 5.1, Ex. 10 **22.** $10\dfrac{3}{4}$; Sec. 5.1 **23.** $-105\dfrac{83}{1000}$; Sec. 5.1, Ex. 11 **24.** $-31\dfrac{7}{100}$; Sec. 5.1 **25.** 1.88; Sec. 5.2, Ex. 11 **26.** -8.8; Sec. 5.2 **27.** 1.012; Sec. 5.3, Ex. 12 **28.** -3.05; Sec. 5.3 **29. a.** 11, 112 **b.** 0, 11, 112 **c.** $-3, -2, 0, 11, 112$ **d.** $-3, -2, 0, \dfrac{1}{4}, 11, 112$ **e.** $\sqrt{2}$ **f.** all numbers in the given set; Sec. 9.1, Ex. 11 **30.** rational numbers, real numbers; Sec. 9.1 **31.** 4, Sec. 9.3, Ex. 5 **32.** 5; Sec. 9.3 **33.** 10; Sec. 9.4, Ex. 2 **34.** 2; Sec. 9.4 **35.** 40 ft; Sec. 9.5, Ex. 2

36. 32 ft; Sec. 9.5 **37.** $\{x \mid x \leq 4\}$ ⟵————•⟶ ; Sec. 9.6, Ex. 7 **38.** $\{x \mid x \geq -10\}$; Sec. 9.6 **39. a.** x^{11} **b.** $\dfrac{t^4}{16}$

c. $81y^{10}$; Sec. 10.1, Ex. 33 **40. a.** y^6 **b.** $\dfrac{8}{27}$ **c.** $64x^6$; Sec. 10.1 **41.** $\dfrac{b^3}{27a^6}$; Sec. 10.2, Ex. 10 **42.** $\dfrac{y^2}{4x^6}$; Sec. 10.2 **43.** $\dfrac{1}{25y^6}$; Sec. 10.2,

Ex. 14 **44.** $\dfrac{1}{27x^{21}}$; Sec. 10.2 **45.** $10x^3$; Sec. 10.3, Ex. 8 **46.** $8x^3$; Sec. 10.3 **47.** $5x^2 - 3x - 3$; Sec. 10.3, Ex. 9 **48.** $-5x^2 + 7x$;

Sec. 10.3 **49.** $7x^3 + 14x^2 + 35x$; Sec. 10.5, Ex. 4 **50.** $-2x^3 + 2x^2 - 2x$; Sec. 10.5 **51.** $3x^3 - 4 + \dfrac{1}{x}$; Sec. 10.7, Ex. 2

52. $2x^6 - 6x + 1$; Sec. 10.7

Chapter 11 Factoring Polynomials

Vocabulary, Readiness & Video Check 1. factors **3.** least **5.** false **7.** $2 \cdot 7$ **9.** 3 **11.** 5 **13.** The GCF of a list of numbers is the largest number that is a factor of all numbers in the list. **15.** When factoring out a GCF, the number of terms in the other factor should have the same number of terms as your original polynomial.

Exercise Set 11.1 1. 4 **3.** 6 **5.** 1 **7.** y^2 **9.** z^7 **11.** xy^2 **13.** 7 **15.** $4y^3$ **17.** $5x^2$ **19.** $3x^3$ **21.** $9x^2y$ **23.** $10a^6b$ **25.** $3(a + 2)$
27. $15(2x - 1)$ **29.** $x^2(x + 5)$ **31.** $2y^3(3y + 1)$ **33.** $2x(16y - 9x)$ **35.** $4(x - 2y + 1)$ **37.** $3x(2x^2 - 3x + 4)$
39. $a^2b^2(a^5b^4 - a + b^3 - 1)$ **41.** $5xy(x^2 - 3x + 2)$ **43.** $4(2x^5 + 4x^4 - 5x^3 + 3)$ **45.** $\dfrac{1}{3}x(x^3 + 2x^2 - 4x^4 + 1)$
47. $(x^2 + 2)(y + 3)$ **49.** $(y + 4)(z + 3)$ **51.** $(z^2 - 6)(r + 1)$ **53.** $-2(x + 7)$ **55.** $-x^5(2 - x^2)$ **57.** $-3a^2(2a^2 - 3a + 1)$
59. $(x + 2)(x^2 + 5)$ **61.** $(x + 3)(5 + y)$ **63.** $(3x - 2)(2x^2 + 5)$ **65.** $(5m^2 + 6n)(m + 1)$ **67.** $(y - 4)(2 + x)$
69. $(2x + 1)(x^2 + 4)$ **71.** not factorable by grouping **73.** $(x - 2y)(4x - 3)$ **75.** $(5q - 4p)(q - 1)$ **77.** $2(2y - 7)(3x^2 - 1)$
79. $3(2a + 3b^2)(a + b)$ **81.** $x^2 + 7x + 10$ **83.** $b^2 - 3b - 4$ **85.** 2, 6 **87.** $-1, -8$ **89.** $-2, 5$ **91.** $-8, 3$ **93.** d **95.** factored
97. not factored **99. a.** 3384 thousand **b.** 3336 thousand **c.** $-3(x^2 - 26x - 968)$ **d.** 2115 thousand **101. a.** 5916 thousand tons
b. 7134 thousand tons **c.** $87(x^2 - 13x + 104)$ **103.** $4x^2 - \pi x^2$; $x^2(4 - \pi)$ **105.** $(x^3 - 1)$ units **107.** answers may vary
109. answers may vary

Vocabulary, Readiness & Video Check 1. true **3.** false **5.** $+5$ **7.** -3 **9.** $+2$ **11.** 15 is positive, so its factors would have to be either both positive or both negative. Since the factors need to sum to -8, both factors must be negative.

Exercise Set 11.2 1. $(x + 6)(x + 1)$ **3.** $(y - 9)(y - 1)$ **5.** $(x - 3)(x - 3)$ or $(x - 3)^2$ **7.** $(x - 6)(x + 3)$
9. $(x + 10)(x - 7)$ **11.** prime **13.** $(x + 5y)(x + 3y)$ **15.** $(a^2 - 5)(a^2 + 3)$ **17.** $(m + 13)(m + 1)$
19. $(t - 2)(t + 12)$ **21.** $(a - 2b)(a - 8b)$ **23.** $2(z + 8)(z + 2)$ **25.** $2x(x - 5)(x - 4)$ **27.** $(x - 4y)(x + y)$
29. $(x + 12)(x + 3)$ **31.** $(x^2 - 2)(x^2 + 1)$ **33.** $(r - 12)(r - 4)$ **35.** $(x + 2y)(x - y)$ **37.** $3(x + 5)(x - 2)$
39. $3(x^2 - 18)(x^2 - 2)$ **41.** $(x - 24)(x + 6)$ **43.** prime **45.** $(x - 5)(x - 3)$ **47.** $6x(x + 4)(x + 5)$ **49.** $4y(x^2 + x - 3)$
51. $(x - 7)(x + 3)$ **53.** $(x + 5y)(x + 2y)$ **55.** $2(t + 8)(t + 4)$ **57.** $x(x - 6)(x + 4)$ **59.** $2t^3(t - 4)(t - 3)$
61. $5xy(x - 8y)(x + 3y)$ **63.** $3(m - 9)(m - 6)$ **65.** $-1(x - 11)(x - 1)$ **67.** $\dfrac{1}{2}(y - 11)(y + 2)$ **69.** $x(xy - 4)(xy + 5)$
71. $2x^2 + 11x + 5$ **73.** $15y^2 - 17y + 4$ **75.** $9a^2 + 23ab - 12b^2$ **77.** $x^2 + 5x - 24$ **79.** answers may vary
81. $2x^2 + 28x + 66$; $2(x + 3)(x + 11)$ **83.** $-16(t - 5)(t + 1)$ **85.** $\left(x + \dfrac{1}{4}\right)\left(x + \dfrac{1}{4}\right)$ or $\left(x + \dfrac{1}{4}\right)^2$ **87.** $(x + 1)(z - 10)(z + 7)$
89. 15; 28; 39; 48; 55; 60; 63; 64 **91.** 9; 12; 21 **93.** $(x^n + 10)(x^n - 2)$

Vocabulary, Readiness & Video Check 1. d **3.** c **5.** Consider the factors of the first and last terms and the signs of the trinomial. Continue to check by multiplying until you get the middle term of the trinomial.

Exercise Set 11.3 1. $x + 4$ **3.** $10x - 1$ **5.** $4x - 3$ **7.** $(2x + 3)(x + 5)$ **9.** $(y - 1)(8y - 9)$ **11.** $(2x + 1)(x - 5)$
13. $(4r - 1)(5r + 8)$ **15.** $(10x + 1)(x + 3)$ **17.** $(3x - 2)(x + 1)$ **19.** $(3x - 5y)(2x - y)$ **21.** $(3m - 5)(5m + 3)$
23. $(x - 4)(x - 5)$ **25.** $(2x + 11)(x - 9)$ **27.** $(7t + 1)(t - 4)$ **29.** $(3a + b)(a + 3b)$ **31.** $(7p + 1)(7p - 2)$
33. $(6x - 7)(3x + 2)$ **35.** prime **37.** $(8x + 3)(3x + 4)$ **39.** $x(3x + 2)(4x + 1)$ **41.** $3(7b + 5)(b - 3)$
43. $(3z + 4)(4z - 3)$ **45.** $2y^2(3x - 10)(x + 3)$ **47.** $(2x - 7)(2x + 3)$ **49.** $3(x^2 - 14x + 21)$ **51.** $(4x + 9y)(2x - 3y)$
53. $-1(x - 6)(x + 4)$ **55.** $x(4x + 3)(x - 3)$ **57.** $(4x - 9)(6x - 1)$ **59.** $b(8a - 3)(5a + 3)$ **61.** $2x(3x + 2)(5x + 3)$
63. $2y(3y + 5)(y - 3)$ **65.** $5x^2(2x - y)(x + 3y)$ **67.** $-1(2x - 5)(7x - 2)$ **69.** $p^2(4p - 5)(4p - 5)$ or $p^2(4p - 5)^2$
71. $-1(2x + 1)(x - 5)$ **73.** $-4(12x - 1)(x - 1)$ **75.** $(2t^2 + 9)(t^2 - 3)$ **77.** prime **79.** $a(6a^2 + b^2)(a^2 + 6b^2)$ **81.** $x^2 - 16$
83. $x^2 + 4x + 4$ **85.** $4x^2 - 4x + 1$ **87.** 18–24 **89.** answers may vary **91.** no **93.** $4x^2 + 21x + 5$; $(4x + 1)(x + 5)$
95. $\left(2x + \dfrac{1}{2}\right)\left(2x + \dfrac{1}{2}\right)$ or $\left(2x + \dfrac{1}{2}\right)^2$ **97.** $(y - 1)^2(4x + 5)(x + 5)$ **99.** 2; 14 **101.** 2 **103.** answers may vary

Vocabulary, Readiness & Video Check 1. a **3.** b **5.** This gives us a four-term polynomial, which may be factored by grouping.

Exercise Set 11.4 1. $(x + 3)(x + 2)$ **3.** $(y + 8)(y - 2)$ **5.** $(8x - 5)(x - 3)$ **7.** $(5x^2 - 3)(x^2 + 5)$ **9. a.** $9, 2$ **b.** $9x + 2x$
c. $(2x + 3)(3x + 1)$ **11. a.** $-20, -3$ **b.** $-20x - 3x$ **c.** $(3x - 4)(5x - 1)$ **13.** $(3y + 2)(7y + 1)$ **15.** $(7x - 11)(x + 1)$
17. $(5x - 2)(2x - 1)$ **19.** $(2x - 5)(x - 1)$ **21.** $(2x + 3)(2x + 3)$ or $(2x + 3)^2$ **23.** $(2x + 3)(2x - 7)$ **25.** $(5x - 4)(2x - 3)$

27. $x(2x + 3)(x + 5)$ **29.** $2(8y - 9)(y - 1)$ **31.** $(2x - 3)(3x - 2)$ **33.** $3(3a + 2)(6a - 5)$ **35.** $a(4a + 1)(5a + 8)$
37. $3x(4x + 3)(x - 3)$ **39.** $y(3x + y)(x + y)$ **41.** prime **43.** $6(a + b)(4a - 5b)$ **45.** $p^2(15p + q)(p + 2q)$
47. $(7 + x)(5 + x)$ or $(x + 7)(x + 5)$ **49.** $(6 - 5x)(1 - x)$ or $(5x - 6)(x - 1)$ **51.** $x^2 - 4$ **53.** $y^2 + 8y + 16$
55. $81z^2 - 25$ **57.** $16x^2 - 24x + 9$ **59.** $10x^2 + 45x + 45; 5(2x + 3)(x + 3)$ **61.** $(x^n + 2)(x^n + 3)$ **63.** $(3x^n - 5)(x^n + 7)$
65. answers may vary

Calculator Explorations

	$x^2 - 2x + 1$	$x^2 - 2x - 1$	$(x - 1)^2$
$x = 5$	16	14	16
$x = -3$	16	14	16
$x = 2.7$	2.89	0.89	2.89
$x = -12.1$	171.61	169.61	171.61
$x = 0$	1	-1	1

Vocabulary, Readiness & Video Check **1.** perfect square trinomial **3.** perfect square trinomial **5.** $(x + 5y)^2$ **7.** false **9.** 8^2
11. $(11a)^2$ **13.** $(6p^2)^2$ **15.** No, it just means it won't factor into a binomial squared. It may or may not be factorable. **17.** In order to recognize the binomial as a difference of squares and also to identify the terms to use in the special factoring formula.

Exercise Set 11.5 **1.** yes **3.** no **5.** yes **7.** no **9.** no **11.** yes **13.** $(x + 11)^2$ **15.** $(x - 8)^2$ **17.** $(4a - 3)^2$ **19.** $(x^2 + 2)^2$
21. $2(n - 7)^2$ **23.** $(4y + 5)^2$ **25.** $(xy - 5)^2$ **27.** $m(m + 9)^2$ **29.** prime **31.** $(3x - 4y)^2$ **33.** $(x + 2)(x - 2)$
35. $(9 + p)(9 - p)$ or $-1(p + 9)(p - 9)$ **37.** $-1(2r + 1)(2r - 1)$ **39.** $(3x + 4)(3x - 4)$ **41.** prime
43. $-1(6 + x)(6 - x)$ or $(x + 6)(x - 6)$ **45.** $(m^2 + 1)(m + 1)(m - 1)$ **47.** $(x + 13y)(x - 13y)$ **49.** $2(3r + 2)(3r - 2)$
51. $x(3y + 2)(3y - 2)$ **53.** $16x^2(x + 2)(x - 2)$ **55.** $xy(y - 3z)(y + 3z)$ **57.** $4(3x - 4y)(3x + 4y)$ **59.** $9(4 - 3x)(4 + 3x$
61. $(5y - 3)(5y + 3)$ **63.** $(11m + 10n)(11m - 10n)$ **65.** $(xy - 1)(xy + 1)$ **67.** $\left(x - \frac{1}{2}\right)\left(x + \frac{1}{2}\right)$ **69.** $\left(7 - \frac{3}{5}m\right)\left(7 + \frac{3}{5}m\right)$
71. $(9a + 5b)(9a - 5b)$ **73.** $(x + 7y)^2$ **75.** $2(4n^2 - 7)^2$ **77.** $x^2(x^2 + 9)(x + 3)(x - 3)$ **79.** $pq(8p + 9q)(8p - 9q)$ **81.**
83. -2 **85.** $\frac{1}{5}$ **87.** $\left(x - \frac{1}{3}\right)^2$ **89.** $(x + 2 + y)(x + 2 - y)$ **91.** $(b - 4)(a + 4)(a - 4)$ **93.** $(x + 3 + 2y)(x + 3 - 2y)$
95. $(x^n + 10)(x^n - 10)$ **97.** 8 **99.** answers may vary **101.** $(x + 6)$ **103.** $a^2 + 2ab + b^2$ **105. a.** 2560 ft **b.** 1920 ft **c.** 13 sec
d. $16(13 - t)(13 + t)$ **107. a.** 2160 feet **b.** 1520 feet **c.** 12 seconds **d.** $16(12 + t)(12 - t)$

Integrated Review **1.** $(x - 3)(x + 4)$ **2.** $(x - 8)(x - 2)$ **3.** $(x + 1)^2$ **4.** $(x - 3)^2$ **5.** $(x + 2)(x - 3)$
6. $(x + 2)(x - 1)$ **7.** $(x + 3)(x - 2)$ **8.** $(x + 3)(x + 4)$ **9.** $(x - 5)(x - 2)$ **10.** $(x - 6)(x + 5)$
11. $2(x - 7)(x + 7)$ **12.** $3(x - 5)(x + 5)$ **13.** $(x + 3)(x + 5)$ **14.** $(y - 7)(3 + x)$ **15.** $(x + 8)(x - 2)$
16. $(x - 7)(x + 4)$ **17.** $4x(x + 7)(x - 2)$ **18.** $6x(x - 5)(x + 4)$ **19.** $2(3x + 4)(2x + 3)$ **20.** $3(2a - b)(4a + 5b)$
21. $(2a + b)(2a - b)$ **22.** $(x + 5y)(x - 5y)$ **23.** $(4 - 3x)(7 + 2x)$ **24.** $(5 - 2x)(4 + x)$ **25.** prime **26.** prime
27. $(3y + 5)(2y - 3)$ **28.** $(4x - 5)(x + 1)$ **29.** $9x(2x^2 - 7x + 1)$ **30.** $4a(3a^2 - 6a + 1)$ **31.** $(4a - 7)^2$ **32.** $(5p - 7)^2$
33. $(7 - x)(2 + x)$ **34.** $(3 + x)(1 - x)$ **35.** $3x^2y(x + 6)(x - 4)$ **36.** $2xy(x + 5y)(x - y)$ **37.** $3xy(4x^2 + 81)$
38. $2xy^2(3x^2 + 4)$ **39.** $2xy(1 + 6x)(1 - 6x)$ **40.** $2x(x - 3)(x + 3)$ **41.** $(x + 6)(x + 2)(x - 2)$ **42.** $(x - 2)(x + 6)(x - 6$
43. $2a^2(3a + 5)$ **44.** $2n(2n - 3)$ **45.** $(3x - 1)(x^2 + 4)$ **46.** $(x - 2)(x^2 + 3)$ **47.** $6(x + 2y)(x + y)$ **48.** $2(x + 4y)(6x - y$
49. $(x + y)(5 + x)$ **50.** $(x - y)(7 + y)$ **51.** $(7t - 1)(2t - 1)$ **52.** prime **53.** $-1(3x + 5)(x - 1)$ **54.** $-1(7x - 2)(x + 3$
55. $(1 - 10a)(1 + 2a)$ **56.** $(1 + 5a)(1 - 12a)$ **57.** $(x + 3)(x - 3)(x - 1)(x + 1)$ **58.** $(x + 3)(x - 3)(x + 2)(x - 2)$
59. $(x - 15)(x - 8)$ **60.** $(y + 16)(y + 6)$ **61.** $(5p - 7q)^2$ **62.** $(4a - 7b)^2$ **63.** prime **64.** $(7x + 3y)(x + 3y)$
65. $-1(x - 5)(x + 6)$ **66.** $-1(x - 2)(x - 4)$ **67.** $(3r - 1)(s + 4)$ **68.** $(x - 2)(x^2 + 1)$ **69.** $(x - 2y)(4x - 3)$
70. $(2x - y)(2x + 7z)$ **71.** $(x + 12y)(x - 3y)$ **72.** $(3x - 2y)(x + 4y)$ **73.** $(x^2 + 2)(x + 4)(x - 4)$
74. $(x^2 + 3)(x + 5)(x - 5)$ **75.** answers may vary **76.** yes; $9(x^2 + 9y^2)$

Vocabulary, Readiness & Video Check **1.** quadratic **3.** $3, -5$ **5.** One side of the equation must be a factored polynomial and the other side must be zero.

Exercise Set 11.6 **1.** $2, -1$ **3.** $6, 7$ **5.** $-9, -17$ **7.** $0, -6$ **9.** $0, 8$ **11.** $-\frac{3}{2}, \frac{5}{4}$ **13.** $\frac{7}{2}, -\frac{2}{7}$ **15.** $\frac{1}{2}, -\frac{1}{3}$ **17.** $-0.2, -1.5$ **19.** $9, 4$
21. $-4, 2$ **23.** $0, 7$ **25.** $0, -20$ **27.** $4, -4$ **29.** $8, -4$ **31.** $-3, 12$ **33.** $\frac{7}{3}, -2$ **35.** $\frac{8}{3}, -9$ **37.** $0, -\frac{1}{2}, \frac{1}{2}$ **39.** $\frac{17}{2}$ **41.** $\frac{3}{4}$ **43.** $-\frac{1}{2},$
45. $-\frac{3}{2}, -\frac{1}{2}, 3$ **47.** $-5, 3$ **49.** $-\frac{5}{6}, \frac{6}{5}$ **51.** $2, -\frac{4}{5}$ **53.** $-\frac{4}{3}, 5$ **55.** $-4, 3$ **57.** $0, 8, 4$ **59.** -7 **61.** $0, \frac{3}{2}$ **63.** $0, 1, -1$ **65.** $-6, \frac{4}{3}$
67. $\frac{6}{7}, 1$ **69.** $\frac{47}{45}$ **71.** $\frac{17}{60}$ **73.** $\frac{7}{10}$ **75.** didn't write equation in standard form; should be $x = 4$ or $x = -2$
77. answers may vary, for example, $(x - 6)(x + 1) = 0$ **79.** answers may vary, for example, $x^2 - 12x + 35 = 0$
81. a. $300; 304; 276; 216; 124; 0; -156$ **b.** 5 sec **c.** 304 ft **83.** $0, \frac{1}{2}$ **85.** $0, -15$

Vocabulary, Readiness & Video Check 1. In applications, the context of the stated application needs to be considered. Each translated equation resulted in both a positive and a negative solution, and a negative solution is not appropriate for any of the stated applications.

Exercise Set 11.7 1. width: x; length: $x + 4$ **3.** x and $x + 2$ if x is an odd integer **5.** base: x; height: $4x + 1$ **7.** 11 units **9.** 15 cm, 13 cm, 22 cm, 70 cm **11.** base: 16 mi; height: 6 mi **13.** 5 sec **15.** width: 5 cm; length: 6 cm **17.** 54 diagonals **19.** 10 sides **21.** -12 or 11 **23.** 14, 15 **25.** 13 feet **27.** 5 in. **29.** 12 mm, 16 mm, 20 mm **31.** 10 km **33.** 36 ft **35.** 9.5 sec **37.** 20% **39.** length: 15 mi; width: 8 mi **41.** 105 units **43.** 2.4 million or 2,400,000 **45.** 2.8 million or 2,800,000 **47.** 2016 **49.** answers may vary **51.** $\frac{4}{7}$ **53.** $\frac{3}{2}$ **55.** $\frac{1}{3}$ **57.** 8 m **59.** 10 and 15 **61.** width of pool: 29 m; length of pool: 35 m **63.** answers may vary

Chapter 11 Vocabulary Check 1. quadratic equation **2.** Factoring **3.** greatest common factor **4.** perfect square trinomial **5.** hypotenuse **6.** leg **7.** hypotenuse

Chapter 11 Review 1. $5(m + 6)$ **2.** $3x(2x - 5)$ **3.** $2x(2x^4 + 1 - 5x^3)$ **4.** $4x(5x^2 + 3x + 6)$ **5.** $(2x + 3)(3x - 5)$ **6.** $(x + 1)(5x - 1)$ **7.** $(x - 1)(3x + 2)$ **8.** $(a + 3b)(3a + b)$ **9.** $(2a + b)(5a + 7b)$ **10.** $(3x + 5)(2x - 1)$ **11.** $(x + 4)(x + 2)$ **12.** $(x - 8)(x - 3)$ **13.** prime **14.** $(x - 6)(x + 1)$ **15.** $(x + 4)(x - 2)$ **16.** $(x + 6y)(x - 2y)$ **17.** $(x + 5y)(x + 3y)$ **18.** $-2(x - 3)(x + 12)$ **19.** $-4(x^2 - 3x - 8)$ **20.** $5y(y - 6)(y - 4)$ **21.** $-48; 2$ **22.** factor out the GCF, 3 **23.** $(2x + 1)(x + 6)$ **24.** $(2x + 3)(2x - 1)$ **25.** $(3x + 4y)(2x - y)$ **26.** prime **27.** $(2x + 3)(x - 13)$ **28.** $(6x + 5y)(3x - 4y)$ **29.** $5y(2y - 3)(y + 4)$ **30.** $3y(4y - 1)(5y - 2)$ **31.** $5x^2 - 9x - 2; (5x + 1)(x - 2)$ **32.** $16x^2 - 28x + 6; 2(4x - 1)(2x - 3)$ **33.** yes **34.** no **35.** no **36.** yes **37.** yes **38.** no **39.** yes **40.** no **41.** $(x + 9)(x - 9)$ **42.** $(x + 6)^2$ **43.** $(2x + 3)(2x - 3)$ **44.** $(3t + 5s)(3t - 5s)$ **45.** prime **46.** $(n - 9)^2$ **47.** $3(r + 6)^2$ **48.** $(3y - 7)^2$ **49.** $5m^6(m + 1)(m - 1)$ **50.** $(2x - 7y)^2$ **51.** $3y(x + y)^2$ **52.** $(4x^2 + 1)(2x + 1)(2x - 1)$ **53.** $-6, 2$ **54.** $-11, 7$ **55.** $0, -1, \frac{2}{7}$ **56.** $-\frac{1}{5}, -3$ **57.** $-7, -1$ **58.** $-4, 6$ **59.** -5 **60.** $2, 8$ **61.** $\frac{1}{3}$ **62.** $-\frac{2}{7}, \frac{3}{8}$ **63.** $0, 6$ **64.** $5, -5$ **65.** $x^2 - 9x + 20 = 0$ **66.** $x^2 + 2x + 1 = 0$ **67.** c **68.** d **69.** 9 units **70.** 8 units, 13 units, 16 units, 10 units **71.** width: 20 in.; length: 25 in. **72.** 36 yd **73.** 19 and 20 **74.** 20 and 22 **75. a.** 17.5 sec and 10 sec; answers may vary **b.** 27.5 sec **76.** 32 cm **77.** $6(x + 4)$ **78.** $7(x - 9)$ **79.** $(4x - 3)(11x - 6)$ **80.** $(x - 5)(2x - 1)$ **81.** $(3x - 4)(x^2 + 2)$ **82.** $(y + 2)(x - 1)$ **83.** $2(x + 4)(x - 3)$ **84.** $3x(x - 9)(x - 1)$ **85.** $(2x + 9)(2x - 9)$ **86.** $2(x + 3)(x - 3)$ **87.** $(4x - 3)^2$ **88.** $5(x + 2)^2$ **89.** $-\frac{7}{2}, 4$ **90.** $-3, 5$ **91.** $0, -7, -4$ **92.** $3, 2$ **93.** $0, 16$ **94.** 19 in.; 8 in.; 21 in. **95.** length: 6 in.; width: 2 in.

Chapter 11 Getting Ready for the Test 1. B **2.** D **3.** A **4.** A **5.** B **6.** B **7.** B **8.** A **9.** C

Chapter 11 Test 1. $3x(3x - 1)$ **2.** $(x + 7)(x + 4)$ **3.** $(7 + m)(7 - m)$ **4.** $(y + 11)^2$ **5.** $(x^2 + 4)(x + 2)(x - 2)$ **6.** $(a + 3)(4 - y)$ **7.** prime **8.** $(y - 12)(y + 4)$ **9.** $(a + b)(3a - 7)$ **10.** $(3x - 2)(x - 1)$ **11.** $5(6 + x)(6 - x)$ **12.** $3x(x - 5)(x - 2)$ **13.** $(6t + 5)(t - 1)$ **14.** $(x - 7)(y - 2)(y + 2)$ **15.** $x(1 + x^2)(1 + x)(1 - x)$ **16.** $(x + 12y)(x + 2y)$ **17.** $3, -9$ **18.** $-7, 2$ **19.** $-7, 1$ **20.** $0, \frac{3}{2}, -\frac{4}{3}$ **21.** $0, 3, -3$ **22.** $-3, 5$ **23.** $0, \frac{5}{2}$ **24.** 17 ft **25.** width: 6 units; length: 9 units **26.** 7 sec **27.** hypotenuse: 25 cm; legs: 15 cm, 20 cm **28.** 8.25 sec

Cumulative Review Chapters 1–11 1. -3; Sec. 2.5, Ex. 13 **2.** -14; Sec. 2.5 **3.** -9; Sec. 2.5, Ex. 14 **4.** -72; Sec. 2.5 **5.** -9; Sec. 2.6, Ex. 3 **6.** -9; Sec. 2.6 **7.** $6y + 2$; Sec. 3.1, Ex. 2 **8.** $-4a - 1$; Sec. 3.1 **9.** $-x + 5$; Sec. 3.1, Ex. 4 **10.** $7x - 6$; Sec. 3.1 **11.** -15; Sec. 3.2, Ex. 6 **12.** -2; Sec. 3.2 **13.** 2; Sec. 3.3, Ex. 4 **14.** 0; Sec. 3.3 **15.** $\frac{31}{2}$; Sec. 6.1, Ex. 5 **16.** -17; Sec. 6.1 **17. a.** $9 \leq 11$ **b.** $8 > 1$ **c.** $3 \neq 4$; Sec. 9.1, Ex. 7 **18. a.** $5 \geq 1$; **b.** $2 \neq -4$; Sec. 9.1 **19.** every real number; Sec. 9.3, Ex. 7 **20.** no solution; Sec. 9.3 **21.** $l = \frac{V}{wh}$; Sec. 9.5, Ex. 5 **22.** $x = \frac{2y + 5}{3}$; Sec. 9.5 **23.** 5^{18}; Sec. 10.1, Ex. 16 **24.** 7^{18}; Sec. 10.1 **25.** y^{16}; Sec. 10.1, Ex. 17 **26.** x^{33}; Sec. 10.1 **27.** x^6; Sec. 10.2, Ex. 9 **28.** y^{22}; Sec. 10.2 **29.** $\frac{y^{18}}{z^{36}}$; Sec. 10.2, Ex. 11 **30.** $\frac{y^8}{x^2}$; Sec. 10.2 **31.** $\frac{1}{x^{19}}$; Sec. 10.2, Ex. 13 **32.** y^{13}; Sec. 10.2 **33.** $4x$; Sec. 10.3, Ex. 6 **34.** $2y$; Sec. 10.3 **35.** $13x^2 - 2$; Sec. 10.3, Ex. 7 **36.** $12y - 2y^2$; Sec. 10.3 **37.** $4x^2 - 4xy + y^2$; Sec. 10.5, Ex. 8 **38.** $9x^2 + 6x + 1$; Sec. 10.5 **39.** $t^2 + 4t + 4$; Sec. 10.6, Ex. 5 **40.** $x^2 - 8x + 16$; Sec. 10.6 **41.** $x^4 - 14x^2y + 49y^2$; Sec. 10.6, Ex. 8 **42.** $x^4 + 14x^2y + 49y^2$; Sec. 10.6 **43.** $2xy - 4 + \frac{1}{2y}$; Sec. 10.7, Ex. 3 **44.** $4ab^2 - 1 + \frac{2}{a}$; Sec. 10.7 **45.** $(x + 3)(5 + y)$; Sec. 11.1, Ex. 9 **46.** $(y - 2)(9 + x)$; Sec. 11.1 **47.** $(x^2 + 2)(x^2 + 3)$; Sec. 11.2, Ex. 7 **48.** $(x^2 + 1)(x^2 - 5)$; Sec. 11.2 **49.** $2(x - 2)(3x + 5)$; Sec. 11.4, Ex. 2 **50.** $5(x + 2)(2x + 1)$; Sec. 11.4 **51.** 3 sec; Sec. 11.7, Ex. 1 **52.** -12 or 10; Sec. 11.7

Chapter 12 Rational Expressions

Vocabulary, Readiness & Video Check 1. rational expression 3. -1 5. 2 7. $\dfrac{-a}{b}, \dfrac{a}{-b}$ 9. yes 11. no 13. Rational expressions are fractions and are therefore undefined if the denominator is zero; if a denominator contains variables, set it equal to zero and solve. 15. You would need to write parentheses around the numerator or denominator if it had more than one term because the negative sign needs to apply to the entire numerator or denominator.

Exercise Set 12.1 1. $\dfrac{7}{4}$ 3. 3 5. $-\dfrac{8}{3}$ 7. $-\dfrac{11}{2}$ 9. $x=0$ 11. $x=-2$ 13. $x=\dfrac{5}{2}$ 15. $x=0, x=-2$ 17. none 19. $x=6, x=-1$ 21. $x=-2, x=-\dfrac{7}{3}$ 23. 1 25. -1 27. $\dfrac{1}{4(x+2)}$ 29. $\dfrac{1}{x+2}$ 31. can't simplify 33. -5 35. $\dfrac{7}{x}$ 37. $\dfrac{1}{x-9}$ 39. $5x+1$ 41. $\dfrac{x^2}{x-2}$ 43. $7x$ 45. $\dfrac{x+5}{x-5}$ 47. $\dfrac{x+2}{x+4}$ 49. $\dfrac{x+2}{2}$ 51. $-(x+2)$ 53. $\dfrac{x+1}{x-1}$ 55. $x+y$ 57. $\dfrac{5-x}{2}$ 59. $\dfrac{2y+5}{3y+4}$ 61. $\dfrac{-(x-10)}{x+8}; \dfrac{-x+10}{x+8}; \dfrac{x-10}{-(x+8)}; \dfrac{x-10}{-x-8}$ 63. $\dfrac{-(5y-3)}{y-12}; \dfrac{-5y+3}{y-12}; \dfrac{5y-3}{-(y-12)}; \dfrac{5y-3}{-y+12}$ 65. correct 67. correct 69. $\dfrac{3}{11}$ 71. $\dfrac{4}{3}$ 73. $\dfrac{117}{40}$ 75. 0.563 77. 0.620 79. 0.595 81. David Ortiz 83. correct 85. incorrect; $\dfrac{1+2}{1+3} = \dfrac{3}{4}$ 87. answers may vary 89. answers may vary 91. a. $403 b. $7 c. decrease; answers may vary 93. 400 mg 95. $C = 78.125$; medium

Vocabulary, Readiness & Video Check 1. reciprocals 3. $\dfrac{a \cdot d}{b \cdot c}$ 5. $\dfrac{6}{7}$ 7. fractions; reciprocal 9. We're converting to cubic feet so we want cubic feet in the numerator. We want cubic yards to divide out, so cubic yards is in the denominator.

Exercise Set 12.2 1. $\dfrac{21}{4y}$ 3. x^4 5. $-\dfrac{b^2}{6}$ 7. $\dfrac{x^2}{10}$ 9. $\dfrac{1}{3}$ 11. $\dfrac{m+n}{m-n}$ 13. $\dfrac{x+5}{x}$ 15. $\dfrac{(x+2)(x-3)}{(x-4)(x+4)}$ 17. $\dfrac{2x^4}{3}$ 19. $\dfrac{12}{y^6}$ 21. $x(x+4)$ 23. $\dfrac{3(x+1)}{x^3(x-1)}$ 25. $m^2 - n^2$ 27. $-\dfrac{x+2}{x-3}$ 29. $\dfrac{x+2}{x-3}$ 31. $\dfrac{5}{6}$ 33. $\dfrac{3x}{8}$ 35. $\dfrac{3}{2}$ 37. $\dfrac{3x+4y}{2(x+2y)}$ 39. $\dfrac{2(x+2)}{x-2}$ 41. $-\dfrac{y(x+2)}{4}$ 43. $\dfrac{(a+5)(a+3)}{(a+2)(a+1)}$ 45. $\dfrac{5}{x}$ 47. $\dfrac{2(n-8)}{3n-1}$ 49. 1440 51. 5 53. 81 55. 73 57. 56.7 59. 1,201,500 sq ft 61. 55.2 miles/hour 63. 1 65. $-\dfrac{10}{9}$ 67. $-\dfrac{1}{5}$ 69. true 71. false; $\dfrac{x^2+3x}{20}$ 73. $\dfrac{2}{9(x-5)}$ sq ft 75. $\dfrac{x}{2}$ 77. $\dfrac{5a(2a+b)(3a-2b)}{b^2(a-b)(a+2b)}$ 79. answers may vary 81. 1912.78 euros

Vocabulary, Readiness & Video Check 1. $\dfrac{9}{11}$ 3. $\dfrac{a+c}{b}$ 5. $\dfrac{5-(6+x)}{x}$ 7. We completely factor denominators—including coefficients—so we can determine the greatest number of times each unique factor occurs in any one denominator for the LCD.

Exercise Set 12.3 1. $\dfrac{a+9}{13}$ 3. $\dfrac{3m}{n}$ 5. 4 7. $\dfrac{y+10}{3+y}$ 9. $5x+3$ 11. $\dfrac{4}{a+5}$ 13. $\dfrac{1}{x-6}$ 15. $\dfrac{5x+7}{x-3}$ 17. $x+5$ 19. 3 21. $4x^3$ 23. $8x(x+2)$ 25. $(x+3)(x-2)$ 27. $3(x+6)$ 29. $5(x-6)^2$ 31. $6(x+1)^2$ 33. $x-8$ or $8-x$ 35. $(x-1)(x+4)(x+3)$ 37. $(3x+1)(x+1)(x-1)(2x+1)$ 39. $2x^2(x+4)(x-4)$ 41. $\dfrac{6x}{4x^2}$ 43. $\dfrac{24b^2}{12ab^2}$ 45. $\dfrac{9y}{2y(x+3)}$ 47. $\dfrac{9ab+2b}{5b(a+2)}$ 49. $\dfrac{x^2+x}{x(x+4)(x+2)(x+1)}$ 51. $\dfrac{18y-2}{30x^2-60}$ 53. $2x$ 55. $\dfrac{x+3}{2x-1}$ 57. $x+1$ 59. $\dfrac{3}{x}$ 61. $\dfrac{3x+1}{5x+1}$ 63. $\dfrac{29}{21}$ 65. $-\dfrac{5}{12}$ 67. $\dfrac{7}{30}$ 69. d 71. answers may vary 73. c 75. b 77. $-\dfrac{5}{x-2}$ 79. $\dfrac{7+x}{x-2}$ 81. $\dfrac{20}{x-2}$ m 83. answers may vary 85. 95,304 Earth days 87. answers may vary 89. answers may vary

Vocabulary, Readiness & Video Check 1. b 3. The exercise is adding two rational expressions with denominators that are opposites of each other. Recognizing this special case can save us time and effort. If we recognize that one denominator is -1 times the other denominator, we may save many steps.

Exercise Set 12.4 1. $\dfrac{5}{x}$ 3. $\dfrac{75a+6b^2}{5b}$ 5. $\dfrac{6x+5}{2x^2}$ 7. $\dfrac{11}{x+1}$ 9. $\dfrac{x-6}{(x-2)(x+2)}$ 11. $\dfrac{35x-6}{4x(x-2)}$ 13. $-\dfrac{2}{x-3}$ 15. 0 17. $-\dfrac{1}{x^2-1}$ 19. $\dfrac{5+2x}{x}$ 21. $\dfrac{6x-7}{x-2}$ 23. $-\dfrac{y+4}{y+3}$ 25. $\dfrac{-5x+14}{4x}$ or $-\dfrac{5x-14}{4x}$ 27. 2 29. $\dfrac{9x^4-4x^2}{21}$ 31. $\dfrac{x+2}{(x+3)^2}$

33. $\dfrac{9b-4}{5b(b-1)}$ **35.** $\dfrac{2+m}{m}$ **37.** $\dfrac{x(x+3)}{(x-7)(x-2)}$ **39.** $\dfrac{10}{1-2x}$ **41.** $\dfrac{15x-1}{(x+1)^2(x-1)}$ **43.** $\dfrac{x^2-3x-2}{(x-1)^2(x+1)}$ **45.** $\dfrac{a+2}{2(a+3)}$

47. $\dfrac{y(2y+1)}{(2y+3)^2}$ **49.** $\dfrac{x-10}{2(x-2)}$ **51.** $\dfrac{2x+21}{(x+3)^2}$ **53.** $\dfrac{-5x+23}{(x-2)(x-3)}$ **55.** $\dfrac{7}{2(m-10)}$ **57.** $\dfrac{2(x^2-x-23)}{(x+1)(x-6)(x-5)}$

59. $\dfrac{n+4}{4n(n-1)(n-2)}$ **61.** 10 **63.** 2 **65.** $\dfrac{25a}{9(a-2)}$ **67.** $\dfrac{x+4}{(x-2)(x-1)}$ **69.** $\dfrac{2}{3}$ **71.** $-\dfrac{1}{2}, 1$ **73.** $-\dfrac{15}{2}$ **75.** $\dfrac{6x^2-5x-3}{x(x+1)(x-1)}$

77. $\dfrac{4x^2-15x+6}{(x-2)^2(x+2)(x-3)}$ **79.** $\dfrac{-2x^2+14x+55}{(x+2)(x+7)(x+3)}$ **81.** $\dfrac{2(x-8)}{(x+4)(x-4)}$ in. **83.** $\dfrac{P-G}{P}$ **85.** answers may vary

87. $\left(\dfrac{90x-40}{x}\right)^\circ$ **89.** answers may vary

Vocabulary, Readiness & Video Check **1.** c **3.** b **5.** a **7.** These equations are solved in very different ways, so we need to determine the next correct step to make. For a linear equation, we first "move" variable terms to one side and numbers to the other; for a quadratic equation, we first set the equation equal to 0. **9.** the steps for solving an equation containing rational expressions; as if it's the only variable in the equation

Exercise Set 12.5 **1.** 30 **3.** 0 **5.** -2 **7.** $-5, 2$ **9.** 5 **11.** 3 **13.** 1 **15.** 5 **17.** no solution **19.** 4 **21.** -8 **23.** 6, -4

25. 1 **27.** 3, -4 **29.** -3 **31.** 0 **33.** -2 **35.** 8, -2 **37.** no solution **39.** 3 **41.** $-11, 1$ **43.** $I=\dfrac{E}{R}$ **45.** $B=\dfrac{2U-TE}{T}$

47. $w=\dfrac{Bh^2}{705}$ **49.** $G=\dfrac{V}{N-R}$ **51.** $r=\dfrac{C}{2\pi}$ **53.** $x=\dfrac{3y}{3+y}$ **55.** $\dfrac{1}{x}$ **57.** $\dfrac{1}{x}+\dfrac{1}{2}$ **59.** $\dfrac{1}{3}$ **61.** answers may vary **63.** $\dfrac{5x+9}{9x}$

65. no solution **67.** 100°, 80° **69.** 22.5°, 67.5° **71.** 5

Integrated Review **1.** expression; $\dfrac{3+2x}{3x}$ **2.** expression; $\dfrac{18+5a}{6a}$ **3.** equation; 3 **4.** equation; 18 **5.** expression; $\dfrac{x-1}{x(x+1)}$

6. expression; $\dfrac{3(x+1)}{x(x-3)}$ **7.** equation; no solution **8.** equation; 1 **9.** expression; 10 **10.** expression; $\dfrac{z}{3(9z-5)}$

11. expression; $\dfrac{5x+7}{x-3}$ **12.** expression; $\dfrac{7p+5}{2p+7}$ **13.** equation; 23 **14.** equation; 3 **15.** expression; $\dfrac{25a}{9(a-2)}$

16. expression; $\dfrac{9}{4(x-1)}$ **17.** expression; $\dfrac{3x^2+5x+3}{(3x-1)^2}$ **18.** expression; $\dfrac{2x^2-3x-1}{(2x-5)^2}$ **19.** expression; $\dfrac{4x-37}{5x}$

20. expression; $\dfrac{29x-23}{3x}$ **21.** equation; $\dfrac{8}{5}$ **22.** equation; $-\dfrac{7}{3}$ **23.** answers may vary **24.** answers may vary

Vocabulary, Readiness & Video Check **1.** c **3.** $\dfrac{1}{x}; \dfrac{1}{x}-3$ **5.** $z+5; \dfrac{1}{z+5}$ **7.** $2y; \dfrac{11}{2y}$ **9.** divided by, quotient **11.** car distance: 325; car rate: $x+7$; car time: $\dfrac{325}{x+7}$; motorcycle distance: 290; motorcycle rate: x; motorcycle time: $\dfrac{290}{x}; \dfrac{325}{x+7}=\dfrac{290}{x}$

Exercise Set 12.6 **1.** 2 **3.** -3 **5.** $2\dfrac{2}{9}$ hr **7.** $1\dfrac{1}{2}$ min **9.** trip to park rate: r; to park time: $\dfrac{12}{r}$; return trip rate: r; return time: $\dfrac{8}{r}$; $r=6$ mph **11.** 1st portion: 10 mph; cooldown: 8 mph **13.** 2 **15.** $108.00 **17.** 20 mph **19.** 5 **21.** 217 mph

23. 8 **25.** 2.2 mph; 3.3 mph **27.** 3 hr **29.** 8 mph **31.** 35 mph; 75 mph **33.** 510 mph **35.** $666\dfrac{2}{3}$ mi **37.** 20 hr **39.** car: 70 mph; motorcycle: 60 mph **41.** $5\dfrac{1}{4}$ hr **43.** 41 mph; 51 mph **45.** $\dfrac{1}{2}$ **47.** $\dfrac{3}{7}$ **49.** faster pump: 28 min; slower pump: 84 min

51. answers may vary **53.** $R=\dfrac{D}{T}$ **55.** 3.75 min

Vocabulary, Readiness & Video Check **1.** $\dfrac{y}{5x}$ **3.** $\dfrac{3x}{5}$ **5.** c **7.** a **9.** a single fraction in the numerator and in the denominator

Exercise Set 12.7 **1.** $\dfrac{2}{3}$ **3.** $\dfrac{2}{3}$ **5.** $\dfrac{1}{2}$ **7.** $-\dfrac{21}{5}$ **9.** $\dfrac{27}{16}$ **11.** $\dfrac{4}{3}$ **13.** $\dfrac{1}{21}$ **15.** $-\dfrac{4x}{15}$ **17.** $\dfrac{m-n}{m+n}$ **19.** $\dfrac{2x(x-5)}{7x^2+10}$ **21.** $\dfrac{1}{y-1}$ **23.** $\dfrac{1}{6}$

25. $\dfrac{x+y}{x-y}$ **27.** $\dfrac{3}{7}$ **29.** $\dfrac{a}{x+b}$ **31.** $\dfrac{7(y-3)}{8+y}$ **33.** $\dfrac{3x}{x-4}$ **35.** $-\dfrac{x+8}{x-2}$ **37.** $\dfrac{s^2+r^2}{s^2-r^2}$ **39.** $\dfrac{(x-6)(x+4)}{x-2}$ **41.** Serena Williams

43. about $9 million **45.** answers may vary **47.** $\dfrac{13}{24}$ **49.** $4\dfrac{1}{4}$ ft or 4.25 ft **51.** $\dfrac{R_1R_2}{R_2+R_1}$ **53.** $\dfrac{2x}{2-x}$ **55.** $\dfrac{1}{y^2-1}$ **57.** 12 hr

Chapter 12 Vocabulary Check 1. rational expression 2. complex fraction 3. $\dfrac{-a}{b}; \dfrac{a}{-b}$ 4. denominator 5. simplifying 6. reciprocals 7. least common denominator 8. unit

Chapter 12 Review 1. $x = 2, x = -2$ 2. $x = \dfrac{5}{2}, x = -\dfrac{3}{2}$ 3. $\dfrac{4}{3}$ 4. $\dfrac{11}{12}$ 5. $\dfrac{2}{x}$ 6. $\dfrac{3}{x}$ 7. $\dfrac{1}{x-5}$ 8. $\dfrac{1}{x+1}$ 9. $\dfrac{(x-2)}{x+1}$ 10. $\dfrac{5(x-5)}{x-3}$ 11. $\dfrac{x-3}{x-5}$ 12. $\dfrac{x}{x+4}$ 13. $\dfrac{x+a}{x-c}$ 14. $\dfrac{x+5}{x-3}$ 15. $\dfrac{3x^2}{y}$ 16. $-\dfrac{9x^2}{8}$ 17. $\dfrac{x-3}{x+2}$ 18. $-\dfrac{2x(2x+5)}{(x-6)^2}$ 19. $\dfrac{x+3}{x-4}$ 20. $\dfrac{4x}{3y}$ 21. $(x-6)(x-3)$ 22. $\dfrac{2}{3}$ 23. $\dfrac{1}{2}$ 24. $\dfrac{3(x+2)}{3x+y}$ 25. $\dfrac{1}{x+2}$ 26. $\dfrac{1}{x-3}$ 27. $\dfrac{2(x-5)}{3x^2}$ 28. $\dfrac{2x+1}{2x^2}$ 29. $14x$ 30. $(x-8)(x+8)(x+3)$ 31. $\dfrac{10x^2y}{14x^3y}$ 32. $\dfrac{36y^2x}{16y^3x}$ 33. $\dfrac{x^2-3x-10}{(x+2)(x-5)(x+9)}$ 34. $\dfrac{3x^2+4x-15}{(x+2)^2(x+3)}$ 35. $\dfrac{4y+30x^2}{5x^2y}$ 36. $\dfrac{-2x+10}{(x-3)(x-1)}$ 37. $\dfrac{-2x-2}{x+3}$ 38. $\dfrac{5(x+1)}{(x+4)(x-2)(x-1)}$ 39. $\dfrac{x-4}{3x}$ 40. $-\dfrac{x}{x-1}$ 41. 30 42. $3, -4$ 43. no solution 44. 5 45. $\dfrac{9}{7}$ 46. $-6, 1$ 47. 3 48. 2 49. faster car speed: 30 mph; slower car speed: 20 mph 50. 20 mph 51. $17\dfrac{1}{2}$ hr 52. $8\dfrac{4}{7}$ days 53. $-\dfrac{7}{18y}$ 54. $\dfrac{6}{7}$ 55. $\dfrac{3y-1}{2y-1}$ 56. $-\dfrac{7+2x}{2x}$ 57. $\dfrac{1}{2x}$ 58. $\dfrac{x(x-3)}{x+7}$ 59. $\dfrac{x-4}{x+4}$ 60. $\dfrac{(x-9)(x+8)}{(x+5)(x+9)}$ 61. $\dfrac{1}{x-6}$ 62. $\dfrac{2x+1}{4x}$ 63. $-\dfrac{3x}{(x+2)(x-3)}$ 64. $\dfrac{2}{(x+3)(x-2)}$ 65. $\dfrac{1}{2}$ 66. no solution 67. 1 68. $1\dfrac{5}{7}$ days 69. $\dfrac{3}{10}$ 70. $\dfrac{2}{3}$ 71. 16.2 72. 5

Chapter 12 Getting Ready for the Test 1. B 2. C 3. D 4. D 5. A 6. D 7. A 8. B 9. B 10. A 11. C 12. A

Chapter 12 Test 1. $x = -1, x = -3$ 2. a. \$115 b. \$103 3. $\dfrac{3}{5}$ 4. $\dfrac{1}{x+6}$ 5. -1 6. $-\dfrac{1}{x+y}$ 7. $\dfrac{2m(m+2)}{m-2}$ 8. $\dfrac{a+2}{a+5}$ 9. $\dfrac{(x-6)(x-7)}{(x+7)(x+2)}$ 10. 15 11. $\dfrac{y-2}{4}$ 12. $-\dfrac{1}{2x+5}$ 13. $\dfrac{3a-4}{(a-3)(a+2)}$ 14. $\dfrac{3}{x-1}$ 15. $\dfrac{2(x+3)(x+5)}{x(x^2+4x+1)}$ 16. $\dfrac{x^2+2x+35}{(x+9)(x+2)(x-5)}$ 17. $\dfrac{4y^2+13y-15}{(y+5)(y+1)(y+4)}$ 18. $\dfrac{30}{11}$ 19. -6 20. no solution 21. no solution 22. $-2, 5$ 23. $\dfrac{xz}{2y}$ 24. $b-a$ 25. $\dfrac{5y^2-1}{y+2}$ 26. 1 or 5 27. 30 mph 28. $6\dfrac{2}{3}$ hr

Cumulative Review Chapters 1–12 1. 35%; Sec. 6.2, Ex. 17 2. 80%; Sec. 6.2 3. $66\dfrac{2}{3}$%; Sec. 6.2, Ex. 18 4. $11\dfrac{1}{9}$%; Sec. 6.2 5. 225%; Sec. 6.2, Ex. 19 6. 375%; Sec. 6.2 7. commutative property of multiplication; Sec. 9.2, Ex. 18 8. commutative property of addition; Sec. 9.2 9. associative property of addition; Sec. 9.2, Ex. 16 10. associative property of multiplication; Sec. 9.2 11. shorter piece, 2 ft; longer piece, 8 ft; Sec. 9.4, Ex. 3 12. 16 ft, 16 ft, 13 ft; Sec. 9.4 13. $\dfrac{y-b}{m} = x$; Sec. 9.5, Ex. 6 14. $b = y - mx$; Sec. 9.5 15. $x \le -10$; ←————•———; Sec. 9.6, Ex. 4 16. $x > -9$; Sec. 9.6 17. x^3; Sec. 10.1, Ex. 24 18. x^8; Sec. 10.1 19. $4^4 = 256$; Sec. 10.1, Ex. 25 20. $7^8 = 5{,}764{,}801$; Sec. 10.1 21. $(-3)^3 = -27$; Sec. 10.1, Ex. 26 22. $(-4)^2 = 16$; Sec. 10.1 23. $2x^4y$; Sec. 10.1, Ex. 27 24. $13ab$; Sec. 10.1 25. $\dfrac{2}{x^3}$; Sec. 10.2, Ex. 2 26. $\dfrac{9}{x^2}$; Sec. 10.2 27. $\dfrac{1}{16}$; Sec. 10.2, Ex. 4 28. $-\dfrac{1}{27}$; Sec. 10.2 29. $10x^4 + 30x$; Sec. 10.5, Ex. 5 30. $12y^3 - 6y$; Sec. 10.5 31. $-15x^4 - 18x^3 + 3x^2$; Sec. 10.5, Ex. 6 32. $-35y^3 + 15y^2 - 5y$; Sec. 10.5 33. $4x^2 - 4x + 6 + \dfrac{-11}{2x+3}$; Sec. 10.7, Ex. 7 34. $3x - 5 + \dfrac{9}{2x+1}$; Sec. 10.7 35. $(x+3)(x+4)$; Sec. 11.2, Ex. 1 36. $(x+7)(x+10)$; Sec. 11.2 37. $(5x+2y)^2$; Sec. 11.5, Ex. 5 38. $4(3a-2b)^2$; Sec. 11.5 39. $11, -2$; Sec. 11.6, Ex. 4 40. $3, -5$; Sec. 11.6 41. $\dfrac{2}{5}$; Sec. 12.2, Ex. 2 42. $\dfrac{6x}{5}$; Sec. 12.2 43. $3x - 5$; Sec. 12.3, Ex. 3 44. 4; Sec. 12.3 45. $\dfrac{3}{x-2}$; Sec. 12.4, Ex. 2 46. $\dfrac{7x+6}{(x-3)(x+3)}$; Sec. 12.4 47. 5; Sec. 12.5, Ex. 2 48. $\dfrac{2}{3}$; Sec. 12.5 49. $2\dfrac{1}{10}$ hr; Sec. 12.6, Ex. 2 50. $7\dfrac{1}{5}$ hr; Sec. 12.6 51. $\dfrac{3}{z}$; Sec. 12.7, Ex. 3 52. $\dfrac{x-3}{9}$; Sec. 12.